HANDBOOK O

ENVIRONMENTAL
FLUID DYNAMICS

SYSTEMS, POLLUTION, MODELING, AND MEASUREMENTS

VOLUME TWO

HANDBOOK OF
ENVIRONMENTAL
FLUID DYNAMICS

SYSTEMS, POLLUTION, MODELING, AND MEASUREMENTS

VOLUME TWO

EDITED BY

H. J. S. FERNANDO

CRC Press
Taylor & Francis Group
Boca Raton London New York

CRC Press is an imprint of the
Taylor & Francis Group, an **informa** business

CRC Press
Taylor & Francis Group
6000 Broken Sound Parkway NW, Suite 300
Boca Raton, FL 33487-2742

First issued in paperback 2018

ISBN-13: 978-1-4665-5601-0 (hbk)
ISBN-13: 978-1-138-37474-4 (pbk)

Library of Congress Cataloging-in-Publication Data

Handbook of environmental fluid dynamics / edited by Harindra Joseph Fernando.
    v. cm.
    Includes bibliographical references and index.
    Contents: v. 1. Overview and fundamentals -- v. 2. Systems, pollution, modeling, and measurements.
    ISBN 978-1-4398-1669-1
    1. Fluid dynamics--Handbooks, manuals, etc. 2. Hydrogeology--Handbooks, manuals, etc. 3. Water quality--Handbooks, manuals, etc. 4. Fluids--Migration--Handbooks, manuals, etc. 5. Earth--Research--Handbooks, manuals, etc. I. Fernando, Harindra Joseph.

QC809.F5H33 2013
620.1'064--dc23

2012015977

Visit the Taylor & Francis Web site at
http://www.taylorandfrancis.com

and the CRC Press Web site at
http://www.crcpress.com

In memory of

Owen M. Phillips

(December 30, 1930–October 13, 2010)

Contents

PART II   Environmental Pollution

PART III   Numerical Modeling of Environmental Flows

PART IV   Laboratory Modeling of Environmental Flows

PART V   Environmental Measurements

Preface

The *Handbook of Environmental Fluid Dynamics* is a culmination of years of work by the contributing authors, publisher, and the volume editor and represents a state-of-the-science exposition of selected subtopics within environmental fluid dynamics (EFD). When I was first contacted by Luna Han (senior editor of Professional & Trade Books, Taylor & Francis Group LLC) in October 2008, regarding the editorship of this volume, I agreed to do so only with a brief reflection of its potential usefulness, perhaps not fully fathoming the task ahead. The original proposal included some 45 papers, and it quickly grew to about 85 topics with a moratorium declared thereafter. Some appealing topics had to be excluded, partly because of space limitations and last minute cancellations, with the hope of including them in a future edition. This two-volume set represents 81 chapters written by a cadre of 135 authors from 20 countries. Without the dedication and commitment of these authors, notwithstanding their regular professional obligations, this volume could not have come to light, and my sincere appreciation goes to all of them.

During the course of the preparation of this volume, two of our accepted contributors, Gerhard H. Jirka and Daewon W. Byun, passed away unexpectedly, and we fondly remember them. Gerhard's chapter, "Jets and Plumes," (Volume One, Chapter 25) was written by his students and collaborators. Daewon wrote his chapter, "Air Pollution Modeling and Its Applications" (Volume Two, Chapter 26), by himself. May their souls rest in peace!

The discipline of EFD descended from the revered geophysical fluid dynamics (GFD), the latter being the platform for fluid dynamics study of natural motions akin to climate, weather, oceanic circulation, and motions of the earth's interior. Conversely, EFD focuses on natural motions influenced by or directly affecting human (anthropogenic) activities. Applications of EFD abound in wind and ocean engineering as well as areas of human health and welfare. Air and water pollution are examples of the latter, which to a large extent sparked the early interest in EFD.

Environmental pollution has been recognized for centuries, as have the environmental regulations that date back to King Edward I of England, who decreed a ban on sea coal burning in 1306 because of its acrid odor. It was not until the 1940s that the pernicious human health effects of air and water pollution were realized beyond their aesthetic and human welfare woes. The appearance of Los Angeles smog or *gas attacks* in the

1940s and London *black fog* in December 1952 that killed more than 4000 clearly indicated the association of air pollution episodes with atmospheric inversions and complex topography. The roles of local pollutant emissions, flow, turbulent mixing, topography, and land use were soon realized in determining the *environmental quality* of urban areas. Thus was born EFD, which has now ramified into a plethora of other problems of the current epoch—the *anthropocene*. The public awareness of human-inflicted permanent damage to the environment started to develop, as exemplified by the 1962 novel by Rachel Carson, *Silent Spring*, and the policymakers became increasingly cognizant of environmental degradation that followed the industrial revolution. The U.S. EPA was established in 1970, and the advent of the Clean Air (1970) and Clean Water (1977) Acts prompted focused, funded research on how air and water flows interact with biogeochemophysical processes, providing a boost to EFD. While the early emphasis was on pollutant emissions, which could be effectively regulated, communities are now faced with additional existential challenges, for example, the hazards of climate change (floods, heat waves, and drainage), rapid urbanization (heat island), and scarcity of water and energy. The quest for economic development without further environmental and resource degradation led, in 1987, to the concept of *sustainability*.

Anthropogenic forcing occurs mainly at scales smaller than the urban scales, and thus are the problems of concern in EFD and for this two-volume handbook. Volume One starts with topics of general relevance (Part I), emphasizing the close relevance of EFD research in societal, policy, infrastructure, quality of life, security, and legal arenas. Some established or rapidly establishing focus areas akin to EFD are discussed next (Part II). Submesoscale flow processes and phenomena that form the building blocks of environmental motions are included in Part III, paying attention to turbulent motions and their role in heat, momentum, and species transport. In Volume Two, engineered structures and anthropogenic activities that affect natural flows and vice versa are discussed in Part I, and in its sequel, Part II, which deals with the most conspicuous anthropogenic effect, environmental pollution. Numerical methodologies that underpin research, predictive modeling, and cyber-infrastructure developments are included in Part III. All quantitative predictions on environmental motions need to be validated by laboratory experiments and field observations, which also help identify new phenomena and processes. These are addressed in Parts IV and V, respectively.

In closing, I would like to acknowledge those who contributed immensely to the preparation of this handbook. Luna Han patiently guided its development despite incessant delays. Jennifer McCulley carried out major administrative chores, assisted in the later stages by Marie Villarreal and Shenal Fernando. The back cover photo was kindly provided by a friend and skillful photographer, Jeff Topping. As always, all anxieties associated with the preparation of this handbook were magnanimously accommodated by my wife, Ravini.

It is my hope that this handbook will be a useful reference to students and professionals in EFD, and the efforts of those who labored to produce it will be handsomely rewarded through readership and dissemination of knowledge.

Harindra Joseph Shermal Fernand
Wayne and Diana Murdy Professor of
Engineering & Geosciences
University of Notre Dame
Notre Dame, Indiana

Editor

Harindra Joseph Shermal Fernando received his BSc (1979) in mechanical engineering from the University of Sri Lanka and his MS (1982) and PhD (1983) in geophysical fluid dynamics from the Johns Hopkins University. He received postdoctoral training in environmental engineering sciences at California Institute of Technology (1983–1984). From 1984 to 2009, he was aDliated with the Department of Mechanical and Aerospace Engineering at Arizona State University, ASU (assistant professor 1984–1987; associate professor 1988–1992; professor 1992–2009). In 1994, Fernando was appointed the founding director of the Center for Environmental Fluid Dynamics, a position he held till 2009, while holding a co-appointment with the School of Sustainability. In January 2010, he joined University of Notre Dame as the Wayne and Diana Murdy Endowed Professor of Engineering and Geosciences, with the primary aDliation in the Department of Civil and Environmental Engineering and Earth Sciences and a concurrent appointment in the Department of Aerospace and Mechanical Engineering.

Among awards and honors he received are the UNESCO Team Gold Medal (1979), Presidential Young Investigator Award (NSF, 1986), ASU Alumni Distinguished Research Award (1997), Rieger Foundation Distinguished Scholar Award in Environmental Sciences (2001), William Mong Lectureship from the University of Hong Kong (2004), and Lifetime Achievement Award from the Sri Lanka Foundation of the USA (2007). He is a fellow of the American Society of Mechanical Engineers, American Physical Society, and American Meteorological Society. He was elected to the European Academy in 2009. In 2007, he was featured in the *New York Times*, *International Herald Tribune*, and other international news media for his work on hydrodynamics of beach defenses. In closing the year 2008, the *Arizona Republic News* honored him by including in "Tempe Five Who Matter"—one of the five residents who have made a notable difference in the life of the city—in recognition of his work on the Phoenix Urban Heat Island.

He has served on numerous national and international committees and panels, including the Sumatra Tsunami Survey Panel (NSF, 2005), Louisiana Coastal Area Science and Technology Board (2006–2011), and the American Geophysical Union Committee on Natural Disasters (2006). He serves on the editorial boards of *Applied Mechanics Reviews* (associate editor), *Theoretical and Computational Fluid Dynamics*, IAHR *Journal of Hydro-environment Research* (associate editor), and *Nonlinear Processes in Geophysics* (editor). He is the editor in chief of *Environmental Fluid Dynamics*. He has published more than 225 papers spanning nearly 50 different international peer-review journals covering basic fluid dynamics, experimental methods, oceanography, atmospheric sciences, environmental sciences and engineering, air pollution, alternative energy sources, acoustics, heat transfer and hydraulics, river hydrodynamics, and fluids engineering. He is also the editor of the books entitled *National Security and Human Health Implications of Climate Change* (2012) and *Double DiTusive Convection* (1994).

Contributors

Sultan lam
SOGREAH
Le Beausset, France

Vincenzo Armenio
Department of Civil and Environmental
    Engineering
University of Trieste
Trieste, Italy

Jong-Jin Baik
School of Earth and Environmental
    Sciences
Seoul National University
Seoul, Korea

Sukanta Basu
Department of Marine, Earth, and
    Atmospheric Sciences
North Carolina State University
Raleigh, North Carolina

F.A. Bombardelli
Department of Civil and Environmental
    Engineering
University of California, Davis
Davis, California

ReE B ritter
Department of Engineering
University of Cambridge
Cambridge, United Kingdom

and

Senseable City Laboratory and Building
    Technology Group
Massachusetts Institute
    of Technology
Cambridge, Massachusetts

Christopher Butler
School of Engineering
University of California, Merced
Merced, California

Daewon W. Byun
Air Resources Laboratory
National Oceanic and Atmospheric
    Administration
Silver Spring, Maryland

Jean-Luc Caccia
Laboratoire de Sondages
    Electromagnétiques de lfEnvironment
    Terrestre
University of the South, Toulon-Var
La Garde, France

Ronald J. Calhoun
School for Engineering
    of Matter, Transport
    and Energy
Arizona State University
Tempe, Arizona

Meredith L. Carr
Cold Regions Research and
    Engineering Laboratory
United States Army Corps
    of Engineers
Hanover, New Hampshire

Olivier Cazaillet
SOGREAH
Echirolles, France

Falin Chen
Institute of Applied Mechanics
National Taiwan University
Taipei, Taiwan, Republic of China

Fei Chen
Research Applications Laboratory
National Center for Atmospheric
    Research
Boulder, Colorado

Jason Ching
Institute for the Environment
University of North Carolina
Chapel Hill, North Carolina

Stuart B. Dalziel
Department of Applied Mathematics and
    Theoretical Physics
University of Cambridge
Cambridge, United Kingdom

Olaf David
Department of Civil and Environmental
    Engineering
and
Department of Computer Science
Colorado State University
Fort Collins, Colorado

M.J. Davidson
Department of Civil and Natural
    Resources Engineering
University of Canterbury
Christchurch, New Zealand

Cecelia Deluca
Cooperative Institute
    for Research in the
    Environmental Sciences
University of Colorado, Boulder
Boulder, Colorado

SilvanaD i Sabatino
Department of Materials Science
University of Salento
Lecce, Italy

P. Diplas
Department of Civil and
    Environmental Engineering
Virginia Polytechnic Institute and State
    University
Blacksburg, Virginia

Marie Farge
LMD–IPSL–CNRS
École Normale Supérieure
Paris, France

Morris R. Flynn
Department of Mechanical
    Engineering
University of Alberta
Edmonton, Alberta, Canada

Christophe C. Frippiat
Division of Post-mining and Geological
    Risks
Scientific Institute for Public Services
Liège, Belgium

Marcelo H. García
Civil and Environmental
    Engineering
University of Illinois,
    Urbana-Champaign
Urbana, Illinois

Jonathañ G oodall
Department of Civil
    and Environmental
    Engineering
University of South Carolina
Colombia, South Carolina

RoG urka
Department of Mechanical
    Engineering
Ben Gurion University of the Negev
Beer-Sheva, Israel

I omas C. Harmon
School of Engineering
University of California, Merced
Merced, California

Chad W. Higgins
Department of Biological
    and Ecological
    Engineering
Oregon State University
Corvallis, Oregon

Tissa H. Illangasekare
Department of Civil and Environmental
    Engineering
Colorado School of Mines
Golden, Colorado

Hong-Ming Jang
Department of Mechanical
    Engineering
Chinese Culture University
Taipei, Taiwan, Republic of China

Ahsan Kareem
Department of Civil Engineering and
    Geological Sciences
University of Notre Dame
Notre Dame, Indiana

Nigel Berkeley Kaye
Glenn Department of Civil Engineering
Clemson University
Clemson, South Carolina

A. Khosronejad
Department of Civil Engineering
University of Minnesota
Minneapolis, Minnesota

Eliezer Kit
School of Mechanical Engineering
Tel Aviv University
Tel Aviv, Israel

G.A. Luz
Luz Social and Environmental Associates
Baltimore, Maryland

Alex Mahalov
Department of Mechanical and
    Aerospace Engineering
and
Department of Mathematics and Statistics
Arizona State University
Tempe, Arizona

J. Ezequiel Martin
Department of Civil and Environmental
    Engineering
University of Illinois,
    Urbana-Champaign
Urbana, Illinois

Mohamed Moustaoui
Department of Mechanical and
    Aerospace Engineering
and
Department of Mathematics and
    Statistics
Arizona State University
Tempe, Arizona

Romain Nguyen van yen
LMD–IPSL–CNRS
École Normale Supérieure
Paris, France

E.T. Nykaza
Engineer Research and Development
    Center
United States Army Corps of Engineers
Champaign, Illinois

C.J. Oliver
Department of Civil and Natural
    Resources Engineering
University of Canterbury
Christchurch, New Zealand

Olivier Pannekoucke
National Centre for Meteorological
    Research
Toulouse, France

Eric R. Pardyjak
Department of Mechanical Engineering
University of Utah
Salt Lake City, Utah

Marc B. Parlange
School of Architecture, Civil and
    Environmental Engineering
Ecole Polychnique Federal de
    Lausanne
Lausanne, Switzerland

L.L. Pater
Engineer Research and Development
    Center
United States Army Corps of Engineers
Champaign, Illinois

I omas Peacock
Department of Mechanical Engineering
Massachusetts Institute of Technology
Cambridge, Massachusetts

ScottD P eckham
Institute of Arctic and Alpine Research
University of Colorado, Boulder
Boulder, Colorado

ChriR R ehmann
Department of Civil, Construction, and
 Environmental Engineering
Iowa State University
Ames, Iowa

Philip J.WR b erts
School of Civil and Environmental
 Engineering
Georgia Institute of Technology
Atlanta, Georgia

Elena Roget
Department of Physics
University of Girona
Catalonia, Spain

M. Roth
Department of Geography
National University of Singapore
Singapore, Singapore

C.A. Sanchez
Department of Soil, Water and
 Environmental Science
University of Arizona
Tucson, Arizona

Sutanu Sarkar
Department of Mechanical and
 Aerospace Engineering
University of California, San Diego
San Diego, California

KennethL S cher(in Memoriam)
Atmospheric Modeling and Analysis
 Division
United States Environmental
 Protection Agency
Research Triangle Park, North Carolina

Kai Schneider
Aix-Marseille University
Marseille, France

Scott A. Socolofsky
Zachry Department of Civil
 Engineering
Texas A&M University
College Station, Texas

F. Sotiropoulos
Department of Civil Engineering
University of Minnesota
Minneapolis, Minnesota

David R. StauJ er
Department of Meteorology
Pennsylvania State University
University Park, Pennsylvania

M.E. Swearingen
Engineer Research and Development
 Center
United States Army Corps
 of Engineers
Champaign, Illinois

J.P.M. Syvitski
Institute of Arctic and Alpine Research
University of Colorado, Boulder
Boulder, Colorado

Josef Tanny
Institute of Soil, Water, and
 Environmental Sciences
Agricultural Research Organization
Bet-Dagan, Israel

Gerhard I eurich
Science Applications International
 Corporation
McLean, Virginia

Akula Venkatram
Department of Mechanical
 Engineering
University of California, Riverside
Riverside, California

JeJ rey C. Weil
Cooperative Institute for Research
 in Environmental Sciences
University of Colorado, Boulder
and
National Center for Atmospheric
 Research
Boulder, Colorado

M.J. White
Engineer Research and Development
 Center
United States Army Corps
 of Engineers
Champaign, Illinois

C.H.K. Williamson
Mechanical and Aerospace
 Engineering
Cornell University
Ithaca, New York

Clinton S. Willson
Department of Civil and Environmental
 Engineering
Louisiana State University
Baton Rouge, Louisiana

D.K. Wilson
Engineer Research and Development
 Center
United States Army Corps of
 Engineers
Hanover, New Hampshire

Poojitha D. Yapa
Department of Civil and Environmental
 Engineering
Clarkson University
Potsdam, New York

D. Zerihun
Department of Soil, Water and
 Environmental Science
University of Arizona
Tucson, Arizona

I

Engineered Systems and Anthropogenic Influence

1

Water Distribution Systems

F.A. Bombardelli
University of California, Davis

1.1 Introduction

Water resources have been essential in the development of mankind and play a fundamental role in the way people relate to the environment. At the individual level, humans need drinking water for daily survival. Water regulates physical and biochemical processes within the human body, ranging from the digestion of food and the control of the body temperature, to the elimination of body wastes. Modern life in the developed world takes for granted the availability of water for basic uses such as drinking, cleaning, and cooking. Water bodies are also exploited for power generation, recreation, irrigation, etc.

Communities have historically developed in close proximity to water bodies. Numerous examples across diverse continents include early villages of Mesopotamia (near the Euphrates and Tigris Rivers), Egypt (in the valley of the Nile River), and Crete (Mays 2007, 2008). Nearby water courses allowed members of the community to obtain fresh water for basic uses, and to discharge their wastes. When communities were of relatively small size, their impact on the environment was relatively low. However, with the historical growth of population and the development of more formal settlements, a strong need for large-scale infrastructure arose. Hydraulic structures were needed to collect water from neighboring water bodies, store it, and distribute it over large distances. According to Mays (2007), rivers and springs were both exploited in ancient Greece, and hydraulic structures such as wells, cisterns, and aqueducts were constructed during the Minoan culture, circa 2900–2300 BC.

Cisterns were sometimes built to store rainfall collected from roofs and courtyards. The city of Tylissos of the Minoan era was built with an aqueduct made of pipes and stone channels, and a sediment tank. Today, it is possible to see relics of these structures (Mays 2007). Figure 1.1 shows portions of the water system in Tylissos. These hydraulic structures can be considered as sample components of what it is known today as water supply systems (WSSs) and water distribution systems (WDSs). One of the milestones in the development of ancient WDSs is the tunnel of Eupalinos on Samos Island, built by the Greek circa 530 BC, the first deep tunnel in history whose construction started from two different openings (Mays 2008). Aqueducts and terracotta pipes became common infrastructure of Greek cities, indicating that water supply was an essential aspect of the welfare of those communities. Greek water systems also used inverted siphons, lead pipes to withstand larger pressures and, for the first time, Archimedes screws (Archimedes, 287–212 BC) and "force pumps" (Mays 2008).

The Romans developed technology to obtain water not only from wells and springs, but also by damming rivers. Large aqueducts were part of massive supply systems built by the Romans, and many of them can still be seen in Spain, France, Rome, and North Africa (Chanson 2002; Mays 2008). Romans built tapped water storage tanks called *castella* that served multiple groups of people at once. The city of Pompeii (located in the Bay of Naples, Italy) developed a WDS that followed a typical Roman design, with pipes to supply water to a few private baths (Mays 2008). Romans also developed diverse types of valves and pumps (Mays 2008).

Handbook of Environmental Fluid Dynamics, Volume Two, edited by Harindra Joseph Shermal Fernando. © 2013 CRC Press/Taylor & Francis Group, LLC. ISBN: 978-1-4665-5601-0.

(a)   (b)

FIGURE 1.1 Images of relics of the water system in Tylissos, in ancient Crete. (a) Aqueduct to transport water from springs. (b) Sedimentation deposit (in foreground) with stone channel connecting to cistern. (Photo courtesy of L.W. Mays.)

Another major contribution of Romans to the water supply of cities was the development of drop shafts within aqueducts. These structures served several purposes: (1) accommodating sharp drops in topography; (2) dissipating kinetic energy; and (3) aerating the flow (Chanson 2002).

Figure 1.2 shows a schematic of water supply and wastewater management systems in a generic, modern city. The left portion of the schematic shows the extraction of water from the river (by an unspecified method in the diagram), as well as from a reservoir, and wells. Water obtained from these sources is transported to a water treatment plant and then pumped toward houses and industries. Historically, WDSs have been associated (in a restricted sense) only with the set of hydraulic structures (mostly tanks and pipes) that convey water to houses and industries. According to this notion, a WDS is composed mainly by pressurized flows. Mays (2004) states that WDSs possess three

main components: pumping systems, storage, and distribution piping. It is possible to add to the description of WDSs discussed previously, the structures that extract water from rivers, reservoirs and wells, and the water treatment plant, obtaining a more general definition of WDSs. Therefore, in this chapter, WDSs are understood in a general sense as a set of canals, pipes, water tower/s and other hydraulic structures designed to provide water to urban areas, *from the source to the consumer*, be it for domestic or industrial use.

In some instances the WDS may cover an area a few kilometers across (say, less than 25 km), while in other situations the pipes, channels, pumping systems, storing tanks, and other structures may cover hundreds of kilometers. Such is the case of the WDS for the Los Angeles area, in California. Water is "exported" south from the lower portion of the Sacramento–San Joaquin Delta (located in the Central Valley of the state), by a

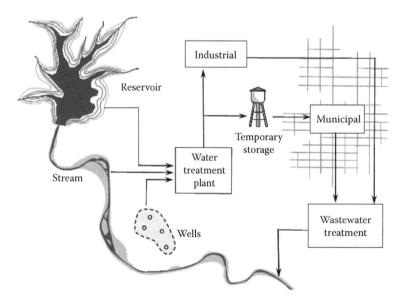

FIGURE 1.2 Schematic of water supply, distribution, and wastewater management systems in a modern city. (Adapted from Masters, G., *Introduction to Environmental Engineering and Science,* Prentice Hall, New York, 1997; Mays, L.W., *Water Supply Systems Security*, McGraw-Hill Professional Engineering, New York, 2004.)

FIGURE 1.3 Pictures of the water supply system of the Los Angeles area. (a) California aqueduct. (Used under the GNU project.) (b) San Luis reservoir (http://www.water.ca.gov/swp). (c) Pyramid Lake (http://www.water.ca.gov/swp). (d) Edmonston pumping plant (http://www.water.ca.gov/swp).

large pumping system. The water transport system is composed of an approximately 500 km long aqueduct that includes several pumping stations along the way. Figure 1.3 shows diverse portions of channels, pipes and pumping systems for the WDS of the Los Angeles area.

A common feature of all hydraulic structures of WDSs is the presence of water flows of *turbulent nature*, which overcome the resistance to flow produced by solid boundaries. Most of these flows can be classified as shear flows, where the velocity is essentially one dimensional, but varies in space in a direction normal to the main direction of motion. Shear flows include wakes, boundary layers, submerged jets, open-channel and pipe flows, etc. (Davidson 2005). This chapter reviews the knowledge required to design hydraulic structures in WDSs, with an emphasis on the quantification of the resistance to flow. It also presents recent theoretical results which connect traditional design formulations to modern analyses of turbulence. These ideas are considered to be very important in everyday engineering as well as in research. The chapter aims at bridging the gap between practitioners and scientists.

1.2 Principles and Methods of Analysis

Current knowledge and understanding of resistance to flow in pipes and open channels is the result of contributions by many scientists and engineers over the last three centuries. These contributions can be broadly classified as coming from the classic field of hydraulics (essentially during the 1700s and 1800s; Dooge 1992), and those coming from the traditional fluid mechanics field (1800s and 1900s; Kundu and Cohen 2008). It is interesting to note that the empirical contributions by engineers in the field of hydraulics were made prior to the seminal

experiments of Reynolds in 1883 (O. Reynolds, 1842–1912), who addressed the existence of two main flow regimes: laminar and turbulent (Pope 2000, p. 5).

1.2.1 Empirical Equations for the Quantification of Flow Resistance

Early systematic developments regarding flow in pipes and open channels were based mainly on empiricism. The equations most widely used to predict the cross-sectionally averaged flow velocity (U) in open channels and pipes include (Yen 1992a,b, 2002; Bombardelli and García 2003) the following:

- Manning's equation (R. Manning, 1830–1920)

$$U = \frac{K_n R_h^{2/3} S^{1/2}}{n} \tag{1.1}$$

- Dimensionally homogeneous Manning formula (Yen 1992a; B. C. Yen, 1935–2001)

$$U = \frac{g^{1/2} R_h^{2/3} S^{1/2}}{n_g} \tag{1.2}$$

- Chèzy's equation (A. de Chèzy, 1718–1798)

$$U = C R_h^{1/2} S^{1/2} \tag{1.3}$$

- Darcy–Weisbach equation (H. Darcy, 1803–1858; J. Weisbach, 1806–1871)

$$U = \left(\frac{2g}{f}\right)^{1/2} D^{1/2} \left(\frac{h_f}{\Delta L}\right)^{1/2} \tag{1.4}$$

ft   Hazen–Williams equation

$$U = K_{HW} C_{HW} R_h^{0.63} S^{0.54} \qquad (1.5)$$

where

> R_h is the hydraulic radius (i.e., the ratio between the wetted area and the wetted perimeter; Chow 1959)
> S denotes the slope of the energy grade line
> D indicates the pipe diameter
> ΔL refers to the length of the pipe segment
> h_f is the energy loss in the pipe segment (expressed per unit weight of fluid)
> g is the acceleration of gravity
> K_n and K_{HW} are both unit conversion factors
> $n, n_g, C, f,$ and C_{HW} indicate the *resistance/conveyance coe cients*

K_n is $1\,\mathrm{m}^{1/2}/\mathrm{s}$ in the International System (SI) or $1.486\,\mathrm{ft}^{1/3}\,\mathrm{m}^{1/6}/\mathrm{s}$ in English units. Using either system, Manning's n should have units of $\mathrm{m}^{1/6}$ (Yen 1992a, 2002). K_{HW} equals 0.849 and 1.318 for the SI and English units, respectively (Jeppson 1977).

It is well known that the Darcy–Weisbach and Manning formulas can both be used for computations in either open-channel or pipe, fully rough turbulent flow, provided the equivalence $D = 4R_h$ (for a circular pipe) is considered, and reliable estimates for the resistance coe cients are available (Yen 1992b, 2002; Bombardelli and García 2003). (Fully rough flow is defined as the flow regime in which the roughness of the boundary controls the flow behavior; see Section 1.2.2.1.) The Chèzy equation can also be employed in either case, also for turbulent, fully rough flow. Based on these ideas, it is possible to state the following equivalence (Yen 1992b, 2002; Bombardelli and García 2003):

$$\sqrt{\frac{8}{f}} = \frac{C}{\sqrt{g}} = \frac{K_n R_n^{1/6}}{\sqrt{g}\,n} = \frac{R_n^{1/6}}{n_g} = \frac{U}{\sqrt{gR_n S}} = \frac{U}{u_*} \qquad (1.6)$$

with u_* indicating the wall-friction (shear) velocity, defined as the square root of the ratio between the shear stress at the wall and the fluid density.

The Hazen–Williams formula, in turn, has quite a restricted range of application that is often overlooked in practice. Diskin (1960) determined the ranges of Reynolds numbers, Re (i.e., the product of the flow velocity, U, and a flow length scale, L, divided by the kinematic viscosity of the fluid, ν) for which the Hazen–Williams formula is applicable, concluding that the expression is *not* valid for fully rough flow. In practical terms, Diskin also found that the use of the formula, when applied in the appropriate range, should result in values of C_{HW} between 100 and 160 (see also Bombardelli and García 2003).

Although many engineers and scientists tend to regard the Darcy–Weisbach formula as "more rationally based than other empirical exponential formulations" (Streeter et al. 1998), such an idea is somewhat misleading. In fact, all expressions can be obtained from dimensional analysis and in the end experiments

are required to measure the resistance coe cients (Yen 1992b, 2002; Gioia and Bombardelli 2002).

Example 1.1 (on the use of the ow resistance equations)

Compute the discharge (the volumetric flow rate) in an open channel of rectangular cross section, with a width of 2.5 m, and a slope of 0.002, for a water depth of 1.193 m, and $n = 0.02\,\mathrm{m}^{1/6}$.

Knowing n, and calculating the hydraulic radius ($R_h = 0.61\,\mathrm{m}$), it is possible to obtain all coe cients from Equation 1.6, as follows: $C = 46\,\mathrm{m}^{1/2}/\mathrm{s}$, $f = 0.037$. The velocity can then be obtained from (1.1) through (1.4), being equal to 1.61 m/s; the discharge is $4.8\,\mathrm{m}^3/\mathrm{s}$.

1.2.2 Semi-Logarithmic Law for Flow Velocities Close to Walls, and Friction Laws for Turbulent Flow in Pipes and Open Channels

1.2.2.1 Flow Regimes in Pipes and Open Channels

Every solid boundary presents a certain degree of roughness, even those that seem smooth to the touch. From a fluid mechanics standpoint, this roughness exists as protrusions or indentations of the solid boundary acting on the flow, and imposing a drag on it. The drag on the flow will depend on the density, distribution, and size of the roughness elements on the surface of the pipe or channel walls (Schlichting 1968). Although the roughness elements are in reality nonuniform in size and location, it is customary to use a unique length scale, k, to represent such roughness. It is certainly remarkable that a relatively small length scale, when compared to other scales in the flow, can play such a tremendous role in defining the flow features close to solid walls under certain flow conditions (Gioia and Bombardelli 2002).

Experiments performed in the 1920s on pipes and open channels under *turbulent* flow conditions allowed for the identification of three main types of flow responses to the roughness elements (Schlichting 1968). In the first one, which occurs for instance in cement channels or cast iron pipes, the resistance to the flow only depends on the *relative roughness*, k/L_R, with L_R denoting either the pipe diameter or the hydraulic radius in open-channel flow. This regime is called *turbulent, fully rough ow*. The second type of response occurs when the protrusions are relatively smaller, or are distributed over larger wall areas, as is the case of the flow in wooden or commercial steel pipes (Schlichting 1968). Under this condition, *both* k/L_R and the Reynolds number characterize the level of the resistance to flow. This regime is called *transitional ow*. The third type of flow is the *smooth regime*, where the flow resistance is dictated solely by the Reynolds number, in a similar fashion to what occurs in the Hagen–Poiseuille (laminar) pipe flow (Schlichting 1968, p. 80).

It turns out that for the same relative roughness, k/L_R, the flow behavior could vary from one regime to another depending on the flow condition. A dimensionless parameter is used to indicate the different flow regimes, obtained by dividing the roughness height, k, with a *viscous length scale*, ν/u_*, the ratio of the kinematic viscosity of water, ν, and the wall-friction (shear) velocity, u_*; i.e., $k\,u_*/\nu$. The different regimes are delimited as follows (Schlichting 1968):

Smooth regime:

$$0 \leq \frac{ku_*}{\nu} < 5 \tag{1.7}$$

Transitional regime:

$$5 \leq \frac{ku_*}{\nu} \leq 70 \tag{1.8}$$

Fully rough regime:

$$\frac{ku_*}{\nu} > 70 \tag{1.9}$$

The viscous length scale also defines the size of the *viscous sublayer*, δ_v, a very thin region close to the wall where viscosity dominates over inertial forces. It must be emphasized that the flow in this region is indeed *turbulent*; not laminar as stated in old books on fluid mechanics (see Davidson 2005, p. 130). Experimental results by Nikuradse (J. Nikuradse, 1894–1979) indicated that the viscous sub-layer can be computed as $\delta_v = 11.6\,\nu/u_*$. The different flow regimes (embedded in Equations 1.7 through 1.9) explain that when the flow is smooth, the viscous sub-layer, δ_v, is larger than k, while it is smaller than k for fully rough flow. However, this explanation has been challenged recently, and is further discussed in Section 1.3.2.

There have been attempts to link the values of the resistance/conveyance coeDcients with the roughness height values, mainly Manningß n. One such relationship was developed in 1923 by Strickler (Strickler 1981; A. Strickler, 1887–1963), as follows:

$$n = \frac{k^{1/6}}{21.1} \tag{1.10}$$

with $k = d_{50}$ computed in meters. Very recently, Travis and Mays (2007) presented an expression connecting the Hazen–Williams C_{HW} with k:

$$k = D\left(3.32 - 0.021C_{HW}D^{0.01}\right)^{2.173} e^{-0.04125C_{HW}D^{0.01}} \tag{1.11}$$

1.2.2.2 Velocity Distribution in Pipes and Open Channels and Expressions for the Friction Factor

It took natural philosophers many centuries to realize that the flow velocity vector at a stationary and impervious wall, right at the wall, was zero. This condition is known today as the *no-slip*

and impermeability conditions (Pope 2000, p. 17). It also took the efforts of many researchers to address the nature and features of the velocity distribution close to a solid wall as a function of the distance from the wall. *The Vbw suﬃciently close to solid boundaries at large Reynolds numbers is very similar for channels, pipes, and Vat-plate boundary layers* in a region where the flow is statistically stationary and does not vary with the coordinate in the direction of motion. This region is called the *fully developed* flow region.

The total shear stress (τ) within the fluid at a given distance z from the wall is the result of stresses coming from molecular origin (the viscous stresses), and the Reynolds stresses ($-\rho\overline{u'w'}$) which arise as a consequence of turbulence (Pope 2000). In this expression for the Reynolds stresses, u' and w' indicate the velocity fluctuations in x and z, respectively, ρ is the density of water, and the overbar indicates average over turbulence. At the wall, the Reynolds stresses are zero, because of the no-slip and impermeability conditions (Pope 2000). Therefore, the contribution from viscosity is the only stress remaining for the total shear stress at the wall (τ_w):

$$\tau_w \equiv \rho\nu\left(\frac{d\overline{u}}{dz}\right)_{z=0} \tag{1.12}$$

Away from the wall, turbulent (Reynolds) stresses become prevalent, and the viscous stresses decay accordingly.

Immediately close to the wall, where viscosity dominates, the distances are better described in viscous coordinates—also called "wall units":

$$z^+ \equiv \frac{zu_*}{\nu} \tag{1.13}$$

where ν/u_* is the viscous length scale defined in the previous section. Far from the wall, $\eta = z/\delta$ provides a better scaling of distances. Here, δ is a measure of the boundary-layer thickness, given by the pipe radius in pipe flow (Schlichting 1968). The distance from the wall relative to the viscous length scale, that is, z^+, and η help identifying different regions of flow behavior close to the wall. First, there is a consensus among researchers that there is a *viscous wall region* for $z^+ < 50$, where viscosity significantly affects the shear stress, and an *outer layer* for $z^+ > 50$, where viscosity effects are negligible. Some researchers have also suggested the existence of subregions within the aforementioned two main regions (Pope 2000; Jiménez 2004). The viscous sub-layer holds for $z^+ < 5$, and there is an *inner layer* which reaches up to $z^+ = 1000$ or up to $\eta = 0.1$. This indicates that there is an *overlap region* of the inner and outer layers from $z^+ = 50$ to 1000 (see Figure 1.4a).

When concerned with the time-averaged velocity distribution in the fully developed flow region, the important variables are: the water density, ρ, the kinematic viscosity of water, ν, the wall-friction (shear) velocity, u_*, and the thickness of the boundary

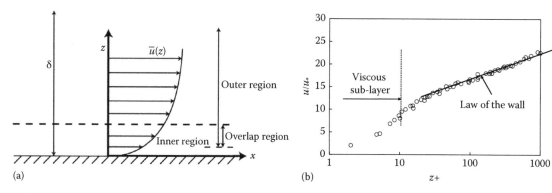

FIGURE 1.4 (a) Schematic of the regions close to a wall. (Adapted from Davidson, P.A., *Turbulence*, Oxford University Press, New York, p. 128, 2005.) (b) Comparison of time-averaged velocity data for fully developed, turbulent channel flow (compiled by Pope from observations by Wei and Willmarth 1989) with the law of the wall. Data pertain to Reynolds numbers ranging from 2,970 to 39,582. (Adapted from Pope, S.B., *Turbulent Flows*, Cambridge University Press, Cambridge, U.K., p. 275, 2000.)

layer, δ. Using dimensionless numbers, the aforesaid variables result in (Pope 2000)

$$\bar{u}(z) = u_* F\left(\frac{z}{\delta}, \text{Re}\right) \quad (1.14a)$$

where $F(\,)$ denotes a universal, nondimensional function. However, it has been shown to be more convenient to work with the time-averaged velocity gradient, because the velocity gradient determines both the viscous stress and the production of turbulence (Pope 2000). Thus,

$$\frac{d\bar{u}(z)}{dz} = \frac{u_*}{z}\phi\left(\frac{zu_*}{\nu}, \frac{z}{\delta}\right) \quad \text{for } \frac{z}{\delta} \ll 1 \quad (1.14b)$$

In the inner region, Prandtl (L. Prandtl, 1875–1953) postulated that the mean velocity gradient is determined by the viscous scale and, thus,

$$\frac{d\bar{u}(z)}{dz} = \frac{u_*}{z}\phi(z^+) \quad \text{for } \frac{z}{\delta} \ll 1 \quad (1.15)$$

with ϕ referring to an unknown function. Integrating (1.15) in the wall-normal direction,

$$u^+ = \frac{\bar{u}(z)}{u_*} = f_w(z^+) = \int_0^{z^+} \frac{1}{z'}\phi(z')dz' \quad \text{for } \frac{z}{\delta} \ll 1 \quad (1.16)$$

As previously mentioned, flow features are *not* dictated by viscosity *above* the inner layer. Therefore, ϕ should not be dependent on δ_ν, which *is* indeed a function of viscosity. Mathematically, this is stated as

$$\phi(z^+) = \frac{1}{\kappa} \quad \text{for } \frac{z}{\delta} \ll 1 \text{ and } z^+ \gg 1 \quad (1.17)$$

where κ is von Kármán constant (T. von Kármán, 1881–1963). This leads to

$$\frac{d\bar{u}(z)}{dz} = \frac{u_*}{z\kappa} \quad \text{for } \frac{z}{\delta} \ll 1 \quad (1.18)$$

which, upon integration gives

$$\frac{\bar{u}(z)}{u_*} = \frac{1}{\kappa}\ln z^+ + C_1 \quad (1.19)$$

Equation 1.19 is known as the logarithmic *law of the wall*, log law, or semi-logarithmic velocity law for a smooth boundary, developed by von Kármán. It is generally accepted that κ is about 0.4 (Davidson 2005), and the constant of integration is $C_1 = 5.2$–5.5. Comparison with experimental data indicates that the predictive capability of the law of the wall is excellent for $z^+ > 30$ and $z/\delta < 0.2$ (see Figure 1.4b). This law has been found to provide accurate predictions for many shear flows (Pope 2000).

Using a Taylor series expansion, it can be shown that, for small values of z^+, $f_w(z^+) = z^+ + O(z^{+2})$, where $O(h)$ indicates the quantity of big order h (Pope 2000). From highly resolved simulation results of Kim et al. (1987), it is possible to conclude that $f_w(z^+) = z^+$ for $z^+ < 5$—the viscous sub-layer region (Pope 2000). This leads to a linear distribution of the velocity along the wall-normal direction in this region (or curved in a semi-logarithmic plot; see Figure 1.4b).

When the flow behavior is fully rough, the roughness height becomes a parameter that *cannot* be disregarded. A combined expression that describes smooth, transitional and fully rough behaviors is as follows (White 1974):

$$\frac{\bar{u}(z)}{u_*} = \frac{1}{\kappa}\ln z^+ + 5.5 - \frac{1}{\kappa}\ln(1 + 0.3k^+) \quad (1.20)$$

where $k^+ = ku_*/\nu$. Equation 1.20 can be immediately interpreted as the result of subtracting a "roughness function" on k^+ from Equation 1.19. Thus, the velocity distribution for a rough boundary has the same slope as the velocity distribution for a smooth one, but it is shifted downward (White 1974, p. 489). An equation that describes solely the fully rough regime is as follows:

$$\frac{\bar{u}(z)}{u_*} = \frac{1}{\kappa}\ln\frac{z}{k} + 8.5 \quad (1.21)$$

Additional information regarding laws of the time-averaged velocity in open channels and pipes can be found in Schlichting (1968), White (1974), Yen (1992b), Pope (2000), and Davidson (2005, p. 130). Some authors have recently questioned the conceptual rigor, accuracy, and universality of the logarithmic law of the wall, and have suggested power laws for the flow velocity instead (Schlichting 1968; Yen 1992b; Barenblatt et al. 2000). Others have focused on the lack of universality of the logarithmic law of the wall with constant parameters, but have stated its standing as a "robust and eﬃcient workhorse for engineering applications" (Buschmann and Gad-el-Hak 2003, 2007, 2009).

Integrating Equations 1.19 through 1.21, or other similar ones, gives the cross-sectionally averaged velocity, U. These results can then be compared to the empirical formulations of flow resistance of Section 1.2.1 (Chen 1992). In fact, integrating Equation 1.21 from k to the water depth, H, results in an equation which provides close results to Manning's power law. The resulting equation is $U = u_*/\kappa \ln (11H/k)$, known as Keulegan's resistance relation for rough flow (García 1999; Bombardelli 2010). This outcome offers an interesting and noteworthy verification of the legitimacy of Manning's empirical expression for fully rough flows.

Integrating Equation 1.19, and applying (1.6), it is possible to obtain, after small corrections,

$$\frac{1}{\sqrt{f}} = 2\log_{10}\left(\sqrt{f}\,\mathrm{Re}\right) - 0.8 \qquad (1.22)$$

Equation 1.22 is known as the Prandtl's law for smooth pipes, and gives an implicit expression for the friction factor f as a function of the Reynolds number. Results obtained with Equation 1.22 agree with experimental data for a large range of values of the Reynolds number. Results obtained with (1.22) are also in close agreement with values obtained with the following empirical equation proposed by Blasius in 1913 (P. R. H. Blasius, 1883–1970), up to Reynolds numbers of approximately 100,000 (Schlichting 1968, p. 573):

$$f = 0.3164\,\mathrm{Re}^{-1/4} \qquad (1.23)$$

beyond which the Blasius equation deviates from the data.

In a paper published in 1937, Colebrook and White proposed a formula for the friction factor, f, characterizing the flow in the smooth, transitional and fully rough regimes. This formula gives f as a function of the relative roughness, k/D, and the Reynolds number, Re, as follows:

$$\frac{1}{\sqrt{f}} = -2\log\left(\frac{k}{D}\frac{1}{3.7} + \frac{2.51}{\mathrm{Re}\sqrt{f}}\right) \qquad (1.24)$$

This is also an implicit equation. Swamee and Jain (1976) developed in turn an *explicit* equation for the friction factor, written as follows:

$$f = \frac{0.25}{\left[-\log\left(\dfrac{k}{D}\dfrac{1}{3.7} + \dfrac{5.76}{\mathrm{Re}^{0.9}}\right)\right]^2} \qquad (1.25)$$

1.2.2.3 Nikuradse, and Moody/Rouse Diagrams for the Friction Factor of Flow in Pipes, and Equivalent Diagram for the Friction Factor of Flow in Open Channels

One of the pioneering and weightier experimental contributions regarding flow resistance in pipes is that of Nikuradse. Nikuradse reported values of the friction factor in a wide range of pipe flow conditions, as a function of the Reynolds number and the relative roughness (Nikuradse 1933). In his distinctive and classic experiments, sand grains of *uniform* size (denoted as k_s), were glued to the walls of pipes (of radius R) in a dense, compact manner, covering a range of ratios k_s/R between 1/15 and 1/500. Using uniform, compact layers of sand grains proved to be crucial in obtaining a simple definition of the roughness height with just *one* parameter (i.e., there was no need for the specification of density or area distribution of the grains, which are factors usually overlooked in works regarding flow in commercial pipes). Figure 1.5 shows the experimental results of Nikuradse for laminar and turbulent flow. Six curves characterize the flow in the right-hand side of the diagram, for different k_s/R values. In that portion of the chart, the curves are horizontal, indicating that the Reynolds number does not exert influence on the friction factor. These curves denote fully rough flow behavior and follow the so-called Strickler scaling, that is, $f \sim (k_s/R)^{1/3}$, which can be obtained from Equations 1.6 and 1.10 (see also Gioia and Bombardelli 2002). (The symbol "~" means "scales with.") Beyond Re $\approx$ 3000 ($\approx$3 in log-scale) within the turbulent regime, the experimental points show higher values of f as the Re increases, where the curves form a "bundle" followed by a "hump" (Gioia and Chakraborty 2006). After the "hump," the bundle curves down toward the regression for smooth flows

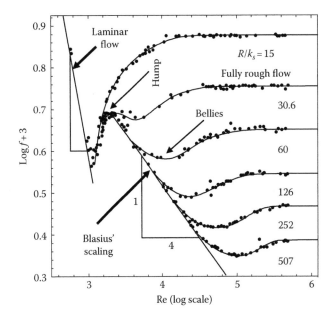

FIGURE 1.5 Nikuradse's diagram for circular pipes. (Adapted from Gioia, G. and Chakraborty, P., *Phys. Rev. Lett.*, 96, 044502, 2006.) In this diagram, the roughness is k_s, obtained as the result of gluing sand grains to the pipe walls.

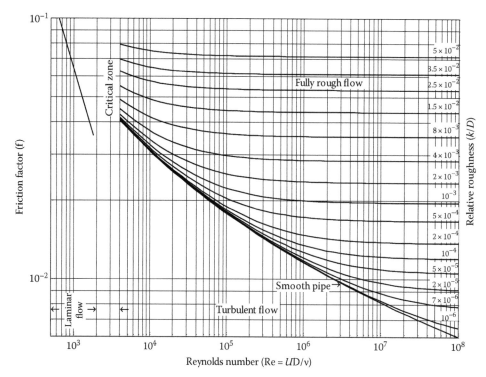

FIGURE 1.6 Moody/Rouse diagram for circular commercial pipes, obtained by plotting Equation 1.25 for the turbulent regime and $f = 64/\text{Re}$ for the laminar regime.

(Blasius scaling, $f \sim \text{Re}^{-1/4}$, Equation 1.23). The diverse curves deviate from the smooth-flow curve forming a set of "bellies" (Gioia and Chakraborty 2006) which continue into the fully rough behavior lines. This region of the diagram represents the transitional regime. Further details about Nikuradse's experiments can be found in Schlichting (1968).

In an effort to characterize the resistance to flow in commercial pipes, Moody (1944) presented a plot of *empirical* relations for turbulent flows developed by Blasius (1913), Prandtl (1935), Colebrook (1939), and Colebrook and White (1937). The idea for such a diagram had been previously suggested by Rouse (H. Rouse, 1906–1996), although he used different variables plotted on each axis. The Moody/Rouse chart is reproduced as Figure 1.6. The transitional regime curves do not show the inflection portion (i.e., neither the "bellies," nor the "hump") seen in the Nikuradse's chart—an intriguing feature. This discrepancy does not have a clear explanation.

Colebrook and White (1937) argued that in commercial pipes with varying roughness heights, the largest roughness elements are responsible for the point of departure of the curves from the smooth line while the point of collapse with the rough curves is determined by the smallest roughness elements (Langelandsvik et al. 2008).

Experimental results on the friction factor for noncircular pipes showed a dependence on shape of the cross section (Schlichting 1968, p. 576). This dependence of the friction factor on the shape of the cross section also appears for open-channel flow.

Yen (1992b, 2002; B. C. Yen, 1935–2001) attempted to produce a diagram for the friction factor for *open-channel flow* for impervious rigid boundaries. To that end, he employed in abscissas the Reynolds number based on the hydraulic radius, Re_R, and included the few existing data on the subject, which were mainly obtained in the 1930s and 1960s. It could be argued that the presence of the free surface would affect the role of the relative roughness in the fully rough regime, but that role of the free surface is currently not completely understood.

Figure 1.7 shows a tentative diagram for open-channel flow (Yen 1992b, 2002). The chart shows that the *shape* of the channel is important both in the laminar regime and the fully rough region. Although all curves are represented by $f = K_L/\text{Re}_R$ in the laminar region, the values of K_L range from 14.2 for an isosceles triangular channel, to 24 for a wide channel. For flow in a semi-circular channel, or a full pipe, $K_L = 16$. In the transitional and fully rough regions, the available information is scarce. Yen (1992b, 2002) recommended using the following modified Colebrook–White equation for $\text{Re}_R > 30,000$:

$$\frac{1}{\sqrt{f}} = -c_1 \log\left(\frac{k}{R_h} \frac{1}{c_2} + \frac{c_3}{4\,\text{Re}_R\,\sqrt{f}} \right) \qquad (1.26)$$

with $c_1 = 2.0$, $c_2 = 14.83$, and $c_3 = 2.52$. Yen noted that the values of K_L and c_3 decrease for decreasing width to depth ratio, while c_2 increases and c_1 slightly increases. In addition, Yen showed that the curves for a wide open channel do not differ significantly

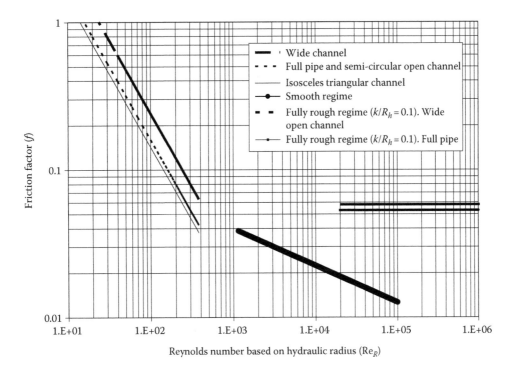

FIGURE 1.7 Diagram for the friction factor for open-channel flow, constructed based on information provided by Yen, B.C., Hydraulic resistance in open channels, in *Channel Flow Resistance: Centennial of Manning's Formula*, B.C. Yen (Ed.), Water Resources Publications, Littleton, CO, 1992a; Yen, B.C., *J. Hydraul. Eng.*, 128(1), 20, 2002.

from those corresponding to full pipes, for small values of the ratio k/R. However, he recommended further experiments to complete and validate the diagram.

1.2.3 Elements of the Phenomenological Theory of Turbulence

Turbulent flow present in WDSs is characterized by a wide range of length scales. This is consistent with all turbulent flows found in nature that have a high Reynolds number. These length scales can be associated with "eddies" ("vortex blobs," Davidson 2005, p. 131) varying from the largest flow length scales (L, determined by the size of the flow) to the smallest one, η_K, given by the Kolmogorov length scale (A. N. Kolmogorov, 1903–1987; see Frisch 1995; Pope 2000; Davidson 2005, p. 17). Although eddies are diﬃcult to define (see discussion in Davidson [2005], p. 52, or Pope [2000], p. 183), the scientific notion of eddies shares remarkable similarities with the scales of flow in the famous sketches of Leonardo Da Vinci (1452–1519) of water falling into a pool (Frisch 1995, cover; Davidson 2005, plate 3). The Kolmogorov length scale is the length scale at which the turbulent fluctuations are dissipated into heat by viscosity (Davidson 2005). In pipe flow, the largest flow length scales are of the order of the pipe radius, and are determined by either the channel width or depth in open-channel flow (Gioia and Bombardelli 2002). The ratio of the large flow length scale and the Kolmogorov length scale can be shown to scale as the flow Reynolds number of the large scales to the ¾ power.

Example 1.2 (on the length scales in a turbulent c ow)

For a pipe with a diameter of 0.4 m with a mean velocity of flowing water of 1 m/s, the ratio $L/\eta_K \sim Re^{3/4} = (1\,\text{m/s}\ 0.4\,\text{m}/10^{-6}\,\text{m}^2/\text{s})^{3/4} \approx 16{,}000$. Usually, the values of η_K are of the order of fractions of millimeters.

The intrinsic problem with the current analysis of turbulent flows is that there is no formal or general theory that is able to explain flow features under a wide range of situations. Most of the existing theories are valid for specific cases only, such as the theories for boundary layers, stratified flows, wall-bounded flows, etc. In spite of this state of affairs, there are two significant tools that can be used to address turbulent flows. These tools are based on the concept of *energy cascade* suggested by Richardson in 1922 (L. F. Richardson, 1881–1953; Pope 2000; Davidson 2005) and the Kolmogorov analysis for small scales (Kolmogorov 1991).

The concept of the energy cascade considers that the largest eddies in the flow suffer from inertial instabilities and that, consequently, they "break up" (i.e., loose identity), splitting their energy into two eddies of approximately half their original size during an overturn time. (The overturn time is defined as the lifespan of an eddy.) These resulting eddies also suffer from further instabilities, evolving into, and simultaneously transferring their energy to smaller eddies. Interestingly, viscosity does not play any role in this transfer of energy (Davidson 2005).

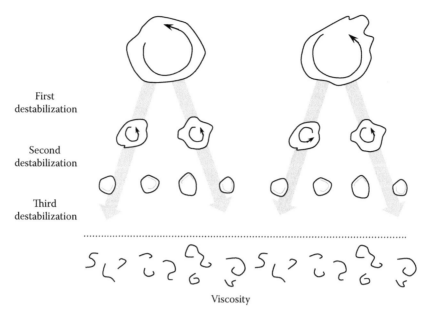

First
destabilization

Second
destabilization

Third
destabilization

Viscosity

FIGURE1 .8 Schematic of the energy cascade. (Adapted from Davidson, P.A., *Turbulence*, Oxford University Press, New York, p. 128, 2005.)

The rate at which the small scales receive energy from the large scales is equal to the rate at which the turbulent kinetic energy is dissipated (i.e., ε) in statistically steady turbulence. The instabilities of eddies and the transfer of energy continues until the size of the eddies becomes so small that viscosity becomes important and the energy is dissipated into heat at the Kolmogorov length scale. Figure 1.8 shows a schematic describing such a cascade.

The Kolmogorov theory for small scales states that these scales are *statistically isotropic* (i.e., the probability density function (PDF) of the velocity field is independent of the point in space considered (local homogeneity), and the PDF is invariant with respect to rotations and reflections of the coordinates axes; see Pope 2000, p. 190) and have also a structure which is *statistically universal* (i.e., a structure which is valid for pipe flow, open-channel flow, boundary layers, wakes, etc.; Davidson 2005). The theory additionally states that for local isotropic turbulence, the statistical properties are a function of only the viscosity and the rate of transfer of energy via the cascade. This description of turbulent flows is summarized in Figure 1.9, which shows the scale ranges in a logarithmic scale. The dissipation range, close to the Kolmogorov length scale, where the turbulent fluctuations are dissipated into heat, extends up to a length scale l_{DI} of about 50–60 η_K (Pope 2000). On the opposite side of the spectrum, there is the energy-containing range,

extending for length scales larger than l_{EI} (approximately, 1/6 L; Pope 2000). The small scales in Kolmogorov's analysis are smaller than l_{EI}. The zone between the energy-containing range and the dissipation range is called the *inertial sub-range*. These tools and ideas constitute the basis for the so-called *phenomenological theory of turbulence*. This theory rests on two main tenets that pertain to the transfer of turbulent kinetic energy: (1) the transfer of energy starts at the length scale of the largest eddies, and (2) the rate of energy transfer is independent of viscosity. Under these tenets, the Taylor–Kolmogorov scaling states that

$$\varepsilon \sim \frac{V^3}{L} \sim \frac{u_l^3}{l} \qquad (1.27)$$

where

V indicates the velocity scale of the large eddies
u_l refers to the velocity scale of a generic eddy located within the inertial sub-range (an eddy of size equal to l)

This scaling was thought to apply in principle only to homogeneous and isotropic turbulence; however, recent research has provided evidence that the range of validity could be extended to flows that do not possess such features (see Section 1.3.1).

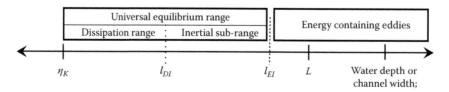

FIGURE 1.9 Schematic of the spectrum of scales in a turbulent flow. (Adapted from Pope, S.B., *Turbulent Flows*, Cambridge University Press, Cambridge, U.K., 2000.)

Although this description of turbulence might be considered a rather simple "cartoon-like" approach, it has been shown to be very useful in providing physical insight into several practical cases (see Sections 1.3.1 and 1.3.2).

Another useful concept is the division of the flow field into a mean flow and a so-called turbulent flow. In describing shear flows (a pipe flow for instance), Davidson (2005) stated: "the mean flow generates, maintains, and redistributes the turbulence, while the turbulence acts back on the mean flow, shaping the mean velocity distribution."

1.2.4 Wave-Number Spectrum of Energy in Turbulent Flows

The velocity spectrum tensor Φ_{ij} is defined for homogeneous turbulence as the Fourier transform of the two-point correlation tensor, R_{ij} (Pope 2000):

$$\Phi_{ij}(\underline{\lambda},t) = \frac{1}{(2\pi)^3} \iiint_{-\infty}^{\infty} e^{-i\,\underline{\lambda}\cdot\underline{r}} R_{ij}(\underline{r},t)\,d\underline{r} \tag{1.28}$$

$$R_{ij}(\underline{r},t) = \iiint_{-\infty}^{\infty} e^{i\,\underline{\lambda}\cdot\underline{r}} \Phi_{ij}(\underline{\lambda},t)\,d\underline{\lambda} \tag{1.29}$$

where

$\underline{\lambda}$ is the wavenumber vector
i and j are indices of the components of the tensor
$\underline{r}$ is the distance vector; in turn, underlines indicate vectors

In Equations 1.28 and 1.29, $d\underline{r}$ and $d\underline{\lambda}$ represent the differentials in the three directions of the coordinate system. At $\underline{r} = \underline{0}$,

$$R_{ij}(0,t) = \iiint_{-\infty}^{\infty} \Phi_{ij}(\underline{\lambda},t)\,d\underline{\lambda} = \overline{u_i'u_j'} \tag{1.30}$$

Equation 1.30 relates the velocity spectrum in terms of the covariance $\overline{u_i'u_j'}$ (see Pope 2000, p. 78). In the definitions (1.28) through (1.30), the variable indicating position, $\underline{x}$, has been removed because the turbulence field has been assumed to be homogeneous.

In addition to the tensorial expression, a simpler scalar version of the velocity spectrum can be defined. This is called the *energy spectrum function*, which removes all information on direction, as follows:

$$E(\|\underline{\lambda}\|) = \oint \frac{1}{2} \Phi_{ii}(\underline{\lambda})\,dS(\|\underline{\lambda}\|) \tag{1.31}$$

Here

$\|\underline{\lambda}\|$ indicates the modulus of the wavenumber vector
$S(\|\underline{\lambda}\|)$ denotes the sphere in the wavenumber space with radius $\|\underline{\lambda}\|$ centered at $\underline{x} = \underline{0}$

It can be shown that (Pope 2000)

$$K = \int_0^{\infty} E(\|\underline{\lambda}\|)\,d\|\underline{\lambda}\| \tag{1.32}$$

where $K\left(= 1/2\,\overline{u_i'u_i'}\right)$ is the flow turbulent kinetic energy (TKE). Several model spectra have been proposed. One example of such models is

$$E(\|\underline{\lambda}\|) = C_s \varepsilon^{2/3} \|\underline{\lambda}\|^{-5/3} f_L(\|\underline{\lambda}\|L) f_\eta(\|\underline{\lambda}\|\eta_K) \tag{1.33}$$

where

f_L is a dimensionless function dictating the shape of the energy-containing range
f_η is associated with the shape of the dissipation range
C_s is a constant of proportionality

The function f_L becomes important when the product $\|\underline{\lambda}\|L$ is large, and f_η becomes large when the product $\|\underline{\lambda}\|\eta_K$ is small. In the inertial sub-range, $f_L = f_\eta = 1$, denoting the dominance of the factor $\|\underline{\lambda}\|^{-5/3}$. This last condition constitutes the well-known Kolmogorov spectrum, i.e., a spectrum in which the slope is −5/3. Experimental evidence obtained from different sources for diverse types of flow (wakes, pipes, grids, jets, etc.; Davidson 2005, p. 226) indicates that *the scaled Kolmogorov spectrum is a universal function of the product of the wavenumber and the Kolmogorov length scale* (Saddoughi and Veeravalli 1994; Pope 2000).

1.3 Analysis and Results

1.3.1 Manning's Formula and the Phenomenological Theory of Turbulence

Manning's formula is an empirical equation obtained as the result of mathematical regressions to several flow datasets that came from flumes and large rivers in North and South America, such as the Mississippi and Paraná Rivers (Dooge 1992; Yen 1992b). In a recent paper, Gioia and Bombardelli (2002) pioneered a derivation of Manning's formula starting from the phenomenological theory of turbulence, a result that connected for the first time two seemingly diverse fields: flow resistance formulations (Section 1.2.1) and flow turbulence (Section 1.2.3). To do this, Gioia and Bombardelli (2002) first applied the momentum balance, i.e., they equaled the component of the weight in the direction of motion, to the force resisting the flow at the boundary in a uniform flow. They then scaled the shear stress effecting the transfer of momentum at the wall as the product of the tangential and normal velocities to a wetted surface tangent to the peaks of the roughness elements, and the density of water (Figure 1.10):

$$\tau \sim \rho v_n v_t \tag{1.34}$$

While the normal velocity was found to scale with the characteristic velocity of the eddies of size k (the size of the roughness elements), u_k, the tangent velocity was found to scale with the cross-sectionally averaged velocity U, leading to

$$u_k U \sim R_h gS \tag{1.35}$$

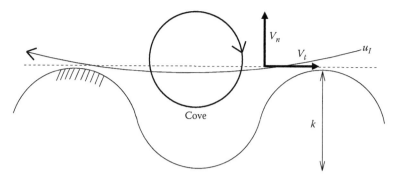

FIGURE 1.10 Schematic of the transfer of momentum by eddies at the wall. (Adapted from Gioia, G. and Bombardelli, F.A., *Phys. Rev. Lett.*, 88(1), 014501, 2002; Bombardelli, F.A. and Gioia, G., Towards a theoretical model of localized turbulent scouring, in *Proceedings of River, Coastal and Estuarine Morphodynamics (RCEM 2005, October 2005)*, Vol. 2, pp. 931–936, G. Parker and M.H. García (Eds.), Taylor & Francis Group, London, U.K., 2005, Gioia, G. and Bombardelli, F.A., *Phys. Rev. Lett.*, 95, 014501, 2005; Bombardelli, FA and Gioia, G., *Phys. Fluids*, 18, 088101, 2006; Gioia, G. and Chakraborty, P., *Phys. Rev. Lett.*, 96, 044502, 2006.)

Next, Gioia and Bombardelli (2002) related u_k and U by making use of Equation 1.27, and assumed that eddies of size k located between roughness elements pertain to the inertial sub-range (Figure 1.10). This gives

$$u_k \sim U \left(\frac{k}{R_h} \right)^{1/3} \qquad (1.36)$$

It is worth mentioning that the scaling embedded in this equation implicitly assumes that the Taylor–Kolmogorov scaling is valid also for flows that are essentially nonhomogeneous and nonisotropic, such as wall-bounded flows. Recent research supports the assumption, including Knight and Sirovich (1990) and Lundgren (2002). Knight and Sirovich (1990) used the "empirical eigenvalue approach" to extend the Kolmogorov spectrum to conditions in which neither translational invariance (homogeneity) nor isotropy holds. They validated the results with data obtained from highly resolved simulations (channel flow, Bénard convection), which show the appearance of an inertial sub-range. Lundgren (2002) employed matching asymptotic expansions to obtain the Kolmogorov two-thirds law (Pope 2000), relating the Kolmogorov theory and the Navier–Stokes equations. In 2003, Lundgren extended his theory to derive the inertial sub-range without imposing homogeneity and/or isotropy (Lundgren 2003). Replacing Equation 1.36 in (1.35) gives

$$U \sim \left(\frac{R_h}{k} \right)^{1/6} \sqrt{R_h g S} \qquad (1.37)$$

When Strickler's regression (Equation 1.10) is applied to Equation 1.37, the standard form of Manning's formula (Equation 1.1) is recovered.

Very interestingly, when $k \to \eta_K$ (the Kolmogorov length scale), Equation 1.37 gives

$$U \sim \mathrm{Re}_R^{1/8} \sqrt{R_h g S} \qquad (1.38)$$

where Re_R is the Reynolds number based on the hydraulic radius. Since $f = 8\,R_h\,gS/U^2$, Equation 1.38 leads to the scaling by Blasius:

$f \sim \mathrm{Re}_R^{-1/4}$ (Equation 1.23)! This finding allows for the connection of well-known but seemingly unrelated scalings: Blasius', Kolmorogov's, and Manning's (Gioia and Bombardelli 2002).

Gioia et al. (2006) employed similar reasoning based on the scaling of eddies to address the issue of intermittency in turbulence, and Bombardelli and Gioia (2005, 2006) and Gioia and Bombardelli (2005) used the same concepts to study the scour due to jets in pools.

1.3.2 Nikuradse's Diagram, the Spectrum, and the Phenomenological Theory of Turbulence

Gioia and Chakraborty (2006) developed a theoretical/mathematical model that advanced the understanding of the physical mechanisms embedded in Nikuradse's chart. The model starts with the following expression for the velocity of eddies of size s, u_s (Pope 2000):

$$u_s^2 = \int_0^s E(\|\underline{\lambda}\|) \|\underline{\lambda}\|^{-2} \, d\|\underline{\lambda}\| \qquad (1.39)$$

where $E(\|\underline{\lambda}\|)$ denotes the turbulence spectrum, including the two correction functions for the energy-containing and the dissipation range (Equation 1.33). Using the Taylor–Kolmogorov scaling (Equation 1.27), and the scaling for the shear stress developed in the previous section (Equation 1.34), Gioia and Chakraborty were able to obtain the following integral expression:

$$f = \Gamma \left[\int_0^{s/R} \Lambda^{-1/3} f_\eta (b\,\mathrm{Re}^{-3/4}/\Lambda) f_L(\Lambda) d\Lambda \right]^{1/2} \qquad (1.40)$$

where $\Lambda \equiv \|\underline{\lambda}\|/R$, and Γ and b are constants. Also, Re is the Reynolds number based on the pipe radius, and f_L and f_η are the dimensionless functions representing the shape of the ranges of energy-containing eddies and of dissipation, respectively (Section 1.2.4). Equation 1.40 is an explicit function of Re and the relative roughness, k/R, of the pipe, because s/R is a function

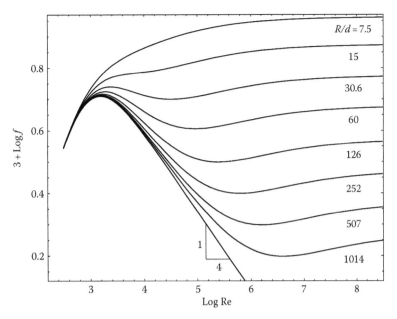

FIGURE 1.11 Numerical results of model by Gioia and Chakraborty (2006) obtained by performing the integral of Equation 1.40. In the diagram, *d* indicates the roughness height, *k*.

of *k*/*R* (Gioia and Chakraborty 2006). When evaluated numerically with the help of certain values of the constants, the integral expression developed by Gioia and Chakraborty provides an impressive *qualitative* agreement with Nikuradse's data, including all aspects of Nikuradse's chart (i.e., the "bundle" of curves, the "bellies," Blasius scaling, and the Strickler scaling) described in Section 1.2.3 (see Figure 1.11). The theoretical model allows the chart to be interpreted as a manifestation of the transfer of momentum close to the wall by eddies located in the (1) energetic, (2) dissipative, and (3) inertial ranges (Figure 1.12). At the low values of the Reynolds number in the turbulent regime of Figure 1.6 (of the order of 3000), the momentum

transfer is dominated by the large (energetic) eddies, which possess a velocity on the order of *U*, i.e., a velocity which scales with Re. This is the reason why the friction factor grows with growing values of the Reynolds number between Re = 3000 and 3500 (approximately). After the peak of the "hump," the curves plunge into the Blasius scaling (Figures 1.6 and 1.11), indicating that the momentum transfer is dominated by scales on the order of the Kolmogorov length scale, which is much larger than the roughness height for the smooth regime. Since the Kolmogorov length scale scales with the inverse of the Reynolds number to the 3/4 power, higher values of the Re lead to smaller eddy sizes and, consequently, the friction factor diminishes within the dissipative range. For larger values of the roughness height, the curves decrease to a lesser extent and start to deviate from Blasius'scaling. This behavior forms the "bellies" seen in Figures 1.6 and 1.11. As the Re increases, the roughness height becomes larger than the Kolmogorov length scale and, thus, the roughness height becomes more important in the transfer of momentum. Finally, when the inertial range is well established, the momentum transfer is effected by eddies of the size *k*, leading to Strickler's scaling, as predicted previously by Gioia and Bombardelli (2002).

Calzetta (2009) discussed Gioia and Chakraborty's (2006) model by using a different spectrum—Heisenberg's spectrum. He found that his results fitted Nikuradse's data well. This demonstrates that the main features of the diagram are independent of the spectrum model employed.

Langelandsvik et al. (2008) reviewed the laws that describe the friction factor of commercial pipes. This was done using data obtained at unique facilities in the Department of Mechanical and Aerospace Engineering at Princeton University. They reported that Gioia extended Gioia and Chakraborty's 2006 theoretical model to consider two roughness heights that differ

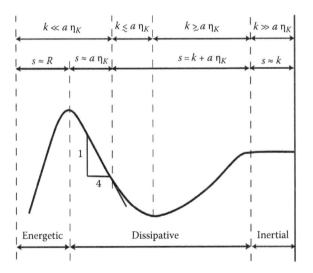

FIGURE 1.12 Schematic of the momentum transfer at the wall according to the eddies effecting the transfer. (Adapted from Gioia, G. and Chakraborty, P., *Phys. Rev. Lett.*, 96, 044502, 2006.) *a* is a constant of proportionality.

in several orders of magnitude. This modified theoretical model predicts a monotonic curve for the friction factor with an abrupt transition from smooth to fully rough (Langelandsvik et al. 2008). Since the roughness heights in commercial steel pipes do not follow such a distribution, it is hard to extrapolate Gioia's theoretical model.

1.3.3 Range of Validity of Formulations for Flow Resistance

Equation 1.6 relates the resistance/conveyance coefficients associated with the different formulas. The details of the equivalence embedded are important to consider, because the diverse formulations have different ranges of applicability. Manning's equation (and the dimensionally homogeneous Manning's equation) should be used for fully rough flows, which is the range for which there are theoretical derivations through integration of the velocity profile (Keulegan's relation; Section 1.2.2) and through the phenomenological theory of turbulence (Section 1.3.1). Chèzy equation should also be used in the fully rough regime. The Hazen–Williams formula's restricted range of applicability was confirmed by Bombardelli and García (2003). Analyzing the original dataset used by Williams and Hazen (1920), they determined that this range of applicability is limited to flows of Reynolds numbers between 10^4 and 2×10^6.

Formulations which relate roughness heights with resistance/conveyance coefficients (such as Equations 1.10 and 1.11, for instance) should also be used with care. Bombardelli and García (2003) compared outcomes from Equation 1.10 against those using (6) and (25). They concluded that Equation 1.10 gives accurate results only when Manning's n is smaller than 0.02.

1.4 Additional Topics of Water Distribution Systems

Other important topics associated with WDSs include (1) the selection of an adequate pumping capacity for a given WDS; (2) hydraulic transients in pipe networks (pressurized flows); (3) water quality in pipe and open-channel flow; (4) flow through gates; (5) design of water towers; (6) design of drop shafts; etc. Information on these and other topics of WDSs can be found in Mays (2005); Larock et al. (2008); Jeppson (1977); Watters (1984); Fischer et al. (1979). Some recent literature on WDSs has been devoted to the description and analysis of software packages used for the computation of flow and water quality in piping systems. One of such software packages is EPANET, developed by the Environmental Protection Agency (see http://www.epa.gov/nrmrl/wswrd/dw/epanet.html; and also Larock et al. 1999; Watters 1984). Another area of increasing interest in recent decades is given by issues associated with intentional spills of pollutants in large pipe systems (Mays 2004). Finally, the issue of proper WDS maintenance (i.e., cleaning of pipes and open channels, maintenance of tanks and buildings, etc.) is important and has received significant attention recently.

Acknowledgment

This chapter is dedicated to the memory of Prof. Ben C. Yen, who was a pioneer in interpreting the design formulations for flow in pipes and open channels in light of the theory of fluid mechanics.

References

Barenblatt, G.I., Chorin, A.J., Prostokishin, V.M. 2000. Self-similar intermediate structures in turbulent boundary layers at large Reynolds numbers. *J. Fluid Mech.*, 410, 263–283.

Blasius, H. 1913. Das Ahnlichkeitsgesetz bei Reibungvorgangen in Flussigkeiten. *Forschg. Arb. Ing.*, 131.

Bombardelli, F.A. 2010. Notes of the course "Urban Hydraulics and Hydrology," Department of Civil and Environmental Engineering, University of California, Davis, CA.

Bombardelli, F.A., García, M.H. 2003. Hydraulic design of large-diameter pipes. *J. Hydraul. Eng.*, 129(11), 839–846.

Bombardelli, F.A., Gioia, G. 2005. Towards a theoretical model of localized turbulent scouring. *Proceedings of River, Coastal and Estuarine Morphodynamics (RCEM 2005, October 2005)*, Vol. 2, pp. 931–936, G. Parker and M.H. García (Eds.). Taylor & Francis Group, London, U.K.

Bombardelli, F.A., Gioia, G. 2006. Scouring of granular beds by jet-driven axisymmetric turbulent cauldrons. *Phys. Fluids*, 18, 088101.

Buschmann, M.H., Gad-el-Hak, M. 2003. Generalized logarithmic law and its consequences. *AIAA J.*, 41(1), 40–48.

Buschmann, M.H., Gad-el-Hak, M. 2007. Recent developments in scaling wall-bounded flows. *Prog. Aerosp. Sci.*, 42, 419–467.

Buschmann, M.H., Gad-el-Hak, M. 2009. Evidence of non-logarithmic behavior of turbulence channel and pipe flow. *AIAA J.*, 47(3), 535–541.

Calzetta, E. 2009. Friction factor for turbulent flow in rough pipes from Heisenberg's closure hypothesis. *Phys. Rev. E*, 79, 056311.

Chanson, H. 2002. An experimental study of Roman dropshaft hydraulics. *J. Hydraul. Res.*, 40(1), 312, 320.

Chen, C. 1992. Power law of flow resistance in open/channels: Manning's formula revisited. In *Channel Flow Resistance: Centennial of Manning's Formula*, B.C. Yen (Ed.). Water Resources Publications, Littleton, 206–240.

Chow, V.T. 1959. *Open-Channel Hydraulics*. McGraw-Hill, New York.

Colebrook, C.F. 1939. Turbulent flow in pipes, with particular reference to the transitional region between smooth and rough wall laws. *J. Inst. Civ. Eng.*, 11, 133–156.

Colebrook, C.F., White, C.M. 1937. Experiments with fluid friction in roughened pipes. *Proc. R. Soc. Lond. A*, 161, 367–378.

Davidson, P.A. 2005. *Turbulence*. Oxford University Press, New York.

Diskin, M.H. 1960. The limits of applicability of the Hazen-Williams formula. *Houille Blanche*, 6, 720–723.

Dooge, J.C.I. 1992. The Manning formula in context. In *Channel Flow Resistance: Centennial of Manning's Formula*, B.C. Yen (Ed.). Water Resources Publications, Littleton, CO, 136–185.

Frisch, U. 1995. *Turbulence*. Cambridge University Press, Cambridge, U.K.

Fischer, H.B., List, E.J., Koh, R.C.Y., Imberger, J., Brooks, N. 1979. *Mixing in Inland and Coastal Waters*. Academic Press, New York.

García, M.H. 1999. Sedimentation and erosion hydraulics. In *Hydraulic Design Handbook*, L.W. Mays (Ed.). McGraw-Hill, New York, 6.1–6.113.

Gioia, G., Bombardelli, F.A. 2002. Scaling and similarity in rough channel flows. *Phys. Rev. Lett.*, 88(1), 014501.

Gioia, G., Bombardelli, F.A. 2005. Localized turbulent flows on scouring granular beds. *Phys. Rev. Lett.*, 95, 014501.

Gioia, G., Chakraborty, P. 2006. Turbulent friction in rough pipes and the energy spectrum of the phenomenological theory. *Phys. Rev. Lett.*, 96, 044502.

Gioia, G., Chakraborty, P., Bombardelli, F.A. 2006. Rough-pipe flows and the existence of fully developed turbulence. *Phys. Fluids*, 18, 038107.

Jeppson, R.W. 1977. *Analysis of Flow in Pipe Networks*. Ann Arbor Science, Ann Arbor, MI.

Jiménez, J. 2004. Turbulent flows over rough walls. *Annu. Rev. Fluid Mech.*, 36, 173–196.

Kim, J., Moin, P., Moser, R. 1987. Turbulence statistics in fully developed channel flow at low Reynolds numbers. *J. Fluid Mech.*, 177, 133–166.

Knight, B., Sirovich, L. 1990. Kolmogorov inertial range for inhomogeneous turbulent flows. *Phys. Rev. Lett.*, 65(11), 1356–1359.

Kolmogorov, A.N. 1991. The local structure of turbulence in incompressible viscous fluid for very large Reynolds numbers. *Proc. R. Soc. Lond. A*, 434, 9.

Kundu, P.K., Cohen, I.M. 2008. *Fluid Mechanics*. Academic Press, New York.

Langelandsvik, L.I., Kunkel, G.J., Smits, A.J. 2008. Flow in a commercial steel pipe. *J. Fluid Mech.*, 595, 323–339.

Larock, B.E., Jeppson, R.W., Watters, G.Z. 2008. *Hydraulics of Pipeline Systems*, CRC Press, Boca Raton, FL.

Lundgren, T.S. 2002. Kolmogorov two-thirds law by matched asymptotic expansion. *Phys. Fluids*, 14(2), 638–641.

Lundgren, T.S. 2003. Kolmogorov turbulence by matched asymptotic expansion. *Phys. Fluids*, 15(4), 1074–1081.

Mays, L.W. 2004. *Water Supply Systems Security*. McGraw-Hill Professional Engineering, New York.

Mays, L.W. 2005. *Water Resources Engineering*. John Wiley & Sons, Inc., Hoboken, NJ.

Mays, L.W. 2007. Water supply systems in arid and semi-arid regions during Antiquity: The use of cisterns. In *Proceedings of Fish Symposium on Environmental Hydraulics*, Tempe, AZ, p. 205 (in CD.)

Mays, L.W. 2008. A very brief history of hydraulic technology during antiquity. *Environ. Fluid Mech.*, 8, 471–484.

Masters, G. 1997. *Introduction to Environmental Engineering and Science*. Prentice Hall, New York.

Moody, L.F. 1944. Friction factors for pipeflow. *Trans. ASME*, 66, 671–684.

Nikuradse, A. 1933. Laws of flow in rough pipes. VDI Forschungshef, 361. (Also NACA TM 1292, 1950.)

Pope, S.B. 2000. *Turbulent Flows*. Cambridge University Press, Cambridge, U.K.

Prandtl, L. 1935 The mechanics of viscous fluids. In *Aerodynamic Theory III*, p. 142, W.F. Durand (Ed.), also Collected Works II, pp. 819–845.

Saddoughi, SG, Veeravalli, SV. 1994. Local isotropy in turbulent boundary layers at high Reynolds number. *J. Fluid Mech.*, 268, 333–372.

Schlichting, H. 1968. *Boundary-Layer Theory*, 6th edn. McGraw-Hill, New York.

Streeter, V., Wylie, E.B., Bedford, K.W. 1998. *Fluid Mechanics*, 9th edn. WCB/McGraw-Hill, New York.

Strickler, A. 1981. *Contribution to the Question of a Velocity Formula and Roughness Data for Streams, Channels and Close Pipelines*. Translation by T. Roesgen and W.R. Brownlie, California Institute of Technology, Pasadena, CA.

Swamee, P.K., Jain, A.K. 1976. Explicit equations for pipe-flow problems. *J. Hydraulic Division*, ASCE, 102 (HY5), 657–664.

Travis, Q.B., Mays, L.W. 2007. Relationship between Hazen-William and Colebrook-White roughness values. *J. Hydraul. Eng.*, 133(11), 1270–1273.

Watters, G.Z. 1984. *Analysis and Control of Unsteady Flow in Pipelines*, 2nd edn. Ann Arbor Science Books, Ann Arbor, MI.

Wei, T., Willmarth, W.W. 1989. Reynolds-number effects on the structure of a turbulent channel flow, *J. Fluid Mech.*, 204, 57–95.

White, F.M. 1974. *Viscous Fluid Flow*. McGraw-Hill, Inc., New York.

Williams, G.S., Hazen, A. 1920. *Hydraulic Tables*, 3rd edn. John Wiley & Sons, Inc., New York.

Yen, BC. 1992a. Dimensionally homogeneous Manning's formula. *J. Hydraul. Eng.*, 118(9), 1326–1332.

Yen, B.C. 1992b. Hydraulic resistance in open channels. In *Channel Flow Resistance: Centennial of Manning's Formula*, B.C. Yen (Ed.). Water Resources Publications, Littleton, CO.

Yen, B.C. 2002. Open channel flow resistance. *J. Hydraul. Eng.*, 128(1), 20–39.

2

Groundwater–Surface Water Discharges

Christopher Butler
University of California, Merced

Thomas C. Harmon
University of California, Merced

2.1 Introduction

Understanding and characterizing groundwater–surface water (gw-sw) interactions has emerged over the past few decades as an important subfield of hydrology and stream ecology. Prior to this, groundwater and surface water were generally assessed as separate domains, except when water resource demands in one domain imparted pressure on the other. For example, groundwater well hydraulics relatively near to streams requires consideration of the stream as a recharging boundary condition, and intensive pumping can actually modify stream flow conditions. More recently, the continuity and complexity of interactions between these hydrologic compartments under natural conditions have become more apparent. In particular, the role of gw-sw hydraulics in streams, often coupled to thermal regimes and biogeochemical cycling, has been the subject of many recent investigations related to water resources management, earth science, and ecosystem assessment (Brunke and Gonser 1997; Kalbus et al. 2006). In this context, it is clear that understanding gw-sw interactions is critical to making well-informed land, water resources, and ecosystem management decisions.

This chapter focuses on methods for assessing gw-sw discharges at multiple observational scales (Figure 2.1). Empirical and analytical approaches differ greatly depending on the spatial and temporal scales of interest. For example, in a water balance approach, the gw-sw discharge component can sometimes be inferred as the difference in surface flow between two gauging stations. However, this discharge is often a relatively small (but significant) fraction of the stream flow, and this approach may be effective only at larger spatial scales. When finer grained assessment is needed, alternative techniques become necessary, such as those involving direct in situ discharge measurements or interpretation of local hydraulic head, temperature, and chemical tracer gradients. Even with these finer-grained approaches, clear delineation of gw-sw interactions can be challenging, depending on both long- and short-term climate factors, stream stage, and geomorphology. Locally gaining and losing zones, for example, are possible within the same river reach, and the distribution of these zones may change seasonally with changes in river stage, groundwater surface elevations, and other hydraulic considerations.

This chapter begins with a review of key hydraulic principles governing gw-sw discharges, followed by a description of some common approaches for assessing these discharges at various observational scales. Several techniques are presented in more detail including example applications in order to provide a deeper understanding of the procedures and potential complications associated with them. As gw-sw interactions are complex and remain an active field of investigation in hydrology, this chapter

Handbook of Environmental Fluid Dynamics, Volume Two, edited by Harindra Joseph Shermal Fernando. © 2013 CRC Press/Taylor & Francis Group, LLC.
ISBN: 978-1-4665-5601-0.

FIGURE 2.1 Gw-sw interactions within a basin, depicting observations at multiple spatial scales: (a) basin, (b) reach, (c) segment, (d) local (hill slope or riparian zone), and (e) point.

concludes with a discussion of some of the key challenges that need to be overcome in the effort to achieve an understanding of these processes and their role in the watershed.

2.2 Principles Governing GW-SW Discharges

This section summarizes the hydraulics of gw-sw interactions, emphasizing those occurring in alluvial-fluvial or fluvial plain systems, where the discharge tends to be diffuse and continuous relative to that associated with crystalline rock settings. The latter, which tends to be associated with discrete fissures along the stream, is important in many settings, such as in mountain hydrology. In fluvial plain systems, the magnitude and direction of gw-sw exchanges depend on the position and geometry of the stream channel relative to the fluvial plain, the hydraulic head difference (river stage vs. water table), and the hydraulic conductivity distribution in the aquifer and stream substrate. In this context, stream reaches are typically categorized in terms of gaining, losing, flow-through, and parallel-flow conditions (Woessner 2000). Conventionally, a gaining stream reach or segment is one into which groundwater discharges positively; the discharge is considered negative for a losing reach or segment. Flow-through conditions exhibit gaining characteristics on one side of the stream, with losing conditions on the other. Parallel-flow systems result when the stream stage is equal to groundwater levels, and the flows roughly parallel one another. Under such conditions, heterogeneity in the streambed and aquifer hydraulic conductivity field can result in local gw-sw exchanges known as hyporheic exchanges. Such exchanges often exhibit complex spatiotemporal patterns, which may bias larger scale groundwater discharge measurements if not well understood. Hyporheic exchanges are important to stream ecology and reviewed elsewhere (Bencala 2000).

The level of detail of the hydraulic model needed to adequately simulate or assess gw-sw interactions depends on the nature and scope of the investigation. For example, when the primary objective is to estimate the average groundwater contribution to base flow for reaches or stream catchments as a whole, then differential gauging (Section 2.2.1) or hydrograph separation (Section 2.2.2) approaches may be appropriate. As objectives shift toward characterizing the magnitude and distribution of local groundwater discharges, groundwater flow and transport principles must be used to quantify these fluxes directly or indirectly (Section 2.2.3). Recent motivation to better understand the coupled gw-sw dynamics at the basin scale has led to the development of coupled gw-sw models governed by the principles of groundwater and surface water flow. This approach has the potential to more accurately simulate the spatiotemporal variability of discharges that is consistently observed in the field. It is still common, however, when the objectives of a hydrologic investigation are more specific and/or localized, to consider gw-sw discharges from either the surface or groundwater perspective. In many such applications, simplified flow and transport models may suﬃce for extracting the desired parameters from the data (Section 2.2.4).

2.2.1 Differential Gauging

Quantifying the flow difference between the upstream and downstream cross sections of a river reach is the most straightforward method for quantifying gw-sw discharges. In theory, this mass balance approach may be applied at multiple scales, but is most often applied at reach to basin scales as stream flow differences at smaller scales become diﬃcult to quantify with adequate precision. This approach averages spatial variability in both the magnitude and the direction of the groundwater, yielding only a net gain or loss over the entire reach. In the absence of tributaries, return flows, and surface runoff between stations, the differential gauging approach can expressed as follows:

$$Q_{in} \pm GW - ET - Q_{out} = 0 \qquad (2.1)$$

where

Q_{in} is the volumetric flow rate entering the reach [L³/T]

ET is the volumetric evapotranspiration rate associated with the riparian vegetation within the reach

Q_{out} is the flow at the downstream end of the reach

GW is the net volumetric groundwater discharge into or out of the reach

The *ET* term is significant when deep-rooted riparian vegetation (phreatophytes) is present in arid and semiarid climates and can create an observable diurnal variation in river stage (Butler et al. 2007). In larger rivers and in more temperate climates, this term may be neglected.

2.2.2 Hydrograph Recession Analysis

The recession curve is the portion of the hydrograph following the peak, where flow typically decreases rapidly and then diminishes rates. Hydrograph recession analysis for assessing gw-sw discharges is appropriate when the watershed's hydrogeology is well characterized as a single unconfined aquifer and when the characteristic response time of the aquifer is less than the period between significant recharge events (Brutsaert and Nieber 1977). Under these conditions, and in the absence of intervening surface flows from rain events or human releases, the recession curve is a function of natural discharges from groundwater. The 2D (horizontal flow) Boussinesq equation has most often been used to describe these discharges, shown here in its linearized form for a homogeneous, isotropic aquifer of uniform thickness (*b*) and where the change in head is modest relative to the aquifer thickness ($h \ll b$):

$$\frac{S_y}{Kb}\frac{dy}{dx} = \frac{\partial^2 h}{\partial x^2} + \frac{\partial^2 h}{\partial y^2} \qquad (2.2)$$

where

S_y is the aquifer specific yield [–]

K is the hydraulic conductivity [LT⁻¹]

Hydrologists have developed several solutions to (2.2), and other forms of the Boussinesq equation, for the purpose of linking hydrograph recession curve analysis to gw-sw discharges, as noted in Section 2.3.2. This approach can be useful under the conditions noted earlier and, like differential gauging, can provide an estimate of the average behavior upstream of the gauging location. Addressing the spatial variability of gw-sw discharges requires a more general theoretical approach, as discussed in the following section.

2.2.3 Hydraulics of GW-SW Flow

Coupled gw-sw models have been developed and are being used to satisfy the need to more accurately simulate watershed-scale responses to changes in climate, land management practices, and other issues (Kollet and Maxwell 2006). The groundwater component of these models generally employs a form of the Richards equation to simulate the movement of water through variably saturated porous media, written here as (Panday and Huyakorn 2004)

$$\nabla \cdot (Kk_{rw}(\psi)\nabla h) - W + q_{gs} + q_{gc} = \phi\frac{\partial S_w}{\partial t} + S_w S_s \frac{\partial h}{\partial t} \qquad (2.3)$$

where

K [LT⁻¹] is the saturated hydraulic conductivity tensor, which is commonly limited to its principal components (K_{xx}, K_{yy}, K_{zz})

$k_{rw}(\psi)$ [–] is the relative permeability, a characteristic of the soil type and a function of the pressure head, ψ [L]

h [L] is the hydraulic head ($h = \psi + z$, where *z* is the elevation)

W [T⁻¹] is a source/sink term expressed as a volumetric flux per unit volume of subsurface domain

q_{gs} [T⁻¹] is the flux per unit volume subsurface domain to/from the overland flow module (see following text)

q_{gc} [T⁻¹] is the flux per unit volume subsurface domain to/from a stream channel

ϕ [–] is the porous medium porosity

S_w is the water saturation [–]

S_s is the specific storage of the porous medium

Overland flow models are typically used to simulate surface conditions in gw-sw models, as expressed by the 2D shallow-flow equation (Panday and Huyakorn 2004):

$$\frac{\partial h_s}{\partial t} = \frac{\partial}{\partial x}\left(dk_x \frac{\partial h_s}{\partial x}\right) + \left(dk_y \frac{\partial h_s}{\partial y}\right) - dq_{sg} - dq_{sc} \qquad (2.4)$$

where

h_s [L] is the water surface elevation

k_x and k_y [LT⁻¹] are the friction-slope-related flow conductance terms, which may be estimated using a Manning, Chezy, or Darcy–Weisbach-type expression

d [L] is the average depth of the surface flow

q_{sg} [T⁻¹] is the flux per unit volume to/from the subsurface domain

q_{sc} [T⁻¹] is the flux per unit volume to/from the stream channel

Solution of (2.3) and (2.4) to simulate real systems is a substantial undertaking that is beyond the scope of this chapter. First, the spatially distributed material properties must be estimated for relevant soil types, including hydraulic conductivity and relative permeability curves. Second, the solution is subject to the appropriate boundary conditions, including precipitation and evapotranspiration, and initial conditions, such as the soil moisture profile. In addition, these equations must be coupled via flux terms (here, q_{gs} and q_{sg}). Numerical coupling of the gw and sw flow equations has been the subject of some research and often involves the use of a surface-subsurface exchange flux (see La Bolle et al. 2003 for review). The exchange parameters are difficult to independently assess and a topic of research. In addition, contrasting overland and groundwater flow timescales can result in numerical instabilities with these coupled models. Numerical methods are emerging to alleviate these problems, such as directly coupling the two models by implementing the overland flow equation as the uppermost boundary to the Richards equation under saturated conditions (Kollet and Maxwell 2006).

2.2.4 Chemical Tracers for Assessing GW-SW Interactions

Many gw-sw assessment techniques are based on the observation of spatiotemporal chemical tracer distributions. Tracers are generally used to attribute sources and partitioning of water between various hydrologic units at basin to reach scales, as in end-member mixing and related diagnostic analyses (Hooper 2003). They may also be used at the stream segment to local (hill slope/riparian zone) scales to identify flow paths and quantify residence times by observing and modeling natural or introduced tracers in time and space.

End-member mixing analysis (EMMA) is based on the assumption that variations in stream chemistry are the result of the mixing of source components from discharging groundwater, precipitation (surface runoff and direct interception), lateral unsaturated zone discharges, and other potential sources (Christophersen and Hooper 1992). EMMA and related diagnostic tools are discussed further in the context of gw-sw discharge assessment in Section 2.3.4.

Solute transport of a conservative tracer in groundwater can be summarized by the advection–dispersion equation:

$$\frac{\partial(nC)}{\partial t} = \nabla \cdot (nD_{HD}\nabla C - nuC) \quad (2.5)$$

where
C [ML^{-3}] is the tracer concentration
n [–] is the aquifer porosity
t [T] is time
D_{HD} [L^{2}T^{-1}] is the hydrodynamic dispersion tensor
u [LT^{-1}] is the pore water velocity vector

The velocity term in (2.5) describes the advective transport of the tracer solute. Hydrodynamic dispersion is used to account for the mixing or dilution of the tracer due to local velocity gradients. Dispersion is generally assumed to increase linearly with velocity, where the scaling factor is referred to as the porous medium dispersivity. At low to negligible groundwater velocities, the dispersion coeDcient becomes equivalent to an effective diffusion coeDcient.

2.2.5 Heat as a Tracer of GW-SW Interactions

Heat is a particularly useful and economical tracer in the context of assessing gw-sw discharges (Stonestrom and Constantz 2003). Due to the disparity of timescales associated with stream flow and groundwater discharge, and given a significant temperature difference between the two, measurable temperature gradients tend to develop in a streambed. The general characteristics for the cases of gaining and losing streams are illustrated in Figure 2.2. Given reasonably stationary conditions and a gaining stream segment, the diurnal variation of stream water temperature caused by solar radiation fails to propagate deeply into the underlying sediments (Figure 2.2a). For a losing segment, the downward flux of stream water transports the diurnal stream temperature cycle into the sediments (Figure 2.2b). The temperature cycles may be delayed and attenuated relative to those observed in the stream, depending on the rate of discharge into the sediments and the observational depths.

Modeling of heat transport is analogous to the advection–dispersion approach discussed earlier for chemical tracers. For interpreting temperature observations in a streambed, for example, the 3D convection–conduction equation can be written as follows (Anderson 2005):

$$\frac{\partial T}{\partial t} = \frac{K_T}{\rho c}\nabla^2 T - \rho_w c_w \nabla \cdot (Tq) \quad (2.6)$$

where
T is temperature [Θ]
K_T [MLT$^{-3}\Theta^{-1}$] is the thermal conductivity
q [LT^{-1}] is the pore water velocity vector
$\rho_w c_w$ [MT$^{-2}\Theta^{-1}$] and ρc [MT$^{-2}\Theta^{-1}$] are the volumetric specific heat of water and solid–water matrix, respectively, where ρ [ML^{-1}] and c [L^{2}T$^{-2}\Theta^{-1}$] are the corresponding density and heat capacities

The first term on the right-hand side of (2.6) describes conductive heat transfer as a function of the thermal conductivity, a thermal property of the solid–fluid matrix. The second term describes convective heat transfer. Note that all thermal properties of the solid–fluid matrix are considered to be constant and homogeneous in this expression.

2.3 Methods and Analysis

This section describes the methods and analysis associated with assessing gw-sw discharges and is intended to complement the preceding discussion on the underlying principles.

2.3.1 Differential Gauging

Groundwater discharge estimation using a mass balance approach entails measuring stream velocity fields using devices such as wading rod/flow meters and acoustic Doppler

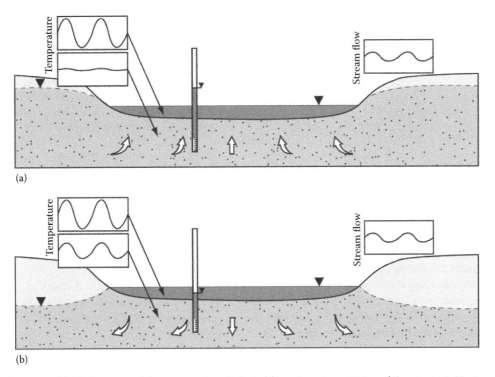

FIGURE 2.2 (a) Gaining and (b) losing perennial stream reaches. (Adapted from Stonestrom, D.A. and Constantz, J., *Heat as a Tool for Studying the Movement of Ground Water Near Streams*, Circular 1260, U.S. Geological Survey, San Francisco, CA, 2003.)

velocimeters (ADVs) and profilers (ADPs) (see Figure 2.3). To obtain flow estimates, the velocity fields are integrated either by the software accompanying the instrument or manually. As noted previously, the target reaches are determined based on the scale and questions being investigated at a given location and must be long enough to yield a significant flow difference and avoid ungauged in/outflows that may bias the mass balance. Specific transect locations are then selected based on channel geometry and site accessibility. With respect to geometry, channel depth is taken into consideration as some sensors require minimum depths in the range of 10–30 cm to avoid excessive

boundary effects. In addition, ADPs provide spatial mapping of the velocity field and require reasonably trapezoidal geometry to capture the majority of the cross section in its effective field of view.

2.3.2 GW-SW Estimates from Recession Analysis

Assuming the conditions are appropriate for using recession analysis to assess gw-sw discharges, a variety of approaches are available. Most of these are based on Brutsaert and Nieber's (1977) approach,

FIGURE2 .3 Stream velocity measurements taken using (a) Wading rod instrument and (b) ADP.

which is based on the Boussinesq equation, as noted in Section 2.2.2. By observing the time derivative of the recession curve (dQ/dt) as a function of Q, hence eliminating time as a dependent variable, Brutsaert and Niebert arrived at the following relationship:

$$-\frac{dQ}{dt} = \alpha Q^b \qquad (2.7)$$

where
  α [T^{-1}] is a function of the physical dimensions and hydraulic properties of the system
  b is an empirical constant [–]

In theory, the recession curve achieves a limiting flow rate or base flow. A variety of expressions for α and b have been obtained using different solutions to the Boussinesq equation (Sujono et al. 2004; Rupp and Selker 2006a,b). Computational tools are available for analyzing recession behavior in long time series (e.g., RECESS and associated routines by the U.S. Geological Survey). As with all models, it is important to understand the underlying assumptions in such programs prior to applying them. For example, RECESS is valid during times when no recharge is occurring, when stream flow consists entirely of groundwater discharge (base flow conditions), and when the profile of the groundwater head distribution is nearly stable (steady state). It is generally applicable to estimating the average discharge to gaining streams at the basin scale on a long-term (interannual) basis.

2.3.3 Groundwater-Flow-Based Methods

Several approaches to assessing gw-sw interactions are based on investigation of groundwater flow conditions near the gw-sw interfacial zone (Cardenas and Zlotnik 2003). Groundwater hydraulics is described by Darcy's Law, a gradient-flux expression relating energy (typically expressed as pressure or hydraulic head) and groundwater discharge:

$$q = \frac{kpg}{\mu}\frac{dh}{ds} \qquad (2.8)$$

where
  q [L^{3}T^{-1}] is the groundwater-specific discharge (or Darcy velocity) [LT^{-1}]
  k [L^2] is the intrinsic permeability of the porous media
  ρ [ML^{-3}] is the fluid density
  μ [ML^{-1}T^{-1}] is the dynamic viscosity
  dh/ds is the pressure gradient along the flow line, with pressure expressed as hydraulic head

If the discharge may be measured directly using devices known as seepage meters (Figure 2.4b), or in the context of point-scale gw-sw discharge observations (Figure 2.4a), then (2.8) is conveniently restated as follows:

$$Q_{gw} = -\frac{kpg}{\mu}A\frac{(h_2 - h_1)}{s} \qquad (2.9)$$

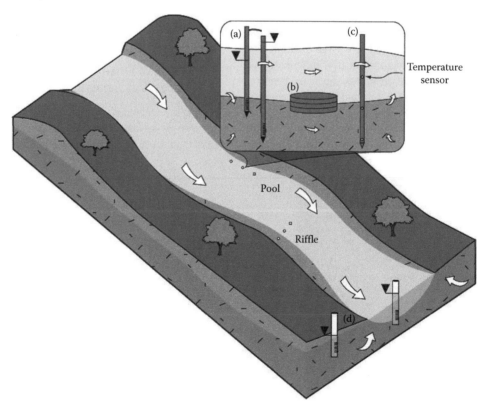

FIGURE 2.4 Reach, segment, and point-scale measurements. (a) Minipiezometer cluster with hyporheic sampling tube, (b) seepage meter, (c) hyporheic temperature probe, and (d) riparian and hyporheic wells.

where Q_{gw} [L^{3}T^{-1}] is the volumetric flux through the porous medium cross-sectional area A [L^2] and is driven by the hydraulic head difference ($h_2 - h_1$) between two measuring points separated by the distance s.

Temperature influences both viscosity and density and therefore the hydraulic conductivity, K [LT^{-1}], of the porous medium underlying a stream:

$$K = \frac{kgp}{\mu} \tag{2.10}$$

Density differences on the gw-sw interface may also affect discharge patterns, particularly if salinity gradients are also present (Langevin et al. 2005). While groundwater temperatures are relatively constant for a given location, stream temperatures can vary diurnally (3°C or more) and seasonally (25°C or more). Across this temperature range, viscosity changes significantly and should be considered when employing gw-sw discharge estimation models (Constantz et al. 1994).

While point-scale discharge assessment is useful, and commonly used to map distributed gw-sw discharges, a larger spatial perspective may be necessary in the context of groundwater flow path delineation and stream segment- or reach-scale investigations. Here, the approach is visualized in a 2D cross section illustrating the connectivity between a shallow, unconfined aquifer and a stream. Given steady groundwater and stream conditions, the Dupuit approximation assumes that (1) the hydraulic gradient is equal to the slope of the water table and (2) groundwater flow is predominantly horizontal, resulting in the following (Figure 2.4d):

$$q'_x = \frac{1}{2} K \frac{\left(h_2^2 - h_1^2\right)}{L} \tag{2.11}$$

where

q'_x [LT^{-1}] is the discharge (per unit width normal to the horizontal flow)

h_2 is the steady-state water table elevation at some distance from the river

h_1 is the steady-state river stage

L is the horizontal distance between these locations

While this model accounts for the primary groundwater flow considerations in this type of problem, it is relatively simplistic and should be implemented only as a first approximation in estimating gw-sw discharges. More detailed models are needed to capture spatial and temporal variations of gw-sw discharges and are discussed in the following section.

2.3.4 Chemical Tracer Methods

This section highlights two important methods for applying chemical tracers in the assessment of gw-sw discharges: mixing-based analyses (EMMA and related diagnostic tools) and flow path analyses. A complete discussion of the application, advantages, and disadvantages of chemical tracers is beyond the scope of this chapter and available elsewhere (Kendall and Jeffery 1998). Common chemical tracers involve the use of natural species, such as chloride (Cl$^-$), sulfate $\left(SO_4^{2-}\right)$, stable (e.g., ^{18}O, ^2H, ^{13}C, ^{15}N), and radioactive isotopes (e.g., ^3H, ^{14}C), and introduced species, such as bromide (Br$^-$) and various dyes (e.g., Rhodamines). The stable oxygen and hydrogen isotopes are particularly useful in identifying water fractions associated with groundwater and surface water sources.

In the context of gw-sw interactions, one or more of the end members would be associated with gw chemistry in wells adjacent to a stream (e.g., shallow versus deep groundwater may have distinctly different compositions). An underlying assumption in EMMA is that the end-member compositions are known and that the tracers are conservative. EMMA effectively fits fractions of the predetermined source components to the observed stream composition (i.e., forward analysis) and is useful when the end members are readily identifiable and reasonable in number (i.e., 2 or 3). Christophersen and Hooper (1992) integrated principal component analysis (PCA) and EMMA to enable the use of stream water composition to identify the potential number, but not the composition, of end members (inverse analysis). In this approach, the end-member criterion is subjectively based on the variance explained by each principal component of the solute correlation matrix. More recently, Hooper (2003) introduced diagnostic tools that not only provide additional guidance on the number of end members, but can also be used to determine whether individual solutes can be used as tracers or not by identifying violations in mixing model assumptions regarding conservativeness and constancy of end-member composition. Liu et al. (2008) presented a recent example of the application of mixing analysis and the associated diagnostic tools in their investigation of snowmelt (including the delineation of subsurface flow contributions) in semiarid mountain catchments.

2.3.5 Temperature-Based Methods

Stallman (1965) used Equation 2.6 to describe temperature redistribution as a function of heat transfer via convection (or advection) and conduction, an approach that has since led to the development of several analytical models for gw-sw exchanges. Many solutions to the Stallman equation have been used to develop both analytical and numerical models. One such analytical model, developed by Turcotte and Schubert (1982), assumes that vertical temperature distributions in groundwater underlying a streambed are developed by steady, upward vertical convection and conduction (i.e., no density-driven flow) and that the porous media properties are homogeneous. This solution is expressed as

$$q_z = \frac{K_T}{\rho_w c_w \cdot z} ln \frac{T(z) - T_L}{T_0 - T_L} \tag{2.12}$$

where
> $T(z)$ is the temperature at a given depth z
> T_L is the deeper (constant) groundwater temperature
> T_0 is the temperature at the upper boundary, which in this case is the stream water temperature (Figure 2.5a)
> the remaining terms are as described for Equation 2.6

Observed temperature gradient data can be evaluated using (2.12) to yield an estimate of the vertical velocities at a given depth given reasonable estimates for the temperature boundary conditions at the streambed water interface ($z = 0$, $T = T_0$), and deep in the riverbed (as $z \to y$, $T = T_L$), as well as for the thermal conductivity value and the volumetric heat capacity for the solid–fluid matrix.

Equation 2.12 requires a discernible vertical temperature gradient within the substrate. If the temperature difference is insufficient, then the uncertainty of the model will be too large to obtain reasonable parameter estimates. Figure 2.5b illustrates conditions where local groundwater is warmer than surface water. For this calculation, groundwater is held constant at 19°C and surface water varies between 5°C and 15°C, which is consistent with observed groundwater and surface water temperatures. It is important to note that this model produces invalid negative flux estimates when the streambed temperature is less than the surface water temperature.

The steady flow assumption may be appropriate in gw-sw systems where pressure head differential (driving the discharge) is not changing appreciably over time. If the system is in a period of significant flow change, as is common in many managed waterways, this assumption may not be suitable and alternative

models should be employed. An example case in which gw-sw temperature data are collected during a seasonal transition period further illustrates this point in Section 2.4.2.

2.4 Applications

This section uses results from several gw-sw investigations to demonstrate the principles and methods described earlier. The examples are presented at several spatial and temporal scales to illustrate the relationship between method selection and investigation-specific objectives.

2.4.1 Differential Gauging: Case Study

The lower Merced River is a heavily managed agricultural drainage, with numerous diversions and inputs. These manipulations pose a problem for most gw-sw discharge questions. However, along a reach near the confluence of Dry Creek and the Merced River, there are no diversions or inputs into the river (see Figure 2.6a), thus making it an excellent site for the differential gauging approach. Dry Creek is an ephemeral stream flowing for short periods after winter storms (December through February).

Three stations were selected on the basis of their historical datasets and geographic location to be used as endpoints in this exercise (Figure 2.6a): Dry Creek near Snelling (station ID = DSN) and Merced River near Snelling (MSN) were used as the upper endpoints, and Merced River at Cressey (CRS) was used as the bottom endpoint of the drainage system. Equation 2.1 was used to assess the mass balance in this system. Flows along the Merced River and Dry Creek converge at their confluence and

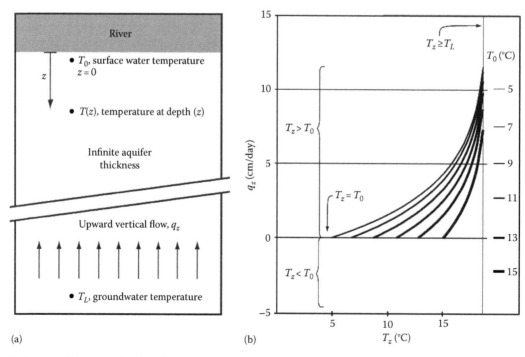

(a)   (b)

FIGURE 2.5 (a) Conceptual diagram of analytical solution Turcotte and Schubert (1982) with boundary conditions. (Adapted from Schmidt, 2007.) (b) Groundwater discharge velocity changes with respect to streambed changes over variable surface water temperatures.

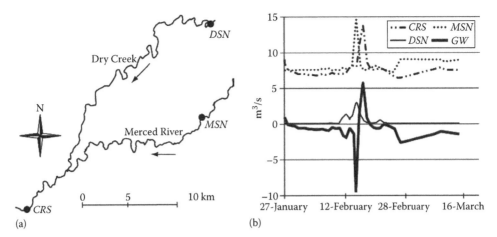

FIGURE 2.6 (a) Confluence of the Merced River and Dry Creek highlighting local gauge stations. (b) Flow rate measurements and calculations of gauge stations and groundwater discharge/recharge.

are monitored at the CRS gauge station. gw-sw exchanges occur throughout the length of the river and are included in the flow measurements at CRS (2.13). Assuming that there are no other inputs or losses of water in this system, and ignoring ET for the wintertime period examined, groundwater discharge/recharge can be calculated using the following equations:

$$CRS = MSN + DSN + GW \qquad (2.13)$$

Figure 2.6b shows reported flow rates from each station, with a peak indicating a significant storm event in mid-February. During this event, the increased flows at *MSN* are evident first and then arrive downstream at *CRS*. A similar but attenuated peak flow is apparent at *DSN*. Estimated groundwater discharge rates mirrors closest to the gauge station flows at *DSN*, during the month of February, with peak recharge values occurring at peak flows along both *DSN* and *MSN* stations. A flow reversal is shown shortly after peak flows occur at *DSN* and *MSN* and is consistent with peak flows at *CRS*. Shallow aquifer levels rise after the February storm event to a level that allows for a flow reversal in groundwater recharge to discharge. After the storm flows pass, the water table levels begin to recede to previous levels as groundwater flow is reversed from discharge back to recharge near February 20. Outside of the storm event, groundwater recharge inversely responds to the Merced River shown by flows at *CRS* and *MSN*.

2.4.2 Heat as a Tracer: Case Study

Downstream from the location of the differential gauging example, a known gaining reach of the Lower Merced River has been the site of an ongoing gw-sw investigation (Zamora 2007; Essaid et al. 2008). Over the winter of 2009–2010, temperature was logged at three depths (0.5, 1.0, and 2.0 m) below the riverbed. In addition, local groundwater and surface water temperature measurements were collected. Groundwater discharge velocities were measured using Thermocron iButtons, small inexpensive self-logging thermisters.

The 1D vertical analytical solution to the conduction-convection equation by Turcotte and Schubert (1982), Equation 2.6, was utilized to calculate the positive vertical groundwater discharge through the substrate into the Merced River at each depth (see Figure 2.7). The river stage remained unchanged throughout the deployment, minimizing the influence of a variable pressure head and maintaining steady hydraulic conditions. During the first 2 days, temperature was reasonably constant, providing good conditions for applying this simple model. The resulting positive (gaining) discharges are consistently around 0.5 cm/day, a number consistent with previous results obtained using a variety of methods (Zamora 2007).

Diurnal temperature fluctuations become more evident in the air and surface water temperatures the last 2 days of the experiment. These cycles in the surface water temperatures are more evident at the shallower depths, resulting in apparent variations in the discharge velocities (Stonestrom and Constantz 2003). However, this apparent diurnal cycling in velocities is likely an artifact of the analytical approach. Velocities resulting from the deepest observation points approach similar values throughout the experiment, casting further doubt upon the cyclic behavior of the estimates based on shallower estimates. While the result here is probably an artifact, temperature-induced gw-sw discharges are possible. Constantz and Thomas (1997) observed diurnal discharge cycles tied to surface water temperature cycling. Fortunately, more robust approaches have been developed that take advantage of diurnal gw-sw temperature fluctuations and the phase shift in these cycles due to the time required for heat to propagate in the substrate (Hatch et al. 2006). Alternatively, numerical modeling approaches can be used to estimate gw-sw discharges under transient conditions.

The use of temperature to calculate groundwater discharge is limited by temperature differences between water sources (surface water and groundwater). Small temperature differences between groundwater and surface water do not allow discharge velocities to be accurately calculated. In instances where temperatures are too similar, the uncertainty associated with the velocity calculation will be large. An additional limitation to this particular method is its inability to calculate

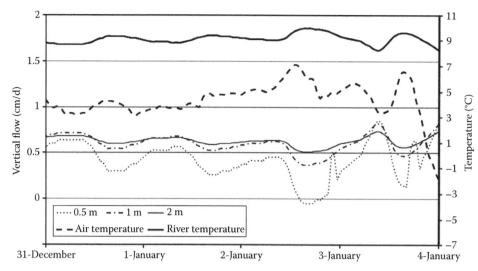

FIGURE 2.7 Calculated vertical velocities at three depths under steady-state stream conditions (gw temperature relatively constant at 19.5°C throughout the experiment).

groundwater recharge. Though negative values may allude to groundwater recharge, the Turcotte and Schubert method cannot accurately calculate recharge values.

2.5 Challenges

Understanding the complex spatiotemporal distributions of gw-sw discharges, as well as their connection to human activities and stream ecology, remains a rich research area. As this chapter strives to demonstrate, gw-sw discharges in a watershed can be assessed at point, stream segment, reach, or across the entire drainage basin. Each observational scale yields information that may be useful in the context of some questions, but which may provide irrelevantly detailed or averaged information with respect to other questions. As with most hydrologic investigations, the resolution of observational scale (both temporally and spatially) is a major factor in the design of a successful field investigation.

Given that the most relevant observational scale has been determined, there is still a need for improved field methods for more eDciently assessing gw-sw discharges and related processes (e.g., hyporheic zone exchanges). The following are some areas where new or modified techniques and instruments would greatly improve our ability to assess gw-sw discharges at various scales:

ft Specialized acoustic Doppler profiler (ADP)-type instruments adept at high-precision flow estimation in shallow, irregularly shaped channel cross sections. In many natural river systems, particularly high-energy streams, cross sections are nonuniform and labor intensive to gauge precisely.

ft Integrated sensing-modeling systems (e.g., for multi-level temperature probes—Figure 2.4d) providing real-time flux estimates. As noted throughout this chapter, instrumentation has been developed for acquiring data by which to determine gw-sw fluxes, via pressure or temperature gradients. Integrated systems would be relatively straightforward to develop and would enable relatively rapid gw-sw discharge mapping campaigns.

ft Additional multiscale modeling efforts and tool development for linking land and water resource management practices at the watershed scale with spatially distributed and coupled gw-sw interactions. Characterizing gw-sw interactions at any scale is a significant undertaking, and better multiscale models would help to identify critical reaches for intensive investigations.

ft Greater application and development of higher resolution gw-sw characterization as a remote-sensing application through the use of high spatial granularity light detection and ranging (LiDAR) and spectral reflectance products. Using these and other products, vegetation classification and physiognomy, subtle water temperature gradients, and geomorphologic features may be useful in the development of eDcient gw-sw discharge mapping tools.

References

Anderson, M. P. (2005). Heat as a ground water tracer. *Ground Water* **3** (6): 951–968.

Bencala, K. E. (2000). Hyporheic zone hydrological processes. *Hydrological Processes* **4** (15): 2797–2798.

Brunke, M. and T. Gonser (1997). The ecological significance of exchange processes between rivers and groundwater. *Freshwater Biology* **3** (1): 1–33.

Brutsaert, W. and J. L. Nieber (1977). Regionalized drought flow hydrographs from a mature glaciated plateau. *Water Resources Research* **3** (3): 637–644.

Butler, J. J., G. J. Kluitenberg et al. (2007). A field investigation of phreatophyte-induced fluctuations in the water table. *Water Resources Research* **4** (2), W02404, doi:10.1029/2005 WR004627.

Cardenas, M. B. and V. A. Zlotnik (2003). A simple constant-head injection test for streambed hydraulic conductivity estimation. *Ground Water* **4** (6): 867–871.

Christophersen, N. and R. P. Hooper (1992). Multivariate-analysis of stream water chemical-data—the use of principal components-analysis for the end-member mixing problem. *Water Resources Research* **8** (1): 99–107.

Constantz, J., C. L. Thomas et al. (1994). Influence of diurnal-variations in stream temperature on streamflow loss and groundwater recharge. *Water Resources Research* 30(12): 3253–3264.

Constantz, J. and C. L. Thomas (1997). Streambed temperatures profiles as indicators of percolation characteristics beneath arroyos in the Middle Rio Grande Basin, USA. *Hydrological Processes* **1** (12): 1621.

Essaid, H. I., C. M. Zamora et al. (2008). Using heat to characterize streambed water flux variability in four stream reaches. *Journal of Environmental Quality* **3** (3): 1010–1023.

Hatch, C. E., A. T. Fisher et al. (2006). Quantifying surface water-groundwater interactions using time series analysis of streambed thermal records: Method development. *Water Resources Research* **42**(10), W10410, doi:10.1029/2005WR004787.

Hooper, R. P. (2003). Diagnostic tools for mixing models of stream water chemistry. *Water Resources Research* **9** (3), 1055, doi:10.1029/2002WR001528.

Kalbus, E., F. Reinstorf et al. (2006). Measuring methods for groundwater—surface water interactions: A review. *Hydrology and Earth System Sciences* **❶** (6): 873–887.

Kendall, C. M. and J. J. McDonnell (1998). *Isotope Tracers in Catchment Hydrology*. Amsterdam, the Netherlands: Elsevier B.V., 839.

Kollet, S. J. and R. M. Maxwell (2006). Integrated surface-groundwater flow modeling: A free-surface overland flow boundary condition in a parallel groundwater flow model. *Advances in Water Resources* **9** (7): 945–958.

La Bolle, E. M., A. A. Ahmed, et al. (2003). Review of the integrated groundwater and surface-water model (IGSM). *Ground Water* **4** (2): 238–246.

Langevin, C., E. Swain et al. (2005). Simulation of integrated surface-water/ground-water flow and salinity for a coastal wetland and adjacent estuary. *Journal of Hydrology* 314(1–4): 212–234.

Liu, F. J., R. C. Bales et al. (2008). Streamflow generation from snowmelt in semi-arid, seasonally snow-covered, forested catchments, Valles Caldera, New Mexico. *Water Resources Research* 44(12), W12443, doi:10.1029/2007WR006728.

Panday, S. and P. S. Huyakorn (2004). A fully coupled physically-based spatially-distributed model for evaluating surface/subsurface flow. *Advances in Water Resources* **2** (4): 361–382.

Rupp, D. E. and J. S. Selker (2006a). Information, artifacts, and noise in dQ/dt-Q recession analysis. *Advances in Water Resources* **9** (2): 154–160.

Rupp, D. E. and J. S. Selker (2006b). On the use of the Boussinesq equation for interpreting recession hydrographs from sloping aquifers. *Water Resources Research* **4** (12), W12421, doi: 10.1029/2006WR005080.

Schmidt, C., B. Conant, Jr., et al. (2007). Evaluation and field-scale application of an analytical method to quantify groundwater discharge using mapped streambed temperatures. *Journal of Hydrology* **4** (3–4): 292–307.

Stallman, R. W. (1965). Steady 1-dimensional fluid flow in a semi-infinite porous medium with sinusoidal surface temperature. *Journal of Geophysical Research* **70**(12): 2821.

Stonestrom, D. A. and J. Constantz (2003). *Heat as a Tool for Studying the Movement of Ground Water Near Streams*, Circular 1260, San Francisco, CA: U.S. Geological Survey.

Sujono, J., S. Shikasho et al. (2004). A comparison of techniques for hydrograph recession analysis. *Hydrological Processes* **8** (3): 403–413.

Turcotte, D. L. and G. Schubert (1982). *Geodynamics Applications of Continuum Physics to Geological Problems*. New York, John Wiley and Sons, Inc.

Woessner, W. W. (2000). Stream and fluvial plain ground water interactions: Rescaling hydrogeologic thought. *Ground Water* **8** (3): 423–429.

Zamora, C. (2007). *Estimating Water Fluxes Across the Sediment-Water Interface in the Lower Merced River*, San Francisco, CA: U.S. Geological Survey.

3

Fluid Mechanics of Agricultural Systems

Josef Tanny
*Agricultural Research
Organization*

3.1 Introduction

Agriculture is the production of food and goods through farming and forestry. In the beginning of the twenty-first century, one-third of the world's workers were employed in agriculture. Nevertheless, agricultural production accounts for less than 5% of the gross world product (Wikipedia, 2010). Since its beginning, roughly 10,000 years ago, agriculture was significantly developed. In particular, during the industrial revolution, which took place from the eighteenth to the nineteenth century, major changes were taking place in agriculture as well as in other fields of life.

Agricultural production is usually classified into two categories: extensive and intensive. Extensive agriculture is crop cultivation using small amounts of labor and capital in relation to the area of land being farmed. The crop yield in extensive agriculture depends primarily on the natural fertility of the soil, the terrain, the climate, and the availability of water. On the other hand, intensive agriculture is associated with high inputs of capital, labor, and technology. The ultimate goal of intensive agriculture is to increase yield (quantity and quality) relative to the farmed land area in an attempt to meet the increasing demand of the growing world population.

One example of intensive agriculture is protected cultivation which became highly popular in agricultural production in the past decades. Here crops are cultivated in some kind of protected environment, usually a greenhouse, screenhouse, or a horizontal shading screen. These structures aim at protecting the crop from external hazards while allowing meticulous supply of resources for optimal production.

Plants need light, water, and carbon dioxide for photosynthesis and carbohydrate production. Besides, optimal production will be achieved by plants under specific conditions of air temperature and humidity. The greenhouse is one of the most sophisticated facilities for agricultural production. Its walls and roof are made of impermeable transparent material like glass or plastic which allows the penetration of radiation to the crop. Due to its relatively high isolation from the outside, climate control is possible and implemented in many greenhouses. Thus, conditions for optimal crop production may be achieved. The screenhouse is a structure which covers the whole plantation or orchard with a porous screen extending over the roof and the sidewalls. The screenhouse protects the crop from invasion of insects, high wind speeds, and supra-optimal solar radiation, usually allowing sufficient ventilation. Screens of different hole sizes and colors have different protection capabilities. Thus, screenhouses may provide passive climate control. Horizontal shading screens deployed over plantations mainly provide wind and radiation protection of the crops. With open sidewalls, much higher ventilation rates are realized and protection level is lower than screenhouses and greenhouses.

In greenhouses, light penetrates through the transparent cover and water is supplied by the irrigation system. However, to supply sufficient CO_2 for photosynthesis, greenhouses are usually ventilated by exchanging the inside air with external fresh air at a certain rate. Ventilation also removes excess moisture

Handbook of Environmental Fluid Dynamics, Volume Two, edited by Harindra Joseph Shermal Fernando. © 2013 CRC Press/Taylor & Francis Group, LLC.
ISBN: 978-1-4665-5601-0.

which may cause certain plant diseases and deteriorate production. Ventilation may be natural, in which case the driving force for airflow is either external wind or buoyancy due to temperature differences between inside and outside. Ventilation may also be forced through the use of mechanical fans which usually suck inside air outside. To keep the inside environment at certain ranges of temperature and humidity, more sophisticated climate control systems are also employed in greenhouses. These may include cooling/heating, humidifying/dehumidifying, and lighting/shading systems.

For ventilation purposes greenhouses are usually equipped with windows. The size and duration of windows openings are controllable and depend on the outside climatic conditions and crop requirements. To avoid the penetration of insects into ventilated greenhouses, openings are usually equipped with insect-proof screens. These screens have very tiny holes with a size smaller than the size of the anticipated insect. While these screens are eDcient in insect exclusion, they inhibit air flow through the opening thus impeding ventilation and affect microclimate (Teitel, 2001). Therefore, knowledge of the properties of flow through openings with and without screens is essential for a proper design of greenhouse ventilation systems.

In recent years, the area of cultivation under screen constructions is steadily increasing mainly in Israel, some Mediterranean countries, and around the world. Major purposes for cultivation under screens are shading from supra-optimal solar radiation, improving the thermal climate (e.g., for frost protection), exclusion of insects (with insect-proof screens, Tanny et al., 2003), changing the solar spectrum for induction of light-mediated processes (e.g., use of colored screens), providing shelter from wind and hail, and saving of irrigation water. The use of insect-proof screenhouses is expanding rapidly because of the increasing demand for produce grown with reduced use of pesticides and the relatively low costs associated with screenhouses compared to fully climate-controlled greenhouses.

In screenhouses, the passive climate control is determined by the screen properties, that is, the screen material, the shape and size of holes, the screen color, and the shading level. In these structures, both sidewalls and roof are covered with a screen. For common horizontal winds over flat terrains, flow through screens will prevail across the screenhouse sidewalls, whereas over the roof the flow will be along the screen (Tanny and Cohen, 2003, Tanny et al., 2009). Hence, these two configurations were studied in past years.

Obviously, the analysis of microclimate, ventilation, or heat and mass exchange is involved with a variety of fluid mechanics principles. Therefore, in this chapter we will review several topics related to fluid mechanics in greenhouse, screenhouse, and shading screen systems. In particular, we will discuss greenhouse natural and forced ventilation, screenhouse ventilation, and flow through and along screens. The chapter will not cover all the relevant literature on these issues. Instead, it will present and highlight certain topics which, according to the view of this author, nicely demonstrate how fluid mechanics principles can be employed for better analysis and design of agricultural systems.

3.2 Greenhouse and Screenhouse Ventilation

3.2.1 General

Greenhouse ventilation may be either natural or forced. Natural ventilation, which may be defined as ventilation driven by the natural forces of wind and temperature, is a reliable, low-maintenance and energy-eDcient method to keep temperature, humidity, and CO_2 concentration inside greenhouses within suitable limits for optimal crop production. The main attraction of natural ventilation is that the airflow within the greenhouse is driven by two naturally occurring forces, namely, the buoyancy force and the wind force. The buoyancy force, also known as the "stack" or "chimney" effect, results from temperature differences between the internal and external environment; warm, less dense air rises and flows out through openings at high levels and draws cooler ambient air in through openings at lower levels. Wind-driven flow through a building depends on the locations and sizes of the openings and may enhance or hinder the stack-driven flow depending upon the wind speed and direction. The disadvantage of natural ventilation is its dependence on natural resources, the availability of which is not always predictable. Also, in certain climates natural ventilation may not be suDcient to induce optimal conditions for crop production, and more sophisticated approaches of climate control are required.

Forced ventilation is generated by operating mechanical fans that suck the air out of the greenhouse, induce small negative internal pressure which causes the inflow of external fresh air. Forced ventilation is controllable in terms of flow rates, operation time and duration, and locations of inlets and outlets. The design of a forced ventilation system should also take into account the drag induced by the canopy elements (stems and leaves) on the airflow through the greenhouse. In many situations in hot and arid climates, forced ventilation is combined with evaporative cooling, supplied either by a wet pad or a fogging system. Wet pads introduce additional airflow resistance and should be taken into account in the design of a forced ventilation system.

3.2.2 Greenhouse Natural Ventilation

Systems for greenhouse natural ventilation consist of openings which can be closed or opened. Greenhouses are usually equipped with either roof openings or side openings or both (Figure 3.1).

An important parameter in greenhouse design is the ventilation rate, G ($m^3 \ s^{-1}$). In naturally ventilated greenhouses, ventilation may be driven by buoyancy and wind. The ventilation rate can be determined either through an energy balance analysis of the greenhouse, or by application of the Bernoulli equation. For a greenhouse equipped with side and roof openings, natural ventilation is governed by the displacement ventilation mode (Linden, 1999) where outflow of less dense air occurs through high-level openings and inflow of external denser air takes place

FIGURE 3.1 Naturally ventilated greenhouse with vertical openings at the roof and sidewalls. The openings are covered by porous screens for insect exclusion.

through low-level openings. For this situation, the ventilation rate is (Katsoulas et al., 2006)

$$G = C_d \sqrt{\left(\frac{A_R A_S}{\sqrt{A_R^2 + A_S^2}}\right)^2 \left(2g \frac{\Delta T_{i-o}}{T_o} h\right) + \left(\frac{A_T}{2}\right)^2 C_w u^2} \quad (3.1)$$

When only roof openings are opened, the governing ventilation mode is mixing ventilation. In this case, exchange flow between inside and outside takes place through a single upper opening. For greenhouses equipped with roof openings only, the ventilation rate is (Katsoulas et al., 2006)

$$G = \frac{A_T}{2} C_d \sqrt{2g \frac{\Delta T_{i-o}}{T_o} \frac{h}{4} + C_w u^2} \quad (3.2)$$

In Equations 3.1 and 3.2, A_R, A_S, A_T are the roof, sides, and total openings/surface area (m²), g is the gravitational acceleration (m s⁻²), u is the outside wind speed (m s⁻¹), C_d is the discharge coefficient of the openings, C_w is the wind effect coefficient, ΔT_{i-o} is the temperature difference between inside and outside (K), T_o is the outside air temperature (K), and h is the vertical distance between the midpoints of side and roof openings (m) in Equation 3.1 and half of the vertical height of the roof opening in Equation 3.2.

The buoyancy effect would be significant in greenhouses equipped with roof and side openings (e.g., Figure 3.1) because of the relatively large vertical distance between openings. In this situation, displacement ventilation would prevail. If only roof or side openings are employed, this effect will be relatively small since the mixing ventilation mode would be dominant and exchange flow will take place through a single opening. In the latter case with an upper opening, relatively high-turbulence intensities were observed in the shear layer

between the inflow and outflow (Tanny et al., 2008), which may decrease the ventilation efficiency.

Wind-induced ventilation is driven by the pressure field developed around the structure. The ventilation rate would depend on the wind and structure properties and the interrelation between them. For example (Hunt and Linden, 2004), openings located at a high level on the windward side of the greenhouse and at low level on the leeward side allow a wind-driven flow from high to low level, opposite to the buoyancy-driven flow. In such a case, the wind and buoyancy-driven ventilation flows will oppose each other and ventilation would be insufficient. Hence, proper orientation of the greenhouse openings is required such that both natural ventilation modes would assist each other. Therefore, commonly, greenhouses are positioned such that high-level openings are leeward and low-level openings are windward, relative to the prevailing wind direction at the site.

Many studies investigated natural ventilation of greenhouses in an attempt to characterize the different ventilation modes under a variety of operating conditions and greenhouse structure configurations. One question of interest is the relative dominance of the wind and stack effects. Kittas et al. (1997) investigated an experimental greenhouse in southern France and have demonstrated that the wind effect predominates on the stack effect for $u/\sqrt{\Delta T_{i-o}} > 1$. For low wind velocities and under typical temperature differences in this study (~5 K), the roof and side ventilation system became more efficient than roof openings only for low wind velocities (<2.5 m s⁻¹). Kittas et al. (1997) have also tested the ventilation models (Equations 3.1 and 3.2) and by fitting measured and modeled ventilation rates obtained $C_d\sqrt{C_w} = 0.2 \pm 0.01$. Boulard et al. (1997a) obtained $C_d\sqrt{C_w} = 0.18$ for a greenhouse with roof openings only, and Teitel and Tanny (1999) reported on $C_d\sqrt{C_w} = 0.11$ also for a roof ventilated greenhouse. The different value obtained in the latter study as compared to earlier literature was explained mainly by different greenhouse configuration and differences in the wind directions relative to the greenhouse openings.

Apart from the overall ventilation rate, research was focused on the detailed flow patterns within and through the openings of naturally ventilated greenhouses. This information is of practical importance in identifying regions that are less ventilated within a large greenhouse, which may cause nonuniform production and even damage to the crop.

Boulard et al. (1996) applied the eddy correlation technique to determine the heat flux through greenhouse roof openings. They deployed a one-axis ultrasonic anemometer and a fine-wire thermocouple at the plane of the vent and measured air velocity and temperature at a frequency of 5 Hz. Moving the system along the 32 m long vent allowed them to identify separate regions of inflow and outflow through the same vent and to verify their measurements by mass balance closure. Boulard et al. (1996) also compared values of $C_d\sqrt{C_w}$ using measured mean and turbulent mass fluxes through the openings and those obtained by ventilation measurements using the tracer gas technique in the same greenhouse. They obtained good agreement between the two approaches over a wide range of wind speeds. Using a three-axis ultrasonic anemometer, Boulard et al. (1997b) have demonstrated that external wind blowing in parallel to a long roof opening gave rise to inflow at the downwind end of the vent and outflow at the windward end. Within the greenhouse, the air followed a spiral flow pattern.

Teitel and Tanny (2005) studied the effect of external wind direction, relative to the greenhouse, on the flow patterns through continuous vertical roof openings. Using sonic anemometers, they measured simultaneously mean and turbulent flow at the two edges of a roof opening. When the wind was not perpendicular to the plane of the openings, there was outflow and inflow at the windward and leeward edges of the openings, respectively (Figure 3.2). However, when the wind was blowing from the back of the openings and nearly perpendicular to the plane of the openings, the mean air velocity through the openings was reduced but the turbulent component was almost unchanged.

Shilo et al. (2004) extended earlier eddy correlation measurements by applying simultaneous high-frequency humidity measurements at the openings using a Krypton hygrometer. This allowed them to measure both sensible and latent heat fluxes through the openings. In addition to the air mass balance, they analyzed the energy balance from which they determined the ventilation rate. For a semi-commercial greenhouse equipped with roof openings covered with insect-proof screens, good agreement was obtained between ventilation rates measured by the tracer gas technique and those deduced from the energy balance. Internal flow patterns measurements showed that under conditions of leeward ventilation (i.e., roof openings facing the leeward direction) the direction of air velocity at plant level was opposite to that of the external wind.

A more recent advancement in the study of greenhouse ventilation is the use of CFD (computational fluid dynamics) for such systems. This can assist in greenhouse design for optimal ventilation depending on external climatic conditions. The effect of ventilation configuration of a tunnel greenhouse with crop on airflow and temperature patterns was numerically investigated

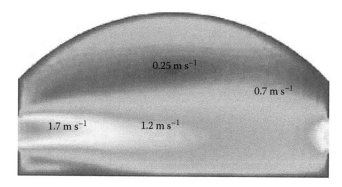

FIGURE 3.3 Computed contours of the air velocities of a tunnel greenhouse with side openings only. Tunnel width (from left to right) is 8 m. Tunnel length (perpendicular to the page) is 20 m. (From Bartzanas, T. et al., *Biosyst. Eng.*, 88(4), 479, 2004, Figure 5. With permission.)

by Bartzanas et al. (2004) using a commercial CFD code. The numerical model was first validated against experimental data with good qualitative and quantitative agreement between the numerical results and the experimental measurements. Then, the CFD model was used to study the consequences of four different ventilator configurations on the natural ventilation system. An interesting finding of the simulations was that while the mean air temperature at the middle of the tunnels varied from 28.2°C to 29.8°C, for an outside air temperature of 28°C, there were regions inside the tunnels 6°C warmer than outside air. An example of velocity contours in one of the configurations studied by Bartzanas et al. (2004) is given in Figure 3.3.

Boulard and Wang (2002) utilized a commercial CFD package (CFD2000) to predict the heterogeneity of plant transpiration in a tunnel greenhouse lettuce crop. The crop was simulated as a porous medium (using the Darcy–Forsheimer equation) exchanging latent and sensible heats with its environment. The radiative and convective heterogeneity in two vertical sections of the tunnel predicted by the CFD model was validated against the experimental results obtained by several solar radiation sensors and sonic anemometers. The validated model was finally used to predict the transpiration flux of a mature lettuce crop in the tunnel. The predicted crop transpiration was in close agreement with the measured value. It was demonstrated that the crop transpiration strongly varied with the location in the tunnel. Specifically, they showed that crop transpiration was lower by about 30% on the northern side of the greenhouse due to lower solar radiation reaching the plants at this region. Such analyses have important implications in the management of greenhouse crops and demonstrate the usefulness in using CFD tools.

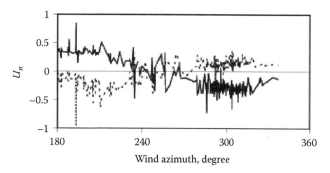

FIGURE 3.2 Air velocity, normalized with respect to external wind speed, through two edges of a greenhouse roof opening as a function of external wind azimuth. Openings spanned 32 m along the north–south direction with their plane facing the east. Azimuth of 270° indicates wind perpendicular to the opening plane. Positive values of velocity indicate outflow and vice versa; solid line—the southern edge of roof opening; dashed line—the northern edge of roof opening. (From Teitel, M. and Tanny, J., *Flow Turbul. Combust.*, 74(1), 24, 2005, Figure 3c. With permission.)

3.2.3 Greenhouse Forced Ventilation

If the natural ventilation system is unable to provide the required inside climatic conditions, forced ventilation systems are applied. These include exhaust fans, usually installed at one side wall with inlet openings at the opposite wall. In warm-arid climates, wet pads, installed at the inlet sidewall, or fogging systems are used to cool down the greenhouse air through evaporative cooling.

In forced ventilation systems, the airflow along the greenhouse crop may generate horizontal gradients in temperature, humidity, and CO_2 concentrations, which induce non-uniform climatic conditions inside the greenhouse. Such gradients were modeled and measured by Kittas et al. (2003) who reported a temperature gradient of up to 8°C along the 60 m distance from pads to fans. Along with the measurements, Kittas et al. (2003) developed a model based on the energy balance of the greenhouse. Components affecting the temperature distribution along the greenhouse crop were: ventilation rate, crop transpiration, percentage of shading (of the downstream half of the greenhouse), water evaporation from the pad, and heat-loss coeDcient of the greenhouse cover. Sensitivity analysis showed that increasing the ventilation rate, that is, fan flow rate, and shading the downstream half of the greenhouse contributed to a smaller horizontal temperature gradient.

Vertical temperature distributions were studied experimentally and using a numerical thermal model by Li and Willits (2008a, b). Measurements were conducted in a fan-ventilated greenhouse equipped with cooling pads that could be turned on or off. Vertical temperature profiles were measured by five thermocouple and relative humidity sensors installed at five levels above the ground. Air velocities were measured by a hand-held anemometer, at 55 sampling points and at three cross sections. In their study, Li and Willits (2008a) distinguished between mean canopy velocity, that is, air velocity within the canopy, and greenhouse mean velocity which is the mean of all measurements at the three cross sections.

Figure 3.4 shows the ratio of mean canopy velocity to mean greenhouse velocity, as a function of the ratio of canopy frontal area to greenhouse cross-section area. An interesting finding of this figure is that with relatively small plants ($A_c/A = 0.2$) and lower ventilation rate, the canopy velocity was higher than mean greenhouse velocity. Li and Willits (2008a) explained that this was possibly because at low ventilation rates, higher velocities were

recorded near the ground since the cooler airstream from outside tended to slide along the ground due to its higher density relative to the warmer inside air. Thermal stratification (defined as the temperature difference between top and bottom temperatures) within the fan-ventilated greenhouse was found by Li and Willits (2008a) to increase with increasing solar radiation and when the wet pad was operating. The stratification decreased with increased ventilation rate and due to the presence of plants, as compared to an empty greenhouse. Li and Willits (2008b) developed a theoretical thermal model which essentially verified the experimental observations of temperature stratification in the greenhouse.

3.2.4 Natural Ventilation of Screenhouses

Unlike greenhouses, screenhouses do not have definite openings. Instead, the whole structure is covered by a porous screen, through which air can flow in and out at various locations. Forced ventilation of screenhouses is therefore impractical and natural ventilation is the only feasible ventilation mechanism.

Tanny et al. (2003) conducted a field study in a commercial fine mesh screenhouse in which pepper was grown. Screenhouses of this type use high mesh screens to mechanically exclude insect penetration and thus avoid potential fruits diseases. Simultaneously, however, these screens inhibit ventilation due to the small holes and high resistance to airflow. Hence, it is of importance to study ventilation rate and microclimatic properties of such screenhouses. Tanny et al. (2003) utilized water vapor as a tracer and applied two physical principles, namely, the flux-gradient ratio and mass conservation of water vapor to estimate the air exchange rate through the screenhouse. The resulting measured air exchange rate increased linearly with the external wind speed (Figure 3.5) and was in good agreement

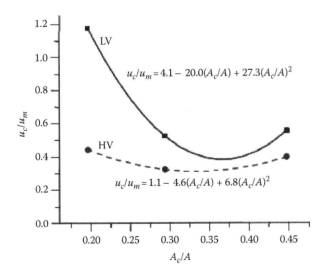

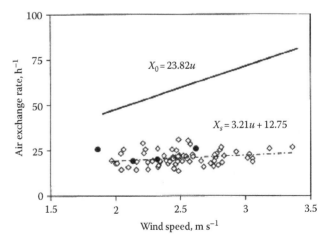

FIGURE 3.4 Ratio of canopy velocity to mean velocity (u_c/u_m) as a function of the ratio of canopy area to greenhouse cross-section area (A_c/A) for two ventilation rates (LV = 0.041 m³ m⁻² s⁻¹; HV = 0.087 m³ m⁻² s⁻¹). The second-order polynomial curves were fitted from three data points. (From Li, S. and Willits, D.H., *Trans. ASABE*, 51(4), 1443, 2008a, Figure 3. With permission.)

FIGURE 3.5 Air exchange rate at the center of the screenhouse (open diamonds) as a function of the external wind speed, *u*, together with the theoretical air exchange rate (solid line) estimated theoretically for an open pepper field. Closed circles: greenhouse exchange rate measured by Fatnassi (2001). X_s and X_0 are air exchange rates in the screenhouse and in the open field, respectively. (From Tanny, J. et al., *Biosyst. Eng.*, 84(3), 331, 2003, Figure 9a. With permission.)

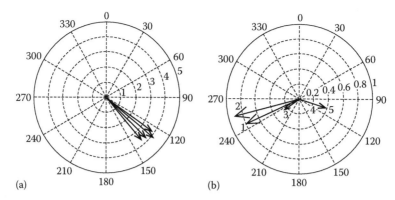

(a)

(b)

FIGURE 3.6 (a) Wind vectors outside and above the screenhouse. Each arrow represents a measuring period of 15 min and corresponds to the times of the inside measurements on the right. The concentric circles indicate the wind speed in m s^{-1}. (b) Wind vectors inside the screenhouse. Each vector corresponds to a location along the central axis of the screenhouse from west to east: 1–13 m from western edge; 2–30, 3–57, 4–78, and 5–102 m from western edge. The concentric circles indicate the wind speed in m s^{-1}. (From Möller, M. et al., *Acta Hortic.*, 614, 445, 2003, Figures 10 and 11. With permission.)

with earlier results by Fatnassi (2001) for a plastic greenhouse fitted with insect-proof screens at its openings.

Measurements of air velocity and direction inside and above the same screenhouse (Möller et al. 2003) revealed a complex flow pattern with a counter-flow (to the atmospheric conditions) in the windward western region of the screenhouse and a concurrent flow along its easterly region, as depicted in Figure 3.6. This flow pattern can be explained by the curvature of external wind streamlines at the leading edge of the structure and the induced lower pressure at this region as compared to the higher pressure at the leeward easterly edge.

In a more recent study, Tanny et al. (2006) applied the eddy covariance technique to investigate turbulent fluxes of water vapor and sensible heat in a large banana screenhouse. They showed that turbulent properties within the screenhouse, in the air gap above the plants and below the screen, were favorable for such measurements. For example, they showed that velocity spectra decay rates were very close to −5/3, the value typical of the inertial sub-range in steady state turbulent boundary layers. Average turbulence intensity for all the data collected was 0.49, a value which marginally supports the validity of Taylor's hypothesis of frozen turbulence. The water vapor flux measured by Tanny et al. (2006) was again used to calculate the ventilation rate of the screenhouse. Results showed a linear increase with the external wind speed and general agreement with literature results.

3.3 Flow along and through Porous Screens

3.3.1 General

The use of porous screens in agriculture became widely popular in recent years. The main two applications are (1) the deployment of screens in greenhouse roof and side openings to exclude insects and (2) crop cultivation inside screenhouses or under horizontal shading screens. In the first case, since

greenhouse openings are usually vertical or slightly inclined, the screen is perpendicular or inclined at some angle to the airflow. In the second application, a flow parallel to the screen takes place over much of the cultivated area. In this chapter, we first discuss flow along screens, mostly related to the screenhouse cultivation practice. Discussion of flow through screens will follow.

3.3.2 Flow along Screens

Covering a crop with a screen reduces the vertical exchange of heat, mass, and momentum between the crop and the free atmosphere. Besides incident radiation, these transport processes are the major factors influencing the crop microclimate. A major goal of recent studies (Tanny and Cohen, 2003, Tanny et al., 2009) was to investigate how screens modify the turbulence characteristics of the wind. Tanny and Cohen (2003) measured vertical profiles of wind and temperature above a small shading screen covering few citrus trees. From these profiles, and by fitting the data to a log-linear profile, they were able to estimate the friction velocity and other boundary layer properties.

When the boundary layer flow along the screen is neutrally stable, the logarithmic wind profile equation is

$$u(z) = \frac{u_*}{k} \ln \frac{(z - d)}{z_0} \qquad (3.3)$$

where
  u is the wind speed (m s^{-1})
  z is the height above the ground (m)
  u_* is the friction velocity (m s^{-1})
  k is the von Karman constant (= 0.41)
  The parameter d is the zero plane displacement (m), that is, the distance that the canopy or the screen "displaces" the wind profile above the ground
  z_0 is the roughness length (m), which is a scaling length for a particular surface

When the boundary layer is not neutral, a modification of the aforementioned equation should be applied to take into account the stabilizing or destabilizing effect of the temperature profile. This modification results with a log-linear profile where the effect of buoyancy is characterized by the bulk Richardson number, Ri, defined as

$$Ri = \frac{(g/T)(\partial T/\partial u)}{\Delta u/\Delta z} \quad (3.4)$$

where

$\Delta u/\Delta z$ is a representative *constant* wind speed gradient, calculated over the boundary layer above the canopy

$\partial T/\partial u$ is the variation of temperature with horizontal mean velocity within this boundary layer

The purpose of past studies was to evaluate the friction velocity from mean wind speed and temperature profile measurements under different conditions of atmospheric stability and to compare the results for flow over orchards covered with screens and over uncovered orchards. Results of such analyses have shown that horizontal screen covers reduced the friction velocity (Figure 3.7) and roughness length and increased the aerodynamic resistance and zero-plane displacement in comparison with an uncovered plantation. Results under unstable conditions were in agreement with a numerical calculation by Louis (1979). The results also exhibited the decay of turbulence at $Ri \cong 0.2$, in agreement with the well-known stability criterion of stratified shear flows (Turner, 1979). The modifications in flow properties along screens may be partially responsible for the reduced transport of water vapor from the crop to the atmosphere which may allow lower irrigation by the growers and thus increase the water saving.

Tanny et al. (2009) further elaborated the aforementioned analysis by comparing the results mentioned earlier for a small shading screen above a citrus orchard with measurements over a larger screenhouse (110 m × 60 m) in which pepper was grown. The analysis showed that under stable conditions the boundary layers in both configurations had almost similar properties. Under unstable conditions, however, more significant differences were observed presumably due to the differences in screen and crop types, and different available fetch in the two experiments.

3.3.3 Flow through Screens

Screens inhibit airflow through increased resistance, that is, increased pressure drop. The pressure drop across a screen, ΔP, can be calculated by the Bernoulli's equation:

$$\Delta P = 0.5\rho K U^2 \quad (3.5)$$

where

ρ is fluid density

K is pressure loss coefficient

U is mean upstream velocity

Many research works were devoted to the characterization of K as a function of screen and flow properties, part of them within the general engineering context, for example, Brundrett (1993) and Turner (1969).

Within the context of insect-proof screens used to protect agricultural crops, Teitel and Shklyar (1998) studied experimentally and numerically the pressure drop across such screens. They conducted pressure drop measurements in a wind tunnel and carried out numerical simulations using a commercial finite element program. Presenting the pressure loss coefficient as a function of the thread Reynolds number for a given screen porosity (defined as open area divided by total screen area) suggested that K also depended on the weave texture. In particular, it was shown that as the distance between adjacent threads increases, the pressure drop becomes less dependent on the detailed weave texture.

Flow patterns of different weave configurations were also investigated by Teitel and Shklyar (1998). An example is presented in Figure 3.8, where the velocity vectors downstream two intersecting threads (designated as cylinders 1 and 2) are shown. It can be observed in the figure that the vertical flow patterns immediately downstream the screen induce a region with backflow; this may assist in the process of insect exclusion.

A recent comprehensive review by Teitel (2007) considered different approaches in characterizing the flow resistance induced by screens. The general equation describing flow through a porous medium can be written as

$$\frac{\mu}{k}u + \rho\left(\frac{Y}{k^{1/2}}\right)|u|u = \frac{\partial P}{\partial x} \quad (3.6)$$

where

μ is the dynamic viscosity

u is upstream fluid velocity

k is permeability

Y is inertial factor of the screen

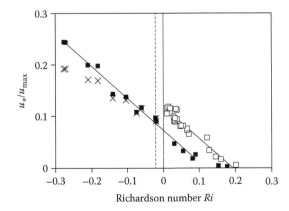

FIGURE 3.7 Friction velocity, normalized by maximum wind speed measured on the tower, u_*/u_{max}, as a function of the bulk Richardson number; open squares, uncovered orchard; solid squares, covered orchard; times symbol, calculations based on Louis (1979); the vertical dashed line is the lower validity limit of the calculation. (From Tanny, J. and Cohen, S., *Biosyst. Eng.*, 84(1), 57, 2003, Figure 6. With permission.)

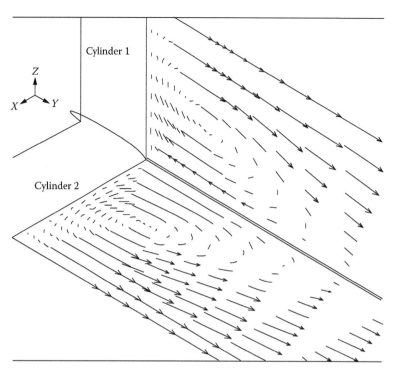

FIGURE 3.8 Velocity vectors downstream an insect-proof screen. Thread Reynolds number = 10. The two orthogonal cylinders represent intersecting screen threads. (From Teitel, M. and Shklyar, A. *Trans. ASAE*, 41(6), 1829, 1998, Figure 7. With permission.)

For very small air velocities (and $Re < 1$), the quadratic term in Equation 3.6 can be neglected, whereas for higher velocities and Reynolds numbers the viscous term can be neglected, and Equation 3.6 is reduced to the Bernoulli equation as was given before in Equation 3.5. Teitel (2007) review found only a few studies that related the permeability (k) and inertial factor (Y) of a screen to its porosity. Comparison among the results of these studies showed significant differences. Differences were also obtained among results relating the porosity and the pressure loss coefficient (K). This indicates the need of a more comprehensive study of these issues.

Teitel and Tanny (2005) investigated the effect of insect-proof screens, deployed at the vertical roof openings of a greenhouse, on various mean and turbulent flow characteristics. High-frequency (5 Hz) measurements of air velocity perpendicular to the screen and air temperature near the screen were used to estimate mean velocity and direction, skewness and flatness, RMS velocity, velocity spectra, velocity-temperature co-spectra, integral scales of turbulent flow, and to carry out a quadrant analysis to identify events of ejections and sweeps.

Results showed that without screens there was inflow at one edge of the opening and outflow at the other edge (see Figure 3.2, Section 3.2.2). With screens, however, in part of the days a flow pattern similar to the case without screens was recorded, while in other days there was an outflow at both edges. This observation, also supported by simultaneous temperature measurements by Teitel and Tanny (2005), suggested a more uniform microclimate near the opening when screens are deployed.

Values of skewness and flatness were significantly different from those for a Gaussian distribution. Skewness was mostly negative, indicating strong negative turbulent motions. Velocity flatness values suggested that the outflow was more intermittent than the inflow for both screened and unscreened openings. Spectral analysis showed that installing screens on the greenhouse openings reduced the turbulence energy density at low frequencies, and increased the decay rate of turbulence, apparently due to the smaller scales generated by the screen. Positive co-spectra of velocity and temperature, estimated for openings with and without screens, indicated turbulent heat transfer from the greenhouse to the external surroundings in both cases. The quadrant analysis showed that with screens, the contribution of the different quadrants to the turbulent heat flux were similar at the two edges of the opening while without screens there was a more pronounced difference between the two edges. This, again, indicated on the more uniform microclimate induced by the screens.

Integral length scales calculated by Teitel and Tanny (2005) for the flow through the roof openings resulted with values of 0.7–0.84 m for the inflow and 0.12–0.31 m for the outflow. The scale for the inflow was commensurate to the height of the roof opening. The calculations showed only a relatively small difference between integral length scales with and without screens deployed at the openings. This result was not surprising since screens are expected to mainly affect the small scales of the flow.

In an attempt to compensate the effect of screens in inhibiting greenhouse ventilation, various screen configurations have been applied in greenhouse openings. The use of inclined or concertina-shaped screens (Bailey, 2003) increases the actual flow area and thus may compensate for the reduced flow due to the higher pressure drop across the screen. An experimental and CFD simulation of flow through vertical and inclined screens was recently conducted by Teitel et al. (2009). In the experiments, a sub-tunnel with a cross section 280 × 280 mm and 1.2 m long was placed on the floor of a larger wind tunnel. Tested screens were mounted on the side of the sub-tunnel facing the upstream flow. Screens were deployed at inclination angles of 45°, 90°, and 135° between the upstream direction and wind-tunnel floor. In addition, three screen porosities were examined at each inclination angle. Vertical profiles of velocity were measured downstream of the screens using an omnidirectional hot-wire sensor. In the simulations, the ANSYS-CFX-11 software package was utilized, with the addition of the pressure drop expression of Equation 3.6 as a source term to the momentum equation. Following Fatnassi et al. (2006), Teitel et al. (2009) used a porous medium with a larger thickness than a real screen, but with similar airflow transmission characteristics, to avoid discontinuities in the meshing of the simulation domain.

Results by Teitel et al. (2009) showed that inclined screens generate a non-uniform downstream velocity profile. Generally, good agreement was obtained between the velocity measurements and the simulations. Positioning the screen at 45° caused more air to penetrate into the sub-channel, thus somewhat increasing the mass flow rate as compared to the 135° inclination (Figure 3.9). With 135° inclination, more air could escape above the sub-channel and mass flow rate was somewhat smaller. The effect of screen porosity on mass flow rate through the screen was more pronounced than the effect of screen inclination. Increasing the porosity from 0.4 to 0.62 nearly doubled the mass flow rate, as shown in Figure 3.9. Teitel et al. (2009) also calculated the forces on the screen induced by the flow, which is an important consideration in the design of the greenhouse structure.

3.4 Summary

A wide variety of fluid mechanics principles and tools are commonly applied in the study of agricultural systems. These tools are mainly aimed at improving the understanding and hence the design and performance of such systems. For example, fluid mechanics studies deal with non-uniform climatic conditions inside agricultural structures which may reduce the total yield. InsuDcient ventilation in greenhouses or screenhouses may result is CO_2 shortage which may deteriorate photosynthesis and production. The reduced friction velocity of the flow along screens indicates smaller exchange between plants and atmosphere which may lead to increased water saving. It must be recalled that the ultimate goal in the design of an agricultural system is increasing the profitability of the growers to facilitate suDcient supply of food in affordable prices for the growing world population.

This chapter focused on some applications of fluid mechanics in protected cultivation since this has been a widespread adopted technology in agricultural production in many regions of the world in the past few decades. The chapter did not cover all topics related to fluid mechanics in agricultural systems. Some topics that were not covered here are fluid mechanics of irrigation systems (e.g., sprinklers, drip irrigation systems), water distribution in the root zone within the soil (i.e., flow in porous media), microclimate of livestock buildings, transport of solutes and nutrients in agricultural soils, and more. A much more comprehensive review is required to cover these topics.

References

Bailey BJ (2003) Screens stop insects but slow airflow. *Fruit and Vegetable Technology*, 3:6–8.

Bartzanas T, Boulard T, Kittas C (2004) Effect of vent arrangement on windward ventilation of a tunnel greenhouse. *Biosystems Engineering*, 88(4):479–490.

Boulard T, Feuilloley P, Kittas C (1997a) Natural ventilation performance of six greenhouse and tunnel types. *Journal ofeAgriculture Engineering Research*, 67:249–266.

Boulard T, Meneses JF, Mermier M, Papadakis G (1996) The mechanisms involved in the natural ventilation of greenhouses. *Agricultural and Forest Meteorology*, 79:61–77.

Boulard T, Papadakis G, Kittas C, Mermier M (1997b) Air flow and associated sensible heat exchanges in a naturally ventilated greenhouse. *Agricultural and Forest Meteorology*, 88:111–119.

Boulard T, Wang S (2002) Experimental and numerical studies on the heterogeneity of crop transpiration in a plastic tunnel. *Computers and Electronics in Agriculture*, 34:173–190.

Brundrett E (1993) Prediction of pressure drop for incompressible flow through screens. *Journal of Fluids Engineering*, 115:239–242.

Fatnassi H (2001) Modelling and characterization of the microclimate and the climatic heterogeneity in large-scale greenhouse fitted with insect-proof netting. PhD thesis, Faculty of Sciences, University Ibn Zohr, Agadir, Morocco.

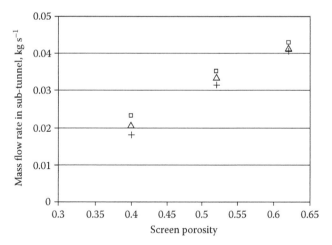

FIGURE 3.9 Mass flow rate (based on simulation results) through the sub-channel as function of screen porosity. Square, 45°; triangle, 90°; plus, 135°. Upstream velocity 0.85 m s⁻¹. (From Teitel, M. et al., *Biosyst. Eng.*, 104, 404, 2009, Figure 10. With permission.)

Fatnassi H, Boulard T, Poncet C, Chave M (2006) Optimization of greenhouse insect screening with computational fluid dynamics. *Biosystems Engineering*, 93(3):301–312.

Hunt GR, Linden PF (2004) Displacement and mixing ventilation driven by opposing wind and buoyancy. *Journal of Fluid Mechanics*, 527:27–55.

Katsoulas N, Bartzanas T, Boulard T, Mermier M, Kittas C (2006) Effect of vent openings and insect screens on greenhouse ventilation. *Biosystems Engineering*, 93(4):427–436.

Kittas C, Boulard T, Papadakis G (1997) Natural ventilation of a greenhouse with ridge and side openings: Sensitivity to temperature and wind effects. *Transactions of the ASAE*, 40(2):415–425.

Kittas C, Bartzanas T, Jaffrin A (2003) Temperature gradients in a partially shaded large greenhouse equipped with evaporative cooling pads. *Biosystems Engineering*, 85(1):87–94.

Li S, Willits DH (2008a) An experimental evaluation of thermal stratification in a fan-ventilated greenhouse. *Transactions of the ASABE*, 51(4):1443–1448.

Li S, Willits DH (2008b) Modeling thermal stratification in fan-ventilated greenhouses. *Transactions of the ASABE*, 51(5):1735–1746.

Linden PF (1999) The fluid mechanics of natural ventilation. *Annual Reviews in Fluid Mechanics*, 31:201–238.

Louis JF (1979) A parametric model of vertical eddy fluxes in the atmosphere. *Boundary Layer Meteorology*, 17:187–202.

Möller M, Tanny J, Cohen S, Teitel M (2003) Micrometeorological characterization in a screenhouse. *Acta Horticulturae*, 614:445–451.

Shilo E, Teitel M, Mahrer Y, Boulrad T (2004) Air-flow patterns and heat fluxes in roof-ventilated multi-span greenhouse with insect-proof screens. *Agricultural and Forest Meteorology*, 122:3–20.

Tanny J, Cohen S (2003) The effect of a small shade net on the properties of wind and selected boundary layer parameters above and within a citrus orchard. *Biosystems Engineering*, 84(1):57–67.

Tanny J, Cohen S, Teitel M (2003) Screenhouse microclimate and ventilation: An experimental study. *Biosystems Engineering*, 84(3):331–341.

Tanny J, Haijun L, Cohen S (2006) Airflow characteristics, energy balance and eddy covariance measurements in a banana screenhouse. *Agricultural and Forest Meteorology*, 139:105–118.

Tanny J, Haslavsky V, Teitel M (2008) Airflow and heat flux through the vertical opening of buoyancy-induced naturally ventilated enclosures. *Energy and Buildings*, 40(4):637–646.

Tanny J, Möller M, Cohen S (2009) Aerodynamic properties of boundary layers along screens. *Biosystems Engineering*, 102(2):171–179.

Teitel M (2001) The effect of insect-proof screens in roof openings on greenhouse microclimate. *Agricultural and Forest Meteorology*, 110:13–25.

Teitel M (2007) The effect of screened openings on greenhouse microclimate. *Agricultural and Forest Meteorology*, 143:159–175.

Teitel M, Tanny J (1999) Natural ventilation of greenhouses: Experiments and model. *Agricultural and Forest Meteorology*, 96:59–70.

Teitel M, Dvorkin D, Haim Y, Tanny J, Seginer I (2009) Comparison of measured and simulated flow through screens: Effects of screen inclination and porosity. *Biosystems Engineering*, 104:404–416.

Teitel M, Shklyar A (1998) Pressure drop across insect-proof screens. *Transactions of the ASAE*, 41(6):1829–1834.

Teitel M, Tanny J (2005) Heat fluxes and airflow patterns through roof windows in a naturally ventilated enclosure. *Flow, Turbulence and Combustion*, 74(1):24–47.

Turner JT (1969) A computational method for the flow through non-uniform gauzes: The general two-dimensional case. *Journal of Fluid Mechanics*, 36:367–383.

Turner JS (1979) *Buoyancy Effects in Fluids*. Cambridge University Press, London, U.K.

Wikipedia (2012) http://en.wikipedia.org/wiki/Agriculture, from: Key Indicators of the Labour Market, International Labour Organization, UN.

4

Desalination and the Environment

M.J. Davidson
University of Canterbury

C.J. Oliver
University of Canterbury

4.1 Introduction

Historically, desalination has played an essential role in providing secure water supplies for ad uent communities where natural water resources were scarce but oil resources were plentiful. This approach is common in the Middle East, where many communities continue to have the financial and energy resources to implement distillation processes to extract fresh water from brackish inland and coastal sources. Modern distillation plants typically employ either multistage flash (MSF) distillation or multi-effect distillation (MED) technologies to extract fresh water, with the former being the more common of the two. While the eDciencies and cost-effectiveness of these processes have improved over time, they still remain prohibitively expensive for most municipalities, and the recent growth in global desalination capacity (approximately 10% per annum between 2004 and 2009; Wangnick/GWI, 2005; GWI/DesalData, 2009) would not have been possible without the development of alternative desalination processes. In particular, the use of membrane technologies, which have improved rapidly in recent years, has provided the basis for more cost-eDcient potable water extraction. Membranes are employed to filter the saltwater under high pressure to produce potable water in a reverse osmosis (RO) process. Based on this alternative technology, desalination has become a feasible option for a broader range of communities throughout the globe, which continue to grapple with water supply and security issues. In 2005, RO provided approximately 50% of the global desalination capacity (Gleick et al., 2007), which reflects the broader appeal of this technology. While the bulk of the capacity and planned expansion remains in the Middle East and North Africa, significant capacity expansions are also being implemented and planned in other parts of the world, such as the United States (notably California); Spain, where desalinated water is also utilized for agricultural purposes; and China. In addition, Singapore and the Australian cities of Perth and Sydney have also recently developed large-scale desalination facilities (extracting well in excess of 100,000 m³ of potable water per day) and other Australian cities are planning or constructing similarly sized facilities; Adelaide and Melbourne are current examples at the time of writing (begining of 2010). It is important to note that this increase in capacity should be part of an integrated solution to water-supply deficiencies, which have essential water conservation and water recycling components. Desalination provides an increased level of independence from less reliable sources of water, and in Australia extended droughts have been a major catalyst for the significant increases in desalination capacity noted earlier. Less reliable traditional water sources are a feature of changes in the global climate and therefore the present rapid increase in global desalination capacity is likely to continue in the foreseeable future.

Large-scale desalination facilities are a feature of this growth in capacity with the largest RO and MSF plants currently extracting 330,000 m³/day (Ashkelon, Israel) and 1.64 M m³/day (Jebel Ali, United Arab Emirates), respectively. With the increased number and capacity of these large-scale plants, there are growing concerns about their potential environmental impacts at global, regional, and local levels. In a global context, desalination plants have significant carbon footprints because of the energy-intensive nature of the processes involved. These energy demands can be mitigated to some extent through energy recovery processes, for example, the high-pressure water in an RO plant can also be utilized to drive turbines to generate electricity; and through the creation of equivalent renewable energy sources, for example, a 272 GWh/a wind farm was constructed in conjunction with the Perth desalination plant that consumes approximately 185 GWh/a (Stover and Crisp, 2007). Regional or transnational problems are evident in areas such as the Arabian Gulf, where desalination plants in coastal nations discharge large quantities of contaminants (copper, chlorine, and others) into a semi-enclosed receiving water, and the cumulative effects are of increasing concern (Latteman and Höpner, 2008). Within each nation there are also local environmental issues associated with land use, plant construction, noise and air pollution, and the impacts on the receiving water of the intake and disposal phases of the desalination process.

Ed uent disposal from desalination plants (including backwash) has the potential to create severe environmental impacts. Elevated salt concentrations associated with this ed uent are known to have adverse effects on corals and sea grasses, but the ed uent also contains chemicals that are utilized to protect the plant and improve the eDciency of the process. The nature of these chemicals depends on the desalination process but in general involve agents for antiscaling and biofouling control, typically chlorination for the latter. RO plants utilize preconditioning coagulants and dechlorination to protect the membranes, whereas with MSF plants dechlorination is not required, but anticorrosion and antifoaming agents are needed. Ed uent from MSF plants typically have elevated copper levels, because of corrosion of the thermally eDcient copper–nickel alloy tubing, that is central to the effectiveness of the distillation process.

Clearly there are significant differences in the chemical composition of the ed uents from the different plant types and detailed epidemiological studies of the impacts of these chemicals are still required in many cases, so that their potential impacts can be adequately assessed. Equally important are the differences in the physical properties of the ed uent, which vary significantly because of the processes involved and because in some instances the ed uent is mixed with wastewater from associated power generation facilities or municipal ed uent. Thus, different configurations can generate ed uents that range from positively buoyant discharges (mixed ed uents from MSF plants which have elevated temperatures), to pure negatively buoyant discharges from RO plants, where there is little or no temperature difference and ed uent salinity is approximately 3% higher than that of the ambient.

Ideally an appropriate regulatory framework should control the environmental impacts of such facilities. While it is outside the scope of this chapter to detail approaches in different parts of the world, it is worthwhile to briefly outline an acceptable structure that potentially provides a basis for addressing the global, regional, and local issues. An appropriate consent process, integrating genuine public consultation, provides an opportunity for the broad range of environmental issues to be addressed. For example, the carbon footprint of the plant can be mitigated through compensatory measures, such as the development of renewable energy sources that are at least equivalent to the energy consumed by the plant. Similarly, issues associated with land use, construction, and operation of the plant can be resolved as part of this process. A critical aspect is the disposal of the very large volumes of ed uent

from these plants, and its impacts should be evaluated in the context of appropriate ed uent standards and ambient standards. Ed uent standards are applied at the discharge point, the source, and are designed to protect the environment from short-term or acute impacts, whereas ambient standards are applied at the edge of the mixing zone and are designed to protect the environment from long-term or chronic impacts. The dilution to be achieved within the mixing zone is the ratio of ed uent to ambient standards for any given contaminant. These standards are set based on the best available scientific information, and in this context there are some diDculties because while some studies exist, a great deal more information is needed for the standards to be set with confidence, as indicated previously.

The definition of mixing zones, within which ambient standards can be exceeded, remains controversial. Assessing the characteristics of the discharge within any proposed mixing zone is of critical importance in designing the disposal system. Here the engineer has a variety of options in terms of the disposal location, number of ports, discharge angle, etc. The design of such systems is commonly informed by the relatively simple integral models, which are employed to predict the mixing that takes place after discharge. These models provide effective predictions of the rapid mixing that occurs as the ed uent moves from its point of release to a point where it impacts on a nearby boundary, either the ocean surface or seabed. This initial dilution region is crucial in assessing the ed uent characteristics within any proposed mixing zone. Initial mixing (or dilution) modeling of brine, and other discharges, represents another application where environmental fluid dynamics provides the basis for developing powerful analysis tools to assist in decision-making processes for important engineering problems. In this chapter, the development and predictive capabilities of this class of models are demonstrated through the formulation of a set of analytical solutions, which are assessed in the context of desalination discharges.

4.2 Principles

Traditionally, predictive models of wastewater discharges have been developed with a primary focus on municipal discharges, where the discharges are typically positively buoyant (e.g., fresh water into a marine environment) and experimental studies have focused on relevant discharge configurations. In this context, horizontal discharges are generally preferred because of the higher dilution levels achieved, while avoiding direct contact with the seabed. Thus, existing models can be directly applied to positively buoyant discharges from desalination facilities with some confidence, although there are potential diDculties where this positive buoyancy is achieved through a combination of elevated temperatures and salt concentrations, typical of some desalination discharges, because of the differing diffusivities of the sources of buoyancy.

In contrast, the discharge configurations employed for negatively buoyant ed uents, such as those from RO plants, differ significantly with the ed uent being released at some inclination to the horizontal (typically 60°). Thus, the wastewater initially

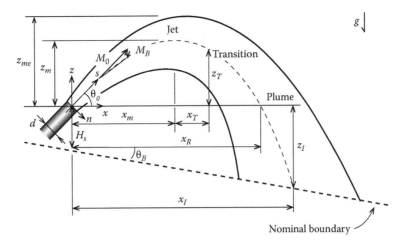

FIGURE4 .1 Schematic diagram of a negatively buoyant discharge that is representative of a submarine disposal of ed uent from an RO plant.

rises toward the ocean surface in the direction of discharge and then falls back to the ocean floor. This configuration is preferred because it enhances the initial mixing of the ed uent, thereby rapidly reducing pollutant concentrations. This mixing is normally further enhanced by the presence of an ambient current, so the more critical cases from the design and environmental impact perspectives are those where there is no motion in the ambient. A schematic diagram of such a discharge is shown in Figure 4.1. Note stratification of the receiving water can also have a significant impact on the mixing, but is not dealt with here.

In Figure 4.1 a source of diameter d is located a height H_S above a sloping seabed, and the slope of this bed is θ_B. The ed u- ent is released at an angle θ_0 to the horizontal, with an initial velocity U_0 and density ρ_0. The ambient density is assumed to be uniform and is represented by ρ_a. Two coordinate systems are useful when describing the flow behavior; one is based on horizontal x- and vertical z-axes with their origin at the source. The second, which also has its origin at the source, makes use of an s coordinate in the direction of the *initial* discharge and an orthogonal n coordinate as shown in Figure 4.1.

Two regions of particular interest when designing these discharge systems are the maximum height and the impact regions. The coordinates of the centerline maximum height are (x_m, z_m), and this is where the minimum cross-sectional dilution is closest to the free surface. Another important parameter in this region is the maximum height of the *edge* of the flow (z_{me}), which is reached at a similar horizontal location of the centerline maximum height. In the impact region, the location where the flow centerline impacts the seabed, (x_I, z_I) is important because this is where the minimum dilution occurs within the cross section. However, another useful reference in this region is location of the return point $(x_R, 0)$, where the flow returns to its original height. Conditions at the return point can be modeled without reference to specific site conditions (H_S and θ_B), and where bed slopes are shallow and source heights small, it is diDcult to distinguish between the return and impact points.

There are several options for modeling flows of this type, which range from sophisticated computational models

(e.g., large-eddy-simulations) to a simple regression analysis based on available experimental data. The resources required to run sophisticated computational models are significant, but they do have the potential to capture the detailed physics of the flows. Conversely, the resources required for a regression analysis are minimal, but much of the physics is lost. Integral models, where the integral forms of the equations of motion are applied to a control volume, provide a useful compromise, and numerical forms of these models are commonly applied in the design of wastewater disposal systems of this nature. These models come in a variety of forms, with some developed in a Lagrangian framework (VISJET—Lee et al., 2000) and others from a Eulerian perspective (CorJet—Jirka, 2004). The examples noted also differ in the selected closure assumptions and cross-sectional profiles. However, both models are generic in nature, modeling key physical characteristics for a broad range of discharge conditions and are reasonably simple to implement with basic computing resources. An alternative involves making use of this integral approach to develop analytical solutions for specific problems. Here there is a further loss of generality, but the analytical relationships provide a relatively simple method for predicting and interpreting the flow behavior for a restricted range of initial conditions. In the context of discharges from RO plants, Kikkert et al. (2007) developed a set of analytical solutions to predict key parameters in the maximum height and return regions, and Oliver et al. (2012a) have extended these solutions to include site-specific parameters, so that conditions at the impact point can be explored in more detail. In addition to their practical value, these solutions represent a very interesting application of two fundamental flows in environmental fluid dynamics: the ubiquitous *jets* and *plumes*. It would seem appropriate therefore to explore the development of these solutions in some depth.

Referring again to Figure 4.1, the flow is released with an initial momentum flux (per unit density) $M_0 = U_0^2 \pi d^2 / 4$. This momentum flux governs the flow behavior near the source, where it behaves in a similar manner to a *jet*. However, the density difference between the source and ambient fluids creates a buoyancy force, and thus a buoyancy-generated momentum flux develops as the flow evolves. When the buoyancy-generated momentum flux exceeds the value of the constant initial momentum flux, the flow behavior changes to that of a *plume*. At the maximum height, the buoyancy-generated momentum flux is equal to the vertical component of the initial momentum flux. Thus, for inclined discharges, the flow reaches maximum height as a *jet* because the transition to a plume occurs after the flow reaches this location. Thus, the conditions at maximum height can be predicted based on standard jet solutions, where it is assumed that the role of the buoyancy-generated momentum flux is to simply deflect the flow. Given this is the simplest of the analytical solutions, it is therefore a convenient starting point.

4.2.1 Conditions at Maximum Height

The bulk behavior in the jet region can be defined in terms of a characteristic spread b_t, velocity u_t, and pollutant concentration c_t at any selected cross section that is perpendicular to the flow centerline. If the velocity and concentration distributions are assumed uniform, then the momentum flux per unit density and pollutant mass flux at a given cross-section can be written as

$$u_t^2 \pi b_t^2 = M_0 \tag{4.1}$$

$$c_t Q = c_t u_t \pi b_t^2 = C_0 Q_0 \tag{4.2}$$

The momentum flux of a *jet* does not change with distance from the source because there is no net force acting (assuming the pressure distributions are hydrostatic). Similarly, the mass flux of the pollutant is unchanged with distance from the source, because the ambient fluid entrained into the jet through turbulent mixing processes is assumed to be unpolluted. Note Q_0 and C_0 are the volume flux and pollutant concentration at the source, respectively, and Q is volume flux at the selected cross-section.

Solutions to the jet problem are then obtained by either assuming a spread relationship or by introducing an entrainment velocity through a mass flux equation and making an assumption about how this velocity relates to the cross-sectional characteristic velocity. The simplicity of the spread assumption appeals and it can be written in the form

$$b_t = k_t s \tag{4.3}$$

where it has been assumed that the flow originates from a point source. This is a reasonable assumption provided predictions are not required close to the source (within $10d$), where there is a zone of flow establishment. The velocity decay and dilution (C_0/c_t) of the jet can therefore be written as

$$u_t = \sqrt{\frac{M_0}{\pi}} \frac{1}{k_t s} \tag{4.4}$$

$$\frac{C_0}{c_t} = \frac{Q}{Q_0} = 2k_t \frac{s}{d} \tag{4.5}$$

The growth of the buoyancy-generated momentum flux per unit density (M_B) within the jet region is based on the conservation of vertical momentum flux, and this can be written as

$$\frac{dM_B}{ds} = g_t' \pi b_t^2 \tag{4.6}$$

where g_t' is the associated reduced gravity. The tracer flux relationship can also be applied to the density excess or buoyancy to give

$$g_t' u_t \pi b_t^2 = g_0' Q_0 \tag{4.7}$$

Combining Equations 4.6 and 4.7 and then making use of Equation 4.4 to define the velocity, because the flow is essentially a jet, gives the following relationship for the development of the buoyancy-generated momentum flux within the jet region:

$$\frac{M_B}{M_0} = k_t \frac{g_0' d}{U_0^2} \left(\frac{s}{d}\right)^2 = \frac{k_t}{F_0^2} \left(\frac{s}{d}\right)^2 \tag{4.8}$$

where F_0 is the densimetric Froude number at the source. Equation 4.8 provides the basis for defining the location of the transition from a jet to a plume and also the location of the maximum height. The transition from a jet to a plume occurs when the momentum flux ratio is approximately equal to 1 and the maximum height occurs when this ratio is equal to the sine of the initial discharge angle. These conditions create the following relationships:

$$\frac{S_{JP}}{d} = \frac{F_0}{\sqrt{k_t}} \tag{4.9}$$

$$\frac{s_m}{d} = \frac{F_0}{\sqrt{k_t}} \sqrt{\sin\theta_0} = S_{JP}\sqrt{\sin\theta_0} \tag{4.10}$$

where
S_{JP} is the value of the s coordinate at the transition point
s_m is the coordinate s value at maximum height

Recall that the s coordinate is measured in the direction of the initial discharge.

The trajectory of the flow is given by

$$\frac{dn_*}{ds_*} = \frac{M_{B*}\cos\theta_0}{1 - M_{B*}\sin\theta_0} \tag{4.11}$$

where the subscript "*" denotes a dimensionless parameter and the selected length and momentum scales are the diameter d and initial momentum flux M_0. Substituting

Equation 4.8 into Equation 4.11 and integrating yields the following relationship for n_*:

$$n_* = \frac{1}{\tan\theta_0} \frac{S_{JP*}}{\sin^{1/2}\theta_0} \left[-\frac{s_* \sin^{1/2}\theta_0}{S_{JP*}} + \frac{1}{2}\ln\left(\frac{1+\left(s_* \sin^{1/2}\theta_0 / S_{JP*}\right)}{1-\left(s_* \sin^{1/2}\theta_0 / S_{JP*}\right)}\right) \right]$$

(4.12)

The transformations between the $n–s$ and $x–z$ coordinate systems are

$$z_* = s_* \sin\theta_0 - n_* \cos\theta_0$$ (4.13)

$$x_* = s_* \cos\theta_0 + n_* \sin\theta_0$$ (4.14)

The location of the centerline maximum height is then determined through substitution of Equation 4.10 into Equation 4.12 and the subsequent transformation to the $x–z$ coordinate system through Equations 4.13 and 4.14 to give

$$\frac{z_{m*}}{F_0} = \frac{1}{\sqrt{k_t}}\sin^{3/2}\theta_0 \left[1 + \frac{1}{\tan^2\theta_0}\left(1 - \frac{1}{2\sin\theta_0}\ln\left(\frac{1+\sin\theta_0}{1-\sin\theta_0}\right)\right) \right]$$

(4.15)

$$\frac{x_{m*}}{F_0} = \frac{1}{2\sqrt{k_t}}\frac{\cos\theta_0}{\sin^{1/2}\theta_0}\ln\left(\frac{1+\sin\theta_0}{1-\sin\theta_0}\right)$$ (4.16)

The cross-sectional average dilution at this location is obtained by combining Equation 4.10 with Equation 4.5:

$$\frac{1}{F_0}\frac{C_0}{c_t} = 2\sqrt{k_t}\sin\theta_0$$ (4.17)

However, it is the minimum cross-sectional dilution that is of particular interest from an environmental perspective, and in addition it is necessary to define a flow boundary to determine the maximum edge height (z_{me}). To obtain these predictions from the model, it is necessary to consider the actual form of the cross-sectional profiles, which are known to have a Gaussian form for a jet. The velocity and mean concentration profiles are commonly represented by the following relationships:

$$\frac{u}{U_m} = \exp\left(-\left(\frac{r}{b}\right)^2\right)$$ (4.18)

$$\frac{c}{C_m} = \exp\left(-\left(\frac{r}{\lambda b}\right)^2\right)$$ (4.19)

In the profile relationships U_m and C_m represent the maximum cross-sectional velocity and concentration, respectively, u and c are the local values of these parameters, r is a radial coordinate,

and finally b and λb are the characteristic spreads of the velocity and tracer profiles, respectively.

A mapping of the so-called top-hat bulk parameters (u_t, b_t, and c_t) to their Gaussian equivalents (b, U_m, and C_m) is obtained by conserving mass and momentum fluxes at a given cross-section. Comparing *mass fluxes* (per unit density), we have

$$u_t\pi b_t^2 = \int_0^\infty u2\pi r\,dr = \int_0^\infty \frac{u}{U_m}2\pi\frac{r}{b}\frac{dr}{b}U_mb^2 = I_Q U_m b^2$$ (4.20)

where

$$I_Q = \int_0^\infty \frac{u}{U_m}2\pi\frac{r}{b}d\left(\frac{r}{b}\right) = \pi$$

Note I_Q is a constant because the functional form of the velocity profiles does not change with distance, that is, the profiles are *self-similar*. Similarly, a comparison of the *momentum fluxes* gives

$$u_t^2\pi b_t^2 = \int_0^\infty u^2 2\pi r\,dr = \int_0^\infty \left(\frac{u}{U_m}\right)^2 2\pi\frac{r}{b}\frac{dr}{b}U_m^2b^2 = I_M U_m^2 b^2$$ (4.21)

where

$$I_M = \int_0^\infty \left(\frac{u}{U_m}\right)^2 2\pi\frac{r}{b}d\left(\frac{r}{b}\right) = 1.7$$

It is also important that the *tracer fluxes* are consistent, and this requires

$$u_t c_t \pi b_t^2 = \int_0^\infty uc2\pi r\,dr = \int_0^\infty \frac{u}{U_m}\frac{c}{C_m}2\pi\frac{r}{b}\frac{dr}{b}U_mC_mb^2 = I_{QC}U_mC_mb^2$$

(4.22)

where

$$I_{QC} = \int_0^\infty \frac{u}{U_m}\frac{c}{C_m}2\pi\frac{r}{b}d\left(\frac{r}{b}\right) = 1.95$$

Note the value of I_{QC} is dependent on the ratio of tracer to velocity spread (λ), and the values of both I_M and I_{QC} have a dependence on turbulent fluxes of momentum and mass, respectively. While these values are known to vary between the jet and plume regions, the variations are generally comparable to experimental error, and it is therefore reasonable to adopt a single set of values, which are normally based on the buoyancy-driven flow because of its practical significance. Recently Wang and Law (2002) carried out a detailed investigation into buoyant jet behavior and their data combined with information from previous studies of a similar nature, Papanicolaou and List (1988) and Hussein et al. (1994), for example, indicate that the turbulent flux contributions for momentum and mass transport are approximately 10% and 15% of the mean fluxes, respectively. These studies also indicate that assuming a Gaussian spreading rate (k) of 0.11 and a spread ratio (λ) of 1.1 is reasonable for these flows.

Combining Equations 4.20 and 4.21 and then substituting the resulting relationships into Equation 4.22 gives the following mappings between the different cross-sectional representations:

$$U_m = \frac{I_Q}{I_M} u_t = 1.85 u_t, \quad b = \sqrt{\frac{\pi I_M}{I_Q^2}} \, b_t = .74 b_t, \quad \text{and}$$

$$C_m = \frac{\pi}{I_{QC}} \frac{u_t}{U_m} \left(\frac{b_t}{b} \right)^2 c_t = 1.61 c_t \qquad (4.23)$$

Noting that $k_t = 0.11/0.74 = 0.15$, the minimum dilution at the maximum centerline height can be written as

$$\frac{1}{F_0} \frac{C_0}{C_m} = k_{mI} \frac{2 k_t^{1/2}}{1.61} \sqrt{\sin \theta_0} = k_{mI} 0.48 \sqrt{\sin \theta_0} \qquad (4.24)$$

Here the coeDcient k_{mI} has been introduced to enable discrepancies between the predicted and measured dilutions to be quantified. A value of 1 for this parameter indicates that the predicted and measured dilutions are in good agreement.

Predictions of the maximum edge height depend on the definition of the flow boundary. Based on measured mean concentration profiles Kikkert et al. (2007) assumed that the flow edge was defined at a radius of $2\lambda b$, which the Gaussian profiles suggest is a 2% mean concentration contour. Equation 4.15 can then be modified to predict the maximum height of the discharge:

$$\frac{z_{me^\star}}{F_0} = \frac{1}{\sqrt{k_t}} \sin^{1/2} \theta_0$$

$$\times \left[\sin \theta_0 \left(1 + \frac{1}{\tan^2 \theta_0} \left(1 - \frac{1}{2 \sin \theta_0} \ln \left(\frac{1 + \sin \theta_0}{1 - \sin \theta_0} \right) \right) \right) + 2\lambda k \right] \qquad (4.25)$$

Thus Equations 4.15, 4.16, 4.24, and 4.25 provide the basis for predicting the conditions at maximum height.

4.2.2 Conditions at the Return Point

As indicated previously, relationships for conditions at the return point can be derived without reference to specific site conditions (bed slope and source height). It is important to note that for shallow discharge angles it is possible for the flow to reach the return point as a jet, because of the deflection due to the buoyancy. Although not of particular practical interest, it is useful

to develop relationships for the flow reaching the return point as a jet in preparation for the more complex situation where it reaches the return point as a plume.

4.2.2.1 Jet at the Return Point

In the jet region, this location can be determined by recognizing that $z_\star = 0$ at this point and by manipulating Equations 4.12 through 4.14 to give the following relationship for the horizontal distance to the return point in the jet region (x_R):

$$\frac{x_{R^\star}}{F_0} = \frac{1}{\sqrt{k_t}} \frac{1}{\tan \theta_0} \frac{1}{\sin^{3/2} \theta_0}$$

$$\times \left[-\frac{x_{R^\star}}{S_{JP^\star}} \sqrt{\sin \theta_0} \cos \theta_0 + \frac{1}{2} \ln \left(\frac{1 + \left(x_{R^\star}/S_{JP^\star} \right) \sqrt{\sin \theta_0} \cos \theta_0}{1 - \left(x_{R^\star}/S_{JP^\star} \right) \sqrt{\sin \theta_0} \cos \theta_0} \right) \right] \qquad (4.26)$$

An approximate solution to Equation 4.26 can be obtained by expressing the logarithmic term as a series and ignoring higher order terms to give

$$\frac{x_{R^\star}}{F_0} \approx \sqrt{\frac{3}{k_t}} \frac{\sqrt{\sin \theta_0}}{\cos^2 \theta_0} \qquad (4.27)$$

At this point, the distance traveled in the source direction ($s_{R^\star}$) = $x_{R^\star} \cos \theta_0$ and the minimum dilution at the return point is therefore given by

$$\frac{C_0}{C_m} = k_{RI} \frac{2\sqrt{3 k_t}}{1.61} \frac{\sqrt{\sin \theta_0}}{\cos \theta_0} = k_{RI} 0.83 \frac{\sqrt{\sin \theta_0}}{\cos \theta_0} \qquad (4.28)$$

The coeDcient k_{RI} plays the same role as k_{mI} in Equation 4.24.

4.2.2.2 Plume at the Return Point

For larger discharge angles, the flow reaches the return point as a plume, and in order to predict key parameters at this location we need a set of plume solutions. A plume spreads at essentially the same rate as a jet, but because it is buoyancy driven it moves predominantly in the vertical (z) direction, so that

$$b_t = k_t z_p \qquad (4.29)$$

where z_p is the vertical distance from the plumeß virtual source. The virtual source is a point source from which a pure plume is released to create the same conditions at the transition from a jet to a plume. To create the same spread at the transition, for example, the virtual source would be located a vertical distance S_{Jp} above the jet to plume transition for the negatively buoyant discharges considered in this chapter. The development of the buoyancy-generated momentum flux within a plume is governed by

$$\frac{dM_B}{dz_p} = g' \pi b_t^2 \qquad (4.30)$$

As with the jet, the buoyancy flux is governed by Equation 4.7 and noting that $M_B = u_t^2 \pi b_t^2$, Equation 4.30 can be integrated to give

$$\frac{M_B}{M_0} = \left(\frac{3k_t}{2F_0^2}\right)^{2/3}\left(\frac{z_p}{d}\right)^{4/3} = \left(\frac{3}{2}\right)^{2/3}\left(\frac{z_{p*}}{S_{JP*}}\right)^{4/3} \quad (4.31)$$

Making use of the definition of M_B noted previously along with the spread assumption, Equation 4.31 can be reorganized to give

$$\frac{u_t}{U_0} = \left(\frac{3}{16k_t^2 F_0^2}\right)^{1/3}\left(\frac{d}{z_p}\right)^{1/3} \quad (4.32)$$

Dilution predictions for the plume can be obtained by combining Equation 4.32 with Equation 4.7 and the spread assumption:

$$\frac{g_0'}{g_t'} = \frac{C_0}{c_t} = \left(12\frac{k_t^4}{F_0^2}\right)^{1/3}\left(\frac{z_p}{d}\right)^{5/3} \quad (4.33)$$

Note these are cross-sectional average dilution predictions and minimum values are obtained by dividing through by a factor of 1.61 (as with the jet predictions).

In the plume region, the horizontal component of the initial momentum flux deflects the flow, so that the trajectory is defined by

$$\frac{dz_p}{dx_p} = \frac{M_B}{M_0 \cos\theta_0} \quad (4.34)$$

where x_p is the horizontal distance from the plume's virtual source. Note the vertical component of the initial momentum flux can also have an influence on the plume trajectory near the transition from a jet to a plume, but this has been neglected for simplicity. Substituting Equation 4.31 into Equation 4.34 and integrating creates a relationship for the plume trajectory beyond the transition point, where $z_p = S_{JP*}$:

$$x_{p*} = 12^{1/3} S_{JP*}\cos\theta_0\left(1 - \frac{1}{\left(z_{p*}/S_{JP*}\right)^{1/3}}\right) \quad (4.35)$$

The described plume model provides the basis for predicting conditions at the return point; however, its implementation requires that the location of the virtual source be defined. Kikkert et al. (2007) assumed $(x_p, z_p) = (0, S_{JP})$ at the transition, which is consistent with the simplistic nature of the model. With this virtual source, there are discontinuities in some of the bulk parameters at the transition point, and although the virtual source location can be adjusted to reduce these discontinuities, such adjustments will increase others and some form of compromise is ultimately required.

With the virtual source defined, the deflected plume solution is implemented beyond the transition point. The coordinates at a particular location are translated between those making reference to the virtual and the real sources. This translation requires the horizontal and vertical distances from the real source to the transition point (x_{T*}, z_{T*}) be determined. Relationships for these coordinates can be derived by noting again that $s_* = S_{JP*}$ at the transition, substituting this condition into Equation 4.12, and then these values of n_* and s_* are substituted into Equations 4.13 and 4.14 to give

$$z_{T*} = S_{JP*}\sin\theta_0\left(1 - \frac{1}{\tan^2\theta_0}\left(-1 + \frac{1}{2\sqrt{\sin\theta_0}}\ln\left(\frac{1+\sqrt{\sin\theta_0}}{1-\sqrt{\sin\theta_0}}\right)\right)\right) \quad (4.36)$$

$$x_{T*} = \frac{S_{JP*}}{2}\frac{\cos\theta_0}{\sqrt{\sin\theta_0}}\ln\left(\frac{1+\sqrt{\sin\theta_0}}{1-\sqrt{\sin\theta_0}}\right) \quad (4.37)$$

The distance z_{T*} can then be added to the vertical distance from the plume virtual source to the transition point (S_{JP*}) to give the vertical distance from the virtual source to the return point. This distance is then substituted into the trajectory relationship for the plume region to give the horizontal location of the return point relative to the transition point (x_p), that is,

$$x_{p*} = 12^{1/3} S_{JP*}\cos\theta_0\left(1 - \left(\frac{1}{K_{R\theta}}\right)^{1/3}\right) \quad (4.38)$$

where for convenience the coeDcient $K_{R\theta}$ is defined as

$$K_{R\theta} = 1 + \sin\theta_0\left(1 - \frac{1}{\tan^2\theta_0}\left(-1 + \frac{1}{2\sqrt{\sin\theta_0}}\ln\left(\frac{1+\sqrt{\sin\theta_0}}{1-\sqrt{\sin\theta_0}}\right)\right)\right)$$

Adding Equations 4.37 and 4.38 gives the horizontal location of the return point relative to the real source:

$$\frac{x_{R*}}{F_0} = \frac{1}{\sqrt{k_t}}\cos\theta_0\left[\frac{1}{2\sqrt{\sin\theta_0}}\ln\left(\frac{1+\sqrt{\sin\theta_0}}{1-\sqrt{\sin\theta_0}}\right) + 12^{1/3}\left(1 - \left(\frac{1}{K_{R\theta}}\right)^{1/3}\right)\right] \quad (4.39)$$

To obtain dilution predictions, the vertical distance from the plume's virtual source is required, and this is simply the sum of z_T and S_{JP}. This summation can be written as the coeDcient $K_{R\theta}$ multiplied by the transition length scale and substituted into Equation 4.33 to give

$$\frac{1}{F_0}\frac{C_0}{C_m} = k_{RI}\frac{12^{1/3}}{1.61}\sqrt{k_t}\,k_{R\theta}^{5/3} = k_{RI}\,0.55k_{R\theta}^{5/3} \quad (4.40)$$

where the factor 1.61 converts the cross-sectional average value to the local minimum dilution.

It remains to determine under what condition the flow returns to the source height in the jet and plume regions. The critical situation in this regard occurs when the transition from jet to plume behavior takes place at the source height, that is, $z_{T*} = 0$. From Equation 4.36, we then have the following condition:

$$1 + \tan^2 \theta_0 = \frac{1}{2\sin^{1/2}\theta_0} \ln\left[\frac{1 + \sin^{1/2}\theta_0}{1 - \sin^{1/2}\theta_0}\right] \qquad (4.41)$$

Expanding the logarithmic function as a series and neglecting higher order terms yields the following approximate condition:

$$\cos^2\theta_0 \approx 3\sin\theta_0 \qquad (4.42)$$

Equation 4.42 indicates that the flow centerline will return to the source height in the jet region for angles less than $\approx 20°$.

4.2.3 Conditions at the Impact Point

The relationships developed for conditions at the return point can be extended to incorporate geometric variations associated with specific site conditions (Oliver et al. 2012b). The source height and slope of the sea bed (Figure 4.1) are two potentially significant variations and these are dealt with in the context a plume impacting on the seabed. A similar approach can be adopted for shallow angles where a jet impacts the seabed, although as the slope of the seabed increases the range of angles for which this occurs diminishes.

The location of the impact point is defined by the coordinates (x_I, z_I) in Figure 4.1, and this occurs on a boundary defined by the relationship

$$z_{I*} - H_{s*} = x_{I*}\tan\theta_B \qquad (4.43)$$

The impact point for a given discharge scenario is defined as the intersection of Equation 4.43 with the plume trajectory relationship defined by Equation 4.35. Noting that $x_{I*} = x_{p*} + x_{T*}$ and that $z_{p*} = S_{JP*} + z_{T*} + z_{I*}$, Equation 4.35 can be substituted into Equation 4.43 to give

$$z_{I*} = H_{s*} + \left(x_{T*} + 12^{1/3}S_{JP*}\cos\theta_0\left(1 - \frac{1}{\left(1 + \left(z_{T*}/S_{JP*}\right) + \left(z_{I*}/S_{JP*}\right)\right)^{1/3}}\right)\right)$$
$$\times \tan\theta_B \qquad (4.44)$$

The solution of Equation 4.44 can be obtained through iteration, and this converges rapidly with a reasonable initial estimate of z_I. One such estimate is based on the gradient of the plume trajectory solution at the return point and has the form

$$z_{I*} = H_{s*} + \left(x_{R*} + z_{R*}\left(\frac{dx_{p*}}{dz_{p*}}\right)_R\right)\tan\theta_B \qquad (4.45)$$

where $z_{R*} = H_{s*} + x_{R*}\tan\theta_B$ and the gradient of the plume trajectory at the return point is given by

$$\left(\frac{dx_{p*}}{dz_{p*}}\right)_R = \left(\frac{2}{3}\right)^{2/3}\cos\theta_0\left(\frac{S_{JP}}{z_p}\right)^{4/3} = \left(\frac{2}{3}\right)^{2/3}\cos\theta_0\left(\frac{1}{k_{R\theta}}\right)^{4/3} \qquad (4.46)$$

With this initial estimate, the solution generally converges within four iterations, and once z_I is known, z_p can be calculated, and the dilution at the impact point can be predicted with Equation 4.33. Minimum dilutions are obtained by dividing the predictions by the factor of 1.61. It is important to note that influences of the impact itself are not modeled with these relationships, although the nature and significance of these influences is not yet clear because of insuᴅcient experimental data.

This section has outlined the development of a system of relatively simple analytical solutions that predict key parameters associated with the near field mixing processes of submarine discharges from RO desalination plants. These solutions are based on well-established integral modeling techniques that effectively predict the bulk parameters of two fundamental flows in environmental fluid dynamics, the jet and the plume. In order to evaluate the effectiveness of the outlined solutions, comparisons are necessary with data from discharges of this type.

4.3 Applications

Unfortunately, comparisons with extensive datasets from field applications are not possible at this point because the data is not yet available. However, increasingly modern implementation of desalination plants requires extensive monitoring of the impacts of the brine discharge on the local environment, and field experiments are an integral part of that process. A recent example of this is the field study conducted on the submarine discharge from Perthᴤ desalination plant (Okely et al., 2007) and an image from that study is shown in Figure 4.2. However,

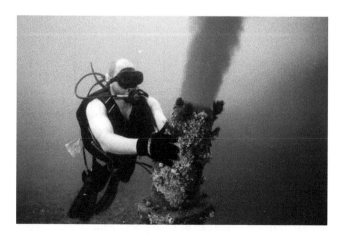

FIGURE 4.2 (See color insert.) Dyed desalination discharge from the Perth field study. (© *The West Australian.*)

data from field studies can be diDcult to interpret, because in general it is not possible to control or measure all the relevant parameters in suDcient detail. In addition, it is not possible to systematically vary important parameters, such as the initial discharge angle, so that effects on the mixing of the discharged fluid can be explored. Thus, while field data provides invaluable insight into the performance of submarine discharge systems in a complex environment, comparisons with data from carefully controlled laboratory experiments provide a more systematic method of exploring model weaknesses.

There is a long history of laboratory investigations into the behavior of jets and plumes, and these have become increasingly sophisticated with the development of nonintrusive optical techniques such as laser-induced fluorescence (LIF), particle image velocimetry (PIV), and particle tracking velocity (PTV). However, these investigations have generally been more relevant to positively buoyant submarine discharges, such as those from municipal treatment plants where it has been noted that a horizontal discharge configuration is generally preferred. Recently, detailed studies into the behavior of inclined negatively buoyant discharges, typical of submarine discharges from desalination plants, have begun appearing in the literature; which is at least in part due to the ongoing rapid growth in global desalination capacity. Valuable data is contained in papers such as Cipollina et al. (2005), Nemlioglu and Roberts (2006), Kikkert et al. (2007), Shao and Law (2010), and most recently Oliver et al. (2012a), where variations in key parameters are explored as a function of initial discharge angle. It is worth noting that the earliest study of this type was published in 1970 (Zeitoun et al., 1970) where, based on trajectory length, it was suggested that a 60° initial discharge angle provides the maximum initial dilution at the return or impact points. This initial discharge angle has been widely adopted by industry for submarine desalination discharges. However, based on predictions from the well-established Corjet model, Jirka (2008) concluded that shallower angles (30°–45°) were preferred because the dilution advantages at higher angles were not significant and the shallower angles

had other advantages, such as the ability to locate the discharge closer to the shore (thereby significantly reducing costs).

In contrast, the most recent data of Oliver et al. (2012a) suggests that the higher initial angles do have a significant dilution advantage and the Corjet predictions are conservative (as are those from other general integral models), particularly for the higher angled discharges. Oliver et al. also compared data from their LIF study with predictions from the analytical solutions outlined earlier. Following Oliver et al., predictions of the minimum dilution at the return point (based on Equations 4.28 and 4.40) are compared with experimental data in Figure 4.3. Predictions are shown with k_{RI} values of 1 and 1.2. Clearly, those based on a k_{RI} value of 1 are conservative, whereas those based on a value of 1.2 are consistent with Oliver et al.ß data. The higher value of k_{RI} indicates that mixing additional to that predicted by the model is taking place. Minimum dilution data from other authors are also presented in Figure 4.3, and there is considerable scatter both within and between the different datasets. Interestingly, there is some consistency for the higher discharge angles of 60° and 75°, but the inconsistencies at lower angles are severe. Unlike Oliver et al.ß data, the remaining studies have a boundary in the vicinity of the return point, and this can have a significant influence on the results. However, such boundaries have the potential to restrict entrainment on the lower side of the flow and thus reduce dilutions, which is clearly not the case at the lower discharge angles. Oliver et al. comment on the need for extended averaging times to obtain repeatability in their data because of the large-scale turbulent motions that occur as the flow descends. Their data were typically averaged over periods of several minutes, as opposed to the 60 s or so that is typical for studies of this type. In addition, they also comment that while most studies consider a range of Froude numbers in determining the dilution coeDcient for each discharge angle, in the case of Nemlioglu and Roberts (2006) these coeDcients were determined based on at most two Froude numbers of approximately 20. However, while the aforementioned issues may be contributing factors, they do not explain the severity of scatter.

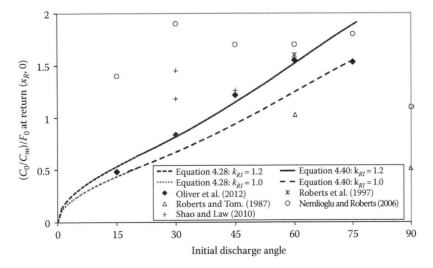

FIGURE 4.3 Dilution at the return point versus initial discharge angle.

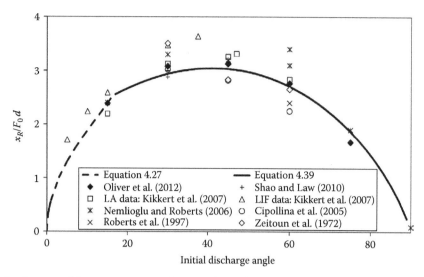

FIGURE 4.4 Dimensionless location of the return point as a function of initial discharge angle.

It is also worth noting the disagreement between the model predictions ($k_{RI} = 1.2$) and Oliver et al.ß data for angles of 75° and above, where re-entrainment effects become significant.

There is also significant scatter in the data for the location of the return point, and this is evident in Figure 4.4. Predictions are shown for flows reaching the return point as a jet, for shallower initial discharge angles, and as a plume for higher initial discharge angles. These predictions fall within the scatter of the data and are particularly consistent with the most recent data.

Oliver et al. (2012b) also show that predictions from the analytical solutions are consistent with data obtained for different source heights and virtual boundaries with slopes ranging from 0° to 20°. The same k_{RI} value of 1.2 is again employed in the dilution relationship to obtain reasonable agreement with the data. This reflects the fact that the selected impact points were in close proximity to the return point and thus the influence of the additional mixing was similar. Therefore, with some adjustment, the relatively simple analytical solutions are able to predict these

parameters with reasonable accuracy. Conditions at the return or impact point are particularly important when assessing potential environmental effects, because this is the region where benthic communities experience the least diluted wastewater and adverse effects are likely to be most significant. In addition, these predictions provide the basis for subsequent modeling of the boundary interactions and formation of the density current which transports the diluted wastewater along the seabed.

Conditions at maximum height are also critical because this is the region where interactions with human activities are more likely to occur, and it is also important to prevent the flow from interacting with the free surface, because such interaction has the potential to restrict the entrainment process and thus the dilution of the ed uent. Two particularly important parameters in this regard are the minimum ed uent dilution at maximum height and the overall height of the edge of the flow. Comparisons of these parameters with analytical solution predictions are shown in Figures 4.5 and 4.6. These figures are modified versions

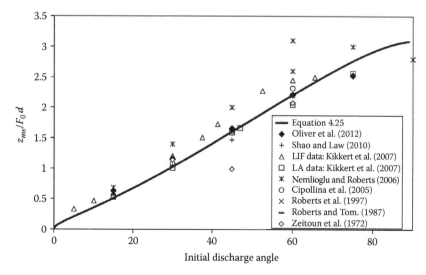

FIGURE 4.5 Dimensionless maximum vertical edge height as a function of initial discharge angle.

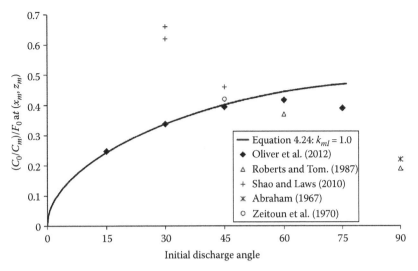

FIGURE4 .6 Minimum dilution at the centerline maximum height.

of those in Oliver et al. (2012a). In Figure 4.5, predictions of the maximum height of the edge of the flow are compared with laboratory data. Note these predictions are based on the deflected jet solutions outlined in Section 4.2. Scatter in the data is at least in part due to difference in the definitions of the edge of the flow, but despite this, the data and model predictions are consistent. Relatively little minimum dilution data has been measured at the centerline maximum height, but comparisons with data from Shao and Law (2010), Oliver et al. (2012a), and others are shown in Figure 4.6. The available data are more consistent at 45° and 60°, but less so at 30°. Model predictions with a k_{ml} value of 1 are in reasonable agreement with most of the experimental data for initial angles less than 70°, suggesting that the centerline mean behavior is consistent with that of a deflected jet. At initial angles above 70°, significant discrepancies are more evident because of re-entrainment processes.

Flow visualization provides additional insight into the near field mixing processes and an example is shown in Figure 4.7. This color-enhanced instantaneous LIF image was recorded on

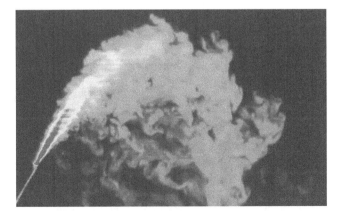

FIGURE 4.7 (See color insert.) Instantaneous uncalibrated LIF image: $d = 4.3\,\text{mm}$, initial angle = 60°, $F_0 = 23$, and initial Reynolds number $R_0 = 3450$.

an x–z plane that passes through the source ($y = 0$; Figure 4.1). The complexities of this turbulent flow are clearly evident in the image, and it is a stark reminder of how much detail is lost with simplistic integral models. Particularly significant in this context is the large quantity of dyed material beneath the main flow in the maximum height region. This fluid has detrained from the jet, and it continues to mix and fall to the bottom of the tank, exterior to the main flow. As noted by Kikkert et al. (2007) and others, detrainment occurs when buoyancy-induced instabilities create velocities in excess of the entrainment velocities that are typical of a jet or a plume. Where the flow's mean motion is predominantly horizontal, these additional induced velocities oppose those associated with the entrainment process on the lower side of jet, and contaminated fluid therefore falls or detrains from the main flow. While the detrainment process itself is essentially independent of discharge angle, the geometry of discharge can enhance the overall significance of the additional mixing. Inclined discharges, such as those discussed here, are an example of this. Here a component of the initial momentum opposes the buoyancy-generated momentum and hence there is an extended region where the motion of the flow is predominantly horizontal, as the flow reverses direction. Thus, detrainment occurs over an extended surface area of the flow, and the influence on the overall mixing of the flow can be significant.

Further evidence of this can been seen in Figure 4.8, where mean centerline concentration contours for a 60° discharge are shown. The presence of the detrained fluid creates significant asymmetries in the contours, and peak concentrations (minimum dilutions) are no longer located at the center of the cross section, but are instead in relatively close proximity to the outer edge of the flow. The asymmetries on the inner side of the flow are not self-similar, and this also makes the application of integral modeling techniques problematic. However, while these processes have a significant influence on the flow behavior, they are not dominant and characteristic jet and plume behavior is still observed. It is therefore possible to develop solutions

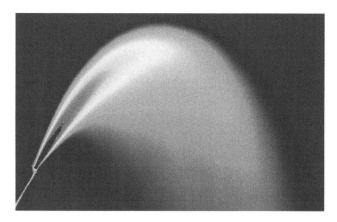

FIGURE 4.8 Mean concentration contours: $d = 4.3\,mm$, initial angle = 60°, $F_0 = 23$, and $R_0 = 3450$. This image was created from data averaged over a period of approximately 9 min.

based on simple jet and plume models that are able to predict with reasonable accuracy key flow parameters for the design of discharges from RO desalination plants and other negatively buoyant discharges. The development of analytical forms of this type of model has been demonstrated in this chapter, where it is evident that predictions of geometric quantities can be made with some confidence. However, reasonable minimum dilution predictions at the return point require the introduction of a coeDcient to account for the additional mixing that takes place because of the detrainment due to the buoyancy-induced instabilities, thus reflecting the simplistic nature of these integral solutions.

4.4 Future Challenges

The scatter in the data present in Figure 4.3 exhibits an immediate challenge. Additional data is needed to confirm the dependence of dilution in the impact region on initial discharge angle. Whether or not higher initial discharge angles offer a significant dilution advantage relative to their shallower counterparts has a direct influence on the preferred initial discharge angle, which in turn determines key design parameters such as the required minimum depth, number of ports, and offshore distance to the discharge location. Shallower discharge angles have the potential to reduce the minimum depth requirement and hence the distance to the offshore discharge location, which has substantial cost implications. Integral models, such as those described in this chapter, provide predictive tools to assess the viability of these shallower discharge options. However, it is essential that their predictive capabilities are validated in context of such discharges before robust decisions can be made. There is clearly a need to improve these models through a physically based model that incorporates the effects of the detrainment processes on the inner side of the flow, rather than simply using a coeDcient to account for the discrepancies. Other challenges from an environmental fluid dynamics perspective include the development and validation of models that account for boundary interactions with a horizontal or inclined seabed. Beyond this a three-dimensional

density current forms and propagates over complex terrain in varying ambient conditions, and the creation of effective physical and numerical models of such flows will be a significant challenge in this field for many years to come.

Our understanding of the environmental fluid dynamics of these problems will undoubtedly continue to develop within the framework of the broader challenges associated with the need for desalination and its potential effects on the environment. There are challenges associated with the processes themselves, in particular with reducing reliance on toxic chemicals to condition the feed water and to protect the process equipment. There are also challenges associated with improving the current understanding of the impacts of key toxins on the marine environment, through extensive epidemiological studies, which will enable ed uent and ambient standards to be determined on a more scientific basis. Political challenges include the implementation of appropriate regulatory frameworks at national and transnational levels so that marine environments are adequately protected from all disposal activities, particularly those that are semi-enclosed with limited natural flushing. Perhaps the most significant challenge is to further improve the energy eDciency of the processes and to increase their reliance on renewable energy resources, so that the global impacts of this increasingly popular means of obtaining a secure supply of potable water are sustainable.

References

Abraham, G. (1967), Jets with negative buoyancy in homogeneous fluid. *J. Hydraul. Res.* **5**(4), 235–248.

Cipollina, A., Brucato, A., Grisafi, F., and Nicosia, S. (2005), Bench-scale investigation of inclined dense jets. *J. Hydraul. Eng.* **ʒ** (11), 1017–1022.

Gleick, P., Cooley, H., and Wolff, G. (2007), With a grain of salt: An update on seawater desalination. In *The Worlds Water 2006–2007—The Biennial Report of Freshwater Resources*, Island Press, Washington, DC.

GWI DesalData/IDA (2009), *22nd Worldwide Desalting Plant Inventory*, Global Water Intelligence, Oxford, U.K.

Hussein, H. J., Cap, S. P., and George, W. K. (1994), Velocity measurements in a high-Reynolds-number, momentum conserving, axisymmetric, turbulent jet. *J. Fluid Mech.* 258, 31–75.

Jirka, G. H. (2004), Integral model for turbulent buoyant jets in unbounded stratified flows. Part 1: Single round jet. *Environ. Fluid Mech.* **4**(1), 1–56.

Jirka, G. H. (2008), Improved discharge configurations for brine ed uents from desalination plants. *J. Hydraul. Eng.* **134**(1), 116–120.

Kikkert, G. A., Davidson, M. J., and Nokes, R. I. (2007), Inclined negatively buoyant discharges. *J. Hydraul. Eng.* **ʒ** (5), 546.

Latteman, S. and Höpner, T. (2008), Environmental impact and impact assessment of seawater desalination. *Desalination*, **ʘ** , 1–15.

Lee, J. W. H., Cheung, V., Wang, W. P., and Cheung, S. K. B. (2000), Lagrangian modeling and visualization of rosette outfall plumes, *Proceedings of Hydroinformatics 2000*, Iowa, LA, July 23–27, 2000 (CD-ROM).

Nemlioglu, S. and Roberts P. J. W. (2006), Experiments on dense jets using three-dimensional laser-induced fluorescence (3DLIF), *MWWD 2006—4th International Conference on Marine Waste Water Disposal and Marine Environment*, DVD, Antalya, Turkey, November 6–10.

Okely, P., Antenucii, J. P., Marti, C. L., and Imberger, J. (2007), The near-field characteristics of the Perth seawater desalination plant, Report WP2175PO, Centre for Water Research, The University of Western Australia, Crawley, WA, Australia.

Oliver, C. J., Davidson, M. J. and Nokes, R. I. (2012a), Removing the boundary influence on negatively buoyant jets. Submitted to *Journal of Environmental Fluid Mechanics.*

Oliver, C. J., Davidson, M. J. and Nokes, R. I. (2012b), The behaviour of desalination discharges beyond the return point. Submitted to *Journal of Hydraulic Engineering, ASCE.*

Papanicolaou, P. N. and List, E. J. (1988), Measurements of round vertical axisymmetric buoyant jets, *J. Fluid Mech.*, **195**, 341–391.

Roberts, P. J. W., Ferrier, A., Daviero, G. (1997), Mixing in inclined dense jets, *J. Hydraul. Eng.*, **3** (8), 639–699.

Roberts, P. J. W. and Toms, G. (1987), Inclined dense jets in flowing current. *J. Hydraul. Eng.* **3** (3), 323–341.

Shao, D. and Law, A. W.-K. (2010), Mixing and boundary interactions of 30° and 45° inclined dense jets. *Environ. Fluid Mech.* **0** (5), 521–553.

Stover, R. and Crisp, G. (2008), Environmentally sound desalination at the Perth seawater desalination plant. *Enviro'08, Australia's Environmental and Sustainability Conference and Exhibition*, DVD, Melbourne, Victoria, Australia, May 5–7.

Wang, H. and Law, A. W. K. (2002), Second-order integral model for a round buoyant jet, *J. Fluid Mech.*, **9** , 397–428.

Wangnick/GWI. (2005), *2004 Worldwide Desalting Plants Inventory*. Global Water Intelligence, Oxford, U.K.

Zeitoun, M. A., Mcllhenny, W. F., Reid, R. O. (1970), Conceptual designs of outfall systems for desalting plants. Research and Development Progress Report No. 550, ODce of Saline Water, U.S. Department of the Interior, Washington, DC, pp. 1–139.

5
Bubble Plumes

Scott A. Socolofsky
Texas A&M University

Chris R. Rehmann
Iowa State University

5.1 Introduction

Bubble plumes, or in general any plumes consisting of an immiscible dispersed phase, appear in many applications important for environmental fluid dynamics. Bubble plumes have been used to contain contaminants, aerate wastewater, prevent ice from forming in harbors, and act as breakwaters for waves. To manage water quality in lakes and reservoirs, bubble plumes are used to oxygenate deep water or to destroy the stratification that can prevent oxygen transport to the deep water. Other applications of multiphase plumes are related to energy use and climate change: The Deepwater Horizon spill of 2010 provides a notorious, spectacular example of an oil-well blowout. Plumes of liquid carbon dioxide are used to sequester carbon in the deep ocean, and recently, researchers have studied the release of methane, a greenhouse gas, from water bodies to the atmosphere. In this chapter, we outline the fluid mechanics of multiphase plumes and experimental and computational approaches toward studying their behavior, and we discuss example applications and challenges to be addressed.

5.2 Principles

Multiphase plumes form when a buoyant dispersed phase of immiscible bubbles, droplets, or particles is discharged locally into a continuous receiving fluid such as a water body or the atmosphere. Because of its buoyancy, the dispersed phase moves away from the discharge source; likewise, the continuous phase fluid is set in motion in reaction to the drag force on each bubble, droplet, or particle. The resulting turbulent flow drives further entrainment of ambient fluid into the mixture, resulting in a buoyant plume flow (Figure 5.1). Due to the local nature of the discharge, the plume is symmetric about the z-axis. The time-averaged lateral profiles of velocity $u(z)$ exhibit a Gaussian shape with maximum velocity $U_m(z)$ and width $b(z)$, typical of single-phase buoyant plumes, while the time-averaged void fraction profile $\chi(z)$ of the dispersed phase spreads at a slower rate, characterized by width $\lambda b(z)$ and centerline magnitude $\chi_m(z)$, and can have a more top-hat shape (Seol and Socolofsky 2008).

Multiphase plumes differ from single-phase plumes because the buoyancy source (dispersed phase) can separate from the induced flow of the continuous phase. In a single-phase plume, the buoyancy arises by changes in the density of the continuous phase through the irreversible addition of heat or dissolved chemical species that alter the equation of state of fluid entrained in the plume. By contrast in multiphase plumes, the dispersed phase remains discrete from the ambient fluid in the form of individual bubbles, droplets, or particles that can follow trajectories different from those of the entrained continuous phase fluid.

Separation can be caused by crossflows or stable ambient density stratification. Consider, for example, the difference between a single-phase wastewater plume and a bubble plume. Crossflow may deflect the wastewater plume and stratification may cause the plume to trap at a level of neutral buoyancy, but all of the wastewater and entrained ambient fluid follow a single trajectory and move together into the far field. In the bubble plume, the entrained fluid may separate from the trajectory of the bubbles. In the case of crossflows, currents may pass through the plume, pulling entrained fluid into the lee wake of the bubble column, thereby continually refreshing the entrained

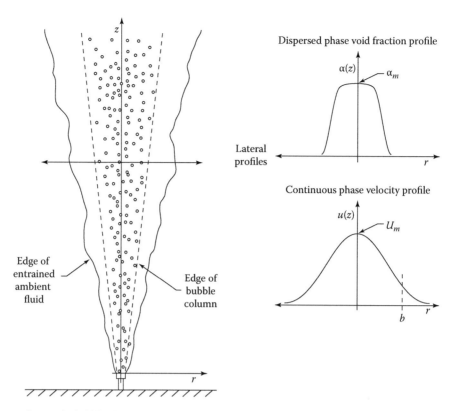

FIGURE 5.1 Schematic of a simple bubble plume in a quiescent fluid. Time-averaged lateral profiles of void fraction and continuous phase velocity are also shown.

fluid in the plume core. Likewise, density stratification may arrest upward rising entrained fluid and cause it to trap, but the bubbles maintain their buoyancy and continue above the trapping level, potentially creating a series of plumes. The dynamic effect of the separation is also most apparent in the stratified case: whereas the wastewater plume intrudes at a level of neutral buoyancy very near the trap height of the plume (because the lighter wastewater flows into the intrusion), the detrained fluid from a bubble plume loses the buoyancy of the bubbles and falls far below the trap height of the plume before intruding.

Because separation may occur to varying degrees and at different locations in different plume configurations, plumes are classified by separation type (Figure 5.2). In a Type 1 plume (Figure 5.2a), separation occurs only at the free surface, where bubbles leave the reservoir and the entrained fluid is arrested and intrudes laterally away from the plume. In an unstratified flow with weak crossflow, entrainment on the downstream side of the plume is suDcient to overcome the suction pressure in the wake of the plume so that no separation occurs (Figure 5.2b). For stronger crossflows, fluid is continually entrained on the front side of the plume, lifted at a characteristic height h_S, and detrained into the lee of the plume (Figure 5.2c); the edge of the plume shown in the figure only tracks fluid discharged with the bubbles at the source.

The remaining types involve the effect of stratification in deep reservoirs in the absence of currents. As the plume rises, the density difference between the entrained fluid and that in the reservoir at the same height increases. Eventually, the buoyancy of the bubbles can no longer support the negative buoyancy of the entrained ambient fluid, and the plume decelerates. Once the entrained fluid stops moving upward, it is ejected, or peeled, from the plume at the height h_P, and forms a downward-flowing outer annular plume that eventually intrudes at a level of neutral buoyancy called the trap height h_T. In Type 1* plumes, the bubbles are small enough that they partially detrain, but because of their buoyancy they ultimately rise out of the downdraught plume to continue above the peel height and form new plumes (Figure 5.2d). As the bubbles increase in size, they become less affected by the peeling fluid: In the Type 2 plume, detrainment is periodic and discrete (Figure 5.2e), and in the Type 3 plume, detrainment is sporadic and nearly continuous (Figure 5.2f). Limited data are available in the literature on the effects of both stratification and crossflow. Socolofsky and Adams (2002) suggested that the crossflow may be neglected when $h_T \ll h_S$ and stratification may be neglected $h_S \ll h_T$.

Several key parameters have been identified to predict plume behavior, including characteristic heights, flow rates, and plume type. In the case of an unbounded, unstratified reservoir without crossflow, the key variables are the elevation z above the source, the terminal rise velocity—or slip velocity—u_s of the dispersed phase, the entrainment coeDcient α (the ratio of the radial speed of fluid entering the plume to the along-axis speed on the centerline), and the buoyancy flux $B = Q_0(\rho_a-\rho_0)g/\rho_r$ at the source, where Q_0 is the volume flow rate at the source, ρ_0 is the density of the

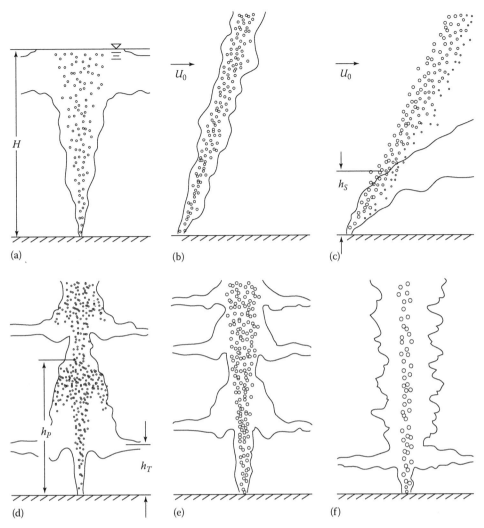

FIGURE 5.2 Characteristic types of bubble plumes in water. (a) Type 1, (b) weak crossflow, (c) strong crossflow, (d) Type 1*, (e) Type 2, and (f) Type 3. (From Asaeda, T. and Imberger, J., *J. Fluid Mech.*, 249, 35, 1993; Socolofsky, S.A. and Adams, E.E., *J. Hydraul. Res.*, 40, 661, 2002; Socolofsky, S.A. and Adams, E.E., *J. Hydraul. Eng.*, 131, 273, 2005.)

dispersed phase at the source, ρ_a is the density of the ambient fluid at the source, ρ_r is a reference density for the continuous phase, and g is the acceleration of gravity. In an unstratified ambient, ρ_r and ρ_a are the same. Although a single-phase plume has no characteristic length scale, including u_s in the parameter set leads to a length scale $D = B/4\pi\alpha^2 u_s^3$. Bombardelli et al. (2007) showed that for $z/D > 5$ the plume loses memory of the initial conditions and dimensionless forms of plume variables such as velocity and densimetric Froude number follow universal behavior.

Other parameters are required to describe more realistic cases. For crossflows, the velocity of the current u_y is also important, and h_S is found by solving (Socolofsky and Adams 2002)

$$\frac{u_\infty}{(B/h_S)^{1/3}} = 6.3\left(\frac{u_s}{(B/h_S)^{1/3}}\right)^{-2.4} \tag{5.1}$$

This equation shows the relationship between the nondimensional crossflow and slip velocities. Alternatively, the equation

may also be solved explicitly for h_s. If the ambient fluid is stratified, the buoyancy frequency $N = [-(g/\rho_0)e\rho/ez]^{1/2}$, where ρ is the density of the continuous phase, must be included. Then, the dominant nondimensional parameter is the nondimensional slip velocity U_N, defined as

$$U_N = \frac{u_s}{(BN)^{1/4}} \tag{5.2}$$

For the bounded reservoir, a length scale of the reservoir depth is needed. Although the reservoir depth H might seem to be a natural choice, the absolute pressure head at the discharge source H_T is used because expansion of compressible dispersed phases is governed by the absolute pressure. Thus, the new parameters that arise are H_T/D in unstratified reservoirs and

$$H_N = \frac{H_T}{(B/N^3)^{1/4}} \tag{5.3}$$

in stratified reservoirs.

The parameters z/D, U_N, and H_N describe the behavior of multiphase plumes. These parameters can be related to the plume number $P_N = H^4N^3/B$ and $M_H = D/H$ used by Asaeda and Imberger (1993). Each of the plume types in Figure 5.2 can be predicted using these governing parameters. Type 1 plumes arise when H_N is of order 1 or smaller. Transition from Type 1* to Type 2 plumes occurs for $U_N = 1.5$, and transition from Type 2 to Type 3 plumes occurs for $U_N = 2.4$. Crossflows transition from weak to strong as h_S/H becomes order 1 or smaller.

5.3 Methods of Analysis

For environmental applications of multiphase plumes, important questions requiring predictive models focus on the trajectory of the plume, concentration and flow rate of the dispersed and continuous phase, and the turbulence generated both within the plume and in the surrounding ambient fluid. The experimental methods applied to these questions include flow visualization and measurement of point and full-field velocity and concentration. Numerical methods span the range from one-dimensional integral plume models to fully three-dimensional computational fluid dynamics (CFD) models. While a comprehensive review of the literature is not possible in this section, we focus on fundamental methods that are applied in most multiphase plume studies and highlight a few of the fundamental results.

5.3.1 Experimental Methods

Early measurements of bubble plumes focused on bulk properties such as plume width, entrainment, and mean velocity profiles (e.g., Milgram 1983), but advances in measurement techniques have allowed details of the turbulence to be investigated. For example, methods for measuring profiles of temperature microstructure in lakes and the ocean have been adapted for bubble plumes to estimate the rate of dissipation of turbulent kinetic energy (Soga and Rehmann 2004) and the eddy diffusivity (Wain and Rehmann 2005). Because the vertical velocities generated by bubble plumes are not small compared to the profiler's fall speed, they must be measured and included in applying Taylor's hypothesis to compute the spectra needed for the turbulence quantities. Another important issue is the finite response of the thermistors, especially in energetic flows. Although Wain and Rehmann (2005) estimated that 95% of the dissipation of temperature variance was resolved in the measurements of Soga and Rehmann (2004), the dissipation was a factor of about 5 smaller than that estimated from measurements with acoustic Doppler velocimeters (ADVs) in the same flow (García and García 2006). Nevertheless, these measurements are useful for evaluating bulk models of the eddy diffusivity and predictions from two-equation turbulence models, such as those used in CFD models discussed in Section 5.3.2.2.

The ADV measurements of García and García (2006) highlight some of the challenges of measuring velocities in flows driven by bubble plumes, especially in large water bodies. The measurements must resolve the small-scale turbulence with a power spectrum including an inertial subrange observed in energetic flows, but they must also capture the large-scale wandering, which can occur on timescales 2 orders of magnitude larger than the integral timescale of the turbulence. Fortunately, the widely different timescales allow filtering to separate the contributions of the two processes to the turbulence parameters; for example, the amount of turbulent kinetic energy generated by wandering decreases as the airflow rate increases. Also, García and García (2006) suggested that the integral length scale of the turbulence grows and reaches a constant value when the length scale D becomes larger than the tank radius.

Although the measurements with temperature microstructure methods and acoustic Doppler velocimetry provide useful information for understanding the flow and for evaluating turbulence models, they focus on the flow outside the bubble plume because of challenges with distinguishing measurements of the dispersed phase and those of the continuous phase. Several methods for removing bubbles are available, and they depend on the type of measurements. Point velocity measurements have been conducted using laser Doppler velocimetry (LDV; e.g., Lance and Bataille 1991) and hot films (e.g., Rensen et al. 2005). Here, bubble signatures can be removed by peak velocity and acceleration thresholding. Typically, these high-resolution velocity measurements are made to evaluate turbulence energy spectra, and results are not very sensitive to the removal method in dilute bubbly flows.

Particle image velocimetry (PIV) is used to obtain full-field velocity data, and three methods are available to remove bubble signatures from the data. The most reliable method is optical separation, in which fluorescent seeding particles are used to track the continuous phase and optical filters are fitted on the PIV imaging cameras, either to block the fluorescent light and allow the laser light reflected from the bubbles to pass or to block the laser light and capture the seeding particles. Because fluorescent light is scattered from the seeding particles and laser light is scattered from the bubbles, signatures of both phases usually appear in each image series; therefore, very careful lighting setup and image processing are necessary to fully remove the undesired phase. Another effective method is to remove the bubble signature by pre-processing images containing both phases, as illustrated in Figure 5.3. First, the image (Figure 5.3a) is converted to binary by assigning pixels above a threshold intensity to white and setting all others to black (Figure 5.3b). Then, the area-open algorithm is applied (Figure 5.3c) to eliminate contiguous white spots below a threshold size. Since bubbles sometimes have hollow areas, the area-close algorithm is applied to create filled bubble masks, and because the binary image is created by clipping the grayscale image at a threshold value, the bubble masks must also be dilated to mask the area that fell below the binary threshold value (Figure 5.3d). Applying this mask to the original image results in an image containing only bubbles (Figure 5.3e) and an image containing only particles (Figure 5.3f). PIV can be

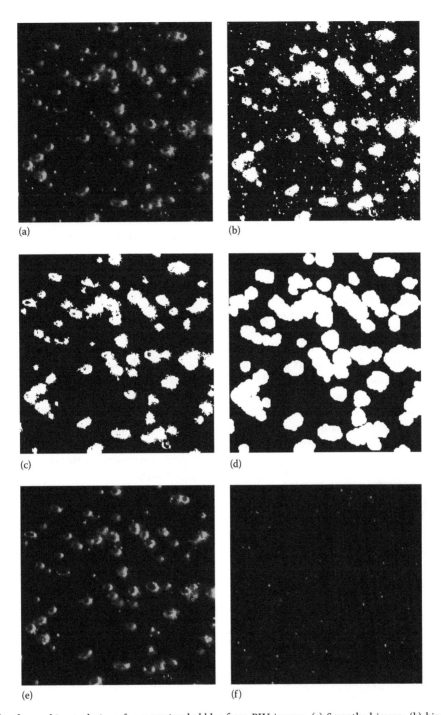

FIGURE 5.3 Example of a masking technique for removing bubbles from PIV images. (a) Smoothed image, (b) binary image, (c) particles removed, (d) bubble masks, (e) bubbles only, and (f) particles only. (With kind permission from Springer Science + Business Media: *Experiments in Fluids,* Vector post-processing algorithm for phase discrimination of two-phase PIV, 45, 2008, 223–239, Seol, D.G. and Socolofsky, S.A.)

applied to the particle image, and depending on the density and size of bubbles, PIV or particle tracking velocimetry can be applied to the bubble image. A final method to remove bubbles from PIV data is to process the mixed-phase image using standard PIV methods and then to try to identify and remove bubble velocity vectors from the mixed-phase velocity field (Seol et al. 2007, Seol and Socolofsky 2008). For dilute bubble plumes, optical separation is generally preferred, and it is matched closely by image masking techniques; vector post processing generally removes more data in order to have a conservative estimate of the continuous phase vectors (Seol and Socolofsky 2008).

Removing bubbles from concentration measurements is much more difficult, and it is often achieved by hand. Optical separation is difficult because of the amount of reflected fluorescent light illuminating the dispersed phase. Separation methods based on image intensity also fail because high-concentration regions make it hard to identify continuous phase fluid from brightly lit bubbles and low-concentration regions make it impossible to distinguish between low-concentration regions and voids caused by non-illuminated bubbles. Figure 5.4 shows a good example for a bubble plume in linear stratification; fluorescent dye is injected at the source to track the continuous phase, and no optical filtering was applied. Shadows are cast behind bubbles in the light sheet. Other bubbles near the light sheet and in the middle of the field of view appear as bright dots, and some bubbles outside the light sheet and near the top of the image, where fluorescent dye concentration is lighter, appear as dark spots. Because the human eye can usually distinguish between bubble signatures and intermittent dye concentration patches, removing the bubbles automatically should be possible, but more sophisticated means than those applied to PIV data are needed. Nonetheless, laser-induced fluorescence of the intrusion in the index-of-refraction matched image is possible, and it yields insight on dilution and the flow rate of the intrusion.

Once bubbles are removed from a multiphase flow, the resulting time series and fields of continuous phase measurements will have voids where the dispersed phase was present. Most authors fill these voids using linear interpolation to obtain a continuous measurement (e.g., Lance and Bataille 1991, Rensen et al. 2005, Seol and Socolofsky 2008, Bryant et al. 2009). In dilute plumes common in the environment (void fractions below 10%), this effect has been shown to be negligible for computing the statistics of the turbulent flow field.

A fundamental result obtained from time-averaged data is an estimate of the entrainment coeDcient. In single-phase plumes, the entrainment coeDcient is a constant and, for Gaussian lateral profiles of velocity, is equal to 0.083. In multiphase plumes, the entrainment is not constant and depends on the local plume properties. Seol et al. (2007) obtained an expression for the entrainment coeDcient for Gaussian velocity profiles consistent with measurements throughout the literature given by

$$\alpha = 0.18 \exp\left(-1.7 \frac{u_s}{(B/z)^{1/3}}\right) + 0.04 \qquad (5.4)$$

for $0.3 < u_s/(B/z)^{1/3} < 25$. Casting Equation 5.4 in terms of z/D (with $\alpha \approx 0.083$ in the definition of D) shows that in the asymptotic region of the plume, $z/D > 5$, the entrainment coeDcient is constant at 0.04. However, the variable entrainment rate is important in the near-source region, $z/D < 5$ because much higher (up to three times greater) entrainment coeDcients apply and because this source adjustment region can be a significant region of the total flow. Indeed, most experimental data on bubble plumes in the literature are for $z/D < 5$. Hence, a variable entrainment rate is an important feature of multiphase plumes.

Measurements of turbulent properties of multiphase plumes also highlight important differences with single-phase plumes. Bryant et al. (2009) performed high-speed PIV of bubble plumes and analyzed both point and full-field turbulence statistics of the flow. Statistics for large-scale (greater than the bubble diameter), individual vortices in the flow were obtained by calculating the swirl strength from the two-dimensional PIV vector fields. Vortex sizes normalized by the local plume radius (distance at which the time-average velocity falls to $1/e$ of the centerline value) collapse well for a range of bubble flow rates (Figure 5.5a). Larger vortices are present on the edges of the plume, where velocity shear in the continuous phase drives vortex production, and smaller eddies are located in the center of the plume, where the dispersed phase continually acts to inhibit growth in vortex size, likely through shear straining of fluid advected around the bubble (Figure 5.5b). The bubbles also affect the slope of the energy spectra (Figure 5.6). The turbulent energy spectra exhibit a reduced slope (−7/6) in the inertial subrange compared to a single-phase flow (−5/3). This reduction, which has been observed in other experiments (e.g., Lance and Bataille 1991, Rensen et al. 2005), is hypothesized to be a result of turbulent energy production by bubbles at a length scale on the order of the Kolmogorov length scale of the continuous phase flow, where dissipation would occur in a single-phase plume. This question is an active area of research and is particularly important for development of reliable numerical models.

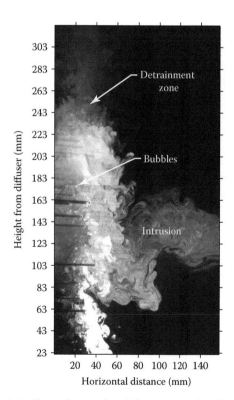

FIGURE 5.4 Planar laser-induced fluorescence (PLIF) image of a Type 2 stratified bubble plume. (Adapted from Seol, D.G. et al., *J. Hydraul. Eng.*, 135, 983, 2009. With permission from ASCE.)

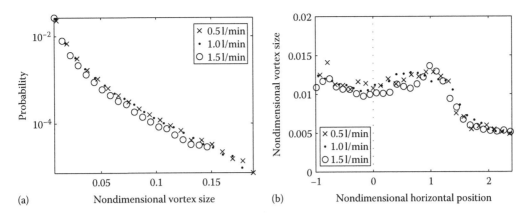

FIGURE 5.5 Measured vortex properties for a simple bubble plume in a quiescent reservoir: (a) occurrence of vortices of various sizes, (b) lateral profile of average vortex size. Length scales are normalized by the local plume radius. (Reprinted with permission from Bryant, D.B. et al., Quantification of turbulence properties in bubble plumes using vortex identification methods, *Phys. Fluids,* 21, 075101, Copyright 2009, American Institute of Physics.)

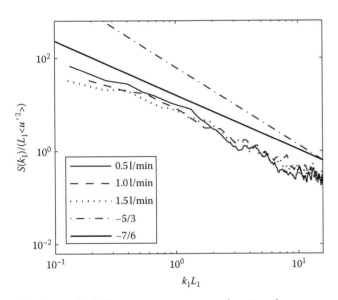

FIGURE 5.6 Turbulent energy spectrum as a function of wavenumber k_1 along the centerline of a simple bubble plume in a quiescent reservoir. The variables are normalized by the integral scale L_1 and the RMS velocity. (Adapted with permission from Bryant, D.B. et al., Quantification of turbulence properties in bubble plumes using vortex identification methods, *Phys. Fluids,* 21, 075101, Copyright 2009, American Institute of Physics.)

5.3.2 Modeling

5.3.2.1 Integral Plume Models

Integral plume models use the entrainment hypothesis and self-similarity to solve the steady equations for conservation of mass and momentum for the variation of laterally averaged quantities along the plume axis. Bubble plumes are not strictly self-similar because the entrainment coeDcient is not constant and the dispersed phase and entrained ambient fluid may not spread laterally at equal rates; however, profiles of continuous phase velocity and void fraction, among others, exhibit similar shapes along the height of the plume (e.g., Milgram 1983,

Seol and Socolofsky 2008), and integral models have been successfully applied to predict flow rates of entrained fluid, rise and trap heights of the plume, and the behavior of detrainment and intrusion formation, among others (e.g., McDougall 1978, Milgram 1983, Asaeda and Imberger 1993, Crounse et al. 2007, Socolofsky and Bhaumik 2008, Socolofsky et al. 2008).

Integral models must account for effects of compressibility and dissolution on the dispersed phase as well as density changes of the entrained fluid caused by entrainment and dissolution. The discrete bubble model of Wüest et al. (1992) accounts for compressibility and dissolution by tracking their effects on a single bubble in the flow and extending the results to all bubbles at a similar height. Density changes of the entrained fluid can be calculated using an equation of state and by solving a conservation equation for each dynamic chemical or thermodynamic component of the flow (e.g., heat, salt, and other dissolved species that affect density). For a comprehensive review of integral models for multiphase plumes, see Socolofsky et al. (2008).

To apply widely, integral models must treat separation events. In the double-plume integral model of Socolofsky et al. (2008), an inner, radial plume of bubbles and entrained fluid rises until the bubbles are fully dissolved (Figure 5.7). Periodically, fluid is lost through detrainment to a descending, outer annular plume of ambient fluid that descends until it eventually intrudes at a level of neutral buoyancy. The detrainment flux for models based on top-hat profiles can be computed with (Crounse et al. 2007)

$$E_p = -\zeta \left(\frac{u_s}{u_i}\right)^2 \frac{B_i}{u_i^2} \qquad (5.5)$$

where
u is the velocity of the continuous phase
the subscript i indicates a quantity calculated for the inner plume
ζ is a calibration parameter, reported in Socolofsky et al. (2008) to be 0.015

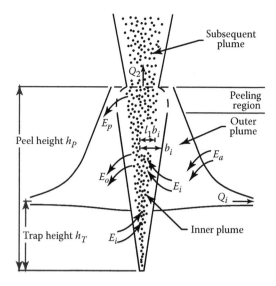

FIGURE 5.7 Diagram of the double-plume integral model showing characteristic length scales and volume fluxes. (From Socolofsky, S.A. et al., *J. Hydraul. Eng.*, 134, 772, 2008. With permission from ASCE.)

Entrainment occurs into the inner plume with a rate E_i, from the inner plume into the outer plume at a rate E_o, and from the ambient fluid into the outer plume with a rate E_a. The general entrainment law for these fluxes is

$$E_i = 2\pi b_i \alpha_i (u_i + cu_o)$$
$$E_o = -2\pi b_i \alpha_o u_o \qquad (5.6)$$
$$E_a = -2\pi b_o \alpha_a u_o$$

where
 the subscript o is for the outer plume
 c is a parameter of value -1 or 0 depending on the entrainment law desired

For the top-hat model, Socolofsky et al. (2008) calibrated the entrainment coeDcients to be $\alpha_i = 0.055$, $\alpha_o = 0.110$, and $\alpha_a = 0.110$; the best model performance was obtained for $c = 0$ though there was not strong sensitivity to the value of c. To close the model equations, the initial conditions required at the plume source and at the initiation of outer plumes are supplied using Froude number criteria, as introduced by Wüest et al. (1992).

In the case of crossflows, Socolofsky and Adams (2002) introduced a simple treatment for separation in the absence of ambient density stratification. The inner plume model described above is solved up to the height h_S given in Equation 5.1. At that point, all entrained water is assumed to be detrained into the wake of the plume. The trajectory of this detrained water is predicted by a second single-phase plume model, which is initiated at the height h_S with the volume, momentum, and buoyancy flux being that predicted by the last iteration of the bubble plume models and with the buoyancy of the bubbles removed. The bubble column continues to rise with a trajectory predicted well by the vector addition of the slip velocity and ambient current velocity.

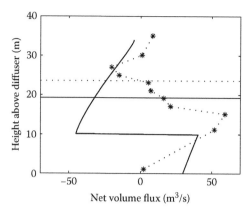

FIGURE 5.8 Validation of the double-plume integral model with the field data of McGinnis et al. (2004). The solid lines present the model results, and the dotted lines with star symbols denote the field data. The solid and dotted horizontal lines indicate the modeled and measured trap heights. (Adapted from Socolofsky, S.A. et al., *J. Hydraul. Eng.*, 134, 772, 2008. With permission from ASCE.)

Despite their simplicity, integral models capture many of the important features of bubble plumes dynamics. Validating the stratified plume model with many length and flow rate measurements from the literature, Socolofsky et al. (2008) found an average root mean square error of 33% for model deviations normalized by the mean of the model results. In the example of a reservoir aeration plume (Figure 5.8), the model predicts very well the height at which the plume dissolves (about 35 m above the diffuser), the flow rate at the top of the plume (about -10 m³/s), the flow rate of the inner plume going into the bottom of the outer plume (about 30 m³/s), and the trap height of the intrusion (about 20 m). The flow rate through the downdraught region of the plume (10–30 m above the source) does not appear accurate in the model since the field data integrate the inner and outer plumes and the model results are presented for the difference between the outer and inner volume fluxes. For the case of the unstratified crossflow (Figure 5.9), the model tracks well the trajectory of the fluid entrained at the injector source. For fluid flowing through the plume, above the height h_S, no laboratory data are available and the behavior is not likely predicted by integral models since self-similarity would be violated.

5.3.2.2 CFD Models

More detailed information on bubble plume parameters can be computed by solving a two-fluid model. As a concrete example, we summarize the work of Buscaglia et al. (2002), who applied the theory of multicomponent fluids to predict mixing and aeration in large wastewater reservoir. Their model includes hydrodynamics, mass transfer, and liquid chemistry. Rather than solve separate equations for the gas and liquid phases, Buscaglia et al. (2002) solved for the density ρ_m, velocity vector $\mathbf{u}_m$, and pressure p_m of the mixture:

$$\rho_m = \chi_\ell \bar{\rho}_\ell + \chi_g \bar{\rho}_g, \quad \rho_m \mathbf{u}_m = \chi_\ell \bar{\rho}_\ell \bar{\mathbf{u}}_\ell + \chi_g \bar{\rho}_g \bar{\mathbf{u}}_g, \quad p_m = \bar{p} \qquad (5.7)$$

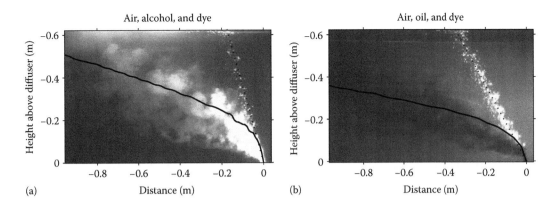

FIGURE 5.9 Validation of the integral model for a bubble plume in strong crossflow. The dotted line tracks the bubble column, and the solid line presents the integral model for the trajectory of fluid discharged at the source, a buoyant solution of alcohol and Rhodamine WT in subplot (a) and a crude oil in subplot (b). (Adapted from Socolofsky, S.A. and Adams, E.E., *J. Hydraul. Res.*, 40, 661, 2002. With permission of Taylor & Francis Ltd.)

where

  p is the pressure

  χ_ℓ and χ_g are the volume fractions of the liquid and gas phases subscripts ℓ and g indicate the liquid and gas phases, respectively

The ensemble average, denoted by the overbar, requires knowledge of the presence of the two phases in space and time. For air–water plumes with small gas fraction [e.g., $\chi_g < O(10^{-2})$], the resulting mixture equations resemble the usual Navier–Stokes equations with the Boussinesq approximation, in which the effect of density differences is neglected except with respect to gravity.

Approaches to further averaging differ. Some researchers take the ensemble average to encompass turbulence scales, while others perform another average in space or time to obtain Reynolds-averaged equations. As Buscaglia et al. (2002) discuss, for dilute flows the chief difference between the approaches involves an extra term, modeled as turbulent dispersion, in the mass balance for the gas, which is considered next. The Reynolds-averaged equations for the mixture variables are then (Bombardelli et al. 2007)

$$\nabla \cdot \hat{\mathbf{u}}_m = 0 \tag{5.8a}$$

$$\frac{\partial \hat{\mathbf{u}}_m}{\partial t} + \hat{\mathbf{u}}_m \cdot \nabla \hat{\mathbf{u}}_m = -\frac{1}{\rho_r} \nabla \hat{p}' + \alpha_g \frac{\hat{\rho}_\ell - \hat{\rho}_g}{\rho_r} g \mathbf{k} + \nabla \cdot [\nu_T (\nabla \hat{\mathbf{u}}_m + \nabla^T \hat{\mathbf{u}}_m)] \tag{5.8b}$$

where

  carets denote the Reynolds average

  $\hat{p}'$ is the dynamic pressure

  $\mathbf{k}$ is the unit vector in the vertical direction

  superscript T indicates a transpose

As in other modeling studies, the hydrodynamic model of Buscaglia et al. (2002) employs a version of the k-ε turbulence model, where k is the turbulent kinetic energy (TKE) and ε is the rate of dissipation of TKE; in the momentum equations,

correcting the Reynolds stress tensor to produce the correct TKE introduces modified dynamic pressure $\hat{p} = p_m - 2k/3$, and the eddy viscosity is $\nu_T = C_\mu k^2/\varepsilon$ with $C_\mu = 0.09$. While Buscaglia et al. (2002) used the standard k-ε model, others have added terms to account for bubble-induced turbulence. Smith (1998) reviews modifications to turbulence models for bubble plumes.

The models for mass transfer and liquid chemistry involve transport equations of the general form

$$\frac{\partial C}{\partial t} + \nabla \cdot (C\mathbf{U}) = \Phi + \nabla \cdot (D_T \nabla C) \tag{5.9}$$

where

  C is a scalar

  $\mathbf{U}$ is the transport velocity

  Φ is a source or sink of the scalar

Various models for the turbulent diffusivity D_T have been proposed; a simple approach is to assume a constant value of the turbulent Schmidt number $Sc_T = \nu_T/D_T$. Modeling the mass transfer of the gas for an air–water plume involves tracking the number of bubbles and the concentrations of gaseous nitrogen and oxygen. For these scalars, Buscaglia et al. (2002) chose $Sc_g = 1$ and set the transport velocity to the gas velocity—that is, $\mathbf{U} = \mathbf{u}_g = \mathbf{u}_m + u_s\mathbf{k}$. The bubble slip velocity can be related empirically to the bubble radius; this approach avoids solving momentum equations for the bubbles, which requires specifying and modeling the forces acting on the bubble. In the balance for the number of bubbles, Buscaglia et al. (2002) ignored breakup and coalescence and set the source to zero, but in the balance for nitrogen and oxygen, they included exchange between the gaseous and dissolved forms.

The liquid chemistry model of Buscaglia et al. (2002) tracks dissolved oxygen, dissolved nitrogen, and—because the application was to wastewater reservoirs—biochemical oxygen demand (BOD). For these scalars, the transport velocity is the liquid velocity ($\mathbf{U} = \mathbf{u} \approx \mathbf{u}_m$). Again the turbulent Schmidt number was assumed to be constant, but a smaller value $Sc = 0.83$ was used. The source term included the exchange between the gaseous form for both dissolved gases and a sink caused by BOD for dissolved oxygen. The sink for BOD follows first-order decay as in standard water quality models. The generality of Equation 5.9 and the examples in Buscaglia et al. (2002) illustrate how a model of a plume with other scalars—for example, temperature, salinity, and suspended sediment—can be constructed.

The two-fluid model can be used to evaluate integral plume models. Buscaglia et al. (2002) obtained an integral plume model by integrating the equations in the two-fluid model over a surface perpendicular to the plume and neglecting effects of dynamic pressure and streamwise turbulent diffusion in the momentum balance. To close the system, they evaluated exchange terms with plume-averaged quantities and specified values of the entrainment coefficient and λ, the ratio of the widths of the bubble core and plume. For a large wastewater reservoir, predictions from the integral plume model regarding gas transfer, fluid velocity, and entrainment agreed well with the predictions from the two-fluid model in most cases (Buscaglia et al. 2002). Furthermore, error due to the approximations in deriving the integral plume model was on the order of 10% or smaller. Therefore, the two-fluid model supports the use of integral plume models in practical situations.

5.4 Applications

5.4.1 Aeration and Destratification of Reservoirs

Bubble plumes have been used for many years to manage water quality in lakes and reservoirs. Summer stratification in deep temperate lakes typically has a well-mixed upper layer and a weakly stratified lower layer separated by a metalimnion with a sharp temperature gradient. Because the strong stratification in the metalimnion reduces vertical transport of oxygen from the surface layer to the bottom, the water quality may suffer. Problems with water quality can develop even in unstratified or weakly stratified reservoirs. For example, McCook Reservoir, an element in a plan to control combined-sewer overflows in Chicago, will be 70 m deep and hold up to $3 \times 10^7 m^3$ of combined sewage (Soga and Rehmann 2004, García and García 2006). Managing the combined sewage and preventing odors will require aeration and mixing.

Improving water quality with bubble plumes uses artificial mixing, oxygenation, or both (Wüest et al. 1992). Artificial mixing involves injecting air near the bottom of the lake and using a coarse-bubble diffuser to create bubbles large enough to reach the lake surface. Schladow (1993) quantified the efficiency of the mixing by comparing the change in potential energy of the water column to the work required to inject the air and plotted it as a function of the parameters $M_S = D(1 + \lambda^2)/H_T$ and $C_S = H_N^4$ (Figure 5.10). Plumes with higher M_S are more efficient because their larger area results in more entrainment. For weak stratification, or low C_S, the efficiency of a bubble plume behaves like that in other stratified flows because it increases as the stratification strength increases. However, as C_S continues to increase, the efficiency of a bubble plume behaves quite differently from

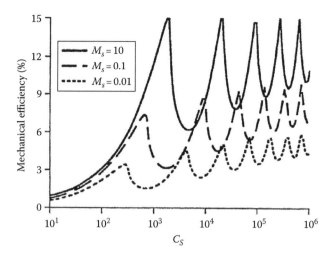

FIGURE 5.10 Mechanical efficiency of a bubble plume as a function of the stratification parameter C_S and the source-strength parameter M_S. (Adapted from Schladow, S.G., *J. Hydraul. Eng.*, 119, 350, 1993. With permission from ASCE.)

efficiencies in other flows. The efficiency reaches a peak when the plume just reaches the water surface: For lower C_S, some of the energy of the plume is wasted, and for higher C_S, the plume detrains below the water surface and another plume continues to the surface (e.g., Figure 5.2d through f). Other peaks in the efficiency as a function of C_S occur when the uppermost plume just reaches the surface. Using these findings to design bubblers, Schladow (1993) recommended choosing values of M_S such that the system starts operating near the second peak in efficiency; the efficiency will remain high as the stratification weakens and the internal detrainment point approaches the water surface.

Designing an effective system for oxygenation involves different considerations. While artificial mixing uses large air bubbles, oxygenation uses oxygen bubbles small enough to dissolve in the hypoxic zone. While the model of Schladow (1993) involves only the hydrodynamics of the plume, a model of oxygenation must also account for mass transfer and liquid chemistry—in particular, the changes in bubble radius caused by mass transfer and compressibility. As discussed in Section 5.3.2, the integral plume model of Wüest et al. (1992) and the two-fluid model of Buscaglia et al. (2002) account for these extra processes. The former model reproduced field measurements well when it used near-field, rather than far-field, conditions because mixing and plume fallback lead to weaker stratification near the plume than farther away (McGinnis et al. 2004). Double-plume models attempt to solve this problem (e.g., Asaeda and Imberger 1993, Socolofsky et al. 2008), but they have not been extensively validated in full-field conditions.

5.4.2 Deepwater Accidental Oil-Well Blowouts

In 2010, the Gulf of Mexico suffered a 2-month long flow of oil and natural gas from the Deepwater Horizon blowout. Estimates for the flow rate of oil have ranged from 5,000 bpd ($9.2 \times 10^{-3} m^3/s$)

early in the disaster to as high as 50,000–100,000 bpd depending on the data source and method for estimating flow. Before this event, two blowout models had been developed, each based on the integral plume modeling method (Johansen 2000, Yapa and Chen 2004), and field validation data were available for only one modeling study (Johansen et al. 2003). These models attempt to account for deflection of the plume and separation of the oil and natural gas by currents, and they treat the oil and entrained fluid as a single-phase fluid so that a double-plume modeling approach is not required—all oil is assumed to intrude in the first stratification- and/or crossflow-arrested detrainment event. Once the oil intrudes into the water column, particle-tracking algorithms predict the oil's rise to the surface. The primary intent of the models is to predict the surface expression of the oil so that clean-up can be adequately deployed.

Several chemical and physical processes acting on the oil and natural gas are included in the models. Natural gas hydrates, which are stable at the temperatures and pressures of a deepwater spill, alter the buoyancy of the gas and reduce the dissolution rate. Although hydrates are stable, they were not observed in the DeepSpill field experiment (Johansen et al. 2003), probably because a threshold concentration of dissolved gas must be present in the water surrounding the bubble before hydrate nucleation can commence; moreover, it is unknown how oil contamination of the bubble surface may affect hydrate formation. Hence, the extent of hydrate formation and its effect on dissolution in a real blowout is an ongoing topic of research.

Other processes affecting the oil include emulsification and droplet formation. The turbulence at the source and within the turbulent plume is expected to have the potential to emulsify the oil (oil in water emulsions). Emulsifications reduce the buoyant effect of the oil by incorporating dense sea water and alter the rise velocity by capturing the oil in a neutrally buoyant water–oil mixture. To predict the initial size distribution of the oil droplets, the properties of the oil (surface tension, viscosity, and density) and the exit (velocity and geometry) are taken into account. Most subsea blowouts are expected to be in or close to the atomization stage, yielding predominantly small oil droplets

in the submillimeter range. Droplet sizes are further reduced by the addition of dispersants at the blowout source, an effect not widely discussed in the literature. Because of the complexity of these competing processes, the droplet size distribution and emulsification state is among the most important and most uncertain parameters describing a deepwater oil-well blowout plume.

Taken together, the small droplet size, the density of oil being close to water (870–930 kg/m^3), and the potential for emulsification work to reduce the rise velocity of the oil and promote subsurface trapping. On the other hand, the plume effect of the natural gas and the oil and water mixture works to bring the oil to the surface. Hence, a complex structure of submerged oil plumes (trapped by stratification or crossflow) and surface oil slicks (generated by oil within the plume and rising out of the submerged intrusions) results, and numerical models capturing all of the physics discussed in this chapter are necessary to make accurate predictions.

The laboratory study of Socolofsky and Adams (2002) presents a scale analysis to determine the regimes of stratification and crossflow dominance for a deepwater accidental oil-well blowout typical of the Gulf of Mexico (Table 5.1). For the smaller spills, crossflow dominates, and the models developed in the literature and cited earlier likely yield accurate results. For the largest spill in the table, stratification is quite important except in the case of strong crossflows. The predictions relevant to the Deepwater Horizon spill of 2010 (for flow rates around 50,000 bpd, or 0.09 m^3/s) are that stratification will cause the oil to trap subsurface and that the first intrusion would be expected around a depth between 200 and 250 m above the sea bottom. Because the plume is stratification dominated (some of the oil will escape the first peel and continue up to the inner plume), multiple subsurface intrusions containing oil would be expected. Measurements from an autonomous underwater vehicle confirm these predictions (Camilli et al. 2010). Hence, even very simple models based on dimensional analysis can provide useful results for designing and analyzing environmental applications of multiphase plumes.

TABLE 5.1 Field-Scale Plume Behavior for Oil-Well Blowouts

In situ Oil Flow Rate[a] (bpd)	In situ Gas Flow Rate (m^3/s)	In situ Hydrate Flow Rate (m^3/s)	h_S (m)	h_T (m)	Critical u_y[b] (cm/s)
540	0	0	4	38	2
540	0.001	0	5	44	2
540	0	0.001	4	41	2
5,400	0.01	0	28	101	5
5,400	0	0.01	43–81	94–104	9–14
54,000	0.1	0	280	235	21
54,000	0	0.1	430–1100	197–216	37–84

Source: Adapted from Socolofsky, S.A. and Adams, E.E., *J. Hydraul. Res.*, 40, 661, 2002. With permission of Taylor & Francis Ltd.

Note: Calculations assume the gas density follows the ideal gas law, and they use a depth of 1000 m, buoyancy frequency of 2×10^{-3} rad/s, crossflow velocities of 15 cm/s, and a crude oil with a density of 900 kg/m^3. Results are presented with and without the formation of natural gas hydrates. For cases with oil or hydrate, the volume flow rate of oil and gas or oil and hydrate are assumed equal.

[a] Flow rate of 540 bpd of petroleum corresponds to 9.9×10^{-4} m^3/s.

[b] Crossflow velocity at which $h_S = h_T$.

5.4.3 Design of Direct Ocean Carbon Sequestration Plumes

Because the capacity of the oceans to store CO_2 is much greater than that of the atmosphere, one strategy to reduce CO_2 emissions to the atmosphere is to pump it directly into the deep ocean, where it will remain sequestered for at least one turnover time of the oceans (about 1000 years). Although not a permanent storage solution, direct ocean carbon sequestration would reduce both the rate of increase of CO_2 in the atmosphere and the peak concentration, two important parameters believed to drive climate change.

Two competing goals for the design of deep-ocean CO_2 sequestration plumes are to dilute the dissolved CO_2 so as to reduce pH changes in the near field and to require that all of the CO_2 is dissolved below the upper mixed layer of the ocean so it will remain adequately sequestered. Both of these goals are met by adjusting the height of rise of the injected liquid CO_2 droplets. Because double-plume integral models predict well the height of rise of dissolving bubble and droplet plumes, they were applied by Socolofsky and Bhaumik (2008) to study the design of a positively buoyant CO_2 sequestration plume at 1000 m depth in the ocean.

An interesting aspect of their study was the nondimensional data reduction that was achieved. The greatest unknown for the design is the degree to which clathrate hydrates may form on the CO_2 droplets, thereby reducing the dissolution rate. To quantify this reduction, Socolofsky and Bhaumik (2008) defined the nondimensional parameter K as the ratio of the hydrate-affected mass transfer rate and the rate of dissolution without hydrates. The data for the nondimensional maximum height of plume rise collapsed to a series of lines when plotted against K (Figure 5.11). Each line in the figure is for a different initial droplet size d_e (in mm). Because the effect of droplet size was to monotonically

increase the rise height as the droplet size increased, Socolofsky and Bhaumik (2008) were able to generate a single design equation, yielding

$$\frac{h_m}{(B/N^3)^{1/4}} = 2.65 d_e K^{-0.08 d_e - 0.63} \tag{5.10}$$

A sensitivity analysis of this equation in the vicinity of the base parameter set showed that the rise height is most sensitive to the initial droplet size and moderately sensitive to the ambient stratification and the factor K; for reasonable buoyancy fluxes, the results were insensitive to B. Errors in the rise height due to variability or uncertainty in these parameters were a factor of about three. Hence, even very simple integral models can provide helpful design information for complicated plumes.

5.5 Major Challenges

For bubble plumes in some environmental applications, the assumption of self-similarity is violated such that the accommodations developed for integral models presented in Section 5.3.2.1 (i.e., double-plume models in stratification and correlation models in crossflow) are inadequate. A good example is the problem of the dissolved gas plume in the wake of a deepwater oil-well blowout in a stratified crossflow. In this case, fluid entrained at all levels of the plume may interact with the gas bubbles and result in elevated concentrations of dissolved gas throughout the wake of the plume. The only applicable numerical tool in such a case would be a fully three-dimensional hydrodynamic model.

The challenge in developing reliable hydrodynamic models of these types of bubble plumes results from the large size of the bubbles and their low void fraction in the plume. Balachandar and Eaton (2010) present an excellent review of turbulent multiphase flow physics and modeling, and highlight two important parameters describing their behavior. To assess the relative independence of the bubble phase to the flow of the carrier phase, the Stokes number St compares the relaxation time of the bubble to the Kolmogorov timescale of the background turbulent flow. For bubble plumes with modest-sized bubbles (2 mm diameter and larger) and relatively low flow rate (1 L/min gas flow rate at the source), St is already of order 1. As either the bubble size or the gas flow rate increases, St also increases; hence, environmental applications of bubble plumes can be expected to have $St \geq 1$. In such cases, the bubble phase cannot be assumed to follow the carrier phase, and either the Lagrangian point particle (LPP) or fully resolved bubble approach to bubble/carrier-phase coupling must be used. A second important parameter is the ratio of the bubble size d to the Kolmogorov length scale η of the carrier-phase turbulence. Most bubbly flows have $d/\eta \geq 1$, in which case the bubbles are expected to modulate the turbulence, as their self-generated turbulence is on a scale that interacts with the carrier-phase turbulence.

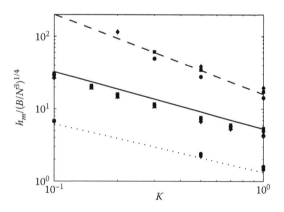

FIGURE 5.11 Variation of the maximum height of plume rise with the mass transfer reduction factor. Symbols are model predictions and lines are the empirical correlation equation. The lines are for initial droplet sizes of 0.5 mm (dotted line), 2 mm (solid line), and 6 mm (dashed line). The symbols are for initial mass fluxes of 0.05 kg/s (circles), 0.5 kg/s (squares), and 1.0 kg/s (diamonds). (From Socolofsky, S. A. and Bhaumik, T., *J. Hydraul. Eng.*, 134, 1570, Copyright 2008. With permission from ASCE.)

While some Lagrangian point particle models have empirical relationships to account for finite-sized bubbles, their application is limited and much new research is needed in this area. Hence, the most reliable simulation tools for environmental scales of bubble plumes are bubble-resolving models. At today's computing power, these models can handle about 100–500 bubbles in a single simulation—clearly not an environmental scale. On the other hand, LPP models can handle about 10^4 to 10^5 bubbles in a single simulation, making environmental scale calculations tenable. Hence, new research is needed to seek LPP models that account for the finite size of bubbles in the flow. These efforts will include both new laboratory experiments designed to validate such models and new theoretical models to describe the physics.

References

Asaeda, T. and J. Imberger. 1993. Structure of bubble plumes in linearly stratified environments. *J. Fluid Mech.* 249: 35–57.

Balachandar, S. and J. K. Eaton. 2010. Turbulent dispersed multiphase flow. *Annu. Rev. Fluid Mech.* 42: 111–133.

Bombardelli, F. A., G. C. Buscaglia, C. R. Rehmann, L. E. Rincón, and M. H. García. 2007. Modeling and scaling of aeration bubble plumes: A two-phase flow analysis. *J. Hydraul. Res.* 45: 617–630.

Bryant, D. B., D. G. Seol, and S. A. Socolofsky. 2009. Quantification of turbulence properties in bubble plumes using vortex identification methods. *Phys. Fluids* 21: 075101, doi:10.1063/1.3176464.

Buscaglia, G. C., F. A. Bombardelli, and M. H. García. 2002. Numerical modeling of large-scale bubble plumes accounting for mass transfer effects. *Int. J. Multiphase Flow* 28: 1763–1785.

Camilli, R., C. M. Reddy, D. R. Yoerger, B. A. S. Van Mooy, M. V. Jakuba, J. C. Kinsey, C. P. McIntyre, S. P. Sylva and J. V. Maloney. 2010. Tracking hydrocarbon plume transport and biodegradation at Deepwater Horizon. *Science* 330: 201–204.

Crounse, B. C., E. J. Wannamaker, and E. E. Adams. 2007. Integral model of a multiphase plume in quiescent stratification. *J. Hydraul. Eng.* 133: 70–76.

García, C. M. and M. H. García. 2006. Characterization of flow turbulence in large-scale bubble-plume experiments. *Exp. Fluids* 41: 91–101.

Johansen, O. 2000. DeepBlow—A Lagrangian plume model for deep water blowouts. *Spill Sci. Technol. Bull.* 6: 103–111.

Johansen, O., H. Rye, and C. Cooper. 2003. DeepSpill—field study of a simulated oil and gas blowout in deep water. *Spill Sci. Technol. Bull.* 8: 433–443.

Lance, M. and J. Bataille. 1991. Turbulence in the liquid phase of a uniform bubbly air-water flow. *J. Fluid Mech.* 222: 95–118.

McDougall, T. J. 1978. Bubble plumes in stratified environments. *J. Fluid Mech.* 85: 655–672.

McGinnis, D. F., A. Lorke, A. Wüest, A. Stockli, and J. C. Little. 2004. Interaction between a bubble plume and the near field in a stratified lake. *Water Resour. Res.* 40: W10206, doi:10.1029/2004WR003038.

Milgram, J. H. 1983. Mean flow in round bubble plumes. *J. Fluid Mech.* 133: 345–376.

Rensen, J., S. Luther, and D. Lohse. 2005. The effect of bubbles on developed turbulence. *J. Fluid Mech.* 538: 153–187.

Schladow, S. G. 1993. Lake destratification by bubble-plume systems: Design methodology. *J. Hydraul. Eng.* 119: 350–368.

Seol, D. G., T. Bhaumik, C. Bergmann, and S. A. Socolofsky. 2007. Particle image velocimetry measurements of the mean flow characteristics in a bubble plume. *J. Eng. Mech.* 133: 665–676.

Seol, D. G., D. B. Bryant, and S. A. Socolofsky. 2009. Measurement of behavioral properties of entrained ambient water in a stratified bubble plume. *J. Hydraul. Eng.* 135: 983–988.

Seol, D. G. and S. A. Socolofsky. 2008. Vector post-processing algorithm for phase discrimination of two-phase PIV. *Exp. Fluids* 45: 223–239.

Smith, B. L. 1998. On the modelling of bubble plumes in a liquid pool. *Appl. Math. Model.* 22: 773–797.

Socolofsky, S. A. and E. E. Adams. 2002. Multi-phase plumes in uniform and stratified crossflow. *J. Hydraul. Res.* 40: 661–672.

Socolofsky, S. A. and E. E. Adams. 2005. Role of slip velocity in the behavior of stratified multiphase plumes. *J. Hydraul. Eng.* 131: 273–282.

Socolofsky, S. A. and T. Bhaumik. 2008. Dissolution of direct ocean carbon sequestration plumes using an integral model approach. *J. Hydraul. Eng.* 134: 1570–1578.

Socolofsky, S. A., T. Bhaumik, and D. G. Seol. 2008. Double-plume integral models for near-field mixing in multiphase plumes. *J. Hydraul. Eng.* 134: 772–783.

Soga, C. L. M. and C. R. Rehmann. 2004. Dissipation of turbulent kinetic energy near a bubble plume. *J. Hydraul. Eng.* 130: 441–449.

Wain, D. J. and C. R. Rehmann. 2005. Eddy diffusivity near bubble plumes. *Water Res. Res.* 41, W09409, doi:10.1029/2004WR003896.

Wüest, A., N. H. Brooks and D. M. Imboden. 1992. Bubble plume modeling for lake restoration. *Water Resour. Res.* 28: 3235–3250.

Yapa, P. D. and F. H. Chen. 2004. Behavior of oil and gas from deepwater blowouts. *J. Hydraul. Eng.* 130: 540–553.

6

Scour around Hydraulic Structures

F. Sotiropoulos
University of Minnesota

P. Diplas
Virginia Polytechnic Institute and State University

A. Khosronejad
University of Minnesota

6.1 Introduction

Rivers often experience changes in bed elevation, which can be caused by the presence of hydraulic structures, other forms of human interference within a stream and/or its watershed, or occur as part of natural *geomorphological* processes. Furthermore, human-induced changes tend to occur at an accelerated pace compared to natural processes. Whatever the cause, there are three generally accepted forms of riverbed elevation change: (1) aggradation or degradation, (2) general scour, and (3) local scour. Aggradation and degradation are long-term processes that take place over long stream reaches. The bed topography evolution associated with these processes is often related to changes in sediment load (e.g., when a braided river changes its planform geometry into a meandering river or downstream of a dam, where the outlet flow contains no sediment load). General or contraction scour is often caused by alteration of the flow patterns and consequently the bed shear stress within a short reach. For example, this type of scour can occur during the passage of a flood when the bed scours on the rising limb and fills during the falling limb. Local scour, on the other hand, is much more complicated than the other two processes leading to bed elevation change. In this case, the removal and transport of bed material occurs in the vicinity of hydraulic structures due to the action of unsteady energetic coherent vortices, which are induced as the approach flow in the reach encounters the hydraulic structure and wraps around it. The classical example of this type of scour is the scour hole that develops around a bridge pier in a river reach due to the so-called horseshoe vortex (see Figure 6.1). Often the effects of general and local scour cannot be readily separated as, e.g., in the case of a rapid flooding event passing through a river reach with hydraulic structures. The process that leads to changes in bed elevation in such cases is the result of the combined action of both scouring processes and will be collectively referred to as scour.

Scour of the streambed at bridge piers and abutments (collectively referred to as foundations) during flooding events poses a major threat to transportation infrastructure as it has resulted in more bridge failures than all other causes combined in recent history (Coleman et al. 2003). Recent examples include hurricanes Katrina and Rita wrecking havoc on numerous bridges in New Orleans; the damage of over 2400 bridge crossings during the 1993 upper Mississippi River basin flooding; the failure of numerous bridges during tropical storm Alberto in central and southwest Georgia (July 3–7, 1994); and the 1987 failure of the I-90 bridge over Schoharie Creek near Albany, New York, which resulted in the loss of 10 lives and millions of dollars for bridge repair/replacement. In addition to the repair and replacement costs, it has been estimated that the total financial loss due to scour-induced damages, including costs due to disruption of local economic and trade activities resulting from bridge closures, is approximately five times higher than the cost of simply repairing and/or rebuilding the damaged bridge. Furthermore, bridge failure may delay significantly access of a distressed area

Handbook of Environmental Fluid Dynamics, Volume Two, edited by Harindra Joseph Shermal Fernando. © 2013 CRC Press/Taylor & Francis Group, LLC. ISBN: 978-1-4665-5601-0.

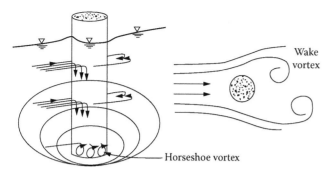

Wake
vortex

Horseshoe vortex

FIGURE 6.1 Schematic view of local scour and flow features around bridge pier showing horseshoe and wake vortices. (From Deng, L. and Cai, C.S., *Pract. Period. Struct. Des. Constr.*, 15(2), 125, 2010.)

by emergency vehicles at a time of great need. As both the frequency and intensity of major storms are forecasted to increase in coming years due to global warming and associated sea-level rise, scour of bridge foundations during extreme flooding events are likely to become more frequent and pronounced. This makes the need for developing effective scour mitigation strategies all that more pressing and critical.

Erosion and failure of in-stream hydraulic structures due to streambed scour is also a major concern in river restoration and infrastructure protection projects (Radspinner et al. 2010). This problem has attracted national attention since efforts to stabilize and restore streams and rivers in the United States have grown dramatically in the last 20 years, with well over $1 billion spent every year since 1990. Most stream restoration projects employ a variety of shallow, in-stream, low-flow structures to assist in stabilizing beds and banks (Radspinner et al. 2010)—e.g., the so-called J-hook structure shown in Figure 6.2. Such structures direct the flow away from the banks, dissipate flow energy, protect stream banks from erosion and scour, create aquatic habitat, and increase habitat diversity. Similar in-stream structures are also employed to protect critical transportation infrastructure. For example, they are used to improve the approach flow at bridge crossings in order to prevent lateral stream migration and suppress local scour near bridge foundations. In spite of the large recent investments in designing and deploying in-stream structures, stream-restoration today is more of an art than science as it is estimated that at least 50% of stream restoration projects will ultimately fail. Streambed erosion due to local scour

in the vicinity of in-stream structures is the major culprit for undermining the stability of such structures and is ultimately one of the leading causes for the failure of stream-restoration and infrastructure protection projects.

A major diDculty for designing hydraulic structures that are not susceptible to scour stems from the complexity of the underlying physical processes and in particular our limited understanding of the interaction of the highly three-dimensional (3D) and unsteady turbulent flow with the sediment bed in the presence of a structure (Dargahi 1990; Melville 1997; Paik et al. 2005; Escauriaza and Sotiropoulos 2011a,b). As a result, design guidelines for hydraulic structures are largely based today either on empirical correlations derived using dimensional analysis arguments with input from field and laboratory experiments (Melville 1997) or on oversimplified *one-dimensional* models that employ cross-sectionally averaged values of the shear stress rather than local values. As such, the task of predicting scour around hydraulics structures remains today as challenging as it has ever been, and many engineers do not trust currently available empirical formulas used for evaluating scour potential (Richardson and Davis 1995).

In this chapter, we seek to (1) provide an overview of the underlying physical processes that give rise to scour around hydraulic structures; (2) review the broad spectrum of empirical and mathematical approaches for modeling flow and transport processes past hydraulic structures; (3) summarize engineering design methods for protecting hydraulic structures from scour; and (4) outline the major challenges that need to be addressed through future experimental and computational modeling research in order to develop reliable scour prediction methodologies.

6.2 Principles

Scour around hydraulic structures may occur under two different flow conditions: (1) *live-bed* scour; or (2) *clear-water* scour (Coleman et al. 2003). Under live-bed conditions, particle entrainment and transport occurs throughout the entire channel bed because the mean channel velocity and the associated bed shear stress are suDciently high to entrain and transport sediment grains. Sediment transport and scour in this case would occur in the channel regardless of the presence of hydraulic structures. The situation is drastically different, however,

FIGURE6 .2 J-hook structures installed in the St. Anthony Falls Laboratory Outdoor StreamLab.

when scour occurs under clear-water conditions. In this case, the water column is devoid of sediment particles since the mean shear stress in the channel is lower than the threshold necessary for initiating sediment motion. Particle entrainment and transport are limited in the vicinity of the hydraulic structure due to instantaneous increments of the bed shear stress at levels above the threshold for initiating motion. Such increments are produced by energetic large-scale vortices that emerge at the junction of the structure with the mobile bed due to 3D separation of the approach turbulent boundary layer induced by the adverse pressure gradients imparted on the flow by the hydraulic structure (Dargahi 1990; Coleman et al. 2003; Escauriaza and Sotiropoulos 2011a).

In what follows, we first discuss the complex hydrodynamic environment in the vicinity of hydraulic structures with main emphasis on the dynamics of the turbulent horseshoe vortex (THSV) past a cylindrical pier mounted on a flat rigid bed. Subsequently, we discuss modes of sediment transport that are relevant to the scour process, present the Lagrangian model for particle transport incorporating the physics of the problem from first principles, and comment on the timescales of the problem.

6.2.1 Large-Scale Hydrodynamics Induced by Hydraulic Structures

The turbulent flow approaching a hydraulic structure mounted on the bed of an open channel experiences a strong adverse pressure gradient and undergoes 3D separation leading to the formation of complex and highly energetic large-scale coherent vortices. The THSV is perhaps the most commonly encountered vortical structure in many applications of practical interest in river hydraulics. Pier- and boulder-like geometries, for instance, give rise to an energetic THSV system in the upstream junction of the structure with the bed, which is known to dramatically increase the production of turbulent stresses and yield instantaneous increments in the wall shear stress that can initiate sediment transport and lead to scour (Dargahi 1990; Devenport and Simpson 1990; Paik et al. 2007; Paik et al. 2009; Escauriaza and Sotiropoulos 2011a,b).

As shown by experiments (Devenport and Simpson 1990) and confirmed and further clarified by recent high-resolution

numerical simulations (Paik et al. 2007, 2009; Apsilidis et al. 2010; Escauriaza and Sotiropoulos 2011a,b), the THSV system upstream of a cylindrical pier is characterized by low-frequency oscillations that produce bimodal probability-density functions (pdfs) of the horizontal and vertical velocities in a region close to the channel bed surface in the vicinity of the hydraulic structure. The presence of bimodal velocity pdfs is the result of the competition between two distinct flow states or modes that occur at the upstream junction between the pier and the bed (see Figure 6.3): (1) the so-called *back-flow mode*, during which the near-bed flow forms a strong wall jet directed upstream (away from the pier) and rolls up to form a well-defined vortex core at the junction region; and (2) the *zero-flow mode* in which the well-defined vortex core is destroyed by strong vertical eruptions of near-wall vorticity flow caused by a pocket of essentially stagnant axial flow and large upward vertical velocities. The flow in the junction region fluctuates continuously in an aperiodic manner between these two flow modes, first identified by Devenport and Simpson (1990): a stable THSV forms during the back-flow mode, it is subsequently destroyed during the zero-flow mode as the result of complex interactions with near-wall vortical structures that engulf and destroy the primary vortex (see Paik et al. (2007) for an explanation of this complex process), and then it forms again as the back-flow mode reemerges and the THSV generation–destruction cycle repeats itself (see Figure 6.4). This continuous, low-frequency interplay between the two flow modes provides the mechanism for producing turbulence stresses in the vicinity of the junction region that are one order of magnitude higher than those produced within the turbulent boundary layer of the approach flow (Devenport and Simpson 1990; Paik et al. 2007). The same mechanism is also responsible for producing large instantaneous increments of the wall shear stress and a pocket of increased wall shear stress root mean square (RMS) values at the bed upstream of the pier (Escauriaza and Sotiropoulos 2011a; Paik et al. 2010). Therefore, the bimodal dynamics of the THSV upstream of a cylindrical pier explains clearly how it is possible for sediment motion to be initiated under clear-water conditions in a region where the mean flow is nearly stagnant, as observed in experiments (Dargahi 1990). Namely, the unsteadiness of the THSV system creates pockets of instantaneous wall shear stress that can exceed

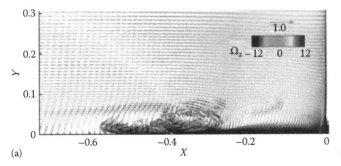 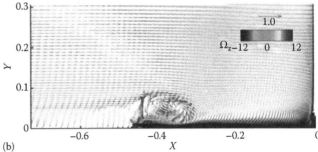

FIGURE 6.3 THSV past a wall-mounted wing-shaped pier at Re = 125,000 for the Devenport and Simpson (1990) case. Calculated (Paik et al. 2007) instantaneous velocity vectors colored with out-of-plane vorticity contours at the vertical plane of symmetry showing the *back-flow* (a) and *zero-flow* (b) modes. The pier leading edge is located at X = 0.

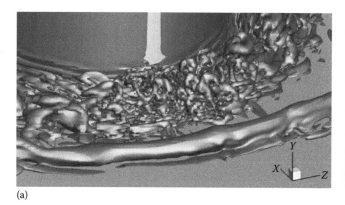

(a)

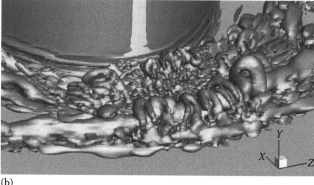

(b)

FIGURE 6.4 THSV past a wall-mounted wing-shaped pier at Re = 125,000 for the Devenport and Simpson (1990) case. Calculated (Paik et al. 2007) instantaneous coherent structures visualized with the q-criterion showing the 3D structure of the THSV during the *back-Wow* (a) and *zero-Wow* (b) modes. The approach flow direction is from right to left.

the critical threshold for dislodging sediment grains, entraining them into suspension and initiating the sediment transport processes that ultimately lead to the creation of a scour hole around the pier (Escauriaza and Sotiropoulos 2011a,b).

Dynamically rich, unsteady, large-scale vortices also emerge and dominate sediment transport processes past other types of hydraulic structures, such as bridge abutments, boulders, groins, etc. (Paik et al. 2005, 2009, 2010). The precise form and dynamics of the large-scale vortices that emerge in such cases are very much dependent on the structure geometry, the Reynolds and Froude numbers of the flow, and the channel geometry within which the structures are embedded. A common feature that is shared between these flows and the flow past the pier-like geometries discussed earlier, however, is that in all cases the emerging coherent vortices are very complex and energetic and are, thus, able to produce significant levels of turbulence stresses near the hydraulic structure. A typical example of the complexity of such flows is shown in Figure 6.5, which depicts the results of recent numerical simulations (Paik et al. 2010) of

high Reynolds number turbulent flow past a rectangular abutment mounted on the side of an open channel. The bed in the simulations was rigid albeit deformed and its shape was obtained from laboratory measurements (Paik et al. 2010) in a flume with a mobile bed after equilibrium was reached.

6.2.2 Flow–Sediment Interactions

Bed material sediment grains in a turbulent open channel flow can in general be transported as suspended load or bed load. Suspended load refers to transport of particles that settle sufficiently slowly to be transported by the flow throughout the entire water column without ever touching or only intermittently coming in contact with the bed. When sediment moves as bed load, on the other hand, particles remain confined in close proximity to the bed within a layer referred to as the bed-load layer, which is typically of the order of few particle diameters thick. Within this layer, particles move in bursts, sliding, rolling, and saltating along the bed in highly intermittent and seemingly chaotic manner (Einstein and El-Samni 1949). Bed-load transport results from the complex interactions between the near-bed turbulent flow structures and the bed particles and depends strongly on the geometrical characteristics of particles, the manner in which they are arranged relative to each other, and the frequency and intensity of interparticle collisions. For the case of scour past hydraulic structures, one would typically anticipate sediment to be transported both as suspended load and bed load under live-bed scour conditions. For clear-water scour, on the other hand, sediment grains would for the most part be transported as bed load in the immediate vicinity of the foundation (Coleman et al. 2003; Escauriaza and Sotiropoulos 2011c).

The fate of a sediment particle lying on the bed in the vicinity of a hydraulic structure is determined by the balance of the various instantaneous forces acting on that particle. These forces can be broadly classified into three categories: (1) flow-induced forces by near-wall turbulent flow structures; (2) gravity; and (3) forces arising due to interactions of the particle both with the bed and/or the hydraulic structure and other particles. Using the

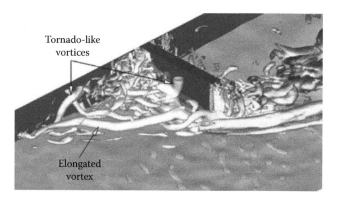

Tornado-like vortices

Elongated vortex

FIGURE 6.5 Flow past a bridge abutment mounted on a mobile bed at Re = 450,000. (From Paik, J. et al., *J. Hydraul. Eng.*, 136(2), 981, 2010.) Calculated instantaneous coherent structures visualized with the q-criterion. Note the multiple tornado-like vortices in the upstream recirculating region and the large elongated vortex that provides the primary mechanism for the creation of the scour hole at the edge of the abutment. The approach flow direction is from left to right.

Lagrangian approach, i.e., considering the motion of individual particles and their interactions with the flow and other particles, the mathematical equations governing the trajectory and momentum of a sediment particle are written as follows:

$$\frac{d\vec{x}}{dt} = \vec{v} \tag{6.1}$$

$$m\frac{d\vec{v}}{dt} = \vec{f}_F + \vec{f}_G + \vec{f}_P \tag{6.2}$$

where

$\vec{x}$ and $\vec{v}$ are the position and the velocity vectors of the sediment particle

m is the particle mass

$\vec{f}_F$ is the resultant force due to the interaction of the instantaneous flow with the particle

$\vec{f}_G$ is the total gravitational force ($\vec{f}_G = m\vec{g}(\rho_s - \rho)/\rho_s$ where $\vec{g}$ is the gravitational acceleration vector and ρ and ρ_s are the fluid and sediment densities, respectively)

$\vec{f}_P$ is the resultant force due to interactions of the particle with solid walls and other particles

The most important forces comprising the flow–particle interactions force $\vec{f}_F$ are the drag, lift, added mass, and fluid stress-induced forces. Assuming spherical nonrotating sediment particles of diameter d, these forces can be expressed as follows (Escauriaza and Sotiropoulos 2011a):

$$\vec{f}_F = \frac{1}{2}\rho C_D \frac{\pi d^2}{4}|\vec{v}_r|\vec{v}_r + \rho C_L \frac{\pi d^3}{6}\left(\vec{v}_r \times \vec{\omega}\right)$$
$$+ \rho C_M \frac{\pi d^3}{6}\left(\frac{D\vec{u}}{Dt} - \frac{d\vec{v}}{dt}\right) + \rho\frac{\pi d^3}{6}\left(-\nabla p + \mu\nabla^2\vec{u}\right) \tag{6.3}$$

where

C_D, C_L, and C_M are the drag, lift, and added mass coeDcients, respectively, obtained from empirical correlations—see Escauriaza and Sotiropoulos (2011a) for details

$\vec{v}_r$ is the particle velocity relative to the local flow velocity $\vec{u}$ ($\vec{v}_r = \vec{v} - \vec{u}$)

$\vec{\omega}$ is the instantaneous vorticity of the flow at the location of the particle ($\vec{\omega} = \nabla \times \vec{u}$)

D/Dt is the material (or Lagrangian) time derivative; p is the pressure; and μ is the dynamic viscosity of the fluid

For more details about the formulas and empirical correlations used to calculate the various terms in the aforesaid equations, the reader is referred to Escauriaza and Sotiropoulos (2011a).

The forces collectively accounted for in the $\vec{f}_P$ term in Equation 6.2 arise as the result of particle–wall and particle–particle interactions. These interactions can induce forces due to (1) frictional resistance when the particle is at rest on the bed; (2) collisions

of the particle with the bed and other solid walls (Schmeeckle and Nelson 2003); and (3) collisions of the particle with other particles (i.e., particle-to-particle interactions). See Escauriaza and Sotiropoulos (2011a) for a more detailed discussion of the various forces included in the $\vec{f}_P$ term.

Equations 6.1 through 6.3 incorporate most of the underlying physics of sediment transport from first principles, and if solved in a coupled manner with the equations governing the flow (the unsteady Navier–Stokes equations) and with the appropriate boundary conditions they can provide an accurate description of scour past hydraulic structures. Such an undertaking, however, is not feasible as it would require computing the trajectories and accounting for the interactions of millions if not billions of individual sediment grains among themselves and with the flow over time suDciently long for scour to reach equilibrium (see the following section). The most challenging case to model using this approach would be the live-bed scour case. Such flows would in general be dense with high concentration of particles and as a result particle–particle collisions as well as changes on the flow field produced by fluid–particle interactions are important and cannot be neglected (Escauriaza and Sotiropoulos 2011a). A significantly simpler case to model, on the other hand, would be clear-water scour since the flow could be treated as dilute (low concentration of particles in the flow) and particle–particle interactions and the impact of particles on the flow can be neglected. Even for such a case, however, using Lagrangian equations to model the initiation and evolution of scour toward equilibrium is not feasible since the computational cost would still be excessive due to the large number of particles and long integration times. A Lagrangian approach could be very effective, however, for gaining insights into the fundamental mechanisms of initiation of motion and sediment transport near hydraulic structures. Escauriaza and Sotiropoulos (2011a) applied a model similar to that described by Equations 6.1 through 6.3 to simulate the motion of 100,000 inertial particles initially placed on the bed upstream of cylindrical pier mounted in a rectangular open channel driven by a simulated flow field that resolves the unsteadiness of the THSV in the junction region (Escauriaza and Sotiropoulos 2011b). They simulated the initial stages of clear-water scour and showed that in accordance to experimental observations the transport of sediment grains is highly intermittent and exhibits essentially all the characteristics of bed-load sediment transport observed in experiments, including random bursting events, sliding, saltation, particle clustering along the bed, etc. (see Figure 6.6). They also showed that the resulting bed-load flux exhibits scale invariance and multifractality as a result of the overall effect of the coherent vortical structures of the flow on sediment transport (Escauriaza and Sotiropoulos 2011a).

To summarize, Lagrangian models readily incorporate and can accurately simulate most of the physics of scour past hydraulic structures but are not feasible to be used as practical engineering tools. Various modeling approaches, which adopt macroscopic (continuum) rather than particle-based

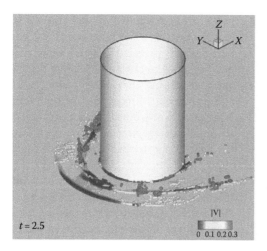
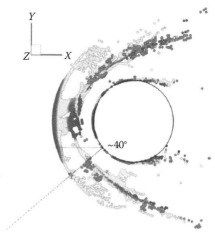

FIGURE 6.6 Calculated instantaneous image of particle transport, showing an ejection event at approximately 40° from the symmetry plane for turbulent flow past a cylindrical pier at Re = 40,000. (Escauriaza, C. and Sotiropoulos, F., *J. Fluid Mech.*, 666, 36, 2011a.) The approach flow direction is from left to right.

formulations and are suitable for carrying out engineering computations of sediment transport and scour, will be discussed in Section 6.4.

6.2.3 Time Evolution of Scour

The equilibrium between the erosive action of forces imparted by the turbulent flow on the sediment and the sediment material resistive forces (e.g., gravity, sediment-wall, and sediment–sediment interactions) is progressively attained through the evolution of the geometry of the mobile bed. The timescale of this evolution, however, is slower than both the timescales of the turbulent flow and the bed-form dynamics observed in the vicinity of the hydraulic structures. The equilibrium scour depth is reached asymptotically with time and could take hours or even days until the final equilibrium condition of local scour topography around a hydraulic structure emerges. In fact, due to the asymptotic nature of the problem the exact time it takes to reach equilibrium is diDcult to determine. For that, an equilibrium time can be defined as the time beyond which the rate of scour depth growth does not exceed 5% of the structure diameter in the subsequent 24 h period (Melville 1997, Melville and Chiew 1999). Several empirical correlations, derived from field and laboratory data, for calculating the so-defined equilibrium time have been proposed in the literature. Among them, the most widely used such formulas are those proposed by Melville and Chiew (1999) and Coleman et al. (2003).

Under clear-water scour conditions, the local scour depth approaches equilibrium asymptotically and in a monotonic fashion as shown in Figure 6.7. An important characteristic of the temporal development of scour also evident in Figure 6.7 is that most of the scour occurs very quickly within the early stages of the scouring process. For example, experimental data for noncohesive material under clear-water condition show that for approach mean-flow velocity U in the range of $0.4 < U/U_{cr} < 1$ (where U_{cr} is the critical value of the approach velocity required

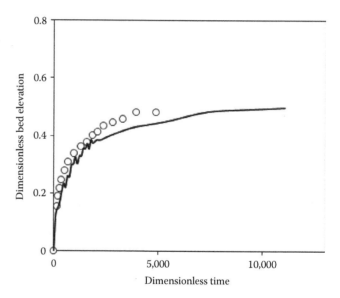

FIGURE 6.7 Variation of measured (circles) and computed (line) dimensionless local scour depth (d/D) with dimensionless timescale (= $t \times U/D$ and t is physical time) under clear-water scour conditions around a laboratory scale diamond bridge pier. (see Khosronejad et al. [2012] for more details).

to initiate sediment motion) 90% of the final scour depth occurs within the first 30% of total equilibrium time (Melville and Chiew 1999).

Under live-bed scour condition, the equilibrium depth is reached faster than in the clear-water case but thereafter the scour depth fluctuates about its equilibrium position due to the influence of bed-form features migrating periodically past the hydraulic structure (Coleman et al. 2003). Melville and Chiew (1999) show that the final local scour depth and related equilibrium time decreases as the U/U_{cr} ratio increases. The live-bed scour typically happens during flood events in which the U/U_{cr} ratio is bigger than one at some part of the flood hydrograph. Hence, the shape and magnitude of the flood hydrograph as well as its duration affect the scour development process. During part of the receding limb of the flood, clear-water scour prevails, which induces additional scour especially if the flow condition remains at near-critical condition over a significant period of time.

6.3 Examples

6.3.1 Scour Past Bridge Foundations: Effects of Geometry and Sediment Type

On the basis of many experimental studies, the shape and alignment of pier and abutment structures have significant influence on the scour process by affecting the dynamics of the coherent vortical structures induced by the foundation (Coleman et al. 2003). For instance, the horseshoe vortex forming at upstream of a blunt nose pier will be quite different than vortices shed by, say, a sharp-nosed pier (see Figure 6.8), and accordingly the spatial distribution and scour and the maximum scour depth for each case will be quite different. The significant influence of the geometry of the hydraulic structure on the properties of scour has led researchers to take into consideration the different geometric characteristics in the formulas they derived for the scour process. As mentioned in the previous section, in the empirical formulas, the predicted depth of scour is corrected according to the geometrical characteristics of the structure. These geometrical characteristics include the length and width of structure, the diameter, if dealing with the cylindrical pier, the angle of attack in horizontal and vertical directions, and the geometrical shape of the structures (Figure 6.8).

The riverbed and bank materials typically encountered in the vicinity of a bridge crossing can range from bedrock to gravel, sand, silt and clay, and many times is represented by a combination of several sediment types. The rate at which the local scour at bridge foundations grows is significantly affected by the type of sediment present. Bedrock erosion is the slowest, requiring

tens or even hundreds of years, while sand is eroded at the fastest rate, reaching the maximum scour depth in a matter of hours. Erosion of cohesive soils will take place at a rate in-between to those encountered in these two extreme cases. While Briaud et al. 1999 have suggested that the final scour hole depth will be similar for both cohesive and noncohesive riverbed materials, several studies with noncohesive sediments have found a dependence of the scour depth size on median bed material size, D_{50}. More specifically, the relative scour depth d/b increases with b/D_{50} for $b/D_{50} \leq 25$ and decreases for b/D_{50} values larger than 25.

While during major floods rivers operate under live bed conditions, it is rather common to perform bridge scour laboratory experiments under clear water settings. Extensive experimental testing indicates that the scour hole dimensions for the live bed case is about the same or marginally larger compared to clear water results (Coleman et al. 2003). However, the scour hole evolution rate is typically higher under live bed conditions. Furthermore, it has been recently suggested that the presence of suspended sediment might have some modest effect on bridge scour dimensions.

Another way the sediment type influences the characteristics and assessment of the scour hole at bridge foundations is through the bed forms present in the channel bed. The most pronounced bed forms are usually encountered in sandy streams. The large variation of the bed elevation, and resulting oscillation in scour hole depth, during the passage of dunes needs to be considered when determining the most critical condition, which coincides with the deepest scour hole and the presence of the bed-form trough in the vicinity of the bridge foundation.

6.3.2 Pressurized Flows

During major flood events, it is possible for the flow free surface at the upstream face of a bridge to reach or exceed the elevation of the low chord of the bridge superstructure, resulting in an orifice-like flow condition. At even more extreme flood events, the bridge can be completely submerged, resulting in a rather complex interaction of weir-like flow over the bridge and orifice-type flow below the bridge superstructure. These are called as pressure flow conditions to distinguish them from the more commonly occurring free surface flows. Laboratory experiments have indicated that during floods having similar flow characteristics, the former mode of operation triggers higher scour within the bridge crossing compared to the latter (Guo et al. 2010). Some of existing studies concluded that the local scour depth around a pier was similar for both pressure and free surface flow conditions. Due to prevailing life threatening conditions, it is diDcult to collect field data on bridge scour under pressure flow and examine the validity of the experimental results.

6.3.3 Scour around Offshore Structures

Scour around offshore structures, such as bridge crossings, platforms, and wind turbines, is a well-known phenomenon. Waves, tides, storm surges, and various combinations of them can cause

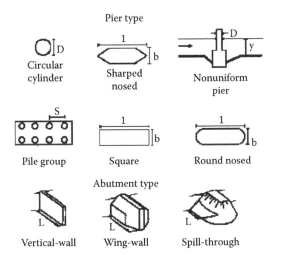

FIGURE 6.8 Typical geometrical shape of hydraulic structures. (From Melville, B.W., *J. Hydraul. Eng.*, 123(2), 125, 1997.)

scour around offshore foundations. Given the large variability associated with the flow types typically encountered in coastal and deep water areas, e.g., wave height and duration, the offshore scour phenomenon is more complicated than its counterpart in rivers. Such scour development may result in uneven settling of the structure and increased stresses which in turn may endanger operational conditions of the corresponding structure. Given the increased popularity of offshore wind turbines to produce aeolian energy, a better understanding of scour around monopiles, and many other foundation combinations and geometries, is of interest. Field measurements from the North Sea and other places have indicated that scour holes around monopiles can have a depth similar to, or even larger than, the pile diameter and cover a horizontal area much larger than the cross-sectional area of the pile. The change in the seabed topography can alter the prevailing local flow characteristics and wave patterns, and increase the hydrodynamic loading upon the pile(s). Considering that the foundation cost of offshore wind turbines can amount up to one-third of the total investment, it is evident that more effort needs to be expended in collecting field data and performing laboratory and numerical studies that will enable us to obtain better predictions of scour in the vicinity of offshore foundations (Harris et al. 2010).

6.3.4 Scour Countermeasures

A countermeasure is defined as a measure incorporated at a bridge stream crossing to monitor, control, inhibit, change, delay, or significantly reduce the severity of scour, and stream and bridge stability problems. Countermeasures have been organized into groups based on their functionality with respect to bridge foundation scour prevention or mitigation. The two main groups of local scour countermeasures are: hydraulic countermeasures and monitoring.

Hydraulic countermeasures are primarily designed either to modify the flow (river training structures) or resist erosive forces caused by the flow (armoring structures). In the latter case, the armoring structures typically act as a resistant layer to hydrodynamic forces providing protection to the more erodible materials underneath. Monitoring describes activities used to facilitate early identification of potential scour problems or survey the scour progress around the bridge foundations. While monitoring does not fix the scour problem, it allows for action to be taken before the potential failure of the bridge. A well-designed monitoring program can be a cost-effective countermeasure.

The results of an extensive survey conducted in the United States indicate that out of more than 220,000 bridges 36,432 sites employed countermeasures (Parker et al. 1998). The study showed that monitoring is by far the most frequently used countermeasure, employed in 76% of the cases. Typical hydraulic countermeasures are, among others, rock gabions, pavements, extended footings, and spurs with riprap being the most widely used. In-stream structures, installed upstream of a highway crossing, can also be used for improving the approach flow and as a result minimize lateral stream migration and scour in the vicinity of the bridge (Radspinner et al. 2010).

Countermeasures may be installed at the time of bridge construction or retrofitted to resolve scour problems. The latter represents good engineering practice because the magnitude, location, and nature of potential scour problems, and associated bridge stability issues, are not always discernible at the design stage and may take several years to develop. The selection of an appropriate countermeasure for scour at a bridge requires an understanding of the erosion mechanism producing the specific scour problem. Furthermore, the installation of such measures can alter the hydraulic conditions to the point that they affect the stability and morphology of the stream. It is worth mentioning that during the design of bridge foundations or implementation of countermeasures, we need to keep in mind that flow direction in some streams can and does change with time (e.g., during the passage of bed forms) and with stage.

6.4 Methods of Analysis

In this section, we discuss a wide spectrum of engineering models for predicting scour at hydraulic structures. We begin with an overview of simple semiempirical equations of scour based on field and laboratory experiments. Subsequently, we discuss 1D, 2D, and 3D numerical models that rely on solving partial differential equations for the flow and sediment transport processes.

6.4.1 Semiempirical Equations of Scour

Semiempirical equations for predicting local scour are among the most widely used equations for designing hydraulic structures (Froehlich 1989). Such equations are based on laboratory and field-scale experiments and attempt to incorporate as many as possible of the geometrical (e.g., channel geometry, hydraulic structure geometry), flow (e.g., mean velocity, approach flow angle, flow depth), and sediment (e.g., sediment size, cohesive vs. noncohesive material) parameters on which scour depends into a simple algebraic equation for calculating the maximum scour depth and the equilibrium timescale. In what follows, we present some of the most popular semiempirical equations for bridge piers and abutments.

One of the most commonly used bridge pier scour equations is the Colorado State University equation which was developed based on laboratory data obtained under both live bed and clear-water scour conditions (Richardson and Davis 1995):

$$d_s = 2.0 h K_1 K_2 K_3 \left[\frac{D}{h} \right]^{0.65} Fr^{0.43} \qquad (6.4)$$

where

d_S is the equilibrium scour depth

K_1, K_2, and K_3 are empirical correction factors (see Richardson and Davis [1995] for more details) accounting for the pier nose shape, the angle of approaching flow toward pier called angle of attack, and the bed material prosperities, respectively

Fr is the Froude number of approaching flow

Another widely used equation for bridge pier scour is that developed by Froehlich (Froehlich 1989) based on laboratory data. This equation incorporates directly the sediment particle characteristics as follows:

$$d_s = 0.32 DKFr^{0.2} \left[\frac{D_e}{D}\right]^{0.62} \left[\frac{h}{D}\right]^{0.46} \left[\frac{D}{D_{50}}\right]^{0.62} \quad (6.5)$$

where

K is a coeDcient related to the shape of the pier nose
D_e is the pier width projected normal to the approach flow
D_{50} is the bed material median grain size.

A bridge-scour equation of somewhat different form is that developed by Melville and Sutherland (1988), which is also based on experimental data:

$$d_s = K_f K_d K_h K_a K_s D \quad (6.6)$$

In the aforementioned equation, K_f, K_d, K_h, K_a, and K_s are coeD-cients representing the effect of approaching flow intensity, sediment size, flow depth, pier alignment relative to flow direction, and pier nose shape, respectively. Detailed expressions for these empirical coeDcients can be found in Melville and Sutherland (1988).

Several semiempirical equations have also been proposed for bridge abutment scour (Ballio and Orsi 2001; Coleman et al. 2003). Among these equations, the equation of Coleman et al. 2003 considers most of the important parameters that determine the equilibrium scour depth and reads as follows:

$$d_s = K_{hL} K_I K_d K_a K_s K_g \quad (6.7)$$

where

K_{hL} is the flow depth-abutment size factor
K_I is the flow intensity factor
K_d is the sediment size factor
K_a is the abutment foundation alignment factor
K_s is the foundation shape factor
K_g is the approach channel geometry factor (which is 1.0 for simple rectangular channel)
K_t is the time factor. Empirical equations for the various factors can be found in Coleman et al. (2003)

Several studies (Melville and Chiew 1999; Coleman et al. 2003) have also been carried out to derive semiempirical equations for the timescale of scour, i.e., the time (t_e) it takes to reach equilibrium for bridge piers (see Section 6.2). These experiments have shown that both the equilibrium time-scale and the equilibrium scour depth are subject to similar influences of geometrical, flow-, and sediment-related parameters as the maximum scour depth. For fine-grained materials (sands and gravels), the equilibrium scour depth d_s is attained much faster under live-bed conditions than under clear-water conditions (Melville and

Chiew 1999). For cohesive sediments, multiple flooding events may be required before the maximum clear-water scour is reached and this may take many years. For bridge abutments, Hoffman and Verheij (1997) studied the time evolution of scour and identified four phases: an initial phase, a development phase, a stabilization phase, and an equilibrium phase. For long abutments, they proposed an exponential equation for describing the temporal evolution of scour. An exponential temporal variation was also identified in the experiments of Ballio and Orsi (2001).

6.4.2 Physics-Based Models of Scour

Semiempirical equations of scour incorporate the physics of the process insofar as they are based on laboratory and field-scale data. Such equations, however, are inherently empirical, and their generality is limited by the basic assumption that a complex and highly nonlinear physical process, such as that of scour, can be described by curve-fitting a limited amount of case-specific experimental data. In this section, we review models of scour that are based on solving numerically differential equations derived from the physical conservation laws (conservation of mass and momentum) for the flow and sediment. Unlike the Lagrangian-based model we discussed in Section 6.2, differential equation models are based on the assumption that the fluid and the sediment phases are interpenetrating continua, which, depending on the degree of model sophistication, may be partially (one-way) or fully (two-way) coupled with each other. In the one-way coupling approach, only the flow impacts the sediment motion directly while the effect of sediment on the flow is indirect via the changes in streambed geometry. In two-way coupling formulations, both phases are interacting with each other. The most complete such model is based on the full 3D equations of mass and momentum conservation. Simpler models that are suitable for expedient engineering calculations can be derived by spatially averaging the 3D equations along the flow depth (2D depth-averaged models) or along the channel cross section (1D cross-sectionally averaged models). In what follows, we present the hierarchy of the most commonly used in engineering practice today models, including 1D, 2D, and the 3D formulations. Due to space considerations, we only discuss one-way coupling models that strictly speaking are suitable for dilute flows.

6.4.2.1 One-Dimensional Models

One-dimensional models simulate flow and sediment transport in the stream-wise direction of a channel in terms of cross-sectionally averaged quantities. Therefore, these models are often applied in the study of long-scale sediment transport processes like degradation and aggradation problems in rivers, reservoirs, estuaries, etc.

The governing equations are the so-called de St. Venant equations, which are written as follows (Wu 2002):

$$\frac{\partial A}{\partial t} + \frac{\partial Q}{\partial t} = 0 \quad (6.8)$$

$$\frac{\partial Q}{\partial t} + \frac{\partial}{\partial x}\left(\beta \frac{Q^2}{A}\right) + gA\frac{\partial z_s}{\partial x} + gAS_f = 0 \qquad (6.9)$$

where
- t is the time
- A is the cross-sectional flow area
- Q is the flow discharge, defined as $Q = AU$, with U being the flow velocity averaged over the flow cross section
- x is the spatial coordinate representing the streamwise distance
- β is momentum correction factor due to the nonuniformity of stream-wise velocity over the flow cross section
- z_s is water surface elevation
- g is gravitational acceleration
- S_f is friction slope which is computed using Manning and/or Chezy equations

Once the flow field is calculated by solving the aforementioned system of equations, the bed change is computed using the following 1D sediment mass balance equation:

$$(1-p')\frac{\partial A_b}{\partial t} + \frac{\partial Q_t}{\partial x} = 0 \qquad (6.10)$$

where
- p' is the sediment material porosity
- A_b is the cross-sectional area of bed above a reference datum
- Q_t is the total sediment transport load at the cross section

This mass balance equation can be solved to compute the total change in bed area, which in turn is used to update the channel geometry before the de St. Venant equations (6.8) and (6.9) are solved again to advance the process in time (Wu 2002).

The presence of hydraulic structures in 1D models is accounted for by specifying internal boundary conditions. By neglecting the storage effect of a hydraulic structure, the flow discharge is conserved across the computational cell where the structure is located while the flow depth across the structure is determined using a stage-discharge relation.

6.4.2.2 Two-Dimensional Models

If the vertical variations of flow and sediment quantities are suDciently small or can be determined analytically, variations in the horizontal plane can be approximately described by a depth-averaged 2D model. The governing equations for 2D depth-averaged turbulent flow are the depth-averaged Reynolds-averaged continuity and momentum equations, which read as follows:

$$\frac{\partial h}{\partial t} + \frac{\partial (hU_x)}{\partial x} + \frac{\partial (hU_y)}{\partial x} = 0 \qquad (6.11)$$

$$\frac{\partial (hU_x)}{\partial t} + \frac{\partial (hU_x^2)}{\partial x} + \frac{\partial (hU_yU_x)}{\partial y}$$
$$= -gh\frac{\partial z_s}{\partial x} + \frac{1}{\rho}\left(\frac{\partial (hT_{xx})}{\partial x} + \frac{\partial (hT_{xy})}{\partial y} - \tau_{bx}\right) \qquad (6.12)$$

$$\frac{\partial (hU_y)}{\partial t} + \frac{\partial (hU_xU_y)}{\partial x} + \frac{\partial (hU_y^2)}{\partial y}$$
$$= -gh\frac{\partial z_s}{\partial y} + \frac{1}{\rho}\left(\frac{\partial (hT_{yx})}{\partial x} + \frac{\partial (hT_{yy})}{\partial y} - \tau_{by}\right) \qquad (6.13)$$

where
- h is the flow depth
- x and y are the horizontal coordinates
- U_x and U_y are the respective depth-averaged horizontal mean velocity components
- ρ is the water density
- τ_{bx} and τ_{by} are the components of the bed shear stress
- T_{xx}, T_{xy}, T_{yx}, and T_{yy} include the viscous and Reynolds stresses

Using the Boussinesq assumption, these stresses can be expressed in terms of the mean flow as follows:

$$T_{xx} = 2\rho(\nu+\nu_t)\frac{\partial U_x}{\partial x} - \frac{2}{3}\rho k \qquad (6.14)$$

$$T_{xy} = T_{yx} = \rho(\nu+\nu_t)\left(\frac{\partial U_x}{\partial y} + \frac{\partial U_y}{\partial x}\right) \qquad (6.15)$$

$$T_{yy} = 2\rho(\nu+\nu_t)\frac{\partial U_y}{\partial y} - \frac{2}{3}\rho k \qquad (6.16)$$

where
- ν is the kinematic viscosity of fluid
- ν_t is the eddy viscosity, which needs to be determined using an appropriate turbulence model
- k is the turbulence kinetic energy

To close the aforementioned equations, the eddy viscosity needs to be determined with an appropriate turbulence model. Such model could range in sophistication from a simple constant eddy-viscosity assumption or an algebraic, mixing-length model, or the most widely used two-equation k–ε model, where ε is the rate of dissipation of k. For an overview of various turbulence models for depth-averaged equations, the reader is referred to Rodi (1993).

By integrating the complete 3D sediment transport equation (next section) over the depth of the suspended-load layer, the 2D depth-averaged suspended-load transport equation is obtained as follows:

$$\frac{\partial (hC)}{\partial t} + \frac{\partial (hU_xC)}{\partial x} + \frac{\partial (hU_yC)}{\partial y}$$
$$= \frac{\partial}{\partial x}\left(\Gamma_{ex}h\frac{\partial C}{\partial x}\right) + \frac{\partial}{\partial y}\left(\Gamma_{ey}h\frac{\partial C}{\partial y}\right) + E_b - D_b \qquad (6.17)$$

where

C is the depth-averaged mean sediment concentration

Γ_{ex} and Γ_{ey} are the effective diffusion coeDcient in horizontal x and y directions, respectively

E_b and D_b are the sediment entrainment flux and the deposition flux at the interface between the bed-load and suspended-load layers, respectively

Therefore, the term $E_b - D_b$ represents the net sediment flux at the interface and needs to be computed by empirical relations. Now by integrating the same 3D sediment transport equation (see the following section) over the bed-load layer, we obtain the sediment mass balance equation for the bed-load layer (Wu 2002):

$$(1 - p')\frac{\partial z_b}{\partial t} + \frac{\partial(\alpha_x q_b)}{\partial x} + \frac{\partial(\alpha_y q_b)}{\partial y} = D_b - E_b \qquad (6.18)$$

where

z_b is the local bed elevation above datum

q_b is the bed-load transport flux

α_x and α_y are direction cosines

Summing Equations 6.17 and 6.18 and neglecting terms of secondary importance leads to the overall sediment balance equation:

$$(1 - p')\frac{\partial z_b}{\partial t} + \frac{\partial q_{tx}}{\partial x} + \frac{\partial q_{ty}}{\partial y} = 0 \qquad (6.19)$$

where $q_{tx} = \alpha_x q_b + h U_x C$ and $q_{ty} = \alpha_y q_b + h U_y C$ are the components of the total sediment load in the x (longitudinal) and y (transverse) directions, respectively.

The so-derived 2D depth-averaged sediment transport governing equations consist of Equations 6.17 and 6.18, but there are three unknowns: $\tilde{C}$, q_b and z_b. Therefore, one more equation is required to close the problem. The third equation is obtained by adopting a local equilibrium assumption for the bed-load transport, which assumes that the bed-load sediment transport rate is equal to the transport capacity under equilibrium conditions. However, this local equilibrium assumption may lead to unrealistic predictions of bed deformation especially in cases of strong erosion and deposition. Another approach to close the governing equations and take into account the nonequilibrium transport effects is to use the following equation to compute the bed-load flux (Wu 2002):

$$\frac{\partial(\alpha_x q_b)}{\partial x} + \frac{\partial(\alpha_y q_b)}{\partial y} + \frac{1}{L_s}(q_b - q_b^*) = 0 \qquad (6.20)$$

where

L_s is the bed-load transport nonequilibrium adaptation length

q_b is the equilibrium bed-load rate

Depth-averaged models can account directly for the contraction and expansion effects the presence of hydraulic structures may have on the flow and sediment transport but are inherently incapable of incorporating the effects of 3D vortical structures induced by hydraulic structures, e.g., horseshoe vortex. Even though depth-averaged models explicitly sensitized to 3D effects have been proposed (Wu 2002), such models are in principle more suitable for situations where contraction scour is the dominant mode of scour. Examples from the application of depth-averaged models to simulate scour past hydraulic structures can be found in Hoffman and Verheij (1997), and Wu (2002).

6.4.2.3 Three-Dimensional Models

In 3D models, there is a dynamic interaction between flow and the sediment transport field. Assuming dilute flows, in which the impact of sediment on the carrier fluid turbulence intensity and structure is negligible, this interaction can be described as follows. First, the flow field induces the sediment transport process and then the sediment transport process impacts the flow by modifying the bed geometry through the formation of scour holes or bed forms, and thus altering the boundary conditions for the flow velocity field. The governing equations for the flow are the unsteady Reynolds-averaged continuity and Navier–Stokes equations, which written in Cartesian tensor notation, where repeated indices imply summation, read as follows (Rodi 1993):

$$\frac{\partial u_i}{\partial x_i} = 0 \qquad (6.21)$$

$$\frac{\partial u_i}{\partial t} + u_j \frac{\partial u_i}{\partial x_j} = -\frac{1}{\rho}\frac{\partial P}{\partial x_i} - \frac{1}{\rho}\frac{\partial}{\partial x_j}(\overline{\rho u_i u_j}) + \nu \frac{\partial^2 u_i}{\partial x_j \partial x_j} + g_i \qquad (6.22)$$

where

u_j (j = 1, 2, 3) are the components of local time averaged flow velocities

P is the pressure

gi are the components of the gravitational acceleration

$-\rho \overline{u_i u_j}$ is the Reynolds stress tensor

In most engineering models of sediment transport, the Reynolds stress tensor is modeled using the Boussinesq hypothesis (Rodi 1993) using turbulence models of varying degree of sophistication as discussed for the depth-averaged model described earlier.

In 3D sediment transport simulations, the sediment concentration profile in the entire water column could be computed using 3D sediment transport equation (repeated indices imply summation):

$$\frac{\partial c}{\partial t} + \frac{\partial}{\partial x_j}[(u_j - w_s \delta_{j3})c] = \frac{\partial}{\partial x_j}\left(\Gamma_c \frac{\partial c}{\partial x_j}\right) \qquad (6.23)$$

where

c is the mean (Reynolds averaged) local sediment concentration

δ_{j3} is Kronecker delta (unit tensor) which is always zero unless in the vertical direction when j = 3

Γ_c is the diffusion coeDcient for sediment concentration

The bed deformation can be calculated by solving Equation 6.19, which is the same in depth-averaged and 3D models since for

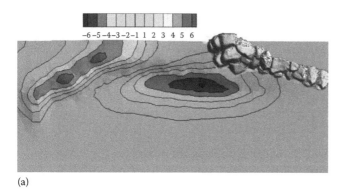

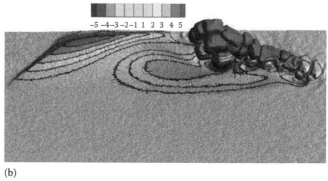

(a)

(b)

FIGURE 6.9 Contours of computed (a) and measured (b) local scour around rock vane structure in cm (flow direction is from right to left).

both cases this equation is derived by averaging across the depth of the bed-load layer. Depending on the relative importance of bed-load and suspended-load transport processes, however, one or both of the two sediment transport Equations 6.18 and 6.19 need to be considered. For instance, for live-bed scour around hydraulic structures both the bed load and the suspended load are significant and need to be modeled. In most clear-water scour problems, on the other hand, suspended load is not important and only the bed-load sediment transport equation needs to be considered.

The bed-load rate is calculated either by adopting a local equilibrium assumption or using the nonequilibrium relation given by (6.20). The suspended-load flux vector, on the other hand, is given as follows ($i = 1, 2$):

$$q_{sx_i} = \int_{\delta_b}^{h} \left[u_i c - \Gamma_c \frac{\partial c}{\partial x_i} \right] dx_3 \qquad (6.24)$$

where $i = 3$ denotes the vertical (gravity) direction. Finally, the total sediment load is determined as follows:

$$q_{tx_i} = \alpha_{x_i} q_b + q_{sx_i} \qquad (6.25)$$

Examples from the application of 3D models to simulate scour past hydraulic structures can be found in Escauriaza and Sotiropoulos (2011a,b,c). More recently, a 3D RANS numerical model has been developed for simulating scour past arbitrarily complex hydraulic structures (Khosronejad et al. 2011, 2012). An example from the application of this model to simulate scour patterns past a rock vane is shown in Figure 6.9. The figure compares the computed bed bathymetry with the results of a flume experiment carried out in the St. Anthony Falls Laboratory after 60 min of scour.

6.5 Major Challenges

6.5.1 Upscaling Issues

Extending the bridge foundation scour measurements obtained from laboratory tests to prototype scale piers and abutments remains problematic because of incomplete geometric and dynamic modeling employed during the experiments. Traditionally, distorted geometric similarity—with $(b/D_{50})_{model} > (b/D_{50})_{prototype}$ and $(H/b)_{model} \neq (H/b)_{prototype}$—and

incomplete dynamic similarity (based on Froude number) are used during model tests. Results from such experiments have been found to over predict field scour data. This discrepancy has been attributed, in part, to the significant difference in pier-based Reynolds number values typically used in model studies and those encountered in prototype structures during the passage of floods. Furthermore, the resulting differences between the two flow patterns are further exacerbated by the lack of geometric similarity—the impact of b/D_{50} on d/b was discussed in Section 6.3, while the free surface effects manifested through the H/D ratio appear to be diminishing for $H/b > 1.4$ (Coleman et al. 2003) or $H/b > 2$. More specifically, the majority of the laboratory experiments are dealing with Reynolds numbers in the tens of thousands up to 1.2×10^5, while prototype values usually exceed 10^6. Experimental work (Devenport and Simpson 1990; Apsilidis et al. 2010) and numerical simulation studies (Paik et al. 2010; Escauriaza and Sotiropoulos 2011b) have demonstrated the strong dependence of the coherent flow dynamics of the THSV on the value of the Reynolds number. Given the dominant role of the THSV in bridge pier scour, both in terms of development rate and maximum depth magnitude, such findings suggest that most laboratory studies do not adequately represent the flow dynamics encountered under field conditions. The lack of complete geometric similarity, combined with the effects of dynamically dissimilar flow patterns, significantly alter the erosional behavior of the laboratory sediments compared to that observed in the field (Paik et al. 2007; Apsilidis et al. 2010). These shortcomings cast a shadow upon the results obtained in traditional laboratory modeling studies and limit their usefulness for predicting prototype scour values. These issues deserve further study by pursuing Reynolds based similarity experiments, numerical simulations at Reynolds numbers representative of field conditions, and by collecting bridge scour foundation data at prototype structures during major floods.

6.5.2 Time-Averaged versus Instantaneous Understanding of Flow and Sediment Transport

A criterion originally proposed by Shields in 1936 (BuDngton and Montgomery 1997) has been used as the standard method

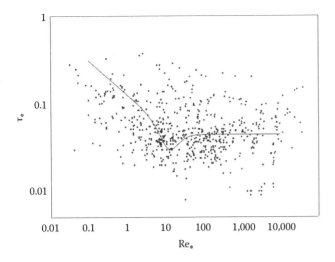

FIGURE 6.10 Shields diagram for the initiation of sediment entrainment. Dimensionless wall shear, τ_*, versus particle (roughness) Reynolds number, R_*. The line, Shields criterion, separates movement from no movement of sediment particles. Data represent a large number of field and laboratory studies on threshold of motion criterion.

for describing threshold conditions of mobile sediment for more than 70 years, even though results from laboratory and field studies have shown more than an order-of-magnitude variability (Figure 6.10, reproduced from BuDngton and Montgomery 1997). This criterion emphasizes the time-averaged boundary shear stress and therefore is incapable of accounting for the fluctuating forces encountered in turbulent flows. The significant variability exhibited by the data shown in Figure 6.10 suggests that Shieldsfl criterion represents a surrogate yet incomplete measure of the turbulent flow processes responsible for particle entrainment. Several researchers have attempted to explain this variability and devised alternative plots for a variety of flow/sediment cases (BuDngton and Montgomery 1997).

In an effort to overcome the limitations of the time-averaged wall shear stress approach, a large number of researchers have explored and advocated the important role that peak turbulent velocity values and resulting hydrodynamic forces play on particle dislodgement, particularly for flow conditions near the threshold of movement. One of the first attempts to highlight the importance of turbulence on particle movement, albeit in an elementary way, is attributed to Varenius (1664). Varenius concluded from observations that the random motion of sediment is related to what we would refer to today as the turbulent motion of the water in the stream. In 1949, based on laboratory experiments, Einstein and El-Samni concluded that near-bed turbulent flow fluctuations are mainly responsible for the dislodgement of particles located at the boundary of a laboratory flume. Since then, a large number of researchers have advocated this point of view based on detailed field, laboratory, and other studies for both smooth (Sumer et al. 2003) and rough boundaries (e.g., Papanicolaou et al. 2001; Schmeeckle and Nelson 2003), as well as in the presence of bed forms (Sumer et al. 2003). It suDces to repeat one of the findings obtained by Sumer et al. (2003) "A 20% increase in the turbulence

level in the bed shear stress induces a factor of 6 increase in the sediment transport at the value of the (dimensionless wall shear stress) Shields parameter $\tau^* = 0.085$ (which represents a flow condition relatively close to threshold)." Similarly, from their experiments, Schmeeckle and Nelson (2003) found that increasing the amplitude of the instantaneous downstream velocity while maintaining the same time-mean value (representing a Shields stress value 50% higher than threshold) resulted in dramatically higher bed-load transport rates.

Regardless of the widespread recognition, and the more recent rather startling results, the attempts to link the characteristics of turbulent flow to particle entrainment have not surpassed qualitative descriptions. Furthermore, the notion that the increase in bed-load transport is solely due to the role of the peak instantaneous velocity values and associated hydrodynamic forces was challenged recently by Diplas and coworkers. Based on detailed laboratory experiments, they demonstrated that peak values in the turbulent record are necessary but not suDcient to explain bed-load transport, especially near threshold of motion conditions (Diplas et al. 2008). Instead, they emphasized that duration is just as important as the magnitude of the hydrodynamic forces in determining threshold conditions. Then, impulse (the product of force magnitude and duration) becomes the relevant parameter characterizing near threshold conditions. Results from laboratory experiments and theoretical formulation of the problem support the concept of impulse as the universal criterion for determining the initiation of particle movement (Diplas et al. 2008; Valyrakis et al. 2010). These recent findings provide fertile ground for research in scour around bridge foundations, a phenomenon dominated by unsteady and turbulent flow phenomena.

6.5.3 Use of High-Resolution Computational Models

As we have repeatedly emphasized at several places in this chapter, a common characteristic of flows past hydraulic structures is that they are dominated by energetic coherent structures that are ultimately responsible for streambed erosion and scour. The most popular example of this situation is encountered in bridge pie scour where sediment erosion, transport, and deposition are dominated by the unsteadiness of the THSV. RANS-based hydrodynamic models, such as those reviewed in Section 6.4, are inherently incapable of resolving the unsteadiness of the THV. Furthermore, such models link the bed-load flux to the mean shear stress and naturally should fail to even qualitatively predict the scour patterns in areas such as the upstream junction of the pier with the streambed where the bed shear stress field fluctuates intensely but is close to zero in the mean (Paik et al. 2010). Therefore, unsteady hydrodynamic models that explicitly resolve unsteady coherent structures, such as detached-eddy simulation (DES) or large-eddy simulation (LES) (see Paik et al. (2010); Kang et al. (2011) for recent overview of such models), are critical prerequisites for developing physics-based models of sediment transport and scour.

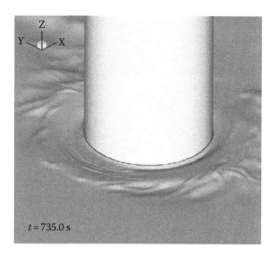

$t = 735.0$ s

FIGURE 6.11 Calculated instantaneous bed surface showing scour geometry development and bed forms in turbulent flow past a bridge pier. (From Escauriaza, C. and Sotiropoulos, F., *J. Geophys. Res.*, 116, F03007, 2011c.)

The potential of coherent-structure-resolving hydrodynamic models insofar as sediment transport calculations are concerned was recently demonstrated by Escauriaza and Sotiropoulos (2011c) who proposed a new unsteady model of bed-load transport, which readily accounts for the impact of the near-bed, fluctuating hydrodynamic forces on the sediment grains. The model was applied to simulate clear-water scour past a cylindrical pier and was shown to yield statistically meaningful bed forms that resemble those observed in experiments. Future work should focus on further refining such models by incorporating in them the previously reviewed recent insights into the role of impulse in determining the initiation of particle movement mechanisms, and demonstrating their predictive capabilities in bridge-scour problems. A major challenge that needs to be addressed in order to be able to use models like that developed by Escauriaza and Sotiropoulos (2011c) in practical computations stems from the large disparity between the timescales of the coherent structures in the flow, which are of the order of seconds, and the timescales of scour, which could be of the order of hours. This challenge can be addressed by drastically increasing the computational eDciency of coherent-structure-resolving hydrodynamic models by taking advantage of massively parallel computational platforms and by developing computational techniques that incorporate the large disparity of time-scales in the algorithms for solving the flow and sediment transport equations (Figure 6.11).

Acknowledgments

This work was supported by NSF Grants EAR-0120914 (as part of the National Center for Earth-surface Dynamics) and EAR-0738726 and the National Cooperative Highway Research Program under Project 24-33. Numerical simulations presented in this paper were carried out in part using computing resources at the University of Minnesota Supercomputer Institute. Figure 6.10 was prepared by Dr. Manousos Valyrakis.

References

Apsilidis, N., Diplas, P., Dancey, C. L., and Sotiropoulos, F. (2010) The effect of Reynolds number on junction flow dynamics. *Sixth International Symposium on Environmental Hydraulics*, Athens, Greece.

Ballio, F. and Orsi, E. (2001) Time evolution of scour around bridge abutments. *Water Eng. Res.*, 2: 243–259.

Briaud, J.-L., Ting, F., Chen, H.-C., Cao, Y., Gudavalli, R., Perugu, S., and Wei, G. (1999) Prediction of scour rate in cohesive soils at bridge piers. *J. Geotech. Geoenviron. Soc. Am.*, 125(4): 237–246.

BuDngton, J. M. and Montgomery, D. R. (1997) A systematic analysis of eight decades of incipient motion studies with special reference to gravel-bedded rivers. *Water Resour. Res.*, 33(8): 1993–2029.

Coleman, S. E., Lauchlan, C. S., and Melville, B. W. (2003) Clear water scour development at bridge abutments. *J. Hydraul. Res.*, 41: 521–531.

Dargahi, B. (1990) Controlling mechanism of local scouring. *J. Hydraul. Eng.*, 116(10): 1197–1214.

Deng, L. and Cai, C. S. (2010) Bridge scour: Prediction, modeling, monitoring, and countermeasures—Review. *Pract. Period. Struct. Des. Constr.*, 15(2): 125–134.

Devenport, W. J. and Simpson, R. L. (1990) Time-dependent and time-averaged turbulence structure near the nose of a wing-body junction. *J. Fluid Mech.*, 210: 23–55.

Diplas, P., Dancey, C. L., Celik, A. O., Valyrakis, M., Greer, K., and Akar, T. (2008) The role of impulse on the initiation of particle movement under turbulent flow conditions. *Science*, 322: 717–720, DOI: 10.1126/science.1158954.

Einstein, H. A. and El-Samni, E. A. (1949) Hydrodynamic forces on a rough wall. *Rev. Mod. Phys.*, 21: 520–524.

Escauriaza, C. and Sotiropoulos, F. (2011a) Lagrangian model of bed-load transport in turbulent junction flow. *J. Fluid Mech.*, 666: 36–76.

Escauriaza, C. and Sotiropoulos, F. (2011b) Reynolds number effects on the coherent dynamics of the turbulent horseshoe vortex system. *J. Flow, Turbul. Combust.*, 86(2): 231–262.

Escauriaza, C. and Sotiropoulos, F. (2011c) Initial stages of erosion and bed-form development in turbulent flow past a bridge pier. *J. Geophys. Res.*, 116: F03007.

Froehlich, D. C. (1989) Local scour at bridge abutments. *Proceedings of ASCE, National Hydraulic Conference*, New York, pp. 13–18.

Guo, J., Kerenyi, K., Pagan-Ortiz, J., Flora, K., and Afzal, B. (2010) Submerged-flow bridge scour under maximum clear-water conditions (I): Experiment. *International Conference on Scour and Erosion (ICSE-5)*, San Francisco, CA, pp. 807–814.

Harris, J. M., Whitehouse, R. J. S., and Benson, T. (2010) The time evolution of scour around offshore structures, *Proc. ICE—Maritime Eng.*, 163(1): 3–17.

Hoffmans, G. J. C. M. and Verheij, H. J. (1997) *Scour Manual.* Rotterdam, the Netherlands: A. A. Balkema.

Kang, S., Lightbody, A., Hill, C., and Sotiropoulos, F. (2011) High-resolution numerical simulation of turbulence in natural waterways. *Adv. Water Resour.*, 34(1): 98–113.

Khosronejad, A., Kang, S., Borazjani, I., and Sotiropoulos, F. (2011) Curvilinear immersed boundary method for simulating coupled flow and bed morphodynamic interactions due to sediment transport phenomena. *Adv. Water Resour.*, 34(7): 829–843.

Khosronejad, A., Kang, S., and Sotiropoulos, F. (2012) Experimental and computational investigation of local scour around bridge piers. *Adv. Water Resour.*, 37: 73–85.

Melville, B. W. (1997) Pier and abutment scour: Integrated approach. *J. Hydraul. Eng.*, 123(2): 125–136.

Melville, B. W. and Chiew, Y. M. (1999) Time scale for local scour at bridge piers. *J. Hydraul. Eng.*, 125(1): 59–65.

Melville, B. W. and Sutherland, A. J. (1988) Design method for local scour at bridge piers. *J. Hydraul. Eng.*, 114(10): 1210–1226.

Paik, J., Escauriaza, C., and Sotiropoulos, F. (2007) On the bi-modal dynamics of the turbulent horseshoe vortex system in a wing-body junction. *Phys. Fluids*, 19: 045107.

Paik, J., Escauriaza, C., and Sotiropoulos, F. (2010) Coherent structure dynamics in turbulent flows past in-stream structures: some insights gained via numerical simulation. *J. Hydraul. Eng.*, 136(2): 981–993.

Paik, J., Sotiropoulos, F., and Porte-Agel, F. (2009) Detached eddy simulation of the flow around two wall-mounted cubes in tandem. *Int. J. Heat Fluid Flow*, 30: 286–305.

Paik, J., Sotiropoulos, F., and Sale, M. J. (2005) Numerical simulation of swirling flow in a complex hydro-turbine draft tube using unsteady statistical turbulence models. *J. Hydral. Eng.*, 131(6), 441–456.

Papanicolaou, A., Diplas, P., Dancey, C. L., and Balakrishnan, M. (2001) Surface roughness effects in near-bed turbulence: Implications to sediment entrainment. *J. Eng. Mech.*, 127(3): 211–218.

Parker, G., Toro-Escobar, C., and Voight, R. L., Jr. (1998) Countermeasures to protect bridge piers from scour Final Report, Vol. 2, Prepared for *National Cooperative Highway Research Program, Transportation Research Board, National Research Council, NCHRP Project 24–07*, St. Anthony Falls Laboratory, University of Minnesota, Minneapolis, MN.

Radspinner, R., Diplas, P., Lightbody, A., and Sotiropoulos, F. (2010) River training and ecological enhancement using in-stream structures. *J. Hydral. Eng.*, 136(12): 967–980.

Richardson, E. V. and Davis, S. R. (1995) Evaluating scour at bridges. *Hydraulic Engineering Circular No. 18*, Federal Highway Administration, Washington, DC.

Rodi, W. (1993) Turbulence models and their application in hydraulics—A state of the art review. *IAHR Monograph*, 3rd edn., Rotterdam, the Netherlands: A. A. Balkema.

Schmeeckle, M. W. and Nelson, J. M. (2003) Direct numerical simulation of bed load transport using a local, dynamic boundary condition. *Sedimentology*, 50: 279–30.

Sumer, B. M., Chua, L. H. C., Cheng, N. S., and Fredsoe, J. (2003) Influence of turbulence on bed load sediment transport. *J. Hydraul. Eng.*, 129(8): 585–596.

Valyrakis, M., Diplas, P., Dancey, C. L., Greer, K., and Celik, A. O. (2010) The role of instantaneous force magnitude and duration on particle entrainment. *J. Geophys. Res. Earth Surf.*, 115: 1–18, DOI:10.1029/2008JF001247.

Varenius, B. (1664) Geographia generalis, in *qua aTactiones generales telluris explicantur*, pp. 748, D. Amstelodami, Elsevier, Amsterdam.

Wu, W. (2002) *Computational River Dynamics.* Taylor & Francis Group, the Netherlands.

7

Flow through Urban Canopies

Rex E. Britter
University of Cambridge

Silvana Di Sabatino
University of Salento

7.1 Introduction

Recently a major milestone was observed in that over half of the worldß population is now living in urban areas and, furthermore, that this urbanization is going to continue. The nature of urban areas and how we create or modify them will become of greater consequence to all our lives. Governments are paying increasing attention to this urbanization phenomenon (United Nations, World population Prospects: The 2006 Revision, 2007). This is leading to new questions for the scientific community concerning more and more densely populated cities.

Because most of us live in cities, many aspects of the city influence us directly or indirectly. For example, the flow of air through the city impacts upon the forces on buildings, urban air quality and the breathability of cities, indoor air quality and building ventilation, pedestrian comfort, the response of the city to climate change, local weather, energy usage, urban mobility, and security concerns. Though air flow in cities has been studied for many years (Vitruvius, 2005, translation), there has been a very marked, recent increase in research on cities driven by these issues. At the same time, this diversity of areas of interest has attracted a wide range of disciplines, technical and nontechnical, all with somewhat different perspectives.

From a fluid dynamical point of view, we are talking about (and living in) a rough-wall turbulent boundary layer flow with changes of surface roughness across the city. In this chapter we are particularly interested in what is happening *within* the surface roughness. Urban morphology affects the flow throughout the atmospheric boundary layer (ABL) which is roughly the lowest 10% of the troposphere. Direct effects are felt within the urban canopy layer (UCL); the layer occupied by the buildings

and indirect, but specific, effects above this. Figure 7.2 describes these layers schematically.

In addition to this, there are effects due to temperature variations throughout the city, these being influenced by conductive, convective, and radiative heat transfers and thermal advection, together with moisture fluxes and phase changes. These have a direct impact on the surface energy balance leading to the possible development of an urban heat island (UHI) effect. These thermal and moisture effects, if large enough, can produce dynamical effects on the air flowing through and over the city.

When first studying the cities as represented in Figure 7.1, we might immediately decide to purchase a very large computer and a suite of computational fluid dynamics (CFD) codes. Experience has shown that, initially, this may be an unwise decision because that type of modeling often obscures the essence (from the physics or thermo-fluid dynamics perspectives) of the processes being modeled. The city comprises a very wide range of spatial inhomogeneities (called scales), for example, a city scale of 10–20 km or even larger for megacities, a neighborhood scale of 1–2 km, a street (canyon) scale of 100–200 m, a building scale, a building façade scale, and so on, all being driven by the flow at a larger meso or regional scale. There are corresponding vertical scales: the lower 10–20 km of the atmosphere, the lower 1–2 km (the boundary layer thickness), and the depth of the roughness sublayer (—two to three times the average building height) for both the neighborhood and the street scale. This representation of scales can be translated into a simplified scheme for describing the flow mathematically. The easier the representation, the better is the development of physical-mathematical models that can be used for different applications. Introduction of the various scales is useful in identifying the dominant forces

Handbook of Environmental Fluid Dynamics, Volume Two, edited by Harindra Joseph Shermal Fernando. © 2013 CRC Press/Taylor & Francis Group, LLC.
ISBN: 978-1-4665-5601-0.

FIGURE 7.1   Rapidly growing cities: Phoenix and Singapore.

or dynamics at each scale that will form the boundary conditions for the next scale (Britter and Hanna, 2003). Thus, some important decisions must be made regarding the phenomena of interest and the larger and smaller scales that must be explicitly modeled, the largest scales being represented by the boundary conditions and the smallest scales by some form of parameterization. Notably, these and other similar choices can be the essence of the solution, and guidance is often provided (for very complex problems) by such a semi-analytical perspective. This chapter will attempt to note and review the descriptions of the various scales, the current problems that have arisen, and what has been done recently to address these problems.

7.2 Principles

7.2.1 Physical Description of the Flow

In the absence of the city, the wind flow pattern close to the surface is determined principally by large-scale pressure variations, surface stresses, thermal effects, orographic forcing, and the Coriolis contribution due to the earth's rotation. Overall, the flow is the result of the interaction among these forces that act on an air parcel. The introduction of a city into this perspective can be thought of as a large perturbation generating a change in the wind flow patterns at various scales.

The focus here is the layer closest to the ground and occupied by the buildings: the UCL. Rather than analyzing the complex flow within and near the UCL, it may be more useful to step back a little and see what we currently know about such flows. These flows have been extensively investigated as a marked reflection of their influence on the environment where we live and that affects our daily life. An alternative approach to what is typically described is to introduce the topic more intuitively by asking ourselves how the air that makes up the wind travels from one side of a city to the other.

It first seems clear that if the city was very densely packed such that there was no space between the buildings, the city would appear to be a mountain and as the flow cannot pass through the mountain it will go around. It could go two ways: it might go horizontally around the city or it might move vertically over

the city (or a combination of the two). If the city was mainly a tight nest of skyscrapers, as occurs in the central business district of U.S. cities and in many Asian cities, then there would be more of the around flow; but if the city has mainly buildings of similar height as in much of Europe, then the flow would rise over the city. Thermal or buoyancy effects may also have a role to play in that surface heating would encourage flow over the city while ground cooling would suppress vertical motion. If there were now some small spaces between the buildings, then some of the air would flow *through* the city. We expect the mass flux of fluid passing through the city to be smaller than the mass flux through a similar cross section in the absence of the city. This would again force flow over and around the city. As the city is made less densely packed, these effects would be less pronounced but would still be evident and form a basic aspect of the flow. Of course it is possible to have regions of the flow within the cities that are locally faster than would have occurred in the absence of the city, but these would be less common. An example of this situation might be when the air flow meets a significant change (increase) in building heights from a suburban area to an urban area. The flow can "accelerate" through the "apparent contractions" created by the tall buildings. This is an effect directly related to the spatial inhomogeneity of the buildings. At the leeward area of the city, there will be a reduction of buildings (both in number and height) and the flow leaving the building regions will experience a reduction in drag force and will be accelerated by the mixing down of high-momentum fluid from above.

The city has imposed a "mass" or "volume" constraint on what the airflow can do. A parcel of air will generally travel at an elevated level within the city and travel slower through the building array. Therefore its transit time across the city will be increased and this might, for example, allow more time for important chemical reactions to occur. The concept of transit time can be linked to the concept of age of air when an air parcel travels from a rural (clean) area into an urban (polluted) area; a concept that can be used to deduce information about pollutant dilution within the UCL as discussed in later sections. Another view of the flow is to note that cities are made up of buildings (basically sharp-edged bluff bodies), and the flow through much of the urban canopy will be strongly influenced by flow separation, flow reattachment, and the presence of recirculating flow regions often called separation bubbles. These will also reduce the bulk wind speed within the city and further increase the transit time for material to cross the city. They may also locally produce substantial vertical exchange of the air. Thus, there is a kinematic displacement of the air due both to physical blockage and to the macroscopic effects of many separated flow regions.

Additionally, and possibly most importantly, we expect the existence of flow channeling, imposed by order (manifesting itself through streets and street canyons), within the building array (see, e.g., Theurer, 1999). A lot of airflow through the city will be along the streets, and there will also be internal adjustments to the flow due to the real inhomogeneity of the city (e.g., downtown versus suburban).

Another approach is to consider the forces and changes in linear momentum within the city. Away from large spatial changes in the geometry of the city, we might assume the flow to be the same everywhere (i.e., homogeneous). The wind flow through the city is resisted by the drag forces on the various buildings, and we might ask why the flow does not slow down continuously as it moves through the city rather than remaining nominally constant. If the flow was within a duct, then a pressure drop along the duct would provide the required force. However, such a pressure drop is not observed in the atmosphere. In fact the drag force on the various buildings must be balanced by a continuous transfer of horizontal momentum into the city from the faster flow above the city. The net momentum transfer into the city from above is crucial to understanding the flow in urban areas and is the key to many of the applications mentioned in the introduction. This transfer of momentum is the result of roughly equal mass fluxes up and down (for a flow that is fully developed) that carry different horizontal momentum from above and below. It is these nominally equal and opposite vertical mass fluxes that drive the horizontal flow within the city. Of importance here is the fact that the UCLs (or the buildings inhabiting that layer) are the sink for the flux of linear momentum into the UCL from above. This sink of linear momentum flux manifests itself as a force on the surface and when divided by the plan area provides a "surface shear stress" τ_w and this is a major parameter describing the flow. It is convenient later to use a slightly modified form of τ_w, that is, $(\tau_w/\rho)^{1/2}$, and this is called the friction velocity $u*$.

Above the roughness sublayer, it is found that that the horizontal shear stress is approximately constant for some height forming the constant stress layer. We might also ask how the shear stress varies through the roughness sublayer. It is clear that the stress must remain constant until a force, such as the drag on a church spire, produces a force on the air flow and thus reduces the spatially averaged shear stress. Going down in height into the UCL, more and more obstacles will provide a retarding force to the wind and further reduce the shear stress toward zero. Eventually it may become close to zero at or above the ground level. Engineers and meteorologists both like to think in terms of nondimensional parameters, for example, the shear stress coefficient $C_f = \tau_w/\tfrac{1}{2}\rho\,U^2$ where U over homogeneous terrain is commonly measured at a 10 m reference height above the surface.

The height or position of the reference velocity is quite uncertain in urban areas. We will return to this complication later. Generally the larger the buildings, the larger the surface shear stress and the larger the shear stress coeDcient. This produces a reduction in the near-surface winds.

A similar interpretation of the energy equation to those for mass and momentum will note the importance of incoming solar radiation to be partially reflected back to space by the buildings and surfaces, partially to impinge on buildings and surfaces and be absorbed, or reflected to other buildings and surfaces. Additionally, the heated buildings and surfaces will emit radiation based on their own temperatures and thermal capacity, emissivity, and other properties. Heat may be absorbed by the buildings and surfaces (the thermal energy storage term) and be partially transferred by heat transfer processes to the air within the urban canopy. There will also be a convective flux of sensible and/or latent heat to or from the region above the urban canopy. Finally, there are likely to be anthropogenic heat sources within the urban canopy such as electric, gas, or fuel oil supplies. Note that if the surfaces within the UCL become too hot, or provide large heat fluxes, then the heat transfer will have an element of free or natural convection leading to a coupling between the momentum and energy equations. In summary, at the simplest descriptive level we can imagine a fluid parcel that will

ﬅ Be advected more slowly through the city

ﬅ Rise and fall or move sideways consistent with the changed advection speed

ﬅ Be caught up in "recirculating regions" within the urban canopy

ﬅ Suffer very locally increases in speed

ﬅ Impinge on buildings reducing its momentum and producing a force on the building

ﬅ Be transported vertically out of the urban canopy and replaced somewhere by an equivalent parcel

ﬅ Be mixed with greater velocity fluid above the canopy and either moved even higher or lower back unto the urban canopy

ﬅ Appear as a momentum increase; and thereby driving the advection in the urban canopy

ﬅ Be heated or cooled, during its transit of the city

Obviously, these provide a fluid mechanics approach to the problem and other thermodynamics, heat transfer, and meteorological processes must also be considered. However, at the scales we are considering in this chapter, these events might be useful in understanding as to what is going on.

7.2.2 Transfer Processes

As suggested in Section 7.2.1, most of the processes occurring within the roughness sublayer can be thought of in terms of transfer processes for mass, momentum, energy, entropy, etc.

The sketch in Figure 7.2 is somewhat misleading in that it accentuates the "horizontal" velocities. While the horizontal velocities are important, particularly at larger scales, when

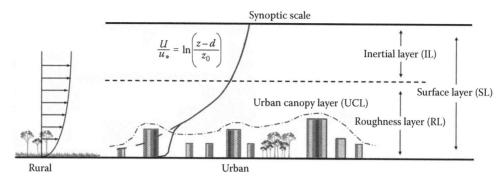

FIGURE 7.2 Schematic of the urban atmospheric layer.

dealing with the UCL, it is the vertical velocities that bring net momentum down into the UCL by mixing high-momentum fluid down and mixing low momentum fluid up. It is this vertical exchange velocity that produces the form of the horizontal flow through the urban canopy. The transfer processes in the vertical direction may have mean (e.g., vertical flow due to macroscopic roughness change and secondary flows around individual buildings) and turbulent (diffusive-like and with intermittent bursts as in vegetative canopies) components. The processes that control transport, mixing, and pollutant dispersion within and above urban areas, span a large range of spatial and temporal scales. Different applications require understanding at different levels of detail. These different scales and applications are traditionally studied by different disciplines.

The transfer processes can be interpreted or modeled in different ways. The use of "exchange velocities" between different regions such as the urban canopy and the part of the roughness sublayer above is broadly consistent with the "turbulent exchange coeDcient" approach used by urban climatologists. However although consistent, there are some crucial differences. The environmental engineer might analyze the transfer process differently from the urban climatologist. The engineer would typically classify the transfer between a buildingß surface or a water surface for a lake and the atmosphere above the urban canopy in two stages: a surface boundary layer approach between the surface and the flow within the urban canopy and a separate "exchange velocity" approach connecting the flow in the urban canopy to that above the urban canopy. Many urban climatologists will consolidate these two processes into one and call it a "turbulent exchange coeDcient." The relevance of these different approaches becomes more evident when one is trying to describe the transfer processes from very complex and inhomogeneous surfaces, such as those found on real buildings. Flow and turbulence that contribute to the transfer processes for momentum, heat, moisture, etc. have several sources including

ft Direct high-momentum fluid transfer—down from above—and low momentum fluid—up out of the UCL. These transfers are greatly enhanced when the urban canopy has buildings with substantial height variations
ft An additional contribution from extensive, ubiquitous regions of 3D separation and reattachment

ft A further contribution from vehicles moving through the city. This contribution becomes more important under very light winds with even a possible contribution from the thermal effects of the heat released from vehicles in very light wind conditions and congested traDc
ft Substantial contributions to these mixing processes due to the heat release from the heated building surfaces, particularly that arising from incident solar radiation
ft Anthropogenic heat sources that will have similar effects

These last two contributions introduce, somewhat belatedly, the importance of the study of the energy distribution and fluxes within and above the UCL. Overall the study and analysis of the surface energy budget (including radiation, sensible and latent heat fluxes, thermal storage and generation) is well established. Less studied is the advection of sensible and latent heat fluxes through the city. The exploitation of the similarity of pollutant transfer and sensible and latent heat fluxes through a city appears to be slight.

The development of parameterizations of the surface processes for the city is of importance in that these can be used as boundary conditions for regional (meso-scale) models, thereby restricting the need for extensive computer resources to be made available for the surface region. These boundary conditions may be in the form of drag coeDcients or, more generally, surface exchange coeDcients. Within the engineering CFD community, similar coeDcients are commonly referred to as "wall functions." The surface exchange processes are significantly different for momentum and, say, heat; the latter having no analog with the forces arising due to the pressure differences.

7.2.3 Approaches

7.2.3.1 Regional or Meso Scale

The regional scale acts as the "background" for the flow on the city scale, while the city scale acts to modify the regional scale flow. As we see in the next section, the city will divert the regional scale flow vertically and laterally both kinematically due to direct flow blockage and dynamically due to the increased drag force provided by the buildings. The city acts as a momentum sink for the regional flow and produces a modification to the velocity profiles. The city acts as a source of turbulent kinetic energy that will enhance vertical mixing over the city. The regional scale flow may also be dynamically affected through a thermally generated "heat island." The heat island produces a flow convergence into the city and vertical motions over the city. The regional scale provides the background chemical composition of the air within which the pollutants released in the city will mix, react, and dilute. The city also provides the pollutant source for an "urban plume" that can be detected for distances of 100–200 km downwind. This scale is often the one for which particular combinations of the city scale processes and the regional scale flows lead to severe pollution episodes. These may be of a generic nature or may be specific to particular cities.

7.2.3.2 City Scale

On the city scale, a common approach is to rely on the city or major parts of it to be represented by a homogeneous rough surface characterized by a small number of semiempirical parameterizations. In the absence of buoyancy effects, it is found that the mean velocity profile (implicitly taken to be a horizontal spatial average) above the urban canopy is logarithmic as sketched in Figure 7.2 and is given by

$$U(z) = \frac{u_*}{\kappa} \ln\left(\frac{z - d}{z_0}\right)$$

based on three main parameters: the roughness length z_0, the displacement height d, and the already introduced friction velocity u_*. Here κ is the von Karman constant typically taken as 0.40. This result should only be valid at heights well above the height of the individual buildings H_r. The parameters, z_0, d, and u_*, summarize much of what is important about the UCL in influencing the substantial region of flow above. Both the mean flow and the turbulence (see Roth, 2000) are determined by these parameters.

This is a conventional view in meteorology and has been used in representing urban areas in regional and mesoscale models. The urban area is viewed as a lower boundary condition on the ABL flow above. Several approaches are in use for determining these parameters including look up tables (see Britter and Hanna, 2003). There exist extensive theory, experiments (laboratory and field), and model types in this area, and there are many studies of this "fully developed" large fetch flow.

Of importance is that the same variables obviously reflect important aspects of the UCL. The displacement thickness d can be thought of as a parameterization of the extent of the region within the UCL of greatly reduced velocities. The friction velocity u_* is a measure of the surface shear stress, and a reflection of the momentum mixing process down from above. Note that the roughness length and the friction velocity are NOT independent for a particular reference velocity at a particular reference height. The friction velocity is a far more physically interpretable parameter than the surface roughness length. This large-scale flow approach has also been extended to accommodate thermal effects and, when these are large, to scenarios where the density variation influences the flow to produce stable or unstable density stratification. In this note we shall discuss what happens within the UCL. This will be addressed in the next section however the results are also directly applicable on the city scale.

7.2.3.3 Neighborhood Scale

The neighborhood is a scale that is large enough for effects of individual streets and buildings to be averaged out, but small enough so that variations across the city in building density and type are resolved. On the other hand, it is also a scale on which a full CFD calculation is feasible. Figure 7.3 provides an example of such a calculation where the flow is predicted at high resolution within and above the inner core of Oklahoma city explicitly resolving more than 700 buildings. This scale impacts most strongly on urban air quality, emergency response, and other urban considerations that affect us.

Overall 768 buildings were modeled using about 14 million cells. Simulations took about 4–5 days on a workstation with 32GB RAM and 8 CPU.

Following the former approach, statistical parameters must be found that are relevant for neighborhoods and that allow determination of flow and possibly turbulence in the urban canopy. These statistical parameters may then be used to obtain spatially averaged velocity profiles and turbulence levels within the urban canopy. Note that it is sometimes unclear whether the spatially averaged velocities are defined in terms of the open space or in terms of open and closed spaces (such as the mass flux). An early approach to urban canopy modeling was based on a porosity model; the porosity providing the drag force of the buildings. This has lost some favor recently, possibly due to

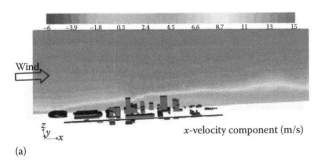

(a)

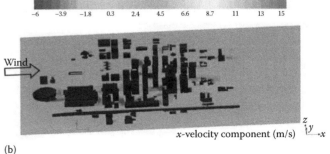

(b)

FIGURE 7.3 **(See color insert.)** CFD simulation for inner core (neighborhood scale) of Oklahoma City, OK. Contours of *x*-velocity component (along the wind direction) at a vertical plane (a) and at a horizontal plane near the ground (b).

the uncertainty of how best to treat the "turbulent length scale" aspect in a very general way.

A more recent approach uses typical urban statistical parameters such as H_r, the average building height, and λ_p and λ_f, the ratio of built area to total area and the ratio of frontal area to total area. Di Sabatino et al. (2010) provide simple methodologies to estimate these parameters and demonstrate their use in determining z_0 and d. Note that the friction velocity u_* is a measure of mixing within the urban canopy and it is likely, though not yet demonstrated conclusively, that turbulence levels within the urban canopy will scale on u_*.

Several approaches have been used to describe the mean flow within the UCL. Cionco (1965) developed a semi-analytical model to produce an exponential velocity profile within the urban canopy. This method was originally developed for vegetative canopies, though the extension to urban canopies is not unreasonable. Of more concern is the fact that a matching between the in-canopy and above canopy flow is made at the "top" of the urban canopy: a position of some uncertainty. Macdonald (2000) produced a similar argument to develop similar results while Bentham and Britter (2003) developed an extremely simple argument to determine a constant velocity prediction for the flow in the urban canopy.

Comprehensive laboratory data for flow is now available (e.g., Macdonald et al., 1998). Field data studies (e.g., Arnold et al., 2004, Rotach et al. 2004) are also more common. It is always a truism that more experimental data is required to assist in the development of models as in the case here. However, it is diDcult to obtain enough field data to assist in model development, though the data is essential for model evaluation and validation. The use of generic CFD studies is an approach that should not be dismissed too quickly. The combination of field studies and generic CFD work is a very attractive option to assist model development.

It is of particular interest to determine the turbulence levels within the urban canopy, especially how the turbulence varied with position; the strong mixing within the urban canopy may produce somewhat uniform levels of turbulence; or at least uniform enough for operational use. It is argued strongly here that turbulence levels within the urban canopy should be nondimensionalized with the u_* for the surface as a whole rather than with a local u_* based on a local Reynolds stress. The latter approach is common in meteorology but seems inappropriate both in using local values and when it is realized that the local Reynolds stress can be zero when turbulence is still very evident.

Analytically, there are many studies on the standard change of roughness problem, that is, the 1D step change problem, only a few on a 1D repeating change of roughness problem, and close to nothing on 2D or gradually varying roughness problems. If we want to treat the city scale problem in terms of parameterizations as previously mentioned, either for the city as a whole or allowing the parameters to vary with different neighborhoods across the city, we need more information on the latter two cases.

There is still great interest on the roles of the turbulence generated mechanically within the urban canopy and the removal of turbulence due to stable conditions within the urban canopy. The importance of this region of urban flows suggests the need for a few comprehensive field campaigns in countries with significantly different climates.

The use of CFD at the neighborhood scale and in a true obstacle-resolving capacity has become very attractive recently partly due to the availability of 3D databases for many cities and the continued increase in computing power at lower cost. This approach can obviously be extended down to encompass the street scale. Such computational studies are a balance between the region to be studied, the level of geometrical detail required, and the computing resources available. The significance of the detail appropriate to ensure adequate computational modeling of the physical processes requires attention as does the development and use of "wall functions" to parameterize the near-wall regions. CFD studies are still only infrequently used to develop phenomenological observations or results. Adequate boundary and initial conditions for a CFD calculation normally require significant experience on the part of the user as this is not "plug and play" territory. Another very real diDculty is that the background flows are only quasi-steady and, for example, pollutant plume meandering may be far more important than diffusive processes, particularly under light or calm wind conditions.

On reflection, it may be that we have a lot of the "wrong kind of information." We need far more information on what happens inside the urban canopy. In particular, a different way of describing the flow is required; one which tells us where the flow goes to and why, rather than looking for more roughness and displacement lengths.

7.2.3.4 Street Scale

The street scale is of particular interest in the sense that it is specifically where we live and work. Using urban air quality as an example, the street scale is the smallest scale that encompasses both the source and the receptor and thus is likely to be the scale that determines the extreme values of concentration.

The street scale flows are mainly straightforward, and it is important to reemphasize that they are essentially driven by the mixing of high-momentum (a vector) fluid down from above into the street canyons where the momentum will be lost to forces (normal and shear) on, for example, the building, vegetation, and street furniture surfaces. For example, consider the flow in a very long street with no intersections and driven by the wind aloft being in the direction of the street. The fully developed bulk wind velocity in the street is linked directly to the momentum mixed down into the street and the skin friction coeDcient of the street canyon walls and floor. Increased mixing due, say, to variability of nearby building heights will tend to increase the wind along the streets while irregularities (building offsets or extensive balconies) will tend to reduce the wind speed. When the wind direction aloft is normal to the street axis, the flow is typically viewed as a recirculating eddy within the street canyon driven by the wind flow at the top of the street canyon with a shear layer separating the above canyon flow from that within. If the street canyon has a large depth to width ratio, then the recirculating region may not reach the canyon floor and the ground level regions may suffer little ventilation. What is often not appreciated is that such flows are rarely as simple as the idealized one presented here. They are nearly always intermittent with the flowing going in the reverse direction for much of the time.

There are many variations on this flow description due to wind directions not parallel or normal to the street axis, the real rather than the simple idealized geometry of the street canyon and the mean flow and turbulence generated by vehicles within the street canyon. There is also the possibility of thermally driven flows arising from incident solar radiation, building heat sources, and the vehicles themselves.

Wind-tunnel visualizations of flow and dispersion at a street canyon intersection (Scaperdas et al., 2000) showed that the flow and dispersion patterns for a symmetrical situation are bistable, the direction of the flow in the street perpendicular to the incident flow periodically switching direction. The symmetrical base case was modified in several ways, principally by varying the incident wind direction or by introducing an offset to produce a staggered road junction. The flow and dispersion were highly sensitive to small changes in both these parameters and less sensitive to subsequent larger changes.

A somewhat distinct but overarching problem that needs to be considered is what to use as the "reference" measurement for any of the urban canopy flow categories. Should it be the geostrophic velocity, velocity above the roughness sub-layer, the velocity at building height, or the number of "reference" measurements required. At present answer to this question is not available. What is known is based on the results of field experiments which in the last 10 years have become more and more numerous. It is hard to identify a common line and a common outcome because all these experiments have been made with different research questions in mind, and they all in some way are far from being comprehensive.

7.3 Methods of Analysis

From the description given in the earlier sections, it is clear that the methods of analysis will follow from the specific approach adopted which might be different according to the specific area of expertise being engineering, meteorology, geography, urban climatology, and so on. In the last 10 years, considerable efforts have been made in all subjects, and with them various methods of analysis have become more sophisticated.

The use of gross parameterizations of the surface and the flow for neighborhoods and cities as a whole was outlined in Section 7.2.3.2, where the urban canopy is replaced with several parameters such as H_r, z_0, d, $u*$, and the skin-friction coefficient. These parameters can be determined on both city and neighborhood scales and are useful for characterizing the flow *above* and within the UCL. Further parameters are introduced for non-adiabatic surfaces such as the Obhukov length, L and a convective velocity scale $w*$. Studies of the surface energy balance in urban areas lead to predictive techniques for the surface heat fluxes (in their various forms), and this is a required input for

determining L. Other dynamically important parameters such as the surface temperature and moisture and heat transfer within the urban canopy arise from the surface energy balances.

A weakness of this approach and most others is the fundamental one; the urban surface characteristics are usually heterogeneous through all length scales and as the surface layer is continuously adjusting, it is not clear that the concepts of roughness length and displacement height hold meaning in this context.

An early approach to describe the urban canopy flow based on porosity is now less preferred than models using a digital elevation models (DEMs); approach based on viewing an urban canopy comprised of buildings as obstacles. Models for the mean flow within the urban canopy such as those developed by Cionco (1965), Macdonald (2000), and Bentham and Britter (2003) are essentially similar in that they all assume that the distribution of drag producing elements are uniformly distributed throughout the UCL. An interesting and very useful extension of this approach by Di Sabatino et al. (2008) was to use the DEMs to provide the vertical distribution of the drag producing elements within the UCL and subsequently predict the spatially averaged mean velocity profile.

A remaining important uncertainty is to investigate whether the "underlying viewpoint" should be one based on an array of buildings as in Bentham and Britter (2003) or an array of streets as in Soulhac et al. (2002). Much work has been done on the former while the latter seems to be essential, at least for European type cities. Figure 7.4 reflects the strong imprint that the streets play on the advection and dispersion problem in urban areas. A release of material at the "yellow dot" clearly shows the tendency of the plume to follow the street system downwind, while some of the released material rises above the building heights and becomes more like a conventional plume. The development of the model Sirane by Soulhac (2000) and co-workers and the

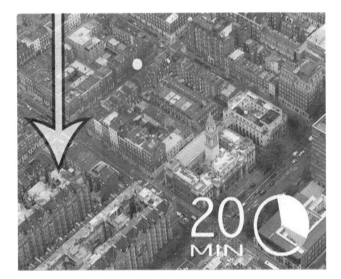

FIGURE 7.4 (See color insert.) Air flow through London with much of the flow being along streets. Note that the marker released at the yellow circle moves upwind a little.

model by Hamlyn and Britter (2005) are based on this perspective. Sirane is in operational use.

Computational fluid dynamics (CFD) has become a major tool for simulating urban canopy flows, particularly over the last decade. Urban canopy flows have been widely investigated considering urban-like building arrays and street networks (e.g., Garbero et al., 2010) and in real urban built-up areas (e.g., Xie and Castro, 2009). Studies have been carried out comparing CFD results with both field and laboratory experiments. Earlier work typically used Reynolds-averaged Navier–Stokes (RANS) turbulence models including commercially available models and more recently using large-eddy simulation (LES) models. Additionally, there is now extensive use of models such as the weather research and forecasting (WRF) model (http://www.wrf-model.org), which is fundamentally a meteorologically based model run at high resolution and incorporating some forms of near surface urban parameterization. It is interesting to note the steady transformation of engineering-based codes to include aspects of meteorology (e.g., inclusion of solar radiation) and meteorological codes to include aspects of engineering studies (e.g., treatment of sharp-edged buildings).

What is most obviously lacking in the area of CFD and more generally for the modeling of urban canopy flows is the undertaking of formal model evaluation studies. Fortunately there is now very clear guidance in this area through COST 732 (2005–2009) for micro-scale meteorological models and through COST 728 (2004–2009) for meso-scale models. These are the results of about 5 years of trans-European research for both studies covering many modeling approaches.

Little effort appears to have been directed at using large CFD computations to determine simpler surface process parameterizations.

It is also clear that what is universally needed is a set of coordinated long-term observations over real urban areas to form a coherent approach to "city condition monitoring," similar in intent to the "engine condition monitoring" common throughout much of engineering. Core requirements of such an observational campaign are that (1) the measurements cover a wide spatial range, (2) are run over a long time, and (3) measure the large-scale meteorological context of the smaller scale flow. Several cities have moved in this direction. It may well be that many cheap instruments are preferable to a few expensive measurements.

7.4 Applications

7.4.1 Urban Breathability

Urban canopy flows impact directly on air quality. Flow entering urban areas from rural environments is a source of clean air that can dilute pollution particularly at street level. The exchange velocity introduced earlier can be interpreted as a surrogate for ventilation within the urban canopy. However, it accounts for bulk vertical exchange of air masses but does not account for the effects of spatial variability. An alternative approach to determine street-level pollutant dilution can be adopted from building ventilation concepts previously used to evaluate indoor exposure. Hang et al. (2009) analyzed the flow being advected into the city from its surrounding rural areas using the concept of the local mean age of air (Etheridge and Sandberg, 1996) which represents the time taken for an air parcel to reach a given place after the clean air enters the city. A poorly ventilated region implies a large mean age and this, in turn, means an accumulation of pollutants in the region and larger concentration. The supplying of clean air into the urban area can be thought of as an *inhale* and pollutant removal reminds us of an *exhale*. Buccolieri et al. (2010) introduced the expression "city breathability" or "urban breathability" to describe these effects. The interaction of the atmospheric approach flow and the city produces complicated flow patterns between buildings, along streets, near stagnant zones, and wake regions. An air mass approaching a city might enter the streets, flow above the buildings, or flow around them. Wind tunnel and numerical simulations suggested a way to classify cities into three groups: sparse, compact, and very compact cities. The sparse city acts as a collection of obstacles, where reversed flow only occurs behind the individual buildings. The compact city behaves as a unique obstacle with respect to the flow. A single wake, whose size scales with the horizontal dimension of the city, forms behind the building array. In the very compact city, the horizontal flow in the streets at the center of the array is negative, that is, opposite to the approaching wind direction. As an example, Figure 7.5 shows the vectors of normalized velocity magnitude U/U_H at $z = 0.5H_r$, the normalized mean age of air at pedestrian level ($\sim$0.06H), where U_H is the undisturbed velocity at the building height H_r, for a sparse case ($\lambda_p = 0.25$) and for a very compact case ($\lambda_p = 0.69$) obtained by CFD simulations. The uniform volume source of pollutants is defined over all gaps within the urban building array, from the ground to the street top. The age of air is normalized using only a portion of the overall gaps volume. The mean age of air is large in poorly ventilated recirculation zones and in downstream regions. The air becomes older in the downstream region of the array as the building packing density increases. Pollutants tend to accumulate in the downstream region along the streets, as a result of the pollutant transport from the street sides in this region and the low vertical air mass through the street top. The local mean age of air is low near the side openings where lower concentrations are found. The figure also shows that the mean age of air is larger close to the middle of the arrays. Moreover, it increases as building packing density increases, and this occurs both in the middle and at the edge of the array. However, some differences can be observed. In general, it can be argued that the mean age of air increases downstream in the sparse case, while for the compact case it reaches a maximum and then decreases close to the end of the array.

The local mean age of air can be considered as a useful indirect way to quantify the breathability of a city. Local mean age of air, estimated for a given city and specific meteorological conditions, can be presented as spatial maps which can be used for a simple and direct evaluation of city breathability and pollutant removal potential.

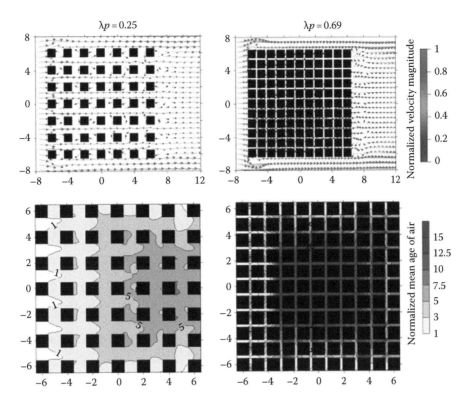

FIGURE 7.5 Vectors of normalized (by the undisturbed velocity U_H at the building height H_r) velocity at $z = 0.5H_r$ (top) and normalized mean age of air at $z = 0.06H_r$ (corresponding to the pedestrian level) for a sparse canopy (left) and a very compact canopy (right).

Neighborhood morphometry affects these results and research is ongoing to evaluate the effects of neighborhood-scale and city-scale models such as round, square, long rectangular cities, etc. The differences in the age of air between round, square, and long cities mainly result from the difference in the wakes behind the city. Ongoing research with a focus on high-packed cities (e.g., typical in Asia), where tall buildings are usually built to provide suDcient residential area, shows that city-scale high-rise urban areas should be avoided. If suDciently wide "urban canyons" are used to separate the city-scale high-rise urban areas into several shorter ones (e.g., less than 1 km or shorter) between which the roughness isolated flow regime exists, the city breathability is improved even in the presence of tall buildings in the urban areas.

7.4.2 Vegetation

How does urban vegetation affect urban canopy flows, the microclimate, and air pollution levels. Vegetation in urban areas has been often used as a design object for urban management. Vegetation is also able to produce continuous "greenways" for recreational walking and the mixing of plant and animal species. The effect of vegetation on urban flows, climate, and sustainable urban growth is infrequently taken into account even though urban vegetation may trigger processes at larger scales, broadly influencing the whole environment. For example, chemical reaction rates that are temperature dependent can form secondary pollutants. Emissions, both from traDc and from vegetation itself, are function of temperature.

Buildings are not the only obstacles offering resistance to the airflow within urban canyons. Trees modify the flow pattern, the turbulent exchange of mass between the UCL and the atmosphere above and consequently affect pollutant dilution. Although particle deposition on plant surfaces removes pollutants from the atmosphere, thus reducing their concentration, trees themselves act as obstacles to airflow decreasing air exchange with the above roof level atmosphere and increasing pollutant concentrations, Litschke and Kuttler (2008).

The presence of trees alter flow distribution according to the specific street canyon aspect ratios, wind direction, crown shape of trees, tree arrangements, and the foliage density, although the dependence on the first two is stronger than the others. The street level concentrations crucially depend on the approaching wind direction and the street canyon aspect ratio rather than on tree planting porosity and configuration. It is usually assumed that larger concentrations are associated with winds perpendicular to the street axis. With oblique winds and tree-free street canyons the larger the aspect ratio, the smaller the street-level concentration. In the presence of trees the reduction of street-level concentration with increasing aspect ratio is not as significant. This is confirmed by observations in wind tunnel experiments (see for instance CODASC Database, 2008) and in numerical simulations using CFD models (Buccolieri et al., 2009). Here trees are modeled as porous media with a pressure loss linked to the specific tree porosity. Typically a modification to the governing equation is done by adding a momentum sink term. The interpretation of concentration results is done by

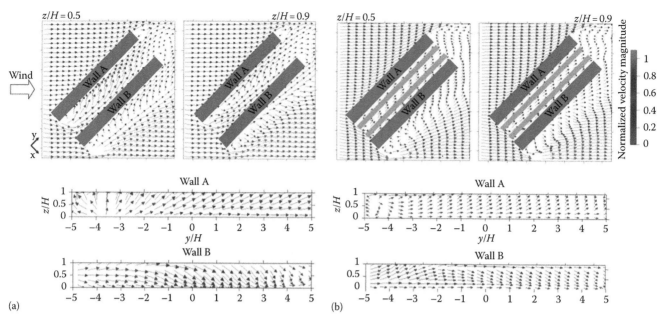

FIGURE 7.6   Isolated street canyon. Vectors of normalized (by the undisturbed velocity U_H at the average building height H_r) velocity magnitude at $z = 0.5H_r$, $0.9H_r$ and at the two canyon walls in the case without (a) and with trees (b).

looking at modifications that the flow undergoes in the presence of trees. Though the wind is highly intermittent which suggests that at least LES type of models should be employed, RANS CFD simulations suggest that the presence of trees in the canyon breaks the main vortex structure expected in the tree-free cases and modifies the corner vortices at the street ends thus reducing mass exchanges with the various openings; therefore, to a first approximation, dilution is much reduced. Figure 7.6 shows an example of flow simulations in a street canyon without trees and with trees. Looking at vertical sections and horizontal sections in Figure 7.6, it can be noted how different the flow structure is in the correspondence of the two walls. Consequently concentration distribution will change within this new flow distribution.

Most current air quality models do not take into account the aerodynamic effect of trees and therefore conclusions made on the basis of such model predictions might be questionable.

7.4.3 Thermal Effects

How do thermal processes influence and change urban canopy flows? The flow and dispersion processes are physically much the same at all the scales although buoyancy- or density-driven flows will manifest themselves more easily if applied over large spatial scales, for example, the urban heat island. Historically the UHI has been defined simply as the air temperature differences $\Delta T_{ut\,r}$ between the city and its surrounding rural area. So, heat island is intended to be a meso-scale type phenomenon. In general terms $\Delta T_{ut\,r}$ depends on a number of features of the urban area. During the day, concrete, asphalt, and building material store larger amounts of incoming solar radiation than bare, grassy or cultivated rural terrains. This heat storage can produce an increase of air temperature. The upward sensible heat flux of urban soils is

larger than that of their rural counterparts. If heat storage is sufficiently large, the upward sensible heat flux can persist during the night, giving rise to nocturnal urban mixed layers. If large-scale (synoptic) winds are absent or weak, airflow associated with the UHI generally consists of a closed circulation characterized by strong updraft motion (thermal plume) at the city downtown, by a horizontal convergent flow above the surface, and by divergent flow at the elevated layers. A slow downdraft far from the city closes the circulation similarly to what happens in the sea breeze circulation. The occurrence of this circulation depends on not only the size of the city and the UHI intensity but also the influence of nearby orographic features such as valleys and the presence of sea. In the context of the interaction between local thermally induced flows, UHI and slope wind interaction is a case of great meteorological and environmental interest because of its influence on dispersion of pollutant emitted into cities located in mountainous areas. Nocturnal (katabatic) and diurnal (anabatic) winds associated with mountain slopes can also change appreciably the characteristics of the urban plume, giving rise to very complex flow structures (Fernando et al., 2001).

Typically, in cities of about hundred thousand inhabitants the heat island causes increases in temperature of a few degrees so that $\Delta T_{u-r}/T$ is small and the associated wind $U = \sqrt{gl_0\Delta T_{u-r}/T}$, where g is the gravitational acceleration and l_0 is a spatial scale, is of the order of 2–3 ms^{-1}. Higher velocities can be found in larger urban areas where the UHI is much more intense. The observation that the UHI circulation is closed indicates that pollutants that are emitted within the UCL may stay trapped for a long time.

At city and smaller scales, thermal processes have been studied in isolation and in a format distinct from the UHI. Recent field experiments such as the one performed in Phoenix in 2008 (see, e.g., Fernando et al., 2010) have shown that the simple definition $\Delta T_{ut\,r}$ given previously is insuDcient to determine in-canopy flow modifications. Thermal processes occur at the street canyon scale though the flow in field campaigns appears to be affected mainly locally.

Further wind tunnel work demonstrated the similarity of the flow fields in a 2D street canyon with CFD calculations for a range of aspect ratios (with width to height ratios between 0.3 and 2.0). The same work demonstrated the influence of solar-induced wall heating on the flow within a street canyon under low speed winds. The Froude number in this case described the transition to a buoyancy-influenced flow. However, it is unclear whether these results will directly carry across to the field; the thermal layers will be relatively larger in the model than at full scale.

7.4.4 Architecture and Urban Planning

Major users of an understanding of urban canopy flows are architects and urban planners. The former are designing the urban canopy on a street scale while the latter are more concerned with the neighborhood or even city scale. Of most importance though is the fact that both groups have many "masters," that is, they are answerable to many disciplines in their activities and there is the common request for simple to use but accurate and robust models; a request that is often diDcult to satisfy.

A major requirement, particularly in an energy-conscious context is knowledge of the local urban climatology rather than that on a flat, unobstructed air field 30 km away. The request also strengthens the argument for pervasive environmental sensing of cities. In the absence of this approach, there is a pressing need for improved surface energy budget models. Generally, numerical models adopt very simplified schemes for the urban surface like bare soil, where the actual presence of the buildings is taken into account by means of empirical laws. The latter are frequently not suited for the real cases, giving rise to large errors in the estimation of the turbulent fluxes and in the underestimation of the turbulent intensity, a key parameter in dispersion processes. For example, Grimmond et al. (1991) proposed the OHM (objective hysteresis model), a scheme based on empirical relationships between the net radiation at the surface with the heat storage within the buildings and the soil. Walko et al. (2000) proposed the LEAF-2 (land ecosystem-atmospheric feedback), a model which is able to take into account several biophysical parameters but neglecting heterogeneous 3D structures and anthropogenic heat sources. New approaches that simulate the city as a 3D environment and not simply based on the bare soil formulation as done in the models mentioned earlier have been proposed by several authors. Martilli et al. (2002) developed the FVM (finite volume model) scheme, an urban soil model in which the UCL is represented by means of a series of street canyons, a 3D entity, ideally straight and with buildings of equal height on either side. The most promising scheme for urban soil modeling seems to be the TEB (town energy balance) model developed by Masson (2000). Starting from the geometrical and physical properties

of the canyons, it is possible to compute the wind velocity, the temperature, and the turbulent fluxes within the canyon. Later applications of the TEB model by several researchers still remain confined to simplified cases in which the cities were assumed spatially homogeneous. The latter hypothesis is not appropriate in many European cities and in some North American and Asian cities where the urban texture is commonly highly inhomogeneous.

Finally, returning to the energy context theme, the architect or urban planner requires that more sophisticated surface energy balance models that include buildings appropriately are made available to them from the urban climatologist and that these models are coupled with building energy models developed by the mechanical engineer in order to make use of improved local urban climatology.

7.5 Major Challenges

The principal and overarching challenge is not being daunted by the importance of understanding and designing our cities with sound physical, chemical, social, cultural, economic, and creative tools that are appropriate to the task. Considerable effort has been made in understanding real urban canopy flows by means of analytical work, laboratory and field studies, and numerical modeling.

Most physical processes are well understood qualitatively by using the spatial scale analysis as a framework. Each scale is characterized by specific dynamical features which dominate over the others. The major challenge is provision of models that can seamlessly deal with a wide range of scales. Interestingly, there is an ongoing competition as the engineer attempts to upscale their models while the meteorologists are downscaling their models. A related challenge that is being actively addressed is the coupling of building energy models with urban canopy models.

Probably the most challenging technical requirement for air quality is the inclusion of appropriate chemistry that is consistent across the scales. Similarly there is a lack of knowledge on the use of vegetation in urban areas.

The introduction of pervasive urban sensing to provide input to city condition monitoring is a challenge that has been taken up by a few cities. Encouraging this activity is a rewarding challenge for it may be essential in the future. The point it may raise is: Is there a role for a wider "applied urban climatology group?"

The last major challenge is to continue to encourage and formalize the evaluation of models to ensure that models are of the right quality, that is, are fit-for-purpose.

References

Arnold S.J., ApSimon H., Barlow J., Belcher S., Bell M., Boddy J.W., Britter R.E., Cheng H., Clark R., Colvile R.N., Dimitroulopoulou S., Dobre A., Greally B., Kaur S., Knights A., Lawton T., Makepeace A., Martin D., Neophytou M., Neville

S., Nieuwenhuijsen M., Nickless G., Price C., Robins A., Shallcross D., Simmonds P., Smalley R.J., Tate J., Tomlin A.S., Wang H., Walsh P., 2004. Introduction to the DAPPLE air pollution project. *Science of the Total Environment* 332, 139–153.

Bentham T., Britter R.E., 2003. Spatially averaged flow within obstacle arrays, *Atmospheric Environment* 37, 2037–2043.

Britter R.E., Hanna S.R., 2003. Flow and dispersion in urban areas, *Annual Review of Fluid Mechanics* 35, 469–496.

Buccolieri R., Gromke C., Di Sabatino S., Ruck B., 2009. Aerodynamic effects of trees on pollutant concentration in street canyons. *Science of the Total Environment* 407, 5247–5256.

Buccolieri R., Sandberg M., Di Sabatino S., 2010. City breathability and its link to pollutant concentration distribution within urban-like geometries. *Atmospheric Environment* 44, 1894–1903.

Cionco R., 1965. A mathematical model for air flow in a vegetative canopy. *Journal of Applied Meteorology* 4, 517–522.

CODASC 2008. CODASC: Concentration data of street canyons. Internet database. Karlsruke Institute of Technology (KIT). http://www.codasc.de

COST 728, 2004–2010. Enhancing mesoscale meteorological modelling capabilities for air pollution and dispersion applications. http://www.cost728.org/home.htm

COST 732, 2005–2009. Quality assurance and improvement of micro-scale meteorological models. http://www.mi.uni-hamburg.de/COST-732-in-Brief.470.0.html

Di Sabatino S., Leo L.S., Cataldo R., Ratti C., Britter R.E., 2010. Construction of digital elevation models for a southern European city and a comparative morphological analysis with respect to Northern European and North American cities. *Journal of Applied Meteorology and Climatology* 49, 1377–1396.

Di Sabatino S., Solazzo E., Paradisi P., Britter R.E., 2008. Modelling spatially averaged wind profiles at the neighbourhood scale. *Boundary-Layer Meteorology* 127, 131–151.

Etheridge D., Sandberg M., 1996. *Building Ventilation: Theory and Measurement*, John Wiley & Sons, Chichester, U.K.

Fernando H.J.S., Lee S.M., Anderson J., Princevac M., Pardyjac E., Grossman-Clarke, S., 2001. Urban fluid mechanics: Air circulation and contaminant dispersion in cities. *Journal of Environmental Fluid Mechanics* 1, 107–164.

Fernando H.J.S., Zajic D., Di Sabatino S., Dimitrova R., Hedquist B., Dallman, A. 2010. Flow, turbulence and pollutant dispersion in the urban atmospheres. *Physics of Fluids* 22, 051301–051320.

Garbero V., Salizzoni P., Soulhac L., 2010. Experimental study of pollutant dispersion within a network of streets. *Boundary-Layer Meteorology* 136, 457–487.

Grimmond C.S.B., Cleugh H.A., Oke T.R., 1991. An objective urban heat storage model and its comparison with other schemes. *Atmospheric Environment* 25B, 311–326.

Hamlyn D., Britter R.E., 2005. A numerical study of the flow field and exchange processes within a canopy of urban-type roughness. *Atmospheric Environment* 39, 3243–3254.

Hang, J., Sandberg M., Li Y., 2009. Age of air and air exchange eDciency in idealized city models. *Building and Environment* 44, 1714–1723.

Litschke T., Kuttler W., 2008. On the reduction of urban particle concentration by vegetation—A review. *Meteorologische Zeitschris* 17, 229–240.

Macdonald R.W., 2000. Modelling the mean velocity profile in the urban canopy layer. *Boundary-Layer Meteorology* 97, 25–45.

Macdonald R.W., GriDths R.F., Hall D.J., 1998. A comparison of results from scaled field and wind tunnel modelling of dispersion in arrays of obstacles. *Atmospheric Environment* 32, 3845–3862.

Martilli A., Clappier A., Rotach M.W., 2002. An urban surface exchange parameterisation for mesoscale models. *Boundary-Layer Meteorology* 104, 261–304.

Masson V., 2000. A physically-based scheme for the urban energy budget in atmospheric models. *Boundary-Layer Meteorology* 98, 357–397.

Rotach M.W., Gryning S.-E., Batchvarova E., Christen A., Vogt R., 2004. Pollutant dispersion close to an urban surface—The BUBBLE tracer experiment. *Meteorology and Atmospheric Physics* 87, 39–56.

Roth M., 2000. Review of atmospheric turbulence over cities. *Quarterly Journal of the Royal Meteorological Society* 126, 941–990.

Scaperdas A., Robins A.G., Colville R.N., 2000. Flow visualisation and tracer dispersion experiments at street canyon intersections. *International Journal of Environment and Pollution* 14, 526–537.

Soulhac L., 2000. Modélisation de la dispersion atmosphérique à lfíntérieur de la canopéeurbaine. PhD thesis, Ecole Centrale de Lyon, Lyon, France.

Soulhac L., Mejean P., Perkins R. J., 2002. Modelling transport and dispersion of pollutant in street canyons. *International Journal of Environment and Pollution* 16, 404–416.

Theurer W., 1999. Typical building arrangements for urban air pollution modelling. *Atmospheric Environment* 33, 4057–4066.

United Nations, Department of Economic and Social Affairs, Population Division, 2007. World population prospects: The 2006 revision—Highlights. Working Paper ESA/P/WP.202, 114 pp. [Available online at http://www.un.org/esa/population/publications/wpp2006/WPP2006_Highlights_rev.pdf]

Vitruvius P., 2005. *Vitruvius. The Ten Books on Architecture.* Translated by M. H. Morgan. Adamant Media Corporation, Boston, MA, 372p.

Walko R.L., Band L.E., Baron J., Kittel T.G.F., Lammers R., Lee T.J., Ojima D. et al., 2000. Coupled atmosphere-biophysics-hydrology models for environmental modeling. *Journal of Applied Meteorology* 39, 931–944.

Xie Z.-T., Castro I.P., 2009. Large-eddy simulation for flow and dispersion in urban streets. *Atmospheric Environment* 43, 2174–2185.

8

Flow through Buildings

Nigel Berkeley Kaye
Clemson University

Morris R. Flynn
University of Alberta

8.1 Introduction

Of all the economic, environmental, and other costs associated with urbanization, perhaps none is more telling than this: for every 1% increase in urban population, energy usage increases by more than 2%. "Energy consumption defines the quality of urban lifek " [17]. One culprit contributing to the said statistic is the energy used by buildings, which, worldwide, consume approximately 40% of the energy generated by mankind. (Buildings are in turn responsible for roughly half of anthropogenic CO_2 emissions [17].) Little wonder that a significant fraction of this energy is consumed by energy-intensive HVAC equipment when, for example, (i) almost three-quarters of Americans consider home air-conditioning a necessity, and (ii) the energy devoted to the cooling of buildings in Los Angeles may, on a hot day, exceed that used by motor vehicles [15].

Despite human efforts to control interior climactic conditions, 30% of the worldﬂs buildings suffer poor air quality. The situation is especially grave in communities that continue to rely on solid fuels; as compared to natural gas, solid fuels produce 10–100 times the volume of particulates, which are associated with a myriad of respiratory ailments [17].

EDcient solutions to the aforementioned problems must harness "natural" factors, e.g., an external wind shear or internal heat gains in facilitating ventilation flows. Consequently, and in keeping with this volumeﬂ emphasis on buoyancy and stratification, we herein focus on natural ventilation, defined as a technology "to improve indoor air quality in urban areas, to protect health, to provide thermal comfort, and to reduce unnecessary energy consumption" [17]. From this general (and optimistic) assessment, we highlight the role played by fluid dynamics and emphasize the connection between flows in the internal and external environments.

Whether or not they employ natural factors, ventilation systems must balance multiple, and often competing, criteria. They are designed to remove stale air from rooms while maintaining agreeable temperatures. These criteria are sympathetic in mild climates, however, in more extreme climates, conditioning, possibly through the use of thermal mass or phase change materials, often requires lower ventilation rates. Air preconditioning and building ventilation may be further complicated by factors such as humidity, condensation, and the need to avoid drafts that may cause thermal discomfort. This necessitates an adequate control system that is responsive to changes in the local internal and external environment.

Our review is organized as follows: In Sections 8.2.1 and 8.2.2, respectively, we summarize the fundamental equations pertinent to buoyancy- and wind-driven flows. We expand upon these results in Section 8.3.1 where we consider details of single- and multichamber ventilation for a variety of design scenarios. The experimental and numerical methods typically used for model validation are briefly summarized in Section 8.3.2. A particular case study of a naturally ventilated building is examined in Section 8.4. Finally in Section 8.5 we identify a number of open questions to be addressed and thereby offer ideas for future research in a field whose significance is expected to grow in lockstep with the environmental consciousness of the public at large.

Handbook of Environmental Fluid Dynamics, Volume Two, edited by Harindra Joseph Shermal Fernando. © 2013 CRC Press/Taylor & Francis Group, LLC. ISBN: 978-1-4665-5601-0.

8.2 Principles

8.2.1 Buoyancy-Driven Flows

Buoyancy-driven (stack-driven) ventilation flows result from temperature-induced hydrostatic pressure differences between the interior and exterior of a building containing upper and lower vents. Consider a single room with a floor-level vent of area A_L and a ceiling-level vent of area A_U that connect the room to a uniform quiescent external ambient of temperature T_A and density ρ_A. Internal heat loads from occupants, equipment, and incoming solar radiation create a thermal stratification within the room, denoted by $T_R(z)$ with corresponding density stratification $\rho_R(z)$ (Figure 8.1a). The thermal stratification drives a flow in which the pressure loss due to flow through the vents is balanced by the hydrostatic pressure difference between the room and the external ambient, i.e.,

$$\frac{c\rho_A U_L^2}{2} + \frac{c\rho_A U_U^2}{2} = \int_0^H g[\rho_A - \rho_R(z)]dz, \qquad (8.1)$$

where

  g is gravitational acceleration
  U is the flow velocity through the appropriate vent
  c is a loss coeDcient
  H is the chamber height [12]

Writing (8.1) in terms of the reduced gravity, $g'(z) = g[\rho_A - \rho_R(z)]/\rho_A$, and the volumetric flow rate, Q, gives

$$Q^2\left(\frac{c}{2A_L^2} + \frac{c}{2A_U^2}\right) = \int_0^H g'(z)dz \Leftrightarrow Q = A^*\left[\int_0^H g'(z)dz\right]^{1/2}, \qquad (8.2)$$

where A^* denotes the effective vent area defined by

$$\frac{1}{A^{*2}} = \frac{c}{2A_L^2} + \frac{c}{2A_U^2}. \qquad (8.3)$$

The room pressure at floor level is less than that of the external ambient. However, the rate of decrease of pressure with height is smaller inside than outside so the pressure difference decreases with height until interior and exterior pressures become equal at the neutral height (Figure 8.1a). Outflow and inflow will occur, respectively, through any vent located above and below this neutral height.

Typical rooms have multiple heat sources including building occupants, mechanical and electrical equipment, lighting and solar radiation. These sources have a broad range of spatial scales. A building occupant occupies only a small percentage of the floor area but can have a height of up to $\frac{2}{3}H$, whereas a patch of sun occupies a larger portion of the floor area but has no height. The buoyancy-driven flow in a typical room is highly turbulent, with Rayleigh numbers, Ra, of the order 10^9 to 10^{10}.

8.2.2 Wind-Driven Flows

Wind-driven (or cross) ventilation results from pressure differences on the outside walls due to wind flow around a building. These differences are due to variations in the wind speed with height and flow separation at the building edges (Figure 8.1b). The distribution of pressure, P, over a building is described in terms of a pressure coeDcient $C_p = P/(\frac{1}{2}\rho U_{ref}^2)$ where U_{ref} is a reference wind speed [6]. Interior flows are calculated by balancing the difference between the windward pressure, P_W, and the leeward pressure, P_L, at the vent locations and the pressure drop due to flow through the vents, i.e.,

$$P_W - P_L = \frac{1}{2}\rho U_{ref}^2(C_{pW} - C_{pL}) = \frac{c\rho}{2}\left(U_W^2 + U_L^2\right). \qquad (8.4)$$

Written in terms of the ventilation flow rate, (8.4) becomes

$$Q = A^* U_{ref}\left(\frac{C_{pW} - C_{pL}}{2}\right)^{1/2}, \qquad (8.5)$$

where A^* is defined by (8.3) with the upper and lower vent areas replaced by the windward and leeward areas.

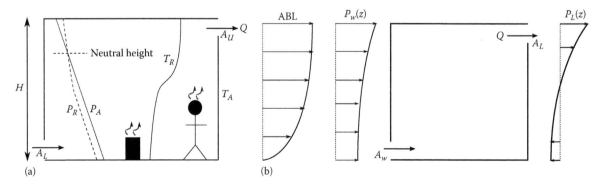

(a)   (b)

FIGURE 8.1 Ventilated room showing (a) the internal temperature stratification $T_R(z)$ and resulting hydrostatic pressure distribution $P_R(z)$ and (b) the incoming ABL and resulting pressure distribution over the windward and leeward walls.

Pressure coeDcients vary across the surface of the building depending on the wind direction, flow disturbances due to surrounding buildings, and the properties of the atmospheric boundary layer (ABL), which is in turn influenced by the upstream topography and weather conditions (see Chapter 7). Details of the local climate and geography therefore affect design. In many instances, a wind tunnel study (as described in Section 8.3.2) can be applied to more accurately determine pressure coeDcients.

8.3 Methods of Analysis

8.3.1 Analytical Description

8.3.1.1 Single Zone

There are two basic models for the ventilation flow through a single zone with upper and lower vents. That of Linden et al. [12] presumes the heat load to be concentrated at a finite number of points; that of Gladstone and Woods [3] presumes the heat load to be evenly distributed over the entire plan area, S, of the room. While these assumptions may be unrealistic, the models provide helpful upper and lower bounds.

Gladstone and Woods [3] further assume that the heat load drives a uniform convection so that the room temperature, T_R, and buoyancy, g', do not vary from point to point (Figure 8.2a). Equation 8.2 therefore yields

$$Q = A^*(g'H)^{1/2}, \quad \text{or} \quad Q = A^{*2/3}(BH)^{1/3}, \tag{8.6}$$

where $B = Qg'$ is the buoyancy flux associated with the heat load $\mathcal{H}$, i.e., $\mathcal{H} = Bc_p\rho_0 T_0/g$ in which c_p is the air specific heat capacity and ρ_0 and T_0 denote a representative air density and temperature, respectively. The aforementioned result prescribes the largest possible stack-driven ventilation flow rate for the given H, A^*, and $\mathcal{H}$. The steady-state buoyancy is specified by

$$g' = \frac{B}{Q} = \left(\frac{B^2}{A^{*2}H}\right)^{1/3}. \tag{8.7}$$

Conversely, Linden et al. [12] assume that the heat load is concentrated at a finite number of points from which issue turbulent plumes that rise and spread laterally along the ceiling. The simplest case is that of a single floor-level source with associated buoyancy flux B. As the plume rises, it entrains ambient fluid; the plume volume flux increases as $Q_p(z) = CB^{1/3}z^{5/3}$ where $C \approx 0.16$ parameterizes the rate of entrainment into the plume. The outflow from the plume impinging on the ceiling forms a buoyant layer that drives a flow through the room. A two-layer stratification forms with a well-mixed upper layer of buoyancy g' and a lower well-mixed layer of buoyancy zero. Thus heterogeneity in the horizontal distribution of buoyancy flux leads to heterogeneity in the vertical distribution of buoyancy. The buoyancy of the upper layer suppresses vertical mixing; the only interfacial transport is through the plume. Hence $Q = Q_p(h)$ where h is the interface height and

$$Q = A^*[g'(H-h)]^{1/2} = CB^{1/3}h^{5/3}. \tag{8.8}$$

Subject to the assumptions of the analytical model, (8.8) prescribes the smallest possible ventilation flow rate for the given H, A^*, and $\mathcal{H}$. Buoyancy conservation requires that whatever buoyancy leaves through the upper vent is balanced by an influx from the plume, i.e.,

$$g' = \frac{B}{Q} = \frac{B^{2/3}}{Ch^{5/3}}. \tag{8.9}$$

Combining (8.8) and (8.9) leads to an expression for the interface height

$$\frac{A^*}{H^2C^{3/2}} = \left(\frac{\zeta^5}{1-\zeta}\right)^{1/2}, \tag{8.10}$$

where $\zeta = h/H$. Note that the interface height is independent of the heat load and is only a function of the room geometry and the mixing characteristics of the plume. Stated differently, simple

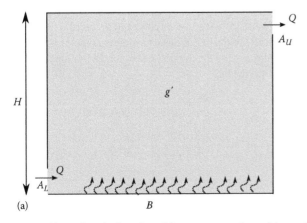

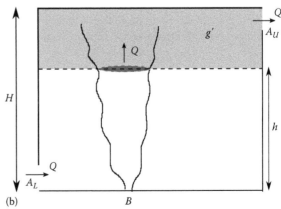

FIGURE 8.2 (a) Uniformly distributed heat source and resulting well-mixed room. (b) A localized heat source with resulting plume and two-layer stratification.

naturally ventilated rooms exhibit an appealing self-correcting behavior: If the thermal forcing is altered, the interface will, after some transient adjustment, return to its original elevation. The duration of this adjustment depends on the timescales of the flow, which are defined in the following subsection.

8.3.1.2 Transient Effects

The time-dependent development of stack-driven ventilation flow is described by writing conservation equations for the volume and buoyancy of the warm layer. For the model of Gladstone and Woods [3], the volume of the warm layer is constant (equal to the volume, $V = HS$, of the room) and therefore only a buoyancy equation is required:

$$\frac{d}{dt}(HSg') = B - Qg' = B - A^*(g'^3 H)^{1/2}. \qquad (8.11)$$

Scaling the buoyancy using (8.7) yields

$$\frac{d\delta_u}{d\tau_u} = 1 - \delta_u^{3/2}. \qquad (8.12)$$

The subscript *u* indicates variables associated with a well-mixed model and $\tau = t/T_u$ where

$$T_u = \frac{H^{2/3} S}{A^{*2/3} B^{1/3}}. \qquad (8.13)$$

For the special case where $g'(t = 0) = 0$, the solution to (8.12) is given implicitly by [10]

$$\tau_u = \frac{2}{3^{1/2}} \tan^{-1} 3^{1/2} - \frac{1}{3} \ln \left| \frac{(1 - \delta_u^{1/2})^2}{1 + \delta_u^{1/2} + \delta_u} \right| - \frac{2}{3^{1/2}} \tan^{-1} \left(\frac{2\delta_u^{1/2} + 1}{3^{1/2}} \right). \qquad (8.14)$$

The aforementioned analysis uses a single timescale for the flow development, namely T_u. By contrast, for the two-layer model of Linden et al. [12], evolution equations are required for the upper layer thickness and buoyancy. The description of Kaye and Hunt [9] assumes that, during the flow development, the upper buoyant layer remains well mixed for all time. Therefore, for the case of a single plume, the conservation equations read as follows:

$$\frac{d}{dt}[S(H - h)] = CB^{1/3} h^{5/3} - A^*[g'(H - h)]^{1/2}, \qquad (8.15)$$

$$\frac{d}{dt}[S(H - h)g'] = B - A^*[g'^3(H - h)]^{1/2}. \qquad (8.16)$$

Two timescales now arise; the filling box time and the draining time, defined as

$$T_f = \frac{S}{CB^{1/3} H^{2/3}}, \quad T_d = \frac{C^{1/2} SH^{4/3}}{A^* B^{1/3}}, \qquad (8.17)$$

respectively. The balance between filling and draining is highlighted by forming the ratio

$$\frac{T_f}{T_d} = \frac{A^*}{H^2 C^{3/2}}. \qquad (8.18)$$

Not coincidentally, the right-hand side of (8.18) is identical to the left-hand side of (8.10).

Equations 8.15 and 8.16 can be integrated to predict the buoyancy and thickness of the warm upper layer and the ventilation flow rate as functions of time. For relatively small effective vent areas ($A^*/H^2 C^{3/2} < 4$), the draining flow development lags behind the filling box flow development and the interface overshoots its steady-state height. For larger effective vent areas no overshoot occurs. In either case, the eventual steady-state stratification and ventilation flow rate are uniquely prescribed by (8.9) and (8.10), i.e., they depend in a one-to-one fashion on the source conditions and room geometry.

The aforementioned analysis assumes that the buoyancy source has zero initial volume flux, i.e., $Q_s = 0$. This is not always the case: Air conditioners and under floor air distribution (UFAD) systems have nonzero source buoyancy and volume fluxes. If the buoyancy source in Figure 8.2b has an associated source volume flux $Q_s > 0$, the outflow through the upper vent is the sum of Q_s and the inflow through the lower vent. Also there is now the possibility of a "blocked" flow regime where $Q_v \le Q_s$ in which Q_v is a characteristic buoyancy-driven ventilation flow rate. Once the chamber becomes blocked, the interface reaches the floor and there is no inflow, and more likely outflow, through the lower vent [19]. The transient development of the blocked regime is complicated by the appearance of an additional replacement time, $T_R = HS/Q_s$, which is the time required for the source volume flux to flush the entire volume of the room. The initial filling is governed by (8.15) and (8.16) with an appropriate virtual origin offset for the plume volume flux. Once the interface reaches the source, a replacement flow is established in which the fluid at the interface is replaced by fluid with source buoyancy, B_s. The room buoyancy then approaches B_s exponentially with a time constant T_R [19].

A similar analytical approach to that applied in deriving (8.15) and (8.16) can be employed in predicting the transport of a neutrally buoyant pollutant through an enclosure [5]. Equations may be written for the pollutant concentration above and below the interface, i.e.,

$$\frac{d}{dt}[K_{lo} hS] = -K_{lo} CB^{1/3} h^{5/3}, \qquad (8.19)$$

$$\frac{d}{dt}[K_{up} S(H - h)] = K_{lo} CB^{1/3} h^{5/3} - K_{up} A^*[g'(H - h)]^{1/2}, \qquad (8.20)$$

where K_{lo} and K_{up} are the pollutant concentrations in the lower and upper layers, respectively. Both layers are assumed to be well mixed at all times. Again, these equations can be solved numerically given the appropriate initial conditions.

By approximating the upper layer as well mixed, it can be shown that the jump of temperature and pollutant concentration across the interface are over- and underestimated, respectively. More sophisticated analyses show that the peak pollutant concentration actually occurs at the interface, i.e., potentially at head height [1]. This implies that the total amount of pollutant being flushed at a given time is less than that predicted by the well-mixed model. Later work by the authors of [1] examined particulate matter pollutants.

8.3.1.3 Multiple Zones

Previously it was argued that transients such as the overshoot of the terminal interface elevation do not alter conditions as $t \rightarrow A$. In practice and for more intricate building geometries, however, the transient approach toward steady state may exert a less ephemeral influence. Consider, for example, a seemingly trivial modification to the canonical geometry of Figure 8.2b whereby an internal partition is placed to the right of the thermal source. The model building is then divided into forced and unforced zones, which may communicate with one another through floor- and ceiling-level openings in the vertical partition. The buoyant layer in the forced zone evolves according to equations akin to (8.15) and (8.16). However, because buoyant fluid is not re-entrained from the unforced zone into the plume, no mechanism exists for eroding the stable stratification of density that develops in this right-hand side room. In other words, a permanent record of the evolution of the plume temperature is preserved [2]. The details of the steady state cannot then be determined from a simple algebraic equation such as (8.10); these are instead resolved by following the path that connects the initial and final states.

A further complication to be examined is that of delayed convergence whereby the approach toward steady state, say in the forced zone, is prolonged due to mass and buoyancy exchange with the adjacent zone. When investigating, for example, a number of small shops connected to a larger atrium, the ramifications of delayed convergence are nontrivial especially when time-varying forcing is considered [2]. Even so, atria are agreed to have a generally positive influence vis-à-vis building cooling: Much like solar chimneys, they serve as deep repositories for escaping buoyant air, which may be further heated by solar radiation resulting in still larger stack pressures. Atria and chimneys are also effective, for better or worse, at providing secondary ventilation to zones with minimal associated heat loads.

8.3.1.4 Wind Forcing and Multiple Steady States

In the previous three subsections, ventilation is driven by buoyancy alone; however, additional factors must be examined when an adverse or assistive wind pressure gradient is included. Consider the flow illustrated in Figure 8.2b. Now assume an imposed pressure difference ΔP between the upper and lower vents. When $\Delta P < 0$, wind and buoyancy act in tandem; correspondingly, we expect the interior interface to rise and the mean interior temperature to fall. When $\Delta P > 0$, buoyant convection is inhibited; the interface and average indoor temperature are

expected to fall and rise, respectively. For suDciently large ΔP, a reversal of flow direction occurs with, for example, inflow through upper vent(s). The flow regime is then wind driven and well mixed.

In the buoyancy-driven flow regime, (8.10) is modified to read [6]

$$\frac{A^\star}{H^2 C^{3/2}} = \frac{\zeta^{5/3}}{\left(\frac{1-\zeta}{\zeta^{5/3}} - \frac{\Delta P}{\rho_0} \cdot \frac{C H^{2/3}}{B^{2/3}} \right)^{1/2}}. \quad (8.21)$$

Moreover, with $\Delta P > 0$ and for both buoyancy- and wind-driven flow regimes, the magnitude of the ventilation flow rate may be estimated from

$$|Q| = |Q_B^2 - Q_W^2|^{1/2}, \quad (8.22)$$

where $Q_W = A^\star (\Delta P/\rho_0)^{1/2}$ and $Q_B = A^\star [g'H(1-\zeta)]^{1/2}$. Therefore

$$\left(\frac{Q}{Q_W} \right)^3 + \frac{Q}{Q_W} - \frac{1}{F^3}(1-\zeta) = 0 \text{ (buoyancy-driven flow with } Q > 0), \quad (8.23)$$

$$\left(\frac{Q}{Q_W} \right)^3 - \frac{Q}{Q_W} - \frac{1}{F^3} = 0 \quad \text{(wind-driven flow with } Q < 0), \quad (8.24)$$

where, in the notation of [6],

$$F = \left(\frac{\Delta P}{\rho_0} \right)^{1/2} \left(\frac{A^\star}{BH} \right)^{1/3}. \quad (8.25)$$

Solutions to (8.23) and (8.24) are exhibited in Figure 8.3a; multiple steady states are predicted, the existence and stability of which have been confirmed by laboratory experiments [6]. Consider, for example, a strong wind forcing that slowly diminishes over time due, say, to a diurnal variation. This corresponds to a continuous transition from points 1 to 2 along the solid curve of Figure 8.3a. From point 2 and for any subsequent decrease of the wind speed, the system jumps discontinuously to a buoyancy-driven flow regime. Provided there are no abrupt changes to the wind velocity, the buoyancy-driven flow regime is maintained whether the external wind speed is increased or decreased. Not until point 4 is reached does the flow regime revert to its original (qualitative) state. Comparable transitions occur when the wind velocity is constant but the heat generated by the internal source varies in time.

Figure 8.3a reveals an intriguing asymmetry: whereas wind-driven flow cannot exist for $F \lesssim 1.37$, buoyancy-driven flow is realizable for all F, provided the source mass flux is vanishingly small [6]. Thus no unique numerical value of F is associated with transition from the buoyancy-driven to the wind-driven flow regime. Rather, as argued by Lishman and Woods [14] and others, the transition value of F depends upon the magnitude and

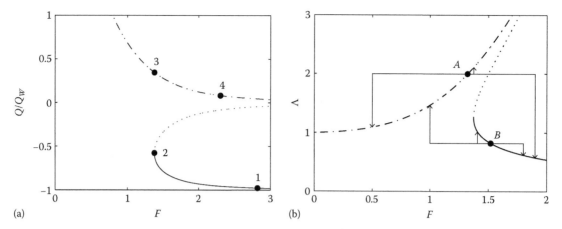

FIGURE 8.3 (a) Ventilation flow rate vs. *F*. When $Q/Q_W < 0$, (8.24) admits stable (solid curve) and unstable (dotted curve) solutions. The dash-dotted curve shows the solution to (8.23) with $\zeta \ll 1$. (b) $\Lambda = g'A^{*2/3}H^{1/3}/B^{2/3}$ vs. *F*. (Adapted from Hunt, G.R. and Linden, P.F., *J. Fluid Mech.*, 527, 27, 2004; Lishman, B. and Woods, A.W., *Build. Environ.*, 44, 666, 2009.)

rate of change of the wind or thermal forcing. To illustrate their argument, consider Figure 8.3b, which shows the nondimensional reduced gravity of the exhaust air as a function of *F*. Line types are as indicated in Figure 8.3a; the upper and lower thin solid lines consider the impact of a sudden increase or decrease in wind forcing. From the point labeled "A," a transition to the stable branch of the wind-driven solution is possible only for sufficient large increases to $\Delta P \propto F^2$. For a small increase (or any decrease) to ΔP, no qualitative change of flow regime will occur. As suggested by the lower thin solid line of Figure 8.3b, a change of flow regime from wind- to buoyancy-dominated also depends on the magnitude (though not in this case the rate of change) of ΔF. What constitutes a "small" or "large" variation in *F* clearly depends upon the flow conditions in the initial state.

More generally, changes to the wind forcing may (i) evolve relatively slowly, i.e., over a timescale comparable to the adjustment timescale of the ventilated building, or (ii) be non-monotonic. Discerning conditions for regime transition in either case is nontrivial. Analytical details and references to other pertinent studies may be found in Lishman and Woods [14].

8.3.1.5 Interior Thermal Mass

When the diurnal temperature variation is fair, thermal mass elements may be used in moderating and phase-shifting outdoor temperature swings. "Night ventilation" is a strategy whereby buildings are cooled by ventilation after sunset; their massive elements then act as heat reservoirs the following day. Despite the nonlinear relationship between the interior temperature and ventilation flow rate in naturally ventilated buildings, a time-periodic variation to T_A results in a time-periodic variation to T_R, here assumed to be spatially uniform. Also, the maximum phase shift is a quarter period [20].

By virtue of their slow response time, however, thermal masses introduce the possibility of quasi-steady states between the buoyancy- and wind-driven flow regimes. In describing this phenomenon, our discussion follows Lishman and Woods [13], who applied the following assumptions: (i) the thermal mass

is isothermal, (ii) the interior space remains well mixed for all time even when the flow is buoyancy driven [3,20], and (iii) the convective heat transfer coeDcient, η, is constant. Under these simplifications, radiative effects may be ignored.

Owing to a slow change in the thermal forcing, say, the interior temperature evolves as

$$\rho_0 c_p V \frac{d\Delta T}{dt} = \mathcal{H} - \rho_0^{1/2} c_p A^* \Delta T \,|\Delta P - \Upsilon g H \Delta T\,|^{1/2} + \Sigma\eta(\Delta T_m - \Delta T). \tag{8.26}$$

Here $\Upsilon = \beta\rho_0$ in which β is the thermal expansion coeDcient and Σ is the thermal mass surface area. Furthermore, ΔT (ΔT_m) is the temperature difference between the interior air (thermal mass) and the exterior air. The thermal mass temperature is determined from

$$\rho_m c_{pm} V_m \frac{d\Delta T_m}{dt} = -\Sigma\eta(\Delta T_m - \Delta T), \tag{8.27}$$

where ρ_m, c_{pm}, and V_m are the density, specific heat capacity, and volume of the thermal mass, respectively. Solutions are obtained by introducing nondimensional (hatted) parameters:

$$\widehat{\Delta P} = \Delta P \cdot \rho_0^{1/3}\left(\frac{c_p A^*}{\Upsilon g H \mathcal{H}_0}\right)^{2/3}, \quad \widehat{\Delta T} = \left(\Upsilon\rho_0 g H\right)^{1/3}\left(\frac{c_p A^*}{\mathcal{H}_0}\right)^{2/3}\Delta T, \tag{8.28}$$

$$\widehat{\mathcal{H}} = \frac{\mathcal{H}}{\mathcal{H}_0} \quad \widehat{\Sigma\eta} = \frac{\Sigma\eta}{(c_p A^*)^{2/3}(\Upsilon\rho_0 g H \mathcal{H}_0)^{1/3}},$$

$$\hat{t} = \left(\frac{\Upsilon g H \mathcal{H}_0}{c_p}\right)^{1/3}\left(\frac{A^*}{\rho_0}\right)^{2/3}\frac{t}{V}, \tag{8.29}$$

where $\mathcal{H}_0 \equiv \mathcal{H}(t = 0)$. Thus (8.26) and (8.27) may be written as

$$\frac{d\Delta T}{dt} = \mathcal{H} - \Delta T\,|\Delta P - \Delta T\,|^{1/2} + \Sigma\eta(\Delta T_m - \Delta T), \tag{8.30}$$

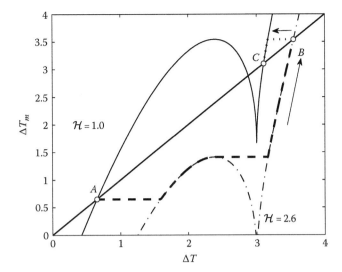

FIGURE 8.4 Trajectories between steady-state solutions in the $(\Delta T, \Delta T_m)$ phase plane. The (thin) solid and dash–dotted curves show steady-state solutions of (8.30) for $\mathcal{H} = 1.0$ and 2.6, respectively. In both cases, $\Delta P = 3.0$ and $\Sigma \eta = 0.75$. The 45 degree line shows steady-state solutions, $\Delta T = \Delta T_m$ of (8.31). The remaining line types are described in the text. (Adapted from Lishman, B. and Woods, A.W., *Build. Environ.*, 44, 762, 2009.)

$$\frac{d\Delta T_m}{dt} = \left(\frac{\rho_0 c_p V}{\rho_m c_{pm} V_m}\right) \Sigma \eta (\Delta T - \Delta T_m). \quad (8.31)$$

To lighten the notation, hats are omitted here. Steady-state solutions to (8.30) and (8.31) must satisfy $\Delta T = \Delta T_m$ and $\mathcal{H} = \Delta T \,|\, \Delta P - \Delta T \,|^{1/2}$. As indicated by the thin solid curve of Figure 8.4, there are, in general, two such (stable) solutions. One corresponds to the buoyancy-driven regime (point C), the other to the wind-driven regime (point A).

Suppose that the thermal forcing is altered abruptly, i.e., $\mathcal{H}$ increases from 1.0 to 2.6. The room air temperature is assumed to adjust much more quickly than that of the thermal mass. From point A of Figure 8.4, therefore, the system progresses along the thick dashed curve. Following an abrupt increase in ΔT, quasi-steady conditions, characterized by slow increases in the thermal mass and indoor air temperatures, are predicted. There is then a second rapid increase in ΔT and a second interval of quasi-steady adjustment. During the quasi-steady states, the thermal mass acts as a heat sink that approximately balances the increase in $\mathcal{H}$. True steady state is not achieved until point B is reached, however.

The flow described by Figure 8.4 shows a familiar hysteresis. If $\mathcal{H}$ is subsequently decreased from 2.6 to 1.0, the system evolves along the thick dotted curve of Figure 8.4 landing ultimately not at point A but rather at point C corresponding to a buoyancy-driven flow.

8.3.1.6 Underfloor Air Distribution

Unfortunately, thermal mass elements are relatively ineffective in establishing hospitable interior conditions in especially humid regions or those where the night temperature remains elevated.

Consequently, some form of conditioning may be necessary. The design basis of traditional air-conditioning systems is a well-mixed interior space. In displacement ventilation systems, cool air is mechanically supplied at low levels and at low velocities. Warm air is displaced to the top of the interior space and is subsequently discharged to the exterior through one or more return vents.

The appearance of an interior temperature gradient is, from the perspective of waste-heat disposal, a hallmark of an energy-eDcient process. Ideally in displacement ventilation only the lower portion of an interior space is cooled. However, sharp vertical variations of temperature, if they descend into the working zone, are potentially disagreeable to building occupants. As a compromise between energy savings and comfort, an under-floor air distribution or UFAD system may be favored. Here cool air is supplied by an underfloor plenum through floor-level diffusers and thereafter takes the form of negatively buoyant jets or fountains. Jet impingement on the interface results in the entrainment of buoyant air from the upper layer, which in turn diminishes the temperature gradient [15].

For illustrative purposes, we shall examine a UFAD system consisting of a single ideal thermal source and multiple diffusers (Figure 8.5). The figure is adapted from Liu and Linden [15] whose discussion guides our exposition. The heat load and total ventilation flow rate, Q, are assumed fixed. The impact of adding more diffusers is to uniformly reduce the volume and momentum fluxes, respectively Q_s and M_s, from any individual diffuser. This in turn reduces interfacial entrainment, whose associated volume flux is determined to be

$$Q_e = \begin{cases} (0.6 \pm 0.1)Q_i & Ri \le 8, \\ (4.8/Ri \pm 0.1)Q_i & Ri > 8, \end{cases} \quad (8.32)$$

where Q_i is the impinging volume flux and the Richardson number is defined as $Ri = (g_2' - g_1')b_i/w_i^2$. Here b_i and w_i are, respectively, the impinging radius and "top-hat" velocity of the fountain and $g_1' = g_2'Q_e/(Q_e + Q/n)$ and $g_2' = B/Q$ are, respectively, the uniform reduced gravity of the lower and upper layer. The number of diffusers is denoted by n.

As n increases, the height of, and density jump across, the interface respectively decreases and increases (Figure 8.5a of Liu and Linden [15]). In the limit as $n \to A$, the lower layer

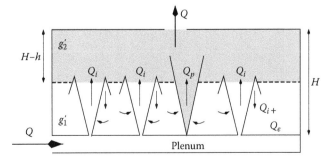

FIGURE 8.5 Schematic illustration of a UFAD system. (Adapted from Liu, Q.A. and Linden, P.F., *J. Fluid Mech.*, 554, 323, 2006.)

temperature approaches the cold air supply temperature, i.e., $g_1' \to 0$. The ventilation regime is then one of pure displacement with vanishing interfacial entrainment.

A second scenario considered by Liu and Linden [15] is that of a fixed plenum pressure. The ventilation rate now grows linearly with n and $g_1'/g_2' = Q_e/(Q_e + Q_s)$. As n increases, the interface rises and the layer temperatures decrease (see Figure 8.5b). When $n \to A$, the stratification disappears and the interior temperature matches the supply air temperature.

8.3.1.7 Heat Source Modeling

Neither the well-mixed conditions of Gladstone and Woods [3] nor the two-layer stratifications of Linden et al. [12] are observed in measurements of real rooms. Given the restrictions of these earlier studies, it is of interest to examine the transition from displacement to mixing ventilation as a function of the heat source geometry. Two scenarios are considered: multiple point sources and sources distributed over a finite area that is still less than S. In the former case, a two-layer stratification forms with the interface height, ζH, determined from

$$\frac{A^*}{nH^2C^{3/2}} = \left(\frac{\zeta^5}{1-\zeta}\right)^{1/2}, \tag{8.33}$$

where n is the number of point sources of buoyancy. The upper layer buoyancy is given by

$$g' = \frac{(B/n)^{2/3}}{C(\zeta H)^{5/3}}. \tag{8.34}$$

As n increases, the interface approaches the floor and

$$\zeta \to \left(\frac{A^*}{nH^2C^{3/2}}\right)^{2/5}. \tag{8.35}$$

By substituting (8.35) into (8.34), (8.7) is recovered. As the heat load is divided among a greater number of equal point sources, the stratification and ventilation flow rate predicted by Linden et al. [12] approach those anticipated by Gladstone and Woods [3] even though the latter model is derived without direct reference to plume entrainment.

Real heat loads are usually distributed over some finite area. A transition from displacement to mixing ventilation is expected as this source area increases. In the near field of a pure source of buoyancy, with source radius b_0, the volume flux increases linearly with height until $z \approx 0.55b_0$. Thereafter the flow behaves like a pure plume rising from a virtual origin a distance $2.2b_0$ below the physical source [11]. Given an expression for the volume flux as a function of z, Linden et al.s analysis can be repeated to predict the stratification in the room, the only complication being the change in volume flux behavior at $z \approx 0.55b_0$ and the introduction of an additional length scale, namely b_0. The associated details were resolved by Kaye and Hunt [11], who showed that the transition from displacement to mixing ventilation is a function of both A^*/H^2 and the ratio of the source radius to the room height $\sigma = b_0/H$. Plots of the interface height and warm-layer buoyancy for different values of σ and A^*/H^2 are shown in Figure 8.6. As σ increases, the interface approaches the floor and the layer buoyancy approaches that of a well-mixed interior.

Another possibility is that the source area tends toward S. Measurements show that the plume breaks down due to shear; mixing ventilation is observed when the source area is only $\sim 0.15S$ (see Kaye and Hunt [11] and the references therein).

Finally, heat sources may have nontrivial vertical extent. Provided B is known and the convective volume flux above the source can be calculated, however, Linden et al.s equations may again be used and the resulting stratification and ventilation flow determined.

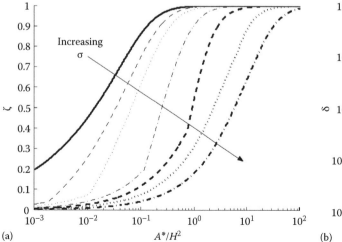
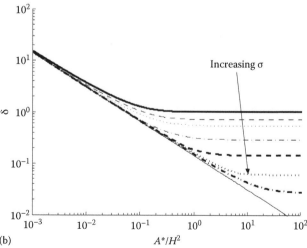

(a) A^*/H^2 (b) A^*/H^2

FIGURE 8.6 (a) Interface height vs. the relative effective vent area and (b) layer buoyancy (scaled on the plume buoyancy at $z = H$) for $\sigma = b_0/H = 0$, 0.1, 0.2, 0.5, 1, 2, and 5. The thin solid line of (b) shows the mixed ventilation limit. (Reprinted from *Build. Environ.*, 45 (4), Kaye, N.B. and Hunt, G.R., The effect of floor heat source area on the induced airflow in a room, 839–847. Copyright 2010, with permission from Elsevier.)

8.3.2 Experimental and Numerical Description

8.3.2.1 Salt Bath Experiments

Owing to the diDculty and expense of model verification at full scale, the water-based similitude experiments pioneered by Dr. Paul F. Linden and others at the University of Cambridge have become increasingly popular over the past two decades. While important exceptions exist (e.g., Gladstone and Woods [3]), experiments are often run "upside down" so that dense saltwater plumes mimic isolated sources of buoyancy [12]. Provided density differences are small, i.e., the Boussinesq approximation is applicable, the resulting flows are dynamically equivalent to those that would be observed if the experiments were instead run "right-side up." Flows in similitude experiments are relatively easy to visualize using food coloring or shadowgraph.

Given the important role of buoyancy in natural ventilation, Reynolds numbers may be defined as $Re = g'^{1/2} H^{3/2}/\nu$ where u is the kinematic viscosity. By tuning salt concentrations, and hence g', appropriately, representative values for Re may be recovered.

One of the criticisms of salt bath experiments is that they predict sharp density gradients that are rarely observed in real buildings. Between full and reduced scale, one cannot simultaneously match Re and the Péclet number, defined as $Pe = Re \cdot u/\kappa$ where κ denotes a diffusion coeDcient for salt or heat. Laboratory values for Pe may exceed their full-scale counterparts by up to 10^3. A recent theoretical study by Kaye et al. [8] quantifies the impact of this disparity and provides, for a canonical ventilation flow, an estimate of the resulting interface thickness for prescribed room geometry and source conditions.

8.3.2.2 Wind Tunnel Experiments

Salt bath modeling is helpful for investigating buoyancy-driven ventilation; however, examining wind-driven flows is more problematic. While studies such as that of Hunt and Linden [6] have explored the interaction of wind and buoyancy in a water flume, such an experimental apparatus is relatively ineffective at relating a given wind field and building geometry to the pressure differences that drive ventilation flows. These questions are more typically addressed using large-scale boundary layer wind tunnels.

Studies of this sort typically assume that Reynolds number effects are minimal. In principle, the appropriate nondimensional groups that must be matched between tunnel and full scale are the ratio of the building height and all significant turbulence length scales in the flow. However, Jensen [7] showed that basic dynamic similarity can be achieved provided (i) the model building is immersed in a fully rough turbulent boundary layer, and (ii) the ratio of the building height to upstream surface roughness is the same for both model and full scale. Wind tunnel studies are therefore conducted in large-scale facilities with a long section of artificial roughness upstream of the test section (Figure 8.7b).

Data giving the pressure distribution over the surface of simple buildings are readily available. Unfortunately this information is generally based on measurements with no upstream buildings, only small roughness elements. The data will therefore lead to an

(a)

(b)

FIGURE 8.7 (a) The DVG oDce building in Hanover, Germany. (b) The wind tunnel model used to assess the design of the ventilation system. (Pictures courtesy of Hascher Jehle Architektur, Berlin.)

overestimate of the force due to wind loading and an underestimate of the ventilation flow rate.

Modeling real-world upwind geometric complexity leads to a bewildering array of parameters that have yet to be fully explored. Consequently, wind tunnel studies are often used for design verification rather than scientific research. An example is shown in Figure 8.7. The DVG oDce building, designed by Hascher Jehle Architektur, Berlin, is a naturally ventilated complex in which buildings are covered by glass airfoils designed to create lift and draw air up through the buildings. The wind tunnel model used to verify the design is presented in Figure 8.7b. It shows the long upstream fetch, as well as models for the surrounding buildings, hedges, and trees.

8.3.2.3 Zonal Models

In multizone buildings, zonal models are often applied in estimating ventilation flow rates and temperatures. The modeling

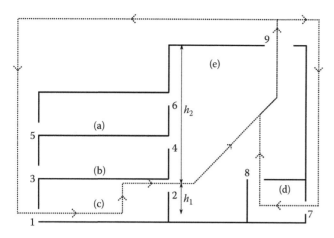

FIGURE 8.8 Schematic of a representative multiroom building with a large atrium. The labels (a)–(e) indicate different possible flow paths.

approach employs a familiar balance between buoyancy and pressure losses due to flow. Equations are written for the pressure change along a path starting near an inlet vent, passing through the building, and then returning to the path origin (Figure 8.8). The net pressure difference along the path is zero.

Equations are written for the pressure along each flow path (for clarity, only two possible paths are indicated in Figure 8.8). For example, the equation for loop $1 - 2 - 9 - 1$ is

$$-\frac{C\rho u_1^2}{2} + \rho g' h_1 - \frac{C\rho u_2^2}{2} + \rho g' h_2 - \frac{C\rho u_9^2}{2} = 0. \quad (8.36)$$

Zones are treated as well mixed, leading to a possible overprediction of the ventilation flow.

A second set of equations expresses volume conservation in each zone. For example,

$$Q_9 - Q_2 - Q_4 - Q_6 - Q_8 = 0 \quad (8.37)$$

quantifies the appropriate balance for zone (e). Finally, equations are written expressing energy conservation. Written in terms of buoyancy, the equation relevant for zone (e) reads

$$g_e' Q_9 - g_c' Q_2 - g_b' Q_4 - g_a' Q_6 - g_d' Q_8 - B_e = 0, \quad (8.38)$$

where B_e is related to the internal heat load in zone (e). The example given yields five equations each for buoyancy and volume and four pressure loop equations, which can be written in terms of the vent areas and flow rates. This yields 14 equations that can be solved for 14 unknowns, such as zone-specific temperatures. This method of analysis is similar to that employed in studying flow in piping networks. A number of zonal modeling software tools are available online including Energy Plus developed by the U.S. Department of Energy and CONTAM/LoopDA developed by the U.S. National Institute of Standards and Technology's Building and Fire Research Laboratory.

8.3.2.4 Computational Fluid Dynamics

Numerical simulations have been used both as a research and a design tool in analyzing airflow in buildings, particularly involving natural ventilation. Research has focused on identifying appropriate turbulence closure schemes to be applied in a given scenario. Many benchmarking studies also compare numerical output to related theoretical results. Using computational fluid dynamics (CFD) rather than salt bath experiments, one can, in principle, consider more complex heat source or room geometries, the effects of thermal mass, and more realistic boundary conditions.

The use of CFD for modeling entire buildings is less common because it requires particular expertise and extensive computational resources. An appealing alternative is to couple CFD simulations applied to part of a building with a zonal model of the entire structure [18]. The CFD simulations may concentrate on those portions of the building where the well-mixed assumption is especially questionable. Such coupling also provides a means of specifying CFD boundary conditions based on the predictions of the coupled zonal model.

8.3.3 Limitations

The analytical, experimental, and numerical models described in Sections 8.3.1 and 8.3.2 provide insights into the physics of airflow in the built environment. By necessity, however, these models typically invoke a series of simplifying assumptions, which we outline in the following subsections.

8.3.3.1 Adiabatic Walls

With the exception of thermal mass models, one often considers perfect insulation and negligible air leakage. Heat transfer at walls and ceilings not only results in energy losses but also creates additional convective boundary layer flows.

8.3.3.2 Mixing by Convection Only

The aforementioned analytical models assume that the motive force is due to pressure differences resulting from wind and buoyancy. Thus, the following influences are ignored: the momentum of the inflowing air, the movement of occupants, diffusion and the action of fans or blowers. Recently published studies (e.g., Hunt and Coffey [4] and Kaye et al. [8]) reveal the extent to which these omissions may jeopardize model predictions, though much further work remains to be done in this area.

8.3.3.3 Negligible Radiant Heat Transfer

As suggested by the discussion in Section 8.4, radiation heat transfer can play an important role in thermal comfort even though it may not significantly alter patterns of air movement. The accurate forecasting of interior wall temperatures will improve the predictive capability of the aforementioned models with respect to thermal comfort.

8.3.3.4 Negligible Humidity and Condensation

Existing architectural fluid dynamics models focus on the forces that drive ventilation flows but often ignore moisture budgets.

Control of humidity is a significant obstacle to the adoption of natural ventilation schemes in hot, humid climates. Humidity in the external ambient can lead to condensation on cool internal or interstitial surfaces. It can also cause occupant discomfort and errors in the energy budget due to latent heat effects. These considerations are often season dependent; in many parts of the world, absolute and relative humidities vary significantly over the course of the year.

8.4 Applications

Examples of naturally ventilated buildings abound; here, we examine one particular case study to illustrate the application of the aforementioned principles, namely the Zion National Park VisitorsflCenter. Constructed in 2000 in a semi-arid region of southwest Utah, the building experiences hot summers, cold winters, and a steady stream of park visitors.

A schematic of the VisitorsflCenter is shown in Figure 8.9. Notable design features include a pair of downdraft cooltowers, clerestory windows, heavy concrete floors, and an equator-facing Trombe wall. Natural ventilation provides the majority of the summer cooling; hot air escapes the building through the high clerestory windows, which are shielded from the summer

sun by optimized overhangs. On especially hot days, water from a nearby river is circulated through the top of the cooltowers. Owing to the associated evaporative cooling, a dense plume of cold air falls, by passive means, through the tower and then spreads as a gravity current along the floor of the Visitorsfl Center. Interior louvers allow park staff to adjust the flux and direction of this draft of cold air. Night ventilation using interior thermal mass elements (e.g., the concrete floor) helps to offset daytime heat gains [16].

During winter, solar energy is exploited for daytime lighting and thermal comfort. In the latter respect, the Trombe wall is especially effective. It consists of masonry coated on its exterior facing side with an absorbent material. A pane of glass is placed between the Trombe wall and the environment; the interstitial air is heated by the greenhouse effect. This raises the temperature of the masonry, making it a stable source of radiative and convective warmth. Solar gains are phase-lagged and take full effect in mid to late afternoon. Electrical radiative ceiling panels supply additional heat on cloudy or unseasonably cold days [16].

By contrasting building energy use against an appropriate baseline, Torcellini et al. [16] estimate a total energy cost savings for the VisitorsflCenter of 67%. Summertime savings are especially pronounced given the absence of air conditioners.

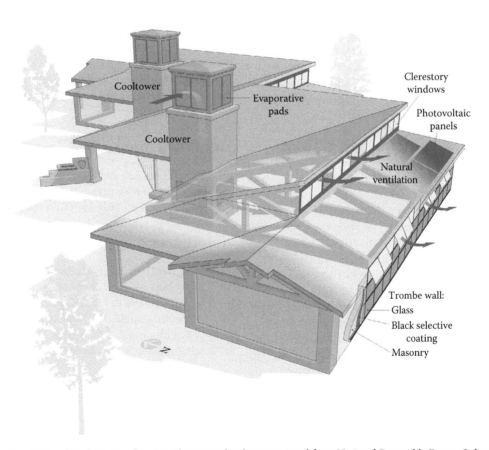

FIGURE 8.9 The Zion National Park VisitorsflCenter. This image has been reprinted from National Renewable Energy Laboratory Conference Paper (NREL/CP-550-36272). (From Torcellini, P. et al., Zion National Park Visitor Center: Performance of a low-energy building in a hot, dry climate, Golden, CO, Preprint, July 2004.)

8.5 Major Challenges

Continuing our discussion initiated in Section 8.3.3, we outline the following topics for future study in the context of science and technology in naturally ventilated building.

8.5.1 Model Input Conditions

Previously, many of the input conditions (e.g., the exterior temperature, wind speed, pollution concentration, insolation) were assumed known and constant. This is rarely true in practice; further research is therefore required to understand the behavior of ventilation systems given oscillatory inputs whether the temporal variation is slow, or, more critically, fast. From the perspective of building design, the number of permutations to consider is often prohibitive. It is therefore important to (i) choose appropriate model inputs for simulation, and (ii) quantify the sensitivity of the relevant dependent variables (such as interior temperature) to variations in the ambient conditions.

8.5.2 Control Strategies

Given the plethora of input conditions to consider, sensitivity studies are most appropriately complemented by robust climate control algorithms. Unfortunately, algorithm development has traditionally been retarded by (i) multiple (and sometimes competing) objectives regarding indoor air quality, thermal comfort, noise reduction, etc., and (ii) the need for multifaceted control that seamlessly accommodates rapid hour-by-hour climactic variations and also those that occur over slow, seasonal timescales. Consequently, many designs continue to rely on building occupants for manual control (e.g., in the adjusting of interior louvers to the downdraft cooltowers at the Zion National Park VisitorsﬂCenter). They may also incorporate auxiliary mechanical or electrical components to be used on a periodic basis. Opportunities for manual control may be couched as user-friendly design. It should be emphasized, however, that real-time adjustments by building occupants are necessary to bridge the chasm that exists between the analytical models described previously and the design and operation of real buildings, which are characterized by complex geometries, flow obstructions, multiple modes of heat transfer, non-isothermal heat sinks, etc.

Very few designers would attempt to eliminate manual control altogether. However, and by virtue of the proliferation of sophisticated control algorithms in the transportation, chemical processing, and related industries, there is now the opportunity to more rigorously couple manual and predictive control strategies, the latter exploiting recent advances of understanding. Obstacles to be overcome in this context include the issue of multiseason design alluded to in earlier sections and, more specifically, hysteretic transitions between multiple steady states and also delayed convergence, whether due to adjacent zones or thermal mass.

8.5.3 Integration into Energy Management Software

When estimating energy operating costs, algorithms such as Energy Plus have traditionally assumed building zones to be well mixed. If a two-layer stratification of the type shown in Figure 8.2b is known to exist, one may instead model the upper and lower layers as their own distinct zones. Even so, there are many further nuances of natural ventilation that have yet to be embedded into Energy Plus and related programs. Appropriately balancing model fidelity and runtime eDciency is an evolving challenge, which will dictate the alacrity with which software is updated.

8.5.4 Pollution Dispersion Modeling

One of the advantages of naturally ventilated buildings is that they are more open to the external environment and therefore avoid many of the problems of sick- or tight-building syndrome. However, this openness makes building occupants more susceptible to external toxins, whether accidentally or maliciously released. Although several studies (e.g., [1,5]) have focused on pollution dispersion within naturally ventilated buildings, the number of such investigations remains small. Also, the task of combining building dispersion, control systems, and large-scale urban dispersion models remains largely unfulfilled: models that resolve the details of flow around a building in the urban canopy coupled with infiltration into, and transport through, that building remain incomplete. Here especially, there are multiple avenues to incorporate ideas, measurements, and results from elsewhere in this volume in advancing the field of architectural fluid mechanics.

Acknowledgments

Drs. Colm P. Caulfield and Diogo T. Bolster kindly reviewed earlier versions of this manuscript. M.R.F. acknowledges the financial support of the NSERC Discovery Grant and SNEBRN programs.

References

1. D. T. Bolster and P. F. Linden. Contaminants in ventilated filling boxes. *J. Fluid Mech.*, 591:97–116, 2007.
2. M. R. Flynn and C. P. Caulfield. Natural ventilation in interconnected chambers. *J. Fluid Mech.*, 564:139–158, 2006.
3. C. Gladstone and A. W. Woods. On buoyancy-driven natural ventilation of a room with a heated floor. *J. Fluid Mech.*, 441:293–314, 2001.
4. G. R. Hunt and C. J. Coffey. Emptying boxes—Classifying transient natural ventilation flows. *J. Fluid Mech.*, 646:137–168, 2010.
5. G. R. Hunt and N. B. Kaye. Pollutant flushing with natural displacement ventilation. *Build. Environ.*, 41:1190–1197, 2006.

6. G. R. Hunt and P. F. Linden. Displacement and mixing ventilation driven by opposing wind and buoyancy. *J. Fluid Mech.*, 527:27–55, 2004.

7. M. Jensen. The model-law for phenomena in natural wind. *Ingenioren, Int. Ed.*, 2(4):121–128, 1958.

8. N. B. Kaye, M. R. Flynn, M. J. Cook, and Y. Ji. The role of diffusion on the interface thickness in a ventilated filling box. *J. Fluid Mech.*, 652, 195–205, 2010.

9. N. B. Kaye and G. R. Hunt. Time-dependent flows in an emptying filling box. *J. Fluid Mech.*, 520:135–156, 2004.

10. N. B. Kaye and G. R. Hunt. Heat source modelling and natural ventilation eD ciency. *Build. Environ.*, 42(4):1624–1631, 2007.

11. N. B. Kaye and G. R. Hunt. The effect of floor heat source area on the induced airflow in a room. *Build. Environ.*, 45:839–847, 2010.

12. P. F. Linden, G. F. Lane-Serff, and D. A. Smeed. Emptying filling boxes: The fluid mechanics of natural ventilation. *J. Fluid Mech.*, 212:309–335, 1990.

13. B. Lishman and A. W. Woods. The effect of gradual changes in wind speed or heat load on natural ventilation in a thermally massive building. *Build. Environ.*, 44:762–772, 2009.

14. B. Lishman and A. W. Woods. On transitions in natural ventilation flow driven by changes in the wind. *Build. Environ.*, 44:666–673, 2009.

15. Q. A. Liu and P. F. Linden. The fluid dynamics of an underfloor air distribution system. *J. Fluid Mech.*, 554:323–341, 2006.

16. P. Torcellini, S. Pless, N. Long and R. Judkoff. Zion National Park Visitor Center: Performance of a low-energy building in a hot, dry climate. Technical Report NREL/CP-550-36272, National Renewable Energy Laboratory, Golden, CO, 2004.

17. M. Santamouris. Energy in the urban built environment: The role of natural ventilation. In C. Ghiaus and F. Allard, (eds), *Natural Ventilation in the Urban Environment*, pp. 1–19. Earthscan, Sterling, VA, 2005.

18. G. Tan and L. Glicksman. Application of integrating multi-zone model with CFD simulation to natural ventilation prediction. *Energy Build.*, 37:1049–1057, 2005.

19. A. W. Woods, C. P. Caulfield, and J. C. Phillips. Blocked natural ventilation: The effect of a source mass flux. *J. Fluid Mech.*, 495:119–133, 2003.

20. J. C. W. Yam, Y. Li, and Z. Zheng. Nonlinear coupling between thermal mass and natural ventilation in buildings. *Int. J. Heat Mass Tran.*, 46:1251–1264, 2003.

Bluff Body Aerodynamics and Aeroelasticity: A Wind Effects Perspective

Ahsan Kareem
University of Notre Dame

9.1 Introduction

The assurance of structural safety and reliability under wind loads requires accurate modeling of wind load effects relying heavily on our understanding of bluff body aerodynamics and aeroelasticity. The intractability of wind–structure interactions amidst complex urban topography has precluded analytical treatment of the subject with the exception of buffeting effects (Davenport, 1967; Zhou and Kareem, 2001; Kwon and Kareem, 2009). Therefore, physical modeling of wind effects in boundary layer wind tunnels has served as a most effective tool for ascertaining these load effects (Cermak, 1975). Accordingly, the last few decades have witnessed significant advances in wind tunnel technology, full-scale monitoring, sensors/transducers, instrumentation, data acquisition systems, data fusion and mining strategies, laser Doppler-based technologies, geographical information and positioning systems (GIS) and (GPS), and information technologies, which have increased our ability to better monitor and process gathered information for improved understanding of the complexities and nuances of how wind interacts with structures, the attendant load effects and their modeling.

On the other hand, in the last few decades, there have been major developments in the computational area to numerically simulate flow fields and their effects on structures. Developments in computational methods, e.g., stochastic computational mechanics, have led to useful tools to further advance the role of numerical analysis. The availability of high-speed computers, individually, in networked clusters, or in the cloud has enhanced the portability and interoperability of computational codes, to most research laboratories and design oﬃces. Rapid advances on all fronts have accordingly led to advances in our understanding of aerodynamics and aeroelasticity and associated load effects and, as a consequence, have improved the prospect of developing the next generation of load simulating facilities and wind tunnels, database-enabled design aids, web-based e-technologies and codes and standards.

These developments have undoubtedly enhanced our abilities to better understand and capture the effects of wind on structures. It is appropriate time, however, to reflect on these developments, reassess their merits and shortcomings, and identify the need for embarking on different modeling philosophies and paradigms as called for by the recent observations. In this context, the rest

Handbook of Environmental Fluid Dynamics, Volume Two, edited by Harindra Joseph Shermal Fernando. © 2013 CRC Press/Taylor & Francis Group, LLC. ISBN: 978-1-4665-5601-0.

of this chapter will identify and discuss a few selected frontiers in bluff body aerodynamics and aeroelasticity with particular reference to wind loads on structures and the challenges these pose to the fluid dynamics and wind engineering communities. These topics include nonstationarity/nonhomogeneous/transient wind events; nonlinearity of structural and aerodynamic origins; non-Gaussianity; mechanical/convective turbulence; unsteady/transient aerodynamics. An expanded version of this chapter is available in Kareem (2010).

9.2 Stationary versus Nonstationary/ Transient Winds

Most extreme wind events are nonstationary in nature and are often highly transient, e.g., wind fields in hurricanes, tornadoes, downbursts, and gust fronts. Therefore, the most critical issue in wind field characteristics concerns the transient wind events, e.g., gust fronts generated by downdrafts associated with thunderstorms. The significance of these transient wind events and their load effects can be readily surmised from an analysis of thunderstorms databases both in the United States and around the world, which suggest that these winds actually represent the design wind speed for many locations (Twisdale and Vickery, 1992; Brooks et al., 2001).

The mechanics of gusts associated with convective gust fronts differs significantly from conventional turbulence (driven by momentum) both in its kinematics and dynamics. A survey of full-scale studies in the meteorological field suggests that winds spawned by thunderstorm, both on the updraft side as tornadoes and on the downdraft side as downburst, fundamentally differ from the synoptic winds in neutrally stable atmospheric boundary layer flows. The key distinguishing attributes are the contrasting velocity profile with height and the statistical nature of the wind field. In gust fronts, the traditional velocity profile does not exist; rather it bears an inverted velocity profile with maxima near the ground potentially exposing low- to mid-rise structures to higher wind loads (e.g., Wood and Kwok, 1998; Letchford and Chay, 2002; Butler et al., 2009). This is compounded by the inherent transient nature of energetic convective gusts, raising serious questions regarding the applicability of conventional aerodynamic loading theories. Although the size of gust fronts may be relatively small and their effects rather local, the fact remains that they can produce significantly damaging winds. The famous Andrews Air force base downburst of 1983 clocked peak gusts of 67 m/s, whereas, ASCE Standard provisions list 50-year recurrence winds of 40–45 m/s in this region (Fujita, 1985). Accordingly, one should question the appropriateness of a design based on conventional analysis frameworks in codes and standards, which generically treat these fundamentally different phenomena in the same manner.

The major challenge in this area is at least twofold, i.e., first, the nature of flow fields in rain bands, the eye wall of hurricanes, downdrafts and gust fronts needs to be better quantified, and second, analysis and modeling tools to capture these features need to be established. Design loads are based on the mean wind speed

for a given site and direction and rely on the assumption that the fluctuations in the mean are characterized by a statistically stationary process, which has led to useful and practical simplifications. The gust fronts generated in thunderstorms/downdrafts differ from the large-scale (extratropical/depressional) storms as the mean wind speed exhibits sharp changes and in some cases changes in wind direction. This leaves the assumption of stationarity open to serious criticism.

Besides this departure in statistical attributes of the wind field, gust fronts are likely to be associated with rapid and substantial changes in the local flow around structures and will likely be correlated over a larger area. These changes in the kinematics and dynamics of the flow field would potentially result in higher aerodynamic loads. These attributes further complicate the concept of "gust factors" which in some forms are central to most wind load assessments. The gust factor concept used for extratropical winds must be revisited as the period used to evaluate average wind speed for thunderstorm winds must be shortened to obtain meaningful results. Longer periods such as an hour, if used for thunderstorm winds, may result in gust factor values almost—two to three times the corresponding values in extratropical winds. Current efforts toward gleaning information regarding the thunderstorm outflow characteristics through modeling nonstationary winds would aid in better capturing the salient features of winds in transient events (Gast et al., 2003; Wang and Kareem, 2004; Chen and Letchford, 2005). In a recent study, Kwon and Kareem (2009) presented a new framework to capture the flow field in a gust front and its attendant load effects.

9.3 Nonlinearity

In the area of wind effects on structures there are three types of nonlinearities that are generally experienced, i.e., geometric, material, and aerodynamic. The geometric nonlinearity is most prevalent in cable-suspended and guyed structures, i.e., suspension and cable stayed bridges and guyed towers and masts, and pneumatic structures. The material nonlinearities may arise from materials of construction, e.g., concrete and composites that do not follow linear constitutive relationships. These effects can be adequately modeled for most structures using finite element models.

The nonlinearity of aerodynamic and aeroelastic origins are prevalent in wind effects on structures. Customarily, in these situations, linearized solutions are invoked, e.g., the widely used gust loading factor in codes and standards is based on the Gaussian framework; thus, the term containing the square of the velocity fluctuations is dropped from the formulation. Kareem et al. (1998) presented a gust loading factor, which included the square velocity term, using the Hermite moment-based distribution in place of the Gaussian distribution. Earlier attempts to capture this effect using Edgeworth series did not adequately represent the tail regions of the distribution.

A linear assumption is often used to compute the response of structures to buffeting forces when nonlinearity in aerodynamic loading arises from the squared velocity term, as alluded to previously and due to dependence of aerodynamic coeDcients on the angle of attack. Therefore, reliance on the quasi-steady theory is currently the method of choice. According to the quasi-steady assumption, aerodynamic forces on structures are expressed as a nonlinear memoryless transformation of the flow-structure relative velocity and the angle of attack, which are variable in time due to oscillation of the body and the presence of turbulence in the incoming flow. The projection of the aerodynamic forces on the shape-functions employed in the finite element or Galerkin discretization of the structure requires approximating the nonlinearities by polynomial expressions. This can be carried out through a Taylor series expansion (Denoel and Degee, 2005) or by an optimization procedure aimed at minimizing some error measure (Carassale and Kareem, 2009).

One of the challenges in aeroelastic analysis remains in the modeling of aerodynamic forces that take into consideration nonlinearities in both structural dynamics and aerodynamics and the ubiquitous issues related to turbulence (Chen and Kareem, 2003a, 2003b). Traditional analysis approaches are not suitable for accommodating these computational challenges. Chen et al. (2000) and Chen and Kareem (2001) proposed a time domain framework incorporating the frequency-dependent characteristics of aerodynamic forces that have been often neglected in most of the previous studies in time domain aeroelastic analysis, potentially impacting the accuracy of the response estimates.

For many innovative bridge sections, even at low levels of turbulence, the effective angle of incidence due to structural motion and incoming wind fluctuations may vary to a level such that the nonlinearities in the aerodynamic forces may no longer be neglected. Current linear force models have proven their utility for a number of practical applications; however, these fail to completely address the challenges posed by aerodynamic nonlinearities and turbulence effects. Diana et al. (1999) proposed a nonlinear aerodynamic force model based on the so-called quasi-static corrected theory, which led them to analytically investigate the turbulence effects on flutter and buffeting response.

An advanced nonlinear aerodynamic force model and attendant analysis framework has been presented by Chen and Kareem (2003a, b) that focused on the needs for modeling of aerodynamic nonlinearity and effects of turbulence on long span bridges. The nonlinear force model separates the forces into the low- and high-frequency components in accordance with the effective angle of incidence corresponding to the frequencies lower than and higher than a critical frequency, e.g., the lowest natural frequency of the bridge. The low-frequency force component can be modeled based on the quasi-steady theory due to its high reduced velocity, while the high–frequency force component is separated into self-excited and buffeting components which are modeled in terms of the frequency-dependent unsteady aerodynamic characteristics at the low-frequency spatiotemporally varying effective angle of incidence. The nonlinear analysis framework is summarized in Figure 9.1 along with the conventional linear scheme. Within this framework, the

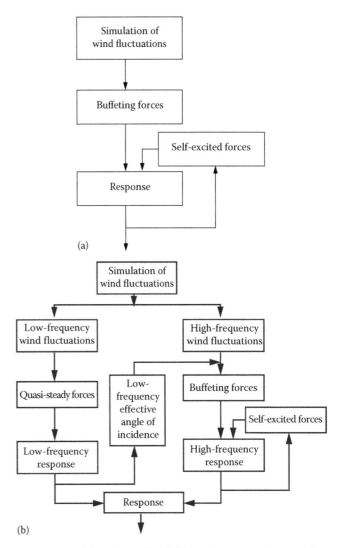

FIGURE 9.1 (a) Traditional and (b) nonlinear aeroelastic analysis framework. (After Chen, X. and Kareem, A., *J. Eng. Mech.*, 129(8), 885, 2003a.)

effects of turbulence on flutter are modeled through the changes in the effective angle of incidence caused by turbulence and its influence on the self-excited forces and the flutter instability. The application of this framework to a long span suspension bridge, with aerodynamic characteristics sensitive to the angle of incidence, revealed a gradual growth in response with increasing wind velocity around the flutter onset velocity, which is similar to the wind tunnel observations of full-bridge, aeroelastic models in turbulent flows (Figure 9.2).

9.4 Non-Gaussianity

Many of the studies encompassing analysis and modeling of wind effects on structures have assumed that the involved random processes are Gaussian. In general, for large structures the assumption of Gaussianity may be valid as a consequence of the central limit theorem. However, the regions of structures under separated flows experience strong non-Gaussian

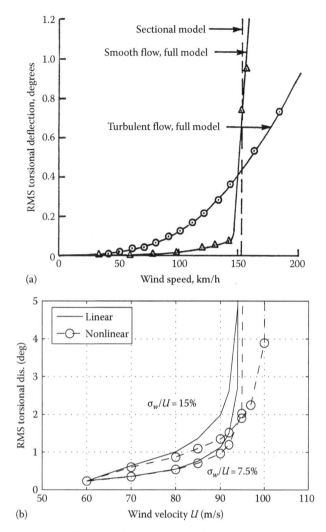

(a)

(b)

FIGURE 9.2 (a) Torsional response of LionsfGate Bridge. (From Irwin, P. Wind tunnel and analytical investigation of the response of Lions Gate Bridge to turbulent wind, NRC of Canada, NAE-LTR-LA-210. 1977.); (b) linear and nonlinear predicted response of a bridge. (After Chen, X. and Kareem, A., *J. Eng. Mech.*, 129(8), 885, 2003a.)

effects in the pressure distribution characterized by high skewness and kurtosis. Figure 9.3 illustrates the nature of pressure fluctuations on the windward and roof regions, where the rooftop region distinctly exhibits non-Gaussian features as noted in the time history of pressure fluctuations. It is also noteworthy that the bending moments (bottom time history) at the supports also exhibit non-Gaussian features despite the fact that the bending moments result from the summation of several force components. The departure from the central limit theorem stems from the fact that the loads over the structures are correlated, which is in contradiction to the premise of the theorem.

The probability density function of pressure fluctuations with large skewness observed in separated flow regions is critical to the modeling of loads on cladding and components, as well as loads on main load-resisting systems (Figure 9.4). Often researchers have attempted to use the lognormal distribution

for pressure fluctuations, but this fails to represent the tail region with high fidelity (Gurley et al., 1997). Among different alternatives, including the maximum entropy-based model, it is noted in Figure 9.4 that the Hermite moment-based distribution provides the best match to the data, especially for the negative tails (Gurley et al., 1997). This model also captures the significance of non-Gaussian pressure fluctuations for determining the equivalent constant pressure for glass design (Gurley et al., 1997), while providing a useful format to account for non-Gaussianity in the estimation of wind-induced fatigue damage, as characterized by a correction factor (Gurley et al., 1997). The non-Gaussian effects result in enhanced local loads and may lead to increased expected damage in glass panels and higher fatigue effects on other components of cladding (Gurley et al., 1997). Progress in quantifying and simulating the non-Gaussian effects of wind on structures has been elusive due to the limitations of traditional analytical tools (Gurley et al., 1996; Grigoriu, 1998).

9.5 Mechanical/Convective Turbulence

The origin of atmospheric turbulence can be classified into two broad categories, i.e., mechanical and convective. While the former results from the shearing action of the flow field by protuberances from ground surface such as urban developments, vegetation, and, in the case of oceans, sea surface roughness. This is the most common type of turbulence studied in the area of wind effects on structures as it manifests different levels of turbulence structure depending on the terrain, topographical and surface features. The level of turbulence is characterized by the turbulence intensity, length scale and spectral and correlation structure. The other type of turbulence evolves from convective origins resulting from atmospheric instabilities associated with atmospheric dynamics. One is more familiar with this type of turbulence while going through cloud cover or around a thunderstorm in a plane. In contrast with the boundary layer type flows, in these storms, the convectively driven flow fields have their own unique character. Thunderstorm winds, downbursts, and hurricane winds have a significant component that results from convective effects. Often the mechanical component is dwarfed by the convective effects due to the transient nature of these flows which do not have time to evolve into well-developed flows and may not reflect the terrain characteristics of the boundary.

The main challenge in this area remains on the structure of turbulence in extreme wind events like hurricanes, tornadoes, downdrafts, and gust fronts. Questions remain unresolved as to the profiles of the mean flow and structure of these events which may play a major role in quantifying wind load effects on structures. Discerning the role of convective turbulence may become equally important as we obtain a better understanding of the overall turbulent structure in these extreme wind events and its ramification on the established loading models that rely on the structure of mechanical turbulence only (e.g., Hogstrom et al., 2002).

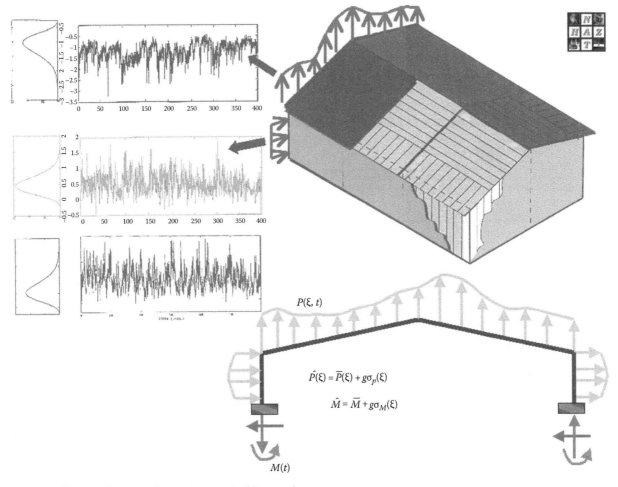

$$\hat{P}(\xi) = \overline{P}(\xi) + g\sigma_p(\xi)$$

$$\hat{M} = \overline{M} + g\sigma_M(\xi)$$

FIGURE 9.3 Schematic of pressure fluctuations on a building envelope.

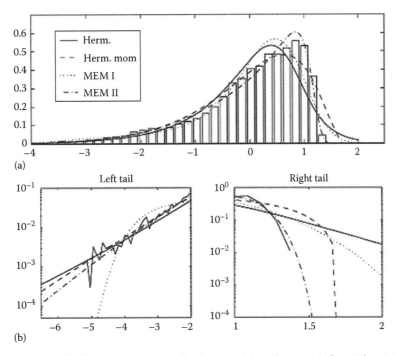

FIGURE 9.4 (a) PDF of pressure fluctuations in a separated region; (b) close-up of the tail regions. (After Gurley, K. et al., Analysis and simulation tools for wind engineering, *Probabilist. Eng. Mech.*, 12(1), 9, 1997.)

9.6 Effects of Turbulence

The role of turbulence in the aerodynamics of stationary bluff bodies has been extensively documented in the literature. Studies have shown that the flow around bluff bodies is governed by the separation and reattachment of the shear layers and by vortex shedding. The shear-layer thickness and the body size are length scales associated with these two phenomena, respectively. Turbulent eddies on the order of these scales are most effective at altering flow structure. The main effect of small-scale turbulence is to cause earlier reattachment of the flow through enhanced mixing in the shear layers. Turbulence in the range of the body scale can enhance or weaken vortex shedding depending on the body geometry. A schematic diagram of the multiscale fluctuations involved in typical bluff body wind interactions is shown in Figure 9.5. Clearly there are three distinct bands of frequency that are characterized by the incident, shear layer or near wake and the wake fluctuations.

9.6.1 Basic Studies

Because bridges and buildings often have rectangular or near rectangular shapes, a notable amount of wind tunnel work has been done studying varieties of rectangular bodies and the shear layers separating from them. Most of this work has been done for stationary bodies, but some has been for prisms in motion. The role of turbulence in the aerodynamics of stationary bluff bodies has been extensively documented in the literature (e.g., Gartshore, 1973; Lee 1975; Kareem and Cermak, 1979; Saathoff and Melbourne, 1989; Nakamura, 1993). While experimental studies focusing on the effects of turbulence on bridge aerodynamics (e.g., Scanlan and Lin, 1978; Huston, 1986; Matsumoto et al., 1991; Larose et al., 1993; Haan et al., 1999) have typically noted an increase in critical wind velocity with added turbulence, several studies have reported destabilizing trends associated with turbulence (Matsumoto et al., 1991; Huston, 1986). The cause of these disagreements remains to be

conclusively determined and signifies the limitations in current understanding of the problem (e.g., Scanlan, 1997; Chen et al., 2000; Haan and Kareem, 2009).

9.6.2 Bridge Aerodynamics Studies

More recent studies investigating the effects of turbulence on rectangular sections have focused on both stationary and/or oscillating prisms with applications to improved understanding of building and bridge aerodynamics to turbulence (Larose, 2003; Haan and Kareem, 2009). This study investigated the distribution of the pressure field around the prism in both chordwise and spanwise directions, including their correlations and in bridge related studies the correlation of the pressure field in comparison with upstream turbulence. In the following, a summary of some of the findings of a recent study, which has not been widely reported in the literature thus far, is presented. It brings out some useful observations and important ramifications for the current state of the art of aerodynamic analysis of long span bridges (Haan and Kareem, 2009). In this study, grid-generated turbulence was used to study the effects of both turbulence intensity and turbulence scale on aerodynamic forces. By examining the unsteady pressure distributions over the bridge model rather than the flutter derivatives alone, a clearer understanding of how turbulence affects the unsteady forces was obtained. Both increasing turbulence intensity and turbulence scale decreased the amplitudes of self-excited pressure fluctuations. The basic shape of the chordwise distributions of pressure amplitude—a single hump shape—is shifted upstream with increasing intensity of the free stream turbulence. This shift, however, was only slightly affected by the turbulence scale. Figure 9.6 shows a space time portrait of the pressure distribution highlighting this observation.

Haan and Kareem (2009) also observed that the phase values of the self-excited pressure (with respect to the body motion) had several regimes in the streamwise direction. Near the leading edge, phase was nearly constant. Downstream of this locale was a region where phase increased rapidly. Beyond this rapidly increasing phase zone was a region where the phase values leveled off and even decreased in some cases. While scale had little discernible effect on this phase value, turbulence intensity shifted the region of rapidly increasing phase toward the leading edge. The stabilizing effects of turbulence observed in the flutter derivatives were related to these turbulence-induced shifts in the pressure amplitudes and phase distributions. By tracking the integrands of expressions for lift and moment, specific changes in pressure amplitude and phase were linked to flutter derivative modifications. In addition, this upstream shifting in the unsteady pressure on the oscillating models was found to be similar to the behavior observed for pressure distributions over stationary models. This suggests that the vast amount of research done in bluff body aerodynamics on stationary bodies may aid in the understanding of oscillating body problems as well.

Prior to this study, no experimental study had justified the conventional analysis technique in which the aerodynamic forces

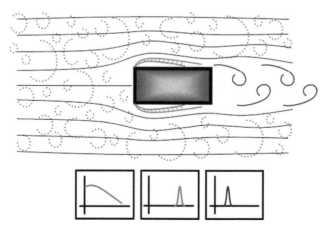

FIGURE 9.5 Schematic of flow around a bluff cross section highlighting three distinct regions of fluctuations.

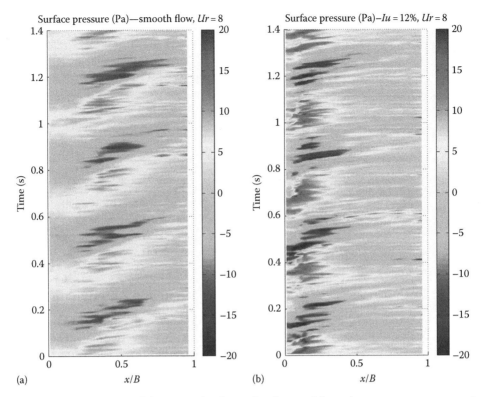

FIGURE 9.6   (See color insert.) Surface plots of the temporal and spatial evolution of the surface pressure on a rectangular prism undergoing torsional oscillations (*x/B*—streamwise direction/deck width: (a) smooth flow; (b) turbulent flow). (After Haan, Jr., F.L. and Kareem, A., *J. Eng. Mech.*, 135(9), 987, 2009.)

are separated into flutter and buffeting components. Pressure measurements made on oscillating models allowed experimental assessment of this assumption. Overall, the assumption is quite close. Examination of the lift and moment spectra showed close agreement throughout the frequency range considered. Where the stationary and oscillating model spectra did not agree, however, the oscillating model values were larger. This oscillation-induced increase in the broadband energy occurred mainly for frequencies above $fD/U = 0.1$ although some differences were observed for lower frequencies as well.

Quantitative analysis of these differences showed that buffeting forces on oscillating model could have RMS values as much as 10% higher than their stationary model counterparts. This difference decreased for increased turbulence intensity and increased turbulence scale. For the flow with the highest intensity and scale considered, these differences were only around 2%–3%. Observation of the streamwise distribution of such differences revealed that the location of oscillation-induced broadband increase was upstream of reattachment—for this case, this meant upstream of the location of the maximum RMS pressure value. This implies that bodies which experience separation over smaller portions of their surface may exhibit less significant differences between stationary and oscillating model buffeting levels.

Through both coherence measurements and correlation calculations of self-excited force components, the self-excited forces were found to have near unity coherence over the entire

spanwise separation range considered. It was noted that cases with larger turbulence length scale showed a slightly lower correlation than those of smaller scales; the estimated 95% confidence intervals of + 0.03 puts all results within the statistical spread of the others suggesting no influence of length scale. The conventional assumption of self-excited forces being fully correlated in the spanwise direction was thus supported by the results of this study. Of course, this also means that the often-suggested hypothesis that a decrease in spanwise correlation of the self-excited forces causes the turbulence-induced increase in the critical flutter velocity is not supported by the current results. A conclusive investigation of self-excited force coherence would require much longer span lengths to observe whether appreciable changes occur for longer spanwise separations. Coherence calculations also showed that the broadband coherence of the oscillating model matched that of the stationary model to within the uncertainty of the experiment.

9.6.3 Impact on Bridge Flutter

While experimental studies focusing on the effects of turbulence on bridge aerodynamics have typically noted an increase of critical wind velocity with added turbulence, several studies have reported destabilizing trends associated with turbulence. The cause of these disagreements remains to be conclusively determined and signifies the limitations in the current understanding of the problem.

Flutter studies utilizing full-bridge models have often shown critical velocity boundaries to be sharply defined in smooth flows and higher or nonexistent in turbulent flows. This has often been attributed to a decrease in spanwise coherence of the self-excited forces due to the turbulence, and a recent study by Scanlan (1997) has shown this to be plausible analytically. Although the stabilizing effect of spanwise correlation loss may be apparent for single-mode torsional flutter (Scanlan, 1997), it is not obvious that this will apply to multimode coupled flutter cases. Correlation loss along the span may stabilize a deck by reducing negative damping effects, and yet it may destabilize a deck by reducing favorable damping (Chen and Kareem, 2003). This issue will become even more important as bridge spans grow longer and multimode flutter becomes more probable.

While spanwise correlation of pressure and buffeting forces on static models in turbulent flows have been measured by a number of researchers, little if any experimental work has been done on the spanwise behavior of the self-excited forces. Full correlation of the self-excited forces is generally assumed in the response analysis, as alluded to earlier by findings in Haan et al. (1999). This suggests, however, that the turbulence-induced behavior of full-bridge models mentioned previously cannot be explained entirely due to a decrease in self-excited force coherence. In view of this, there is a clear need for improved knowledge of the basic physical phenomenon. Analysis framework based on recent developments in bridge aerodynamics may facilitate improved understanding in the subject area (Chen and Kareem, 2006a,b; Wu and Kareem, 2011).

In closing this section, it is very important to note that the preceding anthology of the role of turbulence on the bluff body aerodynamics of prismatic bodies, with applications to buildings and bridges, excluded a vast amount of literature concerning circular cylinders, which is available in a number of review articles. Furthermore, while most of this work has focused on experimental studies, there have been remarkable developments in the area of computational wind engineering centered on the flow around bluff bodies involving 2D and 3D prisms and the role of turbulence (e.g., Yu and Kareem, 1997; Murakami and Mochida, 1999; Tamura and Ono, 2003).

9.7 Changing Dynamics of Aerodynamics

The subject of aerodynamics has been treated traditionally by invoking quasi-steady and strip theories and has been extended to unsteady aerodynamic theories for loads originating from wake induced aeroelastic effects. The current challenges are to address aerodynamics in transient flows. In the following, a brief discussion of these areas is provided with an outlook for their treatment of transient conditions, lessons one may learn from aerodynamics in nature and the role of aerodynamic/aeroelastic tailoring.

9.7.1 Quasi-Steady Aerodynamics

Quasi-steady and strip theories offer reliable estimates of load effects when the dominant mode of loading is attributed to buffeting, e.g., surface pressure responding to large-scale, low-frequency turbulence for the along-wind buffeting load effects. The quasi-steady theory fails to relate the approach flow and the ensuing pressure fluctuations on surfaces in separated flow regions. The shortcoming of the theory stems from the fact that it does not account for the wind–structure interactions at several scales, which may introduce additional components, thus highlighting the need for unsteady aerodynamics. The quasi-steady theory has been successfully applied to the analysis of galloping behavior of structures, often observed in structures exposed to winds at high reduced velocity, e.g., ice-coated transmission line cables and cables experiencing skewed winds.

9.7.2 Unsteady Aerodynamics

Notwithstanding the improved knowledge of wind effects on structures over the past few decades, our understanding of the mechanisms that relate the random wind field to the various wind-induced effects on structures has not developed sufficiently for functional relationships to be formulated. Not only is the approach flow field very complex, the flow patterns generated around a structure are complicated by distortion of the wind field, flow separation, vortex formation, and wake development. Nonlinear interaction between the body motion and its wake results in the "locking in" of the wake to the body's oscillation, resulting in vortex-induced vibrations over a range of wind velocities. The stability of aeroelastic interactions is of crucial importance (Matsumoto, 1999). In an unstable scenario, the motion-induced loading is further reinforced by the body motion, possibly leading to catastrophic failure. Depending on the phase of the force with respect to the motion, self-excited forces can be associated with displacement, velocity, or acceleration. Furthermore, aeroelastic effects can couple modes that are not coupled structurally, leading to more complex issues in bridge aerodynamics as discussed earlier. The intractability of these unsteady aerodynamic features has led to experimental determination of these effects using scale models in wind tunnels. These measurements have led to loading characterizations in terms of spectral distributions of local and integral load effects and aerodynamic flutter derivatives, which, when combined with structural analysis, yield measures of overall structural behavior.

Concerning building aerodynamics, the most widely used wind tunnel technology involving the high-frequency base balance (HFBB) does not include motion-induced loads. Kareem (1982a) noted in a validation study concerning the crosswind spectra derived from statistical integration of surface pressure that the response estimates computed by using the measured spectra began to depart from the estimated values based on aeroelastic model test of the same building at reduced velocities above 6. The damping estimates from the aeroelastic model suggested a constant increase in the negative aerodynamic damping. By including the negative aerodynamic damping, the response predictions provided a better comparison with the aeroelastic tests at higher reduced velocities. It is also important not to simply attribute all motion-induced effects to aerodynamic damping. Indeed the motion of a structure also modifies the flow field around it. Particularly it tends to enhance the spanwise pressure correlation, which may lead to an increased forcing in comparison with that measured by a force balance. A recent study critically evaluates the role of several modeling parameters needed to accurately model aeroelastic effects for base pivoted models (Zhou and Kareem, 2003).

In attempts to relate the incident turbulence to pressure fluctuations in separated regions, higher-order modeling via bi-spectral approach has been utilized, which has identified some correlation although it has not provided a functional relationship. It is noteworthy that that the intermittent nature of the unsteady aerodynamic relationship is vitiated using Fourier-based analysis (Gurley et al., 1997). Higher-order bi-spectral analysis utilizing a wavelet basis offers the promise to capture intermittent relationship between the incident turbulence fluctuations and the attendant pressures under separated flows (Gurley et al., 2003). Thus, challenges remain in establishing transfer functions that could relate, at higher orders in a localized basis, the complexities inherent to bluff body aerodynamics.

9.7.3 Transient Aerodynamics

Both earlier and more recent studies in fluid dynamics have pointed out an overshoot in aerodynamic/hydrodynamics loads on cylinders in unsteady flows (Sarpakaya, 1963). It has also been noted that for the analysis of structures in nonstationary atmospheric turbulence, the traditional stationary analysis fails to account for possible transient overloads, e.g., the sharp changes in gusts were found to cause a transient aerodynamic force on a bridge model, which cannot be explained by a stationary statistical analysis. This clearly points at the need to critically assess the impact of abrupt changes in the wind field magnitudes and associated modifications in aerodynamics of structures and appraises the need for refining the current load descriptions.

9.7.4 Aerodynamics in Nature: Some Lessons

Natural objects with bluff profiles like trees, plants, and jellyfish experience forces in a flowing medium like air or water. Often these objects have the ability to minimize drag by shape reconfiguration, e.g., palm trees bend and through their compliant action survive the fury of hurricanes. This is not an option available to most of the built environment as their shapes are fixed. In architectural and structural communities tapered, twisted, and tilted towers are being built. The fixed shapes used in man-made structures are arrived at on the basis of aerodynamic considerations, but are undertaken a priori and the possibilities of changes in real time during wind storm are not available. Other possible avenues of managing flow around buildings are possible by using biomimetics through introducing scalloped leading edges like whale flippers, which act like vortex generators.

Unfortunately, these fundamental phenomena in nature may not benefit the built environment. As in nature, more flexibility may help survive adverse environmental conditions through excessive bending; however, tall buildings or bridges may not be able to afford this luxury as they may lose their functionality, e.g., occupants of a building experiencing large excursions of motion may get symptoms of sea sickness leading to discomfort and in extreme cases nausea and vertigo. However, such flexible behavior has been utilized in offshore drilling platforms, where the human comfort considerations are not as stringent.

9.7.5 Aerodynamic/Aeroelastic Tailoring

Aerodynamic tailoring can be used to achieve optimal performance of buildings and bridges under winds like it is used routinely in the aerospace and automobile industries. Following the profiles of Japanese pagodas in some respects, recent examples have shown that these considerations can be integrated into the design of tall buildings without sacrificing their appearance and often creating an aerodynamically eDcient signature of the structure, e.g., Taipei 101 in Taipei, Taiwan; Jin Mao Building, Shanghai, PROC; Petronas Towers, Kuala Lumper, Malaysia; International Commerce Center, Hong Kong (Figure 9.7).

Modifications to the corner geometry and building shape, e.g., inclusion of chamfered corners, slotted corners, and softened corners have been found to considerably reduce the response of buildings, in comparison with the performance of a basic square plan, with these improvements becoming more marked as the corners are progressively rounded. Improvements have been noted in buildings that vary their cross-sectional shape with height or which reduce their upper level plans through tapering effects, cutting corners, or progressively dropping off corners with height (e.g., Kareem et al. 1999).

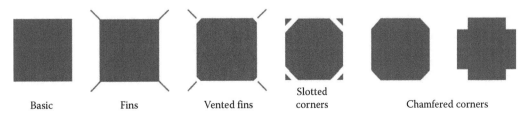

Basic   Fins   Vented fins   Slotted corners   Chamfered corners

FIGURE 9 .7 Aerodynamic modifications to square building shape. (After Kareem, A. et al., *Wind Struct.*, 2(3), 201, 1999.)

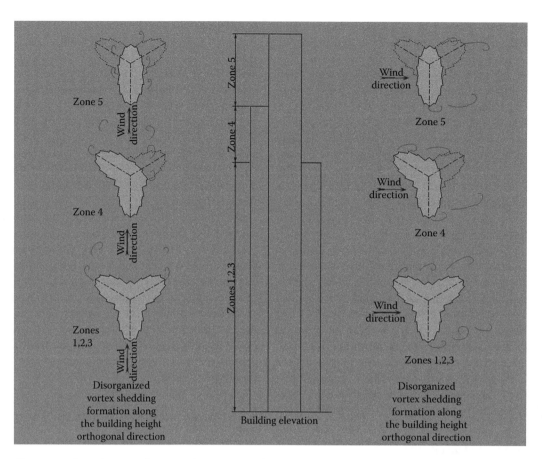

FIGURE 9.8 Illustration of aerodynamic shaping to disrupt vortex shedding along the height. (After Abdelrazak, A.K. et al., Design and full-scale monitoring of the tallest building in Korea: Tower Palace III, *Proceedings of the Sixth Asia-Pacirc Conference on Wind Engineering*, Seoul, Korea, September 12–14, 2005.)

Evidence of the aesthetic benefits of aerodynamic modifications is provided by more recent structures with setbacks along the height including Tower Palace III in Seoul and Burj Khalifa, a 2716 ft plus tall building in Dubai. The influence of vortex-induced excitations was minimized by deterring the formation of a coherent wake structure, as demonstrated conceptually in Figure 9.8. This favorable aerodynamic behavior of the tower was achieved by varying the building shape along the height which offered building cross sections that were aerodynamically less conducive to vortex shedding and lacked dynamic interaction of vortices from both sides of the cross section (Abdelrazak et al., 2005). In the case of Burj Khalifa, the basic footprint of the building represents petals of a desert flower with setbacks that follow a spiral along the height, further vitiating the impact of any potentially coherent formation of vortices along the height.

The inclusion of complete openings through the building, particularly near the top, provides yet another means for improving aerodynamic response by significantly reducing wake-induced forces. Such a design strategy has been utilized in several buildings, most noteworthy of which is the Shanghai World Financial Center, which features a diagonal face that is shaved back with a rectangular aperture at the top of the building. The next generation of high-rise buildings and bridges will heavily rely on implementing fundamental principles of aerodynamics in the design process aided by the wind tunnel and computational fluid dynamics (CFD) as tools for tailoring/sculpturing and shape optimization of building shapes to better manage the demand posed by survivability, serviceability, and human comfort perspectives (e.g., Soobum et al., 2011). Similarly in bridge aerodynamics, the shape of a bridge deck section highly dictates its aerodynamic performance. The decks of planned bridges spanning over large water bodies involve aerodynamically tailored box sections with openings in the middle to modify the flow around the deck in an effort to minimize aerodynamic loads and shift flutter to higher wind speeds.

9.8 Analysis, Computational, Identification, and Modeling Tools and Frameworks

9.8.1 Computational Fluid Dynamics

The large eddy simulation (LES) framework is emerging as a numerical scheme of choice for the solution of Navier–Stokes equations (Murakami and Mochida, 1999). For example, recent studies have shown that the simulated pressure field around prisms has convincingly reproduced experimentally observed

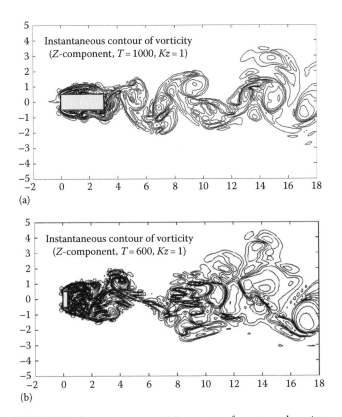

FIGURE 9.9 Instantaneous vorticity contours for rectangular prisms with aspect ratio: (a) 0.3 and (b) 3.0. (After Yu, D. et al., Numerical investigation of the influence of aspect ratio on flow around bluff bodies, *Proceedings of the Computational Wind Engineering*, Yokohama, Japan, July 2006.)

characteristics. This is true with respect to variations in the mean and RMS pressure coeDcients of the drag force and regions of flow reattachment, even as the aspect ratio of the prism is varied (Yu and Kareem, 1997; Murakami, 1998). Other computational schemes like detached eddy and hybrid simulation schemes are offering other opportunities for improved simulations. Figure 9.9 shows instantaneous contours of vorticity for prisms of aspect ratios: 3:1 and 1:3 based on 3D LES schemes. In Figure 9.10 time histories of the drag and lift forces along with their respective PSD and time-frequency scalograms are shown. Examination of the drag and lift force time histories for the selected aspect ratios reveal distinctive features both above and below the critical aspect ratio of 0.62. For aspect ratios below the critical value intermittent behavior (periods of low and high drag force) occurs, while beyond the critical value, the drag force has less intermittence. The corresponding spectral descriptions show a global distribution of energy. The scalograms (Figure 9.11) reaDrm the intermittent bursts of energy for aspect ratio 3:1 and relatively more evenly distributed fluctuations for the 1:3 prism.

One of the challenges remaining in this field entails the inclusion of surrounding structures to capture the influence of flow modification, shielding and interference, as well as the simulation of inflow turbulent boundary layer flow conditions. Recently, advances have been made in these areas, including modeling of aeroelastic instabilities, CFD techniques for practical applications, and modeling of urban roughness and terrain effects (Tamura et al., 2002; Tamura and Ono, 2003). Coupled with computer-aided flow animations, such simulation techniques may in the near future provide "numerical wind tunnels"

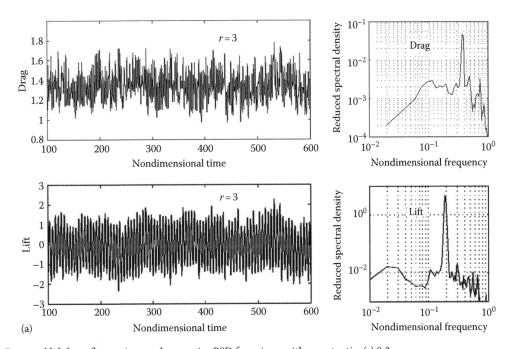

FIGURE 9.10 Drag and lift force fluctuations and respective PSD for prisms with aspect ratio: (a) 0.3;

(*continued*)

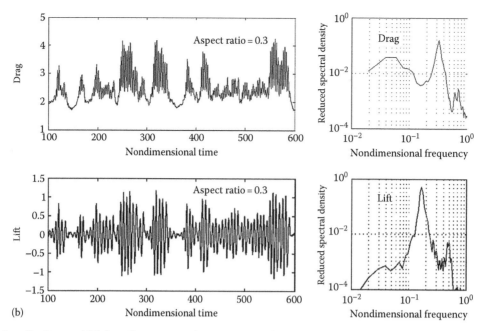

FIGURE 9.10 (continued) Drag and lift force fluctuations and respective PSD for prisms with aspect ratio: (b) 3.0. (After Yu, D. et al., Numerical investigation of the influence of aspect ratio on flow around bluff bodies, *Proceedings of the Computational Wind Engineering*, Yokohama, Japan, July 2006.)

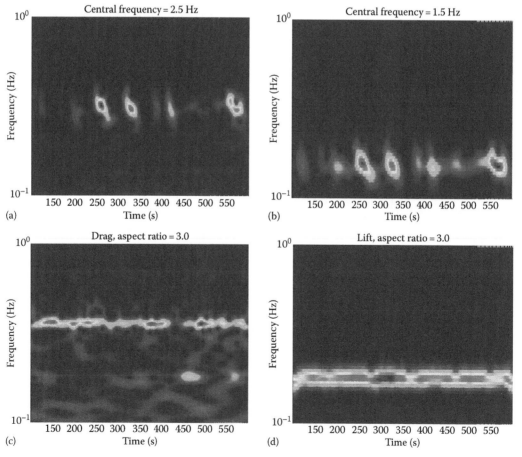

FIGURE 9.11 Wavelet-based scalogram of drag and lift fluctuations for aspect ratio 0.3 (a, b) and 3.0 (c, d), respectively. (After Yu, D. et al., Numerical investigation of the influence of aspect ratio on flow around bluff bodies, *Proceedings of the Computational Wind Engineering*, Yokohama, Japan, July 2006.)

to analyze the evolution of flow around structures and estimate attendant load effects. More recently, developments in CFD on the cloud are evolving (vortex-winds.org).

9.9 Concluding Remarks

This chapter reflected on recent developments in aerodynamics and aeroelasticity of bluff bodies with a viewpoint concerning wind effects on structures. Some of the overarching issues for the future encompass the modeling and impact of imperfect correlation, transient/nonstationary effects, non-Gaussian and nonlinear features, coupled load effects in the formulation of wind loads and their effects on structures. Advances in computational and experimental methods with improved sensing capabilities based on wireless sensors, smart sensors, GPS, laser Doppler-based technologies, and optical and sonic systems would aid in effectively addressing some of the needs highlighted. With foundations steeped in bluff body aerodynamics and aeroelasticity theories and projected research paths imbued in advanced and e-technologies, our abilities will certainly be lifted up to a higher level of understanding, modeling, analysis, design, and monitoring of wind effects on structures.

References

Abdelrazak, A.K., Kijewski-Correa, T., Song, Y.-H., Case, P., Isyumov, N., and Kareem, A. (2005). Design and full-scale monitoring of the tallest building in Korea: Tower Palace III, *Proceedings of the Sixth Asia-Pacirc Conference on Wind Engineering*, Seoul, Korea, September 12–14.

Brooks, H.E., Doswell III, C.A., and Kay, M.P. (2001). Climatology of tornadoes and severe thunderstorm winds in the United States, *Proceedings of the 1st American Conference on Wind Engineering*, Clemson, Lubbock, TX, June 4–6.

Butler, K., Cao, S., Tamura, Y., Kareem, A., and Ozono, S. (2010). Characteristics of surface pressures on prisms immersed in a transient gust front flow field, *J. Wind Eng. Ind. Aerodyn.*, 98, 299–316.

Carassale, L. and Kareem, A. (2010). Modeling nonlinear systems by volterra series, *J. Eng. Mech.*, 136(6), 801–818.

Cermak, J.E. (1975). Application of fluid mechanics to wind engineering—A Freeman Scholar Lecture, *J. Fluids Eng.*, 97, 9–38.

Chen, X. and Kareem, A. (2001). Aeroelastic analysis of bridges under multi-correlated winds: Integrated state-space approach, *J. Eng. Mech.*, 127(11), 1124–1134.

Chen, X. and Kareem, A. (2003a). New frontiers in aerodynamic tailoring of long span bridges: An advanced analysis framework, *J. Wind Eng. Ind. Aerodyn.*, 91, 1511–1528.

Chen, X. and Kareem, A. (2003b). Aeroelastic analysis of bridges: Turbulence effects and aerodynamic nonlinearities, *J. Eng. Mech.*, 129(8), 885–895.

Chen, X. and Kareem, A. (2006a). Revisiting multimodal coupled bridge flutter: Some new insights, *J. Eng. Mech.*, 132(10), 1115–1123.

Chen, X. and Kareem, A. (July 2006b). Understanding the underlying physics of multimode coupled bridge flutter based on closed-form solutions, *Proceedings of Computational Wind Engineering 2006*, Yokohama, Japan.

Chen, L. and Letchford, C.W. (2005). Proper orthogonal decomposition of two vertical profiles of full-scale nonstationary downburst wind speed, *J. Wind Eng. Ind. Aerodyn.*, 93(3), 187–216.

Chen, X., Matsumoto, M. and Kareem, A. (2000). Time domain flutter and buffeting response analysis of bridges, *J. Eng. Mech.*, 126, 7–16.

Davenport, A.G. (1967). Gust loading factor, *J Struct. Div.*, 93(3), 11–34.

Diana, G., Cheli, F., Zasso, A., and Bocciolone, M. (1999). Suspension bridge response to turbulent wind: Comparison of a new numerical simulation method results with full scale data, *Proceedings of the Tenth International Conference on Wind Engineering: Wind Engineering into the 21st Century*, Larsen et al. (eds.), Balkema, Rotterdam, the Netherlands, pp. 871–878.

Fujita, T. (1985). *The Downburst*, University of Chicago, Chicago, IL.

Gartshore, I.S. (1973). The effects of free stream turbulence on the drag of rectangular two-dimensional prisms, *Engineering Science Research Report*, University of Western Ontario, Ontario, Canada, BLWT-4, Vol. 73–74, 24p.

Gast, K., Schroeder, K., and Spercell, J.L. (June 2003). Rear-flank downdraft as sampled in the 2002 Thunderstorm Outflow Experiment, *Proceedings of 11ICWE*, Lubbock, TX.

Grigoriu, M., (1998). Simulation of stationary non-Gaussian translation processes, *J. Eng. Mech.*, 124(2), 121–126.

Gurley, K. and Kareem, A. (1998). A conditional simulation of non-normal velocity/pressure fields, *J. Wind Eng. Ind. Aerodyn.*, 67–68, 673–684.

Gurley, K., Kareem, A., and Tognarelli, M.A. (1996). Simulation of a class of non-normal random processes, *J. Nonlinear Mech.*, 31(5), 601–617.

Gurley, K., Kijewski, T., and Kareem, A. (2003). First and higher-order correlation detection using wavelet transforms, *J. Eng. Mech.*, 129(2), 188–201.

Gurley, K., Tognarelli, M.A., and Kareem, A. (1997). Analysis and simulation tools for wind engineering, *Probabilist. Eng. Mech.*, 12(1), 9–31.

Haan, Jr., F. L. and Kareem, A. (2009). Anatomy of turbulence effects on the aerodynamics of an oscillating prism, *J. Eng. Mech.*, 135(9), 2009, 987–999.

Haan, F.L., Kareem, A., and Szewczyk, A.A. (1999). Influence of turbulence on the self-excited forces on a rectangular cross section. In *Proceedings of the Tenth International Conference on Wind Engineering, Wind Engineering into the 21st Century*, Larsen et al. (eds.), Balkema, Rotterdam, the Netherlands, pp. 1665–1672.

Högström, U., Hunt, J.C.R., and Smedman, A.S. (2002). Theory and measurements for turbulence spectra and variances in the atmospheric neutral surface layer. *Bound. Layer Meteorol.*, 103(1), 101–124.

Huston, D.R. (1986). The effect of upstream gusting on the aeroelastic behavior of long-span bridges, PhD dissertation, Princeton University, Princeton, NJ.

Irwin, P. (1977). Wind tunnel and analytical investigation of the response of Lions Gate Bridge to turbulent wind, NRC of Canada, NAE-LTR-LA-210.

Kareem, A. (1982). Across wind response of buildings, *J. Struct. Eng.*, 108(4), 869–887.

Kareem, A. (January 2010), Bluff body aerodynamics and aeroelasticity: A wind effects perspective, *J. Wind Eng.*, 7(1), 30–74.

Kareem, A., Kijewski, T., and Tamura, Y. (1999). Mitigation of motion of tall buildings with specific examples of recent applications, *Wind Struct.*, 2(3), 201–251.

Kareem, A., Tognarelli, M.A., and Gurley, K. (1998). Modeling and analysis of quadratic term in the wind effects on structures, *J. Wind Eng. Ind. Aerodyn.*, 74–76, 1101–1110.

Kitigawa, M., Shiriashi, N., and Matsumoto, M. (1982). Fundamental study on transient properties of lift force in an unsteady flow, *Proceedings of the 37th Annual Conference of the Japan Society of Civil Engineering*, Kyoto, October 1982 (in Japanese).

Kwon, D., Kijewski-Correa, T., and Kareem, A. (2005), e-Analysis/design of tall buildings subjected to wind loads, *Proceedings of the 10th Americas Conference on Wind Engineering*, Baton Rouge, LA.

Larose, G.L., Davenport, A.G., and King, J.P.C. (1993). On the unsteady aerodynamic forces on a bridge deck in turbulent boundary layer flow, *Proceedings of the 7th U.S. National Conference on Wind Engineering*, UCLA, Los Angeles, CA.

Matsumoto, M. (1999). Recent study on bluff body aerodynamics and its mechanisms. In *Proceedings of the Tenth International Conference on Wind Engineering, Wind Engineering into the 21st Century*, Larsen et al. (eds.), Balkema, Rotterdam, the Netherlands.

Matsumoto, M., Shiraishi, N., and Shirato, H. (1991). Turbulence unstabilization on bridge aerodynamics, *Proceedings of the International Conference Innovation in Cable-Stayed Bridges*, Fukuoka, Japan, pp. 175–183.

Murakami, S. and Mochida, A. (1999). Past, present, and future of CWE: The view from 1999. In *Proceedings of the Tenth International Conference on Wind Engineering, Wind Engineering into the 21st Century*, Larsen et al. (eds.), Balkema, Rotterdam, the Netherlands, pp. 91–104.

Nakamura, Y. (1993). Bluff-body aerodynamics and turbulence, *J.eWind Eng. Ind. Aerodyn.*, 49, 65–78.

Sarpkaya, T. (1963). Lift, drag, and mass coeDcients for a circular cylinder immersed in time dependent flow, *J. Appl. Mech., Ser. E*, 85, 13–15.

Saathoff, P.J. and Melbourne, W.H. (1989). The generation of peak pressure in separated/reattaching flows, *J. Wind Eng. Ind. Aerodyn.*, 32, 121–134.

Scanlan, R.H. (1997). Amplitude and turbulence effects on bridge flutter derivatives, *J. Struct. Eng.*, 123(2), 232–236.

Soobum, L., Andres, T., Renaud, J., and Kareem, A. (July 2011). Topological optimization of building structural systems and their shape optimization under aerodynamic loads, *Proceedings of the 13th International Conference on Wind Engineering*, Amsterdam, the Netherlands.

Tamura, T. and Ono, Y. (2003). LES Analysis on aeroelastic instability of prisms in turbulent flow, *J. Wind Eng. Ind. Aerodyn.*, 91, 1827–1826.

Twisdale, L.A. and Vickery, P.J. (1992). Research on thunderstorm wind design parameters, *J. Wind Eng. Ind. Aerodyn.*, 41, 545–556.

Wang, L. and Kareem, A. (2005). Modeling and simulation of transient winds: Downbursts/hurricanes, *Proceedings of the 10th American Conference on Wind Engineering*, Baton Rouge, LA.

Wood, G.S. and Kwok, K.C.S. (1998). A empirically derived estimate for the mean velocity profile of a thunderstorm downburst, *Proceedings of the 7th Australian Wind Engineering Society Workshop*, Auckland, New Zealand.

Wu, T. and Kareem, A. (2011). Modeling hysteretic nonlinear behavior of bridge aerodynamics via cellular automata nested neural network, *J. Wind Eng. Ind. Aerodyn.*, 99(4), 378–388.

Xu, Y. L. and Chen, J. (2004). Characterizing nonstationary wind speed using empirical mode decomposition, *J. Struct. Eng.*, 130(6), 912–920.

Yu, D. and Kareem, A. (1997). Numerical simulation of flow around rectangular prisms, *J. Wind Eng. Ind. Aerodyn.*, 67–68, 195–208.

Yu, D. and Kareem, A. (1998). Parametric study of flow around rectangular prisms using LES, *J. Wind Eng. Ind. Aerodyn.*, 77–78, 653–662.

Yu, D., Kareem, A., Butler, K., Glimm, J., and Sun, J. (July 2006). Numerical investigation of the influence of aspect ratio on flow around bluff bodies, *Proceedings of the Computational Wind Engineering*, Yokohama, Japan.

Zhou, Y. and Kareem, A. (2003). Aeroelastic balance, *J. Eng. Mech.*, 129(3), 283–292.

10

Wake–Structure Interactions

C.H.K. Williamson
Cornell University

10.1 Introduction

Wake–structure interactions comprise a number of mechanisms; for example, galloping, flutter and vortex-induced vibrations, and the reader is referred to the excellent book by Naudascher and Rockwell (1994) for a clear description of the different types of flow-induced motion of bodies. In this chapter, we are principally focused on vortex-induced vibrations, or VIV. VIV of structures is of practical interest to many fields of engineering affecting the environment. For example, it can cause vibrations in heat exchanger tubes; it influences the dynamics of riser tubes bringing oil from the seabed to the surface; it is important to the design of civil engineering structures such as bridges and chimney stacks, as well as to the design of marine and land vehicles; it can cause large-amplitude vibrations of tethered structures in the ocean; and also the vibrations of rising or falling bodies. These are a few examples out of a large number of problems where VIV is important. The practical significance of VIV has led to a large number of fundamental studies, most of which are referred to in a comprehensive review paper; Williamson and Govardhan (2004), where further significant reviews and references are listed. The reader is also referred to books by Blevins (1990), Naudascher and Rockwell (1994), and Sumer and Fredsøe (1997). Important new results are presented in a recent brief review by Bearman (2011). Here, we focus on the more recent accomplishments of researchers, especially within the last decade.

In this chapter, we are concerned primarily with the oscillations of an elastically mounted rigid cylinder; with forced vibrations of such structures; with bodies in two degrees of freedom; with the dynamics of cantilevers, pivoted cylinders, cables, and tethered bodies. As a *paradigm* for such VIV systems, we shall consider here an elastically mounted cylinder restrained to move transverse to the flow, as Figure 10.1 shows. As the flow speed (U) increases, a condition is reached when the wake vortex formation frequency (f_V) is close enough to the body's natural frequency (f_N) such that the unsteady pressures from the wake vortices induce the body to respond.

Certain wake patterns can be induced by body motion, such as the 2S mode (two single vortices per cycle, like the classic Karman street) and the 2P mode (comprising two vortex pairs formed in each cycle of body motion), following the terminology introduced in Williamson and Roshko (1988). Interestingly, a forced vibration can also lead to other vortex modes including a P + S mode, which is not able to excite a body into free vibration. In essence, a nominally periodic vibration ensues if the energy transfer, or work done by the fluid on the body over a cycle, is positive. This net energy transfer is influenced significantly by the phase of induced side force relative to body motion, which in turn is associated with the timing of the vortex dynamics. The problem of VIV is therefore a fascinating feedback between body motion and vortex motion. In this chapter, we present not only response phenomena, but also the important vortex dynamics modes leading to the response.

Even in the simple case of the elastically mounted cylinder, many challenging questions exist: (1) What is the maximum possible amplitude attainable for a cylinder undergoing VIV, for conditions of extremely small mass and damping? (2) Under what conditions does the classically employed mass-damping parameter collapse peak-amplitude data? What is the functional shape for a plot of peak amplitude versus mass-damping? (3) What modes of structural response exist, and how does the system jump between the different modes? (4) What vortex dynamics give rise to the different body response modes? (5) What generic

Handbook of Environmental Fluid Dynamics, Volume Two, edited by Harindra Joseph Shermal Fernando. © 2013 CRC Press/Taylor & Francis Group, LLC. ISBN: 978-1-4665-5601-0.

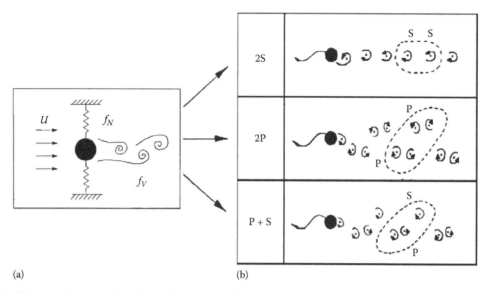

(a)   (b)

FIGURE 10.1 (a) Schematic diagram of an elastically mounted cylinder, restrained to vibrate transverse to the free stream (*Y*-direction). (b) A feedback between the fluid and body motion is now known to lead to several different vortex formation modes, for example, the 2S mode (comprising two single vortices per cycle of motion) or the 2P mode (comprising two vortex pairs per cycle). The P + S mode, ubiquitous in low-*Re* controlled vibration, is never found in free vibration.

features can be discovered that are applicable to all VIV systems? To what extent is the enormous number of studies for bodies restricted to motion transverse to the flow relevant to other, more complex VIV systems? (6) Because almost all of the studies of VIVs are at low and moderate Reynolds numbers, how do these results carry across to high Reynolds numbers? This chapter brings together a number of new phenomena, many of which are related to the earlier-mentioned questions.

10.2 Principles and Methods of Analysis

Numerical simulations are presently used extensively to attack 2D and 3D flows, bodies of long aspect ratio, as well as flexible structures. It is still a challenge to reach Reynolds numbers comparable with many of the experiments ($Re \sim 10{,}000{-}50{,}000$) or full-scale ($Re \sim 10^6$), except in the case where direct numerical simulation (DNS) is replaced by LES computations, for example. On the experimental side, extensive use is made of particle image velocimetry (PIV) technique to determine velocity and vorticity, as well as ingenious means to restrain bodies, with variable damping and springs.

Commonly used in both computation and experiment, here we introduce an equation of motion generally used to represent VIV of a cylinder oscillating in the transverse Y direction (normal to the flow):

$$m\ddot{y} + c\dot{y} + ky = F \qquad (10.1)$$

where
  m is the structural mass
  c is the structural damping
  k is the spring constant
  F is the fluid force in the transverse direction

In the regime where the body oscillation frequency is synchronized with the periodic vortex wake mode (or periodic fluid force), a good approximation to the force and the response is given by

$$F(t) = F_0 \sin(\omega t + \phi) \qquad (10.2)$$

$$y(t) = y_0 \sin(\omega t) \qquad (10.3)$$

where $\omega = 2\pi f$ and *f* is the body oscillation frequency. The response amplitude and frequency may be derived in a straightforward manner from Equations 10.1 through 10.3, yielding equations in terms of a chosen set of nondimensional parameters, as in Khalak and Williamson (1999):

$$A^* = \frac{1}{4\pi^3} \frac{C_Y \sin\phi}{(m^* + C_A)\zeta} \left(\frac{U^*}{f^*}\right)^2 f^* \qquad (10.4)$$

$$f^* = \sqrt{\frac{m^* + C_A}{m^* + C_{EA}}} \qquad (10.5)$$

where
  C_A is the potential added-mass coefficient (taking the value 1.0)
  C_{EA} is an "effective" added-mass coeDcient that includes an apparent effect due to the total transverse fluid force in-phase with the body acceleration ($C_Y \cos\phi$)

$$C_{EA} = \frac{1}{2\pi^3} \frac{C_Y \cos\phi}{A^*} \left(\frac{U^*}{f^*}\right)^2 \qquad (10.6)$$

Quantities in the earlier-mentioned equations are defined in the Appendix. Animated debate often surrounds the definition of added mass (C_{EA}) in these problems. Of course, *it is not a true added mass*, because it has a significant force component due to the vorticity dynamics. Note that the amplitude A^* in Equation 10.4 is proportional to the transverse force component that is in-phase with the body velocity ($C_Y \sin\phi$), and, for small mass and damping, the precise value of the phase angle ϕ has a large effect on the response amplitude.

Feng (1968) contributed some important classic measurements of response and pressure for an elastically mounted cylinder. Figure 10.2a presents his minimum damping case, and it is apparent that there are two amplitude branches, namely, the "initial" branch and the "lower" branch (in the terminology introduced by Khalak and Williamson 1996), with a hysteretic transition between branches. The mass ratio (or relative density) is very large because the experiments were conducted in air

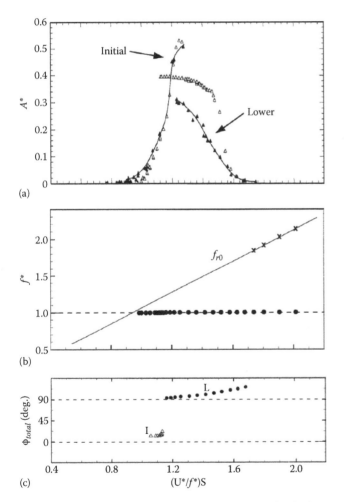

(a)

(b)

(c) $(U^*/f^*)S$

($m^* \sim 250$). Much of the new work we discuss in this chapter comes from the push to explore much smaller mass and damping, over the last 15 years, generally using water as the fluid medium.

Regarding the frequency response, the classical definition of lock-in or synchronization is often perceived as the regime where the frequency of oscillation (f), as well as the vortex formation frequency (f_V), are close to the natural frequency (f_N) of the structure throughout the regime of large-amplitude vibration, so that $f^* = f/f_N \sim 1$ in Figure 10.2b. However, recent studies (shown later) show a dramatic departure from this classical result; bodies can conceivably vibrate with large amplitude, at hundreds of times the natural frequency! Feng also noted that the jump in response amplitude was reflected by a significant jump in the phase of the pressure fluctuations relative to body motion. One might suspect that a jump in phase angle (between transverse force and displacement) through resonance, as shown in Figure 10.2c, will be matched by a switch in the timing of vortex shedding. Zdravkovich (1982) showed this for the first time using selected visualizations from previous studies. An excellent demonstration of this timing switch comes from the comprehensive forced vibration study of Ongoren and Rockwell (1988a), shown in Figure 10.3a, where the switch in timing of vortex formation is evident as the body's frequency is increased through a critical value (roughly $f/f_{VO} = 1.05$, where f_{VO} is the vortex frequency in the absence of vibration). Gu et al. (1994) confirmed this from forced vibrations at small $A^* = 0.2$, in Figure 10.3b, in the groundbreaking first study of this problem using PIV.

It is important to ask what the relationship is between the maximum response amplitude and the system mass and damping. Generally, this information has been plotted as A^*_{max} versus a parameter, S_G, proportional to the product of mass and damping, following the first comprehensive compilation of existing data by Griffin and coworkers in the 1970s, and labeled for convenience as the "Griffin plot" by Khalak and Williamson (1999). Figure 10.3c shows one of Griffin's (1980) original plots, illustrating the characteristic shape whereby the amplitude reaches some limiting value as S_G (reduced damping) becomes small. The logic in choosing a combined mass-damping parameter comes from Equation 10.4 for A^*. For example, Bearman (1984) demonstrated that for large mass ratios ($m^* \gg 1$), the actual cylinder oscillation frequency (f) at resonance will be close to the vortex-shedding frequency for the static cylinder (f_{VO}), and also close to the system natural frequency (f_N), that is, $f \simeq f_{VO} \simeq f_N$, and thus $f^* \simeq 1.0$ (see Equation 10.5 for large m^*). Thus, at resonance, the parameter $(U^*/f^*) = (U/fD) \simeq (U/f_{VO}D) = 1/S$, where S is the Strouhal number of the static cylinder, suggesting a resonance at the normalized velocity, $U^* \simeq 5–6$. Therefore, the assumption is often made that both (U^*/f^*) and f^* are constants, under resonance conditions, giving (from Equation 10.4)

$$A^*_{max} \propto \frac{C_Y \sin\phi}{(m^* + C_A)\zeta} \tag{10.7}$$

FIGURE 10.2 Free vibration of an elastically mounted cylinder at high mass ratios. In (a), we compare the classical response amplitudes of Feng (1968) (triangle symbols), with Brika and Laneville (1993) (open symbols), both at the same ($m^*\zeta$) in air. (b) and (c) show the vibration frequency and phase of the transverse force, as measured in water, but with the same ($m^* + C_A$)$\zeta \sim 0.251$ as used in the air experiments. (From Govardhan, R. and Williamson, C.H.K., *J. Fluid Mech.*, 420, 85, 2000.)

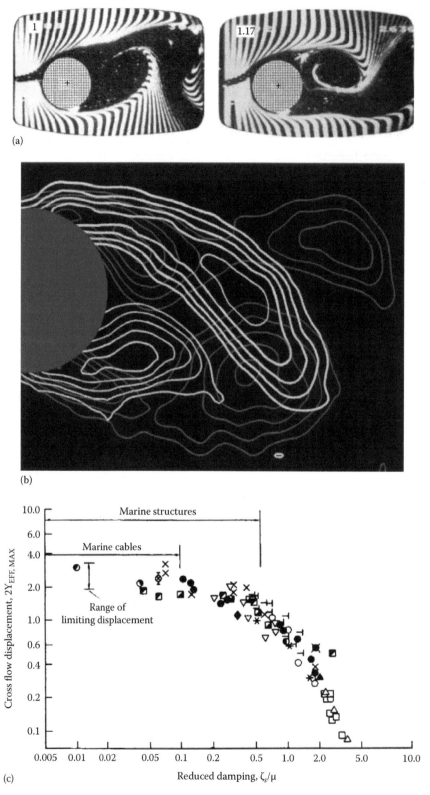

FIGURE 10.3 Some classic results from vortex-induced vibration (VIV). (a) shows the switch in timing of vortex shedding as forced vibration frequency increases. (From Ongoren, A. and Rockwell, D., *J. Fluid Mech.*, 191, 197, 1988a). (b) presents the same phenomenon using the PIV technique, and represents the first use of this technique in the VIV field. (From Gu, W. et al., *Phys. Fluids*, 6, 3677, 1994.) Light vorticity magnitude contours are for low frequency, with the body at the top of its vertical motion. Faded contours are for higher frequencies above a critical value for the switch in timing, again for the body at the top of its motion. (c) presents one of the original GriDn plots, showing peak amplitudes plotted versus the Skop–GriDn parameter (SG). (From GriDn, O.M., *ASME J. Press. Vessel Tech.*, 102, 158, 1980.)

We stress that Equation 10.7 depends on the earlier assumptions remaining reasonable, namely, that $f^* \simeq 1.0$, which is not self-evident.

Aside from studies of elastically mounted structures, one approach to an understanding and possible prediction of vibrations has been to undertake forced vibrations of a structure. A central contribution of Sarpkaya to VIV has been his well-known and much-referenced data set of transverse force coeDcients for controlled sinusoidal vibration of a cylinder transverse to a free stream, at certain fixed amplitudes. Sarpkaya (1978) expressed the transverse force as

$$C_Y = C_{my} \sin \omega t - C_{dy} \cos \omega t \qquad (10.8)$$

where C_{my} and C_{dy} are the inertia (in-phase) and drag (out-of-phase) force coeDcients of the transverse force coeDcient C_Y.

Williamson and Roshko (1988) studied the vortex wake patterns for a cylinder, translating in a sinusoidal trajectory, over a wide variation of amplitudes (A/D up to 5.0) and wavelengths (λ/D up to 15.0). They defined a whole set of different regimes for vortex wake modes, using controlled vibrations, in the plane of $\{\lambda/D, A/D\}$, where a descriptive terminology for each mode was introduced. Each periodic vortex wake pattern comprises single vortices (S) and vortex pairs (P), giving patterns such as the 2S, 2P, and P + S modes, which are the principal modes near the fundamental lock-in region in Figure 10.4. Visualization of the 2P mode is clearly presented in this figure also. Williamson and

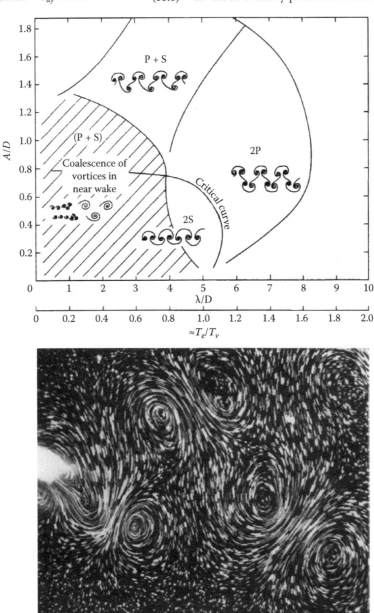

FIGURE 10.4 The map of regimes for vortex wake modes (From Williamson, C.H.K. and Roshko, A. *J. Fluids Struct.* 2, 355, 1988.), showing principally the 2S, 2P, and P + S mode regimes, which are relevant to the fundamental synchronization regime. The 2P mode, comprising two vortex pairs per half cycle, is visualized clearly below the mode map. Cylinder is towed through fluid in a sinusoidal trajectory toward the left.

Roshko (1988) described other patterns, such as those formed by coalescence of vortices, or those formed from more vortices per cycle (e.g., the "2P + 2S," representing a one-third subharmonic mode). The 2P and P + S modes have been found in controlled vibration studies in-line with the flow (GriDn and Ramberg, 1976; Ongoren and Rockwell, 1988b). The P + S mode was also found in GriDn and Rambergß (1974) well-known smoke visualizations for transverse motions. The significance of these modes from controlled vibration is that they provide a map of regimes within which we observe certain branches of free vibration. One deduction from the Williamson and Roshko study was that the jump in the phase ϕ of the transverse force in Bishop and Hassanß (1964) classical forced vibration paper, and also the jump in phase measured in Fengß (1968) free-vibration experiments, were caused by the changeover of mode from the 2S to the 2P mode. This has since been confirmed in a number of free-vibration studies (Brika and Laneville, 1993; Govardhan and Williamson, 2000). Such vortex modes occur for bodies in one or two degrees of freedom, for pivoted rods, cantilevers, oscillating cones, and other bodies. Response data from all of these studies have been correlated with the map of regimes described earlier.

10.3 Applications

10.3.1 Free Vibration of a Cylinder

Brika and Laneville (1993) were the first to show evidence of the 2P vortex wake mode from free vibration, using a vibrating cable in a wind tunnel, with $m^* \sim 100$. However, phenomena at low mass ratios and low mass-damping are quite distinct. A direct comparison is made between the response in water ($m^* \simeq 2.4$) (from Khalak and Williamson, 1997), with the largest response plot of Feng conducted in air (Figure 10.5). The lighter body has a value of ($m^* \zeta$), around 3% of Fengß value, yielding a much higher peak amplitude. The regime of U^* over which there is significant response is four times larger than that found by Feng.

Although these are trends that might be expected, the character of the response for low mass-damping is also distinct. The low-($m^* \zeta$) type of response is characterized by not only the initial branch and the lower branch, but also by the new appearance between the other two branches of a much higher "upper response branch." Khalak and Williamson (1996) showed the existence of these three distinct branches, and using the Hilbert Transform to find instantaneous phase, force, and amplitude, they showed that the transition between the initial and the upper branches is hysteretic, while the upper–lower transition involves instead an intermittent switching. Vorticity measurements for free vibrations, by Govardhan and Williamson (2000), confirmed that the initial and lower branches correspond to the 2S and 2P vortex wake modes (Figure 10.6) while the upper branch also comprises a 2P mode, but the second vortex of each pair (in each half cycle) is much weaker than the first one.

The phenomenon of lock-in, or synchronization, traditionally means that the ratio $f^* = f/f_N$ remains close to unity, as seen in Figure 10.2 for high mass ratio. However, for light bodies in

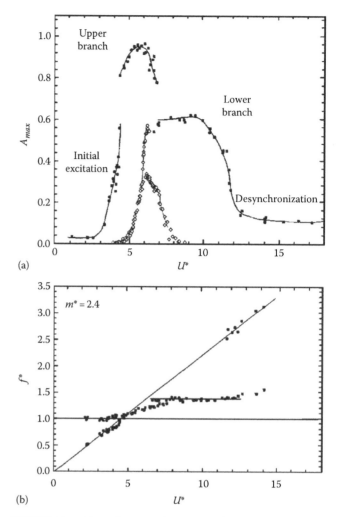

FIGURE 10.5 Free vibration at low mass and damping is associated with the existence of an upper branch of high-amplitude response, which appears between the initial and lower branches. The frequency of the lower branch is not close to the natural frequency, and is remarkably constant in (b). (From Khalak, A. and Williamson, C.H.K., *J. Wind Eng. Ind. Aerodyn.*, 69–71, 341, 1997.) Open symbols in (a) show the contrasting high-$m^* \zeta$ response data of Feng (1968).

water, in this case for $m^* \simeq 2.4$ in Figure 10.5b, the body oscillates at a distinctly higher frequency ($f^* \simeq 1.4$). Therefore, one might define synchronization as the matching of the frequency of the periodic wake vortex mode (or frequency of force on the body) with the body oscillation frequency.

A reduction in vibrating mass, for example, from $m^* \simeq 8.6$–1.2, in Figure 10.7, leads to a wider synchronization regime, in this case yielding a significant increase of the lock-in regime, which reaches $U^* \simeq 17$. The normalized velocity used here, U^*, is the traditional parameter for free-vibration experiments. However, by replotting the data versus the parameter $(U^*/f^*)S$, which is equivalent to (f_{vo}/f) (or the inverse of the ratio of actual oscillating frequency to the fixed-body shedding frequency), the data sets collapse very well. Khalak and Williamson (1999) made the first such collapse of free-vibration data and showed that this collapse cannot a priori be predicted. An equivalent "true" reduced

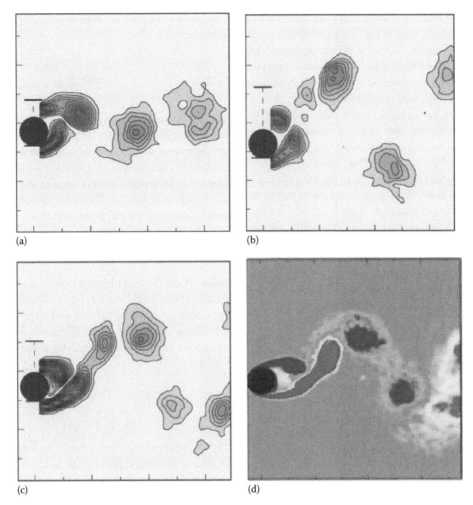

FIGURE 10.6 Evidence from PIV, in the form of vorticity measurements in free vibration, shows that the initial branch corresponds with the 2S vortex wake mode, and the upper and lower branches both reflect the 2P mode. (a) 2S Initial, (b) 2P Upper, (c) 2P Lower (Experiment), (d) 2P Lower (Simulation). (From Govardhan, R. and Williamson, C.H.K., *J. Fluid Mech.*, 420, 85, 2000.) Blackburn et al. (2001) make a good comparison from simulation, computing the 2P mode of the lower branch, which is only possible with 3D computations.

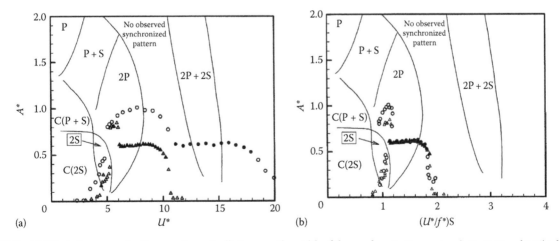

FIGURE 10.7 (a) Effect of a mass reduction can dramatically increase the width of the synchronization regime ($m^* = 8.63$ and 1.19) when plotted with velocity (U^*), used traditionally for free vibration. (b) These response data collapse well if one uses the "true" normalized velocity ($U^*/f^*)S$, yielding a good correspondence between response branches and vortex mode regimes in Williamson and Roshko (1988) map. (From Govardhan, R. and Williamson, C.H.K., *J. Fluid Mech.*, 420, 85, 2000.)

velocity has been used, not only in the numerical simulations, but also in experiment (e.g., Hover et al., 1998). With the renormalization mentioned earlier, there is a good correspondence between the vortex wake modes of the free response branches with the vortex mode regimes deduced from forced vibration in the Williamson and Roshko map (see Figure 10.7b).

One of the more recent debates comes from the *BBVIV-3* Conference in Marseille, in June 2000 (see Bearman et al. 2000), which triggered much-needed clarification. Leonard and Roshko (2001) specifically discussed added mass of an accelerating body, defining it as "the impulse given to the fluid during an incremental change of body velocity, divided by that incremental velocity." They point out that such "properties of the added mass are well known from textbook derivations which are usually obtained for irrotational flow, and so it is not as well known that the resulting definitions are applicable more generally, e.g., in separated flows, such as those that occur in problems of flow-induced vibration. As a result, empirical relations are sometimes introduced into models, unnecessarily." Leonard and Roshko provide a clear proof for the validity of the decomposition of the force in a general viscous flow, which of course includes bluff bodies undergoing VIV. On the other hand, Sarpkaya (2001) stated that such a force decomposition is impossible in the case of the transverse forces acting on bluff bodies undergoing VIV. Of relevance to the preceding text, it is important to note the sometimes misinterpreted use of terminology in these problems, particularly where it is used in practice. For example, it is common in offshore engineering to use the expression "added mass" to simply mean all the fluid force in phase with acceleration (which of course includes a component of force due to vorticity dynamics), which is distinct from the potential added mass.

10.3.2 Existence of a "Critical Mass"

From an applications point of view, it is important to note that, as the structural mass decreases, so the regime of velocity U^*, over which there are large-amplitude vibrations, increases (see, e.g., Figure 10.7). Anthony Leonard indicated the large extent of such regimes for very low mass ratios, based on results related to numerical simulation. We make the deduction here that when mass ratio (m^*) tends to zero, then the extent of the synchronization regime of large-amplitude motion extends to infinity! (We simply deduce this from Equation 10.5.) However, a more surprising result shows that the synchronization regime becomes infinitely wide, not simply when the mass becomes zero, but when the mass falls below a special critical value whose numerical value depends on the shape of the vibrating body, and to a weak extent on the Reynolds number.

The higher end of the synchronization regime for free vibration of a cylinder, with low mass-damping, is generally distinguished by a lower amplitude branch, which has a remarkably constant vibration frequency (f^*_{LOWER}), as typified by Figure 10.5b, and whose frequency level increases as the mass is reduced. Govardhan and Williamson (2000) presented a large data set for the lower branch frequency (f^*_{LOWER}) plotted versus m^*,

yielding a good collapse of data onto a single curve fit based on Equation 10.5:

$$f^*_{LOWER} = \sqrt{\frac{m^* + 1}{m^* - 0.54}} \qquad (10.9)$$

This expression provides a practical and simple means to calculate the highest frequency attainable by the VIV system in the synchronization regime, if one is provided the mass ratio, m^*. An important consequence of Equation 10.12 is that the vibration frequency becomes infinitely large as the mass ratio reduces to a limiting value of 0.54. Therefore, Govardhan and Williamson concluded that a critical mass ratio exists

$$\text{Critical mass ratio, } m^*_{\text{CRIT}} = 0.54 \pm 0.02 \qquad (10.10)$$

below which the lower branch of response can never be reached for finite velocities, U^*, and ceases to exist. These conditions are applicable for finite (U^*/f^*), so when the mass of the structure falls below the critical value, one predicts that large-amplitude vibrations will be experienced for velocities U^* extending to infinity:

$$U^*_{end\ of\ synchronization} = 9.25 \sqrt{\frac{m^* + 1}{m^* - 0.54}} \qquad (10.11)$$

This expression accurately marks the upper boundary of the shaded synchronization regime in Figure 10.8a. The fact that the critical mass turns out to be 54% is significant because it is in the realm of the "relative densities" of full-scale structures in engineering. We note carefully that this unique value of the critical mass is valid under the conditions of low mass and damping, so long as $(m^* + C_A)\zeta < 0.05$. We make the point here that added-mass coefficients having a negative value can be observed in data sets collected from forced vibration (Mercier, 1973, and many subsequent studies) and in recent free-vibration data sets. The implications to free-vibration phenomena, such as the possible existence of a "critical mass," were not deduced in these earlier works.

There is nothing in principle to suggest that an experiment (consider the $m^* \simeq 0.52$ case in Govardhan and Williamson, 2000) cannot reach $U^* = 300$, for example, at which point the system will vibrate vigorously at 32 times the natural frequency. This is far from the classical concept of synchronization, where resonant vibration is expected around $U^* = 5$–6. It is possible, even within a laboratory, to take the normalized velocity (U^*) to infinity simply by removing the restraining springs, as done by Govardhan and Williamson (2002). [Setting damping zeta = 0, can be done more straightforwardly in numerical simulations, as by Shiels et al. (2001) at low Reynolds numbers.] In the experiments, a reduction of mass led to a catastrophic change in response; large-amplitude vigorous vibrations suddenly appear as mass ratio is reduced to below a critical value, $m^* \simeq 0.542$ (see Figure 10.8b). This accurately proves the prediction of the

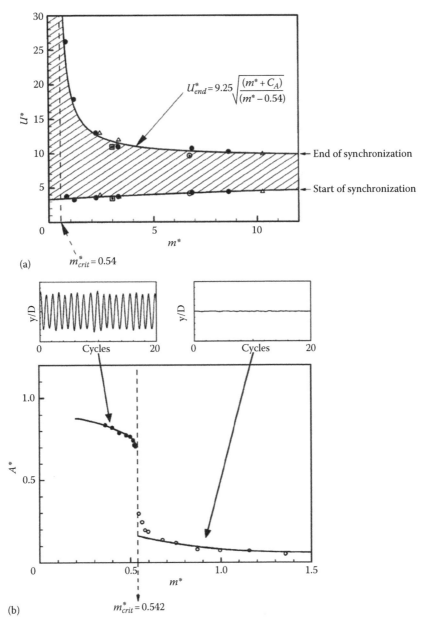

FIGURE 10.8 Discovery of a critical mass. The synchronization regime of high-amplitude vibration (*shaded regime*) extends to infinite velocities as m^* approaches the value 0.54, in (a) (From Govardhan, R. and Williamson, C.H.K., *J. Fluid Mech.*, 420, 85, 2000). The lower plot in (b), from an independent set of experiments at infinite U^*, shows that there is a sudden appearance of large-amplitude response when m^* just falls below 0.54 (Govardhan and Williamson, 2002). Symbols in (a) are: •, Govardhan and Williamson (2000); △, Khalak and Williamson (1999); ▣, Hover et al. (1998); ⊙, Anand (1985).

earlier study; resonant oscillations persist up to infinite (normalized) flow speeds, and in this sense the cylinder "resonates forever."

How generic is the phenomenon of critical mass? Govardhan and Williamson (2002) deduced that it will be a universal phenomenon for all systems of VIV whose induced forces and dynamics are reasonably represented by the Equations 10.1 through 10.3. A critical mass is deduced for laminar vortex formation at low Reynolds numbers in the computations of Ryan et al. (2005). In fact, one finds a critical mass, $m^* = 0.30$, for a tethered sphere system (Govardhan and Williamson, 2003),

a critical mass, $m^* = 0.50$, for a pivoted cylinder (Flemming and Williamson, 2003), as well as $m^* = 0.52$, for an elastically mounted cylinder in two degrees of freedom (Jauvtis and Williamson, 2004). Note that these values are valid for small mass-damping.

10.3.3 Griffin Plot

An important question that has been debated for about 35 years is whether a combined mass-damping ($m^*\zeta$) parameter could reasonably collapse peak-amplitude data A^*_{max} in the GriDn plot.

The use of a mass-damping parameter stems from several studies. Vickery and Watkins (1964), who considered an equation of motion for flexible cantilevers, plotted their peak tip amplitudes versus their *Stability parameter* $= K_S = \pi^2(m^*\zeta)$. Scruton (1965) used a parameter, proportional to K_S, for his experiments on elastically mounted cylinders, which has since been termed the *Scruton number* $= Sc = \pi/2\,(m^*\zeta)$. A slightly different parameter was independently derived from a response analysis involving the van der Pol equation by Skop and Griffin (1973), and they compiled data from several different experiments as a means to usefully predict response amplitudes. The combined response parameter was subsequently termed S_G in Skop (1974), and we have termed this in Williamson and Govradhan (2004) as the *Skop–Grin parameter* (private communication with the late Dick Skop), which is defined here as follows:

$$\text{Skop–Griffin parameter} = S_G = 2\pi^3 S^2(m^*\zeta) \qquad (10.12)$$

Griffin et al. (1975) made the first extensive compilations of many different investigations, using S_G, and subsequently the classical log–log form of the plot (Griffin, 1980), as shown in Figure 10.3c, has become the widely used presentation of peak response data. Multiple problems regarding the validity of this widely used plot were suggested in several papers by, for example, Sarpkaya (1978). He stated that one would be able to use the combined parameter S_G only if $S_G > 1.0$, which would actually rule out most of the Griffin plot, as one can see in Figure 10.3c. On the other hand, there are now a number of results that show the validity of using this Griffin plot over a very wide regime of combined mass-damping, including some of Griffin's own experiments (Griffin and Ramberg, 1982).

As deduced in the review of Williamson and Govardhan (2004), even for the smallest mass-damping, the peak amplitudes were not yet close to saturating at a specific value. They asked: *What is the maximum attainable amplitude that can be reached as $(m^* + CA)\zeta$ gets ever smaller?* The trend of the data suggested no clear limit. However, it subsequently became clear that a missing ingredient in understanding the limit comes from realizing that Reynolds number influences the peak amplitudes of a vibrating body. This realization was also made by Klamo et al. (2005). In these cases, the investigators found a way to control the damping in the VIV system. Govardhan and Williamson (2006) found that the otherwise scattered data in a Griffin plot (see Figure 10.9a) could be collapsed beautifully if one takes into account the effect of Reynolds number, as an extra parameter in a "modified Griffin plot" (see Figure 10.9b). A good best-fit for a wide range of mass-damping and Reynolds number is thus given by the following simple expression, where $A^* = g(\alpha)\,f(Re)$:

$$A^* = (1 - 1.12\alpha + 0.30\alpha^2)\log(0.41 Re^{0.36}) \qquad (10.13)$$

All the earlier results now show that indeed one can collapse very well the peak-amplitude data, using the combined mass-damping parameter (α), down to mass ratios of at least $m^* = 1$, and for S_G far below 1.0.

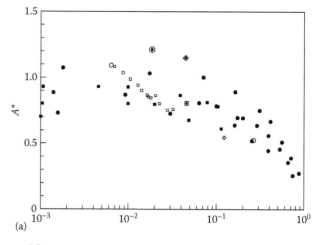

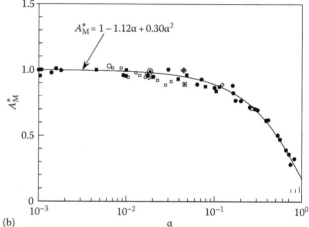

FIGURE 10.9 (a) Collapse of previously scattered data from the Griffin plot, by taking into account Reynolds number. If we replot all of the data in the "modified Griffin plot" in (b), employing our "modified amplitude" $[A_M^* = A^*/f(Re)]$, all of the data collapse beautifully onto a single curve. •, Mechanical "spring" damper; ■, Electromagnetic damper; ⊙, Vikestad (1998); ▣, Hover et al. (1998); □, Khalak and Williamson (1999); ○, Govardhan and Williamson (2000); ◇, Blackburn et al. (2001); ◆, Smogeli et al. (2003);—, present curve fit: $A_M^* = (1 - 1.12\,\alpha + 0.30\alpha^2)$.

10.3.4 Controlled Vibration of a Cylinder

One approach to predicting VIV has been to generate an experimental force database by testing cylinders undergoing forced or controlled sinusoidal oscillations in a free stream. Several investigators have measured the forces on bodies in harmonic, as well as multifrequency motion. As mentioned earlier in conjunction with Sarpkaya's well-known data set, in these experiments the transverse force is generally decomposed into two components, one in phase with the velocity ($C_Y\sin\phi$, which predicts when free vibration should occur) and one in phase with the acceleration ($C_Y\cos\phi$, which yields the "effective added mass").

Response plot predictions using controlled vibration data with an assumed equation of motion have been compared with free-vibration tests using contours of force coefficients by Staubli (1983) and Hover et al. (1998), who developed an ingenious and extremely versatile experiment, namely, a novel force-feedback

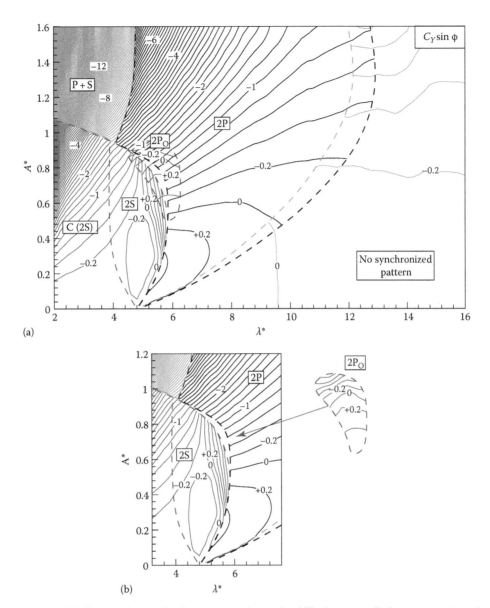

FIGURE 10.10 (a) Contours of the force in phase with velocity, $C_Y \sin\phi$ (normalized "fluid excitation"), for $Re = 4000$. Boundaries between modes are indicated by the dashed lines. Note that contours overlap regions in which multiple vortex-shedding modes are possible. In (b) we pull away the $2P_O$ mode to more clearly show the 2P and 2S mode regions underneath. Contour interval is 0.2.

Virtual Cable Testing Apparatus (VCTA). There are parametric regions where such comparison is successful, and other regions where comparison is not close.

It turns out that accurately predicting free-vibration response from controlled vibration depends on two key aspects: conducting free or forced experiments under almost identical conditions, as well as assembling extremely high-resolution contours. Around 6000 runs were used to construct contour plots of force as a function of amplitude and wavelength by Morse and Williamson (2009, 2010), and an example of the normalized excitation ($C_Y \sin\phi$) is included in Figure 10.10 (See also the discussion in the "Focus on Fluids" article by Bearman, 2009). They were able to clearly identify boundaries between fluid-forcing regimes, which appear remarkably similar to those boundaries

separating vortex formation modes in the Williamson–Roshko map in Figure 10.4. Further, they were able to accurately predict response plots over a range of mass-damping, and to introduce the concept of an "energy portrait," by comparing energy of excitation versus energy dissipated to damping. Stable and unstable solutions were uncovered, which are able to explain the hysteretic jumps or intermittent switching associated with response branch transitions found in free vibration.

10.3.5 *XY* Motion of Bodies

Despite the large number of papers dedicated to the problem of a cylinder vibrating transverse to a fluid flow (*Y* motion), there are very few papers that also allow the body to vibrate in-line with the flow. One principal question that may be posed is as follows: How does the freedom to vibrate in-line with the flow influence the dynamics of the fluid and the structure? Full-scale piles in an ocean current (Wooton et al., 1972), and similar cantilever models in the laboratory (King, 1974), vibrate in-line with the flow with peak amplitudes of the cantilever tip ($A_X^* = 0.15$). Oscillations ensue if the velocity is close to $U^* = 1/2S$. These investigators showed a classical vortex street (antisymmetric) pattern, although they also discovered a second mode where the wake formed symmetric vortex pairs close to the body. One subsequent approach, where these two modes have been observed, is to vibrate bodies in-line with the flow (Ongoren and Rockwell, 1988b).

In most past experimental work with XY vibrations (e.g., Moe and Wu, 1990; Sarpkaya, 1995), the X and Y mass ratios or natural frequencies (or both) were chosen to have different values. Under their chosen special conditions, these studies demonstrated a broad regime of synchronization, but with no evidence of the different response branches. However, in most practical cases, cylindrical structures (such as riser tubes or heat exchangers) have the same mass ratio and the same natural frequency in both the streamwise (X) and transverse (Y) directions. A pendulum setup that ensures such conditions (Jauvtis and Williamson, 2004) has demonstrated a set of response branches. Even down to the low mass ratios, where $m^* = 6$, it is remarkable that the freedom to oscillate in-line with the flow hardly affects the response branches, the forces, and the vortex wake modes. These results are significant because they indicate that the extensive understanding of VIV for Y-only body motions, built up over the last 40 years, remains strongly relevant to the case of two degrees of freedom. However, there is a dramatic change in the fluid–structure interactions when mass ratios are reduced below $m^* = 6$. A new response branch with significant streamwise motion appears in a "super-upper" branch, which yields massive amplitudes of three diameters peak-to-peak ($A^* = 1.5$). This response corresponds with a new periodic vortex wake mode, which comprises a triplet of vortices being formed in each half cycle, defined as a "2T" mode.

10.3.6 "Complex Flows": Flexible, Tapered, Pivoted, Tethered Bodies; Rising and Falling Bodies

As bodies become more directly practical, they generally become more complex, although many of the phenomena discovered for the simpler paradigm of the elastically mounted cylinder carry across to more involved structures, including those whose vibration amplitude varies along the span. Techet et al. (1998) discovered a 2S-2P Hybrid mode (shown in Figure 10.11a), comprising the 2S and 2P modes occurring along different spanwise lengths of their tapered cylinder, with vortex dislocations between the spanwise cells. They showed an excellent correlation and prediction of these modes in the framework of the Williamson and Roshko (1988) map of modes. Vortex-induced vibrations of pivoted cylinders also exhibit similar branches of response to the

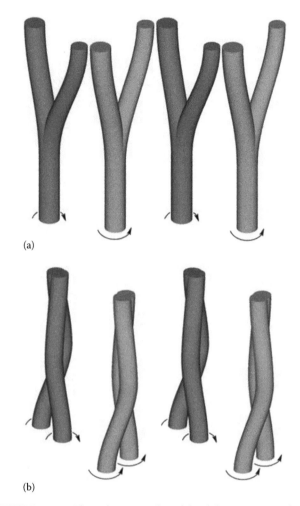

(a)

(b)

FIGURE 10.11 Three-dimensional models of the two new modes of vortex formation, constructed on the basis of the DPIV vorticity layers. The pivot point of the cylinder is toward the bottom of the diagrams and the fluid is moving from the left to the right: (a) 2S-2P hybrid mode and (b) 2C mode.

cantilever and free cylinder. Most studies confine vibrations to transverse motion, but Flemming and Williamson (2003) recently studied the case of a pivoted cylinder free to move streamwise as well as transverse to the flow. Over a range of body inertias I^* (equivalent to m^*), a number of different spanwise modes were discovered. For cases with high I^*, and negligible streamwise motion, either the 2S or 2P modes were observed along the span, but for lighter structures, the Techet et al. 2S-2P Hybrid mode was found, along with a distinct new mode along the span, comprising two corotating vortices formed in each half cycle, namely, the "2C" mode in Figure 10.11b, for the lightest of their structures.

Kim Vandiver at MIT has undertaken extensive field and laboratory experimental studies concerning cable dynamics (see, e.g., Vandiver and Marcollo, 2003), and he has developed a well-known cable VIV prediction program "SHEAR 7" (Vandiver, 2003) that is currently based on data for short laboratory cylinders. The group of George Karniadakis at Brown University has performed extensive computational studies, beginning with their studies to investigate flow past a freely vibrating cable (Blackburn and Karniadakis, 1993). In these cases, they employed a simple wave equation to model the structure and found two possible wake states: one for a traveling wave (oblique vortices), and one for a standing-wave response (Lambda-shaped vortices). Lucor et al. (2001) investigated very long bodies (aspect ratio >500) in uniform and sheared flows to observe vortex dislocations of the kind found for fixed-body flows (see review of Williamson, 1996), which cause substantial modulation of lift forces.

Risers are long flexible circular cylinders exposed to all sorts of oceanographic conditions. They are used to link the seabed to the offshore platforms for oil production. Some of these floating platforms are installed along the continental shelf of the Atlantic Ocean where water depths over 1000 m are common. In such conditions, a better comprehension of the vortex dynamics causing vibration and fatigue of risers is essential. With risers presenting such high aspect ratios and complex flow fields around them, a complete 3D simulation at realistic conditions is unfeasible. With this in mind and aiming at the hydroelastic response of the riser structure, a numerical model in a quasi-3D fashion has been developed by Willden and Graham (2000), Meneghini et al. (2004), and Yamamoto et al. (2004), using a quasi-3D approach, where the hydrodynamic forces are evaluated in 2D strips. An extensive testing of the methods to study riser tube dynamics has involved multiple international groups working to model the same well-defined specific riser problem, and the reader is referred to Chaplin et al. (2005).

Also of relevance to long bodies, de Langre's group (Ecole Polytechnique) has been developing analytical methods to model the dynamics of long flexible structures. Facchinetti et al. (2003) studied the effect of a coupling term between the equation for near-wake dynamics (van der Pol), and the one degree-of-freedom structure oscillator. They found good comparison between their model and recent experimental results, but perhaps the principal conclusion is that the optimal coupling term involves the body acceleration, rather than the displacement or velocity. This is a useful result for future modeling developments. Facchinetti et al. (2003) also looked into the problem of vortex-induced waves (VIW) on flexible structures, using both their modeling approach, and also experiment.

A new impetus to suppress VIV has stemmed from the original work of Tombazis and Bearman (1997) and Bearman and Owen (1998), where they investigated the influence of an imposed spanwise waviness of the flow separation lines around bluff bodies. They achieved a drag reduction of 30% and a suppression of classical vortex shedding. A principal idea is to weaken vortex shedding without the drag increase associated with traditional "helical strakes" (Zdravkovich, 1981). More recently, Bearman's group at Imperial College has achieved some significant results (Assi et al., 2009; Bearman, 2011), where VIV can be practically eliminated by using free-to-rotate control plates, although it is then important to avoid a parameter space where galloping might be initiated.

Finally, we mention the dynamics of tethered structures. In the case of tethered spheres, Govardhan and Williamson (2005) explored a wide range of masses, $m^* = 0.1–1000$, and a range of velocities from $U^* = 0–300$, by using both light spheres in a water channel facility and heavy spheres in a wind tunnel. They found a number of modes of response, analogous to the cylinder VIV problem, yielding amplitudes up to one diameter. The principal vortex structure in Figure 10.12 appears to be a system of streamwise vortex loops, which can be related with

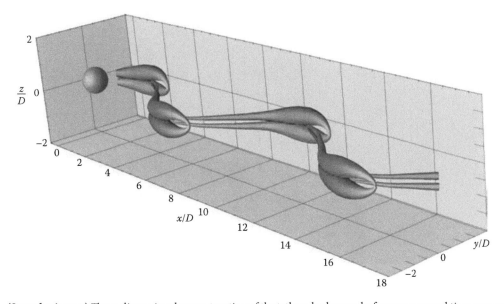

FIGURE 10.12 (See color insert.) Three-dimensional reconstruction of the tethered sphere wake from a measured time sequence of streamwise vorticity. For all tethered sphere modes, the 3D structure clearly shows that the wake comprises a two-sided chain of streamwise vortex loops. Streamwise direction is parallel to the *x*-axis, while the sphere vibrates primarily transverse to the flow (*y*-direction). Blue indicates clockwise vorticity, and red anticlockwise vorticity. $Re \approx 3000$.

the vortex force on the body giving rise to vibration (in a model analogous to the relation between aircraft trailing vortex dynamics and wing lift).

Within the context of VIV, Horowitz and Williamson (2010) have studied the effect of Reynolds number on dynamics and vortex formation of rising and falling spheres. They find that falling spheres always fall rectilinearly, but that there exists a significant regime of m^*, where rising spheres can also rise without vibration. However, below a special critical mass (e.g., below 40%, depending on Re), the sphere suddenly starts to vibrate periodically in a zigzag trajectory within a vertical plane. Helical or spiral trajectories, found readily for rising bubbles (Magnaudet and Eames, 2000), were not found for solid spheres, unless there exists disturbed background fluid motions. Wakes are found for rectilinear paths, which exhibit single-sided (R mode) or double-sided periodic sequences of vortex rings (2R mode), whereas for the zigzag trajectory, a four rings pattern is found (4R mode), all of which are exhibited in Figure 10.13. A map of dynamics and vortex wake modes as a function of mass ratio and Reynolds number $\{m^*, Re\}$ is presented in Figure 10.14. In general, the drag of a vibrating body is higher than that for a fixed body. There is a reasonable collapse of drag measurements, as a function of Re, onto principally two curves, a higher one for the vibrating regime, and one for the rectilinear trajectories. Interestingly, the first person to observe an effect on the drag from the vibration was Isaac Newton (1726), who experimented with falling hog's bladders.

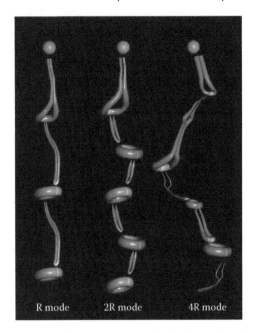

FIGURE 10.13 (See color insert.) The family of periodic wake modes for rising and falling spheres, showing the essential vortex configurations, which are the R mode (single vortex ring per wavelength of wake), 2R mode (two rings per wake wavelength), and 4R mode (four rings per cycle of sphere vibration).

10.4 Challenges

In this chapter, we discussed many of the new fundamental results and methods, but we did not cover all topics fully. Excellent work has been done by many researchers to bring the fundamentals into practical design codes. There is clearly inadequate full-scale data for fluid–structure interactions in a variety

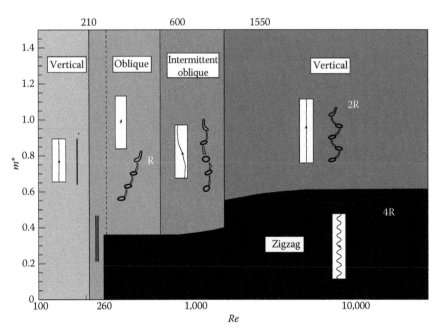

FIGURE 10.14 Map of regimes of sphere motion and associated wake patterns in the $\{m^*, Re\}$ plane. Typical trajectories are shown for each regime. The red regime is essentially the parameter space where the rising sphere exhibits the zigzag dynamics; above the critical mass (between 35% and 60%), the trajectories of rising or falling spheres are rectilinear.

of conditions, including sheared flows in the ocean. VIV behavior at large *Re* is in need of a parallel effort to see which phenomena in this chapter remain relevant to full-scale structures, and to discover what new phenomena appear.

One major challenge is to suppress VIV effectively, while avoiding the usual drag increase that normally accompanies proposed methods, although Bearman's group at Imperial College has achieved some success recently in this direction (Assi et al., 2009). On this and other new results, the reader is referred to the short review of recent progress by Bearman (2011). Further understanding of the modes and regimes for yawed cylinders is needed. There are energetic ongoing efforts to model, compute, and undertake experiments concerning cable and riser tube dynamics in ocean engineering. There are some important efforts underway to explore phenomena at high Reynolds numbers, including the quest to relate vortex modes at the lower *Re* to the higher Reynolds numbers.

One of the most fundamental questions concerning VIV is as follows: What is the maximum attainable amplitude in VIV of an elastically mounted cylinder? We may also ask, what is the functional relationship between peak amplitudes and mass-damping, in the GriDn plot? The recognition that Reynolds number is key to understanding the previously large scatter in the GriDn plot has been a large step forward. Now the question is how does a body vibrate through the critical Reynolds number where a fixed body would undergo boundary layer transition. The status at present is that experiments ($Re \sim 10^5$–10^6) exhibit very large scatter, and no clear pattern exists as yet.

What generic characteristics exist for VIV, which carry across from the paradigm of the elastically mounted cylinder, in transverse vibration, to more complex systems? It is fascinating that the response branches for this "simple" paradigm are found similarly for cylinders in *XY* motion, for flexible cantilevers, for pivoted cylinders, for vibrating cables, for rising and falling bodies, and possibly for other systems. Analogous modes are found also for tethered bodies. Vortex wake modes that are now known to cause free vibration, at moderate *Re*, comprise the following set: {2S; 2P; 2T; 2C}; and in the laminar regime ($Re < 200$), the set comprises only {2S}. For VIV systems with spanwise amplitude variation, we observe the 2S-2P Hybrid mode. The P + S mode, ubiquitous in forced vibrations, apparently does not induce free vibration. Conditions where such free-vibration modes exist in these VIV systems correspond well with the Williamson and Roshko (1988) map of modes in the plane of amplitude-velocity, compiled from controlled vibration experiments.

The concept of a critical mass has been introduced, whereby the regime of synchronization extends to infinite flow velocity—in a sense the body resonates forever! Values of critical mass have been identified for several VIV systems, under conditions of low mass-damping, such as the cylinder in *Y*-only motion, as well as *XY* motion, the pivoted cylinder, the tethered sphere, and so on. In fact, one expects to find a critical mass for all VIV systems.

In essence, we continue to find generic or universal features that are common to all VIV systems.

Some progress has been achieved on certain debates in VIV, for example on the "controversy" that previously existed regarding interpreting added mass, and on the problem of whether accurate free-vibration response can be predicted using data from controlled vibrations. There was also debate about whether results from our paradigm, the *Y*-only free vibration of a cylinder, carry across to two degrees of freedom (*XY* motion). Fortunately, for the hundreds of papers concerned with the paradigm, the results generally carry across very well. However, this similarity breaks down for very low vibrating mass, and for systems having different natural frequencies in different orthogonal directions.

Further ideas have been developed in the last few years. One of these ideas is the concept of utilizing vorticity dynamics to measure the force on bodies, which received a boost from the increased capabilities to simulate flows, and to accurately evaluate vorticity using PIV in experiment. As the tools of analysis, simulation, and experiment are further sharpened, we may expect more fundamentally new contributions to emerge, and further universal or generic characteristics to be discovered, which carry across from one VIV system to another. But perhaps the biggest challenge in these flows is determining whether the phenomena at low to moderate Reynolds numbers (up to $Re \sim 50,000$) carry across to high *Re* or full-scale ocean structures, $Re \sim 10^5$–10^6. Further high Reynolds number experiments are certainly needed for practical design of structures in the environment.

10.A Appendix: Nondimensional Groups

Mass ratio	m^*	$\dfrac{m}{\pi\rho D^2 L/4}$
Damping ratio	Z	$\dfrac{c}{2\sqrt{k(m+m_A)}}$
Velocity ratio	U^*	$\dfrac{U}{f_N D}$
Amplitude ratio	A^*	$\dfrac{y_o}{D}$
Frequency ratio	A^*	$\dfrac{f}{f_N}$
Streamwise force coeDcient	C_X	$\dfrac{F_X}{\frac{1}{2}\rho U^2 DL}$
Transverse force coeDcient	C_Y	$\dfrac{F_Y}{\frac{1}{2}\rho U^2 DL}$
Reynolds number	Re	$\dfrac{\rho UD}{\mu}$

Note regarding these groups: We use f_N as the natural frequency in still water, and correspondingly use ζ as the ratio of (structural damping)/(critical damping in water). The frequency f, used in f^*, is the actual body oscillation frequency during induced vibration. The added mass, $m_A = C_A m_d$, where $m_d = \pi\rho D^2 L/4$ is the displaced mass of fluid, and where L is the cylinder length, $c_A = 1.0$.

References

Assi GRS, Bearman PW, Kitney N. 2009. Low drag solutions for suppressing vortex-induced vibration of circular cylinders. *J. Fluids Struct.* 25:123–130.

Bearman PW. 1984. Vortex shedding from oscillating bluff bodies. *Annu. Rev. Fluid Mech.* 16:195–222.

Bearman, PW. 2009. Understanding and predicting vortex-induced vibration. *J. Fluid Mech.* 634:1–4.

Bearman, PW. 2011. Circular cylinder wakes and vortex-induced vibration. Accepted for *J. Fluids Struct.* 27:648–658.

Bearman PW, Leweke T, Williamson CHK. 2000. *Proceedings of the IUTAM Symposium on Bluff Body Wakes and Vortex-Induced Vibrations BBVIV-2*, Marseille, France, June 13–16, pp. 10–19.

Bearman PW, Owen JC. 1998. Reproduction of bluff-body drag and suppression of vortex shedding by the introduction of wavy separation lines. *J. Fluids Struct.* 12:123–130.

Bishop RED, Hassan AY. 1964. The lift and drag forces on a circular cylinder oscillating in a flowing fluid. *Proc. R. Soc. Lond.* A 277:51–75.

Blackburn H, Karniadakis GE. 1993. Two and three-dimensional simulations of vortex-induced vibration of a circular cylinder. In *3rd International Offshore and Polar Engineering Conference*, Singapore, Vol. 3, pp. 715–720.

Blevins RD. 1990. *Flow-Induced Vibrations.* New York: Van Nostrand Reinhold.

Brika D, Laneville A. 1993. Vortex-induced vibrations of a long flexible circular cylinder. *J. Fluid Mech.* 250:481–508.

Chaplin JR, Bearman PW, Cheng Y, Fontaine E, Heraord K, Huarte FJH, Isherwood M et al. 2005. Blind predictions of laboratory measurements of vortex-induced vibrations of a tension riser. *J. Fluids Struct.* 21:25–40.

Fachinetti ML, DeLangre E, Biolley F. 2003. Vortex-induced travelling waves along a cable. *Eur. J. Mech. B Fluids.* 23:199–208.

Feng CC. 1968. The measurements of vortex-induced effects in flow past a stationary and oscillating circular and D-section cylinders. Master's thesis. University of British Columbia, Vancouver, British Columbia, Canada.

Flemming F, Williamson CHK. 2003. Vortex-induced vibrations of a pivoted cylinder. *J. Fluid Mech.* 522:215–252.

Govardhan R, Williamson CHK. 2000. Modes of vortex formation and frequency response for a freely-vibrating cylinder. *J. Fluid Mech.* 420:85–130.

Govardhan R, Williamson CHK. 2002. Resonance forever: Existence of a critical mass and an infinite regime of resonance in vortex-induced vibration. *J. Fluid Mech.* 473:147–166.

Govardhan R, Williamson CHK. 2005. Vortex-induced vibration of a sphere. *J. Fluid Mech.* 531:11–47.

Govardhan R, Williamson CHK. 2006. Defining the "modified GriDn plot" in vortex-induced vibration: Revealing the effect of Reynolds number using controlled damping, *J. Fluid Mech.* 473:147–180.

GriDn OM. 1980. Vortex-excited cross-flow vibrations of a single cylindrical tube. *ASME J. Press. Vessel Tech.* 102:158–166.

GriDn OM, Ramberg SE. 1974. The vortex street wakes of vibrating cylinders. *J. Fluid Mech.* 66:553–576.

GriDn OM, Ramberg SE. 1976. Vortex shedding from a cylinder vibrating in line with an incident uniform flow. *J. Fluid Mech.* 75:257–271.

GriDn OM, Ramberg SE. 1982. Some recent studies of vortex shedding with application to marine tubulars and risers. *Trans. ASME J. Energy Resour. Tech.* 104:2–13.

GriDn OM, Skop RA, Ramberg SE. 1975. The resonant vortex-excited vibrations of structures and cable systems. In *7th Offshore Technology Conference*, Houston, TX, OTC Paper 2319.

Gu W, Chyu C, Rockwell D. 1994. Timing of vortex formation from an oscillating cylinder. *Phys. Fluids* 6:3677–3682.

Horowitz M, Williamson CHK. 2010. The effect of Reynolds number on the dynamics and wakes of freely rising and falling spheres. *J. Fluid Mech.* 651:251–294.

Hover FS, Techet AH, Triantafyllou MS. 1998. Forces on oscillating uniform and tapered cylinders in crossflow. *J. Fluid Mech.* 363:97–114.

Jauvtis N, Williamson CHK. 2004. The effect of two degrees of freedom on vortex- induced vibration at low mass and damping. *J. Fluid Mech.* 509:219–229.

Khalak A, Williamson CHK. 1996. Dynamics of a hydroelastic cylinder with very low mass and damping. *J. Fluids Struct.* 10:455–472.

Khalak A, Williamson CHK. 1997. Investigation of the relative effects of mass and damping in vortex-induced vibration of a circular cylinder. *J. Wind Eng. Ind. Aerodyn.* 69–71:341–350.

Khalak A, Williamson CHK. 1999. Motions, forces and mode transitions in vortex-induced vibrations at low mass-damping. *J. Fluids Struct.* 13:813–851.

King R. 1974. Vortex-excited oscillations of a circular cylinder in steady currents. In *Proceedings of the Offshore Technology Conference*, Dallas, TX, Paper OTC 1948.

Klamo JT, Leonard A, Roshko A. 2005. On the maximum amplitude for a freely vibrating cylinder in cross flow. *J. Fluids Struct.* 21:429–434.

Leonard A, Roshko A. 2001. Aspects of flow-induced vibration. *J. Fluids Struct.* 15:415–425.

Lucor D, Imas L, Karniadakis GE. 2001. Vortex dislocations and force distribution of long flexible cylinders subjected to sheared flows. *J. Fluids Struct.* 15:651–658.

Magnaudet J, Eames I. 2000. The motion of high Reynolds number bubbles in inhomogeneous flow. *Annu. Rev. Fluid Mech.* 32:659–708.

Meneghini JR, Saltara F, Fregonesi RA, Yamamoto CT, Casaprima E, Ferrari Jr, JA. 2004. Numerical simulations of VIV on long flexible cylinders immersed in complex flow fields, *Eur. J. Mech.* 23:51–63.

Mercier JA. 1973. Large amplitude oscillations of a circular cylinder in a low speed stream. PhD thesis, Stevens Institute of Technology, Hoboken, NJ.

Moe G, Wu ZJ. 1990. The lift force on a cylinder vibrating in a current. *ASME J. Offshore Mech. Arctic Eng.* 112:297–303.

Morse TL, Williamson CHK. 2009. Prediction of vortex-induced vibration response by employing controlled motion. *J. Fluid Mech.* 634:5–39.

Morse TL, Williamson CHK. 2010. Prediction of steady, unsteady, and transient behaviour in vortex-induced vibration from controlled motion. *J. Fluid Mech.* 649:429–451.

Naudascher E, Rockwell D. 1994. *Flow-Induced Vibrations: An Engineering Guide.* Rotterdam, the Netherlands: Balkema.

Newton I. 1726. *Philosophia Naturalis Principia Mathematica*, 3rd edn. Translated by I. B. Cohen and A. Whitman, Berkeley, CA: University of California Press, 1999.

Ongoren A, Rockwell D. 1988a. Flow structure from an oscillating cylinder. Part 1. Mechanisms of phase shift and recovery in the near wake. *J. Fluid Mech.* 191:197–223.

Ongoren A, Rockwell D. 1988b. Flow structure from an oscillating cylinder. Part 2. Mode competition in the near wake. *J. Fluid Mech.* 191:225–245.

Ryan K, Thompson MC, Hourigan K. 2005. Variation in the critical mass ratio of a freely oscillating cylinder as a function of Reynolds number. *Phys. Fluids* 17:038106.

Sarpkaya T. 1978. Fluid forces on oscillating cylinders. *ASCE J. Waterw. Port Coastal Ocean Div.* 104:275–290.

Sarpkaya T. 1995. Hydrodynamic damping, flow-induced oscillations, and biharmonic response. *ASME J. Offshore Mech. Arctic Eng.* 117:232–238.

Sarpkaya T. 2001. On the force decompositions of Lighthill and Morison. *J. Fluids Struct.* 15: 227–233.

Scruton, C. 1965. On the wind-excited oscillations of towers, stacks and masts. In *Proceedings of the Symposium on Wind Effects on Buildings and Structures*, Teddington, U.K.: Her Majesty's Stationery Office, Paper 16, pp. 798–836.

Shiels D, Leaonard A, Roshko A. 2001. Flow-induced vibration of a circular cylinder at limiting structural parameters. *J. Fluids Struct.* 15:3–21.

Skop RA, GriDn OM. 1973. An heuristic model for determining flow-induced vibrations of offshore structures. In *5th Offshore Technology Conference,* Houston, TX, OTC Paper 1843.

Staubli, T. 1983. Calculation of the vibration of an elastically-mounted cylinder using experimental data from forced oscillation. *ASME J. Fluids Eng.* 105:225–229.

Sumer BM, Fredsøe J. 1997. *Hydrodynamics around Cylindrical Structures.* Singapore: World Scientific.

Techet AH, Hover FS, Triantafyllou MS. 1998. Vortical patterns behind a tapered cylinder oscillating transversely to a uniform flow. *J. Fluid Mech.* 363:79–96.

Tombazis N, Bearman PW. 1997. A study of three-dimensional aspects of vortex shedding from a bluff body with a mild geometric disturbance. *J. Fluid Mech.* 330:85–112.

Vandiver JK. 2003. SHEAR 7. *User Guide.* Department of Ocean Engineering, MIT, Cambridge, MA.

Vandiver JK, Marcollo H. 2003. High mode number VIV experiments. In *IUTAM Fully Coupled Fluid-Structure Interaction*, eds. Benaroya H and Wei T. Dordrecht, the Netherlands: Kluwer, pp. 42–51.

Vickery BJ, Watkins RD. 1964. Flow-induced vibrations of cylindrical structures. In *Proceedings of the First Australian Conference on Hydraulics and Fluid Mechanics*, ed. Silvester R. New York: Pergamon Press, pp. 213–241.

Willden RHJ, Graham JMR. 2000. Vortex induced vibration of deep water risers. In *Flow-Induced Vibration*, Lucerne, Switzerland: A.A. Balkema, pp. 29–36.

Williamson CHK. 2004. Vortex dynamics in the cylinder wake. *Annu. Rev. Fluid Mech.* 28:477–539.

Williamson CHK, Govardhan R. 2004. Vortex-induced vibrations. *Annu. Rev. Fluid Mech.* 36:413–455.

Williamson CHK, Roshko A. 1988. Vortex formation in the wake of an oscillating cylinder. *J. Fluids Struct.* 2:355–381.

Wooton LR, Warner MH, Sainsbury RN, Cooper DH. 1972. Oscillations of piles in marine structures. A resume of full-scale experiments at Immingham. *CIRIA Tech. Rep.* 41.

Yamamoto, CT, Meneghini JR, Saltara F, Fregonesi RA. 2004. Numerical simulations of vortex-induced vibration on flexible cylinders. *J. Fluids Struct.* 19:467–489.

Zdravkovich, MM. 1982. Modification of vortex shedding in the synchronization range. *ASME J. Fluids Eng.* 104:513–517.

11
Urban Heat Islands

M. Roth
National University of Singapore

11.1 Introduction

For millions of people living in cities, increased temperatures are a growing fact and concern. The urban heat island (UHI) is a phenomenon whereby urban regions experience warmer temperatures than their rural, undeveloped surroundings. The UHI is the most obvious atmospheric modification attributable to urbanization, the most studied of climate effects of cities and an iconic phenomenon of urban climate. It can be found in settlements of all sizes in all climatic regions and arises from the introduction of artificial surfaces characteristic of those of a city that radically alters the aerodynamic, radiative, thermal, and moisture properties in the urban region compared to the natural surroundings. The heat island is defined on the basis of temperature differences between urban and rural stations, and the isotherm patterns of near-surface air temperatures resemble the contours of an island.

The evaluation of the influence of settlements on the local climate has been an important task for a long time, and the thermal environment in particular has received widespread attention because it has practical implications for energy use, human comfort and productivity, air pollution, and urban ecology. During the winter, heat islands can be beneficial to cities in colder climates by helping to reduce heating costs and cold-related deaths. The negative impacts of summertime heat islands, however, outweigh the benefits of wintertime warming in most cities because the largest population centers are located in (sub)tropical climates. Summertime UHIs increase heat-related illness and mortality, air pollution, energy demand for air conditioning, and indirectly greenhouse gas emissions. Heat waves, for example, which heat islands can exacerbate, are the leading weather-related killer in the United States. Urbanization and UHIs have impacts that range from local to global scales, and cities are an important component of global environmental change research.

Urban climatology has a rich and long history beginning with Luke Howard's pioneering work. Howard was a chemist and amateur meteorologist and is most famous for his classification of clouds. In 1815, he conducted the first ever systematic urban climate study measuring what is now called the UHI effect based on thermometers in the city of London and in the countryside nearby (Howard 1818). It is also remarkable that Howard identified virtually all causes that are responsible for the development of the UHI during this early study. Subsequent urban climate research replicated Howard's findings from urban-rural pairs of thermometers at about 2 m height in many cities and a large number of UHI maps generated from mobile traverses appeared in the literature during the 1930s. Much of the research from this period, with a focus on German work, was summarized by the Benedictine Father Albert Kratzer in a monograph entitled "Das Stadtklima" (The Urban Climate) (Kratzer 1937), which also provides the first systematic review of the influence of settlements on air temperature. A comprehensive summary of UHI studies (maps and statistics) carried out in primarily European and North American cities was subsequently published by Helmut

Landsberg in his book entitled "The Urban Climate" (Landsberg 1981). During the 1970s and 1980s, focus shifted from the largely descriptive early work toward exploring the processes responsible for the urban effect with many fundamental contributions from Tim Oke, laying the foundation for a modern treatment of this topic, much of which will be referred to in the following. The present review also benefits from a recent, comprehensive summary of current knowledge of urban climate to allow cities to become more sustainable (Grimmond et al. 2010).

11.2 Principles

11.2.1 Urban Scales

Urban climates are characterized by a combination of processes occurring at a variety of scales imposed by the biophysical nature of cities and the layered structure of urban atmospheres. Recognition of the impact of scale is key to understand the workings and phenomena of urban climates in general and heat islands in particular. Scale is important when studying the form and genesis and for the measurement and modeling of heat islands. Scale sets the fact that there are not only one but several types of heat islands. The size of the source area from where the thermal influence originates and how it changes over time is determined by scale. Scale also determines the controls and processes affecting and creating each heat island and therefore determines the

conceptual framework required to model each type. In applied studies, the target may be human comfort, building energy use, or city-wide mitigation which all occur at different scales (Oke 2009).

Three basic urban climate scales are usually recognized (Figure 11.1). Individual buildings, trees, and the intervening spaces create an urban "canopy" and define the *microscale*, found inside the roughness sublayer (RSL), which itself is a manifestation of the small-scale variability found close to the urban surface. Typical microscales extend from one to hundreds of meters. Similar houses in a district combine to produce the *local scale*, which is an integration of a mix of microclimatic effects. Typical local scales extend from one to several kilometers. Plumes from individual local scale systems extend vertically and merge to produce the total urban boundary layer (UBL) over the entire city, which is a *mesoscale* phenomenon. During daytime, this layer is usually well mixed due to the turbulence created by the rough and warm city surface, extending to a height of 1 km or more by day, shrinking to hundreds of meters or less at night. The UBL has typical scales of tens of kilometers and can be advected as an urban "plume" downstream from the city by the prevailing synoptic winds.

11.2.2 Conceptual Framework

The formation of the UHI is related to the energy balance of the urban area. Urban structures and materials, land cover change, and human activity alter the individual components of

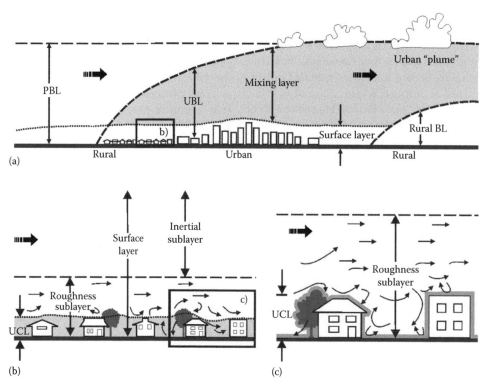

FIGURE 11.1 Idealized vertical structure of the urban atmosphere over (a) an urban region at the scale of the whole city (mesoscale), (b) a land-use zone (local scale), and (c) a street canyon (microscale). Gray-shaded areas (thick line following surface in (c)) show "locations" of the three UHI types corresponding to each scale (see Section 11.2.3 and Table 11.1). (Modified after Oke, T.R., Initial guidance to obtain representative meteorological observations at urban sites, *Instruments and Observing Methods Report No. 81*, WMO/TD-No. 1250, World Meteorological Organization, Geneva, Switzerland, 2006.)

the energy balance in urban areas and subsequently the atmospheric state. The UHI must therefore be the result of urban/rural energy balance differences. The urban energy balance is defined as (Oke 1988)

$$Q^* + Q_F = Q_H + Q_E + Q_S + \Delta Q_A \, (\text{Wm}^{-2}) \quad (11.1)$$

where

Q* is the net all-wave radiation flux

Q_H and Q_E are the turbulent sensible and latent heat fluxes, respectively

Q_S is the net uptake or release of energy by sensible heat changes in the urban ground-canopy-air volume

ΔQ_A is the net horizontal advective heat flux

Q_F is the anthropogenic heat flux from heat released by combustion of fuels (e.g., traDc, building HVAC systems). It is an additional term particular to cities and not present in other ecosystems

The available energy at any location to heat the air or ground or evaporate water depends on the radiation balance

$$Q^* = K^* + L^* = K \downarrow - K \uparrow + L \downarrow - L \uparrow (\text{Wm}^{-2}) \quad (11.2)$$

where

K and L are the shortwave (from the sun) and longwave (or terrestrial) radiation flux, respectively

arrows indicate whether the flow of energy is toward (↓) or away (↑) from the surface

The presence of the city alters all individual components in (11.1) and (11.2). The surface morphology results in a lower albedo for an urban array because of multiple reflections of the incoming solar radiation in street canyons. Street canyons also trap longwave radiation resulting in a lower net longwave loss (L*) at street level. The radiative properties of surface materials influence the albedo, surface temperature and emissivity, and therefore the outgoing short- and longwave radiation and respective net fluxes. The urban atmosphere is both polluted and warm, and the surface is warmer compared to the countryside, which also affects the net longwave radiation balance. Overall the urban effects tend to offset, and Q* in cities is close to that in nearby rural settings. Sometimes the lower albedo of the city can result in positive urban-rural Q* daytime differences, and negative nocturnal differences are often observed as a consequence of extra longwave emissions from the warmer urban surface temperatures (Figure 11.2).

The warm and rough nature of the urban surface promotes turbulent mixing, and unstable conditions prevail during daytime and mildly unstable or neutral conditions at night, especially in summertime and in densely built-up areas. The convective fluxes are therefore usually directed away from the surface at most hours of the day. Reduction of Q_E in the city is common during daytime but evapotranspiration can remain an important energy sink outside densely built urban centers depending on the amount of greenspace, surface wetness, precipitation, or irrigation. Because of lower Q_E in the city, heat during daytime is preferentially channeled into sensible forms (Q_H and Q_S), which results in a warming of the environment. Key characteristics that influence the size of Q_S are the surface materials and the urban structure. The mass of the building fabric presents a large reservoir for heat storage because urban surface materials have good ability to accept, conduct, and diffuse heat. Q_S is therefore significant and often considerably larger in an urban area than its rural surroundings (Figure 11.2). The partitioning of Q* into Q_H is relatively constant across a wide range of cities, geographic locations, and climatic regions during dry and clear daytime conditions (0.35 < Q_H/Q* < 0.45) but is more variable for Q_S (0.2 < Q_S/Q* < 0.55) (Roth 2007). Q_S is usually transported into the building volume in the morning, and by mid to late afternoon heat is transferred back to the surface and released into the atmosphere. This helps to maintain a positive Q_H flux in cities in the evening and at night, contributing to the heat island in the air. At night, the additional radiative drain is entirely supplied from the extra heat stored in the urban fabric during daytime (Figure 11.2).

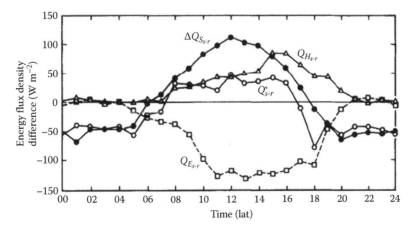

FIGURE 11.2 Differences of hourly ensemble average energy fluxes between a suburban (*s*) and a rural (*r*) site in Vancouver (Canada) for 30 summer days. Values are positive if suburban > rural. (From Cleugh, H. and Oke, T.R., *Bound. Layer Meteorol.*, 36, 351, 1986. With permission.)

11.2.3 Heat Island Types, Characteristics, and Underlying Processes

The UHI commonly refers to the temperature increase measured in the air close to the surface inside a settlement compared to its undeveloped surroundings. Heat islands, however, may be measured as either surface or atmospheric phenomena, and a further distinction is sometimes introduced according to the observation method. Fixed station networks produce different heat islands compared to those measured using car traverses or by the thermal response from true 3-D surface compared to the bird's eye view for surface temperature (Oke 1995). Although they are related, it is essential to distinguish between the different types because the respective processes, observations, and models will differ. Table 11.1 lists the main heat island types (also shown in Figure 11.1) together with their defining characteristics and some of the known impacts.

11.2.3.1 Surface UHI

The *surface UHI* is defined by the temperature of the surface that extends over the entire 3-D envelope of the surface. It is a surface energy balance phenomenon and involves all urban facets (street, vertical walls, roofs, trees, etc.). Urban surface temperatures contain strong microscale patterns that are sensitive to the relative orientation of the surface components to the sun by day and the sky at night, as well as to their thermal (e.g., heat capacity, thermal admittance) and radiative (e.g., reflectivity or albedo) properties. The magnitude and temporal variation of the surface heat island are well known. It is strongest during daytime when solar heating creates large differences between dry/wet and vegetated surfaces and the response is dominated by exposed, horizontal surfaces such as roofs and pavements. During daytime, the warmest surfaces are measured in industrial–commercial zones, especially those with large, flat-topped buildings or extensive open areas of pavement (e.g., airport, shopping malls, and major highway intersections) rather than in the CBD where buildings are tall and roofs are not the principal surface (Figure 11.3). At night, some of the processes are reduced, and urban-rural differences and intra-urban variability of surface temperature are smaller than during the day (Roth et al. 1989). The surface heat island has been less studied compared to its atmospheric counterpart.

The remotely sensed heat island is a surface phenomenon that should not be set equal to its atmospheric counterpart whose magnitude is lower and largest during nighttime, which is the reverse of the surface temperature pattern (Figure 11.3). Also, in daytime,

TABLE1 1.1 Classification of Urban Heat Island Types, Their Scales, Underlying Processes, Timing and Magnitude, and Likely Impacts

UHI Type; Spatial Scale	Processes	Timing: Magnitude	Impacts
Subsurface; Micro (1–100s m)	Subsurface energy balance; heat diffusion into ground	Day/night: small; follows surface heat island	Engineering design for water pipes, road construction, permafrost, groundwater characteristics, and carbon exchange between soil and atmosphere
Surface; Micro (1–100s m)	Day: surface energy balance; strong radiation absorption and heating by exposed dry and dark surfaces	Day: very large and positive	Thermal comfort, planning and mitigation measures, temperature of storm water runoff, and health of aquatic ecosystems
	Night: surface energy balance; roofs—large cooling (large sky view); canyon facets—less cooling (restricted sky view)	Night: large and positive	
UCL; Local (1– <10 km)	Day: strong positive sensible heat flux at surface; sensible heat flux convergence in canyon	Day: small, sometimes negative if shading is extensive	Thermal comfort, building energy use, water use (irrigation), thermal circulation if winds are light, air quality, urban ecology, and ice and snow
	Night: often positive sensible heat flux supported by release of heat from storage in ground and buildings, longwave radiative flux convergence, and anthropogenic heat	Night: large and positive, increases with time from sunset, maximum between a few hours after sunset to predawn hours	
UBL; local-meso (10s km)	Day: bottom-up sensible heat flux through top of RSL, top-down heat entrainment into UBL, and radiative flux divergence due to polluted air	Day/night: small and positive, decreasing with height in UBL	Air quality, photochemical pollutants, local circulation, precipitation and thunderstorm activity downwind, and plant growing season
	Night: Similar to day, but intensity of processes is reduced		
	Anthropogenic heat under special conditions		

Sources: Modified from Oke, T.R., The heat island of the urban boundary layer: Characteristics, causes and effects, in *Wind Climate in Cities*, ed. J.E. Cermak et al., Kluwer Academic Publishers, Dordrecht, the Netherlands, pp. 81–107, 1995; Oke, T.R., The need to establish protocols in urban heat island work. Preprints *T.R. Oke Symposium and Eighth Symposium on Urban Environment*, January 11–15, Phoenix, AZ, 2009.

UCL, Urban canopy layer; UBL, urban boundary layer; RSL, roughness sublayer.

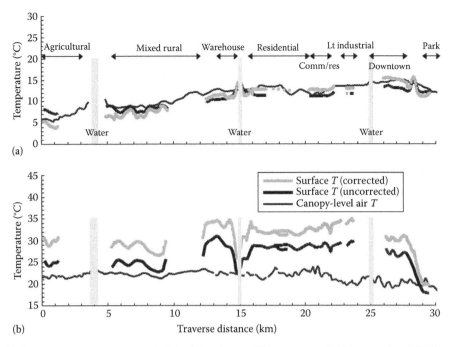

FIGURE 11.3 Heat island transect across Vancouver, BC, for (a) nighttime (YD 238 1992; 9 h after sunset) and (b) daytime (YD 237 1992; 2 h following solar noon) showing canopy-level air temperature and remotely sensed surface temperature before ("raw," directional brightness) and after corrections (for atmospheric effects, variations in emissivity and to account for surfaces not viewed by the remote sensor) have been applied. The plotted results represent values normalized to a single time. The automobile traverse passed through a tunnel (~4 km) and over bridges (~15 and 25 km) along the route. (Adapted from Voogt, J.A. and Oke, T.R., *Remote Sens. Environ.*, 86, 370, 2003.)

the correlation between surface temperature and land use is much stronger than that for air temperature, and the daytime urban-rural surface temperature differences are considerably larger than those measured in the air in the canopy layer (Figure 11.3). Relating surface and air temperatures is complicated and no simple general relation is available. Without considering the differing source areas for the two measurements and atmospheric effects (radiative divergence in the UCL and horizontal heat advection) that affect air temperatures, it is easy to misinterpret canopy-layer UHIs using measurements of surface temperatures.

11.2.3.2 Canopy-Layer UHI

The *canopy-layer UHI* is found within the atmosphere below the tops of buildings and trees (i.e., in the urban canopy). It is an expression of the surface energy balance that influences the air volume inside the canopy, primarily through sensible heat transfer from the surface into the canyon to change the temperature. Some exchange of air between the canyon volume and the air above is also possible during the day. At night, some cold air coming off the roofs may contribute to cooling. The canopy-layer UHI is a local (neighborhood)-scale phenomenon. Given its accessibility and relevance to human activities, it is the most studied of all heat island types.

Urban-rural air temperature differences measured in the canopy-layer show significant spatial and temporal variability within a city but generally form isotherms in an island-shaped pattern resembling height contours of an island on a topographic map, which closely follows the built form of the city. The form

of the pattern varies from city to city, but a peak associated with the city center, a large gradient at the city periphery ("cliff"), and lower values associated with more open areas, parks, and water surfaces are some of the common characteristics found across settlements of different sizes (Figure 11.4). Microscale variations within individual street canyons or courtyards are complex, with large temperature differences possible at small scales.

The diurnal variation of the canopy-layer UHI is very pronounced and its physical basis is understood theoretically and has been confirmed by numerous studies (Figure 11.5). During daytime, the urban-rural difference is relatively small or even negative (i.e., cool island) in city centers or other developments with dense and tall buildings that promote shading at the surface. The heat island intensity increases after sunset and reaches a maximum sometime between a few hours after sunset and before sunrise. The canopy-layer UHI is therefore primarily a nocturnal phenomenon and arises from reduced cooling rates observed in the city in the late afternoon and evening compared to the non-built-up areas resulting in higher urban minimum temperatures. After sunrise, the urban area also warms up more slowly and the heat island is rapidly disappearing. Under ideal conditions for heat island development (calm and clear) and measured in city centers with deep canyons, maximum heat island intensities of up to 12°C have been recorded. On an annual mean basis, including the dampening effects of wind and clouds, a city of about one million inhabitants may have a heat island of about 1°C–2°C (Oke 1997).

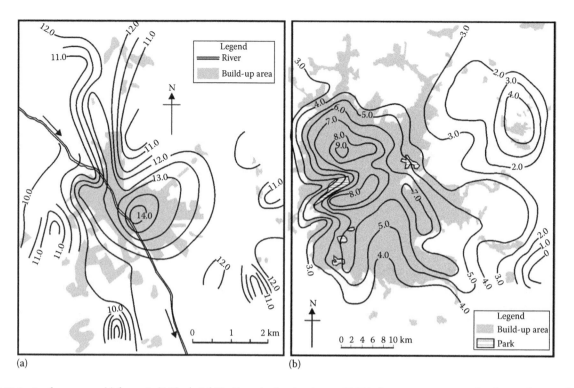

FIGURE 11.4 Isotherm maps (a) for an individual night in Uppsala, Sweden (pop < 20,000; from car traverses under clear and near calm conditions on August 30, 1948), and (b) of mean monthly minimum temperature in Mexico City (pop > 10,000,000; from fixed station network under mostly clear and weak wind conditions in November 1981). (Redrawn from Sundborg, A., *Tellus*, 2, 222, 1950. With permission; Jauregui, E., The urban climate of Mexico city, in *Urban Climatology and Its Applications with Special Regard to Tropical Areas*, ed. T. R. Oke, pp. 63–86, Geneva, Switzerland, World Climate Programme, Publication No. 652, WMO, 1986. With permission.)

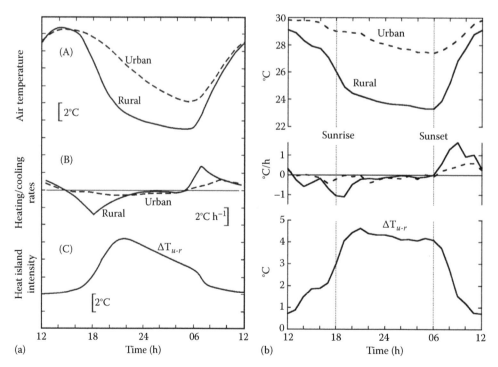

FIGURE 11.5 (a) Idealized diurnal variation of (A) urban and rural air temperatures, (B) heating/cooling rates, and (C) heat island intensity for a midlatitude city under ideal (clear and calm) conditions and (b) the same for measurements in Singapore (commercial site COM; ensemble average of 20 days with clear and calm conditions between May and August 2003). (a: From Oke, T.R., *Q. J. R. Meteorol. Soc.*, 108, 1, 1982. With permission; b: Chow, personal communication, 2010. With permission.)

11.2.3.3 Boundary-Layer UHI

The urban warmth extends into the UBL (above the RSL) through convergence of sensible heat plumes from local scale areas (bottom-up) and the entrainment of warmer air from above the UBL (top-down) to create the *boundary-layer UHI*. Radiative interactions through the polluted boundary layer may also be important. Because of experimental diDculties to probe the air at large heights, the UBL heat island has not received as much attention as its canopy-layer counterpart, but a few airplane, helicopter, remote sensing, balloon, and tower studies have been conducted since the 1960s in a wide range of cities. They provide insight into the vertical structure of the nocturnal UBL and confirm that the UHI extends upward to a depth of several 100 m with magnitudes that decrease almost linearly with height that often become negative near the top of the UBL (so-called crossover effect) (Figure 11.6). The heat island is largest under light winds and when strong rural surface inversions exist, and weaker for strong winds when the vertical temperature distribution is more uniform. The vertical extent of the UBL heat island is limited by the height of the UBL, which during daytime increases in depth with time in accordance with the growth of the mixed layer. At night, the city retains a shallow surface mixed layer because of the canopy-layer heat island. A cross section trough the air above the city shows a simple dome (calm conditions) or plume of warm air advected downwind of the city (more windy conditions), often extending for tens of kilometers (Figure 11.1a). The boundary-layer heat island is a local to mesoscale phenomenon and its intensity is less compared to that measured in the canopy layer (~1.5°C–2°C).

11.2.4 Heat Island Controls

The physical mechanisms through which the UHI effect is driven are well established, and most have been known since the first study of the heat island by Howard (1818) (Table 11.2). The heat island magnitude is related to the size and morphology of the city. It correlates positively with the number of inhabitants and the geometry of the street canyons in the downtown area. Although a positive correlation between population and UHI intensity has been demonstrated in many studies, population should only be considered a surrogate for city size, and the use of canyon geometry is preferred as a more physical parameter that exerts control over radiation access. At night, the amount of surface cooling is related to the net longwave radiation loss. The addition of an urban canyon restricts the ability of the surface to lose heat by radiation because the open sky is replaced by walls that intercept part of the outgoing longwave radiation. The exposure to the sky can be represented by the sky view factor (Ψ_s) and is sometimes approximated by the more readily available canyon height-to-width ratio (H/W). The strong relation between street canyon geometry and nocturnal maximum heat island intensity observed across many cities confirms that urban geometry is a basic physical control (Figure 11.7a).

Another category of surface controls relates to material properties. Urbanization not only changes the 3-D structure of the surface but also introduces radically different materials in terms of their radiative and thermal properties when compared to natural land cover. For example, dark surfaces such as asphalt roads have a low reflectivity to solar radiation and therefore eDciently absorb solar energy during the day. Many urban materials, such as those used for the construction of buildings or roads and parking lots, are dense and have a relatively high heat capacity and large surface thermal admittance. They have the ability to eDciently accept and retain heat during daytime for periods longer than that of natural surfaces and release it at night. In addition, urbanization replaces the natural impervious surfaces with

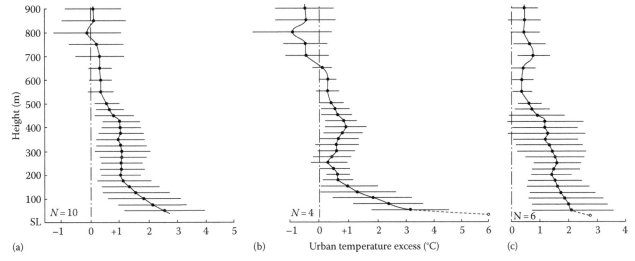

FIGURE 11.6 UBL heat island intensity (Urban Temperature Excess: difference between upwind rural helicopter ascent and the urban sounding corresponding most closely to the center of the heat island) over the metropolitan region of Montreal (Canada) (2.3 million people) in February/March 1968 just after sunrise for (a) all wind speeds, (b) 0–3 m s⁻¹, and (c) 3–6 m s⁻¹. (From Oke, T.R. and East, C., *Bound. Layer Meteorol.*, 1, 411, 1971. With permission.)

TABLE1 1.2 Commonly Hypothesized Causes of the Urban Canopy and Boundary-Layer Heat Islands (Not Rank Ordered)

Altered Energy Balance Terms Leading to Positive Thermal Anomaly	Features of Urbanization Underlying Energy Balance Changes
A. Canopy layer	
1. Increased absorption of shortwave radiation	Canopy geometry—increased surface area and multiple reflection
2. Increased longwave radiation from the sky	Air pollution—greater absorption and reemission
3. Decreased longwave radiation loss	Canyon geometry—reduction of sky view factor
4. Anthropogenic heat source	Building and traDc heat losses
5. Increased sensible heat storage	Construction materials—increased thermal admittance
6. Decreased evapotranspiration	Construction materials—increased "waterproofing"
7. Decreased total turbulent heat transport	Canyon geometry—reduction of wind speed
B. Boundary layer	
1. Increased absorption of shortwave radiation	Air pollution—increased aerosol absorption
2. Anthropogenic heat source	Chimney and stack heat losses
3. Increased sensible heat input—entrainment from below	Canopy heat island—increased heat flux from canopy layer and roofs
4. Increased sensible heat input—entrainment from above	Heat island, roughness—increased turbulent entrainment

Source: Oke, T.R., *Q. J. R. Meteorol. Soc.* 108, 1, 1982.

waterproof materials, which results in a drier urban area where less water is available for evaporation and hence cooling. The remaining vegetation cover imparts substantial spatial variability to the magnitude of the urban temperature. Because the heat island intensity is a difference measure, the surface values of the respective rural surface against which the city is compared are also important. Heat island development is maximized when the city is characterized by dense building materials but surrounded by dry rural areas with low thermal admittance and hence good ability to release heat.

The effects of thermal properties (promoting heat storage) and of street canyon geometry (reducing longwave radiation loss) are almost equally capable of creating a heat island under conditions that maximize UHI generation. This simplified picture is modified by a number of atmospheric factors that contribute to the presence and magnitude of heat islands. Synoptic weather in particular exerts a strong control on the heat island magnitude, which has been observed to be largest during clear (cloudless) and calm (weak wind) conditions (Figure 11.7b). The heat island intensity is most sensitive to wind speed, showing an inverse relationship under cloudless conditions, pointing to the effect of turbulence mixing and advection. Clouds affect the longwave radiation exchange and therefore the potential for the surface to cool. Low, thick clouds (e.g., Stratus) have a larger effect than a similar amount of high, thin ones (e.g., Cirrus).

The specific spatial and temporal form of the canopy-layer UHI is also sensitive to the climate, topography, rural

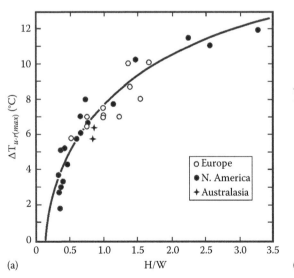

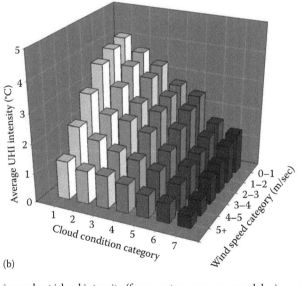

(a)　(b)

FIGURE 11.7 Controls on UHI intensity (I). (a) Relation between maximum heat island intensity (from car traverses measured during calm and clear conditions in dry summers) and canyon geometry expressed as the *H/W* of the street canyon in the center of cities on different continents. Line is empirical relationship (11.4). (b) Average nocturnal UHI intensity in Orlando, Florida, the United States, measured between September 3, 1999, and December 26, 2001, for specific combinations of cloud conditions and wind speed; cloud category 1 represents clear conditions and each successive category represents clouds having greater influence on longwave radiation fluxes. (a: From Oke, T.R., *Boundary-Layer Climates* (2nd edn.), London, U.K., Methuen, 1987. With permission; b: Yow, D.M., *Geogr. Compass*, 1/6: 1227, 2007. With permission.)

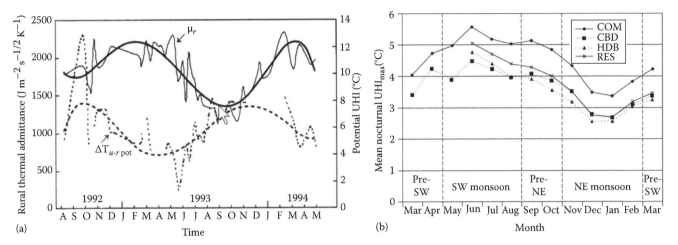

FIGURE 11.8 Controls on UHI intensity (II). (a) Variation of "potential" heat island magnitude (corrected to "ideal" weather value; see (11.7)) and observed rural soil thermal admittance from fixed urban and rural stations in Vancouver, BC, for the period August 7, 1992, to April 27, 1994. Smoothed line is polynomial fit. (b) Seasonal variation of monthly mean UHI maximum under all weather conditions for the period March 2003–2004. NE monsoon season is the wettest of all seasons sampled. (a: From Runnalls, K.E. and Oke, T.R., *Phys. Geogr.*, 21, 283, 2000. With permission; b: Chow, W.T.L. and Roth, M., *Int. J. Climatol.* 26, 2243, 2006. With permission.)

conditions, time of day, season, latitude (length of day and nighttime, which controls the amount of heat storage and cooling), and metabolism of the city. Multiple controls need to be considered when making comparison between observations from different cities. Temporal variability in the moisture content of the rural surrounding, for example, may be responsible for seasonal variations in heat island intensity. Dry-season UHI magnitudes are often higher than those observed during the wet season (Figure 11.8). This variability can be explained by the variability of surface moisture content of the undeveloped rural reference site, which is largely influenced by the seasonal variation of precipitation. Higher (lower) rural thermal admittance during the wet season results in lower (higher) rural cooling rates and hence lower (higher) UHI intensities assuming that the urban surface does not experience a corresponding change in surface moisture.

Seasonal variability is also observed in mid- and high-latitude cities without much seasonal moisture variability. Here, cities often reach their maximum heat island intensities during the cold season in winter, which suggests that space-heating plays a major role. Q_F can be an important term in (11.1) and even exceeds Q^* under these special circumstances when heating demands are large and incoming solar radiation is reduced or where the concentration of sources is large, such as in industrial areas or the center of a large city. Addition of anthropogenic heat may also be important in summer or all seasons in hot climates when air-conditioning use contributes waste heat to the urban canyon air. In these conditions, human activity contributes substantially to the heat balance and can be a main factor in the development of the heat island in the air.

Finally, cities are often located in complex geographic locations (coastal, bottom of valley, confluence of rivers, etc.) where differences in regional land cover and/or topography generate advection and thermal wind systems at scales larger than the city (land/sea breeze or valley/slope flows), which interact with the urban atmosphere and have the potential to modulate the UHI (Fernando 2010).

11.3 Methods of Analysis

Observational and modeling approaches are used to investigate the urban thermal environment. Field observations provide descriptive information about the UHI and the processes causing it, and methods differ for each UHI type. Observations are also essential to evaluate the full range of physical (scale), statistical, or numerical models used to research these processes and predict UHI behavior.

11.3.1 Observations

Observations are supposed to represent the urban effect on the climate; however, determination of the "true" effect is in most cases not possible, because observations from prior to the urban settlement do not exist. Lowry (1977) has investigated to what extent estimates of urban effects can be accepted with confidence unless preurban values are used. He evaluates four surrogate approximations, which are (1) urban-rural differences stratified according to synoptic conditions, (2) upwind-downwind differences (sometimes used to assess precipitation processes), (3) urban-rural ratios, and (4) weekday-weekend differences (measuring the particular weekly anthropogenic cycle that is uniquely related to urbanization). Most UHI studies are based on (1), i.e., the comparison between "urban" and "rural" locations, although it cannot be assumed that the rural observations taken in the vicinity of an urban area represent preurban conditions. This places a particular emphasis on the selection and proper description of the rural reference site during the planning and design of a study to ensure that it is free from urban influences.

11.3.1.1 Surface Temperatures

The spatial and temporal variability of surface temperature in urban areas has been investigated using thermal infrared measurements of upwelling thermal radiance (most often directional radiometric temperatures or directional brightness temperatures) from instruments based at ground level (e.g., pyrgeometer and thermal scanner), on aircrafts (e.g., thermal scanner), and on satellites (e.g., GOES, AVHRR, MODIS, Landsat, and Aster). The capability of remote sensors to provide high spatial resolution thermal imagery is very attractive and valuable to study the spatial and temporal patterns of the UHI across entire urban regions over time.

Temporal coverage and interpretation of remotely sensed data, however, are often restricted to specific times (prescribed by time of satellite overpass), atmospheric conditions (clear or mostly clear skies only), and the viewing geometry of the sensors. Uncritical acceptance of these data can also lead to erroneous conclusions. Recognition of the 3-D surface, possible variations in emissivity across different surface materials, and an appropriate definition and/or control for what defines the "rural" area are important issues (Roth et al. 1989). A representative surface temperature requires averaging an adequate sample of the many surfaces, vertical as well as horizontal, comprising an urban area. In most cases, remote sensors only sample a subset of the complete urban surface or have a nadir view (birds or plan view), which prefers horizontal surfaces and results in temperatures that are different compared to off-nadir views (especially those close to the horizon), which includes some proportion of the vertical surfaces. The proportion of surfaces that are not included, but are part of the active area, as part of the total 3-D area increases with increasing density of the built area.

The error introduced by oversampling the flat surfaces is likely to be largest near midday in the warm season and at low latitudes, i.e., during conditions when solar zenith angles are high. The 3-D nature of urban surfaces, together with solar and sensor geometric considerations, therefore implies that sensors with a narrow field of view see a biased sample of the microscale temperature patterns of the urban surface, and the upwelling longwave radiation received by the sensor shows directional variation as a function of the surface structure, also termed "effective thermal anisotropy" (Voogt and Oke 1998). Observations show that urban areas have strong thermal anisotropy, complicating the interpretation of measurements. Applications need to consider the importance of this effect, and the sum of corrections needed to be applied to the "raw" remotely sensed urban surface temperatures (e.g., directional brightness) can be significant, especially during daytime (Figure 11.3).

11.3.1.2 Air Temperature

Air temperature measurements using direct techniques (thermocouples and thermistors) are relatively simple. The UHI has been well documented using observations from fixed sensor networks and mobile surveys using automobiles. Often simple urban-rural station pairs are used to assess the UHI effect, but they may miss the location of the UHI peak as opposed to car traverses that provide a more complete sample of the full diversity of urban morphologies and land uses. More recently, networks with a large number of stations to provide better spatial resolution have been employed, benefiting from advances made in sensor miniaturization and data transmission technology.

Careful siting and exposure are essential to obtain meaningful observations. The characteristics of the surface and atmosphere within the source determine the measured temperature. The specific requirements on instrument exposure, siting, and metadata requirements for urban temperature measurements are described in Oke (2006). If the objective is to monitor the thermal environment of the canopy layer, the sensors must be exposed so that their microscale surroundings are representative of the local-scale environment representative of the selected neighborhood. The sensor location must be surrounded by "typical" conditions for urban terrain. Ideally, the site should be located in an open space, where the surrounding H/W ratio is representative of the local environment, away from trees buildings or other obstructions. Care should be taken to standardize practice across all sites used in a network regarding radiation shields, ventilation, height (2–5 m is acceptable given that the air in canyons is usually well mixed) and to ensure that sensors are properly calibrated against each other. Locations in urban parks, over open grass areas, or on rooftops should be avoided since they are not representative of the urban canopy.

Much less work is available on the boundary-layer UHI compared to that in the canopy layer. Within the UBL, air temperatures may be measured in situ by sensors on tall towers, tethered or free-flying balloons, aircraft, helicopters, and remote observation methods. Known challenges are the determination of the exact footprint of the measurement, which increases with increasing height and the possibility of large-scale advection of air with different thermal characteristics.

In the absence of a standard methodology to calculate the UHI intensity from urban to nonurban differences, it is necessary to apply strict experimental control to be able to separate the confounding effects of time, weather, relief, and soil moisture on UHI magnitude and obtain meaningful results that can be compared across studies. Basic requirements include the proper classification and reporting of field sites (urban and rural), e.g., using the recently proposed *local climate zones* by Stewart and Oke (2009). This system replaces the traditional and simple descriptors "rural" and "urban" with a more sophisticated one, which takes into account the diversity of real cities and their surroundings. The suggested local climate zones are differentiated according to surface cover (built fraction, soil moisture, albedo), surface structure (sky view factor, roughness height), and cultural activity (anthropogenic heat flux). It is further necessary to ensure that sensor locations are not affected by local thermal gradients (produced by, e.g., coastal areas or mountain-valley effects), synoptic weather is controlled for (in particular wind and clouds), moisture effects on the rural thermal admittance are considered, and to normalize daylength (which controls the amount of daytime heat storage and nocturnal cooling).

11.3.2 Models

11.3.2.1 Physical Scale Models

Physical scale (or hardware) models are not appropriate to study the thermal environment directly but have been useful tools to examine some of the heat island controlling parameters under simplified conditions (e.g., nighttime, negligible turbulent transfer). Nocturnal cooling and heat island intensity have been studied by Oke (1981). Under clear and calm conditions, the nighttime energy balance can be approximated by $L^* = Q_S$, and the resulting cooling is a function of L^* and thermal admittance only (according to the Brunt framework). Using a simple indoor hardware model, Oke (1981) confirmed that thermal admittance and geometry (e.g., expressed through Ψ_s or H/W) are important controls under these ideal conditions, a result that supported the development of the simple empirical models in (11.4) and (11.5) as follows. Another scale modeling study confirmed the influence of geometry and thermal admittance on nocturnal cooling for calm conditions in the case of an urban park (Spronken-Smith and Oke 1999). Parks with significantly lower thermal admittance than their surroundings, and those with large openings to the sky, showed strongest cooling potential. View factor influences also create an edge effect of reduced cooling near the park perimeter within a zone of about 2.2–3.5 times the height of the park border. Other hardware models have investigated the longwave radiation exchange in simple canyon geometries, and outdoor scale models demonstrate that 2-D and 3-D urban geometries can substantially decrease the surface albedo.

11.3.2.2 Statistical Models

Many statistical UHI models have been developed because of their simplicity and ease to use. They generally work well for the city and the limited range of atmospheric conditions for which they were created. Most of these models are simple (multiple) linear regressions of maximum UHI intensity ($\Delta T_{ut\,r(\max)}$) measured during ideal conditions (calm and clear weather) related to population (P), geometric (e.g., H/W ratio) and meteorological variables (wind speed, clouds, rural lapse rate or solar radiation), or use land cover information to model intra-urban temperature variation. One of the best-known models relates maximum heat island magnitude to the population according to a log-linear relation (Oke 1973)

$$\Delta T_{u-r(\max)} = a \cdot \log(P) + b \qquad (11.3)$$

The regression parameters differ for North American, European, and Asian cities and probably for different climatic regions because of diversity in city morphology and rural surroundings (where the reference temperature is measured) across different geographic locations. Equation (11.3) is purely

empirical and dimensionally incorrect, and the use of canyon geometry instead of population is preferred as a more physical parameter that exerts control over the local wind and radiation access. The strong relation between the street canyon geometry and nocturnal maximum heat island intensity measured under ideal conditions is expressed by the following empirical relationships (Oke 1981):

$$\Delta T_{u-r(\max)} = 7.45 + 3.97 \cdot \ln\left(\frac{H}{W}\right) \qquad (11.4)$$

or

$$\Delta T_{u-r(\max)} = 15.27 - 13.88 \cdot \psi_s \qquad (11.5)$$

which are valid for the specific conditions under which the data have been collected (clear, calm, and dry) (Figure 11.7a).

UHI intensity strongly depends on weather conditions, time of observation, and daylength and more advanced models try to predict the canopy-layer UHI intensity for any city for all time periods and weather conditions. Combining already-known, physically based empirical relationships between maximum UHI intensity and known weather controls in a common scheme, Runnalls and Oke (2000) define a weather factor

$$\phi_w = U^{-1/2}(1 - k \cdot N^2) \qquad (11.6)$$

where

U is mean wind speed

N is cloudiness, which allows the prediction of a weather-dependent UHI intensity as

$$\Delta T_{u-r} = \Delta T_{u-r(\max)}\phi_w \qquad (11.7)$$

Equations (11.6) and (11.7) indicate an inverse proportionality of nocturnal $\Delta T_{ut\,r}$ with $U^{-1/2}$ and a linear decrease with N^2 as was first shown by Sundborg (1950). Further refinement of these statistical models is possible, e.g., by applying control to normalize the amplitude of the UHI and length of daytime and nighttime across cities from different latitudes (Oke 1998).

11.3.2.3 Numerical Models

Numerical models offer increased capabilities to simulate the full complexity and diversity of cities and the way they interact with the atmosphere when compared to scale and statistical models. Historically, numerical urban climate models were adapted from vegetation models by modifying surface parameters to better represent particular aspects of the urban surface (e.g., surface albedo, roughness length, displacement height, surface emissivity, heat capacity, and thermal conductivity). They vary substantially according to their physical basis and their spatial and temporal resolution and have been developed to assess impacts of urbanization on the environment and, more recently, provide accurate meteorological information for planning mitigation and adaptation strategies in a changing climate.

Given its central role in understanding and predicting the urban climate, including the UHI, the urban energy balance has received much attention of the modeling community. Because of the complexity of the full surface-atmosphere system and energy processes, such models are often simplified by assuming several parameterizations and simplified geometries. An early example of an energy balance model that has been used to assess UHI causation is the SHIM surface heat island model (Oke et al. 1991). It is a simple energy balance model that can be used to calculate the surface radiation budget and surface temperature of street canyon surfaces on calm and cloudless nights (i.e., during conditions when turbulent exchange can be neglected and radiative transfer is restricted to longwave only). One of the first models to include many of the climate processes typical of the complex building-atmosphere volume to simulate the small-scale interactions between individual buildings, ground surfaces, and vegetation is ENVI-met (Bruse and Fleer 1998). It is a nonhydrostatic model designed for the microscale with very high temporal and spatial (0.5–10 m) resolution and a typical time frame of 24–48 h, capable of simulating the diurnal cycle of temperature, thermal comfort, and many standard atmospheric variables across a realistic building array.

Early models were treating individual building facets or one individual canyon only or were restricted to small domain sizes. Mesoscale meteorological models, on the other hand, are able to simulate the spatial structure and temporal dynamics of the UHI intensity across entire cities. One of the first such studies used CSUMM (CSU mesoscale model), which was modified by incorporating a semiempirical formulation for storage heat flux in urban areas to predict the daytime UHI in Atlanta (United States) (Taha 1999). Masson (2000) further extended the individual canyon approach to larger scales using simplified real city geometry, introducing individual energy budgets for roofs, roads, and walls; refining the radiation budget; and adding anthropogenic heat flux. His town energy balance scheme could be used in a prognostic way in mesoscale models and was successful in simulating many of the observed urban energetics, including the net longwave fluxes and surface temperatures. Extension to larger scales and coupling the urban surface to the atmosphere was an important step forward and demonstrated that mesoscale models that aim to predict weather at the city scale need to include a realistic representation of the urban surface.

Over the last decade, research in urban meteorology has focused on the development of parameterizations for urban canopy models. A number of schemes incorporating urban features have appeared since Masson's pioneering work for a variety of applications that include the assessment of human thermal comfort, prediction of UHI form and magnitude, heat island circulation, numerical weather prediction (NWP), which improves the operational forecasts of screen-level temperatures, or global climate modeling. An example is the new urban scheme MORUSES. The urban geometry is described as a 2-D street canyon and takes into account varying building geometry at the grid scale over the model domain. A complete urban surface energy balance is parameterized for the 2-D geometry,

including multiple reflections within street canyons, transporting heat out of the canyon via a resistance network, and differentiating between the energy balance of canyon and roof facets. This scheme is implemented in the UK Met Office Unified Model (UM) and is capable of producing many of the observed features of the diurnal surface energy balance variation such as reduced Q_E, increased Q_S, increased Q_H, and higher temperature in the city (Bohnenstengel et al. 2009).

To bridge the gap between traditional mesoscale modeling and microscale modeling, the National Center for Atmospheric Research (NCAR), in collaboration with other agencies and research groups, has developed an integrated urban modeling system coupled to the Weather Research and Forecasting (WRF) model (e.g., Chen et al. 2011). This urban modeling community tool includes three methods to parameterize urban surface processes, ranging from a simple bulk parameterization to a sophisticated multilayer urban canopy model with an indoor–outdoor exchange model that directly interacts with the atmospheric boundary layer and procedures to incorporate high-resolution urban land use, building morphology, and anthropogenic heating data. The WRF/urban model has been used, for example, as a high-resolution regional climate model to simulate the summer UHI of Tokyo, demonstrating the improved performance when using an urban canopy model (WRF/Noah/SLUCM) compared to a simple slab (WRF/Slab) approach to parameterize the surface processes (Figure 11.9).

Urbanization is also an important aspect of land-use/land cover change in climate science. Given the increasing resolution of global climate models, it is necessary to properly represent large metropolitan areas in the land-use component of global climate models. Oleson et al. (2008) presented a formulation of an urban parameterization designed to represent the urban energy balance in the Community Land Model (CLM3), which is the land surface component of the Community Climate System Model (CCSM). The model includes urban canyons to simulate the effect of the 3-D geometry on the radiative fluxes, distinguishes between pervious and impervious surfaces, and has separate energy balances and surface temperatures for each canyon facet, and heat conduction into and out of the canyon surfaces. Results indicate that the model does a reasonable job of simulating the observed energy balance of cities and general characteristics of UHIs in a qualitative sense (e.g., importance of Q_S increases with increasing H/W, positive Q_H into the night, and decreased diurnal temperature range in the city).

11.4 Impacts and Applications

Because of the increasingly obvious urban impacts on the environment and society, accurate information about the urban climate is needed for a wide range of users. In terms of consequences for humans, the canopy layer is of greatest interest with the UHI as the predominant phenomenon. Whether this unintended impact is desirable or not depends on the background climate. For cities in cold climates, the UHI may produce positive effects such as less snowfall and frost events, longer growing

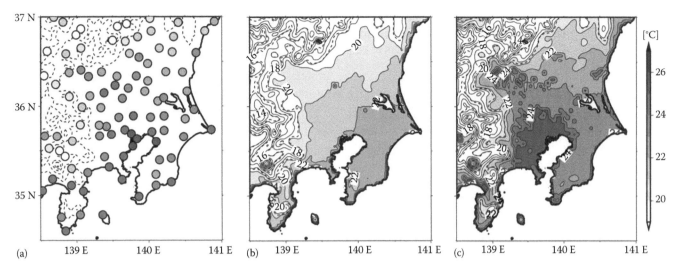

FIGURE 11.9 Monthly mean surface air temperature at 2 m in the Tokyo area at 0500 JST in August averaged for 2004–2007. (a) AMeDAS observations, (b) using WRF/Slab model, and (c) WRF/Noah//SLUCM. When the simple slab model is used, the model is unable to reproduce the observed heat island of Tokyo and the urban areas in the inland northwestern part of the Kanto plain located just to the northwest of Tokyo Bay, which is located in the center of the map. (From Chen, F. et al., *Int. J. Climatol.*, 31, 273, 2011. With permission.)

season, and reduced energy demand because of less domestic heating. In warm/hot climates or temperate climates in summertime, higher temperatures increase the use of energy use for cooling, degrade the air quality (e.g., ozone formation increases with increasing temperature), and may produce stressful conditions and increase in mortality as temperatures rise outside an optimum range. The UHI intensity can be as large as 12°C under certain conditions, posing a threat to the sustainability of an area and calling for the implementation of heat island mitigation strategies from city managers and planners. Strategies have therefore been developed to manage the urban thermal effects that can help to mitigation the UHI, reduce energy demand, and, at the same time, contribute to the reduction of climate change emissions.

11.4.1 Solutions to Mitigate Urban Heat Islands

The UHI effect is a result of urban/rural energy balance differences. A useful starting point is therefore the consideration of the individual terms in (11.1) and (11.2) and their dependence on the urban fabric and geometry (Table 11.2) to decide which terms could be readily altered to have the desired mitigation outcome. Interventions and applications depend on scale, can occur via city function and form (Table 11.3), and have to consider the climate setting of a location.

Many recent building development and design options have concentrated on reducing building energy use and the UHI to improve human comfort in summer in temperate climates or for all seasons in hot, (sub)tropical regions. The scope for change is often limited as the basic morphology (building dimensions and placement, street width, and green areas) is already in place. Consequently, efforts have focused on changing the properties of the urban surface to modify its radiative (by replacing materials with new surface cover to reduce the radiative heat gain

TABLE 11.3 Basis of Selected Mitigation Measures to Control the Urban Heat Island

Scale	Intervention	Control
Building	Roughness	Airflow, ventilation
	Trees, overhangs, narrow spaces	Provide shade and shelter
	Impervious surface fraction	Energy partitioning between sensible (heating) and latent, evaporative (cooling) exchanges
	Porous pavement	Increase surface wetness, evaporative cooling, reduce runoff
	Vegetated roofs	Cool rooftop through shading and evaporative cooling and provide additional insulation to improve building energy performance
	High albedo, light surfaces	Influences surface heat absorption and ensures high reflection of radiation
	Sky view factor	Influences solar access and radiative cooling
	Thermal admittance	Modulates heating and cooling cycles of materials
	Thick walls, roof insulation	Modulates heat storage
Neighborhood	Morphology, building and pavement materials, amount of vegetation, and transport	Influence airflow, ventilation, energy use, anthropogenic heat emissions, pollution and water use via city form and function

in the material) and evaporative properties. Vegetation has been found to be a versatile tool to mitigate the local thermal environment. It can be used to provide shade, thermal insulation to keep the interior cool, and evaporative cooling, and manage noise and air pollution.

Much research has been conducted to evaluate options to mitigate urban heat in low-latitude regions. Given high sun zenith angles in (sub)tropical regions, roofs, and other open surfaces such as large pavements are the most thermally active parts of the urban geometry and prime areas of energy absorption and conduction into the ground calls for high-albedo materials to moderate the conductive heat transfer. Especially in hot, arid areas, protection from solar radiation, which is received in high quantities given the absence of clouds, dry atmosphere, and relative proximity to the equator, is paramount. Compact urban forms with narrow street canyons provide deep shading but promote trapping of shortwave radiation and reduce nighttime longwave cooling. Together with reduced ventilation by the narrow geometry, this may result in elevated canyon air temperatures. Model results show that access to solar radiation and thermal stress can be reduced by high H/W ratios in N-S streets, provided that the building and paving materials have suDciently high heat capacity to store excess heat and radiate it away during the cooler nighttime hours (Pearlmutter et al. 2007). Where water is available, evaporative cooling from strategically placed vegetation can further mitigate the thermal environment. Urban populations in the warm-humid regions of the tropics are exposed to the negative effects of the UHI on an annual basis given the high-incident solar radiation, generally low wind speed, and lack of cool winters. Here, mitigation options need to promote ventilation, e.g., through the manipulation of geometry or street orientation to take advantage of preferential wind patterns (e.g., monsoonal), or provide shade, e.g., through the use of artificial means (overhangs or canopies) or by taking advantage of the abundant natural vegetation, which can also provide evaporative cooling.

11.4.2 Human Comfort and Health

Increasing temperatures decrease human comfort in hot climates and raise mortality rates at temperatures outside an optimum range. People living in urban areas exposed to the UHI are at greater risk than those in nonurban regions. In many parts of the world, heat already has a devastating impact on human health, and excessive heat events (heat waves) have led to a large number of deaths in many large, midlatitude cities (e.g., Athens, Chicago, New York, Paris, Philadelphia, Rome, Shanghai, and Seoul), making it the most important weather-related killer. Higher minimum temperatures due to the UHI and subsequent lack of nighttime relief exacerbate the heat wave process and likely increase heat stress and mortality. Air conditioning seems to have a positive effect in reducing heat-related deaths, but it is possible that waste heat emitted by air conditioning may be actually contributing to the UHI. Other solutions come from the various cooling initiatives to improve indoor climate and mitigate the UHI effect (Section 11.4.1).

11.4.3 Forecasting Urban Weather and Climate and Developing UHI Mitigation Strategies

It is clear that urban planning and architectural design have an important role in regulating thermal comfort, reducing heat-related mortality, and the development of more sustainable cities.

However, before policies can be developed to avoid or adapt to potentially dangerous overheating conditions, tools are required to identify and quantify the effectiveness to mitigation and adaptation strategies. Most operational global weather and climate models still fail to resolve cities and are unable to provide accurate forecasts for large metropolitan areas. Given advances in computer resources, however, it is now possible to resolve urban areas, and several urban models of different complexity are currently available for use in operational NWP models (see also Section 11.3.2.3). Research shows that even a basic representation of urban areas can lead to significant improvement in urban temperature forecasts (Best 2005).

A number of studies are currently in progress to develop climate-sensitive development strategies at the city scale. Using a modified version of the MM5 regional climate model, the UHI and heat island mitigation scenarios, included planting trees in open spaces and along trees, green roofs, high-albedo roofs and surfaces, and a combination thereof, have been simulated for New York City (Rosenzweig et al. 2009). Results from this study indicate that the influence of vegetation (using a combination of tree planting and vegetated roofs) on urban climate is more important than the influence of the albedo of built surfaces for the case of New York City. Applying this strategy reduced simulated city-wide urban air temperature by 0.4°C on average. Because of the relatively large amount of built surfaces and the relative lack of area available for interventions, the temperature reduction is at the lower end of similar mitigation studies carried out in other cities with strongly differing urban geometries that show mitigation potential of up to 3.6°C for average air temperature. UHI mitigation strategies involving vegetation tend to be more expensive per unit area than strategies involving high-albedo surfaces. However, incorporating other benefits, including air quality and public health improvements, and reductions in the cityʻs stormwater runoff and contribution to greenhouse gas emissions might improve the cost effectiveness of strategies involving vegetation (Rosenzweig et al. 2009).

Another application is the use of the new urban scheme MORUSES as part of the UK Met ODce UM to perform long-term climate simulations over London. These two nested models are used as part of the LUCID project (the development of a local urban climate model and its application to the intelligent design of cities; http://www.lucid-project.org.uk/), which brings together meteorologists, building scientists, urban modelers, and epidemiologists in order to investigate the physical processes responsible for the UHI of London and to derive planning strategies for future energy use and green space (Bohnenstengel et al. 2009).

11.4.4 The Urban Heat Island and Global Climate Warming

Urban areas cover less than 3% of the global land area, and the area of extra heat is local and too small to have a direct impact on global climates. However, the UHI has relevance for the study of global climate change because it makes it diDcult to detect the influence of human activity on the global mean temperature. Many of the earliest established observation stations used to construct the globally averaged surface temperature record have initially been located near urban areas and their readings may have become contaminated by the UHI as settlements have grown over time. Some of the research on historical data indicates that global temperature trends are not significantly affected by the UHI and the effect on the global temperature trend is no more than 0.05°C through 1990 and not considered significant (e.g., Parker 2006). Others contend that attempts to adequately remove the urban effect, e.g., by using empirical relations between city size and UHI magnitude from urban-rural pairs together with population data or satellite measurements of night light used to classify urban and rural stations, may be inadequate and underestimate the urban effect (e.g., Kalnay and Cai 2003).

Cities, however, have an indirect responsibility for the observed global warming as the major contributor of greenhouse gases. More than half of the world's population currently lives in cities and, thanks to their intensive metabolism, they release more than 70% of the total emissions of carbon dioxide (CO_2) of anthropogenic origin and a substantial proportion of other known greenhouse gases. In urbanized areas, these emissions have three main causes, which are transport, energy use in households and public buildings, and manufacturing and industry, with each sector contributing about one-third of the total. Energy use is sensitive to temperature, and there is a strong interdependence between the UHI and electricity demand where fossil fuels are used to generate the electricity that is driving air conditioning. Electricity demand for cooling increases 3%–5% for every 1°C increase in air temperature above approximately 23°C " 1°C (Sailor 2002). This implies that a 5°C UHI can increase the rate of urban electric power consumption for cooling by 15%–25% above that used in surrounding rural areas during hot summer months and for cities located in (sub)tropical regions.

UHIs have impacts that range from local to global scales, which emphasize the importance of urbanization to environmental and climate change. Magnitude and rate of urban warming are comparable to that considered possible at the global scale, and any global warming will raise the base temperature on top of which the UHI effect is imposed. Cities are also considered important agents in mitigating global climate change but at the same time their inhabitants and infrastructure are exposed to the effects of climate change. Many of the proposed mitigation and adaptation methods to increase the environmental sustainability of cities and make them more resilient to climate change are related to the urban thermal environment.

11.5 Major Challenges

Cities and their populations will continue to grow, increase stress on local environments, and remain important drivers of global environmental change. A good understanding of the nature of urban warming is important to (1) inform the construction of models to provide predictions to assess impacts on human comfort and mortality, (2) provide sound planning tools to assess the net impact of climate-based interventions for the design of more sustainable cities, and (3) explore the intimate relationship between the UHI and energy use and demand in cities and, hence, GHG emissions, which contribute to anthropogenic climate change. Moreover, much of past urban climate research has focused on developed cities located in midlatitude climates in the Northern Hemisphere. In contrast, only a rudimentary understanding of the physical processes operating in the atmosphere of (sub)tropical climates is available. This is unfortunate because much of the future urban growth will take place in cities located in low latitudes where there is an urgent need to incorporate climatological concerns in their design to provide a better living and working environment for a large segment of the world's inhabitants (Roth 2007). Future research should address the following areas where knowledge gaps exist.

11.5.1 Science

Despite much research, there is still a lack of quantitative analysis on the relative impact of specific contributions from the important controls on the UHI. There is a need to explore the linkages between the different UHI types. For example, the understanding of the coupling of surface and air temperatures and the relationship between the UHI in the canopy layer and the one in the boundary-layer are insuDcient. Satellite-derived UHI data are very valuable, but more research is needed to better assess urban surface emissivity and devise methods to correct for thermal anisotropy to be able to accurately determine surface temperatures in cities. Modeling and prediction of the UHI depend on the ability to properly simulate the surface energy balance of the 3-D urban surface and the transport processes in the UCL and UBL. Realistic geometries and buoyancy effects in microscale models, proper subgrid parameterization of aggregate effects of urban elements in mesoscale models, and the nesting of multiscale models all need more attention.

11.5.2 Observations

There is a growing need for meteorological observations conducted in urban areas, especially in the less developed regions, to support basic research, evaluate the increasing number of urban climate models being developed, building, and urban design or energy conservation measures. Guidelines on how to obtain representative measurements in urban areas are available (Oke 2006). However, no standardized observation protocols exist to estimate the heat island intensity, which has obstructed comparison among studies. A recent review concluded that almost half of 190 primary studies conducted between 1950 and 2007 reported UHI magnitudes that were methodologically unsound or unreliable (Stewart 2011). Weaknesses identified include a lack of experimental control, e.g., weather or relief, inappropriate choice of sites that are not representative of their local surroundings, definition of UHI magnitude is not given, inadequate exposure of sensors, sample sizes are insuDciently large to make statistical interferences, or failure to adequately document the sites. There is an urgent need for the adoption of common protocols in UHI research and its use in applied climatology and to document urban station metadata to improve the present situation and allow assessment of results across cities and weather conditions (see also Section 11.3.1.2).

References

Best, M. J. 2005. Representing urban areas within operational numerical weather prediction models. *Bound. Layer Meteorol.* 114: 91–109.

Bohnenstengel, S. I., Porson, A., Davies, M., and Belcher, S. 2009. Simulations of the London urban climate: The LUCID project. *T. R. Oke Symposium and Eighth Symposium on Urban Environment,* January 11–15, Phoenix, AZ.

Bruse, M. and Fleer, H. 1998. Simulating surface-plant-air interactions inside urban environments with a three dimensional numerical model. *Environ. Model. Sos ware* 13: 373–384.

Chen, F., Kusaka, H., Bornstein, R. et al. 2011. The integrated WRF/urban modeling system: Development, evaluation, and applications to urban environmental problems. *Int. J. Climatol.* 31: 273–288.

Chow, W. T. L. and Roth, M. 2006. Temporal dynamics of the urban heat island of Singapore. *Int. J. Climatol.* 26: 2243–2260.

Cleugh, H. and Oke, T. R. 1986. Suburban-rural energy balance comparisons in summer for Vancouver, B.C. *Bound. Layer Meteorol.* 36: 351–369.

Fernando, H. J. S. 2010. Fluid dynamics of urban atmospheres in complex terrain. *Annu. Rev. Fluid Mech.* 42: 365–389.

Grimmond, C. S. B., Roth, M., Oke, T. R. et al. 2010. Climate and more sustainable cities: Climate information for improved planning and management of cities (Producers/capabilities perspective). Procedia Environmental Sciences 1: 247–274.

Howard, L. 1818. *The Climate of London Deduced from Meteorological Observations.* London, U.K.: W. Phillips.

Jauregui, E. 1986. The urban climate of Mexico city. In *Urban Climatology and Its Applications with Special Regard to Tropical Areas*, ed. T. R. Oke, pp. 63–86. Geneva, Switzerland: World Climate Programme, Publication No. 652, WMO.

Kalnay, E. and Cai, M. 2003. Impact of urbanization and land-use change on climate. *Nature* 423: 528–531.

Kratzer, P. A. 1937. *Das Stadtklima* (in German). Die Wissenschaft, Vol. 90. Braunschweig, Germany : Friedr. Vieweg and Sohn (2nd edn. and English version, 1956).

Landsberg, H. E. 1981. *The Urban Climate.* London, U.K.: Academic Press, Inc.

Lowry, W. P. 1977. Empirical estimation of urban effects on climate: A problem analysis. *J. Appl. Meteorol.* 16: 129–135.

Masson, V. 2000. A physically-based scheme for the urban energy budget in atmospheric models. *Bound. Layer Meteorol.* 94: 357–397.

Oke, T. R. 1973. City size and the urban heat island. *Atmos. Environ.* 7: 769–779.

Oke, T. R. 1981. Canyon geometry and the nocturnal urban heat island: Comparison of scale model and field observations. *J. Climatol.* 1: 237–254.

Oke, T. R. 1982. The energetic basis of the urban heat island. *Q. J. R. Meteorol. Soc.* 108: 1–23.

Oke, T. R. 1987. *Boundary-Layer Climates* (2nd edn.). London, U.K.: Methuen.

Oke, T. R. 1988. The urban energy balance. *Prog. Phys. Geogr.* 12: 471–508.

Oke, T. R. 1995. The heat island of the urban boundary layer: Characteristics, causes and effects. In *Wind Climate in Cities*, ed. J. E. Cermak et al., pp. 81–107. Dordrecht, the Netherlands: Kluwer Academic Publishers.

Oke, T. R. 1997. Urban environments. In *Surface Climates of Canada*, ed. J. Bailey et al., pp. 303–327. Montreal, Quebec, Canada: McGill-Queens University Press.

Oke T. R. 1998. An algorithmic scheme to estimate hourly heat island magnitude, *Proceedings of 2nd Symposium Urban Environment*, November 1998, Albuquerque, NM, pp. 80–83.

Oke, T. R. 2006. Initial guidance to obtain representative meteorological observations at urban sites. *Instruments and Observing Methods Report No. 81*, Geneva, Switzerland: World Meteoreological Organization, WMO/TD-No. 1250.

Oke, T. R. 2009. The need to establish protocols in urban heat island work. Preprints *T.R. Oke Symposium and Eighth Symposium on Urban Environment*, January 11–15, Phoenix, AZ.

Oke, T. R. and East, C. 1971. The urban boundary layer in Montreal, *Bound. Layer Meteorol.* 1: 411–437.

Oke, T. R., Johnson, G. T., Steyn, D. G., and Watson, I. D. 1991. Simulation of surface urban heat islands under idealflconditions at night. Part 2: Diagnosis of causation, *Bound. Layer Meteorol.* 56: 339–358.

Oleson, K. W., Bonan, G. B., Feddema, J., Vertenstein, and M., Grimmond, C. S. B. 2008. An urban parameterization for a global climate model. Part 1: Formulation, evaluation for two cities, *J. Appl. Meteorol. Climatol.* 47: 1038–1060.

Parker, D. E. 2006. A demonstration that large-scale warming is not urban. *J. Clim.* 19: 2882–2895.

Pearlmutter, D., Berliner, P., and Shaviv, E. 2007. Integrated modeling of pedestrian energy exchange and thermal comfort in urban street canyons. *Build. Environ.* 42: 2396–2409.

Rosenzweig, C., Solecki, W. D., Parshall Lily et al. 2009. Mitigating New York cityfl heat island. *Bull. Am. Meteorol. Soc.* 90: 1297–1312.

Roth, M. 2007. Review of urban climate research in (sub)tropical regions. *Int. J. Climatol.* 27: 1859–1873.

Roth, M., Oke, T. R., and Emery, W. J. 1989. Satellite-derived urban heat island from three coastal cities and the utilization of such data in urban climatology. *Int. J. Remote Sens.* 10: 1699–1720.

Runnalls, K. E. and Oke, T. R. 2000. Dynamics and controls of the near-surface heat island of Vancouver, British Columbia. *Phys. Geogr.* 21: 283–304.

Sailor, D. J. 2002. Urban heat islands: Opportunities and challenges for mitigation and adaptation. Sample electric load data for New Orleans, LA (NOPSI, 1995). *North American Urban Heat Island Summit.* Toronto, Ontario, Canada, pp. 1–4.

Spronken-Smith, R. A. and Oke, T. R. 1999. Scale modelling of nocturnal cooling in urban parks. *Bound. Layer Meteorol.* 93: 287–312.

Stewart, I. D. 2011. A systematic review and scientific critique of methodology in modern urban heat island literature. *Int. J. Climatol.* 31: 200–217.

Stewart, I. D. and Oke, T. R. 2009. Newly developed "thermal climate zones" for defining and measuring urban heat island magnitude in the canopy layer. Preprints *T. R. Oke Symposium and Eighth Symposium on Urban Environment,* January 11–15, Phoenix, AZ.

Sundborg, A. 1950. Local climatological studies of the temperature condition in an urban area. *Tellus* 2: 222–232.

Taha, H. 1999. Modifying a mesoscale meteorological model to better incorporate urban heat storage: A bulk-parameterization approach. *J. Appl. Meteorol.* 38: 466–473.

Voogt, J. A. and Oke, T. R. 1998. Effects of urban surface geometry on remotely-sensed surface temperature. *Int. J. Remote Sens.* 19: 895–920.

Voogt, J. A. and Oke, T. R. 2003. Thermal remote sensing of urban climates. *Remote Sens. Environ.* 86: 370–384.

Yow, D. M. 2007. Urban heat islands: Observations, impacts, and adaptation. *Geogr. Compass* 1/6: 1227–1251.

II

Environmental Pollution

12

Atmospheric Dispersion

Jeffrey C. Weil
*University of Colorado, Boulder
and
National Center for Atmospheric
Research, Boulder*

12.1 Introduction

Observations of a smoke plume in the atmosphere show that its instantaneous structure is "puffy" with wide sections interspersed between narrow ones and that the plume meanders in a highly variable manner. Such variability is caused by the randomness of the turbulent wind field in which the plume is embedded. The turbulence is produced by wind gusts or mechanically generated "eddies" due to flow over the rough ground and, during daytime, by warm convective plumes or thermals rising from the heated surface. The turbulence exists primarily in the lowest part of the atmosphere known as the atmospheric boundary layer (ABL), which typically extends from 1 to 2 km above the surface during the day and from a few tens to a few hundreds of meters at night. Since dispersion is driven by turbulence, one needs information on the ABL turbulence properties in order to estimate dispersion.

Dispersion predictions are needed for a range of problems including (1) air quality in both urban and rural areas; (2) toxic and hazardous material releases either planned or unplanned; (3) odor nuisances downwind of feedlots, pulp mills, etc.; (4) agricultural applications such as crop spraying; (5) military uses; and (6) others. These applications encompass a range of release heights; source emissions including both gases and particles; buoyancies that can be positive, negative (e.g. dense gases), or zero; chemically reactive species; particle gravitational settling; surface deposition; etc. However, a common feature to all is turbulent dispersion.

In this chapter, our focus is on dispersion of "point-source" plumes such as those from chimneys or stacks, but the concepts carry over to other release geometries such as area or line sources, for example, freeway traﬃc emissions. Attention is directed to dispersion of passive (nonbuoyant) plumes in an ABL over flat or uncomplicated terrain with a rough surface, distances generally of ~30 km or less, and concentration averaging times of 1 h or less. It is important to recognize the statistical nature of dispersion, that is, the concentration measured at a fixed receptor downwind of a source which varies rapidly and randomly with time and cannot be predicted with precision. To obtain more order or regularity in the observed concentration, one averages it over time or an ensemble (i.e., large number) of observations, and it is this average or mean concentration field that is predicted by most models (see Section 12.3).

In the following, we describe the ABL mean and turbulence properties (Section 12.2) as well as the statistical nature of dispersion and important theories and models of dispersion (Section 12.3). The key features of dispersion and concentration fields as found from theories, field observations, laboratory experiments, and numerical simulations are then presented for both daytime (convective) and nighttime (stable) ABLs (Section 12.4). Section 12.5 gives a brief discussion of the statistical variability in concentration.

12.2 Atmospheric Boundary Layer

Turbulence in the ABL is controlled by the wind speed, surface heat flux, surface roughness, and static stability—the vertical gradient of the potential temperature Θ_a, $a\Theta_a/az$ where z is the height above ground (see Chapter 24 for a discussion of Θ_a and stability as represented by $a\Theta_a/az$). During daytime, the ABL usually takes the form of an unstable or convective boundary layer (CBL) due to strong solar heating of the ground, whereas at night a stable boundary layer (SBL) forms due to the outgoing net radiation (i.e., cooling) at the surface. Neutral conditions

Handbook of Environmental Fluid Dynamics, Volume Two, edited by Harindra Joseph Shermal Fernando· © 2013 CRC Press/Taylor & Francis Group, LLC.
ISBN: 978-1-4665-5601-0.

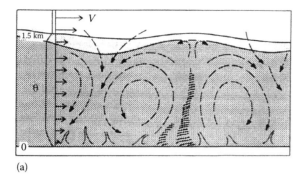

(a)

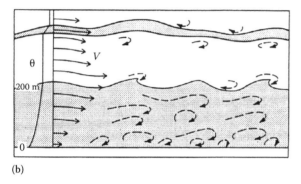

(b)

FIGURE 12.1 Schematic of the instantaneous or unaveraged structure of (a) the CBL showing its large turbulent eddies, thermal plumes or updrafts, wind speed (V) profile, and a thin temperature inversion at its top; and (b) the SBL with its small eddies, wind profile with a jet near 200 m, and wave motion. Note the large difference in the CBL and SBL depths. (Adapted from Wyngaard, J.C., *J. Clim. Appl. Meteorol.*, 24, 1131, 1985.)

($a\Theta_a/az = 0$) are rare at most midlatitude sites but do occur with high cloud cover (i.e., reduced surface heating or cooling) and strong winds.

In the CBL, the surface heating leads to large-scale convective or turbulent motions in the form of "updrafts" and "downdrafts" that extend from the surface to the CBL top, which is capped by an elevated inversion or stable layer, $a\Theta_a/az > 0$ (Figure 12.1a). The inversion defines the CBL height z_i and suppresses upward motion and turbulence due to the stable stratification. The convective elements or "eddies" in the CBL have horizontal dimensions proportional to z_i, with updrafts being narrower and more intense than downdrafts. The CBL has turbulent or root-mean-square (rms) velocities proportional to the convective velocity scale $w_\star$: $w_\star = [gH_o z_i/(\rho_a c_p T_a)]^{1/3}$, where g is the gravitational acceleration, H_o is the turbulent surface heat flux, ρ_a is the air density, c_p is the specific heat of air at constant pressure, and T_a is the absolute temperature of air. At midday, typical values of z_i and $w_\star$ are 1–2 km and 2 m/s, respectively. These large turbulence scales lead to rapid dispersion.

Turbulence arising from mechanical processes or wind shear is generally confined to small heights, $z < -L$, where L is the Monin–Obukhov (MO) length. Here, $L = -\rho_a c_p T_a u_\star^3/(kgH_o)$, where $u_\star$ is the surface friction velocity, and k the von Kármán constant (=0.4). As discussed in Chapter 24, a positive H_o corresponds to an upward heat flux (unstable case) and a negative

L, whereas a negative H_o applies to a downward flux and positive L. Since convection extends to z_i, Deardorff (1972) suggested $-z_i/L$ or $z_i/|L|$ as the stability parameter representing the relative effects of convective and mechanical turbulence over the bulk of the CBL. This can be expressed alternatively by the ratio of velocity scales characterizing the convective and mechanical turbulence—$w_\star/u_\star$, where $w_\star/u_\star = (-z_i/kL)^{1/3}$.

In highly convective conditions ($z_i/|L| > 20$ and best for $z_i/|L| > 50$), the CBL can be considered to have a vertical structure comprising three layers: surface, mixed, and entrainment. In the surface layer ($z < 0.1z_i$), the mean wind and potential temperature exhibit strong gradients that are described by MO similarity theory (Businger et al., 1971). The longitudinal σ_u, lateral σ_v, and vertical σ_w rms turbulence components can be parameterized by (Panofsky et al., 1977)

$$\frac{(\sigma_u, \sigma_v)}{u_\star} = \left[4 + 0.6\left(\frac{-z_i}{L}\right)^{2/3}\right]^{1/2} \quad \text{and} \quad \frac{\sigma_w}{u_\star} = \left[1.6 + 2.9\left(\frac{-z}{L}\right)^{2/3}\right]^{1/2}.$$

(12.1)

The σ_v expression interpolates between a neutral stability limit ($z_i/|L| \ll 1$), $\sigma_v \propto u_\star$, and a highly convective limit ($z_i/|L| \gg 1$), $\sigma_v \propto w_\star$. The σ_w equation interpolates between a neutral limit, $\sigma_w/u_\star = 1.3$, and a "free-convection" limit, $\sigma_w/w_\star = 1.3(z/z_i)^{1/3}$ (Wyngaard et al., 1971), where the latter is valid over the height range $-L < z < 0.1z_i$.

In the mixed layer ($0.1z_i < z < 0.8z_i$), the mean wind, potential temperature, and turbulence components exhibit little height variation, that is, they are well mixed with $\sigma_u, \sigma_v, \sigma_w \sim 0.6w_\star$. In the entrainment layer ($0.8z_i < z < 1.2z_i$), all variables tend toward their values above the CBL.

For the highly unstable regime ($z_i/|L| > 20$ or 50), the convective elements are randomly distributed thermal plumes surrounded by large downdrafts (Figure 12.1a). In this case, the vertical velocity (w) probability density function (PDF) is positively skewed (non-Gaussian) with the downdraft regions comprising 60% of the horizontal area; thus, downdrafts are more probable (Lamb, 1982; Weil, 1988). For $z_i/|L| < 10$ or 20, the structure depends more on $z_i/|L|$ with the effects of shear-generated turbulence extending to greater depths (Deardorff, 1972; Wyngaard, 1988).

In the SBL, turbulence is driven by shear stress and is inherently much weaker than that in the CBL. The weak turbulence coupled with the variable effects of gravity waves, nonstationarity, and local drainage leads to a broad range of potential SBL types. Following Mahrt (1999), it is useful to consider two extreme states: (1) the weakly stable boundary layer (WSBL), which is driven by moderate-to-strong geostrophic winds (≥5 m/s), cloudiness, and weak surface cooling; and (2) the very stable boundary layer (VSBL), which is characterized by light winds (≤3 m/s), clear skies, and strong radiative surface cooling. Turbulence in the WSBL is continuous with height, whereas turbulence in the VSBL is often intermittent—alternating between turbulent and laminar flow (Mahrt, 1999).

The SBL schematic in Figure 12.1b applies mostly to the WSBL and shows the mean wind and temperature profiles as well as the typical eddy size. In the WSBL, the mean profiles follow the MO stable predictions near the surface (Businger et al., 1971), while the shear stress ($\tau = \rho_a u_*^2$) has its maximum value at the surface (near the turbulence source) and decreases with height to near zero at the SBL top, $z = z_i$. The u_* generally ranges from 0.1 to 0.4 m/s in the WSBL. Observations (Caughey et al., 1979; Nieuwstadt, 1984) and large-eddy simulation (LES) (Beare et al., 2006) show that strong vertical inhomogeneity, or height variation, exists in the turbulence, mean wind, and potential temperature Θ.

The typical eddy size or turbulence length l is proportional to z near the surface ($l = kz$) and to a buoyancy length $l_B = \sigma_w/N$ in the upper part of the boundary layer (Brost and Wyngaard, 1978), where $N = [(g/\Theta)(a\Theta/az)]^{1/2}$ is the Brunt–Väisälä frequency. Thus, l_B decreases with an increase in N or the stratification, which makes sense. The N is the natural oscillation frequency of a fluid particle if perturbed from its equilibrium position in a stratified environment. Typical values of N^{-1} and l_B are 1 min and 10–30 m, respectively. Nieuwstadt (1984) derived the following expression for σ_w from his similarity model, $\sigma_w/u_* = 1.4(1 - z/z_i)^{3/4}$, and found that it agreed well with measurements from a 213 m tower in the Netherlands.

The stability of the SBL as a whole can be characterized by z_i/L (Brost and Wyngaard, 1978) or a bulk Richardson number, $Ri_b = (g\Delta\Theta z_i)/(\Theta_a U_i^2)$ (Mahrt, 1999), where $\Delta\Theta$ is the temperature change between the surface and z_i, and U_i is the mean wind speed at z_i. A WSBL exists for $z_i/L < 3$ or 4 and a VSBL for say $z_i/L > 5$ or 10, but these are only very rough estimates.

The VSBL structure is more variable than that of the WSBL due to the weaker winds, more variability in the direction, and the strong stability. The temperature and wind gradients are largest at the surface and decrease monotonically upward to small values at the VSBL top. Recent observations by Banta et al. (2007) and Mahrt and Vickers (2006) show that the VSBL is quite shallow, generally 10 m < z_i < 60 m, and is capped by a quiescent layer of weak and intermittent turbulence. The turbulence forcing or u_* in the VSBL is quite small, typically ranging from 0.02 to 0.1 m/s. Observations by Mahrt and Vickers showed that the σ_w^2 decreased with z in one dataset, but increased with z in another. Thus, the VSBL properties are more variable and much harder to predict than those of the WSBL.

12.3 Dispersion Process and Theories

Dispersion is driven by the ABL mean and turbulence structure and, like turbulence, it is a random or stochastic phenomenon. This means that the concentration observed at a particular time and location downstream of a source cannot be predicted with precision (Chatwin, 1982). Concentration is a *random* variable and should be described statistically by a probability distribution (Chatwin, 1982; Csanady, 1973). The latter can be prescribed by a shape function (e.g., a gamma distribution), the ensemble-mean concentration C, and the rms fluctuating concentration σ_c, which characterizes the width of the distribution. For many dispersion problems, the σ_c is large, $\sigma_c \geq C$, and should be modeled in addition to C. However, most dispersion and air quality models only predict the mean plume spread and C. In Sections 12.3 through 12.4, focus is on the mean dispersion and C, but in Section 12.5, concentration fluctuations and the uncertainty problem are addressed.

An important concept in analyzing a random process such as dispersion is that of an ensemble of observations. The ensemble consists of a set of repeated observations or "realizations" with the same "major" or measurable variables held fixed (e.g., mean wind speed, surface heat flux, and ABL height), while a number of "lesser" variables cannot be controlled and vary randomly. The average or mean value of the concentration (C) and other statistics (e.g., σ_c) can be obtained from the ensemble (Chatwin, 1982; Sykes, 1988; Venkatram, 1988b). In addition, one can construct an ensemble-average plume—an average over many plume "snapshots"—which is smoother in shape and more easily described.

Figure 12.2a illustrates the highly fluctuating concentration profile inside an instantaneous plume as well as the meander. If the concentration is averaged over a finite time period in a fixed or Eulerian reference frame, much of the meandering and internal fluctuations are removed and a smoother concentration profile results (Figure 12.2b). Averaging the concentration over a much longer time period or a large number of realizations

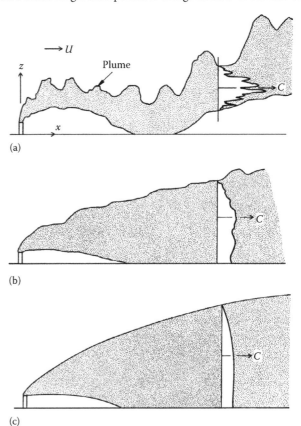

(a)

(b)

(c)

FIGURE 12.2 Schematic of dispersion showing a realization of an (a) instantaneous plume and a (b) time-averaged plume as well as the (c) ensemble-mean behavior.

approximates the ensemble-mean pattern (Figure 12.2c), which is even smoother.

The following theories and models are important in understanding and modeling dispersion.

12.3.1 Statistical Theory

Statistical theory (Taylor, 1921) describes the ensemble-mean or average rms plume spread (dispersion) as a function of time t in a turbulent flow. It is based on the sequential release of passive particles from a source and Lagrangian tracking or following of them in the flow. The turbulence statistics, for example, σ_v or σ_w, measured at fixed points in the flow (i.e., Eulerian measurements) are assumed to be uniform (homogeneous) in space and steady (stationary) in time. The restriction to "passive" particles means those which do not affect the flow (i.e., are neutrally buoyant) and are merely "markers" or "tracers" of fluid motion. The random particle displacement is found from the Lagrangian velocity autocorrelation, which is the correlation of the particle velocity at some t after its release and at a later time $t + \tau$, where τ is a time lag or separation. For the particle lateral (y) velocity component V_y, the autocorrelation is $\overline{V_y(t)V_y(t + \tau)}$, where the overbar denotes an average of the velocity product over many particles at the lag τ. A dimensionless autocorrelation function $f_{Ly}(\tau)$ is defined by

$$f_{Ly}(\tau) = \frac{\overline{V_y(t)V_y(t + \tau)}}{\overline{V_y^2}(\tau = 0)}, \quad (12.2)$$

where the particle velocity variance at zero lag, $\overline{V_y^2}(0)$, equals σ_v^2, the local velocity variance in the fluid.

For short times, the particle velocity is well correlated with its initial velocity, and thus $f_{Ly} \simeq 1$ as $\tau \to 0$. However, with an increase in τ and after many random interactions with the surrounding fluid, the particle loses memory of its initial velocity and $f_{Ly} \to 0$. A measure of the time over which V_y becomes independent of its earlier value at t is the Lagrangian integral time scale or "memory" time:

$$T_{Ly} = \int_0^\infty f_{Ly}(\tau)d\tau. \quad (12.3)$$

From the analysis in Taylor (1921), Csanady (1973), or Pasquill and Smith (1983), the mean square particle displacement or dispersion is found from

$$\sigma_y^2 = 2\sigma_v^2 \int_0^t \left(\int_0^{t'} f_{Ly}(\tau)d\tau \right) dt'. \quad (12.4)$$

Two limiting forms of Equation 12.4 are obtained from the $f_{Ly}(\tau)$ behavior and yield important results for the plume spread regardless of the actual variation of f_{Ly} with τ. For short times ($t \ll T_{Ly}$), $f_{Ly} \simeq 1$, and

$$\sigma_y^2 = \sigma_v^2 t^2, \quad (12.5)$$

where the plume spread is linear with time. This is sometimes referred to as a "ballistic" limit because the particle paths are essentially straight lines. For long times ($t \gg T_{Ly}$), the f_{Ly} approaches zero, but its integral (Equation 12.3) remains finite and Equation. 12.4 leads to

$$\sigma_y^2 = 2\sigma_v^2 T_{Ly} t. \quad (12.6)$$

In this time limit, the σ_y grows parabolically with t, which is a diffusive type of behavior, and the product $\sigma_v^2 T_{Ly}$ is an effective eddy diffusivity, K_y, which has the dimensions of length^2/time, for example, m^2/s. Equation 12.6 also can be written as $\sigma_y^2 = 2K_y t$ and hence $K_y = (1/2)(d\sigma_y^2/dt)$. A simple widely used dispersion parameterization for the full range of times is $\sigma_y = \sigma_v t/(1 + 0.5t/T_{Ly})^{1/2}$, which reduces to Equations 12.5 and 12.6 for short and long times, respectively.

12.3.2 Eddy-Diffusion Theory

In contrast to the Lagrangian treatment in statistical theory, eddy-diffusion theory applies to the concentration flux and mean concentration at fixed points in the fluid; it is a "Eulerian" description. The key assumption in this theory is that the turbulent flux of a scalar is down the mean concentration gradient or from high to low concentration C. The vertical flux F_z is given by

$$F_z = -K_z \frac{\partial C}{\partial z}, \quad (12.7)$$

where the proportionality coeDcient K_z is the turbulent diffusivity. This expression is written in analogy with molecular diffusion, where the proportionality coeDcient is the molecular diffusivity D. In the molecular case, D is a property of the Wuid, whereas in the turbulent case, K_z is a property of the Wow or the *turbulence* and must be parameterized using the turbulence characteristics. In addition, K_z is several orders of magnitude greater than D.

As shown by Csanady (1973) and Venkatram (1988a), the mean concentration field $C(x,y,z)$ given by the Gaussian plume equation is a solution to the diffusion equation, which governs the scalar mass concentration (see Chapter 24). The eddy diffusivities found from the mass conservation equation in the case of homogeneous turbulence, a uniform mean wind U, and the flux–gradient relationship (Equation 12.7) are given by

$$K_y = \frac{U}{2}\frac{d\sigma_y^2}{dx}, \quad K_z = \frac{U}{2}\frac{d\sigma_z^2}{dx}. \quad (12.8)$$

This shows how K_y and K_z are related to the plume dispersion parameters or spreads, σ_y and σ_z. With the substitution $x = Ut$, the diffusivities can be written as

$$K_y = \frac{1}{2}\frac{d\sigma_y^2}{dt}, \quad K_z = \frac{1}{2}\frac{d\sigma_z^2}{dt}, \qquad (12.9)$$

which are consistent with the long-time limit of statistical theory; see Equation 12.6 and the related discussion.

The consistency of Equation 12.9 with statistical theory tells us another property of eddy-diffusion theory—it is applicable only when the characteristic eddy size, l_y $(=\sigma_v T_{Ly})$ or l_z $(=\sigma_w T_{Lz})$, is small compared to the plume width or dispersion (σ_y, σ_z) in the appropriate direction. This can be seen by rewriting Equation 12.6 as $\sigma_y^2 = 2\sigma_y^2 T_{Ly}^2(t / T_{Ly})$ or $(\sigma_y/l_y)^2 = 2t/T_{Ly}$. Since $t / T_{Ly} \gg 1$, the l_y/σ_y is much less than 1. In principle, eddy-diffusion theory does not apply to the "near-field" or "short-time" region of statistical theory, $t < T_{Ly}$. When it is valid, K theory has the advantage of being useful in inhomogeneous conditions where the K is spatially dependent. For most ABL applications, the main inhomogeneity is in the vertical or z direction, and in this case, MO theory provides a formulation for K_z near the surface. The mass diffusivity is essentially the same as for heat, K_h, where $K_h = ku_*z/\phi_h(z/L)$ and ϕ_h is a dimensionless temperature gradient function (Businger et al., 1971); for stable conditions $\phi_h = 1 + \beta z/L$ where $\beta = 4.7$ or 5.

12.3.3 Lagrangian Particle Dispersion Model

In the "real" ABL, dispersion is complicated by the turbulence complexities—the spatial inhomogeneity, nonstationarity, possible non-Gaussian velocity statistics, and a sometimes large timescale T_L. In general, this makes the use of statistical and eddy-diffusion theories invalid except under some special conditions. However, these complexities can be addressed with a Lagrangian particle dispersion model (LPDM), which describes the trajectories of passive particles in a turbulent flow given the random velocity field. The mean concentration C is obtained numerically by simulating thousands of particle trajectories with C being proportional to the PDF of particle position, that is, the number of particles per unit volume or the number density.

There are two main ways of providing the random velocity fields. In the first, the Lagrangian velocities are computed numerically from a stochastic model that requires the velocity statistics (e.g., σ_v, σ_w, third moment of w) as input to an assumed PDF of the velocities (Wilson and Sawford, 1996). In applications, the statistics are provided by parameterized profiles fitted to laboratory data, field observations, or numerical simulations.

In the second, the velocities are obtained from the time-dependent velocity fields of a numerical turbulence model of the ABL—an LES. This approach was pioneered by Lamb (1978), who computed dispersion due to an elevated source in the CBL.

The Lagrangian velocity of a particle was found from the numerically computed LES or "resolved" velocity at the particle position and a random "subgrid-scale" (SGS) velocity that accounts for velocities not resolved by the LES. For the vertical component, the velocity is given by $w_L = w_R(\mathbf{x}_p,t) + w_S(\mathbf{x}_p,t)$, where subscripts L, R, and S denote Lagrangian, resolved, and SGS, respectively, and $\mathbf{x}_p$ (x_p, y_p, z_p) is the vector position of the particle. The particle height z_p can then be found by integrating $dz_p/dt = w_L(\mathbf{x}_p,t)$ and similarly for the x_p and y_p.

12.4 Observed Mean Dispersion Characteristics

The key features of dispersion and the mean concentration field in the CBL and SBL are presented using theory, experiments, observations, and numerical simulations. It will be shown how simple theories—eddy diffusion and statistical—compare with experiments and simulations under some restricted conditions.

12.4.1 Convective Boundary Layer

For a surface source, the diffusion equation with a height-dependent diffusivity, $K_z(z)$, can describe the mean concentration field near the source. This works because the characteristic eddy size l is small and grows with z near the surface—$l \propto z$—thus satisfying a criterion in Section 12.3.2. van Ulden (1978) adopted Roberts's analytical solution to the diffusion equation for the cross-wind-integrated concentration (CWIC), $C^y = \int_{-\infty}^{\infty} C(x,y,z,)dy$:

$$C^y(x,z) = \frac{AQ}{\bar{u}.\bar{z}}\exp\left[-\left(\frac{Bz}{\bar{z}}\right)^s\right]. \qquad (12.10)$$

where

Q is the source emission rate
A and B are constants that depend on the exponent or shape factor s ($1 \leq s \leq 2$)
$\bar{u}$ is the mean wind speed at $\bar{z}$
$\bar{z}(x)$ is the mean plume height at downwind distance x

van Ulden found analytical expressions for $x(\bar{z})$ and $\bar{u}(\bar{z})$ as a function of u_*, L, and the surface roughness length z_o and showed that the predicted C^y along the surface agreed well with measurements from the Prairie Grass experiments. Horst (1979) followed an approach similar to van Ulden's and together their work has been known as the surface layer similarity (SLS) theory for diffusion in the atmospheric surface layer.

Venkatram (1992) derived the limiting behavior of σ_z and the surface CWIC, $C^y(x,0)$, for very unstable conditions, that is, a small L or large $z_i/|L|$, say >50–100. He combined the long-time limit of statistical theory in the form $d\sigma_z^2/dt = 2K_z(t)$ (Section 12.3.1) with K_z given by the MO result for heat, $K_z = K_h(z/L)$,

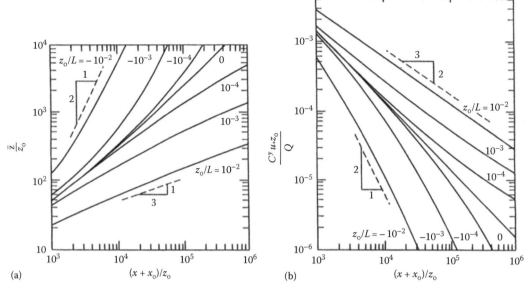

FIGURE 12.3 (a) Dimensionless mean plume height and (b) dimensionless CWIC as a function of dimensionless distance for a surface source in the atmospheric surface layer. (From van Ulden, A.A.P., *Atmos. Environ.*, 12, 2125, 1978, with additions.)

using the Businger et al. (1971) expression for K_h. By evaluating K_h at $z = \sigma_z$, he obtained

$$\sigma_z \propto \frac{x^2}{|L|} \quad \text{and} \quad C^y(x,0) \propto \frac{Q|L|}{u_* x^2} \qquad (12.11)$$

Here, the σ_z dependence on x is greater than the usual statistical theory result ($\sigma_z \propto x$) and is due to the strong vertical inhomogeneity of K_z in the CBL surface layer.

Figure 12.3 shows van Ulden's results (solid lines) for $\bar{z}/z_o$ and $C^y u_* z_o/Q$ as a function of $(x + x_o)/z_o$, where x_o is the distance to a virtual source for an elevated release in the surface layer; results for convective conditions are given by negative values of z_o/L. As can be seen, the $\bar{z}$ and C^y approach an x^2 and x^{-2} dependence, respectively, in the limit of large instability ($z_o/L = -10^{-2}$) and large distances (see dashed lines) consistent with the results of Equation 12.11.

For dispersion throughout the CBL, the concept of "convective scaling" of dispersion (Willis and Deardorff, 1976) is important for analyzing dispersion and comparing laboratory results with field observations. This states that the key length, velocity, and timescales of the turbulence are z_i, w_*, and z_i/w_*, the eddy-turnover time, i.e., $T_L \propto z_i/w_*$. In strong convection, the turbulence, mean wind, etc., can be approximated as nearly uniform or well mixed in the upper 90% of the CBL (Section 12.2) so that $t \simeq x/U$. In addition, the convective eddies are long lived (a large T_L) since z_i/w_* is 500 s or about 8 min for a typical $z_i = 1000$ m and $w_* = 2$ m/s. As shown by experiments and simulations (LPDMs), the mean CWIC can be given by a general expression of the form $C^y = Q f_1(z_s, z_i, U, w_*, t, k)$, where f_1 is some function and the key independent variable is time t for a given set of conditions—z_s,

z_i, U, etc. The dependence on t can be understood since the plume growth or dispersion parameters, σ_y and σ_z, depend on t.

Based on the aforementioned discussion, Willis and Deardorff proposed that the CWIC be scaled or nondimensionalized as

$$\frac{C^y U z_i}{Q} = f_2\left(\frac{z}{z_i}, X, \frac{z_s}{z_i}\right), \qquad (12.12)$$

where

z_s is the source height
f_2 is some function
dimensionless distance X is

$$X = \frac{w_* x}{U z_i}. \qquad (12.13)$$

Here, X is the ratio of travel time $t = x/U$ to the eddy-turnover time or effectively $X \propto t/T_L$. The basis for the scaling of C^y is mass conservation and the observation that far downstream the concentration becomes well mixed in the vertical such that $C^y U z_i = Q$. Thus, the left-hand side of Equation 12.12 is the ratio of the local C^y at (x,z) to the well-mixed CWIC, $Q/(U z_i)$.

The important aspects of dispersion in the CBL were first demonstrated through laboratory experiments (Willis and Deardorff, 1976, 1978, 1981) and numerical simulations (Lamb, 1978, 1982). These studies showed the importance of source height z_s on the dispersion patterns and the utility of convective scaling in organizing the data.

Figure 12.4 shows contours of the dimensionless CWIC (Equation 12.12) in an x-z plane from the laboratory experiments

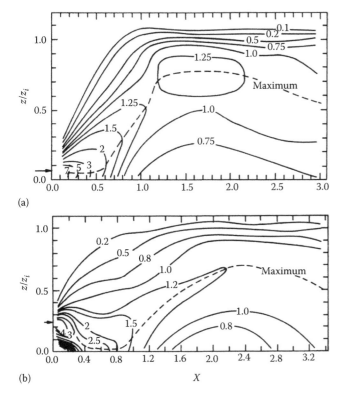

(a)

(b)

FIGURE 12.4 Convection tank results showing contours of dimensionless CWIC as a function of dimensionless height and downwind distance X. Horizontal arrows denote source height: (a) $z_s/z_i = 0.067$, and (b) $z_s/z_i = 0.24$. (From Willis, G.E. and J.W. Deardorff, *Q. J. R. Meteorol. Soc.*, 102, 427, 1976. With permission; Willis, G.E. and J.W. Deardorff, *Atmos. Environ.*, 12, 1305, 1978. With permission.)

as a function of X. For the near-surface source (Figure 12.4a), the average plume centerline as defined by the locus of maximum CWIC ascends after a short distance downwind $X \sim 0.5$, whereas the centerline from the elevated source (Figure 12.4b) descends until it reaches the ground. These behaviors can be understood in terms of the release of material into updrafts and downdrafts. For the surface release, material emitted into the base of an updraft begins rising almost immediately, whereas that

released into a downdraft remains near the ground and moves horizontally. Once a significant amount of material is swept out of downdrafts into neighboring updrafts, the centerline begins rising (Lamb, 1982; Weil, 1988). For the elevated source, the centerline descent is caused by the greater areal coverage by downdrafts (60%) and, hence, the higher probability of material being released into them. In addition, the downdrafts are long lived so that material carried by them persists in its travel to the surface.

In contrast to the behavior in Figure 12.4b, the Gaussian plume model predicts that the elevated plume centerline remains aloft and horizontal until a suDcient number of plume reflections occurs at the surface. The centerline then moves toward the surface (Lamb, 1982).

A number of dispersion observations and comparisons with theory have been made using convective scaling, for example, Briggs (1988, 1993) and Weil (1988). Figure 12.5a shows crosswind dispersion observations based on oil fog and radar chaff releases in the CBL from the CONDORS experiment—dispersion from a 300 m tower near Erie, CO (Briggs, 1993). The measurements covered the following ranges: $x \leq 3.6$ km, 1.6 m/s $\leq U \leq 6.2$ m/s, 520 m $\leq z_i \leq 1600$ m, and typical averaging times of 30–40 min. The results presented in convective scaling form, σ_y/z_i versus X, show reasonable agreement with the line $\sigma_y/z_i = 0.6X$ for $X < 0.7$ or 1. This is the short-time prediction of statistical theory (Equation 12.5) assuming $\sigma_v = 0.6w_*$ (Section 12.2) and $t = x/U$, where U is the vertically averaged wind speed. The good agreement is due to the near uniformity in the σ_v and $\hat{u}$ profiles in the upper part of the CBL and to the relatively large T_{Ly}. The data suggest a reduction in σ_y/z_i from the prediction for $X > 1$ but do not extend far enough to support the long-time prediction, $\sigma_y/z_i \propto X^{1/2}$.

For vertical dispersion (Figure 12.5b), the data support the prediction $\sigma_z/z_i = 0.6X$ for $X < 0.5$, but beyond that distance they converge toward the limit $\sigma_z/z_i = 0.27$. The latter behavior is due to the trapping of the plume by the elevated inversion (at z_i) and the tendency toward a vertically well-mixed concentration by $X = 2$ or 3. The laboratory experiments of Willis and Deardorff (1976, 1978, 1981) and numerical (LPDM) simulations by Lamb

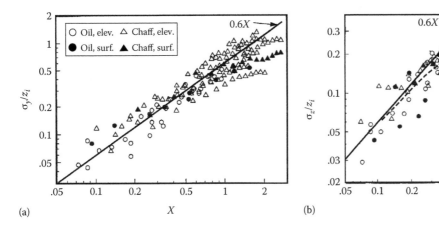

(a)

(b)

FIGURE 12.5 (a) Dimensionless crosswind and (b) vertical dispersion as a function of dimensionless distance X from the CONDORS experiment. (From Briggs, G.A., *J. Appl. Meteorol.*, 32, 1388, 1993. With permission.)

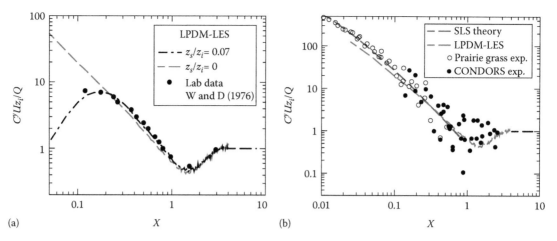

FIGURE 12.6 Dimensionless CWIC at the surface as a function of the dimensionless distance X for (a) a near-surface source ($z_s/z_i = 0.07$) with LPDM results also given for a surface release (for reference), and (b) a surface source. Results from an LPDM–LES model (Weil et al., 2004), SLS theory, laboratory data (Willis and Deardorff, 1976), and field observations. Horizontal dashed line is well-mixed CWIC.

(1978, 1982) generally support the σ_y, σ_z results shown earlier but with a slightly different coeDcient; the prediction is closer to $(\sigma_y, \sigma_z)/z_i = 0.5X$.

In air quality applications, most interest is in the surface concentration distribution $C(x,y,0)$, and this can be determined from the surface CWIC, $C^y(x,0)$ by assuming a Gaussian cross-wind distribution. The ground-level concentration along the plume centerline ($y = 0$) is found from $C = C^y(x,0)/(\sqrt{2\pi}\sigma_y)$ (Weil, 1988). In the following, we present the surface CWIC variation with X as found from experiments, observations, and an LPDM driven by LES fields (Weil et al., 2004).

Figure 12.6a presents the dimensionless CWIC as a function of X for a near-surface source ($z_s = 0.07z_i$). The LPDM shows that the CWIC increases rather quickly with X to reach a maximum value near $X \simeq 0.2$, falls systematically as the plume expands, and attains the well-mixed value ($C^y U z_i/Q = 1$) near $X = 3$. The LPDM is in excellent agreement with the Willis and Deardorff (1976) laboratory data. This C^y behavior is characteristic of that from an elevated source in the CBL, where the maximum CWIC, $C^y_{mx} \propto 1/z_s$, and thus decreases with an increase in the source height. Figure 12.6b shows the dimensionless CWIC behavior for a surface release in the CBL with results given by SLS theory, the LPDM, and field observations from the Prairie Grass and CONDORS experiments. For a surface release, the maximum CWIC occurs at the source, falls systematically with X, and ultimately reaches the well-mixed value near $X = 3$. There is a general agreement among the theory, LPDM, and observations, although there is considerable scatter in the field data especially for $0.4 \leq X \leq 2$.

12.4.2 Stable Boundary Layer

We discuss vertical dispersion first for surface layer sources and then for elevated sources.

For the surface layer, van Ulden (1978) adopted the same eddy-diffusion approach as for the CBL surface layer (Equation 12.10) and found analytical expressions for $x(\hat{z})$ and $\hat{s}(\hat{z})$ using the "stable" similarity forms of $\hat{s}(z)$ and K_h (Businger et al., 1971).

Likewise, Venkatram (1988a, 1992) coupled statistical theory with $K_z = K_h(z/L)$ (Section 12.4.1) using the "stable" forms for $\hat{s}(z)$ and K_h. In the case of small σ_z/L, he obtained $\sigma_z \propto u_*x/U$ and $C^y \propto Q/(u_*x)$, which are consistent with the neutral limit of eddy-diffusion theory (Figure 12.3, $z_0/L = 0$). For very stable conditions ($\sigma_z/L \gg 1$), he found

$$\sigma_z \propto L^{2/3} x^{1/3} \quad \text{and} \quad C^y \propto \frac{Q}{u_* L^{1/3} x^{2/3}}. \tag{12.14}$$

The more rapid falloff of C^y with x than simply $C^y \propto 1/\sigma_z$ or $x^{-1/3}$ is due to the increasing mean transport speed with z in the SBL (Venkatram, 1988a).

Figure 12.3 gives van Ulden's results for $\hat{s}$ and $C^y(x,0)$ for stable conditions, $z_0/L > 0$. The limiting forms for $\hat{s} \propto \sigma_z$ and $C^y(x,0)$ given by (12.14) are shown as dashed lines and can be seen to match the slopes of van Ulden's curves for $z_0/L = 10^{-2}$. Thus, these forms are useful in showing the simple functional dependence of σ_z and C^y on u_*, L, and x.

For elevated sources, the vertical dispersion at long times can vary between the classical diffusion limit, $\sigma_z \propto t^{1/2}$ (Equation 12.6), and a constant vertical thickness or "pancake" limit (Pearson et al., 1983), which has been observed (Hilst and Simpson, 1958). The former is more familiar, and the σ_z over the full range of times can be obtained from an interpolation expression like that given for σ_y (Section 12.3.1):

$$\sigma_z = \frac{\sigma_w t}{(1 + 0.5t/T_{Lz})^{1/2}}. \tag{12.15}$$

This was proposed by Venkatram et al. (1984), who evaluated σ_w at z_s and assumed $T_{Lz} = l/\sigma_w$. They adopted a turbulence length scale l that was interpolated between a neutral limit, $l_n \propto z_s$ valid near the surface, and a stable limit, $l_s \propto l_B$ valid in the upper part of the SBL as suggested by Brost and Wyngaard (1978) (see Section 12.2).

Csanady (1964) and Pearson et al. (1983) proposed models to deal with the constant σ_z regime of pancake-like plumes. For short times, the Pearson et al. model is consistent with statistical theory, but for long times ($t > T_{Lz}$), their model predicts

$$\sigma_z = \frac{\sigma_w}{N}(c_1 + 2\gamma^2 Nt)^{1/2}, \qquad (12.16)$$

where

c_1 is a constant ($\simeq 1.3$)

γ is a parameter determining the degree of "molecular mixing" between fluid elements in a stratified turbulent flow

Note that the σ_z ($\propto \sigma_w/N$ or l_B) increases with the turbulent energy (σ_w) and decreases with increasing N or stratification, that is, the plume is thinner with greater stability as would be expected.

The key idea of the Pearson et al. (1983) model is that if $\gamma = 0$, a fluid particle moving about in a turbulent flow retains its initial density and moves through a vertical distance of $\sim\sigma_w/N$ before the buoyancy force returns it to its initial position. However, for $\gamma > 0$, a particle can become warmed or cooled (by molecular diffusion) if moving up or down; the buoyancy force is then weakened and the particle may be displaced a distance greater than σ_w/N. Their model reduces to the constant σ_z and $t^{1/2}$ regimes of spread depending on the value of γ. For $\gamma \sim 0.1$, σ_z approaches a constant over a considerable range of time, but for $\gamma \sim 0.3$, σ_z follows a parabolic growth with t as in statistical theory.

There are experimental data supporting both of the aforementioned models. For example, Venkatram et al. (1984) compared their model with σ_z measurements downstream of an elevated source, typically 50 m tall, and for distances extending to about 1 km downwind. Figure 12.7 compares a dimensionless form of

Equation 12.15, $\sigma_z/(\sigma_w t)$ versus t/T_{Lz}, with these measurements. In general, the model (solid line) gives a reasonable prediction of the decrease of $\sigma_z/(\sigma_w t)$ with time but tends to overestimate the observations at large travel times, $t/T_{Lz} > 10$. Also shown in Figure 12.7 is an interpolation expression based on Equation 12.16 with the $2\gamma^2 Nt$ term omitted, that is, assuming the constant σ_z limit is followed. As can be seen, this expression provides a lower bound on the data at large times.

The key problem in applying these models is determining the conditions under which each is valid. Intuitively, we expect that the statistical theory limit ($t^{1/2}$) is applicable in a WSBL with continuous turbulence, whereas the pancake limit may apply to the VSBL with weak and intermittent turbulence. Support for the WSBL case has been provided by LPDMs driven by LES through the work of Kemp and Thomson (1996) and Weil et al. (2006). They show that σ_z approximates the short- and long-time limits of statistical theory in WSBLs for $z_i/L < 4$ (Kemp and Thomson, 1996) or $z_i/L = 1.6$ (Weil et al., 2006).

Another issue is knowledge of the SBL height and whether a plume is above or below z_i. In the absence of measurements, the z_i for the WSBL can be estimated from the Zilitinkevich (1972) expression, $z_i = 0.4(u_* L/f)^{1/2}$, where f is the Coriolis parameter (see Chapter 1), but the applicability of this to a VSBL is uncertain. In addition, a better understanding and physical interpretation of the γ parameter and methods for determining it are needed. Finally, the lateral dispersion can be estimated from statistical theory, but the parameterization of σ_v in terms of u_*, L, etc., is questionable due to the contributions from large mesoscale eddies, which are generated by terrain features and the stratification.

12.5 Statistical Variability in Concentration

Turbulence leads to large statistical variability in concentration fields (Chatwin, 1982; Sykes, 1988). In this section, we discuss the importance of concentration fluctuations in point-source plumes, the key processes driving them, and simple models that highlight some of the key physics. Observations are presented that reveal the fluctuations in "real-world" plumes, and a partial diagnosis of them is made based on a simple model. This is only a brief introduction to a complex topic; for further discussion see Sykes (1988).

The random concentration field downwind of a source is best described using a probability distribution, which requires the ensemble-mean (C) and the rms fluctuating (σ_c) concentrations. Laboratory and field experiments show that the fluctuations are significant—$\sigma_c \geq C$—in point-source plumes for short averaging times (≤ 1 h) and relatively short downwind distances (≤ 10 km). The fluctuations are important in a number of problems such as estimating the peak concentration of toxic pollutants and odor thresholds. Knowledge of σ_c also is important in evaluating dispersion models for C because large deviations exist between predicted and observed concentrations when $\sigma_c \sim C$ (Weil et al., 1992).

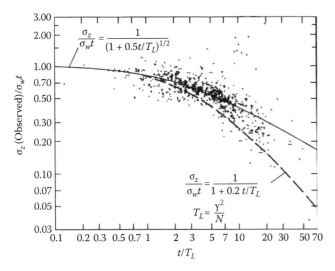

FIGURE 12.7 Vertical dispersion σ_z scaled by $\sigma_w t$ as a function of the dimensionless time t/T_{Lz} in the SBL. Solid line is based on Equation 12.15 and dashed line is an interpolation expression based on Equation 12.16. (From Venkatram, A. et al., *Atmos. Environ.*, 18, 823, 1984, with additions. With permission.)

Concentration fluctuations in plumes arise from two key mechanisms: (1) the instantaneous plume entrains ambient air due to small-scale (inertial-range) turbulence, which leads to concentration fluctuations over distance scales less than the instantaneous plume width (σ_r), and (2) the plume meanders due to the large energy-containing eddies in the environment. Such eddies are of the order of the characteristic turbulence scale l. Meandering causes concentrations at a fixed receptor to vary essentially between "in-plume" peaks and the zero value in the environment. It is an important source of fluctuations as long as the plume width σ_r is less than l, and this condition effectively holds for $t < T_L$.

Two early models for σ_c were developed along the lines of the two aforementioned mechanisms: (1) the "meandering plume" model (Gifford, 1959), and (2) a Eulerian similarity model (Csanady, 1973). As the name suggests, the meandering model describes the fluctuations due to the large eddies alone and applies when $\sigma_r < l$. The similarity model treats the opposite case when $\sigma_r > l$ or $t > T_L$.

In the meandering plume model, the concentration c within an instantaneous plume is assumed to be nonrandom and to have a Gaussian shape about the instantaneous centroid position, y_c and z_c, in the crosswind and vertical directions:

$$c(x,y,z) = \frac{Q}{2\pi U \sigma_r^2} \exp\left[-\frac{(y-y_c)^2 + (z-z_c)^2}{2\sigma_r^2}\right]. \quad (12.17)$$

Although c within the plume is nonrandom, the concentration observed at a fixed location (x,y,z) is random due to the randomness or uncertainty in the plume centroid coordinates, y_c, z_c. In Equation 12.17, it is assumed that (1) the concentration distribution is axisymmetric so that $\sigma_{yr} = \sigma_{zr} = \sigma_r(x)$, and (2) the y_c, z_c are uncorrelated random variables prescribed by a Gaussian PDF. By averaging the aforementioned c over the PDF of y_c, z_c (which have zero means), Gifford obtained the familiar Gaussian plume model for the mean concentration C:

$$C(x,y,z) = \frac{Q}{2\pi U (\sigma_r^2 + \sigma_m^2)} \exp\left[-\frac{y^2 + z^2}{2(\sigma_r^2 + \sigma_m^2)}\right], \quad (12.18)$$

where

$\sigma_m(x)$ is the rms displacement of the meander
$(\sigma_r^2 + \sigma_m^2)^{1/2}$ is the total spread of the mean plume

Using the same averaging procedure, he also found the mean square concentration:

$$\langle c(x,y,z)^2 \rangle = \frac{Q^2}{(2\pi U \sigma_r^2)(\sigma_r^2 + 2\sigma_m^2)} \exp\left[-\frac{y^2 + z^2}{\sigma_r^2 + 2\sigma_m^2}\right], \quad (12.19)$$

where $(\sigma_r^2 + 2\sigma_m^2)^{1/2}$ is the spread of the quantity $\langle c^2 \rangle^{1/2}$. The mean square deviation or variance σ_c^2 was then found from $\sigma_c^2 = \langle c^2 \rangle - C^2$.

A key result from his model is a prediction of the concentration fluctuation intensity σ_c/C or its square, σ_c^2/C^2, in a plume cross section:

$$\frac{\sigma_c^2}{C^2} = \frac{(\sigma_r^2 + \sigma_m^2)^2}{\sigma_r^2(\sigma_r^2 + 2\sigma_m^2)} \exp\left[\frac{\sigma_m^2(y^2 + z^2)}{(\sigma_r^2 + 2\sigma_m^2)(\sigma_r^2 + \sigma_m^2)}\right] - 1, \quad (12.20)$$

which increases in the tails of the plume, that is, as $y^2 + z^2$ gets larger. This occurs because the rms spread of $\langle c^2 \rangle^{1/2}$, $(\sigma_r^2 + 2\sigma_m^2)^{1/2}$, is greater than the spread of C, $(\sigma_r^2 + \sigma_m^2)^{1/2}$.

We note that the similarity model (Csanady, 1973) is based on the diffusion equation and application of that equation to both the mean (C) and variance (σ_c^2) of the concentration. The transport of these quantities is modeled using eddy-diffusion or K theory, which is only valid for $l > \sigma_r$ as noted earlier. However, a key result from Csanady's model is that the fluctuation intensity squared, σ_c^2/C^2, also increases in the tails of the plume as it does for the meandering model. Thus, this aspect of the fluctuations is independent of whether $\sigma_r < l$ or $\sigma_r > l$.

Numerous evaluations of point-source and other dispersion models have been conducted with field data to assess their performance in air quality applications and to understand the model physics (Hanna and Paine, 1989; Hanna et al., 1991; Sykes, 1988; Venkatram, 1988b; Weil et al., 1992). In the following, we analyze the variability in the ratio of predicted C_p to observed C_o concentrations in the plume downwind of an elevated point source. The experiment was conducted in a suburb of Copenhagen (Gryning and Lyck, 1984) using hourly releases of SF_6 from a 115 m tower during daytime convective periods ($1.4 \leq z_i/|L| \leq 14$); surface SF_6 measurements were obtained on arcs at 2, 4, and 6 km downwind. Evaluation (Weil, 2005) was made of the National Atmospheric Release Advisory Center (NARAC) model (Nasstrom et al., 2007), which uses an LPDM for stochastic particle dispersion.

Comparisons between C_p and C_o were made for three categories: (1) predicted centerline ($y = 0$) concentrations versus observed arc-maximum concentrations (Category 1, C1), (2) C_p versus C_o at all monitors (on and off centerline, 309 cases) using the observed plume direction as the model wind direction (C2), and (3) same as (2) but using the predicted wind direction (C3). The model was nearly unbiased for all categories as determined by the geometric mean (GM) of C_p/C_o, which was 0.88 (C1), 0.86 (C2), and 0.78 (C3); the ideal ratio is 1. However, the scatter as measured by the geometric standard deviation (GSD) of C_p/C_o varied dramatically—being 1.5, 4.0, and 5.1 for C1–C3, respectively. That is, there were large deviations between C_p and C_o for the "paired in time and space" groupings (C2, C3), and such deviations are common to all point-source dispersion models, for example, Smith (1984).

For category 2 (C2) mentioned earlier, Figure 12.8 shows the variability in C_p/C_o as a function of y/σ_y across the plume. As can be seen, the scatter is lowest on the plume centerline ($y = 0$), in agreement with the GSD results mentioned earlier, and increases substantially in the plume tails. The lower variability on the centerline occurs because the instantaneous plume spends most of its time there. The behavior in Figure 12.8 is consistent with the meandering

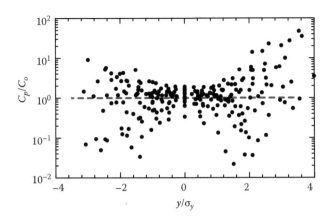

FIGURE 12.8 Variation in the ratio of predicted-to-observed surface concentration and scatter with crosswind position in a point-source plume; dashed line represents a zero bias in predicted concentration. Scatter exhibits a "butterfly" or "bowtie" pattern.

and similarity models for the fluctuation intensity σ_c/C, which is predicted to increase in the plume tails. The "butterfly" or "bowtie" pattern produced in Figure 12.8 helps explain why dispersion models perform poorly (i.e., exhibit large random scatter) in plots of C_p versus C_o when using observations off the centerline.

References

Banta, R., L. Mahrt, D. Vickers, J. Sun, B. Balsley, Y. Pichugina, and E. Williams, 2007. The very stable boundary layer on nights with weak low-level jets, *J. Atmos. Sci.*, **6**, 3068–3090.

Beare, R. J. et al., 2006. An intercomparison of large-eddy simulations of the stable boundary layer, *Boundary-Layer Meteorol.*, **8**, 247–272.

Briggs, G. A., 1988. Analysis of diffusion field experiments, in *Lectures on Air Pollution Modeling*, eds. A. Venkatram and J. C. Wyngaard, pp. 63–117, American Meteorological Society, Boston, MA.

Briggs, G. A., 1993. Plume dispersion in the convective boundary layer. Part II: Analyses of CONDORS field experiment data, *J. Appl. Meteorol.*, **3**, 1388–1425.

Brost, R. A. and J. C. Wyngaard, 1978. A model study of the stably stratified planetary boundary layer, *J. Atmos. Sci.*, **3**, 1427–1440.

Businger, J. A., J. C. Wyngaard, Y. Izumi, and E. F. Bradley, 1971. Flux profile relationships in the atmospheric surface layer, *J. Atmos. Sci.*, **8**, 181–189.

Caughey, S. J., J. C. Wyngaard, and J. C. Kaimal, 1979. Turbulence in the evolving stable boundary layer, *J. Atmos. Sci.*, **6**, 1041–1052.

Chatwin, P. C., 1982. The use of statistics in describing and predicting the effects of dispersing gas clouds, *J. Hazard. Mater.*, **6**, 213–230.

Csanady, G. T., 1964. Turbulent diffusion in a stratified fluid, *J. Atmos. Sci.*, **21**, 439–447.

Csanady, G. T., 1973. *Turbulent Diffusion in the Environment*, Reidel, Dordrecht, the Netherlands, 248p.

Deardorff, J. W., 1972. Numerical investigation of neutral and unstable planetary boundary layers, *J. Atmos. Sci.*, **29**, 91–115.

Gifford, F. A., 1959. Statistical properties of a fluctuating plume dispersion model, in *Advances in Geophysics*, Vol. 6, eds., F. N. Frenkiel and P. A. Sheppard, pp. 117–138, Academic Press, New York.

Gryning, S.-E. and E. Lyck, 1984. Atmospheric dispersion from elevated sources in an urban area: Comparison between tracer experiments and model calculations, *J. Clim. Appl. Meteorol.*, **3**, 651–660.

Hanna, S. R. and R. J. Paine, 1989. Hybrid plume dispersion model (HPDM) development and evaluation, *J. Appl. Meteorol.*, **8**, 206–224.

Hanna, S. R., D. G. Strimaitis, and J. C. Chang, 1991. Evaluation of 14 hazardous gas models with ammonia and hydrogen fluoride field data, *J. Hazard. Mater.*, **6**, 127–158.

Hilst, G. E. and C. I. Simpson, 1958. Observations of vertical diffusion rates in stable atmospheres, *J. Meteorol.*, **15**, 125–126.

Horst, T. W., 1979. Lagrangian similarity modeling of vertical diffusion from a ground-level source, *J. Appl. Meteorol.*, **18**, 733–740.

Kemp, J. R. and D. J. Thomson, 1996. Dispersion in stable boundary layers using large-eddy simulation, *Atmos. Environ.*, **0**, 2911–2923.

Lamb, R. G., 1978. A numerical simulation of dispersion from an elevated point source in the convective planetary boundary layer, *Atmos. Environ.*, **2**, 1297–1304.

Lamb, R. F., 1982. Diffusion in the convective boundary layer, in *Atmospheric Turbulence and Air Pollution Modelling*, eds. F. Nieuwstadt and H. van Dop, Chapter 5, pp. 159–229, Reidel, Dordrecht, the Netherland.

Mahrt, L., 1999. Stratified atmospheric boundary layers, *Boundary-Layer Meteorol.*, **90**, 375–396.

Mahrt, L. and D. Vickers, 2006. Extremely weak mixing in stable conditions, *Boundary Layer Meteorol.*, **9**, 19–39.

Nasstrom, J. S., G. Sugiyama, R. L. Baskett, S. C. Larsen, and M. M. Bradley, 2007. The NARAC modeling and decision support system for radiological and nuclear emergency preparedness and response, *Int. J. Emerg. Manag.*, **4**, 524–550.

Nieuwstadt, F. T. M., 1984. The turbulent structure of the stable nocturnal boundary layer, *J. Atmos. Sci.*, **4**, 2202–2216.

Panofsky, H. A., H. Tennekes, D. H. Lenschow, and J. C. Wyngaard, 1977. The characteristics of turbulent velocity components in the surface layer under convective conditions, *Boundary-Layer Meteorol.*, **1** 355–361.

Pasquill, F. A. and F. B. Smith, 1983. *Atmospheric Diffusion*, Wiley, New York, 437p.

Pearson, H. J., J. S. Puttock, and J. C. R. Hunt, 1983. A statistical model of fluid-element motions and vertical diffusion in a homogeneous stratified turbulent flow, *J. Fluid Mech.*, **129**, 219–249.

Smith, M. E., 1984. Review of the attributes and performance of 10 rural diffusion models, *Bull. Am. Meteorol. Soc.*, **65**, 554–558.

Sykes, R. I., 1988. Concentration fluctuations in dispersing plumes, in *Lectures on Air Pollution Modeling*, eds. A. Venkatram and J. C. Wyngaard, pp. 325–356, American Meteorological Society, Boston, MA.

Taylor, G. I., 1921. Diffusion by continuous movements, *Proc. London Math. Soc.,* **0** , 196–211.

van Ulden, A. A. P., 1978. Simple estimates for vertical dispersion from sources near the ground, *Atmos. Environ.,* **2** 2125–2129.

Venkatram, A., 1988a. Dispersion in the stable boundary layer, in *Lectures on Air Pollution Modeling,* eds. A. Venkatram and J. C. Wyngaard, pp. 229–265, American Meteorological Society, Boston, MA.

Venkatram, A., 1988b. Topics in applied dispersion modeling, in *Lectures on Air Pollution Modeling,* eds. A. Venkatram and J. C. Wyngaard, pp. 267–324, American Meteorological Society, Boston, MA.

Venkatram, A. 1992. Vertical dispersion of ground-level releases in the surface boundary layer, *Atmos. Environ.,* **A** , 947–949.

Venkatram, A., D. Strimaitis, and D. DiCristofaro, 1984. A semi-empirical model to estimate vertical dispersion of elevated releases in the stable boundary layer, *Atmos. Environ.,* **8** , 823–828.

Weil, J. C., 1988. Dispersion in the convective boundary layer, in *Lectures on Air Pollution Modeling,* eds. A. Venkatram and J. C. Wyngaard, pp. 167–227, American Meteorological Society, Boston, MA.

Weil, J. C., 2005. Evaluation of the NARAC modeling system, Technical Report UCRL-AR-217329, Lawrence Livermore National Laboratory, Livermore, CA.

Weil, J. C., E. G. Patton, and P. P. Sullivan, 2006. Lagrangian modeling of dispersion in the stable boundary layer, in *17th Symposium on Boundary Layers and Turbulence,* Number J6.12, San Diego, CA, http://ams.confex.com/ams/pdfpapers/110925.pdf

Weil, J. C., P. P. Sullivan, and C.-H. Moeng, 2004. The use of large-eddy simulations in Lagrangian particle dispersion models, *J. Atmos. Sci.,* **6** , 2877–2887.

Weil, J. C., R. I. Sykes, and A. Venkatram, 1992. Evaluating air-quality models: Review and outlook, *J. Appl. Meteorol.,* **3** , 1121–1145.

Willis, G. E. and J. W. Deardorff, 1976. A laboratory model of diffusion into the convective planetary boundary layer, *Q. J. R. Meteorol. Soc.,* **0** , 427–445.

Willis, G. E. and J. W. Deardorff, 1978. A laboratory study of dispersion from an elevated source within a modeled convective planetary boundary layer, *Atmos. Environ.,* **2** , 1305–1312.

Willis, G. E. and J. W. Deardorff, 1981. A laboratory study of dispersion from an elevated source within a modeled convective planetary boundary layer, *Atmos. Environ.,* **15**, 109–117.

Wilson, J. D. and B. L. Sawford, 1996. Review of Lagrangian stochastic models for trajectories in the turbulent atmosphere, *Boundary-Layer Meteorol.,* **8** , 191–220.

Wyngaard, J. C., 1985. Structure of the planetary boundary layer and implications for its modeling, *J. Clim. Appl. Meteorol.,* **2** , 1131–1142.

Wyngaard, J. C., 1988. Structure of the PBL, in *Lectures on Air Pollution Modeling,* eds. A. Venkatram and J. C. Wyngaard, pp. 9–61, American Meteorological Society, Boston, MA.

Wyngaard, J. C., O. R. Cote, and Y. Izumi, 1971. Local free convection, similarity, and the budgets of shear stress and heat flux, *J. Atmos. Sci.,* **8** , 1171–1182.

Zilitinkevich, S. S., 1972. On the determination of the height of the Ekman boundary layer, *Boundary-Layer Meteorol.,* **3**, 141–145.

Flow and Dispersion in Street Canyons*

Jong-Jin Baik
Seoul National University

13.1 Introduction

A street canyon is a space between buildings, and very complex phenomena are observed within such a space. Street canyon flows are highly turbulent flows, and they depend strongly on building configurations and meteorology. In the daytime, building surface and street bottom are heated by solar radiation. Harmful gases and aerosols are emitted from automobiles or transported from elsewhere into street canyons. These materials interact and are dispersed in street canyons in the presence of turbulent flow. People living in cities are directly exposed to the street canyon environment, which typically has a negative impact on their health because of the poor air quality in street canyons. Accidental or intentional releases of toxic chemicals in street canyons must be dealt with. Urban planning can benefit greatly from information on the climate of street canyons, such as microscale wind and temperature fields.

Flow and dispersion in street canyons have received much attention because of practical as well as fluid dynamical interest. This interest will increase with continued urbanization and increasing concerns about the health, safety, and quality of life of people residing in urban areas. In the past two decades, a great deal of research on street canyon flow and dispersion has been carried out through field experiments, laboratory experiments (wind tunnels and water tanks), and numerical simulations. Several good reviews of this research have been published, including Vardoulakis et al. (2003) and Li et al. (2006). In this chapter, we will use the results of numerical simulations to present some aspects of the basic fluid dynamics of street canyon flow and dispersion.

13.2 Principles

Street canyon flows are governed by the Navier–Stokes equations, which are based on the principles of the conservation for momentum, mass, and energy and the irreversibility principle. There are three approaches to solving the Navier–Stokes equations numerically. The first approach is direct numerical simulation (DNS), in which the model grid size is smaller than the Kolmogorov microscale (η). Hence, all eddies are explicitly resolved, and no turbulence parameterization is needed. The number of grid points required for DNS is roughly given by $(L/\eta)^3 \sim \text{Re}^{9/4}$, where L is the largest eddy size (or the model domain size) and Re is the Reynolds number. For street canyon flow applications, if one takes characteristic ambient wind speed and building height of $5\,\text{m s}^{-1}$ and $10\,\text{m}$, respectively, the calculated Re is $\sim 3.3 \times 10^6$, and the required number of grid points is $\sim 4.7 \times 10^{14}$. Applying the DNS approach to the study of street canyon flows at geometrical scales equivalent to urban canopies is, thus, not possible with current computing facilities. However, DNS studies at much reduced geometrical scales have been attempted (e.g., Coceal et al., 2006). The second approach is large-eddy simulation (LES). LES models perform simulations up to the length scale of inertial subrange. These models represent large eddies explicitly, and small eddies are represented by turbulence parameterizations. Various turbulence parameterizations for use in LES models have been proposed, including the Smagorinsky model and the dynamic model. The use of LES models to study street canyon flows has grown with rapidly increasing computing power (Liu and Barth, 2002; Cui et al., 2004; Kanda et al., 2004; Letzel et al., 2008). The third approach is Reynolds-averaged Navier–Stokes (RANS) simulation. The RANS

Handbook of Environmental Fluid Dynamics, Volume Two, edited by Harindra Joseph Shermal Fernando. © 2013 CRC Press/Taylor & Francis Group, LLC. ISBN: 978-1-4665-5601-0.

* This handbook chapter is largely based on journal papers published by the author.

approach is the approach most commonly used to simulate flow and dispersion in street canyons. The computational demand is much lower in RANS simulations than in LES simulations, and more abundant turbulence statistics can be obtained through LES simulations than through RANS simulations. The DNS, RANS, and LES models used in the field of urban flow and dispersion are computational fluid dynamics (CFD) models.

Of the three approaches mentioned earlier, the RANS approach (a CFD–RANS model) is described here. The CFD–RANS model has been developed to simulate and predict urban flow and dispersion (Kim and Baik, 2004; Baik et al., 2007), and the model description given as follows is based on those works. The model includes the renormalization group (RNG) k–ε turbulence model proposed by Yakhot et al. (1992). The momentum equation, mass continuity equation, thermodynamic energy equation, and transport equation for passive scalar (e.g., any passive pollutant) using the Boussinesq approximation can be written as

$$\frac{\partial U_i}{\partial t} + U_j \frac{\partial U_i}{\partial x_j} = -\frac{1}{\rho_0}\frac{\partial P^*}{\partial x_i} + \delta_{i3} g \frac{T^*}{T_0} + \nu \frac{\partial^2 U_i}{\partial x_j \partial x_j} - \frac{\partial}{\partial x_j}(\overline{u_i u_j}),$$
(13.1)

$$\frac{\partial U_j}{\partial x_j} = 0,$$
(13.2)

$$\frac{\partial T}{\partial t} + U_j \frac{\partial T}{\partial x_j} = \kappa \frac{\partial^2 T}{\partial x_j \partial x_j} - \frac{\partial}{\partial x_j}(\overline{T' u_j}) + S_h,$$
(13.3)

$$\frac{\partial C}{\partial t} + U_j \frac{\partial C}{\partial x_j} = D \frac{\partial^2 C}{\partial x_j \partial x_j} - \frac{\partial}{\partial x_j}(\overline{c u_j}) + S_c.$$
(13.4)

Here,

U_i is the ith (grid) mean velocity component
T is the mean temperature
C is the mean concentration of any passive scalar
u_i is the fluctuation from U_i
T' is the fluctuation from T
c is the fluctuation from C
P^* is the deviation of pressure from its reference value
T^* is the deviation of temperature from its reference value
ρ_0 is the air density
δ_{ij} is the Kronecker delta
g is the gravitational acceleration
ν is the kinematic viscosity of air
κ is the thermal diffusivity of air
D is the molecular diffusivity of scalar
S_h is the source/sink term of heat
S_c is the source term of scalar

The Reynolds stress, turbulent heat flux, and turbulent scalar flux in (13.1), (13.3), and (13.4), respectively, are expressed in terms of mean gradients as

$$-\overline{u_i u_j} = K_m \left(\frac{\partial U_i}{\partial x_j} + \frac{\partial U_j}{\partial x_i} \right) - \frac{2}{3}\delta_{ij} k,$$
(13.5)

$$-\overline{T' u_j} = K_h \frac{\partial T}{\partial x_j},$$
(13.6)

$$-\overline{c u_j} = K_c \frac{\partial C}{\partial x_j}.$$
(13.7)

Here,

K_m is the eddy (or turbulent) viscosity of momentum
K_h is the eddy diffusivity of heat
K_c is the eddy diffusivity of scalar
k is the turbulent kinetic energy

The inclusion of the second term of the right-hand side of (13.5) yields $\overline{u_i u_i}/2 = k$ for $i = j$, which is the expression for turbulent kinetic energy. K_m in the RNG k–ε turbulence model is given by

$$K_m = \nu \left(1 + \left(\frac{C_\mu}{\nu} \right)^{1/2} \frac{k}{\varepsilon^{1/2}} \right)^2,$$
(13.8)

where

C_μ is an empirical constant
ε is the dissipation rate of turbulent kinetic energy

The turbulent Prandtl number Pr ($= K_m/K_h$) and the turbulent Schmidt number Sc ($= K_m/K_c$) are specified for computation of K_h and K_c with K_m. K_m is computed using the prognostic equations of turbulent kinetic energy and its dissipation rate:

$$\frac{\partial k}{\partial t} + U_j \frac{\partial k}{\partial x_j} = -\overline{u_i u_j} \frac{\partial U_i}{\partial x_j} + \frac{\delta_{3j} g}{T_0} \overline{T' u_j} + \frac{\partial}{\partial x_j} \left(\frac{K_m}{\sigma_k} \frac{\partial k}{\partial x_j} \right) - \varepsilon,$$
(13.9)

$$\frac{\partial \varepsilon}{\partial t} + U_j \frac{\partial \varepsilon}{\partial x_j} = -C_{\varepsilon 1} \frac{\varepsilon}{k} \overline{u_i u_j} \frac{\partial U_i}{\partial x_j} + C_{\varepsilon 1} \frac{\varepsilon}{k} \frac{\delta_{3j} g}{T_0} \overline{T' u_j}$$
$$+ \frac{\partial}{\partial x_j} \left(\frac{K_m}{\sigma_\varepsilon} \frac{\partial \varepsilon}{\partial x_j} \right) - C_{\varepsilon 2} \frac{\varepsilon^2}{k} - R,$$
(13.10)

where σ_k, σ_ε, $C_{\varepsilon 1}$, and $C_{\varepsilon 2}$ are empirical constants, and the last term in (13.10) is an additional strain rate term in the RNG k–ε turbulence model.

Equations 13.1 through 13.4, 13.9, and 13.10 can be solved using any numerical method, such as the finite difference method, the finite element method, or the spectral method. For street canyon flow and dispersion simulations, the building configuration setup, the scalar emission source specification, and appropriate boundary/initial conditions are needed. Section 13.3 presents and discusses the results of CFD–RANS model simulations.

13.3 Analysis Results and Applications

13.3.1 Flow in Street Canyons

Flow in real street canyons is quite complex because of complicated building configurations and varying meteorology. To understand such a complex flow and its associated dispersion phenomena, it is first necessary to understand flows in simple building configurations.

Flow regimes in street canyons can be categorized as isolated roughness flow, wake interference flow, and skimming flow (Oke, 1988; Hunter et al., 1992; Sini et al., 1996). Let us consider a building configuration, in which the street canyon and buildings with equal height are infinitely long and the ambient wind blowing perpendicular to the street canyon. In this situation, a crucial factor determining flow regime is the ratio of the building height (H) to the width between two buildings (W), called the aspect ratio. Figure 13.1 schematically depicts flow features in each of the three flow regimes, which can be described as follows (Baik and Kim, 1999). When the buildings are relatively far apart, flow separation at the top edge of the upwind building causes the formation of a low-pressure recirculation vortex behind the upwind building. In addition, an eddy vortex due to flow separation on the downwind building wall is formed. In isolated roughness flow, the flow recovers its upwind profile before it reaches the downwind building, and the interaction between these two corotating vortices is negligible (Figure 13.1a). When the buildings are less far apart, the downwind building disturbs

the recirculation vortex before the flow readjustment takes place, and the two vortices interact with each other. This is a feature of wake interference flow (Figure 13.1b). When the buildings are narrowly spaced, very little of the above-roof flow enters the street canyon, and a stable vortex is formed in the canyon, featuring skimming flow (Figure 13.1c). A transition from isolated roughness flow to wake interference flow occurs at $W/H \sim 8-9$, and a transition from wake interference flow to skimming flow occurs at $W/H \sim 1.5$. Here, we are concerned with skimming flow and its associated dispersion, which are characteristic of flow and dispersion in densely built-up urban areas.

When the ambient wind speed exceeds a certain threshold value for W/H greater than about 1.5, a trapped vortex (or vortices) is formed in the street canyon. This vortex is formed and maintained by the downward transfer of the momentum of the ambient wind into the street canyon. When the ambient wind subsides, the vortex formed in the street canyon disappears. Figure 13.2 shows a streamline field, vertical profiles of the vertical velocity at two locations, and a turbulent kinetic energy field. These (also Figure 13.9) are simulation results obtained using the CFD–RANS model described in Section 13.2. The ambient wind blows toward the right. A primary, clockwise-rotating vortex is formed in the street canyon. A small, counterclockwise-rotating vortex is also noticeable near the bottom corner of the downwind building. The center of the primary vortex is located slightly above and downwind of the street canyon center. The strength of the downward motion near the

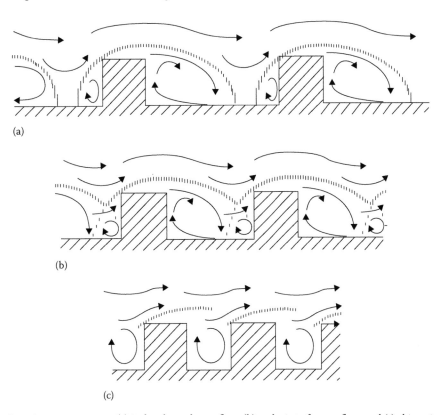

(a)

(b)

(c)

FIGURE 13.1 Flow regimes in street canyons: (a) isolated roughness flow, (b) wake interference flow, and (c) skimming flow. (After Oke, T.R., *Energy Build.*, 11, 103, 1988.)

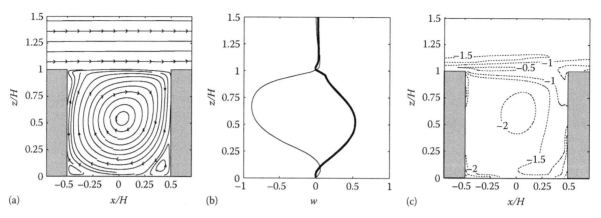

FIGURE 13.2 (a) Streamline field, (b) vertical profiles of the vertical velocity (unit in m s⁻¹) at $x/H = -0.425$ (thick solid line) and $x/H = 0.425$ (thin solid line), and (c) turbulent kinetic energy field (log scale in base 10 with a contour interval of 0.5 and unit in m² s⁻²) for an aspect ratio of 1.

downwind building is stronger than that of the upward motion near the upwind building (Figure 13.2b). Since the wind just below the top level of the street canyon ($z/H = 1$) is much weaker than the ambient wind, a strong shear layer exists across the top level. As wind shear is a source of turbulent kinetic energy, the turbulent kinetic energy is high in the strong shear layer (Figure 13.2c). The turbulent kinetic energy near the rooftops of both the upwind and downwind buildings is high because of the strong wind shear there. Air of relatively high turbulent kinetic energy enters the street canyon near the downwind building by the downward motion there.

The main vortex number in a street canyon varies with the aspect ratio (H/W). Figures 13.3a, 13.4a, and 13.5a show streamline fields for different aspect ratios in a neutral (isothermal) environment (Kim and Baik, 1999). It can be seen that skimming flow regime is characterized by a vortex (vortices) almost completely trapped in the street canyon and that the vortex number increases with increasing aspect ratio. When the aspect ratio is 1, one vortex is formed in the street canyon. When the aspect ratio is 2, two counterrotating vortices appear. When the aspect ratio is 3.5, three counterrotating vortices are formed. The middle vortex has the largest vertical size. The threshold aspect ratio for transition from a one-vortex regime to a two-vortex regime is 1–1.5. The threshold aspect ratio for transition from a two-vortex regime to a three-vortex regime is 3–3.5. Threshold values seem to be dependent to some extent on the model and experimental setup (e.g., Jeong and Andrews, 2002).

In the daytime, building surface and street bottom are heated by solar radiation. This solar heating affects flow in street canyons. Some insights into thermal effects on flow in street canyons can be obtained by performing numerical experiments with different aspect ratios and the heating of individual surfaces. Figures 13.3 through 13.5 are such simulation results, showing streamline fields for aspect ratios of 1, 2, and 3.5 when each of the upwind building wall, street bottom, and downwind building wall is heated (Kim and Baik, 1999). The temperature of the street bottom or building wall remains 5°C higher than the ambient air temperature.

For an aspect ratio of 1 (Figure 13.3), one vortex is formed in the no heating, upwind building-wall heating, and street bottom

heating cases, while two counterrotating vortices appear in the downwind building-wall heating case. In the upwind building-wall heating case, the higher temperature near the upwind building wall induces upward motion there. This thermally induced upward motion combines constructively with the upward motion (Figure 13.3a) that is mechanically induced by the ambient wind, resulting in a strengthened vortex in the street canyon. In the street bottom heating case, the temperature is higher near the street bottom. The temperature is also higher near the upwind building wall because of the advection of heated air near the street bottom toward the upwind building. This induces thermal upward motion near the upwind building wall, consequently enhancing vortex circulation in the street canyon. In the downwind building-wall heating case, the temperature is higher near the downwind building wall, especially in the upper region of the street canyon. At $z \sim 28$ m near the downwind building, the thermally induced upward motion, roughly speaking, balances the mechanically induced downward motion, and the horizontal motion toward the upwind building occurs, forming two counterrotating vortices in the street canyon.

For an aspect ratio of 2 (Figure 13.4), when the upwind building wall or street bottom is heated, only one vortex is formed by combining the two counterrotating vortices (Figure 13.4a) that appear in the no heating case. In the downwind building-wall heating case, two counterrotating vortices appear. The vertical size of the lower (upper) vortex is larger (smaller) in the downwind building-wall heating case than in the no heating case. In the upwind building-wall heating case, the thermally induced upward motion near the upwind building wall overcomes the mechanically induced downward motion in the lower region of the street canyon, eventually producing one vortex. In the street-bottom heating case, the temperature is higher near the upwind building wall as well as near the street bottom, as in the upwind building-wall heating case for an aspect ratio of 1. This results in the formation of one vortex by inducing thermally induced upward motion near the upwind building wall. In the downwind building-wall heating case, the thermally induced upward motion near the downwind building wall combines constructively with the mechanically induced upward motion in the

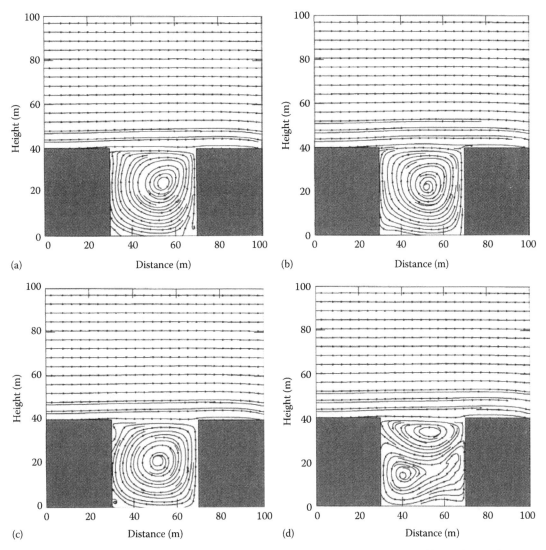

FIGURE 13.3 Streamline fields for an aspect ratio of 1: (a) no heating, (b) upwind building-wall heating, (c) street bottom heating, and (d) downwind building-wall heating cases. (After Kim, J.-J. and Baik, J.-J., *J. Appl. Meteorol.*, 38, 1249, 1999. With permission.)

lower region and combines destructively with the mechanically induced downward motion in the upper region.

For an aspect ratio of 3.5 (Figure 13.5), one vortex appears in the upwind building-wall heating case, as in the cases for aspect ratios of 1 and 2 (Figures 13.3b and 13.4b). The higher temperature near the upwind building wall induces strong upward motion there. This thermally induced upward motion causes the mechanically induced downward motion to disappear in the middle region. When the street bottom is heated, three vortices appear: two counterrotating vortices in the lower-to-middle region and one vortex in the upper region. Below the midlevel of the street canyon, the temperature field is almost symmetric about the center of the street canyon because of very weak flows in the lower region of the deep street canyon. This is associated with the two counterrotating vortices. The street bottom heating has little influence on flow field in the upper region. In the downwind building-wall heating case, the pattern of the streamline field with two counterrotating

vortices is similar to that in the cases for aspect ratios of 1 and 2 (Figures 13.3d and 13.4d).

When the street bottom or building wall is heated, the resultant street canyon flow can be regarded as a combination of the mechanically induced flow and thermally induced flow, as explained earlier. The strength/pattern of mechanically induced flow in a street canyon depends on the ambient wind speed and aspect ratio. The strength/pattern of thermally induced flow in a street canyon depends to a large extent on the heated surface and the intensity of the surface heating. Street canyon flows for wide ranges of heating intensities and aspect ratios are quite diverse when the street bottom or building wall is heated (e.g., Kim and Baik, 2001). In a real urban area, heated urban surfaces and heating intensity by solar radiation are determined by its geographical location, time of day, physical properties of surface materials, and so on.

So far, we have studied street canyon flows in the simplest building configuration (Figure 13.6a). Other prototypes

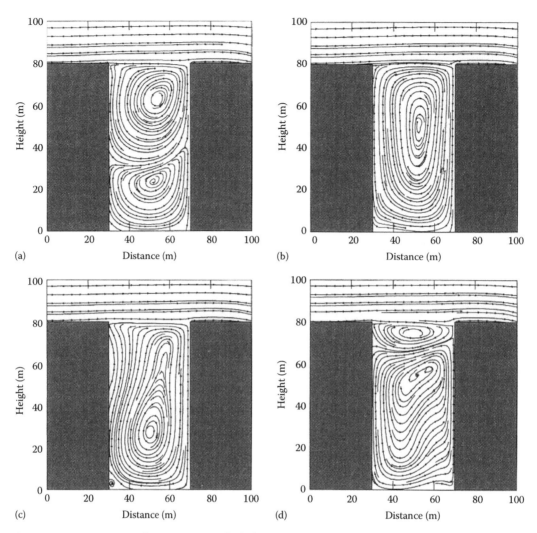

FIGURE 13.4 Same as in Figure 13.3 except for an aspect ratio of 2. (After Kim, J.-J. and Baik, J.-J., *J. Appl. Meteorol.*, 38, 1249, 1999. With permission.)

of building configurations can be employed to gain more insights into street canyon flows. These include a building configuration similar to the simplest one but with a finite building length (Figure 13.6b), a building configuration with a street intersection (Figure 13.6c), and an array of buildings (Figure 13.6d). Some studies employing these simple building configurations have been undertaken. Here, street canyon flows in a cubical building array are briefly described. Figure 13.7 shows the schematic of the mean flow for different ambient wind directions (Kim and Baik, 2004). These are drawn based on analysis of simulated three-dimensional flow fields. Only a basic portion of the cubical building array is shown. Depending on the ambient wind direction, three flow patterns can be identified. In flow pattern I when the ambient wind blows to the right (here eastward), a portal vortex formed behind the east wall of the upwind building is symmetric about the center of the street canyon. In flow pattern II with an ambient wind direction of up to 20°–25°, a portal vortex is also formed behind the east wall of the upwind building but not perpendicular to the ambient wind direction.

In flow pattern III with an ambient wind direction of 25°–45°, the footprints of a portal vortex are located behind the east and north walls of the upwind building. As expected, the mean flow pattern in a cubical building array is quite sensitive to the ambient wind direction.

There are abundant simple cases with various building configurations and ambient flow conditions. For example, when the buildings are not all the same height (e.g., the left building is taller than the right building in Figure 13.6a) and the ambient wind direction is perpendicular to the street canyon, mean flow and turbulence fields can be significantly different from those in the corresponding case for buildings of equal height. Studies in simple frameworks are a prerequisite to understanding complex street canyon flows in urban areas.

Next, we move to a real case. The CFD model can be used to simulate and predict flow (also dispersion) in a densely built-up urban area. Such simulation results for an area of Seoul, Korea, are presented here (Baik et al., 2009). Time-dependent boundary conditions for the CFD model are provided by a mesoscale

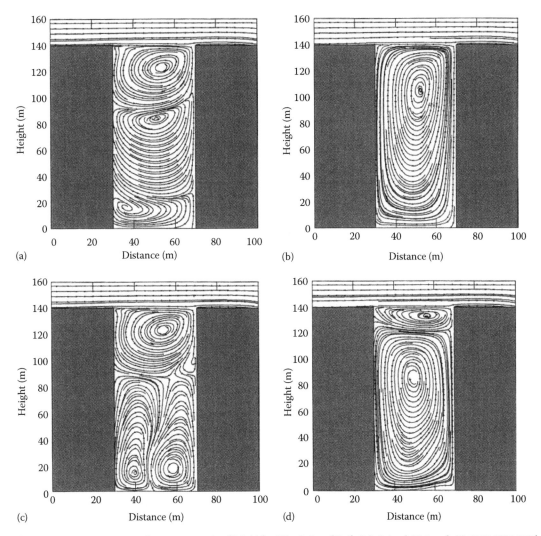

FIGURE 13.5 Same as in Figure 13.3 except for an aspect ratio of 3.5. (After Kim, J.-J. and Baik, J.-J., *J. Appl. Meteorol.*, 38, 1249, 1999. With permission.)

model MM5 (fifth-generation Pennsylvania State University–National Center for Atmospheric Research Mesoscale Model) (Grell et al., 1994). The coupled CFD-MM5 model is integrated for 24 h starting from 0300 LST 2 June 2004. Figure 13.8 shows simulated wind vector fields at 1400 LST 2 June. At $z = 2.5$ m, the fractional area covered by buildings is 0.31, and the MM5-simulated ambient wind is northwesterly. Street canyon flows in this area are very diverse and complex. At $z = 2.5$ m (Figure 13.8a), the CFD model-simulated wind and MM5-simulated ambient wind are generally not in the same direction because of the strong influence of buildings on flows. In most regions, the CFD model-simulated wind is weaker than the MM5-simulated ambient wind. This indicates that buildings reduce wind speed. The wind vector fields in the three small rectangular regions in Figure 13.8a are enlarged in Figure 13.8b through d to show characteristic flow patterns more clearly. In Figure 13.8b, the westerly flow toward the building elongated in the north–south direction (labeled 1) at $y = 510$ m diverges northward and southward in front of the building. The northward flow combines with the incoming flow near the northern

edge of the building, and a clockwise-rotating vortex is formed behind the building. The southward flow combines with the northwesterly inflow, and a counterclockwise-rotating vortex is formed. These two counterrotating vortices are produced mainly by the influence of the three buildings on the left and resemble a double-eddy circulation formed behind an isolated building (e.g., Zhang et al., 1996). The buildings on the right corners affect the size and intensity of the vortices to some extent. Figure 13.8c shows a typical channeling flow pattern. Flows converge and strengthen near the entrance of the channel-like street canyon, pass through the channel, and leave. The flow near the entrance and in the channel region is stronger than the MM5-simulated ambient wind. Figure 13.8d shows a recirculation vortex in the vertical plane, which is formed behind the tall buildings on the left. The lower portion of the recirculation vortex is affected by the two relatively low buildings on the right. At $z = 2.5$ m, the northeasterly flow enters this subregion (Figures 13.8a and d) and interacts with the recirculation vortex. As the MM5-simulated ambient wind speed and direction vary with time, flow in street canyons changes significantly.

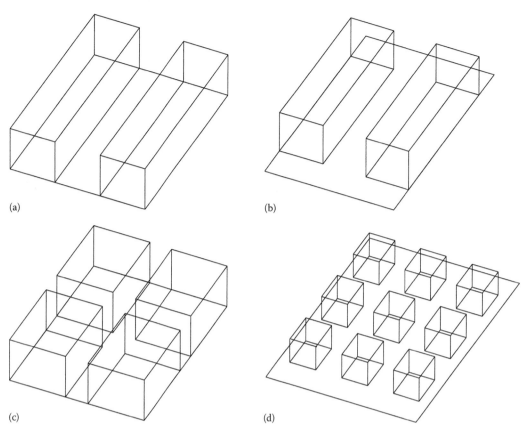

(a)

(b)

(c)

(d)

FIGURE1 3.6 Simple building configurations employed to study flow and dispersion in street canyons. See text for details.

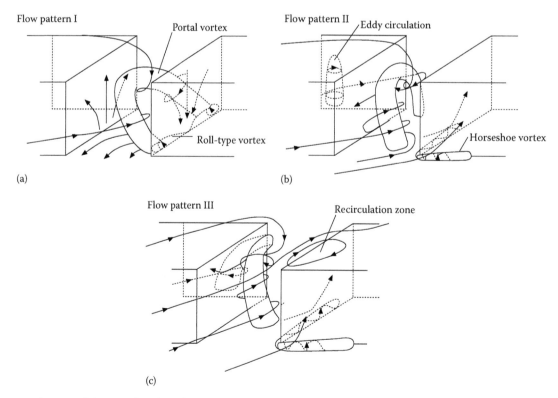

Flow pattern I — Portal vortex — Roll-type vortex

(a)

Flow pattern II — Eddy circulation — Horseshoe vortex

(b)

Flow pattern III — Recirculation zone

(c)

FIGURE 13.7 Schematic of the mean flow for different ambient wind directions. The dimensions of the portal vortex are reduced for the clarity of figure. See text for details. (After Kim, J.-J and Baik, J.-J., *Atmos. Environ.*, 38, 3039, 2004. With permission.)

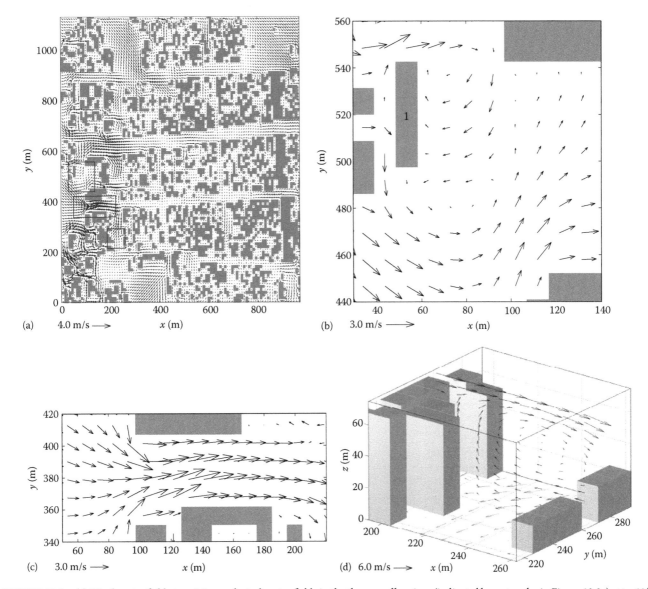

FIGURE 13.8 (a) Wind vector field at $z = 2.5$ m and wind vector fields in the three small regions (indicated by rectangles in Figure 13.8a) at $t = 11$ h (corresponding to 1400 LST 2 June 2004) in a densely built-up area of Seoul. (b) and (c) are at $z = 2.5$ m, and (d) is wind vector field in a horizontal plane of $z = 2.5$ m and that in a vertical plane of $y = 254.4$ m. (After Baik, J.-J. et al., *J. Appl. Meteorol. Climatol.*, 48, 1667, 2009. With permission.)

13.3.2 Dispersion in Street Canyons

To study scalar dispersion patterns in street canyons, let us consider a situation in which passive pollutants are continuously emitted near the street bottom, imitating emissions from automobiles. Figure 13.9 shows pollutant concentration fields for aspect ratios of 1 and 2. For an aspect ratio of 1, pollutant concentration is low near the downwind building in the middle-to-upper region of the street canyon because ambient air of relatively very low concentration enters the street canyon by downward motion on the downwind side. Pollutant concentration is also low near the vortex center. Pollutant concentration is high near the upwind building wall because highly polluted air passing through the street-level source is transported upward on the upwind side. Pollutant concentration near the bottom corner

of the downwind building is high. This is because a small, counterclockwise-rotating vortex (see Figure 13.2a) traps emitted pollutants. For an aspect ratio of 2, the pollutant concentration field pattern in the upper region of the street canyon is similar to that for an aspect ratio of 1. In the lower region of the street canyon, pollutant concentration is higher near the downwind building wall than near the upwind building wall. The highly polluted air passing through the street-level source is transported upward on the downwind side of the lower region. Near the midlevel, pollutants are transported from the downwind side to the upwind side. On the upwind side, pollutants are transported either upward by the upward motion of the upper vortex or downward by the downward motion of the lower vortex. Ambient air of relatively very low pollutant concentration enters the street canyon by the downward motion of the upper vortex on the downwind side.

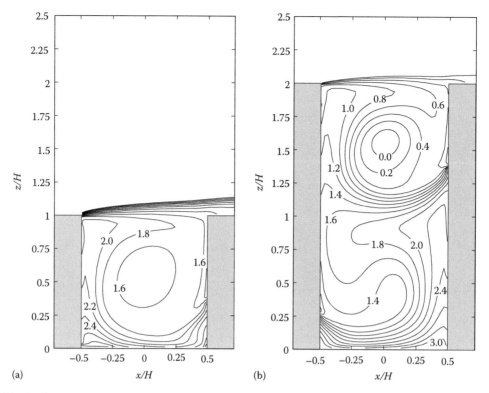

FIGURE 13.9 Fields of pollutant concentration (log scale in base 10 and unit in ppb) for aspect ratios of (a) 1 and (b) 2 for the street-level emission source. The contour interval is 0.2.

These explain the pollutant concentration field pattern. As explained earlier, the dispersion of pollutants in a street canyon is directly related to vortex circulation.

For the same ambient wind speed and pollutant emission rate, pollutant concentration, for example, near the pedestrian level is higher for larger aspect ratios. Thus, air quality can be very poor in a deep street canyon. For a given street canyon and pollutant emission rate, meteorology greatly affects air quality in the street canyon. As the ambient wind speed increases, pollutant concentration in the street canyon becomes low, hence improving the air quality. The ambient wind direction, ambient turbulence intensity, and surface heating by solar radiation also influence pollutant concentration in the street canyon.

Pollutants in a street canyon escape from the canyon, and some of the escaped pollutants can reenter the canyon. The question to which process (mean or turbulent) is more responsible for pollutant escape from a street canyon remains to be answered. We answer this question by calculating the vertical flux of pollutants by mean flow (F_m) and the vertical flux of pollutants by turbulent flow (F_t):

$$F_m = CW, \qquad (13.11)$$

$$F_t = \overline{cw} = -K_c \frac{\partial C}{\partial z}. \qquad (13.12)$$

The horizontal distributions of U (mean horizontal velocity), W (mean vertical velocity), k (turbulent kinetic energy), F_m, and F_t

at the top level of a two-dimensional street canyon with an aspect ratio of 1 ($z/H = 1$) are shown in Figure 13.10 (Baik and Kim, 2002). Pollutants are assumed to be continuously emitted near the street bottom. A positive F_m (F_t) value means that pollutants escape from the street canyon, while a negative F_m (F_t) value means that escaped pollutants reenter the street canyon. The vertical mean flux (F_m) and vertical velocity have similar horizontal patterns. The vertical mean flux is positive from the upwind building edge ($x = 30\,\text{m}$) to $x = 52\,\text{m}$, negative up to $x = 68\,\text{m}$, and then positive up to the downwind building edge ($x = 70\,\text{m}$). The location of the maximum downward mean flux is very close to that of the maximum downward motion. The vertical turbulent flux (F_t) is upward at all positions with its maximum at $x = 38\,\text{m}$. In most regions, the magnitude of the vertical turbulent flux is much larger than that of the vertical mean flux. These results indicate that pollutants escape from the street canyon mainly by turbulent process.

Furthermore, the net vertical mean and turbulent pollutant fluxes across the top level of the street canyon are calculated by integrating F_m and F_t in the horizontal from $x = 30$ to $70\,\text{m}$ at $z/H = 1$:

$$\beta_m = \int_l F_m dx, \qquad (13.13)$$

$$\beta_t = \int_l F_t dx. \qquad (13.14)$$

The calculated net fluxes in this particular simulation are $\beta_m = -6.57\,\text{ppb m}^2\,\text{s}^{-1}$ and $\beta_t = 70.63\,\text{ppb m}^2\,\text{s}^{-1}$. β_m is negative,

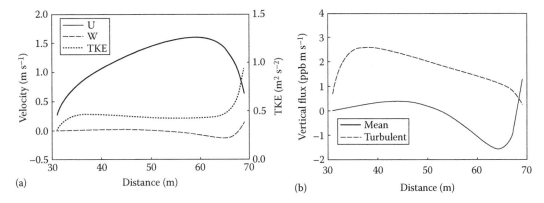

FIGURE 13.10 Horizontal distributions of (a) mean horizontal and vertical velocities and turbulent kinetic energy and (b) the vertical fluxes of pollutants by mean flow and turbulent flow at the top level of a street canyon with an aspect ratio of 1. (After Baik, J.-J. and Kim, J.-J., *Atmos. Environ.*, 36, 527, 2002. With permission.)

implying that the net effect by mean flow is to cause some escaped pollutants to reenter the street canyon. The magnitude of β_t is much larger than that of β_m, again indicating that the turbulent process is dominantly responsible for pollutant escape from the street canyon. Simulations with different inflow turbulence intensities, ambient wind speeds, and street aspect ratios confirm this finding (Baik and Kim, 2002). When the street bottom or building wall is heated, the relative importance of mean flow and turbulent flow in expelling pollutants from the street canyon can differ depending on the heating intensity.

Automobiles are one of the main pollutant sources in urban areas. Pollutants emitted from automobiles, such as nitric oxide (NO) and nitrogen dioxide (NO_2), are chemically reactive. Complex photochemical processes in densely built-up urban areas with heavy traDc often cause serious air pollution problems when meteorological conditions are favorable. To further understand dispersion in street canyons, reactive pollutants need to be taken into account (Baker et al., 2004; Baik et al., 2007). The reactive pollutants we will introduce are NO, NO_2, and ozone (O_3). It is assumed that NO and NO_2 are emitted from automobiles into a street canyon in the presence of background O_3 and sunlight. The chemical reactions involved are

$$NO_2 + h\nu \rightarrow NO + O, \tag{13.15}$$

$$O + O_2 + M \rightarrow O_3 + M, \tag{13.16}$$

$$O_3 + NO \rightarrow NO_2 + O_2, \tag{13.17}$$

where M represents a molecule (N_2 or O_2 or another third molecule) that absorbs excess energy. Sunlight of wavelength shorter than 420 nm photodissociates NO_2 to NO, and NO_2 is reproduced by the reaction of O_3 with NO.

Then, the transport equations for NO, NO_2, and O_3 can be written as

$$\frac{\partial [NO]}{\partial t} + U_j \frac{\partial [NO]}{\partial x_j} = D \frac{\partial^2 [NO]}{\partial x_j \partial x_j} + \frac{\partial}{\partial x_j}\left(K_c \frac{\partial [NO]}{\partial x_j}\right)$$
$$+ J_{NO_2}[NO_2] - k_1[O_3][NO] + S_{NO}, \tag{13.18}$$

$$\frac{\partial [NO_2]}{\partial t} + U_j \frac{\partial [NO_2]}{\partial x_j} = D \frac{\partial^2 [NO_2]}{\partial x_j \partial x_j} + \frac{\partial}{\partial x_j}\left(K_c \frac{\partial [NO_2]}{\partial x_j}\right)$$
$$- J_{NO_2}[NO_2] + k_1[O_3][NO] + S_{NO_2}, \tag{13.19}$$

$$\frac{\partial [O_3]}{\partial t} + U_j \frac{\partial [O_3]}{\partial x_j} = D \frac{\partial^2 [O_3]}{\partial x_j \partial x_j} + \frac{\partial}{\partial x_j}\left(K_c \frac{\partial [O_3]}{\partial x_j}\right)$$
$$+ k_2[O][O_2][M] - k_1[O_3][NO]. \tag{13.20}$$

Here,

J_{NO_2} is the photolysis rate of NO_2 in (13.15)
k_1 is the rate constant for the reaction in (13.17)
k_2 is the rate constant for the reaction in (13.16)
S_{NO} and S_{NO_2} are the source terms of NO and NO_2, respectively

The oxygen atom is highly reactive, so its depletion rate in (13.16) is equal to its formation rate in (13.15). This is called the pseudo-steady-state approximation for a highly reactive chemical species. Thus, given $k_2[O][O_2][M] = J_{NO_2}[NO_2]$, (13.20) can be rewritten as

$$\frac{\partial [O_3]}{\partial t} + U_j \frac{\partial [O_3]}{\partial x_j} = D \frac{\partial^2 [O_3]}{\partial x_j \partial x_j} + \frac{\partial}{\partial x_j}\left(K_c \frac{\partial [O_3]}{\partial x_j}\right)$$
$$+ J_{NO_2}[NO_2] - k_1[O_3][NO]. \tag{13.21}$$

Using the aforementioned chemistry model coupled to the CFD–RANS model described in Section 13.2, we can study the role and relative importance of individual processes in dispersing reactive pollutants in a street canyon. This can be accomplished by performing a budget analysis for each transport equation.

Figure 13.11 shows the advection, turbulent diffusion, and chemical reaction terms in the transport equation for each reactive pollutant in the case of street bottom heating (Baik et al., 2007). In the NO budget, the magnitude of the advection term, $-U_j \partial [NO]/\partial x_j$ (also the turbulent diffusion term,

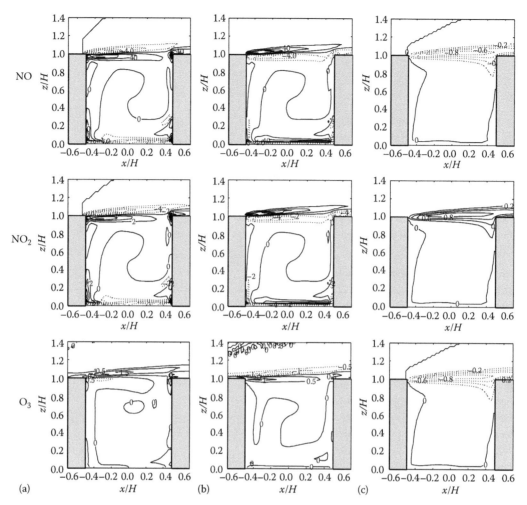

FIGURE 13.11 Advection, turbulent diffusion, and chemical reaction terms in the transport equations for the reactive pollutants (unit in ppb s^{-1}). Figures in the upper, middle, and lower rows are for NO, NO$_2$, and O$_3$, respectively. Figures in the (a), (b), and (c) columns are for advection, turbulent diffusion, and chemical reaction terms, respectively. The contour interval in each figure is uniform, and the aspect ratio is 1. (From Baik, J.-J. et al., *Atmos. Environ.* 41, 934, 2007. With permission.)

$\partial/\partial x_j (K_c \partial[NO]/\partial x_j))$, is much larger than that of the chemical reaction term, $J_{NO_2}[NO_2] - k_1[O_3][NO]$. The advection term largely balances the turbulent diffusion term. Just below the top level of the street canyon ($z/H = 1$), positive NO advection increases NO concentration, and the turbulent diffusion process reduces NO concentration. In the region close to the street bottom, negative NO advection reduces NO concentration. The turbulent diffusion process reduces NO concentration in the region very close to the street bottom and increases NO concentration above it. Near $z/H = 1$, the NO depletion rate is larger than its formation rate by photolysis of NO$_2$. Hence, the chemical reaction process reduces NO concentration near $z/H = 1$, but its contribution to the total NO concentration is minor. In the NO$_2$ budget, the spatial distribution pattern of the advection term (also the turbulent diffusion term) is very similar to that in the NO budget, but its magnitude is much smaller. The chemical reaction term near $z/H = 1$ increases NO$_2$ concentration there. In the O$_3$ budget, all three terms are small in the street canyon but large

near $z/H = 1$. The magnitude of the chemical reaction term is comparable to that of the advection term (also the turbulent diffusion term). Near $z/H = 1$, the chemical reaction term reduces O$_3$ concentration. These budget analysis results indicate that for certain chemical species, the chemical reaction process can play an important role in determining the concentrations of chemical species in a street canyon. Among many factors, the relative importance of the chemical reaction process for any chemical species depends on the location and strength of its source.

Dispersion of aerosols in street canyons is another interesting problem. In urban areas, automobiles emit aerosol particles, mostly ultrafine particles (particles smaller than 100 nm), into the atmosphere. Dispersion of aerosols in street canyons can be studied using an aerosol dynamics model coupled to a CFD model. An aerosol dynamics model may include transport equations for aerosols with nucleation, condensation, coagulation, and deposition processes. Some research along this line has been undertaken (Kim, 2010).

13.4 Major Challenges

Progress in modeling and understanding flow and dispersion in street canyons has been made in the past two decades. Studies performed so far have provided very valuable insights into the basic fluid dynamics of street canyon flow and dispersion and also have helped to resolve many of the environmental problems related to street canyons.

Many challenging problems remain in the field of street canyon flow and dispersion. Some problems are as follows. (1) Flow and dispersion in the simplest building configuration (Figure 13.6a) have been investigated extensively. However, there have been fewer studies using other simple building configurations, such as those depicted in Figure 13.6b through d. Since understanding flow and dispersion in simple building configurations is the cornerstone of understanding complex flow and dispersion in built-up urban areas, systematic studies with various prototypical building configurations are needed. (2) Compared with the characteristics of mean flow in street canyons, those of turbulent flow in and above street canyons (coherent structures, intermittency, and so on) are less well understood. Fundamental studies, especially using DNS and LES models, are needed to gain a better understanding of the nature of urban turbulent flows. (3) The majority of street canyon flow and dispersion studies have focused on neutral stability conditions. Street canyon flow and dispersion can change significantly depending on stability changes caused by building surface/street bottom heating, ambient flow, and ambient thermal stratification. Studies specifically designed to investigate stability (thermal) effects on flow and dispersion in street canyons are needed. (4) A street canyon is a space, in which aerosols and chemically reactive species exist, interact, and are dispersed in the presence of turbulent flow. To understand such complex phenomena through numerical simulations, the development of a coupled CFD-chemistry-aerosol model with reasonable levels of complexity in photochemistry and aerosol dynamics is required. This is a very challenging task. (5) Development of a coupled CFD-mesoscale model in two-way interactions is necessary to predict flow and dispersion in built-up urban areas reliably. The coupled mesoscale model needs to be urbanized. Coupling the CFD-chemistry-aerosol model to a regional air quality model is also a demanding task.

References

Baik J-J, Kang Y-S, Kim J-J. 2007. Modeling reactive pollutant dispersion in an urban street canyon. *Atmos. Environ.* 41:934–949.

Baik J-J, Kim J-J. 1999. A numerical study of flow and pollutant dispersion characteristics in urban street canyons. *J. Appl. Meteor.* 38:1576–1589.

Baik J-J, Kim J-J. 2002. On the escape of pollutants from urban street canyons. *Atmos. Environ.* 36:527–536.

Baik J-J, Park S-B, Kim J-J. 2009. Urban flow and dispersion simulation using a CFD model coupled to a mesoscale model. *J. Appl. Meteorol. Climatol.* 48:1667–1681.

Baker J, Walker HL, Cai X. 2004. A study of the dispersion and transport of reactive pollutants in and above street canyons—A large eddy simulation. *Atmos. Environ.* 38:6883–6892.

Coceal O, Thomas TG, Castro IP, Belcher SE. 2006. Mean flow and turbulence statistics over groups of urban-like cubical obstacles. *Boundary Layer Meteorol.* 121:491–519.

Cui Z, Cai X, Baker CJ. 2004. Large-eddy simulation of turbulent flow in a street canyon. *Q. J. R. Meteorol. Soc.* 130:1373–1394.

Grell GA, Dudhia J, Stauffer DR. 1994. A description of the fifth-generation Penn State/NCAR Mesoscale Model (MM5). NCAR/TN-398+STR. 138pp.

Hunter LJ, Johnson GT, Watson ID. 1992. An investigation of three-dimensional characteristics of flow regimes within the urban canyon. *Atmos. Environ.* 26B:425–432.

Jeong SJ, Andrews MJ. 2002. Application of the kt ε turbulence model to the high Reynolds number skimming flow field of an urban street canyon. *Atmos. Environ.* 36:1137–1145.

Kanda M, Moriwaki R, Kasamatsu F. 2004. Large-eddy simulation of turbulent organized structures within and above explicitly resolved cube arrays. *Boundary Layer Meteorol.* 112: 343–368.

Kim J-J, Baik J-J. 1999. A numerical study of thermal effects on flow and pollutant dispersion in urban street canyons. *J.eAppl. Meteor.* 38:1249–1261.

Kim J-J, Baik J-J. 2001. Urban street-canyon flows with bottom heating. *Atmos. Environ.* 35:3395–3404.

Kim J-J, Baik J-J. 2004. A numerical study of the effects of ambient wind direction on flow and dispersion in urban street canyons using the RNG k–ε turbulence model. *Atmos. Environ.* 38:3039–3048.

Kim M-W. 2010. CFD modeling of aerosol dispersion in a street canyon. MS thesis. Seoul National University, Seoul, South Korea, 73pp.

Letzel MO, Krane M, Raasch S. 2008. High resolution urban large-eddy simulation studies from street canyon to neighbourhood scale. *Atmos. Environ.* 42:8770–8784.

Li X-X, Liu C-H, Leung DYC, Lam KM. 2006. Recent progress in CFD modelling of wind field and pollutant transport in street canyons. *Atmos. Environ.* 40:5640–5658.

Liu C-H, Barth MC. 2002. Large-eddy simulation of flow and scalar transport in a modeled street canyon. *J. Appl. Meteorol.* 41:660–673.

Oke TR. 1988. Street design and urban canopy layer climate. *Energy Build.* 11:103–113.

Sini J-F, Anquetin S, Mestayer PG. 1996. Pollutant dispersion and thermal effects in urban street canyons. *Atmos. Environ.* 30:2659–2677.

Vardoulakis S, Fisher BEA, Pericleous K, Gonzalez-Flesca N. 2003. Modelling air quality in street canyons: a review. *Atmos. Environ.* 37:155–182.

Yakhot V, Orszag SA, Thangam S, Gatski TB, Speziale CG. 1992. Development of turbulence models for shear flows by a double expansion technique. *Phys. Fluids.* A4:1510–1520.

Zhang YQ, Arya SP, Snyder WH. 1996. A comparison of numerical and physical modeling of stable atmospheric flow and dispersion around a cubical building. *Atmos. Environ.* 30:1327–1345.

14

Air Flow through Tunnels

Hong-Ming Jang
Chinese Culture University

Falin Chen
National Taiwan University

14.1 Introduction

A traDc transportation tunnel is a highly enclosed man-made domain. In the domain not only the transfers but also the transformations of constituents, momentum, and energy can happen. In that context, the air flow is primarily for maintaining this complex and changeable environment at an acceptable condition for human beings to pass through, or to be free from any hazards. To achieve this purpose, some ventilation systems have been developed. In this chapter, the ventilation systems, flow driving units, flow models, examples, methods of analysis, and major challenges have been introduced.

14.1.1 Tunnels and the Ventilation Concerns

Up to probably 2017, the longest transportation tunnel in this world would be as long as 57,072 m (Gotthard Base Tunnel in Switzerland) whose dimension is still far smaller than the dimension of its foundation, the lithosphere of the earth. Even when the sum of the volume of all the transportation tunnels in this world just occupies a tiny portion of the volume of this lithosphere, their operation is not likely to have minor impact on the entire ecosystem. However, studies (Berglund et al. 1993, Tornqvist and Ehrenberg 1994, Barrefors 1996) and tunnel catastrophic events (Chen 2000) have indicated that the environment in a tunnel has been crucial to the health and even the life of tunnel users; two of the most annoying situations in a tunnel are traDc jams and fires. Since no measure can completely prevent the accident from happening in a tunnel and that the number of tunnels built in the past half century has increased faster than ever at the same time (as shown in Figure 14.1), the trend which seems to remain, tunnel ventilation issues will be more

intimate to our daily life and its ventilation technology will have to face more challenges.

Tunnels can effectively shorten the transportation distance and therefore reduce the associated travelling time and energy consumption. Although tunnels are categorized into many types, such as road tunnels, rail tunnels, metro tunnels, subsea tunnels, mountain tunnels, or immersed tube tunnels based on a specific feature, they are primarily built for road and rail vehicles. Road tunnels constructed are generally shorter than the rail ones (Figure 14.1b); by probably 2017 there will be about 70 rail tunnels but only 16 road tunnels longer than 10 km and till then the longest rail tunnel (57,072 m) will be more than twice that of the longest road tunnel (24,510 m, the Laerdal Tunnel in Norway). One of the major factors that influence the development paces of the road tunnel and the rail tunnel is the ventilation concern. Vehicles fleet needs to continuously consume oxygen from the surrounding air and in the meantime expel gaseous and particulate pollutants into it. In a tunnel, the cumulated gaseous exhausts can cause discomfort and illness to the passengers and the particulates that are mostly emitted from diesel engines can reduce road visibility. Moreover, a fire may occur due to mechanical failure of vehicles, vehicle collisions, or even vandalism attack. Being able to achieve a well dilution of vehicle exhausts and being able to properly control smoke flow are commonly the main concerns in designing and operating the ventilation system of a tunnel. Generally, the traDc complexity in a road tunnel is higher than that in a rail tunnel, so that traDc jam and vehicle collision are more likely to occur there. The ventilation requirements are therefore more rigorous for road tunnels and the technologies for ventilating a road tunnel can almost be applied on ventilating a rail tunnel. This article will aim at introducing the road tunnel ventilation systems and

Handbook of Environmental Fluid Dynamics, Volume Two, edited by Harindra Joseph Shermal Fernando. © 2013 CRC Press/Taylor & Francis Group, LLC.
ISBN: 978-1-4665-5601-0.

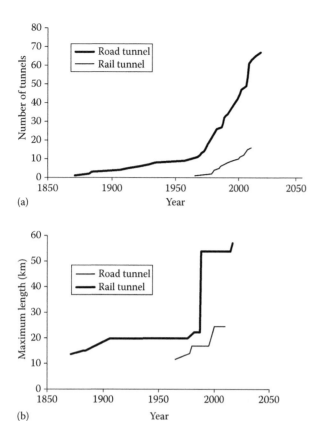

FIGURE 14.1 Development trend of long tunnels in the world. (a) Number of tunnels longer than 10 km. (b) The maximum tunnel length. (Based on the data published in 2009 on the World's Longest Tunnel Page, http://www.lotsberg.net/data/rail.html)

the principles commonly employed to analyze the flow behaviors and the ventilation performances of these systems.

14.1.2 Road Tunnel Ventilation Systems

No matter how short or long road tunnels are, there have been three types of ventilation systems developed to deal with the ventilation problems: longitudinal ventilation, transverse ventilation, and semi-transverse ventilation. For a particular traDc status, the number of vehicles in a tunnel tube normally increases linearly with the length of the tunnel tube and so does the air flow rate required to maintain an acceptable air quality. For any existing system, the air flow rate is proportional to the air flow speed and the pressure loss of the flow is proportional to the square of the air flow speed; as a result the power required to support the air flow is proportional to the cube of this air flow speed. Therefore, the ventilation system of a long tunnel would become very expensive in terms of equipment and energy costs if it were to be ventilated as a whole unit. Consequently, a long tunnel, regardless of the type of ventilation system equipped, is divided into separate ventilation sections using intermediate shafts or air-cleaning plants.

In the following subsections, the configurations, flow patterns, and ventilation characteristics of these three basic types of ventilation systems will be introduced and their one-dimensional (1D) flow models described in Section 14.2.

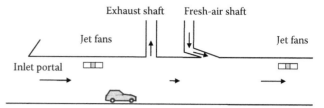

FIGURE 14.2 Typical flow pattern in a longitudinal ventilation system that is divided into ventilation sections by intermediate shafts.

14.1.2.1 Longitudinal Ventilation System

The longitudinal ventilation system is primarily designed for a uni-directional traDc tunnel. In a ventilation section of a long longitudinally ventilated tunnel, fresh air is entrained from the inlet portal or is injected into the tunnel section through a duct that brings fresh air from the intermediate shafts, as shown in Figure 14.2, and meanwhile the pollutants and smoke are carried to exhaust shafts and outlet portal by the air flowing along this tunnel section.

Since the traDc in a longitudinally ventilated tunnel is uni-directional, in case of normal ventilation the pollutant concentrations are linearly growing in the traDc direction and the maximum concentration is near the intermediate shafts and outlet portal. The level of pollution in this system can be lowered by increasing the air flow speed, but due to the considerations of comfort and the motor bike safety there is usually a limitation for the air flow speed in the tunnel. For example, Permanent International Association of Road Congresses (PIARC) recommends that 10 m/s is the limitation of air flow speed. As a result, this speed limitation confines the air volume available for dilution in this system.

In case of fire, the air flow carries the smoke from the fire to the nearest exhaust shaft or portal with a possible smoke filling of the cross section in the downstream. As a result, if the traDc is so low that vehicles downstream smoothly leave the tunnel, there will be no safety concern. If there is traDc jam in the downstream, those vehicles that are trapped may be exposed to smoke. This is mostly likely to occur in urban tunnels. Therefore in order to ensure user's safety in tunnels with a risk of traDc jam, installation of smoke exhaust system is necessary.

Most of short tunnels in the world are equipped with longitudinal ventilation system. Two examples of long longitudinally ventilated tunnels that are separated into ventilation sections with vertical shafts are Hsuehshan tunnel in Taiwan and Kan Etsu tunnel in Japan. Laerdal tunnel in Norway is an example of long longitudinally ventilated tunnel that is separated into three ventilation sections by an air cleaning plant and a shaft.

14.1.2.2 Transverse Ventilation and Semi-Transverse Ventilation Systems

The transverse and semi-transverse ventilation systems are primarily designed for bi-directional traDc tunnels. A transverse ventilation tunnel is equipped with fresh-air ducts that intake

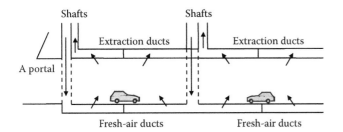

FIGURE 14.3 Typical flow pattern in a transverse ventilation system that is divided into ventilation sections by shafts.

fresh air from portals or fresh-air shafts and provide fresh air directly inside the tunnel. This tunnel is also equipped with extraction ducts that extract polluted air and smoke directly inside the tunnel and discharge them at portals or through exhaust shafts, Figure 14.3. Dampers are distributed along the ducts, through which fresh air is injected into and polluted air is extracted from the tunnel. In order to sustain the air-exchange flow through the entire system, local ventilation plants are usually built at portals and shafts.

In case of normal ventilation, the volume of fresh air supplied to the tunnel is equal to the volume of polluted air extracted in a full transverse ventilation system, while in a semi-transverse ventilation system, the volume of fresh air supplied to the tunnel is not equal to (mostly more than, to be explained in Example 14.3) the volume of polluted air extracted. In a semi-transverse ventilation system, part of vehicle exhausts will be carried by the air flow along the tunnel to portals or exhaust shafts. Therefore, the pollutant distributions along the tunnel equipped with full transverse ventilation would be quite uniform and those in a semi-transverse ventilation tunnel would increase in the axial flow direction. The air quality in these ventilation systems would depend on the traDc pattern, traDc volume, and the volume rate of air supplied and extracted.

In case of fire, the remote-controlled system would open the dampers near fire and in the meantime close the dampers elsewhere so that smoke would be extracted through exhaust ducts. Ideally the smoke could be confined near the fire and prevented from attacking the vehicles and people trapped in both sides of the fire. When the meteorology makes the pressure difference between portals so high that induces the air flow speed to a certain magnitude beyond the ventilation system can manage, the smoke would be partially extracted through ducts and partially carried by air flow from the fire to the portals. Therefore, there is still a risk of smoke-filling of the cross section for the transverse and semi-transverse ventilation systems. This is most likely to occur in tunnels located under very tall mountains on a windy day. In order to ensure userß safety in these tunnels, banks of jet fans are commonly added to suppress the possible longitudinal flow in case of a fire.

Full transverse ventilation system is less common than semi-transverse ventilation system due to more expansive costs of construction and operation, e.g., bigger tunnel cross section for installing larger ducts, more powerful ventilation plants and energy consumption. The Gotthard tunnel in Switzerland is one

of the examples that adopted full transverse ventilation system. There are more examples of tunnels equipped with semi-transverse ventilation system in the world, such as the Frejus tunnel connecting France and Italy, the Arlberg tunnel in Austria, and the Mont Blanc tunnel connecting France and Italy. The ventilation systems of Mont Blanc tunnel and Frejus tunnel have been modified to reinforce their smoke extraction capacities after the 1999 and 2005 fatal fire that revealed the smoke extraction capability of a semi-transverse system may be severely weakened by the wind pressure difference at portals.

14.1.3 Tunnel Flow Driving Units

Due to viscosity, the air flow in a tunnel is subjected to friction that continuously transforms its mechanical energy into heat. Therefore, mechanical energy must be continuously supplied into the flow so that the flow can be sustained or accelerated. The flow in a tunnel can be driven by vehicles fleet, jet fans, large axial flow fans, and Saccardo nozzles. Their working properties and principles are to be introduced next.

14.1.3.1 Vehicle Piston Effect

As a vehicle passes through a tunnel, it will be subjected to a drag force due to the aerodynamic response of air. Conversely, on the entire air column of the tunnel each vehicle in the tunnel will exert a driving force, F_{di}, that is equal in magnitude but reverse in direction with the drag force, Figure 14.4. This is the so-called vehicle piston effect in a tunnel and its quantity can be expressed in terms of the following equation:

$$F_{vehicle} = \frac{1}{2}\rho \sum_{i=1}^{N_{vehicle}} C_{di} \left| V_{vehicle\,i} - V \right| \left(V_{vehicle\,i} - V \right) A_{vehicle\,i} \quad (14.1)$$

where

ρ is the air density
V is the airflow longitudinal mean velocity
$N_{vehicle}$ is the total number of vehicles in the tunnel section
$V_{vehicle\,i}$, $A_{vehicle\,i}$, and C_{di} denote the speed, frontal area, and drag coeDcient of vehicle i, respectively

Note that drag coeDcient is normally evaluated for a vehicle moving in an open area. However, after a vehicle enters a tunnel, its body more or less narrows down the air flow passage.

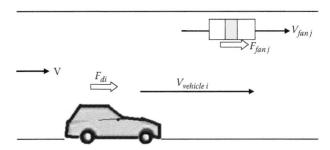

FIGURE 14.4 Dynamic effects of a jet fan and a vehicle.

The blockage raises the pressure in the field in front of the vehicle, lowers the pressure in the field behind it, and increases the flow speed between the vehicle and the tunnel wall (Jang and Ma 2009). As a result, the vehicle is subjected to a higher frictional coeDcient and forms drags and has higher drag coeDcient as it passes through a tunnel than an open area. Another factor that may affect the "piston effect" of vehicle is the tail gating effect or shadow effect (Katz 1995), which needs to be taken into consideration if vehicles are very close to each other.

Equation 14.1 shows that if $V_{vehicle\,i} > V$, the vehicle i will add momentum to the flow and facilitate ventilation, and if $V_{vehicle\,i} < V$ the existence of the vehicle will slow down the flow.

14.1.3.2 Jet Fans

In a short tunnel or a properly divided ventilation section of a long tunnel, piston effect of traDc flow is suDcient to dilute vehiclesﬂemissions and air quality can be maintained during normal operation. But as vehicles are slowed down by unusual traDc situations, their piston effect may be too weak to support the flow speed to meet the ventilation requirement. In these situations, another source of momentum input is needed and jet fans are the most common machine installed for this purpose. Normally banks of jet fans are installed at intervals in the space between the tunnel roof and the traDc envelope. They intake a small portion of the tunnel air and inject it back into the tunnel with a high momentum and then the propulsion jet transfers its momentum to the air column of the tunnel, Figure 14.4. The equivalent total thrust of jet fans applied on the column of tunnel air can be written as

$$F_{jet\,fan} = \sum_{j=1}^{N_{fan}} F_{fan\,j} = \rho A_{fan} \sum_{j=1}^{N_{fan}} \left(V_{fan\,j} - V \right) \left| V_{fan\,j} \right| \eta_{fan\,j} \quad (14.2)$$

where

N_{fan} is the total number of jet fans started
A_{fan} is the jet fan outlet cross-sectional area
$V_{fan\,j}$ is the outflow velocity of jet fan j
$\eta_{fan\,j}$ denotes the eDciency of jet fan j, which accounts for the losses due to friction (Costeris 1991, Betta et al. 2009)

In order to save energy, jet fans are normally not activated until the air quality deteriorates suDciently to require ventilation assistance. Under normal operation, the jet fans are used to create a speed of the air column so that permissible pollution limits can be remained. However, in case of fire the jet fans are primarily used for controlling the smoke flow. Therefore jet fans can be operated to push the air flow forward as well as backward. Since jet fans have shown good performance on controlling smoke flow, they are employed in not only the longitudinal but also the transverse ventilation systems. However, the driving force of a jet fan in downstream of a large fire may be reduced to a large extent due to the decrease in the density of air.

14.1.3.3 Large Axial Flow Fans and Saccardo Nozzles

For a tunnel located in a densely populated city, the highly polluted air in the tunnel may need to be extracted from the tunnel before reaching the outlet portal so that the air quality near the outlet portal can be maintained within permissible limits. The polluted air is mostly extracted with large axial flow fans and distributed via an exhaust duct or shaft into the atmosphere. For a long tunnel of longitudinal ventilation, it is usually divided into several ventilation sections. At the end of each section, the polluted air is extracted through exhaust shaft or duct and meanwhile the fresh air is injected into the next section through Saccardo nozzles, see Figure 14.5. Both the powers required to extract and inject air are provided by large axial flow fans.

As the fresh air is injected through a Saccardo nozzle into the tunnel tube, it will add momentum to the main stream of the tunnel and raise the inlet pressure of the next ventilation section of the tunnel. According to the conservation of mass and Newton﬒ 2nd law of motion, the ideal pressure increment made by the Saccardo nozzle can be estimated using the following equation:

$$P_2 - P_1 = \rho V_s^2 \left(\frac{A_s}{A} \right) \left[\cos \propto - \left(\frac{A_s}{A} \right) \left(1 + \frac{2V_1}{V_s} \right) \right] \quad (14.3)$$

where

A_s is the Saccardo nozzle cross-sectional area
V_1 is the mean longitudinal velocity of the air flow in ventilation Section 1

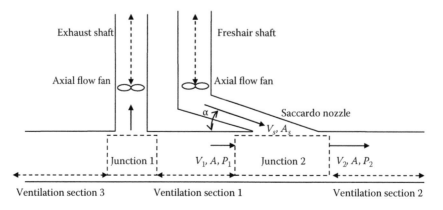

FIGURE14.5 Saccardo nozzle and ventilation sections.

Equation 14.3 shows that the pressure increment of a Saccardo nozzle can be increased by increasing the discharge velocity V_s and reducing the injection angle $\propto$. Actual pressure increment made by a specific nozzle depends on the nozzle configurations (Tabarra et al. 2000).

14.2 Principles

The flow in a real road tunnel is a transient, 3D, turbulent flow driven by ventilation machines and the dynamic activity of vehicles fleet (Jang and Chen 2000). Normally, vehicles fleet continuously release exhaust heat, various components of gaseous pollutants, and particulates into the flow; and then these components conduct 3D dispersion. In case of fire, the flow will suddenly turn into another phase with the addition of chemical reaction, thermal radiation, and buoyancy-induced convection around the fire and with the abrupt change of traDc and a possible strong impulse of ventilation machines. However, a tunnel has a unique geometric feature; its crosswise dimensions are normally much smaller than its longitudinal one. In addition, there is strong local mixing caused by both the turbulent nature of the flow and the pass-by disturbances of vehicles. As a result, most of the time, all the physical quantities in the air flow are crosswise uniform and vary predominantly along the tunnel axis, regardless of the ventilation systems adopted. Based on the characteristics described earlier and for practical purpose, 1D models have been commonly employed for analyzing general ventilation performances of all kinds of existing ventilation systems (te Velde 1988, Chang and Rudy 1990, Bring et al. 1997). The air flow through a tunnel can be complex or simple as described earlier. On this basis, a variety of models and approaches have been developed for studying or solving different problems. This section will focus on introducing the principles most often adopted for engineering analysis purposes.

There are a variety of harmful substances in vehicle exhaust, such as monoxide (*CO*), hydrocarbons (*HC*), nitrogen oxides (*NO_x*), particulates, etc. One of them would be the most critical pollutant concerned, depending on many factors such as the average age of vehicles fleet, percentage of diesel-powered vehicles, percentage of vehicles with catalytic converter, ascent rate of the tunnel, government air quality regulations, and so on. A road tunnel ventilation system is designed and operated to maintain the most critical pollutant concentration below permissible limit. Mathematical models that practically satisfy this need for the three basic types of ventilation systems are to be introduced.

Consider one of the ventilation sections of a long tunnel. The section has length L and cross-sectional area A and is equipped with a distribution of inflow and outflow of air along the tunnel, Figure 14.6. Under 1D assumption and with negligence of chemical reactions and longitudinal diffusion, the concentration $C(x)$ of the most critical pollutant satisfies the following transient mass-balance equation which is written on

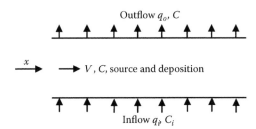

FIGURE 14.6 Pollutant flow in a ventilation section.

the basis of a unit volume of tunnel air (Chang and Rudy 1990, for example).

$$\frac{\partial C}{\partial t} + \frac{\partial VC}{\partial x} = e_C + q_i C_i - q_o C - kC \qquad (14.4)$$

where

t (s) denotes time

x (m) indicates the longitudinal coordinate

e_C (μg s^{-1} m^{-3}) is the vehicle emission strength of the most critical pollutant

q_i (s^{-1}) and q_o (s^{-1}) are the inflow and outflow rates (both are zero for longitudinal ventilation system)

C_i (μg m^{-3}) is the pollutant concentration in the inflow

k (s^{-1}) is the deposition rate of the component concerned

The longitudinal component of mean velocity V (m s^{-1}) is determined from the equation of continuity

$$\frac{dV}{dx} = q_i - q_o \qquad (14.5)$$

In case that q_i and q_o are independent of x, Equation 14.5 yields

$$V(t,x) = V(t,0) + (q_i - q_o)x \qquad (14.6)$$

For full transverse and semi-transverse systems, $V(t,0)$ is generally a complex function of all the longitudinal flow driving forces, tunnel layout, and meteorological condition and needs to be derived numerically. However, the following equation, representing a superposition property, provides a good approximation of $V(t,0)$ for many situations (Chang and Rudy 1990):

$$V(t,0) \approx V_{natural}(t) - (q_i - q_o)\frac{L}{2} \qquad (14.7)$$

where $V_{natural}$ is the longitudinal velocity induced normally by vehicle piston effect and portal pressure difference, which could be calculated based on the principle described next for longitudinal ventilation system.

For a longitudinal ventilation system the mean flow velocity V is independent of x, and can be calculated by the following momentum equation written on the basis of a tunnel ventilation section of air

$$F_{vehicle} + F_{jet\ fan} + F_{\Delta p} + F_{wall\ friction} + F_{entrance\ loss} = \rho A L \frac{dV}{dt} \quad (14.8)$$

where

$F_{vehicle}$ (Equation 14.1) and $F_{jet\ fan}$ (Equation 14.2) are the thrusts exerted on the air column by the vehicles fleet and jet fans, respectively

$F_{\Delta p}$ and $F_{wall\ friction}$ represent the forces applied on the air column due to pressure difference between both ends of the tunnel section and the wall friction with fully developed flow assumed all over the section

Both forces can be expressed by the following equations, respectively.

$$F_{\Delta p} = (p_1 - p_2)A \quad (14.9)$$

$$F_{wall\ friction} = -f \frac{\rho}{2} \frac{L}{D_h} V |V| A \quad (14.10)$$

where p_1 and p_2 are the static pressures on both ends of the tunnel section. Both pressures vary with the meteorology condition if both ends connect to the atmosphere. If this tunnel is an intermediate section of the system, p_1 and p_2 will depend on the pressures at the ends of adjacent ventilation sections and the pressure raised by the Saccardo nozzle in between. In this case, the pressures at the end of the ventilation section and the velocity in the ventilation section are unknown and need to be solved simultaneously from the governing equations of the entire ventilation system, described in Section 14.4. As air flows through the tunnel, the tunnel wall will exert a friction force on it, which is expressed as Equation 14.10. In this equation D_h is the hydraulic diameter of the tunnel and f is the friction factor that is a function of the relative wall roughness and Reynolds number (Fox et al. 2009, for example). In tunnel ventilation the normal value of f is 0.020 (Glerum et al. 1991, Bring et al. 1997).

$F_{entrance\ loss}$ in Equation 14.8 represents the extra friction in the entrance region due to flow development. If this end is an inlet portal of the entire system, this extra friction force can be taken into account by the following equation:

$$F_{entrance\ loss} = -K_{port} \frac{\rho}{2} V |V| A \quad (14.11)$$

As this end connects to a fresh air shaft with Saccardo nozzle and connects to another ventilation section through a junction, as shown in Figure 14.5, this extra friction force in the entrance region of this ventilation section will reduce the ideal pressure rise of the Saccardo nozzle expressed in Equation 14.3. Then, both the pressure rise of the Saccardo nozzle and the pressure loss due to flow development in the entrance region of this

ventilation section are combined together and expressed in terms of the loss coefficients of the junction (Ward-Smith 1980).

Besides gaseous pollutants and particulates, temperature is another factor that might be of interest in tunnel ventilation. For the analysis of this factor, a simplified 1D energy equation must be added to the flow model. This equation depicts appropriately the generation and the transfer of heat in a tunnel. Generally, in a road tunnel the heat generated from equipments and the vehicles fleet, the heat conducted through the tunnel wall, the heat brought in by inflow, and the heat taken out by outflow must be taken into account, but the heat conduction in the longitudinal direction is negligible compared to the heat advection in the same direction. As a result, the energy equation is the first-order partial differential equation (Chen and Jang 2000, for a longitudinal system).

In case of fire in a longitudinal ventilation system, the air temperature, density, and velocity would have an abrupt change over the fire site. Therefore the wall friction on the flow and the performance of jet fans in the downstream of the fire would be greatly affected. The 1D model needs to be modified accordingly (te Velde 1998).

For a long tunnel that includes several ventilation sections, the 1D ventilation model of the whole system can be set up via the combination of the 1D governing equations for the flow in each ventilation section as described earlier, the 1D governing equations for the flows in exhaust-air and fresh-air ducts or shafts, and the continuity equations and the correlations of the pressure losses through every junction that joins these ventilation sections and shafts. For the modeling details, the reader can refer to te Velde (1998), for example. For the ventilation system of a long road tunnel, the 1D flow model is attractive because it passes over the complex process of meshing components and has very low demand on computation resources and time.

14.3 Examples

In order to demonstrate the uses of the principles described in Section 14.2 and to reveal the characteristics of the three basic ventilation systems, three examples are included, in which, CO, a high toxicity component in vehicle exhaust, is assumed as the most critical component in vehicle exhausts. The tunnels to be studied in these three examples are equipped with different ventilation systems but all have length (L = 1000 m), cross-sectional area (A = 56 m²), hydraulic diameter (D_h = 7.8 m), and two traffic lanes. Other basic data regarding the aerodynamic properties of air flow through the tunnels, the dimensions and performance of jet fans, traffic flows, vehicle mean frontal area, vehicle drag coefficient, vehicle CO emission rate, and environmental CO concentrations are listed in Tables 14.1 and 14.2. Example 14.1 shows the approach of calculating the natural velocity and CO concentration distribution in a longitudinally ventilated tunnel under normal and congested traffic statuses. The effect of operation of jet fans on air quality improvement is also shown. Example 14.2 is to demonstrate the solution process for the CO concentration in a full-transverse ventilation system.

TABLE1 4.1 Basic Data of TraDc Flow and Vehicle *CO* Emission Rate for the Cases of Examples 14.1–14.3

	Normal TraDc	Congested TraDc
TraDc volume N (vehicles h^{-1})	2800	1600
Vehicle speed V_v (km h^{-1})	70.0	8.0
Vehicle category	100% sedans with petro engines with catalytic converter	100% sedans with petro engines with catalytic converter
Mean drag coeDcient of sedans, C_d	0.25	0.25
Mean frontal area of sedans, $A_{vehicle}$ (m^2)	2.5	2.5
Vehicle *CO* emission rate E ($g\ h^{-1}vehicle^{-1}$)	40.0	40.0

TABLE1 4.2 Values of Flow Parameters and Environmental *CO* Concentrations for the Cases of Examples 14.1–14.3

Air density ρ ($kg\ m^{-3}$)		1.18
Friction factor f		0.02
Tunnel entrance loss coeDcient K_{port}		0.6
Jet fan	Diameter, d (m)	1.0
	Outlet velocity V_{fan} ($m\ s^{-1}$)	30.0
	EDciency η_{fan}	0.8
CO concentration at tunnel entrance $C(0)$ ($\mu g\ m^{-3}$)		100.0
CO concentration in inflow of transverse ventilation systems C_i ($\mu g\ m^{-3}$)		10.0
CO deposition rate k (s^{-1})		0.0

Uni-directional traDc flows and bi-directional traDc flows are considered in the example, so that the vehicle piston effect on the performance of this type of ventilation system can be revealed. The intrinsic different characteristics of both the outflow and the inflow types of semi-transverse ventilation systems are discussed in Example 14.3. For simplicity, the effect due to meteorological effect is ignored and quasi-steady state is assumed.

Examp 4.1

Suppose the tunnel is equipped with a longitudinal ventilation system and the traDc in the tunnel is uni-directional. Estimate the *CO* concentration distributions under the normal traDc and congested traDc statuses. If the maximum *CO* concentration in the tunnel to be maintained is below 8 mg m^{-3}, how many jet fans must be started for the congested traffic case.

Solution

For quasi-steady state, Equation 14.8 can be reduced to the following algebraic equation of V:

$$aV^2 + bV + c = 0 \qquad (14.12)$$

where

$$a = \frac{1}{2}\rho C_d A_{vehicle} N_{vehicle} - \frac{1}{2}\rho A K_{port} - \frac{1}{2}\rho A f \frac{L}{D_h}$$

$$b = -\rho N_{vehicle} C_d A_{vehicle} V_{vehicle} - \rho N_{fan} A_{fan} V_{fan} \eta_{fan}$$

$$c = \frac{1}{2}\rho C_d A_{vehicle} N_{vehicle} V^2_{vehicle} + \rho N_{fan} A_{fan} V^2_{fan}\eta_{fan} + (P_1 - P_2)A$$

Total number of vehicles in tunnel, $N_{vehicle}$ is equal to $NL/(1000V_V)$, which is 40 and 200 for normal and congested traDc flows, respectively. The mean velocity of vehicles $V_{vehicle} = 1000V_V/3600$, which is 19.44 and 2.22 ms^{-1} for normal and congested traDc flows, respectively. Substitute these traDc flow data, $N_{fan} = 0$, and other parameters listed in Tables 14.1 and 14.2 for normal and congested traDcs into the aforementioned equations of coeDcients a, b, and c then solve Equation 14.12 to obtain $V = 5.31$ and 1.01 ms^{-1} for normal traDc and congested traDc, respectively.

Assume that the vehicle emission strength of the component *CO*, e_C, is uniform in the tunnel, then Equation 14.4 will provide the following solution for the *CO* concentration along the tunnel:

$$C(x) = \frac{e_C}{V}x + C(0) \qquad (14.13)$$

where the emission strength of *CO* in the tunnel can be calculated according to the equation

$$e_C = \frac{N}{1000V_vA}\frac{10^6 E}{3600} \ (\mu g\,s^{-1}\,m^{-3}) \qquad (14.14)$$

Substitute the traDc flow and vehicle *CO* emission rate data listed in Table 14.1, we will find that e_C is 7.94 and 39.68 $\mu g\ s^{-1}m^{-3}$, for normal traDc and congested traffic, respectively.

Therefore,

$$C(x) = \frac{7.94}{5.31}x + 100\,(\mu g\,m^{-3}),$$

for normal traffic without any jet fan being started

$$(14.15)$$

$$C(x) = \frac{39.68}{1.01}x + 100\,(\mu g\,m^{-3}),$$

for congested traffic without any jet fan being started

$$(14.16)$$

Under congested traDc, the *CO* concentration distribution along the tunnel can be expressed as $C(x) = 39.68x/V$ 100, which indicates that the maximum concentration occurs at the outlet, which is equal to 1000 m. To satisfy the requirement $C(1000\,m) \le 8\,mg\ m^{-3}$, the mean air flow

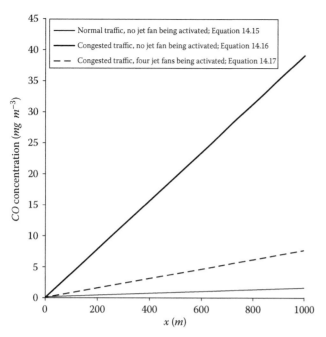

FIGURE 14.7 Effects of traDc status and jet fans on the distribution of *CO* in a longitudinally ventilated tunnel.

velocity *V* must be greater than 5.02 ms^{-1}. By substituting this velocity value into Equation 14.12, we will find $n_{fan} > 4$, i.e., at least four jet fans must be started. For congested traDc with four jet fans being started, the mean velocity *V* would be 5.24 ms^{-1} at steady state, and the *CO* concentration distribution in the tunnel would be

$$C(x) = \frac{39.68}{5.24}x + 100 \, (\mu g \, m^{-3}) \qquad (14.17)$$

Equations 14.15 through 14.17 are plotted in Figure 14.7. The results show the following features.

1. Generally in a longitudinal ventilation system the pollutant concentration increases linearly along the axis of the tunnel.
2. The air quality under congested traDc is much worse than under normal traDc.
3. Jet fans can effectively improve the air quality for the congested traDc situation; therefore, they are commonly equipped in road tunnels.

Examp▶ 4.2

Estimate the *CO* concentration distributions again for the same uni-directional normal and congested traDc cases as in Example 14.1 with the ventilation system now changed into a full transverse ventilation system which is operated under the total volume of ventilation flow rate $Q_i = Q_o = 280 (m^3 s^{-1})$. If the total traDc volume and vehicle speed remain the same but the traDc flow pattern turns bi-directional,

estimate the *CO* concentration distributions for both the normal and congested traDc situations.

Solution

In Example 14.1, the longitudinal air flow velocities we obtained under the piston effect of uni-directional normal traDc and congested traDc are 5.31 and 1.01 ms^{-1}, respectively. Since the inflow rate q_i (=$Q_i/(AL)$) and the outflow rate q_o(=$Q_o/(AL)$) have the same value of 0.005 s^{-1}, Equations 14.7 and 14.6 indicate that the velocity at the inlet portal as well as inside the full transverse ventilation system would approximately equal to the natural velocity; $V(x) = V_{natural} = 5.31$ and 1.01 ms^{-1} for normal traDc and congested traDc, respectively.

Under quasi-steady state assumption, Equation 14.4 becomes a linear nonhomogeneous first-order ordinary differential equation, $V(dC/dx) + 0.005C = e_C + 0.05$, which has the following analytic solution:

$$C(x) = \frac{e_C + 0.05}{0.005} + \left(C(0) - \frac{e_C + 0.05}{0.005} \right) e^{-0.005x/V} \, (\mu g \, m^{-3})$$

$$(14.18)$$

or

$$C(x) = \frac{7.94 + 0.05}{0.005} + \left(100 - \frac{7.94 + 0.05}{0.005} \right) e^{-0.005x/5.31} \, (\mu g \, m^{-3}),$$

for normal traffic $\qquad (14.19)$

$$C(x) = \frac{39.68 + 0.05}{0.005} + \left(100 - \frac{39.68 + 0.05}{0.005} \right) e^{-0.005x/1.01} \, (\mu g \, m^{-3}),$$

for congested traffic $\qquad (14.20)$

If the traDc flow pattern changes to bi-directional with the same traDc volume in both directions, then the piston effect of the traDc flow is canceled and there is no longitudinal velocity, i.e., $V_{natural} = V(x) = 0$ ms^{-1}. In this case, $C(x)$ is the solution of Equation 14.4 with the ignorance of the transient term and the advection term.

$$C(x) = \frac{e_C + q_i C_i}{q_o} \qquad (14.21)$$

or

$$C(x) = \frac{7.94 + 0.005 \times 10}{0.005} = 1598 \, (\mu g \, m^{-3}),$$

for normal traffic $\qquad (14.22)$

$$C(x) = \frac{39.68 + 0.005 \times 10}{0.005} = 7946 \, (\mu g \, m^{-3}),$$

for congested traffic $\qquad (14.23)$

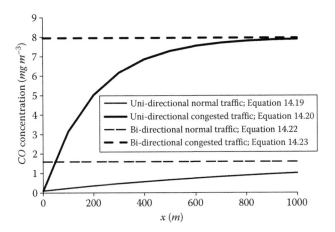

FIGURE 14.8 Effects of traffic flow pattern and traffic status on the distribution of CO in a full transverse ventilation tunnel.

Equations 14.19, 14.20, 14.22, and 14.23 are plotted in Figure 14.8. The results show the following features.

1. In a full transverse ventilation system operated for a bi-directional traffic, the pollutant concentration is normally uniform.
2. If the traffic flow is turned into uni-directional, the air quality could be improved to a certain extent. This reveals that the piston effect produces additional ventilation capacity for the system, which is larger at normal traffic than the congested traffic.

Example 4.3

Analyze all the cases in Example 14.2 again by changing the full transverse ventilation system into an inflow semi-transverse ventilation system with the total volume of inflow rate $Q_i = 280$ m^3 s^{-1} and the total volume of outflow rate $Q_o = 0$ m^3 s^{-1}. Perform the analysis for all the cases again if the ventilation system is changed into a outflow semi-transverse ventilation system with the total volume of outflow rate $Q_o = 280$(m^{3}s^{-1}) and the total volume of inflow rate $Q_i = 0$ (m^{3}s^{-1}).

Solution

For the inflow semi-transverse ventilation system, the inflow rate q_i is 0.005 s^{-1} and the outflow rate q_o is 0 s^{-1}. According to Equations 14.6 and 14.7, the longitudinal velocity in the ventilation system can be written as

$$V(x) \approx V_{natural} + q_i\left(x - \frac{L}{2}\right) \quad (14.24)$$

For uni-directional normal traffic flow, the longitudinal velocity is

$$V(x) \approx 5.31 + 0.005\left(x - \frac{1000}{2}\right) = 2.81 + 0.005x \text{ (m s}^{-1}) \quad (14.25)$$

The velocity at the inlet is $V(0) = 2.81$ ms^{-1} > 0; therefore, the CO concentration at the inlet would be equal to the given value, i.e., $C(0) = 100 \mu$g m^{-3}.

Substituting $V(x)$, Equation 14.24, into Equation 14.4 and solving the differential equation by analytical method, we can obtain the CO concentration in the tunnel

$$C(x) = \frac{e_C + q_i C_i}{q_i} + \left(C(0) - \frac{e_C + q_i C_i}{q_i}\right)\left(1 + \frac{q_i x}{V_{natural} - q_i L/2}\right)^{-1} \quad (14.26)$$

or

$$C(x) = 1598 - \frac{4209}{2.81 + 0.005x} \text{ } (\mu\text{g m}^{-3}) \quad (14.27)$$

which is obtained by substituting $V_{natural} = 5.31$ ms^{-1}, $e_C = 7.94 \mu$g s^{-1}m^{-3}, $q_i = 0.005$ s^{-1}, $C_i = 10 \mu$g m^{-3}, $L = 1000$ m, and $C(0) = 100 \mu$g m^{-3} into Equation 14.26.

By substituting $V_{natural} = 1.01$ ms^{-1} into Equation 14.24, the longitudinal velocity for uni-directional congested traffic flow can be expressed as

$$V(x) \approx V_{natural} + q_i\left(x - \frac{L}{2}\right) = -1.49 + 0.005x\text{(m s}^{-1}) \quad (14.28)$$

where $V(0) \approx -1.49$ ms^{-1} < 0, which means that the CO concentration at $x = 0$ is not equal to 100μg m^{-3}, the given value. However, the solution process that derived Equation 14.26 still holds for this case. Equation 14.28 implies that there is a stagnation point at $x = x_s = \frac{L}{2} - \frac{V_{natural}}{q_i} = 298$ m. Equation 14.26 implies that the CO concentration at the stagnation point, $C(x_s)$, would be infinity, if $C(0)$ was not equal to $\frac{e_C + q_i C_i}{q_i}$. Therefore, in this case the CO concentration in the tunnel is $C(x) = 7946 \mu$g m^{-3}, which is calculated using the following equation with $e_C = 39.68 \mu$g s^{-1} m^{-3}, $q_i = 0.005$ s^{-1}, and $C_i = 10 \mu$g m^{-3}.

$$C(x) = \frac{e_C + q_i C_i}{q_i} \quad (14.29)$$

For bi-directional normal and congested traffic flows, $V_{natural}$ is 0 m s^{-1} and $V(x) \approx q_i\left(x - \frac{L}{2}\right)$. Since $V(0)$ is -2.5 m s^{-1}, less than 0 m s^{-1}, there is a stagnation point at, $x_s = L/2 = 500$ m, inside the tunnel, where the CO concentration must be finite. Therefore, the CO concentration in the tunnel is uniform and can also be calculated according to Equation 14.29.

$$C(x) = \frac{e_C + q_i C_i}{q_i} = \frac{7.94 + 0.005 \times 10}{0.005} = 1600 \, (\mu g \ m^{-3}),$$

normal traffic   (14.30)

$$C(x) = \frac{e_C + q_i C_i}{q_i} = \frac{39.68 + 0.005 \times 10}{0.005} = 7946 \, (\mu g \ m^{-3}),$$

congested traffic   (14.31)

The longitudinal velocity and *CO* concentration distributions for all the cases analyzed for the inflow semi-transverse ventilation system are plotted in Figure 14.9. The results show the following features.

1. If the total piston effect of the traffic flow is weak, such as under uni-directional congested flow or most of the bi-directional traffic flows, there is normally a stagnation point in the tunnel and then the *CO* concentration is uniform in the tunnel.
2. Generally, for the inflow semi-transverse ventilation system, the vehicle piston effect makes no contribution on lowering pollutant concentration level if it is not strong enough to move the stagnation point out of the tunnel.

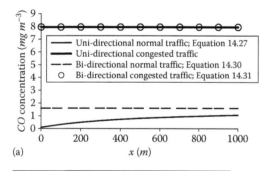

(a)

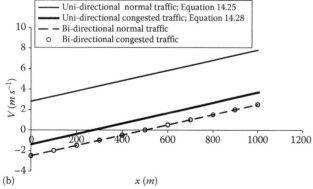

(b)

FIGURE 14.9 (a and b) Effects of traffic flow pattern and traffic status on the distribution of *CO* and longitudinal velocity in an inflow semi-transverse ventilation tunnel, respectively.

For the outflow semi-transverse ventilation system, the inflow rate q_i is $0 \, s^{-1}$ and the outflow rate q_o is $0.005 \, s^{-1}$. According to Equations 14.6 and 14.7, the longitudinal velocity in the ventilation system can be written as

$$V(x) \approx V(0) - q_o x \qquad (14.32)$$

$$V(0) \approx V_{natural} + q_o \left(\frac{L}{2} \right) \qquad (14.33)$$

There is a stagnation point at $x_s = \frac{V(0)}{q_o} = \frac{L}{2} + \frac{V_{natural}}{q_o}$.

The governing equation of *CO* concentration, Equation 14.4, can be simplified as

$$[V(0) - q_o x] \frac{dC}{dx} = e_C \qquad (14.34)$$

The aforementioned equation has the following analytical solution.

$$C(x) = C(0) + \frac{e_C}{q_o} \ln \left| \frac{V(0)}{V(0) - q_o x} \right| \qquad (14.35)$$

Under uni-directional normal traffic flow, $V_{natural} = 5.31 \, m \ s^{-1}$, $V(0) \approx 5.31 + 0.005 \times \frac{1000}{2} = 7.81 \, m \, s^{-1}$, and the velocity and *CO* concentration in the tunnel are

$$V(x) \approx 7.81 - 0.005x \, (m \, s^{-1}) \qquad (14.36)$$

$$C(x) = C(0) + \frac{e_C}{q_o} \ln \left| \frac{V(0)}{V(0) - q_o x} \right|$$

$$= 100 + \frac{7.94}{0.005} \ln \left| \frac{7.81}{7.81 - 0.005x} \right| (\mu g \ m^{-3}) \qquad (14.37)$$

Under uni-directional congested traffic flow, $V_{natural} = 1.01 \, ms^{-1}$, $V(0) \approx 1.01 + 0.005 \times \frac{1000}{2} = 3.51 m \, s^{-1}$, and the velocity and *CO* concentration in the tunnel are

$$V(x) \approx 3.51 - 0.005x \, (m \, s^{-1}) \qquad (14.38)$$

$$C(x) = C(0) + \frac{e_C}{q_o} \ln \left| \frac{V(0)}{V(0) - q_o x} \right|$$

$$= 100 + \frac{39.68}{0.005} \ln \left| \frac{3.51}{3.51 - 0.005x} \right| (\mu g \ m^{-3}) \qquad (14.39)$$

Under bi-directional normal flow, $V_{natural} = 0 \, m \ s^{-1}$, $V(0) \approx 0.005 \times \frac{1000}{2} = 2.5 \, m \, s^{-1}$, and the velocity and *CO* concentration in the tunnel are

$$V(x) \approx 2.5 - 0.005x \, (m \, s^{-1}) \qquad (14.40)$$

$$C(x) = C(0) + \frac{e_C}{q_o} \ln \left| \frac{V(0)}{V(0) - q_o x} \right|$$

$$= 100 + \frac{7.94}{0.005} \ln \left| \frac{2.5}{2.5 - 0.005x} \right| (\mu g\ m^{-3}) \qquad (14.41)$$

Under bi-directional congested flow, $V_{natural} = 0\ m\ s^{-1}$, $V(0) \approx 0.005 \times \frac{1000}{2} = 2.5\ m\ s^{-1}$, and the velocity and CO concentration in the tunnel are

$$V(x) \approx 2.5 - 0.005x\ (m\ s^{-1}) \qquad (14.42)$$

$$C(x) = C(0) + \frac{e_C}{q_o} \ln \left| \frac{V(0)}{V(0) - q_o x} \right|$$

$$= 100 + \frac{39.68}{0.005} \ln \left| \frac{2.5}{2.5 - 0.005x} \right| (\mu g\ m^{-3}) \qquad (14.43)$$

The longitudinal velocity and CO concentration distributions for all the cases analyzed for the outflow semi-transverse ventilation system are plotted

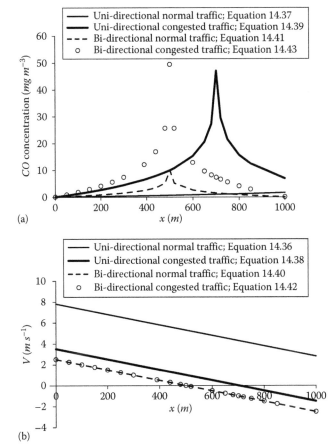

(a)

(b)

FIGURE 14.10 (a and b) Effects of traDc flow pattern and traDc status on the distribution of CO and longitudinal velocity in an outflow semi-transverse ventilation tunnel, respectively.

in Figure 14.10. The results show the following features:

1. If the total piston effect of the traDc flow is weak, such as under uni-directional congested flow or the bi-directional traDc flow, there is normally a stagnation point in the tunnel and the CO concentration is extraordinarily high around this point. This is why the outflow semi-transverse ventilation systems are rarely adopted.
2. The vehicle piston effect can lower pollutant concentration level in this outflow semi-transverse ventilation system, if it is strong enough to move the stagnation point out of the tunnel.

14.4 Methods of Analysis

Most of the time, the air flow through a tunnel is to maintain the air quality for passengers. A quasi-steady 1D flow model is appropriate for evaluating the air quality in a ventilation section. In these cases, the governing equations consist of an algebraic equation for momentum transport and a first-order ordinary differential equation for pollutant transport. They can be solved using analytical methods in a sequential manner, first the momentum equation and then the pollutant transport equation, as shown in the examples of Section 14.3.

The 1D model can also simulate some transient phenomena, for example, the pulsating behavior of the air flow in a road tunnel (Jang and Chen 2000). If the air quality in a specific longitudinal ventilation section during transitional periods is interested, the momentum equation, Equation 14.8, a nonlinear first-order ordinary differential equation with variable coeDcients, and the pollutant transport equation, Equation 14.4, a first-order partial differential equation coupled with Equation 14.8 can be solved sequentially or simultaneously using numerical methods. However, besides Equations 14.4 and 14.8, Equations 14.5 and 14.7 also have to be included in the solution process if a transverse system is equipped in this ventilation section.

In case of a long tunnel, the ventilation system is normally divided into multiple ventilation sections. The flow in each ventilation section can be depicted by a set of governing equations as described in Section 14.2. In addition to these equations, equations for the flows through each junction that connect different ventilation sections are required to form a complete set of governing equations for the entire ventilation system of the long tunnel. The governing equations for the flows through each junction include a continuity equation of the air flow, a continuity equation of the most critical pollutant component, and pressure-loss correlations for the flows through each junction. All the governing equations of the entire ventilation system must be solved simultaneously by numerical methods. For details of the work the reader can refer to the work of te Velde

(1988), for example. Although the pressure-loss correlations are normally obtained from steady flow experimental test data, they can perform well for unsteady flow; references 28 and 43 in Bassett et al. (2003) have demonstrated this point for pipe networks.

The 1D flow model provides a reasonable compromise between modeling accuracy and computational speed. It therefore is usually employed by engineers to analyze tunnel ventilation performances. However, there are certain tunnel ventilation problems, such as the pressure losses of the flow through a junction in a tunnel ventilation system, the eDciency of a Saccardo nozzle, the installation eDciency of a jet fan in a tunnel, the piston effect of a vehicle in different tunnels, the back-layering phenomena and the smoke propagation in a tunnel fire (Chen 2000), etc. that must be studied using multi-dimensional models. In the past two decades, computational fluid dynamics (CFD) has been employed to analyze these local flow behaviors, particularly in tunnel fires.

14.5 Major Challenges

The purpose of the air flow in a tunnel is to provide a comfortable, healthy, and safety environment for people to pass through. The existing tunnel ventilation systems in the world can attain this purpose most of the time and the situation may even be better in the future if the vehicle exhausts are reduced via the improvement of its power systems. However, this does not ensure that a fetal fire can be prevented or properly controlled once it occurs. If a fatal fire occurs in a ventilation section of a long tunnel, two critical parameters, the flow direction and the flow speed, in this ventilation section during the initial stage are decisive to property damage and casualties. The flow direction can be determined according to the traDc pattern and status near the fire. The flow speed depends on the fire size and tunnel cross-sectional dimensions (see Chen 2000, for example). The size of a fetal fire, depending primarily on the chemical property of the goods carried by the vehicle on fire and the trapped neighboring vehicles, is actually hard to estimate at the critical moment. To start up ventilation machines to make a quick and proper response to this ventilation requirement during the initial dynamical period is really a challenge in practical operation.

To analyze the air flow during such a crucial and dynamical period using mathematical models is also a big challenge. The prediction accuracy of a 1D model for a long tunnel with multiple ventilation sections highly depends on not only the proper inputs of the realistic conditions but also on the accuracy of the pressure-loss correlations for the flow through each junction. The pressure-loss correlations cannot fully capture the complex frictional effects of the geometry of the junctions; the junctions in long road tunnels normally have much more complex in geometry than those in the usual fluid transport

networks. Therefore they have major influence on the prediction accuracy of the flow direction and speed in every ventilation section. On the other hand, a 3D model though including relatively less assumptions still also has its limitation on the prediction accuracy of the two critical parameters. The limitation comes from some elements; for example, the incomplete turbulence modeling that influences the flow separation in a junction and therefore the pressure losses and other results, the big technical challenges to input the realistic dynamical conditions into the model and to properly meshing the entire flow domain of a long road tunnel under dynamical conditions.

In the past two decades, the analytical technologies and computation powers have had great advancements which will carry on to the future. However, by looking at the development trend of the tunnel in the world, we can see there will be more new long tunnels built and new ventilation systems installed. Since no ventilation system fits for all situations, particularly for a highly enclosed public domain, tunnel-ventilation challenges may never be stopped by the advancement of technologies.

References

Barrefors G. 1996. Air pollutants in road tunnels. *Science of the Total Environment*. 189–190: 431–435.

Bassett MD, Pearson RJ, and Fleming NP. 2003. A multi-pipe junction model for one-dimensional gas-dynamic simulations. Society of Automotive Engineers, Inc. Paper No. 2003-01-0370, Detroit, MI, March 3–6.

Berglund M, Bostrom C-E, Bylin G, Ewetz L, Gustavsson L, Moldeus P, Norberg S, Pershagen G, and Victorin K. 1993. Health risk evaluation of nitrogen oxides. *Scandinavian Journal of Work Environment and Health*. 19(Suppl. 2): 1.

Betta V, Cascetta F, Musto M, and Rotondo G. 2009. Numerical study of the optimization of the pitch angle of an alternative jet fan in a longitudinal tunnel ventilation system. *Tunnelling and Underground Space Technology*. 24: 164–172.

Bring A, Malmstrom TG, and Boman CA. 1997. Simulation and measurement of road tunnel ventilation. *Tunnelling and Underground Space Technology*. 12: 417–424.

Chang TY. and Rudy SJ. 1990. Roadway tunnel air quality models. *Environmental Science and Technology*. 24: 672–676.

Chen F. and Jang H-M. 2000. Temperature rise in Ping-Lin tunnel. *Journal of the Chinese Society of Mechanical Engineers*. 21: 325–340.

Costeris N. 1991. Impulse fans. Ventilatoren Stork Howden B.V. 7550 AZ Hengelo, the Netherlands.

Chen F. 2000. Smoke propagation in road tunnels. *Applied Mechanics Review*. 53: 207–218.

Fox RW, Pritchard PJ, and McDonald AT. 2009. *Introduction to Fluid Mechanics*, 7th edn, Wiley, New York.

Glerum A, Swart L, ⸺Hooft EN, Costeris NP, Franken A, Hartman PF, Rijkeboer RC, Speulman H, and den Tonkelaar WAM. 1991. Ventilation of road tunnels. Royal Institute of Engineers (KIVI), Department of Tunnel Technology and Underground Engineering Works Working Party "Ventilation of Road Tunnels."

Jang H-M. and Chen F. 2000. A novel approach to the transient ventilation of road tunnels. *Journal of Wind Engineering and Industrial Aerodynamics*. 86: 15–36.

Jang H-M. and Ma C-K. 2009. Investigation of the drag-coeDcient correction factor for a tractor traveling through Hsuehshan Tunnel using CFD method. *Proceedings of the 26th National Conference on Mechanical Engineering*. Nov. 20–21, Tainan, Taiwan. Paper no. A06–A019.

Katz J. 1995. Race car aerodynamics designing for speed, Bentley, Cambridge, MA.

Tabarra M, Matthews RD, and Kenrick BJ. 2000. The revival of Saccardo ejectors–history, fundamentals, and applications. *10th International Symposium on the Aerodynamics and Ventilation of Vehicle Tunnels*, Boston, MA, November 1–3.

Tornqvist M. and Ehrenberg L. 1994. Cancer risk: Estimation of urban air pollution. *Environmental Health Perspectives*. 102(Suppl. 4): 173.

te Velde K. 1988. A computer simulation for longitudinal ventilation of a road tunnel with incoming and outgoing slip roads. *6th International Symposium on the Aerodynamics and Ventilation of Vehicle Tunnels*, Durham, U.K., September 27–29.

Ward-Smith AJ. 1980. *Internal Fluid Flow: The Fluid Dynamics of Flow in Pipes and Ducts*. Clarendon Press, New York.

15

Sound Outdoors and Noise Pollution

D.K. Wilson
United States Army Engineer Research and Development Center

E.T. Nykaza
United States Army Engineer Research and Development Center

M.J. White
United States Army Engineer Research and Development Center

M.E. Swearingen
United States Army Engineer Research and Development Center

L.L. Pater
United States Army Engineer Research and Development Center

G.A. Luz
Luz Social and Environmental Associates

15.1 Introduction

Humans have relied on sound and the sense of hearing for communication, detection, navigation, and many other purposes important to our survival. But in modern life, we often become overwhelmed by the many competing and intrusive sounds in the acoustic environment surrounding us. Noise is usually described subjectively as an unwanted sound; still, to regulate and reduce noise, we must attempt to objectively quantify noise and the effects it has on us.

This chapter provides an overview of outdoor noise and sound. First, in Section 15.2, the general principles of sound are discussed: How it is produced, how sources are characterized, and how the sound travels through the atmosphere to the listener. Methods for measuring noise exposure and predicting sound are described in Section 15.3. Some practical applications

to preserving natural quiet, managing noise impacts, and using natural and man-made barriers to attenuate noise are discussed in Section 15.4. Lastly, in Section 15.5, we offer a perspective on current major challenges, which include sound propagation in complex environments, statistical sampling of noise, and understanding noise annoyance in the context of soundscapes.

15.2 Principles

15.2.1 General

Sound waves consist of oscillating compressions in a fluid. The compressions are initially produced by fluids and surfaces undergoing relative acceleration, such as when vibrating or rotating objects alternately compress the surrounding air. Sound can

Handbook of Environmental Fluid Dynamics, Volume Two, edited by Harindra Joseph Shermal Fernando. © 2013 CRC Press/Taylor & Francis Group, LLC. ISBN: 978-1-4665-5601-0.

also be produced by unsteady fluid flow such as in explosions or jet exhaust. The waves propagate through the air and thus can reach a sensor, such as the human ear or a microphone, which converts the sound energy to electrical signals and processes the information. In noise control, we refer to the sound produced by the source as noise *emission*, whereas the sound reaching the sensor is referred to as noise *immission* or *exposure*.

Most measurements and predictions of sound waves describe the associated pressure fluctuations, although sometimes the particle velocity or wave intensity is measured. The pressure of a single-frequency (harmonic) sound wave may be written as $p(t) = A \cos(\omega t - \phi)$, where A is the amplitude, $\omega = 2\pi f$ is the angular frequency, f is the frequency, t is time, and ϕ is the phase. The wavelength of the sound is $\lambda = c/f$. Usually, measurements actually refer to the root-mean-square (rms) average pressure, p_{rms}, which equals $A/\sqrt{2}$ in the single-frequency case. As in other engineering fields, acousticians customarily represent harmonic signals using a complex notation. The pressure is thus written as $p(t) = \text{Re}[\hat{p}e^{-i\omega t}]$, where $\hat{p} = Ae^{i\phi}$ is the complex pressure phasor.

For harmonic spherical or planar sound waves, the intensity (average rate of energy flow per unit area normal to the propagation direction) can be shown as $I = p_{rms}^2/\rho_0 c_0$, where ρ_0 is the ambient air density, and c_0 is the sound speed. In practice, this equation is often assumed to be a good approximation at distances far from a source, where the wave fronts may be nearly planar.

15.2.2 Sound Sources and Their Characterization

Some sound sources emit steady signals, like a generator's "hum" or airflow noise created by drag forces on a moving vehicle. Other sources are unsteady, such as intermittent noise at a railway crossing or raindrops falling on a rooftop. Unsteady, impulsive sources include thunder, explosions, and the splash of a rock thrown into a pond. Most sources radiate sound more intensely in certain directions. For example, the human voice is significantly louder directly in front of the head of the speaker than behind. Different loudness and frequency characteristics are observed as one walks around an idling car. Some sources, such as a balloon popping, may be considered omnidirectional, which means that sound is emitted equally in all directions. At sufficient distances, many sources can be approximated as a point or a line source. For example, a chirping bird can be regarded as a point source when observed from a distance. Noise from a steady flow of road traffic is often approximated as a line source.

Sound sources are usually characterized by the sound pressure or intensity they would produce if they happened to be radiating into *open space* and the measurement were made in the *far field*. The open-space condition is important because observed sound loudness is affected by the presence of acoustically reflective objects. Outdoors, even far from buildings, trees, and other objects, the ground will produce significant reflections. When possible, source levels are thus measured in *anechoic chambers* built with walls made of an acoustically absorptive material such as open-cell foam. Alternatively, *reverberant chambers*, which have near-perfectly

reflecting walls, may be used. However, outdoor noise emissions often involve sources that cannot practically be moved into a controlled chamber for measurement, and this introduces some additional uncertainty into the process of characterizing a sound level.

The far-field condition means that sound pressure amplitude decays as $1/r$ along radial lines extending outward from the source, where r is the radial distance. (The far-field pressure amplitude may still depend on the azimuth and elevation angles of observation.) Far-field conditions occur, in general, when (1) the distance is many wavelengths from the source ($r \gg \lambda$), and (2) the distance is much larger than the size of the source ($r \gg a$, where a is a characteristic length dimension of the source) (Beranek 1986). Again, such conditions may be challenging to duplicate in practice.

Finally, presuming one has characterized the open space, far-field sound pressure or intensity, normal practice is to adjust the received value to a reference distance of $r_0 = 1$ m from the source. Pressure values are compensated to the reference distance by multiplying by r/r_0.

Acousticians almost always indicate sound pressure and intensity on a decibel (dB) scale. Because the decibel scale is logarithmic, it conveniently compresses the large range of sound-wave amplitudes encountered in practice. This compression also mimics human perception of sound, which is nearly logarithmic. Sound pressure level (SPL) is defined as

$$L_p = 20 \log_{10}\left(\frac{p_{rms}}{p_{ref}}\right) \tag{15.1}$$

where p_{ref}, the reference sound pressure, is $20\,\mu$Pa in air. The reason for selecting $20\,\mu$Pa as a reference is that it very roughly corresponds to the hearing threshold of an acute young listener at the most sensitive frequency (near 1000 Hz), which means that the quietest audible sound would be 0 dB.

Similarly, sound intensity level is defined as

$$L_I = 10 \log_{10}\left(\frac{I}{I_{ref}}\right) \tag{15.2}$$

TABLE 15.1 Example Sound Pressure Levels of Common Noises

Source	Level (dB(A))[a]
Train horn	96–110
727 aircraft, 6500 m from takeoff roll	100
Heavy truck, 65 mph, 50 ft away	88
Heavy truck, 35 mph, 50 ft away	82
Shouting, 3 ft away	80
Auto, 65 mph, 50 ft away	75
Auto, 35 mph, 50 ft away	64
Normal speech, 3 ft away	62
Quiet urban nighttime	40
Quiet suburban nighttime	35
Quiet rural nighttime	25

[a] All Levels are in dB(A) referenced to $20\,\mu$Pa.

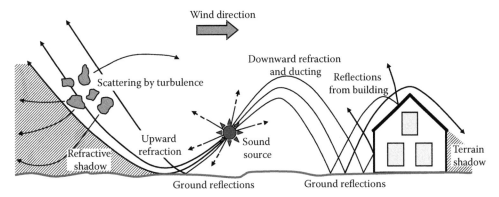

FIGURE1 5.1 Main influences of the atmosphere and terrain on sound propagation outdoors.

where the reference level $I_{\text{ref}} = 10^{-12}$ W m^{-2}. Since $I = p_{\text{rms}}^2/\rho_0 c_0$ for a planar or spherical sound wave, and $\rho_0 c_0 \simeq 410$ kg s m^{-2} at 20°C, a consequence is that $I_{\text{ref}} \simeq p_{\text{ref}}^2/\rho_0 c_0$. Thus L_p and L_I are usually close in value (e.g., within 0.5 dB for a 12% change of $\rho_0 c_0$).

When the sound energy propagates between two points in the far field, from $r = r_1$ to $r = r_2$, the L_p and L_I will change by $10\log(r_2/r_1)^2 = 20\log(r_2/r_1)$ dB. The sound level thus decreases by $20\log 2 = 6.02$ dB for a doubling of distance.

Table 15.1 provides sound pressure levels for some common sources and environments. The levels are actually indicated in A-weighted decibels. The purpose of the weighting, which will be discussed further in Section 15.3.1.1, is to account for the response of human hearing to the sound.

15.2.3 Sound Propagation Outdoors

Propagation of sound waves outdoors is strongly dependent on the atmosphere and terrain. As such, it is subject to many dynamic, complex, and interacting phenomena. The near-ground atmosphere and terrain states exhibit temporal variability driven by turbulence, the diurnal cycle, synoptic weather variations, and seasonal forcing. Spatial variability results from turbulence, gravity waves, and from natural and man-made variations in terrain elevation and material properties.

The main influences on outdoor sound propagation, which will be discussed in this subsection, are illustrated in Figure 15.1. We begin with a description of the sound speed and absorption in air. Next, reflections from the ground are introduced. Lastly, various atmospherically induced propagation effects are described. Our emphasis is on near-ground propagation out to distances of several kilometers, which is usually most relevant to noise control problems.

15.2.3.1 Sound Speed and Absorption in Air

Fundamentally, the atmosphere affects sound propagation by introducing variations in the acoustic index of refraction; these variations lead to refraction and scattering of sound.* Atmospheric

variations in the acoustic index of refraction are very strong, in the range of 10^{-3}–10^{-2}. This is several orders of magnitude greater than typical variations in the optical refractive index.

The dependence of sound speed on temperature and humidity is given by (Ostashev 1997, Wilson 2003)

$$c(T,q) = \sqrt{\gamma_d R_d T(1 + \eta q)} \simeq c_0\left[1 + \frac{(\Delta T/T_0 + \eta \Delta q)}{2}\right], \quad (15.3)$$

where

$\gamma_d = 1.402$ is the ratio of specific heat capacities in dry air
$R_d = 287.04$ J K^{-1} kg^{-1} is the gas constant in dry air
$\eta = 0.511$
T is the temperature (K)
q is the water vapor mixing ratio
$c_0 = c(T_0, q_0)$

Here a subscript zero indicates constant reference values (not necessarily ensemble averages), whereas ΔT and Δq indicate small perturbations of T and q about their corresponding reference values.

The velocity of sound waves is also affected by the wind. A wavefront travels with velocity $\mathbf{v}_{\text{ray}} = c\mathbf{n} + \mathbf{v}$, where $\mathbf{n}$ is the normal to the wavefront and $\mathbf{v}$ is the local wind velocity. Since the velocity depends on the wavefront orientation, and the orientation varies over a refracted (bent) path, equations for sound propagation can be complicated considerably in comparison to their counterparts for a nonmoving medium (Pierce 1989, 1990, Ostashev 1997). Hence, a heuristic known as the *effective sound speed* is often employed. This quantity can be defined as (Ostashev 1997)

$$c_{\text{eff}} = c + \mathbf{e} \cdot \mathbf{v} = c + \upsilon \cos\theta, \quad (15.4)$$

where

$\mathbf{e}$ is a constant, nominal direction of propagation, usually specified as the horizontal direction from the source to the receiver
$\upsilon = |\mathbf{v}|$
θ is the angle between the wind and $\mathbf{e}$

---

* The index of refraction is defined as $n = c_0/c$, where c_0 is the reference speed, and c is the actual speed. We define a variation as $\Delta n = c_0/c - 1 \simeq -\Delta c/c_0$. The second (approximate) form is valid when the sound speed variation $\Delta c \ll c_0$.

The effective sound speed is then substituted for the actual sound speed in equations derived for a nonmoving medium. The validity of such a substitution can be demonstrated for propagation confined to a narrow angular beam (roughly " 20°" about the normal propagation direction, but in many other cases it is used without rigorous justification (Ostashev 1997).

Sound energy is *absorbed* (dissipated into heat) as the waves propagate through air. Absorption may be considered to have four primary component processes (ISO 1993): the "classical" (viscosity and heat conduction), rotational relaxation of air molecules, vibrational relaxation by oxygen molecules, and vibrational relaxation by nitrogen molecules. All four processes depend on air temperature, humidity, and pressure. The classical and rotational contributions are both proportional to f^2. In most of the audible frequency range, the vibrational mechanisms also scale as f^2. However, at higher frequencies (greater than about 10 kHz), the vibrational mechanisms become independent of frequency, and the classical and rotational mechanisms thus eventually dominate at very high frequencies.

Standard methods are available to calculate the various contributions (ISO 1993). Absorption typically increases with increasing temperature. The humidity dependence is more complicated: for very dry air, absorption increases with increasing humidity, but as humidity increases further, the absorption decreases. At 0.0°C and 40% relative humidity, absorption is 2.3 dB/km at 500 Hz, and 74.3 dB/km at 4000 Hz. At 20.0°C and 40% relative humidity, absorption is 2.8 dB/km at 500 Hz, and 33.7 dB/km at 4000 Hz (ISO 1993). In response to temperature and humidity changes, absorption exhibits substantial diurnal, synoptic, and seasonal variability.

15.2.3.2 Ground Interaction

Sound waves are reflected at interfaces between the air, ground, and other obstacles such as buildings. As a starting point for discussing ground reflections, let us idealize the ground surface as perfectly flat, and the atmosphere and sub-surface as half spaces exhibiting no variations in space or time. Conceptually, we can consider the sound field to be the sum of sound emitted by the actual source and by an image (virtual) source. The image source is positioned exactly below the actual source, at the same distance below the ground plane as the actual source is above. The complex pressure above the ground can thus be written as

$$\hat{p}(r) = \hat{p}_0 r_0 \left(\frac{e^{ikr}}{r} + R_s \frac{e^{ikr_i}}{r_i} \right), \tag{15.5}$$

where

- $k = \omega/c$ is the wavenumber
- $\hat{p}_0$ is the complex pressure at the reference distance $r_0 = 1$ m from the source (in a free field condition)
- R_s is the spherical wave reflection factor ($R_s = 1$ for a perfectly rigid, reflecting surface)
- r is the distance from the actual source to the observer
- r_i is the distance from the image source to the observer

Here $r = \sqrt{(z-h)^2 + x^2 + y^2}$ and $r_i = \sqrt{(z+h)^2 + x^2 + y^2}$, where h is the height of the source above the ground, and (x, y, z) is the receiver position.

The functional dependence of R_s becomes rather complicated when non-planar wave fronts and surface waves along the boundary are considered. To accommodate partially absorbing ground surfaces, the image source method must be modified to allow the reflection coeDcient to be complex and dependent on the source and receiver positions. Often, R_s is written as $R_p + (1 - R_p)F(w)$, where

$$R_p = \frac{Z_s \sin\phi - \rho_0 c_0}{Z_s \sin\phi + \rho_0 c_0} \tag{15.6}$$

is the plane-wave reflection coeDcient, Z_s is the specific acoustic impedance of the ground surface, and $F(w)$ is a function accounting for the spherical nature of the waves (Attenborough et al. 1980).

Usually, the ground surface may be considered a perfect reflector ($R_p = 1$) if it is much denser than air and has a very low porosity. Some surfaces possessing these characteristics are water, ice, and solid rock. But, even some seemingly hard materials such as asphalt, as well as many common outdoor surfaces such as soil, sand, and gravel, are imperfect reflectors of sound. Sound waves enter and are rapidly dissipated within the pores in these materials. Snow is a particularly eDcient absorber of sound energy.

Emissions from the actual and image sources may interfere constructively or destructively, depending on the reflection coefficient and relative lengths of the paths, r and r_i. An appearance of lobes emanating from the source position results, as illustrated in Figure 15.2a. The calculation is for a unit-amplitude source ($|\hat{p}_0| = 1$ Pa) with frequency 400 Hz, and positioned at a height of 1 m.

15.2.3.3 Atmospheric Interactions

Spatial and temporal variations in the atmosphere introduce considerable complications into the idealized, homogeneous half-space model just described. *Refraction*—the bending of the propagation paths by wind, temperature, and humidity (usually mean vertical) gradients—directs the sound energy into certain regions, as shown in Figure 15.1. When the vertical gradient of the effective sound speed is positive, sound is refracted downward. Downward refraction near the ground may combine with ground reflections to create a duct. If the ground surface has a high reflection coeDcient ($|R_p| \simeq 1$), sound may propagate over long distances with very little loss. When the vertical gradient of the effective sound speed is negative, sound is refracted upward. A refractive shadow, characterized by very low sound levels, may form.

Figure 15.2c shows a calculation of sound levels downwind and upwind of a 400 Hz source. (The parabolic equation, or PE, method used for the calculation will be discussed later.) Formation of a refractive shadow is evident in the upwind direction.

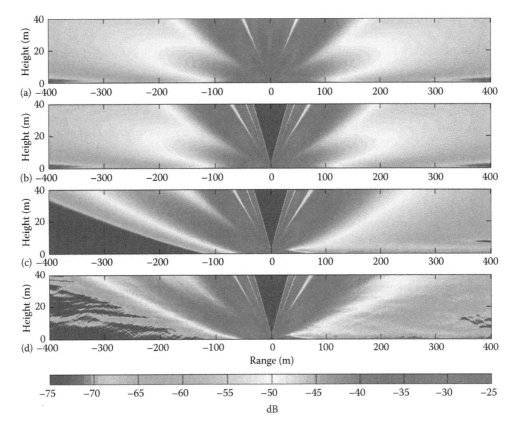

FIGURE 15.2 **(See color insert.)** Calculations of the sound field above an impedance ground surface. A vertical plane, oriented along the wind direction, is shown: upwind propagation is to the left, downwind to the right. The sound source, which is positioned at zero range, has unit strength, a frequency of 400 Hz, and height of 1 m. The ground impedance, $Z_s/\rho_0 c_0 = 10 + i8.5$, is typical of grass-covered soil. For (c) and (d), a logarithmic wind profile with friction velocity of 0.5 ms^{-1} was used. (a) Calculation with no wind and no turbulence, based on an image source model. (Attenborough, K. et al. *J. Acoust. Soc. Am.*, 68, 1493, 1980.) (b) Calculation with no wind and no turbulence, using a PE method. (c) Calculation with wind but no turbulence, using a PE method. (d) Calculation with wind and turbulence, using a PE method and a von Kármán spectrum for the turbulence.

Downwind, some enhancement of the sound level due to ducting is evident between about 3 and 5 m. However, the sound level is actually depressed at a height of about 1 m. This phenomenon, which is due to mode ducting above an impedance ground, was identified and named the "quiet height" by Waxler et al. (2006).

Since the effective sound speed incorporates both wind and temperature (with a lesser effect from humidity), wind and temperature gradients combine to determine the net refractive characteristics. Four distinctive refractive conditions may be identified:

1. *Weak, upward refraction prevalent in all directions.* These conditions often occur on days or nights with thick cloud layers. Stratification is close to neutral, thus producing a weak negative temperature gradient. Similar stratification may also occur in (usually thin) layers during the morning and evening transitions in fair weather.

2. *Strong, upward refraction prevalent in all directions.* Typically, this condition occurs during low wind, clear, daytime conditions in which solar heating produces a positive heat flux from the ground surface to the overlying air. Upward refraction may occur in other situations in which cool air overlies a relatively warm surface, such as above a lake during the fall.

3. *Downward refraction prevalent in all directions.* This condition is characteristic of strong temperature inversions. Occurrence is typical during clear nights and other low-wind conditions involving a ground surface that is relatively cooler than the air.

4. *Refraction dependent upon the propagation direction.* The distinguishing characteristic of this condition is a high wind speed, strong enough to keep the air well mixed near the ground. This condition may occur in clear or cloudy conditions, day or night. Refraction is usually downward in the downwind direction, and upward in the upwind direction.

Sound energy in shadow regions and other locations of relatively low levels (such as the quiet height) is often increased by scattering processes. In atmospheric acoustics, *scattering* generally refers to redirection of sound into multiple directions by dynamic atmospheric variations, such as turbulence, and by small roughness elements, such as vegetation. This process is illustrated in Figure 15.1 and by a numerical calculation in Figure 15.2d.

15.3 Methods of Analysis

15.3.1 Sound Measurements

15.3.1.1 Sound-Level Metrics

A great variety of noise metrics (measures) are used for describing noise. Because such a variety increases the likelihood of comparing "apples to oranges," care should be taken to distinguish between the metrics. The following is an essential subset of noise metrics in use. More complete discussions can be found in standards documents (ISO 2003, ISO 2007).

The instantaneous sound pressure may be spectrally filtered by *frequency weighting*, usually with A-weighting or C-weighting filters. These filters correspond somewhat to perception of loudness of tones for barely audible (A-weighting) and intense sound levels (C-weighting) compared to measurements. These frequency weightings are implemented in most sound-level meters, and they are commonly applied to broadband noise.

The rms average of the frequency-weighted sound pressure is used in assessing the *equivalent level*. Such levels are indicated as L_{AeqT} or L_{CeqT}, where T is an averaging time. Common choices for the averaging time interval are 1 h, 8 h, 1 day, and 1 year. However, T may be chosen to correspond to any time period, such as the operating time of noisy machinery. The equivalent level is most useful for describing steady noise.

Sound levels are often adjusted to account for quiet and the increased potential for disturbance during the evening and nighttime. The most widely adopted such adjustment is the day–night average sound level (L_{dn}, or DNL), which is a time-averaged, A-weighted sound level of all noise within a 24 h period, with a 10 dB penalty added for noise occurring between 10 PM and 7 AM local time.

Received sound levels may vary due to variations of the source power, or variations of the transmission properties of the medium between source and receiver. Consequently, the received sound may vary by several dB over short time intervals. For unsteady sounds, a useful metric is the *N percent exceedance level*, L_{NT}, the level exceeded for N percent of the time interval. The exceedance level is sometimes used to distinguish relatively loud (e.g., L_1 or L_5) periods from quiet (e.g., L_{95} or L_{99}) ones, or for distinguishing on-off periods.

For rapidly varying or *impulsive sound*, the (possibly frequency-weighted) sound can be further characterized by applying *time weighting*. Time weighting employs a moving average applied to the recent signal history with damping time constant of either 1 s (slow response) or 1/8 s (fast response). The fast response approximates the 200 ms integration time of human hearing. For time-weighting metrics, the maximum value within some time interval is usually of chief interest (e.g., $L_{A,F,max}$, $L_{C,S,max}$). In the absence of any time weighting, the maximum is referred to as the *peak level*, L_{peak}.

Steady sound may be characterized by its narrow-band Fourier spectrum, as from the output of an FFT (fast Fourier transform) analyzer. Such an analysis, which characterizes the sound in fixed-bandwidth intervals, is useful for identifying tones and their harmonics. Alternatively, the signal may be partitioned into proportional frequency bands, meaning that the bandwidth is proportional to the band's center frequency. Proportional bandwidth filtering is convenient for identifying coarse spectral features of signals. Filters based on fractional octave-band, particularly full- and one-third-octave bands, are widely used in acoustics. The one-third-octave filter has band center frequencies displaced in progressive multiples of $2^{1/3}$ (approximately $10^{1/10}$). Unsteady sounds require additional methods to evaluate spectra such as the *short-time Fourier transform*, which uses a sliding time window and is sometimes presented in a "waterfall spectrum."

15.3.1.2 Instrumentation

Microphones are used to sense sound pressure. Capacitive microphones rely on the difference between the exterior and interior pressures of a sealed volume to deflect the position of a flexible, charged diaphragm relative to a conducting backplate. The voltage between the diaphragm and backplate changes roughly in direct proportion to the instantaneous sound pressure. The constant of proportionality may be determined through calibration by exposing the microphone to a sound of known pressure and measuring its output voltage. High-quality microphones have nearly constant sensitivity as a function of frequency, a trait determined by the microphone's resonance frequency. The interior of the capacitive volume will resonate according to its dimensions, thus limiting the useable bandwidth to frequencies well below the resonances. Generally, microphones must be larger to be more sensitive at lower frequencies, but smaller to decrease directivity, increase bandwidth, and avoid self-diffraction at high frequencies. Amplifiers with high input impedance are needed to sense the diaphragm voltage, but amplifiers are a source of low-level electronic noise. The *dynamic range* encompasses the range of sound levels over which the microphone output is linear yet remains above the electronic noise. A variety of microphone designs are in use, and each offers some combination of cost, size, linearity, frequency range, dynamic range, sensitivity, directivity, and environmental tolerance.

Sound-level meters (SLMs) employ a microphone along with several stages of signal processing in order to provide the noise metrics outlined in Section 15.3.1.1. Simple analog SLMs may display selectable noise metrics on a d'Arsonval meter. Digital SLMs are capable of continuously processing several noise metrics at once, logging events according to trigger levels or external controls, and storing signals for replay or further processing.

Microphone signals are voltages, and these can be digitally sampled via analog-to-digital (A/D) conversion. Doing so involves some important considerations when choosing the sampling rate and the number of sample bits of the digitizer. The microphone may sense high-frequency sounds above the sampling capability of the A/D converter. The Nyquist frequency, equal to one-half of the digital sampling rate, is an upper limit above which signals will be aliased into lower bands. Aliasing may be avoided by increasing the sampling rate until it exceeds twice the highest microphone response frequency, or by filtering microphone output to remove the high frequencies prior to A/D sampling. The number of sample bits b should be large enough to adequately span the dynamic range D. The requirement is $20b \log_{10} 2 > D$, where D is in decibels. Finally, the maximum input amplitude of the digitizer should exceed the maximum output of the microphone, to avoid signal clipping.

Other instrumentation for environmental noise sampling includes various configurations of microphone arrays for separating sounds by their directions of arrival. Two microphones separated by some small distance may be processed to enhance signals arriving from one direction (relative to their common axis) and reduce signals arriving from other directions. Such an array is not very directional when compared to a directional microphone, but it may be adjusted to select any angle, and additional microphones dramatically increase the number of baseline axes. At higher frequencies, larger separations, and longer propagation paths the microphone arrays lose correlation due to turbulent scattering of the sound.

A specialized two-microphone array is used for evaluating the acoustic impedance of ground surfaces in ANSI S1.18 (2004). In this method, a loudspeaker broadcasts a sequence of tones, which are received a short distance away ($\sim$2 m) by a vertical two-microphone array. The spectral complex pressures are evaluated by FFT for each microphone, and the spectral elements are combined in a ratio. Using putative values of the surface impedance, equations similar to (15.5) and (15.6) are used to evaluate the complex pressure ratio. In an iterative fashion, the estimated ratio is compared to the measured ratio, and the surface impedance is adjusted to resolve a mismatch between them.

15.3.2 Predictive Methods

Often, direct measurement of noise levels is infeasible. For example, we may wish to estimate the noise impacts of a facility prior to its construction or modification. Or, we may wish to explore the cost effectiveness of potential mitigation measures, such as barriers and vegetative screens, prior to their installation. Predictive methods can be valuable in such situations.

15.3.2.1 Rigorous Approaches

Many numerical approaches have been developed to predict sound propagation. These approaches vary greatly in their fidelity and computational requirements. The following first-order differential equations (Ostashev et al. 2005), which apply to small acoustical fluctuations in an ideal gas moving at a low Mach number, are usually an appropriate starting point:

$$\frac{\partial p}{\partial t} + \upsilon_i \frac{\partial p}{\partial x_i} + \rho c^2 \frac{\partial w_i}{\partial x_i} = \rho c^2 Q, \tag{15.7}$$

$$\frac{\partial w_j}{\partial t} + \upsilon_i \frac{\partial w_j}{\partial x_i} + w_i \frac{\partial \upsilon_j}{\partial x_i} = \frac{F_j}{\rho}. \tag{15.8}$$

In the preceding, υ_i is a component of the wind vector, and w_i is a component of the acoustic particle velocity. The terms on the right sides of Equations 15.7 and 15.8 are mass and force source terms, respectively. From these equations, various wave equations and geometric (ray) approximations can be derived (Ostashev et al. 2005). However, these equations are not intended for gravity waves or high-amplitude (nonlinear) acoustic propagation.

Let us consider several common numerical methods for solving Equations 15.7 and 15.8, or related approximations thereof.

- *Ray tracing* is based on solution of a high-frequency approximation to the wave equation. In effect, the trajectories of particles moving through the fluid at the speed of sound are traced. Amplitude is related to the spacing between adjacent rays. Unlike optical ray tracing, the motion in the fluid (atmospheric wind and turbulence) is quite important in acoustics. The fluid motion makes the ray tracing equations (Pierce 1989, Ostashev 1997) and their solution rather more involved. Ray tracing can readily accommodate many complications, such as a spatially and temporally varying atmosphere and irregular terrain features. One of the main, inherent difficulties with ray tracing is determining the sound field behind buildings or in refractive shadows. For such situations, the ray tracing may be supplemented by the geometric theory of diffraction.

- In *wavenumber integration* (also called the *fast-field program* or *FFP*, see Salomons 2001, and references therein) one solves a wave equation to which a Fourier (or Hankel) transform has been applied in the horizontal directions and in time. The resulting equation is an ordinary differential equation with respect to the vertical coordinate. After solving this equation, e.g., with a finite-element method, the result can then be transformed back to the spatial domain with a fast Fourier transform. Wavenumber integration approaches generally assume a *layered* propagation medium. These approaches are thus valid when the atmosphere and ground are horizontally stratified. Therefore these approaches cannot be applied to turbulence or variations in terrain elevation.

- The *parabolic equation* (or *PE*, see Salomons 2001, and references therein) method solves a one-way approximation to the full wave equation in which the wave energy propagates within a confined beam. The angular extent of the beam depends on the type of parabolic approximation. Narrow-angle approximations (less than 20°–30°) are simplest and most commonly used, but wider angle approximations, including those that do not depend on an effective sound-speed approximation, are available. PEs can be applied to non-layered (horizontally varying) environments and can therefore handle turbulence and uneven terrain. PE calculations are normally performed in two-dimensional (2D), vertical planes. The primary disadvantage of PEs is that they do not include reflections or scattering in directions outside of the beam. Hence they are unsuitable for propagation modeling in urban environments, in particular.

ft *Finite-diTbrence, time-domain* (FDTD) methods directly solve discretized versions of coupled, first-order differential equations such as Equations 15.7 and 15.8. FDTD methods can readily incorporate many complications that the previously described methods cannot, including transient source mechanisms, sound-wave interactions with uneven ground, dynamic turbulent scattering, and backscattering in complex terrain such as urban environments. Their main disadvantages are computational intensiveness and the diDculty of implementing an impedance (frequency-domain) boundary condition in the time domain.

Except for ray tracing, the computational burden of the previously described methods increases with frequency (decreasing wavelength). This is because the solution is calculated on a spatial grid having resolution proportional to the wavelength. FDTD solution also involves a temporal grid with resolution inversely proportional to frequency. The PE and FFP are generally implemented in the frequency domain.

Figure 15.2b through d illustrates calculations with the PE method. Each image actually combines two separate calculations; one upwind and one downwind from the source. As an artifact of the finite-angle characteristic of the PE, no sound energy is present in "fans" pointing upward and downward from the source. Also, in comparing Figure 15.2a and b, some inaccuracies in the PE calculation are evident when the propagation angle deviates substantially from the horizontal. Figure 15.2c and d introduce refraction and turbulence, respectively. Scattering into the upwind refractive shadow region is clearly evident.

15.3.2.2 Heuristic Approaches

Heuristic approaches are usually based on analysis of empirical and/or simulation data for particular noise sources and mitigation measures of interest. These approaches are adopted to speed predictions or to enable predictions in situations that are impractical to treat by more rigorous approaches, such as those described in the previous subsection. The diDculty with heuristic approaches, in general, is that they may provide poor results when applied to situations for which they were not specifically designed or tested. Existing heuristic approaches often assume that adjustments associated with various propagation effects (such as refraction, shielding by barriers, and ground attenuation) are independent and sum linearly.

Many of the heuristic schemes employed for outdoor sound propagation are based on classifying atmospheric and ground conditions into a number of defined categories. Different adjustments to the sound-level predictions are then applied for each category. At the simplest, ground categories may specify whether the surface is acoustically hard (such as asphalt or frozen ground) or soft (such as loose soil or snow). Atmospheric categories may specify whether refraction is upward or downward,

and whether it is weak or strong. Some of the most widely used schemes for refractive categories have been based on heuristic surface-layer classes familiar elsewhere in micrometeorology. For example, Marsh (1982) employed Pasquill's stability classes for atmospheric diffusion (based on wind speed and solar radiation) to partition sound propagation conditions into six refractive classes.

As an alternative to Pasquill's classes and similar heuristic schemes, stability can be quantified using the Monin–Obukhov similarity theory (MOST). An analysis of the effective soundspeed gradient based on MOST (Wilson 2003) and Equation 15.3 leads to a parameterization involving a refractive strength parameter $c^*_{eff} = c_* + u_* \cos\theta$ and a shape ratio c_*/u_*, where $c_* = (c_0/2)(T_*/T_0 + \eta q_*)$, u_* is the friction velocity, T_* is the surface-layer temperature scale, and q_* is the surface-layer humidity scale. Heimann and Salomons (2004) approximate the MOST profiles with log-linear profiles and then develop refractive classes based on the log-linear profile parameters. Parabolic equation calculations are used to determine sound levels within each class.

Complex heuristic approaches have been formulated to incorporate the effects of noise-reducing barriers, ground materials, and refraction. Exemplifying such approaches is Nord2000, which was designed for near-ground propagation out to 3 km. Arbitrary terrain profiles and ground surfaces are supported, and refraction is handled heuristically by a ray-based approach. The Harmonoise model incorporates a number of the numerical methods described in Section 15.3.2.1, including ray tracing and the PE. The numerical calculations are used to calibrate a more eDcient engineering (heuristic) model, which has been widely used in Europe to map noise levels. Attenborough et al. (2007) provide more information and references on Nord2000, Harmonoise, and other models, along with a comparative discussion.

15.4 Applications

Noise exposure can affect the physiology and psychology of humans and animals. As a result, various regulations, standards, and engineering practices have been developed to reduce the potentially harmful effects of noise exposure, alleviate annoyance, and preserve quiet. This section gives an overview of noise regulations and discusses five applications in which the assessment and mitigation of noise exposure is affected by meteorology:

ft Preservation of natural quiet in national parks
ft Low-frequency noise from airports
ft Management of the impact of high-intensity impulsive sound
ft Threatened or endangered species
ft Noise reduction by barriers and vegetation

15.4.1 Overview of Noise Regulations

Local, state, and national noise regulations exist for a variety of residential, industrial, and military noise sources. On the national level, in the United States, responsibility for noise regulation and assessment is currently partitioned among many agencies, including the Occupational Safety and Health Administration (OSHA, which is responsible for occupational/industrial noise), the Department of Transportation (DOT, responsible for road and traffic noise), the Federal Aviation Administration (FAA, responsible for aircraft noise), and the Federal Railroad Administration (FRA, responsible for train noise).

Technical standards describe professionally sanctioned methodologies, procedures, and practices. Organizations that develop standards related to acoustics and noise include the American National Standards Institute (ANSI) and the International Organization for Standardization (ISO). Standards are not mandatory unless adopted by government agencies.

Research studies, which serve as the foundation of many noise regulations and standards, seek to quantify the aspects of the stimulus that best predict response. Human and animal responses can be quantified with physiological measures, such as temporary/permanent threshold shifts or physical damage to the eardrum, and with psychological measures, such as annoyance or complaints. Establishment of precise dose-response relationships between noise stimuli and various response metrics is complex, and often imprecise, because of the subjective nature of psychological response measures (discussed further in Section 15.5.3) and the ethical considerations of exposing humans to potentially damaging noise levels. Physiological and psychological response metrics also vary between and within species, as well as among different types of noise sources. These factors hinder the formulation and widespread adoption of common practices for regulating noise.

Noise regulations and standards are often based on single or cumulative event "not to exceed" thresholds, which seek to protect a certain percentage of the affected population. For example, in the United States, the Occupational Safety and Health Administration mandates that no worker shall be exposed to single event greater than a peak level of 140 dB and a cumulative noise dose of 90 dB over an 8 h time period (OSHA 1983) to protect workers from hearing damage.

15.4.2 Preservation of Natural Quiet

Preservation of natural quiet in national parks has proven to be a particularly interesting case study in noise regulation, as a paradigm shift in normal regulatory practice was found necessary. Concern with preservation of natural quiet in the United States began with the National Parks Overflights Act of 1987, Public Law 100–91. The initial impetus for this law was the proliferation of helicopter tours in particularly quiet parkland.

From the beginning of national regulation of outdoor sound levels, which Congress tasked to the U.S. Environmental Protection Agency (U.S. EPA) in 1972, the goal of noise assessments was to predict community response and/or annoyance. Pursuant to this goal, the EPA determined that public health and welfare is protected with an adequate margin of safety if

outdoor sound levels in residential areas do not exceed an average A-weighted equivalent level (L_{Aeq}) of 55 dB between 7 AM and 10 PM and 45 dB between 10 PM and 7 AM. However, given that existing ambient A-weighted levels in the Grand Canyon and Haleakala National Parks can be as low as 10 dB, adoption of the EPA thresholds would have severely disturbed the natural quiet of these settings. The National Park Service (NPS) thus chose the percentage of time that the noise is *audible* as its preferred regulatory measure.

15.4.3 Low-Frequency Noise at Airports

Historically, assessment of the impact of airport noise on nearby residential areas has been conducted in terms of the A-weighted average sound levels. Under the FAA's Part 150 regulations, homeowners may apply for the funding of noise insulation and/or sound-attenuating windows if their day–night average sound level (DNL) exceeds 65 dB on an annual basis. However, recent experience indicates that this policy does not adequately address all problems associated with airport noise. In particular, at some locations in the immediate vicinity of airports, residents may be subjected to considerable low-frequency sound energy from takeoff rolls and reverse-thrust landing operations. Research conducted at the San Francisco and Minneapolis-Saint Paul International Airports (Fidell et al. 2002) has demonstrated that low-frequency sound levels in excess of 75–80 dB are likely to result in house rattle and vibration. In some situations, the residential occupant's perception of rattle and vibration may be a more important determinant of annoyance than the perceived loudness as reflected by the A-weighted sound level.

15.4.4 High-Energy Impulsive Sound

Time-averaged noise metrics used to assess typical environmental and transportation noises do not meaningfully characterize high-energy impulsive sounds. A short high-energy impulsive event with a high unweighted peak sound level (e.g., greater than 115 dB) will barely affect the time-averaged noise level, despite the fact that most community members find the noise annoying.

Within the United States, three common sources of high-intensity impulsive sound are quarry blasts, supersonic military flights, and military weapons noise from testing and training activities. Federal agencies responsible for regulating each of these sources have conducted source-specific research that has focused on (1) the importance of low-frequency sound as the primary cause of annoyance; (2) the propagation of high-energy impulsive sound over long distances (often several kilometers); (3) the importance of meteorology in predicting the sound levels at distant noise-sensitive receptors; and (4) the response to impulsive noise events.

Noise regulations for each of these impulse sources vary. Quarry blasts are subject to state and local noise laws; however, only a few states have imposed legal limits. Noise from military operations is exempt from regulation but subject to disclosure under the National Environmental Protection Act (NEPA) and often is a major problem for military installations. High-energy impulsive noise generated from military testing and training activities negatively affects residents and communities near installations and often results in lawsuits and training curfews/restrictions (Nykaza et al. 2009).

Increasingly, computer models (such as those described in Section 15.3.2) are used to generate noise exposure contour maps for large military guns, bombs, and explosions. The noise exposure maps are used by installations to manage the day-to-day and long-term impacts of their testing and training operations, and by local planning and zoning boards to properly zone areas for compatibility with residential development.

15.4.5 Threatened and Endangered Species

The potential impact of noise on wildlife is a topic of substantial concern because of federal mandates such as the Endangered Species Act and the National Environmental Policy Act. While the noise exposure response criterion for humans is annoyance, the critical issue for endangered species is survival. Noise can potentially impact habitat use, reproductive success, predation risk, and communication. Immediate responses, such as alert and flushing behavior at nests or roosts and changes in activity patterns, are of special interest when associated with foraging and rearing offspring.

Human response depends on the type noise source and is highly variable among individuals. Animal response may be unique not only for a particular type of noise but also for each significantly different species (Pater et al. 2009). Frequency-weighting algorithms designed for humans are tenuous for animals with significantly different audiograms. For example, bats use ultrasonic echoes, usually well above the range of human hearing, to locate their prey and navigate. Many other animals, including cats and dogs, also hear high-frequency sound much better than humans do. Elephants use infrasound (below the range of human hearing) to communicate over distances of many kilometers at favorable times in the evening and early morning. Research investigations have shown that animals often acclimate to a specific type of noise after only a few exposures, which may suggest that they quickly determine the noise does not represent a threat.

15.4.6 Noise Reduction by Barriers and Vegetation

Trees can benefit a community in many ways, including aesthetics, air pollution reduction, and visual screening. Trees can also be considered for their beneficial effects in reducing noise levels. However, research indicates that special care must be taken when planting a vegetative barrier in order to maximize its effectiveness for noise reduction. Fang and Ling (2003) found strong correlations between the density of plant material, as indicated by visibility, and attenuation. Their research with tropical plantings indicates that 6 dB of attenuation of typical traffic noise

can be attained by planting a 5 m width planting with 1 m visibility, or a 18 m width planting with 10 m visibility. They also found that dense shrubs provided the best benefit, so long as the shrubs were at least as high as the intended receiver. Cook and Haverbeke (1974) recommended that plantings have a high density near the ground and utilize a graded height structure. Plantings are most effective if they are near the noise source as opposed to closer to the receiver. Swearingen and White (2005) found that requiring the acoustic signal to pass from the open into the forest provides additional attenuation, thus indicating that belts of trees are more effective than continuous forests.

When the amount of space available for noise mitigation devices is limited, it is often most effective to construct a barrier. For example, busy roads with nearby residential areas frequently have walls constructed to reduce noise in the communities and act as a visual and safety screen. In order to be effective, these barriers must be higher than the source and receiver heights. This is because the primary purpose of a barrier is to block the direct sound, limiting the propagated sound to only the diffracted part of the wave. The effectiveness of the barrier varies greatly with construction, area ground surfaces, and atmospheric conditions. Kurze (1974) provides an excellent synopsis of noise barriers. Recent innovations include various modifications to reduce diffraction over the barrier's upper edge.

15.5 Major Challenges

15.5.1 Modeling the Complexity of Real-World Scenarios

The discussion in Section 15.3.2 on predictive methods indicated many of the challenges involved in realistically modeling outdoor sound propagation. While incorporation of realistic non-uniform flow, turbulence, variations in ground topography, noise barriers, and vegetation has recently been demonstrated with some of the mentioned numerical approaches, most notably FDTD simulation, extensive computational resources and high-resolution environmental data are often required for accurate calculations. Atmospheric fields and acoustical properties of the natural and man-made materials comprising the environment are not usually available at the high spatial resolution (often submeter) needed for propagation calculations.

To a large extent, these challenges parallel or are rooted in related challenges in the atmospheric sciences. It is increasingly common to incorporate data from atmospheric computational fluid dynamics (CFD) simulations, such as Reynolds-averaged Navier–Stokes calculations, large-eddy simulation, and mesoscale models (e.g., Hole et al. 1999), into acoustical models. Advances in atmospheric CFD of complex terrain can thus help to improve sound propagation modeling. Alternatively, kinematic methods (Wilson et al. 2009, and references therein) for synthesizing turbulence fields with realistic, second-order statistics are often employed to rapidly provide high-resolution inputs to acoustical and other wave propagation calculations.

15.5.2 Statistical Sampling of Variability in Sources and Propagation

Noise emissions are rarely steady. TraDc density and vehicle types on a highway, and thus noise emissions, vary over time. Construction, airports, sporting and performing arts events, and military testing all produce intermittent and/or impulsive noises. As mentioned earlier, propagation is also highly variable, due to variations in the environment on a variety of scales in space and time.

Given the resulting, random nature of sound levels, sampling strategies are needed to ensure accurate characterization of exposures and uncertainties. The problem of estimating a noise level from a limited number of data samples parallels many others in the environmental sciences, including sampling of air pollutants, groundwater contamination, temperature trends over a broad area (climate change), coastal erosion, etc. A thorough strategy should sample variations occurring at different times of day, different days of the week, different weather conditions, etc. Nonetheless, in current practice sound levels are quite commonly characterized from a very limited set of measurements or predictions. Cost and time limitations are primary factors. Also, because the main goal of a measurement or prediction is often to determine compliance (or will comply) with a particular regulatory threshold, extensive measurements or predictions may not be necessary when a value is well above or below the threshold.

To alleviate the need for collecting large datasets, the ISO 9613-2 standard (ISO 1996) prescribes collection during *moderately* downward refracting conditions, which are considered the likeliest circumstances to produce noise annoyance; downward refraction tends to concentrate the sound near ground-based listeners, but *strong* downward refraction conditions should be infrequent. But, given the complex and interacting factors affecting sound propagation, the standard procedure perhaps places more confidence in the ability to identify the "right" sampling conditions than is warranted.

Statistical sampling issues are relevant for model predictions as well as for observational studies. Model predictions are compromised by epistemic uncertainties (limited knowledge) of the atmospheric and terrain representations. Methods should therefore be employed to quantify and statistically sample the impacts of these uncertainties.

15.5.3 Soundscapes

Recent research indicates that perceived annoyance to sound is not entirely linked to loudness and other typically measured characteristics; cultural and emotional context also play important roles (Schulte-Fortkamp et al. 2006). The noise environment should thus be considered holistically as a *soundscape*, analogous to a visual landscape. Hence, good physical models for the generation and propagation of the sound must be combined with an understanding of the preferences and perceptions of a community to accurately predict noise annoyance. Addressing these considerations within the context of practical noise regulation will be challenging.

References

American National Standards Institute. 2004. ANSI S1.18-2004, American National Standard Template Method for Ground Impedance. New York.

Attenborough, K., Hayek, S. I., and Lawther, J. M. 1980. Propagation of sound over a porous half-space. *J. Acoust. Soc. Am.* 68: 1493–1501.

Attenborough, K., Li, K. M., and Horoshenkov, K. 2007. *Predicting Outdoor Sound*. Oxon, U.K.: Taylor & Francis Group.

Beranek, L. L. 1986. *Acoustics*. New York: American Institute of Physics.

Cook, D. I. and Haverbeke, D. F. V. 1974. *Trees and Shrubs for Noise Abatement*. Lincoln, NE: University of Nebraska, College of Agricultural Experimental Station Bulletin, RB246.

Fang, C. F. and Ling, D. L. 2003. Investigation of the noise reduction provided by tree belts. *Landscape Urban Plan.* 63: 187–195.

Fidell, S., Pearson, K., Silvati, L., and Sneddon, M. 2002. Relationship between low-frequency aircraft noise and annoyance due to rattle and vibration. *J. Acoust. Soc. Am.* 111: 1743–1750.

Heimann, D. and Salomons, E. M. 2004. Testing meteorological classifications for the prediction of long-term average sound levels. *Appl. Acoust.* 65: 925–950.

Hole, L. R. and Mohr, H. M. 1999. Modeling of sound propagation in the atmospheric boundary layer: Application of the MIUU mesoscale model. *J. Geophys. Res. Atmos.* 104: 11891–11901.

International Organization for Standardization. 1993. ISO 9613-1, Acoustics—Attenuation of sound during propagation outdoors—Part 1: Calculation of the absorption of sound by the atmosphere. Geneva, Switzerland.

International Organization for Standardization. 1996. ISO 9613-2, Acoustics—Attenuation of sound during propagation outdoors—Part 2: General method of calculation. Geneva, Switzerland.

International Organization for Standardization. 2003. ISO 1996-1, Acoustics—Description, measurement and assessment of environmental noise—Part 1: Basic quantities and assessment of procedures. Geneva, Switzerland.

International Organization for Standardization. 2007. ISO 1996-2, Acoustics—Description, measurement and assessment of environmental noise—Part 2: Determination of environmental noise levels. Geneva, Switzerland.

Kurze, U. J. 1974. Noise reduction by barriers. *J. Acoust. Soc. Am.* 55: 504–518.

Marsh, K. J. 1982. The CONCAWE model for calculating the propagation of noise from open-air industrial plans. *Appl. Acoust.* 15: 411–428.

Nykaza, E. T., Pater, L., Luz, G., and Melton, R. 2009. Minimizing sleep disturbance from blast noise producing training activities for residents living near a military installation. *J.eAcoust. Soc. Am.* 125(1): 175–184.

Occupational Safety and Health Administration. 1983. Occupational noise exposure: Hearing conservation amendment; Final Rule, CFR 1910.95. *Fed. Regist.* 46(162).

Ostashev, V. E. 1997. *Acoustics in Moving Inhomogeneous Media.* London, U.K.: E & FN SPON.

Ostashev, V. E., Wilson, D. K., Liu, L. et al. 2005. Equations for finite-difference, time-domain simulation of sound propagation in moving inhomogeneous media and numerical approximation. *J. Acoust. Soc. Am.* 117: 503–517.

Pater, L.L., Grubb, T.T., and Delaney, D.K. 2009. Recommendations for improved assessment of noise impacts on wildlife. *J.&Wildl. Manage.* 73(5): 788–795.

Pierce, A. D. 1989. *Acoustics—An Introduction to its Physical Principles and Applications.* New York: American Institute of Physics.

Pierce, A. D. 1990. Wave equation for sound in fluids with unsteady inhomogeneous flow. *J. Acoust. Soc. Am.* 87: 2292–2299.

Salomons, E. M. 2001. *Computational Atmospheric Acoustics.* Dordrecht, the Netherlands: Kluwer Academic.

Schulte-Fortkamp, B. and Dubois, D. 2006. Recent advances in soundscape research. *Acta Acust. Acust.* 92: v–viii.

Swearingen, M. E. and M. J. White. 2005. Effects of forests on blast noise, Technical Report TR-05–29. U.S. Army Engineer Research and Development Center Champaign, IL.

Waxler, R., Talmadge, C. L., Dravida, S., and Gilbert, K. E. 2006. The near-ground structure of the nocturnal sound field. *J.&Acoust. Soc. Am.* 119: 86–95.

Wilson, D. K. 2003. The sound-speed gradient and refraction in the near-ground atmosphere. *J. Acoust. Soc. Am.* 113: 750–757.

Wilson, D. K., Ott, S., Goedecke, G. H., and Ostashev, V. E. 2009. Quasi-wavelet formulations of turbulence and wave scattering. *Meteorol. Z.* 18: 237–252.

16

Riverine Transport, Mixing, and Dispersion

J. Ezequiel Martin
*University of Illinois,
Urbana–Champaign*

Meredith L. Carr
*United States Army Engineer
Research and Development Center*

Marcelo H. García
*University of Illinois,
Urbana–Champaign*

16.1 Introduction

The transport of pollutants is a fundamental part of the study of riverine flows (e.g., Garcia, 2001). The problem can be summarized in two questions: How long until a contaminant reaches a certain location, and how concentrated will the cloud be when it arrives? These two questions have fundamental importance in the field, as the determination of remediation actions, in the case of accidental spills, and the design of ed uent discharges both depend on providing an accurate description of the situation. Several factors affect the mixing and dilution of contaminants in rivers, including the type of release, the chemical and physical interactions of the contaminant and the river system, and the hydrodynamic conditions of the river. The following sections focus on the instantaneous release of dilute, passive, conservative contaminants for which the characteristics of the river regulate the dilution process.

The main parameter regulating the long-term fate of a pollutant in a channel is the longitudinal dispersion coeDcient, which is typically orders of magnitude larger than other coeDcients affecting the mixing of a contaminant, such as molecular and turbulent diffusivities. The dispersion coeDcient synthesizes the effect of velocity gradients in the flow as the pollutant is advected; it will be shown that mean spatial variations (with respect to the average velocity) rather than the turbulent fluctuations are responsible for dispersion of the contaminants. Turbulence plays an indirect role in the process, by redistributing the pollutant across the cross-section rather than longitudinally. As the mixing process depends mostly on the advection capacity of the flow, the existence of regions in the channel where advection is reduced directly affects the dispersion capacity of the stream.

The classical theory of dispersion based on the work of Taylor (1921, 1954) and extended by others is covered in Section 16.2. The existence of separation, or dead, zones which are responsible for the discrepancy between the one-dimensional (1D) theory and field measurements is discussed in Section 16.3. Field and empirical methods for the determination of dispersion coefficients are presented in Section 16.4. Numerical codes available for the simulation of the transport equations are discussed briefly in Section 16.5. A particular case, the transport of non-conservative substances, is discussed in Section 16.6.

16.2 Classical Theory: One-Dimensional Advection-Dispersion Equation

Many processes contribute to the mixing of a contaminant in a channel, from the molecular diffusion of the species to turbulent processes and the differential transport due to velocity differences across the cross-sectional area of the stream. As with many turbulent flows, the contribution of molecular diffusion can be neglected. The relative importance of each mechanism

Handbook of Environmental Fluid Dynamics, Volume Two, edited by Harindra Joseph Shermal Fernando. © 2013 CRC Press/Taylor & Francis Group, LLC. ISBN: 978-1-4665-5601-0.

to the overall transport of pollutants can be assessed through scaling considerations of the terms of the general 3D Eulerian advective-diffusion equation (which expresses the mass conservation of the solute). It can be shown (e.g., Fischer, 1973) that the complete equation can be reduced to

$$\frac{\partial c}{\partial t} + \vec{u} \cdot \nabla c = \frac{\partial}{\partial y}\left(\varepsilon_y \frac{\partial c}{\partial y}\right) \qquad (16.1)$$

where
  $\vec{u}$ is the velocity vector
  c is the pollutant concentration
  y is the transverse coordinate
  t is time
  ε_y is the transverse turbulent mixing coeDcient

Equation 16.1 can be integrated over the cross-section to yield a 1D advection-dispersion equation (ADE). This approach highlights the relationship between the transverse mixing and the dispersion coeDcient and will be considered after describing the properties of the 1D ADE. Alternatively, a general expression can be given as

$$\frac{\partial C}{\partial t} + U\frac{\partial C}{\partial x} = \frac{1}{A}\frac{\partial}{\partial x}\left(KA\frac{\partial C}{\partial x}\right) \qquad (16.2)$$

where
  A is the cross-sectional area
  C is the cross-sectional average concentration of the pollutant
  U is the cross-sectional average velocity
  K is the longitudinal dispersion coeDcient
  x is the longitudinal coordinate

For the case of a constant area, Equation 16.2 reduces to

$$\frac{\partial C}{\partial t} + U\frac{\partial C}{\partial x} = K\frac{\partial^2 C}{\partial x^2} \qquad (16.3)$$

This expression was first presented by Taylor (1954) in the context of transport of passive contaminants in pipe flow, generalized by

Aris (1956) to arbitrary cross-sections, and extended to turbulent open channel flow by Elder (1959). Wallis (2007) provides a review and discussion of the numerical methods in use to solve the ADE.

It can be shown that the dispersion coeDcient K is given by

$$K = \overline{e_x} - \overline{u''c''}\left(\frac{\partial C}{\partial x}\right)^{-1} \approx -\overline{u''c''}\left(\frac{\partial C}{\partial x}\right)^{-1} \qquad (16.4)$$

with e_x the longitudinal turbulent or eddy diffusivity, u'' the cross-sectional velocity variation defined as

$$u''(x,y,z) = u(x,y,z) - U(x)$$

$$= u(x,y,z) - \frac{1}{A}\int_A u(x,y,z)dydz = u(x,y,z) - \overline{u(x,y,z)} \qquad (16.5)$$

where
  y and z are the transverse and vertical coordinates, respectively
  the operator $=$ indicates a cross-sectional average

Equivalently c'' represents the cross-sectional variation of the concentration. Typically the dispersion coeDcient K is much larger than the mean longitudinal turbulent diffusivity, and the contribution from $\overline{e_x}$ can be neglected (as it was shown in Equation 16.1), indicating that dispersion is driven by cross-sectional variations of the concentration and velocity fields.

For the case of uniform flow, the equation has a simple Gaussian solution

$$C(x,t) = \frac{M}{A\sqrt{4\pi Kt}}\exp\left[-\frac{(x-Ut)^2}{4Kt}\right] \qquad (16.6)$$

where M is the total mass of solute. The mean position of the cloud $\bar{x} = Ut$ is controlled by advection, while the variance $\sigma_x^2 = 2Kt$ increases linearly with time and shows that the spread of the cloud is directly proportional to the dispersion coefficient. While the complete solution presents no skewness, it will appear skewed when the temporal evolution of the concentration is observed at a fixed point. This is important for field

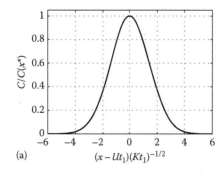

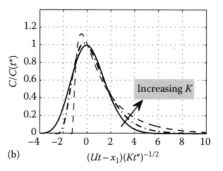

FIGURE 16.1 Solution of Equation 16.6. (a) the spatial distribution at time t_1 is shown; (b) the temporal distribution at position x_1 is given. The dimensionless curve on the left is unique regardless of the value of K, while the curves on the right become increasingly skewed for increasing K. The scaling parameters t^* and x^* are defined as $t^* = x_1/U$ and $x^* = Ut_1$.

applications where the concentration might be sampled at a fixed location as the temporal signal will differ from a normal distribution. Figure 16.1 shows the differences between solutions of Equation 16.6 sampled spatially and temporally. Notice that the maximum concentration at a fixed location occurs before the advective time scale $t^* = x_1/U$.

Taylor's analysis is only exactly valid for asymptotically large times, but it can be applied for finite times if certain conditions are met. Traditionally the dispersion process has been described as consisting of two periods: an initial advective regime followed by a dispersive, Taylor or Fickian (in reference to Fick's law that states that the flux of a quantity is proportional to its gradient), regime for which the previous description is valid. Fischer (1967) established that a sufficient condition for the validity of Equation 16.2 is that the cross-sectional variations c'' become much smaller than the average concentration C. Sayre (1968) suggested that this condition might not always occur even for large times, particularly at the ends of the distribution. He suggested that a necessary condition is instead that the flux $u''c''$ should be proportional to the concentration gradient along the channel. Notice that this condition is equivalent to requiring a uniform value for K (from Equation 16.4). Fischer (1967) gives the following three conditions to identify the dispersive period:

- Linear growth of variance with time
- Decay of concentration following a $t^{-1/2}$ law, in agreement with Equation 16.6
- The function $c''(dC/dx)^{-1}$ becomes steady and uniform (i.e., it is only a function of y and z, the transverse and vertical coordinates)

The previous conditions as well as those from Sayre (1968) are given as operational conditions to evaluate measured or simulated results, but do not establish the transition between the two regimes. Different authors have tried to establish a criterion to determine the duration of the initial period, t_I. Fischer (1967, 1973) considered two possible time scales for this quantity: the transverse mixing time scale T, or Eulerian time scale, and the Lagrangian integral time scale, T_L. The time scales can be defined by

$$T = \frac{\ell^2}{\varepsilon} \tag{16.7}$$

and

$$T_L = k\frac{\ell^2}{\varepsilon} = kT \tag{16.8}$$

with a characteristic length scale, ε a characteristic mixing coefficient of the section, and k a variable depending on the cross-sectional distribution of the mixing coefficient and the velocity. Conceptually, the Lagrangian time scale is defined as the integral time scale for which there is no correlation between a fluid particle velocity and its initial value, and can be computed

directly from the velocity profile, while the Eulerian time scale arises from dimensional analysis. Moreover, Fischer (1967) shows that the Lagrangian time scale is related to the dispersion coefficient by

$$K = \overline{(u'')^2}T_L \tag{16.9}$$

By assuming a logarithmic profile in an infinitely wide channel, as proposed by Elder (1959), Fischer (1967) showed that the coefficient relating the two time scales is $k = 0.067$. While not universal, this estimate can be used, in conjunction with the operational definitions of the mixing scale as the lateral turbulent diffusivity and the length scale as the distance between the position of maximum surface velocity and the most distant bank, to determine the Lagrangian time scale for other flows. From his experimental work in laboratory open channel flumes, Fischer (1967) concluded that the time for the validity of the dispersive description of the mixing is about $6T_L$, while the variance presents a linear growth as soon as $t \sim 3T_L$. While other authors (e.g., Sayre, 1968; Yotsukura, 1968) have found some discrepancies with these estimates for other numerical and experimental studies, the order of magnitude of the estimates appears to be adequate. An important point made by Fischer (1967) regarding the time scaling is that it implies that for small rivers the convective length is rather small and the dispersive description can be used relatively close to the injection site, while in a large river these conditions might not be achieved for tens or hundreds of kilometers. Chatwin and Allen (1985) review some of the methods used to address the initial advective period, but these will not be discussed here.

Estimates of the dispersion coefficient can be calculated using Equation 16.9 for given velocity and diffusivity fields; for instance, for a logarithmic profile, Elder (1959) estimated $K = 5.93Hu^*$, with H the flow depth and u^* the shear velocity. Values of K for other flows have also been calculated, but this approach is of limited use in the field as it requires prescribed profiles. It is also possible to derive a general expression for K as a function of the velocity and transverse mixing by integrating Equation 16.1, under the assumptions for the validity of the dispersive description of the flow. Fischer (1967) presents such expression

$$K = -\frac{1}{A}\int_0^B dy\, h(y)u''(y)\int_0^y dy'\,\frac{1}{\varepsilon_y(y')h(y')}\int_0^{y'} dy''\,h(y'')u''(y'') \tag{16.10}$$

where

 B is the top width of the channel
 h is the depth

Equation 16.10 gives an operational definition of the dispersion coefficient which depends exclusively on the velocity field, the depth and the transverse mixing, that has been used successfully in the field to estimate K, as is shown in Section 16.4.

16.3 Other Models of Dispersion

Many experimental studies in natural streams have shown that contrary to the normal distribution predicted by Equation 16.6, profiles of concentration have more persistent skewness even well into the dispersion period. Several mechanisms have been considered to explain this behavior.

In a review of over 50 field studies, Sabol and Nordin (1978) found "systematic departures from Fickian behavior" including non-Gaussian concentrations distributions whose long tail increases in length with time, a non-linear increase in variance with time, and a decrease in peak concentration that is faster than the predicted Fickian rate of $t^{-1/2}$. Jobson (1997), using a set of over 400 cross-sections from more than 60 rivers, found that the best fit for peak concentration as a function of time followed a rate of $t^{-0.887}$. These shortcomings of the advection-dispersion equation have been addressed by developing alternative models to ADE, with different degrees of success. Some of these alternative models are described next, with particular emphasis on the conceptualization of the models rather than in their implementation; in many cases there are variations in the implementation of the method depending on the authors, while at the same time these models are numerically more complex than ADE.

16.3.1 Dead Zone and Transient Storage Models

Many authors have argued that the missing element in Fickian dispersion theory is demonstrated most clearly by tails of low concentration seen in field concentration distributions (e.g., Valentine and Wood, 1977; Sabol and Nordin, 1978; Nordin and Troutman, 1980). These authors attribute this tail to mass trapped in stagnant areas, referred to as dead zones or storage zones, which include small side pockets, aquatic vegetation zones, and roughness areas on the bottom and banks. To add this missing element, Thackston and Schnelle (1970) added a term representing a well-mixed channel bottom dead zone to the advection dispersion equation and coupled it with a dead zone equation.

$$\frac{\partial C}{\partial t} + U \frac{\partial C}{\partial x} = \frac{1}{\hat{A}} \frac{\partial}{\partial x}\left(\hat{A} K \frac{\partial C}{\partial x}\right) - \frac{\hat{f}\theta_m}{\hat{H}}(C - C_d) \quad (16.11)$$

$$\frac{\partial C_d}{\partial t} = \frac{\theta_m}{d}(C - C_d) \quad (16.12)$$

where
  C_d is the dead zone concentration
  $\hat{A}$ is the cross-section area without the dead zones
  $\hat{H}$ is the depth without the dead zones
  $\hat{f}$ is the fraction of the bed that is the dead zone
  θ_m is an mass exchange term
  d is the depth of the dead zone

Pedersen (1977) derived a similar equation, relying on Q_E, an exchange discharge per unit length, to produce the exchange of mass between the stream and dead zone as follows:

$$\frac{\partial C}{\partial t} + U \frac{\partial C}{\partial x} = \frac{\partial}{\partial x}\left(K \frac{\partial C}{\partial x}\right) - \frac{Q_E}{v}(C - C_d) \quad (16.13)$$

$$\Omega \frac{\partial C_d}{\partial t} = Q_E(C - C_d) \quad (16.14)$$

where
  v is the stream volume per unit length
  Ω is the dead zone volume per unit length

Most of these storage zone models require calibration of three parameters: a longitudinal dispersion coefficient, a parameter that relates dead zone residence time to cross-sectional mixing time, and a parameter that relates dead zone volume to stream volume. These parameters are typically determined from field measured concentration distributions and there is usually difficulty in finding a unique solution (Rutherford, 1994). Though some work has been done to relate these parameters to measurable physical parameters (e.g., Pedersen, 1977), the reliance on concentration distributions for determination of parameters removes these models from the physical understanding of the dead zone. The dead zone that is modeled may not have a volume or an exchange discharge that can be observed or measured in the field, though some transient storage models (Bencala and Walters, 1983; Seo and Cheong, 2001) have focused on the alternate geometry of pool and rid es present in natural streams to try to relate the parameters of their model with the actual physical conditions of the channel. As well as presenting difficulties for calibration and explanation of dead zones, these models do not solve the problem of predicting the persistent skewness observed in the field. They do produce larger skewness and better fits, but do not decrease the peak concentration as quickly as seen in the field (Nordin and Troutman, 1980).

Recently, it has been proposed that the persistent skewness may be a factor of both surface dead zones and hyporheic exchange (Harvey et al., 1996). Thus, several two- and multi-zone storage models have been proposed. Deng and Jung (2009) proposed a conceptual model with two layers: a more traditional bed dead zone where advective pumping dominates the tracer exchange over a short time scale, and a longer scale where diffusion dominates and exchange occurs between the bed surface layer and the hyporheic zone. While a standard transient storage zone model has an exponential residence time distribution (RTD), these two-zone and multi-zone models generally use a combination of RTDs with some success, particularly at producing the more persistent skewness. For example, STAMMT-L (Gooseff et al., 2007) uses a combination of exponential and power law RTDs, while STIR (Marion et al., 2008) uses a combination of exponential and advective pumping RTDs and produces physically relevant parameters, as the two different time scale processes are treated separately. However, as with the simple transient storage model, the combination of processes involved (here subsurface and surface mixing) leads to non-unique parameters and the definition of RTDs for different types of storage zones produces yet another part of the model that must be specified and may only apply to specific situations. However, a variable residence time model (VART, Deng and Jung, 2009) has recently been proposed which does not require a user-specified RTD and has shown success in reproducing field concentration curves.

There are also some models which do not represent dead zones by coupling an additional equation as in 16.13 and 16.14 but use a modified version of the advection dispersion equation itself. Deng et al. (2004) described the use of the fractional advection-dispersion equation (FRADE) in streams, which can be written as

$$\frac{\partial c}{\partial t} + u\frac{\partial c}{\partial x} = \frac{\partial}{\partial x}\left(D_F \frac{\partial^{F-1} c}{\partial x^{F-1}}\right) \tag{16.15}$$

where F, the factor, produces the skewed tails attributed to storage zones and is in the range 1.4–2.0, where $F = 2$ is the special case of the standard ADE. This equation must be solved numerically and parameters are found by optimization methods, such as the fractional Laplace transform-based parameter estimation method (Deng et al., 2006). Alternatively Singh (2003) proposed the modified advection dispersion model (MADE),

$$(1 - \eta + k)\frac{\partial c}{\partial t} + u\frac{\partial c}{\partial x} = D\frac{\partial^2 c}{\partial x^2} \tag{16.16}$$

where η represents the effect of storage zones. This model has a simple analytical solution and limited parameters which are related to physical criteria (Singh, 2008).

16.3.2 Cells-in-Series Model

The cells-in-series (CIS; e.g., Banks, 1974) model departs from the Fickian model in that the model separates the river into a series of cells in which the concentration is assumed to be well mixed (Rutherford, 1994, p. 222). A mass balance conducted on these cells can be used to determine concentration profiles. For a conservative tracer in a stream with constant velocity, a simple analytical solution is

$$C_{n+1}(t) = \frac{t^n}{T_r^n n!}\frac{M_0}{V}\exp\left(-\frac{t}{T_r}\right) \tag{16.17}$$

where

n is the cell number
$T_r = V/Q$ is the cell retention time, V is the cell volume and Q is the discharge

The resulting concentration distribution shows that the variance in time σ_t depends only on the cell retention time and the number of cells.

$$\sigma_t^2 = nT_r^2 \tag{16.18}$$

The variance in time increases linearly with time; thus this model does not solve the problems with the Fickian model. Other models, such as the aggregated dead zone model, described next, and the hybrid-CIS (Ghosh et al., 2004), are variations of this model that attempt to resolve some of these problems by including a decoupled advection term into the cell model.

16.3.3 Aggregated Dead Zone Model

A complete deviation from the advection dispersion equation was suggested by Beer and Young (1983) who claimed that mixing in and out of dead zones, not shear dispersion, was the dominant mechanism of longitudinal dispersion. They proposed an "aggregated dead zone" (ADZ), which included identifiable stagnant areas as well as mixing due to other mechanisms such as eddies from bottom roughness (Wallis et al., 1989). Similar to the CIS model, the ADZ model assumes a series of well-mixed dead zones, but it accounts for advection using a time delay. Discretization of the individual dead zones results in an aggregate dead zone. Calibration of the model requires determining the number of ADZs in the reach, the cell residence time, and the time delay. As with the single dead zone models, the parameters are diDcult to fit and have little physical meaning (Rutherford, 1994), though the model has been quite successful in practice.

16.4 Calculation of Dispersion Coefficients

In the previous sections, a number of models used to estimate the transport of pollutants in rivers have been discussed. All of these models require the estimation of parameters for their application to particular cases, and the different methods to estimate the dispersion coeDcient are considered in this section. Parameters for other models are not considered explicitly but can usually be obtained from the same data sets used to evaluate K.

Broadly speaking, the methods to determine K can be divided into three categories: empirical relations based on flow variables, tracer studies, and methods based on the evaluation of Equation 16.10.

16.4.1 Tracer Studies

The most common approach used to measure the longitudinal dispersion coeDcient has been to conduct tracer studies. Tracer studies are regarded as the most reliable method to determine dispersion coeDcients in a stream. This direct method consists of injecting a passive tracer into the flow and recording concentrations downstream (Wilson et al., 1986). The benefits of such a method include the direct measurement of the dispersion to be parameterized and the relative simplicity of the tracer study in small streams. The disadvantages include tracer loss, the complex planning and often high expense in large rivers, and the assumptions necessary in analyzing the results and determining a dispersion coeDcient. Risks with dyes involve sorption to sediments and vegetation, photo-degradation or reaction with other materials, and the possibility of the tracer already being present in the system. Fluorescent dyes can be particularly tricky, since their concentration is related to reflected light of a specific wavelength, which may also be caused by the presence of organic components in the water (Smart and Laidlaw, 1977). The planning and expense necessary for a tracer study involve optimizing the method of injection to create the proper initial conditions, significant manpower and equipment to take and analyze water samples at a variety of locations, and the expense of the dye. In particular, a large amount of dye is usually needed for a single injection that results in only a single value for the dispersion coeDcient at the current conditions. A recent tracer in use has been SF_6, a synthetic gas. SF_6 is significantly less expensive than fluorescent dyes and more conservative, but it requires discrete sampling which often has less spatial resolution than continuous fluorometric sampling, leaving a potential for loss of fine-scale features of the concentration cloud (Clark et al., 1996).

The method of moments is a simple method to determine the dispersion coefficient from tracer concentration data based on the relationship between the longitudinal variance and the dispersion coefficient. The accuracy of this method can be poor because of its sensitivity to errors in measuring low concentrations, which are found in the tail of the tracer clouds and are difficult to measure (Rutherford, 1994). Also, spatial concentration distributions are rarely measured, since most data are taken as a time series of concentrations at a set of measurement sites. In this case, the frozen cloud approximation is necessary to convert from temporal to spatial variances. The approximation neglects transport due to dispersion and assumes the main mechanism is advection, which introduces error (Koussis et al., 1983).

Another method, referred to as a routing method, uses evaluation of the ADE with a dispersion coeDcient and knowledge or modeling of the flow. The best coeDcient is chosen based on a fit to field concentration distributions. Though this method reduces the impact of low concentration errors and conversion from spatial to temporal values, results can differ based on the details of conditions or how they are modeled. For example, using a Muskingum routing method, Ponce and Theurer (1982) found that the maximum spatial grid size is a function of the flow, velocity, depth, and slope of the system. Routing methods can be time-consuming and require more data than the method of moments, but they are generally accepted as more accurate (Rutherford, 1994).

16.4.2 Empirical Relations

Empirical equations, based on flow parameters such as mean velocity, cross-sectional average depth H, width and mean shear velocity can be reduced to the general form (Seo and Baek, 2004).

$$\frac{K}{u^* H} = a \left(\frac{U}{u^*} \right)^b \left(\frac{B}{H} \right)^c \tag{16.19}$$

Table 16.1 (Carr, 2007) summarizes some of the relations that have been proposed. These types of relationships work best for those rivers for which they were calibrated, while for other rivers the values from the different expressions can vary by a factor of 10 or more (Rutherford, 1994; see also Figures 16.2 and 16.3). They do, however, have a significant advantage over other methods, as they can provide a quick estimate based on a few flow parameters. Figures 16.2 and 16.3 show some of the available measured coeDcients for K (Carr, 2007) compared to the values from the different predictors. It is important to emphasize that while most empirical relations rely on tracer studies to determine the dispersion coeDcient, the empirical method only requires the knowledge of general flow variables for the estimation of K.

The proposed formulas of Deng et al. (2001) and Seo and Baek (2004) can be considered intermediate methods between truly empirical methods and those based on Equation 16.10, as they use this equation in combination with cross-sectional models of bathymetry and velocity to determine an empirical formulation. The resulting estimations cannot be readily extended to previous data as they require this additional information on the cross-section and results of other models can only be compared to the authorsflused data (Figures 16.4 and 16.5).

TABLE1 6.1 Empirical Estimates of the Longitudinal Dispersion CoeDcient (Equation 16.19)

Reference	a	b	c	Sample	Method
Elder (1959)	5.9	0	0	—	Log law
McQuivey and Keefer (1974)	$0.058S_o^{-1}$	1	0	18 rivers	Regression of tracer study
Fischer (1975)	0.011	2	2	Lab	Regression of tracer study
Liu (1977)	0.18	0.5	2	14 rivers	Inspectional analysis and regression of tracer study
Iwasa and Aya (1991)	2	0	1.5	62 lab, 79 rivers	Regression of tracer study
Seo and Cheong (1998)	5.92	1.43	0.62	26 rivers	Regression of tracer study
Koussis and Rodriguez-Mirasol (1998)	0.6	0	2	16 rivers	von Karmanß defect law velocity profile
Deng et al. (2001)	$0.019\varepsilon_{to}^{-1}$	2	5/3	29 rivers	Power law velocity profile, stable channel shape
Seo and Baek (2004)	γ	2	2	—	γ Velocity profile fit to data

Note: S_o is the slope and ε_{to} is the transverse mixing coeDcient as defined by Deng et al. (2001).

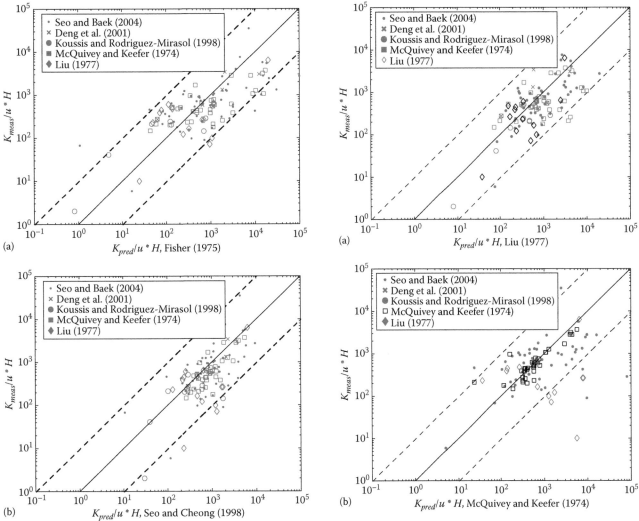

FIGURE 16.2 Observed longitudinal dispersion coeDcient from different studies, compared to the predicted value of K using Fischer (1975) empirical relation (a) and Seo and Cheong (1998) relation (b), as given in Table 16.1. Lines at $K_{pred}/K_{meas} = 1$ (solid), 0.1, and 10 (dashed) are also included.

FIGURE 16.3 Observed longitudinal dispersion coeDcient from different studies, compared to the predicted value of K using Liu (1977) empirical relation (a) and McQuivey and Keefer (1974) relation (b), as given in Table 16.1. Data from their studies are highlighted. Lines at $K_{pred}/K_{meas} = 1$ (solid), 0.1, and 10 (dashed) are also included.

16.4.3 Theoretical Equation

The derivation of the ADE from the depth-averaged turbulent diffusion equation (Equation 16.1) led to a theoretical equation for the shear dispersion coeDcient (Fischer, 1967) shown earlier in Equation 16.10 and verified by Fischer in the laboratory. In the past, the use of Equation 16.10 for evaluation of the longitudinal dispersion coeDcient has been impeded by the challenge of measuring detailed velocities and bathymetry and of determining the transverse dispersion coeDcient. An example of its early application in the field is the survey of a very small stream (about 20 m wide by 1.5 m deep), the Green-Duwamish River in Washington, where eight subsections were discretely integrated to estimate K (Fischer et al., 1979, p. 130). Methods of measuring depth and velocity at the time limited the sampling and thus

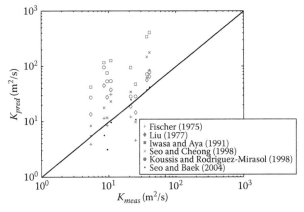

FIGURE 16.4 Observed and predicted dispersion coeDcient for data from Seo and Baek (2004). (Data from Seo, I.W. and Baek, K.O., *J. Hydraul. Eng.-ASCE*, 130, 227, 2004.)

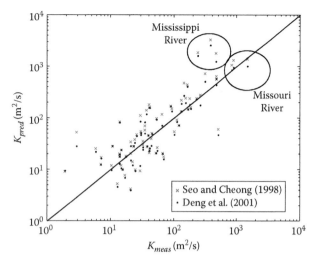

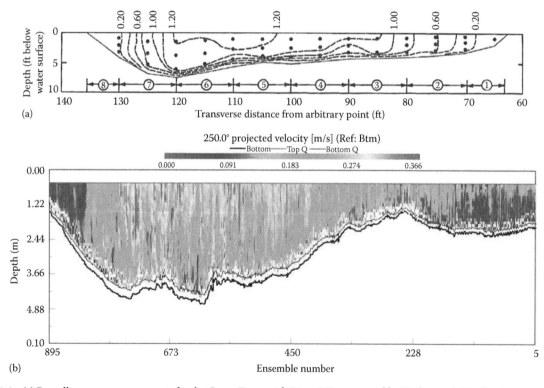

FIGURE 16.5 Observed and predicted dispersion coeDcient for data from Deng et al. (2001). Data for large rivers is highlighted. (Data from Deng, Z.-Q. et al., *J. Hydraul. Eng.-ASCE*, 127, 919, 2001.)

the detail of the velocity profile obtainable within the time that velocities might remain constant.

It was not until fast and accurate acoustic instruments were used in the field that it was possible to extend this method to rivers of much larger magnitude. Acoustic instruments, in particular the acoustic Doppler current profiler (ADCP), increase the density of velocity measurements about a hundred times

when compared with traditional propeller meter measurements, while at the same time drastically reducing the sampling time and providing the bathymetry of the channel as part of the data set. An example of measurements from an ADCP in the Illinois River compared to the data for the Green-Duwamish River experiment is shown in Figure 16.6.

Bogle (1997) used ADCP velocity profiles to calculate K from Equation 16.10, but he compared them only with a single empirical estimate of dispersion and not with more thorough dye study estimates, thus shedding little light on the validity of the ADCP method. Bogle (1997) did, nevertheless, address some of the concerns of using the ADCP for these measurements, including dealing with the ADCP̆s inability to measure near the shoreline and surface and estimating the transverse mixing coeDcient. More recent studies (Carr and Rehmann, 2007; Shen et al., 2010) have shown the viability of using ADCP data to estimate longitudinal dispersion in rivers, in particular in medium to large size rivers where tracer studies can be prohibitive.

Use of the theoretical equation 16.1 is limited by assumptions made by Fischer in his derivation. The main assumptions made during this derivation are as follows:

ft Measurements are made downstream of the advective zone. The velocity of any tracer particle must not be correlated with its initial velocity, and the deviation of the concentration from the cross-section average must be small compared to the average concentration (Fischer, 1967).

FIGURE 16.6 (a) Propeller meter measurements for the Green-Duwamish River, WA, measured by Fischer et al. (1979) using conventional propeller meters. Dots indicate locations of velocity measurement. (b) Illinois River, USGS measurement (unpublished) using an ADCP. (Reproduced from Carr, M.L., An eDcient method for measuring longitudinal dispersion in rivers, PhD thesis, University of Illinois at Urbana-Champaign, Urbana, IL, 2007.)

ft Secondary currents must be negligible, with flow only in the streamwise direction.

ft Turbulent diffusion can be represented by a flux-gradient diffusion relationship, and it must be much larger than molecular diffusion. The turbulent diffusion tensor must be aligned with the principal axes of flow.

ft All parameters are turbulent statistics of the flow (ensemble averaged to remove random turbulent fluctuations).

ft Vertical turbulent diffusion can be neglected, usually because mixing occurs more quickly than in the transverse direction since the aspect ratios in rivers are usually large.

If measurements are made within the advective zone, tracer measurements of shear dispersion will be less than the value in the shear dispersion zone (Fischer et al., 1979). In an analysis of K by the theoretical equation, velocity profiles analyzed within the advective zone may not be representative of the correct value of K downstream and may be biased high or low dependent on the variation in velocity profiles in the streamwise direction.

The assumption of negligible secondary currents has been shown to be valid for situations where secondary currents are less than 4%–5% of the mean velocity (Shiono and Feng, 2003), which is common in most straight or mildly bending rivers. The presence of secondary currents produced by bends can alter the equilibrium between transverse turbulent mixing and the transverse velocity gradient, which is responsible for shear dispersion (Rutherford, 1994). In the case of long bends, the secondary currents may be small enough to allow the 1D ADE to hold. In the case of repeated meanders, Fischer (1969) found that if the bend is long enough, a steady-state velocity profile would be reached, where shear dispersion would hold. If the sinuosity of the channel is too large, the longitudinal dispersion coeDcient will be underestimated. Another cause of secondary currents can be the presence of dead zones. If the system is primarily composed of dead zones, the shear dispersion equation will not apply. However, if a storage zone model, which separates the effects of shear dispersion and dead zones into two mechanisms, is used, evaluation of K using the theoretical equation may be more valid than tracer study values, which will be skewed by the presence of the dead zone.

Fischer (1967) suggested that for $B/H > 6$, the mixing in the vertical direction can be neglected. For smaller aspect ratios, vertical mixing occurs on comparable time scales to transverse mixing, which make the vertical and transverse velocity gradients important in shear dispersion.

16.4.4 Estimation of the Transverse Mixing Coefficient

A challenge with using Equation 16.10 with any velocimetry technique is that the transverse mixing coeDcient ε_y must be estimated. This transverse mixing may occur in bursts along the river at sharp bends or at a divergence or a confluence of flow. The validity of using a reach averaged transverse mixing coeDcient

in Equation 16.10 which applies to a single cross-section has not been discussed, primarily since the only known way to measure the transverse mixing coeDcient is over a reach using tracer studies. Empirical methods to estimate the transverse mixing coeDcient are used but often depend on reach averaged bulk parameters of the flow (e.g., shear velocity, average velocity). It should be kept in mind that transverse mixing will vary along and across the river and that these properties may not be represented by the methods discussed in the following.

The transverse mixing coeDcient can be determined directly by measuring the transverse mixing length from a point injection of dye, yielding an average value only applying to that length of the river (Rutherford, 1994). In larger rivers, where the transverse mixing length may be diDcult to measure, the generalized method of moments (Holley et al., 1972) can be used to determine the transverse mixing coeDcient from a steady source. Since these traditional methods are as costly and diDcult as the tracer study methods for determining longitudinal dispersion, empirical estimates have also been developed. Recent interest in estimating the transverse diffusion coeDcient has resulted in simpler tracer study methods (Boxall and Guymer, 2001), relations between longitudinal and transverse dispersion coeDcients (Gharbi and Verrette, 1998), and empirical estimates that include the effects of secondary currents that cause transverse shear dispersion in addition to turbulent mixing (Deng et al., 2001).

Usually empirical estimates of the transverse mixing coeDcient are written as

$$\varepsilon_y = \theta u^* h \tag{16.20}$$

where θ is a coeDcient.

Different estimates of the shear velocity based on the logarithmic law and other methods can be found in the literature and will not be discussed here. It is worth mentioning that, however, that shear velocity can be assessed from ADCP data (Nystrom, 2001), if using a suDciently large data set, typically larger than what is measured to determine mean flow.

The value of θ is expected to vary from 0.15 to 0.30 in straight channels and from 1 to 3 in curved channels. The effects of turbulent diffusion, or transverse turbulent diffusivity, is well understood and is usually taken as $\bar{\varepsilon}_y/u^* H = 0.145$, while the remainder is usually attributed to secondary currents and/or transverse shear dispersion. Many studies have tried to relate the variation of θ to bulk parameters of the flow: Bansal (1971), based his relation on the aspect ratio B/H; Yotsukura and Sayre (1976), on the ratio of the width and the radius of curvature of the channel; Rathbun and Rostad (2004), found a relation between ε_y and the flow rate. Other authors have tried to estimate θ using simple models of the transverse mixing process: Smeithlov (1990) attributed transverse mixing to a combination of turbulent diffusivity and transverse shear dispersion; Boxall and Guymer (2003) used a method similar to a streamtube model to estimate ε_y directly from transverse velocities which are assumed to transfer materials by an interzone mixing coeDcient set equal to

TABLE1 6.2 Formulas for Transverse Mixing CoeDcient ε_y and Parameter θ

Reference	Formula
Bansal (1971)	$\theta = 10^{-2.698}\left(\dfrac{B}{H}\right)^{1.498}$
Yotsukura and Sayre (1976)	$\theta = 0.4\left(\dfrac{B}{R_c}\right)^2\left(\dfrac{U}{u^*}\right)^2$
Smeithlov (1990)	$\theta = 0.145 + \dfrac{1}{3250}\left(\dfrac{B}{H}\right)^{1.38}\left(\dfrac{U}{u^*}\right)$
Deng et al. (2001)	$\theta = 0.145$
Albers and Stede r (2007)	$\theta = 0.15 + \dfrac{0.31}{\kappa^5}\left(\dfrac{H}{R_c}\right)^2\left(\dfrac{U}{u^*}\right)^2$
Rathbun and Rostad (2004)	$\ln \varepsilon_y = -5.61 + 0.554\ln Q$
Gharbi and Verrette (1998)	$\varepsilon_y = 0.0035\left(\dfrac{(Q/H)^{1.75}(B/H)^{0.25}}{K^{0.75}}\right) + 0.0005$

Note: R_c is the radius of curvature, κ is von Karman constant, $\kappa = 0.41$. Units in Rathbun and Rostad (2004) formula are in m^3/s (Q) and m^2/s (ε_y).

the turbulent diffusivity between well-mixed zones; Albers and Sted er (2007) proposed that much of transverse mixing is due to secondary current shear dispersion. Formulas for the different models are given in Table 16.2; Boxall and Guymer (2003) are dependent on multi-zone parameters which are not included. All of these models present significant scatter, particularly once applied to data not used to fit a particular model, and represent the largest sources of uncertainty for the ADCP estimate for K.

16.5 Numerical Simulation of Transport Processes

A variety of public domain codes are available for solving contaminant transport in rivers. The enhanced stream water quality model (QUAL2E; Brown and Barnwell, 1987) developed by the Environmental Protection Agency (EPA) solves the ADE using a finite difference solution at steady state. The model can be used to simulate both point and non-point sources, but requires the definition of a longitudinal dispersion coeDcient in each reach for both types of sources. HSCTM-2D, a model of sediment and contaminant transport also developed by the EPA (Hayter et al., 1995), allows a dynamic solution of the depth-averaged ADE which requires specification of the dispersion coeDcients. The USGS-developed OTIS is a 1D solution to the ADE coupled with transient storage zones (Runkel, 1998). This model allows incorporation of transient storage, lateral inflow, first-order decay, and sorption.

Another approach to solving the ADE has been to evaluate it in a Lagrangian frame of reference moving with the flow (Jobson and Schoellhamer, 1987). The branched Lagrangian transport model is a public code developed by the USGS that solves the dynamic Lagrangian ADE and has been more economical to solve numerically compared to the Eulerian ADE models.

This model requires the definition of a longitudinal dispersion coeDcient for each "branch" or reach of the system. It also allows incorporation of both zero- and first-order decay/production and integration of tributary inflows as source terms.

Commercial codes are also available, such as MIKE 11 (DHI, 2002), a code that solves the hydrodynamic St. Venant Equations and allows an ADE add-on. RMA4 is a code developed by the U.S. Army Corps of Engineers to solve the depth-averaged ADE using finite element analysis (Donnel and Letter, 2003). This code is available through a commercial code SMS (surface water modeling system). These and most commercial codes require definition of the longitudinal dispersion coeDcient throughout the system and recommend the use of coeDcients determined from physical measurements. Two- and three-dimensional codes are seldom used to simulate the transport of pollutants without the specification of longitudinal dispersion coeDcients, as the required size of meshing to correctly describe the dispersion process based only in turbulent mixing results in prohibitively large numerical domains. These codes can be used, however, to analyze the near-field distribution of a contaminant; it is important to notice that for this type of study, momentum from the pollutant discharge as well as stratification effects due to large initial concentrations of the contaminant can play a role in the dynamics of the flow. An example of this type of models is given by McGuirk and Rodi (1978) for side discharge into a river. Many other models and applications can be found in the literature.

16.6 Non-Conservative Species

The general theory of dispersion has been discussed in the previous sections. Many particular cases where the dispersion process is relevant could be discussed. Some, such as the effects of stratification in mixing (Brocard and Harleman, 1976; Smith, 1976), transport processes in estuaries (Fischer et al., 1979; Geyer and Signell, 1992), pollution of gravel spawning grounds by fine sediments (Huang and Garcia, 2000), the effect of vegetation on transport and mixing (Nepf et al., 1997; López and Garcia, 1998; Perucca et al., 2009), and the transport of biological matter and species (Carr et al., 2004; Wolter and Sukhodolov, 2008) constitute areas of research on their own and will not be covered here. The case of a non-conservative species is considered as an example of how the characteristics of the pollutant or the features of the hydraulic system may require modifications of the basic theory.

Some substances of interest in river transport and mixing are non-conservative, as they react chemically, are retained by vegetation or the bottom of the stream or are exchanged to the boundaries of the system. In many cases, the complex mechanisms involved in the process of the decrease of concentration of the substance of interest can be modeled as a first-order decay of the concentration (Fischer et al., 1979).

$$\frac{\partial C}{\partial t} + U\frac{\partial C}{\partial x} = \frac{1}{A}\frac{\partial}{\partial x}\left(KA\frac{\partial C}{\partial x}\right) - \lambda C \tag{16.21}$$

with λ the coeDcient of solute decay. Neglecting the local temporal variation of C (i.e., assuming a quasi-steady spatial variation), Fischer et al. (1979) showed that the solution of Equation 16.21 for a constant release, and after the initial period takes the form

$$C(x) = C_0 \exp\left[-\left(\frac{Ux}{2K}\right)\left(\sqrt{\alpha+1}-1\right)\right]; \quad \text{with } \alpha = 4\frac{K\lambda}{U^2} \quad (16.22)$$

C_o is in general an unknown, as the amount of solute decayed in the initial period needs to be accounted for. Fischer et al. (1979) noted that in most cases of interest Equation 16.22 is of limited applicability, as either the rate of decay is too fast and most of the solute disappears during the initial advective period or else the effect of dispersion is negligible and the solution decays exponentially only as a function of λ. Numerical simulation of the general advection-diffusion equation with a reactive term is necessary in the first case, while in the second case the solution can be approximated by $C = \dot{M}/Q \exp(-\lambda x/U)$.

More recent works (Schmid, 1995; Runkel et al., 1996; Wörman, 1998; Jonsson et al., 2004) have included reactive terms to other models of dispersion. For example, Schmid (1995) presents the following formulation:

$$\frac{\partial C}{\partial t} + U\frac{\partial C}{\partial x} = K\frac{\partial^2 C}{\partial x^2} + \frac{\varepsilon}{T}(C - C_d) - \lambda C \quad (16.23)$$

$$\frac{\partial C_d}{\partial t} = \frac{1}{T}(C - C_d) - \lambda_d C_d \quad (16.24)$$

where

ε is the dead zone ratio

T is a constant of proportionality (a measure of the kinetics of the solute exchange)

λ and λ_d are the coeDcients of solute decay in the channel and dead zone, respectively

This equation does not take into account sorption, exchange with the hyporheic zone and assumes the decay is a first-order reaction. Wörman (1998) introduced a model with an analytical solution which includes the effects of first-order reaction, sorption to the bed and suspended sediments based on a partitioning ratio, and exchange with storage zones. This model does not include longitudinal shear dispersion, assuming that the storage zone exchange will capture the effect. Jonsson et al. (2004) propose that the sorption effect must be represented by both an instantaneous and a kinetic step, which yield better fits with observed break-through curves.

The models described so far are limited in that they only represent reactions that are kinetic and sorption that is based on simple partitioning (Runkel et al., 1996). Many inorganic reactions cannot be represented using these models, so models which couple hydrodynamic transport with chemical reactions, such as equilibrium reaction (Runkel et al., 1996), were introduced.

Such models generally required numerical solutions and some consideration must be given to the method of coupling. An extensive description of the available models will not be given here; the reader is referred to Dzombak and Ali (1993) for a review of available hydrochemical models.

References

Albers, C. and P. Sted er: 2007, Estimating transverse mixing in open channels due to secondary current-induced shear dispersion. *J. Hydraul. Eng. ASCE*, **3** , 186–196.

Aris, R.: 1956, On the dispersion of a solute in a fluid flowing through a tube, *Proc. R. Soc. Lond. Ser. A*, **3** , 67–77.

Banks, R. B.: 1974, A mixing cell model for longitudinal dispersion in open channels. *Water Resourc. Res.*, **0** , 357–358.

Bansal, M. K.: 1971, Dispersion in natural streams. *J. Hydraul. Eng. Div. ASCE*, **9** (11), 1867–1886.

Beer, T. and P. C. Young: 1983, Longitudinal dispersion in natural streams, *J. Environ. Eng. ASCE*, **0** (5), 1049–1067.

Bencala, K. E. and R. A. Walters: 1983, Simulation of solute transport in a mountain pool-rid e stream: A transient storage model. *Water Resourc. Res.*, **9** , 718–724.

Bogle, G. V.: 1997, Stream velocity profiles and longitudinal dispersion. *J. Hydraul. Eng. ASCE*, **3** , 816–820.

Boxall, J. B. and I. Guymer: 2001, Estimating transverse mixing coeDcients. *Water Mar. Eng.*, **8** , 263–275.

Boxall, J. B. and I. Guymer: 2003, Analysis and prediction of transverse mixing coeDcients in natural channels. *J. Hydraul. Eng. ASCE*, **9** , 129–139.

Brocard, D. N. and D. R. F. Harleman: 1976, One-dimensional temperature predictions in unsteady flows. *J. Hydraul. Div. ASCE*, **0** , 227–240.

Brown, L. C. and T. O. Barnwell: 1987, The enhanced stream water quality models QUAL2E and QUAL2E-UNCAS: Documentations and users manual. EPA/600/3-81/007 edition.

Carr, M. L.: 2007, An eDcient method for measuring longitudinal dispersion in rivers. PhD thesis, University of Illinois at Urbana-Champaign, Urbana, IL.

Carr, M. L. and C. R. Rehmann: 2007, Measuring the dispersion coeDcient with acoustic Doppler current profilers. *J. Hydraul. Eng. ASCE*, **3** , 977–982.

Carr, M. L., C. R. Rehmann, J. A. Stoeckel, D. K. Padilla, and D. W. Schneider: 2004, Measurements and consequences of retention in a side embayment in a tidal river. *J. Mar. Syst.*, **49**, 41–53.

Chatwin P. C. and C. M. Allen: 1985, Mathematical models of dispersion in rivers and estuaries. *Annu. Rev. Fluid Mech.*, **7** , 119–149.

Clark, J. F., P. F. Schlosser, M. Stute, and H. J. Simpson: 1996, SF_6-^{3}He tracer release experiment: A new method of determining longitudinal dispersion coeDcients in large rivers. *Environ. Sci. Technol.*, **0** , 1527–1532.

Deng, Z.-Q., L. Bengtsson, and V. P. Singh: 2006, Parameter estimation for fractional dispersion model for rivers. *Environ. Fluid Mech.*, **6**, 451–475.

Deng, Z.-Q. and H.-S. Jung: 2009, Variable residence time–based model for solute transport in streams. *Water Resourc. Res.*, **4**, W03415.

Deng, Z.-Q., V. P. Singh, and L. Bengtsson: 2001, Longitudinal dispersion coeDcient in straight rivers. *J. Hydraul. Eng. ASCE*, **1**, 919–927.

Deng, Z.-Q., V. P. Singh, and L. Bengtsson: 2004, Numerical solution of fractional advection-dispersion equation. *J. Hydraul. Eng. ASCE*, **3** 0, 422–431.

DHI: 2002, MIKE 11—A modeling system for rivers and channels. DHI software.

Donnel, B. P. and J. V. Letter: 2003, Users Guide for RMA4 Version 4.5. U.S. Army, Engineer Research and Development Center, Waterways Experiment Station.

Dzombak, D. A. and M. A. Ali: 1993, Hydrochemical modeling of metal fate and transport in freshwater environments. *Water Pollut. Res. J. Can.*, **8**, 7–50.

Elder, J. W.: 1959, The dispersion of marked fluid in turbulent shear flow. *J. Fluid Mech.*, **5**, 544–560.

Fischer, H. B.: 1967, The mechanics of dispersion in natural streams. *J. Hydraul. Eng. Div. ASCE*, **9**, 187–216.

Fischer, H. B.: 1969, The effect of bends on dispersion in streams. *Water Resourc. Res.*, **5**, 496–506.

Fischer, H. B.: 1973, Longitudinal dispersion and turbulent mixing in open-channel flow, *Annu. Rev. Fluid Mech.*, **5**, 59–78.

Fischer, H. B.: 1975, Discussion of "Simple method for predicting dispersion in streams by R.S. McQuivey and T.N. Keefer." *J. Environ. Eng. Div. ASCE*, **0**, 453–456.

Fischer, H. B., E. J. List, R. C. Y. Koh, J. Imberger, and N. H. Brooks: 1979, *Mixing in Inland and Coastal Waters*. Academic Press, New York.

Garcia, M. H.: 2001, Modeling sediment entrainment into suspension, transport, and deposition in rivers, in *Model Validation in Hydrologic Science*, Paul D. Bates and M. G. Anderson (Eds.). Wiley & Sons, Chichester, U.K.

Geyer, R. W. and R. P. Signell: 1992, A reassessment of the role of tidal dispersion in estuaries and bays. *Estuaries*, **5**, 97–108.

Gharbi, S. and J. Verrette: 1998, Relation between longitudinal and transversal mixing coeDcients in natural streams. *J. Hydraul. Res.*, **6**, 43–53.

Ghosh, N. C., G. C. Mishra, and C. S. P. Ojha: 2004, A hybrid-cells-in-series model for solute transport in a river. *J. Environ. Eng. ASCE*, **0**, 10, 1198–1209.

Gooseff, M. N., R. O. Hall Jr. and J. L. Tank: 2007, Relating transient storage to channel complexity in streams of varying land use in Jackson Hole, Wyoming. *Water Resourc. Res.*, **4**, W01417.

Harvey, J. W., B. J. Wagner and K. E. Bencala: 1996, Evaluating the reliability of the stream tracer approach to characterize stream-subsurface water exchange, *Water Resourc. Res.*, **3**, 2441–2451.

Hayter, E. J., M. A. Bergs, and R. Gu: 1995, HSCTM-2D, a finite element model for depth-averaged hydrodynamics, sediment and contaminant transport. Technical report, EPA National Exposure Research Laboratory, Athens, GA.

Holley, E., J. Siemons, and G. Abraham: 1972, Some aspects of analyzing transverse diffusion in rivers. *J. Hydraul. Res.*, **10**, 27–57.

Huang, X. and M. H. García: 2000, Pollution of gravel spawning grounds by deposition of suspended sediment, *J. Environ. Eng. ASCE*, **6** (10), 963–967.

Iwasa, Y. and S. Aya: 1991, Predicting longitudinal dispersion coeDcient in open-channel flows. *Environmental Hydraulics: Proceedings of the International Symposium on Environmental Hydraulics*, Hong Kong, China, December 16–18, 1991, J. Lee and Y. Cheung (Eds.), pp. 505–510.

Jobson, H. E. 1997, Predicting travel time and dispersion in rivers and streams. *J. Hydraul. Eng. ASCE*, **3**, 971–978.

Jobson, H. E. and D. H. Schoellhamer: 1987, Users manual for a branched Lagrangian transport model. Water-Resources Investigation Report 87-4163, U.S. Geological Survey, Denver, CO.

Jonsson, K., H. Johansson, and A. Wörman: 2004, Sorption behavior and long-term retention of reactive solutes in the hyporheic zone of streams. *J. Environ. Eng.*, **0**, 573–584.

Koussis, A. D. and J. Rodriguez-Mirasol: 1998, Hydraulic estimation of dispersion coeDcient for streams. *J. Hydraul. Eng. ASCE*, **2**, 317–320.

Koussis, A. D., M. A. Saenz, and I. G. Tollis: 1983, Pollution routing in streams. *J. Hydraul. Eng. ASCE*, **0**, 1636–1651.

Liu, H.: 1977, Predicting dispersion coeDcient of streams. *J. Environ. Eng. ASCE*, **0**, 59–69.

López, F. and M. H. García: 1998, Open-channel flow through simulated vegetation: Suspended sediment transport modeling, *Water Resourc. Res.*, **4** (9), 2341–2352.

Marion, A., M. Zaramella, and A. Bottacin-Busolin: 2008, Solute transport in rivers with multiple storage zones: The STIR model. *Water Resourc. Res.*, **4**, W10406.

McGuirk, J. J. and W. Rodi: 1978, A depth-averaged mathematical model for the near-field of side discharges into open channel flow. *J. Fluid Mech.*, **6**, 761–781.

McQuivey, R. S. and T. N. Keefer: 1974, Simple method for predicting dispersion in streams. *J. Environ. Eng. ASCE*, **0**, 997–1011.

Nepf, H. M., C. G. Mugnier, and R. A. Zavistoski: 1997, The effects of vegetation on longitudinal dispersion. *Estuar. Coast. Shelf Sci.*, **4**, 675–684.

Nordin, C. F. and B. M. Troutman: 1980, Longitudinal dispersion in rivers: The persistence of skewness in observed data. *Water Resourc. Res.*, **6**, 123–128.

Nystrom, E. A.: 2001, Applicability of acoustic Doppler profilers to measurement of mean velocity and turbulence parameters. Master's thesis, University of Illinois at Urbana-Champaign, Urbana, IL.

Pedersen, F. B.: 1977, Prediction of longitudinal dispersion in natural streams. Series paper no. 14, Institute of Hydrodynamics and Hydraulic Engineering, Technical University of Denmark, Kongens Lyngby, Denmark.

Perucca, E., C. Camporeale, and L. Ridolfi: 2009, Estimation of the dispersion coeDcient in rivers with riparian vegetation. *Adv. Water Resourc.*, **3**, 78–87.

Ponce, V. M. and F. D. Theurer: 1982, Accuracy criterion in diffusion routing. *J. Hydraul. Eng. ASCE*, ❶ , 747–757.

Rathbun, R. E. and C. E. Rostad: 2004, Lateral mixing in the Mississippi River below the confluence with the Ohio River. *Water Resourc. Res.*, ❶ , W05207.

Runkel, R. L.: 1998, One dimensional transport with inflow and storage (OTIS): A solute transport model for streams and rivers. Water-Resources Investigation Report 98-4018, U.S. Geological Survey, Denver, CO.

Runkel, R. L., K. E. Bencala, R. E. Broshears, and S. C. Chapra: 1996, Reactive solute transport in streams. 1. Development of an equilibrium-based model. *Water Resourc. Res.*, ❸ , 409–418.

Rutherford, J.: 1994, *River Mixing*. Wiley, Chichester, U.K.

Sabol, G. V. and C. F. Nordin: 1978, Dispersion in rivers as related to storage zones. *J. Hydraul. Eng. Div. ASCE*, ❶ (5), 695–708.

Sayre, W.W.: 1968, Discussion of "The mechanics of dispersion in natural streams by H.B. Fischer." *J. Hydraul. Div. ASCE*, ❹ , HY6, 1549–1556.

Schmid, B. H.: 1995, Simplification in longitudinal transport modeling: Case of instantaneous slug release. *J. Hydrol. Eng.*, 9, 319–324.

Seo, I. W. and K. O. Baek: 2004, Estimation of the longitudinal dispersion coeDcient using the velocity profile in natural streams. *J. Hydraul. Eng. ASCE*, ❶ , 227–235.

Seo, I. W. and T. S. Cheong: 1998, Predicting longitudinal dispersion coeDcient in natural streams. *J. Hydraul. Eng. ASCE*, ❹ , 45–32.

Seo, I. W. and T. S. Cheong: 2001, Moment-based calculation of parameters for the storage zone model for river dispersion, *J. Hydraul. Eng. ASCE*, ❶ (6), 453–465.

Shen, C., J. Niu, E. J. Anderson and M. S. Phanikumar: 2010, Estimating longitudinal dispersion in rivers using acoustic Doppler current profilers. *Adv. Water Res.*, ❸ 615–623.

Shiono, K. and T. Feng: 2003, Turbulence measurements of dye concentration and effects of secondary flow on distribution in open channel flows. *J. Hydraul. Eng. ASCE*, 129, 373–384.

Singh, S. K.: 2003, Treatment of stagnant zones in riverine advection-dispersion. *J. Hydraul. Eng. ASCE*, ❷ (6), 470–473.

Singh, S. K.: 2008, Comparing three models for treatment of stagnant zones in riverine transport. *J. Irrig. Drain. Eng. ASCE*, ❹ , 853–856.

Smart, P. and I. Laidlaw: 1977, An evaluation of some fluorescent dyes for water tracing. *Water Resourc. Res.*, ❸ , 15–33.

Smeithlov, B. B.: 1990, Effect of channel sinuosity on river turbulent diffusion. *Yangtze River*, ❷ , 62.

Smith, R.: 1976, Longitudinal dispersion of a buoyant contaminant in a shallow channel. *J. Fluid Mech.*, ❽ , 677–688.

Taylor, G. I.: 1921, Diffusion by continuous movements. *Proc. R. Soc. Lond. Ser. A*, ❶ , 196–211.

Taylor, G. I. 1954, The dispersion of matter in turbulent flow through a pipe. *Proc. R. Soc. Lond. Ser. A*, ❸ , 446–468.

Thackston, E. L. and K. B. Schnelle: 1970, Predicting the effects of dead zones on stream mixing. *J. Sanit. Eng. Div. ASCE*, ❾ , 319–331.

Valentine, E. M., and I. R. Wood: 1977, Longitudinal dispersion with dead zones, *J. Hydraul. Div., Am. Soc. Civ. Eng.*, ❶ , 975–990.

Wallis, S.: 2007. The numerical solution of the Advection-Dispersion Equation: A review of some basic principles. *Acta Geophys.*, 5 , 85–94.

Wallis, S. G., P. C. Young, and K. J. Beven: 1989, Experimental investigation of the aggregated dead zone model for longitudinal solute transport in stream channels. *Proc. Inst. Civ. Eng.*, 2, 1–22.

Wilson, J. F., E. D. Cobb, and F. A. Kilpatrick: 1986, Fluorometric procedures for dye tracing. *Techniques of Water-Resources Investigations*. Book 3, Chapter A12, U.S. Geological Survey, Washington, DC.

Wolter, C. and A. Sukhodolov: 2008, Random displacement versus habitat choice of fish larvae in rivers. *River Res. Appl.*, ❷ , 661–672.

Wörman, A.: 1998, Analytical solution and timescale for transport of reacting solutes in rivers and streams. *Water Resourc. Res.*, ❸ 2703–2716.

Yotsukura, N.: 1968, Discussion of "The mechanics of dispersion in natural streams by H.B. Fischer." *J. Hydraul. Div. ASCE*, ❹ , HY6, 1556–1559.

Yotsukura, N. and W. Sayre: 1976, Transverse mixing in natural channels. *Water Resourc. Res.*, ❷ , 695–704.

17

Ocean Outfalls

Philip J.W. Roberts
Georgia Institute of Technology

17.1 Introduction

"Outfall pipes into the ocean are analogous to chimneys in the atmosphere: they are each intended for returning contaminated fluids to the environment in a way that promotes adequate transport and dispersion of the waste fluids." This is the opening sentence in a review published more than 35 years ago (Koh and Brooks, 1975), and nothing has changed in the purpose of outfalls since. Much else has changed, however, such as public awareness of the environment, and the science and technology of marine wastewater disposal in myriad ways including understanding of hydrodynamics, refined modeling, and increased use of advanced instrumentation in the laboratory and field. Ocean outfalls dispose of many types of wastewater: domestic sewage, industrial ed uents, heated water for power plants, and, increasingly, brine from desalination plants. Each has specific complexities and features, so here we will limit ourselves to sewage outfalls.

A typical marine wastewater system, which includes a treatment plant and an outfall, is shown in Figure 17.1. The outfall is a pipeline or tunnel, or combination of the two, which terminates in a diffuser whose purpose is to ed ciently mix the ed uent in the receiving water. Outfalls generally discharge onto coastal shelves with a gradual slope, typically less than 1%, so the horizontal flow extents are much greater than vertical ones. Most outfalls range from 1 to 5 km long and discharge into waters 20–70 m deep. Some may lie outside these ranges, for example shorter outfalls when the seabed slope is very steep and deep water is readily available, or longer when the slope is very gradual. The overall disposal system can be thought of as the treatment plant, outfall, diffuser, and also the region round the diffuser (known as the near field) where rapid mixing and dilution occurs.

Domestic sewage contains many contaminants that could potentially cause pollution problems, particularly biochemical oxygen demand (BOD), bacteria and pathogens, suspended solids, toxics, and nutrients. Of these, the National Research Council (NRC, 1993) states that the highest priorities are nutrients, pathogens, and toxic materials. They rate BOD as low priority because the high initial dilution, large surface area available for reaeration, and eD cient coastal flushing means that dissolved oxygen depletion is not usually a problem. This is different from discharges into inland waters where advanced treatment, usually secondary, is often used whose primary purpose is to reduce BOD. Although the NRC report placed nutrients as a high priority, they are also actually not usually of great concern. Coastal waters are productive due to heavy infusion of nutrients from continents and upwelling. Inputs from outfalls are often minimal in comparison and their effects are small due to high dilution and flushing. This may not be true in regions of poor flushing, however, such as bays and estuaries, where more advanced treatment may be required for nutrient removal. Finally, domestic sewage is degradable, so potential problems are, at most, local, and not regional or global.

Because of the high dilution capabilities of diffusers and the highly dispersive nature of coastal waters, the outfall is frequently the major element in eliminating or reducing environmental impacts, and the level of treatment of lesser importance. This is why the near field is included in the "system" in Figure 17.1. Consider, for example, a dilution of 100:1. This corresponds to a reduction in the concentration of any contaminant by 99%, which is far greater than that can be achieved by any conventional wastewater treatment plant. The eD cacy of outfalls is not widely known however, and advanced treatment is frequently advocated as "the solution" to marine wastewater disposal problems. Outfalls are effective because they accomplish high dilution, separate the ed uent from environmentally sensitive regions, and move it from regions of poor mixing to where it can

Handbook of Environmental Fluid Dynamics, Volume Two, edited by Harindra Joseph Shermal Fernando. © 2013 CRC Press/Taylor & Francis Group, LLC. ISBN: 978-1-4665-5601-0.

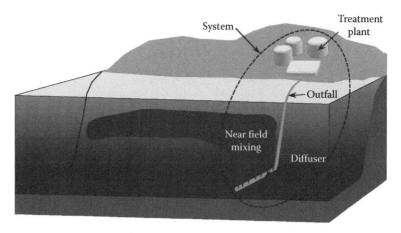

FIGURE1 7.1 Schematic depiction of a marine wastewater disposal scheme.

be eDciently mixed and dispersed. This mixing and dispersion is the subject of this chapter.

Outfalls are usually governed by water quality regulations that are set by regulatory authorities. They often include an initial dilution requirement, typically 100:1. Toxics are controlled by specification of limiting values, for example, Table B of the California Ocean Plan (SWRCB, 2005). These dilution and toxic concentration values apply close to the diffuser, at the boundaries of a region known as the mixing zone. Bacteria limits typically apply at water contact regions that are farther from the diffuser, especially the shoreline. Predicting these water quality impacts requires knowledge of complex hydrodynamics, many of them unique to marine discharges. The most important of these are buoyancy and density stratification effects, mechanics of jets and plumes, plume merging, boundary effects, coastal oceanography and circulation, and tidal effects. In this chapter, we review some of these essential processes, and present some simple methods to analyze them.

17.2 Principles

Figure 17.1 illustrates the essential features of a typical ocean outfall discharge. The wastewater is usually ejected horizontally as round turbulent jets from a multiport diffuser. The ports may be spaced uniformly along both sides of the diffuser or clustered in risers attached to the outfall pipe. Buoyancy and oceanic density stratification play fundamental roles in determining the fate and transport of marine discharges. Because the density of domestic sewage is close to that of fresh water, it is very buoyant in seawater. The jets therefore begin rising to the surface and may merge with their neighbors as they rise. The turbulence and entrainment induced by the jets causes rapid mixing and dilution in a region called the "near field" or "initial mixing region." If the water is deep enough, oceanic density stratification may trap the rising plumes below the water surface where they stop rising and begin to spread laterally. The wastefield then drifts with the ocean current and is diffused by oceanic turbulence in a region called the "far field," where the rate of mixing, or increase of dilution, is much slower than in the near field. Finally, large-scale flushing

and chemical and biological decay processes remove contaminants and prevent long-term accumulation of pollutants.

It is instructive to think of these processes in terms of their length and time scales. These scales increase as the ed uent moves away from the diffuser, in a way that Brooks (1984) has referred to as a "cascade of processes at increasing scales." The most important of these processes, illustrated graphically in Figure 17.2, are as follows.

- ft Near field mixing: Mixing caused by the buoyancy and momentum of the discharge; it occurs over distances of 10–1000 m and times of 1–10 min
- ft Far field mixing: Transport by ocean currents and diffusion by oceanic turbulence; it occurs over distances of 100 m to 10 km and times of 1–20 h
- ft Long-term flushing: Large-scale flushing, upwelling or downwelling, sedimentation; it occurs over distances of 10–100 km and times of 1–100 days and longer

Further complicating the situation is that these processes can overlap, and there may be transitions between various phases. Although these transitions are gradual, for predictive purposes it is convenient to model them as if they were instantaneous. The mixing mechanisms in each of the three major phases and methods to predict them are discussed next. Near field mixing is the only one of these phases that is under the control of the designer but it can also have a substantial effect on all of the subsequent transport phases. Because of the wide range of time and length scales involved, it is not usually feasible to model them in one overall "omnibus" model. Instead, sub-models that are linked together are often used.

17.3 Methods of Analysis

17.3.1 Introduction

The three major phases of mixing, near field, far field, and long-term flushing, are discussed separately in the following sections. Some of the major principles involved are discussed and simple analytical equations are presented to predict them. Each of

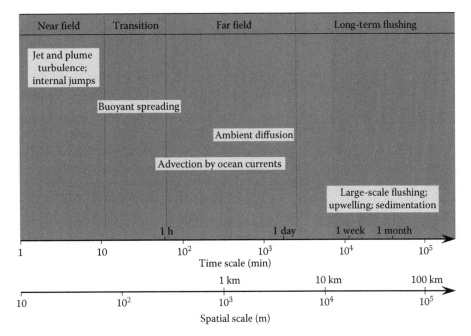

FIGURE17.2 Main processes of outfall discharge fate and transport and their approximate length and time scales.

these topics is complex, with an extensive literature; for a more detailed discussion, see Roberts et al. (2010).

17.3.2 Near Field

Mixing in the near field involves the mechanics of jets and plumes. Fischer et al. (1979) define jets as flows driven by the source momentum flux only and plumes as flows driven by the source buoyancy flux only. Flows with both momentum and buoyancy fluxes are called buoyant jets. Because of the large density difference between domestic sewage (which is essentially fresh water) and seawater, the flows issuing from marine outfalls are dominated by buoyancy forces, in other words they are plumes. The literature on buoyant jets is extensive and they are discussed elsewhere in this book; here we discuss only some of their features that are particularly relevant to outfalls.

Further, we will limit ourselves to the common case of discharge from a multiport diffuser (e.g. Figure 17.1). A definition sketch for this case is shown in Figure 17.3. The effluent issues horizontally at a velocity u_j as buoyant jets from ports of diameter d that are spaced a distance s apart. A current of speed u

flows at an angle Θ to the diffuser axis. The problem is to predict the near field plume properties, particularly the rise height and dilution, as a function of the diffuser and oceanic variables.

Length-scale analyses of jets and plumes based on their source volume, momentum, and buoyancy fluxes (Fischer et al., 1979) have proven to be very productive. The fluxes can be defined either as coming from point or line sources. For isolated sources, such as chimneys in the atmosphere or discharge from an open-ended pipe, point source definitions are obviously prescribed. But for long multiport ocean sewage outfalls, line source parameters are more relevant. They include source volume, buoyancy, and momentum fluxes per unit diffuser length which are defined as follows:

$$q = \frac{Q}{L}; \quad m = u_j q; \quad b = g'_o q \quad (17.1)$$

where

Q is the total flow discharged by the outfall

L is the diffuser length

$g'_o = g(\Delta\rho/\rho)$ is the modified acceleration due to gravity, g the acceleration due to gravity, $\Delta\rho$ the density difference between the effluent and seawater, and ρ a reference density

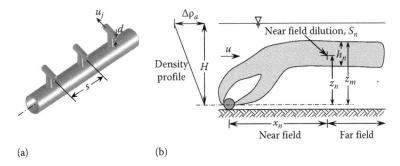

FIGURE17.3 Plumes from a multiport diffuser into an unstratified current. (a) Perspective view and (b) side view.

If the density stratification is linear, it can be characterized by the buoyancy frequency, $N = \sqrt{(-g/\rho)(d\rho_a/dz)}$, where ρ_a is the ambient density at height z.

Any dependent variable can be expressed in terms of these variables. For example, the plume rise height in a stratified environment, z_m is given by

$$z_m = f(q, m, b, s, N, u, \Theta) \qquad (17.2)$$

For most sewage outfalls, it can be shown that the dynamic effect of the source volume flux is small and can be neglected. In that case, a dimensional analysis of Equation 17.2 yields the following:

$$\frac{z_m}{l_b} = f\left(\frac{l_m}{l_b}, \frac{s}{l_b}, F, \Theta\right) \qquad (17.3)$$

where $l_b = b^{1/3}/N$ and $l_m = m/b^{2/3}$ are length scales. The ratios l_m/l_b and s/l_b can be thought of as diffuser parameters that express the relative importance of the source momentum flux and port spacing. If these ratios are small, the ed uent has small momentum flux relative to the buoyancy flux, and the ports are so close that the individual plumes rapidly merge; this is called a "line plume." The experiments of Daviero and Roberts (2006) suggest that it occurs when $l_m/l_b < 0.2$ and $s/l_b < 0.5$ which is satisfied by many ocean wastewater outfalls. Most of the major plume characteristics of interest are determined by their line plume properties, with the effect of port spacing and momentum flux as (small) perturbations from the line plume values. $F = u^3/b$ is a type of Froude number that describes the relative strength of the ambient current speed to the source buoyancy flux.

The flows of Equation 17.3 encompass a wide range of possibilities: jets and plumes and buoyant jets into flowing and stationary currents, stratified and unstratified environments, merging and non-merging plumes, boundary interactions, etc. Many of these situations are discussed elsewhere in this book, and Roberts et al. (2010) summarize many equations for limiting cases that are applicable to ocean outfalls. Here we only discuss some results for line plumes as they are particularly relevant. Note that we have not included viscous, or Reynolds number, effects in Equation 17.3. This is because the flows are assumed to be fully turbulent, in which case Reynolds number effects can be neglected (Fischer et al., 1979).

The plume sketched in Figure 17.3 is trapped by the oceanic stratification with a rise height z_m. Stratification is often important, especially for water depths greater than about 20 m. Even a weak stratification will submerge a typical plume (indeed submergence is usually a desirable design objective because the plume will probably remain submerged and the probability of it reaching shore is very low) so we concentrate on the stratified case here. If the water is unstratified, or weakly stratified, the plume will surface and the water depth H then becomes a relevant length scale of the problem.

For the line plume case in a stratified environment l_m/l_b and s/l_b drop out of Equation 17.3 which becomes

$$\frac{z_m}{l_b} = f(F, \Theta) \qquad (17.4)$$

Similarly, other variables such as the thickness of the wastefield h_n, the length of the near field x_n, and the near field dilution S_n (i.e., the dilution at the end of the near field) can also be shown to be the functions of F and Θ only.

$$\frac{z_m}{l_b}, \frac{h_n}{l_b}, \frac{x_n}{l_b}, \frac{S_n q N}{b^{2/3}} = f(F, \Theta) \qquad (17.5)$$

Photographs of multiport diffuser discharges that approximate a line plume in density-stratified perpendicular currents of various speeds are shown in Figure 17.4. The Froude number F is the most important dynamical parameter that expresses the effect of the current, and different flow regimes occur depending on the value of F. For zero current speed ($F = 0$) the two plumes from each side of the diffuser partially merge before entering the horizontally spreading layer; they overshoot their equilibrium rise height before collapsing back to form a wastefield which occupies a substantial fraction of the total rise height. As the current speed is

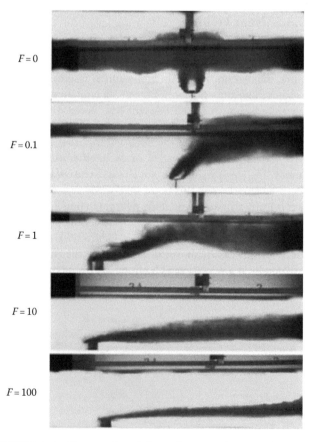

FIGURE 17.4 Photographs of plumes from a multiport diffuser into a stratified current. (From Roberts, P.J.W. et al., *J. Hydraul. Eng.*, 115(1), 26, 1989. With permission.)

increased slightly to $F = 0.1$, the upstream layer is expelled and all of the flow is swept downstream but dilution and rise height are not substantially changed. Increasing the current speed to $F = 1$ causes the plumes from the opposite sides of the diffuser to merge, and the merged wastefield oscillates in an internal wave-like form before reaching its equilibrium level. For these current speeds the flow has the normal plume-like pattern with the plume bent downstream. At higher current speeds (e.g., $F = 10$), however, the plume cannot entrain all of the oncoming flow while maintaining the free plume pattern and the wastefield bottom stays at the nozzle level. This is sometimes called the forced entrainment regime, and it occurs when the Froude number exceeds a value that lies somewhere between 1 and 10 for the stratified case. The dilution increases and the rise height and thickness decrease substantially with increasing current speed in the forced entrainment regime.

For zero current speeds, the right-hand sides of Equation 17.5 become simple constants, whose values are approximately (Roberts et al., 2010)

$$\frac{S_n q N}{b^{2/3}} = 0.86 \qquad (17.6a)$$

$$\frac{z_m}{l_b} = 2.5 \qquad (17.6b)$$

$$\frac{h_n}{l_b} = 1.5 \qquad (17.6c)$$

In other words, all the geometrical plume properties such as z_m and h_n scale in proportion to l_b.

The expression for dilution, Equation 17.6a, has important practical design consequences. g_o' is essentially constant for marine discharges with a value of about $0.25\,\text{m/s}^2$, and the density stratification depends on oceanic conditions. Therefore, the only design variables are Q and L which affect dilution as $S_n \propto (Q/L)^{-1/3}$. But probably the flow to be disposed of Q is also fixed, so the dependency on diffuser length, essentially the only variable under the control of the designer is $S_n \propto L^{1/3}$ Thus, dilution increases only slowly with diffuser length; for example, to double dilution requires a diffuser eight times as long! This insensitivity is partially due to the reduction in rise height (Equation 17.6b) that occurs as the diffuser length is increased. For an unstratified environment, the rise height is constant and equal to the water depth so the dependency of dilution on diffuser length is less.

This can lead to very long diffusers in stratified coastal waters to achieve high dilutions. Consider, for example, a discharge of $5\,\text{m}^3/\text{s}$, typical of a city of about 2 million people. Suppose the oceanic density stratification is linear with a density difference of $0.7\,\sigma_t$-unit over a depth of $25\,\text{m}$. (This small density difference corresponds to a temperature difference of $\sim 3^\circ\text{C}$ or a salinity difference of $\sim 0.9\,\text{ppt}$, or some combination of the two.) Then the buoyancy frequency $N = 0.017\,\text{s}^{-1}$. For a near field dilution of 100:1, Equation 17.6a yields a discharge per unit diffuser length of $0.0081\,\text{m}^2/\text{s}$, so for a total flow of $5\,\text{m}^3/\text{s}$, a diffuser length of $620\,\text{m}$ is required. Even smaller values of "diffuser loading" q are required to produce high dilution when strong stratifications

FIGURE 17.5 Shadowgraph of multiport diffuser discharge into stratified current, $F = 10$. (From Roberts, P.J.W. et al., *J. Hydraul. Eng.*, 115(1), 26, 1989. With permission.)

are expected. For example, the San Francisco outfall has $q \approx 0.005\,\text{m}^2/\text{s}$ (Isaacson et al., 1983) because of the strong stratification caused by freshwater discharges from San Francisco Bay.

The length of the near field is determined by the strength of the density stratification. Consider, for example, the photograph for $F = 10$ in Figure 17.4. Figure 17.5 is a shadowgraph of this flow which shows, qualitative at least, that the plume near to the diffuser is turbulent. It entrains ambient seawater and mixes with it so that dilution increases with distance from the source. Eventually, however, the ambient stratification causes this turbulence to collapse, at which point near field mixing essentially ceases. Following the ideas of Thorpe (1982), Roberts et al. (1989b) argued that this occurred when an internal Froude number of the turbulence $F_i = w'/lN$ where l is a turbulence macroscale ($\sim$size of the largest eddies) and w' is their velocity scale ($\sim$rms value of the vertical velocity fluctuations) falls to some critical value which is of order unity. This is equivalent to saying that the time scale of the vertical velocity fluctuations l/w' is approximately equal to the buoyancy period N^{-1} at collapse. Applying further dimensional analysis for the forced entrainment regime, it can be shown that these arguments lead to

$$\frac{x_n}{l_b} = 8.5 F^{1/3} \qquad (17.7)$$

where the value of the constant is obtained experimentally (Roberts et al., 1989b). As for the other geometrical parameters of Equation 17.6, x_n is proportional to l_b for fixed current speed. Note that x_n is now synonymous with the length of the near field, which is defined as the location where mixing due to the turbulence generated by the discharge itself ceases. An interesting implication of Equation 17.7 is that it can be written as $x_n = 8.5 u N^{-1}$ which does not contain any diffuser variables; in other words, the length of the near field depends only on the current speed and stratification! It also shows that the length of the near field is directly proportional to current speed, and decreases as the stratification becomes stronger. Although Equation 17.7 was derived for the forced entrainment regime, it works reasonably well for lower current speeds also.

What is the length of the near field for typical oceanic conditions? Suppose for the previous example, we have $u = 0.20\,\text{m/s}$. Then $x_n = 8.5 \times 0.20 \times (0.017)^{-1} = 100\,\text{m}$. The buoyancy flux per unit length, $b = 0.25 \times 5/620 = 0.0020\,\text{m}^3/\text{s}^3$ and the Froude number, $F = u^3/b = (0.20)^3/0.002 = 4$. So the flow is in the forced entrainment regime, somewhere between the photographs of $F = 1$ and 10 on Figure 17.4.

These are fairly typical values for coastal outfalls. The near field usually extends tens but not more than a few hundreds of meters from the diffuser. In an unstratified environment, the length of the near field scales with the water depth but would be expected to be similar order of magnitude.

The previous discussion, for example the plumes of Figure 17.4, is primarily for currents flowing perpendicular to the diffuse. As would be anticipated, this gives the highest near field dilution, and diffusers are usually oriented that way, if possible. Even for this case, however, the flow is three-dimensional (3D) because diffusers are not infinitely long. There are end effects, so diluting flow can be supplied from the ends, and the wastefield will spread laterally as a density current. Also, it may not be possible to orient the diffuser perpendicular to the current, and in an ocean environment, currents can flow in all directions. Lowest dilution occurs when the current is parallel to the diffuser axis which also results in strong lateral gravitational spreading and a wide wastefield. Three-dimensional effects are not discussed further here due to space limitations; for a full discussion, see Roberts et al. (1989a).

Further knowledge about the details of the near field mixing is now being provided by advanced laboratory instrumentation, especially 3D laser-induced fluorescence (3DLIF). For experiments using this technique for ocean outfalls, see Tian et al. (2004a,b, 2006).

17.3.3 Far Field

17.3.3.1 Introduction

So the near field extends for distances of the order of 100 m from the diffuser, beyond which the plume drifts with the current and is diffused by ambient oceanic turbulence in a region called the far field. As we will see, the rate of dilution and mixing in this region is much slower than in the near field. Further complications arise from the unsteady nature of the currents, particularly in a tidal environment. The instantaneous tidal current speeds can be much larger than mean current speeds. These cause the plume to wander in the vicinity of the diffuser for some time before being flushed away by the mean current.

There are two main ways to predict contaminant fate and transport in the far field, an Eulerian or Lagrangian approach.

In the Eulerian approach, the pollutant concentration field is obtained by solving the 3D advective-diffusion equation for conservation of constituent mass for the substances of interest (bacteria, nutrients, dissolved oxygen, etc.) on a fixed grid.

$$\frac{\partial c}{\partial t} + u\frac{\partial c}{\partial x} + v\frac{\partial c}{\partial y} + w\frac{\partial c}{\partial z} = \frac{\partial}{\partial x}\left(\varepsilon_h \frac{\partial c}{\partial x}\right) + \frac{\partial}{\partial y}\left(\varepsilon_h \frac{\partial c}{\partial y}\right)$$
$$+ \frac{\partial}{\partial z}\left(\varepsilon_v \frac{\partial c}{\partial z}\right) - kc \tag{17.8}$$

where
  c is the concentration of the substance
  u, v, and w are the mean velocities in the x, y, and z directions
  ε_h and ε_v are horizontal and vertical turbulent (eddy) diffusion coeDcients

kc represents loss of bacteria due to mortality, settling, etc., which is usually assumed to follow a first-order process with k the decay constant. In order to solve Equation 17.8, the values of the current speeds, u, v, and w must be known. This usually requires a hydrodynamic model.

In the Lagrangian approach, fate and transport is predicted on a moving coordinate system either by a particle tracking model in which many particles representing contaminants are released, or by a puff model in which puffs of ed uent are released and tracked over time. The current field can either be supplied by hydrodynamic simulations or from current meters.

The Eulerian approach has several disadvantages for bacterial modeling. First is numerical diffusion due to the often large grid sizes. This can lead to overestimation of bacterial transport to sensitive locations where none actually occurs. It also leads to smooth spatial gradients of contaminants, as opposed to the patchy concentration distributions that are usually observed in nature, and cannot reproduce the sharp gradients that occur at the edge of plumes.

Lagrangian models (particle tracking models, random walk models, or puff models) resolve many of these deficiencies. Lagrangian models have a moving coordinate system. In a particle tracking model, the contaminants are represented as particles that are advected (transported) by the local current with a random velocity added to the deterministic velocity to simulate turbulent diffusion. The particles can be assigned properties, such as mass and age, which makes the method particularly well suited to bacterial predictions. Other properties can also be readily incorporated such as a settling velocity for heavy particles or an upward velocity for bacteria attached to, for example, buoyant particles such as grease and oil. Lagrangian models are particularly well suited to address whether transport to any particular location, such as water intakes or the shoreline, occurs. They are less well suited to addressing more complex water quality issues such as dissolved oxygen, BOD, and nutrients, however. For a thorough discussion of Lagrangian models, see Israelsson et al. (2006).

In the following, we first address turbulent diffusion in steady flows, then outline a Lagrangian model to assess the effects of temporally varying currents, and finally propose a simple box model to estimate long-term contaminant buildup.

17.3.3.2 Diffusion Due to Ambient Turbulence

Consider a line source approximating a multiport diffuser diffusing into an ocean current as shown in Figure 17.6. Assuming uniform steady flow and diffusion in the lateral direction only, Equation 17.8 becomes

$$u\frac{\partial c}{\partial x} = \frac{\partial}{\partial y}\left(\varepsilon \frac{\partial c}{\partial y}\right) - kc \tag{17.9}$$

where ε is the turbulent diffusion coeDcient for transverse mixing. Equation 17.9 is a 2D steady-state equation that neglects diffusion in the vertical and longitudinal directions.

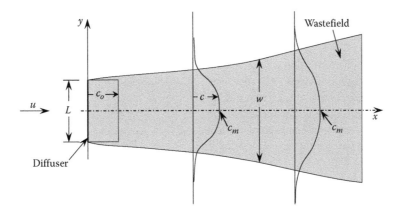

FIGURE1 7.6 Diffusion from a continuous line source of finite length.

A diDculty in solving Equation 17.9 is specification of the turbulent diffusion coeDcient, ε. In contrast to molecular diffusion in laminar flows, and even in some turbulent flows such as rivers, the diffusion coeDcient in large-scale geophysical flows like atmospheres or oceans is not constant. It depends on the size of the diffusing cloud or plume and increases as the cloud size increases. This is known as relative diffusion because the value of ε is "relative" to the cloud size. It is often assumed that the diffusion coeDcient follows the "4/3 power law" originally due to Richardson (1926) where the diffusion coeDcient is proportional to some measure of the plume width raised to the 4/3 power. For this case, Brooks (1960) obtained a solution to Equation 17.9 for the centerline (maximum) tracer concentration, c_m.

$$c_m(x) = c_o e^{-kx/u} \operatorname{erf} \sqrt{\frac{3/2}{\left(1 + \frac{2}{3}\beta\frac{x}{L}\right)^3 - 1}} \tag{17.10}$$

where
  c_o is the initial concentration of the contaminant of interest contained in the wastewater after completion of near field dilution
  $\beta = 12\varepsilon_o/uL$
  erf() is the standard error function
  ε_o is the initial value of the lateral diffusion coeDcient given by

$$\varepsilon_o = \alpha L^{4/3} \tag{17.11}$$

where α is a constant whose value depends on the rate of energy dissipation. Values of α for oceanic diffusion are given in Figure 3.5 in Fischer et al. (1979), and range (in cgs units) from 0.002 to 0.01 cm$^{2/3}$/s.

For a conservative substance (i.e., zero decay rate, $k = 0$), Equation 17.10 can be expressed in terms of a "far-field dilution" $S_f = c_o/c_m$

$$S_f = \left[\operatorname{erf}\left(\frac{3/2}{\left(1 + 8\alpha L^{-2/3}t\right)^3 - 1} \right)^{1/2} \right]^{-1} \tag{17.12}$$

TABLE 17.1 Far Field Dilutions for Diffusers of Various Lengths Assuming 4/3 Power Law

| | Far Field Dilution S_f | |
| | Diffuser Length, L (m) | |
Travel Time t (h)	$L = 35$	$L = 700$
1	2.4	1.0
3	7.4	1.4
10	35.5	3.2
20	95.9	6.9

The wastefield width is given by

$$w = L\left(1 + 8\frac{\varepsilon_o t}{L^2}\right)^{3/2} \tag{17.13}$$

Therefore, the far field dilution and width depend only on Lagrangian travel time $t = x/u$ of the ed uent to any location.

To show the magnitudes of far field dilution for ocean outfalls, some typical values (using Equation 17.12), are given in Table 17.1. Dilutions are given for a short and long diffuser, assuming an upper value for $\alpha = 0.01$ cm$^{2/3}$/s. It can be seen that dilution by oceanic turbulence can be quite effective for short diffusers, but is relatively minor for long diffusers. This is because the time needed for the centerline concentration to be reduced is the time required for eddies at the plume edges to "bite" into it and reach the centerline. For a wide field produced by a long diffuser, the eddies have farther to go so it takes them longer to get there. Consequently, because far field dilution can be quite slow, it is more important to know where the wastefield goes, rather than the exact magnitude of the far field dilution, and ways to estimate this are discussed in the following.

17.3.3.3 Statistical Description of Far Field

The variability of coastal currents and the slow diffusion of outfall plumes make analysis of their far field behavior particularly amenable to Lagrangian analyses with a moving coordinate system. Lagrangian techniques either follow puffs as they move and diffuse, or employ a particle tracking model with a random walk.

Puff models are computationally more eDcient, and we use one in the following to illustrate some of the essential features of dispersion in an unsteady environment.

The preceding analysis for turbulent diffusion was for steady-state conditions. Currents in coastal waters often fluctuate widely, however, with a fluctuating component, u', whose magnitude is much larger than the mean drift, U. The wastefield can then wander near the diffuser for some time before being flushed away. As shown earlier, the centerline concentrations may not be substantially reduced by turbulent diffusion for travel times of a few hours after release. Pollutant concentrations at any location will then be quite intermittent with relatively high values when the plume is present interspersed with lengthy periods of near-background levels.

Csanady (1983a) considers that a measure of environmental impact for this situation consists of computations of background concentration, the maximum concentration, and the frequency of immersion of any point in the plume. Csanady suggests that the modeling approach be based on a division of the plume into contaminant puffs of distinct "ages." "Young" puffs are those which have traveled for a few hours after release and are advected by local currents; "old" puffs have traveled for days or more and contribute to what may be called the background concentration. Csanady defines a dividing time t_d to distinguish between "young" and "old" puffs. In a tidal environment, it would be expected that t_d is of the order of the tidal period. The modeling of young puffs, which we identify with the far field, is discussed in this section, and modeling of old puffs by means of a box model in the following section.

The frequency of immersion of any point in the plume is termed the "visitation frequency" by Csanady. He presents methods to compute this quantity from the statistics of currents measured by a meter at a fixed location. A somewhat similar approach is given by Koh (1988), who refers to "advective transport probabilities."

A model based on this procedure, illustrated in Figure 17.7, was developed by Roberts (1999). This is a statistical model in which the wastefield is discretized as a series of puffs released at regular intervals. Each puff grows by turbulent diffusion and is followed up to a maximum time horizon. If a puff overlays a target point, this is counted as a "visit." The number of visits by a puff of age younger than the maximum time horizon is summed and divided by the total number of releases to compute the visitation frequency.

The location of the puff center at time t is

$$\vec{x}_c(t) = \int_0^t u_L(t')dt' \qquad (17.14)$$

where u_L is the Lagrangian velocity of the puff center. The fundamental problem in air and water pollution, however, is that the Lagrangian velocities of the puffs are not generally known. Instead, what are commonly available are Eulerian velocity measurements at a fixed point such as from a moored current meter. The usual assumption, often applied in air pollution (Pasquill and Smith, 1983), is to infer Lagrangian displacement from an Eulerian record by the approximation

$$\vec{x}_c(t) \approx \int_0^t u_E(t')dt' \qquad (17.15)$$

where u_E is the fixed point (Eulerian) record of velocity. The trajectory computed for various travel times t by Equation 17.14 is a path line; the trajectory computed by Equation 17.15 is known as a progressive vector diagram in oceanography. Clearly, the displacement predicted by Equation 17.15 becomes increasingly unreliable as the distance from the source increases. Furthermore, Zimmerman (1986) has pointed out that regularly varying tidal flows over irregular topography can produce "Lagrangian chaos," that is, unpredictable and non-repeatable trajectories. Therefore, even if we had perfect Lagrangian information for an individual release, we could not use this to predict future trajectories even under identical forcing functions. For these reasons individual plume trajectories should not be inferred by these methods, but it is usually assumed that statistical inferences of the pattern and scale of the dispersion can be made. Support for this assumption is provided by List et al. (1990) who found good general agreement between diffusivities

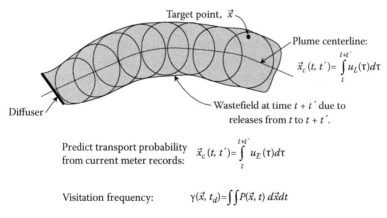

FIGURE1 7.7 Statistical far field transport model.

in coastal waters computed from drogues and from fixed current meters. Because of the complexity of the dispersion process and the relatively poor spatial resolution of coastal measurements, it is clearly not possible to pretend for great accuracy in any predictive method. But the statistical approach is very useful in assessing the probability of exceedance of some threshold concentration at particular locations such as near the coastline.

Outfall studies usually produce current meter records that consist of discrete measurements at a fixed point with a fixed sampling interval Δt, typically 10–30 min. To use these data, we discretize the plume as a series of puffs released at a rate equal to the sampling frequency $(\Delta t)^{-1}$. The location of a puff released at time $t_0 = n\Delta t$ after travel time $T = m\Delta t$ is then assumed to be given by the discrete form of Equation 17.15

$$\vec{x}\langle t_o | T \rangle = \sum_{i=n}^{i=n+m} \vec{u}_i \Delta t \qquad (17.16)$$

where $\vec{u}_i(t)$ is the local measured velocity at time $t = i\Delta t$. This computation is repeated for all releases during the whole data record, and each puff is followed up to the maximum travel time or "time horizon," t_d. This procedure would typically involve thousands of releases, each of which is tracked at each time step as they travel. The area around the diffuser is overlain with a grid, and if a puff is within $\pm \vec{w}/2$ of a grid node, this is counted as a "visit." The number of visits by a puff of age younger than t_d is summed and divided by the total number of releases to obtain the visitation frequency at that location.

As the plume travels it is diffused and grows due to oceanic turbulence. The decay of peak concentration is assumed to be given by Equation 17.12 and the puff size by Equation 17.13. These equations apply to a continuous line source whose concentration is reduced by lateral diffusion only. For an isolated puff growing in three-dimensions the diffusion rate will be greater and the plume dilution will increase away from the puff center. We neglect these effects and use the conservative assumptions

that the dilution in the puff is given by the continuous solution and is constant across the puff. This approach is computationally much more eDcient than a particle tracking model due to its simplicity and the diDculties of keeping track of the huge number of particles that would be necessary for an extended simulation over many months.

The actual dilution S at any location when the plume is present is the product of the near field dilution S_n and the far-field dilution S_f:

$$S = S_n \times S_f \qquad (17.17)$$

where S_n is the dilution at the end of the near field (computed, for example, by Equation 17.6a). The corresponding contaminant concentration is then given by

$$c = \frac{c_{oo}}{S} \qquad (17.18)$$

where c_{oo} is the contaminant concentration in the ed uent leaving the treatment plant. The concentration estimated by Equation 17.18 is the maximum expected at any location. This will occur very infrequently, however, and time-average concentrations will be much lower.

This method of analysis is particularly well suited to situations where extensive current measurements are available. Such data are increasingly being gathered for ocean outfall studies, especially since the development of low-cost acoustic Doppler current profilers (ADCPs) in recent years.

An example of visitation frequencies computed using this approach is shown for Mamala Bay, Hawaii, in Figure 17.8 for travel times up to 12 h. The plots give a good visualization of wastefield impact. The contours elongate in the East/West directions in accordance with the direction of the main current components. These currents are strongly tidal, with the result that the wastefield is swept back and forth in the vicinity of the diffuser. This causes the visitation frequency to diminish rapidly with

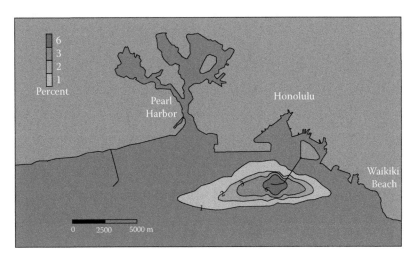

FIGURE17.8 Visitation frequencies for the Sand Island, Hawaii, outfall. (After Roberts, P.J.W., *J. Hydraul. Eng.*, 125(6), 574, 1999.)

distance from the diffuser with very low probability of shore-ward impaction. The wastefield is spread over an area whose dimensions are equal to the maximum tidal excursion within a time scale of about 6 h. The visitation frequency decreases rapidly with distance from the diffuser to about 1% at ~5 km. Put another way, the plume is *not* present for about 99% of the time at this distance. Note that a visitation frequency of 1% is about 7 h per month. Expansion of the contours for travel times longer than about 6 h is primarily caused by the mean drift.

Of more importance than visitation frequency, at least in the regulatory sense, are bacterial predictions. These regulations are usually written in statistical terms, such as the frequency with which a particular bacterial level may be exceeded or geometric mean values. The statistical model outlined earlier is well suited to such predictions.

An example is shown in Figure 17.9, for an outfall proposed for Cartagena, Colombia. In this example, predictions of the fecal indicator organisms total and fecal coliforms are shown. The advective transport is computed from a moored ADCP near the proposed diffuser. Far field diffusion is computed from Equations 17.12 and 17.13, and the bacteria are allowed to decay at a time-variable rate. The near field dilution is computed from a near field model that depends on current speed so is also temporally variable. The near field model is coupled to the far field model by Equations 17.17 and 17.18; in other words, the source concentrations for the far field model are the outputs from the near field model. Figure 17.9 represents thousands of releases

over a 1 year period. It shows the decay of bacteria from the diffuser as a result the combined effects of near and far field dilution and decay. As for the visitation frequencies in Figure 17.8, the contours expand rapidly in the principal current direction, which is essentially parallel to the shoreline, but only slowly in the onshore direction.

A number of very important conclusions can be derived from Figures 17.8 and 17.9. The impact of an ocean outfall decreases very rapidly with distance from the diffuser. The rapid decrease of visitation frequency and consequent increase in average dilution and decreases in contaminant concentrations with distance is typical of coastal discharges and should result in confinement of any impacts to an area extending for at most a few kilometers around the diffuser. Furthermore, many of the assumptions made are conservative, for example, shear dispersion and vertical mixing are neglected.

The level of treatment proposed for the Cartagena wastewater is preliminary, i.e., milliscreening only, and that was assumed for the simulations shown. A very important conclusion from Figure 17.9 therefore is that shoreline bacterial standards can be met with a comfortable safety factor with only preliminary treatment. The same conclusion applies to the water quality regulations for toxic concentrations at the edge of the mixing zone, which will be met through near field dilution alone.

These conclusions are expected to be generally true for coastal discharges. Coastal waters constitute a very dispersive environment. The models used here simultaneously keep track

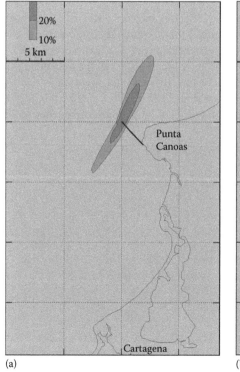
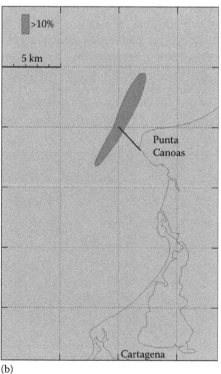

(a)   (b)

FIGURE 17.9 Predicted coliform exceedance frequencies (area where California water contact standard may be exceeded is shown by inner contour). (a) Total coliforms = 1000 per 100 mL. (Fecal coliforms = 200 per 100 mL). (b) Fecal coliforms = 400 per 100 mL.

of currents, stratification, ed uent flow, and bacterial mortality in real time. The results indicate that worst-case conditions are extremely improbable, and their use would lead to overly conservative outfall designs and treatment levels. The methods outlined previously should help to make rational choices of treatment levels, including consideration of the relative public health risks involved and the benefits and costs expected.

17.3.3.4 Long-Term Flushing

Finally, we consider the long-term buildup of contaminants in the vicinity of the discharge, or coastal "flushing" which occurs on long time scales (Figure 17.2). In the previous section, we divided the plume into "young" and "old" puffs. Young puffs, whose travel times are of order a day or less, contribute most of the local bacterial impacts and can be analyzed by means of the statistical model presented earlier. "Old" puffs are subject to considerable decay and diffusion and generate a concentration that can be considered to be a "background" mean concentration field in the vicinity of the diffuser. The level of this concentration is governed primarily by flushing due to the mean drift, horizontal diffusion, and, for non-conservative substances, chemical and biological decay. One approach to predicting the physical dilution caused by these processes is to estimate it from a solution to the 2D diffusion equation (Csanady 1983a; Koh 1988). We here consider a simpler method, however, which is particularly useful for comparing the relative orders of magnitude of the various processes. This is a mass-balance box model (Csanady, 1983b) as shown in Figure 17.10.

Tidal currents distribute the ed uent over an area, or "box" whose dimensions are approximately equal to the tidal amplitude. These dimensions are approximately $X = u_t T/2$ and $Y = v_t T/2$, in the alongshore and cross-shore directions, respectively, where u_t and v_t are the amplitudes of the tidal currents, and T is the tidal period. Csanady (1983b) calls this area the "extended source region." It would be comparable to the outer edge of the visitation frequency contours shown in, for example, Figure 17.8.

Long-term average current speeds are usually much slower than instantaneous values. They lead to an average dilution equal to UhY/Q, where Q is the total ed uent flow rate, h the average depth of the plume over the extended area, and U the long-term average "flushing velocity."

This can be extended to include the other processes by applying a mass balance to the box. This yields a "long-term average dilution" S_p.

$$S_p = \frac{UhY}{Q} + \frac{v_e hX}{Q} + \frac{khXY}{Q} \qquad (17.19)$$

The first term on the right-hand side is the dilution due to flushing by the mean current. The second is dilution due to cross-shore mixing. This is parameterized by v_e, a mass transfer "diffusion velocity," which can be assumed equal to the standard deviation of the cross-shore tidal fluctuations (probably an underestimate). The third term is "dilution" due to chemical or biological decay, where k is a first-order decay rate. The total effective dilution is the sum of these individual dilutions.

Consider a typical outfall problem. Suppose we have the discharge $Q = 5\,\mathrm{m^3/s}$ considered previously into a tidal current whose alongshore amplitude is $u_t = 0.25\,\mathrm{m/s}$, cross-shore amplitude is $v_t = 0.08\,\mathrm{m/s}$, and cross-shore rms velocity is $v_e = 0.04\,\mathrm{m/s}$. Suppose the average current speed (the flushing velocity) is $U = 0.06\,\mathrm{m/s}$. For a semi-diurnal tide, the period T is about 12 h. Suppose further that the average depth (thickness) of the wastefield is 15 m, and the average bacterial decay rate over 24 h corresponds to $k \approx 6 \times 10^{-5}\,\mathrm{s^{-1}}$.

Then the extended source area (size of the box in Figure 17.10) is as follows:

$$X = \frac{u_T T}{2} = \frac{0.25 \times 12 \times 3600}{2} \approx 5400\,\mathrm{m} \approx 5.4\,\mathrm{km}$$

and

$$Y = \frac{v_T T}{2} = \frac{0.08 \times 12 \times 3600}{2} \approx 1700\,\mathrm{m} \approx 1.7\,\mathrm{km}$$

and the dilutions are (Equation 17.19):

Due to the mean current: $\quad \dfrac{UhY}{Q_T} = \dfrac{0.06 \times 15 \times 1700}{5} \approx 300$

Due to cross-shore exchange: $\quad \dfrac{v_e hX}{Q_T} = \dfrac{0.04 \times 15 \times 5400}{5} \approx 650$

Due to decay: $\dfrac{khXY}{Q_T} = \dfrac{6 \times 10^{-5} \times 15 \times 5400 \times 1700}{5} \approx 1650$

The total effective dilution, the sum of these dilutions, is about 2600.

These are obviously only approximate order of magnitude calculations, but they are very useful for estimating long-term impacts. They can be applied to other substances such as toxic materials to estimate their potential accumulation. It is clear that accumulation will not generally be a problem for open coastal sites.

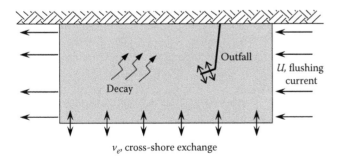

FIGURE 17.10 Box model for estimating long-term buildup of contaminants. (After Csanady, G. T., Advection, diffusion, and particle settling, *Ocean Disposal of Municipal Wastewater: Impacts on the Coastal Environment*, E. P. Myers and E. T. Harding, eds., MIT Sea Grant College Program, Cambridge, MA, pp. 177–246, 1983b.)

17.4 Major Challenges

The fate and transport of ocean outfall discharges is discussed here in terms of processes whose length and time sales increase with time after their discharge (Figure 17.2). It could also be said that the predictive uncertainty increases with travel time. Near field mixing is fairly well understood, and modern laboratory instrumentation is further elucidating it, and mathematical models of near field plume behavior are well established. But predicting far field mixing for travel times up to a day is less reliable. In particular, the dependence of the diffusion coeDcient on travel time or plume size (for example, Equation 17.11) does not have much verification for coastal outfall plumes, especially submerged plumes. The effects of current shear over depth (speed as well as direction) and the quantification of the resulting shear dispersion are not well known. At times longer than a few days, field observations (for example, Hunt et al., 2010) often show the plume breaking up into patches, for which there are presently no reliable models.

There are many other predictive uncertainties that are not related to hydrodynamics. For example, the decay rates of bacteria, and chemical and biological reactions involving dissolved oxygen and nutrients, are not well known. Again, this is especially true for submerged plumes which could have low rates of bacterial decay due to reduction of solar radiation with depth.

More field experiments are needed to address these issues and can shed much light on mixing at all scales. For the near field, improved laboratory instrumentation, such as 3DLIF, enables high spatial resolution measurements of the plume in complex situations. This is not possible in the field, but more measurements should be undertaken with fluorescent tracers added to the wastewater (for example, Hunt et al., 2010) to further evaluate near field mixing processes and laboratory and mathematical models. This is diDcult, however, especially in stratified situations with a submerged plume that cannot be seen. Further diDculties arise from the large spatial extents covering many square kilometers that must be mapped, especially considering the low speeds at which survey vessels must travel. And due to temporal variability in ocean conditions, the plume can change substantially in the time required to map it making it diDcult to get instantaneous synoptic plume maps. Field measurements are also diDcult because many variables must be monitored simultaneously in addition to plume dilution, such as wastewater flow rate, oceanic stratification, current speed and direction over depth, winds, etc. Field campaigns must therefore be carefully planned to ensure that the data obtained are useful and relevant.

Just as in the laboratory, new oceanographic field instrumentation has considerable promise for spatial mapping of submerged plumes. For example, the use of gliders combined with submersible fluorometers, and even smaller sensors deployed in swarms of miniature vehicles. All of this instrumentation is well suited to observations of far field transport also.

Surfacing plumes in unstratified or weakly stratified oceans have different challenges, especially wind effects on the plume and the transport of floatable contaminants such as grease balls. It is not well understood how wind can transport them, especially to shore, or the variation with depth of wind effects. Surface current radar that can map out surface currents in real time and at high resolution over large areas could be very useful for the developments of models that incorporate wind effects.

Because we have focused in this chapter on discussion of the main dispersion mechanisms, we have not addressed the use of hydrodynamic models to predict coastal circulation and far field transport. They are being increasingly used, and this trend will presumably continue. They have considerable uncertainties, however, due to the complexity of many coastal environments. These arise from density stratification, 3D effects, and unsteady forcing, especially wind effects and tides, and complex bathymetry. And a major problem with coastal environments is specifying the model boundary conditions over depth and time at the open ocean boundaries. New coastal observing systems are now coming into use, especially in the United States, that provide extensive measurements of currents, winds, surface currents, etc., often in real time. A challenge here is to combine the data with models to predict plume behavior in real time. This could make it possible to optimize treatment plant performance, for example, chlorination, leading to improved environmental protection at lower cost.

References

Brooks, N. H. (1960). Diffusion of sewage ed uent in an ocean current. *Proceedings of the First International Conference on Waste Disposal in the Marine Environment*, Pergamon Press, New York, pp. 246–267.

Brooks, N. H. (1984). Dispersal of wastewater in the ocean—A cascade of processes at increasing scales. *Proceedings of the Conference on Water for Resource Development*, Coeur dfl Alene, ID, August 14–17.

Csanady, G. T. (1983a). Dispersal by randomly varying currents. *J. Fluid Mech.*, 132, 375–394.

Csanady, G. T. (1983b). Advection, diffusion, and particle settling. *Ocean Disposal of Municipal Wastewater: Impacts on the Coastal Environment*, E. P. Myers, and E. T. Harding, eds., MIT Sea Grant College Program, Cambridge, MA, pp. 177–246.

Daviero, G. J. and Roberts, P. J. W. (2006). Marine wastewater discharges from multiport diffusers III: Stratified stationary water. *J. Hydraul. Eng.*, 132(4), 404–410.

Fischer, H. B., List, E. J., Koh, R. C. Y., Imberger, J., and Brooks, N. H. (1979). *Mixing in Inland and Coastal Waters*, Academic Press, New York.

Hunt, C. D., Mansfield, A. D., Mickelson, M. J., Albro, C. S., Geyer, W. R., and Roberts, P. J. W. (2010). Plume tracking and dilution of ed uent from the Boston sewage outfall. *Mar. Environ. Res.*, 70(2), 150–161.

Isaacson, M. S., Koh, R. C. Y., and Brooks, N. H. (1983). Plume dilution for diffusers with multiple risers. *J. Hydraul. Eng.*, 109(2), 199–220.

Israelsson, P. H., Kim, Y. D., and Adams, E. E. (2006). A comparison of three Lagrangian approaches for extending near field mixing calculations. *Environ. Model. Sos w.*, 21(12), 1631–1649.

Koh, R. C. Y. (1988). Shoreline impact from ocean waste discharges. *J. Hydraul. Eng.*, 114(4), 361–376.

Koh, R. C. Y. and Brooks, N. H. (1975). Fluid mechanics of waste water disposal in the ocean. *Annu. Rev. Fluid Mech.*, 7, 187–211.

List, E. J., Gartrell, G., and Winant, C. D. (1990). Diffusion and dispersion in coastal waters. *J. Hydraul. Eng.*, 116(10), 1158–1179.

NRC (1993). *Managing Wastewater in Coastal Urban Areas*, National Research Council, Committee on Wastewater Management for Coastal Urban Areas, National Academy Press, Washington, DC.

Pasquill, F. and Smith, F. B. (1983). *Atmospheric DiThsion*, E. Horwood, New York.

Richardson, L. F. (1926). Atmospheric diffusion shown on a distance-neighbor graph. *Proc. R. Soc. Lond.*, A110, 709–739.

Roberts, P. J. W. (1999). Modeling the Mamala Bay plumes. II: Far field. *J. Hydraul. Eng.*, 125(6), 574–583.

Roberts, P. J. W., Salas, H. J., Reiff, F. M., Libhaber, M., Labbe, A., and Thomson, J. C. (2010). *Marine Wastewater Outfalls and Treatment Systems*, International Water Association, London, p. 493.

Roberts, P. J. W., Snyder, W. H., and Baumgartner, D. J. (1989a). Ocean Outfalls. I: Submerged Wastefield Formation. *J.dHydraul. Eng.*, 115(1), 1–25.

Roberts, P. J. W., Snyder, W. H., and Baumgartner, D. J. (1989b). Ocean outfalls. II: Spatial evolution of submerged wastefield. *J. Hydraul. Eng.*, 115(1), 26–48.

SWRCB. (2005). *Water Quality Control Plan, Ocean Waters of California*, State Water Resources Control Board, Sacramento, CA.

Tian, X., Roberts, P. J. W., and Daviero, G. J. (2004a). Marine wastewater discharges from multiport diffusers I: Unstratified stationary water. *J. Hydraul. Eng.*, 130(12), 1137–1146.

Tian, X., Roberts, P. J. W., and Daviero, G. J. (2004b). Marine wastewater discharges from multiport diffusers II: Unstratified flowing water. *J. Hydraul. Eng.*, 130(12), 1147–1155.

Tian, X., Roberts, P. J. W., and Daviero, G. J. (2006). Marine wastewater discharges from multiport diffusers IV: Stratified flowing water. *J. Hydraul. Eng.*, 132(4), 411–419.

Thorpe, S. A. (1982). On the layers produced by rapidly oscillating a vertical grid in a uniformly stratified fluid. *J. Fluid Mech.*, 124, 391–409.

Zimmerman, J. T. F. (1986). The tidal whirlpool: A review of horizontal dispersion by tidal and residual currents. *Neth. J. Sea Res.*, 20, 133–156.

18

Modeling Oil Spills to Mitigate Coastal Pollution

Poojitha D. Yapa
Clarkson University

18.1 Introduction

Oil spills in water continue to pose a risk to the coastal environment. Such spills impact fishery, wild life, coastal vegetation, water supplies, recreational facilities, and floating equipment. Oil transport and oil storage near water bodies are major sources for surface and near surface spills. Oil drilling and underwater oil pipelines are the source for oil spills that originate well below the water surface. Modern technology makes it commercially viable to drill and produce from 1500 m deep or even deeper in water. Well head blowouts, pipeline bursts, and damages to facilities due to extreme weather are common sources of leaks. In underwater discharges it is common to have gases mixed with oil in their release. The presence of gases in the oil mix changes the transport and fate of oil. The presence of gas either alone or in the oil as a mix further complicates predicting the environmental risks because of a number of physico-chemical processes it undergoes during its travel from the deep water to the surface. The focus of this chapter is only on oil releases from the water surface and near surface.

Considering the amount of oil used worldwide, the risk of a spill is low. However, when an accident happens the consequences and damage to the environment can be high and long lasting (e.g., Amoco Cadiz, Exxon Valdez, Horizon). Smaller spills are more frequent but they get less attention because they do not always make dramatic news. But a series of smaller spills can also do lasting damages to a coastal area.

The main strategies for minimizing the environmental damage from an accidental spill are as follows. (1) Prepare emergency response plans that include contingency planning and conducting of regular training exercises; (2) decision making related to emergency response and cleanup; (3) decision making related to toxic effects, fire hazards, and buoyancy altering effects on floating vessels and equipment, so that people can be evacuated and equipment can be moved.

All these strategies require the use of comprehensive spill models. Furthermore, models are used in post spill damage assessments as the data collected alone is not suDcient to estimate the damage. Another use of these models is for guiding the data collection crews so that the efforts can be optimized and better quality data can be collected.

Past 30 years have seen a significant improvement in our understanding of the physical, chemical, biological processes that oil undergo when spilled in a water environment. A comprehensive review paper (ACSE 1996) reports that over 50 models are available. The actual number of models could easily exceed 100.

Handbook of Environmental Fluid Dynamics, Volume Two, edited by Harindra Joseph Shermal Fernando. © 2013 CRC Press/Taylor & Francis Group, LLC.
ISBN: 978-1-4665-5601-0.

Most of these are completely out of date. Only a few of these are extensively in use today. They vary from two-dimensional (2D) trajectory types to 3D models that include the oil spill processes governing the fate and effects of spilled oil. The usefulness of each model type depends on the objectives for which it is going to be used: from short-term control and contain of the spill to long-term impact assessment. Several papers appeared in the recent decade explaining new models developed. However, these models are an assembly of ad hoc theories or empirical methods developed nearly 20 years ago for describing the physico-chemical processes in oil spills. The paper by Guo and Wang (2009) is a good example of the state of oil spill models with related to how the physico-chemical aspects are modeled.

This chapter discusses various physico-chemical processes that oil undergoes during a spill and how the models are formulated but does not list or compare all the available models. The focus here is on the physico-chemical processes that take place after an oil spill and the relative importance of different processes for a given objective. The modeling techniques and what to expect from a model are also discussed. Hence it is hoped that with the knowledge gained from this chapter, the reader will be able to assess the suitability of a given model for their objectives or embark on developing their own model.

18.2 Oil Spill Processes

Oil spilled on or near the water surface is subjected to many physico-chemical processes which are referred to as oil spill processes. However, even the oil spilled in deepwater once they come near the water surface is subjected to many of the same processes. Some of these processes are unique to oil (Galt 1994).

Figure 18.1 shows pictorially some of the processes that oil undergoes after a spill. Many of these are inter-dependent as described by Xie et al. (2007). For example, the wind affects evaporation through two ways: evaporation rate and oil slick area. Evaporation and dissolution are competitive processes. Spreading due to diffusion and transport affects the oil slick area and hence affects the evaporation and dissolution. Oil consists of a mixture of hydrocarbons. Many of the hydrocarbon components that are highly volatile are also the ones easily dissolved. Emulsification is highly dependent on the hydrocarbon composition. This composition is affected by evaporation, photo-oxidation, and dissolution. Oil that does not emulsify at the time of spill may start forming stable emulsions after the oil is weathered for some time. Breaking up of oil into droplets and getting submerged in the water column and resurfacing change the rates at which the aforementioned processes take place. Oil density changes due to all these processes. The change in oil density may affect how the oil is transported. The processes discussed in this chapter are advection and diffusion, mechanical spreading, vertical mixing, Langmuir circulation, evaporation, dissolution, emulsification, photo-oxidation, oil-sediment interaction, sedimentation, and shoreline deposition, and re-entrainment of oil.

18.2.1 Mechanical Spreading

When oil is spilled on the water surface under relatively calm conditions, it initially spread rapidly due to different forces acting on

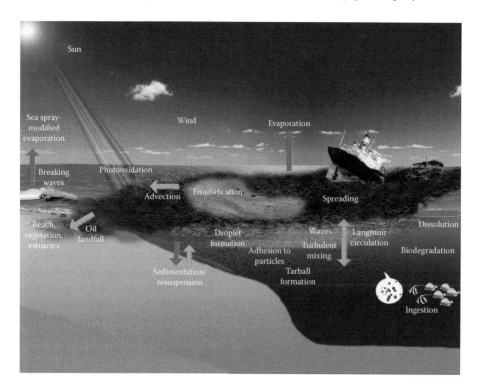

FIGURE 18.1 Physico-chemical processes after an oil spill. (From Poojitha D. Yapa, Clarkson University, Potsdam, NY; Peter Y. Sheng, University of Florida, Gainesville, FL.)

the oil slick. This spreading is much more rapid than the one due to turbulent diffusion under calm conditions. To differentiate it from turbulent spreading this is referred to as "mechanical spreading." The forces acting on a slick during this process are inertia, gravity, viscous, and interfacial tension. While some of these forces act to spread the oil slick, other forces retard such spreading. Initially gravity acts to spread while inertia tries to reduce spreading. As the slick gets somewhat thinner, viscous forces dominate over inertia while gravity continues to act as spreading forces. Further thinning causes the interfacial tension forces to become more dominant and become the main spreading force while viscosity continues to retard growth. At some point, the slick growth reaches an equilibrium condition. However, that thickness is so small, that in field conditions, it may never be achieved.

Fay (1971) and Hoult (1972) shed light on the physics of this spreading process. However, it is now evident that in open water turbulent sea conditions this process is not important compared to spreading by turbulence. Hence, more recent works have paid less attention to it. Many models do not include it. A detailed discussion of the mechanical spreading including the governing equations can be found in Yapa (1994). However, when oil is spilled under ice covers this process is extremely important and the research on it is much more current (e.g., Yapa and Chowdhury 1990; Yapa and Weerasuriya 1997; Izumiyama et al. 1998).

18.2.2 Vertical Mixing and Resurfacing

Turbulence caused by water current, wind, and waves can break up the oil slick into smaller droplets and disperse them into the water column. Dispersed droplets move toward the surface at varying speeds depending on the size of the droplets. This process is important because oil on the water surface travels differently from oil in water column. Hence, the transport and fate of oil depends on how much oil is on the surface and water column, respectively. Despite the importance, studies on this process are limited. In the 1970s, there were some fundamental studies on the physics of the oil break-up and vertical mixing, which are too detailed to be listed here. However, these studies did not aid in modeling field situations and were not used by subsequent researchers. The most common method in use is the empirical method developed by Delvigne and Sweeney (1988) and Delvigne (1994). The amount of oil dispersed into the water column can be calculated by

$$Q(d) = C_o D^{0.57} d^{0.7} \Delta d\, S_{cov} F_{wc} \tag{18.1}$$

where

$Q(d)$ is the entrained mass rate of oil droplets with droplet sizes in the interval Δd around diameter d, per unit surface area and per unit time (kg/m²s)

D is the energy dissipation of the breaking wave per unit surface area (J/m²)

C_o is a constant

F_{wc} is the fraction of sea surface area hit by breaking waves per unit time (s⁻¹)

S_{cov} is the fraction of sea surface covered by oil

Details can be found in Delvigne (1994). Once the break-up mass is determined by Equation 18.1, the oil droplet sizes and their distribution are determined by empirical methods established through limited experiments by Delvigne (1994). Recently Bandara and Yapa (2011) developed a non-empirical method to calculate oil droplet sizes and their distribution in deepwater plumes. In their computations, they not only calculated initial oil droplet sizes, but allowed oil droplet breakup and coalescence as they move about in the water column. Their results compared very well with experimental data. Their method is very promising and can be adopted to calculate oil droplet size distribution in surface spills as well.

18.2.3 Langmuir Circulation

Langmuir circulation (LC) results from the interaction between wind-driven surface currents and waves. LC results in the formation of helical cells that are aligned mainly in the direction of the wind. Wind blowing across the water surface generates a surface shear force which varies due to minor variations in wind. These create vertical cells with adjacent cells rotating in opposite directions. Stokes drift shift the cells generating the helical cells. The net effect of all this results in oil on the water surface being formed into long streaks, called windrows, aligned mainly in the direction of the wind instead of a spread out coherent slick as many would imagine. Figure 18.2 shows an oil slick that has undergone Langmuir effect. LC would increase the dispersion of oil into the water column altering the oil transport significantly. The presence of LC may attribute to the diDculties the modelers have had in accurately predicting the movement of oil. Very few oil transport models have included LC. The proceedings of a special workshop on LCs available as a special issue (Simecek-Beatty et al. 2000) would provide the reader more detailed information.

18.2.4 Evaporation and Dissolution

In many spills, evaporation causes the major loss of oil mass. Evaporation has more significance beyond mass loss. It can affect

FIGURE 18.2 Oil slick that has undergone Langmuir effect. (Photo courtesy of NOAA/Hazmat.)

emulsification, oil sedimentation, and photo-oxidation. Oil dissolution accounts only for a small fraction of the oil mass lost. However, it is important unless the simulations are for very short term, because of its impact on oil toxicity, and adsorption of oil dissolved to sediments. Evaporation and dissolution are discussed together because the chemistry of the processes is similar and can be modeled using similar equations. Furthermore, they are competing mechanisms that affect each other. Most studies have focused on the dissolution of oil from the surface slick. Oil dispersed into the water column rises to the sea surface in a short time unless the dispersed droplets are very fine. Therefore, accounting for evaporation from the surface slick accounts for most of the oil lost due to evaporation.

Oil is considered to be a mix of basic hydrocarbons of differing characteristics (molecular weight, vapor pressure, etc.). Evaporation amount can be calculated by (Mackay et al. 1980a)

$$\Delta M_{Ei} = \frac{A \Delta t K_i^e X_i P_i^s}{RT} \qquad (18.2)$$

where

ΔM_{Ei} is the amount of component i lost by evaporation (mol)
A is the area of oil slick (m²)
Δt is the time increment (s)
K^e is the mass transfer coefficient of evaporation (m/s)
R is the gas constant (atm-m³/mol·K)
P_i^s is the pure vapor pressure of component i (atm-m³)
$X_i P_i^s$ is the partial vapor pressure (VP) of component i (atm-m³)
T is temperature (K)

Yang and Wang's (1977) method is known to give reasonable values for VP of components (Yapa and Zheng 1998). However, it has three limitations: (1) oil must be classified into the eight specific subcomponents; (2) VP is only related to the temperature; and (3) the parameters are empirical. In reality, a method that can take a variety of classifications is very useful. An improved method for calculating VP P_i^s is recommended by API (Daubert and Danner 1989). In this method, equations are set up to calculate the VPs based on the specific gravities, molecular weights of the components, and oil temperature. Therefore, the model can take many kinds of classifications for oil subcomponents.

Many previous models used Mackay and Matsugu's (1973) equations to estimate the evaporation mass transfer coefficient (MTC). The calculated value is a function of the wind speed and other parameters but does not depend on the hydrocarbon type or oil temperature. Riazi and Al-Enezi (1999) derived an equation based on their data in which the MTCs are related to wind speed, oil temperature, and molecular weight of oil component given as

$$K_i^e = 1.5 \times 10^{-5} W_s^{0.8} \left(\frac{T}{M_i} \right)^2 \qquad (18.3)$$

in which the unit of evaporation MTC is m/s.

Yapa and Xie (2006) compared the results from the two methods and found that using Mackay and Matsugu's (1973) method always gives at least 1 order higher values than the other method. They concluded, however, that there is insufficient data to decide which one is better.

An equation similar to that used for evaporation can also be used for dissolution.

$$\Delta M_{Di} = A \Delta t K_i^d \left(e_i X_i C_i^s - C_i^w \right) \qquad (18.4)$$

where

subscript i refers to the component
ΔM_{Di} is the amount lost by dissolution
K_i^d is the dissolution MTC (m/s)
e_i is the solubility enhancement factor
X_i is the oil phase mole fraction
C_i^s is the pure solubility (mol/m³)
C_i^w is the bulk water phase concentration (mol/m³)

Typically it is assumed that the water phase concentration is zero because it is far less than the solubility.

Leinonen and Mackay (1973) obtained a value of $K^d = 2.36 \times 10^{-4}$ cm/s for MTC and was used by others (e.g., Yapa and Zheng 1998). Riazi and Edalat (1996) proposed a way to calculate K^d as

$$K_i^d = \frac{4.18 \times 10^{-9} T^{0.67}}{V_{Ai}^{0.4} A_i^{0.1}} \qquad (18.5)$$

where

V_{Ai} is the molar volume of oil at its normal boiling point (m³/mol), calculated from Rackett equation given by Reid et al. (1987)
A_i is the surface area of component i (m³)

Yapa and Xie (2006) calculated K_i^ds and found the value to be in the same order as the value by Leinonen and Mackay (1973) for the oil as a whole.

Most past research on dissolution focused on the surface oil slick. Studies on oil dissolution in the water column have been very limited. The behavior of oil dissolution in water column is a kinetic process. Shiu et al. (1990) proposed an equation to fit their experimental data as follows:

$$C_i^s = \frac{C_{0i}^s}{(1 + Q/K_i)} \qquad (18.6)$$

where

K is the oil–water partition coefficient of component i, which depends on the oil matrix
Q is the water-to-oil volume ratio
C_{0i}^s is the solubility of component i in the water phase at $Q = 0$

18.2.5 Emulsification

Forming of stable emulsions can take days to weeks. Once the emulsions are formed they significantly impact the fate and transport of oil, hence should be included in longer term models. This is not an important process to include in short-term tactical model.

Although in other research areas emulsions have been studied for a long time, it is important to realize that emulsion-related works in areas such as food industry and beauty products do not necessarily help to model the emulsions related to oil spills. Oil emulsified under turbulent ocean conditions can take various forms. Figure 18.3a and b shows two examples. Figure 18.3a shows oil emulsion balls formed in turbulent water conditions created in the laboratory. Similar oil balls can also be formed in the open ocean conditions as well. Figure 18.3b shows oil emulsions formed in the Gulf of Mexico during the IXTOC spill in 1979. In this case, oil was subjected to photo-oxidation as well.

Emulsion formation is dependent on ocean turbulence and the oil chemistry. Many studies over the last 20 years shed sufficient insight to the chemistry of oil emulsions. Many of these studies originated at Environment Canada, under the leadership of Dr. Merv Fingas. The details of resulting work can be found in the proceedings of AMOP conferences held every year. Much of the work was summarized in the review paper. We now know that certain oils form stable emulsions while others do not, no matter what the turbulence level is. Emulsified oil can contain a large amount of water as much as 80% (Fingas 1994). Emulsions have a higher density than the parent oil and sometimes can be slightly higher than that of ambient water. The viscosity of emulsion can be a few orders of magnitude higher than the parent oil, making it even semi-solid. Evaporation and photo-oxidation (weathering) impact emulsification. Oil that originally does not emulsify may emulsify after a few days of weathering. Emulsification complicates oil spill cleanup. Interrelation between emulsification and other processes makes modeling very complex.

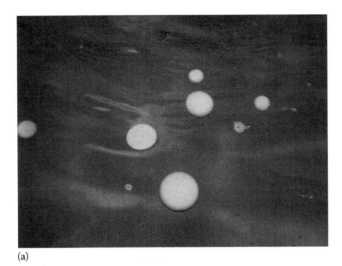

(a)

(b)

FIGURE 18.3 Different forms of emulsified oil: (a) formed in turbulent conditions created in the laboratory; (b) formed in the open ocean with oil subject to the photo-oxidation. (Photos courtesy of NOAA/Hazmat).

Xie et al. (2007) presented a model to simulate emulsification by combining the chemistry of emulsification with ocean turbulence and is the only known model of this type. This model can be refined in the future, if more data become available. Xie et al. (2007) computed the turbulence energy using an expression given by Zhang and Cui (1999)

$$E = \sqrt{\frac{\rho_w \mu_T \varepsilon}{C_\mu}} \quad (18.7)$$

where

E is the turbulence energy in a unit volume of water (J/m^3)
ρ_w is the water density (kg/m^3)
ε is the dissipation rate of turbulence energy in a unit weight of water (m^2/s^3)
μ_T is the eddy viscosity (kg/m · s)
C_μ is a constant = 0.09 for water

Equation 18.7 requires a series of equations to determine the parameters in it. Xie et al. (2007) found them as follows: μ_T from Neumann and Pierson (1996), ε from Terray et al. (1996). In doing so they considered the wind conditions, energy under breaking waves, and the variation of turbulence with depth in the water column.

Water uptake was determined using Mackay et al.'s (1980b) equations given as

$$\ln\left[(1 - K_2 W)\exp\left(\frac{-2.5\,W}{(1 - K_1 W)}\right)\right] = -K_3 t \quad (18.8)$$

where

W is the water content fraction of the emulsion
t is time
K_1, K_2, K_3 are rate constants

Xie et al. (2007) established an emulsion stability index, S_A, based on the work of Mackay and Zagorski (1982) which included the effects of asphaltene and wax content, and temperature. In a series of studies, Fingas et al. (1997, 2001, 2002) established that the stable emulsions must have at least 7% asphaltenes in weight. Meso-stable conditions are possible if the combined asphaltene and resin content is over 3%. These conditions were used to find values of S_A corresponding to stable and meso-stable conditions. This can be considered one method.

The second method is based on chemical composition index (CCI) introduced by Zeidan et al. (1997) which took into account the asphaltene, resin, and wax content in oils. CCI in combination with an experimental chart they provided could be used to determine the conditions for oil emulsification. Recognizing that both the methods mentioned previously have advantages and the fact there is not enough data to find which one is more reliable, Xie et al. (2007) combined the two methods to find the conditions for formation of stable emulsions.

18.2.6 Photo-Oxidation

As the name implies the photo-oxidation is a result of solar radiation and is important to be included only in long-term simulations. Forming stable emulsions from spilled oil is known to be affected by photo-oxidation. Thus, it is worthwhile to include this process in long-term simulation models determining oil fate. Photo-oxidation is known to depend on the amount of sunlight received, hence on the location, time of the year, and the cloud cover. The amount of solar radiation absorbed by different hydrocarbon components is different; therefore, photo-oxidation is dependent on the type of oil. Given that oil has multiple components and solar irradiance contains a broad range of wave lengths, it is diDcult to know the molar absorptivities for all these components under solar irradiance with different wave lengths. This combined with other complexities and dearth of studies on this topic makes it diDcult to model.

Yapa and Xie (2006) formulated a model to simulate photo-oxidation of spilled oil. The model can be considered preliminary, developed based on existing studies of the physics and chemistry of the process. A discussion of their approach will help future researchers attempting to model the process. Yapa and Xie (2006) only considered the direct photo-oxidation as it is nearly impossible to model the indirect photo-oxidation at the present time. They considered oil to consist of different hydrocarbons and modeled photo-oxidation as a first-order process based on the previous work by Zepp and Cline (1977), Miller et al. (1981), Payne and Phillips (1985), and Literathy et al. (1991). In doing so, they considered how the sun changes the position based on the day of the year and location. These were calculated based on the work of Iqbal (1983). Further details of the model formulation are beyond the scope of this chapter. For the cases they simulated the amount of oil lost due to photo-oxidation is in the order of 1% or less for Venezuelan crude with a PAH of 2%, but it can be slightly higher for some other oils. Their study showed that from the mass balance point of view photo-oxidation is not that critical during the first 10 days. However, the oil compositional change may have a dramatic impact on the fate of oil which they did not study. Their results showed that, the location of the spill, time of year, and cloud cover can make a significant impact on the changes in percentage of mass loss.

18.2.7 Oil and Sediment Interaction

Oil sedimentation is one of the main processes of removing oil from a water body after oil is spilled, yet it is among the least studied processes. This lack of understanding of the process is due to: (1) relative inattention to it partly because of the complexity of the process; and (2) in the early development of models, the researchers focused on short-term tactical forecast for control and containment of the spill. For long-term simulations to assess the coastal pollution, this process has been identified as a very important one.

The oil-sediment interaction includes adsorption and desorption of oil by sediment. There are two processes that take place: *sinking* and *sedimentation*. In the literature, at times one can find that these two processes were not clearly identified. The mechanism that transports oil that is denser than the ambient water to the bottom is *sinking* (NRC 2003). Oil may have become denser than water due to weathering or because it accumulated sediment. Sedimentation is the sorption of oil to sediments in a water column followed by their eventual transport to the seafloor (NRC 2003). The relative amount of oil carried by the two processes can be significantly different. Sinking oil may contain only a small percent of sediment, whereas sedimentation will carry only a low percent of oil to the seafloor.

Oil is adsorbed to sediments at the surface as well as deeper due to capillarity (Zhao et al. 1997). At the surface, the adsorption force for ions is weak and therefore can be easily desorped. The capillary force is much stronger and the oil is not easily desorped. Zhao et al.ß (1997) experimental results and the mathematical model showed that as the sediment concentration increases, the oil adsorption decreases. Chao et al. (1997a) divided desorption into equilibrium desorption and complete desorption. Experimental results of Chao et al. (1997a,b) and Li et al. (1997) showed that adsorption and desorption followed Langmuir–Freundlich equation. These observations allowed them to derive a key equation for modeling oil adsorption and desorption:

$$Q_s = \frac{1}{2}\left[\left(\frac{C_0}{S} + Q_0 + \frac{1}{kS} + b + k_m d_s^{\ m}\right) - \sqrt{\left(-\frac{C_0}{S} - Q_0 - \frac{1}{kS} + b + k_m d_s^{\ m}\right)^2 + \frac{4b}{kS}}\right] \quad (18.9)$$

where

   Q_s is the adsorbed oil at equilibrium adsorption (mL/g)
   C_0 is the initial volume concentration of dissolved oil
   S is sediment concentration (g/mL)
   k and b are empirical adsorption constants; they found through experiments and are functions of sediment size d_s, k_m, and m = constants (8.14×10^{-6} and -0.772)

Chao et al. (2003) used this equation in their oil spill model that included many other processes as well to compute the adsorption of oil by sediments.

Yapa and Xieß (2006) model to simulate the oil sediment interaction recognizes that oil exists in a water column as dispersed oil droplets and dissolved oil. Hence the interaction of oil and sediment occurs through two primary mechanisms: collisions of oil droplets and sediments, and the partition of dissolved oil compounds between water and sediment. Bandara et al. (2011) improved on Yapa and Xieß (2006) work to model the fate and transport of oil in sediment laden coastal waters. The model simulates oil sediment transport, oil-sediment aggregate formation, oil partitioning into sediments, and sediment flocculation. This model although can be improved in the future took into account oil sediment dynamics and kinetics into account better than any other known model. The models show that the amount of oil partitioned into sediments is 4 to 5 orders of magnitude smaller than the amount of oil-sediment aggregates.

18.2.8 Shoreline Deposition and Re-Entrainment of Oil

When oil reaches the shore, at least a portion of it is removed from the water. While this is good for the biota in water, it can damage the beaches and destroy the associated biota. In cases of highly volatile spills, residents near beaches may have to be evacuated. After a major oil spill, it may take many years for the affected beaches to be back to normal. Some endangered species may be lost forever. The monitory loss to the recreational industry is significant. This process is important for short- and long-term simulations and depends on how far the spill is from the shore.

Two pictures are shown here, to highlight the impact of oil depositions on the beaches and to show that even a small spill can impact a beach. The spill in Figure 18.4 was a very small one that was not even recorded. Oil washing ashore lasted for only about an hour, yet the damage to the beach is visible. The cleanup consisted of manual removal of sand using hand shovels. The oil deposition on the shore shown in Figure 18.5 is from a much larger oil spill that required a major effort to clean up the shore.

COZOIL model (Reed et al. 1986, 1989) has a significant component to deal with the oil shore interaction. The model identifies the complex mechanisms that take place in oil

FIGURE1 8.4 Effect to shoreline due to a small spill (a beach at Bentota, Sri Lanka).

FIGURE 18.5 Effect to shoreline due to a larger spill (during Amoco Cadiz Spill). (Photo courtesy of NOAA/Hazmat.)

shore interaction: tidal effects; oil penetration into underlying sediments which depends on sediment size and oil viscosity; porosity of the beach; the re-entrainment; wind and wave effect; existence of three categories as oil on surface, subsurface, and associated with suspended particulate matter; and different types of beaches from rocky cliffs to marsh land. COZOIL identified that during a rising tide, some fraction of the oil was re-suspended and the amount was determined by

$$M_i = M_{i0}e^{-k_f t} \qquad (18.10)$$

where
  M_i is the oil mass in the beach i
  M_{i0} is oil mass originally deposited on the beach
  k_f is a removal coeDcient to characterize the oil loss rate for
    each beach

Gundlach and Hayes (1978) classified coastal areas on a scale of 1 to 10 based on their potential vulnerability that included the duration the oil stayed on a shoreline and the extent of damage to shoreline biota due to the contact with oil. This index later evolved into what is now known as environmental sensitivity index (ESI). Environment Canada and the U.S. Coast Guard sponsored extensive surveys to produce ESI atlases for many shorelines. A correlation between k_f values and ESI can be established (Yapa et al. 1993).

Gundlach (1987) proposed to use two components to simulate oil deposition along shorelines: the oil holding capacity and a removal coeDcient (half-life method). The oil holding capacity is the maximum amount of oil that a shoreline can retain and some values were given by Gundlach (1987). Humphrey et al. (1993) modeled the oil deposition in coarse sediments and re-entrainment. Their study reinforced the identity of some of the parameters that were known to affect oil shoreline interaction. A key conclusion of the rare field study of oil interaction with the shoreline by Sergy et al. (1998) and Sergy (1999) was that the mechanical relocation of oiled sediments from the upper inter-tidal zone to the lower intertidal zone significantly accelerated

the rate of oil removal. The sediment relocation did not elevate the toxicity in the near shore environment to unacceptable levels.

Maximum holding capacity concept (Humphrey 1993) was implemented by Chao et al. (2003). However, in their approach beaches were characterized only by the porosity of the beach sediments. While this may work for a smaller range of beach types, it will not characterize wider range of beaches. Therefore, Gundlach₿ (1987) method appears more suitable.

18.2.9 Governing Equations for Advection-Dispersion of Oil

The transport of oil on the water surface and in water column is governed by advection diffusion equation with modifications to account for oil buoyancy and spill processes. The equations for water surface and water column are slightly different and can be stated as

$$\frac{\partial C_s}{\partial t} + \frac{\partial}{\partial x}(u_s C_s) + \frac{\partial}{\partial y}(v_s C_s) = \frac{\partial}{\partial x}\left(D_x \frac{\partial C_s}{\partial x}\right) + \frac{\partial}{\partial y}\left(D_y \frac{\partial C_s}{\partial y}\right)$$
$$+ \left[\alpha_1 V_b C_v - w C_v + D_z \frac{\partial C_s}{\partial z}\right]_{z=0} - \gamma C_s - S_{ws} \quad \text{(water surface)}$$
$$(18.11)$$

$$\frac{\partial C_v}{\partial t} + \frac{\partial}{\partial x}(u C_v) + \frac{\partial}{\partial y}(v C_v) + \frac{\partial}{\partial z}(w C_v)$$
$$= \frac{\partial}{\partial x}\left(D_x \frac{\partial C_v}{\partial x}\right) + \frac{\partial}{\partial y}\left(D_y \frac{\partial C_v}{\partial y}\right) + \frac{\partial}{\partial z}\left(D_z \frac{\partial C_v}{\partial z}\right)$$
$$+ \frac{\partial}{\partial z}(\alpha V_b C_v) - \frac{\partial}{\partial z}(\gamma' C_v) - S_{wc} \quad \text{(water column)} \quad (18.12)$$

where
  x, y, z, and t are space and time variables
  z is the vertical coordinate measured downward from the
    water surface
  C_s is the oil concentration on the water surface (kg/m² or m³/m²)
  C_v is the oil concentration in water column (kg/m³ or m³/m³)
  u_s, v_s are components of the local surface drift velocity
  u, v, w are x, y, and z components of the local velocity in water
    column
  D_x, D_y, and D_z are dispersion coeDcients in x, y, and z direc-
    tions, respectively (these may have different values for
    water surface and water column, they can also be depth
    dependent)
  α_1, α are parameters to account for the probabilities of oil
    droplets reaching the surface and upper water layer,
    respectively
  V_b is the buoyant velocity of oil (Zheng and Yapa 2000)
  γ, γ' are coeDcients describing the rates at which the surface
    oil is dispersed into the water column
  S_{ws}, S_{wc} are source and sink terms (i.e., the processes that change
    the oil mass) for water surface and water column, respectively

Examples are that evaporation and dissolution reduce mass, but emulsification increases the mass. When oil is broken up from the surface slick and entrained into the water column, oil mass is reduced from the surface and increased by equal amount in the water column.

18.3 Methods and Analysis

Hydrodynamics of the water body is important to modeling oil transport and fate. They are almost always used as given input from a different model or data. This inherently assumes that the presence of oil does not affect the hydrodynamics of the water body to any significant degree. This is approximately true, although the main reason for this decoupled handing is that there has not been much research on modeling the water hydrodynamics in an integrated way with wind shear stress transfer through the oil layer.

The wind affects the transport and fate of oil on the water surface. Generally this effect is taken as a percentage of the wind velocity at 10 m above the surface (wind factor method). More elaborate treatment of the wind effect has not improved the results in a significant way. Most common approach is to consider the surface drift as 3% of wind vector plus the surface water current vector. For more details on how the wind affects the oil transport on the water surface the reader is referred to other papers (e.g., Huang and Monastero 1982; Yapa 1994; ASCE 1996). There are no models that can consider how the wind effect on water hydrodynamics changes as result of the presence of large oil slick on the surface. The same is true about any chances to water hydrodynamics as a result of bulk density changes due to the presence of oil in the water column.

Equations 18.11 and 18.12, although look straightforward, are actually very complex to solve when oil spill processes are involved. The source/sink terms are a result of many transformations that oil undergoes. These transformations change the character of oil which will result in a species different from the parent one. This may mean density or droplet size changes resulting in a change in species. Oil may absorb water resulting in change in species and an increase in mass to be transported. Mass loss can occur due to the processes discussed. Equations 18.11 and 18.12 have to be applied to each species separately which will make modeling using Eulerian fixed grid methods very diDcult if not nearly impossible. It is a multi-species problem. The solution is further complicated by the fact that each species may consist of multiple sizes of droplets creating sub-species.

18.3.1 Lagrangian Parcel Method

In the Lagrangian parcel (LP) method oil is represented as a large number of LPs, each containing a mass of oil. This mass of oil may be in the form of oil droplets. In water column different LPs may contain different droplet sizes. Emulsified, photooxidized oil may be assigned as different species. The same is done for oil that has undergone other transformations. The location, mass, and the characteristics (i.e., type, density, water content, sediment content, etc.) are time-dependent. LPs are first introduced into the water at the spill locations at rates to represent the spill rate. Later new LPs may be introduced or removed based on the processes at locations corresponding to the existing LPs. During each time step the LPs are displaced according to the drift velocity at their respective locations and a fluctuation component representing the turbulent diffusion. The fluctuation component is computed using the random walk method (Fischer et al. 1979) to satisfy the aforementioned advection-diffusion equations. How LP method can be used in multi-species applications is given in Yapa et al. (1996).

LP method is favored for the transport of oil due to a number of reasons: (1) LP method is faster when the domain is large but the oil slick occupies a smaller portion of the domain but moves with time, which is typically the case with oil spills; (2) LP method has less numerical diffusion; (3) dealing with multiple species and change of mass from surface to water column and vice versa is easier with LP method; (4) dealing with multiple mass loss and gain of different species is easier with LP method. Typical implementation of the method is first-order explicit. However, higher order methods can be used as in Yapa and Zheng (1995) to improve the accuracy in high vorticity areas. For each time step, LP location is computed by

$$\vec{s}_{n+1} = \vec{s}_n + \vec{v}_n \Delta t \qquad (18.13)$$

where

$\vec{s}$ is the position vector

$\vec{v}$ is the total advective velocity at the point of LP

Δt is the advection and diffusion time step

n and $n+1$ correspond to the time $n\Delta t$ and $(n + 1)\Delta t$, respectively

The total advective velocity is the sum of mean velocity plus a velocity component V' to account for the turbulent diffusion. The mean velocity is typically obtained from an external source. For the water surface, the wind component needs to be added to the surface velocity. The x, y, and z components of V' for the 3D case are given by

$$u_{ran} = R_{nx}\sqrt{\frac{2D_x}{\Delta t}}, \quad v_{ran} = R_{ny}\sqrt{\frac{2D_y}{\Delta t}}, \quad w_{ran} = R_{nz}\sqrt{\frac{2D_z}{\Delta t}}, \qquad (18.14)$$

where

u_{ran}, v_{ran}, and w_{ran} are random walk velocities in x, y, and z direction, respectively

R_{nx}, R_{ny}, and R_{nz} are normally distributed random numbers with a mean = 0 and a standard deviation = 1

Using the LP method, the location of each LP containing different species must be traced at specified time steps. This time step may or may not be the same as advection-diffusion time step. At each time step, each of the oil spill processes must be separately calculated to find the mass or character change for different species. This may also result in some of the oil changing from one type of species to another (e.g., from a surface LP to water column LP or a density change). All oil must be accounted for in applying the oil spill processes. Each process may be applied sequentially although in reality they may occur simultaneously. An often misunderstood point when using LP method is that one LP may contain many droplets; therefore, the volume of a LP is not correlated to the diameter of the oil droplets.

18.3.2 Hydrocarbon Components and the Orders of Magnitude of Evaporation

When simulating oil evaporation and dissolution, oil is considered to consist of a mix of hydrocarbon components such as paraDn (C_6–C_{12}), paraDn (C_{13}–C_{25}), cycloparaDn (C_6–C_{12}), cycloparaDn (C_{13}–C_{23}), aromatic (C_6–C_{11}), aromatic (C_{12}–C_{18}), naphtheno aromatic (C_9–C_{23}), and residue. The percentages of components vary depend on the oil used. Calculations (e.g., Yapa and Xie 2006) show that the lighter components like paraDn (C_6–C_{12}) and cycloparaDn (C_6–C_{12}) have very high evaporation rates. They are completely evaporated in 10–20 h respectively. The medium component aromatic (C_6–C_{11}) is evaporated 10% after 144 h. The components aromatic (C_{12}–C_{18}), naphtheno aromatic (C_{12}–C_{18}), and residue evaporate only by small amounts.

18.4 Applications and Discussion

18.4.1 Use of Oil Spill Models for Different Purposes

There is no known model that comprehensively simulates all the processes described in this chapter. Most known models have the processes incorporated to be generally consistent with the objectives of the model. Among the main uses of models is to use them in real emergencies and contingency planning that includes training exercises.

For surface or near surface spills, many organizations find that a 2D trajectory model would be suDcient for emergency response. NOAA/ORR (oDce of response and restoration) is responsible for providing modeling support for oil spills in the United States. They use a 2D trajectory model (GNOME) to stay ahead of the spill to know where the oil may be heading. This type of model provides good useful information on where and how widespread the oil is. It also provides information on areas of thick oil slicks and thinner oil slicks. Since one cannot rely entirely on a model for simulations that lasts for many days to be accurate under complex ocean turbulence and wind conditions, they uses overflights to regularly update the model conditions. In cases where the spilled substance is volatile, they will try to assess the exposure hazards and flammability using other models available.

In general, the models at the most basic level can provide information on the location and extent of the oil slick and where it is heading. This is typically what is referred to as the trajectory and spread. Such simulation results can be quite useful in making decisions related to oil boom placement. Most oil booms are

not effective in high turbulent areas. Model simulations can be also used for decision making beyond basic response. For example, if dispersant application is considered models can be used to assess the relative difference in impact. A main impact on oil due to dispersants is that the oil will be broken into smaller droplets and may even disperse more oil into the water column. Both of these effects significantly alter the transport and fate of oil. Oil in the water column will travel differently from the surface oil. Therefore, oil will end up reaching a different shoreline compared to no dispersant application. Such a model will need to be 3D. Oil broken up by the dispersants or turbulence into droplets of various sizes and entrained in the water column takes time to resurface and it needs to be accounted in the model. Models that include the physico-chemical processes can also be used to estimate the oil budget as a function of time (e.g., Yapa et al. 1999). In all cases involving models and the use of their results, it must be emphasized that model results must be interpreted by an expert. To rely on the numerical values that come out of the model without the use of a human expert is unwarranted.

18.4.2 Fate of Oil During Spills: Some Qualitative Information

Major mass loss from spilled oil for many oils is evaporation. This happens in a very short time scale in the order hours to days. If the spill occurs nearer to the land, shoreline deposition also accounts for a major portion and acts in short time scale. If the spill occurs in the deep sea, oil does not reach the shoreline for weeks and may account for a smaller portion of the total oil spilled. Sinking, sedimentation, and bio-degradation will remove oil from the water body. These occur on a relatively longer time scale of days to weeks and could be months for bio-degradation. It should be noted that oil undergoing one process may affect how the same oil may undergo a subsequent process.

The amounts of oil evaporated and dissolved heavily depend on the oil type, in addition to a variety of parameters like wind, slick size, and temperature. To provide a rough idea to the reader, for oil like Nigerian crude the amount evaporated in 72 h can be 30% for calm conditions to 60% under a moderate wind of 6 m/s. It is worth noting that Nigerian crude also may have many variations in composition. For some other crudes the equivalent percentages can be 7–15 (Yapa and Xie 2006). For oil like Bunker C, the percentages will be much lower. Computations have shown that dissolution amounts are 1–2 orders of magnitudes smaller than evaporation. Typical dissolution amount is around 3%. Evaporation and dissolution competes against each other with evaporation being more dominant. Evaporation has been shown to cut the dissolution rate by about half compared to pure hydrocarbon dissolution (Yapa and Xie 2006). Computing the dissolved amount is not critical in determining the fate of the oil. However, the dissolved concentration is very important in determining the toxicity to biological life in water.

The amounts of oil dissolved from the surface slick and the oil droplets in the water column also depend on the turbulence level. A spill occurring in relatively calm conditions results in less dispersed oil. Hence the amounts of dissolved oil in the water column will be small. Reverse is true for high turbulence levels.

Oil deposition along a shoreline and its re-entrainment are very important complex processes associated with oil spills which involve many parameters. Understanding of the physics of the process is still far from complete and need further research. Oil interaction with shoreline may account for a significant portion of oil budget. This amount depends on the beach type and a variety of other factors. COZOIL is the only known model to have taken many of the complex parameters into consideration. However, due to the complexity of implementation and the lack of reliable values for many of the parameters almost no one else has adopted that method. Most other models use a simplified approach of half life method (e.g., Shen and Yapa 1986) or maximum holding capacity (e.g., Chao et al. 2003) or a combination of both (Yapa et al. 1994). Despite the problems mentioned, including this process in the oil spill model at least in a simplified way is important. It allows the model to estimate relative oil deposition amounts on the beaches of different types (high and low vulnerability) under different scenarios such as dispersant application.

Although LPs are Lagrangian, the velocities are defined in a fixed grid system. They also contain a single value for a grid. This can be improved by linear interpolation between the grids. The spread due to velocity shear within a grid is embedded into D_x, D_y, and D_z, and they represent the dispersion coefficients because of the large grid sizes used in computing the velocities in ocean modeling. Values for these coefficients are difficult to find because they depend on the sea state. In the absence of reliable data, the modelers use their experience to estimate the values. Reported values for the water surface vary from 1 to 100 m^2/s (ASCE 1996).

18.5 Major Challenges

Verification of oil spill models through controlled field experiments is a nearly impossible task because it is not possible to spill oil in the ocean. Surface or near surface oil spill experiments have never been conducted. For deep water spills, a set of limited large-scale field experiments, *Deepspill*, were conducted in Norway (Johansen et al. 2003). Laboratory experiments do not result in the same quality results because of the inability to scale oil properties. Most of the data for surface oil spills are sparsely collected data during accidents. Because of the much variability of parameters these data are difficult to use although some have been extremely useful. Data collected during accidental spills remain very limited in nature due to the emergency nature of the problem, logistical issues, and concerns on litigation. Therefore, lack of high quality data for verifying models continue to be an issue. Lack of reliable methods to provide good estimates on dispersion coefficients also continues to be a problem.

Oil breakup from the water surface and its entrainment into water column are key processes that govern the transport and fate of oil. When oil slick breaks up and entrains oil into the water column, the size distribution of oil bubbles is equally important.

However, for modeling all these processes the modeling community entirely depends on empirical methods developed two decades ago and have only been partially tested. Newer and better methods are badly needed. No research has been done on how the oil droplets dispersed in the water column coalesce or further break up due to ocean turbulence during their travel upward. It can be reported from the author's own research that a breakthrough in this area is imminent.

One of the main processes of oil removal from water is its interaction with sediment and then eventual sedimentation. Modeling this process has hardly been attempted. NOAA has identified this as a process important to oil spill modeling (personal communications). The knowledge on it is highly rudimentary at best. More research is needed in this area.

Modeling the transport and fate of oil spilled in waters, at a minimum, requires solid knowledge of the following disciplines: fluid mechanics, oceanography, and oil spill chemistry. It is a truly multi-disciplinary subject. So far the integration of expertise from these different disciplines has not been the best. Let us hope for better integration in the future.

Acknowledgments

The summary of oil spill modeling knowledge expressed here was possible due to various sponsored projects. The author would like to thank Professor Hung Tao Shen with whom the author collaborated in the earliest project. Professor Kisaburo Nakata was instrumental in making several of these projects possible and engaged in many discussions. The author would also like to thank his graduate students. Special thanks to Li Zheng, Fanghui Chen, Hao Xie, Lalith K. Dassanayaka, and Uditha C. Bandara.

References

ASCE Task Committee on Oil Spills (1996). The state-of-the-art of modeling oil spills, *Journal of Hydraulic Engineering*, 122, 11, 594–609.

Bandara U. C. and Yapa, P. D. (2011). Bubble sizes, breakup, and coalescence in deepwater gas/oil plumes, *Journal of Hydraulic Engineering*, 137, 7, 729–738.

Bandara U. C., Yapa, P. D., and Xie, H. (2011). Fate and transport of oil in sediment laden marine waters, *Journal of Hydro-Environment Research*, 5, 145–156.

Chao, X., Shankar, N. J., and Wang, S. S. Y. (2003). Development and application of oil spill model for Singapore coastal waters, *Journal of Hydraulic Engineering*, 129 (7), 495–503.

Chao, X., Zhao, W., and Qiu, D. (1997b). Study on the adsorption properties of oil by sediment applying fractal theory, *Shuili Xuebao* (9).

Chao, X., Zhao, W., and Zhang, P. (1997a). Desorption properties of oil on sediment, *Acta Scientiae Circumstantiae*, 17(4), 434–438.

Daubert, T. E. and Danner R. P. (Eds.) (1989). *API Technical Data Book—Petroleum Rerning*, 5th edn, American Petroleum Institute (API), Washington, DC.

Delvigne, G. A. L. (March 1994). Natural and chemical dispersion of oil, *Journal of Advanced Marine Technology*, 11, 23–63.

Delvigne, G. A. L. and Sweeney, C. E. (1988). Natural dispersion of oil, *Oil and Chemical Pollution*, 4(4), 281–310.

Fay, J. A. (1971). Physical processes in the spread of oil on water surface, *Proceedings of the Joint Conference on Prevention and Control of Oil Spills*, Washington, DC, June, American Petroleum Institute, Washington, DC, pp. 463–467.

Fingas, M. (1994). Chemical processes of oil spills, *Journal of Advanced Marine Technology*, 5 (1).

Fingas, M. and Fieldhouse, B. (2006). A review of the knowledge on water-in-oil emulsions, *Proceedings of the 29th Arctic Marine Oilspill Program Technical Seminar,* Vancouver, British Columbia, Canada, Environment Canada, Ottawa, Ontario, Canada, pp. 1–56.

Fingas, M., Fieldhouse, B., Lerouge, L., Lane, J., and Mullin, J. (2001). Studies of water-in-oil emulsions: Energy and work threshold as a function of temperature, *Proceedings of the 24th Arctic Marine Oilspill Program Technical Seminar*, Edmonton, Alberta, Canada, Environment Canada, Ottawa, Ontario, Canada.

Fingas, M., Fieldhouse, B., and Mullin, J. (1997). Studies of water-in-oil emulsions: Stability Studies, *Proceedings of the 20th Arctic and Marine Oil Spill Program (AMOP) Technical Seminar*, Vancouver, British Columbia, Canada, Environment Canada, Ottawa, Ontario, Canada, pp. 21–42.

Fingas, M., Fieldhouse, B., Noonan, J., and Lambert, P. (2002). Studies of water-in-oil emulsions: Testing of emulsion formation in OHMSETT, year II, *Proceedings of the 25th Arctic Marine Oil Spill Program Technical Seminar*, Calgary, Alberta, Canada, Environment Canada, Ottawa, Ontario, Canada.

Fischer H. B., List E. J., Koh R. C. Y, Imberger, J., and Brooks, N. (1979). *Mixing in Inland and Coastal Waters*, Academic Press, New York.

Galt, J. A. (1994). *Personal communications*.

Gundlach, E. R. (1987). Oil-holding capacities and removal coefficients for different shoreline types to computer simulate spills in coastal waters, *Proceedings of the 1987 Oil Spill Conference*, Baltimore, MD, pp. 451–457.

Gundlach, E. R. and Hayes, M. D. (1978). Vulnerability of coastal environments to oil spill impacts, *Marine Technology Society Journal*, 12(4), 18–27.

Guo, W. J. and Wang, Y. X. (2009). A numerical oil spill model based on a hybrid method, *Marine Pollution Bulletin*, 58, 726–734.

Hoult D. P. (1972). Oil spreading on the sea, *Annual Review of Fluid Mechanics*, 4, 341–367.

Huang, J. C. and Monastero, F. C. (June 1982). Review of the state-of-the-art of oil spill simulation models, Final Report submitted to the American Petroleum Institute, Raytheon Ocean Systems Company, East Providence, RI.

Humphrey, B., Owens, E., and Sergy, G. (1993). Development of a stranded oil in coarse sediment (SOCS) model, *Proceedings of the 1993 Oil Spill Conference*, Tampa, FL, pp. 575–582.

Iqbal, M. (1983). *An Introduction to Solar Radiation*. Academic Press, Toronto, Ontario, Canada.

Izumiyama, K., Uto, S., Narita, S., and Tasaki, R. (1998). Effect of interfacial tension on the spreading of oil under an ice cover. In *Ice in Surface Waters*, Shen, H. (ed.), Balkema, Rotterdam, the Netherlands, ISBN 9054109718.

Johansen, O., Rye, H., and Cooper, C. (2003). Deepspill: Field study of a simulated oil and gas blowout in deepwater. *Spill Science & Technology Bulletin*. 8(5–6), 433–443.

Leinonen, P. J. and Mackay, D. (April 1973). The multi-component solubility of hydrocarbons in water, *Canadian Journal of Chemical Engineering*, 51, 230–233.

Li, C., Zhao, W. and Luo, L. (1997). Study on characteristics and effect factors of the absorption, desorption of oil by sediment in rivers, *China Environmental Science*, 17 (1).

Literathy, P., Morel, G., and Al-Bloushi, A. (1991). Environmental transformation, photolysis of fluorescing petroleum compounds in marine waters, *Water Science and Technology*, 23, 507–516.

Mackay, D. I., Buistt, I. A., Mascarenhas, R., and Paterson, S. (1980a). Oil spill processes and models, Manuscript Report No. EE-8, Environment Canada, Ottawa, Ontario, Canada.

Mackay, D. and Matsugu, R. S. (August 1973). Evaporation rates of liquid hydrocarbon spills on land and water, *Canadian Journal of Chemical Engineering*, 51, 434–439.

Mackay, D., Paterson, S., and Trudel, K. (1980b). *A Mathematical Model of Oil Spill Behavior*, Environmental Protection Service, Fisheries and Environment, Ottawa, Ontario, Canada.

Mackay, D. and Zagorski, W. (1982). Water-in oil emulsions: A stability hypothesis, *Proceedings, 5th Arctic Marine Oil Spill Program Technical Seminar*, Environment Canada, Ottawa, Ontario, Canada, pp. 61–74.

Miller, T., Mabey, W. R., Lan, B. Y., and Baraze, A. (1981). Photolysis of polycyclic aromatic hydrocarbons in water, *Chemosphere*, 10(11/12), 1281–1290.

National Research Council (2003). *Oil in the Sea III: Inputs, Fates, and EThcts*, Joseph Henry Press, Washington, DC.

Neumann G. and Pierson, W. Jr. (1966). *Principles of Physical Oceanography*, 196, 351 and 420, Prentice-Hall Inc., Englewood Cliffs, NJ.

Payne, J. R. and Phillips, C. R. (1985). Photochemistry of petroleum in water, *Environmental Science & Technology*, 19(7), 569–579.

Reed, M., Gundlach, E. R., and Kana, T. W. (1989). A coastal zone oil spill model : Development and sensitivity studies, *Oil and Chemical Pollution*, 5, 441–449.

Reed, M., Spaulding, M. L., Gundlach, E. R., Kana, T. W., and Siah, S. J. (1986). Formulation of a shoreline/oil spill interaction model, *Proceedings of the 9th Annual Arctic and Marine Oil Spill Program*, Edmonton, Alberta, Canada, pp. 77–100.

Riazi, M. R. and Al-Enezi, G. A. (1999). Modeling of the rate of oil spill disappearance from seawater for Kuwaiti crude and its products, *Chemical Engineering Journal*, 73, 161–172.

Riazi, M. R. and Edalat, M. (1996). Prediction of the rate of oil removal from seawater by evaporation and dissolution, *Journal of Petroleum Science and Engineering*, 16, 291–300.

Sergy, G. A. (1999). Key conclusions of the Svalbard shoreline experiment, *Proceedings of the 22nd Arctic and Marine Oil Spill Program Technical Seminar*, AMOP, Calgary, Alberta, Canada, *p*. 845.

Sergy, G. A., Guénette, C, C., Owens, E. H., Prince, R. C., and Lee, K. (1998). The Svalbard shoreline oil spill field trials, *Proceedings of the 21st Arctic and Marine Oil Spill Program Technical Seminar*, AMOP, Edmonton, Alberta, Canada, pp. 873–889.

Shen, H. T. and Yapa, P. D. (1986). Oil slick transport in rivers, *Journal of Hydraulic Engineering*, 114(5), 529–543.

Shiu, W. Y., Bobra, M., Bobra, A. M., Maijanen, A., Suntio, L., and Mackay, D. (1990). The water solubility of crude oils and petroleum products, *Oil and Chemical Pollution*, 7, 57–84.

Simecek-Beatty, D., Lehr, W. J., Lai, R., and Overstreet, R. (Eds.) (2000). Langmuir circulation and oil spill modeling, *Spill Science and Technology: Special Issue*, 6(3/4).

Terray, E., Donelan, M., Agrawal, Y., Deennan, W., Kahma, K., Williams III, A., Hwang, P., and Kitaigorodskii, S. (1996). Estimates of kinetic energy dissipation under breaking waves, *Journal of Physical of Oceanography*, 26, 792–807.

Xie, H., Yapa, P. D., and Nakata, K. (2007). Modeling emulsification after an oil spill in the sea, *Journal of Marine Systems*, 68(2007), 489–506.

Yang, W. C. and Wang, H. (1977). Modeling of oil evaporation in aqueous environment, *Water Research*, 11, 879–887.

Yapa, P. D. (March 1994). Oil spill processes and model development, *Journal of Advanced Marine Technology*, 11, 1–22.

Yapa, P. D. and Chowdhury, T. (December 1990). Spreading of oil under ice covers, *Journal of Hydraulic Engineering*, 1468–1483.

Yapa, P. D., Shen, H. T., and Angammana, K. (March 1994). Modeling oil spills in a river-lake system, *Journal of Marine Systems*, 4, 453–471.

Yapa, P. D., Shen, H. T., and Angammana, K. (1993). Modeling oil spills in a River-Lake system?, Report No. 93-1, Department of Civil and Environmental Engineering, Clarkson University, Potsdam, New York, pp. 1–95.

Yapa, P. D. and Weerasuriya, S. A. (1997). Spreading of oil spilled under floating broken ice, *Journal of Hydraulic Engineering*, 123(8), 676–683.

Yapa, P. D. and Xie, H. (January 2006). Modeling of chemical processes in oil spills (final report), Report submitted to CCTI and MBRIJ (Japan), Report 06-01, Department of Civil and Environmental Engineering, Clarkson University, Potsdam, New York, pp. 1–220.

Yapa, P. D. and Zheng, L. (1995). Review of oil spill progress, Department of Civil and Environmental Engineering, Clarkson University, New York, Report No. 95-1.

Yapa, P. D. and Zheng, L. (1998). A three-dimensional model for simulating the behavior of oil and gas released from deep water, Report No. 98-11, Department of Civil and Environmental Engineering, Clarkson University, Potsdam, New York, 13699.

Yapa, P. D., Zheng, L., and Kobayashi, T. (January 1996). Application of linked-list approach to pollutant transport models, *Journal of Computing in Civil Engineering*, 10(1), 88–90.

Yapa, P. D., Zheng, L., and Nakata, K. (May 1999). Modeling underwater oil/gas jets and plumes, *Journal of Hydraulic Engineering*, 481–491.

Zeidan, E., Zahariev, K., Li, M., and Garrett, C. (1997). The breakup of oil spills in the marine environment, *Proceedings of the 20th Arctic Marine Oilspill Program Technical Seminar*, Vancouver, British Columbia, Canada, Environment Canada, Ottawa, Ontario, Canada.

Zepp, R. G. and Cline, D. M. (1977). Rates of direct photolysis in aquatic environment, *Environmental Science & Technology*, 11(4), 359–366.

Zhang, Z. and Cui, G. (1999). *Fluid Mechanics*, Tsinghua University Publisher, Beijing (in Chinese), pp. 1–317.

Zhao, W., Chao, X., and Huang, Q. (1997). Mathematical model and experimental study on adsorption of oil by sediment, *Shuili Xuebao* (12).

Zheng, L. and Yapa, P. D. (2000). Buoyant velocity of spherical and non-spherical bubbles/droplets, *Journal of Hydraulic Engineering*, 126 (11), 852–854.

19

Miscible and Immiscible Pollutants in Subsurface Systems

Tissa H. Illangasekare
Colorado School of Mines

Christophe C. Frippiat
*Scientific Institute for
Public Services*

19.1 Introduction

Subsurface soil and water contamination is a result of chemicals and other waste products that are introduced into the environment through improper disposal, accidental spills, and industrial and agricultural activities. The risk to human health is associated with the chemical dissolved in groundwater appearing at receptors such as drinking water wells or streams. The chemical may enter the subsurface directly dissolved in water or as a separate phase. An engineering classification system presented by La Grega et al. (1994) grouped hazardous waste that has the potential to contaminate soil and groundwater under six broad groups: inorganic aqueous waste, organic aqueous waste, organic liquids, oils, inorganic sludges/solids, and organic sludges/solids. Pathogens and nuclear wastes also act as sources of groundwater contamination (Kreider et al. 1998). As these chemicals are fully or partially soluble in water, they will eventually enter the saturated zone of the aquifers as contaminants that are hazardous to humans and ecological life.

In general, dissolved chemicals are referred to as miscible contaminants and separate phase contaminants are referred to as non-aqueous phase liquids or NAPLs. NAPLs can get entrapped both in the unsaturated zone of the subsurface (or vadose zone) and in the saturated zone. When introduced into the saturated zone, the solubility of NAPLs in flowing groundwater is so low

that they remain entrapped as a separate phase for long periods of time. NAPLs in the unsaturated zone act as source of vapor that has the potential to intrude into subsurface structures. Depending on the specific gravity, NAPLs are categorized as lighter than water NAPLs (or LNAPLs) or denser than water NAPLs (or DNAPLs). Common LNAPLs are petroleum hydrocarbons that are introduced into the subsurface at refineries and from leaking underground tanks and pipes. Chlorinated solvents, such as wood preservatives or coal tar, with specific gravities that are larger than one, are classified as DNAPLs. After reaching the water table, LNAPLs and DNAPLs behave in fundamentally different ways. DNAPLs that are heavier than water will penetrate the water table and move through the saturated zone. A heavier and less viscous fluid staying on top of water in the saturated zone creates an unstable situation resulting in the formation of fingers that will move the DNAPL through preferential channels through the saturated zone. Under certain geologic conditions, these fingers will carry the DNAPL into deeper formations and form highly saturated pools at the bedrock.

The design of effective remediation schemes to mitigate impacts or to reduce exposure risks, requires the use of models that have the ability to predict the fate and transport of both miscible and immiscible contaminants in the subsurface. In Section 19.2, a short summary of the fundamental principles behind the movement of fluids and miscible and immiscible

Handbook of Environmental Fluid Dynamics, Volume Two, edited by Harindra Joseph Shermal Fernando. © 2013 CRC Press/Taylor & Francis Group, LLC.
ISBN: 978-1-4665-5601-0.

contaminants in the subsurface is presented. When contaminants move in the subsurface, they go through physical, chemical and biological transformations, and radioactive decay. Two of these, sorption processes and dissolution resulting from mass transfer between phases, are discussed. In field applications, governing equations based on the fundamental principles are generally solved using numerical models, but to obtain insights into the processes, a few analytical solution methods are presented in Section 19.3. Finally, in Section 19.4, a set of examples is presented to demonstrate these solution methods. Some of the key challenges linked to the characterization of subsurface systems and their associated uncertainty are summarized in Section 19.5.

19.2 Principles

19.2.1 Principle of Water and Immiscible Fluid Flow in Porous Media

Rocks and soils naturally contain a certain amount of voids and cavities, called pores, which can be either filled with water, with gas, or with other immiscible fluids. The ratio of the volume of void per unit volume of bulk soil is called total porosity and is noted here ϕ [$L^3 L^{-3}$]. When multiple fluid phases are present in the pore space, fluid content is defined as the volume of fluid per unit volume of bulk soil. For a given phase p, fluid content is noted here θ_p [$L^3 L^{-3}$] and saturation S_p [$L^3 L^{-3}$] is defined as

$$S_p = \frac{\theta_p}{\phi}. \tag{19.1}$$

When multiple fluids compete for the pore space, their respective degrees of saturation sum up to one.

Saturation is one of the key variables that govern the physical processes controlling fluid flow in porous media. In the saturated zone of the soil, in the absence of immiscible pollutants, water saturation is equal to one and flow is described by Darcy's law (Section 19.2.1.1). If immiscible fluids are present, each phase is characterized by its own pressure state that depends on fluid saturation, and each phase is described by its own governing equation (Section 19.2.1.2). However, it is often hypothesized that, in the absence of other liquid phase than water, gas phase flow can be neglected and water flow in the unsaturated zone of the soil can be described by a simplified governing equation called *Richards equation* (Section 19.2.1.3).

19.2.1.1 Flow of Water in Saturated Porous Media

Pressure gradients and gravitational forces govern flow of a single fluid in a fully saturated porous media according to Darcy's law

$$q = -\frac{K_i}{\mu}(\nabla P - \rho g), \tag{19.2}$$

where

bold lowercase and uppercase symbols refer to vector and matrix variables, respectively

q [$L T^{-1}$] is the specific discharge

ρ [$M L^{-3}$] and μ [$M L^{-1} T^{-1}$] are the volumetric weight and the dynamic viscosity of the fluid, respectively

The gravity vector is noted g [$L T^{-2}$]. K_i [L^2] is a geometric property of the medium called *intrinsic permeability*

It is a second-order tensor that describes the ability of the porous medium to conduct fluids. In a soil, it is linked to total porosity and to the distribution of grain sizes.

Hydraulic head h [L] is a measure of a fluid's total energy

$$h = \frac{P}{\rho g} - z, \tag{19.3}$$

where

P [$M L^{-1} T^{-2}$] is the pressure of the fluid

z [L] is depth from a reference level (e.g., the ground surface)

Kinetic energy is usually not included in Equation 19.3 as fluid velocities are naturally low in subsurface systems. Combining Equations 19.2 and 19.3, Darcy's law can be expressed as

$$q = -K_i \frac{\rho g}{\mu} \nabla h = -K \nabla h, \tag{19.4}$$

where $K = K_i(\rho g/\mu)$ [$L T^{-1}$] is called permeability or saturated hydraulic conductivity. Under the form of Equation 19.4, Darcy's law expresses that head losses are linear with respect to velocity. It is thus valid provided the flow regime is laminar.

Mass balance considerations and the use of Equation 19.4 allow the derivation of a governing equation for head distributions in saturated porous subsurface systems in the absence of source or sink

$$S_s \frac{\partial h}{\partial t} = -\nabla \cdot (-K \nabla h), \tag{19.5}$$

where S_s [L^{-1}] is called specific storage. It expresses that changes in the mass of water within a reference volume of soil depend only on soil and water compressibility.

19.2.1.2 Flow of Immiscible Fluids in Porous Media

In the presence of multiple fluid phases, each phase is characterized by its own pressure state. The differences in pressure arise from imbalances of molecular forces at fluid interfaces (Corey 1994). Such pressure differences are best understood when considering water rising in a capillary tube. At equilibrium, the height of water h_c [L] can be computed from Jurin's law

$$h_c = \frac{2\sigma \cos \alpha}{\rho g r}, \tag{19.6}$$

where

    ρ [M L^{-3}] is the volumetric weight of water

    r [L] is the radius of the tube

    σ [M T^{-2}] is called the *surface tension*

    α is called the *contact angle*

Surface tension is a property of both fluids, while the contact angle depends on the fluids and on the solid they are in contact with. The air–water interface will exhibit a downward curvature that indicates that the pressure in the air phase is larger than that in the water phase. The curvature also indicates that the water phase is more prone to be attracted to the surface of the tube as compared to the air phase. *Wettability* is defined as the tendency of one fluid to being attracted to solid particles in preference to another. In natural subsurface systems, the gas phase is usually the non-wetting phase and water is the wetting phase. Immiscible pollutants usually have an intermediate wettability.

The capillary pressure P_c is defined as the pressure difference between wetting and non-wetting phases

$$P_c(S_w) = P_{nw} - P_w,\qquad (19.7)$$

where

    P_{nw} is the non-wetting phase pressure

    P_w is the wetting phase pressure

According to this definition, the capillary pressure is a positive quantity. In a capillary tube, the capillary pressure is $P_c = \rho g h_c$. In natural subsurface systems, capillary pressure depends on fluid saturation. At high fluid content, only the largest pores are empty and fluid interfaces are characterized by large radii. According to Jurin's law, the corresponding capillary pressure values are thus low. Reciprocally, at low fluid content, only the smallest pores still contain fluid, fluid interfaces are characterized by small radii

and the capillary pressures are high. The distribution of capillary pressure as a function of fluid content is called the *retention function*. As indicated in Equation 19.7, it is usually expressed as a function of the saturation of the wetting fluid.

Since it depends on the pore sizes and their distribution, the retention function for a pair of wetting–non-wetting fluids is an intrinsic property of a given porous medium and the two-fluid system. Data on capillary pressure as a function of saturation are obtained experimentally and fitted with mathematical functions to obtain constitutive models for multiphase flow. Two commonly used constitutive models for capillary pressure used in groundwater applications are by Brooks and Corey (1964) and van Genuchten (1980). The Brooks–Corey model for the retention function is

$$P_c(S_w) = P_d S_e^{-1/\lambda}\quad \text{for } P_c \geq P_d,\qquad (19.8)$$

where

    λ [–] is a fitting parameter

    P_d [M L^{-1} T^{-2}] is the pressure at which the non-wetting fluid enters the pores and is called *displacement pressure*

The effective saturation S_e is defined as

$$S_e = \frac{S_w - S_{wr}}{1 - S_{wr}},\qquad (19.9)$$

where S_{wr} [L^3 L^{-3}] is the residual or minimum wetting fluid saturation. Figure 19.1a shows typical air–water retention functions for sandy, silty, and clayey soils, expressed as a function of water content.

When multiple fluid phases are present in the pore space, the ability of the medium to conduct a given fluid will not depend on the geometry of the pore space only, but also on geometry of the fluid-filled part of the pore space. This additional dependency on

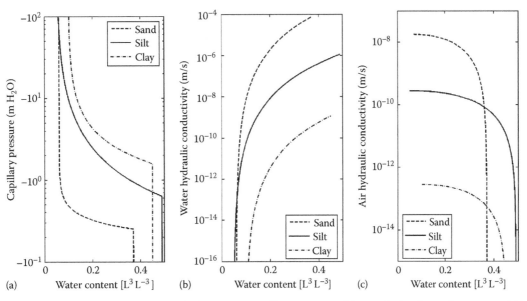

FIGURE 19.1 Typical Brooks–Corey distributions of (a) air–water capillary pressures, (b) water hydraulic conductivity, and (c) air hydraulic conductivity, as a function of water content for sandy, silty, and clayey soils.

fluid content is introduced into Darcy's law using the concept of relative permeability (Mayer and Hassanizadeh 2005)

$$q_p = -K_i \frac{k_{rp}(S_p)}{\mu_p}(\nabla P_p - \rho_p g).$$ (19.10)

The relative permeability k_{rp} [–] of fluid p ranges between zero and one and depends on the fluid saturation S_p. By extension, the hydraulic conductivity also becomes a function of saturation $K = K(S_p) = K_i \, k_{rp}(S_p) \, \rho_p g/\mu_p$ [L T^{-1}]. As for the retention function, data on relative permeability as a function of saturation are obtained experimentally and fitted with mathematical functions. The Brooks–Corey models for the relative permeabilities of the wetting and the non-wetting phases are

$$k_{rw} = S_e^{(2+3\lambda)/\lambda},$$ (19.11a)

$$k_{rnw} = (1 - S_e)^2 \left(1 - S_e^{(2+\lambda)/\lambda}\right).$$ (19.11b)

Figure 19.1b and c shows typical ranges of water and air hydraulic conductivity as a function of water content for sandy, silty, and clayey soils.

Darcy's law (Equation 19.10) and mass balance for each phase are combined to derive the governing equations for multiphase flow. The governing flow equation for a fluid phase p, with pressure P_p and saturation S_p as dependent variables, is written as

$$\phi \frac{\partial(S_p)}{\partial t} = \nabla \cdot \left[K_i \frac{k_{rp}(S_p)}{\mu_p}(\nabla P_p + \rho_p g \nabla z) \right],$$ (19.12)

where it is assumed that the vertical axis points downward. Equation 19.12 written for each phase, in combination with the constitutive models for capillary pressures and relative permeabilities provides the full formulation of the mathematical model to solve for the phase pressures and saturations for given initial and boundary conditions and source terms.

19.2.1.3 Flow of Water in Unsaturated Porous Media

In unsaturated soils, water and air flow are coupled through their respective saturations. Since air viscosity is orders of magnitude smaller than water viscosity, it is of common practice to neglect air flow with respect to water flow and to assume that the air phase is at constant zero atmospheric pressure (gage pressure). In that case, capillary pressure can be equated to water pressure and the governing equation obtained by simplifying Equation 19.12 is usually referred to as *Richards equation*

$$\phi \frac{\partial S_w}{\partial t} = \nabla \cdot \left(K(S_w) \nabla \left(\frac{-P_c(S_w)}{\rho_w g} + z \right) \right)$$

$$= \nabla \cdot \left(K(S_w) \nabla \left(\frac{-P_c(S_w)}{\rho_w g} \right) \right) + \frac{\partial K(S_w)}{\partial z},$$ (19.13)

where subscript w refers to the wetting (water) phase and replaces subscript p in previous equations.

19.2.2 Transport of Miscible Pollutants

At low concentration, the movement of inert miscible pollutants is mainly controlled by three physical processes: advection, diffusion, and dispersion. The governing equation for solute transport is called the *advection-dispersion equation* (ADE) or the *convection-dispersion equation* (Domenico and Schwartz 1997, Fetter 1999)

$$\frac{\partial \theta C}{\partial t} - Q = \nabla \cdot (\theta D_e \cdot \nabla C + \theta D_m \cdot \nabla C - C \cdot q(\theta))$$

$$= \nabla \cdot (\theta D \cdot \nabla C - C \cdot q(\theta)),$$ (19.14)

where
- θ [L^3 L^{-3}] is the fluid content
- C [M L^{-3}] is the concentration of miscible pollutant in the fluid phase

$D_e = \omega(\tau) \times D_{mol}$ [L^2 T^{-1}] is an effective diffusion coefficient, $\omega(\tau)$ [–] being a coefficient that depends on soil tortuosity τ [–], and D_{mol} [L^2 T^{-1}] being the coefficient of molecular diffusion in free phase. D_m [L^2 T^{-1}] is a second rank tensor called *tensor of mechanical dispersion*. $D = D_e + D_m$ [L^2 T^{-1}] is called *tensor of hydrodynamic dispersion* and includes a characterization of both diffusion and mechanical dispersion. Q [M L^{-3} T^{-1}] is a sink-source term and $q(\theta)$ [L T^{-1}] is the specific discharge of the fluid.

Pollutant transport caused by bulk movement of flowing groundwater is called advection. If no other process occurs, then plumes of miscible contaminants are simply translated at the velocity of groundwater. The average transport velocity, called seepage velocity, is the ratio of the specific discharge $q(\theta)$ to the average fluid content θ in the soil. Diffusion is the net flux of solutes from zones of higher concentration to zones of lower concentration. As the driving forces are spatial gradients in concentration, diffusion is described by Fick's law adapted to porous media with the introduction of an effective diffusion coefficient that accounts for tortuosity of the molecular pathways at the pore scale. Mechanical dispersion is the tendency for fluids to spread out from the flow lines that they would be expected to follow according to the advective hydraulics of the flow system. This spreading process results from microscopic variations in velocity. As pollutant particles move at various velocities through the tortuous paths of the medium, the initial plume is distorted and becomes elongated. Although the driving force is under no circumstance linked to concentration gradients, mechanical dispersion is described using a Fickian law. The fictitious diffusion coefficient D_m that characterizes the distribution of pore-scale velocities is a second-order tensor.

It has been found both from simple theoretical models and from experimental evidence that coefficients of mechanical dispersion vary linearly with seepage velocity. When the principal directions of the tensor of mechanical dispersion are aligned with the principal directions of the flow field, it takes the following form (Bear 1972)

$$\theta D = \theta D_e + \begin{bmatrix} \alpha_L & 0 & 0 \\ 0 & \alpha_{TH} & 0 \\ 0 & 0 & \alpha_{TV} \end{bmatrix} q(\theta), \qquad (19.15)$$

where α_L [L], α_{TH} [L], and α_{TV} [L] are characteristic mixing lengths called longitudinal dispersivity, horizontal transverse dispersivity, and vertical transverse dispersivity, respectively. At the pore scale, dispersivities mainly depend on the distribution of grain sizes. In the unsaturated zone of the soil, dispersivity additionally depends on water content (see Illangasekare et al. 2010). However, there exists great debate about the values of dispersivities that are applicable to field-scale problems. Longitudinal dispersivity has been found to vary greatly with the mean travel distance of a plume of miscible pollutant, as a result of natural heterogeneity of subsurface systems. As a very bulk rule of thumb, one can generally assume (Gelhar et al. 1992, Fetter 1999)

$$\alpha_L = \frac{L}{10}, \quad \alpha_{TH} = \frac{\alpha_L}{6 \dots 20}, \quad \text{and} \quad \alpha_{TV} = \frac{\alpha_{TH}}{1 \dots 10}, \qquad (19.16)$$

where L [L] is the mean travel distance. Fundamentals of dispersion in porous media are discussed by Bear (1972). Further details on the estimation of dispersion coeDcients, either empirically or from laboratory or field tests, are given by Illangasekare et al. (2010).

19.2.3 Reactive Processes

In subsurface systems, a chemical species can undergo a wide variety of reactions through which it can be removed or added to the groundwater. Two types of reactive processes that are encountered in most field settings with dissolved chemicals and NAPL contaminated source zones are discussed in this section: (1) sorption, which involves mass transfer between the flowing groundwater and the soil matrix (Section 19.2.3.1), and (2) NAPL dissolution, which involves mass transfer from NAPL sources to the flowing groundwater (Section 19.2.3.2). Discussions of other reactive processes occurring in subsurface systems can be found e.g., in Domenico and Schwartz (1997) or Fetter (1999).

19.2.3.1 Sorption

Sorption commonly refers to the multiple processes through which a miscible pollutant is attracted and retained near or to the surface of a solid particle. These include adsorption, chemisorption, absorption, or ion exchange (Fetter 1999). Defining solid concentration C^* [M M^{-1}] as the ratio of mass retained by sorption to the mass of soil within a reference volume, the associated sink term in Equation 19.14 can be expressed as

$$Q = -\rho_B \frac{\partial C^*}{\partial t}, \qquad (19.17)$$

where ρ_B [M L^{-3}] is the bulk volumetric weight of the soil. Injecting this expression into the advection-dispersion equation yields for left-hand side of the equation

$$\frac{\partial \theta C}{\partial t} - Q = \frac{\partial \theta C}{\partial t} + \rho_B \frac{\partial C^*}{\partial t} = \frac{\partial \theta C}{\partial t}\left(1 + \frac{\rho_B}{\theta}\frac{dC^*}{dC}\right) = R\frac{\partial \theta C}{\partial t},$$
$$(19.18)$$

where $R \geq 1$ [–] is called retardation coeDcient or retardation factor. If the groundwater flow is suDciently slow, sorption can be considered comparatively fast and at equilibrium. In that case, the capacity of the solid matrix to retain a solute will mainly depend on its concentration in the groundwater. The three major models used to characterize equilibrium solute sorption in subsurface systems are the linear model, the Freundlich model, and the Langmuir model.

The linear model is the simplest one. It assumes that the solid concentration C^* [M M^{-3}] is a linear function of solute concentration C [M L^{-3}]

$$C^* = K_{d,l}C, \qquad (19.19)$$

where $K_{d,l}$ [L^3 M^{-1}] is a constant called *partitioning coe cient*. The corresponding retardation coeDcient does not depend on solute concentration and remains constant. This model is therefore computationally advantageous, since the governing equation for miscible transport remains linear in concentration. However, the use of a linear sorption model is appropriate only at low concentration, when the number of sites available at the surface of soil grains does not represent a limiting factor for sorption.

Freundlich and Langmuir sorption models are nonlinear models that require an additional empirical parameter to characterize solid concentration as a function of solute concentration. The Freundlich model takes the form

$$C^* = K_{d,f}C^N, \qquad (19.20)$$

where

N [–] is called Freundlich exponent

$K_{d,f}$ is a partitioning coeDcient with units depending on N

The Langmuir model has an improved physical base as compared to the Freundlich model, as it incorporates a parameter C^*_{lim} [M M^{-1}] being the maximum solid concentration

$$C^* = \frac{K_{d,lang}C^*_{lim}C^*}{1 + K_{d,lang}C^*}, \qquad (19.21)$$

and $K_{d,lang}$ [L^3 M^{-1}] is a partitioning coeDcient.

If groundwater flow is slow as compared to sorption kinetic, an equilibrium sorption model is not adequate and one has to resort to using a rate-limited model similar to those described in the following section.

19.2.3.2 Mass Transfer from Entrapped NAPL Sources

The process by which the NAPL becomes a source for groundwater contamination is through dissolution of NAPL entrapped in soil pores. In naturally heterogeneous subsurface systems, due to unstable behaviors (Held and Illangasekare 1995) and capillary barrier effects at the lithological interfaces of soil layers, NAPLs get entrapped in complex spatial configurations (Illangasekare et al. 1995a,b). The distribution of entrapped NAPL in space is referred to as entrapment architecture. It is usually characterized by isolated zones of NAPL residuals and ganglia entrapped in multiple pores, and pools of high saturations formed at the texture interfaces. The mass transfer that occurs at the NAPL/water interface produces a groundwater contaminant plume downstream of the source area. In this section, phenomenological-based mass transfer models used to predict the up-scaled effective dissolution rate coeDcients are presented.

In porous media applications, a simple conceptual model for NAPL dissolution is based on the stagnant film theory. It assumes that the mass transfer from the NAPL phase to the water phase occurs through a stagnant film adjacent to the interface. Based on this theory, the rate at which mass transfer occurs is represented through a linear driving model that includes a lumped mass transfer rate coeDcient K_c [T^{-1}] defined as (Pankow and Cherry 1996)

$$Q = K_c(C_s - C), \qquad (19.22)$$

where

Q [M L^{-3} T^{-1}] is the flux of mass, being the source term in Equation 19.14
C_s [M L^{-3}] is the aqueous phase concentration under conditions when the NAPL is in thermodynamic equilibrium with the solute in the aqueous phase (i.e., the solubility limit of NAPL in water)

Phenomenological models that are used to predict the mass transfer rate coeDcient K_c are referred to as Gilland–Sherwood models. In chemical engineering applications for simple systems such as dissolution from the surface of a tube, analytical expressions can be derived for the mass transfer rate coeDcients. However, for porous media applications, the expressions that are used to predict the overall mass transfer rate coeDcients are empirically based. Such a Gilland–Sherwood model developed and validated by Saba and Illangasekare (2000) is given as

$$Sh' = aRe^{\beta}Sc^{\alpha}\left(\frac{\theta_N d_{50}}{\tau L}\right)^{\eta}, \qquad (19.23)$$

where Sh' is a modified form of Sherwood number used for porous media applications defined as

$$Sh' = \frac{K_c d_{50}^2}{D_{mol}}, \qquad (19.24)$$

and Re and Sc are the Reynolds and the Schmidt number, respectively. The terms a, β, α, and η are empirical coeDcients. τ is the tortuosity factor of the flow path, L [L] is the dissolution length, d_{50} [L] is the mean grain size, θ_N is the volumetric NAPL content,

and D_{mol} [L^2 T^{-1}] is the molecular diffusion coeDcient for the soluble constituent. The corresponding model that was fitted by Saba and Illangasekare (2000) to dissolution data obtained in a 2D flow configuration in a small test tank, is

$$Sh' = 11.34Re^{0.28}Sc^{0.33}\left(\frac{\theta_N d_{50}}{\tau L}\right)^{1.037}. \qquad (19.25)$$

Similar expressions for other test systems have been reported in literature (see Illangasekare et al. 2010).

The prediction of effective dissolution from the source zone with complex entrapment architecture requires up-scaling of the mass transfer from the NAPL/water interfaces that occur at the pore scale to field scale characterized by spatially varying soil properties. The up-scaling method proposed by Saenton and Illangasekare (2007) is based on the recognition that when water passes through a source zone with complex NAPL entrapment architecture, the net mass generation is a result of the mass contribution from the isolated pockets of NAPL mixing with the water that bypasses the NAPL zones. The up-scaled Gilland–Sherwood type model uses a geostatistical characterization of soil heterogeneity and the statistics that describe how the NAPL saturation is distributed in the source zone. The up-scaled mass transfer correlation is given by (Saenton and Illangasekare 2007)

$$\overline{Sh} = Sh_0(1+\sigma_Y^2)^{\phi_1}\left(1+\frac{\Delta z}{\lambda_z}\right)^{\phi_2}\left(\frac{\hat{M}_{II,z}}{\hat{M}_{II,z}^*}\right)^{\phi_3}, \qquad (19.26)$$

where

$\overline{Sh}$ is the up-scaled Sherwood number containing the effective mass transfer rate coeDcient
Sh_0 is the Sherwood number defined at the laboratory measurement scale (Equation 19.25)
σ_Y^2 is the variance of the ln K field
Δz is the vertical dimensions of the simulation grid
λ_z is the vertical correlation length
ϕ_1, ϕ_2, and ϕ_3 are empirical fitting parameters and the last set of terms is the dimensionless second moment of the vertical saturation distribution determined from the NAPL entrapment architecture

19.3 Methods of Analysis

19.3.1 Solution to Miscible Transport Problems

Space-time distributions of miscible pollutant concentration are obtained by solving the coupled initial and boundary value problems of flow and transport, either sequentially or simultaneously. When pollutant concentration is low, the bulk density of the flowing fluid remains virtually constant and the flow problem can be solved independently to obtain space-time distributions of seepage velocity that is required as input to the advection-dispersion equation. At higher solute concentration, the density of the flowing fluid becomes a function of concentration.

The aforementioned flow partial differential equations (PDEs) have to be complemented by terms accounting for this additional dependency and the flow and transport problems have to be solved simultaneously. This case is not considered here.

Two example solutions of potentially high interest are presented here: (1) a 1D solution applicable to solute infiltration in the soil coupled to transient unsaturated flow, and (2) the widely used approximate solution of Domenico (1987), which describes the 3D migration of a solute plume in a uniform flow field. There exists a large variety of other analytical solutions to the advection-dispersion equations documented elsewhere. Kreft and Zuber (1978) present a variety of solutions to the 1D form of the advection-dispersion equation under the assumption of constant specific discharge and water content. They also discuss their corresponding ranges of applications. Several solutions to solute migration in non-uniform flow fields, usually applied to analyze to movement of miscible tracers in the vicinity of one or multiple wells, are listed by Domenico and Schwartz (1997) and Illangasekare et al. (2010), among others.

19.3.1.1 One-Dimensional Vertical Infiltration in the Soil

Richards equation is a nonlinear equation and, even when relatively simple functional descriptions of the retention function and of relative permeability are adopted, it usually does not lend itself to the derivation of closed-form or analytical solutions. When transient unsaturated flow has to be coupled with transient pollutant migration, the problem often becomes so complex that one will have to resort to using numerical finite-difference or finite-element codes.

Analytical solutions for transient flow in the vadose zone are only available in certain simplified situations. As an example, the solution by Broadridge and White (1988) and extended by Nachabe et al. (1995) is presented here. Initially, the soil is assumed to be at a constant water content θ_n, with a corresponding hydraulic conductivity K_n. At time 0, infiltration at a constant rate q_{in} [L T^{-1}] is initiated. The problem is 1D: only downward vertical flow is considered and the water table is assumed to be deep enough that it does not influence water content. The soil is assumed to be homogeneous with a dependence of capillary pressure and relative permeability to water content described as

$$p(\theta) = -\rho g \lambda_s \left(\frac{1-\theta_*}{\theta_*} + \frac{1}{C} \log\left(\frac{C-\theta_*}{\theta_*(C-1)} \right) \right), \quad (19.27)$$

$$K(\theta) = (K_s - K_n)\left(\frac{C-1}{C-\theta_*}\theta_*^2 \right) + K_n, \quad (19.28)$$

where
$\theta_* = (\theta - \theta_n)/(\theta_s - \theta_n)$ [–] is an effective water saturation
θ_s [L^3 L^{-3}] is the water content at saturation
K_s [L T^{-1}] the saturated hydraulic conductivity

The subscript w used previously was dropped here for the sake of brevity. $C > 1$ [–] controls the shape of the retention function and λ_s [L] can be interpreted as the capillary height. The solution by Broadridge and White (1988) requires the introduction of an additional parameter ζ. The depth of the point under consideration and its corresponding water content are found from

$$z = \frac{\lambda_s}{C}(\rho^2(1+\rho^{-1})\tau + \rho(2+\rho^{-1})\zeta - \ln u), \quad (19.29)$$

$$\theta = (\theta_s - \theta_n)C\left(1 - \left(2\rho + 1 + u^{-1}\frac{du}{d\zeta}\right)^{-1}\right) + \theta_n, \quad (19.30)$$

where

$$\rho = \frac{1}{4C(C-1)}\left(\frac{q_{in}-K_n}{K_s-K_n} \right), \quad (19.31)$$

$$\tau = 4C(C-1)\frac{t}{\lambda_s}\left(\frac{K_s-K_n}{\theta_s-\theta_n} \right), \quad (19.32)$$

and

$$u = \frac{1}{2}\exp\left(\frac{-\zeta^2}{\tau} \right)\left(2\exp\left(\frac{(\zeta+\rho\tau)^2}{\tau} \right) + \left(f_1^+ + f_1^-\right) - \left(f_2^+ + f_2^-\right) \right). \quad (19.33)$$

The additional functions appearing in Equation 19.33 are

$$f_1^\pm = f\left(\frac{\zeta \pm \rho\sqrt{1+\rho^{-1}}\tau}{\sqrt{\tau}} \right), \quad f_2^\pm = f\left(\frac{\zeta \pm \rho\tau}{\sqrt{\tau}} \right), \quad \text{and}$$

$$f(x) = \exp(x^2)\text{erfc}(x). \quad (19.34)$$

An analytical expression for $du/d\zeta$ can be easily found from Equations 19.33 and 19.34. The solution is fully analytical and can be easily evaluated using e.g., a spreadsheet. An example application will be presented in Section 19.4.1.

Nachabe et al. (1995) extended the analytical solution to describe the migration of a conservative miscible species that moves downward with the infiltrating front described by Equations 19.29 and 19.30. The full solution includes the effect of downward advection and longitudinal dispersion and would be rather lengthy to report here. However, if initial water content and hydraulic conductivity are zero and only advection is considered, the depth of the front of pollutant z_f at time t can be computed from

$$z_f = \frac{\lambda_s}{C}(\zeta_f + \rho\tau), \quad (19.35)$$

where ζ_f is a solution of

$$(\rho^2 C_1 + \rho\theta_n)\tau + (2\rho C_1 + \theta_n)\zeta - C_1 \ln(u) = 0, \qquad (19.36)$$

and $C_1 = C\,(\theta_s-\theta_n) + \theta_n$. The velocity of the front is therefore not constant and decreases with depth and time. An evaluation of Equation 19.35, together with the full solution by Nachabe et al. (1995) will be presented in Section 19.4.1.

19.3.1.2 Three-Dimensional Transport in a Uniform Flow Field

Tools for evaluating the 3D movement of a solute plume are required in a variety of applications, including tracer tests analysis, modeling pollutant migration from a point source or from a source of finite extent, or even risk assessment, and evaluation of natural attenuation.

Domenico (1987) proposed an approximate analytical solution to the advection-dispersion equation when solute transport occurs in a homogeneous aquifer from a source of finite extent. For an inert non-decaying solute continuously released at a concentration C_0, the solution takes the following form

$$\frac{C}{C_0} = \frac{1}{8}\,\mathrm{erfc}\left(\frac{x-vt}{2\sqrt{D_L t}}\right) \times \left(\mathrm{erf}\left(\frac{y+W/2}{2\sqrt{\alpha_{TH} x}}\right) - \mathrm{erf}\left(\frac{y-W/2}{2\sqrt{\alpha_{TH} x}}\right)\right)$$

$$\times \left(\mathrm{erf}\left(\frac{z+H/2}{2\sqrt{\alpha_{TV} x}}\right) - \mathrm{erf}\left(\frac{z-H/2}{2\sqrt{\alpha_{TV} x}}\right)\right), \qquad (19.37)$$

where W [L] and H [L] are the width and the height of the rectangular source zone. Although of approximate nature, Domenico's solution is relatively widely used. As the solution is symmetrical with respect to the vertical axis, it can be adopted to model solute transport originating either from a finite source located at middle depth in a infinitely thick aquifer, or from a finite source located at the top of a semi-infinite aquifer.

19.3.2 Solution to Immiscible Flow Transport Problems

In field applications involving the prediction of NAPL migration and recovery, the nonlinear and coupled governing system of equations that describe immiscible flow in porous media have to be solved using numerical codes. A number of such codes exist and they have been extensively used in a variety of applications involving multiphase flow in porous media that includes (e.g., NAPL flow or carbon sequestration). The most commonly used modeling codes that are referenced in the literature are the TOUGH family of codes developed at Lawrence Berkeley Laboratory (Pruess et al. 1999), STOMP developed by the Pacific NorthWest Laboratory (White and Oostrom 2006), FEHM developed by Los Alamos National Laboratory (Zyvoloski 2007)

and UTCHEM developed at the University of Texas at Austin (Pope et al. 1999).

Solutions to the multiphase phase flow equations for simplified systems have been developed and reported in many text books and papers (e.g., Morel-Seytoux 1973, Marle 1981). These solutions are typically useful to estimate front travel times and effectiveness of recovery. One of these solutions is referred to as the Buckley–Leverett solution. This solution has been widely used in petroleum engineering applications to analyze water flooding problems (water displacing oil in 1D systems). The same solution can be used to analyze problems involving displacement of a NAPL by water, simulating a recovery operation. The derivation and the solution of the Buckley–Leverett equation, first presented by Buckley and Leverett (1942), have been documented in a number of text books and publications (e.g., Morel-Seytoux 1973, Marle 1981, Helmig 1997). The derivation and solution presented by Marle (1981) is reproduced here for a two-phase water-NAPL system.

For a two-phase flow in a 1D column where a wetting fluid w is displaced by a non-wetting NAPL phase N, Equation 19.12 can be reduced to a second-order partial differential equation that solves for the wetting fluid (water) saturation (Marle, 1981)

$$\frac{v}{\phi}\left[\frac{d\phi_w}{dS_w}\frac{\partial S_w}{\partial z} + \frac{\partial}{\partial z}\left(\psi_w \frac{\partial S_w}{\partial z}\right)\right] + \frac{\partial S_w}{\partial t} = 0. \qquad (19.38)$$

The parameters that appear in Equation 19.38 depend on the constitutive models adopted for the porous media and fluid properties. First, for an incompressible porous medium, using the equation of continuity it can be shown that the total velocity v remains constant along the length of the column and is defined as

$$v = v_w + v_N, \qquad (19.39)$$

where

v_w is the Darcy velocity of water
v_N is the velocity of the NAPL phase

As v is space invariant, it can be defined using the flux boundary conditions at the entrance of the column.

The functional parameters ϕ_w and ψ_w depend on multiphase fluid and porous media system properties. They are defined as

$$\phi_w = \frac{\dfrac{\mu_N}{k_{rN}} + \dfrac{K_i(\rho_w - \rho_N)g}{v}}{\dfrac{\mu_w}{k_{rw}} + \dfrac{\mu_N}{k_{rN}}}, \qquad (19.40)$$

and

$$\Psi_w = \frac{K_i}{v}\frac{1}{(\mu_w/k_{rw}) + (\mu_N/k_{rN})}\frac{dP_c}{dS_w}, \qquad (19.41)$$

K_i [L^2] being the isotropic intrinsic permeability of the soil. The basic assumption that was made by Buckley–Leverett is that capillary gradients are small and hence ψ_w can be neglected. In that case, Equation 19.38 reduces to

$$\frac{v}{\phi}\left[\frac{d\phi_w}{dS_w}\frac{\partial S_w}{\partial z}\right]+\frac{\partial S_w}{\partial t}=0. \tag{19.42}$$

Moreover, the parameter ϕ_w becomes equal to the fractional flow function defined as

$$f_w=\phi_w=\frac{v_w}{v}. \tag{19.43}$$

Equation 19.42, referred to as Buckley–Leverett equation, is a first-order partial differential equation with unknown $S_w(z,t)$. The solution to this equation for one fluid displacing another along the length of a column can be obtained using the method of characteristics (MOC). The MOC converts the partial differential equation to a system of ordinary differential equations. Following the approach presented by Marle (1981), the three ordinary differential equations that are referred to as the "characteristic" system of equations are written as

$$\frac{dz}{d\lambda}=\frac{v}{\phi}\frac{d\phi_w}{dS_w}, \tag{19.44}$$

$$\frac{dt}{d\lambda}=1, \tag{19.45}$$

$$\frac{dS_w}{d\lambda}=0, \tag{19.46}$$

where λ is a parameter. The characteristics as defined by Equations 19.45 and 19.46 are straight lines parallel to a plane on which $S_w = 0$. The slope of these characteristic lines dz/dt defines the rate of propagation of any specified value of S_w, and is given by

$$\left(\frac{dz}{dt}\right)_{S_w}=\frac{v}{\phi}\frac{df_w}{dS_w}. \tag{19.47}$$

This equation allows one to create the saturation profile of S_w along the column as a function of time by tracking the rate at which a given S_w value propagates. In the implementation of this scheme, the profile will develop into a physically unrealistic saturation profile that results in two saturation values for at the same spatial location. Methods have been developed to correct this situation by defining a sharp saturation front that maintains mass balance. This method will be demonstrated through an example in the next section.

A few other methods to solve Equation 19.38 without neglecting the capillary terms have been reported. McWhorter and

Sunada (1990) presented a quasi-analytical solution without neglecting the capillary effects. The solution method is restricted to a horizontal column with a very specific flux boundary condition. More recently, this solution method was extended using quasi-numerical methods to model flow in a column for more general boundary conditions (Fulík et al. 2007), a medium with discontinuity (Fulík et al. 2008) and allowing for dynamic effects in the two-phase retention function (Fulík et al. 2010).

19.4 Applications

19.4.1 Vertical Infiltration in the Vadose Zone

Large quantities of radioactive and chemical pollutants are present in the vadose zone at the Hanford site, north of Richland, WA, as a result of long-lasting production of plutonium for nuclear weapons. While the subsurface at the Hanford site is characterized by highly stratified sediments that result in complex subsurface-flow paths, simplified analyses can be conducted to obtain rough estimates of pollutant travel distances, e.g., during a rain event.

The solutions of Broadridge and White (1988) and Nachabe et al. (1995) are used to make a rough estimate of travel distances of Tritium from the ground surface downward in the vadose zone. Tritium can be considered as an inert miscible pollutant. Infiltration rate will be varied from $K_s/100$ to $K_s/2$. At the Hanford site, transport of a mobile species such as Tritium is usually linked to elevated recharge rates that occur primarily where land surfaces are void of vegetation. The high values used here are for illustrative purpose only and could be associated to an uncontrolled spill e.g., from a pond.

Typical hydraulic properties of the Hanford sand are summarized in Table 19.1. The soil is assumed to be homogeneous and initially at residual water content, with a corresponding zero hydraulic conductivity. Vertical profiles of water content after 1, 3, and 6h. are shown in Figure 19.2a. It can be seen that steady-state water content is highly dependent on the infiltration rate. When q_{in} increases, the front of water propagates faster and the maximum water content is higher. It must be noted that a value of infiltration rate larger that the saturated hydraulic conductivity is not acceptable within the framework developed by Broadridge and White (1988) as it will lead to ponding at the surface. Such a surface

TABLE 19.1 Flow and Transport Parameters at the Hanford Site

Parameter	Symbol	Value	Units
Initial water content	θ_n	0.027	—
Maximum water content	θ_s	0.349	—
Saturated hydraulic conductivity (assumed isotropic)	K_s	2.27×10^{-5}	m s^{-1}
Initial hydraulic conductivity	K_n	0	m s^{-1}
Parameter of the Broadridge and White model	C	1.51	—
Parameter of the Broadridge and White model	λ_s	0.14	m
Longitudinal dispersivity	α_L	0.203	m

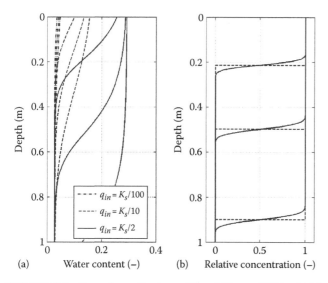

(a) Water content (−)   (b) Relative concentration (−)

FIGURE 19.2 Example applications of Broadridge and White (1988) and Nachabe et al. (1995) solutions. (a) Water content profiles after 1, 3, and 6 h for different infiltration rates, and (b) concentration distributions after 1, 3, and 6 h under an infiltration rate $q_{in} = K_s/2$, with and without considering mechanical dispersion.

TABLE 19.2 Example Properties of Sand and Fluids for One-Dimensional Infiltration in a Laboratory Column

Parameter	Symbol	Value	Units
Porosity of sand	ϕ	0.20	—
Absolute or intrinsic permeability (assumed isotropic)	K_i	10^{-6}	m^2
Brooks–Corey constitutive model pore index	λ	2.5	—
Density of water	ρ_w	1000	$kg\ m^{-3}$
Density of NAPL	ρ_N	1460	$kg\ m^{-3}$
Viscosity of water	μ_w	0.001	$kg\ m^{-1}\ s^{-1}$
Viscosity of NAPL	μ_n	0.0009	$kg\ m^{-1}\ s^{-1}$

horizontal column of length 250 m filled with sand. The column is initially filled with a NAPL with a constant residual water saturation of 0.2 along the length of the column and a water flood is applied at the end of the column at a rate of 1.5×10^{-4} kg m^{-2} s^{-1} while maintaining the water saturation at 0.8. The properties of the sand and the fluids are given in Table 19.2.

In a horizontal column, the fractional flow function in Equation 19.40 can be written as

boundary condition requires a different mathematical treatment. Figure 19.2b shows the concentration profiles corresponding to the highest infiltration rate. The piston flow solution (Equation 19.35) is compared to the full solution. It can be seen that the concentration profile is much sharper than the water content profile: The shape of water distributions is controlled by the retention function and the relative permeability, while the shape of concentration distributions is controlled by dispersion.

$$\phi_w = f_w = \frac{\dfrac{\mu_N}{k_{rN}}}{\dfrac{\mu_w}{k_{rw}} + \dfrac{\mu_N}{k_{rN}}} = \frac{1}{1 + \dfrac{\mu_w}{k_{rw}}\dfrac{k_{rN}}{\mu_N}}. \qquad (19.48)$$

19.4.2 Modeling of Immiscible Flow in a Column

An example is presented to demonstrate how the Buckley–Leverett method is used to solve a problem of water displacing a NAPL in a one-dimensional horizontal column. Consider a

Figure 19.3a shows the relative permeabilities and the fractional flow function for varying water saturation. The relative permeability functions that appear in the expression are calculated using the Brooks–Corey constitutive models given in Equations 19.11a and 19.11b.

Since capillary effects are neglected, simulating water saturation distributions along the column using Equation 19.47 results in physically unrealistic situations where two water saturations occur at the same longitudinal location along the column.

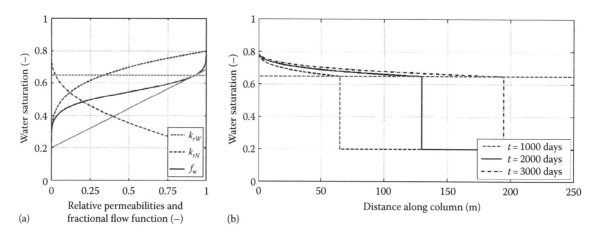

(a) Relative permeabilities and fractional flow function (−)   (b) Distance along column (m)

FIGURE 19.3 (a) Relative permeability and fractional flow function as a function of water saturation. (b) Water saturation distribution simulated using Buckley–Leverett solution along the column at 1000, 2000, and 3000 days, respectively.

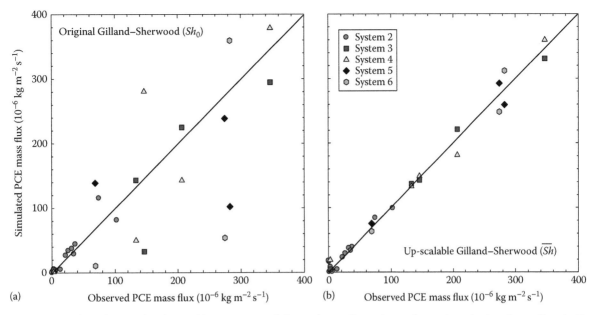

FIGURE 19.4 Up-scaling of NAPL dissolution. (a) Comparison of observed mass flux and mass flux estimated using the small-scale dissolution model, (b) comparison of observed mass flux and mass flux estimated using the up-scaled dissolution model (system 2–5 refer to different grid sizes used in the simulations).

Welge (1952) suggested a graphical procedure to determine the saturation of a shock front of that maintains mass balance, using the fractional flow function. This saturation at the shock front is determined by constructing a tangent to the fractional flow function from the initial water saturation in the column as shown on Figure 19.3a. The front saturation in this case becomes 0.65. Figure 19.3b shows the saturation profile of the water front propagating along column. This simple example illustrates how it is possible to determine the rate of propagation of the wetting fluid displacing a NAPL.

19.4.3 Up-Scaling of NAPL Dissolution

The data from a laboratory intermediate scale study is used to demonstrate how the up-scaling method that was presented in Section 19.2.3.2 was validated. Natural dissolution experiments were conducted within a large-scale tank having dimensions of 4.88 m × 1.22 m × 0.05 m. The tank was packed with five sands to represent a spatially correlated random field. The geostatistical parameters that characterize the test aquifer are mean of log transformed hydraulic conductivity (m day^{-1}) of 4.18, variance of 1.22, horizontal correlation length of 0.50 m, and vertical correlation length of 0.05 m.

A DNAPL source zone was created by injecting 587 mg of PCE and the saturation distribution was mapped using gamma attenuation method. Controlling the heads at the upstream and downstream ends of the tank created a steady groundwater flow field along the length of the tank. Down-gradient distribution of dissolved PCE was monitored by extracting aqueous samples at three vertical arrays. The observed DNAPL source zone was discretized into a very fine grid and observed dissolution data were used to calibrate the mass transfer model

by fitting the empirical parameters of the small-scale Gilland–Sherwood correlation (similar to Equation 19.23). The source zone was then discretized into five different grid systems consisting of different number of columns and layers. The mass transfer simulations were made for each grid systems using the small-scale and up-scaleable Gilland–Sherwood correlations (Equation 19.26). For simulations with the up-scaleable correlation, an inverse modeling algorithm was used to determine the empirical parameters ϕ_1, ϕ_2, and ϕ_3 in Equation 19.26. Simulated concentrations and mass fluxes are plotted against observed values for both small-scale and up-scaleable Gilland–Sherwood correlations in Figure 19.4a and b. These figures show that the small-scale Gilland–Sherwood correlation was not able to simulate dissolution in the larger grid systems. However, as shown in Figure 19.4b, concentration and flux obtained with the up-scaleable correlation matched relatively well with observed values.

19.5 Major Challenges

19.5.1 Site Characterization

Natural subsurface systems are intrinsically heterogeneous and 3D variations of physical and chemical properties of the soil influence pollutant transport drastically. Pore-scale processes occurring at the centimeter scale and below are usually well understood through laboratory investigation, and classical field methods generally allow a proper characterization of large geologic units at the kilometer scale. As soil properties will usually vary in complicated ways from the decimeter to the kilometer scale, the main challenge in predicting pollutant transport for field applications lies in finding a proper characterization of the processes occurring at such intermediate scales. The challenge of site characterization is currently being tackled from two opposite sides: (1) by improving field methods to collect data and (2) by improving models and methods to analyze field data and characterize soil and pollutant transport processes.

The identification of the variations of physical and chemical properties of soils, including spatial distributions of pollutant concentrations, using improved field characterization methods is constantly receiving a growing attention. Field methods can be categorized into static and dynamic methods, the latter involving a solicitation of the subsurface system and requiring the follow up of its response in time. Pumping tests and tracer tests are common dynamic tools applied to the characterization of subsurface systems. The advancing edge of geophysics also provides new methods to collect spatially distributed data, such as time-domain reflectometry, ground penetrating radar or electrical resistivity tomography, that are currently being applied to the characterization of subsurface transport problems. Other geophysical approaches include electromagnetic surveys, crosshole, downhole, and surface tomography. Adequate combinations of multiple characterization techniques also allow gaining significant insights into subsurface systems. For example, combining dynamic resistivity imaging with a tracer test should permit the tracking of a saline tracer plume in time, and infer a 3D map of the variations of subsurface hydraulic properties.

Characterization of source zones with entrapped DNAPLs to estimate mass and saturation distribution remains a major challenge as the unstable behavior in combination with the geologic heterogeneity results in the separate phase migrating deep into the formation to locations that are not easy to detect. Intrusive methods such as soil boring, has the potential for suspended pools of DNAPL at soil texture interfaces to be mobilized, thus moving the DNAPL to deeper locations that are again diDcult to detect. Non-intrusive methods such as the use of partitioning tracers (Jin et al. 1995) have been developed. A partitioning tracer test is conducted by injecting several tracers with different water-NAPL partition coeDcients up-gradient of a NAPL source zone and monitoring tracer migration at down-gradient locations. In most applications, this technique has been only successful to assess remediation performance (Rao et al. 1997). Partitioning tracer tests also are expected to underestimate DNAPL volume in certain conditions where the hydrodynamic accessibility is constrained or where non-equilibrium mass transfer occurs (Rao et al. 2000, Dai et al. 2001). Methods that combine tracer methods with the inversion of the plume concentration data collected using multi-level samples have the potential to better characterize source zone architecture (Saenton and Illangasekare 2004, Moreno-Barbero and Illangasekare 2005, 2006, Moreno-Barbero et al. 2007). However, field implementation of this technique still requires further development. Geophysical methods have the potential to characterize DNAPL source zones non-intrusively, but they again have not been developed adequately for field applications.

Since only a partial characterization of the soil can be obtained from current field methods, a second way of improving the prediction of pollutant transport relies on the use of up-scaling methods. Up-scaling methods link parameters characterizing pollutant transport, such as dispersivity or mass transfer coefficients, to a simplified representation of soil heterogeneity. Most of these up-scaling methods are generally stochastic methods based on a geostatistical description of heterogeneity (Gelhar 1993), in which soil properties are assumed to vary in a given limited range and within a given spatial scale. Among others, such up-scaling methods have been shown to be powerful tools when characterizing solute dispersion in mildly heterogeneous media (Gelhar 1993).

19.5.2 Dealing with Uncertainty

While geophysical methods contribute largely to the improvement of soil characterization at intermediate scales, their results are usually affected by larger uncertainties and the fine spatial resolution needed to completely solve field-scale pollutant transport is currently out of reach of other field characterization methods. Economical costs also limit the range of field investigation techniques used at a given site. As a result 3D variations of physical and chemical properties of subsurface materials are currently never fully characterized in a deterministic way. However, as subsurface heterogeneity usually results from formation processes (such as e.g., sedimentation), it is not fully random either. Stochastic analysis enables then the variability and the uncertainty in flow and pollutant transport to be related to variability and the uncertainty associated to physical and chemical properties of the heterogeneous medium under consideration.

In applications related to flow through porous formations, uncertainty can also arise from other simplifying assumption adopted to conceptualize pollutant transport. Most of the time, source zone and intensity are not fully mastered. Uncertainty can also arise from other assumptions, such as those linked e.g., to the position of geologic layer boundaries, to far-field boundary conditions, or initial conditions, can be addressed through scenario analysis, or multi-model analysis. In multi-model analysis several alternative hydrologic models are constructed for a given site. Model selection criteria are used to (1) rank these models, (2) eliminate some of them, and/or (3) weigh and average predictions and statistics generated by multiple models (Poeter and Anderson 2005).

References

Bear, J. 1972. *Dynamics of Fluids in Porous Media*. New York: American Elsevier Publishing Company.

Broadridge, P. and I. White. 1988. Constant rate rainfall infiltration: A versatile nonlinear model. 1. Analytical solution. *Water Resources Research* 24: 145–154.

Brooks, R. H. and A. T. Corey. 1964. *Hydraulic Properties of Porous Media*. Hydrological paper No. 3, Fort Collins, CO: Colorado State University.

Buckley, S. E. and M. C. Leverett. 1942. Mechanism of fluid displacement in sand. *Transactions of the AIME* 146: 107–116.

Corey, A. T. 1994. *Mechanics of Immiscible Fluids in Porous Media.* Highlands Ranch, CO: Water Resources Publications.

Dai, D., F. T. Barranco Jr., and T. H. Illangasekare. 2001. Partitioning and interfacial tracers for differentiating NAPL entrapment configuration: Column-scale investigation. *Environmental Science and Technology* 35: 4894–4899.

Domenico, P. A. 1987. An analytical model for multidimensional transport of a decaying contaminant species. *Journal of Hydrology* 91: 49–58.

Domenico, P. A. and F. W. Schwartz. 1997. *Physical and Chemical Hydrogeology.* New York: John Wiley & Sons.

Fetter, C. W. 1999. *Contaminant Hydrogeology.* Upper Saddle River, NJ: Prentice Hall.

Fulík, R., J. Miyska, M. Benes, and T. H. Illangasekare. 2007. An improved semi-analytical solution for verification of numerical models of two-phase flow. *Vadose Zone Journal* 6: 93–104.

Fulík, R., J. Miyska, M. Benes, and T. H. Illangasekare. 2008. Semianalytical solution for two-phase flow in porous media with a discontinuity. *Vadose Zone Journal* 7: 1001–1007.

Fulík, R., J. Mikyska, T. Sakaki, M. Benes, and T. H. Illangasekare. 2010. Significance of dynamic effect in capillary during drainage experiments in layered porous media. *Vadose Zone Journal* 9: 697–708.

Gelhar, L. W. 1993. *Stochastic Subsurface Hydrology.* Englewood Cliffs, NJ: Prentice Hall.

Gelhar, L. W., C. Welty, and K. R. Rehfeldt. 1992. A critical review of data on field-scale dispersion in aquifers. *Water Resources Research* 28: 1955–1974.

van Genuchten, M. Th. 1980. A closed-form equation for predicting the hydraulic conductivity of unsaturated soils. *Soil Science Society of America Journal* 44: 892–898.

Held, R. J. and T. H. Illangasekare. 1995. Fingering of dense nonaqueous phase liquids in porous media. 1. Experimental investigation. *Water Resources Research* 31: 1213–1222.

Helmig, R. 1997. *Multiphase Flow and Transport Processes in the Subsurface.* Berlin, Germany: Springer.

Illangasekare, T. H., C. C. Frippiat, and R. Fulík. 2010. Dispersion and mass transfer coeDcients in groundwater of near-surface geologic formations. In *Handbook of Chemical Mass Transport in the Environment*, eds. L. J. Thibodeaux and D. Mackay. Boca Raton, FL: Taylor & Francis Group.

Illangasekare, T. H., J. L. Ramsey, K. H. Jensen, and M. Butts. 1995a. Experimental study of movement and distribution of dense organic contaminants in heterogeneous aquifers. *Journal of Contaminant Hydrology* 20: 1–25.

Illangasekare, T. H., D. N. Yates, and E. J. Armbruster. 1995b. Effect of heterogeneity on transport and entrapment of nonaqueous phase waste products in aquifers: An experimental study. *ASCE Journal of Environmental Engineering* 121: 572–579.

Jin, M. et al. 1995. Partitioning tracer tests for detection, estimations, and remediation performance assessment of subsurface nonaqueous phase liquids. *Water Resources Research* 31: 12011–1211.

Kreft, A. and A. Zuber. 1978. On the physical meaning of the dispersion equation and its solutions for different initial and boundary conditions. *Chemical Engineering Science* 33: 1471–1480.

Kreider, J. F., N. Cook, T. H. Illangasekare, and R. H. Cohen. 1998. Source of pollutions and regulations. In *The CRC Handbook of Mechanical Engineering*, ed. F. Kreith. Boca Raton, FL: CRC Press.

La Grega, M. D., P. I. Buckingham, and J. C. Evans. 1994. *Hazardous Waste Management*, New York: McGraw-Hill.

Marle, C. M. 1981. *Multiphase Flow in Porous Media.* Houston, TX: Gulf Publishing Company.

Mayer, A. and S. M. Hassanizadeh. 2005. *Soil and Groundwater Contamination: Nonaqueous Phase Liquids: Principles and Observations.* Washington, DC: American Geophysical Union.

McWhorter, D. B. and D. K. Sunada. 1990. Exact integral solutions for two-phase flow. *Water Resources Research* 26: 399–413.

Morel-Seytoux, H. J. 1973. Two-phase flow in porous media. In *Advances in Hydrosciences*, ed. V. T. Chow. New York: Academic Press, pp. 119–202.

Moreno-Barbero, E. and T. H. Illangasekare. 2005. Simulation and performance assessment of partitioning tracer tests in heterogeneous aquifers. *Environmental and Engineering Geoscience* 11: 395–404.

Moreno-Barbero, E. and T. H. Illangasekare. 2006. Influence of pool morphology on the performance of partitioning tracer tests: Evaluation of the equilibrium assumption. *Water Resources Research* 42: W04408.

Moreno-Barbero, E., Y. Kim, S. Saenton, and T. H. Illangasekare. 2007. Intermediate-scale investigation of nonaqueous-phase liquid architecture on partitioning tracer test performance. *Vadose Zone Journal* 6: 725–734.

Nachabe, M. H., A. L. Islas, and T. H. Illangasekare. 1995. Analytical solutions for water flow and solute transport in the unsaturated zone. *Ground Water* 33: 304–310.

Pankow, J. F. and J. A. Cherry. 1996. *Dense Chlorinated Solvents and Other DNAPLs in Groundwater.* Rockwood, TN: Waterloo Educational Services.

Poeter, E. and D. Anderson. 2005. Multi-model ranking and inference in ground-water modeling. *Ground Water* 43: 597–605.

Pope, G. A., K. Sepehrnoori, M. M. Sharma, D. C. McKinney, G. E. Speitel, Jr. and R. E. Jackson. 1999. Three-dimensional NAPL fate and transport model, Report EPA/600/R-99/011. Cincinnati, OH: National Risk Management Research Laboratory, U.S. Environmental Protection Agency.

Pruess, K., C. Oldenburg, and G. Moridis. 1999. *TOUGH2 User's Guide*, Version 2.0, Report LBNL-43134. Berkeley, CA: Lawrence Berkeley National Laboratory.

Rao, P. S. C. et al. 1997. Field-scale evaluation of in situ cosolvent flushing for enhanced aquifer remediation. *Water Resources Research* 33: 2673–2686.

Rao, P. S. C., M. D. Annable, and H. Kim. 2000. NAPL source zone characterization and remediation technology performance assessment: Recent developments and applications of tracer techniques. *Journal of Contaminant Hydrology* 45: 63–78.

Saba, T. and T. H. Illangasekare. 2000. Effect of groundwater flow dimensionality on mass transfer from entrapped nonaqueous phase liquids. *Water Resources Research* 36: 971–979.

Saenton, S. and T. H. Illangasekare. 2004. Determination of DNAPL entrapment architecture using experimentally validated numerical codes and inverse modeling, In *Proceedings of XVth Computational Methods in Water Resources 2004 International Conference*, eds. C. T. Miller, M. W. Farthing, W. G. Gray and G. F. Pinder. Amsterdam, the Netherlands: Elsevier.

Saenton, S. and T. H. Illangasekare. 2007. Upscaling of mass transfer rate coeDcient for the numerical simulation of dense nonaqueous phase liquid dissolution in heterogeneous aquifers. *Water Resources Research* 43: W02428, doi:10.1029/2005WR004274.

Welge, H. J. 1952. A simplified method for computing oil recovery by gas or water drive. *Transactions of the AIME* 195: 91–98.

White, M. D. and M. Oostrom. 2006. STOMP subsurface transport over multiple phases, Version 4.0, Report PNNL-15782. Richland, WA: Pacific Northwest National Laboratory.

Zyvoloski, G. A. 2007. FEHM: A control volume finite element code for simulating subsurface multi-phase multi-fluid heat and mass transfer, Los Alamos Unclassified Report LA-UR-07–3359. Los Alamos, NM: Los Alamos National Laboratory.

FIGURE 4.2 Dyed desalination discharge from the Perth field study. (© *The West Australian*.)

FIGURE 4.7 Instantaneous uncalibrated LIF image: $d = 4.3\,\text{mm}$, initial angle $= 60°$, $F_0 = 23$ and initial Reynolds number $R_0 = 3450$.

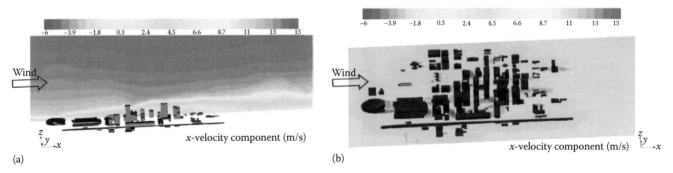

FIGURE 7.3 CFD simulation for inner core (neighborhood scale) of Oklahoma City, OK. Contours of x-velocity component (along the wind direction) at a vertical plane (a) and at a horizontal plane near the ground (b).

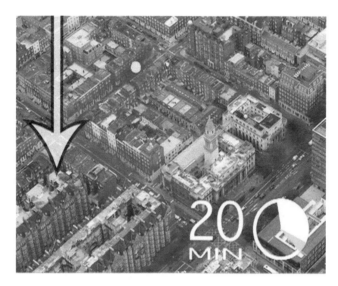

FIGURE 7.4 Air flow through London with much of the flow being along streets. Note that the marker released at the yellow circle moves upwind a little.

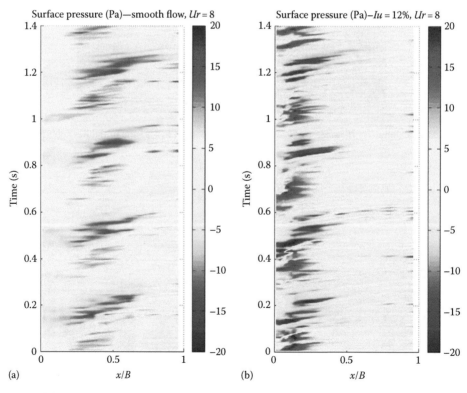

FIGURE 9.6 Surface plots of the temporal and spatial evolution of the surface pressure on a rectangular prism undergoing torsional oscillations (x/B—streamwise direction/deck width: (a) smooth flow; (b) turbulent flow). (After Haan, Jr., F.L. and Kareem, A., *J. Eng. Mech.*, 135(9), 987, 2009.)

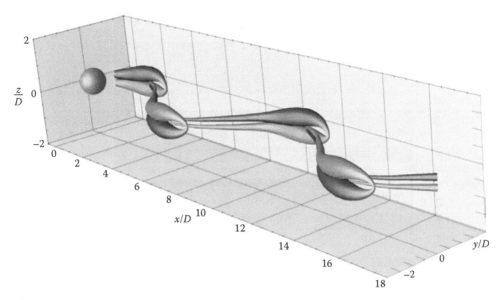

FIGURE 10.12 Three-dimensional reconstruction of the tethered sphere wake from a measured time sequence of streamwise vorticity. For all tethered sphere modes, the 3D structure clearly shows that the wake comprises a two-sided chain of streamwise vortex loops. Streamwise direction is parallel to the x-axis, while the sphere vibrates primarily transverse to the flow (y-direction). Blue indicates clockwise vorticity, and red anticlockwise vorticity. $Re \approx 3000$.

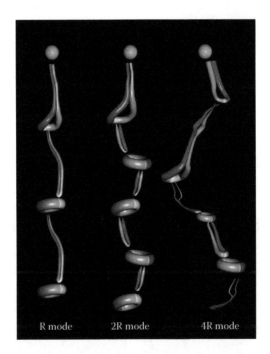

FIGURE 10.13 The family of periodic wake modes for rising and falling spheres, showing the essential vortex configurations, which are the R mode (single vortex ring per wavelength of wake), 2R mode (two rings per wake wavelength), and 4R mode (four rings per cycle of sphere vibration).

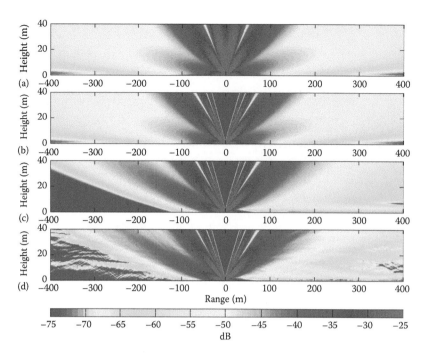

FIGURE 15.2 Calculations of the sound field above an impedance ground surface. A vertical plane, oriented along the wind direction, is shown: upwind propagation is to the left, downwind to the right. The sound source, which is positioned at zero range, has unit strength, a frequency of 400 Hz, and height of 1 m. The ground impedance, $Z_s/\rho_0 c_0 = 10 + i8.5$, is typical of grass-covered soil. For (c) and (d), a logarithmic wind profile with friction velocity of 0.5 ms^{-1} was used. (a) Calculation with no wind and no turbulence, based on an image source model. (Attenborough, K. et al. *J. Acoust. Soc. Am.*, 68, 1493, 1980.) (b) Calculation with no wind and no turbulence, using a PE method. (c) Calculation with wind but no turbulence, using a PE method. (d) Calculation with wind and turbulence, using a PE method and a von Kármán spectrum for the turbulence.

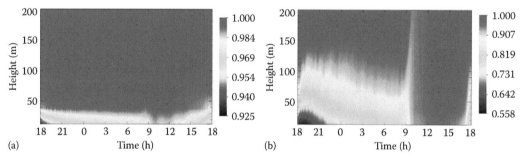

FIGURE 20.4 Time-height plots of activity ratios of ^{218}Po (a) and ^{214}Pb (b) with respect to ^{222}Rn. The S_1/S_0 activity ratios are significantly different from one during nighttime conditions. The time axis denotes local time. (From Vinuesa, J.-F., *Atmos. Chem. Phys.*, 7, 5003, 2007. With permission.)

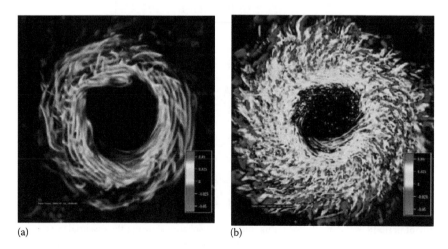

FIGURE 20.7 Vertical vorticity resulting from lower (0.185 km grid; a) and higher (0.062 km grid; b) resolution WRF simulations. Region shown (in both panels) is approximately 35 km in diameter. Note that the internal structure of the higher-resolution simulation could not be anticipated from the lower-resolution results. (From Norton, A. and Clyne, J., Analysis and visualization of high-resolution WRF hurricane simulation using VAPOR, *10th WRF Users' Workshop*, Boulder, CO, 2009. With permission.)

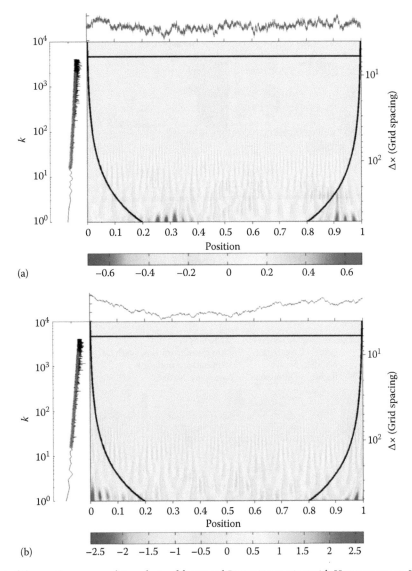

FIGURE 23.13 Real part of the continuous wavelet analysis of fractional Brownian motion with Hurst exponent $H = 0.25$ (a), and of classical Brownian motion (b).

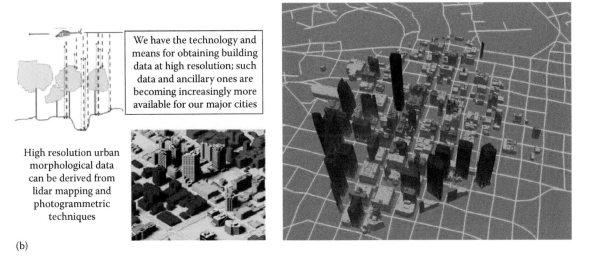

FIGURE 25.2 (b) Advanced high-resolution monitoring of morphological feature and advanced data processing tools provide a base for developing urban canopy parameters.

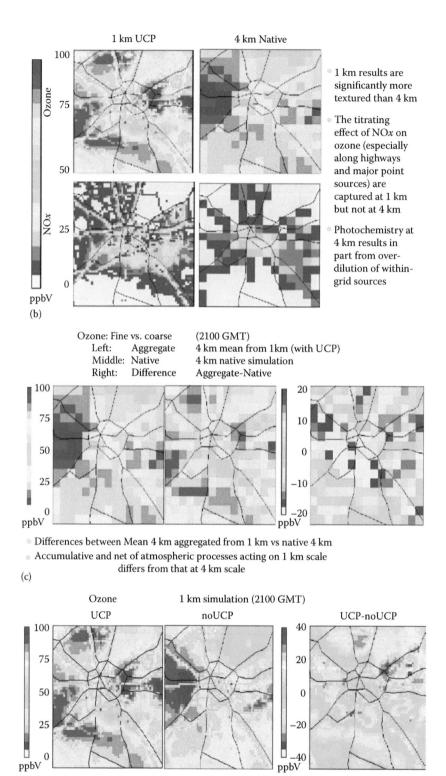

1 km UCP   4 km Native

- 1 km results are significantly more textured than 4 km

- The titrating effect of NOx on ozone (especially along highways and major point sources) are captured at 1 km but not at 4 km

- Photochemistry at 4 km results in part from over-dilution of within-grid sources

(b)

Ozone: Fine vs. coarse   (2100 GMT)
Left: Aggregate   4 km mean from 1km (with UCP)
Middle: Native   4 km native simulation
Right: Difference   Aggregate-Native

- Differences between Mean 4 km aggregated from 1 km vs native 4 km
- Accumulative and net of atmospheric processes acting on 1 km scale differs from that at 4 km scale

(c)

Ozone   1 km simulation (2100 GMT)
UCP   noUCP   UCP-noUCP

- Significant differences in the spatial patterns between UCP and noUCP runs (titration effect occurs in both sets)

- Meteorology (flow and thermodynamic) and turbulent fields differ between the simulations

(d)

FIGURE 25.4 (b) Nested CMAQ 4 and 1 km simulations. (c) Difference between aggregated fine versus coarse grid CMAQ simulations. (d) CMAQ results based on MM5 simulations run with and without UCPs.

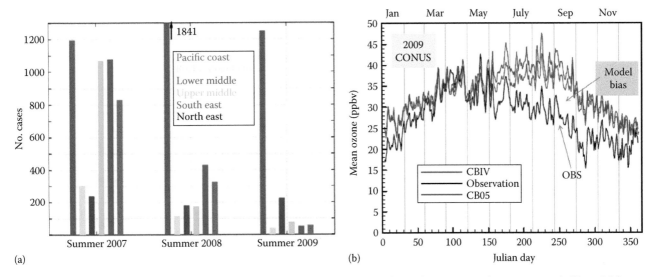

(a)

(b)

FIGURE 26.5 (a) Number of ozone exceedances in summer 2007–2009. (Courtesy of J. Gorline, NOAA, Silver Spring, MD.); (b) model forecasting biases over CONUS domain in 2009 with the CB-IV and CB05 mechanisms.

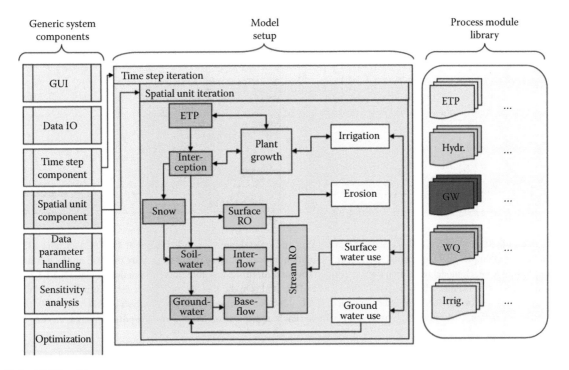

FIGURE 28.3 OMS3 architecture.

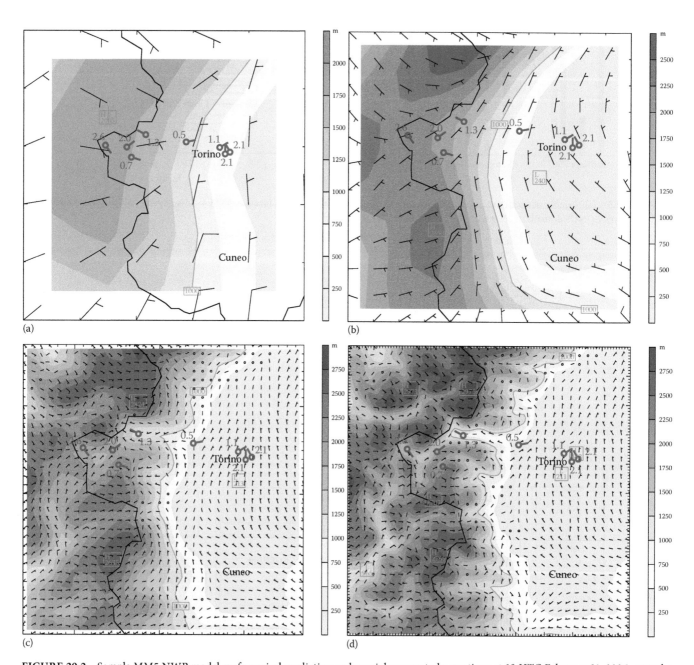

FIGURE 29.2 Sample MM5 NWP model surface wind prediction and special mesonet observations at 18 UTC February 21, 2006, over the 1.3-km Torino Winter Olympics domain area for each model resolution forecast. Surface winds (m s⁻¹) are overlaid on the terrain field (m, color code on right of figure) for each model resolution domain. (a) 36-km domain, (b) 12-km domain, (c) 4-km domain, and (d) 1.3-km domain. One full barb is 10 m s⁻¹. Dark line is France-Italy border. (After Stauffer, D.R. et al., On the role of atmospheric data assimilation and model resolution on model forecast accuracy for the Torino Winter Olympics, 11A.6, *22nd Conference on Weather Analysis and Forecasting/18th Conference on Numerical Weather Prediction*, Park City, UT, June 25–29, 2007b, available at http://ams.confex.com/ams/pdfpapers/124791.pdf).

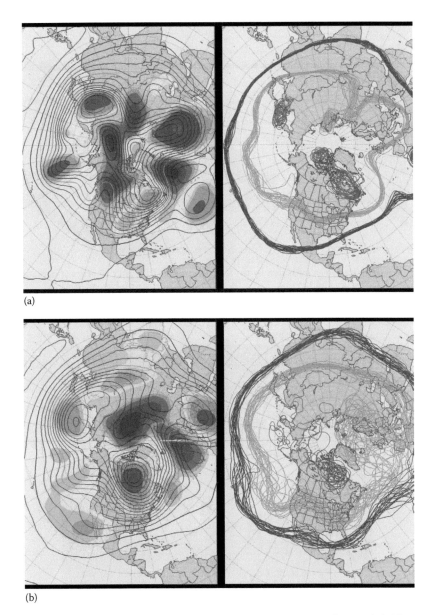

(a)

(b)

FIGURE 29.7 Sample mean and variance ("spaghetti plot") of select 500 hPa height contours from a global forecast ensemble system for the Northern Hemisphere. (a) $t = 5$ days and (b) $t = 10$ days.

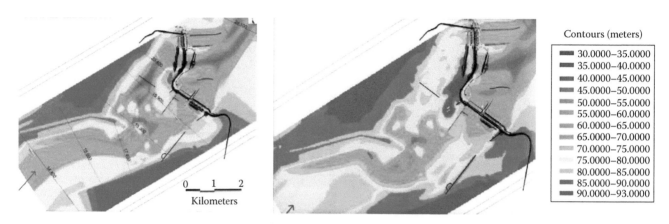

Contours (meters)

- 30.0000–35.0000
- 35.0000–40.0000
- 40.0000–45.0000
- 45.0000–50.0000
- 50.0000–55.0000
- 55.0000–60.0000
- 60.0000–65.0000
- 65.0000–70.0000
- 70.0000–75.0000
- 75.0000–80.0000
- 80.0000–85.0000
- 85.0000–90.0000
- 90.0000–93.0000

0   1   2
Kilometers

FIGURE 30.12 Upper pool bed configuration as measured by the scanner: initial (left); after 10 years of operation (middle); and on the left is contour color code indicating sedimentation of the left arm to powerhouse two up to elevations 70–75 m.

FIGURE 30.14 General view of the model during verification tests at SOGREAH.

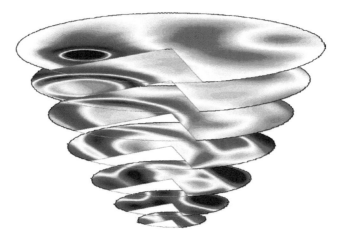

FIGURE 31.6 Density perturbations due to internal gravity waves in a paraboloidal tank. The full three-dimensional density field was reconstructed using tomographic synthetic schlieren. (From Hazewinkel, J. et al., *Exp. Fluids*, 2011.)

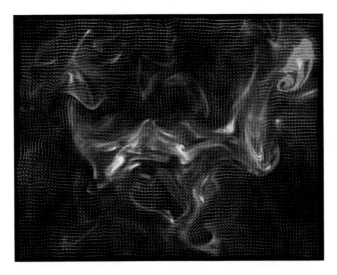

FIGURE 34.6 Simultaneous measurement of velocity and scalar using PLIF and PIV.

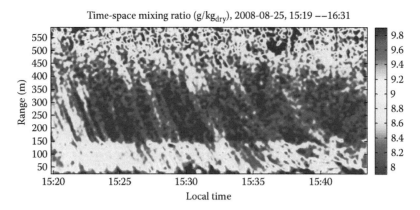

FIGURE 37.8 Contour plot of the mixing ratio of water vapor (g/kg) as a function of distance and time over a small lake in Switzerland using the EPFL Raman LIDAR.

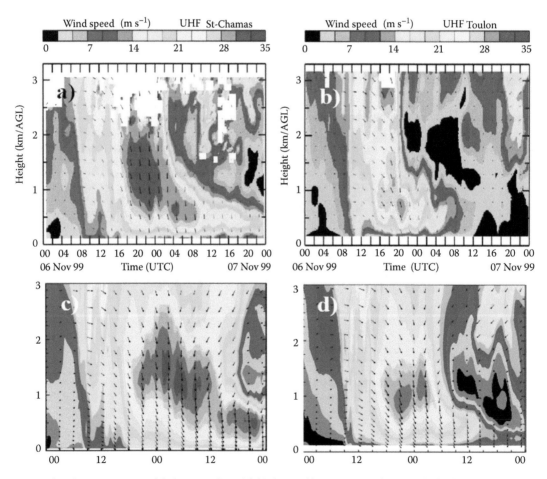

FIGURE 38.4 Time–height cross sections of the horizontal wind field observed by two UHF radars installed 90 km apart (a) at St-Chamas (EDF/ LA/CRA) and (b) at Toulon (Degréane/LSEET), and the corresponding Meso-NH simulations (c) and (d), respectively, during the MAP campaign. The wind vector is given in colored levels with superimposed arrows indicating the wind direction so that an arrow toward the bottom corresponds to a northerly wind. The same wind speed scale, i.e., 0–35 m s^{-1}, timescale, and altitude range, i.e., 0–3 km, are used in (a), (b), (c), and (d). (After Caccia, J.-L. et al., *Metorol. Zeit.* 10, 469, 2001.)

Numerical Modeling of Environmental Flows

20

Turbulent Flow Modeling

Sukanta Basu
North Carolina State University

20.1 Introduction

To this date, we neither have a rigorous definition of turbulence nor a universal theory for it. Nevertheless, there is a consensus in the scientific community that a set of partial differential equations, known as the Navier–Stokes (N–S) equations, likely contains all the ingredients to accurately model turbulence. It is also agreed upon that most of the turbulent flows we encounter in our natural environment and also in engineering practice portray a few generic characteristics including: three-dimensionality; strong unsteadiness; propensity toward intense mixing and dissipation; occurrence of fine-scale intermittency; co-existence of coherent structures and random fluctuations; chaoticity and associated (un)predictability; and presence of a wide range of spatio-temporal scales.

Even though modeling turbulent flows merely involves numerically solving the N–S equations, it is a rather challenging task and numerous scientists and engineers have been confronting it for the past 100 years or so. The persistence is not at all surprising given the immense importance of turbulence in a multitude of scientific fields. Currently, turbulence modeling activities in both academia and the industry are expanding at an incredible rate. The ongoing exponential growth in the (high-performance) computing capacity is at the root of this expansion, without any doubt.

Thus, at the advent of petascale computing, the time is opportune to document the past achievements, the present status, and the future challenges of turbulent flow modeling. This chapter attempts to provide such information in a succinct manner. In contrast to other excellent reviews on this subject (e.g., Reynolds 1976;

Rogallo and Moin 1984; Pope 2000), this chapter is focused on environmental flows rather than engineering flows.

While the fundamental methodologies for modeling engineering and environmental turbulent flows are essentially the same, the latter one is often more diﬃcult to model due to its high Reynolds number (*Re*) and also because of complex interactions among several physical processes (e.g., interactions among turbulence, radiative transfer, and cloud microphysics in atmospheric flows). Figure 20.1 visually bears witness to this assertion. These wake vortex streets over Alaska's Aleutian Islands are strikingly similar to von Karman vortex streets observed in several engineering flows involving bluff bodies. Over the years, several successful simulations of von Karman vortex streets have been reported in the engineering literature. In contrast, realistic simulations of island wake vortex streets do not exist in the atmospheric science literature,[K] because they require the faithful representation of a multitude of flow phenomena, including: unstable atmospheric boundary layer turbulence; marine stratocumulus clouds; stably stratified turbulence in capping inversion layer; and mountain waves (Young and Zawislak 2006).

There is no dearth of equally intriguing turbulent flows in various fields of environmental fluid mechanics (e.g., meteorology, limnology, and oceanography). Thus, a comprehensive account of modeling of all these diverse types of turbulent flows is beyond the scope of this chapter. Instead, we describe the basics of turbulent flow modeling by drawing numerous illustrations from the field of atmospheric boundary layers (ABLs).

[K] To the best of our knowledge, only a handful of idealized simulations have been performed so far.

Handbook of Environmental Fluid Dynamics, Volume Two, edited by Harindra Joseph Shermal Fernando. © 2013 CRC Press/Taylor & Francis Group, LLC.
ISBN: 978-1-4665-5601-0.

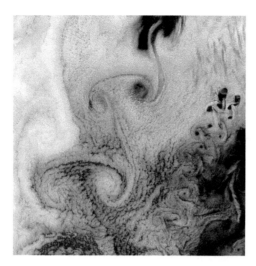

FIGURE 20.1 Formation of wake vortices over Alaska's Aleutian Islands. The cloud patterns clearly emphasize the underlying multi-scaling traits of atmospheric flows. This satellite image was acquired by Landsat 7's Enhanced Thematic Mapper plus sensor. (Image courtesy of USGS National Center for EROS, Sioux Falls, SD, and NASA Landsat Project Science Office, Greenbelt, MD.)

Owing to its high Re, ABL plays a critical role in advancing fundamental turbulence research. For decades, it has been a favorite playground for the theoretical physics community for testing a variety of universal scaling and similarity hypotheses. At the same time, ABL has immense practical importance as wide ranges of industrial (e.g., stack gas dispersion; wind energy generation), biological (e.g., pollen transport and deposition), natural (e.g., dust devil formation), and meteorological (e.g., cloud formation) activities take place in this turbulent layer.

20.2 Principles

The N–S equations for incompressible flows can be written as (using Einstein's summation notation) follows:

$$\frac{\partial u_i}{\partial t} + \frac{\partial (u_i u_j)}{\partial x_j} = -\frac{1}{\rho}\frac{\partial p}{\partial x_i} + \nu \frac{\partial^2 u_i}{\partial x_j \partial x_j} + F_i \qquad (20.1)$$

where
   t is the time
   x_j is the spatial coordinate in the j direction
   u_j is the velocity component in that direction
   p is the dynamic pressure
   ρ is the density
   ν is the kinematic viscosity
   F_i is a forcing term (e.g., geostrophic wind or imposed mean
      pressure gradient)

The individual terms of Equation 20.1, from left to right, represent inertia, advection, pressure gradient force, viscous stress, and forcing, respectively. For high Re flows, the influence of molecular viscosity is quite insignificant (except very close to the surface), so the viscous stress term is often neglected. For ABL simulations, terms describing buoyancy and rotational influences are usually included in Equation 20.1.

Direct numerical simulation (DNS) is a method to solve the N–S equations (Equation 20.1) without any averaging, any filtering, or any other approximations. Owing to its precise representation of the N–S equations, DNS-generated datasets are considered to be of extremely high fidelity and are utilized to provide a better physical understanding of various types of turbulent flows (see <http://turbulence.pha.jhu.edu> for a recent example). Quite often, these datasets are used as complement to (or even substitute of) laboratory measurements. Furthermore, these datasets play a key role in the development, improvement, and validation of turbulence parameterizations.

In DNS, all the scales of turbulent motions, from the energy-containing integral scale (L) to the energy-dissipating Kolmogorov scale (η), are computed. It is straightforward to show that $L/\eta \sim Re^{3/4}$, where Re is referenced to L. For geophysical turbulence (like ABL turbulence; $Re \sim 10^8$), L is on the order of 1 km, whereas η is on the order of 1 mm. Therefore, to perform the DNS of ABL turbulence, one would need to have on the order of 10^6 (or more) grid points in each of the three co-ordinate directions. There is little doubt that the amount of computational resources needed to perform this kind of high-performance simulation will not be available to us in the foreseeable future. So, for the time-being, the only options we have are (1) to be complacent with the DNS of low/medium Re flows (i.e., L/η on the order of 10^3) and extrapolate the results to high Re number flows via heuristic arguments (e.g., the Reynolds number similarity hypothesis); or (2) to employ computationally less expensive methods (e.g., large-eddy simulation) which solve the N–S equations with certain approximations.

In large-eddy simulation (LES), the larger scales of motion (more energetic; called resolved scale) are solved explicitly and the smaller ones (less energetic; called subgrid scale or SGS) are modeled. The separation of large and small scales is achieved via spatial filtering. When Equation 20.1 is filtered, one arrives at

$$\frac{\partial \tilde{u}_i}{\partial t} + \frac{\partial (\tilde{u}_i \tilde{u}_j)}{\partial x_j} = -\frac{1}{\rho}\frac{\partial \tilde{p}}{\partial x_i} + \nu \frac{\partial^2 \tilde{u}_i}{\partial x_j \partial x_j} + \tilde{F}_i - \frac{\partial \tau_{ij}}{\partial x_j} \qquad (20.2)$$

where tilde denotes a spatial filtering operation using a filter of characteristic width Δ_f (considerably larger than Kolmogorov scale η). The effects of the SGS (smaller than Δ_f) on the evolution of $\tilde{u}_i$ appear in the rightmost term of Equation 20.2. The SGS stress term, τ_{ij} is defined as

$$\tau_{ij} = \widetilde{u_i u_j} - \tilde{u}_i \tilde{u}_j \qquad (20.3)$$

This term arises due to the inherent nonlinearity of the advection term of Equation 20.1, which does not commute with the linear filtering operation. During the past decades, numerous SGS models have been proposed in the literature (Geurts 2003; Sagaut 2004). Each SGS model is based on distinct assumption(s). However, they share one common trait: they all attempt to mimic the forward energy cascade process by artificially draining energy from the resolved scales.

Instead of only modeling the smaller scales of turbulence (as in LES), one can opt for a further parsimonious representation by modeling the entire spectrum of scales (from integral scale to Kolmogorov scale). This approach is known as Reynolds averaged Navier–Stokes (RANS) modeling. In this approach, one is solely interested in statistically averaged flow fields and not the instantaneous ones.

In RANS modeling, every dependent variable in the N–S equations (i.e., Equation 20.1) is first decomposed into (ensemble or temporal) mean and turbulent fluctuations around the mean. This step is known as Reynolds decomposition. Then, the Reynolds averaging operation is invoked, which leads to the following RANS equations:

$$\frac{\partial \bar{u}_i}{\partial t} + \frac{\partial (\bar{u}_i \bar{u}_j)}{\partial x_j} = -\frac{1}{\rho}\frac{\partial \bar{p}}{\partial x_i} + \nu \frac{\partial^2 \bar{u}_i}{\partial x_j \partial x_j} + \bar{F}_i - \frac{\partial (\overline{u_i' u_j'})}{\partial x_j} \qquad (20.4)$$

where overbar denotes the Reynolds averaging operation. $\overline{u_i' u_j'}$, called the Reynolds stress, is parameterized by a turbulence closure model. As with the SGS models in LES, numerous RANS closure formulations exist in the literature (see Wilcox 1998).

In the case of LES, the SGS models are only applied to scales which are much smaller than the flow-dependent integral scales. Thus, in theory, these models should be universally applicable for different types of turbulent flows without much modification. However, we are yet to discover any such universal SGS model. The hope still lingers on. On the other hand, since, by design, the RANS closure models are supposed to capture all the turbulent scales, they cannot be universal. As a result, these models are always tuned for different applications.

There are some who believe that RANS is a methodology of the past and that LES will soon replace it. A recent article by Hanjalic (2005), aptly titled: "Will RANS survive LES? A view of perspectives," elaborates on this contentious topic.

20.3 Methods of Analysis

20.3.1 Direct Numerical Simulation

Key references
The review paper by Moin and Mahesh (1998) provides a comprehensive account of the direct numerical simulation (DNS) of different types of turbulent flows (e.g., homogeneous isotropic turbulence; shear-free boundary layers) and also discusses associated computational issues. The readers are also encouraged to consult the recent report by Coleman and Sandberg (2009) for further technical details on DNS.

Salient computational features

- Spectral methods or high-order finite difference schemes are typically used for DNS.
- The required computational domain for DNS is a few times larger than the integral scale of the flow under consideration.

- The required grid-resolution for DNS is on the order of the Kolmogorov scale (η). The temporal resolution is always commensurate with the spatial resolution to avoid any numerical instability issues.
- In DNS, more than 99.98% of the computational effort is usually devoted to resolving the dissipation range of turbulence (Pope 2000). In other words, less than 0.02% of the computational cost is attributed to the inertial range.
- Prescribing realistic turbulent inflows for DNS is a challenging task. For this reason, periodic boundary conditions are used whenever feasible.
- No-slip boundary conditions are used at solid walls.

Illustrative example
In this section, we will briefly describe the DNS of neutrally stratified Ekman layer flows. In the following sections, we will use the same flow in order to compare different turbulence modeling methods.

Over the past two decades, Coleman and his co-workers simulated neutrally stratified Ekman layer flows with increasing *Re*. Coleman et al. (1990) considered *Re* = 400 and 500 (based on geostrophic wind speed and the laminar Ekman layer depth). In 1999 and 2006, they conducted follow-up runs with *Re* = 1000 and 2000, respectively. This increasing trend is more or less consistent with the turbulence modeling community's current belief that the *Re* achievable by DNS can be doubled every 5–6 years.

For *Re* = 500, 1000, and 2000 runs, Coleman et al. used 128 × 128 × 50, 384 × 384 × 85, and 1024 × 1792 × 200 grid points, respectively. As theoretically expected, with increasing *Re*, the logarithmic region of the velocity profile was more pronounced (Figure 20.2a). Based on Figure 20.2b, Coleman et al. (2006) claimed that the value of the von Karman constant should be equal to 0.38 and not 0.4. Based on the high-fidelity DNS runs, they were also able to formulate *Re*-dependent similarity relationships for geostrophic drag coefficient (the ratio of surface friction velocity and geostrophic wind) and surface shear angle. Even though most of these results are rigorously valid only for low *Re* Ekman flows, their heuristic extrapolations (if hold) could have serious ramifications for ABL turbulence.

20.3.2 Large-Eddy Simulation

Key references
The following references are recommended for a thorough description of large-eddy simulation (LES) and its various attributes Sagaut (2004), Geurts (2003), and Meneveau and Katz (2000).

Salient computational features

- The numerical methods and the lateral boundary conditions used for LES are quite similar to those used in DNS.
- In LES, the dynamical complexity of turbulent flows under consideration is reduced by spatial low-pass filtering. The most commonly used filters are spectral cutoff, Gaussian, and box filters. The coarser the filter width (Δ_f), the higher the dependence of the simulated results on the employed SGS model.

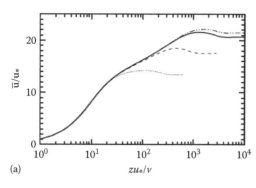
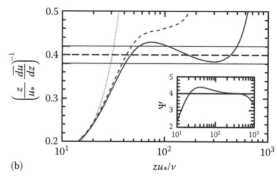

FIGURE 20.2 (a) Inner scaling of mean velocity component aligned with surface shear stress. (b) Effective logarithmic parameters. The dotted, dashed, and solid lines represent $Re = 400$, 1000, and 2000, respectively. In (a), normalized velocity magnitude for $Re = 2000$ is also shown (dash-dotted line). The subplot in the right panel represents $\psi = \bar{u}/u_* - (1/0.38) \ln (zu_*/\nu)$. (From Coleman, G.N. et al., DNS of the turbulent Ekman layer at $Re = 2000$, *Direct and Large-Eddy Simulation VI*, Springer, Berlin, Germany, Part IV, pp. 267–274, 2006. With permission.)

ft The rule of thumb is that an LES should resolve at least 80% of total energy anywhere in the flow. Sometimes (e.g., near wall region, under stably stratified condition), it is not feasible to resolve 80% of the total energy. This type of simulation is then called a very large-eddy simulation or VLES. At present, all the documented LES results of ABL flows are actually VLES results.

ft It is common practice to take the ratio of filter width (Δ_f) to grid spacing (Δ_g) to be equal to 1 or 2. Please refer to Chapter 9 of Geurts (2003) for detailed discussion on this ratio and its impact on error dynamics.

ft Quite often, uniform grids are used in LES because the effects of grid non-homogeneity and anisotropy on SGS modeling are yet to be fully understood.

ft In addition to forward cascade of energy (large to small scales), some SGS models do allow for local reverse energy transfer (known as backscatter). It has been established that certain filters (specifically the spectral cutoff filter) can spuriously enhance the backscatter of energy.

ft In contrast to DNS, no-slip lower boundary conditions are rarely used in LES. In order to utilize this boundary condition, one needs to resolve the small-scale, energy-containing eddies near the wall. This would require very high grid-resolution (comparable to DNS) near the wall and would be computationally very expensive. An alternate approach is to use a wall function (an empirical relationship between the wall shear stress and the velocity at wall-adjacent grid points) as the lower boundary condition.

20.3.2.1 Subgrid-Scale Models

The eddy-viscosity models are the most commonly used subgrid-scale (SGS) models in the environmental fluid mechanics community. These models use the gradient hypothesis and formulate the deviatoric part of the SGS stress tensor as follows:

$$\tau_{ij} - \frac{1}{3}\tau_{kk}\delta_{ij} = -2\nu_t \tilde{S}_{ij} \tag{20.5}$$

where

$$\tilde{S}_{ij} = \frac{1}{2}\left(\frac{\partial \tilde{u}_i}{\partial x_j} + \frac{\partial \tilde{u}_j}{\partial x_i}\right) \tag{20.6}$$

is the resolved strain rate tensor and ν_t denotes the eddy viscosity. From dimensional analysis, the eddy viscosity can be interpreted as the product of a characteristic velocity scale and a characteristic length scale. Different eddy-viscosity formulations use different velocity and length scales. The most popular eddy-viscosity formulation is the Smagorinsky model (Smagorinsky 1963):

$$\nu_t = (C_S \Delta_f)^2 |\tilde{S}| \tag{20.7}$$

where C_S is the so-called Smagorinsky coeDcient, which is adjusted empirically or dynamically to account for shear, stratification, and grid resolution, and

$$|\tilde{S}| = (2\tilde{S}_{ij}\tilde{S}_{ij})^{1/2} \tag{20.8}$$

is the magnitude of the resolved strain rate tensor. In contrast to the Smagorinsky-type eddy-viscosity model, the one-equation eddy-viscosity model utilizes

$$\nu_t = C_K l E_{SGS}^{1/2} \tag{20.9}$$

where

C_K is a modeling coeDcient

l is a length scale (related to Δ_f)

E_{SGS} is the SGS turbulent kinetic energy (TKE)

This modeling approach involves solving a prognostic equation for E_{SGS} in addition to the filtered momentum equations.

It is well known that the eddy-viscosity SGS models give a poor prediction of the SGS stresses on a local level. Their underlying assumption of strain rates being aligned with the SGS stress tensor is unrealistic. Furthermore, these SGS models are purely dissipative; that is, they do not allow backscatter. Recognizing the inherent deficiencies of the eddy-viscosity SGS models, turbulence modelers proposed numerous non-eddy-viscosity-type SGS models (e.g., the similarity model). An extensive review of these SGS formulations is given by Sagaut (2004). Unfortunately, the non-eddy-viscosity-type SGS models have their own set of limitations (e.g., the lack of adequate dissipation in the case of the similarity model), and thus, they did not succeed to substantially reduce the usage of the eddy-viscosity SGS models.

20.3.2.2 Dynamic SGS Modeling

The values of the eddy viscosity-type SGS coeDcients (e.g., C_S) are well established for homogeneous, isotropic turbulence. However, these parameters are expected to be dependent on mean shear and stratification. Traditionally, various types of wall-damping functions (e.g., van Driestß model) and stability correction functions are used to account for these physical effects. An alternative approach is to use the dynamic SGS modeling approach of Germano et al. (1991). In this approach, one computes the value of the unknown SGS coeDcients (e.g., C_S) dynamically at every time and position in the flow. By analyzing the dynamics of the flow at two different resolved scales and assuming scale similarity as well as scale invariance of the model coeDcient, one can optimize its value. Thus, the dynamic model avoids the need for a priori specification and tuning of the SGS coeDcients because they are evaluated directly from the resolved scales in an LES. Given the generic nature of the dynamic modeling approach, it has been extended for non-eddy-viscosity-type SGS models as well.

20.3.2.3 Testing of SGS Models

The SGS models and their underlying hypotheses can be evaluated by two approaches: a priori testing and a posteriori testing (Geurts 2003). In a posteriori testing, LES computations are actually performed with specific SGS models and validated against reference solutions (in terms of mean velocity, scalar and stress distributions, spectra, etc.). However, owing to the multitude of factors involved in any numerical simulation (e.g., numerical discretizations, time integrations, averaging, and filtering), a posteriori tests, in general, do not always provide much insight about the detailed physics of the SGS model. A complementary and perhaps more fundamental approach is to use data from direct numerical simulations, laboratory experiments or field observations to compute the "real" and modeled SGS tensors directly from their definitions and compare them subsequently. This approach, widely known as the a priori testing, does not require any actual LES modeling and is theoretically more tractable. In the recent past, the ABL community has organized several field campaigns (e.g., horizontal array turbulence study—HATS) exclusively for a priori testing.

Illustrative example

Andren et al. (1994) conducted an intercomparison of LES models for a neutrally stratified Ekman layer flow (similar to the DNS studies by Coleman et al.). Here, we only report results related to the non-dimensional velocity gradient (ϕ_M). According to the similarity theory, in the surface layer, ϕ_M should be equal to one (the dashed line in the right panel of Figure 20.3). Andren et al. (1994) found that the non-dynamic SGS models (with the exception of Mason and Brownß SGS model with backscattering option) were highly dissipative near the surface and were unable to reproduce the correct behavior of ϕ_M (left panel of Figure 20.3). The contemporary dynamic SGS models (and their variants) offer major improvements over the non-dynamic models (right panel of Figure 20.3).

20.3.3 Reynolds-Averaged Navier–Stokes Modeling

Key references

The book by Wilcox (1998) is an excellent resource for RANS modeling. Alfonsi (2009) provides a recent review of this topic.

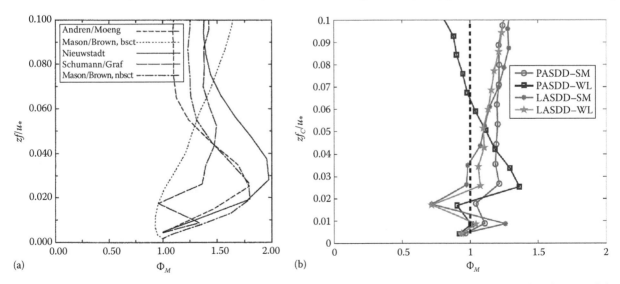

FIGURE 20.3 (a and b) Panels show simulated non-dimensional velocity gradients reported by Andren et al. (1994) and Anderson et al. (2007), respectively. The dashed line in the right panel corresponds to the value of 1.

Salient features

ﬅ In contrast to DNS and LES, RANS can be utilized for both steady and unsteady simulations.

ﬅ Typically, RANS modeling does not involve more than second-order accurate numerical schemes.

ﬅ Since, in RANS, only averaged quantities are needed for inflow boundary conditions, their prescription is of a much lesser challenge compared to DNS and LES. The prescription of lower boundary conditions is also straight-forward in RANS.

ﬅ In general, RANS closure models have diﬃculties in simulating complex flows with pronounced vortex shedding.

ﬅ RANS closure models can be used as wall models in LES.

20.3.3.1 RANS Closure Models

The Reynolds averaged Navier–Stokes (RANS) closure models can be broadly classified into four classes: (1) algebraic models, (2) one-equation models, (3) two-equation models, and (4) Reynolds stress equation models.*

Qualitatively, the RANS algebraic models are similar to the Smagorinsky SGS model used in LES. These models parameterize the deviatoric part of the Reynolds stress tensor as a function of the stress-rate tensor of the mean field. The most popular RANS algebraic models are Cebeci–Smith model and Baldwin–Lomax model.

In one-equation models, one uses a prognostic equation for TKE. Again, this approach is qualitatively similar to the one-equation modeling approach for LES. In this approach, the dissipation term is parameterized by making use of scaling arguments.

In two-equation models, one solves two prognostic equations related to turbulence length and time scales. One of the most widely used models of this type is the $K - \varepsilon$ model. In this case, one solves for TKE and mean energy dissipation rate (ε). The turbulence length and time scales are then proportional to $TKE^{3/2}/\varepsilon$ and TKE/ε, respectively.

The Reynolds stress equation models involve a large number of prognostic equations (corresponding to all the Reynolds stress terms and energy dissipation rate) and a long list of tuning coeﬃcients. These models outperform simpler algebraic and $K - \varepsilon$ models for turbulent flows which are far from equilibrium. However, the complexity of these models has deterred their wide usage in various practical applications.

Illustrative example

Xu and Taylor (1997) simulated an Ekman layer using the $K - \varepsilon$ model. They reported that this widely used RANS closure model was unable to reproduce the well-known Ekman spiral. To the best of our knowledge, all the DNS and LES studies of Ekman layers have always reproduced the Ekman spiral. The inability of the $K - \varepsilon$ model to reproduce this key ABL feature was rather surprising and underlines the limitations of certain popular RANS closure schemes for environmental flows. A new RANS closure scheme proposed by Xu and Taylor (1997) was able to remedy this problem.

20.4 Applications

In this section, a mélange of turbulent flow modeling examples are presented.

20.4.1 Atmospheric Dispersion

In order to protect public health and environmental assets from potentially harmful air pollutants, one needs detailed spatio-temporal characterizations of these pollutants in the ABL. Given the scarcity of field observations of pollutants, atmospheric dispersion models traditionally play critical roles in determining compliance with National Ambient Air Quality Standards and other regulatory requirements. It is needless to point out that the accuracy of these models strongly depends on the quality of the ABL turbulence simulations. Here, as an illustrative example, we discuss the atmospheric dispersion of Radon (^{222}Rn) and its progeny (e.g., ^{218}Po, ^{214}Pb). According to the Environmental Protection Agency (EPA), Radon is the number one cause of lung cancer among non-smokers, with 21,000 victims each year in the United States.

Vinuesa et al. (2007) studied the diurnal evolution of ^{222}Rn decaying family using a dynamic LES model. They reported several interesting results related to the transport and mixing of Radon and its daughters. For example, they found that a departure from secular equilibrium[K] between Radon and its short-lived daughters prevailed in the stably stratified nighttime boundary layer (see Figure 20.4). This disequilibrium was attributed to the proximity of the Radon source and the weak vertical transport during stable condition. Since a significant fraction of Radon was fresh, the Radon and progeny mixture was deficient in daughters. The mixing induced by convective turbulence induced a fast restoration of the secular equilibrium during the morning transition.

20.4.2 Wind Energy

Energy production is one of the critical issues facing the world. Fortunately, the wind energy industry has witnessed a significant surge in recent years. However, to make wind a cost-effective and reliable alternative energy source, the industry will need to make significant advancements on several fronts, including turbine micro-siting, wind resource assessment, short-term wind forecasting, and turbine inflow generation. We believe that the state-of-the-art turbulent flow modeling frameworks have the potential to make noteworthy

---

* Please note that the ABL community follows a different classification of RANS models, originally proposed by Mellor and Yamada (1982).

K The secular equilibrium is the condition according to which the ratio of the activities of the nuclide participating to the decay chain is equal to one.

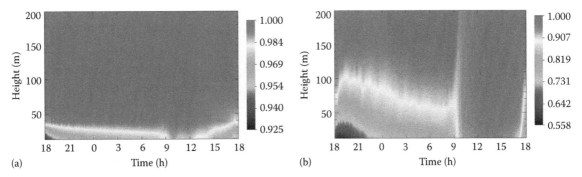

FIGURE 20.4 **(See color insert.)** Time-height plots of activity ratios of ^{218}Po (a) and ^{214}Pb (b) with respect to ^{222}Rn. The S_1/S_0 activity ratios are significantly different from one during nighttime conditions. The time axis denotes local time. (From Vinuesa, J.-F., *Atmos. Chem. Phys.*, 7, 5003, 2007. With permission.)

contributions to some of these areas. In the following, we elaborate on the problem of turbine micro-siting.

For turbine micro-siting projects over complex topography, the wind industry is progressively opting for computational fluid dynamics (CFD) models instead of the traditional mass-consistent and linear models. All of these commercial CFD models (e.g., Meteodyn, WindSim) essentially utilize RANS closures. An alternative and arguably better approach would be to use LES models. At present, due to high computational costs, LES models have not gained popularity in the wind industry. With the increasing accessibility to high-performance computing resources, it is plausible to envisage that the LES models will eventually find their place in a variety of micro-siting projects.

A recent study by Bechmann and Sørensen (2009) compared the abilities of a RANS model and a newly proposed hybrid RANS/LES model to capture the observed features of atmospheric flows over a hill. They showed that both the RANS and the hybrid models were able to simulate the relatively simple windward flow quite accurately (see Figure 20.5). However, in the complex wake region, these models gave very different results.

In terms of mean velocity, the RANS model's performance was superior (Figure 20.5a). On the other hand, the hybrid RANS/LES model outperformed the RANS model in the case of TKE estimation (Figure 20.5b). From these results, it is quite evident that there is still room for improvements for the RANS and LES models for high *Re* atmospheric flows.

20.4.3 Evapotranspiration

Evapotranspiration (ET; directly related to latent heat flux) represents the loss of water from the Earth's surface through the combined processes of evaporation (from soil and plant surfaces) and plant transpiration. Its reliable estimation at multiple spatio-temporal scales is of paramount importance for a multitude of fields, including (but not limited to) agriculture, forestry, ecology, and meteorology. Albertson et al. (2001) developed a coupled framework, involving a LES model, a land-surface model, and satellite remote sensing data, which showed great promise in the estimation of ET (and other surface turbulent fluxes) over heterogeneous land surfaces (Figure 20.6).

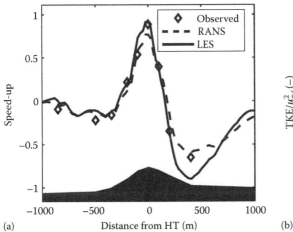
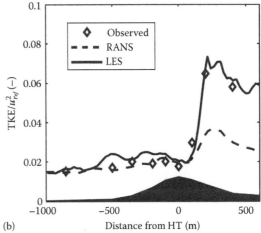

FIGURE 20.5 Comparison of observed velocity speed-up (a) and turbulence kinetic energy (b) to RANS and hybrid RANS/LES simulations. The dashed line shows the RANS results while the full lines are for the hybrid RANS/LES. Measurements are marked with diamonds. (From Bechmann, A. and Sørensen, N.N., *Wind Energy*, 13, 36, 2009. With permission.)

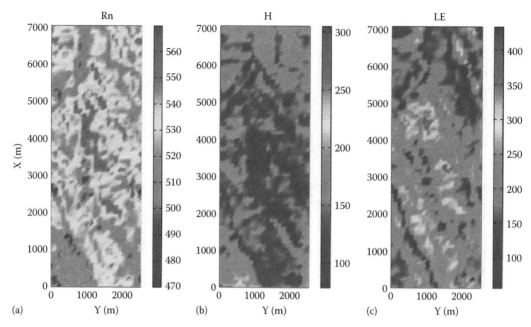

FIGURE 20.6 Large-eddy model simulated net radiation (a), surface sensible heat flux (b), and surface latent heat flux (c) fields. (From Albertson, J.D. et al., *Water Resour. Res.*, 37, 1939, 2001. With permission.)

20.4.4 Tropical Cyclones

Given that almost every landfalling tropical cyclone (TC; e.g., hurricane) causes tremendous amount of physical destructions, massive economic impacts, and staggering loss of lives, their accurate forecasting and simulation is of utmost importance for meteorologists and atmospheric scientists. At present, numerical weather prediction (NWP) models can reasonably capture some of the structural features of TCs, such as eyewall or rainbands. However, these NWP models always parameterize the effects of small-scale turbulence (utilizing the RANS approach).

Recently, Rotunno et al. (2009) provided improved understanding of the effects of small-scale turbulence on TC dynamics, by performing LES of an idealized TC using a new-generation NWP model known as the weather research and forecasting (WRF) model. They used six nested domains of grid resolutions: 15, 5, 1.67, 0.556, 0.185, and 0.062 km. Turbulence was only (partially) resolved in the finest grid ($\Delta = 0.062$ km). In the coarser domains, it was almost entirely SGS. In Figure 20.7, the simulated flow structures resulting from two different grid resolutions are shown (taken from Norton et al. 2009). The structural differences in the vertical vorticity fields near the TC eyewall are

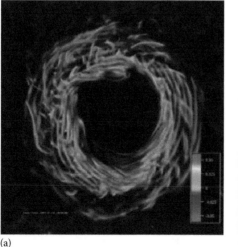
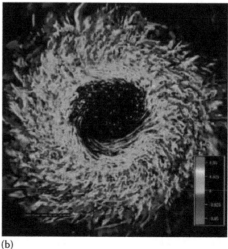

FIGURE 20.7 (See color insert.) Vertical vorticity resulting from lower (0.185 km grid; a) and higher (0.062 km grid; b) resolution WRF simulations. Region shown (in both panels) is approximately 35 km in diameter. Note that the internal structure of the higher-resolution simulation could not be anticipated from the lower-resolution results. (From Norton, A. and Clyne, J., Analysis and visualization of high-resolution WRF hurricane simulation using VAPOR, *10th WRF Users' Workshop*, Boulder, CO, 2009. With permission.)

simply stunning. More importantly, Rotunno et al. (2009) found strong dependence of simulated TC intensity on the amount of resolvable turbulence. This crucial result highlights the need for high-resolution turbulent flow modeling for accurate forecasting and simulation of TC.

20.5 Major Challenges

All the turbulent flow modeling approaches suffer from different technical challenges, which need to be addressed. For example, in order to apply DNS over complex geometries, we need to find a better way to prescribe inflow conditions. In LES, we need to find improved wall models and more robust SGS models which are less sensitive to grid resolution. Several higher-order RANS closure models are not realizable. All these challenging issues and more are extensively covered by Wilcox (1998), Moin and Mahesh (1998), Piomelli (1999), Pope (2004), and others. For brevity, these challenges will not be reiterated here.

The author believes that the most diDcult challenge confronting the environmental turbulent flow modelers is in the arena of turbulent model validation. There is a general consensus that we do not have rigorous theoretical results for environmental flows. As mentioned earlier in this chapter, we will not be able to perform DNS of high Re environmental flows in the foreseeable future. At the same time, environmental flow observations will (almost) always be corrupted by non-turbulent motions, measurement noises, sensor limitations, etc. So, in the absence of "truth" (i.e., rigorous theories or high Re DNS data or precise observations), how should one conduct rigorous model validation studies? The author sincerely hopes that the new generation of scientists and engineers will be able to answer this question.

References

Albertson JD, Kustas WP, Scanlon TM. 2001. Large-eddy simulation over heterogeneous terrain with remotely sensed land surface conditions. *Water Resour. Res.* 37: 1939–1953.

Alfonsi G. 2009. Reynolds-averaged Navier-Stokes equations for turbulence modeling. *Appl. Mech. Rev.* 62: 040802.

Anderson WC, Basu S, Letchford CW. 2007. Comparison of dynamic subgrid-scale models for simulations of neutrally buoyant shear-driven atmospheric boundary layer flows. *Environ. Fluid Mech.* 7: 195–215.

Andren A, Brown AR, Graf J, Mason PJ, Moeng C-H, Nieuwstadt FTM, Schumann U. 1994. Large-eddy simulation of a neutrally stratified boundary layer: A comparison of four computer codes. *Q. J. R. Meteorol. Soc.* 120: 1457–1484.

Bechmann A, Sørensen NN. 2009. Hybrid RANS/LES method for wind flow over complex terrain. *Wind Energy* 13: 36–50.

Coleman GN. 1999. Similarity statistics from a direct numerical simulation of the neutrally stratified planetary boundary layer. *J. Atmos. Sci.* 56: 891–900.

Coleman GN, Ferziger JH, Spalart PR. 1990. A numerical study of the turbulent Ekman layer. *J. Fluid Mech.* 213: 313–348.

Coleman GN, Johnstone R, Ashworth M. 2006. DNS of the turbulent Ekman layer at $Re = 2000$. *Direct and Large-Eddy Simulation VI*, Springer, Berlin, Germany, Part IV, pp. 267–274.

Coleman GN, Sandberg RD. 2009. A primer on direct numerical simulation of turbulence—Methods, procedures and guidelines, Technical Report AFM-09/01, University of Southampton, Southampton, U.K.

Germano M, Piomelli U, Moin P, Cabot WH. 1991. A dynamic subgrid-scale eddy viscosity model. *Phys. Fluids A* 3: 1760–1765.

Geurts BJ. 2003. *Elements of Direct and Large-Eddy Simulation.* R. T. Edwards, Inc., Philadelphia, PA, 329pp.

Hanjalic K. 2005. Will RANS survive LES? A view of perspectives. *J. Fluids Eng.-Trans. ASME* 127: 831–839.

Mellor GL, Yamada T. 1982. Development of a turbulence closure model for geophysical fluid problems. *Rev. Geophys.* 20: 851–875.

Meneveau C, Katz J. 2000. Scale-invariance and turbulence models for large-eddy simulation. *Annu. Rev. Fluid Mech.* 32: 1–32.

Moin P., Mahesh K. 1998. Direct numerical simulation: A tool in turbulence research. *Annu. Rev. Fluid Mech.* 30: 539–578.

Norton A, Clyne J. 2009. Analysis and visualization of high-resolution WRF hurricane simulation using VAPOR. *10th WRF Users' Workshop*, Boulder, CO.

Piomelli U. 1999. Large-eddy simulation: Achievements and challenges. *Prog. Aerosp. Sci.* 35: 335–362.

Pope SB. 2000. *Turbulent Flows.* Cambridge University Press, Cambridge, U.K. 771pp.

Pope SB. 2004. Ten questions concerning the large-eddy simulation of turbulent flows. *New J. Phys.* 6, doi:10.1088/1367–2630/6/1/035.

Reynolds WC. 1976. Computation of turbulent flows. *Annu. Rev. Fluid Mech.* 8:183–208.

Rogallo RS, Moin P. 1984. Numerical simulation of turbulent flows. *Annu. Rev. Fluid Mech.* 16: 99–137.

Rotunno R, Chen Y, Wang W, Davis C, Dudhia J, Holland GJ. 2009. Large-eddy simulation of an idealized tropical cyclone. *Bull. Am. Meteorol. Soc.* 90: 1783–1788.

Sagaut P. 2004. *Large Eddy Simulation for Incompressible Flows* (2nd edn). Springer, Berlin, Germany, 426pp.

Smagorinsky J. 1963. General circulation experiments with the primitive equations. *Mon. Wea. Rev.* 91: 99–164.

Vinuesa J-F, Basu S, Galmarini S. 2007. The diurnal evolution of ^{222}Rn and its progeny in the atmospheric boundary layer during the Wangara experiment. *Atmos. Chem. Phys.* 7: 5003–5019.

Wilcox DC. 1998. *Turbulence Modeling for CFD* (2nd edn). DCW Industries, Inc., La Cãnada, CA, 540pp.

Xu D, Taylor PA. 1997. An $E - \varepsilon - l$ turbulence closure scheme for planetary boundary-layer models: The neutrally stratified case. *Boundary-Layer Meteorol.* 84: 247–266.

Young GS, Zawislak J. 2006. An observational study of vortex spacing in island wake vortex streets. *Mon. Wea. Rev.* 134: 2285–2294.

21

Direct and Large Eddy Simulation of Environmental Flows

Sutanu Sarkar
University of California, San Diego

Vincenzo Armenio
University of Trieste

21.1 Introduction

Numerical simulation is a powerful tool to both understand and predict environmental flows. These flows are multiscale, e.g., a mesoscale eddy in the ocean has scales of motion that range from the eddy's horizontal extent of say 100 km to the 1 cm scale at which molecular mixing is accomplished. Different spatial scales are governed by qualitatively different physical processes that include geostrophic flow, quasi-geostrophic turbulence, linear and nonlinear gravity waves, shear instabilities, convective instabilities, and three-dimensional (3D) turbulence. Time scales are similarly disparate, even if we exclude the interest in local variability of climate, and could range from months to hours to seconds.

In response to the multiscale and multiphysics nature of environmental flows and to the differences in user-set objectives, different types of numerical models have been developed. These include 2D (in space) or quasi-2D models, hydrostatic 3D models, and non-hydrostatic 3D models. Turbulent transport in the environment is parameterized or resolved in numerical models. Parameterization may be as simple (and correspondingly inaccurate) as a constant eddy viscosity/diffusivity or considerably more complicated such as additional differential equations that govern turbulence quantities, e.g., turbulent kinetic energy

(TKE) or turbulence length scales. *Turbulence resolving* simulations can be of the direct numerical simulation (DNS) type or large eddy simulation (LES) type. When a simulation resolves all the dynamically important scales of motion, it is termed as DNS. Simulations with the order of 500 million grid points have been performed by several research groups in environmental fluid dynamics (EFD) including a recent simulation of a stratified wake by Brucker and Sarkar [7] that employed a grid with approximately 2 billion grid points. The largest DNS to date is the simulation of homogeneous turbulence on a 4096^3 grid by Ishihara et al. [23]. Although DNS with current computing resources can be at Reynolds numbers lower than in application and is customarily performed in non-complex geometry, it does result in detailed and high-accuracy description of complex nonlinear dynamics and turbulence in environmental flows. Also, the Reynolds number in several simulations has been larger than in corresponding laboratory experiments. LES, when employed with sufficient resolution and an appropriate subgrid model, resolves most of the scales responsible for turbulent transport but does not resolve the fine scales at which molecular dissipation occurs.

This chapter reviews DNS and LES of 3D environmental flows. The emphasis is on canonical flows which provide high-fidelity descriptions of one or more of the following complications

Handbook of Environmental Fluid Dynamics, Volume Two, edited by Harindra Joseph Shermal Fernando. © 2013 CRC Press/Taylor & Francis Group, LLC.
ISBN: 978-1-4665-5601-0.

peculiar to the natural environment: density stratification, rotation, oscillation (as in an ocean tide), or topography. References to related experimental and analytical studies of importance will be kept sparse since they will likely be covered elsewhere in this handbook. Numerical simulation of atmospheric flows is discussed in other chapters and, therefore, will not be considered here. Subgrid models for LES are discussed in another chapter and will also not be covered here.

21.2 Computational Methods

The three-dimensional, unsteady Navier–Stokes equations (NSE) under the Boussinesq approximation for density variation are solved without any additional model in the DNS approach, while the LES approach requires an additional model for the unknown subfilter (or subgrid) fluxes. There are many texts on computational fluid dynamics which can be consulted for numerical algorithms for the unsteady, multidimensional Navier–Stokes equations. Most fundamental DNS and LES studies of turbulent mixing in flows of interest to EFD have been carried out in simple geometrical conditions. In these cases, the Cartesian grid formulation of the NSE has been used and the equations have been numerically integrated using highly eDcient algorithms. In this section, we will briefly discuss methods for flows in simple geometry followed by methods for complex geometry.

DNS of density-stratified turbulent flows started with the study of turbulence that is statistically homogeneous, i.e., statistics of velocity and density fluctuations do not vary in space but may vary in time. Although statistically homogeneous, the fluctuations are strongly anisotropic owing to the direct effect of buoyancy in the vertical direction. The first DNS was performed by Riley et al. [38] who simulated unforced turbulence, also called box turbulence, subject to uniform stratification. Periodic boundary conditions were applied in all directions and a pseudo-spectral algorithm was adopted. This first DNS at a micro-scale Reynolds number, $Re_\lambda = 27.2$ utilized only 32^3 points (!) but clearly showed suppression of certain types of nonlinear transfers as well as the existence of wave-like behavior. Both observations were consistent with the evolution of stratified turbulent flows in the laboratory. The next problem with statistically homogeneous turbulence that was simulated corresponded to uniform mean shear and mean stratification. This enabled study of the competing roles of mean shear and stratification on turbulent fluctuations. A pseudo-spectral algorithm and resultant spectral accuracy was possible by transforming to a coordinate system advected with the mean flow.

In shear flows, turbulence statistics often have one direction of strong variation, e.g., the cross-stream direction in flows such as a finite-thickness shear layer, a plane jet or a plane far wake. Periodic boundary conditions in the horizontal directions are natural for such flows and, therefore, Fourier collocation can be used in the horizontal. Finite difference (possibly with high accuracy such as Padé derivatives) is used in the vertical. Solution of the pressure equation, equivalently the

enforcement of the divergence-free constraint on the velocity, is necessary. A projection method or fractional step method is customarily utilized for this purpose. Time advancement can be accomplished by a Runge–Kutta (RK) scheme. A low-storage, third-order RK scheme has proved popular. Wall-bounded flows such as the Ekman boundary layer and channel flow also have a single direction of inhomogeneity in the vertical. The additional complexity is the small wall-normal scale in the vertical direction that is required to resolve near-wall turbulence. An implicit method, for instance the Crank–Nicolson method, is then necessary for time advancement of the viscous derivatives in order to avoid the stringent restriction on time step of an explicit method.

Spatially evolving flows require inflow and outflow conditions in the streamwise directions which are problem-specific. If the domain is infinite in the vertical, boundary conditions are necessary at the boundary of the truncated domain. Stratified flows admit propagating waves. Spurious reflection of these waves from the boundaries must be minimized to avoid unphysical generation of turbulence within the computational domain. Non-reflecting boundary conditions that have been devised are exact only in simplified situations such as 1D problems or linear problems. In practice, sponge regions are utilized whereby fluctuations in variables are relaxed to known ambient conditions through additional forcing terms in the governing equations.

Peculiar to EFD problems is the presence of geometrical complexities. Anabatic and katabatic flows are examples where topographic features affect the atmospheric boundary layer, whereas the presence of coastline and bathymetry affect circulation in marine applications. In case of geometrical complexities, two main approaches have been followed in the literature. The first one involves solution of the Cartesian form of the NSE and incorporation of geometrical complexities through the imposition of a set of body forces, properly set to mimic the presence of boundaries not aligned with the cell faces. Due to its simplicity, this technique known as *immersed boundary method* (IBM) has become popular in the DNS/LES scientific community. An alternative strategy consists in solving the curvilinear form of the NSE, and to fit the non-Cartesian boundaries of the physical domain using curvilinear coordinate lines. Among others, the curvilinear-coordinate fractional step algorithm of Zang et al. [61] has been widely employed in the EFD community. Subsequently, this formulation has been used by Armenio and Piomelli [1] to supply a LES subgrid scale (SGS) model recast onto a contravariant form. Finally, very recently the advantages of curvilinear coordinates have been merged with the flexibility of the IBM in Ref. [41] to give a methodology able to treat problems of water mixing in coastal, shallow water, areas [42]. In order to deal with practical high Reynolds number flows, a wall-layer approach has been used in conjunction with LES methodology. The classical wall-layer model which reconstructs the wall-shear stress assuming the first velocity point to lie in a log-region has been adapted to work satisfactorily in conjunction with immersed boundaries.

21.3 Stratified Free Shear Flow

Shear in the vertical direction can overcome the stabilizing effect of density stratification. The evolution of shear instabilities and turbulence in the presence of buoyancy is illustrated here using primarily examples where numerical simulations have been extensively validated against laboratory experiments or applicable theory. Indeed, successful simulations of these benchmark flows have significantly contributed to our confidence in numerical simulation as a tool for predicting and understanding environmental turbulence.

21.3.1 Uniform Shear

The simplest case with mean shear is homogeneous shear flow that corresponds to unbounded shear flow with spatially uniform mean shear, $a\langle U\rangle/az = S$, and constant buoyancy frequency, N. The gradient Richardson number, $Ri_g = N^2/S^2$, is also a constant. This flow is asymptotically stable to exponentially growing modes according to linear analysis; nevertheless, initial fluctuations can build up to turbulence through algebraic growth followed by nonlinear evolution. The flow is spatially homogeneous in a reference frame that moves with the mean flow so that periodic boundary conditions are allowable in that frame. The flow statistics are not a function of space but do depend on time. The first DNS of uniform shear flow in a fluid with uniform stratification was performed by Gerz et al. [19] on a 64^3 grid. A second-order finite difference method was used and time advancement was performed using an Adams–Bashforth scheme. Subsequent simulations of this flow by [20] and [24] adapted the Rogallo algorithm which is a spectral collocation scheme that solves the governing equations in a non-orthogonal frame moving with mean flow to achieve periodicity in all directions.

The TKE is found to evolve exponentially, $K = K_0\exp(\gamma St)$ with γ a problem-dependent coeDcient. Structural turbulence parameters, namely non-dimensional parameters such as the shear number, SK/ϵ, the Reynolds stress anisotropy, b_{ij}, and the ratio of turbulence production to dissipation, P/ϵ, asymptote to *constant* values which are of the same order as in more common shear flows such as a jet or a mixing layer. Indeed, the ability to predict the values of these structural parameters in unstratified uniform shear flow is a key test of turbulence models. As discussed by Jacobitz et al. [24], the value of TKE growth rate γ depends on Ri_g which is a constant, on the micro-scale Reynolds number, Re_λ, whose influence is small in the high-Reynolds number limit and the shear number, SK/ϵ, which is the ratio of a characteristic turbulence time scale to mean flow time scale. It appears that, although SK/ϵ asymptotically approaches a constant during the turbulence evolution, the value of the asymptotic constant has a sensitive dependence on its initial value in stratified shear flow [24] as well as unstratified shear flow.

Stratification inhibits turbulence growth. Thus, with increasing stratification, i.e., Ri_g of the flow, the TKE growth rate, γ, which is positive in unstratified flow, decreases, becomes zero at a critical value, $Ri_{g,cr}$, of Richardson number and eventually becomes negative. The value of $Ri_{g,cr}$ in uniform shear flow becomes independent of Reynolds number but does depend on the shear number, spanning a range of 0.17–0.25. It is interesting that the values of $Ri_{g,cr}$ observed in the fully nonlinear simulations of uniform shear flow are less than 0.25 and, thus, do not contradict the Miles-Howard result derived from linear theory that a mean flow, $U(z)$, is linearly stable if $Ri_g > 0.25$ everywhere in the flow.

21.3.2 Shear Layer of Finite Thickness

The flow corresponds to a stratified shear layer of finite thickness between two currents, each with a different velocity and a different density. The typical configuration corresponds to an inflectional velocity profile, $\langle u_s\rangle = -(\Delta U/2)\tanh(2z/\delta_{\omega,0})$, and a similar inflectional density profile. ΔU and $\Delta\rho$ are the constant velocity difference and density difference across the layer while the vertical extent of the layer is measured by the vorticity thickness, $\delta_\omega = \Delta U/S_{max}$. Both mean shear, $S(z)$, and buoyancy frequency, $N(z)$, have peak values in the middle of the shear layer that decrease to zero at the top and bottom boundaries. This flow has been the subject of vigorous study using simulations, e.g., [8, 11, 48].

The life-cycle of a stratified shear layer is well-understood as summarized in the following. The evolution of the shear layer from a state of small-amplitude initial fluctuations to turbulence occurs through a sequence of instabilities. The fundamental wavelength, given by linear analysis, grows exponentially and the vorticity rolls up to form coherent Kelvin Helmholtz (KH) billows that are connected by thin braids. KH billows pair through 2D interactions. Three-dimensional secondary instabilities with coherent streamwise vortices develop. In the stratified case, these streamwise vortices appear to originate at the periphery of the KH billows and are triggered by a convective instability associated with heavy fluid being transported over light fluid during the KH rollup. Strong 3D turbulence results when streamwise vortices that are swept into the braid region collide. The turbulence evolution depends on Reynolds number, Re. For instance, increasing Re based on vorticity thickness from 1200 to 5000 in our recent simulations, caused the breakdown of KH billows before pairing could occur and, furthermore, intense turbulence developed in the KH billows independent of turbulence in the braids. Although 2D dynamics of KH billows leads to chaotic mixing, it is only after the development of 3D secondary instabilities that broadband turbulence accompanied by a rapid increase in turbulent dissipation rate is found. This is an example of the fact that 3D simulations are necessary to capture turbulent mixing events.

The control parameter that determines the strength of buoyancy in a shear layer is the bulk Richardson number,

$$Ri_b = \frac{\Delta\rho g\delta(t)}{\rho_0\Delta U^2}. \tag{21.1}$$

The thickness δ is customarily taken to be the vorticity thickness, δ_ω or, equivalently, $4\delta_\theta$, based on the momentum thickness.

The characteristic gradient Richardson number, $Ri_g = N^2/S^2$, at the centerline of the shear layer is equal to Ri_b. The shear layer thickens by turbulent entrainment up to a point after which turbulence collapses. Correspondingly, the bulk Richardson number reaches a final value $Ri_{b,f}$. The critical value in experiments shows a spread, $Ri_{b,f} = 0.32 " 0.06$. Recent simulations [8] show that the spread has a systematic trend: $Ri_{b,f}$ increases with increasing Reynolds number and decreasing Prandtl number. Unlike the uniform shear case, the shear layer eventually relaminarizes, however small is the choice of initial stratification. This can be understood by taking the centerline value of Ri_g, equivalently Ri_b defined by Equation 21.1, as the characteristic value of gradient Richardson number in the shear layer. Since Ri_g increases with time, eventually it exceeds a critical value whereby, in analogy with the uniform shear case, turbulence must decay. The critical value of $Ri_g = 0.32 " 0.06$ appears to be larger in the case of the shear layer relative to uniform shear flow, $Ri_g = 0.17$–0.25.

Turbulent flow is a multiscale phenomenon. The fluctuations span a wide range of length scales whose upper boundary is characterized by a large-eddy scale, l, and lower boundary by 10η where $\eta = (v^3/\epsilon)^{1/4}$ is the Kolmogorov scale. Eddies with a vertical velocity scale w and a vertical length scale l are directly affected by buoyancy if l is suDciently large as may be surmised by comparing the potential energy required for overturning motion at that scale to vertical kinetic energy or by comparing the characteristic acceleration w^2/l of turbulent motion with the restoring gravitational acceleration, $(\delta\rho/\rho_0)g = -N^2 l$. In turbulent flow, motion with vertical length scales above the Ozmidov length scale, $L_O = \sqrt{\epsilon/N^3}$, is directly affected by buoyancy. However, there is a range of length scales between L_O and the Kolmogorov scale that is not directly affected by buoyancy and it is motion at these length scales that leads to effective transport and mixing. Stratified turbulence is said to be active when $L_O/10\eta$ is suDciently large or equivalently the buoyancy Reynolds number, $Re_b = \epsilon/vN^2$, is suDciently large.

The evolution of turbulence length scales, l and 10η, relative to the Ozmidov scale, L_0, in the stratified shear layer is instructive [48]. A case with mild stratification allows the formation of strong turbulence, i.e., L_O increases while η decreases to give a relatively wide range of turbulent scales unaffected by buoyancy. During this time period, the shear layer thickness grows. Nevertheless, the largest scales of turbulence are affected by buoyancy since L_O is smaller than the large-eddy scale, l, and therefore the thickness growth rate is smaller than that in the corresponding unstratified case. Once the shear layer thickens so that $Ri_b \sim 0.25$, the flow attains a stage of decaying turbulence wherein L_O decreases and 10η increases so that a progressively larger range of turbulence is affected by buoyancy. Ultimately, $L_O \simeq \eta$ so that turbulence at all scales is damped by buoyancy and the flow approaches a laminar state. Smyth and Moum [49] follow the Thorpe method based on resorted density profiles to calculate the large-eddy scale, l, but other estimates of l have been used in the past such as the Ellison scale $\rho_{rms}/(d\langle\rho\rangle/dz)$, a length

estimated based on geometry in near-boundary flows, and an integral scale using two-point correlations.

21.3.3 Shear Layer with Continuous, Nonuniform Background Stratification

In nature, the background fluid outside the sheared region may also be stratified with non-zero buoyancy frequency and thus support propagating internal gravity waves. The importance of background stratification stems from the fact that there is an additional *nonlocal* pathway to mixing: fluctuation energy from shear instabilities and turbulence in a localized shear region is transported away to a remote location where there may be mixing under the right circumstances, for example, in a critical layer where the wave phase speed is equal to the local fluid velocity.

In a laboratory experiment [54], internal waves with a narrow band of propagation angles between 45° and 60° were observed. The Reynolds stress was measured and it was found that approximately 7% of the average momentum across the shear depth was lost due to wave transport. Internal wave propagation was also observed in the experiments of Strang and Fernando [51]. When the lower layer was linearly stratified, "interfacial swelling" in the shear layer was observed and argued to be responsible for internal wave excitation.

Three-dimensional simulations [36] have allowed the examination of internal wave dynamics in the presence of realistic turbulence. The wave generation process is conceptualized as being analogous with that in flow over a corrugated surface with the corrugations being forced by shear layer instabilities and turbulence. Figure 21.1 shows internal waves generated by KH rollers in part (a) and by broadband turbulence in part (b). The distinct angle of 32°–38° of the phase lines in Figure 21.1a can be explained using linear wave theory. Briefly, the horizontal wavenumber is given by the most unstable wavelength, i.e., the KH mode, the temporal frequency is that due to the bottom freestream moving over the approximately stationary KH rollers, and the linear dispersion relationship for internal gravity waves predicts the observed angle of the phase lines (equivalently group velocity). At later time, in the turbulence regime, internal waves continue to be generated by the shear layer. The wave field shows a broad range of phase lines and also develops variations in the spanwise direction. Nevertheless, the phase lines tend to cluster into a narrow band around 45° with increasing depth. The TKE equation provides a useful diagnostic to analyze the impact of internal waves on flow energetics. The wave energy flux also called pressure transport, $\langle p'w' \rangle$, integrated over time at a location in the region of linear propagation is found to be approximately 17% of the spatially integrated production, 33% of the integrated dissipation, and 75% of the integrated buoyancy flux in simulations [36] of a shear layer with $Re = 1200$ based on initial vorticity thickness. In recent work, the Reynolds number was increased to $Re = 5000$. The mechanism of internal wave excitation did not change. The internal wave flux remained significant but about half of that at the lower Reynolds number.

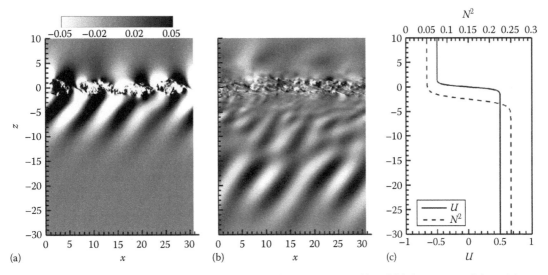

FIGURE 21.1 Shear layer with strong background stratification in the deep region. Parts (a) and (b) show a vertical slice of the vertical velocity field. (a) At early time, Kelvin–Helmholtz (KH) rollers are seen and associated with them are phase lines with a distinct wavelength and a distinct angle with the vertical, (b) at a later time, the KH rollers break down into turbulence and the adjacent internal waves span a wide range of angles, (c) initial profiles of mean velocity and buoyancy frequency.

21.3.4 Jet with Continuous, Nonuniform Background Stratification

A jet with asymmetric stratification that allows shear instability in a limited region and supports waves in an adjacent portion with stronger stratification may lead to a complex field of turbulence and internal waves in a region which is nominally stable, $Ri_g > 0.25$. Internal waves associated with unstable shear in the Equatorial Undercurrent (EUC), whose vertical structure is similar to an asymmetrically stratified jet, have been suggested to be the dominant source of mixing in the thermocline, e.g., see [31]. Internal wave signals observed near the jet stream are thought to originate from stratified turbulence at the upper portion of the jet stream adjacent to the tropopause. The early numerical simulations of the problem were 2D and helped understand the generation of internal waves. Tse et al. [60] performed three-dimensional simulations where the velocity and temperature profiles of the base flow, constructed to model a jet in the atmospheric tropopause, are forced. Turbulence was observed in the

jet core while patchy turbulence and nonlinear evanescent waves were seen at the edges without significant propagation of internal waves in the jet far field.

Recently, Pham and Sarkar [35] used 3D simulations to examine variability in a stratified jet with shear and stratification profiles chosen to model the vertical structure of the EUC. As shown in Figure 21.2b, there is a thin region at the top with shear and low stratification. This flow is unstable since $Ri_g < 0.25$ and, therefore, vortical structures develop which lead to internal waves that propagate into the stratified jet below. Although KH billows form, internal waves of the corresponding wavelength are not seen as can be explained by linear theory. Briefly, the Taylor–Goldstein equation, specialized to the chosen shear and stratification profiles and to the horizontal wave number of the KH mode, leads to an imaginary vertical wave number in the jet region showing that wave propagation is not supported. Later in time, internal waves with a larger wavelength, a little more than twice the KH wavelength and allowable according to the Taylor–Goldstein equation, are observed in the jet. Figure 21.2a shows

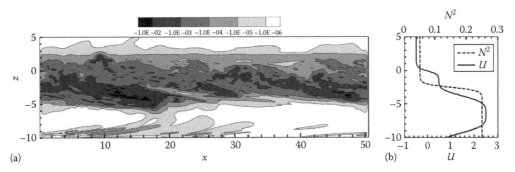

FIGURE 21.2 Asymmetrically stratified jet: (a) turbulent dissipation in a vertical slice of the jet, contour level increases from 10^{-6} (light gray) to 10^{-2} (dark black), (b) initial profiles of mean velocity and buoyancy frequency.

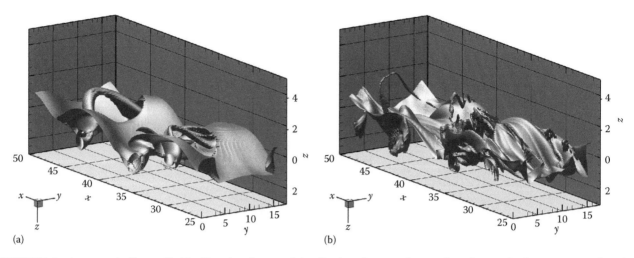

(a)   (b)

FIGURE 21.3 Asymmetrically stratified jet. Note that, for ease of visualization, the vertical axis is flipped: up in the figure corresponds to down in physical space. (a) Isopycnal that shows horseshoe-like density structures that originate from the shear layer, (b) isopycnal at later time. The prominent horseshoe at the left in part (a) has been stretched by the mean shear.

the turbulent dissipation rate, ε. In the nominally stable upper flank of the jet, there are patches with strong dissipation, 3 orders of magnitude larger than the background value. These patches of dissipation are spatially coherent and are convected with the local mean velocity of the jet.

Highly resolved simulations allow examination of mixing events in detail using a variety of diagnostics and visualizations. For example, Pham and Sarkar [35] have determined that the coherent patches of dissipation in their simulation of a stable jet are associated with 3D horseshoe-like structures visualized in Figure 21.3 using the density field. These structures which, in 2D cuts of the density field, appear as "hot pockets" that penetrate into the jet are created by horseshoe vortices that originate from the unstable shear layer. The vortices are stretched and convected by the jet mean shear to break down into turbulence and lead to local mixing.

21.3.5 Horizontal Shear

Shear in the horizontal direction, $a\langle U\rangle/ay$, is abundant in coastal flows, at fronts and in coherent vortices that form in the ocean or the atmosphere. The effect of stratification on horizontal (lateral) mixing processes and the possibility of vertical transport by three-dimensionalization of horizontal shear flow are two important questions in stratified flow with horizontal shear that can be addressed only through 3D simulations. DNS of a flow with uniform horizontal shear (HS) component [25] showed that, in a stratified fluid, vertical transport was larger when a given shear is horizontal rather than vertical. The reason is that turbulence production in the HS case is by horizontal eddies not directly inhibited by gravity and, in the absence of phase-locking in the vertical by rotation, the associated pressure fluctuations have a vertical gradient leading to vertical velocity. Furthermore, the decoupling between layers with horizontal shear leads to vertical shear, au/az, that promotes shear instabilities if the local Ri_g is small. In simulations of a

finite-thickness shear layer [5] with HS, rich dynamics is found including vortical instabilities, internal wave emission and intermittent lateral intrusions of mixed fluid into the interior. There is a nonlinear buoyancy instability that results in a staggered system of vortices with $Fr = \Delta U/Nl_v = O(1)$ and a vertical scale, l_v, of the vortices which is smaller than the horizontal scale. In addition, there are dislocations between the vortices whose vertical thickness is small, order of the diffusive scale, $\sqrt{D/N}$, with D the kinematic mass diffusivity. This is enhanced buoyancy flux as well as diapycnal mixing associated with the vortex dynamics. The observed buoyancy instability is reminiscent of the zig-zag instability [6, 33] of idealized vortex pairs in a stratified fluid.

21.4 Neutral Turbulent Boundary Layers

Rotation and tidal forcing are fundamental ingredients of environmental flow fields. Earth's rotation is known to dominate large scales of motion in geophysical flows which have values of the Rossby number, $Ro = U_0L/f \ll 1$. At environmental scales, ranging from several hundreds of meters to few kilometers, the Rossby number may attain values of the order one, such that rotational effects are still relevant and inertial effects cannot be neglected. The presence of a solid wall modifies rotating flow by introduction of a rotating boundary layer (BL) and the turbulent penetration depth is proportional to $u*/f$ where $u* = \sqrt{\tau_w/\rho}$ is the friction velocity related to the wall shear stress, τ_w. This flow is known as the Ekman BL and it is archetypal of the planetary boundary layer induced on the ground by the geostrophic wind. It is also representative of the ocean BBL. Regarding marine applications, rotation may affect the wind-driven boundary layer, giving rise to the so called wind-driven Ekman layer. In marine applications, the presence of a tide is a rule more than an exception, and the boundary layer thus often develops under unsteady forcing. Numerical simulations have been performed for different conditions representative of atmospheric or marine environments. Specifically, here we discuss the following classes: a class of simulations carried out in neutral stratification under the action of Coriolis force; a second class with wind forcing; and a third class of simulations where harmonic forcing has been considered. Finally, we discuss simulations where a combination of different forcing terms is considered.

21.4.1 Wall-Bounded Ekman Layer

The turbulent Ekman layer developing over a solid surface has been extensively simulated in the atmospheric turbulence community as part of LES modeling efforts. ABL LES is covered in another chapter of the present book. Here, we discuss DNSs of the bottom Ekman layer that have been performed over the past 20 years. The characteristics of the turbulent flow depend on the geostrophic wind $\mathbf{G}$, fluid viscosity ι and rotation vector Ω. Note that the rotation vector has two components, namely the vertical one Ω_V and the horizontal one Ω_H oriented along the south–north direction. The classical Ekman layer refers to the *polar case* where $\Omega_H = 0$ and Ω_V coincides with the Earth rotation rate Ω_E. Conversely the equatorial case is characterized by the presence of the horizontal background vorticity $\Omega_H = \Omega_E$ and the absence of vertical background vorticity. In the general cases when the latitude is $0° < \phi < 90°$ both components are present and affect the flow field. In the seminal paper by Coleman et al. [13], a low Reynolds number ($Re = G\delta_E/\nu = 400$, where $\delta_E = \sqrt{2\nu/f}$ and $f = 2\Omega_E\sin\phi$) DNS has been carried out considering different latitudes and also different directions of the geostrophic wind. In spite of the smallness of Re (i.e., absence of a log-layer profile), the study confirmed the role of latitude in the dynamics of the turbulent field as well as that of the direction of the geostrophic wind. In nearly parallel flows the presence of a mean velocity shear *parallel* to the axis of rotation may have either a stabilizing or a destabilizing effect, depending on whether the angular velocity and the mean shear vorticity have the same or opposite signs. The parameter S, defined as the ratio between the background vorticity and the mean shear vorticity, governs the importance of horizontal component of rotation in shear flows. If $S > 0$ the background vorticity and the shear vorticity are parallel, if $S < 0$ they are anti-parallel. The effect of suppression/enhancing of turbulence is quantified by the Bradshaw number $B = S(S + 1)$. Horizontal rotation enhances turbulence when $B < 0$, that is when $-1 < S < 0$, and the maximum occurs for $B = -1/4$, that means $S = -1/2$. Both $S < -1$ and $S > 0$ are associated with stabilized (turbulence suppression) flow. Turbulence enhancement/suppression occurs because the horizontal component of rotation directly affects the Reynolds stresses through production/destruction terms proportional to Ω_H. Due to the substantial influence played by the horizontal component, Ω_H, on turbulence activity, Coleman et al. [13] concluded that the "f-plane" approximation cannot be adopted for numerical simulations of the non-polar Ekman layer.

Simulations up to $Re = 2828$ of the *polar* case were carried out and discussed in Ref. [50] and in other papers therein cited.

These studies were mainly driven by the need to model the effect of finite Re on similarity theory which is strictly applicable in the limit of infinite Re. The study evidenced that, although a log-layer profile was absent in the simulations up to $Re = 500$, at $Re = 1000$ and larger values, an approximate log region is present in the flow. The results have shown that low-Re number effects are present for all turbulent quantities up to the value of Re investigated. Similarity laws of relevance to the atmospheric community were also tested and modified to fit the low-Re regime. Specifically in the following similarity laws

$$\kappa\frac{G}{u_*^2}\cos\theta - \ln\frac{u_*^2}{f\nu} = -A, \quad \kappa\frac{G}{u_*^2}\sin\theta = B \qquad (21.2)$$

where

$$\frac{u_*^2}{f\nu} = \frac{Re^2}{2(G/u_*)^2},$$

the authors utilized the angle θ instead of the angle α of deviation of the surface stress from the geostrophic wind direction. The angle θ is defined as the angle which satisfies the mean momentum conservation in the layer $0 < zf/u_* < 1$:

$$\theta = \alpha + \frac{C_5}{u_*^2/f\nu}.$$

The coefficient C_5 is defined by the shape of the velocity profile below the bottom of the log-layer. Note that the high-Re limit of $\theta = \alpha$ is obtained in the limit $\nu \to 0$, whereas low-Re behavior is recovered by the corrective term. The importance of the similarity laws reported earlier stems from the fact that, once the universal constants κ, A and B are known, the friction velocity and the angle α can be determined starting from knowledge of the geostrophic wind. The authors found a good fit between their DNS results and similarity laws using $A = 0.0$, $B = 2.1$, $\kappa = 0.41$, and $C_5 = -52$. In order to extend the use of the similarity law to more practical cases with roughness, the constant A was replaced with $A_0 = \alpha - \ln z_0/(\nu/u_*)$ where the ratio between the roughness to the smooth-wall length-scale was introduced. The study also showed the difficulty to identify a unique value of the von Karman constant in the whole range of Re numbers considered, also highlighting the absence of a perfect log- profile (in all cases κ varied from about 0.38 to about 0.42 along the log-region). This can indicate either that a power law might be better suited to represent the velocity profile or that the range of Reynolds number investigated via DNS so far is not sufficiently high.

21.4.2 Wind-Driven Ekman Layer

The wind-driven Ekman layer is a boundary layer created by the tangential stress supplied by the action of the wind over the free surface of the ocean. The Coriolis force causes a rotation of the velocity profiles and thus generates a component in the spanwise direction. It is important for mixing in the upper ocean along with other processes such as breaking of surface waves and Langmuir cells. In laminar conditions, the penetration depth of the Ekman layer (the thickness on the boundary layer) is proportional to $\sqrt{2\nu/f}$. In the turbulent regime the penetration depth is found to be proportional to u_*/f where u_* is the friction velocity associated to the imposed wind stress. A detailed discussion on the Ekman layer and its relevance in environmental fluid mechanics is in [37]. Although important, detailed DNSs or wall-resolving* LESs have not been carried out. Worth mentioning is the $Re \rightarrow y$ LES of Zikanov et al. [62]. In this study the wind stress was imposed directly at the free surface as a wall-layer model and the SGS stresses were modeled through a standard eddy-viscosity dynamic model. The authors focused on two main aspects: first, they checked the classical Ekman model where the analytical solution is preserved through replacement of the molecular viscosity with a constant eddy viscosity; second, they investigated the role of latitude and of the angle γ between the wind stress and the south–north direction. The study showed that under real turbulent conditions, the eddy viscosity is not constant along the vertical direction and this gives rise to substantial variations of the velocity field with respect to the classical Ekman model. The angle α between the wind stress and the velocity at the free surface obtained in the polar simulations is of about 28.5° clockwise, substantially smaller than the value of 45° predicted by the Ekman model. As in the stratified bottom Ekman layer case (see the subsection mentioned earlier for a discussion), the horizontal component of rotation and the angle γ play an important role in turbulent mixing for the non-polar cases. Since the Coriolis force does not produce energy in the system, as already discussed in the previous subsection, it distributes energy back and forth from the mean field to the turbulent one. The decrease of latitude produces a variation of the spiral and consequently a noticeable variation of the angle α. In the northern hemisphere, wind coming from the eastern quadrants produces an increase of α with respect to the polar case, whereas the opposite is true for wind coming from the western quadrants. This analysis is consistent with that of Ref. [13] for the bottom Ekman layer. Moreover, the angle γ governs the redistribution of energy from the mean field to the turbulent one. Specifically, in the northern hemisphere when the wind comes from the south–east quadrant ($0° < \gamma < 90°$) energy is extracted from turbulence to feed the mean field, whereas the opposite is true for wind direction from the north–west quadrant $180° < \gamma < 270°$. This phenomenon is related to the sign of the additional terms proportional to Ω_H present in the transport equations of the Reynolds stresses.

---

* A simulation where the near-wall viscous layer is directly resolved, implying the use of about 8–10 grid points within the first 10 wall units in the wall normal direction, a grid spacing of the order of 40–50 wall units in the streamwise direction and a grid spacing of about 15 wall units in the spanwise direction.

21.4.3 Oscillatory Boundary Layer

A turbulent field of great interest in environmental fluid mechanics is that developing under an oscillatory forcing around a bluff body or simply over a solid wall. In the first case, oscillatory motion induces periodic loads and their 3D modulation that can cause severe damage to pipelines and offshore structures. For a discussion on 3D loads over circular cylinders placed in an oscillatory current, see Ref. [32]. Here we focus on turbulence dynamics developing over a solid surface. This flow has several implications: among them it is important for understanding sediment dynamics under wave action, i.e., for the formation of the sand ripples in the near shore environment; this flow is also relevant for comprehension of turbulent mixing in the ocean under tidal forcing. From a fundamental point of view, the boundary layer developing under an unidirectional oscillatory external pressure gradient is known in literature as the Stokes boundary layer (SBL), and, in the laminar regime, its thickness is proportional to $\delta = \sqrt{2\nu/\omega}$, where ω is the frequency of the oscillatory pressure gradient, $dP/dx(t) = -\omega U_0 \cos(\omega t)$, with U_0 the amplitude of the free stream velocity $U(t) = U_0 \sin(\omega t)$ induced by the pressure gradient. A Reynolds number is usually defined as $Re_\delta = U_0 \delta/\nu$. Depending on the value of Re_δ the flow field undergoes different regimes: the laminar regime, the disturbed laminar regime, the intermittent turbulent regime, and the fully developed turbulent regime. At small values of Re_δ, turbulence activity appears at the beginning of the decelerating phase, associated to the appearance of explosive near-wall bursts; with the increase of Re_δ, turbulence occurs at earlier phases of the cycle. The laminar regime (L) holds for Re_δ smaller than around 100; for $100 < Re_\delta < 500$–550 the disturbed laminar regime is present, characterized by the presence of small-amplitude perturbations, over-imposed to the base Stokes flow. For $Re_\delta > 550$ the 2D pattern typical of the disturbed laminar regime turns into a 3D one and the flow moves to the intermittent regime, where explosive turbulence bursts are created during the decelerating phases of the cycle. Fully developed turbulence is present during the cycle only for higher values of Re_δ. In the exhaustive experimental work of Jensen et al. [28] it is clearly shown that at $Re_\delta = 1790$ fully developed turbulence is already present in most of the cycle of oscillation, whereas fully developed turbulence was recorded throughout the cycle for $Re_\delta \geq 3460$. Recently the turbulent regime has been investigated *via* LES by Salon et al. [44] and by Radhakrishnan and Piomelli [39]. The wall-resolving LES of Salon et al. [44] was performed at $Re_\delta = 1790$ corresponding to test 8 of the experiments of Jensen et al. [28]. The numerical simulation was able to correctly reproduce the alternating phases of acceleration and deceleration as well as the sharp transition to turbulence observable at a phase angle between 30° and 45°, and its climax between 90° and 105°. At this value of Re_δ, from the mid acceleration to the late deceleration phases, fully developed equilibrium turbulence was observed in the central part of the cycle, and the boundary layer is similar to that of a canonical, steady, wall-bounded flow. The phases of low turbulent activity are characterized by the presence of two separate regions: a near-wall one, where the vertical turbulent velocity fluctuations vary much more rapidly than the other two components, thus giving rise to the formation of horizontal, pancake-like turbulence; an outer region where both vertical and spanwise velocity fluctuations vary much faster than the streamwise ones, hence producing cigar-like turbulence. The simulation used a dynamic mixed SGS model, which proved to be effective in reproducing the alternate re-transition and decay of turbulence, at least for the turbulent regime therein investigated. The authors also performed additional simulations in the intermittent regime ($Re_\delta = 990$) to check the performance of the plane-averaged dynamic mixed model (DMM) employed in strongly non-equilibrium conditions. A coarse DNS simulation was able to reasonably reproduce the time-varying wall shear stress, whereas noticeable differences were obtained using LES, which strongly delays transition and appreciably underpredicts the maximum value of the wall shear stress. The underprediction of the wall stress obtained with LES was attributed to the following possible reasons: absence of equilibrium turbulence during the oscillatory cycle; the fact that at $Re_\delta = 990$ turbulence appears strongly intermittent and in spots and consequently plane averaging of the constant may not be suitable for reproducing such physics.

In order to make LES suitable for the study of oscillatory flows of practical interest to the oceanographic community, wall-layer modeling is required, with the aim to skip the direct resolution of the near-wall structures whose dimensions scale with fluid viscosity and, at the same time, to solve properly the inertial large scale structures responsible for mixing in the water column. The study of Radhakrishnan and Piomelli [39] focused on this aspect. The authors reproduced test 10 of Ref. [28] at $Re_\delta = 3600$ characterized by equilibrium turbulence along the whole oscillatory cycle, and test 13, equivalent to test 10 as regards the flow conditions but carried out over a rough wall. Four different SGS models were used in conjunction with two different wall-layer treatments. The main conclusion of this study is that turbulence dynamics developing in an oscillatory flow can be better reproduced by low-dissipative subgrid scale (SGS) models, like those belonging to the family of dynamic ones. This excessive dissipation may make the response of the model too slow with respect to the variation imposed by the subsequent acceleration-deceleration phases of the flow. As regards wall-layer modeling, the classical log-reconstruction gave satisfactory results, although a slightly better agreement was obtained with a more recent model which introduces a shift in the velocity field at the first grid point and a better reconstruction of the wall stress fluctuations. Finally, the quality of the results was maintained also in the rough wall case, of more practical interest in the environmental fluid mechanics community.

To conclude this section we mention flow fields of interest to the environmental fluid mechanics community, where the harmonic driving pressure gradient appears together with additional forcing terms. Among them, the case of the presence of an oscillatory current superimposed to a mean current, making the oscillatory cycle nonsymmetric along the two half-periods is of interest. This flow is controlled by three free parameters: the friction velocity associated to the mean flow component u_*, the frequency of the oscillatory component of the motion ω, and the ratio between the oscillating and mean centerline velocity a_{uc}. A numerical investigation was recently carried out by Scotti and Piomelli [47]. The authors have investigated the current-dominated regime ($a_{uc} < 1$), of interest to the oceanographic community, spanning a wide range of frequencies, from the quasi-steady case to the quasi-Stokes flow case. The simulations were carried out using both DNS and wall-resolving LES with a standard eddy viscosity dynamic model. The study showed that at high frequencies the log profile remains little affected by the oscillation and modulated back and forth by the oscillation itself. At low frequencies the whole flow is dominated by the oscillatory component and resembles a turbulent Stokes boundary layer as that studied in Ref. [44] (although the oscillation is not symmetric due to the presence of the mean field), with phases of relaminarization and phases of re-transitions to turbulence. At intermediate frequencies the log-profile is present only at certain phases of the oscillatory cycle and relaminarization and successive retransition are observed along a cycle. These behaviors were found to be controlled by a length scale $l_t = \sqrt{2(\nu + \nu_T)/\omega}$ where ν_T is the turbulent eddy viscosity. This length scale is a measure of how far the oscillation generated in the near wall layer propagates into the core region. The high frequency regime is characterized by $2l_t/H \ll 1$ meaning that the disturbance produced by the oscillation is not able to propagate in the core region; the low frequency regime is such that $2l_t/H > 1$ and consequently the oscillation fully controls the flow; in the intermediate regime with $2l_t/H < 1$ the disturbances generated near the wall propagate into the core region and interact with turbulence present there, leading to local departure from equilibrium. Another important result of this research is the proof of the reliability of the dynamic model to simulate unsteady flow characterized by cyclic relaminarization and retransition to turbulence.

Another combination of interest is when rotation effects are not negligible and interact with a tidal motion. In practical marine applications, the M_2 tide having a period of about 12.4 h often drives the large amplitude components of the tidal flow. For a non-polar case, the Coriolis parameter is $f = 2\Omega_E sin\theta$. The Rossby number, defined in this case as $Ro = \omega_t/f$, where ω_t is the oscillatory frequency of the tidal motion can be of the order 1. For example, for the M_2 tide at a latitude of 45, $Ro \sim 1.4$. This value decreases with increasing latitude. Classical theory suggests rotation effects to be negligible in case $Ro \gg 1$, whereas rotation dominates when $Ro \ll 1$. In cases characterized by $Ro \sim 1$ both rotation and inertial terms concur to the development of turbulent mixing. This is the case when one is interested to study mixing related to tidal motion. In spite of the importance of this topic, to the best of our knowledge the only LES study devoted to the comprehension of modification induced by earth rotation over a tidal flow is in the recent paper by Salon et al. [45]. The authors extended the study of Salon et al. [44] considering $\mathfrak{R}_\delta = 1790$, and including the effect of rotation for a mid-latitude ($\theta = 45$) shallow-water environment. As regards the modifications with respect to the purely oscillatory case, for the mid-latitude condition therein analyzed, the evolution along the period of the wall shear stress appeared little influenced by rotation, whereas the turbulent penetration length appeared much larger and extended up to the free surface region. Growth and decay of TKE were observed to maintain similar characteristics with respect to the pure oscillatory case: turbulence is generated by shear instability during the accelerating phases, whereas the decelerating ones are characterized by the velocity inversion in proximity to the bottom that causes a splitting of the fluid column in two separate layers. A rapid growth of a laminar boundary layer is then observed close to the bottom while, in the outer layer, turbulence decays due to the absence of production rate and few large-scale structures coming from the previous, productive phases populate the upper part of the water column. Such a history effect is a peculiar feature of the pure oscillatory case and is still observable in the presence of rotation. Rotation redistributes the turbulent energy among the three directions, with larger intensity along the horizontal planes near the bottom surface during the first accelerating phases. As a consequence, the shape of turbulence appears modified, with a tendency to isotropization observed in the core region. This produces an increase of the eDciency of mixing along the water column and weakens the two-layer structure already observed in the purely oscillating turbulent boundary layer. The horizontal background vorticity breaks the symmetry of the two half-cycles with turbulence activity in the decelerating phases of the second semi-cycle substantially larger than that of the first one. The results of the study were in agreement with those of [13], who observed an east/west enhancement/reduction trend associated with the stabilizing/destabilizing asymmetry. It is worth noting that, although the simulations were carried out at a value of Re substantially smaller than that encountered in practical problems, relevant turbulent quantities obtained in this laboratory-scale numerical experiment were in good qualitative agreement with data taken from field measurements in two mid-latitude tidal channels. This lends justification to the extension of results obtained in fundamental studies at a laboratory scale to full scale flows.

21.5 Stratified Turbulent Boundary Layers

A stratified water column generally bounds upper ocean turbulence from below and, similarly, bottom ocean turbulence is bounded from above by the interior stratification. Therefore, stabilizing buoyancy effects must be taken into account in ocean bottom boundary layer (BBL) dynamics with some exceptions such as shallow coastal regions and energetic tidal channels. The atmospheric boundary layer (ABL) can be stratified by strong surface cooling. Such a situation corresponds to a stable ABL

which has important differences with respect to the neutral or convective ABL. The present section presents results from numerical simulations which investigate stabilizing buoyancy effects in the ocean.

21.5.1 Stratified Channel Flow

Turbulent flow in a channel is the canonical flow in DNS and LES investigations of wall-bounded turbulence. The first simulations of stratified turbulent channel flow were at a friction Reynolds number of $Re_* = 180$ and performed by Garg et al. [17] who simulated the initial buoyancy-dominated transient and Armenio and Sarkar [2] whose simulation continued into the final statistically steady state. Both sets of authors employed high-resolution LES with grids that were fine enough so that the effect of the subgrid model on the turbulence statistics was small. Constant temperature boundary conditions at the lower and upper walls were imposed by Armenio and Sarkar [2]. The high shear near the wall allowed a low-Ri_g region which was found to support active wall turbulence with low-speed streaks, albeit with reduced turbulent stresses. The central low-shear region with high Ri_g exhibited strongly suppressed Reynolds shear stress and internal wave activity. The two regions were approximately separated at the location of $Ri_g \simeq 0.2$. Taylor et al. [55] simulated open channel flow with different boundary conditions: an adiabatic bottom and stratification forced by a surface heat flux. This choice of boundary conditions, more representative of marine applications, leads to a well-mixed turbulent bottom layer that is capped by a stable pycnocline. The parameter, Ri_g, is found to be less useful in this situation since the near-wall region, where shear production of turbulence dominates, is relatively unaffected by stratification. Instead, it is found that turbulence at any distance from the lower wall can be classified into three regimes based on the magnitude of the Ozmidov length scale, L_O, relative to a vertical length, l_z, characterizing the large scales of turbulence, and to the Kolmogorov scale, η. The three regimes are as follows: (a) turbulence near the boundaries that is essentially unstratified when $L_O > l_z$ and $L_O > \eta$, (b) *buoyancy-aThcted* turbulence when $L_O < l_z$ but $L_O > 9\eta$, and (c) *buoyancy-dominated* turbulence when $L_O >$ both l_z and 9η. Regimes (b) and (c) are clearly analogous with the relative ordering of length scales observed in a stratified shear layer [49] at early time and late time, respectively, as discussed in Section 21.3.2. Regime (a) could potentially occur at early time in a stratified shear layer if the initial Ri_b is chosen to be suDciently small. A stratified channel flow with vertical walls was simulated by Armenio and Sarkar [3], who found that vertical mixing was higher than in typical channel flow with vertical shear.

21.5.2 Stratified Bottom Ekman Layer

The ocean bottom can be taken to be nominally adiabatic leading to $d\langle\rho\rangle/dz = 0$ as the bottom boundary condition on the density. Therefore, adjacent to the bottom, there is generally a mixed layer where buoyancy effects are weak. Nevertheless, the overlying stratification with buoyancy frequency, N_A, can affect the outer layer of turbulence. The environmental parameter, N_A/f, determines the strength of buoyancy effects. Taking a mid-latitude value of $f = 10^{-4}$ rad/s and taking the buoyancy frequency to vary from $N_A = 10^{-2}$ rad/s in a typical thermocline to $N_A = 10^{-3}$ rad/s in weakly stratified deep ocean gives $10 < N_A/f < 100$. The thickness of Ekman layer decreases with increasing N_A and is proportional to $u_*/\sqrt{N_\infty f}$ for $N_A/f \gg 1$. Therefore, the observed thickness of the oceanic bottom Ekman layer is often much smaller than that given by turbulent Ekman scaling, $0.5u_*/f$, and ranges between 2 and 20 m.

Turbulence in a bottom Ekman layer under an outer geostrophic flow has been studied recently over a wide range of Reynolds numbers using DNS [57], resolved-scale LES [56] and LES with a near-wall model [58]. The following summary of buoyancy effects in a stratified Ekman layer is adapted from the extensive discussion in Ref. [58] of results obtained over a range of background stratifications, $0 < N_A/f < 75$, in LES of a high-*Re* Ekman layer with wall-roughness. The case with $N_A/f = 75$ has external stratification and current velocity chosen to correspond to the BBL measured by Perlin et al. [34] in an 81 m water column on the Oregon shelf. The flow, initialized with a linear temperature profile in the simulations, develops a three-layer structure: a bottom mixed layer, a capping pycnocline, and a uniformly stratified external layer with propagating internal gravity waves. At $N_A/f = 75$, the pycnocline is thin, 1.65 m, which is a small fraction of the mixed layer thickness of approximately 8 m. Over the thickness of the pycnocline, the density gradient increases from zero in the mixed layer to its peak value of about four times the background value and then decreases back to the background value. The height, h, of the turbulent boundary layer is found to approximately follow the following scaling law proposed by Zilitinkevich and Esau [63] based on numerical simulations and observations of a stable ABL: $h = C(u_*/f)(1 + \alpha N_A/f)^{-1/2}$ with $C = 0.5$ and $\alpha = 0.15$. Although the turbulent boundary layer thickness decreases substantially, the Ekman velocity component increases so that the effect on net Ekman transport is found to be insignificant. The Ekman veering angle, α_0, made by the velocity at the bottom surface with the geostrophic velocity direction, increases from its unstratified value of 15.4° to a value of 24.8° at the largest stratification, $N_A/f = 75$. The veering angle, α, exhibits a small decrease in the mixed layer followed by a rapid decrease to zero at the top of the capping pycnocline. The drag exerted by the bottom on mean currents and mesoscale eddies is an important input into operational ocean models that typically do not resolve BBL dynamics. A quadratic drag law is usually employed with the surface stress magnitude computed as $\tau_w = CU^2$ where $C = 0.002$ is a model coefficient and U is the near-bottom velocity in the ocean model. The LES results show that C does not change significantly with increasing stratification.

A classical log law (or law of the wall) in an unstratified turbulent Ekman layer is approximately realized in numerical simulations. However, in the stratified cases, the mean velocity is larger than that given by the log law with the deviation starting from within the mixed layer. An overshoot of mean shear in the outer turbulent layer with respect to the log law has been found in field observations [29] and is likely related to a decrease in vertical mixing length scale with increasing stratification. In order to represent this buoyancy effect on the mean velocity, Taylor and Sarkar [58] modified the length scale, $l = \kappa z$, in the expression for mean shear as follows:

$$\frac{dU}{dz} = \frac{u_*}{l}, \quad \frac{1}{l} = \frac{1}{\kappa z} + \frac{N(z)}{C_b u_*}. \tag{21.3}$$

Integration of the first part of Equation 21.3 using the second part for l leads to

$$\frac{U(z)}{u_*} = \frac{1}{\kappa}\log\left(\frac{z}{z_0}\right) + \frac{1}{C_b u_*}\int_0^z N(z')dz'. \tag{21.4}$$

If $N(z)$ is taken to be constant and equal to the external frequency N_A, the following *log-linear* profile is obtained for the mean velocity:

$$\frac{U(z)}{u_*} = \frac{1}{\kappa}\log\left(\frac{z}{z_0}\right) + z\frac{N_\infty}{C_b u_*}. \tag{21.5}$$

The mean profile given by Equation 21.4 or Equation 21.5 can be viewed as the oceanic boundary layer analog to the log-linear profile, widely used to represent the mean velocity in stable ABLs, that follows from Monin-Obukhov (MO) theory. Note that MO theory cannot be directly used in the oceanic BBL since the surface buoyancy flux is zero.

Turbulence levels in the mixed layer exhibit a small reduction with respect to values in the inner layer of an unstratified Ekman boundary layer. The main effect of stratification is the reduction of mixed layer height and the overall boundary layer height. Internal waves are observed to propagate in the background. The internal wave flux, $\langle p'w' \rangle$, is a small fraction, approximately 2%, of the integrated dissipation. This fraction is about an order of magnitude smaller than in the free shear layer discussed earlier. Nevertheless, the internal wave flux is about the same magnitude as the integrated buoyancy flux. The properties of these turbulence-generated waves have been characterized by Taylor and Sarkar [56]. The wave amplitude is small and their propagation is governed by linear, viscous theory. The spectrum of the vertical strain, $\partial w/\partial z$, is broadband at the edge of the boundary layer with the maximum frequency (in a frame moving with the current velocity) larger than N. Such super-N waves are evanescent and are also present in field observations. As the waves propagate into the interior, the frequency spectrum progressively becomes narrower and tends to cluster in a range of 35°–60°. Turbulence-generated internal waves are often observed to eventually propagate at a narrow band of angles around 45°, although they might span a wide frequency range at the region of generation. This phenomenon has been observed in laboratory experiments of a shear layer by Sutherland and Linden [54] and grid turbulence by Dohan and Sutherland [15]. The following explanation for the narrow band of propagation is offered by Taylor and Sarkar [56] and verified in the boundary layer case: both, high- and low-frequency waves, have low vertical group velocity, larger time of flight to a given vertical level, and larger viscous attenuation leaving behind mid-frequency waves clustered around 45°.

21.5.3 Stratified Wind-Driven Ekman Layer

The stratified wind-driven Ekman layer has strong implications in the analysis of mixing in the upper ocean. Indeed, it is well known that heat exchange with the atmosphere occurring at the free surface always produces stratified conditions in the upper surface layer of the ocean [37]. A LES study of the stable stratified Ekman layer has been recently carried out by Inghilesi et al. [22]. In this LES study, the viscous layer and the diffusive heat transfer at the free surface have not been directly resolved, rather the free-surface shear stress and the heat flux have been directly imposed in the fashion of a *wall-layer model*. This allowed the study of a relatively high value of Reynolds number ($Re = 10,000$, based on the friction velocity and the turbulent penetration depth u_*/f). A mid-latitude case was investigated meaning that the horizontal component of the background vorticity was taken into account for two stratification levels, respectively $Ri = (N^2)_{fs}\delta^2/u_*^2 = 0$ and $Ri = 40$. The $Ri = 40$ case gives a strongly stratified flow so that there is turbulence in a *buoyancy-dominated* regime [2,55]. The results showed that the stratification strongly reduces the turbulent penetration depth, increases the veering angle between the velocity vector at the free surface and the wind stress, strongly reduces the Reynolds stresses in the free surface region, and induces internal waves in the pycnocline region. In the paper, dispersion of buoyant (fresh water) particles in the background (salt water) environment was also investigated by means of a Lagrangian technique. In order to take into account that the particles move into a variable-density field, the equation of motion of the single particle of the swarm was modified by introduction of an additional term proportional to Ri. The results showed that very inaccurate prediction of particle dispersion is obtained using the standard particle motion equation which assumes a constant background density field.

21.5.4 Stratified Oscillatory Boundary Layer

The response of an oscillating boundary layer to stratification is of interest in tidal flows over the ocean bottom. DNS and LES studies of this situation are scarce. DNS of a stratified tidal bottom Ekman layer for various values of the Rossby number is reported in Ref. [43]. The case of pure oscillation was simulated at a Stokes Reynolds number, $Re_\delta = 1000$, and a stratification, $N_A/\omega \simeq 7$. The authors studied the growth of the mixed layer and also the exchange between kinetic and potential energy.

A recent resolved LES by Gayen et al. [18] simulated a stratified, boundary layer with external velocity $U_A\sin(\omega t)$ and studied the dependence of boundary layer properties on the phase of the external velocity. A range of stratifications, $0 < N_A/\omega < 50$, was investigated. The tidal frequency, e.g., 12.4 h for the M_2 tide, is much lower than characteristic turbulence frequencies. Nevertheless, as discussed in Section 21.4.1, turbulence is strongly affected by acceleration and deceleration. The LES of Gayen et al. [18] was performed at $Re_\delta = 1790$ to enable comparison with the unstratified case where detailed LES [44] and laboratory experimental data [28] are available. Stratification has a strong influence on many flow statistics. The duration of the log-law is observed to shrink, e.g., from 40°–140° in the unstratified case to 80°–125° in the case with $N_A/\omega = 50$. The boundary layer height, $\delta_t(\phi)$, is computed using the height of the 10% of maximum TKE contour. The quantity $\delta_t(\phi)$ increases from its minimum value in the early acceleration stage to reach a maximum in the early deceleration stage followed by a decrease in the late deceleration stage. In the stratified cases, the decrease of $\delta_t(\phi)$ is abrupt and to a value of almost zero. The peak value of $\delta_t(\phi)$ at $N_A/f = 50$ decreases to about 0.6 of the corresponding value in the unstratified case. The phase lead of the wall shear stress increases from 17°–25°; note that the laminar case has a phase lead of 45°. The friction coeDcient, computed with the cycle-averaged wall stress, exhibits a slight increase with increasing stratification. The turbulent boundary layer is found to excite internal gravity waves that propagate upward into the ambient. The tilt of the phase lines of the internal waves with respect to the vertical depends of the relative velocity of the eddies with respect to the external background current which changes with phase. Similar to the steady Ekman layer, the internal waves span a narrow band clustered around 45° in the far field although, at the edge of the boundary layer, the internal wave signature is broadband.

21.6 DNS and LES Studies of Turbulent Flows with Geometrical Complexities

The discussion of the previous sections pertained to flows with simple geometry but complex physics. Geometric complexities introduce additional complications especially when stratification is involved. As an example, lee waves in flow over an obstacle and the anabatic/katabatic flows developing along the inclined walls of mountains are typical examples of effects introduced by topography in the presence of stratification. Here we first discuss studies of turbulent fields of interest to EFM developing over topography in the neutral case as well as in the stable stratified cases. Subsequently, we discuss results of very recent studies aimed at understanding the mechanisms of interaction between turbulence and sediments in local scour processes.

A widely investigated complex-geometry flow is that developing over an infinite train of sinusoidal waves. Analysis of turbulence dynamics in this geometry is of importance in many applications, for example, to understand the interaction between surface waves and the low atmosphere, to quantify vertical mixing over terrains and to understand turbulence-sediment interaction over sand ripples or dunes. When the wave height to wavelength ratio (wave slope parameter λ) is smaller than 0.05, the flow does not separate and it resembles a canonical boundary layer. Conversely, when λ > 0.05, a recirculation bubble appears in the downslope part of the wavy surface. In the latter case, the flow field appears split into two separate main regions: an outer region is observable well above the wave crest, for which the wavy wall appears as a wall roughness, and an inner region directly affected by the wavy geometry. The latter is composed of a shear layer extending from the crest of the wave downstream above the recirculation region, the recirculation bubble itself, and a thin boundary layer that is regenerated downstream of the reattachment point over the upslope part of the wave. DNS of this flow was first carried out by De Angelis et al. [14] considering two values of λ belonging to the non-separated regime and to the separated one, respectively. The authors also observed the presence of *splat-like* phenomena, consisting of high-speed fluid patches that impact the upslope wall of the wave to produce large fluctuations in the spanwise direction. The analysis also showed that the recirculation region exhibits large intermittency which causes strong fluctuations of the wall shear stress and of the pressure. These effects may be responsible for local erosion phenomena in the case of an erodible bed.

The DNS study by Sullivan and McWilliams [52] was devoted to the analysis of the interaction between a train of sinusoidal waves and a stratified Couette flow. The wave slope was λ = 0.015, belonging to the regime without flow separation. In this study, a wide range of levels of stratifications (from convective to stable conditions), was considered. The bulk Reynolds number was set equal to 8000 corresponding to values of the friction Reynolds number ranging from 100 to 164 depending on the stratification level and the phase speed of the waves. Finally, different values of the wave phase speed were considered in the range $0 \leq c \leq 0.7$. Isothermal conditions were prescribed at the top and bottom surfaces of the computational box. The study addressed an important question, namely the influence of stratification on the pressure acting over the wave train. It was shown that the surface pressure stress depends on the relative velocity of the waves compared to the freestream velocity and on stratification. For fast-moving waves, stratification has a negligible effect on the surface form stress, which is very small or negative. On the other hand, for slow-moving waves the pressure stress is significant and it is dramatically affected by stratification.

LES of the flow over 2D sinusoidal wavy terrains has been performed [1, 10, 21, 30]. In Refs. [1, 21], the separated regime was investigated for values of λ up to 0.20. The physics, already identified by De Angelis et al. [14] in their DNS, was correctly reproduced in all the LES studies and the main qualitative features of the flow field appeared not to be influenced by the Reynolds number. Interestingly, all numerical studies gave very similar values of the mean separation and reattachment points for given values of Re and λ, whereas some disagreements were detected when comparing with reference experimental data. This feature might be attributed to some differences between the numerical setup and the experimental one.

The wavy-wall LES analysis of Chang and Scotti [12] and of Marchioli et al. [30] were, respectively, driven by the need to understand sediment motion and dynamics of inertial Lagrangian particles in the presence of surface waviness. The LES of Chang and Scotti [12] was mainly devoted to clarify the role of the coherent structures in suspension and entrainment of sediments in water. The study, carried out at a bulk Reynolds number $Re = 6500$ showed that the process of suspension can be divided into two phases: the formation of a cloud of particles in the downslope part of the wave and the subsequent ejection of the particles across the shear layer present above the sediment cloud. The vortical structures developing on the upslope part of the wave govern the amount of sediment in the cloud.

In the LES study of Marchioli et al. [30], the bulk Reynolds number was $Re = 3108$ and the particle density ratio was representative of solid particles in air. The authors were interested in the role of coherent structures in particle dispersion and on the analysis of particle segregation and deposition over the wavy surface. According to classical literature, particles were observed to accumulate in high-strain low-vorticity near-wall regions. The study also showed that particle deposition, segregation, and resuspension strongly depend on particle inertia, and these phenomena are controlled by the particle Stokes number. In particular, since inertia acts as a low-pass filter, it reduces the effect of coherent wall structures on the particle motion and thus reduces the degree of particle segregation. Finally, the study showed that the wall corrugation increases the deposition rate by several orders of magnitude with respect to the case of a plane channel flow, with a non-dimensional deposition coeDcient monotonically increasing with the Stokes number.

The effect of stratification on the flow developing over wavy walls was studied by Calhoun et al. [10] whereas the neutral case was investigated by the same authors in a previous paper reported in their references. A large amplitude sinusoidal bottom wall with λ = 0.1 was considered at $Re = 7000$ (based on the mean velocity and the depth of the fluid column) for two levels of stable stratification. The molecular Prandtl number was chosen equal to unity. Isothermal boundary conditions were considered in this study. The investigation showed that the mean velocity field is affected by the presence of stable stratification which reduces the extent of the downstream recirculation region. The increase of the friction Richardson number from 31 to 62 did not alter appreciably the mean field, probably due to the fact that both simulations resulted with turbulence in the buoyancy-affected regime rather than the buoyancy-dominated regime. An increase of turbulent anisotropy related to the presence of stable stratification was also found. In particular, anisotropy appeared more pronounced over the wave crest, whereas it appeared to be less affected by stratification in the recirculation region.

Finally this study did not show the presence of lee waves probably due to the weakness of the stratification levels investigated.

Another flow of interest is that which develops along longitudinal ridges. This flow, of great interest to the environmental hydraulics community, has been little investigated compared to the wavy wall case. Differently from the wavy wall case, this flow field exhibits a 3D mean field, due to the development of secondary Prandtlß second-kind cellular motion. This motion consists of the presence of longitudinal mean vortices generated by the imbalance between the cross-sectional Reynolds stresses, superposed on a mean streamwise flow. These secondary structures are known to generate sediment motion thus creating longitudinal bars in river applications, and to increase heat transfer in industrial applications. The flow over longitudinal ridges is also archetypal of along-canyon flows that can develop in oceans or in the low atmosphere.

The LES of Falcomer and Armenio [16] studied turbulent mixing in a channel flow with walls equipped with longitudinal ridges with height equal to 1/8 the half-height of the domain. The friction Reynolds number was set equal to 580 with the aim of comparing the numerical results with available experimental data. The agreement between numerical and experimental data was satisfactory thus proving the ability of LES to predict flow fields where classical RANS solvers with algebraic or $k - \epsilon$ model are known to fail. The ridge generates a large cross-sectional secondary motion, whose mean velocity is of the order of 3%–5% of the bulk streamwise velocity and a spanwise modulation of the primary wall stress. The wall stress decreases going from the trough to the bottom corner of the ridge, then it rapidly increases in the upslope wall and then slowly decreases from the top corner to the symmetry axis of the ridge. Near-wall, small secondary recirculations were observed close to the ridge. Unlike the large secondary current that extends up to the centerline of the channel, these small cellular flows are highly intermittent and cause a spotty distribution of the secondary wall stress, the latter responsible for erosion phenomena and generation of sand bars in river applications. The analysis of the coherent structures showed that the longitudinal ridges cause their re-organization and alignment along the streamwise direction.

A stratified along-ridge flow was considered in a later paper [4]. The authors used a no-slip condition over the bottom surface and a free-slip condition at the top-boundary, thus mimicking the presence of a zero-Froude number free surface. Isothermal conditions were prescribed at the top/bottom boundaries and the molecular Prandtl number was set equal to 5, corresponding to thermally stratified water. The ridge height was equal to one-eighth of the channel depth. The SGS dynamic model described in Ref. [2] was used in the simulations. The study showed that stable stratification affects the secondary recirculations in the cross-sectional plane. These recirculations constitute an effective means for the overall mixing in the fluid column, driving warm (light) fluid from the top to the bottom surface and cold (heavy) fluid from the bottom to the free-surface. The increase of stratification produces the formation of a strong density gradient in the vicinity of the free surface creating a barrier of

potential energy and reducing the vertical extension of the main secondary circulation. Conversely, the small cellular flow at the trough of the ridge increases with stratification. The intensity of these small recirculations increases with stratification due to the increased anisotropy of the cross-sectional normal Reynolds stresses.

The analysis of the stable stratified flow over an isolated hill has important implications in oceanic as well as in atmospheric applications. The flow over a hill is considered to be strongly stratified if the ratio $U_0/Nh < 1$ (where h is the height of the hill, U_0 is the free stream velocity and N the Brunt–Vaisala frequency), whereas it is weakly stratified if the ratio exceeds unity. Traditionally, the analysis of the stable stratified flow around a hill has received considerable attention, due to the implications to the dispersion of pollutants. In fact, depending on the location of a pollutant source point, completely different characteristics of dispersion can be observed. A LES study was carried out by Li et al. [27] considering an axisymmetric hill that has the same shape as that used in a previous experimental study. The study was carried out at a laboratory-scale Reynolds number for several levels of stratification, in the range $1 \geq U_0/Nh \geq 0.2$. Consistent with prior results, the study showed that the increase of stratification suppresses the vertical motion. In the strongly stratified case ($U_0/Nh = 0.2$) the streamlines move around the obstacle horizontally, as the vertical rise of the upstream streamline is inhibited by stratification. Since, under these conditions, horizontal motion is predominant, lee-waves are not observed. As U_0/Nh increases, vertical motion develops and the presence of lee-waves is clearly detected. The results of the simulation were in reasonably good agreement with those from previous experimental studies. The numerical results were also used for the verification of analytical theories used for the prediction of the streamline patterns around a hill in stable stratified conditions.

Very recently, LES has been employed for assessing the role of turbulence in erosion phenomena. This is a very important field of classical and environmental hydraulics. These studies have implications to determine physically based formulas for the prediction of the maximum scour depth in local scour phenomena occurring around bridge structures or to find new formulas for the evaluation of homogeneous sediment transport in river or marine applications. In Ref. [26] (and in a companion paper therein cited) the turbulent field around a spur-dike placed over a sidewall of a straight channel was investigated through LES at $Re = 18,000$ (based on the bulk velocity and the flow depth in the channel). First the authors discussed the flat bed case, successively they moved to the scoured condition. The first study confirmed the literature findings about the main structure of the flow field. The necklace main vortex structure was detected in all cases examined. These structures were found to be highly fluctuating in time thus inducing a high level of TKE and pressure fluctuations. The main necklace vortex was found to fluctuate between two states, one in which the structure is close to the spur-dike, the second in which a strong near-bed jet convects the vortex far from the spur-dike. The study also showed that the pressure and shear stresses at the bottom walls appear amplified with respect to the straight channel case. The second study, which considered scoured conditions, showed that the scour has a stabilizing effect on the main turbulent structures and this produces a reduction of the stresses at the erodible wall.

The LES study of Teruzzi et al. [59] was aimed at the analysis of the stresses induced by turbulence near an obstacle representative of a bridge abutment. The Reynolds number of the simulation was set equal to 7000 based on the bulk velocity and the channel height and a trapezoidal abutment was considered placed over the side wall of a closed duct. The numerical experiment was designed to reproduce a prior laboratory experiment. In the study, a flat bed condition was considered, representative of the initial stage of the scour process, and the first and second order statistics of the stresses over the bottom wall were analyzed and discussed. The study showed that the abutment causes a strong 3D flow in the obstacle region which modifies the structure of turbulence nearby. As a consequence, the shear and pressure stress distributions appear dramatically modified with respect to the classical channel flow case. The mean shear stress was observed to increase by more than one order of magnitude at the abutment upstream edge, whereas the increase of the horizontal pressure gradient was as large as two orders of magnitudes at the same location. This explains why, under flat bed conditions, erosion first takes place at the upstream abutment edge. The analysis of the second order statistics clearly showed the presence of large fluctuations of the wall shear stress and of the pressure gradients in a wide area around the abutment. In the study, a *reference sediment* was considered having diameter and density so as to be in condition of incipient motion (based on the classical Shields theory of sediment transport). The analysis showed that the local pressure field may produce uplifting force comparable with the sediment weight and additional horizontal destabilizing forces. Based on the results of this study the authors concluded that local scour models should incorporate parameterization of the first- and second-order statistics of the shear and pressure stresses over the erodible wall.

References

1. Armenio, V. and Piomelli, U., 2000. A Lagrangian mixed subgrid-scale model in generalized coordinates. *Flow Turbul. Combust.*, **6** , 51–81.
2. Armenio, V. and Sarkar, S., 2002. An investigation of stably stratified turbulent channel flow using large-eddy simulation. *J. Fluid Mech.*, **9** , 1–42.
3. Armenio, V. and Sarkar, S., 2004. Mixing in a stably stratified medium by horizontal shear near vertical walls. *Theor. Comput. Fluid Dyn.*, **7** , 331–349.
4. Armenio, V., Falcomer, L. and Carnevale, G. F., 2003. LES of a stably stratified flow over longitudinally ridged walls. In *Direct and Large-Eddy Simulation V*, R. Friedrich, B.J. Geurts and O. Metais eds., Dordrecht, the Netherlands, Kluwer Academic Publishers, pp. 299–306.
5. Basak, S. and Sarkar, S., 2006. Dynamics of a stratified shear layer with horizontal shear. *J. Fluid Mech.*, **568**, 19–54.
6. Billant, P. and Chomaz, J.-M., 2000. Experimental evidence for a new instability of a vertical columnar vortex pair in a strongly stratified fluid. *J. Fluid Mech.*, **418**, 253–272.
7. Brucker, K. A. and Sarkar, S., 2010. A comparative study of self-propelled and towed wakes in a stratified fluid. *J. Fluid Mech.*, **8** , 253–272.
8. Brucker, K. A. and Sarkar, S., 2007. Evolution of an initially turbulent stratified shear layer. *Phys. Fluids*, **9** , 101105.
9. Burchard, H., Craig, P. D., Gemmrich, J. R., Haren van, H., Methieu, P.-P., Meier, H. E. M., Smith, W. A. M. N., Prandke, H., Rippeth, T. P., Skyllingstad, E. D., Smyth, W. D., Welsh, D. J. S. and Wijesekera, H. W., 2008. Observational and numerical modeling methods for quantifying coastal ocean turbulence and mixing. *Prog. Oceanogr.*, **6** , 399–442.
10. Calhoun, R. J., Street, R. L. and Koseff, J. R., 2001. Turbulent flow over a wavy surface: Stratified case. *J. Geophys. Res.-Oceans*, **6** , 9295–9310.
11. Caulfield, C. P. and Peltier, W. R., 2000. The anatomy of the mixing transition in homogeneous and stratified free shear layers, *J. Fluid Mech.*, **4** , 147.
12. Chang, Y. S. and Scotti, A., 2003. Entrainment and suspension of sediments into a turbulent flow over ripples. *J.éTurbul.*, 4, 22.
13. Coleman, G. N., Ferziger, J. H. and Spalart, P. R., 1990. A numerical study of the turbulent Ekman layer. *J. Fluid Mech.*, **3** , 313–348.
14. De Angelis, V., Lombardi, P. and Banerjee, S., 1997. Direct numerical simulation of turbulent flow over a wavy wall. *Phys. Fluids*, 9, 2429–2442.
15. Dohan, K. and Sutherland, B. R., 2003. Internal waves generated from a turbulent mixed region. *Phys. Fluids*, **5** , 488–498.
16. Falcomer, L. and Armenio, V., 2002. Large-eddy simulation of secondary flow over longitudinally ridged walls. *J.éTurbul.*, 3, 008.
17. Garg, R. P., Ferziger, J. H., Monismith, S. G. and Koseff, J. R., 2000. Stably stratified turbulent channel flows. I. Stratification regimes and turbulence suppression mechanism. *Phys. Fluids*, **2** , 2569–2594.
18. Gayen, B., Sarkar, S. and Taylor, J. R., 2010. Large eddy simulation of a stratified boundary layer under an oscillatory current. *J. Fluid Mech.*, **6** , 233–266.
19. Gerz, T., Schumann, U. and Elghobashi, S. E., 1989. Direct numerical simulation of stratified homogeneous turbulent shear flows. *J. Fluid Mech.*, **0** , 563–594.
20. Holt, S. E., Koseff, J. R. and Ferziger, J. H., 1992. A numerical study of the evolution and structure of homogeneous stably stratified sheared turbulence. *J. Fluid Mech.*, 237, 499–539.
21. Henn, D. S. and Sykes, R. I., 1999. Large-eddy simulation of flow over wavy surfaces. *J. Fluid Mech.*, **8** , 75–112.

22. Inghilesi, R., Stocca, V., Roman, F. and Armenio, V., 2008. Dispersion of a vertical jet of buoyant particles in a stably stratified wind-driven Ekman layer. *Int. J. Heat Fluid Flow*, **9**, 733–742.

23. Ishihara, T., Kaneda, Y., Yokokawa, M., Itakura K., and Uno, A., 2007. Small-scale statistics in high-resolution direct numerical simulation of turbulence: Reynolds number dependence of one-point velocity gradient statistics. *J. Fluid Mech.*, **9**, 335–366.

24. Jacobitz, F. G., Sarkar, S. and VanAtta, C. W., 1997. Direct numerical simulations of the turbulence evolution in a uniformly sheared and stably stratified flow. *J. Fluid Mech.*, **3**, 231–261.

25. Jacobitz, F. G. and Sarkar, S., 1998. The effect of nonvertical shear on turbulence in a stably stratified medium. *Phys. Fluids*, **0**, 1158–1168.

26. Koken, M. and Constantinescu, G., 2008. An investigation of the flow and the scour mechanisms around isolated spur dikes in a shallow open channel: 1. Conditions corresponding to the final stage of the erosion and deposition process. *Water Resour. Res.*, **4**, W08407.

27. Ding, L., Calhoun, R. J. and Street, R. L., 2003. Numerical simulation of strongly stratified flow over a three-dimensional hill. *Bound. Layer Meteorol.*, **0**, 81–114.

28. Jensen, B. L., Sumer, B. M. and Fredsøe, J., 1989. Turbulent oscillatory boundary layers at high Reynolds numbers. *J. Fluid Mech.*, **0**, 265–297.

29. Johnson, G., Sanford, T. and Baringer, M., 1994. Stress on the Mediterranean outflow plume: Part I. Velocity and water property measurements. *J. Phys. Oceanogr.*, 24, 2072–2083.

30. Marchioli, C., Armenio, V., Salvetti, M. V. and Soldati, A., 2006. Mechanisms for deposition and resuspension of heavy particles in turbulent flow over wavy interfaces. *Phys. Fluids*, **8**, 025102.

31. Moum, J. N., Hebert, D., Paulson, C. A. and Caldwell, D. R., 1992. Turbulence and internal waves at the equator. Part I: Statistics from towed thermistors and a microstructure profiler. *J. Phys. Oceanogr.*, 2, 1330–1345.

32. Nehari, D., Armenio, V. and Ballio, F., 2004. Three-dimensional analysis of the unidirectional oscillatory flow around a circular cylinder at low Keulegan-Carpenter and beta numbers. *J. Fluid Mech.*, **0**, 157–186.

33. Otheguy, P., Chomaz, J.-M. and Billant, P. 2006. Elliptic and zigzag instabilities on co-rotating vertical vortices in a stratified fluid. *J. Fluid Mech.*, **3**, 253–272.

34. Perlin, A., Moum, J. N., Klymak, J. M., Levine, M. D., Boyd, T. and Kosro, P. M., 2005. A modified law-of-the-wall applied to oceanic bottom boundary layers. *J. Geophys. Res.* **0** (C10S10 doi:10.1029/2004JC002310).

35. Pham, H. T. and Sarkar, S., 2010. Internal waves and turbulence in a stable stratified jet. *J. Fluid Mech.*, **648**, 297–324.

36. Pham, H. T., Sarkar, S. and Brucker, K. A., 2009. Dynamics of a stratified shear layer above a region of uniform stratification. *J. Fluid Mech.*, **6**, 191–223.

37. Price, J. F. and Sundermeyer, M. A., 1999. Stratified Ekman layers. *J. Geophys. Res.*, **0**, 20467–20494.

38. Riley, J. J., Metcalfe, R. W. and Weissman, M. A., 1981. Direct numerical simulations of homogeneous turbulence in density-stratified fluids, *AIP Conf. Proc.*, **6**, 79–112.

39. Radhakrishnan, S. and Piomelli, U., 2008. Large-eddy simulation of oscillating boundary layers: Model comparison and validation. *J. Geophys. Res. C: Oceans*, **113** (2), C02022.

40. Riley, J. J. and Lindborg, E., 2008. Stratified turbulence: A possible interpretation of some geophysical turbulence measurements. *J. Atmos. Sci.*, **6**, 2416–2424.

41. Roman, F., Napoli, E., Milici, B. and Armenio, V., 2009. An improved immersed boundary method for curvilinear grids. *Computers Fluids*, **8**, 1510–1527.

42. Roman, F., Armenio, V., Stipcich, G., Inghilesi, R. and Corsini, S., 2010. Large Eddy simulation of mixing in coastal areas. *Int. J. Heat Fluid Flow*, **3** (3), 327–341.

43. Sakamoto, K. and Akitomo, K., 2009. The tidally induced bottom boundary layer in the rotating frame: Development of the turbulent mixed layer under stratification. *J. Fluid Mech.*, **0**, 235–259.

44. Salon, S., Armenio, V. and Crise, A., 2007. A numerical investigation of the Stokes boundary layer in the turbulent regime. *J. Fluid Mech.*, **0**, 253–296.

45. Salon, S., Armenio, V. and Crise, A., 2009. A numerical (LES) investigation of a shallow-water, mid-latitude, tidally-driven boundary layer. *Environ. Fluid Mech.*, **9**, 525–547.

46. Sarkar, S. and Gayen, B., 2010. Numerical modeling of turbulence in a stratified tidal boundary layer on a slope, EOS Trans. AGU, 91(26), *Ocean Sci. Meet. Suppl.*, Abstract PO23D-02, *2010 Ocean Sciences Meeting*, Portland, OR.

47. Scotti, A. and Piomelli, U., 2001. Numerical simulation of pulsating turbulent channel flow. *Phys. Fluids*, **13** (5), 1367–1384.

48. Scotti, A. and Sarkar, S., 2010. A multiscale approach to boundary layer mixing, EOS Trans. AGU, 91(26), *Ocean Sci. Meet. Suppl.*, Abstract PO23D-02 (invited), *2010 Ocean Sciences Meeting*, Portland, OR.

49. Smyth, W. and Moum, J., 2000. Length scales of turbulence in stably stratified mixing layers. *Phys. Fluids*, **2**, 1327–1342.

50. Spalart P. R., Coleman, G. N. and Johnstone, N., 2008. Direct numerical simulation of the Ekman layer: A step in Reynolds number and a cautious support for a log law with a shifted origin. *Phys. Fluids*, **0**, 101507.

51. Strang, E. J. and Fernando, H. J. S., 2001. Entrainment and mixing in stratified shear flows. *J. Fluid Mech.*, **8**, 349–386.

52. Sullivan, P. P. and McWilliams, J. C., 2002. Turbulent flow over water waves in the presence of stratification. *Phys. Fluids*, **4**, 1182–1195.

53. Sullivan, P. P., McWilliams, J. C. and Melville, W. K., 2007. Surface gravity wave effects in the oceanic boundary layer: Large-eddy simulation with vortex force and stochastic breakers. *J. Fluid Mech.*, **9**, 405–452.

54. Sutherland, B. R. and Linden, P. F., 1998. Internal wave excitation from stratified flow over a thin barrier. *J. Fluid Mech.*, 377, 223–252.

55. Taylor, J. R., Sarkar, S. and Armenio, V., 2005. Large eddy simulation of stably stratified open channel flow. *J. Fluid Mech.*, **7**, 1–18.

56. Taylor, J. R. and Sarkar, S., 2007. Internal gravity waves generated by a turbulent bottom Ekman layer. *J. Fluid Mech.*, **9**, 331–354.

57. Taylor, J. R. and Sarkar, S., 2008. Direct and large eddy simulations of a bottom Ekman layer under an external stratification. *Int. J. Heat Fluid Flow*, **9**, 721–732.

58. Taylor, J. R. and Sarkar, S., 2008. Stratification effects in a bottom Ekman layer. *J. Phys. Oceanogr.* **8**, 2535–2555.

59. Teruzzi, A., Ballio, F. and Armenio, V., 2009. Turbulent stresses at the bottom surface near an abutment: Laboratory-scale numerical experiment. *J. Hydraul. Eng.*, **135**, 106–117.

60. Tse, K. L., Mahalov, A., Nicolaenko, B. and Fernando, H. J. S., 2003. Quasi-equilibrium dynamics of shear-stratified turbulence in a model tropospheric jet. *J. Fluid Mech.*, **8**, 73–103.

61. Zang, Y., Street R. L. and Koseff, J. R., 1994. A non-staggered grid, fractional step method for time-dependent incompressible Navier-Stokes equations in curvilinear coordinates. *J. Comput. Phys.*, **4**, 18–33.

62. Zikanov, O., Slinn, D. N. and Dhanak, M. R., 2003. Large-Eddy simulations of the wind-induced turbulent Ekman layer. *J. Fluid Mech.*, **9**, 343–368.

63. Zilitinkevich, S. and Esau, I., 2002. On integral measures of the neutral barotropic planetary boundary layer. *Bound. Layer Meteorol.*, **8**, 371–379.

22

Multiscale Nesting and High Performance Computing

Alex Mahalov
Arizona State University

Mohamed Moustaoui
Arizona State University

22.1 Introduction

The challenge of multi-scale nesting and high performance computing (HPC) simulations for the dynamically evolving limited area atmospheric environments is to achieve robust real-time predictability and ensemble forecasting of high impact events. Significant advances in computation of atmospheric and environmental flows have been achieved over the last few decades. The dramatic increase in computer power has facilitated developments of nonhydrostatic mesoscale numerical weather prediction (NWP) codes that have capabilities to resolve small-scale atmospheric processes. This was achieved by implementation of nesting techniques with multiple domains resolving horizontal scales ranging from few to 100 km, and by the improvement of sub-grid scale parameterizations. Among these models, the advanced research version of the weather research and forecasting model (WRF-ARW) is a next generation mesoscale NWP model (Skamarock and Klemp, 2008). It is the first fully compressible conservative-form nonhydrostatic atmospheric model suitable for both research and weather prediction applications. The WRF-ARW model represents the latest developments following a particular modeling approach that uses time-splitting techniques to eDciently integrate the fully compressible nonhydrostatic equations of motion. The integration scheme uses a time-split method to circumvent the acoustic-mode time-step restriction, where the meteorologically significant modes are integrated by using a third-order Runge–Kutta (RK) scheme (Wicker and Skamarock, 2002). The spatial discretization typically uses a fifth-order difference for advection, and the vertical coordinate is based on terrain following hydrostatic pressure coordinate.

Nesting options are implemented in WRF-ARW. Nevertheless, as in many nonhydrostatic mesoscale atmospheric models, nesting is allowed only in the horizontal direction and all nests use the same grid distribution in the vertical (i.e., MM5: Dudhia, 1993; COAMPS: Hodur, 1997; WRF: Skamarock and Klemp, 2008). For real applications, NWP models still use a limited number of grid points in the vertical that is typically well below 100. Usually, grid stretching is implemented to increase the vertical resolution in the boundary layer and lower levels at the expense of the upper troposphere and lower stratosphere (UTLS). The extended region consisting of the bulk of the troposphere and the lower stratosphere represents a significant challenge for numerical prediction. The collusion between the stratification and shear in the UTLS region leads to many complex multi-scale physics phenomena. The lack of vertical resolution in the UTLS region as well as at lower atmospheric levels may present a severe limitation in resolving small-vertical-scale processes such as motion associated with clear air turbulence and thin adiabatic layers characterized by sharp vertical gradients at the edges. These small-scale processes are particularly sensitive to the vertical resolution, implying that the vertical grid spacing typically used in operational models is likely insufficient to resolve these vertical scales.

22.2 Principles

There are two main techniques that are used in atmospheric and oceanic models to improve resolution over limited areas. In dynamically adaptive methods, the spatial resolution is constantly changing with time by coarsening or refining the grid

Handbook of Environmental Fluid Dynamics, Volume Two, edited by H. J. S. Fernando. © 2013 CRC Press/Taylor & Francis Group, LLC. ISBN: 978-1-4665-5601-0.

spacing depending on local conditions (Dietachmayer and Droegemeier, 1992; Skamarock and Klemp, 1993). The adaptive methods are not well established in the atmospheric modeling systems for several reasons (Behrens, 2005): (1) adaptive techniques can incur massive overhead due to indirect data addressing and additional efforts for grid handling which increase the cost of real-time forecasting; (2) physical parameterizations of sub-grid processes are usually optimized for a specific grid resolution, making it diDcult to use dynamically and temporally refined or coarsened grids. The other method uses nesting to improve spatial resolution over a limited area. Nesting is widely used in atmospheric (Dudhia, 1993; Hodur, 1997; Skamarock and Klemp, 2008) and oceanic (Shchepetkin and McWilliams, 2005) models. Large domain models with coarse resolution are used to predict large-scale dynamics, while limited area models with boundary conditions interpolated from coarse grids are used over small domains with finer resolution. The improvement allowed by nesting techniques is that small-scale processes which are not resolved in a coarse grid model, and therefore need to be represented by using sub-grid-scale parameterizations, may be explicitly resolved in the nested model.

One of the main diDculties faced in atmospheric as well as oceanic nested modeling is the specification of the lateral boundary conditions. Usually, the prognostic fields at the lateral boundaries of the nested grid are specified from the large domain. These fields have coarse resolution, and are interpolated in space and time to the nested grid. The inconsistencies between the limited and the large domain solutions create spurious reflections that may propagate and affect the solution in the interior of the nested domain. Several approaches are used to handle the lateral boundary conditions. The flow relaxation scheme (Davies, 1983) is the most frequently used for atmospheric mesoscale forecasting models over a limited domain. Lateral open boundary conditions are often used in limited area ocean modeling. These conditions include radiation condition, combined radiation, and prescribed condition depending on the inflow and outflow regime at the boundary, and a scale selective approach. A review of these methods is given in Oddo and Pinardi (2008).

The coupled mesoscale WRF with microscale vertical nesting model simulations to be illustrated here are produced by conducting mesoscale simulations with several nests interacting in a two way mode, with a finest WRF nest that uses a horizontal grid spacing of 1 km. The number of vertical sigma pressure levels used for these nests is 150, and they can be adjusted to achieve an improved resolution at any vertical level. The pressure at the top is $p_{top} = 10$ mb, and the vertical resolution is about $\delta z = 150$ m WRF simulations are initialized with high resolution global datasets provided by the European Centre for Medium-Range Weather Forecasts (ECMWF) analysis with the spectral truncation T799L91. These data have a resolution that uses a horizontal grid spacing of 25 km and are distributed on 91 vertical levels.

Horizontal and vertical nesting is implemented by coupling the finest resolution nest of WRF with a sequence of microscale nests. The microscale nests are constructed with increased resolution in both the horizontal and the vertical, with refined vertical gridding to resolve small-scale processes. A new scheme of implicit relaxation, to be described in the next section, is used for both upper and lateral boundary conditions. The microscale fields are relaxed within relaxation zones to the finest mesoscale WRF nest fields. The WRF fields are interpolated in time and space in these relaxation zones. The vertically relaxed upper boundary conditions prevent the formation of spurious wave reflection at the top of the domain. WRF code run first, the output of the finest grid mesoscale nest is used as a coarse grid input for the microscale nest. WRF outputs are interpolated in time, in the horizontal and in the vertical, to provide initial and boundary conditions for the microscale code, and finally, the microscale nest run is made alone.

22.3 Example of the Analysis

The WRF model was used as a base to develop the microscale code. This code solves the 3D fully compressible nonhydrostatic Navier–Stokes equations for atmospheric dynamics. To insure consistency with WRF, these equations are cast in conservative form and are formulated using a terrain-following pressure coordinate denoted by η and defined as: $\eta = (p_{dh} - p_{dht})/\mu_d$ where $\mu_d = p_{dhs} - p_{dht}$, and μ_d represent the mass of the dry air in the column and p_{dh}, p_{dht}, and p_{dhs} represent the hydrostatic pressure of the dry atmosphere and the hydrostatic pressure at the top and the surface of the dry atmosphere. The formulated moist equations are

$$\partial_t U + (\nabla \cdot \mathbf{V}u)_\eta + \left(\frac{\alpha}{\alpha_d}\right)(\alpha_d \partial_x p + \partial_\eta p \partial_x \phi) = F_U \quad (22.1)$$

$$\partial_t V + (\nabla \cdot \mathbf{V}v)_\eta + \left(\frac{\alpha}{\alpha_d}\right)(\alpha_d \partial_y p + \partial_\eta p \partial_y \phi) = F_V \quad (22.2)$$

$$\partial_t W + (\nabla \cdot \mathbf{V}w)_\eta - g\left[\left(\frac{\alpha}{\alpha_d}\right)\partial_\eta p - \mu_d\right] = F_W \quad (22.3)$$

$$\partial_t \Theta + (\nabla \cdot \mathbf{V}\theta)_\eta = F_\theta \quad (22.4)$$

$$\partial_t \mu_d + (\nabla \cdot \mathbf{V})_\eta = 0 \quad (22.5)$$

$$\partial_t \phi + \mu_d^{-1}[(\mathbf{V} \cdot \nabla \phi)_\eta - gW] = 0 \quad (22.6)$$

$$\partial_t Q_m + (\nabla \cdot \mathbf{V}q_m)_\eta = F_{Q_m} \quad (22.7)$$

where

$\mathbf{v}(u, v, w)$ is the physical velocity vector
θ is the potential temperature
p is the pressure
g is the acceleration of gravity
$\phi = gz$ is the geopotential
$(U, V, W, \Omega, \Theta) = \mu_d(u, v, w, \omega, \theta)$, where $\omega = d\eta/dt$ is the vertical velocity in the computational space
$\mathbf{V} = (U, V, \Omega)$ is the coupled velocity vector

The right-hand-side terms F_U, F_V, F_W, and F_θ represent forcing terms arising from model physics, turbulent mixing, spherical projections, and the earth's rotation.

The aforementioned governing equations are solved together with the diagnostic equation for coupled dry inverse density,

$$\partial_\eta \phi = -\alpha_d \quad (22.8)$$

and the diagnostic relation for the full pressure (vapor plus dry air) $p = p_0(R_d\Theta_m/p_0\alpha_d)^\gamma$. In these equations, α_d is the coupled inverse density of the dry air (μ_d/ρ_d) and α is the coupled inverse density taking into account the full parcel density $\alpha = \alpha_d(1 + q_v + q_c + q_r + q_i + {}^k)^{-1}$ where q_* are the mixing ratios (mass per mass of dry air) for water vapor, cloud, rain, ice, etc. Additionally, $\Theta_m = \Theta(1 + (R_v/R_d)q_v) \approx \Theta(1 + 1.61q_v)$, and $Q_m = \mu_d q_m$; $q_m = q_v, q_c, q_i, k$.

The main difference between the Equations 22.1 through 22.8 and those solved in WRF is the use of a coupled inverse density α_d and α rather than the inverse density as in WRF. This results in a different formulation for the pressure gradient terms in the momentum equations, and in the acoustic time steps. The pressure gradient terms in the momentum equations have the following forms:

$$P_x = \left(\frac{\alpha}{\alpha_d}\right)\left(\alpha'_d \bar{p}_x + \alpha_d p'_x + \bar{\mu}\phi'_x + p'_\eta \phi_x\right)$$

$$P_y = \left(\frac{\alpha}{\alpha_d}\right)\left(\alpha'_d \bar{p}_y + \alpha_d p'_y + \bar{\mu}\phi'_y + p'_\eta \phi_y\right)$$

$$P_z = -g\left(\left(\frac{\alpha}{\alpha_d}\right)p'_\eta - \mu' + (\alpha/\alpha_d - 1)\bar{\mu}\right)$$

and Equation 22.8 takes the following form:

$$\partial_\eta \phi' = -\alpha'_d. \quad (22.9)$$

The prime variables indicate the deviation with respect to the reference state (i.e., $p' = p - \bar{p}$).

The details of the solver used are described in Mahalov and Moustaoui (2009) and Skamarock and Klemp (2008). Here, a brief summary is given and the differences between the microscale and the WRF models are discussed. The time-split scheme developed in Wicker and Skamarock (2002) is used as the basis for the integration scheme. In this method, low-frequency modes that are meteorologically significant are integrated using a third-order RK time integration scheme; while high-frequency acoustic modes are handled by using an implicit scheme in the vertical with smaller time steps to maintain stability. Note that the RK scheme is implemented in a way that is different from the standard RK schemes. This scheme is easily implemented and it does not require the storage of all evaluated steps until the final step. It has many advantages compared with the leap-frog scheme which is used in other time-split nonhydrostatic NWP models. It is more accurate, stable for both diffusion and advection equations and allows to use centered and upwind biased spatial discretizations. It does not require time filtering that degrades the accuracy of the solution as in the leap-frog scheme, where time filtering is necessary to damp the computational modes. The spatial discretization uses a C grid staggering: normal velocities are staggered one-half grid length from the thermodynamic variables. Advection of vector and scalar fields is in the form of flux divergence; the advection uses a fifth- or a third-order accurate up-winded spatial discretization. A technique that consists of integrating perturbation equations is used as in WRF. In these equations, each variable ψ is decomposed as $\psi = \psi^t + \psi''$, where ψ^t is the most updated value in the RK step for ψ, and ψ'' is the perturbation with respect to this value.

In the formulation presented here, the pressure gradient terms in the horizontal momentum perturbation equations have the following form:

$$P''_x = \left(\frac{\alpha}{\alpha_d}\right)\left(\alpha''_d \bar{p}_x + \alpha_d p''_x + \bar{\mu}\phi''_x + p''_\eta \phi_x\right)$$

$$P''_y = \left(\frac{\alpha}{\alpha_d}\right)\left(\alpha''_d \bar{p}_y + \alpha_d p''_y + \bar{\mu}\phi''_y + p''_\eta \phi_y\right)$$

Equation 22.9 becomes in the acoustic integration

$$\partial_\eta \phi'' = -\alpha''_d \quad (22.10)$$

and the linearized form of the equation of state is given by

$$p'' = \gamma \frac{\Theta''}{\Theta} - \frac{\alpha''}{\alpha} \quad (22.11)$$

By introducing a coupled inverse density, the number of arithmetic operations required for the pressure gradient evaluations in the horizontal momentum equations is reduced by 20%, and the linearized equation of state (22.11) does not include the mass of the dry air μ''.

For the initial state, WRF fields are interpolated horizontally; then the mesoscale fields are also interpolated in the vertical to adjusted levels that can be chosen independently from the WRF vertical levels. The interpolation approach follows a class of high-order monotone shape preserving schemes (Smolarkiewicz, 1992). This interpolation preserves monotonicity and avoids generation of numerical noise or local extrema. Once the fields are interpolated the coupled inverse density is calculated by using Equation 22.8. Computation of this variable instead of interpolating it from WRF ensures that (22.8) is satisfied for the initial conditions. This avoids generation of large transients at the initial time if (22.8) is not satisfied.

The relaxation toward coarse grid fields is implemented using operator splitting method applied in the acoustic steps where a prognostic variable is updated first without any relaxation. The relaxation is applied as an implicit correction step after each acoustic step. The corrected variable is then used to update the other prognostic variables.

For the lateral boundary conditions, the normal velocity component located at the boundary is treated differently from the tangential velocities and the thermodynamic variable which are located half grid point inside the domain and adjacent to the boundary. The relaxation boundary scheme consists of progressively constraining the main prognostic variables of the limited area model to match the corresponding values from the coarse grid model in a buffer zone next to the boundary called "relaxation" zone. The flow relaxation scheme used in this study is a combination of Newtonian and diffusive relaxation that has the following form:

$$\partial_t \psi = -N(x)(\psi - \psi^c) + D(x)\partial_{xx}(\psi - \psi^c)$$

where

  ψ is a prognostic variable of the limited area model that needs to be relaxed to the corresponding variable from the coarse grid model ψ^c
  x denotes the direction normal to the boundary
  $N(x)$ and $D(x)$ are the Newtonian and the diffusive relaxation factors

The choice of the profiles and the values of the coeDcientsÂ $N(x)$ and $D(x)$ in the relaxation zones control reflections at the boundary.

In the microscale model, the relaxation is implemented after each acoustic step as an implicit correction. Let $\tilde{\psi}''^{,n+1}$ denote the perturbation of the updated variable after each acoustic time step. The relaxation is then applied as a correction in a subsequent step using the following implicit flow relaxation equation:

$$\frac{\psi_i''^{,n+1} - \tilde{\psi}_i''^{,n+1}}{\delta\tau} = -N_i\left(\psi_i''^{,n+1} + \psi_i^t - \psi_i^{c,n+1}\right)$$
$$+ \frac{D_i}{\delta x^2}\left\{\left(\psi_{i+1}''^{,n+1} + \psi_{i+1}^t - \psi_{i+1}^{c,n+1}\right)\right.$$
$$- 2\left(\psi_i''^{,n+1} + \psi_i^t - \psi_i^{c,n+1}\right)$$
$$\left. + \left(\psi_{i-1}''^{,n+1} + \psi_{i-1}^t - \psi_{i-1}^{c,n+1}\right)\right\}$$

where

  $\delta\tau$ is the acoustic time step
  δx is the grid spacing
  $\psi_i^{c,n+1}$ is the coarse grid value interpolated in space and to the time step $(n + 1)$
  $\psi_i''^{,n+1} = \psi_i^{n+1} - \psi_i^t$, where ψ_i^{n+1} is the total fine grid value, ψ_i^t is the most updated value in the RK step, and $\psi_i''^{,n+1}$ is the perturbation with respect to ψ_i^t

For the prognostic variables located at half grid points adjacent to the lateral boundary, this implicit equation is solved for $\psi_i''^{,n+1}$ along the relaxation zone, subject to the boundary conditions:

$$\psi_{s+1}''^{,n+1} = \tilde{\psi}_{s+1}''^{,n+1} \quad \text{and} \quad \psi_1''^{,n+1} = \psi_1^{c,n+1} - \psi_1^t$$

where s is the index of the last relaxed point in the interior of the domain.

For the normal velocities at the boundary, the same system is solved except that consistency between the coarse and the limited area mass fluxes in the continuity Equation 22.5 is imposed, that is

$$\nabla \cdot \mathbf{V} = \nabla \cdot \mathbf{V}^c$$

where $\mathbf{V}$ and $\mathbf{V}^c$ are the velocity vectors from the limited area and the coarse grid models, coupled with the mass of the dry air per unit area within the column. Since the tangential velocities adjacent to the boundary are imposed by the coarse grid model, the aforementioned relation reduces to

$$\frac{\partial U}{\partial x} = \frac{\partial U^c}{\partial x},$$

where U and U^c are the normal components of $\mathbf{V}$ and $\mathbf{V}^c$, respectively. This relation is imposed implicitly at the lateral boundaries, and the implicit equation is solved for $U''^{,n+1}$ along the relaxation zone as mentioned earlier, but with the following implicit boundary conditions:

$$U_{s+1}''^{,n+1} = \tilde{U}_{s+1}''^{,n+1} \text{ and } U_2''^{,n+1} - U_1''^{,n+1} = \left(U_1^t - U_1^{c,n+1}\right) - \left(U_2^t - U_2^{c,n+1}\right).$$

In the vertical, a constant pressure is imposed at the top of the domain. All the prognostic variables are computed by the model including at the top. After updating these variables in the acoustic step, the same relaxation is applied in a buffer zone near the top of the domain, except that the implicit Newtonian relaxation is used alone. The geopotential, however, is not relaxed in the vertical. The Newtonian and diffusive relaxation times are fixed by the choice of the coeDcients N_i and D_i. Optimal profiles for the Newtonian relaxation coeDcient N are proposed in Davies (1983), Lehmann (1993), and Marbaix et al. (2003). They are constructed in such a way that, under idealized conditions, the unwanted partial reflection of outgoing waves leaving the domain is minimized. The numerical code is written in Fortan90 which facilitate its portability to different platforms. The code is fully parallelized using the message passing interface (MPI), and the memory used by the code is optimized.

22.4 Example of Applications

For the purpose of illustration here, the finest microscale limited area is defined as a 3D domain centered on a geographical region 100 km × 100 km horizontally and extended to 30 km altitude. In real case forecasting, initial and boundary conditions for WRF nested simulations are obtained from the high resolution global model of the ECMWF T799L91 analysis. Global atmospheric data are uploaded every 6 h and then zoom into a specific limited area with a sequence of horizontally and vertically nested domains until the required resolution is obtained. In this chapter limited area microscale simulations are presented. The microscale model is nested to the mesoscale WRF model.

The goal is to improve techniques for identifying, forecasting, and detecting areas of clear air turbulence (CAT), and modeling the related phenomenon optical turbulence (OT) under extreme environmental conditions (Mahalov and Moustaoui, 2010). High resolution nested WRF mesoscale/microscale code simulations for real atmospheric conditions are carried out to demonstrate the ability of the microscale nests to resolve small-scale processes. Effective resolution and prediction of multi-scale physical phenomena and laminated structures in high impact atmospheric environments present a significant challenge for real-time operational forecasting. The terrain-induced rotor experiment (T-REX) campaign (Grubišin et al., 2008), which took place in Owens Valley, California, represents an important benchmark for real case simulations of strongly nonlinear multi-scale dynamics near the tropopause and in the lower stratosphere associated with high activity of topographic gravity waves. Real case simulations with resolutions that are higher than those used in standard mesoscale codes are required to adequately resolve these multi-scale processes. Vertical nesting is also required to resolve nonlinear processes involving small vertical scales. Examples of benchmark and validation of the model simulations during the TREX campaign of measurements are given in the following:

The nested mesoscale WRF with microscale vertical nests simulations are conducted for the period from March 31, 2006 00 UTC to April 2, 2006 00 UTC of the T-REX campaign of measurements. WRF domains are centered over (36.49 N, 118.8 W). Figure 22.1 shows the global distribution of the wind field near the tropopause (320 K) at 12 UTC March 31, 2006 from the ECMWF analyses. This field shows the jet stream with strong anomalies near the Pacific Coast in Southern California. In Figure 22.1 four telescoping domains that are used for the limited area simulations

are superimposed. Each limited area domain uses initial and boundary conditions from the adjacent parent domain. The largest domain uses the global data for initial and boundary conditions. The three mesoscale domains (Figure 22.2a through c) are used with horizontal resolutions of 15, 3 and 1 km, and 150 vertical sigma pressure levels from the ground up to 10 mb (~30 km altitude). These levels are adjusted for better resolution of the tropopause and the lower stratosphere. The grid spacing is about $\delta z \sim$ 150 m. Mesoscale WRF simulations are initialized with high resolution ECMWF T799L91 analysis data, 25 km horizontal resolution and 91 vertical levels.

The microscale domain (Figure 22.2d) uses 300×300 in the horizontal and much higher vertical resolution with 450 staggered vertical levels, and a vertical grid spacing of $\delta z \sim 50$ m. It is nested both horizontally and vertically, with initial and boundary conditions from the finest WRF mesoscale domain. Figure 22.2d shows topography for the microscale nest (100 km $\times$ 100 km horizontal dimensions). The black curve superimposed in Figure 22.2 presents a trajectory of a balloon launched on April 1, 2006 at 7:50 UTC from the location (36.49 N, 118.84 W) in the upstream side of the highest elevation. The wind vector field at $z = 12$ km from the microscale simulation on April 1, 2006 at 8:00 UTC is also superimposed in Figure 22.2d.

The wind directions are dominated by south-westerlies in agreement with observations. Note that the relaxation of the wind field is very smooth at the boundaries. Regions with strong turbulent flow are well resolved in the microscale nest and are found above the valley, toward the south of the balloon trajectory. In these regions, the horizontal wind field shows strong drag, and the direction of the wind is complex.

Figure 22.3 is a vertical cross section of Figure 22.2 showing the mountainous terrain at the bottom of the figure. Strongly laminated multi-scale structures with finer details compared to the mesoscale simulations can be seen in the upper troposphere/lower stratosphere region in Figure 22.3d (altitudes 8–20 km). The earlier mentioned multi-scale simulations demonstrate that higher resolution and finer topography bring up new details and resolve inversion layers, which are characterized by steep vertical gradients and are located at the edges of relatively well-mixed regions produced by shear instabilities and wave breaking. High impact inversion layers in urban atmosphere may cause trapping of pollution resulting in a serious degradation of air quality. Nested multi-scale simulations with adequate vertical resolution are required to resolve such inversion layers. In the context of regional flows, while mesoscale models resolve dynamics induced by topography, high resolution microscale models are needed to resolve essential surface heterogeneity at horizontal scales below 1 km in limited area environments.

Figure 22.4a through d shows potential temperature, and the three wind components obtained from the microscale domain along a portion of the NSF/NCAR G-V aircraft flight path on March 25. The selected portion of the flight track was 3 h long and was flown between 12 and 13 km in the lower stratosphere along the racetrack pattern shown of the aircraft. High fluctuations, which are signatures of mountain waves, are found in all fields.

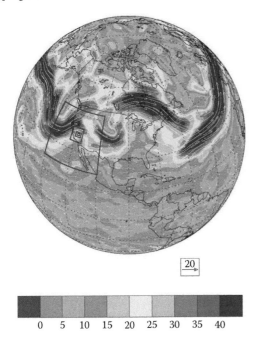

$$\boxed{20}\rightarrow$$

| | | | | | | | |
|0|5|10|15|20|25|30|35|40|

FIGURE 22.1 Wind speed with wind vector field on 320 K isentropic surface at 12 UTC March 31, 2006 from the global data. Four telescoping domains used for limited area simulations are also superimposed (rectangles).

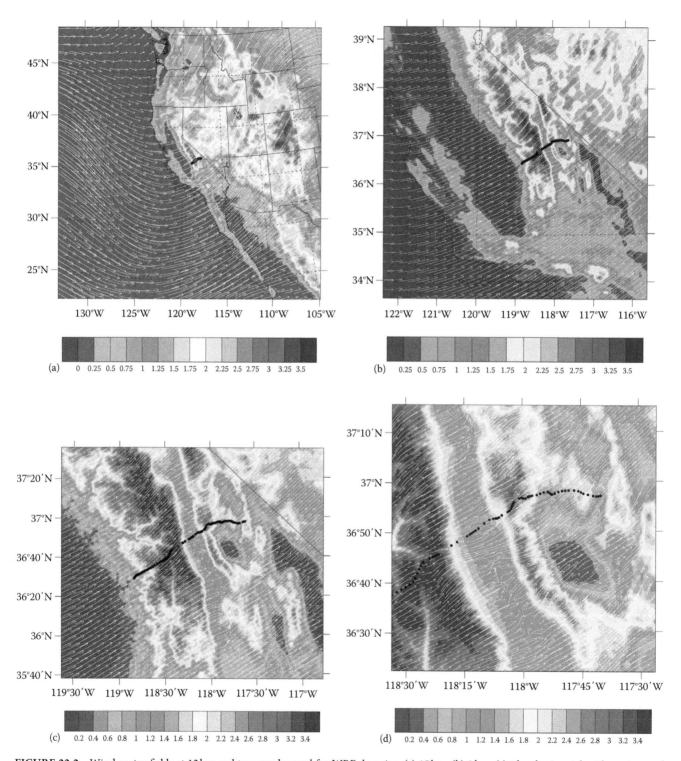

FIGURE 22.2 Wind vector fields at 12 km and topography used for WRF domains: (a) 15 km, (b) 3 km, (c) 1 km horizontal grid spacing, and (d) for the microscale nest. The black curve shows the balloon trajectory. The time is April 1, 2006 at 8:00 UTC.

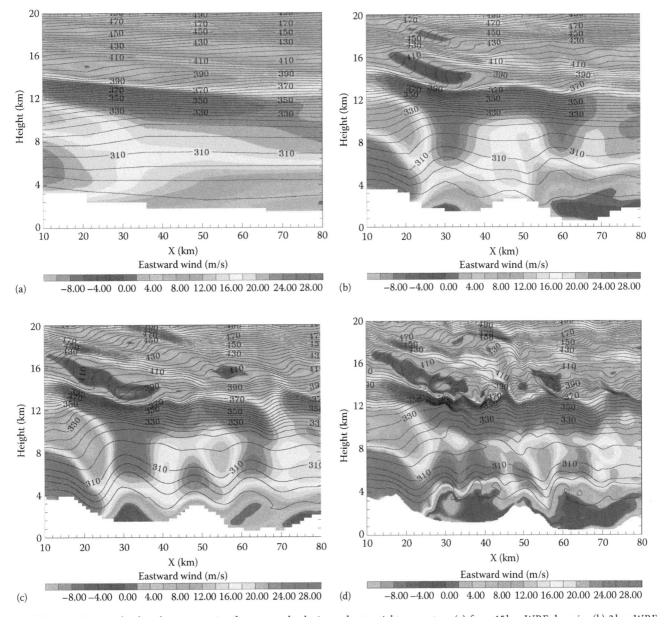

FIGURE 22.3 Longitude-altitude cross-section for eastward velocity and potential temperature (a) from 15 km WRF domain, (b) 3 km WRF domain, (c) 1 km WRF domain, and (d) from the microscale nest. The time is April 1, 2006 at 8:00 UTC.

The vertical velocity shows patches with very high values that reach 12 m s^{-1}. These fields agree well with the observations shown in Figure 22.4e through h, where in addition to the potential temperature and velocity components, the CO and O$_3$ measured by the aircraft are also shown. The time in this figure is relative to 15:00 UTC.

22.5 Summary and Challenges

Numerical weather prediction (NWP) systems employ a range of parameterizations to model the effects of unresolved subgrid processes on the large-scale dynamics. Resolving explicitly these small-scale processes is a significant challenge and requires computations using fine mesh in both the vertical and the horizontal to encompass all pertinent multi-scale phenomena in the UTLS region.

By using advanced and massively parallel computer systems integrating a new generation of many-core processors with revolutionary HPC capabilities and multi-scale atmospheric physics codes, which use improved sub-grid scale parameterizations and micro-scale vertical nesting with refined vertical gridding, recent advances in HPC hardware and software developments will give a real-time multi-dimensional view of dynamically evolving limited area atmospheric environments.

The rapid introduction of many-core microprocessors is the harbinger of a seismic shift in HPC and computing in general. This technology shift requires changes in application codes as well to maintain maximum performance. While the existing

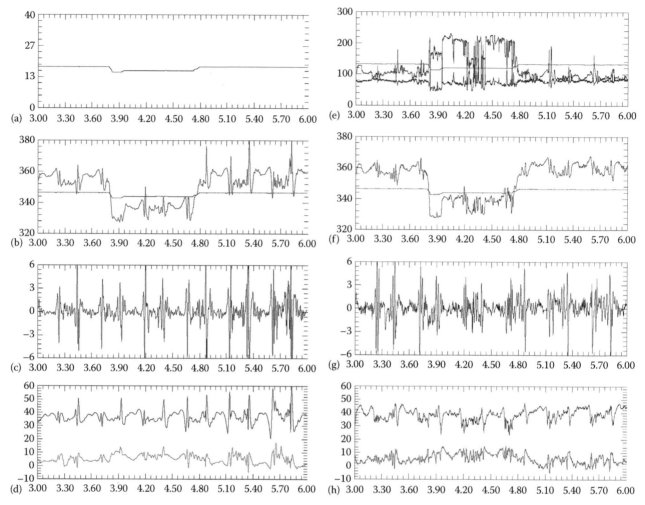

FIGURE 22.4 (a) Profiles of the altitude of the aircraft, (b) potential temperature, (c) vertical velocity, and (d) the wind components simulated by the microscale code on March 25, 2006. The time is relative to 15 UTC. (e) Profiles of CO and O_3 measured by the aircraft; f, g and h are the same as b, c and d but from NCAR GV HIAPER aircraft measurements.

atmospheric physics codes support the MPI standard, and are therefore well suited for application to HPC clusters, the introduction of many-core chips requires changes to the codes and additional performance tuning. The multi-scale atmospheric physics codes need to be adapted to a hybrid multi-threaded, multi-task programming models, which exploits through threading the shared memory capabilities of multi-core CPUs, and through multi-tasking the clustering of many of these multi-core CPUs.

For the study of multi-scale atmospheric dynamics over a limited geographic area, nested micro-scale models need to be created and used to generate high-resolution data in real time to produce accurate predictions and forecasts. Real-time high resolution simulations over limited area atmospheric environments are needed to address complex research and environmental forecasting problems, from identifying the optical effects of the jet stream over observatories to understanding the effects of environmental transport on global and regional scales. This will be

useful for the aviation industry, optical communications, remote sensing applications, environmental, and air quality forecasting.

In the real case, the coupled WRF mesoscale/microscale simulations presented in this study are conducted for 48 h of physical time. The micro-scale code runs for 3 h covering TREX campaign observational period and aircraft measurements. The combined computational time of both simulations in a platform using 512 processors is less than 48 h. Thus, these simulations could be conducted in real-time forecasting. The computational time could be even faster if the number of vertical levels used in WRF is reduced to a typical operational NWP setups (~50), with a refined vertical grid in the micro-scale code (~150). A large speedup can also be achieved if WRF and the micro-scale codes run in a concurrent mode. Coupling the mesoscale and the microscale codes in a two-way concurrent mode is a major challenge. This will eliminate the large slow down caused by the frequent input/output required when these two codes run separately.

Acknowledgments

During the period of preparation of this chapter, the authors were supported by NSF, AFOSR, and the Fulton High Performance Computing Center at ASU.

References

Behrens, J. 2005. Adaptive atmospheric modeling: Scientific computing at its best. *Comput. Sci. Eng.* **7**: 76–83.

Davies, H. C. 1983. Limitations of some common lateral boundary schemes used in regional NWP models. *Month. Weather Rev.* **1**: 1002–1012.

Dietachmayer, G. S. and K. K. Droegemeier. 1992. Application of continuous dynamic grid adaptation techniques to meteorological modeling, Part I: Basic formulation and accuracy. *Month. Weather Rev.* **●**: 1675–1706.

Dudhia, J. 1993. A nonhydrostatic version of the Penn State NCAR mesoscale model: Validation tests and simulation of an Atlantic cyclone and cold front. *Month. Weather Rev.* **121**: 1493–1513.

Grubišin, V., Doyle, J. D. Kuettner, J. et al. 2008. The Terrain-induced Rotor Experiment: An field campaign overview with some highlights of special observations. *Bull. Am. Meteor. Soc.*, **89**: 1513–1533.

Hodur, R. M. 1997. The Naval Research Laboratory Coupled Ocean/Atmosphere Mesoscale Prediction System (COAMPS), *Month. Weather Rev.* **1**: 1414–1430.

Lehmann, R. 1993. On the choice of relaxation coeDcients for Davies lateral boundary scheme for regional weather prediction models. *Meteorol. Atmos. Phys.* **1**: 1–14.

Mahalov, A. and M. Moustaoui. 2009. Vertically nested nonhydrostatic model for multi-scale resolution of flows in the upper troposphere and lower stratosphere. *J. Comp. Phys.* **8**: 1294–1311.

Mahalov, A. and M. Moustaoui. 2010. Characterization of atmospheric optical turbulence for laser propagation, *Laser Photonics Rev.* **4**: 144–159.

Marbaix, P., Gallee, H., Brasseur, O., and J. P. Van Ypersele. 2003. Lateral boundary conditions in regional climate models: A detailed study of the relaxation procedure. *Month. Weather Rev.* **1**: 461–479.

Oddo, P. and N. Pinardi. 2008. Lateral open boundary conditions for nested limited area models: A scale selective approach. *Ocean Model.,* **●**: 134–156.

Shchepetkin, A. and J. C. McWilliams. 2005. Regional ocean model system: A split-explicit ocean model with a free surface and topography-following vertical coordinate. *Ocean Model.* **9**: 347–404.

Skamarock W. C. and J. B. Klemp. 1993. Adaptive grid refinement for two-dimensional and three-dimensional nonhydrostatic atmospheric flow. *Month. Weather Rev.* **121**: 788–804.

Skamarock, W. C. and J. B. Klemp. 2008. A time-split nonhydrostatic atmospheric model for weather research and forecasting applications. *J. Comp. Phys.* **2**: 3465–3485.

Smolarkiewicz, P. K. and G. A. Grell. 1992. A class of monotone interpolation schemes. *J. Comp. Phys.* **●**: 431–440.

Wicker, L. J. and W. C. Skamarock. 2002. Time splitting methods for elastic models using forward time schemes. *Month. Weather Rev.* **●**: 2088–2097.

23

Multiscale Representations: Fractals, Self-Similar Random Processes and Wavelets

Marie Farge
École Normale Supérieure

Kai Schneider
Aix-Marseille University

Olivier Pannekoucke
National Centre
for Meteorological Research

Romain Nguyen van yen
École Normale Supérieure

23.1 Introduction

Many growth processes that shape the human environment generate structures over a wide range of scales, e.g., trees, rivers, lightning bolts. Likewise, most geophysical flows happen on a wide range of scales, e.g., winds in the atmosphere, currents in the oceans, seismic waves in the mantle. In general, both kinds of phenomena are governed by nonlinear dynamical laws that give rise to chaotic behavior, and it is thus very difficult to follow their evolution, let alone predict it. Only in the last few decades could the systems of nonlinear equations modeling environmental fluid flows be solved, thanks to the development of numerical methods and the advent of supercomputers. Although the present computer performances still remain insufficient to simulate from first principles, i.e., by direct numerical simulation (DNS), many environmental fluid flows, especially those which are turbulent, appropriate multiscale representations may contribute to the success of that ongoing enterprise. The goal of this review is to present three of them: fractals, self-similar random processes, and wavelets.

A fractal is a set of points that presents structures that look essentially the same at all scales. When only its large-scale features are considered, a certain shape is observed, which does not become simpler when zooming toward small scales but on the contrary remains quite similar to which it is at large scale. This goes on from one scale to the other, up to the point that one cannot tell what is the scale of observation. When measuring the length, surface, or volume of a fractal object, it is found that, in contrast to classical geometrical objects, e.g., circle or polygons, a definite answer cannot be obtained since the measured value increases when the scale of observation decreases. Let us now consider a simple example of a drop falling into water, an experiment that can be easily done with a glass of water, a drop of oil, and a drop of ink. While falling, the shape of the oil drop becomes more and more spherical, therefore more regular than it was at the instant of impact. Since oil is hydrophobic, the drop tends to minimize the interface between oil and water for a given volume. In contrast, the shape of the ink drop becomes more and more convoluted, since the drop is unstable and splits into smaller drops. In absence of surface tension and of dissipation, the interface between ink and water would then become fractal in the limit of long times. Indeed, since ink is hydrophilic the drop tries to maximize the interface for a given volume. Both systems satisfy the same equations and only one parameter, the surface tension, differs, which implies either minimization or maximization of the interface. The solution of the former

Handbook of Environmental Fluid Dynamics, Volume Two, edited by Harindra Joseph Shermal Fernando. © 2013 CRC Press/Taylor & Francis Group, LLC. ISBN: 978-1-4665-5601-0.

exists and is smooth, while the maximum does not exist. John Hubbard, who suggested this example, concludes:

> The world is full of systems which are trying to reach an optimum which does not exist, and consequently they evolve towards structures which are complicated at all scales. This happens for trees, which try to maximize their exposure to light, for lungs and capillaries, which try to maximize the interface between tissue and blood. The great work of Mandelbrot has been to tell, very loudly and in a very convincing way, that the world is full of complicated phenomena, of complicated objects having structure at all scales. [28]

Fractals can be traced back to the discovery of continuous nondifferentiable functions, e.g., the Weierstrass function, and nonrectifiable curves, e.g., the Sierpinski gasket. Measure theory, as developed in particular by Felix Hausdorff at the end of the nineteenth century, and integration theory, as redesigned by Henri Lebesgue and others at the beginning of the twentieth century, together with the study of recursive sequences in the complex plane, by Pierre Fatou and Gaston Julia, were all precursors of fractals, although a different terminology was used those years. Only when computer graphics became widely available in the 1960s was one able to visualize fractals and wonder about their apparent complexity. Although the mathematical tools were already there, it is Benoît Mandelbrot, while working for the IBM Research Center in Yorktown (USA), who popularized fractals and named them in the 1970s. Actually, before he started talking about fractals, Mandelbrot was a specialist of the theory of Brownian motion that he had learned about during the time he was at Ecole Polytechnique in Paris, where he studied under the French probabilist Paul Lévy [38]. It was Mandelbrot who gave in 1968 the name "fractional Brownian motion" [42] to the self-similar stochastic processes proposed by Kolmogorov in 1940 [33], which are generalizations to long-range correlated increments of the classical Brownian motion.

The mathematical foundation of wavelets is more recent, since the continuous wavelet transform has been introduced only in the 1980s by Jean Morlet and Alex Grossmann. Jean Morlet was researching on oil exploration for the French company Elf, while Alex Grossmann was a specialist of coherent states in quantum mechanics and a member of the CPT (Centre de Physique Théorique) in Marseille (France) (see also [21] for more on the early history of wavelets). From their work, Ingrid Daubechies, Pierre-Gilles Lemarié, and Yves Meyer constructed several orthogonal wavelet bases. Soon after, Stéphane Mallat and Yves Meyer introduced the concept of multiresolution analysis (MRA), which led to the Fast Wavelet Transform (FWT). Without the FWT, the wavelet transform would have remained confined to text books and theoretical papers. The same was true for the Fourier transform that would not have entered our everyday life without the combination of computers and FFT (Fast Fourier Transform), invented by Gauss around 1805 and rediscovered by Cooley and Tukey in 1965.

The aim of this chapter is to give researchers working in environmental fluid dynamics some mathematical tools to study the multiscale behavior of many natural flows. For the sake of clarity, we propose to divide what is presently named "fractals" into two classes: deterministic fractals and self-similar random processes. We will keep the terminology "fractals" to designate the former, which are constructed following some deterministic procedure iterated scale by scale. For the latter we propose to return to the "pre-fractal" terminology of "self-similar random processes," which are ensembles of random realizations whose statistics exhibit some scaling behavior. In this paper we will present several multiscale methods developed from three different view points: fractals, self-similar random processes, and wavelets. All of them are mathematical tools that do not have any explanatory power per se. They require the scientist who uses them to have enough physical insight to interpret the results and decide if this tool is actually appropriate to his problem. If a new technique is not mastered well enough, it induces an a priori interpretation, built in within it without the user being aware of that. To avoid such a drawback, we will here limit ourselves to give definitions, expose methods, and illustrate their use on academic examples rather than from applications. We will justify this choice in the conclusion by showing how such misinterpretation already happened while applying either fractals or wavelets to study turbulence.

23.2 Principles

23.2.1 Fractals

23.2.1.1 Definition and History

To define what "fractal" means is quite a diﬃcult endeavor since one finds in the literature different definitions. Here we use the following definition: a fractal is a geometrical object that is so convoluted, irregular, or fragmented that its length, surface, volume, as well as higher-dimensional generalizations of these measures, all equal either zero or infinity. Its boundary is a set of points, either connected or disconnected, which looks the same at different scales and tends to be space-filling. For instance, a fractal curve is not rectifiable, i.e., its length is infinite. If the points remain connected the boundary can be parametrized by a continuous but nondifferentiable function. Otherwise, the fractal is a dust of disconnected points that can only be parametrized by a measure. A fractal shape may look complicated although it is not, since it may have been generated by a simple iterative procedure. The diﬃculty is, given an observed complicated shape, can we infer the simple rule that has generated it? In most cases the answer is no and this is why methods developed under the trademark "fractals" are rather descriptive than predictive.

Benoît Mandelbrot introduced the word "fractal" in 1975, in a book first published in French [44] and 2 years later in English [45], but he managed to keep the definition vague and varied them throughout his books. The first definition he gave is: *N'fractal object' and 'fractal,' terms that I have formed for* this book from the Latin adjective 'fractus' which means irregular or broken [44]. Subsequently, Mandelbrot succeeded in gathering under the same name different mathematical objects that were proposed before but were considered by most mathematicians as surprising, anecdotic, or weird. Poincaré recalled that

we have seen a rabble of functions arise whose only job, it seems, is to look as little as possible like decent and useful functions. No more continuity, or perhaps continuity but no derivatives [k] Yesterday, if a new function was invented it was to serve some practical end, today there are specially invented only to show up the arguments of our fathers, and they will never have any other use" [7].

An example of such entertaining mathematical object was the fractal curve known as the "snow flake," see Figure 23.1a, published in 1904 by Helge von Koch in the Swedish journal "Arkiv for Matematik" [32].

In 1918, the French Academy set for its "Grand Prix des Sciences Mathématiques" the iteration of fractional functions and Gaston Julia won that prize. Independently, Gaston Julia and Pierre Fatou were studying rational maps in the complex plane by iterating polynomials, e.g. quadratic maps. In 1977, Adrien Douady and John Hubbard used Newton's method to solve the quadratic map $f_c(z) = z^2 + c$, with $z \in \mathbb{C}$, $c \in \mathbb{C}$ a parameter. This quadratic map is the simplest nonlinear dynamical system one can think of in the complex plane and they studied the set K_c of z for which the n-th iterate of f, $f_c^n(z)$, converges. The frontier of K_c is now called the Julia set of f_c. Benoît Mandelbrot, who worked for IBM and had thus access to large computers, graphical facilities, and good programmers, made visualizations to help understanding that problem. In a paper published in 1982, Douady and Hubbard [15] showed that the set of all cs for which $0 \in K_c$ is connex, and they baptized it the Mandelbrot set M in order to pay tribute to Mandelbrot for his visualizations. They commented as follows: *Benoît Mandelbrot has obtained on a computer a very beautiful picture of M, exhibiting small islands which are detached from the principal component. These islands are in fact connected by*

r laments which escape the computer [15]. Without any doubt computer visualization has played an essential role in the dissemination of fractals outside mathematics.

The main contribution of Mandelbrot has been to widely popularize fractals, thanks to computer visualization. His argument is that fractals are more appropriate to describe natural phenomena than the classical objects geometers have been using for centuries, namely rectifiable curves (e.g., circle and other ovals) or piecewise regular curves (e.g., triangle and other polygons). He illustrated that with many examples [44,45] such as the length of the coast of Britain, fluctuations of stock exchange, flood data, etc.

23.2.1.2 Fractal Dimension

The box-counting dimension d of a simple geometrical object A is defined by

$$N(l) \underset{l \to 0}{\sim} l^{-d}, \tag{23.1}$$

where $N(l)$ is the minimal number of boxes of side length l required to cover the whole set of points A. For instance, if A is a regular curve (i.e., everywhere differentiable), like a segment, then $d = 1$. If A is as simple surface (respectively a simple volume), then $d = 2$ (respectively $d = 3$). In those cases, d corresponds to the topological dimension of the manifold. The definition of d given by Equation 23.1 can be extended to more general sets, for which d is in general no more an integer. These sets are thus fractal sets and d is called their fractal dimension. A more rigorous definition of the fractal dimension relies on the Hausdorff dimension [24]. But the latter is less easy to compute from data, and, in all the examples we shall consider thereafter, the box-counting dimension d and the Hausdorff dimension are equal. Hence, we consider thereafter that the Hausdorff dimension is equivalent to the fractal dimension as defined by Equation 23.1.

Classical illustrations of fractal sets of points are given by the Cantor dust and the von Koch curve. The former is a set of points obtained by dividing recursively a segment into three parts, where only the first and the third subsegment are retained; this construction is illustrated in Figure 23.1a. Since each step of the algorithm doubles the number of segments while their length is divided by three, after n iterations there are 2^n segments of length 3^{-n}. Since each segment includes all the subsegments of the following iterations, it results that one can cover this ensemble of segments with 2^n balls of radius 3^{-n}. The fractal dimension of the Cantor set as defined by the box-counting method is as follows:

$$d_C = \lim_{n \to \infty} \left(-\frac{\ln 2^n}{\ln 3^{-n}} \right) = \frac{\ln 2}{\ln 3}. \tag{23.2}$$

Therefore, the fractal dimension of the Cantor set is between 0 and 1, which implies that the set is neither an ensemble of isolated points nor a line.

The second example, the von Koch curve, is also obtained using a recursive process where in this case each segment of length l is

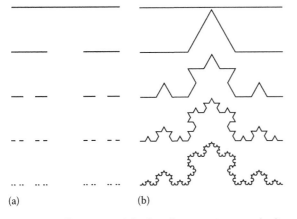

(a)   (b)

FIGURE 23.1 Illustration of the first four iteration steps leading to the Cantor dust (a) and to the von Koch snow flake (b).

replaced by four segments of length $l/3$ as illustrated in Figure 23.1b. Starting from the unit length segment, after n iterations there are 4^n segments of length 3^{-n}. The fractal dimension of the von Koch curve, as defined by the box-counting method, is as follows:

$$d_K = \lim_{n \to \infty} \left(-\frac{\ln 4^n}{\ln 3^{-n}} \right) = \frac{\ln 4}{\ln 3}. \qquad (23.3)$$

The fractal dimension is hence contained between 1 and 2, implying that the length of the von Koch curve is infinite while its surface is zero.

23.2.1.3 Hölder Exponent and Singularity Spectrum

The fractal dimension was defined earlier as a geometrical property that characterizes a set of points, but it can also be used to analyze the regularity of functions or distributions as detailed now. Complex signals, like those encountered in environmental data analysis, can be seen as superpositions of singularities. One way of detecting a singularity of a function f at a point x is to measure its Hölder regularity. The function f is said to be α-Hölder in x if there exists a polynomial P_n of degree n and a constant K such that for suDciently small l

$$|f(x+l) - P_n(l)| \le K|l|^\alpha, \qquad (23.4)$$

where n is the integer part of α (i.e., $n \le \alpha < n+1$). The Hölder regularity of f in x is the maximum α such that f is α-Hölder in x. Note that for $\alpha = 1$ the function is called Lipschitz-continuous in x. If f is $n+1$ times differentiable in x, then $P_n(l) = \sum_{k=0}^{n} \frac{f^{(k)}(x)}{k!} l^k$, the Taylor expansion of f in x. The smaller the Hölder exponent, the stronger the singularity (as illustrated by the examples in Figure 23.2).

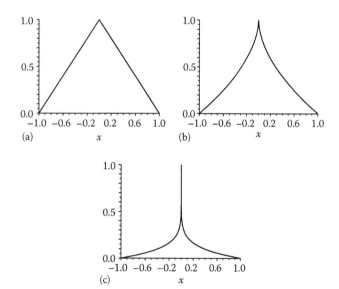

(a)

(b)

(c)

FIGURE 23.2 Illustrations of singularities at point $x = 0$ with the graph of the function $f(x) = 1 - |x|^\alpha$, with $\alpha = 1$, 5/9, and 1/9, respectively, for (a), (b), and (c).

Some functions, sometimes called multifractal functions [48], have a Hölder regularity that varies from one point to the other. It is thus interesting to analyze the set of points A_α where a function has Hölder regularity α, for example, by computing its fractal dimension $d(\alpha)$. The singularity spectrum is the function that associates $d(\alpha)$ to each value of the Hölder regularity α. It is not easy to compute directly, but a trick can be used to estimate it. We briefly sketch the idea without giving a rigorous demonstration.

If we consider a covering $\mathcal{B}_l$ of the support of the function f by boxes of the form $B_{x,l} = [x, x+l]$, then, by definition of the regularity, we obtain that

$$|f(x+l) - f(x)| \sim l^{\alpha_x}, \qquad (23.5)$$

where $\sim$ stands for the magnitude order. Hereafter l is assumed to be small ($l \ll 1$). By definition of the fractal dimension, the minimal number of balls needed to recover the support of A_α is

$$N_{A_\alpha}(l) \sim l^{-d(\alpha)}. \qquad (23.6)$$

The moment function $Z_q(l)$ associated to the cover $\mathcal{B}_l$ of the domain is defined by $Z_q(l) = \sum_{B_l \in \mathcal{B}_l} |f(x+l) - f(x)|^q$. Note that it is sometimes called partition function by analogy with statistical physics. Contributions of boxes containing an α-singularity are given by $|f(x+l) - f(x)|^q \sim l^{q\alpha}$, while the number of such boxes is given by Equation 23.6. Hence, the moment function can be approximated by $Z_q(l) \sim \sum_h l^{q\alpha - d(\alpha)}$. Since l is assumed to be small, the leading contribution in Z_q is given by the term of minimum exponent $q\alpha - d(\alpha)$. It follows that the moment function is approximated by $Z_q(l) \approx l^{\tau(q)}$, where $\tau(q) = \inf_\alpha \{q\alpha - d(\alpha)\}$ is the multiscale exponent. Hence, as shown in [48] the singularity spectrum $d(\alpha)$ appears as being the Legendre–Fenchel inverse transform of the multiscale exponent $\tau(q)$

$$d(\alpha) = \inf_\alpha \{q\alpha - \tau(q)\}. \qquad (23.7)$$

For instance, the singularity spectrum of the Riemann function $f(x) = \sum_{n=1}^{\infty} \frac{\sin n^2 x}{n^2}$, is $d(\alpha) = 4h - 2$ if $\alpha \in [1/2, 3/4]$ and $d(3/2) = 0$. Another example is given by the Devil\$ staircase, related to the Cantor set as follows. In the Cantor set generation algorithm that we have described earlier, each interval was split into two pieces in a symmetric fashion at each iteration. Denoting by μ the characteristic function of the set obtained after n iterations of the procedure, one can easily show that the limit $f(x) = \lim_{n \to \infty} \int_0^x \chi_n(u) du$ exists, and the resulting function f, shown in Figure 23.3a, is called the Devil\$ staircase. It can be shown that each singularity of f is of the same Hölder regularity $\alpha = \ln 2/\ln 3$ and the support of these singularities is the Cantor set. Therefore, in that case the singularity spectrum is reduced to the point $d(\ln 2/\ln 3) = \ln 2/\ln 3$.

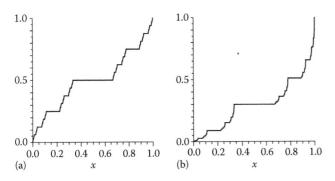

FIGURE 23.3 Illustration of the Devil's staircase with a homogeneous repartition of mass (a) and a heterogeneous repartition of mass where each left subsegment receives 30% of the mass (b).

More complex singularity spectra can be obtained by considering more general fractals similar to the Devil's staircase (see Figure 23.3b), which we do not detail here.

23.2.2 Self-Similar Random Processes

23.2.2.1 Definition and History

Stochastic fractals, sometimes also called fractal noise, are self-similar random processes, which yield models for many applications, e.g., turbulent velocity fields. The self-similarity of a stochastic process is only satisfied in the statistical sense and hence a given realization is not necessarily self-similar. One can distinguish between scalar- or vector-valued random processes in one or higher space dimensions. For the sake of simplicity, we restrict ourselves in the following to scalar-valued processes in one space dimension, which typically corresponds to time t or space x. The simplest ones are Gaussian random processes.

Denoting by $\xi(t)$ a Gaussian random process that we assume to be stationary (i.e., all its statistics are invariant by translation), its one-point probability distribution function (pdf) is given by

$$p(\xi) = \frac{1}{\sqrt{2\pi\sigma^2}} \exp\left(-\frac{(\xi - \mu)^2}{2\sigma^2}\right), \tag{23.8}$$

where

μ is the mean

σ the standard deviation

In the following we suppose that the mean vanishes since we are only interested in the fluctuations. The process $\xi(t)$ is then characterized by its autocovariance function, defined as $\langle \xi(\tau)\xi(0) \rangle$, where $\langle \cdot \rangle$ denotes the expectation, computed either from ensemble, time, or space averages. Equivalently, it can be characterized by its energy spectrum defined as the Fourier transform of its autocovariance function:

$$E(f) = \int_{\mathbb{R}} \langle \xi(\tau)\xi(0) \rangle e^{-\imath 2\pi\tau f} d\tau = \left\langle |\hat{\xi}(f)|^2 \right\rangle \text{ with}$$

$$\hat{\xi}(f) = \int \xi(t) e^{-\imath 2\pi tf} dt \quad \text{and} \quad \imath = \sqrt{-1}. \tag{23.9}$$

The energy spectrum yields the spectral distribution of energy and summing over all frequencies thus yields the total energy.

A simple example of a Gaussian process is the Wiener process, also called Brownian motion, which was proposed in 1900 by Louis Bachelier as a model to describe market price fluctuations [4]. Its mathematical properties were studied in 1923 by Norbert Wiener who called it the fundamental random function. The nomenclature "Brownian" is due to Paul Lévy who named the Wiener process Brownian motion in memory of the Scottish botanist Richard Brown, who in the beginning of the nineteenth century observed the random motion of pollen suspension in water [37]. An extension of Brownian motion has been introduced by Kolmogorov in 1940 [33], of which a spectral representation was given by Hunt in 1951 [29] and that Mandelbrot proposed in 1968 to call fractional Brownian motion [42].

23.2.2.2 Brownian Motion

For Brownian motion the variance of the increments scales as

$$\langle |B(t) - B(\tau)|^2 \rangle = |t - \tau| \tag{23.10}$$

and the Hölder regularity of the trajectories is 1/2. The formal derivative of a Wiener process is called a Gaussian white noise. It is stationary and uncorrelated, i.e., its autocovariance function is $\langle \xi(\tau)\xi(0) \rangle = \delta(\tau)$, where δ is the Dirac distribution or equivalently its energy spectrum is constant, $E(f) = 1$. The constant spectrum means that all frequencies f have the same weight, and hence the noise is called white by analogy with white light. Correlated Gaussian processes have nonconstant spectra and they are called colored noise. Power-law spectra $E(f) \propto f^\beta$ are of particular interest as the processes are statistically self-similar, i.e., $\langle \xi(\lambda\tau)\xi(0) \rangle = \lambda^\alpha \langle \xi(\tau)\xi(0) \rangle$. However, such processes are not necessarily stationary and, in order to recover stationarity, we consider their increments. Due to nonstationarity the energy spectrum can only be defined formally and can no more be integrated (due to infrared divergence). For example, the generalized energy spectrum of Brownian motion satisfies the power law $E(f) \propto 1/f^2$. Brownian motion thus belongs to the class of so-called $1/f$ processes, which have been studied for many applications.

23.2.2.3 Fractional Brownian Motion

Fractional Brownian motion is a kind of self-similar Gaussian process that is nonstationary and whose energy spectrum follows a power law. A given realization of such a noise is almost everywhere singular and has the same Hölder regularity at all points, i.e., it is mono-fractal.

The fractional Brownian motion $B_H(t)$ is the Gaussian process with zero mean such that

$$B_H(t = 0) = 0 \tag{23.11}$$

and

$$\langle |B_H(t) - B_H(\tau)|^2 \rangle = |t - \tau|^{2H}, \tag{23.12}$$

where $0 < H < 1$ is an additional parameter called Hurst exponent [30]. Here H determines the regularity of the trajectories.

The smaller H, the lower the regularity. For $H < 1/2$ the increments of the process are correlated, while for $H > 1/2$ they are anti-correlated. For $H = 1/2$ we get the classical Brownian motion. In all cases the process is said to be long-range dependent.

The covariance function of B_H is given by

$$\langle B_H(t)B_H(\tau)\rangle = \frac{1}{2}(|t|^{2H} + |\tau|^{2H} - |t-\tau|^{2H}) \qquad (23.13)$$

Note that one given realization of fractional Brownian motion is not a fractal: the self-similarity is only fulfilled in the statistical sense. Indeed, Equation 23.12 implies that

$$\langle |B_H(\lambda t) - B_H(\lambda\tau)|^2 \rangle = \lambda^{2H}\langle |B_H(t) - B_H(\tau)|^2\rangle. \qquad (23.14)$$

However, it can be shown that a given trajectory has the pointwise Hölder regularity $H = \alpha$ almost surely and is almost (besides for a set of measure zero) nowhere differentiable.

The self-similarity of the fractional Brownian motion $B_H(t)$ implies for the energy spectrum a power-law behavior with exponent $2H + 1$,

$$E(f) = \frac{C_H}{f^{2H+1}}. \qquad (23.15)$$

Gaussian processes, and thus also fractional Brownian motion, can be represented in Fourier space using the Cramer representation

$$B_H(t) = \int_{\mathbb{R}} \sqrt{E(f)}(e^{\iota 2\pi ft} - 1)d\xi(f), \qquad (23.16)$$

where $d\xi(f)$ is an orthogonal Gaussian increment process with $\langle d\xi(f)d\xi(f')\rangle = \delta(f - f')$, which means that the measure corresponds to Gaussian white noise. The term $(e^{\iota 2\pi s} - 1)$ instead of $e^{\iota 2\pi s}$ guarantees that $B_H(0) = 0$.

23.2.2.4 Multifractional Brownian Motion

Allowing for time (or space) varying Hurst exponents generalizes fractional Brownian motion, which is mono-fractal, to introduce stochastic multifractals. Such multifractional Brownian processes can be defined by generalizing the spectral representation, Equation 23.16, as follows:

$$B_\theta(t) = \int_{\mathbb{R}} \frac{e^{i2\pi ft} - 1}{|f|^{\theta(t)+1/2}} d\xi(f), \qquad (23.17)$$

where θ is a function $\theta: [0,1] \rightarrow]0,1[$ which can be seen as a local Hurst exponent of the process. Indeed, the pointwise Hölder regularity of $B_\theta(t)$ is almost surely equal to $\theta(t)$ and the Hausdorff dimension of the graph of B_θ is $2 - \inf\{\theta(t), 0 \le t \le 1\}$.

Methods for synthesizing fractional Brownian motion are presented in Section 23.3.2.

23.2.3 Wavelets

23.2.3.1 Definition and History

In a signal the useful information is often carried by both its frequency content and its time evolution, or by both its wavenumber content and its space evolution. Unfortunately the spectral analysis does not give information on the instant of emission of each frequency, or on the spatial location of each wavenumber. This is due to the fact that, since the Fourier representation spreads time or space information among the phase of all Fourier coefficients, the energy spectrum (i.e., the modulus of the Fourier coeDcients) does not carry any structural information in time or space. This is a major limitation of the classical way to analyze nonstationary signals or inhomogeneous fields. A more appropriate representation should combine these two complementary descriptions.

From now on we will consider a signal $f(x)$, which will only depend on space. The theory is the same for a signal $f(t)$, which depends on time, except that the wavenumber k should in that case be replaced by the frequency ν and the spatial scale l by the time scale or duration τ. Any function $f \in L^2(\mathbb{R})$ also has a spectral representation $\hat{f}(k)$ defined as

$$\hat{f}(k) = \int_{-\infty}^{\infty} f(x)e^{-2\pi kx}dx, \qquad (23.18)$$

where $\iota = \sqrt{-1}$.

However, there is no perfect representation due to the limitation resulting from the Fourier's uncertainty principle (also called Heisenberg's uncertainty principle when it is used in quantum mechanics). One thus cannot perfectly analyze the signal f from both sides of the Fourier transform at the same time, due to the restriction $\Delta x \cdot \Delta k \ge C$, where Δx is the spatial support of $|f(x)|$ and Δk the spectral support of $|\hat{f}(k)|$, with C a constant that depends on the chosen normalization of the Fourier transform. Due to the uncertainty principle, there is always a compromise to be made in order to have either a good spatial resolution Δx at the price of a poor spectral resolution Δk or a good spectral resolution Δk while losing the space resolution Δx, as it is the case with the Fourier transform. These two representations, in space or in wavenumber, are the most commonly used in practice because they allow to construct orthogonal bases onto which one projects the signal to be analyzed and processed.

In order to try to recover some space locality while using the Fourier transform, Gabor [23] has proposed the windowed Fourier transform, which consists of convolving the signal with a set of Fourier modes $e^{2\pi kx}$ localized in a Gaussian envelope of constant width l_0. This transform allows then a space-wavenumber (or time-frequency) decomposition of the signal at a given scale l_0, which is kept fixed. But unfortunately, as shown by Balian [5], the bases constructed with such windowed Fourier modes cannot be orthogonal. In 1984, Grossmann and Morlet [25] have proposed a new transform, the so-called wavelet transform, which consists of convolving the signal with a set of wave packets, called wavelets, of different widths l and locations x. To analyze the signal $f(x)$, we generate the family of analyzing wavelets $\psi_{l,x}$ by dilation (scale parameter l) and translation (position parameter x) of a given function ψ, which oscillates with a characteristic wavenumber k_ψ in such a way that its mean remains zero. The wavelet transform thus allows a space-scale decomposition of the signal f given by its wavelet coeDcients $\tilde{f}_{l,x}$. The wavelet representation yields the best compromise in view of the Fourier uncertainty principle, because the product $\Delta x \cdot \Delta k$ remains constant in the process of dilating and translating ψ. In fact it gives for the large scales a good spectral resolution Δk but a poor spatial resolution Δx, while, on the contrary, it gives a good spatial resolution Δx with a poor spectral resolution Δk for the small scales.

In 1989, the continuous wavelet transform has been extended to analyze and synthesize signals or fields in higher dimensions [1,47]. In 1985, Meyer, while trying to prove the same kind of impossibility to build orthogonal bases as done by Balian [5] in the case of the windowed Fourier transform, has been quite surprised to discover an orthogonal wavelet basis built with spline functions, now called the Meyer–Lemarié wavelet basis [36]. In fact the Haar orthogonal basis, which was proposed in 1909 in the PhD thesis of Haar and published in 1910 [26], is now recognized as the first orthogonal wavelet basis known, but the functions it uses are not regular, which limits its application. In practice one often likes to build orthogonal wavelet bases in which the expansion of some signals of interest (depending on the application) are sparse, i.e., involve as few large wavelet coefficients as possible, while the rest are negligible. In particular, following Meyerß work, Daubechies has proposed in 1988 [12] orthogonal wavelet bases built with compactly supported functions defined by discrete Quadrature Mirror Filters (QMFs) of different lengths. The longer the filter, the sparser is the expansion of smooth signals, thanks to the higher number of vanishing moments of the wavelet, as detailed later. In 1989, Mallat has devised a fast algorithm [39] to compute the orthogonal wavelet transform using wavelets defined by QMF. Later Malvar [41] and Coifman and Meyer [8] have found a new kind of windows of variable width that allows the construction of orthogonal adaptive local cosine bases, which have then been used to design the MP3 format for sound compression. The elementary functions of such bases, called Malvarß wavelets, are parametrized by their position x, their scale l (width of the window), and their wavenumber k (proportional to the number of oscillations inside each window). In the same spirit, Coifman, Meyer, and Wickerhauser [9] have proposed the so-called wavelet packets, which, similarly to compactly supported wavelets, are wavepackets of prescribed number of vanishing moments, defined by discrete QMFs, from which one can construct orthogonal bases.

The Fourier representation is well suited to solve linear equations, for which the superposition principle holds and whose generic solutions either persist at a given scale or spread to larger scales. In contrast, the superposition principle does not hold anymore for nonlinear equations, for example, the Navier–Stokes equations, which are the fundamental equations of fluid dynamics. In this case, the equations can no more be decomposed as a sum of simpler equations, which can be solved separately. Generically, the time evolution of their solutions involves a wide range of scales and could even lead to finite-time singularities, for example shocks. The "art" of predicting such nonlinear evolutions (a generic case being turbulent flows) consists in disentangling the nonlinear from the linear dynamical components: the former should be deterministically computed while the latter could either be discarded or their effect be statistically modeled. A review of the different types of wavelet transforms and their applications to analyze and compute turbulent flows is given in [19,52].

23.2.3.2 Continuous Wavelet Transform

The only condition a real function $\psi(x) \in L^2(\mathbb{R})$, or a complex function $\psi(x) \in L^2(\mathbb{C})$, should satisfy to be called a wavelet is the admissibility condition:

$$C_\psi = \int_0^\infty |\hat{\psi}(k)|^2 \, dk < \infty, \tag{23.19}$$

where $\hat{\psi} = \int_{-\infty}^\infty f(x)e^{-2\pi kx} dx$ is the Fourier transform of ψ. From (23.18), we see that for ψ to be admissible it should satisfy in particular $\psi(0) = 0$, i.e., the space average of ψ should vanish and only then can the wavelet transform be invertible. The wavelet ψ may also have other properties, such as being well-localized in physical space $x \in \mathbb{R}$ (fast decay of f for $|x|$ tending to y) and smooth, i.e., well localized in spectral space (fast decay of $\psi(k)$ for $|k|$ tending to y). For several applications, in particular to study deterministic fractals or random processes, one also wishes that $\psi(k)$ decays rapidly near 0 or equivalently that the wavelet has enough cancellations such that

$$\int_{-\infty}^\infty x^m \psi(x)dx = 0 \quad \text{for } m = 0, \ldots, M-1, \tag{23.20}$$

namely that its first M moments vanish. In this case the wavelet analysis will enhance any quasi-singular behavior of the signal by hiding all its polynomial behavior up to degree m.

One then generates a family of wavelets by dilatation (or contraction), with the scale parameter $l \in \mathbb{R}^+$, and translation, with the location parameter $x \in \mathbb{R}$, of the so-called mother wavelet and obtains

$$\psi_{l,x}(x') = c(l)\psi\left(\frac{x'-x}{l}\right) \tag{23.21}$$

where $c(l) = l^{-1/2}$ corresponds to all wavelets being normalized in the L^2-norm, i.e., they have the same energy, while for $c(l) = l^{-1}$ all wavelets are normalized in the L^1-norm.

The continuous wavelet transform of a function $f \in L^2(\mathbb{R})$ is the inner product of f with the analyzing wavelets $\psi_{l,x}$, which yields the wavelet coeDcients

$$\tilde{f}(l,x) = \langle f, \psi_{l,x} \rangle = \int_{-\infty}^{\infty} f(x')\psi_{l,x}^*(x')dx', \qquad (23.22)$$

with ψ^* denoting the complex-conjugate of ψ. The continuous wavelet coeDcients measure the fluctuations of f at scale l and around position x. If the analyzing wavelets are normalized in L^2-norm ($c(l) = l^{-1/2}$), then the squared wavelet coeDcients correspond to the energy density of the signal whose evolution can be tracked in both space and scale. If the wavelets are normalized in L^1-norm ($c(l) = l^{-1}$), the coeDcients are related to the previous ones by the following relation:

$$\tilde{f}_{L^1} = l^{-1/2} \tilde{f}_{L^2}. \qquad (23.23)$$

Note that to study the Hölder regularity of a function and estimate its singularity spectrum, one typically uses wavelet coeDcients in L^1-norm (see Section 23.2.1.3).

The function f can be reconstructed without any loss as the inner-product of its wavelet coeDcients $\tilde{f}$ with the analyzing wavelets $\psi_{l,x}$:

$$f(x') = C_\psi^{-1} \int_{\infty}^{\infty} \int_{0^+}^{\infty} \tilde{f}(l,x)\psi_{l,x}(x') \frac{dl}{l^2} dx \qquad (23.24)$$

with C_ψ the constant of the admissibility condition given in Equation 23.19, which only depends on the chosen wavelet ψ.

Like the Fourier transform, the wavelet transform is linear, i.e., we have

$$\widetilde{\beta_1 f_1(x) + \beta_2 f_2(x)} = \beta_1 \widetilde{f_1(x)} + \beta_2 \widetilde{f_2(x)} \qquad (23.25)$$

with $\beta_1, \beta_2 \in \mathbb{R}$, and it is also an isometry, i.e., it conserves the inner product (Plancherel's theorem) and in particular the energy (Parseval's identity). The continuous wavelet transform is also covariant by translation and by dilation, both properties that are partially lost by the orthogonal wavelet transform. Let us also mention that, due to the localization of wavelets in physical space, the behavior of the signal at infinity does not play any role. In contrast, the nonlocal nature of the trigonometric functions used for the Fourier transform does not allow us to locally analyze or process a signal with it.

Figure 23.4 shows six examples of wavelet analyses of academic signals using the complex-valued Morlet wavelet: a Dirac spike (a), a step function (b), a superposition of two cosine functions having different frequencies (c), a succession of two cosine functions having different frequencies (d),

a chirp (e), and a Gaussian white noise (f). The modulus of the wavelet coeDcients is plotted as a function of position x on abscissa and the log of the scale l on ordinate. The curved black lines delimitate the region where the coeDcients are not influenced by left and right boundaries, which correspond to the spatial support of the wavelets localized in $x = 0$ and $x = 1$. The horizontal straight black line indicates the scale below which the wavelet coeDcients are aliased, due to undersampling of the wavelets at small scales. Note in particular that three signals, namely Figure 23.4a, e, and f, have similar flat Fourier and wavelet spectra (see Section 23.3.3.2), although the space-scale representation of the energy density in wavelet space exhibit very different behaviors.

The extension of the continuous wavelet transform to analyze signals in d dimensions is made possible by replacing the aDne group by the Euclidean group including rotations. One thus generates the d-dimensional wavelet family $\psi_{l,r,\vec{x}}$ with l the dilation factor, R the rotation matrix in $\mathbb{R}^d$, and $\vec{x}$ the translation such that

$$\psi_{l,\vec{x},r}(x') = \frac{1}{l^{d/2}} \psi\left(r^{-1}\left(\frac{\vec{x}' - \vec{x}}{l}\right)\right) \qquad (23.26)$$

where the wavelet ψ should satisfy the admissibility condition, which becomes in d-dimensions:

$$C_\psi = \int_0^\infty |\hat{\psi}(k)|^2 \frac{d^d k}{|k|^d} < \infty \qquad (23.27)$$

If we consider $d = 2$ then the rotation matrix $R(\theta)$ is

$$\begin{pmatrix} \cos\theta & -\sin\theta \\ \sin\theta & \cos\theta \end{pmatrix}. \qquad (23.28)$$

The wavelet analysis of a two-dimensional scalar field $f(\vec{x})$ is

$$\widetilde{f(l,\vec{x},\theta)} = \int_{-\infty}^{\infty} \int_{-\infty}^{\infty} f(\vec{x}')\psi_{l,\vec{x},\theta}^*(x')d\vec{x}', \qquad (23.29)$$

and the wavelet synthesis is

$$f(\vec{x}') = \frac{1}{C_\psi} \int_0^\infty \int_{-\infty}^\infty \int_{-\infty}^\infty \int_0^{2\pi} \tilde{f}(l,\vec{x},\theta)\psi_{l,\vec{x},\theta}^*(\vec{x}') \frac{dl\,d\vec{x}\,d\theta}{l^3}. \qquad (23.30)$$

In dimensions larger than two, one needs $(d-1)$ angles to describe the rotation operator R.

23.2.3.3 Orthogonal Wavelet Transform

Wavelets can also be used to construct discrete representations of various function spaces, called frames [11], by selecting a discrete

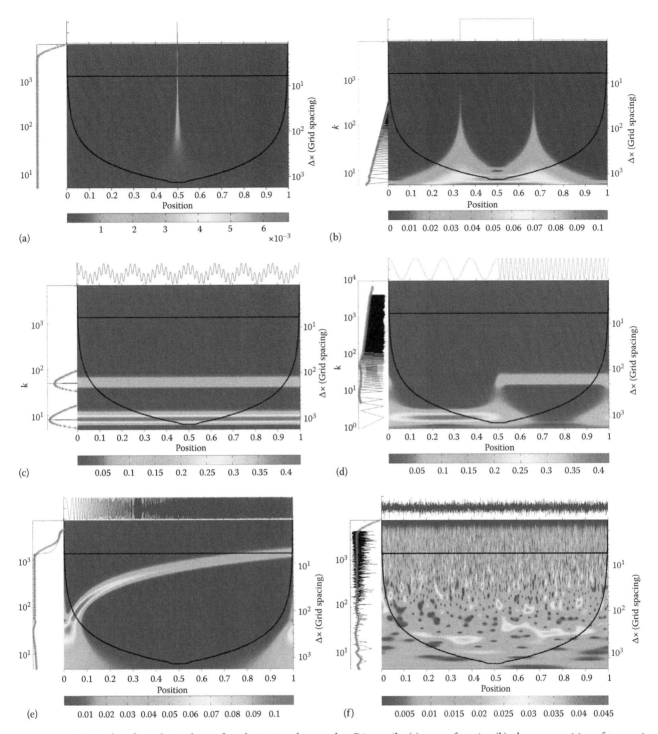

FIGURE 23.4 Examples of wavelet analyses of academic signals, namely a Dirac spike (a), a step function (b), the superposition of two cosine functions having different frequencies (c), the succession of two cosine functions of very different frequencies (d), a chirp (e), and finally one realization of a Gaussian white noise (f). The moduli of the complex-valued Morlet wavelet coeDcients are plotted as a function of position and scale. The original signal is plotted on the top. The Fourier spectrum (black curve) and the wavelet scalogram (grey crosses), as defined in Section 23.3.3.2, are also shown on the left, with the axes rotated by 90°.

subset of all their translations and dilations. Some special frames sampled on a dyadic grid $\lambda = (j, i)$, i.e., for which the scale l has been discretized by octaves j and the position x by spatial steps $2^{-j}i$, constitute orthogonal wavelet bases. The main difference between the continuous and the orthogonal wavelet transform is that all

orthogonal wavelet coeDcients are decorrelated. This is not the case for the continuous wavelet coeDcients, which are redundant and correlated in both space and scale. These correlations can be visualized by plotting the modulus of the continuous wavelet coefficients of one realization of a white noise computed with a Morlet

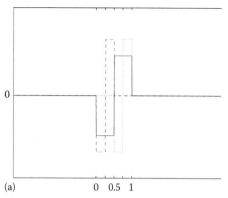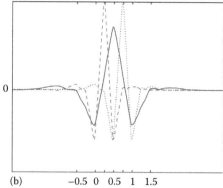

(a)   0   0.5   1   (b)   −0.5   0   0.5   1   1.5

FIGURE 23.5 Orthogonal wavelets: Haar wavelet (a) and Coifman 12 wavelet (b). We have superposed one wavelet at scale $j = 0$ (solid line) and position $x = 0.5$) and two wavelets at the next smaller scale $j = 1$, located at position $x = 0.25$ (dashed line) and $x = 0.75$ (dotted line). They are mutually orthogonal, which can be directly seen for the Haar wavelet and which is much less obvious for the Coifman wavelet.

wavelet, see Figure 23.9b. The patterns one thus observes are due to the reproducing kernel of the continuous wavelet transform, which corresponds to the correlation between all the analyzing wavelets themselves. Note that the redundancy of the continuous wavelet transform is actually useful for algorithms such as edge and texture detection. Moreover, its translation and dilation invariance eliminates some artifacts one encounters when denoising with the orthogonal wavelet transform, which does not preserve those invariances (Figure 23.5).

As a tutorial example, we explain the orthogonal wavelet decomposition of a three-dimensional vector field. For this we consider a square integrable vector-valued field $\vec{x} \to \vec{f}(\vec{x}) \in L^2(\mathbb{T}^3)$, where $\mathbb{T}^3 = (\mathbb{R}/\mathbb{Z})^3$ is the 3D torus and $\vec{x} = (x_1, x_2, x_3) \in \mathbb{T}^3$. Note that in practice the fact that f is defined on a torus simply means that periodic boundary conditions are assumed. The input data consist in discrete values of f sampled with a resolution $N_k = 2^J$ in each direction. N_k is thus the number of grid points and J is the number of octaves in each of the three directions, and the total number of grid points is thus $N = N_1 \times N_2 \times N_3 = 2^{3J}$. The mother wavelet is denoted ψ as earlier and we assume that it satisfies all the necessary conditions (see, e.g., [13]) so that the wavelets $\psi_{l,i}$ defined by Equation 23.21 are pairwise orthogonal if (l,x) is sampled on the dyadic grid $\{(2^{-j}, 2^{-j}i) \mid j = 0, k, J - 1, i = 0, k, 2^j - 1\}$. We also assume that the wavelet has been suitably periodized. To expand the components f_d of $\vec{f}$ (with $d = 1, 2, 3$) into an orthogonal wavelet series from the largest scale $l_{max} = 2^0$ to the smallest scale $l_{min} = 2^{-J+1}$, we need to construct a 3D MRA as follows [13,19]:

For λ_0 belonging to the index set

$$\Lambda_0 = \{(j, \vec{\mu}, \vec{\tau}) \hat{u} j = 0, \dots, J - 1, \quad \vec{\mu} \in \{0,1\}^3, \quad \vec{\tau} \in \{0, \dots, 2^j - 1\}^3\},$$

define the 3D wavelet ψ_λ by

$$\psi_\lambda(x_1, x_2, x_3) = 2^{3j/2} \prod_{\substack{1 \le k \le 3 \\ \mu_k = 0}} \phi(2^j x_k - i_k) \prod_{\substack{1 \le k \le 3 \\ \mu_k = 1}} \psi(2^j x_k - i_k),$$

where ϕ is the scaling function (also called father wavelet) associated to ψ [13]. Here, the parameters j and $\vec{\tau}$ are the 3D

equivalent to the scale and position parameters that we are already familiar with from the preceding discussion of the one-dimensional continuous wavelet transform. The new parameter, $\vec{\mu}$, provides an additional degree of freedom, which is necessary to represent 3D data without loss of information. It controls the directions of oscillation of the wavelet. For example, if $\vec{\mu} = (1, 0, 0)$, the wavelet is oscillatory (i.e., it has vanishing mean) in the first direction, whereas it has nonvanishing mean in the two others directions. If $\vec{\mu} = (0, 0, 0)$, ψ_λ is the 3D equivalent to a scaling function, in which case we shall denote it ϕ_λ, following the classical convention. The wavelets are thus indexed by the subset of Λ_0 whose elements satisfy $\vec{\mu} \neq 0$, which we denote Λ. The wavelet coeDcients and scaling coeDcients of f_d are then simply defined by

$$\tilde{f}_\lambda^d = \langle f_d, \psi_\lambda \rangle$$
$$\overline{f}_\lambda^d = \langle f_d, \phi_\lambda \rangle,$$

where $\langle \cdot, \cdot \rangle$ denotes the inner product in $L^2(\mathbb{R}^3)$.

Now we have all the ingredients to write down the wavelet series of f_d:

$$f_d = \overline{f}_{(0,0,0)} + \sum_{\lambda \in \Lambda} \tilde{f}_\lambda^d \psi_\lambda. \tag{23.31}$$

The first term is a constant, which is in fact the mean value of f, and the sum over λ contains all the oscillations of f at finer and finer scales, $j = 0, k, J - 1$, while preserving some amount of space-locality thanks to the position index $\vec{\tau}$ and also some amount of directionality thanks to $\vec{\mu}$. Hence the expansion coefficients appearing in Equation 23.31 can be used to compute directional and/or scale-wise statistics of $\vec{f}$, as we shall see further down. Importantly, there exists a fast wavelet algorithm with $O(N)$ complexity, where N denotes the number of wavelet coefficients used in the computation. It is thus computationally even faster than the FFT, whose complexity is $O(N \log_2 N)$.

23.3 Methods of Analysis

23.3.1 Fractals

23.3.1.1 Estimation of the Fractal Dimension

The box-counting algorithm is a simple method to compute the fractal dimension of a given object (a set of points S in Euclidean space $\mathbb{R}^d$, for example, a curve in two dimensions or an iso-surface in three dimensions) by counting the number of boxes (squares in two dimensions, cubes in three dimensions, k) which cover the object. First the object is overlaid with an equidistant Cartesian grid of size l. Then the number of boxes with side length l covering the object is counted, which yields $N(l)$. Subsequently the grid size l is reduced (e.g., by a factor 2), a refined grid is overlaid, and the number of boxes covering the object is counted again. The earlier procedure is repeated until the finest resolution of the object is obtained. Finally, the number of boxes $N(l)$ covering the object is plotted against the inverse grid size $1/l$ in log–log representation. A straight line is fitted to the curve thus obtained and the slope of the curve yields the fractal dimension of the set S as defined by Equation 23.1.

For a regular smooth curve (e.g., a straight line in two or three dimensions) we can observe that the number of boxes covering the curve is proportional to the inverse of the grid size and hence its dimension is 1, which is equal to its topological dimension. For a smooth surface (e.g., the surface of a sphere in three dimensions) we find that the number of boxes increases quadratically with the inverse grid size, which yields its topological dimension of two. For fractals the obtained dimension differs from its topological one.

Besides pathological cases, e.g., singular sets, the limit obtained with the box-counting algorithm corresponds to the Hausdorff dimension (box counting dimension t Hausdorff dimension) and thus this technique is an eDcient way for computing it.

23.3.1.2 Synthesis of Fractal Sets

Now we discuss a method to generate a fractal set of points based on iterated functions, recursively applied. An iterated function system (IFS) is a set of functions $\{f_i\}_{i\in[1,N]}$ from $\mathbb{R}^d$ into itself which are contractions, i.e., such that there exists for each i a constant c_i such that $0 < c_i < 1$ with $|f_i(x) - f_i(y)| \leq c_i|x - y|$. The Hutchinson function F associated to the IFS is the transformation from $C(\mathbb{R}^d)$ to itself, where $C(\mathbb{R}^d)$ denotes the set of all compact subsets of $\mathbb{R}^d$, defined by

$$F(A) = f_1(A) \cup \cdots \cup f_N(A), \tag{23.32}$$

with $A \in C(\mathbb{R}^d)$. It can be shown that F itself is also a contraction defined into $C(\mathbb{R}^d)$ for the Hausdorff distance δ_H, that is $\delta_H(F(A),F(B)) \leq c\delta_H(A,B)$, where $\delta_H(A,B) = \max\{\sup_{x\in A}\inf_{y\in B}|x - y|,\sup_{y\in B}\inf_{x\in A}|x - y|\}$ and $c = \max\{c_i\}$. Because of the completeness of the metric space $(C(\mathbb{R}^d), \delta_H)$, F admits a fixed point in $C(\mathbb{R}^d)$, and this fixed point is a compact limit ensemble A_F, obtained as $A_F = \lim_{n\to y} F^n(A)$, where A is an arbitrary initial compact set and A_F verifies $A_F = F(A_F)$.

As illustration for an IFS, we consider the IFS $\{f_1,f_2\}$ defined on the real line $\mathbb{R}$ by $f_1(x) = x/3$ and $f_2(x) = x/3 + 2/3$. These functions are contractions with ratio 1/3. When applying these two contractions to the segment [0, 1], we obtain the algorithm for generating the Cantor set, as illustrated in Figure 23.1. The Cantor set is thus the limit ensemble of the IFS $\{f_1,f_2\}$. In the particular case where the IFS is made of disconnected or just-touching aDne functions $f_i(x) = c_iR_ix + b_i$ where $0 < c_i < 1$ is the magnitude, R_i the rotation matrix, and b_i the translation, then the fractal dimension d of the limit set is linked to the similitude magnitude c_i by the relation

$$\sum_i c_i^d = 1. \tag{23.33}$$

By applying this relation to the Cantor set, we obtain the equation $2(1/3)^d = 1$, whose solution is the fractal dimension $d = \ln 2/\ln 3$ already found earlier. Similarly, the von Koch curve can be obtained from an IFS of four similitudes of magnitude 1/3, so that its fractal dimension satisfies $4(1/3)^d = 1$, leading to the known result $d = \ln 4/\ln 3$.

To construct the limit ensemble, a direct solution is to start from a simple compact set and to make it evolve by using the Hutchinson function associated to the IFS. However this solution is computationally costly, since we have to deal with sets. A more eDcient alternative is to use a random procedure as we will describe now. From a single point $A = \{x_0\}$, which is a compact set, a recursive process is generated so that $x_{n+1} = w_n$ where w_n is randomly chosen within the list $\{f_i(x_n)\}$ where $f_i(x_n)$ is sampled with probability p_i. If $f_i(x) = A_ix + b_i$, where A_i is a matrix, then p_i can be defined as $p_i = \dfrac{|\det A_i|}{\sum_k|\det A_k|}$. The intuitive reason of this choice for p_i is that the volume of the unit square transformed by f_i is $|\det A_i|$. When the determinant is zero, p_i is set to a small value compared to the other nonzero determinants, and then normalized to ensure the probability normalization $\sum_i p_i = 1$.

Another possibility to construct a fractal set of points from an existing set of points is given by the collage theorem [6]. We consider a compact ensemble S of $\mathbb{R}^d$ and $\varepsilon > 0$. The idea is to be able to reconstruct this ensemble from an IFS strategy, which would be easy if an IFS, generating the pattern was known exactly. However, in practical applications the generating system is unknown. The collage theorem states that, if one finds an IFS $\{f_i\}_{i\in[1,N]}$ such that the Hutchinson function F leaves S invariant up to a tolerance ε, i.e., $\delta_H(S, F(S)) \leq \varepsilon$, then the limit ensemble A_F associated to the IFS satisfies

$$\delta_H(S,A_F) \leq \frac{\varepsilon}{1-c}, \tag{23.34}$$

where c is the contraction ratio of F.

Even if this theorem does not lead to a constructive method to determine an appropriate IFS, it provides a useful way for building fractal sets from a given set of points. In practice, the IFS can be looked for within a reduced class of contractions. For instance, one can try to estimate the smallest set of similitudes required to ensure a given tolerance ε.

23.3.1.3 Singularity Spectrum

As an illustration of the singularity spectrum and its limitations, we compute the singularity spectrum of a function f and we compare its singularity spectrum when noise is added.

In Figure 23.6b, we show the singularity spectrum of the function f plotted in Figure 23.6a. The support of the spectrum is the whole interval (0, 1) and the fractal dimension of the Hölder exponent close to $\alpha = 1$ is about $d = 0.7$. It is larger than the fractal dimension of stronger singularities (having small Hölder exponents). Hence, the support where the signal is regular is larger than the one where it is irregular, as seen in Figure 23.6a. If a white noise with a weak standard deviation of $\sigma = 0.01$ is added, see Figure 23.6c, then the signal becomes more irregular leading to a singularity spectrum truncated at a Hölder exponent close to $\alpha = 0.5$, as seen on Figure 23.6d. Moreover, the support of the singularities becomes larger since the fractal dimension $d(\alpha)$ for $\alpha = 0.5$ for the noise-free signal, in Figure 23.6a, is close to $\alpha = 0.5$,

see Figure 23.6b, while for the noisy signal in Figure 23.6c it is close to $\alpha = 1$, see Figure 23.6d. This effect is reinforced with a more intense noise of standard deviation $\sigma = 0.1$, see Figure 23.6e and f.

This illustrates that the computation of the singularity spectrum is sensitive to the amount of noise present in the signal. Thus adding white noise to a signal reduces the regularity since large Hölder exponents disappear as the amount of noise increases, as seen in Figure 23.6.

23.3.2 Self-Similar Random Processes

23.3.2.1 Analysis

The Hurst exponent H of a stochastic process can be estimated by considering the quadratic variation of a given realization, e.g., observed data. For fractional Brownian motion $B_H(t)$ with $t \in [0, 1]$ the quadratic variation V_N associated to

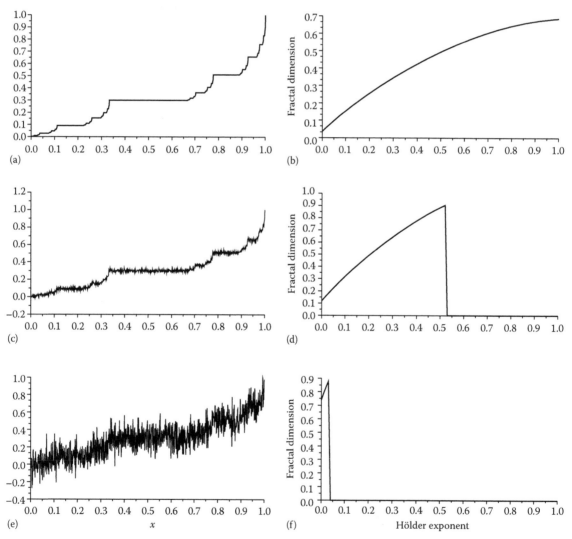

FIGURE 23.6 Singularity spectrum of a function f (a) (the Devil's staircase) and its noisy versions perturbed with a white noise of standard deviation $\sigma = 0.01$ (c) and $\sigma = 0.1$ (e). The corresponding singularity spectra are shown on the right column. Without noise (b), with noise of standard deviation $\sigma = 0.01$ (d) and $\sigma = 0.1$ (f).

the step size $\delta t = 1/N$, N being the number of sampling points, is given by

$$V_N = \sum_{k=0}^{N-1} \left[B_H\left(\frac{k}{N}+\frac{1}{N}\right) - B_H\left(\frac{k}{N}\right) \right]^2. \qquad (23.35)$$

This quadratic variation can be related to the Hurst exponent by

$$V_N = c N^{1-2H}, \qquad (23.36)$$

where c is a constant. Moreover the quadratic variation of the dyadically subsampled data, taking only one out of two values of $B_H(k/N)$, is $V_{N/2}$. It follows that

$$\frac{V_N}{V_{N/2}} = 2^{1-2H}, \qquad (23.37)$$

which leads thus to the Hurst exponent

$$H = \frac{1}{2}\left(1 - \log_2 \frac{V_N}{V_{N/2}}\right). \qquad (23.38)$$

Hence this relation can be used to estimate H from the data. It only requires to compute the quadratic variation of both the data and the dyadically subsampled data.

23.3.2.2 Synthesis

Different approaches are available for the synthesis of self-similar random processes, which are typically either based on the spectral representation of stochastic processes or construct the process in physical space using a decomposed covariance matrix. Additionally wavelet techniques have been developed that allow the eDcient generation of realizations with long-range dependence and with many scales without imposing a cutoff scale thanks to the vanishing moment property of the wavelets.

For synthesizing fractional Brownian motion numerically one can either discretize the Cramer representation in a suitable way or generate it directly in physical space by applying the decomposed covariance matrix to Gaussian white noise.

For the latter the discrete covariance matrix $\Gamma_{i,j} = \langle B_H(t_i)B_H(\tau_j)\rangle$ for $i, j = 1, k, N$, where N denotes the number of grid points, is first assembled. Then a Cholesky decomposition $\Gamma = LL^t$ is computed (where L is a lower triangular matrix with positive diagonal entries and L^t is its transpose). Then, a vector of length N is constructed by taking one realization of Gaussian white noise with variance 1, i.e., $\xi(t_i)$ for $i = 1, k, N$. A realization of fractional Brownian motion is then obtained by multiplication of ξ with L,

$$B(t_i) = L_{ij}\xi(t_j)$$

where summation over j is assumed. Finally, let us recall that the Hausdorff dimension of the graph of B_H is $2 - H$.

Different wavelet techniques for synthesizing fluctuating fields using self-similar random processes with a wide range of scales have been proposed. Elliot and Majda [16,17] proposed a wavelet Monte-Carlo method to generate stochastic Gaussian processes with many scales for one-dimensional scalar fields and for two-dimensional divergent-free velocity fields. The fields thus obtained have a $k^{-5/3}$ scaling of the energy spectrum (which means that the increments grow as $l^{2/3}$) and thus correspond to fractional Brownian motion with a Hurst exponent $H = 2/3$. Applications were dealing with the simulation of particle dispersion (Elliot and Majda) [17]. A related construction was proposed by Tafti and Unser [53].

An interesting technique from image processing, which was originally developed for generating artificial clouds in computer animations was proposed in [10]. Therewith intermittent scalar valued processes in two space dimensions can be eDciently generated, which have a given energy distribution, that could be self-similar. The resulting process is strictly band-limited.

23.3.2.3 Application to Fractional Brownian Motion

To illustrate the fractional Brownian motion, we show in Figure 23.7b three realizations of different fractional Brownian

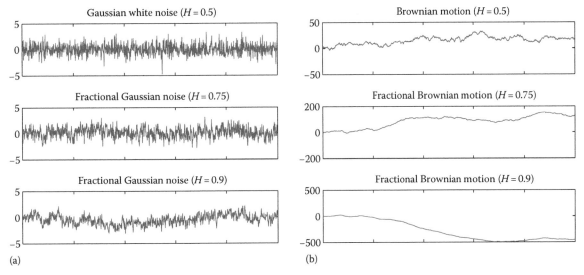

FIGURE 23.7 Sample trajectories of Gaussian fractional noise (a), and of fractional Brownian motion (b) for three different values of the Hurst exponent H. The Gaussian fractional noise (a) corresponds to increments of the fractional Brownian motion (b). The resolution is $N = 1024$.

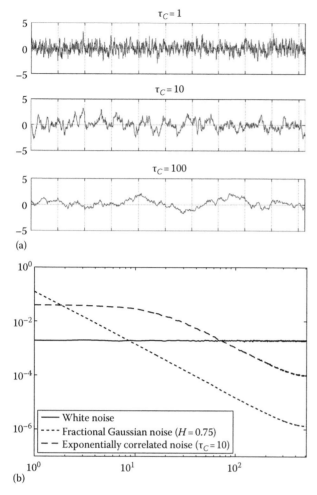

(a)

(b)

FIGURE 23.8 (a) Sample trajectories of Gaussian noise with exponentially decaying covariance. (b) Spectra averaged over 1000 realizations for three types of noise with identical variances, sampled on $N = 2^{14}$ points.

motion for $H = 0.5$ (corresponding to classical Brownian motion), $H = 0.75$, and 0.9. The corresponding increments, which are fractional Gaussian noise with different correlations, are shown in Figure 23.7a. We can observe that the regularity of the curves increases for larger values of H.

To model random process with short-range correlation we can suppose that the covariance function decays exponentially $\propto \exp(-t/\tau_c)$, with τ_c being the correlation time. The corresponding spectral density decays $\propto 4\tau_c/(1 + (f\tau_c)^2)$. Figure 23.8 shows examples for different values of τ_C (a) and different spectral densities (b). For increasing $\propto$ the apparent regularity of the trajectory increases, although the actual regularity of the underlying function remains the same.

23.3.3 Wavelets

23.3.3.1 Wavelet Analysis

The choice of the kind of wavelet transform one needs to solve a given problem is essential. Typically if the problem has to do with signal or image analysis, then the continuous wavelet

transform should be preferred. The analysis benefits from the redundancy of the continuous wavelet coeDcients, which thus allows to continuously unfold the information content into both space and scale. The best is to choose a complex-valued wavelet, e.g., the Morlet wavelet, since from the wavelet coeDcients one can directly read off the space-scale behavior of the signal and detect for instance frequency modulation laws or quasi-singularities, even if they are superimposed. For this one plots the modulus and the phase of the wavelet coeDcients in wavelet space, with a linear horizontal axis corresponding to the position x, and a logarithmic vertical axis corresponding to scale l, with the largest scale at the bottom and the smallest scale being at the top.

A classical real-valued wavelet is the Marr wavelet, also called "Mexican hat," which is the second derivative of a Gaussian,

$$\psi(x) = (1 - x^2)e^{\frac{-x^2}{2}} \tag{23.39}$$

and its Fourier transform is

$$\hat{\psi}(k) = k^2 e^{\frac{-k^2}{2}} \tag{23.40}$$

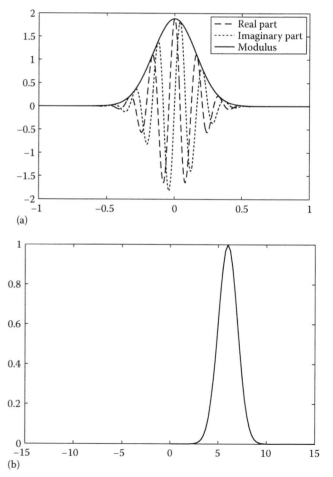

(a)

(b)

FIGURE 23.9 Localization of the Morlet wavelet in physical space (a) and in spectral space (b).

The most useful complex-valued wavelet is the Morlet wavelet (Figure 23.9):

$$\psi(x) = e^{ik_\psi x} e^{-\frac{x^2}{2}} \qquad (23.41)$$

with the wavenumber k_ψ denoting the barycenter of the wavelet support in Fourier space given by

$$k_\psi = \frac{\int_0^\infty k \,|\,\hat{\psi}(k)\,|\,dk}{\int_0^\infty |\,\hat{\psi}(k)\,|\,dk}. \qquad (23.42)$$

The wavenumber k_ψ controls the number of oscillations inside the wavelet. Actually the Morlet wavelet does not *stricto sensus* respects the admissibility condition as defined in Equation 23.19 since its mean is not zero. One should take $k_\psi > 5$ to insure that it vanishes up to the computer round-off errors. A better solution is to define the Morlet wavelet in Fourier space and enforce the admissibility condition by putting its mean, i.e., $\hat{\psi}(0)$, to zero, which gives

$$\hat{\psi}(k) = \begin{cases} e^{-\frac{(k-k_\psi)^2}{2}} & \text{for } k > 0, \\ 0 & \text{for } k \leq 0. \end{cases}$$

If the problem one would like to solve requires filtering or compressing a signal, an image or a vector field under study, then one should use the orthogonal wavelet transform to avoid the redundancy inherent to the continuous wavelet transform. In this case there is also a large collection of possible orthogonal wavelets and their choice depends on which properties one prefers, e.g., compact-support, symmetry, smoothness, number of cancellations, and computational eDciency.

From our experience, we recommend the Coifman 12 wavelet, which is compactly supported, has four vanishing moments, is quasi-symmetric, and is defined by a filter of length 12, which leads to a computational cost of the FWT in $24N$ multiplications (since two filters are needed for the wavelet and the scaling function).

To analyze fluctuating signals or fields, one should use the continuous wavelet transform with complex-valued wavelets, since the modulus of the wavelet coeDcients allows to read the evolution of the energy density in both space (or time) and scales. If one uses real-valued wavelets instead, the modulus of the wavelet coeDcients will present the same oscillations as the analyzing wavelets and it will then become diDcult to sort out features belonging to the signal or to the wavelet. In the case of complex-valued wavelets the quadrature between the real and the imaginary parts of the wavelet coeDcients eliminates these spurious oscillations and this is why we recommend to use complex-valued wavelets, such as the Morlet wavelet. If one wants to compress turbulent flows, and a fortiori to compute their evolution at a reduced cost compared to standard methods (finite difference, finite volume, or spectral methods), one should use orthogonal wavelets. In this case there is no more redundancy of the wavelet coeDcients and one has the same number of wavelet coeDcients as the number of grid points and one uses the FWT [13,19,40]. The first application of wavelets to analyze turbulent flows has been published in 1988 [18]. Since then a long-term research program has been developed for analyzing, computing, and modeling turbulent flows using either continuous wavelets or orthogonal wavelets and also wavelet packets (one can download the corresponding papers from http://wavelets.ens.fr in "Publications").

As an example we show the continuous wavelet transform, using the complex-valued Morlet wavelet, of several signals: a deterministic fractal, which is the DevilB staircase (Figure 23.10), and two self-similar random signals, which are fractional Brownian motions (FBM) having different Hurst exponent, i.e., $H = 0.25$ and $H = 0.75$ (Figure 23.11).

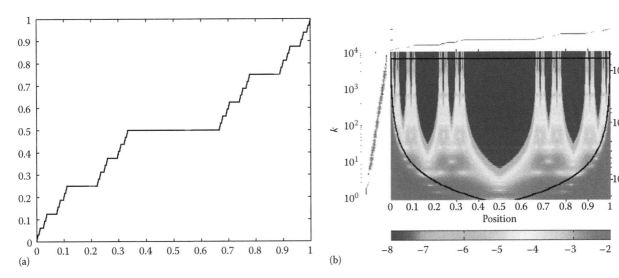

FIGURE2 3.10 DevilB staircase (a) and its continuous wavelet analysis (b).

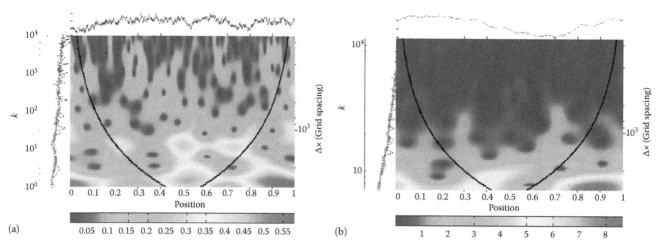

FIGURE 23.11 Continuous wavelet analysis of fractional Brownian motion with Hurst exponent $H = 0.25$ (a) and $H = 0.75$ (b).

23.3.3.2 Wavelet Spectrum

Since the wavelet transform conserves energy and preserves locality in physical space, one can use it to extend the concept of the energy spectrum and define the local energy spectrum of the function $f \in L^2(\mathbb{R})$, such that

$$\tilde{E}(k,x) = \frac{1}{C_\psi k_\psi} \left| \tilde{f} \left(\frac{k_\psi}{k}, x \right) \right|^2 \text{ for } k \geq 0, \quad (23.43)$$

where

k_ψ is the centroid wavenumber of the analyzing wavelet ψ
C_ψ is defined by the admissibility condition given in Equation 23.19

By measuring $\tilde{E}(k, x)$ at different instants or positions in the signal, one estimates which elements in the signal contribute most to the global Fourier energy spectrum that might suggest a way to decompose the signal into different components. One can split a given signal or field using the orthogonal wavelet transform into two orthogonal contributions (see Section 23.3.3.5) and then plot the energy spectrum of each to exhibit their different spectral slopes and therefore their different correlation.

Although the wavelet transform analyzes the flow using localized functions rather than complex exponentials as for Fourier transform, one can show that the global wavelet energy spectrum approximates the Fourier energy spectrum provided the analyzing wavelet has enough vanishing moments. More precisely, the global wavelet spectrum, defined by integrating Equation 23.43 over all positions,

$$\tilde{E}(k) = \int_{-\infty}^{\infty} \tilde{E}(k,x)dx \quad (23.44)$$

gives the correct exponent for a power-law Fourier energy spectrum $E(k)$ scaling as $k^{-\beta}$ if the analyzing wavelet has at least $M > \frac{\beta-1}{2}$ vanishing moments. Thus, the steeper the energy spectrum one would like to study, the more vanishing moments the analyzing wavelet should have. In practice one should choose first a wavelet with many vanishing moments and then reduce this number until the estimated slope varies. This will give the optimal wavelet to analyze the given function.

23.3.3.3 Relation to Fourier Spectrum

The wavelet energy spectrum $\tilde{E}(k)$ is related to the Fourier energy spectrum $E(k)$ via,

$$\tilde{E}(k) = \frac{1}{C_\psi k_\psi} \int_0^\infty E(k') \left| \hat{\psi} \left(\frac{k_\psi k'}{k} \right) \right|^2 dk', \quad (23.45)$$

which shows that the wavelet spectrum is a smoothed version of the Fourier spectrum, weighted with the square of the Fourier transform of the wavelet ψ shifted at wavenumbers k. For increasing k, the averaging interval becomes larger, since wavelets are filters with constant relative bandwidth, i.e., $\frac{\Delta k}{k} =$ constant. The wavelet energy spectrum thus yields a stabilized Fourier energy spectrum.

Considering, for example, the Marr wavelet given in Equation 23.39, which is real-valued and has two vanishing moments only, the wavelet spectrum can estimate exponents of the energy spectrum for $\beta < 5$. In the case of the complex-valued Morlet wavelet given in Equation 23.43, only the zeroth-order moment is vanishing. However higher mth-order moments are very small $\left(\propto k_\psi^m e^{(-k_\psi^2/2)} \right)$, provided that k_ψ is suﬃciently large. For instance choosing $k_\psi = 6$ yields accurate estimates of the exponent of power-law energy spectra for at least $\beta < 7$.

There also exists a family of wavelets with an infinite number of vanishing moments

$$\hat{\psi}_n(k) = \alpha_n \exp\left(-\frac{1}{2} \left(k^2 + \frac{1}{k^{2n}} \right) \right), \quad n \geq 1, \quad (23.46)$$

where α_n is a normalization factor. Wavelet spectra using this wavelet can thus correctly measure any power–law energy spectrum. This choice enables, in particular, the detection of the difference between a power–law energy spectrum and a Gaussian energy spectrum such that $E(k) \propto e^{-(k/k_0)^2}$. This is important in turbulence to determine the wavenumber after which the energy spectrum decays exponentially. The end of the inertial range, dominated by nonlinear interactions, and the beginning of the dissipative range, dominated by linear dissipation, can thus be detected.

23.3.3.4 Relation to Structure Functions

Structure functions, classically used to analyze nonstationary random processes, e.g., turbulent velocity fluctuations, have some limitations that can be overcome using wavelet-based alternatives. Structure functions are defined by moments of increments of the random process. The latter can be interpreted as wavelet coeDcients using a special wavelet, the difference of two Diracs (called DoD wavelet), which is very singular and has only one vanishing moment, namely its mean value. This unique vanishing moment of the DoD wavelet limits the adequacy of structure functions to analyze suDciently smooth signals. Wavelets having more vanishing moments do not have this drawback.

For second-order statistics, the classical energy spectrum, defined as the Fourier transform of the autocorrelation function, is naturally linked to the second-order structure function. Using the earlier relation of the wavelet spectrum to the Fourier spectrum, a similar relation to second-order structure functions can be derived. For structure functions yielding a power-law behavior the maximum exponent can be shown to be limited by the number of vanishing moments of the underlying wavelet.

The increments of a function $f \in L^2(\mathbb{R})$ are equivalent to its wavelet coeDcients using the DoD wavelet

$$\psi^\delta(x) = \delta(x+1) - \delta(x).\tag{23.47}$$

We thus obtain

$$f(x+a) - f(x) = \tilde{f}_{x,a} = \langle f, \psi^\delta_{x,a} \rangle,\tag{23.48}$$

with $\psi^\delta_{x,a}(y) = 1/a\left[\delta(\frac{y-x}{a+1}) - \delta(\frac{y-x}{a})\right]$, where the wavelet is normalized with respect to the L^1-norm. The pth-order moment of the wavelet coeDcients at scale a yields the pth-order structure function:

$$S_p(a) = \int \left(\tilde{f}_{x,a}\right)^p dx.\tag{23.49}$$

As already mentioned earlier the drawback of the DoD wavelet is that it has only one vanishing moment, its mean. Consequently the exponent of the pth-order structure function in the case of a power-law behavior is limited by p, i.e., if $S_p(a) \propto a^{\zeta(p)}$ then $\zeta(p) < p$. The detection of larger exponents necessitates

the use of increments with a larger stencil, or wavelets with more vanishing moments.

We now focus on second-order statistics, the case $p = 2$. Equation 23.45 yields a relation between the global wavelet spectrum $\tilde{E}(k)$ and the Fourier spectrum $E(k)$ for a given wavelet ψ. Taking the Fourier transform of the DoD wavelet, we get $\hat{\psi}^\delta(k) = e^{\imath k} - 1 = e^{\frac{\imath k}{2}}(e^{\frac{\imath k}{2}} - e^{-\imath k2})$ and therefore we have $|\hat{\psi}^\delta(k)|^2 = 2(1 - \cos k)$. The relation between the Fourier and the wavelet spectrum thus becomes

$$\tilde{E}(k) = \frac{1}{C_\psi k} \int_0^\infty E(k')\left(2 - 2\cos\left(\frac{k_\psi k'}{k}\right)\right) dk',\tag{23.50}$$

and the wavelet spectrum can be related to the second-order structure function by setting $a = k_\psi/k$

$$\tilde{E}(k) = \frac{1}{C_\psi k} S_2(a).\tag{23.51}$$

Using now the result of Section 23.3.3.2 that for a Fourier spectrum that behaves like $k^{-\alpha}$ for $k \to y$, the wavelet spectrum only yields $\tilde{E}(k) \propto k^{-\alpha}$ if $\alpha < 2M + 1$, where M denotes the number of vanishing moments of the wavelet, we find for the structure function $S_2(a)$ that $S_2(a) \propto a^{\zeta(p)} = \left(\frac{k_\psi}{k}\right)^{\zeta(p)}$ for $a \to 0$ if $\zeta(2) \le 2M$.

For the DoD wavelet we have $M = 1$, which explains why the second-order structure function can only detect slopes smaller than 2, which corresponds to wavelet energy spectra with slopes being shallower than -3. This explains why the usual structure function gives spurious results for suDciently smooth signals.

23.3.3.5 Detection and Characterization of Singularities

The possibility to evaluate the slope of the energy spectrum is an important property of the wavelet transform, related to its ability to characterize the regularity of the signal and detect isolated singularities [27,31]. This is based on the fact that the local scaling of the wavelet coeDcients is computed in L^1-norm, i.e., with the normalization $c(l) = l^{-1}$ instead of $c(l) = l^{1/2}$ in Equation 23.21.

If the function $f \in C^m(x_0)$, i.e., if f is continuously differentiable in x_0 up to order m, then

$$\left[\tilde{f}(l,x_0)\right]_{l \to 0} \le l^{m+1/2}l^{1/2} = l^m,\tag{23.52}$$

The factor $l^{1/2}$ comes from the fact that to study the scaling in x_0 of the function f we compute its wavelet coeDcients in L^1-norm, instead of L^2, i.e., with the normalization $c(l) = l^{-1}$ instead of $c(l) = l^{1/2}$ in Equation 23.21. More generally if f has Hölder regularity α at x_0 (see Section 23.2.1.3), then

$$\left[\tilde{f}(l,x_0)\right]_{l \to 0} \approx Ce^{\imath \Phi}l^{1/2}\tag{23.53}$$

where Φ is the phase of the wavelet coeDcients in x_0. The phases of the wavelet coeDcients $\Phi(l, x)$ in wavelet coefficient space allow to localize the possible singularities of f since the lines of constant phase converge toward the locations of all the isolated singularities when $l \to 0$. If the function f presents few isolated singularities, their position x_0, their strength C, and their scaling exponent α can thus be estimated by the asymptotic behavior of $\tilde{f}(l, x_0)$, written in L^1-norm, in the limit l tending to zero. If, on the contrary, the modulus of the wavelet coeDcient becomes zero at small scale around x_0, then the function f is regular at x_0. This result is the converse of Equation 23.52, but it only works for isolated singularities since it requires that in the vicinity of x_0 the wavelet coeDcients remain smaller than those pointing toward x_0. Consequently its use is not applicable to signals presenting dense singularities. The scaling properties presented in this paragraph are independent of the choice of the analyzing wavelet ψ. Actually we recommend to use complex-valued wavelets, since one thus obtains complex-valued wavelet coeDcients whose phases locate the singularities while their moduli estimate the Hölder exponents of all isolated singularities, as illustrated in Figure 23.10. We can then compute the singularity spectrum (see Section 23.2.1.3).

23.3.3.6 Intermittency Measures

Localized bursts of high-frequency activity define typically intermittent behavior. Localization in both physical space and spectral space is thus implied, and a suitable basis for representing intermittency should reflect this dual localization. The Fourier representation yields perfect localization in spectral space but global support in physical space. Filtering a fluctuating signal with an ideal high-pass Fourier filter implies some loss of spatial information in physical space. Strong gradients are smoothed out and spurious oscillations occur in the background. This comes from the fact that the modulus and phase of the discarded high-wavenumber Fourier modes have been lost. The artifacts of Fourier filtering lead to errors in estimating the flatness and hence the signal's intermittency.

An intermittent quantity (e.g., velocity derivative) contains rare but strong events (i.e., bursts of intense activity), which correspond to large deviations reflected in "heavy tails" of the probability distribution function of that quantity. Second-order statistics (e.g., energy spectrum, second-order structure function) are not very sensitive to such rare events whose spatial support is too small to play a role in the integral. For higher-order statistics, however, these rare events become increasingly important, may eventually dominate, and thus allow to detect intermittency. Of course, not for all problems intermittency is essential, e.g. second-order statistics are suDcient to measure dispersion (dominated by energy-containing scales), but not to calculate drag or mixing (dominated by vorticity production in thin boundary or shear layers).

Using the continuous wavelet transform we have proposed the local intermittency measure [19,50], which corresponds to

the wavelet coeDcients renormalized by the space-averaged energy at each scale, such that

$$I(l, \vec{x}) = \frac{|\tilde{f}(l, \vec{x})|^2}{\int_{-\infty}^{\infty} |\tilde{f}(l, \vec{x})|^2 \, d^2\vec{x}}. \tag{23.54}$$

It yields information on the spatial variance of energy as a function of scale and position. For regions where $I(l, \vec{x}) \approx 1$ the field is nonintermittent, while regions of larger values are intermittent.

Similar to the continuous wavelet transform the orthogonal wavelet transform allows to define intermittency measures, either local as shown earlier or global as illustrated in the following text. The space-scale information contained in the wavelet coefficients yields suitable global intermittency measures using scale-dependent moments and moment ratios [51]. For a signal f the moments of wavelet coeDcients at different scales j are defined by

$$M_{p,j}(f) = 2^{-j} \sum_{i=0}^{2^j-1} \left(\tilde{f}_{j,i}\right)^p. \tag{23.55}$$

The scale distribution of energy, i.e., the scalogram, is obtained from the second-order moment of the orthogonal wavelet coeDcients: $E_j = 2^{j-1} M_{2,j}$. The total energy is then recovered by the sum: $E = \sum_{j \geq 0} E_j$ thanks to the orthogonality of the decomposition.

Ratios of moments at different scales quantify the sparsity of the wavelet coeDcients at each scale and thus measure the intermittency:

$$Q_{p,q,j}(f) = \frac{M_{p,j}(f)}{(M_{q,j}(f))^{\frac{p}{q}}}, \tag{23.56}$$

which correspond to quotient of norms computed in two different sequence spaces, l^p- and l^q-spaces. Typically, one chooses $q = 2$ to define statistical quantities as a function of scale. For $p = 4$ we obtain the scale-dependent flatness $F_j = Q_{4,2,j}$, which equals 3 for a Gaussian white noise at all scales j and indicates that a signal is not intermittent. Scale-dependent skewness, hyperflatness, and hyperskewness are defined for $p = 3, 5$, and 6, respectively. Intermittency of a signal is reflected in increasing $Q_{p,q,j}$ for increasing j (smaller scale) supposing $p > q$.

23.3.3.7 Extraction of Coherent Structures

To study fluctuating signals or fields, we need to separate the rare and extreme events from the dense events and then calculate their statistics independently for each one. For this we cannot use pattern recognition methods since there is no simple patterns to characterize them. Moreover there is no clear scale separation between the rare and the dense events and therefore a Fourier filter cannot disentangle them. Since the rare events are well localized in physical space, one might try to use an on–off filter defined in physical space to extract them. However, this approach changes the spectral properties by introducing spurious discontinuities, adding an artificial scaling (e.g., k^{-2} in one dimension) to the energy spectrum. The wavelet representation can overcome these problems since it combines both physical and spectral localizations (bounded from below by the uncertainty principle).

We have proposed in 1999 [20] a better approach to extract rare events out of fluctuating signals or fields, which is based on the orthogonal wavelet representation. We rely on the fact that rare events are localized while dense events are not and we assume that the later are noise-like. From a mathematical viewpoint a noise cannot be compressed in any functional basis. Another way to say this is to observe that the shortest description of a noise is the noise itself. Note that one often calls "noise" what actually is "experimental noise," i.e., something that one would like to discard, although it may not be noise-like in the earlier mathematical sense. The problem of extracting the rare events has thus become the problem of denoising the signal or the field under study. Assuming that they are what remains after denoising, we need a model, not for the rare events, but for the noise. As a first guess, we choose the simplest model and suppose the noise to be additive, Gaussian and white, i.e., uncorrelated.

We now describe the wavelet algorithm for extracting coherent structures out of a signal corrupted by a Gaussian noise with variance σ^2 and vanishing mean, sampled on N equidistant grid points. The noisy signal $f(x)$ is projected onto orthogonal wavelets using Equation 23.31 to get $\tilde{f}_\lambda$. Its wavelet coeDcients are then split into two sets, those whose modulus is larger than a threshold ε that we call "coherent," and those remaining that we call "incoherent." The threshold value, based on minmax statistical estimation [14], is $\varepsilon = (2/d\sigma^2 \ln N)^{1/2}$. Note that besides the choice of the wavelet there is no adjustable parameter since σ^2 and N are known a priori. In case the variance of the noise is unknown, one estimates it recursively from the variance of the incoherent wavelet coefficients, as proposed in [3]. The convergence rate increases with the signal-to-noise ratio, namely if there is only noise it converges in zero iteration. The coherent signal f_C is reconstructed from the wavelet coeDcients whose modulus is larger than ε and the incoherent signal f_I from the remaining wavelet coeDcients. The two signals thus obtained, f_C and f_I, are orthogonal.

To illustrate the method we choose an academic signal (Figure 23.12a), which is a superposition of several quasi-singularities having different Hölder exponents, to which we have superimposed a Gaussian white noise yielding a signal-to-noise ratio of 11.04 dB (Figure 23.12b). Applying the extraction method we recover a denoised version of the corrupted signal, which preserves the quasi-singularities (Figure 23.12c). It could be checked a posteriori that the incoherent contribution is spread, and therefore does not compress and has a Gaussian probability distribution.

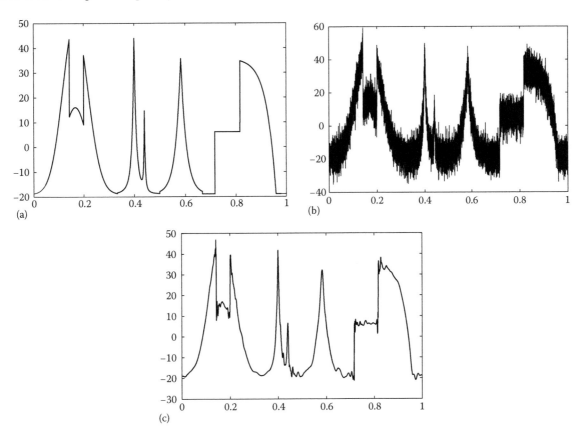

FIGURE 23.12 Academic example of denoising of a piecewise regular signal using the algorithm for coherent structure extraction. Original signal (a), same signal plus Gaussian white noise giving a signal to noise ratio (SNR) of 11.04 dB (b), denoised signal with SNR of 27.55 dB (c).

23.4 Recommendations

In the introduction we stated cautious remarks about the risk of misusing new mathematical tools, if one has not first gained enough practice on academic examples. The problem is the following. When doing research, the questions one addresses are still open and there exist several competing theories, models, and interpretations. Nothing being clearly fixed yet, neither the comprehension of the physical phenomenon under study, nor the practice of the new techniques, one runs the risk to perform a Rorschach's test rather than a rational analysis. Indeed, the interpretation of the results may reveal one's unconscious desire for a preferred explanation. Although it is a good thing to rely on one's intuition and have a preferred theory, one should be conscious of that risk and make sure to avoid bias. Moreover, when a new technique is proposed, most of referees do not master it yet and therefore are not able to detect flaws in a submitted paper.

Let us take as example the case of turbulence, which has applications in everyday life and plays an important role in environmental fluid dynamics. For centuries, turbulence has been an open problem and thus a test ground for new mathematical techniques. Let us focus here on the use of fractals and wavelets, as they were applied to study turbulence. Kolmogorov's statistical theory of homogeneous and isotropic turbulence [34] assumes that there exists an energy cascade from large to small scales, which is modeled as a self-similar stochastic process whose spectrum scales as $k^{-5/3}$, where k is the wavenumber. Although this prediction only holds for an ensemble average of many flow realizations, many authors interpret the energy cascade as caused by the successive breakings of whirls into smaller and smaller ones, as if they were stones. This interpretation was inspired by a comment Lewis Fry Richardson made in 1922:

> When making a drawing of a rising cumulus from a fixed point, the details change before the sketch was completed. We realize thus that: big whirls have little whirls that feed on their velocity, and little whirls have lesser whirls and so on to viscosity–in the molecular sense [49].

We think that Richardson's quote has been misunderstood and turbulence misinterpreted. Indeed, his remark concerns the interface between a cumulus cloud and the surrounding clear air, which is a very convoluted two-dimensional surface developing into a three-dimensional volume. Such an interface may develop into a fractal since its topological dimension is lower than the dimension of the space that contains it. But keeping such a fractal picture to describe three-dimensional whirls that evolve inside a three-dimensional space does not make sense since both have the same topological dimension. In 1974, Kraichnan was already suspicious about this interpretation, when he wrote:

> The terms ~~scale of motion~~ or ~~eddy of size l~~ appear repeatedly in the treatment of the inertial range. One gets an impression of little, randomly distributed whirls in the fluid, with the fission of the whirls into smaller ones, after the fashion of Richardson's poem. This picture seems to be drastically in conflict with what can be inferred about the qualitative structures of high Reynolds numbers turbulence from laboratory visualization techniques and from plausible application of the Kelvin's circulation theorem [35].

Unfortunately Kraichnan's viewpoint was not taken into account and, on the contrary, the picture of breaking whirls was even reinforced by the terminology fractals due to its Latin root *fractare* (to break). This gave rise to numerous models of turbulence, which were based on fractals and later on multifractals (for a review of them see [22]).

Let us now consider the use of wavelets to analyze turbulent flows and illustrate the risk of misinterpretation there too. If one performs the continuous wavelet analysis of any fluctuating signals, for example, the temporal fluctuations of one velocity component of a three-dimensional turbulent flow, one should be very cautious, especially when using a real-valued wavelet. Indeed, for this class of noise-like signals one observes a tree-like pattern in the two-dimensional plot of their wavelet

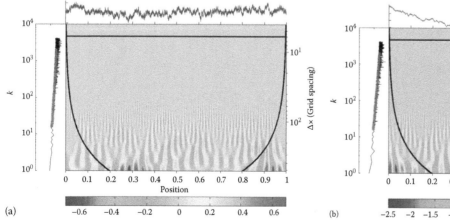
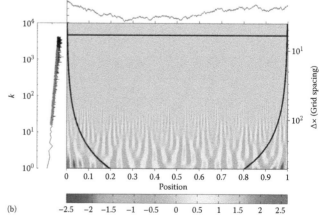

FIGURE 23.13 **(See color insert.)** Real part of the continuous wavelet analysis of fractional Brownian motion with Hurst exponent $H = 0.25$ (a) and of classical Brownian motion (b).

coeDcients, which is generic to the continuous wavelet transform and corresponds to its reproducing kernel [19]. When one performs the continuous wavelet transform of one realization of a Gaussian white noise, one observes such a pattern (see Figure 23.13), which proves that the correlation is among the wavelets but not within the signal itself. Unfortunately, in the case of turbulent signals this pattern has been misinterpreted as the evidence of whirls breaking in a paper published by *Nature* in 1989 under the title "Wavelet analysis reveals the multifractal nature of the Richardsonß cascade" [2]. If the authors had used an orthogonal wavelet transform instead of a real-valued continuous wavelet transform they would not have observed and correlation.

Later Benoît Mandelbrot concludes: "In the domain I know of, there are many words which are meaningless, that do not have any content, which have been created just to impress, to give the feeling that a domain exists when actually there is none. If one gives a name to a science, this science maybe does not exist. And, once more, due to the fierce discipline I was imposing to myself, I avoided that [k]. Therefore I have created the word -fractalfl with much reflection. The idea was that of objects which are dispersed, which are broken into small pieces"[43]. The question remains for us: are fractals a new science or only consist of refurbishing older concepts to launch a new fashion? In the same vein, Yves Meyer wrote:

"Wavelets are fashionable and therefore excite curiosity and irritation. It is amazing that wavelets have appeared, almost simultaneously in the beginning of the 1980s, as an alternative to traditional Fourier analysis, in domains as diverse as speech analysis and synthesis, signal coding for telecommunications, (low-level) information, extraction process performed by the retinian system, fully-developed turbulence analysis, renormalization in quantum field theory, functional spaces interpolation theoryk But this pretention for pluridisciplinarity can only be irritating, as are all "great syntheses" which allow one to understand and explain everything. Will wavelets soon join "catastrophe theory" or "fractals" in the bazaar of all-purpose systems?" [46]

Let the future tell us the answerk

Acknowledgments

We are very thankful to Barbara Burke, John Hubbard, and Rodrigo Pereira for useful comments. We thank CEMRACS 2010 (Centre dfEté de Recherche Avancée en Calcul Scientifique) and CIRM (Centre International de Rencontres Mathématiques), Marseille, France, for their hospitality while writing this chapter. M.F., R.N.V.Y. and O.P. acknowledge financial support from ANR (Agence Nationale de la Recherche) under the grant Geo-FLUIDS (ANR- 09-SYSC-005-01). M.F. is grateful to the Wissenschaftskolleg zu Berlin for its hospitality while writing this chapter. M.F., K.S., and R.N.V.Y. acknowledge financial support from the PEPS program of INSMI-CNRS. We also thank the Association CEA-EURATOM and the FR-FCM (Fédération de Recherche Fusion par Confinement Magnétique—ITER) for supporting our work within the framework of the EFDA (European Fusion Development Agreement) under contract V.3258.001. The views and opinions expressed herein do not necessarily reflect those of the European Commission.

References

1. J. P. Antoine, R. Murenzi, P. Vandergheynst, and S. T. Ali, 2004. *Two-Dimensional Wavelets and Their Relatives*. Cambridge University Press, Cambridge, U.K.
2. F. Argoul, A. Arnéodo, G. Grasseau, Y. Gagne, E. J. Hopfinger, and U. Frisch, 1989. Wavelet analysis of turbulence reveals the multifractal nature of the Richardson cascade. *Nature*, **8** , 51–53.
3. A. Azzalini, M. Farge, and K. Schneider, 2005. Nonlinear wavelet thresholding: A recursive method to determine the optimal denoising threshold. *Appl. Comput. Harm. Analysis*, **8** (2), 177–185.
4. L. Bachelier, 1900. *Théorie de la Spéculation*. Gauthier-Villars, Paris, France.
5. R. Balian, 1981. Un principe dfincertitude en théorie du signal ou en mécanique quantique. *C. R. Acad. Sci. Paris*, **9** , série 2, 1357–1361.
6. M. Barnsley, 1988. *Fractals Everywhere*. Academic Press, Boston, MA.
7. B. Burke Hubbard, 1998. *The World According to Wavelets*. A. K. Peters, Wellesley, MA.
8. R. R. Coifman and Y. Meyer, 1991. Remarques sur lßanalyse de Fourier à fenêtre. *C. R. Acad. Sci. Paris*, **312**, série I, 259–261.
9. R. R. Coifman and V. M. Wickerhauser, 1992, Entropy based algorithms for best basis selection. *IEEE Trans. Inf. Theory*, **3** , 712–718.
10. R. L. Cook and T. DeRose, 2005. Wavelet noise. *ACM Trans. Graph.*, **2** (3), 803–811.
11. I. Daubechies, A. Grossmann and Y. Meyer, 1986. Painless nonorthogonal expansions. *J. Math. Phys.*, **2** , 1271–1283.
12. I. Daubechies, 1988. Orthonormal bases of compactly supported wavelets. *Commun. Pure Appl. Math.*, **4** (7), 909–996
13. I. Daubechies, 1992. *Ten Lectures on Wavelets*. SIAM, Philadelphia, PA.
14. D. Donoho and I. Johnstone, 1994. Ideal spatial adaptation via wavelet shrinkage. *Biometrika*, **8** (3), 425–455.
15. A. Douady and J. H. Hubbard, 1982. Iterations des polynomes quadratiques complexes, *C. R. Acad. Sci. Paris*, **9** , 123–126.
16. F. W. Elliott and A. J. Majda, 1994. A wavelet Monte-Carlo method for turbulent diffusion with many spatial scales. *J.ßComput. Phys.*, **3** (1), 82–111.
17. F. W. Elliott and A. J. Majda, 1995. A new algorithm with plane waves and wavelets for random velocity fields with many spatial scales. *J. Comput. Phys.*, **7** (1), 146–162.

18. M. Farge and G. Rabreau, 1988. Transformée en ondelettes pour détecter et analyser les structures cohérentes dans les écoulements turbulents bidimensionnels. *C. R. Acad. Sci. Paris*, 2(307), 1479–1486.

19. M. Farge, 1992. Wavelet transforms and their applications to turbulence. *Ann. Rev. Fluid Mech.*, 2 , 395–457.

20. M. Farge, K. Schneider and N. Kevlahan, 1999. Non-Gaussianity and coherent vortex simulation for two-dimensional turbulence using an orthonormal wavelet basis. *Phys. Fluids*, 1 (8), 2187–2201.

21. M. Farge, A. Grossmann, Y. Meyer, T. Paul, J.-C. Risset, G. Saracco, and B. Torrésani, 2012. Les ondelettes et le CIRM. *Gazette Math.*, 3 , 47–57.

22. U. Frisch, 1995. *Turbulence: The Legacy of A. N. Kolmogorov*. Cambridge University Press, Cambridge, U.K.

23. D. Gabor, 1946. Theory of Communication. *J. Inst. Electr. Eng.*, 9 (3), 429–457.

24. J.-F. Gouyet, 1996. *Physics and Fractal Structures*. Springer-Verlag, Berlin, Germany.

25. A. Grossmann and J. Morlet, 1984. Decomposition of Hardy functions into square integrable wavelets of constant shape. *SIAM J. Appl. Anal.*, 5 (4), 723–736.

26. A. Haar, 1910. Zur Theorie der orthogonalen Funktionensysteme. *Math. Ann.*, 69(3), 331–371.

27. M. Holschneider, 1988. On the wavelet transform of fractal objects. *J. Stat. Phys.*, 8 (5/6), 963–996.

28. Interview de John Hubbard, *Science Publique, France-Culture*, 19 October 2010.

29. G. A. Hunt, 1951. Random Fourier transforms. *Trans. Am. Math. Soc.*, 7 , 38–69.

30. H. E. Hurst, 1951. Long-term storage capacity of reservoirs. *Trans. Am. Soc. Civil Eng.*, 6 , 770–808.

31. S. Jaffard, 1989. Construction of wavelets on open sets. *First International Conference on Wavelets, Marseille*, 14–18 December 1987 (eds. J.M. Combes, A. Grossmann and P. Tchamitchian), Springer, pp. 247–252.

32. H. von Koch, 1906. Sur une courbe continue sans tangente, obtenue par une construction géomètrique élémentaire pour l'étude de certaines questions de la théorie des courbes planes. *Acta Math.*, 8 , 145–174.

33. A. N. Kolmogorov, 1940. Wienersche Spiralen und einige andere interessante Kurven im Hilbertschen Raum. *C. R. Dokl. Acad. Sci. SSSR*, 8 , 115–118.

34. A. N. Kolmogorov, 1941. The local structure of turbulence in incompressible viscous fluid for very large Reynolds numbers. *Proc. USSR Acad. Sci.*, 8 , 299–303 (Russian), translation in *Proc. Roy. Soc. Ser. A Math. Phys. Sci.*, 4 , 9–13 (1991).

35. R. H. Kraichnan, 1974. On Kolmogorov's inertial-range theories. *J. Fluid Mech.*, 8 , 305–330.

36. P. G. Lemarié and Y. Meyer, 1986. Ondelettes et bases Hilbertiennes. *Rev. Mat. Ibero Am.*, 2, 1–18.

37. P. Lévy, 1954. *Le mouvement Brownien*. Fascicule CXXVI of Mémorial des Sciences Mathématiques. Gauthier-Villars, Paris, France.

38. P. Lévy, 1965. *Processus stochastique et mouvement Brownien*. Fascicule CXXVI of Mémorial des sciences mathématiques. Gauthier-Villars, Paris, France.

39. S. Mallat, 1989. Multiresolution approximations and wavelet orthonormal bases of $L^2(\mathbb{R})$. *Trans. Am. Math. Soc.*, 5 , 69–87.

40. S. Mallat, 1998. *A Wavelet Tour of Signal Processing*. Academic Press.

41. H. Malvar, 1990. Lapped transforms for multiresolution signal decomposition: the wavelet decomposition. *IEEE Trans. Pattern Anal. Mach. Intell.*, 1 , 674–693.

42. B. B. Mandelbrot and J. W. van Ness, 1968. Fractional Brownian motions, fractional noises and applications. *SIAM Rev.*, 0 (4), 422–437.

43. Interview de Benoît Mandelbrot. *A voix nue, France-Culture*, October 1990.

44. B. B. Mandelbrot, 1975. *Les Objets Fractals*. Flammarion, Paris, France.

45. B. B. Mandelbrot, 1977. *Fractals: Form, Chance and Dimension*. Freeman, San Fransisco, CA.

46. Y. Meyer, 1990. Ondelettes et applications. *J. Ann. Soc. Math. Société Mathématique de France (SMF)*, Paris, 1–15.

47. R. Murenzi, 1990. Ondelettes multidimensionnelles et application à l'analyse d'images. Thèse de Doctorat de l'Université Catholique de Louvain, Louvain-la-Neuve, Belgium.

48. G. Parisi and U. Frisch, 1985. On the singularity structure of fully developed turbulence. *Turbulence and Predictability in Geophysical Fluid Dynamics* (eds. M. Ghil, R. Benzi and G. Parisi), pp. 84–87, North-Holland, Amsterdam, the Netherlands.

49. L. F. Richardson, 1922. *Weather Prediction by Numerical Process*. Cambridge University Press, Cambridge, U.K.

50. J. Ruppert-Felsot, M. Farge and P. Petitjeans, 2009. Wavelet tools to study intermittency: Application to vortex bursting. *J. Fluid Mech.*, 6 , 427–453.

51. K. Schneider, M. Farge and N. Kevlahan, 2004. Spatial intermittency in two-dimensional turbulence. *Woods Hole Mathematics, Perspectives in Mathematics and Physics* (eds. N. Tongring and R.C. Penner), World Scientific, Hackensack, NJ, pp. 302–328.

52. K. Schneider and O. Vasilyev, 2.010. Wavelet methods in computational fluid dynamics. *Annu. Rev. Fluid Mech.*, 2 , 473–503.

53. P. Tafti and M. Unser, 2010. Fractional Brownian vector fields. *SIAM Multiscale Model. Sim.*, 8(5), 1645–1670.

24

Dispersion Modeling

Akula Venkatram
University of California, Riverside

24.1 Introduction

Air pollution models play an important role in the implementation of air pollution regulations. For example, before an industrial plant can be constructed, its impact on air quality is determined through an air pollution model to show that emissions from the plant do not lead to ambient concentrations above the regulated level. In the United States, the AMS/EPA Regulatory Model (AERMOD, Cimorelli et al., 2005) is used to make such permitting decisions. U.S. regulations that govern air toxics recommend the use of AERMOD to quantify risk associated with emissions of toxic chemicals in urban areas. Air pollution models that include chemistry are used to make decisions to control emissions that are precursors of ozone and acidifying pollutants. Such decisions can have multimillion-dollar implications associated with installing equipment to reduce emissions, or delaying, or even disallowing the construction of the industry responsible for the emissions.

This chapter examines commonly used air pollution models applicable to scales of the order of tens of kilometers. The effects of chemistry are assumed to be negligible at these scales, although this might not be always true. The models described in this chapter are designed to estimate concentrations averaged over an hour. They cannot be used to estimate instantaneous concentrations, which are relevant to odor (see Weil in this handbook for more on instantaneous concentrations). The chapter also

provides the background necessary to understand the approach used in the formulation of dispersion models. This includes the essentials of the micrometeorology used to construct the inputs for the model.

24.2 Point Source in the Atmospheric Boundary Layer

Most short-range dispersion models are based on the assumption that meteorological conditions are spatially homogeneous and vary little with time during the period of interest, which is typically 1 h. This is equivalent to saying that the timescale governing the variation in meteorology is greater than the time of travel between source and receptor. If the meteorological timescale is 1 h and the wind speed is 5 m/s, the assumption of steady state is not likely to be valid for distances much greater than 10 km. At lower wind speeds, the "valid" distances become smaller. In spite of these limitations, steady-state plume models are often applied beyond their range of applicability with the justification that the concentration at the receptor is representative of that when the plume eventually reaches the receptor. In principle, dispersion during unsteady and spatially varying conditions can be treated with puff or particle models, which attempt to model the dispersion of puffs or particles as the unsteady wind field

Handbook of Environmental Fluid Dynamics, Volume Two, edited by Harindra Joseph Shermal Fernando. © 2013 CRC Press/Taylor & Francis Group, LLC. ISBN: 978-1-4665-5601-0.

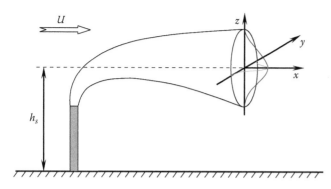

FIGURE 24.1 Gaussian distribution used to model plume from point source. For the time being, we have ignored the effects of the impermeable ground on the concentration field.

carries them along their trajectories. This chapter will not discuss models based on puff dispersion.

Models such as AERMOD are based on the steady-state Gaussian dispersion equation. If the release point is taken to be the origin ($z = 0$), with the x-axis of the coordinate system aligned along the wind direction at the source, empirical evidence indicates that the time-averaged (typically 1 h) concentration field can be described in terms of the Gaussian distribution (see Figure 24.1):

$$C(x,y,z) = \frac{Q}{2\pi\sigma_y\sigma_z U} exp\left[-\frac{z^2}{2\sigma_z^2} - \frac{y^2}{2\sigma_y^2}\right] \quad (24.1)$$

where

  y is the crosswind coordinate
  Q is the source strength (mass/time)
  U is the time-averaged wind speed at source height
  σ_y and σ_z are the plume spreads normal to the mean wind direction

Equation 24.1 assumes that along-wind dispersion is much smaller than transport by the mean wind. This assumption breaks down when the mean wind is comparable to the turbulent velocity along the wind, σ_u. The form of the dispersion model under such low wind speed conditions is discussed in a later section.

The effect of the ground on concentrations is accounted for by making sure that there is no flux of material through the ground, which we now take to be $z = 0$. The mathematical trick to achieve this is to place an "image" source at a distance $z = -h_e$, where h_e is the effective height of the source aboveground. The upward flux from this image source essentially cancels out the downward flux from the real source without affecting the mass balance. Then, the concentration becomes

$$C(x,y,z) = \frac{Q}{2\pi\sigma_y\sigma_z U} exp\left[-\frac{y^2}{2\sigma_y^2}\right]$$

$$\times \left\{ exp\left[-\frac{(z-h_e)^2}{2\sigma_z^2}\right] + exp\left[-\frac{(z+h_e)^2}{2\sigma_z^2}\right]\right\} \quad (24.2)$$

In the real atmosphere, dispersion in the upward direction is limited by the height of the atmospheric boundary layer. This limitation of vertical mixing is incorporated into the Gaussian formulation by "reflecting" material off the top of the mixed layer. Then, Equation 24.2 can be modified to account for the infinite set of "reflections" from the ground and the top of the mixed layer (see Csanady, 1973). When the pollutant is well mixed through the depth of the boundary layer, z_i the expression for the concentration becomes

$$C(x,y) = \frac{Q}{\sqrt{2\pi}\sigma_y z_i U} exp\left(-\frac{y^2}{2\sigma_y^2}\right) \quad (24.3)$$

Equation 24.2 can be modified to account for dry deposition of material at the surface using approaches described in Horst (1984).

The Gaussian formulation for a point source can be used to model volume, area, and line sources by discretizing them into point sources; the associated concentrations are simply the sums of the contributions from these point sources. We can illustrate this approach by applying Equation 24.2 to model a line source.

24.3 Line Source Model

The line source is a basic building block of air pollution models. It is generally used to model emissions from roads, the major source of air pollution in urban areas. In addition, an area source of pollution can be represented as a set of line sources to facilitate the computation of its impact.

The concentration at a receptor can be calculated by representing the line source as a set of point sources. Figure 24.2 illustrates the coordinate systems used to derive the expression for concentration. The finite line source, Y_1Y_2, lies along the Y axis of the fixed X–Y coordinate system. If the length of the line source is L, the ordinates of the ends of the source are Y_1 and $Y_2 = Y_1 + L$.

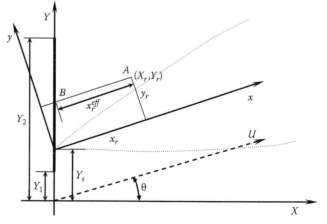

FIGURE 24.2 Coordinate systems used to calculate contribution of point source at Y_s to concentration at (X_r, Y_r). The system $xt\,y$ has the x-axis along the mean wind direction, which is at an angle θ to the fixed X-axis.

The wind blows at an angle θ to the normal to the line source. We want to calculate the concentration at a receptor (X_r, Y_r) caused by emissions from the line source with an emission rate of q (mass/(time × length)). We first represent the line source by a set of point sources of strength $qd\,Y$. To compute the impact of a point source at Y_s, we use a new coordinate system, x–y, with its origin at $(0, Y_s)$ and the x-axis along the direction of the mean wind; the x-axis is rotated by an angle θ relative to the fixed X-axis. In this coordinate system, the coordinates of the receptor (x_r, y_r) are given by

$$x_r = X_r \cos\theta + (Y_r - Y_s)\sin\theta$$
$$y_r = (Y_r - Y_s)\cos\theta - X_r \sin\theta \qquad (24.4)$$

Then, the contribution of the elemental point source to the concentration at (X_r, Y_r) is given by

$$dC = \frac{qd\,Y_s}{\pi U \sigma_y(x_r)\sigma_z(x_r)} exp\left(-\frac{y_r^2}{2\sigma_y^2(x_r)}\right), \qquad (24.5)$$

assuming that the concentration is given by the Gaussian model and the source and the receptor are at ground level. The following analysis can be modified for an elevated source or receptor. Then, the concentration due to the point sources along the Y axis from Y_1 to $Y_2 = Y_1 + L$ becomes

$$C(X_r, Y_r) = \int_{Y_1}^{Y_1+L} \frac{qd\,Y_s}{\pi U \sigma_y(x_r)\sigma_z(x_r)} exp\left(-\frac{y_r^2}{2\sigma_y^2(x_r)}\right). \qquad (24.6)$$

Equation 24.6 requires numerical integration because both σ_y and σ_z depend on x_r, which in turn is a function of the variable of integration, Y_s. However, we can approximate the integral with a fair degree of accuracy when the source and receptor heights are zero (see Calder, 1973; Venkatram and Horst, 2006).

If we define an effective downwind distance, x_r, independent of Y_s,

$$x_r^{eff} = \frac{X_r}{\cos\theta}, \qquad (24.7)$$

and evaluate σ_z at this effective distance, the integral in Equation 24.6 can be evaluated to yield (Venkatram and Horst, 2006)

$$C(X_r, Y_r) = \frac{q}{\sqrt{2\pi}} \frac{1}{U \cos\theta \sigma_z^{eff}}\left[erf(t_1) - erf(t_2)\right], \qquad (24.8)$$

where *erf* is the error function and the limits are given by

$$t_i = \frac{(Y_r - Y_i)\cos\theta - X_r \sin\theta}{\sqrt{2}\sigma_y(x_i)}, \qquad (24.9)$$

where $i = \{1,2\}$. Here σ_y is evaluated at $x_i \equiv x_r(Y_s = Y_i)$ so that σ_y in the definitions of t_1 and t_2 correspond to downwind distances, x_r, from the endpoints Y_1 and Y_2 of the line to the receptor at (X_r, Y_r).

The expression for an infinite line source is obtained by setting $Y_1 = -y$ and $Y_2 = y$ in Equation 24.8 to yield Calder's (1973) formula:

$$C(X_r, Y_r) = \sqrt{\frac{2}{\pi}}\frac{q}{U\cos\theta\sigma_z^{eff}}. \qquad (24.10)$$

Notice that the wind direction has two effects. The dilution wind speed becomes $U \cos\theta$, which is the component of the wind that is normal to the line source. At the same time the vertical and horizontal plume spreads are evaluated at an oblique distance $X_r/\cos\theta$. We note Calder's observation that the concentration given by Equation 24.10 is relatively insensitive to the wind direction because the decrease in the normal wind speed, with θ, is compensated by the increase in the effective σ_z, which increases with θ. In fact, if σ_z grows linearly with downwind distance, the concentration in Equation 24.10 becomes independent of the wind direction.

It turns out that Equation 24.8 is exact when the wind is perpendicular to the line source, $\theta = 0$, for all source and receptor heights (Venkatram and Horst, 2006). This allows the use of the expression as the kernel of the integral used to compute concentrations associated with an area source. The two-dimensional integral for an area source can be computed by representing the area as a set of line sources perpendicular to the wind.

We see that estimates from the dispersion models described here depend on the plume spread parameters, σ_y and σ_z. Most of the currently used regulatory dispersion models use expressions derived empirically from field experiments. The new generation of regulatory dispersion models, such as AERMOD (Cimorelli et al., 2005), estimate dispersion using information, measured or modeled, on the mean and turbulent structure of the atmospheric boundary layer. The next section on the atmospheric boundary layer provides the background necessary to understand the formulation of these dispersion curves, described later.

24.4 Atmospheric Boundary Layer

Turbulence in the atmospheric boundary layer is generated by wind shear and buoyancy associated with radiative heating at the ground. During the daytime, sensible heating at the surface results in parcels of air that are warmer, and hence less dense than their surroundings. These parcels are subject to buoyancy forces that accelerate them upwards. The mixing induced by these moving parcels gives rise to the boundary layer or mixed layer, whose growth is inhibited by a layer in which the rising parcels are denser than their surroundings. This layer, referred to as an inversion, is characterized by increasing temperature with height. This inversion usually develops when there is large-scale downward motion or subsidence of the air. It can be shown that at heights below about a tenth of the mixed layer height, z_i, buoyancy generates vertical turbulent velocities given by (Stull, 1988)

$$\sigma_w = 1.3u_f; \quad z \leq 0.1z_i, \qquad (24.11)$$

where the free convection velocity scale, u_f is defined by

$$u_f = \left(\frac{g}{T_s}Q_o z\right)^{1/3}. \tag{24.12}$$

where

Q_o is the surface kinematic heat flux, which is the sensible heat flux (W/m²) divided by the product of the density and the specific heat of air

T_s is the surface temperature (K)

The heat flux, Q_o, is taken to be positive when it is directed away from ground and into the atmosphere as during the daytime, and is negative when it is into the ground as during most nights.

Between 0.1 z_i and close to the top of the mixed layer, σ_w associated with buoyancy production of turbulence is proportional to the so-called convective velocity scale given by

$$w_\star = \left(\frac{g}{T_s}Q_o z_i\right)^{1/3}. \tag{24.13}$$

In this region, we find that (Stull, 1988)

$$\sigma_w = \sigma_v = \sigma_u \cong 0.6w_\star. \tag{24.14}$$

It is found that σ_u and σ_v are also proportional to $w_\star$, even below 0.1 z_i.

Where turbulence production is dominated by wind shear, σ_w close to the ground is roughly proportional to the surface friction velocity, $u_\star$

$$\sigma_w = 1.3u_\star, \tag{24.15}$$

where $u_\star$ is related to the shear stress at the ground, τ_o, through

$$u_\star = \sqrt{\frac{\tau_o}{\rho_a}}, \tag{24.16}$$

where ρ_a is the density of air. The absolute value of the Monin–Obukhov length, L, is roughly the height at which the turbulent velocity generated by buoyancy is equal to that produced by shear

$$L = -\frac{T_s u_\star^3}{gkQ_o}, \tag{24.17}$$

where the Von Karman constant $k = 0.4$.

Thus, shear production of turbulence dominates that by buoyancy at heights below the Monin–Obukhov length. L is usually negative during the daytime when the heat flux is into the atmospheric boundary layer and positive during nighttime when the heat flux is directed toward the ground.

In describing the structure of the atmospheric boundary layer, it is convenient to define a potential temperature at given height with temperature T and pressure, p:

$$\theta = T\left(\frac{p_0}{p}\right)^{R_a/C_p}, \tag{24.18}$$

where

$p_0 = 1000\,\text{mb}$ is a reference pressure
R_a is the gas constant
C_p is the specific heat of air

The potential temperature, θ, represents the temperature that a parcel with temperature, T, would acquire if it is moved adiabatically from p to p_0.

The potential temperature definition allows us to make statements about the stability of a parcel of air when it is displaced adiabatically without worrying about the effects of pressure changes in the atmosphere. It can be shown that a parcel resists vertical motion in an atmosphere in which the potential temperature increases with height; it is in stable equilibrium. A decreasing potential temperature denotes an unstable atmosphere, while a potential temperature that is constant with height characterizes an atmosphere that is neutral to parcel motion.

In the daytime boundary layer, the potential temperature decreases with height near the surface. Above a tenth of the mixed layer height, the potential temperature and the horizontal velocity are relatively uniform because of vigorous vertical mixing. The mixed layer is usually capped by a sharp temperature inversion and the velocity can also change rapidly across the inversion.

The height of the daytime boundary layer is primarily determined by the surface heat flux. The simplest model for its height variation assumes that the surface energy input into the boundary layer modifies the early morning temperature profile to result in a near uniform potential temperature within the boundary layer, and the unmodified profile above it. This model results in a boundary layer that grows with time from its initial value at sunrise to its maximum when the surface heat flux goes to zero.

When the sun sets, turbulence energy production by buoyancy ceases. Over a period of an hour, the turbulence in the mixed layer collapses and shear becomes the primary mechanism for the production of turbulence. Because the ground is initially warmer than the atmosphere, the thermal radiation leaving the ground exceeds that being supplied by the atmosphere. This deficit leads to a cooling of the ground.

Initially, both the sensible heat flux and the ground heat flux are directed away from the earth's surface. The surface cools rapidly and a point is reached at which the ground becomes colder than the layers above in the atmosphere. At this stage, the heat flux from the atmosphere is directed toward the earth's surface, and the surface boundary layer becomes stable with the potential temperature increasing with height.

The stable potential temperature gradient in the nighttime boundary layer suppresses the production of turbulence because it opposes vertical motion. Under these circumstances, shear production of turbulence is matched by the destruction associated with the stable temperature gradient and viscous dissipation. This balance between these processes of production and destruction leads to relatively small levels of turbulence in the nocturnal boundary layer. We know that turbulence levels in the stable boundary layer are of the order of the surface friction velocity. However, estimating the height of the stable boundary layer or the variation of turbulence levels with height is an uncertain exercise. The height is assumed to proportional to some power of the surface friction velocity (Venkatram, 1980). The horizontal turbulent velocities in the stable boundary layer do not appear to be related to micrometeorological variables. They are affected by mesoscale flows and local topography, which are diDcult to characterize using models.

Dispersion in the atmospheric boundary layer is governed by the turbulent and mean flow fields. Most short-range dispersion models assume that these fields vary only in the vertical, so that vertical profiles of the relevant variables can be constructed using a one-dimensional boundary layer model. The inputs to the boundary layer model generally consist of (1) the near surface wind speed, U; (2) the near surface temperature, (3) Cloud cover, and (4) Upper air sounding. The surface heat flux, Q_0, is estimated using a surface energy balance, which in turn is based on estimates of the incoming solar radiation, the incoming and outgoing thermal radiation, and the ground heat flux. The heat flux estimate is then combined with information on the surface wind speed and roughness length to compute the surface friction velocity. Details of one of the most popular schemes used to estimate these parameters are provided in Holtslag and Van Ulden (1983).

The next section describes how regulatory models use information on the turbulent and mean flow fields in the atmospheric boundary layer to estimate plume spreads.

24.5 Dispersion in the Atmospheric Boundary Layer

Until recently, plume spread formulations were based on those derived empirically by Pasquill (1961) from observations made during the Prairie Grass dispersion experiment conducted in Nebraska in 1956 (Barad, 1958). These formulations were modified subsequently by Gifford and Turner and are commonly referred to as the Pasquill-Gifford-Turner (PGT) curves. For dispersion in urban areas, Industrial Source Complex (ISC, predecessor of AERMOD) uses the McElroy-Pooler curves that are derived from experiments conducted in St. Louis, Missouri (McElroy and Pooler, 1968).

The dispersion curves are keyed to stability classes that are related to ranges in the wind speed and incoming solar radiation. The wind speed, measured at 10 m, is an indicator of turbulence produced by shear, while the incoming solar radiation

is a surrogate for the sensible heat flux, which generates turbulence. Thus, the stability classes contain information on shear and buoyancy produced turbulence.

Classes A, B, and C correspond to unstable conditions when buoyancy production of turbulence adds to that due to shear. The sensible heat flux under these conditions is upward. Class A, the most unstable, is associated with the most rapid dispersion rates; the plume spread parameters (σ_y, σ_z) for a given distance decrease as we go from class A to C. Class D corresponds to neutral conditions when turbulence production is dominated by shear. Classes E and F are associated with stable conditions. Class F corresponds to the lowest dispersion rates. Thus, six dispersion curves, which are only functions of distance from the source, are used to describe the entire range of possible dispersion conditions.

The major advantage of the PGT curves is that they are based on observations, and thus provide realistic concentration estimates under a variety of meteorological conditions. Their shortcoming is that they are derived from dispersion of surface releases, and are thus not applicable to elevated releases. Furthermore, their formulation does not allow the use of on-site turbulence levels to describe dispersion more accurately than the "broad brush" PGT curves.

In the more recently formulated models such as AERMOD (Cimorelli et al., 2005), the expressions for plume spread are based on theoretical analysis first proposed by Taylor (1921). His equation describes the variance of particle positions as a function of travel time from a fixed point of release in a flow that is steady and the turbulence statistics do not depend on location. Rather than presenting all of his analysis, we will highlight the major results using the asymptotic behavior of horizontal plume spread (Csanady, 1973):

$$
\begin{aligned}
\sigma_y &= \sigma_v \tau && \text{for } \tau \ll T_{Lv} \\
\sigma_y &= \sigma_v \left(2\tau T_{Lv} \right)^{1/2} && \text{for } \tau \gg T_{Lv},
\end{aligned}
\tag{24.19}
$$

where

τ is the travel time from the source, given by $\tau = x/U$

σ_v is the standard deviation of the horizontal turbulent velocity fluctuations

A similar expression applies to the vertical spread of the plume.

In Equation 24.19, T_{Lv} is the Lagrangian timescale, which can be formally defined in terms of the statistics of the turbulent flow. For our purposes, it is suDcient to interpret the timescale as roughly the time over which a particle retains its initial velocity. For small travel times, a particle's velocity remains essentially unchanged from its value at the release point, and the particle trajectory is a straight line. This explains the result that, for small travel times, the spread of particles is proportional to the travel time from the source (Equation 24.19). On the other hand, when the travel time is large compared to the Lagrangian timescale, the plume spread is proportional to the product of the "average" step size, $\sigma_v T_{Lv}$, and the square root of the number of steps, τ/T_{Lv}, taken by the particle.

The new generation of dispersion models, such as AERMOD, relates dispersion to atmospheric turbulence using the theoretical framework described earlier. The problem in doing so is that the theory applies to a boundary layer in which the mean and turbulent properties are constant in space and time. To apply it to a real boundary layer in which the properties are highly inhomogeneous, we can use one of two approaches. The first is to average the turbulence and mean properties over the region of interest and use the average properties in the (homogeneous) formulations discussed earlier. This is not as straightforward as it seems because the limits of the average requires an estimate of the plume dimensions, which in turn depends on the average properties. Furthermore, the averaging procedure is necessarily arbitrary. The validity of the method needs to be established by comparing the results obtained from the formulations with observations or theory that accounts for inhomogeneity more explicitly. In general, empirical knowledge derived from observations plays a major role in the development of practical models of dispersion. As in most turbulence research, theory can suggest plausible forms for a dispersion model, but the model almost always contains parameters that have to be estimated from observations.

Even if we could treat the boundary layer as vertically homogeneous, the presence of boundaries, such as the ground and the top of the mixed layer, makes it diDcult to estimate the Lagrangian timescale, T_{Lv}, from a priori considerations. Thus, the timescale is often treated as an empirical parameter that is derived by fitting plume spread expressions to observations. Let us illustrate this by using an expression that is often used to describe horizontal plume spread:

$$\sigma_y = \frac{\sigma_v \tau}{(1 + \tau/2T_{Lv})^{1/2}} \qquad (24.20)$$

Note that Equation 24.20 satisfies the asymptotic limits given by Equation 24.19. We then postulate an expression for T_{Lv} in terms of a length scale l as $T_{Lv} = l/\sigma_v$. The length scale is taken to be proportional to a length characterizing the eddies responsible for transport, and the constant of proportionality is obtained by fitting estimates of plume spread from Equation 24.20 to observations. In AERMOD, the vertical spread for elevated releases in the stable boundary layer is given by an expression similar to Equation 24.20.

The second approach to accounting for inhomogeneity in the boundary layer is based on the solution of the species conservation equation:

$$\frac{\partial C}{\partial t} + \frac{\partial}{\partial x_i}(U_i C) = \frac{\partial}{\partial x_i}\left(K^i \frac{\partial C}{\partial x_i}\right), \qquad (24.21)$$

where K^i is the so-called eddy diffusivity, and the superscript negates the summation convention. The eddy diffusivity is defined as the negative of the ratio of the turbulent mass flux to the local mean concentration gradient. The concept, which is based on an analogy with molecular transport, cannot be justified rigorously for turbulent transport. However, it has heuristic value and is useful for developing semiempirical models of turbulent transport.

It can be shown that the eddy diffusivity concept is most applicable when the scale of concentration variation, the plume spread, is larger than the scale of the eddies responsible for plume spreading. In the surface boundary layer, plume spread in the vertical direction is comparable to the length scale of the eddies responsible for vertical transport. It turns out that that the eddy diffusivity concept is useful in the surface boundary layer, where semi-empirical theories, referred to as Monin–Obukhov similarity, provide useful relationships between velocity and temperature gradients and the corresponding heat and momentum fluxes. These relationships can be used to derive eddy diffusivities for heat and momentum, which can be used in the solution of Equation 24.20.

Existing regulatory models for short-range dispersion do not use the eddy diffusivity based mass conservation equation to avoid the associated numerical effort and to make the most eDcient use of observations of plume spread. However, the eddy diffusivity concept can be useful in deriving expressions for plume spread in the inhomogeneous surface layer. For example, AERMOD's expressions for plume spread are based on Venkatram's (1992) formulations:

$$\sigma_z = \sqrt{\frac{2}{\pi}\frac{u_* L}{U}}\,\bar{x}\,; \qquad\qquad \text{for } \bar{x} \le 1.4$$

$$= \sqrt{\frac{2}{\pi}\frac{u_* L}{U}}\,1.12\bar{x}^{2/3}; \qquad \text{for } \bar{x} > 1.4,\ L > 0,$$

$$= \sqrt{\frac{2}{\pi}\frac{u_* |L|}{U}}\frac{\bar{x}}{(1+0.006\bar{x}^2)^{-1/2}};\ \text{ for } L < 0 \qquad (24.22)$$

where $\bar{x} = x/|L|$, and the wind speed U corresponds to an average over the surface layer. In practice, the ground-level concentration is insensitive to the choice of U, because the dilution is determined by the combination $\sigma_z U$. These expressions provide a good description of the crosswind integrated concentrations observed during the Prairie Grass experiment (see Van Ulden, 1978 for a listing of the data).

An expression for the vertical spread of releases from an arbitrary height in the boundary layer can be formulated by combining the σ_z for a surface release (Equation 24.22) with that from an elevated release:

$$\sigma_z = (1-f)\sigma_{z,surface} + f\sigma_{z,elevated}, \qquad (24.23)$$

where the interpolating factor f is taken to be the ratio of the effective release height to the mixed layer height to ensure the correct limiting behavior. The expression for σ_z for an elevated release is based on Equation 24.20, where σ_v is replaced by σ_w and the Lagrangian timescale incorporates an appropriate length scale (see Weil in this handbook for the formulation in the stable boundary layer).

This approach to deriving an expression for the plume spread is based on a general principle that underlies most of the formulation of AERMOD: the formulation describing a physical quantity should be constrained by observations at the known limits of the quantity. This ensures that the model does not produce results that are outside the range of possibility. The horizontal plume spread is based on a Lagrangian timescale that interpolates between an empirical fit to the data from the Prairie Grass experiment (Barad, 1958), and a formulation for an elevated release that assumes that the T_{Lv} is proportional to the height of release/σ_v.

In AERMOD, the horizontal spread, σ_y is based on an equation similar to Equation 24.20. The Lagrangian timescale was derived by fitting the equation to observations of plume spread from the Prairie Grass experiment.

In order to use the dispersion model represented by Equation 24.2, we need to estimate the effective stack height, h_e, which is the sum of the physical stack height and vertical rise of the plume associated with momentum and buoyancy of the plume. The next section describes the calculation of plume rise in practical dispersion models.

24.6 Plume Rise

The effective stack height, h_e, is given by

$$h_e = h_s + h_p, \qquad (24.24)$$

where
h_s is the physical stack height
h_p is the plume rise

Most practical plume rise models are based on the model that assumes that the plume rises in a neutral atmosphere with a constant wind speed, U (Weil, 1988):

$$h_p = \left(\frac{3}{\beta^2} \frac{F_m}{U} t + \frac{3}{2\beta^2} \frac{F_b}{U} t^2 \right)^{1/3}, \qquad (24.25)$$

where
β is an entrainment parameter taken to be 0.6
$t = x/U$ is the time of travel from the source
the momentum flux, F_m, and the buoyancy flux, F_b, of the plume are given by

$$F_m = v_s^2 r_s^2 \left(\frac{T_s}{T_a} \right)$$

$$F_b = \frac{g}{T_s} v_s r_s^2 (T_s - T_a), \qquad (24.26)$$

where
v_s is the velocity of the exhaust gases
r_s is the stack inner radius
T_s is the exhaust gas temperature
T_a is the ambient temperature

Equation 24.25 is a useful approximation when the plume is rising in the unstable boundary layer in which the potential temperature is relatively uniform in the vertical. The formulation assumes that the growth of the plume is dominated by turbulence induced within the plume by the vertical motion of the rising plume. Atmospheric turbulence is important in the later stages of plume rise when the rise velocity becomes comparable to the standard deviation of the vertical velocity fluctuations, σ_w. Although the effects of atmospheric turbulence can be explicitly accounted for when the plume is rising, it is generally assumed that atmospheric turbulence only affects the final plume rise, when the plume levels off. The plume is assumed to reach its final height when the plume rise rate is comparable to σ_w:

$$\frac{dh_p}{dt} = \sigma_w. \qquad (24.27)$$

In most plumes, plume buoyancy dominates initial plume momentum by the time the plume reaches its final height. Then, Equations 24.25 and 24.27 yield

$$h_f = 1.8 \frac{F_b}{U \sigma_w^2}. \qquad (24.28)$$

The equations governing the rise of a plume in a stable atmosphere (gradient of potential temperature is positive) account for the fact that plume buoyancy relative to the atmosphere decreases as the plume rises. The solution for the path of the plume differs from that given by Equation 24.25. However, practical dispersion models assume that the Equation 24.25 holds until the plume reaches its final height given by

$$h_f = 2.6 \left(\frac{F}{UN^2} \right)^{1/3}, \qquad (24.29)$$

where N is the Brunt–Vaisala frequency of the atmosphere given by

$$N = \left(\frac{g}{T_a} \frac{d\theta}{dz} \right)^{1/2}, \qquad (24.30)$$

where $d\theta/dz$ is the average potential temperature gradient that the plume rises through.

As indicated earlier, the plume model assumes a vertically homogeneous atmosphere. The formulation is applied in the real atmosphere by using the properties at the stack height to represent the entire boundary layer. This approximation can be improved iteratively by first calculating the plume rise using stack height properties and then using averages over the plume rise height. The errors in such approximations become critical near the ground where wind speed and temperature change substantially with height. Some practical dispersion models avoid these errors by numerical solution of the governing equations, which explicitly account for the variation of atmospheric properties with height.

Most dispersion models account for the interaction of the rising plume with the inversion capping the mixing layer height through a "plume penetration factor," which is a function of the plume buoyancy and the temperature difference across the capping inversion. This factor multiplies the emission rate of the source. When the plume completely penetrates the inversion and is trapped within the elevated stable layer, the factor is zero. When the plume does not have suDcient buoyancy to penetrate the inversion, the factor is taken to be unity. All the material within the atmospheric boundary layer is "reflected" from the top of the boundary layer.

24.7 Building Downwash

Buildings and other structures near a relatively short stack can have a substantial effect on plume transport and dispersion and on the resulting ground-level concentrations that are observed. The "rule of thumb" is that a stack should be at least 2.5 times the height of adjacent buildings to avoid the effects of the buildings. Much of what is known of the effects of buildings on plume transport and diffusion has been obtained from wind tunnel and field studies.

When the airflow meets a building (or other obstruction), it is forced around, up, and over the building. On the lee side of the building the flow separates, leaving a closed circulation containing lower wind speeds (see Figure 24.3). Farther downwind, the air flows downward again. In addition, there is more shear and, as a result, more turbulence. This is the turbulent wake zone.

If a plume gets caught in the cavity, concentrations next to the building can be relatively high. If the plume escapes the cavity, but remains in the turbulent wake, it may be carried downward and dispersed more rapidly by the turbulence. This can result in either higher or lower concentrations than would occur without the building, depending on whether the reduced plume height or increased turbulent diffusion has the greater effect. The height to

which the turbulent wake has a significant effect on the plume is generally considered to be about the building height plus 1.5 times the lesser of the building height or width. This results in a height of 2.5 building heights for cubic buildings, and less for tall, slender buildings. Since it is considered good engineering practice to build stacks taller than adjacent buildings by this amount, this height is called the "good engineering practice" (GEP) stack height.

Most treatments of building effects on dispersion are based on incorporating two effects: (1) the effective reduction of source height associated with the trapping of pollutants in the cavity, and (2) the increased turbulence in the building wake. If the emissions are entrained into the cavity, the source is assumed to be at ground level, but the plume is assigned initial values to account for the fact that the emissions originate from a cavity whose size scales with the dimensions of the building. For example, the initial spreads of the plume can be taken to be

$$\begin{aligned} \sigma_{yo} &= \alpha w \\ \sigma_{zo} &= \beta h \end{aligned} \text{,} \tag{24.31}$$

where

w and h are the width and height of the building
α and β are constants

Alternatively, these initial spreads can be modeled in terms of a "virtual" source at ground level at an upwind distance that results in these spreads.

The fraction of the emissions that is entrained into the building cavity is taken to be a function of the stack height and the building height. The fraction that is not entrained into the cavity is treated as a conventional point source, except that plume dispersion is enhanced to account for the increased turbulence levels in the building cavity. The concentration at a downwind receptor is then a sum of the concentrations from the elevated source and the ground-level source, corresponding to the emissions from the cavity. Current models, such as the PRIME algorithm (Schulman et al., 2000), use approaches based on these ideas.

24.8 Terrain Treatment

Several complicated processes govern dispersion in complex terrain. Under unstable conditions, the plume is depressed toward the surface of the obstacle as it goes over it. The implied compression of the streamlines is associated with a speed-up of the flow and an amplification of vertical turbulence. Under stable conditions, part of the flow flowing toward an obstacle tends to remain horizontal, while the other part climbs over the hill. Experiments show (Snyder et al., 1985) that this tendency for the flow to remain horizontal can be described using the concept of the dividing streamline height, denoted by H_c. Below this height, the fluid does not have enough kinetic energy to surmount the

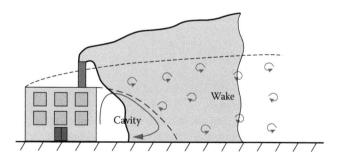

FIGURE 24.3 Plume dispersion affected by cavity and wake behind building.

top of the hill; a plume embedded in the flow below H_c either impacts on the hill or goes around it. On the other hand, the flow and hence the plume above H_c can climb over the hill.

Terrain features can rise toward the plume, deflecting its flow over or around, or allowing the plume to come in contact with the terrain. In convective (unstable) conditions, the airflow, and thus the plume, will be forced over the terrain obstacle. On the lee side of the obstacle, a wake or cavity may occur in the flow, resulting in high concentrations on that side of the terrain feature.

The alignment of ridges and valleys can channel the flow. This can result in high concentrations appearing in areas quite different than would be expected if this effect were not accounted for. The presence of hills and valleys can also help to create local wind flows. These flows may alter the transport of low-level plumes. Modeling these flows using wind data from above or distant from the site may result in incorrect modeling results. Conversely, wind measurements are influenced by these local flows, and if used to model a tall stack source that emits above the local flow, can result in incorrect modeling results. One example is the case of a narrow valley with a north to south orientation. In the morning, the sun will first heat the west wall of the valley. This warmer air will rise, creating a cross-valley flow from east to west (in the absence of strong winds aloft). Conversely, in the evening, the east wall will be heated more, resulting in a cross-valley flow from the west.

Accounting for these effects in air quality models presents a significant challenge. The effects cannot be ignored in regulatory modeling, since terrain effects generally contribute to higher concentrations than would be observed in flat terrain situations. On the other hand, representing terrain effects accurately may require the use of computational fluid dynamics models or other modeling approaches that require extensive computer resources.

24.9 Treatment in Short-Range Models

The Complex Terrain Dispersion Model (CTDMPLUS, Perry, 1992) accounts for the major effects associated with the concept of the dividing streamline height described in the previous section. AERMOD incorporates a semiempirical model (Venkatram et al., 2001) that mimics the major features of CTDM. It assumes that the concentration at a receptor, located at a (x, y, z), is a weighted combination of two concentration estimates: one assumes that the plume is not affected by the terrain and the other assumes that the plume follows the terrain.

The concentration associated with the plume unaffected by the terrain dominates during stable conditions, while that caused by the terrain-following plume is more important during unstable conditions. These assumptions allow us to write the concentration, $C(x, y, z)$, as

$$C(x,y,z) = fC_f(x,y,z) + (1-f)C_f(x,y,z_e) \qquad (24.32)$$

The first term on the right-hand side of Equation 24.32 represents the contribution of the plume that ignores the terrain, while the second term is the contribution of the terrain-following plume. The concentration, $C_f(x, y, z)$, is that associated with a plume that is unaffected by the terrain. The receptor height, z, in the first term on the right is measured relative to the stack bottom. In the second term, the concentration is evaluated at the height of the receptor $z_e = zt\,z_h$ above the local terrain, z_h, to simulate the plume following the terrain contour.

The weighting factor, f, is a function of the fraction of the plume below the dividing streamline height, H_c, defined by the implicit equation

$$\frac{1}{2}U^2(H_c) = \int_{H_c}^{z}(z-\varsigma)N^2(\varsigma)d\varsigma, \qquad (24.33)$$

where

z is the height of the receptor
N is the local Brunt–Vaisala frequency (Equation 24.30)

The left-hand side of the equation is the kinetic energy of the fluid at H_c and the right-hand side is the potential energy gained in climbing from H_c to the receptor at z. If N is constant, the expression for H_c becomes

$$H_c = z - \frac{U(H_c)}{N}. \qquad (24.34)$$

The fraction, φ, of the plume mass below H_c at the downwind receptor distance and the weighting factor, f, are

$$\phi = \frac{\int_0^{H_c} C_f(x,y,z)dz}{\int_0^{\infty} C_f(x,y,z)dz}.$$

$$\qquad (24.35)$$

$$f = \frac{1}{2}(1+\phi).$$

When the entire plume lies below H_c, f goes to unity and the concentration corresponds to a plume that does not see the hill. When the dividing streamline height goes to zero under unstable conditions, f becomes ½. This means, that under unstable conditions, the concentration at an elevated receptor is the average of the contributions from the flat terrain plume and the terrain-following plume (Figure 24.4).

This formulation of the complex terrain model ensures that the model estimates are sensible in that they range between values corresponding to two limits of plume behavior. This simple semiempirical model has been tested at several complex terrain sites and it performs at least as well as CTDM in the limited task of describing observed concentration statistics.

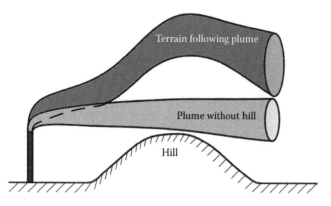

FIGURE 24.4 The two plumes used to formulate the complex terrain model.

24.10 Modifications to the Gaussian Framework

New models, such as AERMOD, incorporate physics that cannot be readily accommodated within the framework of the Gaussian distribution of the concentration. One example is dispersion in the unstable boundary layer. In the unstable boundary layer, both the mean wind and turbulence levels are relatively uniform above a height of about 1/10th of the boundary layer height. In principle, this should allow a straightforward application of Taylor's equations for plume spread in the Gaussian expression. However, the Gaussian equation is not appropriate because the turbulent vertical velocities in the middle of the convective boundary layer do not follow a Gaussian distribution; the distribution has a negative mode and has a long positive tail as shown in Figure 24.5. This implies that material released in the middle of the boundary layer has a greater probability of being caught in downdrafts than in updrafts. This leads to the descent of the plume centerline, which cannot be described with a symmetric Gaussian model. Several approaches have been used to capture this feature of dispersion of elevated releases in the convective boundary layer.

AERMOD uses what is commonly referred to as the probability density function (pdf) approach, which assumes that a particle does not forget its velocity at release. This implies that the crosswind-integrated concentration at ground level is determined by the probability density function of vertical velocities at the source:

$$\bar{C}^y = \frac{2Q}{U\sigma_z}\sigma_w P\left(w = -\frac{uh}{x}\right) \qquad (24.36)$$

where $P(w = t\, uh/x)$ is the probability density function evaluated at the vertical velocity that brings plume material from the elevated release to the receptor at x in a straight line. The factor 2 accounts for reflection at the ground.

That the Gaussian formulation is recovered if the pdf is Gaussian. AERMOD uses a skewed pdf that allows for the plume centerline to descend toward the ground and leads to concentrations that can be over 30% higher than that associated with a Gaussian pdf (see Venkatram, 1993). The actual formulation in AERMOD combines plume rise with dispersion and mimics the non-Gaussian pdf in Equation 24.34 as a sum of two Gaussian distributions, which results in the required mode and skewness.

24.11 Other Features in Practical Models

Regulatory models also need to account for special features of urban areas. In ISC, dispersion in urban areas is treated using empirical dispersion curves derived from tracer experiments conducted in St. Louis (McElroy and Pooler, 1968). These so-called McElroy–Pooler curves, which are keyed to stability classes, lead to enhanced dispersion in urban areas.

AERMOD treats urban dispersion by accounting for the processes that lead to the enhancement of turbulence in urban areas. When rural air flows into a warmer urban area, the boundary layer becomes convective because of surface heating. Thus, when the rural boundary layer is stable during the night, the urban boundary layer can be convective. AERMOD accounts for this effect by formulating an upward heat flux and a boundary layer height in terms of the urban–rural temperature difference, which in turn is parameterized in terms of the population of the urban area. Then, a convective velocity scale is calculated using this heat flux, and the associated boundary layer height. This convective velocity scale is then used to calculate a turbulence profile, which is then added to that from the rural area. The increased roughness over an urban area is included in the calculation of the rural turbulence profile.

When the wind speeds become comparable to the turbulent velocities, it becomes necessary to account for the fact that the plume spreads horizontally over large angles relative to the mean wind, and upwind receptors can be affected by the source. Such conditions are common in urban areas, where buildings can enhance turbulence and reduce the mean flow.

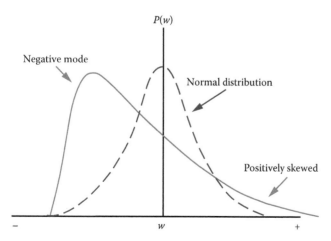

FIGURE2 4.5 Vertical velocity distribution in the CBL.

AERMOD accounts for low wind speed conditions by assuming that the concentration is a weighted average of concentrations in two possible states: a random spread state and plume state. In the random spread state, the release is allowed to spread equally in all directions. Then, the weighted horizontal distribution is written as

$$H(x,y) = f_r \frac{1}{2\pi r} + (1 - f_r) \frac{1}{\sqrt{2\pi}\sigma_y} \exp\left(-\frac{y^2}{2\sigma_y^2}\right), \qquad (24.37)$$

where the first term represents the random state, the second term is the plume state, and $r = (x^2 + y^2)^{1/2}$. The plume is transported at an effective velocity given by

$$U_e = \left(2\sigma_v^2 + U_m^2\right)^{1/2}, \qquad (24.38)$$

where U_m is the mean vector velocity. The weight for the random component in Equation 24.37 is taken to be

$$f_r = \frac{2\sigma_v^2}{U_e^2} \qquad (24.39)$$

This ensures that the weight for the random component goes to unity when the mean wind approaches zero.

Modeling dispersion of plumes from stacks on the shoreline needs to account for features governed by the horizontal inhomogeneity associated with the flow of air from the water to the land surface. In an area close to water, the land surface is warmer because the water heats up less rapidly than land in response to solar heating during the day. These essentially two-dimensional effects, especially those related to the temperature differences between urban and rural areas, are not treated reliably in models such as AERMOD.

As the stable air from the water flows onto the warmer land, the resulting upward heat flux gives rise to an internal boundary layer that has a significant effect on the ground-level impact of elevated power plant sources. Elevated emissions, even when initially released into a stable layer, can be brought down to the ground when the elevated plume intersects the growing thermal boundary layer, as shown in Figure 24.6. The concentration close to the point of fumigation is given by Equation 24.3 corresponding to the well-mixed boundary layer, where z_i is now the height of the boundary layer where the elevated plume intersects the internal boundary layer.

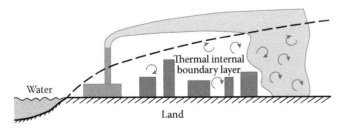

FIGURE 24.6 Growth of the thermal internal boundary layer (TIBL) at the land–water interface.

24.12 Concluding Remarks

Air quality models used in practice for source–receptor distances of a few kilometers assume that emissions from a source can be described by a plume in which the concentration distributions in the horizontal and vertical follow the Gaussian distribution; as we have seen, other distributions can be used to describe the vertical concentration profile. This framework allows the incorporation of several processes that affect ground-level concentrations. We have demonstrated how it can accommodate building effects, terrain effects, and dispersion in shoreline and urban areas.

The plume framework can be readily used to interpret data from field studies, and thus can be improved empirically to provide better descriptions of dispersion. These features, coupled with its computational simplicity, explain its popularity in applications that require realism as well as transparency.

References

Barad M. L. (Ed.) (1958) *Project Prairie Grass*. A field program in diffusion, Geophysical Research Paper No. 59, Vols. I (300pp.) and II (221pp.), AFCRF-TR-58-235, Air Force Cambridge Research Center, Bedford, MA.

Calder, L. K. (1973) On estimating air pollution concentrations from an highway in an oblique wind. *Atmos. Environ.*, 7, 863–868.

Cimorelli, J. A., Perry, G. S., Venkatram, A., Weil, C. J., Paine, J. R., Wilson, B. R., Lee, F. R., Peters, D. W., and Brode, W. R. (2005) AERMOD: A dispersion model for industrial source applications. Part I: General model formulation and boundary layer characterization. *J. Appl. Meteorol.*, 44, 682–693.

Csanady, G.T. (1973) *Turbulent Diffusion in the Environment*, Reidel, Dordrecht, the Netherlands, p. 248.

Holtslag, A. A. M. and Van Ulden, A. P. (1983) A simple scheme for daytime estimates of the surface fluxes from routine weather data. *J. Clim. Appl. Meteorol.*, 22, 517–529.

Horst, T. W. (1984) The modification of plume models to account for dry deposition. *Boundary Layer Meteorol.*, 30, 413–430.

McElroy, J. L. and Pooler, F., (1968) The St. Louis dispersion study-volume II-analysis. *National Air Pollution Control Administration*, Pub. No. AP-53, U.S. DHEW, Arlington, TX, 50pp.

Pasquill, F. (1961) The estimation of the dispersion of windborne material. *Meteorol. Mag.*, 90, 33–49.

Perry, S. G. (1992) CTDMPLUS: A dispersion model for sources in complex topography. Part I: Technical formulations. *J. eAppl. Meteorol.*, 31, 633–645.

Schulman L. L., Strimaitis D. G., and Scire J. S. (2000) Development and evaluation of the PRIME Plume Rise and building downwash model. *J. Air Waste Manag. Assoc.* 50, 378–390.

Snyder, W. H., Thompson, R. S., Eskridge, R. E., Lawson, R. E., Castro, I. P., Lee, J. T., Hunt, J. C. R., and Ogawa, Y. (1985) The structure of the strongly stratified flow over hills: Dividing streamline concept, *J. Fluid Mech.*, 152, 249–288.

Stull, R. B. (1988) *An Introduction to Boundary Layer Meteorology.* Kluwer Academic, Dordrecht.

Taylor, G. I. (1921) Diffusion by continuous movements. *Proc. Lond. Math. Soc. Ser. 2,* 20(1), 196–211.

Van Ulden, A. P. (1978) Simple estimates for vertical dispersion from sources near the ground. *Atmos. Environ.,* 12, 2125–2129.

Venkatram, A. (1980) Estimating the Monin-Obukhov length in the stable boundary layer for dispersion applications. *Boundary Layer Meteorol.,* 19, 481–485.

Venkatram, A. (1992) Vertical dispersion of ground-level releases in the surface boundary layer, *Atmos. Environ.,* 26, 947–949.

Venkatram, A. (1993) Estimates of the maximum ground level concentration in the convective boundary layer: The error in using the Gaussian distribution. *Atmos. Environ.,* 27A: 2187–2191.

Venkatram, A., Brode, R., Cimorelli, A., Lee, R., Paine, R., Perry, S., Peters, W., Weil, J., and Wilson, R. (2001) A complex terrain dispersion model for regulatory applications. *Atmos. Environ.,* 35, 4211–4221.

Venkatram, A. and Horst, T. W. (2006) Approximating dispersion from a finite line source. *Atmos. Environ.,* 40, 2401–2408.

Weil, J. C. (1988) Plume rise. In: *Lectures on Air Pollution Modeling,* eds. A. Venkatram and J. C. Wyngaard, pp. 119–166, American Meteorological Society, Boston, MA.

Modeling the Urban Boundary and Canopy Layers*

Jason Ching
University of North Carolina

Fei Chen
*National Center for
Atmospheric Research*

25.1 Background

Today, we are confronted with increasingly more sophisticated application requirements for urban modeling. These include those that address emergency response to acute exposures from toxic releases, health exposure assessments from adverse air quality, energy usage, and characterizing and potential amelioration of heat islands relevant to human sustainability issues. We contend that their requirements cannot be adequately met with simplistic one size fits all atmospheric models primarily due to combination of wide range of scales and complexities of urban structures and land uses. In this context, this chapter provides an overview of modeling scale dependent meso-to-urban scale flows and air quality. Typical modeling methods and approaches for urban modeling in current use are reviewed, the implications and limitations are presented, and the unique features of the urban canopy layer are highlighted. Various new advanced modeling approaches and a generalized model framework able to take advantage of detailed parameterizations from emerging databases with gridded fields of canopy parameters

and anthropogenic heating to support current canopy implemented simulation schemes follow. The urbanized versions of the Weather Research and Forecasting (WRF) model are then described and illustrative examples of both meteorology and air quality modeling are provided to demonstrate a range of urban focused applications followed by discussion of challenges and opportunities for urban modeling.

Mesoscale models are now capable of providing accurate meteorological forecasts at grid sizes of order 10 km. Modeling tools such as CFD (Coirier et al., 2005) are being tested for fine-scale simulations, and when run at grid sizes of order of meters, can provide detailed flow fields around obstacles such as buildings. Significant model advancements including refinements to model science and computational optimization and capacity are currently underway at grid scales of the order of 1 km or less. Such efforts address the need to resolve the important spatial concentration resolution details commensurate to urban areas and support the increasingly varied type and sophistication levels of urban assessment applications (Taha 2008a–c). As will be seen, a new multilayer canopy modeling framework is being advanced

Handbook of Environmental Fluid Dynamics, Volume Two, edited by Harindra Joseph Shermal Fernando. © 2013 CRC Press/Taylor & Francis Group, LLC.
ISBN: 978-1-4665-5601-0.

* The U.S. Environmental Protection Agency through its ODce of Research and Development collaborated in the research described here. It has been subjected to Agency review and approved for publication.

for implementation at finer scales to overcome limitations of the more traditional roughness approach which serves as the basis for coarser scale atmospheric models. Further, this survey will serve to indicate that the user community can now select from a range of science options that are either currently available or emerging to support the within-canopy modeling framework. The implication is that a user will have the task of selecting the appropriate modeling options/approach that most aptly suits the application. The concept of "fitness for purpose" (Baklanov et al., 2009) guidance is thus a critical tactical requirement and a part of the model design framework. An important model design issue is the requirement to determine the level and, if possible, the grid-by-grid specificity of horizontal detail of the underlying surfaces and also the vertical resolution for the specific problem of interest. One important new advancement is the extension of mesoscale models to simulate on a grid mesh, the wind, thermodynamics, and turbulence within the urban canopy (e.g., below the tops of buildings and trees) with location-specific descriptors of the underlying canopy. It is this aspect, its implications, both its capability and its drawbacks, which will be the major consideration of this chapter. A framework for such a system is described in Chen et al. (2011).

25.2 Principles and Framework for Modeling Urban Boundary Layers

25.2.1 Traditional Boundary Layer Formulations

The Reynolds Averaged Navier–Stokes (RAN) approach is the basis of roughness boundary layer model formulations. It is based upon a paradigm where the flow comes into adjustment with the turbulence generated by the roughness elements within the sublayer containing the roughness elements. With the constraint that some relative degree of homogeneity of the roughness elements in this sublayer exists, the boundary layer can be formulated in terms of scaling lengths (roughness, displacement, and mixing heights), scaling velocities, u_*, w_*, and other scaling variables for temperature and moisture. Typically, these properties are described in the similarity scaling framework of Monin–Obukhov similarity theory (MOST) where turbulent fluxes of momentum, heat, and moisture are approximately height invariant. One of the key methods for the scaling parameters used in modeling urban areas utilize land-use classification schemes in which the modeled flow is governed primarily by the underlying roughness elements for land-use class that is dominant of the modeled grid. Of course, this brings into consideration the property of the scale of the dominant land-use class. However, for urban areas, such land-use classification schemes do not typically conform to the uniformity of numerical grid meshes and the assumption of dominant land-use class is typically imposed and applied at the modeled grid scale.

Other implications of RAN schemes are as follows: First, the scheme applies to flows starting above the roughness sublayer. Second, there are constraints between surface roughness, displacement heights, building heights, Z_b, and stability based on MOST. In practical terms, this imposes a constraint on the minimum

thickness of its lowest model layer, Z_1. In the WRF model, the minimum thickness for this layer is roughly $Z_1 = 0.6Z_b + 4$. Using smaller values for the first layer will cause the simulations to become unstable. This suggests that model grid cells with significant density of tall buildings will either require employing relatively thick value of Z_1 or an arbitrary, artificial, and thereby unrealistic reduction of Z_b values. In urban areas, such roughness lengths and displacement heights can range to several meters for high-density commercial and industrial land-use sectors. For such areas, this imposes a limit of several tens of meters for the first simulation layer. Typically, the model vertical layer structure is preset so as not to violate this condition, and so is set for the largest scaling lengths in the model domain. This is not typically a problem for mesoscale grid sizes of order 10 km or so where even with the presence of tall buildings, the grid averaged roughness length should still be reasonably small enough (or can be modified so as to be small enough) to satisfy RANs roughness length constraints. The problem becomes significant for fine-scale grid simulations where, for many cells, the density and height of tall buildings dominate and the characteristic roughness element exceeds the RANs constraints. For modeling photochemically active air pollutants, this creates a constraint of artificially enhanced dilution of photochemically active emission precursors into overly large layers prior to the photochemistry simulation step, and thus, a reduction in model accuracy. Notwithstanding these implementation issues, such formulations apply with reasonable success to models based on grid sizes of order 10 km or more and for the region above the roughness sublayer and the so-called blending layer where the requirements for constancy of fluxes is generally applicable. As applications require smaller grid sizes of order 1 km, the flow adjustments to the underlying surface morphological features cannot be ignored since buildings in urban areas can be very large relative to typical roughness lengths associated with mesoscale flows. There is now ample evidence that the vertical structure within the building canopy is quite different from the boundary layer structure assumed in RANS or MOST formulations. Observations by Rotach (1999) and wind tunnel simulations (Kastner-Klein and Rotach, 2004) as examples amply demonstrate that momentum fluxes and turbulence profiles within the urban roughness (canopy) sublayer do not follow the MOST, and thus other treatments are required to model the canopy regime. It is thus instructive to turn to studies of forest canopies as a surrogate for buildings as discussed next.

25.2.2 Urban Canopy Model Formulations

The RANs (or aerodynamic) approach applies to the region above the canopy elements (called the displacement height) and thus cannot capture the flow dynamics and turbulence within the canopy layers. For details of flows and turbulence within such layers, models must be able to capture the drag, turbulence, and thermal effects attributed to the distribution of canopy elements. Brown (2000) provides a history of the model development and an urban modeling framework along these lines. Much of the development was based on modeling of forest canopies; they introduced model formulations for the momentum conservation equation and turbulence closure schemes that describe form drag, turbulence production and dissipation by the forest elements (Wilson and Shaw, 1977; Yamada, 1982). Current urban canopy models are extensions of this framework. However, new terms (parameterizations) for modeling the drag effects and descriptions of urban canopy elements are required in such implementation.

Models have also treated the heat equation for forest canopies. The effect of radiative properties by trees include recognition of vertical treatment for albedo and emissivity, and in turn, require modeling the vertical profile of leaf (and non-leaf) surface area density in treatments for the attenuation of solar and long wave radiative heat fluxes. For urban modeling, additional treatments for modeling the heat storage and surface energy budget terms in urban areas will require more complex and specialized treatments for the roof energy budget and multiple reflection and scattering of radiation within street canyons. Further, the proportional introduction of anthropogenic heating differentially to the surface energy budget or directly to the atmosphere must be specified, and in the context of the volumetric distribution of buildings.

25.2.3 Characteristics and Uniqueness Aspects Regarding Modeling of Urban Boundary Layer Structure and Dynamics

As the grid size of mesoscale models decreases, the modeling of the meteorology and air quality in urban areas require greater recognition and treatments of uniquely urban surface features that impact the structure of the flow, turbulence, thermodynamic fields, and their concomitant upscale and downscale dependencies (Roth, 2000). Figure 25.1 illustrates the unique scale dependent aspects of typical urban boundary layers. When encountering an urbanized area, an urban boundary layer (UBL) forms within the air mass. At the local scale, the underlying UBL includes the urban canopy layer (UCL) whose vertical extent is from the surface to the tops of the building, and overlying blending layer above the tops of building where the flow is adjusting to the properties of the UCL; both such layers form the roughness sublayer (RSL). The bulk effects of the RSL sets up approximately constant flux layer and the overlying inertial sublayer. At the microscale, the RSL and UCL the details and complexities of the flow induced by individual buildings are in a state of constant adjustment as the flow proceeds from the upwind edge of the urban area.

This paper identifies a number of important modeling aspects that require enhanced model treatments from that for more simple surfaces, many of which have been described in Brown (2000). Key features contributing to the need for modeling considerations include (1) variability in building morphology, construction materials, and vegetations impacting radiation and surface energy budgets, (2) human activity resulting in spatial variability in anthropogenic heating rates and moisture sources, emissions of photoactive pollutants from traDc and industry. Under this broad class of features that can contribute to spatial heterogeneity are a myriad of mixtures of subgrid features, the combination of which must somehow be from a modeling perspective, made

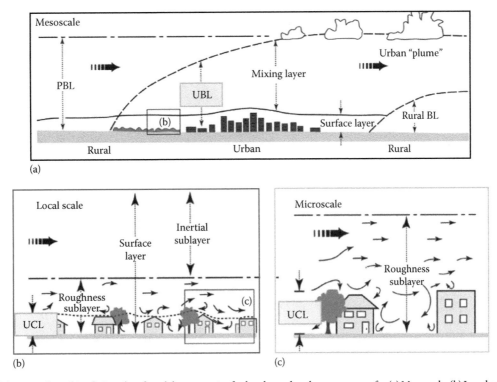

FIGURE 25.1 Schematic describing heirarchy of model treatments of urban boundary layer processes for (a) Mesoscale. (b) Local scale. (c) Microscale perspectives. (Courtesy of C.S.B. Grimmond and M. Piringer et al.) UBL and UCL refer to urban boundary and urban canopy layers, respectively.

to vary from one urban area to another (Cionco and Ellefson, 1998). These include but are not limited to

1. Spatial land-use heterogeneity in industrial, commercial, dwellings, mix of natural and man-made surfaces, and recreational areas
2. High variability in (a) building density, (b) distribution of building heights, (c) their sizes and horizontal position distributions, (d) architectural characteristics (e) construction material
3. Content and distribution of streets and highways, block sizes, parking areas and densities variations in street widths and sky view factors
4. Orientation of street canyons relative to sun angle
5. Depth of street canyons
6. Horizontal distributions, groupings, density, and genus of vegetation and trees

7. Distribution of nonconstructional subareas such as parks and lakes
8. Mix and distribution of high rise, high building density and low rise, low building density areas
9. Mix and density of industrial/commercial use

25.2.4 Model Input Databases

The strategic goal of developing improved models for the UCL typically involve modifying the governing equations of motion, energy, and turbulence with additional terms and their supporting science descriptions prescribed with varying degrees of complexity. They typically involve introducing within-grid details of the urban morphological features, including building and other pertinent urban features.

Figure 25.2a illustrates the tactical strategy being adopted for practical implementation. In an urban area it is given that

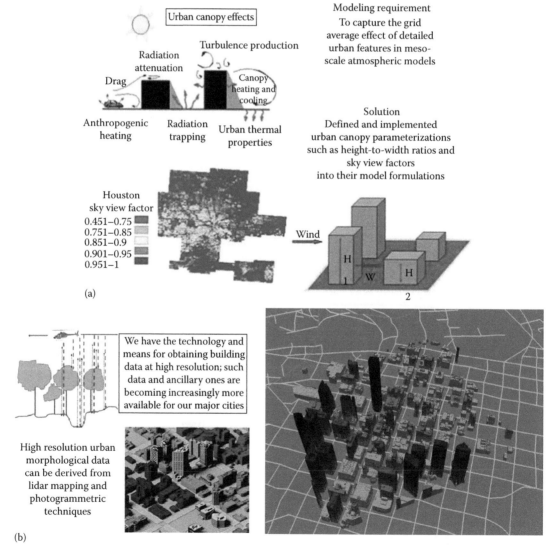

FIGURE 25.2 (See color insert.) (a) Schematic illustrating the implementation approaches to local scale modeling of the UCL. (b) Advanced high-resolution monitoring of morphological feature and advanced data processing tools provide a base for developing urban canopy parameters.

for any and every model grid, there is a unique set of building distributions and other urban features; however, it is the aggregated effect of morphology features that need to be implemented. These features are captured with a variety of urban canopy parameters (UCPs) such as frontal and building plan areas density, sky view factors, and others, and are discussed in further detail in Ching et al. (2009). The detailed urban features are now able to be mapped in great detail and at high resolution using photogrammetric techniques, from airborne lidar surveys or from other techniques and data from satellites (Figure 25.2b). Such data are being collected routinely. With GIS processing tools it is now possible to automate the processing of these high-resolution data into the desired UCP for grids of any size and geographic orientation reported in Burian et al. (2007) and Burian and Ching (2010). Thus, given the specification of, and the means to provide, the requisite set of UCPs, we now have the means to proceed with modeling the UCL for a user-specified grid. Based on the need for advanced treatments of high-resolution urban morphological features (e.g., buildings, trees) in meteorological, dispersion, air quality, and human exposure modeling systems for future urban applications, a prototype project was launched called the National Urban Database and Access Portal Tool (NUDAPT, Ching et al., 2009). NUDAPT was designed to produce and provide gridded fields of UCPs for various new and advanced descriptions of model physics to improve urban simulations given the availability of new high-resolution data of buildings, vegetation, and land use. Additional information, including gridded anthropogenic heating and population data, is incorporated to further improve urban simulations and to encourage and facilitate decision support and application linkages to human exposure models. NUDAPT is a "community"-based system utilizing web-based portal technology to facilitate customizing of data handling and retrievals (Ching et al., 2009).

We now review and discuss the various aspects and issues associated with the local scale modeling of the UBL and its UCL in the subsequent chapters.

25.3 Framework for Modeling the UCL at the Local Scale

Given the modeling requirement to resolve features of the UCL, the user currently has a single or multilayer option. While both require additional UCL parameters from that of the traditional RANs approach, they each have a different set of strengths and weaknesses; thus, the choice between single or multilayer approach will depend upon the application and mainly involve the trade-offs between computational cost and improved science requirements. With single-layer urban canopy parameters (SLUCP) systems (Masson, 2000; Kusaka et al., 2001) the emphasis is on the thermodynamics but they retain the vertical resolution limitation of traditional RAN. Compared to SLUCP systems, the use of multilayer approach (Martilli et al., 2002) has a significantly higher computational requirement but has

no practical limitation on layer thickness and accommodates improved model description of the effects of building drag on the momentum and generation of turbulence fields. In both approaches common requirements include but are not limited to (1) extension of the governing equations to incorporate morphological impacts on the flow, thermodynamics, and turbulence generation; (2) land surface energy budget and boundary layer schemes, (3) land-use classification and grid apportionment schemes as a framework for grid-size downscaling from meso-to-urban/local scale resolutions, (4) provisions for model operational options, (5) data on urban canopy parameters (albeit different for the two approaches and (6) other uniquely urban type data such as anthropogenic heating. The following sections will briefly explore each of these topics.

25.3.1 Generalized Formulations with Urban Canopy Parameterizations

The fundamental grid-size-dependent modeling framework requires the following: (1) the numerical solution of the momentum, thermodynamic, and turbulence equations and appropriate initial and boundary conditions usually handled with model nesting procedures; (2) horizontal grid-size-appropriate physics, inputs regarding a host of critical parameters that represent the underlying surfaces, and energy inputs appropriate to the grid size and, finally, given the presence of a roughness sublayer that includes the urban canopy of high building structures, (3) the introduction of multilayers within such canopies.

Clearly, the mesoscale approaching flow are modified horizontally and vertically as it traverses urban areas. We know from a large body of studies that the subsequent adjustments are complex, profound, and scale dependent. Thus, it is appropriate that the system of governing equations be capable of producing simulations that can accurately describe such adjustments. As a practical matter, we rely on (one- or two-way) grid nesting techniques to establish the required input initial and boundary conditions for the more finely resolved meteorological fields over urban areas. For the urban domain, we envision nesting down to approximately 1 km grid size (or less). This can set up the conditions for even finer scale modeling approaches needed for building scale flows. For the urban scale, the governing equations is typically modified to introduce additional terms such as added drag, heating, and turbulence, reflecting canopy influence on the UBL. Brown (2000) and Masson (2006) provide reviews of proposed approaches along these lines. Subsequently, several such approaches have been described and are in various stages of implementation and applications including single-layer systems such as TEB Masson (2000) and Kusaka et al. (2001), and multilayer within urban canopy treatments (Martilli et al., 2002; Dupont et al., 2004; Otte et al., 2004; Martilli and Schmitz, 2007; Taha 2008a,c). The first scheme does not take into account the within-cell building volume as do the other schemes. The multilayer system implementations all require explicit expressions and solutions for TKE and are typically based on scaling mixing lengths appropriate to the urban canopy (Bougeault and Lacarrere, 1989).

25.3.2 Land-Surface Boundary Layer Schemes

A variety of land surface models have been applied to urban modeling and the Noah (Chen et al., 1996, 1997; Chen and Dudhia, 2001; Ek et al., 2003; Mitchell et al., 2004), PX (Xiu and Pleim, 2001; Pleim and Xiu, 2003) and submeso (Dupont and Mestayer, 2006) schemes have been modified for urban applications. Each of their model formulations account for the drag effect of buildings, for considerations of each of the different factors of the energy budgets and moisture from the top of the canopy to the surface, especially as regards the albedo effects and the trapping and attenuation of both long and short wave radiation from buildings. For instance, the WRF model includes a bulk urban parameterization in Noah by modifying several parameters to represent the zero-order effects of urban surfaces (Liu et al., 2006): (1) increasing the roughness length to 0.8 m to represent enhanced drag due to buildings; (2) reducing surface albedo to 0.15 to represent the shortwave radiation trapping in urban canyons; (3) increasing the volumetric heat capacity to $3.0 \, J \, m^{-3} \, K^{-1}$ for the urban surfaces (walls, roofs, and roads); (4) increasing the thermal conductivity to $3.24 \, W \, m^{-1} \, K^{-1}$ to represent the large heat storage in the urban buildings and roads; and (5) reducing the green-vegetation fraction over urban areas to decrease evaporation.

25.3.3 Land-Use Classification Schemes

Mesoscale models make extensive use of land-use and land-cover (LULC) classification schemes as a means of grouping the wide range of values (in table lookup format) for the various model parameters such as the roughness length, soil moisture content, thermodynamic properties of different surfaces, albedo, and others. For instance, the WRF modeling system includes two 1 km global LULC data sets: (1) U.S. Geological Survey (USGS) 24-category date based on satellite Advanced Very High Resolution Radiometer (AVHRR) images, and (2) International Geosphere–Biosphere Programme (IGBP) 20-category data based on more recent moderate-resolution imaging spectroradiometer (MODIS) images. Common in these types of global LULC data sets, there is a generic urban land-use category. Furthermore, for many urban regions, high-resolution urban land-use maps, derived from in situ surveying (e.g., urban planning data) and remote-sensing data (e.g., Landsat 30 m images) are available. One example is the USGS National Land Cover Data (NLCD) classification, available for the United States, with three urban land-use categories: (1) low-intensity residential with a mixture of constructed materials and vegetation (30%–80% covered with constructed materials), (2) high-intensity residential with highly developed areas such as apartment complexes and row houses (usually with 80%–100% covered with constructed materials), and (3) commercial/industrial/transportation including infrastructure (e.g., roads, railroads). Figure 25.3a shows the spatial distribution of urban land use in NLCD data for the Houston region, United States.

25.3.4 Operational Model Technology Implementations

Grid size and time steps are part of the consideration of computational requirements for fine-scale modeling for meteorological and chemistry models. Since simulations are grid-size dependent, a common approach for downscaling employs the use of embedded nests with smaller modeling domains that employ smaller grid-size elements. Using nesting, there is yet an additional choice of using one versus two-way nesting, each producing uniquely different simulation outcomes given that simulations are grid-size dependent and will in general require application of a set of science options that is more appropriate to the grid size. Alternative approaches could invoke the use of adaptive grid techniques.

One way nesting with regular downscaling grid intervals is typically employed to provide assurance of reasonable and accurate initial and boundary condition. However, this approach has the computation advantage of not being bound to strict rules of downscaling in order to model at fine scales, as long as the input IC/BC are reasonably accurate. Two-way nesting, however, has the advantage of providing a means for including feedback between the coarse and finer grid size simulations, and in practice, does require a more formalize downscaling structure (cf Section 25.4).

Another consideration regards the closeness of the coupling between the meteorology and chemistry model systems. Feedbacks between meteorology and chemistry are optimally captured with a coupled systems approach; e.g., WRF-Chem (Grell et al., 2005). With meteorological simulations used as od ine to atmospheric chemistry models, this feedback is ignored.

25.3.5 UCP Databases

Urban canopy parameters are required for SLUCP and the multilayer canopy models (Masson, 2000; Kusaka et al., 2001; Ching et al., 2009). Here, there is a range of options for such information, mainly differing on grid specificity. Options range from use of table lookup values applicable to different land-use classes, to assigned unique grid-by-grid values obtained directly from high-resolution, detailed building observations, or from generalized extrapolation from more limited data, e.g., Britter and Hanna (2003).

25.3.6 Anthropogenic Heating

This is a contribution that varies widely over a city and can be a major term of the surface energy and water budgets. It is typically introduced as a flux term. However, there is a range of methods for introducing this term in models from table lookup (Lin et al., 2008) to more detailed approaches as in Sailor and Lu (2004), and Salamanca et al. (2009).

Figure 25.3b is a schematic from Dupont et al. (2004) which specifically illustrates an implementation for modeling the UBL in which detailed features of UCL are explicitly recognized and their specific contribution to models are implemented through prescription of a number of UCPs. As shown, some major details of the UCL are ignored or unresolved when using roughness based approach for grids containing tall and varied morphological elements. Moreover, the thickness of the first such layer above ground for such grids must necessarily be quite large and thus are inadequate for many applications where the UCL is well defined. The design basis for urban canopy models is that it handles the dynamic, thermodynamic, and turbulence structure on local scale grid sizes, but in a practical manner based upon model formulations using grid-specific UCPs as surrogates for the impact of the aggregated effects of individual morphology elements. Some important features included in canopy models are the aggregated effects of the surface and wall energy budget based on descriptions and appropriate aggregations of morphological elements such as characterizations and distribution of buildings and their roofs as isolated features or as organized street canyons, the distribution of tree and other green areas, existence and

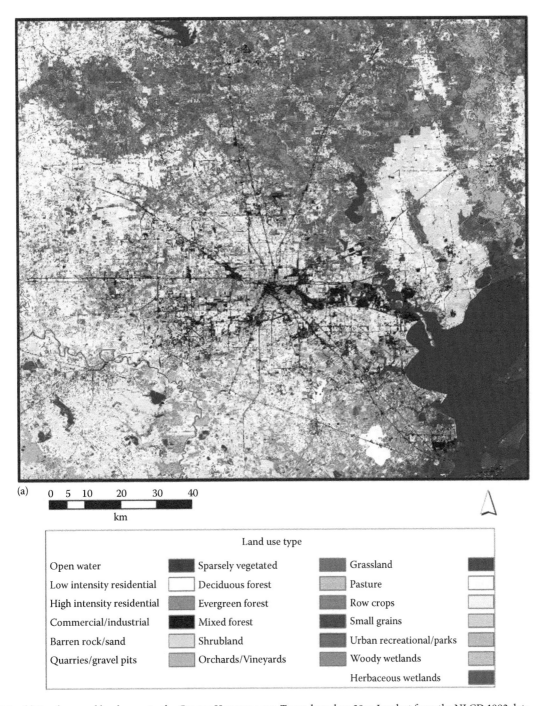

Land use type

Open water	Sparsely vegetated	Grassland	
Low intensity residential	Deciduous forest	Pasture	
High intensity residential	Evergreen forest	Row crops	
Commercial/industrial	Mixed forest	Small grains	
Barren rock/sand	Shrubland	Urban recreational/parks	
Quarries/gravel pits	Orchards/Vineyards	Woody wetlands	
		Herbaceous wetlands	

FIGURE 25.3   (a) Land use and land cover in the Greater Houston area, Texas, based on 30 m Landsat from the NLCD 1992 data.

(continued)

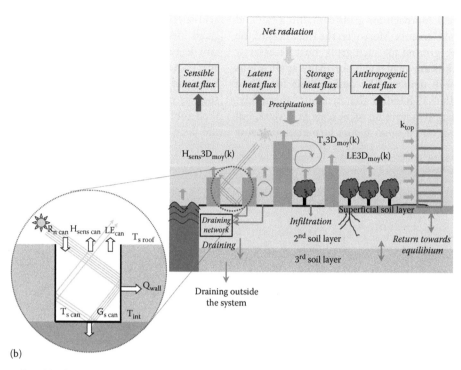

FIGURE 25.3 (continued)   (b) Schematic illustrating the introduction of urban morphological features of the urban canopy layer with a multi-layer scheme into mesoscale models.

extent of water bodies, fraction of impervious versus exposed surfaces, and others. Their definitions will prompt response in terms of model outputs of the gridded energy budget that includes effects of capturing fluxes of short and long wave radiative downwelling and long wave radiative upwelling and changes to albedo, and of alterations to the momentum and spectrum of turbulence given momentum drag and turbulence generation at the surface and throughout the canopy layer.

25.4 Specific Implementations/ Applications

25.4.1 Urbanization of the WRF System

The WRF modeling system, a community based framework, developed at the National Center for Atmospheric Research (NCAR) and its companion chemistry model (WRF-Chem.), with its cross scale modeling capability have been widely used for operations and research in the fields of numerical weather prediction, regional climate, emergency response, air quality, and regional hydrology, and water resources. The WRF model at its core is a nonhydrostatic, compressible model with mass coordinate system. It contains a number of options for describing various physical processes, e.g., a nonlocal closure and Mellor-Yamada based PBL schemes, the land surface schemes including the community Noah LSM (e.g., Chen et al., 1996; Chen and Dudhia, 2001), and a land data assimilation system including the North America Land Data Assimilation System (Mitchell et al., 2004). One basic function of the Noah LSM is to provide surface

sensible and latent heat fluxes and surface skin temperature as lower boundary conditions for coupled atmospheric models.

An integrated urban modeling system coupled to the Noah LSM was developed for WRF and WRF-Chem to improve WRF weather forecasts for cities and thereby air quality prediction and to establish a modeling tool for assessing impacts of urbanization on environmental problems by providing accurate meteorological information for planning mitigation and adaptation strategies in the changing climate. The urbanizations aspect of this system consists of (1) a suite of urban parameterization schemes with varying degrees of complexities, (2) the capability of incorporating in situ and remotely sensed data of urban land-use, building characteristics, anthropogenic heating, and moisture sources, (3) companion fine-scale atmospheric and urbanized land data assimilation systems, and (4) the ability to couple WRF/urban with fine-scale urban transport and diffusion (T&D) models and with chemistry models. It is anticipated that in the future this modeling system will interact with human response models and be linked to an urban decision system.

This framework is distinguished by the availability of modeling choices capable of representing effects of urban environments on local and regional weather and the cross-scale modeling ability ranging from continental scale, to city scale, and to building scale. Through the WRF preprocessing system (WPS), the user is able to specify all the requisite model control parameters to generate the data inputs and values controlling variables for performing specific model applications, including the science options, specifications of nested model domain, their vertical structure, grid size definitions, and initial and boundary conditions needed to run WRF. The urban modeling tools are in the core physics of the urbanized version of WRF, and Chen et al. (2011) describes the system and its current three urban parameterization schemes (i.e., bulk parameterization, single-layer urban canopy model (SLUCM) and building environment parameterization (BEP) for treating urban surface processes. Science questions and the choice of specific applications will dictate careful selection of different sets of science options and available databases. For instance, the bulk parameterization and SLUCM may be more suitable for real-time weather and air quality forecasts than the resource-demanding BEP that needs much higher vertical resolution in the atmospheric boundary layers and more input data. Figure 25.4a shows a schematic comparison of SLUCM and BEP models. The system also has the capability and flexibility to introduce adaptation development strategies such as, the impact of air conditioning by invoking the sophisticated BEP coupled with indoor-outdoor exchange models or by invoking changes to the albedo or vegetative cover by modifying their respective input values.

The problem of providing input data is challenging to most modeling systems. This problem is most acute for urban models and addressing this will continue to be a challenge. Currently the WRF-urban system has been testing several innovative methods and emerging new datasets as part of its initial implementation. These include an approach for initializing UCM state variables using the u-HRLDAS system; this replaces the need to assign a uniform temperature profile within building roofs and walls from air temperature for model grid points across a city. Specification of over 20 UCPs in WRF is a challenge due to large disparity in data availability and methodology for mapping fine-scale, highly variable data for a WRF modeling grid. Currently through WPS one can ingest high-resolution urban

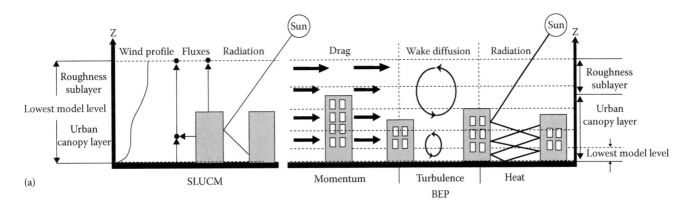

(a)

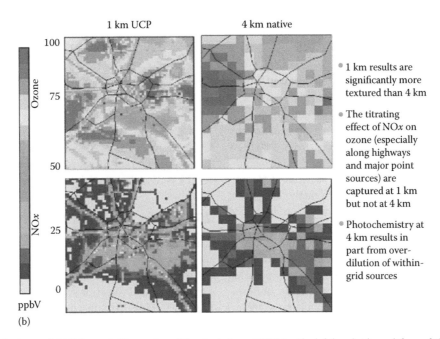

(b)

- 1 km results are significantly more textured than 4 km

- The titrating effect of NOx on ozone (especially along highways and major point sources) are captured at 1 km but not at 4 km

- Photochemistry at 4 km results in part from over-dilution of within-grid sources

FIGURE 25.4 **(See color insert.)** (a) Schematic illustration of the single-layer UCM (on the left-hand side, and the multilayer BEP models on the right-hand side (Chen et al., 2011). (b) Nested CMAQ 4 and 1 km simulations.

(continued)

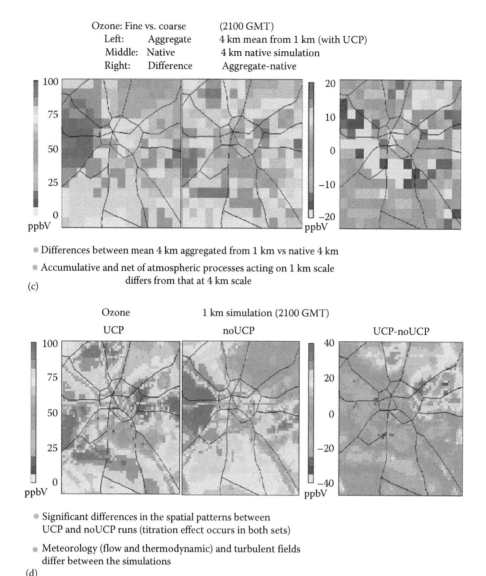

Ozone: Fine vs. coarse (2100 GMT)
Left: Aggregate 4 km mean from 1 km (with UCP)
Middle: Native 4 km native simulation
Right: Difference Aggregate-native

- Differences between mean 4 km aggregated from 1 km vs native 4 km
- Accumulative and net of atmospheric processes acting on 1 km scale differs from that at 4 km scale

(c)

Ozone 1 km simulation (2100 GMT)
UCP noUCP UCP-noUCP

- Significant differences in the spatial patterns between UCP and noUCP runs (titration effect occurs in both sets)
- Meteorology (flow and thermodynamic) and turbulent fields differ between the simulations

(d)

FIGURE 25.4 (continued) **(See color insert.)** (c) Difference between aggregated fine versus coarse grid CMAQ simulations. (d) CMAQ results based on MM5 simulations run with and without UCPs.

land-use maps and then assign UCPs based on a parameter table, or ingest gridded UCPs such as those from NUDAPT (Ching et al., 2009). It would be useful to blend these two methods whenever gridded UCPs are available. It is useful to optimize UCMs with observations, based upon a methodology that identifies a set of UCPs to which the performance of the UCM is most sensitive, and eventually to define optimized values for those UCPs for a specific city (Loridan et al., 2010). Another data issue is to improve the inputs and means for incorporating anthropogenic heating rates. Methods for incorporating this parameter varies from merely assigning specific, but arbitrary values to very sophisticated methods involving aggregating available heating rates buildings, industry/manufacturing, and transportation sectors (e.g., Sailor and Lu, 2004; Sailor and Hart, 2006; Torcellini et al., 2008) albeit obtaining and processing these data are far from automated tasks. Nevertheless, it has

been demonstrated that data are currently available for some U.S. cities in NUDAPT in which data were combined from all anthropogenic heating sources to create gridded, hourly input for the WRF-urban model.

25.4.2 Fine-Scale Air Quality Simulations

Air quality and multimedia simulation models provide tools for assessments and management of air quality and sustainability of ecosystems. Emission-based models such as the U.S. Environmental Protection Agency/Community Multiscale Air Quality (CMAQ) (Byun and Ching, 1999; Byun and Schere, 2006), represent a genre of tools capable of simulating a wide range of pollutant species for different grid sizes and at hourly time intervals. The CMAQ depends on pre-processors that provide the requisite gridded meteorological and emissions fields. The emissions preprocessor also require the inputs from the meteorological preprocessor for its hourly gridded fields. The concomitant simulations at increasingly finer scales for the meteorological and emissions preprocessors are achieved through application of successive nests with increasingly finer grid sizes for both preprocessor and CMAQ as shown in Figure 25.4b. The actual gridded outputs from meteorological simulations for a given event will differ for different grid sizes and for different science options. The air quality simulations will be impacted in a variety of complex ways by these differences. For one, the results of increasingly finer grid size resolution will cause large differences in the gridded precursor mixtures leading to differences in photochemical productions rates and concentration outcomes. For example, the aggregation value of the 16 concentration outputs from 1 km simulation will be different from the simulation result of comparable 4 km simulation, even using the same total aggregated emissions. An example of this is shown in Figure 25.4c. Additionally, even when using exactly the same size grids, different meteorological outcomes will occur using different input data and physics packages, e.g., difference in distribution of cloud amounts and types will impact the incoming radiation and resultant pollutant photolysis rates; differences in simulating transport winds and dilution parameters such as mixing heights will have a definite effect on the emissions distributions and the subsequent modeled concentration (Figure 25.4d). Another as yet to be determined impact will be the ability to simulate pollutant concentrations within the urban canopy using multilevel modeling capability. The inability to simulate within-canopy profiles is anticipated to have an important role in producing accurate results, and this is especially so in urban areas where the canopy height can be significantly larger that the height of the first layer of the atmosphere. Likewise, the ability to perform simulations with a prescribed thinner layer structure such as using the WRF multilayer urban-canopy model, BEP, together with fine-scale LES models will lead to improved simulations during times when the atmosphere is stable stratified as during a typical nighttime period.

25.5 Discussion and Summary

The advent of recent multilayer within-canopy model simulation capability provides an improved basis for fine-scale modeling and thus an important new advancement for urban meteorological and air quality simulation modeling. It is thus expected to improve both accuracy and realism of the modeled pollutant fields. Envisioned applications include detailed forecasts, emergency response and guidance, air quality and human health exposure assessments, urban heat island and mitigation strategies, urban development impacts and planning, and urban hydrology. Resources required for processing and analyzing simulations can vary considerably depending on model configuration implementations on domain and grid sizes, vertical layer structure, time steps, choices of science options and input data selected for an application. Some important modeling considerations on fine-scale grid simulations are summarized in the next stage:

25.5.1 Model Evaluation, Grid Size, and the Presence of Subgrid Variations

Simulations vary considerably in concentration values and in its degree of accuracy and precision for different modeling systems. The inherent "change of support" (Ching et al., 2006) constraint (typically unmet) due to the inherent subgrid scale characteristics of grid models, require the assumption that point measurements represent the simulation at the chosen grid size. Representativeness requirement of simulations to measurements is exacerbated for simulations over urban areas. Standard comparisons between model and observation require the set of observations (typically point measurements to be representative of the grid average prediction), a requirement that cannot be met with traditional sampling techniques for such cells where buildings dominate. Technically, it is by no means assured that the most sophisticated and elaborate urban models will yield the most accurate results. At this time, there is no clear and acceptable sampling protocol and methodology by which model evaluation can be performed given that inherent flow distortions by buildings are not simulated. Siting of monitors for model evaluation in urban areas is problematic; siting guidance for monitors is usually to be at some distance from obstacles. Thus, such measurements cannot represent the "virtual" drag and canopy heating effects of the grid simulations and thus grid simulations for wind speeds are likely to be biased low since the monitors are exposed to open field conditions.

The application of standard Model-Observation evaluation techniques from within canopy measurements is not possible without a dense network of observations within the canopy, and in theory, may not be possible since the model is predicting a virtual grid mean value representing an infinite degree of within-grid variability. Given that evaluation of within-canopy predictions is problematic, alternative nonstandard inductive evaluation methods are desirable. For temperature predictions, these might employ use of high-resolution skin temperature observations from satellite platforms. Such data are typically at a resolution commensurate with grid size of urban models. Also, it may be feasible to perform a process type evaluation of predicted rates of change of temperatures using satellite based measurements. Evaluation of the within canopy transport and turbulence fields are more problematic, but perhaps the use of lidar, sodars, or sounders to determine mixing height measurements may be usefully employed.

25.5.2 Fit-for-Purpose Guidance

As discussed in Baklanov et al. (2009) there are growing numbers of science options, model configurations and input data sets available or emerging. With no "one size fits all" circumstance, it is highly recommended to invoke the "Fit for Purpose" guidance as a critical design step. As discussed below, there are at least four major trade-off cost–benefit considerations when choosing among urban modeling alternatives including but not limited to (albeit related) technical considerations, model infrastructure, data requirements, and computational issues.

25.5.3 Model Data Requirements

Urban canopy models require a set of canopy parameters that represent the aggregated drag, thermal, turbulence generation, and vertical exchanges between the canopy and the overlying roughness sublayer. Current model formulations utilize a host of such parameters as well as ancillary information such as material conductive and radiative properties, anthropogenic heat and moisture source information either as table lookup or as gridded fields. In the future, it is a desired goal to optimize the model formulations so as to reduce the number and type of UCPs to as few in number as possible, to achieve reduction in data requirements and computational burdens (Martilli and Santiago, 2009).

25.5.4 Computational Issues

Model testing has already demonstrated that the CPU time will be significantly increased (order factor of 2) for simulations between one parent grid mesh size and a nested grid meshed daughter domain. This situation is exacerbated when a number of additional grid layers are specified for within-canopy simulations as, e.g., between a WRF RANS and the WRF-urbanized version. The WRF system offers the option of one- versus two-way nesting. The application of a two-way nesting implementation serves to minimize the aggregation differences between the parent and the nested daughter simulations, which is scientifically preferred. However, this type solution is not possible when additional vertical layers are introduced to resolve canopy features. For such simulations, the downscaling mode is restricted to one way nesting protocol. Thus, while advanced modeling options now being introduced that can alleviate certain fundamental limitations to a RANs type simulations, and while as yet not fully proven, they should yield an increase in the prediction accuracy and have the scientific benefit of simulations with thinner vertical layers and the ability to resolve the virtual conditions of the canopy. However, such models may be too computationally impractical to satisfy forecast timeliness requirements.

25.5.5 Resolving Within-Canopy Flow Details

This chapter has focused on fine-scale grid modeling and has not addressed the important subgrid canopy flow details. For example, current canopy implementations do not provide explicit treatment for specialized flows that have building height to street width dependency. Flow features induced by buildings include jetting, flow separation, recirculation zones, high shear, and production of turbulent eddies. Such idealization of flows in an urban canopy or within the roughness sublayer show flow affected by obstacles are roughly dependent on building height to street width ratios; leading to isolated roughness flows, wake interference flows, and skimming flows (Oke, 1978). However, for any given model grid cell, urban areas may contain

a wide range of distribution of building height to street width ratios. Modeling for this genre of applications has begun which involves introduction and testing of building resolving flow models such as with Computational Fluid Dynamics as in Coirier et al. (2005) coupled with grid simulations as shown in Chen et al. (2011), Martilli and Santiago (2009), and Tewari et al. (2010). Additionally hybrid and subgrid treatments such as plume or puff-in-grid methods can be introduced with grid models to provide greater within-grid cell detail to meteorological, dispersion models and also to air quality simulation models (Karamchandani et al., 2007).

References

Baklanov, A. et al. (eds.) 2009. *Meteorological and Air Quality Models for Urban Areas*, Springer-Verlag, Berlin, Germany.

Bougeault, P. and P. Lacarrere, 1989. Parameterisation of orography -induced turbulence in a mesobeta-scale model, *Monthly Weather Review*, 117, 1872–1890.

Britter R.E. and S. Hanna, 2003. Flow and dispersion in an urban area, *Annual Review of Fluid Mechanics*, 35, 469–496, doi:10.1146/annul rev.fluid.35.101101.161147.

Brown, M.J. 2000. Urban parameterizations for mesoscale meteorological models. In Z. Boybeyi (ed.), *Mesoscale Atmospheric Dispersion*, Advances in Air Pollution Series Volume 9, Wessex Press.

Burian, S., N. Augustus, I. Jeyachandran, and M. Brown, 2007. National Building Statistics Database: Version 2, LA-UR-08–1921, 82p.

Burian, S.J. and J. Ching, 2010. Development of gridded fields of urban canopy parameters for advanced urban meteorological and air quality models. EPA/600/R-10/007.

Byun, D.W. and J.K.S. Ching, 1999. Science Algorithms of the EPA Models-3 Community Multiscale Air Quality (CMAQ) Modeling System, March 1999. EPA/600/R-99/030. See also http://www.epa.gov/asmdnerl/models3/doc/science/science.html

Byun, D. and K.L. Schere, 2006. Review of the governing equations, computational algorithms, and other components of the Models-3 Community Multiscale Air Quality (CMAQ) Modeling System. *Applied Mechanics Reviews*, 59, 51–77.

Chen, F. and J. Dudhia, 2001. Coupling and advanced land surface-hydrology model with the Penn State-NCAR MM5 modeling system. Part I: Model implementation and sensitivity. *Monthly Weather Review*, 129(4), 569–585.

Chen, F., Z. Janjic, and K. Mitchell, 1997. Impact of atmospheric surface layer parameterization in the new land-surface scheme of the NCEP mesoscale Eta numerical model. *Boundary Layer Meteorology*, 85, 391–421.

Chen, F., H. Kusaka, R. Bornstein, J. Ching, C.S.B. Grimmond, S. Grossman-Clark, T. Loridan et al. 2011. The integrated WRF/urban modeling system: Development, valuation, and applications to urban environmental problems, *International Journal of Climatology*, 31, 273–288, doi: 10.1002/joc.2158.

Chen, F. et al. 1996. Modeling of land-surface evaporation by four schemes and comparison with FIFE observations. *Journal Geophysical Research*, 101, 7251.

Ching, J., M. Brown, S. Burian, F. Chen, R. Cionco, A. Hanna, T. Hultgren, T. McPherson, D. Sailor, H. Taha, and D. Williams, 2009. National Urban Database and Access Portal Tool, NUDAPT, *Bulletin of the American Meteorology Society*, 908, 1157–1168.

Ching, J., J. Herwehe, and J. Swall, 2006. On joint deterministic grid modeling and subgrid variability conceptual framework for model evaluation. *Atmospheric Environment*, 40(26), 4935–4945.

Cionco, R.M. and R. Ellefsen, 1998. High resolution urban morphology data for urban wind flow modeling. *Atmospheric Environment*, 321, 7–17.

Coirier, W.J., D.M. Fricker, M. Furmanczyk, and S. Kim, 2005. A computational fluid dynamics approach for urban area transport and dispersion modeling. *Environmental Fluid Mechanics*, 5(5), 443–479.

Dupont, S. and P.G. Mestayer, 2006. Parameterisation of the urban energy budget with the sub mesoscale soil model. *Journal Applied Meteorology and Climatology*, 45(12), 1744–1765.

Dupont, S., T.L. Otte, and J.K.S. Ching, 2004. Simulation of meteorological fields within and above urban and rural canopies with a mesoscale model (MM5). *Boundary Layer Meteorology*, 113, 111–158.

Ek, M.B. et al. 2003. Implementation of Noah land surface model advances in the national Center for Environmental Prediction operational mesoscale ETA model. *Journal of Geophysical Research*, 108(D22), 8851, doi: 10.1029/2002JD003296.

Grell, G. et al. 2005. Fully coupled "online" chemistry within the WRF model. *Atmospheric Environment*, 39, 6957–6975.

Grimmond C.S.B. and T.R. Oke, 1999. Aerodynamic properties of urban areas derived from analysis of surface form. *Journal of Applied Meteorology*, 38, 1262–1292.

Karamchandani, P., K. Lohman, and C. Seigneur, 2007. Sub-grid scale modeling of air toxics concentrations near roadways. *Extended Abstract at 6th CMAS Conference*, Chapel Hill, NC, October 1–3, 2007.

Kastner-Klein, P. and M. Rotach, 2004. Mean flow and turbulence characteristics in an urban roughness sublayer, *Boundary Layer Meteorology*, 111, 55–84, 2004.

Kusaka, H., H. Kondo, Y. Kikegawa, and F. Kimura, 2001. A simple single-layer urban canopy model for atmospheric models: Comparison with multi-layer and slab models. *Boundary Layer Meteorology*, 101, 329–358.

Lin, C-Y. et al. 2008. Urban Heat Island effect and its impact on boundary layer development and land-sea circulation over northern Taiwan, *Atmospheric Environment*, 42, 5635–5649, doi:10.1016.

Liu, Y., F. Chen, T. Warner, and J. Basara, 2006. Verification of a mesoscale data-assimilation and forecasting system for the Oklahoma City area during the Joint Urban 2003 Field Project. *Journal Applied Meteorology*, 45, 912–929.

Loridan, T., C.S.B. Grimmond, S. Grossman-Clarke, F. Chen, M. Tewari, K. Manning, A. Martilli, H. Kusaka, and M. Best, 2010. Trade-offs and responsiveness of the single-layer urban canopy parameterization in WRF: An od ine evaluation using the MOSCEM optimization algorithm and field observations. *Quarterly Journal of the Royal Meteorological Society*, 136 997–1019, doi: 10.1002/qj.614.

Martilli, A., A. Clappier, and M. Rotach, 2002. An urban surface exchange parameterization for mesoscale models. *Boundary Layer Meteorology*, 104, 261–304.

Martilli, A. and J.L. Santiago, 2009. How to use Computational Fluid Dynamics model for urban canopy parameterizations. In A. Baklanov, S. Grimmond, A. Mahura, and M. Athanassiadou, (eds.), *Meteorological and Air Quality Models for Urban Areas*, Springer-Verlag, Berlin, Germany, 2009.

Martilli A. and R. Schmitz, 2007. Implementation of an Urban Canopy Parameterization in WRF-chem. Preliminary results. *Seventh Symposium on the Urban Environment of the American Meteorological Society*, San Diego, CA.

Masson, V. 2000. A physically-based scheme for the urban energy budget in atmospheric models. *Boundary Layer Meteorology*, 94, 357–397.

Masson, V. 2006. Urban surface modeling and the meso-scale impact of cities. *Theoretical and Applied Climatology*, 84, 35–45.

Mitchell, K.E. and Coauthors, 2004. The multi-institution North American Land Data Assimilation System (NLDAS): Utilizing multiple GCIP products and partners in a continental distributed hydrological modeling system. *Journal Geophysical Research*, 109, D07S90, doi:10.1029/2003JD003823.

Oke, T.R. 1978. *Boundary Layer Climates*. Methuen, London, U.K., 372p.

Otte, T.L., A. Lacser, S. Dupont, and J.K.S. Ching, 2004. Implementation of an urban canopy parameterization in a mesoscale meteorological model. *Journal Applied Meteorology*, 43, 1648–1665.

Piringer, M., C.S.B. Grimmond, S.M. Joffre, P. Mestayer, D.R. Middleton, M.W. Rotach, A. Baklanov et al. 2002. Investigating the surface energy balance in urban areas—Recent advances and future needs. *Water, Air & Soil Pollution: Focus*, 2(5–6), 1–16.

Pleim, J.E. and A. Xiu, 2003. Development of a land surface model. Part II: Data assimilation. *Journal Applied Meteorology*, 42, 1811–1822.

Rotach, M. 1999. On the influence of the urban roughness sublayer on turbulence and dispersion, *Atmospheric Environment*, 33, 4001–4008.

Roth, M. 2000. Review of atmospheric turbulence over cities. *Quarterly Journal Royal Meteorological Society*, 126, 941–990.

Sailor, D.J. and M. Hart, 2006. An anthropogenic heating database for major U.S. cities, *Sixth Symposium on the Urban Environment*, Atlanta, GA, January 28–February 2, American Meteorological Society, Boston, MA, paper 5.6.

Sailor, D.J. and L. Lu, 2004. A top-down methodology for developing diurnal and seasonal anthropogenic heating profiles for urban areas. *Atmospheric Environment*, 38(17), 2737–2748.

Salamanca, F., A. Krpo, A. Martilli, and A. Clappier, 2009. A new building energy model coupled with an Urban Canopy Parameterization for urban climate simulations—Part I. Formulation, verification and sensitivity analysis of the model. *Theoretical and Applied Climatology*, doi: 10.1007/s00704-009-0142-9.

Taha, H. 2008a. Episodic Performance and Sensitivity of the Urbanized MM5 (uMM5) to Perturbations in Surface Properties in Houston, TX, *Boundary Layer Meteorology*, 127(2), 193–218, doi: 10.1007/s10546-007-9258-6.

Taha, H. 2008b. Meso-urban meteorological and photochemical modeling of heat island mitigation. *Atmospheric Environment*, 42(38), 8795–8809. doi:10.1016/j.atmosenv.2008.06.036.

Taha, H. 2008c. Urban surface modification as a potential ozone air-quality improvement strategy in California: A mesoscale modeling study, *Boundary Layer Meteorology*, 127, 219–239, doi: 10.1007/s10546-007-9259-5.

Tewari, M., H. Kusaka, F. Chen, W.J. Coirier, S. Kim, A. Wyszogrodzki, and T.T. Warner, 2010. Impact of coupling a microscale computational fluid dynamics model with a mesoscale model on urban scale contaminant transport and dispersion. *Atmospheric Research*, 96(2010), p. 656–664.

Torcellini, P., M. Deru, B. GriDth, K. Benne, M. Halverson, D. Winiarski, and D.B. Crawley, 2008. *DOE Commercial Building Benchmark Models*, Preprint of Conference paper NREL/CP-550-4329, ACEEE Summer study on Energy EDciency in Buildings, Pacific Grove, Ac, August 17–22, 2008, 15p.

Wilson, R.N. and R.H. Shaw, 1977. A higher order closure model for canopy flow, *Journal of Applied Meteorology*, 16, 1197–1205.

Xiu, A. and J.E. Pleim, 2001. Development of a land surface model. Part I: Application in a mesoscale meteorological model. *Journal Applied Meteorology*, 40, 192–209.

Yamada, T. 1982. A numerical model study of turbulent airflow in and above a forest canopy, *Journal Meteorological Society of Japan*, 60(1), 439–453.

26

Air Pollution Modeling and Its Applications

Daewon W. Byun*
National Oceanic and Atmospheric Administration

26.1 Introduction

Earth's dry air consists of about 78% nitrogen, 21% oxygen, and 1% other noble and trace gases. Variable trace constituents include water vapor and a variety of chemical compounds such as carbon dioxide, methane, ozone, carbon monoxide, nitrogen species, volatile organic compounds (VOCs), ammonia, and particulate matter (PM) that contribute to air pollution. Some of the pollutants are directly emitted from anthropogenic activities such as industrial and agricultural practices and traﬃc or through natural occurrences such as forest fires, wind-driven suspensions of surface materials, and volcanic eruptions. Air also carries water droplets (clouds) and small particles such as dust, smoke, soot, organic matter, and sea salts. Once released in the atmosphere, these gases and particles go through chemical reactions and physical processes, forming other pollutants, are transported and diffused by air motions, and are removed from the atmosphere via dry deposition and wet scavenging by precipitation. Among these substances, ground-level ozone and fine PM (particulate matter of aerodynamic diameter less than $2.5\,\mu m$, $PM_{2.5}$) are of concern because of their direct harmful effects on human health. Ozone, a major constituent of smog, is not emitted directly, but is formed in a complex series of gas-phase reactions involving oxides of nitrogen and VOCs in the presence of sunlight. When inhaled, ground-level ozone damages lung tissue, worsens asthma, and increases mortality rate. It can also damage the leaves of plants and trees. Fine PM is produced from combustion processes (burning of fossil fuels, residential fireplaces, agricultural burning, and both prescribed and natural fires), volcanic emissions, windblown dust, and as a result of chemical reactions in the atmosphere. Exposure to fine PM is harmful to most people but much more damaging to children and people with respiratory and cardiovascular diseases.

Air quality models help to expand our understanding of the science of chemical and physical interactions of such pollutants in the atmosphere and forecast the quality of the air we breathe. Modeling of air quality also provides a means of evaluating the effectiveness of air pollution and emission control policies and regulations that influence energy management and agricultural practices to protect public health. The primary objectives of such air quality models are to (1) improve the ability of environmental managers to evaluate the impact of air quality management practices for multiple pollutants at multiple scales, (2) enhance scientific ability to understand and model chemical and physical atmospheric interactions, (3) guide the development of air quality regulations and standards, and (4) assist in the creation of state implementation plans. They have been also used to evaluate longer-term pollutant climatology as well as short-term transport from localized sources,

Handbook of Environmental Fluid Dynamics, Volume Two, edited by Harindra Joseph Shermal Fernando. © 2013 CRC Press/Taylor & Francis Group, LLC. ISBN: 978-1-4665-5601-0.

* Dr. Daewon Byun passed away unexpectedly while this volume is in preparation and after submitting his manuscript. A brief reflection of his life follows this chapter, written by his former colleague, Kenneth L. Schere of the United States Environmental Protection Agency. Daewon will be sorely missed by the atmospheric sciences community and his friends.

and they can be used to perform simulations using downscaled regional climate from global climate change scenarios. Various observations from the ground and in-situ and from aircraft and satellite platforms can be used at almost every step of the processing of this decision support tool (DST) for air quality.

In this article, I present (1) basic principles of air quality modeling, in particular as an integrated part of atmospheric modeling; (2) methods of air quality modeling and analysis, including numerical approaches focusing on ozone problems and environmental data used; and (3) application examples of using modeling tools for daily air quality forecasting and for an emissions uncertainty study to guide pollution control strategies.

26.2 Principles

To simulate weather and air quality phenomena realistically, adaptation of a one-atmosphere perspective based mainly on a first principle description of the atmospheric system is necessary. This perspective emphasizes that the influence of interactions at different dynamic scales and among multi-pollutants cannot be ignored. For example, descriptions of processes critical to producing oxidants, acid and nutrient depositions, and fine particles are too closely related to treat separately from meteorological changes.

The governing set of equations describing atmospheric motion and thermodynamics for dry air is established from first principles of mass, momentum (velocity multiplied by density), and energy conservation. With the addition of an equation of state that relates atmospheric pressure with air density and temperature, one can derive a set of Navier–Stokes equations describing instantaneous relations among the six state variables representing air density, temperature, pressure, and three components of momentum. Previously, many weather models have been built with limited atmospheric dynamics assumptions and ranges of time and spatial scales of applicability. To simplify the model development process, the governing equations were first simplified to focus narrowly on the target problems before computer codes were implemented. This approach enabled rapid development of models but extensive developmental efforts were needed to expand the applicability of the models to a wide range of atmospheric phenomena. Therefore, newer weather models, e.g., WRF/ARW (Klemp et al., 2007) and the NOAA/NWS Weather Research and Forecast Non-hydrostatic Mesoscale Model (WRF/NMM) (Janjic, 2003) which is the operational North American Mesoscale Model (NAM) that replaced the Eta model, utilize a nearly full-form of Navier–Stokes equations for describing the fully compressible atmosphere.

In most weather prediction models, temperature and pressure, as well as moisture variables, are used to represent the thermodynamics of the system. Often, these thermodynamic parameters are represented with the advective form equations in meteorological models. Most of the time, density is diagnosed as a by-product of the simulation, usually through the ideal gas law. For multiscale air quality applications where strict mass conservation is required, prognostic equations for the thermodynamic

variables are preferably expressed in a conservative form similar to the continuity equation. Ooyama (1990) has proposed the use of prognostic equations for entropy and air density in atmospheric simulations by highlighting the thermodynamic nature of pressure. Entropy is a well-defined state function of the thermodynamic variables such as pressure, temperature, and density. Therefore, entropy is a field variable that depends only on the state of the fluid. The principle he uses is the separation of dynamic and thermodynamic parameters into their primary roles. An inevitable interaction between dynamics and thermodynamics occurs in the form of the pressure gradient force.

A set of governing equations for the fully compressible atmosphere is presented (Byun, 1999a). Here, density and entropy are used as the primary thermodynamic variables. For simplicity, a dry adiabatic atmosphere is considered. As the density and temperature vary significantly in the vertical direction, variables representing altitude are often replaced with those in a pressure-dependent coordinate $s(x, y, z, t)$.:

$$\frac{\partial(\rho J_s \hat{\mathbf{V}}_s)}{\partial t} + (\nabla_s \cdot \rho J_s \hat{\mathbf{V}}_s)\hat{\mathbf{V}}_s + \frac{\partial \rho J_s \hat{\mathbf{V}}_s \dot{s}}{\partial s} + f\hat{\mathbf{k}} \times \rho J_s \hat{\mathbf{V}}_s$$
$$+ J_s\left[\nabla_s p - \frac{J_s}{g}\left(\frac{\partial s}{\partial z}\right)\frac{\partial p}{\partial s}\nabla_s \Phi\right] = \rho J_s \hat{\mathbf{F}}_s \tag{26.1}$$

$$\frac{\partial(\rho J_s w)}{\partial t} + \nabla_s \cdot (\rho J_s w \hat{\mathbf{V}}_s) + \frac{\partial(\rho J_s w \dot{s})}{\partial s}$$
$$+ J_s\left(\frac{\partial p}{\partial s} + \rho \frac{\partial \Phi}{\partial s}\right)\left(\frac{\partial s}{\partial z}\right) = \rho J_s\left(F_3 + \frac{w Q_\rho}{\rho}\right) \tag{26.2}$$

$$\frac{\partial(\rho J_s)}{\partial t} + \nabla_s \cdot (\rho J_s \hat{\mathbf{V}}_s) + \frac{\partial(\rho J_s \dot{s})}{\partial s} = J_s Q_\rho \tag{26.3}$$

$$\frac{\partial(\varsigma J_s)}{\partial t} + \nabla_s \cdot (\varsigma J_s \hat{\mathbf{V}}_s) + \frac{\partial(\varsigma J_s \dot{s})}{\partial s} = J_s Q_\varsigma \tag{26.4}$$

$$\frac{\partial(\varphi_i J_s)}{\partial t} + \nabla_s \cdot (\varphi_i J_s \hat{\mathbf{V}}_s) + \frac{\partial(\varphi_i J_s \dot{s})}{\partial s} = J_s Q_{\varphi_i} \tag{26.5}$$

where

J_s is the Jacobian of coordinate transformation

ρ is the dry air density

p is the atmospheric pressure

Φ is geopotential

$\hat{\mathbf{V}}_s = u\hat{\mathbf{i}} + v\hat{\mathbf{j}}$ (contravariant horizontal wind) and $\dot{s} = ds/dt$ (contravariant vertical wind)

The Q terms represent sources and sinks of each conservative property

φ_i represents each trace species concentration for different phases of water vapor and individual pollutant species

The thermodynamic variable ς is the entropy per unit volume (entropy density) defined as

$$\varsigma = \rho C_{vd} \ln\left(\frac{T}{T_{oo}}\right) - \rho R_d \ln\left(\frac{\rho}{\rho_{oo}}\right) \qquad (26.6)$$

where

C_{vd} is the specific heat at constant volume
R_d the gas constant for dry air
T_{oo} and ρ_{oo} are reference air temperature and density, respectively

Depending on the atmospheric models, different vertical coordinates may be used. For example, WRF/ARW utilizes a terrain following hydrostatic pressure coordinate (note that $\sigma_\pi = 1 - s$):

$$s = \frac{(\pi - \pi_s)}{\mu}, \qquad (26.7)$$

where

π is the hydrostatic pressure
$\mu = \pi_s - \pi_t$ represents the difference in hydrostatic pressure between the base (π_s) and the top (π_t) of the model column

WRF/NMM utilizes a terrain-following hybrid pressure-sigma vertical coordinate, with the lower part below a cutoff hydrostatic pressure level (say 420 hPa) using σ_π and the upper part using hydrostatic pressure levels. According to Janjic (2003), this hybrid coordinate restricts possible inaccuracies due to the sloping coordinate surfaces to only the lower half of the mass of the atmosphere while the sigma-p coordinate has large errors due to topography in the stratosphere.

An air quality model should be viewed as an integral part of atmospheric modeling and the governing equations and computational algorithms should be consistent and compatible with the host weather prediction model. For successful one-atmosphere simulations, it is imperative to have consistent algorithmic linkage between meteorological and chemical transport models (CTMs). In the set of governing equations presented earlier, Equation 26.5 represents conservation of both water vapor and trace species. Both are subject to injection of materials from the Earth's surface as well as removal processes such as precipitation. However, the origins of the source terms and the eventual impacts on the atmospheric motions and thermodynamics are distinctively different. The water vapor budget is affected greatly by phase changes and microphysical processes. As the amount of water vapor is substantial (about 4%–7% by mass), it is directly involved in the determination of atmospheric stability and thus atmospheric motion. In actuality, water vapor cannot be considered as a passive trace species and therefore, the governing set of equations presented earlier should be rewritten utilizing the actual moist air density which provides consistent dynamics and thermodynamics connected with the parameterization of physical processes. Continuity equations for trace species like air pollutants are expressed in conservative form as the entropy density.

Proper modeling of air pollutants requires that the broad range of temporal and spatial scales of multi-pollutant interactions be considered simultaneously. Depending on the required level of specificity, the instantaneous equations are averaged over appropriate time intervals and grid sizes to form a set of deterministic conservation equations for the averaged state variables before they are solved numerically. Ideally, such averaging must be done over the ensemble of realizations; however, we will replace such ensemble averaging with time averaging and further apply volume averaging over a spatial discretization (mesh) to derive equations suitable for atmospheric modeling with practical grid resolutions.

For dry air density and species concentrations, we apply the Reynolds decomposition, i.e., separating each into mean and fluctuating components:

$$\rho = \bar{\rho} + \rho'; \quad \varphi_i = \bar{\varphi}_i + \varphi_i' \qquad (26.8)$$

And similarly, the mean contravariant wind components and their fluctuations, defined as

$$\bar{u} = \overline{\rho u}/\bar{\rho}; \quad \bar{v} = \overline{\rho v}/\bar{\rho}; \quad \bar{s} = \overline{\rho \dot{s}}/\bar{\rho}, \qquad (26.9)$$

are used to write the continuity equation for the mean dry air density to keep the original conservation form as

$$\frac{\partial(J_s\bar{\rho})}{\partial t} + \nabla_s \cdot (J_s\bar{\rho}\,\bar{\mathbf{V}}_s) + \frac{\partial(J_s\bar{\rho}\bar{\dot{s}})}{\partial s} = J_s\bar{Q}_\rho \qquad (26.10)$$

If the weather model simulates mean dry air density and mean wind that are mass consistent, the density error term $\bar{Q}_\rho$ should vanish. With an imperfect weather model, this term could be non-zero and the impact of such an error must be handled in the trace species transport equation. In a meteorological model that utilizes a kind of hydrostatic pressure coordinate, the continuity equation for the mean air density is utilized to estimate the pressure-tendency term as $J_s\bar{\rho}$, which is constant in space and time, assuming the atmosphere is confined by the rigid model top π_t and the base π_s.

Moisture in the air as well as pollutant concentrations can be accounted for as a mass of trace species per unit mass of mean dry air (i.e., mixing ratio). With the use of the mean mixing ratio, its fluctuation component, and the fluctuation wind components defined as

$$\bar{q}_i = \bar{\varphi}_i/\bar{\rho}; \quad q_i' = \varphi_i'/\bar{\rho}; \quad \hat{\mathbf{V}}_s' = \hat{\mathbf{V}}_s - \bar{\hat{\mathbf{V}}}_s; \quad \dot{s}' = \dot{s} - \bar{\dot{s}}, \qquad (26.11)$$

the mean continuity equation for the trace species becomes a pollutant dispersion equation:

$$\frac{\partial(J_s\bar{\rho}\bar{q}_i)}{\partial t} + \underbrace{\nabla_s \cdot (J_s\bar{\rho}\bar{q}_i\,\bar{\hat{\mathbf{V}}}_s) + \frac{\partial(J_s\bar{\rho}\bar{q}_i\bar{\dot{s}})}{\partial s}}_{(a)}$$

$$+ \underbrace{\nabla_s \cdot (J_s\bar{\rho}\overline{q_i'\hat{\mathbf{V}}_s'}) + \frac{\partial(J_s\bar{\rho}\overline{q_i'\dot{s}'})}{\partial s}}_{(b)} = J_s Q_{\varphi_i} \qquad (23.12)$$

Key assumptions applied in the derivation of the partial differential equation (PDE) for air pollution modeling are as follows:

1. Pollutant concentrations are suDciently small, such that their presence would not affect the meteorology to any detectable extent; thus, species conservation equations can be solved independently of the Navier–Stokes and energy equations.

2. The velocities and concentrations of the various species in atmospheric flow are turbulent quantities and undergo turbulent diffusion, while molecular diffusion for most trace species can be ignored.

3. The ergodic hypothesis holds for the ensemble average and thus can be substituted with the time average.

4. The turbulent atmospheric motion is assumed stationary for the averaging time period of interest (say 30 min to 1 h).

5. The source functions (i.e., emissions of pollutants and chemical reaction source/sink terms) are deterministic for all practical purposes and contributions of covariance terms among tracer species are neglected.

Although not presented here, the equations for describing atmospheric dynamics, Equations 26.1 and 26.2, should undergo the same averaging processes and the mean entropy equation should take a similar form as Equation 26.12. It should be noted here that the averaging processes applied produce additional unknowns, such as covariance terms $\overline{q_i' \hat{\mathbf{V}}_s'}$ and $\overline{q_i' s'}$, known as the Reynolds fluxes. To match the number of unknowns with the number of independent equations, additional principles have been sought to close the equation set. The turbulent closure problem has been the focus of much research over the years and remains a great unsolved problem of physics. The terms in the left-hand side (l.h.s.) of Equation 26.12 other than the rate of the change term represent (a) horizontal and vertical advection and (b) horizontal and vertical diffusion. The right-hand side (r.h.s.) of Equation 26.12 can be expanded to represent contributions from various sources and sinks such as (c) chemistry that includes gaseous, aerosol, and aqueous-phase reactions, (d) dry deposition, and (e) emissions;

$$J_s Q_{\varphi_i} = \left. \frac{\partial (J_s \overline{\varphi}_i)}{\partial t} \right|_{\text{chemistry}} + \left. \frac{\partial (J_s \overline{\varphi}_i)}{\partial t} \right|_{\text{drydep}} + J_s S_{\varphi_i} \big|_{\text{emis}} \quad (26.13)$$

$$\underset{(c)}{} \qquad \underset{(d)}{} \qquad \underset{(e)}{}$$

Description of the key science processes and their numerical algorithms for the computer modeling, together with the initial and boundary conditions necessary for solving the PDE is provided in the following section.

26.3 Method of Modeling: Numerical Algorithms and Modeling Inputs

To build a computer model simulating air pollution phenomena, accurate but fast numerical algorithms must be applied to provide a solution to the governing PDE. Also, the source and sink terms need to be parameterized with known quantities in the model as well as inputs to the model. Highlights of numerical algorithms used for the terms in the pollutant species equation in the U.S. EPA's Community Multiscale Air Quality (CMAQ) modeling system (Byun and Ching, 1999; Byun and Schere, 2006) are described in the following. Similar approaches are applied to other contemporary air quality models.

26.3.1 Advection Algorithms

There have been many studies on the numerical advection algorithms used in atmospheric models. Because the equation is hyperbolic in nature and spatial discretization of the solution generates a finite number of Fourier modes that travel at different speeds, numerical approximation leads to constructive and destructive interference. If the high-wave-number Fourier modes are damped significantly, then numerical diffusion becomes prevalent. Requirements for numerical advection algorithms for pollutant transport are somewhat different from those used for the nonlinear momentum transport in the meteorological models. In particular, algorithms are sought with minimal mass conservation errors to account for pollutant mass, small numerical diffusion to minimize artificial spread of concentrations, and small phase errors to keep the ratios of pollutant species in the mixture accurate. Furthermore, the algorithm should be positive definite and monotonic so that new local extreme values due to the numerical computation are not produced.

To maintain mass conservation characteristics in the numerical solution, the advection process is preferably expressed with $\varphi_i{}^* = J_s \overline{\rho} \overline{q}_i$ in conservative (flux) form, as follows:

$$\frac{\partial (\varphi_i{}^*)}{\partial t} = -\nabla \cdot \left(\varphi_i{}^* \hat{\overline{\mathbf{V}}}_s \right) - \frac{\partial (\varphi_i{}^* \overline{s})}{\partial s} \quad (26.14)$$

Because of large-scale differences in the vertical and horizontal motions in the atmosphere, it is customary to solve the horizontal and vertical components separately. Although 2-D schemes may be more desirable in this regard, they are often more diDcult to implement and less computationally eDcient than 1-D schemes. The 1-D flux-form advection equation, in the Cartesian x-coordinate, is given as

$$\frac{\partial \varphi_i{}^*}{\partial t} = -\frac{\partial (u \varphi_i{}^*)}{\partial x} \quad (26.15)$$

A flux-form discretization with first-order accuracy in time results in

$$(\varphi_i{}^*)_j^{n+1} = (\varphi_i{}^*)_j^n - \frac{\Delta t}{\Delta x_j} \left[F_{j+1/2}^n - F_{j-1/2}^n \right] \quad (26.16)$$

where $F_{j+1/2}^n$ and $F_{j-1/2}^n$ denote advective fluxes ($F_x = u \varphi_i{}^*$) through the faces of cells j in the x-direction, $\Delta x_j = x_{j+1/2} - x_{j-1/2}$, over the time step Δt, which must satisfy $\Delta t < \max_j \left\{ \Delta x_j / |u_{j+1/2}| \right\}$ to maintain numerical stability. Depending on the methods used to estimate the advective fluxes, the 1-D flux-form algorithms may be classified as finite difference, finite volume, flux corrected, finite-element schemes, or spectral schemes.

The practice of applying 1-D numerical schemes in 3-D pollutant transport will lead to some diDculties, such as the representativeness of the mass continuity and the formulation of proper boundary conditions for non-orthogonal 2-D horizontal and vertical directions when simulating a region with complex topography or strong vertical motion in the atmosphere. Compared to engineering fluid dynamics in which density variation can often be ignored, advection algorithms for atmospheric trace species need to consider accurate treatment of air density variations, which must satisfy the dynamic continuity equation in 3-D. By combining the continuity equations for mean air density and trace species (Equations 26.10 and 26.14), one can readily show that (Byun, 1999b)

$$\frac{d\overline{q}_i}{dt} = \overline{q}_i \left(\frac{Q_{\varphi_i}}{\varphi_i} - \frac{Q_\rho}{\overline{\rho}} \right) \tag{26.17}$$

When the meteorological data are mass consistent (i.e., $Q_\rho = 0$), Equation 26.17 shows that the pollutant mixing ratio follows the Lagrangian conservation form preserving the quantity along the atmospheric motion responding to the amount of pollutant injected, i.e., Q_{φ_i}. This provides the basis of the Lagrangian numerical description of pollutant mixing ratio conservation. Because the meteorological inputs may not be always mass consistent due to various reasons (i.e., $Q_\rho \neq 0$), 3-D numerical advection can result in mass conservation errors of pollutants. Various "mass correction" approaches have been developed to mitigate the impact of spurious pollutant sources in air quality modeling. One way to avoid this diDculty is to compensate the mass consistency error in the air density and wind fields by insisting that at minimum, pollutant mixing ratios be conservative (i.e., ratios of mixtures be maintained during the advection in spite of mass consistency error). This amounts to adding an additional correction process that makes the r.h.s. of Equation 26.17 vanish by setting $Q_{\varphi_i} = \overline{\varphi}_i Q_\rho / \overline{\rho}$. Another approach is to modify wind velocity components used for pollutant advection to maintain total pollutant mass conservation despite the fact that the air mass may not be conserved. When the 1-D numerical algorithm is utilized in sequence to simulate 3-D advection, the 1-D version of the continuity equation for air density for each direction is also applied to modify wind components (mostly vertical velocity) slightly. The latter approach may conserve both pollutant mass and mixing ratio, but at the expense of utilizing a modified wind field. For meteorological conditions with serious divergence and convergence flows as they happen during frontal passages, such modification of velocity components can sometimes seriously distort pollutant movement.

26.3.2 Boundary Layer Diffusion Algorithms

Mixing of pollutants in the atmospheric boundary layer (ABL) is primarily due to the turbulent motion of the air, which is represented by the gradient of Reynolds flux terms in the turbulent diffusion equation:

$$\frac{\partial(\varphi_i^*)}{\partial t} = -\nabla_s \cdot \left(J_s \overline{\rho} \, \overline{q_i' \hat{\mathbf{V}}_s'} \right) - \frac{\partial(J_s \overline{\rho} \, \overline{q_i' s'})}{\partial s} \tag{26.18}$$

The Reynolds flux terms need to be parameterized (i.e., expressed as a function of known quantities in the model) to solve the equation. In meteorological models, the number of equations that include the Reynolds flux terms is further reduced by utilizing an equation for the turbulent kinetic energy (TKE) instead of three equations for individual variances of momentum components. In an approach called one-and-a-half order closure, one can retain only equations for the variances while reducing all other covariance terms with parameterized formulas. The meteorological part of the model can utilize either the equations for TKE and variances of temperature and moisture, or the first-order formulations. Including the variance equations increases the number of unknowns that need to be parameterized compared to the first-order closure approach. In air quality models that deal with photochemical problems, adding variance equations for tracer species can be very expensive computationally. It doubles the number of chemical equations and the covariance terms among the pairs of chemical species involved must be parameterized.

The closure approximation is further classified either as local, in which only mean quantities nearby the location (such as first order gradient theory) are used in the parameterization, or nonlocal, in which the parameterization accounts for the fact that larger-size eddies can transport fluid across finite distances before the smaller eddies have a chance to cause mixing. In nonlocal closure, the mass exchange coeDcients are provided by the meteorological part of the model.

If first-order gradient theory is utilized to parameterize the Reynolds terms, Equation 26.18 becomes

$$\frac{\partial(\varphi_i^*)}{\partial t} = -\nabla_s \cdot \left(J_s \overline{\rho} K_H \nabla_s \overline{q}_i \right) - \frac{\partial}{\partial s} \left[\left(J_s \overline{\rho} K_s \frac{\partial \overline{q}_i}{\partial s} \right) \right] \tag{26.19}$$

where K_H and K_s represent contravariant eddy diffusivities in horizontal and vertical directions, respectively. Because of the relatively large numerical diffusion of the advection process, regional-scale model predictions are much less sensitive to the treatment of horizontal diffusion. Therefore, only the vertical diffusion algorithm is discussed later. The 1-D vertical diffusion kernel solves a parabolic diffusion equation:

$$\frac{\partial \varphi_i^*}{\partial t} = -\frac{\partial}{\partial s} \left(K_s J_s \overline{\rho} \frac{\partial \overline{q}_i}{\partial s} \right) \tag{26.20}$$

The vertical component of the contravariant eddy diffusivity K_s can be related to the Cartesian z-coordinate eddy diffusivity K_z with $K_s \approx J_s^{-2} K_z$. K_z is estimated using formulations that are usually parameterized with the planetary boundary layer (PBL) height (h_{PBL}), friction velocity (u_*), and a stability parameter from the meteorological modeling.

Similar to the advection equation, a numerical solver for the diffusion equation should preserve mass, the amplitude should decay monotonically (without generating local peaks), no phase errors should be introduced, and positivity preserved. The popular semi-implicit Crank-Nicholson algorithm, in which both the current and future time steps are used in the numerical solver, is stable in the sense that the amplitude of the signal either decays or stays the same. The positivity condition is kept by utilizing a short enough integration time step that satisfies the Courant-type condition for the diffusion:

$$\Delta t_{vdiff} \leq \min\left\{\Delta s_j \left(\frac{K_{j-1/2}}{\Delta s_{j-1/2}} + \frac{K_{j+1/2}}{\Delta s_{j+1/2}}\right)^{-1}\right\} \quad (26.21)$$

A limitation of local gradient models for representing convective conditions is that eddy diffusion assumes that the turbulent eddies are much smaller in size than the vertical grid spacing of the model. In the convective boundary layer (CBL), the assumption of eddy diffusion fails to represent turbulent fluxes that are often counter to the gradients. To describe the algorithms based on a nonlocal closure, it is useful to express the vertical mixing in flux form as

$$\frac{\partial \varphi_i^*}{\partial t} = -\frac{\partial F_D}{\partial s} \quad (26.22)$$

where $F_D = J_s \overline{\rho} \overline{q_i' s'}$ is the turbulent flux. Nonlocal closure recognizes that larger-size eddies can transport fluid across distances longer than the grid increment before the smaller eddies

have a chance to cause mixing. Nonlocal closure schemes have been successful at simulating CBL fluxes and profiles. Many PBL schemes simply add a gradient adjustment term γ_{q_i} to the eddy diffusion equation (e.g. Deardorff, 1972):

$$\frac{\partial \varphi_i^*}{\partial t} = -\frac{\partial}{\partial s}\left[K_s J_s \overline{\rho}\left(\frac{\partial q_i}{\partial s} - \gamma_{q_i}\right)\right] \quad (26.23)$$

The difficulty with such an approach is that the individual adjustment term needs to be defined for each species. In general, the mass exchange in the boundary layer can be represented with the transilient turbulent expression in which the magnitudes of mixing coefficients are prescribed depending on the numerical integration time step across all the model layers.

A popular nonlocal closure algorithm used in air quality modeling is the asymmetrical convective model (ACM) (Pleim and Chang, 1992), which is a simple transilient model with a very sparse, efficient semi-implicit matrix solver. The original ACM was developed through direct modification of the Blackadar convective model (Blackadar, 1978), which prescribes convective transport as originating in the lowest model layer(representing the surface layer), and rising directly to all other layers within the CBL with symmetrical return flow from each layer back to the lowest layer. ACM replaces the symmetrical downward transport with an asymmetrical layer-by-layer downward transport. Thus, mass fluxes are represented in the ACM by rapid upward transport in convectively buoyant plumes and gradual downward transport. A drawback of the original ACM (ACM1) is that it has no facility for local upward diffusion, resulting in an unrealistic step function mass exchange from the first to the second layer of the model. A new version of the ACM (ACM2) (Pleim, 2007) has been developed that includes the nonlocal scheme of the original ACM combined with an eddy diffusion scheme. Thus, the ACM2 is able to represent both the supergrid- and subgrid-scale components of turbulent transport in CBL (Figure 26.1).

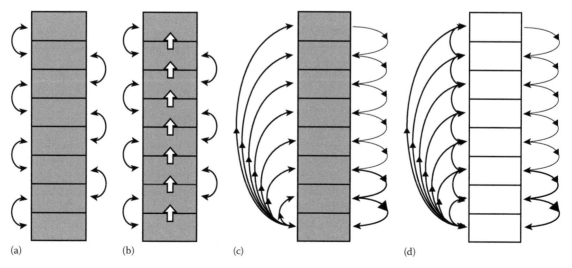

(a)   (b)   (c)   (d)

FIGURE 26.1 Local and nonlocal models of the atmospheric turbulent mixing schemes. (a) Eddy diffusion, (b) nonlocal formulation with constant upward flux originating from the surface, (c) asymmetric convective model (ACM), and (d) a combined local and nonlocal model (ACM2). Top of the column represents the height of the PBL (h_{PBL}). (Adapted from Pleim, J.E., *J. Appl. Meteorol. Climatol.*, 46, 1383, 2007.)

26.3.3 Atmospheric Chemistry

Representation of chemical processes involving various atmospheric constituents is an essential element of an air quality model. Although air pollution modeling must include heterogeneous reactions for aqueous- and aerosol-phase substances; for the sake of simplicity needed for the compendium's focus, only tropospheric gas-phase reactions affecting urban- and regional-scale air pollution are considered here. Important chemical reactions that determine tropospheric atmospheric chemical states are, for convenience, divided into inorganic reactions among the inorganic species, such as nitrogen oxides, atomic and molecular oxygen, hydroxyl radical compounds, and sulfur species, etc., and organic reactions that involve a variety of VOC compounds.

26.4 Gas-Phase Reactions in the Troposphere

All important chemical reactions must be included to make accurate predictions of ambient pollutant concentrations. Ozone in the stratosphere is formed from the photodissociation of molecular oxygen by intense ultraviolet radiation ($\lambda < 205$ nm) and the subsequent recombination of atomic and diatomic oxygen in the presence of a third body. Ground-level ozone is accumulated from photochemical oxidation of oxides of nitrogen and reactive VOCs involving a series of chain reactions. In the presence of sunlight, ozone photodissociates to generate an excited oxygen atom, which reacts with water to form a hydroxyl radical ($^{\bullet}OH$):

$$O_3 + h\nu \rightarrow O_2 + O(^1D)\,(\lambda < 320\,\text{nm})$$

$$O(^1D) + H_2O \rightarrow 2\,{}^{\bullet}OH$$

The hydroxyl radical reacts with VOC in a series of chain reactions to create alkylperoxy radicals ($RO_2^{\bullet}$) that react with NO to generate alkoxy radicals ($RO^{\bullet}$) and NO_2. $RO^{\bullet}$ reacts further with O_2 by forming carbonyls (ketones $RR'CO$ and aldehydes $RCHO$) and $HO_2^{\bullet}$, whose reaction with NO recreates $^{\bullet}OH$ and nitrogen dioxide (NO_2), propagating the cycle:

$$VOC + {}^{\bullet}OH \rightarrow RO_2^{\bullet}$$

$$RO_2^{\bullet} + NO \rightarrow RO^{\bullet} + NO_2$$

$$RO^{\bullet} + O_2 \rightarrow HO_2^{\bullet} + \text{Carbonyl}$$

$$HO_2^{\bullet} + NO \rightarrow {}^{\bullet}OH + NO_2$$

Increase in ground-level ozone under sunlight is due to the photolysis of NO_2 that creates atomic oxygen without consuming ozone:

$$NO_2 + h\nu \rightarrow NO + O(^3P)\,(\lambda < 420\,\text{nm})$$

$$O(^3P) + O_2 \rightarrow O_3$$

$$NO + O_3 \rightarrow NO_2 + O_2$$

For this reason, tracking fluxes of $RO_2^{\bullet}$ and $HO_2^{\bullet}$ reactions with NO can be used as a measure of the total photochemical production of ozone. In the dark, with sufficiently high NO, ozone is titrated, forming NO_2, which can further react with other reactive nitrogen species. Other sources of $^{\bullet}OH$ and $HO_2^{\bullet}$ are the photolysis reactions of nitrous acid and formaldehyde:

$$HONO + h\nu \rightarrow NO + {}^{\bullet}OH\,(300\,\text{nm} < \lambda < 405\,\text{nm})$$

$$HCHO + h\nu \begin{cases} \rightarrow H + HCO\,(< \lambda < 370\,\text{nm}) \rightarrow \cdots \\ (\text{additional reactions})\cdots \rightarrow HO_2^{\bullet} \\ \rightarrow H_2 + CO \end{cases}$$

$^{\bullet}OH$ and NO_x are removed from the atmosphere through the following radical termination reactions:

$$HO_2^{\bullet} + HO_2^{\bullet} \rightarrow H_2O_2$$

$$RO_2^{\bullet} + HO_2^{\bullet} \rightarrow ROOH$$

$$NO_2 + {}^{\bullet}OH \rightarrow HNO_3$$

$$RO_2^{\bullet} + NO \rightarrow RONO_2$$

$$RO_2^{\bullet} + NO_2 \rightarrow ROONO_2$$

where

$ROOH$ represents hydroperoxides
$RONO_2$ represents organic nitrates
$ROONO_2$ represents peroxyacyl nitrates (PANs)

Both organic nitrates and PAN species can thermally decompose to release fresh NO_x after being transported for some distance from sources. Photolysis of hydrogen peroxide (H_2O_2) can reproduce hydroxyl radicals but the efficiency is quite low. Formation of nitric acid (HNO_3) is a prominent sink of nitrogen from the atmosphere. The mechanism for ozone formation described earlier can be summarized as interactions of reactions involving odd hydrogen species $\left(\text{i.e., } HO_x = H + {}^{\bullet}OH + HO_2^{\bullet} + \cdots\right)$, and odd nitrogen ($NO_x = NO + NO_2$).

In a polluted environment, like urban and industrial areas, reactions of highly reactive VOCs (HRVOCs), in particular anthropogenic alkenes such as ethylene ($CH_2 = CH_2$) and biogenic isoprene (ISOP, C_5H_8) with chemical structure $CH_2 = C(CH_3)CH = CH_2$ (2-methyl-1,3 butadiene), can significantly contribute to local ozone problems. An alkene is an unsaturated hydrocarbon containing at least one carbon-to-carbon double bond ($> C = C <$), which can donate a pair of electrons, facilitating reactions with radicals. An alkane is a saturated hydrocarbon with exclusively single-bond linkage between hydrogen and carbon atoms. The simplest alkane, methane (CH_4), reacts slowly with hydroxyl radical, so it is often not considered in a photochemical reaction system.

For this reason, the EPA has traditionally not accounted for this in the emissions inventories as part of the VOC species, leading to the term "non-methane hydrocarbons (NMHCs)."

In the troposphere, reactions of ethylene with hydroxyl radical and ozone further enhance ozone production through the following steps:

$$C_2H_4 + {}^{\cdot}OH \rightarrow \{HCHO, RO_2^{\cdot}\} \xrightarrow{NO} NO_2 \xrightarrow{h\nu} O_3$$

$$C_2H_4 + O_3 \rightarrow \{CO, HCHO, {}^{\cdot}OH, HO_2^{\cdot}\} \xrightarrow{NO} NO_2 \xrightarrow{h\nu} O_3$$

Also, mostly after dark, ethylene reacts with nitrate radicals (NO_3) to form nitroxyethyl radical, a PAN species:

$$C_2H_4 + NO_3 \rightarrow CH_2CH_2ONO_2$$

Alkenes with a higher number of carbons than ethylene, e.g., propylene and butadiene, react with radicals to form carbonyls such as acetaldehydes (CH_3CHO) or higher aldehydes ($RCHO$) and alkyl radicals contributing to the extended cycles of HO_x radicals and increased ozone formation. At lower temperatures, carbonyls further react with hydroxyl radicals to form PAN species through

$$RCHO + OH \rightarrow RCO^{\cdot} + H_2O$$

$$RCO^{\cdot} + O_2 \xrightarrow{M} RC(O)O_2^{\cdot}$$

$$RC(O)O_2^{\cdot} + NO_2 \xrightarrow{M} RC(O)OONO_2$$

where $RC(O)OONO_2$ is peroxyacetyl nitrate (called PAN for short). Although the formation of PAN sequesters nitrogen temporarily, it is not eDciently removed from the atmosphere unlike nitric acid. When heated, it thermally decomposes to release NO_2 and peroxyacyl radicals. Therefore, PAN species produced at night can be transported away from the source region and provide an additional source of NO_x in locations downwind of primary sources.

Similarly, isoprene from biogenic sources reacts with ${}^{ft}OH$ to give (e.g., Zhang et al., 2000)

$$ISOP + {}^{\cdot}OH \rightarrow ISOP'OH \xrightarrow{O_2} ISOP'O_2^{\cdot} \xrightarrow{NO}$$

$$\{HCHO, MVK, MACR, NO_2\} \xrightarrow{h\nu} O_3$$

Like alkenes, ${}^{ft}OH$ addition to the $> C = C <$ bonds results in isoprene ${}^{ft}OH$ adduct ($ISOP'OH$), yielding peroxy radicals. In turn, the adduct forms methyl vinyl ketone (MVK) and methacrolein (MACR) and new ozone is created through photolysis of NO_2 in the presence of NO emissions. Zhang et al. (2000) summarized the products of the isoprene–ozone reactions as follows:

$$ISOP + O_3 \rightarrow PO^* \xrightarrow{M} PO \rightarrow \cdots \rightarrow$$

$$\{CI, HCHO, MVK, MACR, RO_2^{\cdot}\} \rightarrow \cdots \rightarrow {}^{\cdot}OH$$

The isoprene–ozone reaction, i.e., ozonolysis, leads to an excited primary ozonide, (PO^*), a five-member oxygen and carbon ring formed from addition of O_3 to a double bond in the isoprene. The ozonide eventually loses extra energy through collision with air molecules (M) and decomposes into the Crigee intermediate (CI) and carbonyls, including MVK and MACR. The formation of ${}^{ft}OH$ radical supplies fresh ${}^{ft}OH$, enhancing photochemical reactions of HRVOCs and producing ozone through the interaction with the NO_x photolytic cycle. Isoprene also reacts with nitrate, after sunset, to form

$$ISOP + NO_3 \rightarrow CH_2CH_2ONO_2,$$

which can be a source of fresh NO_x at a downwind location in the following morning when the isoprene nitrate dissociates.

26.4.1 Tropospheric Chemical Mechanisms

As seen earlier, VOCs present in the atmosphere are diverse and their oxidation products cascade through degrading reactions with other pollutants and radicals, leading to an unwieldy total number of reactions that need to be tracked. Because of the need for eDcient computation, the set of chemical reactions in air quality models must be simplified instead of keeping track of the minute details of their reaction products. A concise yet accurate set of reactions among primary and secondary air pollutants suitable for capturing key chemical processes affecting certain air pollution problems is called a chemical mechanism. This is accomplished by lumping VOC species and their derivatives with similar properties together to represent their overall effects on the formation of ozone and secondary particulates. A chemical mechanism must provide a collection of reactions that coherently describes the transformation of reactants into products, including all important intermediates. In principle, all reactions in the mechanism are treated as if they are elementary, and the stoichiometric coeDcients for all reactants must be one. Since all reactions are assumed to be elementary, a reaction can have no more than three reactants. These conventions permit the reaction rate to be derived directly from the stoichiometric equation, thereby simplifying the mathematical representation of the reactions. To reduce the number of reactions and intermediate species in the mechanism, products of a chain of complex reactions may assume non-integer stoichiometric coeDcients. Two popular methods of chemical abstraction within photochemical mechanisms are the structural approach and the molecular approach to lumping organic reactions.

Overall, species in an atmospheric chemical mechanism can be divided into inorganic and organic categories. Inorganic reactions are almost always represented explicitly in chemical mechanisms because the number of important inorganic species and their reaction products is relatively small and manageable. Ozone, nitric oxides (NO), nitric acid, nitrous acid, hydrogen peroxide, sulfur dioxide, and several non-carbon radicals are inorganic species important for air pollution study. Among mechanisms commonly used in air quality modeling, there are only small differences in the chemical reactions involving inorganic species, except for occasional omission of less important reaction pathways and use of different rate constants, in particular photolytic rates. The manner in which the grouping of organic compounds is carried out typically distinguishes one mechanism from another. Some species in the mechanism may represent real organic compounds, but in others they represent a mixture of several different compounds with similar properties. Designers of chemical mechanisms endeavor to include a number of species representing primary (i.e., emitted) hydrocarbons and their oxidation products that involve photochemistry. In particular, they attempt to represent ozone productivity as closely as possible with the results of smog chamber experiments. Chemical mechanisms may include artificial species and operators and many of the mechanism reactions are parameterizations of a large set of actual elementary atmospheric reactions. To simplify the product pathways, the mechanism reactions may include non-integer stoichiometric coeDcients and even have elements with no physical significance (e.g., products with negative stoichiometry).

Although it is diDcult to cleanly categorize mechanisms due to their subtle differences, there are three common representations adapted for urban and regional air quality modeling: the lumped structure technique, the surrogate species approach, and the lumped species method. In the lumped structure approach, organic compounds are apportioned to one or more mechanism species on the basis of chemical bond type associated with individual carbon atoms (Whitten et al., 1980), as shown earlier as the single- or double-bond structure between carbon and hydrogen atoms. The carbon bond (CB) mechanism, a lumped-structure type, utilizes the reactivity differences manifested by the different carbon–hydrogen atomic bond.

The CB-IV mechanism, the fourth in a series of the CB mechanisms, differs from its predecessors notably in the detail of the organic compound representation (Gery et al., 1989). The CB-IV uses nine primary lumped organic species (i.e., species emitted directly), including the carbon atoms that contain only single bonds (PAR), highly reactive terminal double-bonded carbon atoms (OLE), 7-carbon ring structures represented by toluene (TOL), 8-carbon ring structures represented by xylene (XYL), the carbonyl group and adjacent carbon atom in acetaldehyde and higher-molecular-weight aldehydes represented by acetaldehyde (ALD2), and nonreactive carbon atoms (NR). A more complex organic compound can be apportioned to these CB species simply on the basis of molecular structure. For example, propane is represented by three PARs because all three carbon atoms have only single bonds, and propylene is represented as one OLE (for its one double bond) and one PAR (for one single-bond carbon). Some apportionments are based on reactivity considerations. For example, alkenes with internal double bonds are represented as ALD2s and PARs rather than OLEs and PARs. Further, the reactivity of some compounds may be lowered by apportioning some of the carbon atoms to the nonreactive class NR (e.g., ethane is represented as 0.4 PAR and 1.6 NR). The CB-IV mechanism also includes explicit representations for

the inorganic and carbonyl species, ethylene (ETH), isoprene (ISOP), and formaldehyde (FORM). The base mechanism has 33 species and 81 reactions. Apportioning rules have been established for hundreds of organic compounds, and are built into the emissions processing subsystems to produce the appropriate emission rates for the CB-IV mechanism species.

In 2005, an updated version of the CB mechanism called CB05 was developed (Yarwood et al., 2005). It is a condensed mechanism of atmospheric oxidant chemistry that provides a basis for computer modeling studies of ozone, PM, visibility, acid deposition, and air toxics issues. Compared to earlier CB versions, CB05 includes photolysis and reaction rate updates, NO_x recycling reactions over multiple days, and improved atmospheric chemistry representations of ethane (ETHA). It further introduces internal olefin ($R - HC = CH - R$) species (IOLE), higher aldehyde species (ALDX) rendering ALD2 solely for acetaldehyde, higher peroxyacyl nitrate species from ALDX called PANX, and lumped terpene species (TERP). The base CB05 mechanism has 51 species and 156 reactions.

In the surrogate species method, the chemistry of a single species is used to represent compounds of the same class. SAPRC-07 (Carter, 2010) is the latest version of Statewide Air Pollution Research Center (SAPRC) mechanism by Dr. William Carter, which keeps the general structure of SAPRC-99 (Carter, 2000) and SAPRC-90 (Carter, 1990). The organic species are lumped based on the similarity of reactivity with ^fOH. Each lumped species is defined with adjusted mechanism parameters for a particular surrogate to account for variations in the composition of the compounds. The base mechanism of SAPRC-07 is the portion of the mechanism that represents the reactions of the inorganic species, the common organic products, and the intermediate radicals leading to these products, including those formed from the initial reactions of the represented VOCs not in the base mechanism. Formaldehyde, acetaldehyde, acetone, ethylene, isoprene, and in particular benzene and acetylene for the 2007 version, are represented explicitly in the base mechanism due to their importance in emissions inventories and because representation with lumped model species can be inadequate. VOCs that are not in the base mechanism can be added to the mechanism either as explicit reactions for individual VOCs or as lumped model species whose parameters are derived from the mixture of detailed model species they represent, as needed in the model application. The recent revision of the SAPRC-07 includes updates to the reaction rates, more explicit aromatic mechanisms, and a simplified chlorine chemistry.

An important component of the SAPRC mechanism is that it provides the mechanism generation and estimation tools to derive the mechanism representations for most of the nonaromatic VOCs of interest. Because air quality model applications require simulations of highly complex mixtures of large numbers of VOCs, the flexible lumping approach and recommended set of lumped model species allows for a computationally economical mechanism with an appropriate number of lumped chemical species that represent reactions of a large number of species with similar reaction rates. The SAPRC mechanism has the option to vary the lumping approach in terms of the number of model species used and how they are lumped, and also to vary the mixture of compounds used to derive the parameters for the lumped model species, based on the emissions in the particular model scenario or other considerations (Carter, 2000). This is referred to as the "adjustable parameter" mechanism for air quality models.

In practice, most air quality models (e.g., CMAQ) have used a version of the "fixed parameter" SAPRC mechanism. It involves use of a fixed set of lumped model species, with the values for the adjustable parameter lumped species being derived using a predefined mixture. These include the lumped model species that are added to the base mechanism whose parameter can be adjusted based on the emissions mixture being represented, the organic product model species already in the base mechanism that are also used to represent various types of primary emitted organics, and the model species for compounds that are explicitly represented. The lumped or explicit species represent the emitted compound on a mole-for-mole basis without considering reactivity weighting. The mechanism includes reactions of inorganics, CO, formaldehyde, acetaldehyde, PAN, propanal, peroxypropionyl nitrate (PPN), glyoxal and its PAN analog, methylglyoxal, and several other product compounds are represented explicitly. A "chemical operator" approach is used to represent peroxy radical reactions. Generalized reactions with variable rate constants and product yields are used to represent the primary emitted alkane, alkene, aromatic, and other VOCs. Most of the higher-molecular-weight oxygenated product species are represented with lumped species, where simpler molecules such as propanal or 2-butanone (MEK) are used to represent the reactions of higher-molecular-weight analogues that are assumed to react similarly. To limit the number of free radical species, a highly condensed mechanism of the peroxy reactions is employed with organic peroxy operators similar to those used in CB05.

26.4.1.1 Chemistry Solvers

As seen earlier, a chemical mechanism is composed of a set of limited number of elementary reactions, for which the rate of a chemical reaction l (r_l) can be expressed as the product of a rate constant and concentrations of reactants:

$$r_l = \begin{cases} k_l C_1 \\ k_l C_1 C_2 \\ k_l C_1 C_2 C_3 \end{cases} \quad (26.24)$$

where C_1, C_2, and C_3 refer to the concentration of reactants 1, 2, and 3, respectively. When a species reacts with itself, the concentration is squared. Several important termolecular reactions involve air molecules ($M = O_2 + N_2$). Since their concentrations are dependent on atmospheric pressure, some mechanism developers convert second- or third-order reactions that involve air molecules to a reaction one order lower by multiplying the higher-order rate coeDcient by the concentration of M, O_2, or N_2.

By using the kinetics laws for elementary reactions and applying a mass balance to each species, an equation for the local time rate of change of each species concentration in a single cell can be derived:

$$\frac{\partial C_i}{\partial t} = P_i(\mathbf{c}, t) - L_i(\mathbf{c}, t) C_i = f_i(\mathbf{c}, t) \quad i = 1, 2, \ldots, N_s \quad (26.25)$$

where N_s is the total number of species in the system of the chemical mechanism, as well as production:

$$P_i = \sum_{j=1}^{j_i} v_{i,j} r_j \quad (26.26)$$

and loss:

$$L_i C_i = \sum_{l=1}^{l_i} r_l \quad (26.27)$$

C_i is the concentration of species i, $v_{i,l}$ is the stoichiometric coeDcient for species i in reaction j, and r_j is the rate of reaction j. $\mathbf{c}$ is the vector of species concentrations of vector length L (number of total species). The sum $j = 1$ k j_i is over all reactions in which species i appears as a product, and the sum $l = 1$ k l_i is over all reactions in which species i appears as a reactant. In this discussion, the emission source term is not included as part of the production term P_i. In vector form, Equation 26.25 is rewritten as

$$\mathbf{c}' = \frac{\partial \mathbf{c}}{\partial t} = \mathbf{f}(\mathbf{c}, t) \quad \text{with initial condition } \mathbf{c}_o = \mathbf{c}(t^o) \quad (26.28)$$

where $\mathbf{c}'$ is the first-order local time derivative of $\mathbf{c}$, and $\mathbf{f}$ is the vector of production and loss function.

The ordinary differential equation (ODE) system is nonlinear because the rate of reactions r_j involves products of concentrations and is highly stiff due to the large differences in the reaction rates. A stiff system can be described mathematically as one in which all the eigenvalues of the Jacobian matrix of Equation 26.28 are negative, and the ratio of the absolute values of the largest-to-smallest real parts of the eigenvalues is much greater than one. Systems are typically termed stiff if the latter ratio is greater than 10^4 but for typical tropospheric photochemical conditions, it is 10^{10}. Analytic solutions of this system are not usually available because of the nonlinearity in the reaction rates and destruction terms. General numerical solution methods for non-stiff ODE systems are readily available but they become ineDcient for stiff systems due to the stability concerns. The stiffness problem coupled with the fact that these equations must be solved for hundreds of thousands of cells in a typical air quality modeling application calls for eDcient numerical schemes. The accuracy and stability of explicit schemes are of great concern. Classical implicit methods that are both accurate and stable have also been found to be inadequate because of high computational demands stemming from the need for suDciently small time steps to constrain the relative errors. A few of the popular numerical schemes that have been utilized in air quality models are discussed as follows.

The explicit *Quasi-Steady-State Approximation* (QSSA) method assumes integration time steps suDciently small such that in Equation 26.25 the production and loss rate terms P_i and L_i can be considered constant. The Euler finite difference equation for species i is

$$C_i^n - C_i^{n-1} = h\left[P_i(\mathbf{c}^{n-1}) - L_i(\mathbf{c}^{n-1})C_i^{n-1}\right] \quad (26.29)$$

where $C_i^n = C_i(t^n)$, $C_i^{n-1} = C_i(t^{n-1})$, and $h = \Delta t = t^n - t^{n-1}$, is the time step that may vary with time step n. Although the method makes no a priori assumptions about the reaction timescales, numerical solutions are separated into either a Euler step, a fully explicit integration, or an asymptotic evaluation based on photochemical lifetimes estimated from an initial, predictor calculation of P_i and L_i:

Euler step for slow reactions ($L_i h \leq 0.01$):
$C_i^n = C_i^{n-1} + (P_i + L_i C_i^{n-1})h$

Fully explicit ($0.01 \leq L_i h \leq 10$): $C_i^n = C_i^\infty + (C_i^{n-1} - C_i^\infty)Exp[-L_i h]$

Asymptotic for fast reactions ($L_i h > 10$): $C_i^n = C_i^\infty$ \quad (26.30)

where $C_i^\infty = P_i/L_i$. Employment of separate solutions depending on the reaction timescales can introduce a discontinuity in the computer codes and cause large numerical errors. Although the QSSA is computationally eDcient and can maintain stability of the solution in common tropospheric photochemical conditions, it is lower order and produces less accurate solutions than other methods that are introduced later.

Studies have shown that implicit numerical solvers perform much better for stiff equations than explicit solvers. In the *Euler Backward Iterative* (EBI) solver (e.g. Hertel et al., 1993), which is the simplest implicit solver, the solution is approximated by extrapolating an as-yet-unknown target point along the tangent line at each time step:

$$\mathbf{c}^n = \mathbf{c}^{n-1} + h\mathbf{f}(\mathbf{c}^n, t^n) \quad (26.31)$$

Solving Equation 26.31 in general requires an iterative solution method as the production and loss function $\mathbf{f}$ is nonlinear. Utilizing $(C_i^n)' = f_i(\mathbf{c}^n, t^n)$ a solution is obtained with the iterative Newton method:

$$[\mathbf{I} - h\mathbf{J}(t^n)]\left[(\mathbf{c}^n)_{m+1} - (\mathbf{c}^n)_m\right] = \mathbf{c}^{n-1} - \mathbf{c}^n + h\mathbf{f}((\mathbf{c}^n)_m, t^n) \quad (26.32)$$

where m is the iteration number, $\mathbf{I}$ is the identity matrix, and $\mathbf{J}(t^n)$ is the Jacobian matrix of partial derivatives at time step n whose entries are defined as follows:

$$J_{il}(t^n) = \frac{\partial f_l(\mathbf{c}^n, t^n)}{\partial C_i^n} \quad (26.33)$$

Each Newton iteration step requires the solution of a system of linear algebraic equations that involves the formation of the Jacobian matrix, thus making it computationally expensive. The Jacobian matrix can be computed using approximate formulas based on finite differences with small perturbations of the arguments of *f*. In certain air quality models, the implicit expression is relaxed only to its own species to give

$$\left(C_i^n\right)_{m+1} = \frac{C_i^{n-1} + P_i\left(\mathbf{c}_m^n\right)h}{1 + L_i\left(\mathbf{c}_m^n\right)h} \quad (26.34)$$

A Gauss–Seidel procedure is used to estimate P and L terms by using the concentrations of the previous time step as a first guess and then updating the concentrations in sequence through the iterative process that may be stopped either by a fixed number of operations or based on a set of convergence criteria. Further, to reduce stiffness of the system, i.e., to allow larger time steps, customization of the solver, such as grouping fast-reacting species together, can be applied for specific chemical mechanisms. For example, coupled species groupings are solved analytically together, such as {NO, NO$_2$, O$_3$, O(^{3}P)}, {OH, HO$_2$, HONO, HNO$_4$}, {NO$_3$, N$_2$O$_5$}, and {PAN, C$_2$O$_3$}; then the EBI method is applied for other species individually.

The most popular stiff ODE solver by Gear (1971) utilizes the *Backward DiTherentiation Formula* (BDF), which is implicit and stable with automatic time step size and error control. The generalized BDF that forms the basis for Gears method can be expressed as follows:

$$\mathbf{c}^n = h\beta_o\mathbf{f}\left(\mathbf{c}^n, t^n\right) + \sum_{j=1}^q \alpha_j\mathbf{c}^{n-j} \quad (26.35)$$

where n refers to the time step, h is the size of the time step, q is the assumed order, and $\beta_o \geq 0$ and α_j are scalar parameters. The method is implicit since concentrations at the desired time step n depend on values of the first derivatives contained in $\mathbf{f}(\mathbf{c}^n, t^n)$ that are functions of the concentrations at the same time. The order of the method q ($1 \leq q \leq 5$) corresponds to the number of concentrations at previous time steps that are incorporated in the summation. The multistep method is modified to multi-value differentiation that relies on information from only the previous time step. The solution is first approximated by predicting higher-order concentration derivatives at the end of a time step for each species with the following matrix equation:

$$z_i^{n_o} = \mathbf{B}z_i^{n-1} \quad (26.36)$$

where $z_i = \left[C_i, hC_i', \ldots, (hq)C_i^{(q)}/q!\right]^T$ and $\mathbf{B}$ is the Pascal triangle matrix. Superscript n_o refers to the prediction at the end of time step n. The prediction is then corrected by solving for z_i^n with

$$z_i^n = z_i^{n_o} + \mathbf{r}\left[hf_i(\mathbf{c}^n, t^n) - h\left(C_i^{n_o}\right)'\right] \quad (26.37)$$

Utilizing $\left(C_i^n\right)' = f_i(\mathbf{c}^n, t^n)$, solution is obtained with the iterative Newton's method:

$$[\mathbf{I} - h\beta_o\mathbf{J}(t^n)][(\mathbf{c}^n)_{(m+1)} - (\mathbf{c}^n)_{(m)}]$$

$$= -\mathbf{c}^n + h\beta_o\mathbf{f}((\mathbf{c}^n)_{(m)}, t^n) + \sum_{j=1}^{q}\alpha_j\mathbf{c}^{n-j} \qquad (26.38)$$

The subscript (m) stands for the number of applied iterations that meet the prescribed convergence criteria. The linear implicit systems that arise are solved by direct methods (LU factor/solve). In LSODE (Livermore Solver for Ordinary Differential Equations), which is one of the most popular general solver packages for both stiff and non-stiff systems, the Gear method is used as the stiff system solver. Although computationally expensive, it is very accurate and therefore serves as a "gold standard" to evaluate other faster solvers for accuracy. Computational eDciency is of major importance for air quality modeling as the more complex chemistry and high spatial resolution are required to study multispecies interactions of pollutants that affect human health. In particular, for air quality forecasting, timely model output is a necessity. The *Sparse-Matrix Vectorized Gear* (SMVGEAR) algorithm from Jacobson and Turco (1994) is a modified gear algorithm to incorporate additional computational eDciencies that can achieve speedups on the order of 100 on vector computers. About half of the improvement is attributed to enhanced vectorization and half to improved matrix operations. Because of the improved matrix operations, SMVGEAR also runs faster than traditional Gear solvers due to the use of faster cache memory. Instead of applying the solver for each grid individually, single calls for SMVGEAR for a vector machine are made for a block of several hundreds of cells, which could be substantially larger than the number of species. However for nonvector computers, users must experiment with the block size because the speedup factor can vary dramatically depending on the architecture of the CPUs and communication speed and size of the memory cache.

The *Rosenbrock method* (e.g., Sandu et al., 1997) originated from the one-step implicit Runge-Kutta formulas (RKF). An s-stage implicit RKF is given as

$$\mathbf{c}^{n+1} = \mathbf{c}^n + h\sum_{l=1}^{s}b_l\mathbf{k}_l \qquad (26.39)$$

with

$$\mathbf{k}_l = \mathbf{f}\left(\mathbf{c}^n + h\sum_{j=1}^{s}a_{lj}\mathbf{k}_j\right), \quad l=1,\dots,s, \qquad (26.40)$$

where a_{lj} and b_l are parameters that determine consistency and stability of the numerical solver. Because the nonlinear equation

$$\mathbf{g}(\mathbf{k}_l) = \mathbf{k}_l - \mathbf{f}\left(\mathbf{c}^n + h\sum_{j=1}^{s}a_{lj}\mathbf{k}_j\right) = 0, \quad l=1,\dots,s, \qquad (26.41)$$

of dimension $s \times N_s$ must be solved at each integration time step, the fully implicit RKF method is even more expensive than the BDF described earlier. Again, the iterative Newton's method is used to solve the nonlinear equation

$$[\mathbf{I} - ha_{ij}\mathbf{J}][(\mathbf{k}^n)_{(m+1)} - (\mathbf{k}^n)_{(m)}] = -\mathbf{g}(\mathbf{k}_l)^n \qquad (26.42)$$

Even though the cost of solving the preceding equation can be reduced by exploiting the special structure of the iteration matrix, the implicit RKF is too expensive. Further reduction of computational cost can be accomplished by forcing the system to be only diagonally implicit. Then $\mathbf{k}_l$ can be solved successively for s-stages with only one inversion of $[\mathbf{I} - h\gamma\mathbf{J}]$ of dimension N_s by equating the diagonal components a_{ij} to a single value γ as in the singly diagonally implicit Runge-Kutta (SDIRK) method:

$$\mathbf{c}^{n+1} = \mathbf{c}^n + h\sum_{l=1}^{s}b_l\mathbf{k}_l \qquad (26.43)$$

where

$$\mathbf{k}_l = \mathbf{f}\left(\mathbf{c}^n + h\left(\sum_{j=1}^{l-1}a_{lj}\mathbf{k}_j + \gamma\mathbf{k}_l\right)\right), \quad l=1,\dots,s, \qquad (26.44)$$

This approach has a cost of evaluation of function s-times per iteration and one LU-factorization of $[\mathbf{I} - h\gamma\mathbf{J}]$ of dimension N_s. The Rosenbrock method replaces the iteration by introducing the Jacobian term directly into the integration formula as

$$\mathbf{c}^{n+1} = \mathbf{c}^n + \sum_{l=1}^{s}b_l\hat{\mathbf{k}}_l \qquad (26.45)$$

where

$$\hat{\mathbf{k}}_l = h\mathbf{f}\left(\mathbf{c}^n + \sum_{j=1}^{l-1}a_{lj}\hat{\mathbf{k}}_j\right) + h\mathbf{J}\sum_{j=1}^{l}\gamma_{lj}\hat{\mathbf{k}}_j, \quad l=1,\dots,s, \qquad (26.46)$$

Note that now $\hat{\mathbf{k}}_l = h\mathbf{k}_l$, which has a unit of concentration. As before, a formula with constant diagonal components, i.e., $\gamma_{ll} = \gamma$ for all $l = 1, \dots, s$ is of interest because the same LU factorization of $[\mathbf{I} - h\gamma\mathbf{J}]$ can be utilized. Like in the RKF, the Rosenbrock method forms an intermediate result for each stage l:

$$\hat{\mathbf{c}}_l = \mathbf{c}^n + \sum_{j=1}^{l-1}a_{lj}\hat{\mathbf{k}}_j, \quad l=1,\dots,s \qquad (26.47)$$

The intermediate time points (t_l) can be related to the stage l as

$$t_l = t^n + h\sum_{j=1}^{l}a_{lj} \qquad (26.48)$$

Although it can vary depending on the testing conditions, the Rosenbrock solver is in general faster and more accurate than the QSSA solver, faster but less accurate than the SMVGEAR solver, and slower but more accurate than the EBI solver when comparable convergence tolerance levels are used. In air quality models, it is common practice to specify both the absolute tolerance parameters around 10^{-6} ppb and the relative tolerance at 0.001 unless more accurate solutions are desired.

26.4.1.2 Dry Deposition

Dry deposition refers to the process wherein gaseous and aerosol species are removed through contact with the Earth's surface elements such as leaves, soil, asphalt, and water bodies as well as gravitational settling of the heavier particles. Because contact with the surface is necessary, the deposition process affects the concentrations in the lowest layer directly. In air quality modeling, it is often implemented as a boundary flux condition of the vertical diffusion by introducing the concept of a dry deposition velocity:

$$\left.\frac{\partial(\varphi_i^*)}{\partial t}\right|_{drydep} \approx -\frac{v_{d_i}}{h_{dep}} \cdot \varphi_i^* \Big|_{layer1} \tag{26.49}$$

where h_{dep} is the thickness of the deposition layer (usually the middle of first model layer, $h_{dep} = z_1$) and v_{d_i} the dry deposition velocity for species i. The magnitudes of the dry deposition velocities, which represent the effective-removal timescales when divided by the distance from the surface, are determined by surface conditions and chemical characteristics of species (surface resistance components), as well as strength of turbulence (aerodynamic resistance). They actually cannot be measured directly and therefore are inferred through the measurement of pollutant gradients, atmospheric stability, physical characteristics of the surface, and chemical affinity of the surface with target chemicals. Figure 26.2 shows the relationship among concentrations at different heights during the deposition process. F is the flux of pollutant available for dry deposition by turbulent atmospheric transport while F_s is the flux actually taken out of the air.

In the lowest model layer, the available flux is approximated with

$$F = v_{da}(z_1) \cdot C(z_1) \tag{26.50}$$

where the aerodynamic resistance (r_a) and deposition velocity (v_{da}) is computed with the vertical eddy diffusivity K_z as follows:

$$r_a(z_1) = \frac{1}{v_{da}} = \int_{z_0}^{z_1} \frac{dz}{K_z} \tag{26.51}$$

The actual deposition flux (F_s) must be smaller than the available flux

$$F_s = v_{da}(z_1) \cdot [C(z_1) - C_o] \tag{26.52}$$

The deposition velocity that takes into account the overall physical and chemical processes (v_d) can be estimated based on the following relations for the monotonically increasing concentration profiles (i.e., $C(z_1) > C_b > C_c > C_g$) in the atmospheric surface layer:

$$F_s = v_d(z_1)[C(z_o) - C_g] = v_d(z_1)[(C(z_o) - C_o) + (C_o - C_b)$$

$$+ (C_b - C_c) + (C_c - C_g)]$$

$$= v_d(z_1)\left[\frac{F_s}{v_{da}} + \frac{F_s}{v_{db}} + \frac{F_s}{v_{dc}} + \frac{F_s}{v_{dg}}\right]$$

$$= F_s v_d(z_1)[r_a + r_b + r_c + r_g] \tag{26.53}$$

This leads to

$$v_d(z_1) = \frac{1}{[r_a + r_b + r_s]} \tag{26.54}$$

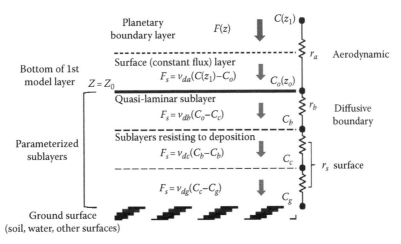

FIGURE 26.2 Schematic of dry deposition process representing relationship among concentrations at different heights in the deposition process. $C(z_1)$ and C_o stand for concentrations at height z_1 (middle of lowest model layer) and z_o (cell average roughness length). C_g is concentration at the deposition surface which is assumed to be negligible. See the text for the definitions of other symbols in the picture. (Adapted from Byun, D.W. and Dennis, R.L., *Atmos. Environ.*, 29 (1), 105, 1995.)

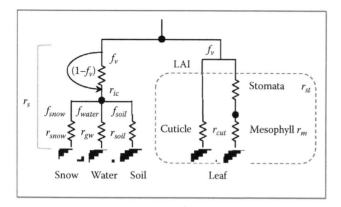

FIGURE 26.3 Schematic of components of surface resistance. See the text for the definitions of other symbols in the picture. (Adapted from Byun, D.W. and Ching, J.K.S., eds. Science algorithms of the EPA models-3 community multiscale air quality (CMAQ) modeling system, EPA Report, EPA/600/R-99/030, NERL, Research Triangle Park, Durham, NC, 1999.)

where the surface resistance represents some of the resistances in the parameterized layers (i.e., $r_s = r_c + r_g$). Diffusive boundary layer resistance (r_b) depends on the land-use-specific friction velocity and molecular diffusivity of the gaseous species. Most of the uncertainty in the dry deposition velocity originates from the diﬃculties in the estimation of the overall surface resistance.

As illustrated in Figure 26.3, various elements of the Earth's surface offer multiple pathways with different deposition eﬃciencies due to their distinct physical, biological, and chemical characteristics. Leaves show distinct physiological behavior resisting to the removal of the air pollutants. Stomata found in the leaf and stem are used for exchange of water vapor, CO_2, and other gases affecting the deposition process. Plants open and close their stomata in response to ambient conditions such as availability of photosynthetically active solar radiation (PASR) (visible radiation with wavelengths between 400 and 700 nm), and ambient temperature (T_a), moisture, and CO_2 concentrations. Assuming the CO_2 concentration is nearly constant, most formulations for the stomatal resistance (r_{st}) are parameterized with PASR, T_a, relative humidity (RH), leaf area index (LAI), and soil moisture. The mesophyll resistance (r_m) takes into account the fate of gases that entered the stomata and therefore can be added to the stomatal resistance directly. It is commonly ignored for the soluble gases but a nonzero value should be used for others. The in-canopy aerodynamic resistance (r_{ic}) is often parameterized with the LAI, friction velocity, and canopy height. Although resistance by cuticles (r_{cut}), which is covered with soluble waxes, is not well understood, but as their eﬃciency for removing species will be different for dry and wet cuticles, the canopy wetness fraction (f_w) is estimated from the land surface model. The surface resistance for the wet cuticle (r_{cw}) is estimated with the nondimensional Henry's law constant and aqueous dissociation factor. These non-stomatal vegetation resistances as well as the ground surface resistances (r_g) for other gases are estimated by applying scaling factors that depend on the solubility and reactivity to the observed values for ozone and SO_2.

The total bulk surface resistance for a cell must be estimated based on the fractional amounts of snow cover, water and soil areas (i.e., $f_{snow} + f_{water} + f_{soil} = 1$), as well as the vegetation fraction (f_v) as follows (personal communication with J.E. Pleim):

$$r_g = \left[\frac{f_{snow}}{r_{snow}} + \frac{f_{water}}{r_{gw}} + \frac{f_{soil}}{r_{soil}} \right]^{-1} \tag{26.55}$$

$$r_s = \left[\frac{f_v}{r_{st} + r_m} + LAI \left(\frac{f_v(1 - f_w)}{r_{cut}} + \frac{f_v f_w}{r_{cw}} \right) + \frac{1 - f_v}{r_g} + \frac{f_v}{r_{ic} + r_g} \right]^{-1} \tag{26.56}$$

26.4.1.3 Emissions

Air quality models depend on preprocessed emission inputs, represented as $S_{\varphi_i}\big|_{emis}$ in Equation 26.13. Historically, emission data are compiled by State and local air pollution control agencies to be used for regulatory or scientific assessment of the spatial distribution, temporal patterns, and trends of air pollution problems such as surface ozone, acid deposition, PM, and air toxics. The EPA compiles the speciated emissions data into national annual inventories for the United States. The inventories for studying ozone and $PM_{2.5}$ commonly include sulfur dioxide (SO_2), nitrogen oxides, $PM_{2.5}$, and VOCs divided into (1) area and (2) point source data, (3) on-road mobile, (4) non-road mobile source emissions, and (5) inputs necessary for the biogenic emissions processing.

Area source inventories: Emissions are reported by categories and their processing methods vary depending on the type of data available for each category. Included are emissions from stationary source fuel combustion (e.g., residential burning), solvent use, product storage and transportation distribution, small industrial/commercial sources, agricultural activities (feedlots, crop burning, etc.), waste management, and other miscellaneous sources. Usually, the amount of fuel consumed and materials data are used for emission estimates, but emission factors for certain categories in association with county population data can be used instead. Because area emissions inventories are often available on a county-wide basis, they needed to be disaggregated for the model grids utilizing additional information available in the geographical information system (GIS).

Point source inventories: Industrial and nonindustrial stack emissions from stationary equipment having significant impact on air pollution (i.e., facilities with about 1 T or more emissions per year) are included in these inventories. Point sources include industrial and commercial boilers, electric-utility boilers, turbine engines, refineries, petrochemical plants, oil rigs, petroleum storage tanks, wood and pulp processers, industrial solvent, and surface coating facilities. In addition to speciated emission amounts, stack parameters such as stack height, diameter, eﬄuent velocity (or fluxes) are compiled to estimate vertical distribution of emissions after plume rise. The NO_x and SO_2 emissions data from the electricity generating utilities (EGUs) are obtained with continuous emissions monitoring (CEM). Emissions from other point sources (non-EGUs) are estimated following the U.S. Environmental Protection Agency standard methods known as AP-42.

On-road mobile sources: This category accounts for emissions from highway and other on-road vehicles, such as cars, trucks, and buses. To estimate emissions, vehicles are segregated into different engine types and fuels used (light-duty vs. heavy-duty and gasoline vs. diesel), as well as vehicle weight. Emissions factors for different mobile source categories are estimated using a sophisticated mobile emission model that utilizes a set of complex mathematical equations and user inputs describing vehicle operating and environmental conditions (e.g., average vehicle speed, ambient temperature, and humidity). Contributions from mobile sources can be estimated with the emissions factor and vehicle miles traveled (VMT) data compiled by the states and local governments usually at county basis or the link-node basis. The county-based mobile emissions are then spatially allocated for the grid cells intersecting with road networks represented in the GIS shapefiles and distributed over time, utilizing the temporal profiles. Several mobile emissions processors such as EPA's Mobile 6 and MOVES are available for the United States.

Non-road mobile sources: They include a wide variety of internal combustion engines not associated with highway vehicles, such as commercial and military aircraft, locomotives, ships and barges, as well as small engines for lawn and garden, airport support vehicles, recreational marine, light commercial generators, construction equipments, and agricultural tractors. Different emissions calculation methods, either computer models or actual fuel usage data, are utilized for different emission categories.

Biogenic emissions: Emissions from trees and vegetation are also important and contribute to air pollution problems. Trees emit significant amount of photochemically sensitive biogenic VOC, carbon monoxide (CO) as well as monoterpenes which are precursors of secondary organic aerosols (SOAs). Also soil microbial processes emit NO. The rates of the biogenic emissions are heavily dependent on the environmental factors that affect tree and vegetation physiology, such as ambient temperature, moisture, and in particular the photosynthetically active solar radiation (PASR). Isoprene, the most important biogenic emission affecting surface ozone, comes mostly from oaks, sweet gums, and poplars, while terpenes come from pines, sycamores, and eucalypti. For the purpose of photochemical modeling, biogenic emissions are not directly measured but are usually estimated with meteorological data, such as PASR, surface temperature, land type, and distribution of vegetation. The Biogenic Emissions Inventory System (BEIS) was first developed in 1988 to estimate VOC emissions from vegetation and NO emissions from soils. The latest version is BEIS 3 (BEIS3). The Model of Emissions of Gases and Aerosols from Nature (MEGAN), developed at the National Center for Atmospheric Research (NCAR), can also be used instead of BEIS, which relies on the Biogenic Emissions Land-Use Database, version 3 (BELD3) (http://www.epa.gov/ttn/chief/emch/ biogenic/) that contains land-use and land-cover as well as demographic and socioeconomic information.

Temporal allocation: In addition to the spatial allocation, emissions data need to be temporally allocated to hourly resolution and speciated for the chemical mechanism species used in a gas-phase chemistry mechanism, which is discussed later. Allocation of emission data from time periods greater than hourly down to hourly resolution is accomplished by use of source category-specific seasonal, monthly, weekly, and daily temporal allocation factors. The daily values are then transformed into emission values for each hour of a typical day by using user-supplied or default temporal allocation profiles to differentiate, e.g., weekday and weekend effects. Mechanism-dependent emission inputs are prepared first by associating the source categories with the speciation database for emissions species and groups to split emissions for mechanism species by applying mixture profiles.

Chemical speciation for emissions assignments: As seen earlier, the VOC species in a chemical mechanism may or may not be consistent with the chemical species included in the inventories. Depending on the inventories, such emissions are reported as VOC, reactive organic gases (ROG), or total organic gases (TOG). USEPA initially suggested the term VOC for organic species that have the hydroxyl reaction rates higher than that of ethane to account for atmospheric oxidation processes (i.e., use ethane as the "bright line" of the VOC emission control). However, in the national inventories, VOC excludes only methane from the TOG as the ROG used by the California Air Resources Board (CARB). To convert emissions of aggregated organic species for emissions of model chemical species in the mechanisms, speciation profiles are used. Because the speciation profiles that are used to segregate (i.e., speciate) organic species are based on TOG, the inventory VOC (or ROG) data must be modified to reflect the contributions from methane and ethane. The conversion factors are available either by FIPS (Federal Information Processing Standards) Source Classification Codes (SCC) or by Speciation Profile ID. For emissions processing, both mole-based and mass-based speciation matrices are created. While, for real chemical species, the conversion can be done simply using a molecular weight, the conversion between the two in the emission processing is not straightforward because the molecular weight of a chemical species is different for each speciation profile. The speciation process does not necessarily conserve mass: reduced mass occurs when some pollutants do not map to chemically reactive species while increased mass happens when some VOC compounds may be associated with two model species to represent required chemical reactivity. Since one part of the VOC could be mapped to two model species, its mass appears to be double counted when summing the model-species mass.

26.5 Applications

Air quality models are used to understand atmospheric processes involving air pollution problems, to test suggested emission control strategies for improving air quality, to provide air quality forecasting to the public, and to perform climate impact studies under different scenarios. Application examples for daily air quality forecasting over the conterminous U.S. (CONUS) domain and the impact of emissions uncertainties on ozone prediction for an industrialized area are provided in the following.

26.5.1 National Air Quality Forecasting Capability

The National Air Quality Forecasting Capability (NAQFC) system is based on the National Weather Service (NWS) National Center for Environmental Predictions (NCEP) operational North American Mesoscale (NAM) model called the Nonhydrostatic Mesoscale Model (NMM) and the U.S. EPAs Community Multiscale Air Quality (CMAQ) modeling system. An initial version of the system was deployed in June 2004 with the NCEPs ETA meteorological model to forecast surface-level O₃ pollution over the northeast United States. Since that time, the domain coverage has expanded incrementally, now to the entire CONUS domain, and developmental PM₂.₅ forecasts are also now produced.

The NMM model generates a meteorological forecast over the North American continent at 12 km resolution on a rotated latitude–longitude grid with the staggered Arakawa E-grid, which must be interpolated for CMAQs C-grid. Meteorological observations from satellite, surface, radiosondes, and profilers are assimilated into NMM using the Gridpoint Statistical Interpolation (GSI) scheme. Lateral boundary conditions for NAM forecasts come from the NCEPs Global Forecasting System (GFS), which consists of a global model with a coarser resolution and a global version of the GSI to assimilate meteorological and ozone observations. Recently, NCEP has launched development of weather models such as the new NMM dynamic core using B-grid staggering (hereafter called NMMB) with the NOAA Environmental Modeling System (NEMS), which uses standard Environmental Systems Modeling Facility (ESMF) compliant software. Work is on the way to utilize the NMMB meteorology data, directly for CMAQ simulations without redundant interpolations.

The operational NAQFC for ozone is currently based on CMAQ version 4.6 with the Carbon Bond-IV (CB-IV) chemical mechanism. The experimental and developmental versions of NAQFC include aerosol components and they utilize the 2005 version of the CB chemical mechanism called CB05, which is considered to be suitable for ozone, PM, visibility, and acid deposition at urban and regional scales. Since 2004, ozone forecasting with NAQFC has been made utilizing EPAs national emissions inventories (NEIs) for base year 2001, 2002, and 2005. In the near future, the 2008 NEI will be utilized as it becomes available. In the NEIs, emissions of sulfur dioxide (SO₂), nitrogen oxides (NOₓ), PM₂.₅, and VOCs are divided into anthropogenic (area, point source, on-road mobile, non-road mobile source emissions) and biogenic emissions. For the NAQFC operation, emissions that are independent from meteorological conditions are processed first using the Sparse Matrix Operator Kernel Emissions (SMOKE) modeling system (Houyoux et al., 2000). Emission components that vary with meteorological conditions

are processed later in PREMAQ in real time using NMM meteorological fields. Currently, NAQFC utilizes constant boundary conditions (BCs) and initial conditions from the previous days forecasting results.

At present, the initial conditions are not specified using observed data even for those species routinely measured as part of the controlled criteria species listed in the National Clean Air Act and its Amendments (CAAA) in an urban area with a dense measurement network. This is because of the diĐculty in specifying multispecies conditions that satisfy chemical balance in the system, which is subject to the diurnal evolution of radiative conditions and the ABL as well as temporal changes in the emissions that reflect constantly changing human activities. With the development of the coupled NAMB-CMAQ core, the GFS ozone profiles will be reintroduced to improve BCs and a multi-cycle-based data assimilation method will be utilized to improve ICs. The results will be compared with that of other global chemical transport models that include stratospheric chemistry. Currently, development of the NAMB-CMAQ online/od ine linkage is in progress to enable NAQFC simulations over the entire CONUS domain at 4 km resolution. Examples of air quality forecast guidance for ozone and PM are presented in Figure 26.4.

To improve model performance, the results of daily forecasting are routinely evaluated with available measurements. Recent analysis of several years of forecasting results shows that fewer cases of ozone exceedances occurred in the eastern United States in 2009 as compared to 2007 and 2008 (Figure 26.5a). The ozone mean time series for 2009 (Figure 26.5b) reveals that the summertime bias increased in the summer for both CB-IV and CB05 chemical mechanisms. Some regional NAQFC focus group members have suggested that not only the NMM temperature and cloud biases but also changes in anthropogenic NOₓ emissions in 2009 due to economic conditions may be the cause of high ozone bias in the model. Others have suggested that diĐculties in simulating the interannual meteorological changes due to the prevailing El Niño may have affected predictions.

Differences in ozone predictions from the two mechanisms are currently under investigation and may be due to specific reaction differences as well as reactivity differences in the VOC precursor emissions as they are split differently for each mechanism. Trends in observed ozone concentrations are thus a result of interannual variation of meteorological conditions as well as actual changes in the emissions. To isolate effects of emission uncertainties on ozone predictions, the meteorological variations must be identified and decoupled from other model biases. Possible factors causing seasonal differences are annual variations in emissions and meteorological parameters such as temperature, precipitation, and solar radiation. Note that although these statistics are affected mostly by the biases in the middle-range O₃ concentrations, CB05 performs better for peak O₃ values. Similar to the analysis for summer 2007, NAQFC simulations with both mechanisms show overpredictions for concentrations below 70 ppb with CB05 showing the worst bias

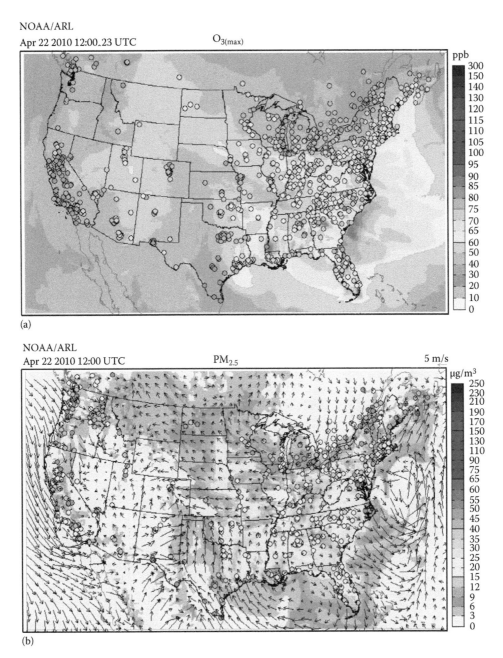

FIGURE 26.4 Air quality forecast guidance by NAQFC: (a) daily maximum ozone for April 22–23, 2010 and (b) hourly PM2.5 for 12:00 UTC, April 22, 2010. NAQFC forecasts for ozone and PM2.5 were verified with the EPAs AirNOW surface measurements.

in southeastern U.S. rural sites for July 2009 (Figure 26.6a), but the system underpredicts urban site concentrations that exceed 70 ppb (Figure 26.6b).

26.5.2 Retrospective Air Quality Simulation of Houston Metropolitan Area

This example studies effects of uncertainties of the industrial emissions of highly reactive VOC (HRVOC) emissions on the severe ozone problem in Houston, Texas. Under the Clean Air Act, the Houston–Galveston–Brazoria (HGB) area and several other metropolitan regions in the United States are classified as

ozone non-attainment areas. Houstons bad air quality is mainly due to a very large amount of VOC emissions from the regional petrochemical industries and NO_x emissions from mobile sources and power plants. Studies of the high ozone episodes of the Texas Air Quality Study in 2000 (TexAQS 2000) based on aircraft measurements and VOC reactivity analysis of gas chromatography observations at multiple surface sites revealed that the reported emissions of light olefins were not consistent with measured atmospheric concentrations. Therefore, the state agency "imputed" the HRVOC emissions (i.e., selectively increased ethylene and propylene) from the regular inventory values. The base VOC emissions inventory from point sources

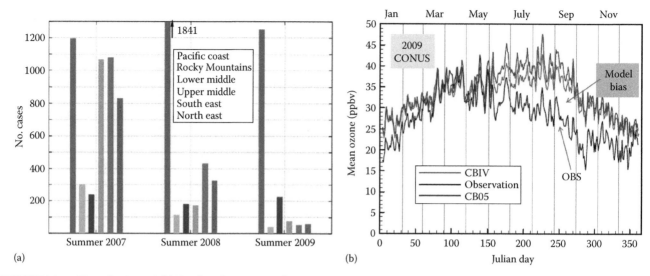

(a)   (b)

FIGURE 26.5 (See color insert.) (a) Number of ozone exceedances in summer 2007–2009. (Courtesy of J. Gorline, NOAA, Silver Spring, MD.); (b) model forecasting biases over CONUS domain in 2009 with the CB-IV and CB05 mechanisms.

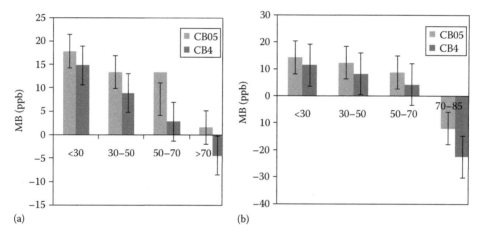

(a)   (b)

FIGURE 26.6 CB-IV and CB05 performance differences of maximum 8 h ozone for (a) rural and (b) urban sites in the southeastern United States, where biogenic emissions are high.

were 325 T/day for the HGB 4 km modeling domain, but ~170 T/day of HRVOC emissions from the sources were added.

Subsequently, the Second Texas Air Quality Study (TexAQS-II) was conducted in 2006 to evaluate effectiveness of local emissions controls implemented. University of Houston (UH) has been performing air quality forecasting simulations since 2004, utilizing CMAQ with the MM5 meteorological model. As a part of the retrospective evaluation of model performance with the TexAQS-II measurements to assess effects of emission uncertainty, UH regenerated meteorological inputs utilizing 4-D data assimilation and then conducted air quality simulations for the TexAQS-II ozone episode with two different emission inputs:

1. The 2006 Eastern Texas Air Quality (ETAQ) forecast emissions obtained by projecting the 2000 Texas emissions inventory (TEI) utilizing growth and control factors. The inventory VOC emissions for 2006 were reduced to ~230 T/day from the 325 T/day of 2000 case

but the same ~170 T/day of imputed HRVOC emissions were added. The mobile emission inventory used was for the base year 2003.

2. A best-effort model-ready (BEMR) 2006 Texas emission data set was prepared from the 2006 special point source emission inventories for summer 2006 TexAQS-II with reported hourly and speciated VOC and NO_x emissions from point sources. Total point source VOC emissions in the BEMR emissions were around 210 T/day for the HGB area. The mobile emission inventory used was for the base year 2005 with ~40 T of NO_x per day less than that of 2003.

Figure 26.7 shows model-predicted surface ozone distributions for the high ozone episode of August 31, 2006. With the ETAQ forecast emission inputs, daily peak ozone values were well simulated, although there was a slight southwestward shift in the peak area because the simulated wind field had a slightly stronger northeasterly wind component around 300 m above

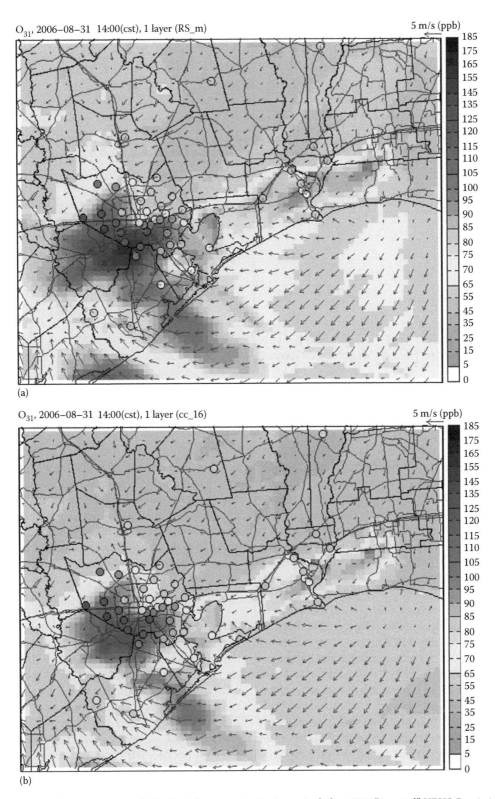

FIGURE2 6.7 Simulated high ozone event with (a) ETAQ forecast emission inputs including 2000 "imputed" HRVOC emissions and (b) BEMR emissions without imputation for 14:00 CST, August 31, 2006.

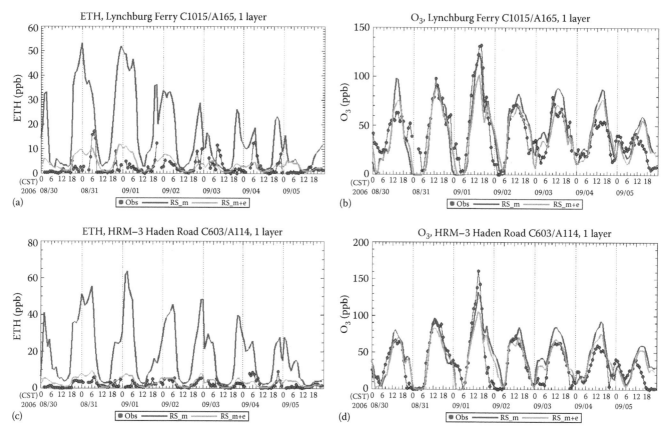

FIGURE 26.8 Time series plots of ethylene (ETH) and ozone (O3) concentration at the Lynchburg (a) and (b), and HRM-3 (c) and (d) site with two different emissions for the August 30–September 5, 2006 period. Simulations with the imputed HRVOC emissions are denoted as "RS_m" and with the BEMR as "RS_m + e."

ground level as compared to the wind profiler measurements at the La Porte site near the Houston Ship Channel (HSC). For the simulation with the BEMR emissions, predicted ozone concentrations were lower with less overprediction in southwestern Houston but missed high ozone observed in the HSC area.

While the peak ozone episodes were predicted more accurately with the imputed HRVOC emissions, comparison with the Automated Gas Chromatograph (AutoGC) monitoring data showed serious overprediction of ethylene (one of the key anthropogenic HRVOCs responsible for ozone formation in the HGB area) at the industrial locations (Figure 26.8). Both the Houston Regional Monitoring (HRM) network site (C603) and Lynchburg Ferry site (C1015) are located in the HSC. Simulated ETH concentrations with the BEMR emissions for both sites (Figure 26.8a and c, for RS_m + e) match well with observations at both sites although on some days fail to reach observed peak values. On the other hand, the simulations with base AQF emissions (Figure 26.8a and c, for RS_m) overpredict ETH by a factor of 4 or more, even during daytime hours. The predicted peak ozone concentrations with the AQF emissions are higher than those with the BEMR but not by much, considering the significant difference in the emissions inputs used. Although simulations with the BEMR emissions could not predict the peaks for the episodic days, their performances were better for other days than those with the imputed emissions. While detailed analysis

is underway, we hypothesize that these are due to the large variability in the actual industrial emissions over those reported in the BEMR emissions, but the amount of underestimation in the annual inventory is not as large as the 2000 imputed emissions. Further investigation on the industrial emission uncertainty is required to develop appropriate emission control strategies that protect public health with minimal burden to the industry.

26.6 Major Challenges

There is no one perfectly correct environmental model since models are only abstractions of the real world with different details and accuracy. Model outputs do not represent reality but are only the consequences of the simplifications and assumptions applied in the model formulations and limitations in the inputs, as well as inevitable bugs in the computer codes. Models are useful only when the results provide answers relevant to the questions asked. The validity of each model is therefore not necessarily an inherent quality of the model but is a characteristic feature that must be determined through the use of the measurements that are appropriate to the questions being asked.

Air quality modeling practices as currently operated have several limitations. The challenge is how to improve air quality modeling tools and data to provide useful information with suDcient accuracy in time and space, appropriate for tracking public health implications due to bad air quality. To utilize air quality modeling for forecasting and in health effects studies, improvements in several modeling components are essential. First, the science parameterizations representing different physical and chemical processes, including two-way interactions with meteorology, must be improved to represent the natural collective behavior of air pollutants more accurately. Second, as regional air quality models employ limited modeling domains, proper dynamic boundary conditions must be provided. Satellite data products can be used directly or assimilated in global chemical transport models to provide better dynamic lateral boundary conditions for regional air quality modeling. One diDculty in implementing coarse-to-high resolution model linkage is that differences in the algorithms and data among the models at different scales must be harmonized. Third, the basic model inputs, including land use and vegetation cover descriptions and emissions inputs, must be improved. Common diDculties of first principle-based environmental models are how to represent accurately "forcing terms" in the system, particularly those due to the influence of the Earth's surface, long-range transport, and uncertainties in the model inputs such as daily emissions changes due to anthropogenic and natural events. Fourth, model representativeness issues, including grid-resolution problems, compensating errors among the model components, and incommensurability of the model results compared with the dimensionality of the measurements (i.e., inherent differences in the modeled outputs that represent volume- and time-averaged quantities to the point or path-integrated measurements) need to be addressed. In addition, there is ample opportunity to reduce some uncertainties associated with air quality modeling through evaluation and verification with observed meteorological and atmospheric chemistry data. Although current state-of-science air quality models are far from perfect, it is clear that with appropriate use, accounting for their weaknesses as well as their strengths, these tools can provide invaluable assistance to forecasters, policymakers, and scientists in understanding and mitigating the effects of air pollution on public health and well-being.

References

Blackadar AK. 1978. Modeling pollutant transfer during daytime convection, Preprints, *Fourth Symposium on Atmospheric Turbulence, DiThsion, and Air Quality*, American Meteorological Society, Reno, NV, pp. 443–447.

Byun DW. 1999a. Dynamically consistent formulations in meteorological and air quality models for multiscale atmospheric applications: Part I. Governing equations in generalized coordinate system. *Journal of Atmospheric Science*, 56, 3789–3807.

Byun DW. 1999b. Dynamically consistent formulations in meteorological and air quality models for multiscale atmospheric applications: Part II. Mass conservation issues. *Journal of Atmospheric Science*, 56, 3808–3820.

Byun DW, Ching JKS, eds. 1999. Science algorithms of the EPA models-3 community multiscale air quality (CMAQ) modeling system, EPA Report, EPA/600/R-99/030, NERL, Research Triangle Park, Durham, NC.

Byun DW, Dennis RL. 1995. Design artifacts in Eulerian air quality models: Evaluation of the effects of layer thickness and vertical profile correction on surface ozone concentrations. *Atmospheric Environment*, 29 (1), 105–126.

Byun DW, Schere KL. 2006. Review of the governing equations, computational algorithms, and other components of the models-3 community multiscale air quality (CMAQ) modeling system. *Applied Mechanics Reviews*, 59 (2), 51–77.

Carter WPL. 1990. A detailed mechanism for the gas-phase atmospheric reactions of organic compounds. *Atmospheric Environment*, 24A, 481–518.

Carter WPL. 2000. Documentation on the SAPRC-99 chemical mechanism for VOC reactivity assessment. Final Report to California Air Resources Board Contract No. 92-329 and 95-308, Sacramento, CA, May 2000.

Carter WPL. 2010. Development of a condensed SAPRC-07 chemical mechanism, *Atmospheric Environment*, 27A, 2591–2611. Doi:10.1016/j.atmosenv.2010.01.024.

Deardorff JW. 1972. Theoretical expression for the countergradient vertical heat flux. *Journal Geophysical Research*, 77, 5900–5904.

Gear CW. 1971. *Numerical Initial Value Problems in Ordinary DiThrential Equations*. Prentice-Hall, Englewood Cliffs, NJ.

Gery MW, Whitten GZ, Killus JP, Dodge MC. 1989. A photochemical kinetics mechanism for urban and regional scale computer modeling. *Journal of Geophysical Research*, 94 (D10), 12925–12956.

Hertel O, Berkowicz R, Christensen J, Hov O. 1993. Test of two numerical schemes for use in atmospheric transport-chemistry models. *Atmospheric Environment*, 27A, 2591–2611.

Houyoux MR, Vukovich JM, Coats Jr. CJ, Wheeler NM, Kasibhatla PS. 2000. Emission inventory development and processing for the seasonal model for regional air quality (SMRAQ) project. *Journal of Geophysical Research*, 105, 9079–9090.

Klemp JB, Skamarock WC, Dudhia J. 2007. Conservative split-explicit time integration methods for the compressible nonhydrostatic equations. *Monthly Weather Review*, 135, 2897–2913.

Janjic ZI. 2003. A nonhydrostatic model based on a new approach. *Meteorology and Atmospheric Physics*, 82, 271–285. http://dx.doi.org/10.1007/s00703-001-0587-6

Jacobson M, Turco RP. 1994. SMVGEAR: A sparse-matrix, vectorized gear code for atmospheric models. *Atmospheric Environment*, 28, 273–284.

Ooyama KV. 1990. A thermodynamic foundation for modeling the moist atmosphere. *Journal of Atmospheric Science*, 47, 2580–2593.

Pleim JE. 2007. A combined local and nonlocal closure model for the atmospheric boundary layer. Part I: Model description and testing. *Journal of Applied Meteorology and Climatology*, 46, 1383–1395.

Pleim JE, Chang J. 1992. A nonlocal closure model for vertical mixing in the convective boundary layer. *Atmospheric Environment*, 26A, 965–981.

Sandu A, Verwer JG, Blom JG, Spee EJ, Carmichael GR, Potra FA. 1997. Benchmarking stiff ODE solvers for atmospheric chemistry problems II: Rosenbrock solvers, *Atmospheric Environment*, 31, 3459–3472.

Whitten GZ, Hogo H, Killus JP. 1980. The carbon-bond mechanism: A condensed kinetic mechanism for photochemical smog. *Environmental Science Technology*, 14, 690–700.

Yarwood G, Rao S, Yocke M, Whitten GZ. 2005. Updates to the carbon bond chemical mechanism: CB05. Final Report to the US EPA, RT-0400675, December 8, 2005 /http://www.camx.com/publ/pdfs/CB05_Final_Report_120805.pdf

Zhang R, Suh I, Lei W, Clinkenbeard AD, North SW. 2000. Kinetic studies of OH-initiated reactions of isoprene, *Journal of Geophysical Research*, 105, 24627–24635.

In Memoriam—Daewon W. Byun (1956–2011)

Dr. Daewon W. Byun, the author of this chapter, passed away suddenly and unexpectedly in February 2011 at 55 years of age. Even though his life and career ended much too soon, Daewon had achieved remarkable success as an accomplished atmospheric scientist and numerical modeler. He was born in South Korea and received a B.S. degree in meteorology from Seoul National University in 1977. After serving as a meteorological oDcer in the South Korean Navy, Daewon came to the United States for graduate training and received M.S. (1983) and Ph.D. (1987) degrees in atmospheric sciences from the North Carolina State University. His professional career started with service in the private sector with ERT/ENSR and Computer Sciences Corporation (1988–1992). Daewon then joined the National Oceanic and Atmospheric Administration in 1992 (NOAA, on assignment at U.S. EPA: 1992–2001), and then returned to NOAA in 2009. In the intervening years, Daewon moved to academia at the University of Houston (2001–2009). Daewon's research interests in the atmospheric sciences were quite broad, and included air pollution meteorology and chemistry; atmospheric boundary layer turbulence and diffusion; numerical algorithms and software development for atmospheric transport and chemistry models; model evaluation, sensitivity, and uncertainty analysis; and numerical air quality forecasting. While at U.S. EPA, Daewon helped implement and evaluate the Regional Acid Deposition Model, one of the first air quality models applicable to 1000 km and larger domains, and he was the principal designer and a key developer of the successor Community Multiscale Air Quality (CMAQ) model, a model now being used by thousands of modelers in the United States and worldwide. The great success and widespread adoption of the CMAQ modeling system is a direct testament to Daewon's vision, attention to detail, and perseverance. His modeling achievements and notable scientific publications while at EPA were recognized with many awards from NOAA and EPA. At the University of Houston, Daewon helped establish the Institute for Multidimensional Air Quality Studies (IMAQS) and he successfully attracted new faculty members, postdoctoral research fellows, and graduate students to this growing program. Daewon and the IMAQS Team developed a real-time air quality forecast modeling system applied over national to local scales, with a focus on Texas and the Houston–Galveston region. Using this operational system, Daewon and his colleagues at IMAQS studied the sensitivity of the forecast results to different meteorological models, emissions and land-use data, and choice of physics parameterization schemes. He later carried the lessons from that experience with him as he led the NOAA research team supporting the NWS air quality forecast modeling capability, with the CMAQ model as the operational air quality component. Daewon is fondly remembered by his many colleagues and friends as a brilliant and inspirational scientist, patient mentor and friend, and for his caring and friendly nature. He is survived by his wife, Chin Ree Byun, and daughters, Christine and Lydia.

Kenneth L. Schere
Atmospheric Modeling and Analysis Division
United States Environmental Protection Agency
Research Triangle Park, North Carolina

27

Mathematical Models to Improve Performance of Surface Irrigation Systems

D. Zerihun
University of Arizona

C.A. Sanchez
University of Arizona

27.1 Introduction

Surface irrigation is the most widely used method of irrigation water application to croplands worldwide. Rising concerns about the negative environmental impacts of irrigation, changing climatic patterns that have resulted in unusually long dry spells, as well as increasing demands for land and water resources from urban and industrial interests are putting pressure on the share of freshwater supply and land resources available for irrigation. Obviously, if agricultural irrigation is to remain sustainable, it needs to evolve into a more eDcient, cost-effective, and eco-friendly technology.

Proper system design and management is essential for the eDcient application of water and nutrients with surface irrigation. In theory, hydraulic and plant nutrient transport data derived through field studies can be used in the development of design and management recommendations. However, the scale of the time and effort needed to collect suDciently varied field data for the evaluation of a wide range of irrigation/fertigation design and management scenarios is prohibitive. On the other hand, mathematical models offer a much more flexible and inexpensive alternative for developing optimal surface irrigation system design and management recommendations.

This chapter presents a brief review of the literature on surface irrigation hydraulic and solute transport modeling and the application of models in surface irrigation system characterization, design, and management.

27.2 Principles

27.2.1 Surface Irrigation System and Processes

An irrigation system can be conceived as an assemblage of four subsystems: the water source/abstraction, conveyance and control, on-farm water/nutrient application, and drainage. This chapter discusses the mathematical modeling and design and management of the on-farm water and nutrient application subsystem. Surface irrigation methods utilize gravity as the driving force to convey and distribute water over the field surface to be irrigated. The primary types of surface irrigation methods are basins, borders, and furrows. In general, basins are completely closed and can have level or sloping beds. Borders are sloping channels with a free-draining downstream boundary. Furrows can be graded and free draining or can have a level bed and a closed-end downstream boundary. An irrigated field may consist of tens or hundreds of furrows or it can be irrigated with

Handbook of Environmental Fluid Dynamics, Volume Two, edited by Harindra Joseph Shermal Fernando. © 2013 CRC Press/Taylor & Francis Group, LLC.
ISBN: 978-1-4665-5601-0.

one or more basins or borders, typically running along the main slope of the irrigated field. Slight cross slope can be tolerated in furrow-irrigated fields, but no cross slope is allowed in basins and borders. The basins, borders, and furrows in a field receive their water supply from a canal or a pipeline that runs along the head end of the irrigated field (across the main field slope). Local variations and adaptations to this description of the surface irrigation system exist; however, this is the configuration typical of modern surface-irrigated farming systems that are amenable to mathematical modeling–based hydraulic analysis.

In general, the surface irrigation process can be conceived as being composed of four phases (Figure 27.1a): (1) The *advance phase* covers the initial wetting process during which a positive surge propagates down a dry bed at a progressively decreasing rate (assuming constant inflow and uniform field characteristics). (2) The *wetting/ponding* phase begins at the end of the advance phase and ends when inflow is cut. During this phase, the downstream boundary of the flow domain is either free-drainage or a no-flow boundary, and at the upstream end there is a constant or variable inflow. Note that the term "wetting phase"

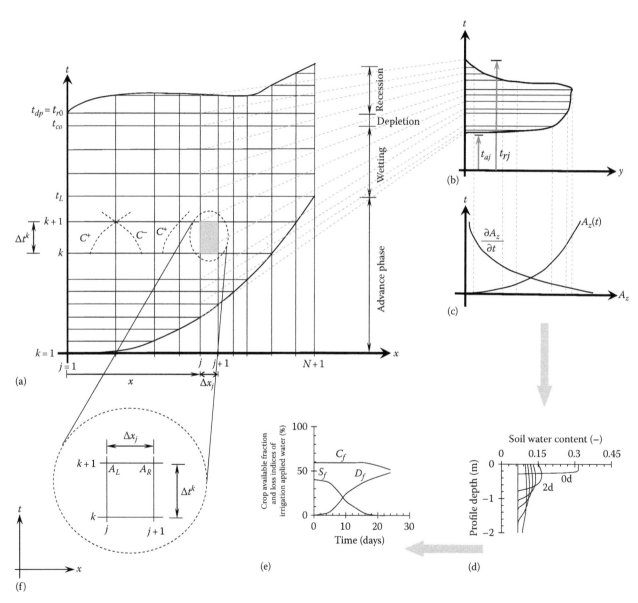

FIGURE 27.1 Schematics of the surface irrigation process: (a) discretization of the *x-t* space for numerical solution, (b) flow depth hydrograph at node *j*, (c) infiltration function at node *j*, (d) post-irrigation soil water redistribution (From Zerihun, D. et al., *J. Irrig. Drain. Eng.*, 131(2), 111, 2005a), (e) crop available fraction of irrigation applied water and irrigation water loss indices (From Zerihun, D. et al., *J. Irrig. Drain. Eng.*, 131(2), 111, 2005a), and (f) a computational cell in the *x-t* plane (*k*, time step index; *j*, distance step index; t_{aj}, advance time to node *j*; t_{rj}, recession time to node *j*; t_L, advance time to downstream end of channel; t_{dp}, depletion time; t_{r0}, recession time at channel inlet; t_{co}, cutoff time; C^+ and C^-, forward and backward characteristics, respectively; C_f, crop available fraction of irrigation applied water; D_f, deep percolation fraction; S_f, surplus fraction of irrigation-applied water; and A_L and A_R, cross-sectional area on the left- and right-hand-side nodes, respectively, of a computational cell at the current time level).

is used in association with free-draining systems and "ponding phase" is preferably used with closed-end systems. The wetting phase does not exist in surface irrigation events in which inflow cutoff precedes completion of advance (advance and recession taking place simultaneously). (3) The *depletion* phase is generally defined as the time interval between inflow cutoff and the time that water depth falls to zero at the inlet end. Often the definition of the depletion phase assumes that it is an irrigation phase sandwiched between the wetting phase and the recession phase. However, from the computational point of view, an important characteristic feature of the depletion phase is that at the channel inlet, flow rate is zero and there is a finite flow depth. The corresponding downstream end boundary condition could be free-drainage, no-flow boundary, or zero depth and zero flow rate. (4) The *recession* phase covers the time interval between the end of depletion and the time that the channel dewaters completely. In sloping channels, during the recession phase, the receding tip (i.e., the zero depth upstream boundary of the flow domain) moves from the inlet end to the downstream end of the channel. However, in level basins or closed-end level furrows, if water surface becomes level subsequent to inflow cutoff, the entire channel dewaters simultaneously; hence the recession phase and the depletion phase are not distinct.

The preceding discussion on the hydraulic phases of an irrigation event is important in terms of the following: (1) specification of boundary conditions for the solution of the governing equations of flow and transport; (2) identification of the main processes, important at any given stage of irrigation, the interaction of which determines the evolution of surface irrigation hydraulics and solute transport processes in time and distance; and (3) determination of applicable forms (and necessary simplifications) of the governing equations at different stages of the irrigation process. Note that Figure 27.1a also depicts the computational domain for the solution of surface irrigation flow and transport equations. Figures 27.1b through e show the wetting and drying processes, infiltration, subsequent post-irrigation redistribution, and the time evolution of crop available, surplus, and deep percolation fractions of irrigation-applied water at a computational node along a channel, respectively. Figure 27.1f shows a typical computational cell, bounded by nodes j and $j + 1$ and time levels k and $k + 1$, in the two-dimensional *x-t* plane.

27.2.2 System Characteristics

At any given time during an irrigation season, a surface irrigation system can be characterized in terms of two sets of physical quantities: (1) *System parameters*: physical quantities that reflect the inherent, but transient, properties of the system at any given point in time, namely, the hydraulic resistance coefficient and infiltration parameters, as well as the dispersion coefficient, dispersivity, or some related constant if solute transport is modeled, and (2) *System variables*: physical quantities whose magnitude can be selected by the system designer or operator at the system design/management phase and/or at the time of

system operation. Generally, physical dimensions (length [L], width [W], or cross-sectional geometry), inflow rate (Q_0), inflow cutoff criteria (cutoff time [t_{co}] or cutoff length [L_{co}]), volumetric net irrigation requirement (A_{zr}), bed slope (S_o), and solute concentration at the channel inlet (C_0), if solute transport is modeled, can be considered as system variables. The interaction of these factors (as governed by pertinent physical principles) determines the outcomes of an irrigation/fertigation event, defined in terms of flow depth and flow rate hydrographs, solute breakthrough curves, post-irrigation longitudinal infiltration profile, and irrigation/fertigation performance. From the perspective of system design, management, and evaluation, the most important category of dependent irrigation variables are the performance indices and are described subsequently.

27.2.3 System Performance

Irrigation performance measures how close an irrigation event is to a reference or an ideal irrigation (e.g., Zerihun et al., 1997). A complete characterization of the performance of an irrigation event can be obtained using application eDciency (E_a, %), water requirement eDciency (E_r, %), and a uniformity index (e.g., distribution uniformity based on minimum depth, DU_{min}, (–)):

$$E_a = \frac{\int_0^L A_z dx - \int_0^{L_{ov}} A_z dx + A_{zr}L_{ov}}{\int_0^{t_{co}} Q_0 dt} 100,$$

$$E_r = \frac{\int_0^L A_z dx - \int_0^{L_{ov}} A_z dx + A_{zr}L_{ov}}{A_{zr}L} 100, \qquad (27.1)$$

$$DU_{min} = \frac{A_{z\,min}}{A_{zav}} = \frac{A_{z\,min}L}{\int_0^L A_z dx}$$

where

L is the channel length (L)

A_z is the volumetric cumulative intake function (L^3/L)

x is the distance (L)

L_{ov} is the over irrigated length (L)

A_{zr} is the required amount of application (L^3/L)

t_{co} is the cutoff time (T)

Q_0 is the inflow hydrograph (unit inflow rate for a basin/border or inflow rate into a furrow) (L^3/T)

A_{zmin} is the minimum infiltrated amount (L^3/L)

A_{zav} is the average infiltrated amount (L^3/L)

A more comprehensive formulation, compared to Equation 27.1, was developed for fertigation performance evaluation by Zerihun et al. (2003) and can readily be adapted for irrigation applications. The application of these performance functions is limited to the spatial scale of an irrigated field and the time scale of an irrigation event. Evaluation of system performance at a larger temporal and spatial scale requires taking into account off-site consequences of a field-scale design/management recommendation (e.g., Burt et al., 1997).

Hydraulic and solute transport equations describing physical laws that govern the interaction of surface irrigation system parameters and variables and their effects on irrigation/fertigation performance are presented subsequently.

27.2.4 Hydraulic Equations

The surface irrigation process is hydraulically described as a gradually varied unsteady open channel flow over a porous bed with variable intake rate (Figure 27.2a). At the most fundamental level, the governing equations of the surface irrigation process are the instantaneous forms of the fluid dynamics equations for incompressible flow, consisting of expressions for mass continuity and linear momentum conservation (Granger, 1995). However, the time and cross-section-averaged (1D) forms of the continuity and momentum conservation equations (Strelkoff, 1969) are generally deemed adequate to describe the hydraulics of surface irrigation with suDcient detail. Noting that infiltration is an important component of surface irrigation hydraulics, the generic equations of 1D open channel hydraulics (e.g., Cunge et al., 1980) are appended with sink terms accounting for infiltration effects on mass and momentum conservation, leading to the form:

$$\frac{\partial Q}{\partial x} + \frac{\partial A}{\partial t} + \frac{\partial A_z}{\partial t} = 0 \qquad (27.2)$$

$$gA\left(\frac{\partial z}{\partial x} - \frac{\partial H}{\partial x} - \frac{1}{A}\frac{\partial P}{\partial x}\right) - \frac{Q}{A}\frac{\partial A_z}{\partial t} = \frac{\partial Q}{\partial t} + \frac{\partial}{\partial x}\left(\frac{Q^2}{A}\right) \qquad (27.3)$$

where

   Q is the flow rate (L^3/T)
   x is the longitudinal dimension (L)
   t is the time (T)
   A is the flow cross-sectional area (L^2)
   g is the acceleration due to gravity (L/T^2)
   P is the ratio of hydrostatic pressure force to weight density of water (L^3) (Figures 27.2a and b)
   z is elevation referenced from datum (L)
   H is the total energy per unit weight of water ($z + y + v^2/2g$) (L)
   y is the flow depth (L)
   v is the cross-sectional average velocity (L/T)
   ez/ex is channel bed slope, S_o (–)
   eH/ex is friction slope, S_f (–) (Figure 27.2a)

Derivation of Equations 27.2 and 27.3 using the control volume approach is described by several authors (e.g., Cunge et al., 1980).

Equation 27.2 is a statement of mass continuity. At any given point in the computational domain (Figure 27.1a), the sum of the spatial gradient of the longitudinal mass flux and the sink term (infiltration flux) is balanced by the time rate of change of surface storage. Equation 27.3 is the momentum equation, also known as the dynamic equation (Cunge et al., 1980). It is a statement of the second law of motion. Each of the terms in parenthesis, on the left-hand side, represent one of the three forces acting on the stream (component of weight in the direction of irrigation, friction force, and pressure force), expressed per unit weight and per unit length of stream. The second term on the left-hand side is a sink term accounting for momentum removal from the surface stream by the infiltrating water. Often this term is considered negligible. The two terms on the right-hand side of Equation 27.3 represent inertial reactions to the net unbalanced force acting on a stream of unit length. The first term represents the local acceleration, which is a measure of the unsteadiness of the flow and the second term is the convective acceleration (the

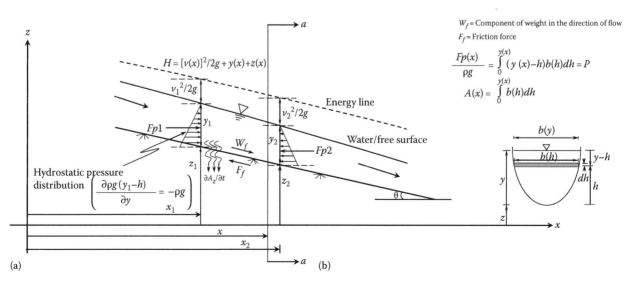

FIGURE2 7.2 (a) Definition sketch for the description of elements of Equations 27.2 and 27.3 and (b) Section at *a–a*.

spatial gradient of the velocity head), a measure of the spatial variation of the flow.

The following limiting assumptions apply to Equations 27.2 and 27.3: (1) Channel is prismatic. (2) Velocity variation and water surface slope in a cross section are negligible, hence flow can be considered one-dimensional. (3) Density of water is constant. (4) Streamline curvature is insignificant; hence vertical acceleration is negligible leading to linear variation of pressure head with depth (hydrostatic pressure distribution at a section of the stream). (5) A hydraulic resistance equation for steady uniform flow can be used to quantify the energy loss due to boundary friction and form drag. (6) The angle of inclination of the channel bed measured from the horizontal, θ, is suĐciently small that its cosine can be taken to be 1.0; hence the vertical distance between channel bottom and free-surface and flow depth can be considered equal.

Equations 27.2 and 27.3 are applicable to channels of arbitrary shape; hence, as such they can be used to model furrow irrigation processes and can be simplified for basins/borders (wide rectangular channels). One-dimensional hydraulic analysis of irrigation systems is conducted on the scale of unit characteristic width of the channel. For furrows, the unit characteristic width is the furrow spacing; for a basin and a border, it is a strip with transverse dimension of 1 m or 1 ft (considering the most common units used). Furrows are small (shallow and narrow) channels with cross-sectional dimensions much smaller than the longitudinal dimension. Velocity variation in a cross section can be assumed negligible relative to the variation in the longitudinal direction; hence, a 1D description of the flow process is realistic (see assumption 1 mentioned earlier). On the other hand, basins and borders are wide rectangular channels with widths orders of magnitude larger than flow depths. Although vertical velocity variations in a basin/border stream can generally be considered insignificant; the existence of cross-currents cannot be totally ignored, unless additional restrictive assumptions are satisfied. The additional assumptions are as follows: (1) Variation in system parameters such as surface roughness, infiltration, and bed slope in the transverse direction are too small to cause transverse water surface gradient. (2) The total basin inflow is uniformly distributed over the width of the basin or at least the total inflow rate and aspect ratio of the basin are sufficiently large to induce complete lateral coverage and the development of a laterally uniform advancing front within a short distance from the inlet end relative to the length of the basin.

Equations 27.2 and 27.3 have two dependent variables, flow rate, Q, and cross-sectional area, A, expressed as a function of two independent variables—distance, x, and time, t. The volumetric infiltration rate, eA_z/et ($L^3/L/T$), is computed in one of the three following ways: (1) with an empirical infiltration equation as a function of infiltration opportunity time, τ (T), and a characteristic infiltration width, w; (2) based on numerical solutions of the governing equation for variably saturated porous media flow (Richardsflequation); or (3) using some simplified analytical solutions of the Richards equation. Discussion on recent developments related to coupled surface-subsurface flow

models will be deferred to a later section in this chapter, where we will briefly touch on this topic.

Because of their simplicity and minimal data need for parameter estimation, empirical infiltration functions are widely used to express infiltration rate and cumulative infiltration in surface irrigation hydraulic modeling. The most widely used type being the Kostiakov function and variants thereof, such as the modified Kostiakov function and the branch infiltration function (Clemmens, 1981). Arguably, the most common form used in surface irrigation modeling is the modified Kostiakov equation. In general, the volumetric intake rate over an infiltration characteristic width, eA_z/et, and the instantaneous intake rate at a point, i, are expressed with the modified Kostiakov function as

$$\frac{\partial A_z}{\partial t} = \bar{i}w \quad \text{and} \quad i = ka\tau^{a-1} + b \quad (27.4)$$

where

$\bar{i}$ (L/T) is the average intake rate over a characteristic infiltration width, w (L)

k (L/T^a), a (–), b (L/T) are empirical constants of the modified Kostiakov infiltration equation

For basins/borders, w is equal to the unit width and $\bar{i} = i$. For furrows, at any given cross-section and at any given time, $\bar{i}$ represents the 2D intake rate averaged over the characteristic infiltration width, w, and the characteristic infiltration width is the nodal wetted perimeter, which is variable with time. Infiltration has a significant effect on flow depth, flow velocity, and rates of advance and recession, and hence on irrigation performance.

Although any steady uniform flow hydraulic resistance law can be used to calculate S_f for gradually varied flows with the same depth and velocity, the most widely used equation in surface irrigation applications is the Manning equation:

$$S_f = \frac{Q|Q|n^2}{C_u^2 A^2 R^{4/3}} \quad (27.5)$$

where

n is the Manningß roughness coeĐcient ($L^{1/6}$)

C_u is the dimensional constant, the equivalent of 1 m$^{1/2}$/s in an appropriate unit ($L^{1/2}/T$)

R is the hydraulic radius (L)

The Manning n is used as an aggregate measure of the resistance effects that the flowing water in a channel is subjected to due to boundary friction and form drag, the latter induced by tillage and vegetative growth. Manning n has a significant effect on flow depth, flow velocity, and rates of advance and recession.

Referring to Figures 27.2a and b, the ratio of hydrostatic pressure force to the weight density of water, P, is

$$P(x) = \int_0^{y(x)} (y(x) - h)b(h)dh \quad (27.6)$$

Based on commonly used empirical approximations of the geometric elements of a furrow cross section (Walker and Skogerbore, 1987)

$$y = \gamma_1 A^{\gamma_2} \quad \text{and} \quad A^2 R^{4/3} = \rho_1 A^{\rho_2} \qquad (27.7)$$

The ratio of the hydrostatic pressure force to the weight density of water, P, can be expressed as

$$P = \beta_1 A^{\beta_2} \quad \text{where} \ \beta_1 = \frac{\gamma_1 \gamma_2}{\beta_2} \quad \text{and} \quad \beta_2 = \gamma_2 + 1 \qquad (27.8)$$

In Equation 27.7, $\gamma_1 (L^{(1-2\gamma_2)})$, γ_2 (–), $\rho_1 (L^{(16-6\rho_2)}/3)$, and $\rho_2 (–)$ are empirical furrow geometry parameters and R is the hydraulic radius (L). Substituting Equation 27.8 in Equation 27.3, and noting that the second term on the left-hand side of Equation 27.3 is negligible compared to the other terms, yields the momentum equation for a prismatic furrow of arbitrary cross section:

$$\frac{1}{g}\frac{\partial Q}{\partial t} + \frac{\partial}{\partial x}\left(\beta_1 A^{\beta_2} + \frac{Q^2}{gA}\right) = A\left(S_o - \frac{Q|Q|n^2}{C_u^2 \rho_1 A^{\rho_2}}\right) \qquad (27.9)$$

A unit width basin/border is a special case of a furrow, with the geometry parameters set at $\gamma_1 = \gamma_2 = \rho_1 = 1$ and $\rho_2 = 10/3$; hence Equation 27.9 is applicable to basins and borders as well. The hydrodynamic model (the name often associated with Equations 27.2 and 27.9 in surface irrigation hydraulics) consists of nonlinear first-order partial differential equations of the hyperbolic type and are solved numerically in all but the most trivial cases. Numerical solutions of the hydrodynamic model, as applied to surface irrigation, were developed, among others, by Katopodes and Strelkoff (1977a), Bautista and Wallender (1992), and Walker (2003).

Field conditions, often, do not satisfy assumptions made in the derivation of Equations 27.2 and 27.9 in full or in a rigorous sense. Typically, the average deviation in the elevation of a basin/border surface from the assumed surface is significant compared to average flow depths. Soil and crop hydraulic properties (infiltration, hydraulic resistance) exhibit random spatial and temporal variations, hence diDcult to predict. However, surface irrigation models can reproduce the essential features of a surface irrigation event in an average sense, provided the system is properly characterized, generally based on past data. Such results can be most useful in system evaluation if model limitations relative to actual field conditions are taken into account in the interpretation and usage of the results. Most importantly, however, due to the substantial advantages of cost and flexibility that hydraulic modeling–based surface irrigation analysis offers compared to field studies, these limitations of the governing equations have not impeded the use of surface irrigation hydraulic models as research, design, and management tools.

The hydrodynamic model is the most complete expression of the two physical laws as applied to the description of 1D open channel flow in the context of surface irrigation, but it is also

relatively the most complex. Thus, any realistic simplification can be useful. Under special flow conditions, some of the terms in the momentum equation (Equation 27.9) are negligible compared to others; hence can be ignored, leading to simper forms. The following three levels of simplification of the governing equations of flow can be discerned for surface irrigation applications.

The zero-inertia model: Surface-irrigated systems are characterized by flat bed slopes and high hydraulic drag, which leads to low velocities that vary slowly in time and space, with typical characteristic Froude number, F, well below 1.0 (Katopodes and Strelkoff, 1977b). Under such flow conditions, the inertial/acceleration terms are much smaller than bed slope and friction slope and appreciably smaller than the depth gradient term. Hence, the acceleration terms can be ignored without loss of accuracy, leading to a force equilibrium relationship:

$$\frac{\partial(\beta_1 A^{\beta_2})}{\partial x} = A\left(S_o - \frac{Q|Q|n^2}{C_u^2 \rho_1 A^{\rho_2}}\right) \qquad (27.10)$$

It can indeed be shown (through formal dimensional analysis) that for the case where $F \ll 1.0$, the acceleration terms dropout leaving the three forces acting on the stream in essential balance. Equations 27.2 and 27.10 are the zero-inertia equations. It can also be shown that the zero-inertia equations constitute an advection-diffusion equation, a parabolic partial differential equation (e.g., Cunge et al., 1980). Hence, compared to the hydrodynamic model, the zero-inertia model is computationally more stable. In addition, numerical solution of the zero-inertia equations is simpler than that of the hydrodynamic equations. The zero-inertia model has successfully been used to simulate irrigation in borders, furrows, and basins (Strelkoff and Katopodes, 1977; Strelkoff, 1985; Walker and Skogerboe, 1987; Walker, 2003; Zerihun et al., 2005a). In general, the zero-inertia model is most suitable for simulating flows in channels with relatively flat slopes, including channels with backwater: basins and closed-end level furrows.

Kinematic wave model: In channels with relatively steep bed slopes (often $S_o = 0.001$ is considered as a cutoff value for practical surface irrigation modeling purposes), the pressure gradient term is typically much smaller than the bed and friction slopes. This leads to further simplification of Equation 27.10 into the uniform flow equation (channel bed slope, S_o, equals the friction slope, S_f), which can be expressed as

$$Q = \alpha_1 A^{\alpha_2}, \quad \text{where} \ \alpha_1 = C_u \sqrt{\frac{\rho_1 S_o}{n^2}} \quad \text{and} \quad \alpha_2 = \frac{\rho_2}{2} \qquad (27.11)$$

Equations 27.2 and 27.11 constitute the kinematic wave model. This model is applicable to sloping free-draining channels only and is most suitable for simulating low Froude number flows in long and steep channels. The kinematic wave equations were solved numerically for both borders and sloping free-draining furrows (e.g., Strelkoff, 1985; Walker and Skogerboe, 1987).

The volume balance model: Ignoring the dynamic equation altogether and integrating the continuity equation over time and distance, assuming a constant inflow, Q_0, results in an equation that can be used to compute stream advance over a dry channel with a sufficiently steep bottom slope (Lewis and Milne, 1938):

$$Q_0 t_X = \int_0^X A(s,t)ds + \int_0^X A_z(s,t)ds \qquad (27.12)$$

where

Q_0 is the inflow rate (L^3/T)
t_X is the time elapsed since the start of irrigation, advance time to point X (T)
X is the advance distance (L) [$0 \le X \le L$]
s is the distance from inlet end (L) [$s \le X$]

Although the volume balance model is an exact statement of mass balance for a constant inflow, it cannot predict the time and distance variation of velocity (which requires some form of the dynamic equation). In general, the use of the volume balance model implies that the modeling objective is finding estimates of the overall system response in terms of surface and subsurface volumes as a function of applied impulse (unit inflow rate) over a specified time. Equation 27.12 is valid only for the advance phase of the surface irrigation process. However, it is often coupled with other algebraic expressions, derived based on additional simplifying assumptions or, empirically, to simulate the complete surface irrigation process (e.g., Walker and Skogerboe, 1987). Volume balance models are robust and execute very fast on digital computers, hence often used in applications that require the evaluation of a wide range of alternative irrigation scenarios (design and management computations) and for parameter estimation purposes (e.g., Elliott and Walker, 1982; Walker and Skogerboe, 1987; Walker, 2003). There are several approximate solutions of Equation 27.12 and we will touch upon them briefly in subsequent sections.

27.2.5 Solute Transport Equations

Currently, mathematical modeling of solute transport processes in surface irrigation context is limited to the transport of solutes that are neutrally buoyant, nonsorbing, nonvolatile, and nonreactive at the temporal scale of a typical irrigation/fertigation event. The cross section–averaged (1D) advection-dispersion equation that describes the transport of such a solute in a surface irrigation stream (in a unit width basin/border or a furrow) can be derived using either a rigorous approach that involves progressive averaging over time, depth, and width of the instantaneous form of the continuity equation (Holly, 1985; Rutherford, 1994), or with a simpler approach based on the application of mass balance principles directly to a control volume in a 1D flow-field where dispersion is fully developed as a gradient-diffusion process

$$\frac{\partial(AC)}{\partial t} + \frac{\partial(QC)}{\partial x} + C\frac{\partial A_z}{\partial t} = \frac{\partial}{\partial x}\left(AK_x\frac{\partial C}{\partial x}\right) \qquad (27.13)$$

where

C is the solute concentration (M/L^3)
K_x is the longitudinal dispersion coefficient (L^2/T)

Note that A, Q, and $\partial A_z/\partial t$ are obtained from the solution of the hydraulic model. Equation 27.13 can be recast in its nonconservative form, a form commonly used for numerical solutions, by expanding it and combining the resulting equation with an expression for $\partial A_z/\partial t$, obtained from the continuity equation of the hydrodynamic model (Equation 27.2), resulting in

$$A\frac{\partial C}{\partial t} + Q\frac{\partial C}{\partial x} = \frac{\partial}{\partial x}\left(AK_x\frac{\partial C}{\partial x}\right) \qquad (27.14)$$

Equations 27.13 or 27.14 can be applied to furrows as well as borders and basins. The following limiting assumptions apply to Equations 27.13 and 27.14: (1) Hydrodynamic dispersion can be modeled as a gradient diffusion process using Fick's law. (2) Mass transport due to molecular diffusion is much less significant than is due to turbulent diffusion, hence can be readily absorbed in the dispersion term. (3) No significant loss/addition of solute occurs through chemical reaction, volatilization, sorption/desorption, and dissolution/precipitation in the course of a typical irrigation. Assumption 3 is reasonable in sofar as the model is used to simulate the transport of solutes that are not susceptible to the aforementioned transformation mechanisms in the time scale of an irrigation event (e.g., NO_3^- and salt). The second assumption is valid provided the first assumption holds. The first assumption is valid only in the channel segment where differential advection is in equilibrium with turbulent diffusion (i.e., complete mixing can be assumed in a cross section). For a nonsteady solute input, complete mixing can be assumed only after the passage of an initial advective period, the length of which is directly proportional to the width of the channel. Note that this is an important constraint that needs to be taken into account when applying Equations 27.13 and 27.14 to basins/borders (wide rectangular channels), under a nonsteady solute input. For furrows, however, complete mixing of solutes in a furrow cross section could be assumed at a short distance from the furrow inlet.

The solution of Equation 27.14 requires that the longitudinal dispersion coefficient, K_x, be known. Approximate values of K_x can be computed with formulas proposed by various authors (e.g., Rutherford, 1994). At the end of the chapter we will briefly touch upon coupled surface-subsurface solute transport models for simulating field-scale solute influx into and subsequent movement through the soil profile as well as post-fertigation redistribution.

27.3 Methods of Analysis

27.3.1 Solutions of the Governing Equations of Flow

In general, the flow equations described earlier are solved after having been coupled with pertinent initial and boundary conditions (Table 27.1). A brief review of the most widely used solution

TABLE2 7.1 Initial and Boundary Conditions, Hydraulic Equations

Type of Condition	Description	Range of Application
Initial condition	$Q(x,0) = A(x,0) = 0$	$0 \leq x \leq L$
Upstream BC	$Q(0,t) = Q_0(t)$	$t \leq t_{CO}$ [a]
	$Q(x_r,t) = 0$	$0 \leq x_r \leq L$ and $t_{CO} < t$
	$A(x_r,t) = 0$	$t_{r0} \leq t \leq t_{rL}$ [b]
Downstream BC	$Q(x_a,t) = A(x_a,t) = 0$	$0 < x_a \leq L$ and $0 < t \leq t_{aL}$ [c]
	$Q(L,t) = 0$	$t_{aL} \leq t \leq t_{rL}$ [d]
	$Q(L,t) = f(A_L)$	$t_{aL} < t \leq t_{rL}$ [e]

x_r, distance of the receding tip from upstream end; t_{r0}, recession time at the inlet end; t_{rL}, recession time at the downstream end; x_a, advance distance; t_{aL}, advance time to downstream end; A_L, flow depth at the downstream end of the computational domain.

[a] For the kinematic wave model A_0 is a function of Q_0 in accord with Equation 27.5 (with $S_f = S_O$).
[b] Applicable to graded channels.
[c] Not applicable to kinematic wave model.
[d] Applicable to closed-end channels.
[e] Applicable to free-draining channels.

techniques to the hydrodynamic, zero-inertia, kinematic wave, and volume balance equations is presented subsequently. The hydrodynamic and kinematic wave equations (hyperbolic partial differential equations) are amenable to solutions by the method of characteristics, but have also been solved with finite differencing (e.g., Walker and Skogerboe, 1987; Strelkoff and Clemmens, 2007). Implicit finite difference schemes were, typically, used to solve the zero-inertia equations (parabolic partial differential equations) and a variety of approximate solutions were proposed for solving the volume balance equation.

1. Method of characteristics: The method of characteristics has been used to solve the hydrodynamic and kinematic wave equations of border irrigation (Katopodes and Strelkoff, 1977a; Strelkoff, 1985; Walker and Skogerboe, 1987). The method transforms the partial differential equations of continuity and momentum into an equivalent set of ordinary differential equations—four equations for the hydrodynamic model and two equations for the kinematic wave model. The resulting equations define the time and space variation of the dependent variables as a function of flow perturbations propagating in the x-t plane along defined paths called characteristic lines (or simply characteristics) (Figure 27.1a). At any given point in the x-t plane, the state of flow is defined by two intersecting characteristics for the dynamic wave model and by perturbations propagating along the forward characteristics in the case of the kinematic wave model (Figure 27.1a).

The characteristic equations can be discretized through finite differencing in the framework of an Eulerian grid (e.g., Katopodes and Strelkoff, 1977a). The resulting nonlinear algebraic equations (four equations with four unknowns for the dynamic wave model and two equations with two unknowns in the case of the kinematic wave model) can be solved simultaneously

for the unknowns. For each time step, solution starts at the upstream end taking into account the boundary condition there and marches sequentially down the channel. At any given time step, each simultaneous solution determines the unknowns: distances measured from channel inlet of the intersection(s) of the back trajectories with the previous time line and values of depth (and velocity for the dynamic wave model) at a node on the current time level. An appropriate interpolation scheme is used to compute flow depth (and velocity in case of the dynamic wave model) at the intersection(s) of the back trajectories and the preceding time line based on known values at adjacent nodes (e.g., Katopodes and Strelkoff, 1977a). For the dynamic wave model, at the downstream end of the computational domain as well the boundary condition, either to compute incremental advance distance or flow depth and velocity needs to be taken into account. For the kinematic wave model, however, flow depth at the downstream end and incremental advance distance (during advance) emerge from the solution. The advantages of the characteristic transformation are as follows: (1) The resulting equations are more tractable than the original forms, yet equivalent to the original equations. (2) It shows the wave nature of unsteady flow, with physical interpretations useful in the identification and definition of required boundary conditions for solving the flow equations and in the physical interpretation of the Courant stability criteria for difference schemes.

Numerical discretization of the differential equations along the characteristics were shown to violate mass balance and for the dynamic wave model, the characteristic equations may represent a poorly posed problem in the flow region where the variables change rapidly, hence the iterative solution may fail to converge (Strelkoff and Falevy, 1993). On the other hand, implicit finite differencing methods have inherent advantages of being (1) unconditionally stable, (2) accurate if the time weighing coeDcient is properly selected and the solution domain is discretized with prudence, and (3) mass conservative at all times when coupled with iterative numerical solutions. Hence, with the widespread availability of high-speed digital computers for solving a large system of coupled algebraic equations resulting from implicit finite differencing, such schemes have become the most widely used techniques in discretizing the flow equations for surface irrigation applications.

2. Implicit rnite diThrencing: The four-point implicit finite difference scheme (Ligget and Cunge, 1975) is the most commonly used numerical technique to discretize the governing equations of surface irrigation hydraulics—typically the hydrodynamic and zero-inertia models. With this scheme, the governing partial differential equations (Equations 27.2 and 27.9 or 27.2 and 27.10) are converted into a pair of nonlinear algebraic equations, representing the residuals of mass and momentum conservation over a computational cell. Time and space discretization of the solution domain and a typical computational grid and cell are shown in Figures 27.2a and f, respectively. At a given time step, say the *kth* time step, application of the pair of algebraic equations to all the computational cells (e.g., *N* cells, Figure 27.2a) and imposing pertinent boundary conditions (Table 27.1), with concomitant simplifications as needed at the boundaries as well as at the initial time step, yields a coupled system of *2N* nonlinear equations with *2N* unknowns, which can be solved iteratively with the Newton-Raphson method (Walker and Skogerboe, 1987; Zerihun et al., 2005a). During each iterative step, a linear system of equations are solved with the double-sweep algorithm (Cunge et al., 1980; Strelkoff, 1992).

The four-point implicit finite difference scheme was also used to discretize the kinematic wave equation. Substituting Equation 27.11 in Equation 27.2 reduces the kinematic wave equation to a single partial differential equation with one unknown (*A* or *Q*, often *A* is used). Application of the finite differencing scheme to the kinematic wave equation will then result in a nonlinear algebraic equation with two unknowns for each computational cell (Walker and Skogerboe, 1987): the cross-sectional areas (at the current time level) at the left- and right-hand nodes of the cell, A_L and A_R, respectively (Figure 27.1f). However, the resulting nonlinear algebraic equations need not be coupled, because in each computational cell (beginning with the upstream end), A_L is known, leaving only A_R as unknown. Hence, the nonlinear equation for each cell can be solved explicitly for A_R, with the explicit solution marching sequentially down the channel. At the downstream end of the flow domain cross-sectional area, and during the advance phase the incremental advance distance, are computed as part of the solution. If, on the other hand, distance step is fixed and time step is to be computed as part of the numerical solution during advance, then the system of equations (arising from the application of the finite difference expression over all the computational cells in the flow domain) can be coupled and solved iteratively with the Newton method and the double-sweep algorithm (Strelkoff, 1992).

27.3.2 Solutions of the Volume Balance Equation

The Lewis and Milne (1938) integral equation (Equation 27.12) has been solved using different techniques, involving different assumptions and levels of approximation (Walker and Skogerboe, 1987; Strelkoff and Clemmens, 2007). All solutions of the volume balance equation are based on the assumption that flow depth at the inlet is at normal depth throughout the advance phase (i.e., flow depth at the inlet rises to normal depth instantaneously). In order for this assumption to be valid in the approximate sense, an important constraint is that the bed slope be suDciently steep. It is also a standard assumption to consider the average flow depth to be a constant irrespective of time, in which case it can be computed as a product of a constant surface shape factor and the inlet flow depth. Given this assumption, the volume balance equation reduces to

$$Q_0 t_X = \sigma_y A_0 X + \int_0^X A_z(s,t)\,ds \qquad (27.15)$$

where

σ_y is the surface shape factor expressed as the ratio of the surface volume to the product $A_0 X$ (–)

A_0 is the inlet cross-sectional area of flow corresponding to normal depth (L^2)

X is the advance distance (*L*)

In theory, σ_y can vary between 0.5 and 1. Often σ_y is set at 0.77 (Walker and Skogerboe, 1987). This is equivalent to assuming flow depth variation in accord with a power-law monomial of distance with a constant exponent of about 0.3. Equation 27.15 is often recast in a form amenable to analytical solutions (the convolution integral form):

$$Q_0 t_X = \sigma_y A_0 X + \int_0^{t_X} A_z(t_X - t_s)\frac{dx}{dt}\Big|_{t=t_s} dt_s \qquad (27.16)$$

where *x* is the longitudinal dimension (*L*). The solutions of the volume balance model can be grouped into five major categories:

1. *The Lewis-Milne solution*: Lewis and Milne (1938) solved Equation 27.16 analytically for the special case where the infiltration function, $A_z(t)$, can be expressed as a solution of a linear differential equation of the *n*th order with constant coeDcients. One such form, for which Lewis and Milne developed analytical solution, is the Horton equation, which expresses cumulative intake as an exponential function of time.

2. *Numerical approximation*: Considering a homogeneous soil and assuming that over a time step the subsurface profile varies linearly between consecutive nodes along the border, Hall (1956) solved Equation 27.15 numerically for the incremental advance distance down an irrigation border over a sequence of constant time steps. Recognizing that the linear subsurface profile assumption leads to underestimation of infiltration volume, corrections were proposed subsequently.

3. *The Kernel function approach*: Hart et al. (1968) presented a comprehensive analysis of the kinematics of the advance phase of the surface irrigation process based on a dimensionless form of the volume balance equation (Equation 27.16) for a unit width border, in which infiltration is described with the Kostiakov function. Some of the important results stemming from the analysis by Hart et al. (1968) are as follows: (a) For small times ($t^* \ll 1$, where t^* = dimensionless time), infiltration is less of a factor in terms of its effect on advance compared to the surface irrigation hydraulic variables (inflow rate, roughness, and bed slope); (b) for large times ($1 \ll t^*$), infiltration plays a dominant role in terms of its effect on advance relative to the surface irrigation hydraulic variables; (c) establishing the theoretical limits for the range of variation of the subsurface shape factor (defined as the ratio of the subsurface volume to the product of advance distance and cumulative intake

at channel inlet); and (d) the specification of the shape of the surface profile, the form of the infiltration function, and the advance function all at the same time results in an over conditioned problem (mathematically speaking), in which case the problem may not have suDcient degrees of freedom to be considered a physically valid description of mass balance. However, numerical evidence (e.g., Hart et al., 1968) suggests that mass balance error from coupling power law description of advance with the other aforementioned assumptions could be negligible, if the channel bed slope is suDciently steep and irrigation time is suDciently large.

4. *Power law advance*: In addition to the shape of the water surface profile and the form of the infiltration function, this approach assumes a specific form for the advance function. The most commonly used form is the power law advance ($s = pt_s^r$, where p [L/T'] and r [–] = empirical constants dependent on inflow rate, infiltration, hydraulic resistance, bed slope, and cross-sectional geometry for furrows). Using the modified Kostiakov infiltration function, for instance, the second term on the right-hand side of Equation 27.16 can be integrated over the wetted distance, following series expansion of the power term of the infiltration equation. Often, however, the sum of the power series terms is approximated with satisfactory accuracy by a more compact expression, Equation 27.17 (Walker and Skogerboe, 1987):

$$Q_o t_X = \left(\sigma_y A_0 + \sigma_z k(t_X)^a + \sigma_z' b t_X \right) X,$$

$$\text{where } \sigma_z = \frac{a + (1-a)r + 1}{(1+a)(1+r)} \quad \text{and} \quad \sigma_z' = \frac{1}{1+r} \quad (27.17)$$

where σ_z and σ_z' are subsurface shape factors (–). Equation 27.17 suggests that the subsurface shape factors are constants, dependent only on infiltration and the exponent of the power law advance function. However, it has been shown that the subsurface shape factor may not be constant with time.

5. *Solution of the convolution integral (Equation 27.16) with Laplace Transform*: Philip and Farrell (1964) used Laplace transformation to solve Equation 27.16 without making any assumption about the form of the advance function. Using the Faltung theorem of Laplace transform, the general solution of the irrigation advance problem is given as

$$\frac{X}{Q_0} = L^{-1} \left(\frac{1}{\bar{A} s'^2 + L[A_z] s'^3} \right) \quad (27.18)$$

where
$\bar{A} = \sigma_y A_0$ and $L[\ ]$ and $L^{-1}[\ ]$ are the Laplace transform and the inverse Laplace transform of [ ], respectively
s' is the a complex variable of the transformed function

Using the Kostiakov function, for example, the particular solution to the irrigation advance problem is given by Equations 27.19 and 27.20, for small and large times, respectively:

$$X = \frac{Q_o t_X}{\bar{A}} \sum_{m=0}^{M} \frac{\left(-\beta t_X^a \right)^m}{\Gamma(2 + ma)} \quad (27.19)$$

$$X = -\frac{Q_o t_X}{\bar{A}} \sum_{m=1}^{M} \frac{\left(-\beta t_X^a \right)^{-(m+1)}}{\Gamma(2 - a(m+1))} \quad (27.20)$$

where
M is the number of terms required for convergence (–)
$\beta = k\Gamma(1+a)/\bar{A}$, and Γ [.] are the gamma function of [ ]

The first term in Equation 27.19 represents advance on an impervious surface, with the volume balance assumption of constant average surface depth in force. Subsequent terms contain infiltration effects as well. Hence, Equation 27.19 shows that for small times, surface irrigation hydraulic variables are dominant in terms of their effect on advance relative to infiltration. For large times (Equation 27.20), however, infiltration parameters appear in all terms including the first term, underscoring their significance on advance relative to the hydraulic variables. Philip and Farrell (1964) developed particular solutions to Equation 27.18 using other infiltration functions, including Horton and Philips two-term equations.

27.3.3 Solution of the Solute Transport Equation

For the case in which flow is steady and uniform, Equation 27.14 can be reduced to a form that can be solved analytically after being coupled with a specific set of initial and boundary conditions. Although such flow conditions are rarely realized in a surface irrigation context, the analytical solutions are useful in testing numerical models under limiting assumptions and in parameter estimation applications. The more comprehensive form of the cross-section-averaged advection-dispersion equation (Equation 27.14) can only be solved numerically. Equation 27.14 can be discretized using an Eulerian scheme, in which case solution is sought at a discrete number of grid points within the solution domain (Burguete et al., 2009). Often, however, numerical solutions are obtained with procedures that combine a semi-Lagrangian advective scheme (in which a finite number of "particles" are tracked along a set of characteristics over a time step) with an Eulerian discretization for the diffusion step (Garcia-Navarro et al., 2000; Zerihun et al., 2005b; Perea et al., 2010).

While the diffusion term in Equation 27.14 can be discretized effectively (accurately and stably) with a wide variety of Eulerian schemes, the satisfactory discretization of the advection term with these numerical schemes is a considerable challenge (Leonard, 1991). In advection-dominated mass transport, the numerical treatment of the advection term often leads to artificial diffusion much larger than the physical diffusion and to nonphysical oscillations, especially near the vicinity of large concentration gradients. A classic method for solving the advection equation (a hyperbolic partial differential equation) is the method of characteristics. Hence, the differing requirements, with respect to effective numerical schemes, between the advection and dispersion terms of Equation 27.14 means that solution accuracy and eDciency can be enhanced by decoupling (in the mathematical sense) the advective transport from the mechanism of dispersion. The resulting pair of equations are then solved in two separate but consecutive steps, using numerical techniques most appropriate to each subproblem. This procedure, commonly known as the operator-splitting method, is widely used in solute transport modeling in a free-surface, turbulent flow-field (Garcia-Navarro et al., 2000; Zerihun et al., 2005b; Perea et al., 2010). With this method, Equation 27.14 is treated as a combination of two subsystems: pure advection ($K_x = 0$) described by

$$\frac{\partial C}{\partial t} + v\frac{\partial C}{\partial x} = 0 \qquad (27.21)$$

and pure diffusion ($v = 0$), expressed as

$$A\frac{\partial C}{\partial t} + \frac{\partial}{\partial x}\left(AK_x\frac{\partial C}{\partial x}\right) = 0 \qquad (27.22)$$

TABLE2 7.2 Initial and Boundary Conditions, Solute Transport Equation

Type of Condition	Description	Range of Application
Initial condition	$C(x,0) = 0$	$0 \leq x \leq L^{\text{I and II}}$
	$C_x(x,0) = 0$	$0 \leq x \leq L$
	$C(x,0) = C_i$	$0 \leq x \leq L^{\text{III and IV}}$
Upstream BC	$C(0,t) = C_0$	$t \leq t_r$
	$C_x(0,t) = F_d$	$t \leq t_r$
Downstream BC	$C_x(x_s,t) = 0$	$0 \leq x_s < L$
	$C_x(L,t) = C_x(L,t)^a$	$t_{aL} \leq t \leq t_{rL}$

L, channel length; C_0, solute concentration at the inlet; C_x, eC/ex; F_d, time averaged backward differencing approximation of C_x at the inlet end (Zerihun et al., 2005b); x_r, distance of receding tip from inlet; x_s, length of solute plume in the surface stream, t_{aL}, water advance time to downstream end, t_{rL}, water recession time at the downstream end; C_i, initial concentration at each of the computational nodes.[I and II] applicable only to fertilizer injection configurations I and II;[III and IV] applicable only to fertilizer injection configurations III and IV (where[I] solute is applied throughout the irrigation application time;[II] solute application begins at the start of irrigation but concentration drops to zero before inflow is stopped;[III] solute application begins after the start of irrigation and continues until inflow is stopped;[IV] solute application begins some time after the start of irrigation and solute concentration drops to zero before inflow is cut; $C_x(L,t)^a$, spatial concentration derivative calculated at the end of the advective step. Note that specification of upstream boundary conditions during the recession phase is not a trivial issue and some discussion on this is provided by Zerihun et al. (2005b).

Equations 27.21 and 27.22 are solved in two separate but consecutive steps, using numerical techniques most appropriate to each subproblem. With the operator-splitting method, Equation 27.21 is, first solved, for the intermediate solute concentrations, C^a, using a semi-Lagrangian integration scheme (Holly and Preissmann, 1977). The advected concentrations, C^a, are then diffused longitudinally by solving Equation 27.22 with a finite differencing scheme (e.g., Garcia-Navarro et al., 2000; Zerihun et al., 2005b; Perea et al., 2010). Pertinent initial and boundary conditions for Equations 27.21 and 27.22 are given in Table 27.2.

27.4 Applications

27.4.1 Surface Irrigation System Characterization

Surface irrigation system variables such as basin/border width, basin/border/furrow length, inflow rate, and cutoff time/distance are physical quantities that can be measured accurately. A second set of variables, consisting of furrow cross-sectional geometry, required amount of irrigation application, and bed slope, can be determined through direct/indirect measurements; however, they are spatially varied and hence often approximated with field-scale averages. A third group of physical quantities represent the inherent, but transient, properties of a surface irrigation system at any given time, namely: the soil and crop hydraulic parameters (hydraulic resistance coeDcient and infiltration parameters). Soil and crop hydraulic parameters are often estimated through the solution of the inverse problem of surface irrigation hydraulics. Since coupled surface-subsurface flow and solute transport modeling of surface irrigation systems are limited to the direct problem, subsequent discussion will only briefly review some selected soil and crop hydraulic parameter estimation methods that are based on surface irrigation hydraulic models with empirical infiltration equations. The methods described here are selected, either because they are widely used or because of the availability of operational public domain software tools developed based on these concepts.

A simple approach to the surface irrigation parameter estimation problem involves inverting an approximate mass balance equation (e.g., Elliott and Walker, 1982; Strelkoff et al., 1999; Bautista et al., 2009) to determine the infiltration parameters and deriving the hydraulic resistance coeDcient (Manning *n*) from literature or computing it with a simplified procedure (e.g., Strelkoff et al., 1999).

Perhaps the most widely used method for estimating the parameters of the Kostiakov and modified Kostiakov infiltration functions in surface irrigation applications is the two-point method of Elliott and Walker (1982) and subsequent variants. The approach involves the formulation of two volume balance equations corresponding to two instants of time during the advance phase and solving the pair of algebraic equations simultaneously to obtain estimates of the two parameters of the Kostikaov function. If the modified Kostiakov function is used to describe infiltration, instead of the Kostiakov equation, the steady state tailwater runoff rate (for a free-draining channel) can be used in conjunction with the inflow rate to estimate the basic intake rate. The method was originally developed for sloping free-draining furrows, but it can be used for sloping borders, provided inflow is not cut prior to completion of advance.

Strelkoff et al. (1999) developed an interactive graphical surface irrigation parameter estimation model—EVALUE—which estimates infiltration parameters by matching the growth with time of infiltration volume, calculated based on measured inflow hydrograph and surface flow depth profiles, with the temporal growth of the infiltration volume computed based on measured opportunity times and a user-selected empirical intake function. EVALUE calculates the Manning roughness coeDcient with the Manning equation, Equation 27.5, based on measured water surface gradients, flow rates, and flow depths.

Bautista et al. (2009) described an adaptation of the post-irrigation mass balance approach of Merriam and Keller as a parameter estimation component of WinSRFR (USDA-ARS-ALARC, 2009). Given the inflow and outflow hydrographs, the infiltration volume, V_z, is known at the end of an irrigation event. A post-irrigation mass balance equation can, then, be formulated by equating V_z with the distance integral of the cumulative infiltration function. Considering the Kostiakov-type infiltration functions, the resulting expression can be solved for the coeDcient of the power term, k, if the other infiltration parameters are determined independently.

Parameter estimates based on inverse solution of the flow equations are not pure expressions of the physical characteristics of the system being studied as such, but they also account for the inadequacies of the inverted hydraulic model in terms of its ability to reproduce the totality of the essence of the physical phenomenon being modeled (flow depth and flow rate hydrographs and longitudinal infiltration profile). Hence, infiltration and roughness parameters are often correlated. The implication being a simultaneous estimation of both parameters within the same modeling framework is desirable and, in principle, can lead to a consistent definition of the state of the system. Insofar as the methods described earlier are based on mass balance approximation only (a simplified description of the physics of the surface irrigation process) and that they cannot be used to estimate infiltration and roughness parameters simultaneously, they have limitations. In addition, the methods described earlier and most other simplified mass balance–based approaches cannot be used to estimate spatially distributed parameter sets. They also lack the capability to estimate infiltration parameters for a condition in which nodal volumetric infiltration from a furrow is modeled as a function of nodal intake opportunity times and wetted perimeter.

A physically rigorous method for a simultaneous estimation of infiltration and roughness parameters requires the coupling of an optimization model with a surface irrigation hydraulic model. Yost and Katopodes (1998) reported a globally convergent algorithm for the simultaneous estimation of the Kostiakov infiltration parameters and the Manning *n*. Walker (2005) coupled a hydrodynamic model with a multilevel parameter estimation approach that combines heuristics with a one-dimensional

optimization algorithm to estimate the parameters of the modified Kostiakov infiltration function and the Manning *n*. In general, these approaches are relatively complex as they require the coupling of a numerical hydraulic model with an optimization subroutine. As a result, they have not yet been integrated into an operational public domain surface irrigation model (e.g., SIRMOD III, Walker, 2003; WinSRFR, USDA-ARS-ALARC, 2009). Hence, the calibration of surface irrigation hydraulic models is often performed with the mathematically more tractable approaches (e.g., Elliott and Walker, 1982; Strelkoff et al., 1999; Bautista et al., 2009).

Generally, the type and volume of data used, the time scale over which the data is collected as related soil type, and the duration of the irrigation event as well as the functional form used to model infiltration and the friction slope are important determinants of the accuracy of parameter estimates (Clemmens, 2009) in terms of their ability to reproduce the totality of the essential features of an irrigation event (flow depth and rate hydrographs and post-irrigation longitudinal infiltration profile). The physics of infiltration suggests that among the different dependent irrigation variables used as an input in infiltration parameter estimation, only depth hydrographs provide a complete physical description of the infiltration process. At a point in a basin/border or at a furrow cross section, the depth hydrograph there provides data on intake opportunity time and the boundary condition for infiltration and its evolution with time. These data—coupled with furrow geometry (in case of furrows), soil physical properties, and initial and bottom boundary conditions—constitute a complete set needed to fully describe the physics of infiltration from a basin/border or a furrow cross section. If depth hydrographs from a series of points, spanning the length of a channel, are taken together, a full physical description of the basin/border or furrow infiltration problem at a field-scale follows.

Although flow depth hydrographs allow a complete physical description of the infiltration process, that does not necessarily guarantee solution uniqueness for the inverse problem of surface irrigation hydraulics, because nonunique solutions can be numerical/mathematical artifacts of the parameter estimation approach. Furthermore, unlike infiltration parameters, the effect of the Manning roughness coeDcient cannot be fully taken into account by depth hydrographs only, because the effect of the Manning roughness coeDcient on flow depth is closely related to its influence on flow velocity. Hence at the physical level, at least, a unique characterization of field-scale average roughness and infiltration parameter set may require the use of depth hydrographs in conjunction with flow rate hydrographs.

27.4.2 Surface Irrigation System Design and Management

The authors are not aware of any comprehensive study aimed at developing a system design and management procedure for a surface fertigation system. Hence, subsequent discussion on system design and management will be limited to the hydraulic component of the on-farm water and nutrient application subsystem.

Design is a decision making process in which both the physical configuration of the system and a set of tentative operational criteria (inflow rate and inflow cutoff time) are selected such that a certain measure of merit (gross/net economic benefit or irrigation performance) is maximized or economic cost is minimized. Design recommendations are developed based on anticipated average field conditions (soil and crop hydraulic properties) over an irrigation season. Typically, irrigation performance is used as a system design criteria in preference to economic benefit or cost—due mainly to limitations in data availability.

The geometry of a surface irrigation system is defined in terms of channel length, width (for basins and borders), cross-sectional geometry and spacing (for furrows), average longitudinal bed slope, and downstream boundary condition. In acknowledgment of the limitations of available design and management tools (1D hydraulic models), surface irrigation system design is based on unit width analysis. The implication is that, although soil and crop hydraulic properties and cross-sectional geometry (for furrows) may vary in the direction of irrigation, they are considered uniform over the entire width or over a section of the width of an irrigated farm. In general, average channel bed slopes are set based on considerations of land grading costs and drainage requirements (relative to weather and regional hydrology). Width of basins and borders are selected based on considerations of land grading requirements (as related to topography and depth of top soil), width of available machinery, crop spacing, and available flow rate. Furrow spacing and cross-sectional geometry is set taking into account such factors as type and size of available farm machinery, soil type, and crop spacing. To the extent that bed slope, basin and border width, furrow spacing, and cross-sectional geometry are selected based on considerations that are not explicitly related to irrigation performance, they are not considered here as design variables.

Hence, system design, in the context of surface irrigation, is defined in a more restrictive and limited sense, as a decision making process in which a certain measure of merit (often irrigation performance) is maximized with respect to field length, L, and the operational criteria of inflow rate, Q_0, and cutoff time, t_{co}. Considering a feasible set, for system design, consisting of irrigation scenarios satisfying the condition that the minimum applied irrigation amount equals the requirement, water requirement eDciency becomes constant (100%). In which case, spatial uniformity of applied irrigation water is a redundant quantity as an irrigation performance index (Zerihun et al., 1997, 2003). This leaves application eDciency, E_a, as the primary design criterion for surface irrigation systems. With E_a as the design criterion, the design problem can be formulated as (Zerihun et al., 2005c):

$$
\left.
\begin{array}{l}
\text{Max} \quad E_a(L, Q_0, t_{co}) \\[6pt]
St. \\[6pt]
A_{zr} - A_{z\min}(L, Q_0, t_{co}) = 0, \\[6pt]
DU_{\min^l} - DU(L, Q_0, t_{co}) \leq 0, \\[6pt]
\text{and } C_i(L, Q_0, t_{co}) \leq 0
\end{array}
\right\} \qquad (27.23)
$$

where

$DU_{\min^l}$ is the lowest allowable level of distribution uniformity based on minimum depth (–)

C_i represents a set of constraints that can be categorized as variable bounds, conservation-like, and management related

The first constraint imposes a restriction on the minimum cumulative infiltration and target E_r (note that this constraint needs to be relaxed for numerical solutions). These constraints can be implicit in the hydraulic simulation model or explicitly taken into account by the optimization algorithm, depending on whether a physically based model or explicit empirical functions are used to evaluate the terms in the constraint functions. In Equation 27.23, the cutoff time, t_{co}, instead of cutoff distance, L_{co}, is used as an inflow cutoff criterion, because t_{co} is a more general cutoff criterion than L_{co}.

For a given field condition, the cutoff time, t_{co}, needed to apply a minimum depth just equal to the requirement is dependent on Q_0 and L, hence t_{co} cannot be treated as an independent system design variable (Zerihun et al., 2005c). This reduces the surface irrigation design variables to just two: Q_0 and L. For free-draining channels, Zerihun et al. (2001, 2005c) showed that, for the condition $A_{z\min} = A_{zr}$ (Equation 27.23), the E_a function is unimodal with respect to L and Q_0. Based on these results, simple yet highly accurate optimality conditions were derived for the $E_a(Q_0)$ and $E_a(L)$ functions. Furthermore, these authors also showed that for practical design purposes, the solution of Equation 27.23 can be reduced to the solution of a series of one-dimensional problems, simplifying the problem significantly. For level basins and closed-end level furrows, the $E_a(Q_0)$ function is monotonic increasing and approaches a limiting E_a value asymptotically. On the other hand, the $E_a(L)$ function is monotonic decreasing with a minimum E_a corresponding to the maximum advance distance or the field length, whichever is smaller. Hence, given a basin, the optimum unit inflow rate is the threshold Q_0 value beyond which $E_a(Q_0)$ function does not show appreciable increase with an increase in Q_0. However, optimum Q_0 can be limited by the available flow rate and soil erosion considerations. The optimum length is the minimum acceptable length based on considerations of system cost accruing from short lengths. For graded basins, the $E_a(L)$ and $E_a(Q_0)$ functions can be bimodal (with a local maximum and a local minimum point), hence design and management computations need to take that into account. Note that the preceding discussion assumes Q_0 is constant with time.

Design decisions select an optimal system configuration for an irrigation season based on anticipated seasonal average system parameters. Considering the fact that the transient system characteristics (infiltration and hydraulic roughness) exhibit significant variability through the irrigation season, seasonal averages often contain significant uncertainties. Hence, high irrigation performance can be sustained through an irrigation season, only if the operational criteria (Q_0 and t_{co}) set at the design phase are adjusted in the course of the irrigation season to match the changing field conditions. Given the system configuration, including its length, the process of selecting an optimal set of operational criteria prior to the initiation of every irrigation event is often described as system management. In which case, the objective criterion (measure of merit) reduces to the maximization of E_a with respect to Q_0 only, subject to the constraints listed in Equation 27.23. In general, the robustness of both design and management prescriptions need to be evaluated based on sensitivity analysis to uncertainties in parameter estimates. While accounting for the effects of system design and management decisions at a larger spatial and temporal scale (seasonal, annual, or any convenient time horizon larger than an event scale) is desirable, currently data requirement and technical complexity of the required analytical tools are limiting factors.

A design and management approach developed at the USDA-ARS-ALARC, as part of the WinSRFR modeling system, involves the discretization of a two-dimensional solution domain (defined in terms of the operation variables or the physical dimensions) into a finite number of grid points (Clemmens, 2007). Simulations are then conducted at each of the grid points and the resulting irrigation performance and application depth data are summarized in a series of contours (that can potentially be superimposed as different layers of information on a single graph and) viewed dynamically. Computational eDciency is enhanced through the use of the volume balance model (calibrated through limited simulations with a more rigorous hydraulic model) to conduct the large number of simulations required to generate the performance contours. Although earlier versions of this approach were based on dimensionless variables (Clemmens et al., 1995), recent solutions use dimensional variables. The advantage of the method is that it presents an overview of the available management options in a set of graphs, from which regions of high performance along with associated sensitivity and robustness can be gleaned with comparative ease and in a relatively short time. Often, however, a static database of irrigation performance indices based on a seasonal average parameter set is used in practical irrigation management. For example, Sanchez et al. (2009) described the development of management guidelines for closed-end level furrows using a combination of field and modeling studies: (1) field experimental studies to generate the data needed for a modeling study; (2) model calibration, verification, and selection; (3) simulation experiment and development of management tools (i.e., performance tables and charts); and (4) development of management guidelines to accompany the management tools.

27.5 Potential Research Areas

The established approach to surface irrigation hydraulic modeling is based on numerical solutions of the governing equations of 1D unsteady gradually varied flow, described in a preceding section. In general, infiltration is modeled using empirical functions. Subsurface flow dynamics is often ignored if the objective is to study the surface irrigation hydraulics. On the other hand,

if performance characterization is needed for system design, management, and evaluation purposes, the physics of subsurface flow dynamics is often replaced with simplifying assumptions as regards the vertical distribution of irrigation-applied water in the soil profile at the end of an irrigation event. The surface irrigation/fertigation process is complex and multifaceted, hence a comprehensive and accurate physics-based description of the process during and subsequent to an irrigation/fertigation event (or over an irrigation cycle or an irrigation season) requires a capability to model subsurface flow dynamics, constituent transport, sequestration, and fate; constituent transport in the surface stream; as well as higher dimensioned (e.g., 2D) surface irrigation hydraulic modeling. Some of the most important potential research areas in surface irrigation/fertigation system modeling, design and management, and system characterization are outlined as follows:

1. Although assumptions are routinely made to the effect that flow in basins and borders can be modeled as a one-dimensional process, in practice flow in irrigation basins and borders is often too complex to be modeled accurately as a 1D process. A satisfactory description of basin and border hydraulics may require a numerical solution of the depth-averaged shallow water equations, with a capability for modeling interacting overland and channel flow processes. Advances have been made in 2D numerical modeling of basin and border irrigation hydraulics (e.g., Playan et al., 1994; Bradford and Katopodes, 2001; Strelkoff et al., 2003). However, further progress is required in such areas as computational eDciency and robustness as well as in the application of statistical techniques for generating reliable high-resolution, spatially distributed input data from limited field measurements, before such models can be made available for routine system design, management, and evaluation applications.

2. Surface irrigation is a coupled surface-subsurface flow process. Water flow on the surface and through the subsurface domains are linked by the process of infiltration. Physics-based description of infiltration and subsurface flow dynamics, both during and subsequent to an irrigation event, requires the coupling of a surface irrigation hydraulic model with a variably saturated porous media flow model. Infiltration and subsurface flow dynamics in a basin/border can be modeled with 1D variably saturated porous media flow model, while a 2D model (e.g., HYDRUS-2D, Simunek et al., 1999) is needed for furrows. The advantages of a coupled physically based surface-subsurface flow model are as follows: (a) Unlike the parameters of empirical infiltration functions, soil hydraulic parameters are independent of initial and boundary conditions; hence they are not event specific. (b) Physics-based models can track the time evolution of the vertical/cross-sectional distribution of irrigation-applied soil water during and subsequent to an irrigation event, hence allowing an accurate and complete characterization

of irrigation performance and water loss indicators over any desired time scale (Zerihun et al., 2005a). When crop water uptake is taken into account, coupled models can be used as seasonal irrigation management tools (Wöhling and Schmitz, 2007). However, limitations as related to required computational time and robustness and to a certain extent data needs (i.e., parameter estimation for the coupled model) must be overcome before such models can be available for routine system design and management applications (Bautista et al., 2010).

3. Given the significance of irrigated soils as important sources and sinks of agricultural chemicals and sediments, modeling constituent transport in surface irrigation systems is important. Progress has been made in mathematical modeling of solute transport in surface irrigation streams (Garcia-Navarro et al., 2000; Zerihun et al., 2005b; Burguete et al., 2009; Perea et al., 2010). However, some significant challenges remain before constituent transport models evolve to the level that they can be integrated into established surface irrigation modeling systems. Some of the most important research areas in constituent transport modeling are as follows: (a) Existing numerical solutions to the 1D advection-dispersion equation in surface irrigation applications can be evaluated for possible incorporation into surface irrigation hydraulic models. (b) Numerical solutions of the depth-averaged advection dispersion equation need to be developed for use in solute transport modeling in irrigation basins and borders. (c) Research on erosion and sediment transport modeling in surface irrigation systems is important in itself, but it is also an important component of solute transport modeling in a surface irrigation stream wherein sorption is a significant source and sink mechanism (note that WinSRFR has a 1D erosion and sediment transport modeling capability). (d) While most agricultural chemicals can be treated as nonreactive during the time scale of an irrigation event, some of them, however, are volatile and many can have significant sorptive characteristics. Hence, there is a need to incrementally expand modeling capabilities to include sorption and volatilization in the mass transport component of surface irrigation/fertigation models. (e) Simulation of plant nutrient uptake and chemical reaction in surface irrigation context is, generally, important in the post-irrigation/fertigation redistribution phase. Such a capability is needed for seasonal irrigation/fertigation management. Note that subsurface flow and constituent transport models, commonly used in conjunction with surface irrigation hydraulic models (e.g., HDYRUS-2D) have functionalities for simulating plant water and nutrient uptake, chemical reactions, and physical transformations (such as sorption, volatilization, precipitation/dissolution) in variably saturated porous media.

4. Given the widespread availability of increasingly powerful computational resources (computer hardware and software), holistic modeling of the surface irrigation processes is becoming common. The coupling of surface irrigation hydraulic and subsurface flow models as well as flow and transport models (both within the surface and subsurface domain) can be accomplished at different levels of integration of the physics of the component processes. Three different categories can be discerned: (a) fully coupled models: consisting of models in which the differential equations describing the component processes are coupled and solved simultaneously (e.g., Burguete et al., 2009); (b) internally coupled models: consisting of standalone models (designed to simulate the component processes independently) that are programmatically coupled, such that they interact at the level of the duration of a computational time step (e.g., Zerihun et al., 2005b; Wöhling and Schmitz, 2007); (c) externally coupled models: consisting of standalone models that may or may not be programmatically coupled, but interact at the level of the duration of an irrigation/fertigation event (e.g., Bautista et al., 2010; Perea et al., 2010). The broad attributes of these model coupling approaches in the context of surface irrigation have been described in the aforementioned references, but more complete studies may be needed to evaluate the advantages and limitations in terms of numerical stability, accuracy, computational eDciency, and complexity. In addition, often surface irrigation models are designed to simulate flow and transport processes along a furrow and a unit width basin/border. However, a capability to simulate field-scale surface irrigation/fertigation processes requires coupling these models with models describing the flow and transport processes in the field water supply mechanism: which could be a pipe with multiple outlets or a canal with sluice-gates or siphons. Such models could be more useful practical field-scale surface irrigation/fertigation system design and management tools.

5. There exists a large volume of published work describing various surface irrigation system design and management procedures. These approaches differ in the design and management objective criterion, the decision variables, problem formulation, solution techniques, and input data requirements. There is a need for (a) standardization of the definition of terms as regards system design and management; (b) standardization of alternative objective criterion, associated design variables, constraints, and computational procedures; and (c) a clear statement of pertinent assumptions, limitations, and advantages. In addition, an important research area in surface irrigation system design and management would be the development of a practically useful procedure for the propagation of uncertainty in system parameter estimates (quantified in probabilistic terms) to the output side of a design and management problem.

6. Perhaps the most significant limitation, to the practical application of surface irrigation mathematical models, is related to the diDculty associated with accurate system characterization. The limitations are threefold: (a) Solution nonuniqueness (the inability of the computed

system parameter set to reproduce the totality of the essential features of an irrigation event within a reasonable error margin) is an important problem. Nonuniqueness of the parameter vector could be the result of the type and volume of data used in parameter estimation, the parameter estimation approach, and the functional form used to express infiltration and friction slope (Clemmens, 2009). Physics-based analysis coupled with the theory of parameter identification and inverse problems of physical systems can be used to determine the most rigorous and complete formulation of the surface irrigation parameter estimation problem as well as possible simplifications and applicable conditions. (b) Deterministic treatment of essentially stochastic system properties is another limitation of widely used practical parameter estimation approaches (i.e., there is a lack of specification of the amount of uncertainty and associated confidence level implicit in the parameter estimates). (c) Implicit in surface irrigation system design and management prescriptions is that the system characteristic (soil and crop hydraulic parameter set) is known at some future time. However, a capability to forecast the state of a surface irrigation system at some future time does not exist; hence system design and management recommendations are based on past data. Since soil and crop hydraulic properties of surface-irrigated systems are highly dynamic with time, system design and management prescriptions based on past data typically contain a high degree of uncertainty, often leading to underperforming or infeasible irrigation scenarios.

References

Bautista, E., Clemmens, A.J., and Strelkoff, T.S. (2009). Optimal and postirrigation volume balance infiltration parameter estimates for basin irrigation. *J. Irrig. Drain. Eng.*, 135(5), 579–587.

Bautista, E. and Wallender, W.W. (1992). Hydrodynamic furrow irrigation model with specified space steps. *J. Irrig. Drain. Eng.*, 118(3):450–465.

Bautista, E., Zerihun, D., Clemmens, A.J., and Strelkoff, T.S. (2010). An external iterative coupling strategy for surface-subsurface flow calculations in surface irrigation. *J. Irrig. Drain. Eng.*, 136(10):692–703.

Bradford, S.F. and Katopodes, N.D. (2001). Finite volume model for nonlevel basin irrigation. *J. Irrig. Drain Eng.*, 127(4):216–223.

Burguete, J. Zapata, N., Garcia-Navarro, P., Maikak, M., Playan, E., and Murillo, J. (2009). Fertigation in furrows and level furrow systems. I: Model description and numerical tests. *J.drrig. Drain. Eng.*, 135(4):401–412.

Burt, C.M., Clemmens, A.J., Strelkoff, T.S., Solomon, K.H., Bliesner, R.D., Hardy, L.A., Howell, T.A. and Eisenhauer, D.E. (1997). Irrigation performance measures: EDciency and uniformity. *J. Irrig. Drain. Eng.*, 123(6):423–442.

Clemmens, A.J. (1981). Evaluation of infiltration measurements for border irrigation. *Agric. Water Manage.*, 3:251–267.

Clemmens, A.J. (2007). Simple approach to surface irrigation design: Theory. *e-J. Land Water* 1:1–19.

Clemmens, A.J. (2009). Errors in surface irrigation evaluation from incorrect model assumptions. *J. Irrig. Drain. Eng.*, 135(5):556–565.

Clemmens, A.J., Dedrick, A.R., and Strand, R.J. (1995). BASIN: A computer program for the design of level-basin irrigation systems. V.2.0. WCL Report #19. USDA-ARS, U.S. Water Conservation Laboratory, Phoenix, AZ.

Cunge, J.A., Holly, F.M., and Verwey, A. (1980). *Practical Aspects of Computational River Hydraulics*. Pitman Publishing Limited, London. U.K.

Elliott, R.L. and Walker, W.R. (1982). Field evaluation of furrow infiltration and advance functions. *Trans. ASAE*, 25(2):396–400.

Garcia-Navarro, P., Playan, E., and Zapata, N. (2000). Solute transport modeling in overland flow applied to fertigation. *J. Irrig. Drain. Eng., ASCE*, 126(1):33–40.

Granger, R.A. (1995). *Fluid Mechanics*. Dover Publications, Inc., New York.

Hall, W.A. (1956). Estimating irrigation border flow. *Agric. Eng.*, 37(4):262–265.

Hart, W.E., Bassett. D.L., and Strelkoff, T.S. (1968). Surface irrigation hydraulics-kinematics. *J. Irrig. Drain. Div.*, 94(4):419–440.

Holly, F.M. (1985). Dispersion in rivers and coastal waters—I. Physical principles and dispersion equations. In *Developments in Hydraulic Engineering*, Vol. 3 (ed. Novak, P.). Elsevier Applied Science Publishers, New York.

Holly, F.M. and Preissmann, A. (1977). Accurate calculation of transport in two dimensions. *J. Hydraul. Div.*, 103(11):1259–1277.

Katopodes, N.D. and Strelkoff, T.S. (1977a). Hydrodynamics of border irrigation. *J. Irrig. Drain. Div.*, 103(3):309–323.

Katopodes, N.D. and Strelkoff, T.S. (1977b). Dimensionless solution of border irrigation advance. *J. Irrig. Drain. Div.*, 103(4):401–407.

Leonard, P.B. (1991). The ULTIMATE conservative difference scheme to unsteady one-dimensional advection. *Comput. Meth. Appl. Mech. Eng.*, 88(1991):17–74.

Lewis, M.R. and Milne, W.E. (1938). Analysis of border irrigation. *Agric. Eng.*, 19:267–272.

Ligget, J.A. and Cunge, J.A. (1975). Numerical methods of solution of the unsteady flow equations, pp. 89–182, In *Unsteady Flow in Open Channels*, Vol. I (eds., Mahmood, K. and Yevejevich, V.), Water Resources Publications, Fort Collins, CO.

Perea, H, Strelkoff, T.S., Adamsen, F.J., Hunsaker, D.J., and Clemmens, A.J. (2010). Nonuniform and unsteady solute transport in furrow irrigation I. Model development. *J. Irrig. Drain. Eng.*, 136(6):365–375.

Philip, J.R. and Farrell, D.A. (1964). General solution of the infiltration advance problem in irrigation hydraulics. *J. Geophys. Res.*, 69(4):621–631.

Playan, E., Walker, W.R., and Merkley, G.P. (1994). Two-dimensional simulation of basin irrigation. I. Theory. *J. Irrig. Drain. Eng.*, 120(5):837–856.

Rutherford, J.C. (1994). *River Mixing.* John Wiley & Sons Ltd., New York.

Sanchez, C.A., Zerihun, D., and Farrell-Poe, K.L. (2009). Management guidelines for eDcient irrigation of vegetables using closed-end level furrows. *Agric. Water Manage.*, 96:43–52.

Simunek, J., Sejna, M., and van Genuchten, M.Th. 1999. The HYDRUS-2D software package for simulating two-dimensional movement of water, heat, and multiple solutes in variably saturated media. Ver. 2.0. Rep. IGWMC-TPS-53. Int. Ground Water Model. Ctr., Golden, CO.

Strelkoff, T.S. (1969). One-dimensional equations of open channel flow. *J. Hydraul. Div.*, 95(3):861–876.

Strelkoff, T. (1985). BRDRFLW: A mathematical model of border irrigation. U.S. Water Conservation Laboratory, USDA-ARS, Phoenix, AZ.

Strelkoff, T.S. (1992). EQSWP: Extended unsteady-flow double-sweep equation solver. *J. Hydraul. Eng.*, 118(5):735–742.

Strelkoff, T.S. and Clemmens, A.J. (2007). Hydraulics of surface irrigation, pp. 437–498, In *Design and Operation of Farm Irrigation Systems* (eds. Hofmann, G.J., Evans, R.G., Jensen, M.E., Martin, D.L., Elliott, R.L). Chapter 13, 2nd edn., ASABE, St. Joseph, MI.

Strelkoff, T., Clemmens, A. J., El-Ansary, E., and Awad, M. (1999). Surface-irrigation evaluation models: Application to level basin in Egypt. *Trans. ASAE* 42(4):1027–1036.

Strelkoff, T.S. and Falevy, H.T. (1993). Numerical methods used to model unsteady canal flow. *J. Irrig. Drain. Eng.*, 119(4):637–655.

Strelkoff, T. and Katopodes, N.D. (1977). Border irrigation hydraulics with zero-inertia. *J. Irrig. Drain. Eng.*, 103(3):325-342.

Strelkoff, T.S., Tamimi, A.H., and Clemmens, A.J. (2003). Two-dimensional basin flow with irregular bottom configuration. *J. Irrig. Drain. Eng.*, 129(6):391–401.

USDA-ARS-ALARC. (2009). *WinSRFR Version 3.1 User Manual,* U.S. Arid Land Agricultural Research Center, Maricopa, AZ.

Walker, W.R. (2003). SIRMOD III—Surface irrigation simulation, evaluation and design. Userß Guide and Technical Documentation. Department of Biological and Irrigation Engineering, Utah State University, Logan, UT.

Walker, W.R. (2005). Multilevel calibration of furrow infiltration and roughness. *J. Irrig. Drain. Eng.*, 131(2):129–136.

Walker, W.R. and Skogerboe, G.V. (1987). *Surface Irrigation: Theory and Practice.* Prentice-Hall, Inc., Englewood Cliffs, NJ.

Wöhling, Th. and Schmitz, G.H. (2007). Physically based coupled model for simulating 1D surface–2D subsurface flow and plant water uptake in irrigation furrows. I: Model development. *J. Irrig. Drain. Eng.*, 133(6):538–547.

Yost, S.A. and Katopodes, N.D. (1998). Global identification of surface irrigation parameters. *J. Irrig. Drain. Eng.*, 124(3):131–139.

Zerihun, D., Furman, A., Warrick, A.W., and Sanchez, C.A. (2005a). A coupled surface-subsurface flow model for improved basin irrigation management. *J. Irrig. Drain. Eng.*, 131(2):111–128.

Zerihun, D., Furman, A., Warrick, A.W., and Sanchez, C.A. (2005b). A coupled surface-subsurface solute transport model for irrigation borders and basins, I. Model development. *J. Irrig. Drain. Eng.*, 131(5):396–406.

Zerihun, D., Sanchez, C.A., and Farrell-Poe, K.L. (2001). Analysis and design of furrow irrigation systems. *J. Irrig. Drain. Eng.*, 127(3):161–169.

Zerihun, D., Sanchez, C.A., Farrell-Poe, K.L., Adamsen, F.J., and Hunsaker, D. J. (2003). Performance indices for surface N fertigation. *J. Irrig. Drain. Eng.*, 129(3):173–183.

Zerihun, D., Sanchez, C.A., Farrell-Poe, K.L., and Yitayew, M. (2005c). Analysis and design of border irrigation systems. *Trans. ASAE*, 48(5): 1751–1764.

Zerihun, D. Wang, Z., Suman, R., Feyen, J., and Reddy, J.M. (1997). Analysis of surface irrigation performance terms and indices. *Agric. Water. Manage.*, 34(1):25–46.

28

Cyberinfrastructure and Community Environmental Modeling

J.P.M. Syvitski
University of Colorado, Boulder

Scott D. Peckham
University of Colorado, Boulder

Olaf David
Colorado State University

Jonathan L. Goodall
University of South Carolina

Cecelia Deluca
University of Colorado, Boulder

Gerhard Theurich
*Science Applications
International Corporation*

28.1 Introduction

The development of cyber-infrastructure in the field of Environmental Fluid Dynamics has been long, and tortuous. With the proliferation of acceptable core university courses in the sciences and engineering, in the 1960s and 1970s, and the reduction of core credits needed for graduation, a knowledge gulf has been growing between computational scientists and software engineers (Wilson and Lumsdaine 2008). Yet Computational Science and Engineering (CSE) has been growing rapidly at most research universities, particularly in the 1990s and 2000s, penetrating to some extent most of science and engineering disciplines (McMail 2008). Large codes by their nature involved more than one environmental domain, for example, wind-driven currents and wave dynamics in oceanography, or channelized flow overland flow and groundwater flow in hydrology. As these codes grew in source lines, so did diversity of experts needed in their development, and thus the birth of community modeling. Models require readily available data and data systems, and community-modeling efforts both supported and directed

new field campaigns and observational systems (e.g., satellites, ships, planes, telecommunication). Inevitably, when the codes reached a certain size (say 50,000 lines of code), they became modeling frameworks. Too large for individuals to understand the rigid framework, developers would pass on their process modules to be implemented by a master(s) of the code. As these large framework models were to be linked to other framework models, such as between an ocean general circulation model (GCM) and an atmospheric GCM, they were so rigid that flux couplers unique to the domains being coupled were required. Great attention to the details to grid meshing, time stepping, computational precision, and data I/O made these couplers rigid themselves. Meanwhile the field of software engineering had been changing rapidly, developing new standards for data exchange, model interfaces, and ways to employ varied computational platforms (laptop, server, high-performance computing clusters, distributed and cloud computing). This chapter is written by a combination of software engineers and scientists behind four representative cyberinfrastructure projects involved in the field of environmental community modeling. Each project faced

Handbook of Environmental Fluid Dynamics, Volume Two, edited by Harindra Joseph Shermal Fernando. © 2013 CRC Press/Taylor & Francis Group, LLC.
ISBN: 978-1-4665-5601-0.

different cyber requirements, different but overlapping scientific communities, different funding histories, and to some extent different customers. The projects each support one another, a necessity given limited funds.

28.2 Principles

Concepts like direct numerical simulation (DNS), large-eddy simulation (LES), and Reynolds-averaged Navier Stokes simulation (RANS), as defined in other chapters of this book, are familiar to most working in the field of CSE. But CSE experts are less familiar with the jargon of cyberinfrastructure, even though they underpin the third pillar of CSE. In the following we provide a window into the key principles behind modern community modeling and cyberinfrastructure efforts (see Section 28.3), including frameworks, architectures, component-based modeling, interface standards, drivers, ports, and protocols.

28.2.1 Frameworks and Architectures

Driving forces for framework adoption are (1) saving time and reducing costs, (2) providing quality assurance and control, (3) repurposing model solutions for new business needs, (4) ensuring consistency and traceability of model results, and (5) mastering computing scalability to solve complex modeling problems. The model developer should be able to eDciently develop and deliver a simulation model, and look forward to increasing modeling productivity.

Environmental modeling frameworks support developing and deploying environmental simulation models. Functions provided may include support for coupling of models into functional units (e.g., components, classes, or modules), component interaction and communication, time stepping, regridding of arrays, up or downscaling of spatial data, multiprocessor support, and cross language interoperability. A framework may also provide a uniform method of trapping or handling exceptions (i.e., errors).

An architecture is a set of standards that allow components to be combined and integrated for enhanced functionality, for instance, on high-performance computing systems. The standards are necessary for the interoperation of components developed in the context of different frameworks. Software components that adhere to these standards can be ported with relative ease to another compliant framework.

28.2.2 Modularity, Components, and Component-Based Modeling

Modularity applies to the development of individual software modules, often with a standardized interface to allow different modules to communicate. Typically, partitioning a system into modules helps minimize coupling and leads to an "easier-to-maintain" code.

Components are functional units that once implemented in a particular framework are reusable in other models within the same framework, with little migration effort. One advantage of using a modeling framework is that preexisting components can be reused to facilitate model development. Component-based modeling brings about the advantages of "plug-and-play" technology. Component programming builds upon the fundamental concepts of object-oriented programming, with the main difference being the introduction or presence of a framework. Components are generally implemented as classes in an object-oriented language, and are essentially "black boxes" that encapsulate some useful bit of functionality. They are often present in the form of a library file, that is, a shared object (.so file, Unix) or a dynamically linked library (.dll file, Windows; .dylib file, Macintosh). A framework then provides for an environment in which components can be linked together to form applications. One thing that typically distinguishes components from ordinary subroutines, software modules, or classes, is their capability to communicate with other components written in a different programming language.

28.2.3 Interfaces and Drivers

Components typically provide one or more interfaces by which a caller can access their functionality. The word "interface" refers to a boundary between two things and what happens at the boundary. An interface can be between a human and a computer program, such as a command-line interface (CLI), or a graphical user interface (GUI). In the context of plug-and-play components, the word interface refers to a named set of member functions (also called methods), each defined completely with regard to argument types and return types but without any actual implementation. An interface is a user-defined type, similar to an abstract class, with member function "templates" but no data members.

If a component does "have" a given interface, then it is said to expose or implement that interface, meaning that it contains an actual implementation for each of those member functions. It is perfectly fine if the component has additional member functions beyond the ones that comprise a particular interface. Because of this, it is possible and often useful for a single component to expose multiple, different interfaces. This allows it to be used in a greater variety of settings.

Most surface dynamics models advance values forward in time on a grid or mesh and have a similar internal structure. This structure consists of lines of code before the beginning of a time loop (the initialize step), lines of code inside the time loop (the run step), and finishes with additional lines after the end of the time loop (the finalize step). Virtually all component-based modeling efforts (e.g., ESMF, OpenMI, OMS, CSDMS) recognize the utility of moving these lines of code into three separate functions, with names such as initialize, run, and finalize. These three functions constitute a simple model-component interface that we refer to as an "IRF interface." Such an interface provides a calling program with a fine-grained access to a model's capabilities and the ability to control its overall time stepping so that it can be used in a larger application. The calling program "steers" a set of components and so is referred to as a driver.

28.2.4 Provides Ports and Uses Ports

A meaningful linkage of components often requires more than just the IRF functions; most linkages also require data exchange. Therefore, a model's interface must also describe functions that access data that it wishes to provide (getter functions) as well as methods that allow other components to change its data (setter functions). With getter and setter interface functions, connected components can query generated data as well as alter data from the other model.

Within a general component framework, a component will have two types of connections with other models. These connections are made through ports that come in two varieties. These ports are called "provides-ports" and "uses-ports" within a common component architecture framework (Armstrong et al. 1999). The first provides an interface to the component's own functionality (and data). The second specifies a set of capabilities (or data) that the component requires from another component to complete its task. A provides-port presents to other components an interface that describes its functionality. For instance, a provides-port that exposes an IRF interface allows another component to gain access to its initialize, run, and finalize steps. Any interface can be exposed through a port, but it can only be connected to another port with a similar interface.

The uses-port of a component presents functionality that it lacks and therefore requires from another component. Any component that provides the required functionality is able to connect to it. Thus, the component is not able to function until it is connected to a component that has the required functionality. This allows a model developer to create a new model that uses the functionality of another component without having to know the details of that component or to even have that component exist at all.

This style of plug-and-play component programming benefits both model programmers and users. Within a framework model, developers are able to create models within their own areas of expertise and rely on experts outside their field to fill in the gaps. Models that provide the same functionality can easily be compared to one another simply by unplugging one model and plugging in another, similar model. In this way, users can easily conduct model comparisons and more simply build larger models from a series of components to solve new problems.

28.2.5 Model Protocols

The procedures or system of rules governing contributed community software are referred to as its "protocols," and provide both technical and social recommendations to model developers. The protocols for software contribution to the CSDMS Model Repository, for example, are as follows:

1. Software should hold an open-source license (e.g., GPL2 compatible; OSI approved).
2. Software should be widely available to the community of scientists (e.g., CSDMS Model Repository; Computers & Geosciences Repository).

3. Software should receive some level of vetting (e.g., a colleague; a CSDMS Working Group). The software should be determined to do what it says it does.
4. Software should be written in an open-source language (C, C++, any FORTRAN, Java, Python), or have a pathway for use in an open-source environment (e.g., IDL & MATLAB® code can be made compatible).
5. Code should be written or refactored to be a component with an interface (IRF), with specific I/O exchange items (getters, setters, grid information) documented.
6. Code should be accompanied with metadata information, for example, http://csdms.colorado.edu/wiki/Form: Module_questionnaire, and test files (input files to run the model, output files to verify the model run).
7. Code should be clean and documented, and annotated using keywords within comment blocks to provide basic metadata for the model and its variables.

28.3 Methods and Applications

In the following, we provide details about four community projects that can be considered representative of the large suite of community cyberinfrastructure efforts. The CUAHSI HIS offers a service-oriented architecture, unique interface specification, data-exchange standard, and data model, with successful penetration of its hydrological offerings into academic and government research centers. OMS is a noninvasive modeling framework employing metadata annotations that employs a large library of legacy (FORTRAN) code developed originally by the U.S. Department of Agriculture, while being supported by modern computational platforms. ESMF is a computational-efficient modeling framework for mostly scalable FORTRAN-operational codes (NOAA, NASA, NCAR, DoD), offering standards for gridded components and coupler components. CSDMS is a complete modeling environment involving a model repository of numerous research-grade code, offering model coupling; language interoperability; unstructured, structured, and object-oriented code; framework interoperability; structured and unstructured grids; by employing the common component architecture and open modeling interface standards, with augmented services and tools.

28.3.1 Hydrological Information System

The Consortium of Universities for the Advancement of Hydrologic Science, Inc. (CUAHSI) Hydrologic Information System (HIS) is a collection of standards and software components designed to enhance access to water data (Maidment 2008). A multi-university team (U. Texas at Austin, Utah State U., San Diego Supercomputer Center, U. South Carolina, Drexel U., and Idaho State U.) supports the development and design of CUAHSI HIS (http://his.cuahsi.org).

The CUAHSI HIS addresses the common problem of data access and integration in hydrologic analysis and modeling.

Because hydrologic models, particularly those at the catchment or watershed-scale, require a significant amount of data to digitally describe the natural environment (e.g., observations, terrain, soil, vegetation), hydrologists devote a substantial amount of time to locating, reformatting, and integrating these data to support modeling and analysis activities. Even simple analyses that require cross-site integration of data are hindered by inconsistent data formats and data access protocols used in the hydrologic science community. New remote and in situ sensing technologies promise to amplify this challenge with exponential increases in scientific data generated over the coming decades. For this reason, standardization of how water data is digitally communicated must not only focus on allowing humans to more easily find and leverage databases, but also on enabling machines to access and effectively synthesize data across sites, organizations, and disciplinary boundaries.

The core design of CUAHSI HIS follows a service-oriented architecture (SOA) approach for organizing complex, geographically distributed software systems. In SOA design, system components are loosely coupled and communicate through web services that have standardized interfaces with clearly defined protocols and data exchange specifications (Erl 2005). An important innovation of CUAHSI HIS is the web service standards for communicating water data between software components within a distributed computing environment. These web service standards include an interface specification termed WaterOneFlow that defines how a client application requests data when using the web service, and a data exchange standard termed the Water Markup Language (WaterML) that specifies how water data is encoded when transferred from server to client applications (Figure 28.1). Both WaterOneFlow and WaterML standards extend core web service standards established by the

World Wide Web Consortium, making these general standards more specific and useful to the hydrologic community. Although an end user may never directly interact with the WaterOneFlow and WaterML standards, these core design specifications are a critical innovation that pervades the entire HIS cyberinfrastructure and allow the distributed software components to function as an integrated system.

The CUAHSI HIS development team has created a set of open source software components based on the WaterOneFlow and WaterML standards to support publishing, discovering, and accessing water data. The software components are organized into five primary applications: HIS Server and HIS Central are server-side applications, while HydroSeek, HydroDesktop, and HydroModeler are client-side applications (Figure 28.1). HIS Server is designed as a tool to support scientific investigators in storing and publishing observational data (Tarboton et al. 2009). The core of the HIS Server is the Observation Data Model (ODM), a database schema for representing point observational data that is being used to publish and archive field data collected at environmental observatories (Horsburgh et al. 2008). HIS Central provides the ability to catalog data stored within multiple HIS Servers and third-party data servers that have adopted the CUAHSI HIS web service standard (Tarboton et al. 2009). It consists of a central registry of CUAHSI HIS web services, a metadata catalog harvested from registered web services, and semantic mediation and search capabilities that enable discovery of information distributed across multiple web services.

On the client-side, HydroSeek is a web application for discovering data indexed within HIS Central. It allows for keyword searches and provides a map-based view for discovering relevant data. While HydroSeek is a web-based "thin-client" application that does not require the user to install software

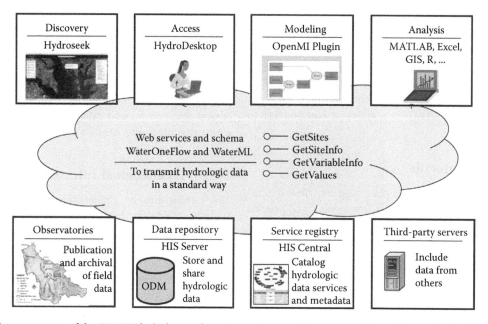

FIGURE 28.1 The components of the CUAHSI hydrologic information system are interconnected through a standardized web service protocol called WaterOneFlow and a standardized communication language called WaterML.

on their personal computer, HydroDesktop (formally named HIS Desktop) is a "thick-client" that must be installed on a personal computer, but provides a larger feature set compared to HydroSeek. HydroDesktop provides the ability to both discover and access data indexed within HIS Central. Downloaded data can be visualized in a map view that leverages Geographic Information System (GIS) technology, a graph view designed specifically for visualizing time series data, and a data view that provides a user interface to the raw data stored in the underlying database. While HydroDesktop's core functionality is intended for data management and visualization, the core software can be extended through the development of third-party software extensions (i.e., plug-ins) to perform more complex tasks. An example of a plug-in that extends the core HydroDekstop functionality is an integrated modeling environment called HydroModeler that adopts the Open Modeling Interface (OpenMI) standard for linking model components into workflows. Finally, it is important to note that users are not limited to HydroSeek or HydroDesktop as client applications for performing analysis on HIS data; many third-party analysis systems (e.g., MATLAB, Excel, and R) have the ability to directly ingest data served using web services that are based on industry standards. This feature keeps HIS open and extensible beyond the core tool set being developed by the HIS team.

From a systems integration perspective, the CUAHSI HIS automates the flow of data from sensors to scientists (Figure 28.2). Data enters the HIS either by investigators loading observations collected in the field into an ODM Database or from existing databases that have been made compliant to the HIS web service standards. Utility software such as ODM Tools, ODM Data Loader, and the Streaming Data Loader are provided as part of the HIS to assist investigators in loading their data into an ODM Database. Once the data has been loaded into an ODM database, software in the HIS Server stack makes that data accessible to client applications using the HIS web service standard. When a new HIS Server is registered in HIS Central, a software harvester uses the web service to extract and store metadata associated with that server within a metadata catalog. By design, only the metadata associated with observations (the site locations, the variable names, measurement units, etc.) are stored within HIS Central to support search and discovery applications; the data values themselves are not stored within HIS Central to reduce data storage volumes and to maintain data ownership with data collectors. Semantic mediation occurs within HIS Central by requiring data providers to perform a semantic tagging operation to relate their own variable names to variable concepts within the HIS ontology. Software tools such as the Hydrotagger support the user in the semantic mediation process. Once the dataset has been related to the HIS ontology, it becomes searchable through various client applications that access HIS data.

While the CUAHSI HIS is designed and has been implemented as a tool to aid research investigators in their efforts to advance hydrologic science and education in the United States,

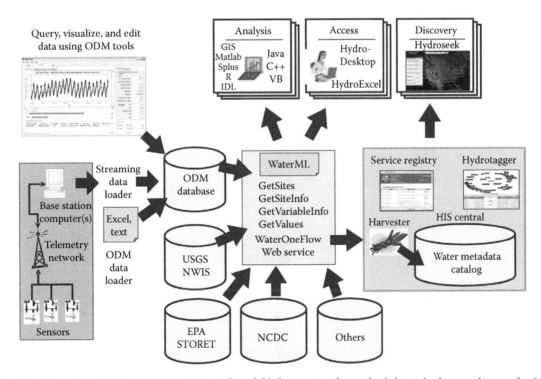

FIGURE 28.2 Data flows through the CUAHSI HIS starting from field observations that are loaded into databases and exposed as WaterML data using the WaterOneFlow web services. Metadata for these databases is automatically harvested and organized into a metadata catalog as part of HIS Central. Finally, client applications are able to use the web services and HIS Central catalog to analyze, access, and discover data distributed across multiple databases.

it has also been used by governmental agencies to assist in water resources management. Many federal datasets are registered in HIS Central and, in doing so, the CUAHSI HIS is providing integrated access to national-scale datasets describing stream-flow, groundwater, water quality, and climate conditions. Some state government agencies have also adopted CUAHSI HIS approaches for integrating water data to support water management mandates, a prime example being work done in Texas as part of the Texas Water Data Services (http://data.crwr.utexas.edu). The CUAHSI HIS is also being implemented outside of the United States, for example, in Australia, to integrate data collected and maintained by different agencies across the country. The integration of water data across such a wide range of groups speaks to the generality of the data integration and management challenge addressed by the CUAHSI HIS.

For further information about the CUAHSI HIS including any of the software tools discussed in this section, please visit the HIS website (http://his.cuahsi.org).

28.3.2 Object Modeling System

The Object Modeling System (OMS) is an integrated environmental modeling framework. There are four foundations identified for OMS3 (Figure 28.3): modeling resources, the system knowledge base, development tools, and the modeling products. OMS3 consists of an internal knowledge base and development tools for model and simulation creations. The system derives information out of various modeling resources, such as data bases, services, version control systems, or other repositories, and transforms it into a framework knowledge base that the OMS3 development tools use to create modeling products.

Products include model applications; simulations that support calibration, optimization, and parameter sensitivity analysis; output analyses; audit trails; and documentation. Implementing OMS3 requires a commitment to a structured model and simulation development process, such as the use of a version control system for model source code management, or a simulation-run database to store audit trails. Such features are important for institutionalized implementation of OMS3; however, a single modeler may not be required to adhere to it.

OMS3 adheres to the notion of objects as the fundamental building blocks for a model and to the principles of component-based software engineering for the model development process. The models within the OMS3 Framework are objects or components. However, the design of OMS is unique in that it is considered noninvasive and sees models and components as plain objects with metadata by means of annotations. Modelers do not have to learn an extensive object-oriented application programming interface (API), nor do they have to comprehend complex design patterns. Instead, OMS3 plain objects are perfect fits as modeling components as long as they communicate the location of their (1) processing logic and (2) data flow. Annotations do this in a descriptive, noninvasive way.

Most agro-environmental modelers, at least early in the development life cycle, are natural resource scientists with experience in programming (often self-taught), but not software architecture and design. Most modeling projects do not have the luxury employing an experienced software engineer or computer scientist. Software engineers understand and apply complex design patterns, UML diagrams, advanced object-oriented techniques such as parameterized types, or higher-level data structures and composition. A hydrologist or other natural resource scientist

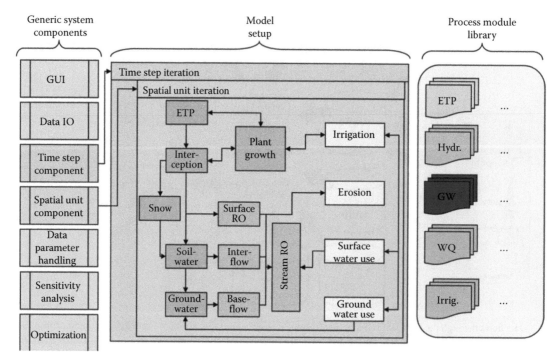

FIGURE 28.3 (See color insert.) OMS3 architecture.

may lack these skills. The targeted use of object-oriented analysis and design principles for modeling could be productive for a specific model having limited expectation for reuse and extensibility. However, for a framework, the extensive use of object-oriented features for models puts an undesirable burden on the scientist.

The agro-environmental modeling community maintains a large number of legacy models. Some methods and equations still in use were developed as long as 60 years ago. What has changed and will continue to change is the infrastructure around them that delivers the output from these models. A lightweight framework adjusts to an existing design as opposed to define its own specification or API. The learning curve is small, as there is no complex API to learn or new data types to manage. This has some very practical implications for a modeler, since there is no major paradigm shift in using existing modeling codes and libraries.

Since OMS3 is a noninvasive modeling framework, the modeler does not need an extensive knowledge of object-oriented principles to make the model-framework integration happen. Creating a modeling object is very easy. There are no interfaces to implement, no classes to extend and polymorphic methods to overwrite, no framework-specific data types to replace common native language data types1, etc. OMS3 uses metadata by means of annotations to specify and describe "points of interest" for existing data fields and methods for the framework.

There are several operational and research-focused OMS3 model applications to date. The National Water and Climate Center (NWCC) of the U.S. Department of Agriculture (USDA) Natural Resources Conservation Service (NRCS) is moving to augment seasonal, regression-equation-based water supply forecasts with shorter-term forecasts based on the use of distributed-parameter, physical process hydrologic models and an ensemble streamflow prediction (ESP) methodology. The primary model base is built using OMS3 and the PRMS hydrological watershed Model. The model collection will be used to assist in addressing a wide variety of water-user requests for more information on the volume and timing of water availability, and improving forecast accuracy. This effort involves developing and implementing a modeling framework and associated models and tools, to provide timely forecasts for use by the agricultural community in the western United States where snowmelt is a major source of water supply.

At the USDA-ARS Agricultural Systems Research Unit, an OMS3-based, component-oriented hydrological system for fully distributed simulation of water quantity and quality in large watersheds is being developed. The AgES-WS (AgroEcosystem Watershed) model was evaluated on the Cedar Creek Watershed (CCW) in northeastern Indiana, USA, 1 of 14 benchmark watersheds in the USDA-ARS Conservation Effects Assessment Project (CEAP) watershed assessment study. Model performance for daily, monthly, and annual stream flow response using non-calibrated and manually calibrated parameter sets was assessed using Nash-Sutcliffe model eDciency (ENS), coeDcient of determination (R2), Root Mean Square Error (RMSE), relative absolute error (RAE), and percent bias (PBIAS) model evaluation coeDcients. The results show that the prototype AgES-WS watershed model was able to reproduce the hydrological dynamics with suDcient quality, and more importantly should serve as a foundation on which to build a more comprehensive model to better quantify water quantity and quality at the watershed scale. The study is unique in that it represents the first attempt to develop and apply a complex natural resource system model under the OMS3.

OMS3 represents an easy to use, transparent, and scalable implementation of an environmental modeling framework. In OMS3, the internal complexity of the framework itself was vastly reduced while allowing models to implicitly scale from multicore desktops to clusters to clouds, without burdening the model developer with complex technical details.

28.3.3 Earth System Modeling Framework

The Earth System Modeling Framework (http://www.earthsystemmodeling.org) is an open-source software for building model components, and coupling them together to form applications. ESMF was motivated by the desire to increase collaboration and capabilities, and reduce cost and effort, by sharing codes. It was initiated in 2002 under NASA funding and has evolved to multiagency support and management. The project is distinguished by its strong emphasis on community governance and distributed development and by a diverse customer base that includes modeling groups from universities, major U.S. research centers, the National Weather Service, the Department of Defense, and NASA. Some of the major codes that have implemented ESMF coupling interfaces include the Community Climate System Model (CCSM4), the NOAA National Environmental Modeling System (NEMS), the NASA GEOS-5 atmospheric general circulation model, the Weather Research and Forecast (WRF) model, and the Coupled Ocean/Atmosphere Mesoscale Prediction System (COAMPS). The ESMF core development team is based at the NOAA Earth System Research Laboratory.

ESMF was originally designed for tightly coupled models (Hill et al. 2004). Tight coupling is exemplified by the data exchanges between the ocean and atmosphere components of a climate model: a large volume of data is exchanged frequently, and computational eDciency is a primary concern. Such models usually run on a single computer with hundreds or thousands of processors, low-latency communications, and a Unix-based operating system. Almost all components in these domains are written in FORTRAN, with just a few in C or C++. There are a large number of hardware vendors, compilers, and configurations that must be supported, and ESMF is regression tested nightly on >24 platforms.

As the ESMF customer base has grown to include modelers from other disciplines, such as hydrology and space weather, the framework has evolved to support other forms of coupling. For these modelers, ease of configuration, ease of use, and support for heterogeneous components may take precedence over performance. Heterogeneity here refers to programming language (Python, Java, etc., in addition to FORTRAN and C), function (components for analysis, visualization, etc.), grids and algorithms, and operating systems. In response, the ESMF team has introduced support for the Windows platform, and is exploring an approach to language interoperability through a simple switch that converts components to web services. It has introduced more general data structures, including an unstructured mesh, and several strategies for looser coupling, in which components may be in separate executables, or running on different computers.

The ESMF architecture is based on the concept of component software elements. Components are ideally suited for the representation of a system comprised of a set of substantial, distinct, and interacting elements, such as the atmosphere, land, sea ice and ocean, and their subprocesses. Component-based software is also well suited for the manner in which Earth system models are developed and used. The multiple domains and processes in a model are usually developed as separate codes by specialists. The creation of viable applications requires integration, testing, and tuning of these pieces, a scientifically and technically formidable task. When each piece is represented as a component with a standard interface and behavior, that integration, at least at the technical level, is more straightforward. Interoperability of components is a primary concern for researchers, since they are motivated to explore and maintain alternative versions of algorithms (such as different implementations of the governing fluid equations of the atmosphere), whole physical domains (such as oceans), parameterizations (such as convection schemes), and configurations (such as a standalone version of the atmosphere).

There are two types of components in ESMF: gridded components and coupler components. Gridded components (ESMF_GridComps) wrap the scientific and computational functions in a model, and coupler components (ESMF_CplComps) wrap the operations necessary to transform and transfer data between them. ESMF components can be nested, so that parent components can contain child components with progressively more specialized processes or refined grids.

As an Earth system model steps forward in time, the physical domains represented by gridded components must periodically transfer interfacial fluxes. The operations necessary to couple gridded components together may involve data redistribution, spectral or grid transformations, time averaging, and/or unit conversions. In ESMF, a coupler component encapsulates these interactions. Coupler components share the same standard interfaces and arguments as gridded components. A key data structure in these interfaces is the ESMF_State object, which holds the data to be transferred between components.

Each gridded component is associated with an import state, containing the data required for it to run, and an export state, containing the data it produces. Coupler components arrange and execute the transfer of data from the export states of producer gridded components into the import states of consumer gridded components. The same gridded component can be a producer or consumer at different times during model execution.

Both gridded and coupler components are implemented in the FORTRAN interface as derived types with associated modules. ESMF itself does not currently contain fully prefabricated gridded or coupler components—the user must connect the wrappers in ESMF to scientific content. Tools available in ESMF include methods for time advancement, data redistribution, online and off-line calculation of interpolation weights, application of interpolation weights via a sparse matrix multiply, and other common modeling functions.

Coupler component arrangements can vary for ESMF applications. Multiple couplers may be included in a single modeling application. This is a natural strategy when the application is structured as a hierarchy of components. Each level in the hierarchy usually has its own set of coupler components. Figure 28.4 shows the arrangement of components in the GEOS-5 model.

Design goals for ESMF applications include the ability to use the same gridded component in multiple contexts, to swap different implementations of a gridded component into an application, and to assemble and extend coupled systems easily.

A design pattern that addresses these goals is the mediator, in which one object encapsulates how a set of other objects interact (Gamma 1995). The mediator serves as an intermediary, and keeps objects from referring to each other explicitly. ESMF coupler components are intended to follow this pattern; this important as it enables the gridded components in an application to be deployed in multiple contexts without changes to their source code. The mediator pattern promotes a simplified view of intercomponent interactions. The mediator encapsulates all the complexities of data transformation between components. However, this can lead to excessive complexity within the mediator itself. ESMF has addressed this issue by encouraging users to create multiple, simpler coupler components and embed them in a predictable fashion in a hierarchical architecture, instead of relying on a single central coupler. This approach is useful for modeling complex, interdependent Earth system processes, since the interpretation of results in a many-component application may rely on a scientist's ability to grasp the flow of interactions system wide.

Computational environment and throughput requirements motivate another set of design strategies. ESMF component wrappers must not impose significant overhead (generally less than 5%) and must operate eDciently on a wide range of computer architectures, including desktop computers and petascale supercomputers. To satisfy these requirements, the ESMF software relies on memory-eDcient and highly scalable algorithms (e.g., Devine 2002). ESMF has proven to run eDciently on tens of thousands of processors in various performance analyses.

How the components in a modeling application are mapped to computing resources can have a significant impact on performance. Strategies vary for different computer architectures, and ESMF is flexible enough to support multiple approaches. ESMF components can run sequentially (one following the other, on the same computing resources), concurrently (at the same time, on different computing resources), or in combinations of these execution modes. Most ESMF applications run as a single executable, meaning that all components are combined into one program. Starting at a top-level driver, each level of an ESMF application controls the partitioning of its resources and the sequencing of the components of the next lower level.

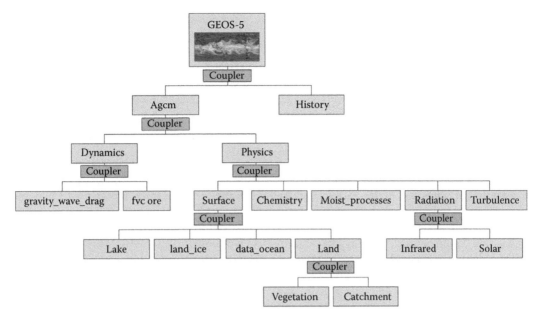

FIGURE 28.4 ESMF-based hierarchical architecture of the GEOS-5 atmospheric general circulation model. Each box is an ESMF component. Every component has a standard interface to facilitate exchanges. Hierarchical architecture enables the systematic assembly of many different systems.

It is not necessary to rewrite the internals of model codes to implement coupling using ESMF. Model code attaches to ESMF standard component interfaces via a user-written translation layer that connects native data structures to ESMF data structures. The steps in adopting ESMF are summarized by the acronym *PARSE*:

Prepare user code. Split user code into initialize, run, and finalize methods and decide on components, coupling fields, and control flow.

Adapt data structures. Wrap native model data structures in ESMF data structures to conform to ESMF interfaces.

Register user methods. Attach user code initialize, run, and finalize methods to ESMF components through registration calls.

Schedule, synchronize, and send data between components. Write couplers using ESMF redistribution, sparse matrix multiply, regridding, and/or user-specified transformations.

Execute the application. Run components using an ESMF driver.

Immediate ESMF plans include enabling the framework to generate conservative interpolation weights for a wide variety of grids and improving handling of masking and fractional areas. Implementation of a high-performance I/O package is also a near-term priority. Looking to the longer term, the ESMF team is exploring the use of metadata to broker and automate coupling services and to enable self-describing model runs.

28.3.4 Community Surface Dynamics Modeling System

The Community Surface Dynamics Modeling System (CSDMS) develops, supports, and disseminates integrated software modules that predict the movement of fluids (wind, water, and ice) and the flux (production, erosion, transport, and deposition) of sediment and solutes in landscapes, seascapes, and their sedimentary basins. CSDMS is an integrated community of experts to promote the quantitative modeling of earth-surface processes. The CSDMS community comprised >810 members from 133 U.S. Academic Institutions, 22 U.S. Federal labs and agencies, 225 non-U.S. Institutes from 60 countries, and an industrial consortia of 20 companies. CSDMS operates under a cooperative agreement with the National Science Foundation (NSF) with other financial support through industry and agencies. CSDMS serves this diverse community by promoting the sharing and reuse of high-quality, open-source modeling software. The CSDMS Model Repository comprises a searchable inventory of >200 contributed models (>6 million lines of code). CSDMS employs state-of-the-art software tools that make it possible to convert stand-alone models into flexible, "plug-and-play" components that can be assembled into larger applications. The CSDMS project also serves as a migration pathway for surface dynamics modelers toward high-performance computing (HPC). CSDMS offers much more than a national Model Repository and model-coupling framework (e.g., CSDMS Data Repository, CSDMS Education & Knowledge Transfer Repository), but this paper focuses on the former only.

The CSDMS Framework is predicated on the tools of the common component architecture (CCA) that is used to convert member-contributed code into linkable components (Hutton et al. 2010). CCA is a component architecture standard adopted by the U.S. Department of Energy and its national labs and many academic computational centers to allow software components to be combined and integrated for enhanced functionality on high-performance computing (HPC) systems (Kumfert et al. 2006). CCA fulfills an important need of scientific, high-performance, open-source computing (Armstrong et al. 1999). Software tools developed in support of the CCA standard are often referred to as the "CCA tool chain," and include those adopted by CSDMS: Babel, Bocca, and Ccaffeine.

Babel is an open-source, language interoperability tool (and compiler) that automatically generates the "glue code" that is necessary in order for components written in different computer languages to communicate (Dahlgren et al. 2007). It currently supports C, C++, FORTRAN (77, 90, 95, and 2003), Java, and Python. Almost all of the surface dynamics models in the contributed CSDMS Model Repository are written in one of these languages. Babel enables passing of variables with data types that may not normally be supported by the target language (e.g., objects, complex numbers). To create the glue code that is needed in order for two components written in different programming languages to "communicate" (or pass data between them), Babel only needs to know about the interfaces of the two components. It does not need any implementation details. Babel can ingest a description of an interface in either of two fairly "language neutral" forms XML (eXtensible Markup Language) and SIDL (Scientific Interface Definition Language) that provide a concise description of a scientific software component interface. This description includes the names and data types of all arguments and the return values for each member function. SIDL has a complete set of fundamental data types to support scientific computing, including booleans, double precision complex numbers, enumerations, strings, objects, and dynamic multidimensional arrays. SIDL syntax is very similar to Java.

Bocca helps create, edit, and manage CCA components and ports associated with a particular project, and is a key tool that CSDMS software engineers use when converting user-contributed code into plug-and-play components for use by CSDMS members. Bocca is a development environment tool that allows rapid component prototyping while maintaining robust software engineering practices suitable to HPC environments. It operates in a language-agnostic way by automatically invoking the lower-level Babel tool. Bocca can be used interactively at a Unix command prompt or within shell scripts. Once CCA-compliant components and ports are prepared using Bocca, CSDMS members can assemble models into applications with the CSDMS Modeling Framework that employs Ccaffeine (Figure 28.5).

Ccaffeine is the most widely used CCA-compliant frameworks. Ccaffeine has a simple set of scripting commands that are used to instantiate, connect, and disconnect CCA-compliant components. There are at least three ways to use Ccaffeine: (1) at an interactive command prompt, (2) with a "Ccaffeine script," or (3) with a GUI that creates a Ccaffeine script. The GUI is especially helpful for new users and for demonstrations and simple prototypes, while scripting is often faster for programmers and provides them with greater flexibility. The GUI allows users to select components from a palette and drag them into an arena. Components in the arena can be connected to one another by clicking on buttons that represent their ports (Figure 28.6). The component with a "config" button allows its parameters to be changed in a tabbed dialog with access to HTML help pages. Once components are connected, clicking on a "run" button on the "driver" component starts the application. The CSDMS modeling tool offers significant extensions and improvements to the basic Ccaffeine GUI, including a powerful, open-source

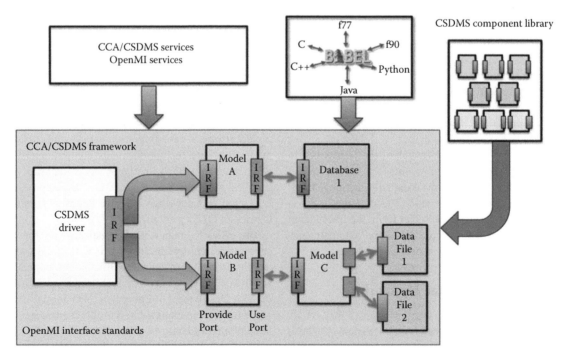

FIGURE 28.5 CSDMS modeling architecture reliant on componentization of models, a library of component models, a CCA/CSDMS framework, OpenMI interface standards, a suite of CCA/CSDMS and OpenMI services, and the language neutral tool/compiler Babel.

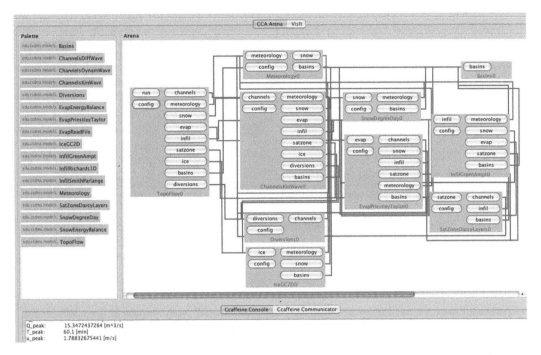

FIGURE 28.6 A "wiring diagram" for a CSDMS application (TopoFlow) project. The CCA framework called Ccaffeine provides a "visual programming" GUI for linking components to create working applications.

(DOE) visualization package called VisIt, which is specifically designed for HPC use with multiple processors.

CSDMS employs some of the OpenMI interface standards (OpenMI version 2.0 to be released soon is more closely aligned with the needs of CSDMS than version 1.4). CSDMS components have the member functions required for an OpenMI-style "IRF interface" as well as the member functions required for use in a CCA-compliant framework such as Ccaffeine. OpenMI provides for an open-source interface standard for environmental and hydrologic modeling (Gregersen et al. 2007). An IRF interface lies at the core of OpenMI, with additional member functions for dealing with various differences between models (e.g., units, timestepping, and dimensionality). While OpenMI is primarily an interface standard that transcends any particular language or operating system, its developers have also created a software development kit (or SDK) that contains many tools and utilities that make it easier to implement the OpenMI interface. Both the interface and the SDK are available in C# and Java. The C# version is intended for use in Microsoft's .NET framework on a PC running Windows. CSDMS has incorporated key parts of the OpenMI interface (and the Java version of the SDK) into its component interface.

In summary, the suite of services, architecture, and framework offered by a mixture of OpenMI and CCA tools and standards form the basis of the CSDMS Modeling Framework (Figure 26.5). By downloading the CSDMS GUI, users are able to (1) rapidly build new applications from the set of available components, (2) run their new application on the CSDMS supercomputer, (3) visualize their model output, and (4) download their model runs to their personal computers and servers. The CSDMS GUI supports

(1) platform-independent GUI (Linux, OSX, and Windows), (2) parallel computation (via MPI standard), (3) language interoperability, (4) legacy (non-protocol) code and structured code (procedural and object-oriented), (5) interoperability with other coupling frameworks, (6) both structured and unstructured grids, and (7) offers a wide range of open-source tools.

28.4 Major Challenges

Voinov et al. (2010) recently outlined technical challenges that community-modeling efforts that address earth-surface dynamics face: (1) fundamental algorithms to describe processes; (2) software to implement these algorithms; (3) software for manipulating, analyzing, and assimilating observations; (4) standards for data and model interfaces; (5) software to facilitate community collaborations; (6) standard metadata and ontologies to describe models and data; and (7) improvements in hardware. Voinov et al. (2010) wisely noted that perhaps the most diDcult challenges may be social or institutional. Among their many recommendations we emphasize the following:

- ft As a requirement for receiving federal funds, code has to be open source and accessible through model repositories.
- ft The production of well-documented, peer-reviewed code is worthy of merit at all levels.
- ft Develop effective ways of for peer review, publication, and citation of code, standards, and documentation.
- ft Recognize at all levels the contributions to community modeling and cyberinfrastructure efforts.

ft Use/adapt existing tools first before duplicating these efforts; adopt existing standards for data, model input and output, and interfaces.

ft Employ software development practices that favor transparency, portability, and reusability, and include procedures for version control, bug tracking, regression testing, and release maintenance.

References

Armstrong, R., D. Gannon, A. Geist et al. 1999. Toward a common component architecture for high-performance scientific computing. In *Proceedings of the 8th International Symposium on High Performance Distributed Computing*, Washington, DC, pp. 115–124.

Collins, N., G. Theurich, C. DeLuca et al. 2005. Design and implementation of components in the Earth System Modeling Framework. *International Journal of High Performance Computing Applications* 19: 341–350.

Dahlgren, T., T. Epperly, G. Kumfert, and J. Leek. 2007. *Babel User's Guide. 2007 edition*. Center for Applied Scientific Computing, U.S. Department of Energy and University of California Lawrence Livermore National Laboratory, 269pp.

Devine, K., E. Boman, R. Heaphy, B. Hendrickson, and C. Vaughn. 2002. Zoltan: Data management services for parallel dynamic applications. *Computing in Science and Engineering*, 4(2): 90–96.

Erl, T. 2005. *Service-Oriented Architecture: Concepts, Technology, and Design*, Prentice Hall PTR, Upper Saddle River, NJ.

Gamma, E., R. Helm, R. Johnson, and L. Vlissides. 1995. *Design Patterns: Elements of Reusable Object-Oriented Sos ware*, Addison-Wesley, Boston, MA.

Gregersen, J.B., P.J.A. Gijsbers, and S.J.P. Westen. 2007. OpenMI: Open modeling interface. *Journal of Hydroinformatics* 9: 175–191.

Hill, C., C. DeLuca, V. Balaji, M. Suarez, and A. da Silva. 2004. The architecture of the earth system modeling framework. *Computing in Science and Engineering* 6: 18–28.

Horsburgh, J.S., D.G. Tarboton, D.R. Maidment, and I. Zaslavsky. 2008. A relational model for environmental and water resources data. *Water Resources Research* 44: W05406, 12pp. doi:10.1029/2007WR006392.

Hutton, E.W.H., J.P.M. Syvitski, and S.D. Peckham. 2010. Producing CSDMS-compliant morphodynamic code to share with the RCEM community. In *River, Coastal and Estuarine Morphodynamics RCEM 2009*, eds. C. Vionnet et al. Taylor & Francis Group, London, U.K., pp. 959–962. ISBN 978-0-415-55426-CRC Press.

Kumfert, G., D.E. Bernholdt, T. Epperly et al. 2006. How the common component architecture advances computational science. *Journal of Physics: Conference Series* 46: 479–493.

Maidment, D.R. 2008. Bringing water data together. *Journal of Water Resources Planning and Management*, 134: 95. doi:10.1061/(ASCE)0733–9496(2008)134:2(95).

McMail, T.C. 2008. Next-generation research and breakthrough innovation: Indicators from US academic research. *Computing in Science and Engineering* 11: 76–83.

Moore, R.V. and I. Tindall. 2005. An overview of the open modelling interface and environment (the OpenMI). *Environmental Science and Policy* 8: 279–286.

Tarboton, D.G., J.S. Horsburgh, D.R. Maidment et al. 2009. Development of a community hydrologic information system. In *18th World IMACS/MODSIM Congress*, Cairns, Australia.

Voinov, A., C. DeLuca, R. Hood, S. Peckham, C. Sherwood, and J.P.M. Syvitski. 2010. Community modeling in earth sciences. *EOS Transactions of the AGU* 91: 117–124.

Wilson, G. and A. Lumsdaine. 2008. Software engineering and computational science. *Computing in Science and Engineering* 11: 12–13.

29

Uncertainty in Environmental NWP Modeling

David R. Stauffer
Pennsylvania State University

29.1 Introduction

Environmental numerical weather prediction (NWP) is expanding its role in science and society, to provide new understanding of and more actionable information about our environment (e.g., Holt et al. 2006, Seaman et al. 2012). Better weather guidance helps decision makers save lives and increase personal safety, protect property, enhance economic eDciency, and strengthen national defense (e.g., Stauffer et al. 2007a, Deng et al. 2012). It is therefore critical to identify, understand, quantify, and reduce as many sources of uncertainty in environmental models as possible. Because systematic model error, or bias, can be reduced by post-processing techniques, model uncertainty is often defined as that part of the model error that cannot be corrected by such methods, (e.g., Eckel and Mass 2005). In actuality, it is very diDcult, if not impossible, to remove all bias. Model uncertainty, therefore, is defined here in the broadest sense to include (1) model error, whether systematic or non-systematic, in the mean or variance of the model solution, and (2) variability or sensitivity of a model solution to observational data or internal attributes of the model itself.

29.1.1 Types of Models

There are many types of models that could be considered environmental models, including diagnostic and prognostic models,

meteorological, oceanographic, climate and ecosystem models, and models representing scales from the global scale (Leutbecher and Palmer 2008) to the turbulent microscale (Wyngaard 2004). Restricting our discussion to the atmosphere, it is valuable to give an overview of the different types of weather models and then define what is meant here by an environmental NWP model. We note that weather describes the atmosphereʼs short-term variations over minutes to days, while climate refers to the slowly varying or average atmospheric state over longer periods such as a month or more.

29.1.1.1 Diagnostic Models

Diagnostic models rely heavily on observations to describe atmospheric states, rather than using a set of nonlinear time-varying (prognostic) hydrodynamic equations. Although most can be run quickly and eDciently on very modest computing platforms, diagnostic models have very limited application to environmental problems in general. They typically provide horizontally mass-consistent wind fields over some local region where the model skill is directly tied to an objective analysis of observations (Kalnay 2003) and thus is very sensitive to the availability and density of the data (Wang et al. 2008). Generally, there is significant model uncertainty in areas where data are sparse, such as in mountainous regions or coastal zones, or when conditions are rapidly changing. The vertical motion fields produced by diagnostic models are generally so poor as to be unusable, and

Handbook of Environmental Fluid Dynamics, Volume Two, edited by Harindra Joseph Shermal Fernando. © 2013 CRC Press/Taylor & Francis Group, LLC.
ISBN: 978-1-4665-5601-0.

thermodynamic fields may not be provided (Wang et al. 2008). Lacking prognostic hydrodynamic equations, stand-alone diagnostic models can only be used for current or historical case studies. For predictive applications, the growing availability of faster and more affordable computing power favors wider use of sophisticated prognostic models in which only the initial states are defined from observations. As with diagnostic models, the accuracy of predictive models will always be limited in some way by the quality of these data.

29.1.1.2 Prognostic Models

A prognostic model generally can provide much greater value since it compensates somewhat for data-sparse conditions by using time-varying interdependent relationships for mass, momentum, and energy, usually in the form of nonlinear partial differential equations that can be solved numerically by computer. All NWP models require an initial state defined over a three-dimensional (3D) domain by meteorological observations, while geophysical data are used to specify the lower boundary (terrain, soil wetness, land use, etc.). Thus, given an initial atmospheric state, the NWP model solves an initial-value problem by integrating forward in time a set of nonlinear hydrodynamic equations, using relatively small time steps, to produce a forecast ranging from minutes to perhaps 1 week into the future.

Prognostic model error, or uncertainty, has many sources and is free to grow in time and space. Note that uncertainty is not the same as predictability, since our knowledge of the atmospheric state is always uncertain, even at the initial time. Predictability refers to a deterministic forecast of a nonlinear system being very sensitive to its initial state such that even a very small perturbation to the initial state can produce a very different large-scale forecast at 1–2 weeks in the future (e.g., Yoden 2007). Initial-condition error and model error are inseparable since the model is often used to create the initial state from the observations (e.g., Leutbecher and Palmer 2008). Observations used to define the initial state are always incomplete and also contain measurement and representativeness errors. Stochastic turbulent subgrid-scale (unresolvable) energy also adds to the uncertainty in the model forecast (e.g., Hanna and Yang 2001). NWP models, based on the Reynolds-averaged Navier–Stokes (RANS) equations, do not explicitly reflect uncertainty due to turbulent motions, but rather the unpredictability and randomness introduced by the nonlinearities of the equations and extremely large number of spatial degrees of freedom needed to describe the motion (Yoden 2007).

29.1.1.3 Data-Assimilating Models

Another type of analysis model is based on a variation of the prognostic models, making use of their full set of time-varying nonlinear hydrodynamic equations to analyze current and past atmospheric states. Like the simpler diagnostic models, data-assimilating models (DAMs) reduce analysis uncertainty by making direct use of observations. However, they do so by assimilating data throughout a four-dimensional time-space application, rather than at only a single time, as in the diagnostic models. The DAM is especially attractive for dynamic analysis (Stauffer

and Seaman 1994), where the observations are applied to constrain the model solution throughout the analysis period, while the dynamic consistency of time-varying model equations adds value by propagating observational information into data-sparse regions and also producing fine-scale realistic local structure not present in the data. Together these allow the DAM to produce a high-quality, dynamically consistent data set describing states throughout the assimilation period. The DAM fields can be used for scientific analysis, initialization of a prognostic model (e.g., Dixon et al. 2009, Reen and Stauffer 2010), or as input into other types of models such as air-chemistry or atmospheric transport and dispersion (AT&D) models (e.g., Tanrikulu et al. 2000, Deng et al. 2004, Deng and Stauffer 2006, Stauffer et al. 2009a).

29.1.1.4 Mesoscale NWP Models

For brevity, the scope of discussion of prognostic models in this chapter will be restricted to limited-area, nonhydrostatic, fully compressible mesoscale NWP models with grid lengths (Δx) from ~1 to 50 km (e.g., Grell et al. 1995, Steppeler et al. 2003, Skamarock et al. 2008). This type of model is chosen because it is generally applicable to a large range of environmental problems over spatial scales ranging from the meso-alpha scale (~200–2000 km) down to the meso-gamma scale (~2–20 km, Orlanski 1975), and a host of atmospheric processes including the effects of complex terrain, internal gravity waves, and thunderstorms. Thus, these models fall between the global prognostic models, which are generally hydrostatic, without any lateral boundary conditions and capable of representing global-scale circulations (e.g., Leutbecher and Palmer 2008), and large eddy simulation (LES), where the largest energy-containing turbulent eddies are explicitly resolved with $\Delta x \sim 100$ m or less (e.g., Wyngaard 2004). Note that mesoscale NWP models are not "building-aware"; that is, they are not capable of explicit simulation of fine-scale flow around buildings and small obstacles, as in an urban setting.

29.1.2 Observations, Models, and Filter Scales

Both observations and models contain information on many different scales of motion (see Chapter 1). Inconsistent representation of these scales when comparing models and observations can be another source of uncertainty. For example, Figure 29.1 compares observed and model-predicted time series of near-surface wind speeds based on 1 min averaged intervals (raw) and similar series to which a 2 h running vector mean filter has been applied (filtered). The model time series are derived from a mesoscale prediction by the Weather Research and Forecast (WRF, Skamarock et al. 2008) model with $\Delta x = 0.44$ km and 10 layers within the lowest 50 m. Both the model and observations apply within a stable nocturnal boundary layer on the night of October 7, 2007 (e.g., Stauffer et al. 2009b). The 1 min time series reveals weak observed winds exhibiting many scales of motion ranging from nonturbulent, submeso motions (Mahrt et al. 2010) with periods of ~2 min to meso-gamma/meso-beta scale motions due to interactions between inertia-gravity waves and drainage flows from the nearby mountains and having

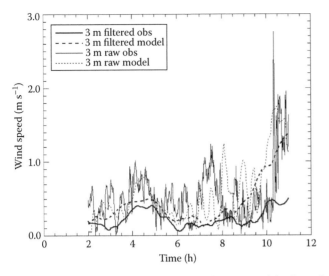

FIGURE 29.1 Sample time series of 3-m above ground level wind speed (m s^{-1}) through the night from a high-resolution WRF NWP model forecast and observations in the Nittany Valley of central PA, U.S. raw (filtered) model and observation data represent 1 min (2 h) temporal averages.

periods from tens of minutes to 1–2 h. Larger meso-alpha-scale and synoptic-scale influences lasting many hours to days also may appear (e.g., end of the series).

Instantaneous sampling of the modeled and observed series can lead to large apparent error because the smallest scales are not able to be predicted by the model. It is also much more diDcult for the model and observations to match at the finer temporal and spatial scales, and this can greatly affect a given quantitative measure of uncertainty (e.g., Casati et al. 2008). The model and observations must be properly sampled and consistently filtered (averaged) for comparison. Note that the filtered time series clearly exhibit higher correlation between the model and observations. To summarize, the model must be configured to represent the energy-containing scales of motion that are important to a given application, and then the model output and observations must be properly sampled and filtered to be consistent with each other. This general topic of "representativeness" will be discussed and demonstrated in further detail in the following section.

29.2 Principles: Sources of Model Uncertainty and Their Reduction

Environmental model uncertainty depends on both the observations (e.g., how many and what type of observations are available?) and NWP models (e.g., are the modelß domain, resolution, physics, and initialization configured correctly for the given application?). Observations may contain instrument errors due to either calibration or measurement precision, as well as representativeness errors. Instrument errors reflect the accuracy of the sensors, while representativeness errors account for how well a measurement represents relevant scales resolved by the model grid. Observations undergo a variety of quality control (QC)

checks to assess both measurement and representativeness error before they are used in a model (e.g., Kalnay 2003). Figure 29.2 shows forecasts from the MM5 mesoscale NWP model (Grell et al. 1995) at 36-, 12-, 4-, and 1.3-km horizontal resolutions for winds in complex terrain in the Italian Alps during the Torino 2006 Winter Olympics (Stauffer et al. 2007b). The finer resolution domains better resolve local drainage and channeled flows and show much better agreement with the special surface-mesonet wind data (in red). Note how these same data, when overlaid on the smoother 12- and 36-km wind fields, appear nearly random because the model cannot resolve the fine-scale terrain influences affecting the observations. Although the observations are generally point measurements and the model predictions are grid-volume averages, there is less model error or uncertainty when using the higher-resolution model fields, especially in complex terrain.

Figure 29.3 shows how model-output frequency can produce sampling errors that misrepresent the temporal evolution of vertical velocities at two locations over the eastern United States on a 4-km MM5 NWP model domain. The full time series (every time step shown) in Figure 29.3a is typical of weakly forced conditions in the warm sector ahead of an approaching frontal system, while the similar series in Figure 29.3b includes the passage of the cold front. In this case, the sign and magnitude of the modelß vertical motions, including the amplitude and timing of convective updrafts during the cold frontal passage, appear quite different depending on whether the output frequency is at every time step (12 s), every 10 min, every hour or every 3 h. (Note that output fields from many operational models are made available to most users at relatively coarse 3 or 6 h temporal resolutions (e.g., Schroeder et al. 2006, Stauffer et al. 2007a, Penn Stateß electronic map wall at http://www.meteo.psu.edu/%7Egadomski/ewall.html). Of course the predicted weather (clouds, precipitation) and planetary boundary layer (PBL) structure are largely dependent on the modelß vertical motion fields. So representativeness errors in either space or time can be a major source of environmental model uncertainty.

In addition to spatial and temporal representativeness of the models and observations, environmental NWP model solutions generally have three broad sources of model uncertainties or errors: (1) initial and lateral boundary conditions, (2) numerics, and (3) physics.

29.2.1 Initial and Lateral Boundary Conditions

The equations and algorithms comprising a limited-area mesoscale NWP model represent a highly complex initial-value, boundary-value problem (Kalnay 2003), the solution of which requires definition of accurate initial conditions and lateral boundary conditions.

29.2.1.1 Model Initialization and Data Assimilation

The ability of a high-resolution NWP model to produce accurate predictions, especially in the nowcast time frame (i.e., 0–6 h following initialization, Schroeder et al. 2006, Dixon et al. 2009,

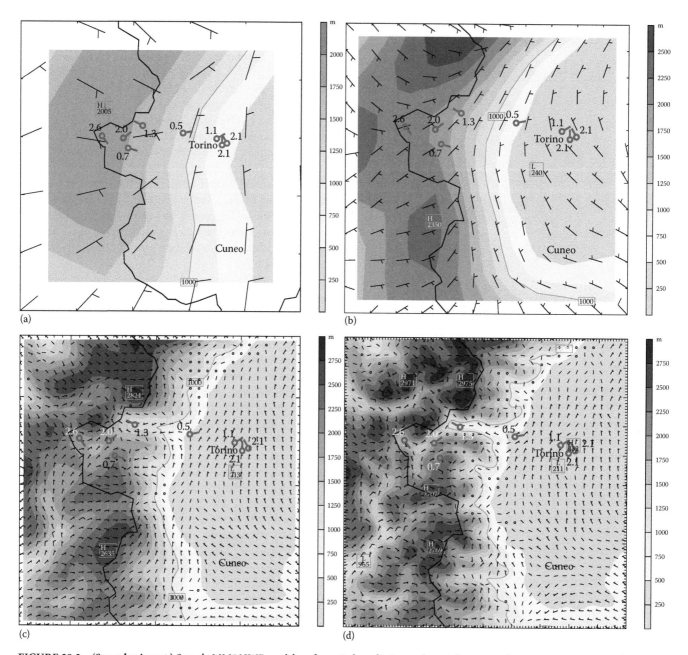

FIGURE 29.2 (See color insert.) Sample MM5 NWP model surface wind prediction and special mesonet observations at 18 UTC February 21, 2006, over the 1.3-km Torino Winter Olympics domain area for each model resolution forecast. Surface winds (m s⁻¹) are overlaid on the terrain field (m, color code on right of figure) for each model resolution domain. (a) 36-km domain, (b) 12-km domain, (c) 4-km domain, and (d) 1.3-km domain. One full barb is 10 m s⁻¹. Dark line is France-Italy border. (After Stauffer, D.R. et al., On the role of atmospheric data assimilation and model resolution on model forecast accuracy for the Torino Winter Olympics, 11A.6, *22nd Conference on Weather Analysis and Forecasting/18th Conference on Numerical Weather Prediction,* Park City, UT, June 25–29, 2007b, available at http://ams.confex.com/ams/pdfpapers/124791.pdf).

Deng et al. 2012), depends critically on the accuracy of the model's initial moisture and vertical motion fields. Because direct observations of these variables are usually either unavailable or incomplete at the model's grid scale, it must "spin up" its own cloud, precipitation, and vertical motion fields. Inconsistencies among the initial states of these crucially linked variables contribute to large forecast uncertainty in the early part of a statically initialized ("cold-start") model forecast and may amplify over time. As an example, Figure 29.4 shows the spinup of

domain-wide grid-maximum and domain-average precipitation rates after a cold-start initialization for the 36-, 12-, 4-, and 1.3-km forecasts computed over the area of the innermost 1.3-km Torino Olympics domain shown in Figure 29.2. Note that both precipitation rates generally increase faster and become greater in the first few hours (shorter spinup period) with finer grid resolution (Figure 29.4). For example, the magnitudes of domain-averaged precipitation rate on the 1.3- and 4-km domains after ~2 h in Figure 29.4b are not reached on the 12-km domain until

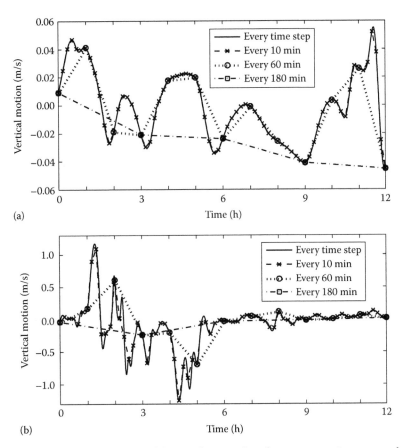

FIGURE 29.3 Sample time series of 4-km MM5 NWP model vertical motion (m s⁻¹) outputs at various temporal sampling frequencies (every time step [12 s], every 10 min, every 60 min, and every 180 min) for (a) location in the warm sector and (b) location experiencing a frontal passage.

after 4 h and not on the 36-km domain until after 6 h. Thus, an NWP forecast may contain much greater uncertainty during the first 6–12 h while the model is first generating mesoscale structure (Skamarock 2004) and spinning up its moist processes.

Generally, the model spinup problem can be reduced by replacing the "cold-start" static initialization with an effective "running-start" dynamic initialization, where observations in space and time are inserted or assimilated throughout a "pre-forecast" period (say $t = -3$ to 0 h) to produce dynamically balanced initial conditions at $t = 0$ h for a subsequent model forecast. The length of the pre-forecast period increases with grid size (e.g., Stauffer et al. 2007b) since the spinup period is resolution dependent (e.g., Figure 29.4). The goal is to create an initial state at the end of the pre-forecast period with spunup clouds, precipitation and local circulations, thereby improving the accuracy of all meteorological variables in subsequent short-term (24 h) forecasts (e.g., Stauffer et al. 2007b).

Nearly all objective analysis and four-dimensional data assimilation (FDDA, representing three space dimensions and time) methods are generally based on combining a model-generated "first guess" or background (x^b) and observations (y^o) to produce a gridded analysis of the observations (x^a) as described by the analysis equation

$$x^a = x^b + W[y^o - H(x^b)] \qquad (29.1)$$

where

$y^o - H(x^b)$ is the "observation increment" or "innovation"

H is the observation operator that performs the necessary interpolation and transformation from model variables to observation space

Details on how the model and observations are combined vary among the methods (Kalnay 2003). For example, the weight, W, applied to the innovation can be a simple function based on distance between the observation and model grid cell (e.g., Cressman successive scan or traditional nudging), or based on minimizing the analysis error at each grid point (e.g., optimal interpolation, OI), or based on model and observation error covariances (e.g., ensemble Kalman filter, EnKF). The x^a may also be determined by minimizing a cost function that measures the "distance" between the analysis and both the observations and model background, scaled by the observation error covariance and model background error covariance, respectively (e.g., three-dimensional variational, 3D-VAR). An innovation-like term can also be included directly within the model itself (nudging), or inserted in a derivative (adjoint) model used to compute the model initial state that produces a forecast that "best fits the observations" (minimizes the cost function) through the forecast/assimilation period (e.g., four-dimensional variational, 4D-VAR). Hybrid data-assimilation

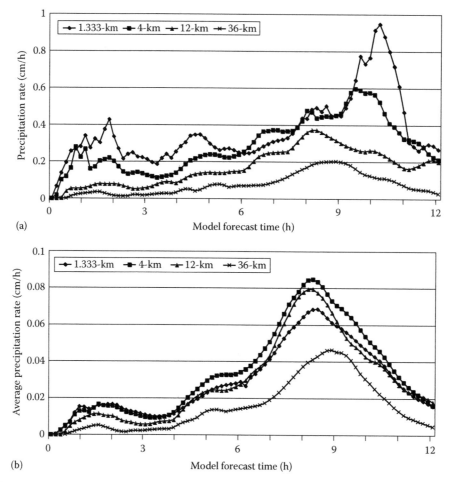

(a)

(b)

FIGURE 29.4 Sample MM5 NWP model time series of precipitation rates (cm h⁻¹) following a cold- start initialization on a 1.3-, 4-, 12-, and 36-km domain over the innermost 1.3-km domain area shown in Figure 29.2. (a) domain-wide grid-cell maximum precipitation rate and (b) domain-average grid-cell precipitation rate.

methods combining different strategies are also being developed (e.g., Warner 2011). In all of these methods, the role of the model is central since it propagates observation information into data-sparse regions and provides dynamic constraints.

Data assimilation strategies can be generally divided into intermittent and continuous methods, describing the frequency of data ingest into the model. Preference should be given to those FDDA methods that generate the least amount of spurious insertion "noise" while enhancing the spinup and accuracy of the model forecast fields. Intermittent methods use a prior model forecast as the background for data insertion, then run the model without further data ingest for some time period (say 1 h), and stop the model for another data ingest step before running the model again. This process is repeated throughout the assimilation period. These intermittent methods can produce large corrections to the model fields, even when using more sophisticated techniques such as the EnKF, and produce error/noise "spikes" related to the data insertion (e.g., Fujita et al. 2007). Digital filters are often used to reduce insertion errors (Kalnay 2003), but they can also remove realistic high-frequency atmospheric modes that may play a role in the forecast.

Continuous FDDA methods such as nudging (e.g., Stauffer and Seaman 1994, Dixon et al. 2009) add relaxation terms to the model's prognostic equations to gradually and continuously assimilate data at every time step, which minimizes insertion noise. The innovations can be computed in grid space (analysis nudging) or observation space (observation nudging). In a nested-grid framework where multiscale FDDA is common, the latter is more attractive for finer-scale grids and asynoptic data, while the former is more suitable for coarser-scale grids and synoptic data (Stauffer and Seaman 1994, Stauffer et al. 2007b). Hybrid methods that combine, for example, the advantages of EnKF (flow-dependent error covariance and observation weighting) and nudging (time-continuous, gradual corrections) are especially attractive and are also being developed (e.g., Lei et al. 2012).

A cold-start initialization is simply inappropriate for short-term forecasting (nowcasting). Available observational data (surface, rawinsonde, satellite, radar, etc.) are best applied through a pre-forecast period, rather than at only the initial time because static analyses usually have larger uncertainties in their vertical motion, vapor, cloud, and precipitation fields that take time to spinup in the model. Initialization of clouds and precipitation based on observations may be possible, but effective use of these data requires consistency with the 3D motion field (e.g., Dixon et al. 2009). Analyzed cloudy areas must correspond to the initial divergence field and regions of upward motion to be retained by the model. Nudging toward observed cloud and rain-rate-derived latent heating profiles can be very effective (e.g., Dixon et al. 2009) and is currently more practical than applying a full 4D-VAR.

In the absence of suDcient data for direct analysis of cloud, divergence, and vertical motion fields, short-term forecast improvements are generally obtained with time-continuous nudging and 4D-VAR methods that assimilate asynoptic data at the times they are observed, rather than at discrete synoptic "analysis times." Nudging is more computationally eDcient and adaptable to very fine grids, while 4D-VAR may provide a better framework for assimilating non-state variables (e.g., satellite radiance data) on large domains. Observation nudging has been successful in fine-scale model applications, where, for example, a snow squall was predicted at the initial time of a dynamically initialized forecast during the Torino Winter Olympics (Stauffer et al. 2007b). The localized squall was intense enough to postpone a women's downhill skiing event, while a similar "cold-start" model forecast could not spin up the precipitation. Caution is advised when applying intermittent FDDA methods (e.g., 3D-VAR) to short-term forecasting or dynamic analysis, especially without proper filtering, since insertion noise can create spurious vertical motion, cloud, and precipitation fields. However, such after-the-fact filtering may also suppress meteorologically realistic modes that can adversely affect forecast accuracy and increase model uncertainty. Nudging is also widely used for DAM applications where seamless dynamically consistent meteorological fields are needed for input into air-chemistry models, tactical decision aids, or AT&D models (e.g., Tanrikulu et al. 2000, Stauffer et al. 2007a,b).

29.2.1.2 Model Lateral Boundary Conditions

The demand for more precise NWP guidance from high horizontal resolution models, and hence limited-area nested-grid domains, requires definition of lateral boundary conditions (LBCs), which also contribute to the total model uncertainty (Warner 2011, Warner and Hsu 2000). Lateral boundary errors can propagate through the domain interior by advection (speeds ~5–50 m s⁻¹), as internal (~20–50 m s⁻¹) or external (~300 m s⁻¹) gravity waves. Ideally, the lateral boundaries should be located as far from the region of chief interest as possible to reduce their effects and to include important upstream forcing. Care is warranted to avoid placing an artificial lateral boundary in areas of strong physiography (e.g., very steep terrain, coastal zones). To the extent possible, consistent numerics and physics methods should be used between outer and inner nested domains. Communication between grids should be frequent (ideally, every time step), while the ratio of grid resolution between domains should be no more than 3:1 or 4:1 (Warner and Hsu 2000).

29.2.2 Model Numerics

29.2.2.1 Model Resolution and Time Step

Errors due to insuDcient model resolution (e.g., Figure 29.2) or inaccurate numerical methods also contribute to model uncertainty. For limited-area grid-point models, resolution is often defined in terms of the minimum number of grid lengths, Δx, needed to simulate the amplitude, speed and group velocity of wave-like phenomena with "suDcient" accuracy to be useful. Generally, features having spatial scales less than 6–10 Δx are considered too small to be resolvable on the grid. However, different numerical solution methods can have very different accuracy for a given grid size. Kalnay (2003) demonstrates the impact of several numerical methods on phase speed and group velocity for homogeneous linear waves (Figure 29.5). For each numerical method, fewer Δx per wavelength cause phase speeds to be artificially slowed. At $2^*\Delta x$, the phase speed becomes stationary, while group velocity is in the opposite direction! In general, higher-order numerical methods and finer grid resolution reduce these spatial truncation errors for a given wave, but at increased computational cost. Linear terms in spectral models are unaffected by spatial truncation errors since spatial derivatives are computed analytically. Thus, the linearly advected waves in the spectral model in Figure 29.5 have perfect phase speed at all wavelengths. However, the nonlinear advection terms in these models, as well as physics calculations, still involve calculations in grid-point space (Kalnay 2003).

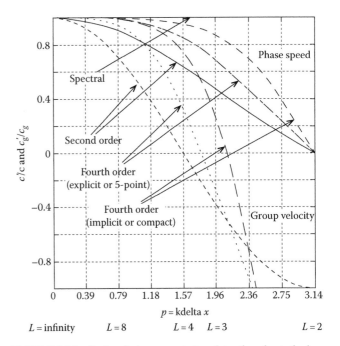

FIGURE 29.5 Ratio of the computational to the physical phase speed and group velocity for a simple linear wave equation, neglecting time truncation errors, as a function of wavelength (*L* in grid points) for second-order, fourth-order explicit and implicit, and spectral schemes. Perfect speed ratio is 1.0 along the top of the figure. Top curves represent phase speed; bottom curves represent group velocity. (After Kalnay, E., *Atmospheric Modeling, Data Assimilation and Predictability*, Cambridge University Press, Cambridge, MA, 2003.)

Generally, the finite time increment, Δt, used in a grid point NWP model to integrate the dynamical equations forward in time is subject to a stability limit beyond which wave amplitudes in the model solutions will grow exponentially in a nonphysical manner. Advanced numerical schemes that allow much larger time steps have been developed (e.g., see Steppeler et al. 2003), but they should be used with caution. For example, semi-implicit methods allow larger time steps by artificially slowing the high-speed sound and gravity modes. Sound waves have no effect on the meteorological modes, but slowing the speed of gravity waves may adversely affect solutions in some cases. Hence, for non-wavelike phenomena like thermodynamic diffusion, implicit methods can be very attractive. Semi-implicit methods are often combined with semi-Lagrangian spatial techniques to lengthen the time step while maintaining the accuracy of advective phase speed (Kalnay 2003). However, the longer time steps still introduce temporal truncation errors (Figure 29.3), although they are normally assumed smaller than spatial truncation errors. Finally, Teixeira et al. (2007) found that changing the model time step in a global model using semi-Lagrangian advection greatly affected the model solutions.

29.2.2.2 Model Diffusion

Numerical diffusion may also adversely affect a model's ability to simulate mesoscale structures. Numerical diffusion is a necessity in numerical models since nonlinear interactions between waves of certain wavelengths cause transfer of energy to shorter unresolved wavelengths. Without wavelength-selective diffusion, wave energy will accumulate in the $2-4\,\Delta x$ range, where it is misrepresented ("aliased"). If unchecked, nonlinear interactions can lead to unbounded amplification of energy in the $2-4\,\Delta x$ range. However, caution must be exercised so that the horizontal or vertical diffusion is not so large that it also damps otherwise resolvable atmospheric modes. Diffusion can be defined explicitly via a ∇^{2n} term or appear implicitly within the model's numerical solution method (Kalnay 2003). Higher-order diffusion methods are preferred since they are more scale selective, removing less wave energy at longer wavelengths. Skamarock (2004) defined a model's "effective resolution" as the smallest scale that preserves a $k^{-5/3}$ slope in the model's mesoscale kinetic energy spectrum, found to be $\sim 7-10\,\Delta x$. Note that operational NWP models often use somewhat larger numerical diffusion than comparable research models because they must remain numerically stable for the widest possible range of meteorological scenarios.

29.2.2.3 Model Conservation Properties

The conservation properties of a model's dynamic equations and the positive definiteness of its numerics (i.e., assuring that certain model fields, such as moisture, remain positive during integration) are also sources of uncertainty. Ideally, the model should conserve mass, momentum, and energy, but most current nonhydrostatic models ignore this issue (Steppeler et al. 2003). One exception is the Weather Research and Forecast (WRF) model that conserves mass and scalars (Skamarock et al. 2008). Conservation properties are being given more emphasis today, especially for climate models, where accumulation of errors becomes more significant with longer integration times. Mass conservation is especially important for climate simulations and in models using advanced model physics or where advection of small scalar concentrations is important (e.g., Seaman 2000, Li et al. 2009). Energy conservation is more complex, and conservation issues for atmospheric model dynamic cores should be addressed in the context of realistic atmospheric flows rather than idealized test cases (Thuburn 2008).

29.2.2.4 Grid Nesting

Grid nesting is an effective means to increase horizontal resolution in an area of special interest while limiting overall computer resource requirements (e.g., Warner 2011). As motivation, increasing the horizontal resolution for a given area by a factor of three generally increases total computation time by a factor of 27, since the time step must also be reduced by a factor of three to maintain numerical stability, as expressed by the Courant number (Kalnay 2003). Similarly, an increase of vertical resolution may require smaller time steps to maintain the numerical stability of vertical advection. Vertical grid nesting is rarely used in NWP, mostly because of the atmosphere's strong vertical stratification compared to its horizontal gradients.

29.2.3 Model Physics Parameterizations

Model physics are often treated in a modular fashion based on a sometimes-artificial separation of scales and processes (e.g., Arakawa 2004, Warner 2011). Physics parameterizations should respond smoothly over different spatial scales and include interactions among different processes (e.g., land-surface, PBL, convection, radiation), especially at small scales.

A physical parameterization (submodel) basically enables an atmospheric model to adjust back to a quasi-equilibrium state when forced by an unresolvable process. For example, in response to solar heating of the land-surface, the model atmosphere adjusts primarily through the PBL submodel, but also simultaneously through the land-surface and longwave radiation submodels. When there is deep unstable stratification, a convective parameterization scheme (CPS) exchanges the temperature and moisture within the vertical column to remove convective available potential energy (CAPE) and restore a neutrally stable state. Thus, a model physics "parameterization" represents a subgrid-scale process (e.g., turbulent eddies, convective clouds) in terms of its net effect on grid-resolved fields, which requires using one or more closure assumptions (e.g., K-theory, CAPE-removal). Different physical processes tend to be highly interactive, so separate submodels for the surface energy balance, PBL turbulence, deep moist convection, shallow clouds, moist microphysics, longwave and shortwave radiation, ozone, aerosol, etc., effectively comprise a complex model physics "suite." A well-designed physics suite is defined here as a broadly tested consistent set of parameterizations spanning the range of physical processes significant for the model's intended scale and application, using common physical assumptions and parameters among its various physics components. For example, atmospheric radiation schemes do not usually take into account the effects of subgrid (partial) cloud as represented by a CPS.

The ability of a model to represent diurnally forced resolved-scale circulations, such as land-sea breezes or mountain-valley circulations and their associated convection, depends not only on model resolution (see Figure 29.2) but on the accuracy of the diurnal surface temperature cycle, which is greatly affected by the land-surface parameterization and the effects of soil moisture, soil temperature, vegetation, and other surface characteristics (e.g., Holt et al. 2006). Use of more sophisticated land surface models (LSMs) can be an example of how using more model input data and parameters can actually increase model uncertainty (e.g., Reen et al. 2006). For example, a sophisticated multilayer LSM with time-varying soil temperature and soil moisture profiles that respond to surface fluxes and precipitation effects also introduces more risk when necessary input data (initial 3D soil moisture and temperature, soil type characterization, vegetation coverage, root depth, leaf area index, etc.) are not well known. In this case, a simpler force-restore "slab" model using seasonal-average land-use information and lookup tables for fewer surface parameters may provide a more robust solution (e.g., Stauffer et al. 2007a).

The use of a CPS versus grid-resolved moist physics (i.e., explicit microphysics parameterization) represents another potentially important issue for environmental NWP model accuracy. A model using a CPS will produce convective precipitation when certain "trigger" conditions (e.g., based on CAPE or moisture convergence into the column) are met, while a model exclusively using explicit microphysics will usually produce precipitation only when upward motion leads to grid-scale supersaturation. However, many CPSs have underlying assumptions, which if not met, add uncertainty to the model solutions. Convective parameterizations are appropriate for grid lengths of 50 km or larger but are often applied down to grid lengths of ~10 km (e.g., Deng and Stauffer 2006, Figure 29.6). An underlying assumption, in this case, is that the convective updrafts should cover only a small percentage of the grid cell area. Thus, as the model grid size decreases, this assumption may no longer be valid and there is risk of "double counting" or erroneous CPS-generated tendencies feeding back to the resolved grid scale. In many applications, models contain both a CPS for subgrid-scale deep convection and an explicit microphysics parameterization that represents grid-resolved cloud and precipitation processes. These two types of precipitation schemes are able to differentiate conditions favoring deep convection versus larger-scale stratiform precipitation, and so normally work very well together.

Storm-scale models are often defined as those having $\Delta x \sim 4$ km or less, where an explicit microphysics scheme is assumed to resolve both deep convective precipitation (at least partially) and stratiform precipitation, eliminating the need for a CPS (Warner and Hsu 2000). However, grid-resolved explicit microphysics is best applied without a CPS at very fine resolutions ($\Delta x \sim 1-2$ km), where convective thunderstorm updrafts can begin to be resolved by the grid (e.g., Deng and Stauffer 2006). NWP models typically use single-moment bulk microphysics schemes where the

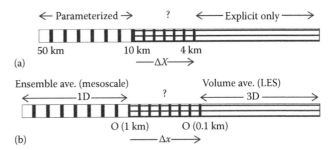

FIGURE 29.6 A schematic illustrating the horizontal grid-scale (Δx) dependence of parameterizations for (a) convection (CPS) and (b) turbulence (PBL). The crosshatched regions indicate where the underlying assumptions of the physical parameterizations are not strictly valid, and there is currently no robust, satisfactory solution in these "no man's land" scale ranges: convection ($\sim 4-10$ km Δx) and turbulence ($\Delta x \sim$ depth of mixed layer). (After Deng, A. and Stauffer, D.R., *J. Appl. Meteorol. Climatol.*, 45, 361, 2006.)

particle size distributions of various hydrometeor types are specified and there are prognostic equations for the mixing ratio for each type (e.g., Grell et al. 1995, Skamarock et al. 2008). More advanced are the two-moment bulk schemes, which usually also predict particle number concentrations and can be used to study aerosol-cloud interactions (e.g., Saleeby and Cotton 2005), but still require assumptions regarding the cloud droplet activation process and shapes of particle size distributions that can affect model uncertainty. Cloud-resolving models (CRMs) applied at 1 km or higher horizontal resolutions can use much more expensive bin microphysics, where particle bins are used to represent the actual size spectra of cloud condensation nuclei as well as different hydrometeor particles (e.g., Li et al. 2009). Aerosol-cloud-precipitation interactions are best represented by bin schemes and are therefore missing in most environmental NWP models.

Although there is still no clear consensus regarding the finest grid scale that can safely use a CPS (e.g., Lee and Hong 2005, Deng and Stauffer 2006), treatment of deep convection in models having $\Delta x \sim 4-10$ km is problematic because neither the CPS nor explicit-microphysics approach is entirely satisfactory (see Figure 29.6). That is, in this range the grid is too coarse to even begin to resolve convective updrafts explicitly, while it is too fine to satisfy the CPS assumption that updraft area is very small compared to the mesh size. Prediction of precipitation at these scales can contain much greater uncertainty. However, the smaller spatial scale of shallow convection (non-precipitating cumulus) still allows it to be parameterized in storm-scale models (e.g., Dixon et al. 2009). In nested-grid mesoscale models, grid resolutions are often prescribed to jump over these "no man's land" uncertainties by defining a 12-km outer domain (with a CPS) and a 4-km nest (with only explicit microphysics) (e.g., Schroeder et al. 2006).

The PBL and its turbulent processes often contribute to help trigger convection. Similar to a CPS and most atmospheric radiation schemes, current turbulence parameterizations in mesoscale models are 1D in the vertical direction (e.g., see appendix of Zhang and Zheng 2004). That is, based on mesoscale scaling arguments, only the vertical fluxes and turbulent mixing processes are parameterized, while lateral mixing is considered negligible compared to advection and other resolved processes. However, for very high-resolution mesoscale models (or very coarse LES), Δx may be similar to the PBL depth, $Z_i \sim 1–2\,\mathrm{km}$. Hence, Δx (or, simply Δ, called the model filter scale) approaches the scale of the largest energy-containing turbulent eddies, l (Wyngaard 2004). When $\Delta \sim l$, neither the mesoscale model closure assumption ($\Delta \gg l$) nor the LES closure ($l \gg \Delta$) is appropriate. LES makes explicit 3D flux calculations based on volume averages, while mesoscale model closures parameterize the vertical flux calculations based on the assumption of "ensemble averages" over many realizations. (Note that definition of ensemble average in the context of turbulence modeling is not the same as that described in Section 29.3.2.) As in the case of deep convection, a substantially greater risk and model uncertainty are incurred when using intermediate grid resolutions in these "no man's land" scales ($l \sim \Delta$), where phase and amplitude errors of advected waves may be improved (see Figure 29.5), but the physics parameterization assumptions are not strictly valid (Figure 29.6). Mesoscale PBL closures can still be valid for sub-kilometer horizontal resolutions in stable conditions, where the scale of the largest turbulent eddies is still very small compared to the grid length (e.g., Figure 29.1, Stauffer et al. 2009a). Hybrid mesoscale-LES turbulence parameterizations that adapt smoothly across these no man's land scales are currently unavailable, but would be of great advantage to allow general use of environmental NWP models at these very high resolutions.

Model physics (PBL, microphysics, radiation, etc.), whether explicitly resolved or parameterized, can also be very sensitive to the model's vertical resolution. Model tendencies, especially vertical fluxes and flux divergences, often become larger in magnitude with finer vertical layer spacing. Thus attention is needed to make sure that the vertical resolution, and the aspect ratio of horizontal to vertical grid length, is reasonable for the physics expected to affect the important feature(s) to be simulated. For example, modeling of processes associated with a shallow stable PBL requires not only very fine horizontal resolution, but much higher than normal vertical resolution near the surface than is typically used for deep daytime PBLs (e.g., Hanna and Yang 2001, Zhang and Zheng 2004, Mahrt et al. 2010, Stauffer et al. 2009a).

29.3 Methods of Analysis: Quantification of Model Uncertainty

Knowing what we do not know is always important in anything we do. The development of better verification methods for our environmental NWP models has been motivated by our higher-resolution models producing somewhat misleading statistical results using traditional grid-point spatial verification methods (e.g., mean square error, MSE), and the need to quantify probabilistic weather information, such as that represented by a probability density function (PDF). A PDF allows one to represent meteorological information in terms of a mean and its variance (another measure of uncertainty).

29.3.1 Model Verification

High-resolution environmental NWP models generally produce higher-amplitude, finer-scale atmospheric structures (e.g., Figures 29.2 and 29.4), and traditional grid-point spatial verification methods can produce much larger errors due to mislocation of the higher-intensity embedded structures (e.g., Roebber et al. 2004). These methods do not account for phase errors or the intrinsic spatial correlation existing within these fields. Casati et al. (2008) reviews new methods more appropriate for higher-resolution models including feature-based and neighborhood-based (fuzzy) approaches. The feature-based methods identify features in the forecast and observation fields and then assess different attributes associated with each individual pair of forecast-observed features (such as position, orientation, size, intensity). This is more similar to a subjective evaluation of a modeled feature's existence, location, shape, size, etc. Neighborhood-based approaches consider values nearby in space and time in the forecast-observation matching process and relax the requirements for perfect time-space matching. This metric assesses whether the model predicts some event within a given radius of the observed location.

For example, higher horizontal model resolution and FDDA were generally (but not always) found to improve traditional short-term forecast skill, especially in the mountainous areas including the Appalachians (Schroeder et al. 2006) and the Alps (Stauffer et al. 2007b). In the Alps, there was also a clear subjective (e.g., Figure 29.2) and statistical advantage (Stauffer et al. 2007b) of 4-km resolution over 12-km resolution forecasts, but a somewhat reduced statistical advantage between the 4- and 1.3-km domains. Careful subjective evaluation still suggested some added value to the 1.3-km domain forecasts while the statistical improvement between the 4- and 1.3-km domains remained quite small. Despite the similar mean absolute error (MAE) grid-point statistics in this case, even when using vector wind difference (VWD) that measures both speed and direction error and is less sensitive to larger wind direction errors in weak flows, the AT&D model's plume predictions based on the 4- and 1.3-km NWP model forecasts were quite different! Therefore, it is very important to consider using the newer verification methods described earlier, and develop more user-oriented statistical scores to isolate the meteorological variables, levels, etc., most important to the user (e.g., see Hanna and Yang 2001 for AT&D, Seaman 2000 for air quality, Stauffer et al. 2007a for field artillery NWP systems). For example, the U.S. Army field artillery and users of the Army MMS-P mobile NWP system (Schroeder et al. 2006, Stauffer et al. 2007a) are most concerned about the circular error probable (CEP), a measure of the accuracy of the field artillery weapons systems. The CEP is defined by the radius of a circle centered about the mean, whose boundary is expected to include 50% of the rounds. Thus weather information is often of intermediate importance to what matters most to a specific user of environmental NWP models, and obtaining probabilistic information regarding the likely accuracy of specific model-forecast variables and critical derived parameters is also very important to users of environmental NWP models.

29.3.2 Model Ensembles

The verification of probabilistic weather information involves a PDF, and a very practical way to estimate this PDF is with an NWP model ensemble that creates flow-dependent forecast uncertainty. The ensemble is typically comprised of a set of NWP model forecasts (typically at least 10–30 members, although ten times more members are really needed (Kolczynski et al. 2011)), designed to represent model uncertainty by perturbing the model initial conditions in some members and the model itself in others and using the spread or variance of the ensemble to quantify the uncertainty.

The method used to construct the ensemble is more of an "art" than a science, and it is expected to be different for increasing ensemble spread and representing uncertainty for, say, quantitative precipitation forecasting (QPF) versus AT&D forecasts focused more on surface and PBL processes (e.g., Kolczynski et al. 2009). There are many methods to perturb the model initial state, and they vary for global models whose spread depends more on large-scale dynamic instabilities, as compared to limited-area mesoscale models whose spread is based more on dry and moist convective disturbances and lateral boundary conditions (Kalnay 2003). Mesoscale model ensembles often involve also perturbing surface properties, sea-surface temperature, etc. (e.g., Eckel and Mass 2005), while global model ensembles use techniques that include isolating the fastest growing dynamic modes (e.g., Leutbecher and Palmer 2008). Methods to account for model error, which is still poorly understood, include defining ensemble members by using multiple NWP models, and perturbing model physics parameters or using alternate physics parameterizations. One of the more interesting methods involves using stochastic model physics where model physical tendencies include random or structured errors over some range (Leutbecher and Palmer 2008). The Grell-Devenyi ensemble CPS is innovative in this respect since it uses an ensemble of CPS closures within a single CPS (Lee and Hong 2005, Skamarock et al. 2008). Many studies have demonstrated the value of using both initial-condition and model-diversity ensemble members (e.g., Eckel and Mass 2005, Fujita et al. 2007).

Probabilistic verification methods originated with applications to statistical regression methods such as model output statistics (MOS), and are now being revisited and further developed for ensemble forecasting (Casati et al. 2008). In addition to verifying the ensemble mean forecast against observations, there are two popular measures for verifying the skill and usefulness of the uncertainty information within the ensemble: the statistical consistency of the predicted probabilities and the width of the predicted distribution. These two aspects for verifying the ensemble PDF are referred to as reliability (the ability of the forecast probability to match the observed relative frequency of an event over many cases) and resolution (the degree of sharpness of the PDF to better distinguish events from nonevents). Reliable probability forecasts with a narrow distribution will be more useful than those with a broad distribution. A popular metric that measures both reliability and resolution and is a single number used to assess the quality of probabilistic forecasts for a scalar variable is the continuous ranked probability score (CRPS) (Leutbecher and Palmer 2008). The rank histogram (Talagrand diagram) is used to determine the extent to which the ensemble spread matches the spread of the distribution of verifying observations (Casati et al. 2008). If the ensemble members are equally likely, unbiased and uncorrelated, they produce the desired "flat" rank histogram. However, they are usually under-dispersive (ensemble PDF does not include all the observations) and the rank histogram is then "U-shaped," and attention is needed to add a "good spread." A good spread is defined as an increase in ensemble variance that simultaneously improves statistical consistency; that is, the ensemble variance matches the MSE of the ensemble mean (Eckel and Mass 2005). The Brier score and related methods (Casati et al. 2008) are used to measure probabilistic skill of dichotomous variables ("yes/no" events). Finally, economic value and cost/loss models can be used to help decision makers take maximum advantage of probabilistic information (Casati et al. 2008). Familiarity with all of these statistics is strongly advised when using and evaluating ensemble forecast systems.

29.4 Applications: Use of Model Uncertainty Information

29.4.1 Subjective Use of Model Uncertainty Information

A visualization of uncertainty information from a forecast PDF, the ensemble mean and variance from a global ensemble forecast system's 500 hPa height fields at 5 and 10 days, is shown in Figure 29.7. The ensemble mean is a forecast with generally lower MSE than that of an unperturbed forecast because the unpredictable details have been filtered out in the ensemble mean and only the predictable features remain (e.g., Leutbecher and Palmer 2008). This can be used to predict those regions that will be colder/moister versus warmer/drier. However, note how the uncertainty of the system, as seen in the "spaghetti plots" in Figure 29.7 for this period, increases dramatically from day 5 to day 10, and from Asia to the United States. The time scale and region over which the ensemble mean outperforms a single high-resolution forecast also depends on forecast variable. For example, the precipitation field, especially heavy precipitation in complex terrain, is likely to be better predicted by a higher-resolution model (e.g., Figures 29.2 and 29.4). It can be used to refine the intensity and location of a feature in the "smoother" ensemble mean forecast. Note that ensemble-averaged mass and wind fields will not necessarily obey dynamic constraints such as geostrophic balance, and the ensemble mean wind fields may likely cross the ensemble mean 500 hPa height contours. If a dynamic balance is important for a model application, at least for shorter forecast times, a best member can be chosen to be the one closest to the ensemble mean, and still retain dynamic consistency among its fields.

29.4.2 Objective Use and Calibration of Model Uncertainty Information

Uncertainty in AT&D or similar models that use environmental model data is doubly uncertain, so NWP ensemble models are now being used to drive the AT&D models and incorporate their

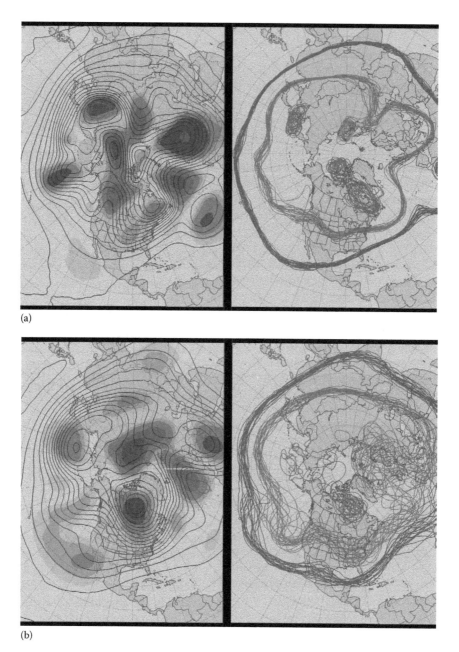

(a)

(b)

FIGURE 29.7   (See color insert.) Sample mean and variance ("spaghetti plot") of select 500 hPa height contours from a global forecast ensemble system for the Northern Hemisphere. (a) $t = 5$ days and (b) $t = 10$ days.

meteorological uncertainty information directly into the AT&D predictions (e.g., Kolczynski et al. 2009). The NWP-ensemble mean wind/stability fields are used along with the ensemble wind variance/covariance fields as inputs to an AT&D model to account for meteorological uncertainty and better predict the plume hazard area, orientation, etc., and it was shown to produce results similar to those based on an ensemble of AT&D models running off the NWP ensemble. Furthermore, a new linear variance calibration (LVC) of the NWP ensemble variances to actual error variances was shown to further improve the skill of the single AT&D forecasts (Kolczynski et al. 2009).

Statistical methods can also be used to reduce systematic errors in the NWP model outputs (e.g., Eckel and Mass 2005),

and a weighted mean of the ensemble can also be computed based on each member's performance over some recent training period (e.g., Roebber et al. 2004). It is important to note that there are many methods for post-processing environmental NWP ensemble model data, and post processing can further improve the value of probabilistic weather forecasting (e.g., Eckel and Mass 2005, Kolczynski et al. 2011).

29.5 Major Challenges

It is important to realize first that any aspect of environmental NWP that can be improved will contribute toward reducing environmental model uncertainty and also improving environmental ensemble model forecasting. We can always potentially benefit from obtaining more observations for initialization (and developing effective FDDA techniques to use these data), especially vertical-profile mass and wind data, cloud and precipitation fields, and also the specialized data needed for developing and evaluating new model physics schemes that are more unified and vary smoothly across the no man's land scales. Improved verification and post-processing methods using the new data sources can also reduce model uncertainty.

The next biggest challenge involves determining the best way to use available/expanding computer resources in terms of running the highest-resolution models possible versus the use of lower-resolution ensemble members with more simplified physics (e.g., Roebber et al. 2004). There is always a trade-off between model resolution and the number of ensemble members. High-resolution models better allow atmospheric scientists to compare model results against their conceptual understanding of phenomena. Ensembles, on the other hand, are a practical way to estimate the forecast PDF, although defining an appropriate ensemble for a given application is not easy. Since higher-resolution ensembles may be traded for a larger number of ensemble members, another related major challenge will be effective dynamic or statistical "downscaling" of the coarser-model resolution data (e.g., Warner 2011). Nesting a higher-resolution model from a best member or subset of ensemble members would be one solution. The other is statistically based where MOS-type equations are used to define finer-scale structure from coarser-resolution model outputs (Roebber et al. 2004). This, however, requires freezing the model configuration and running the model(s) over the same domain, every day, for a sufficiently long training period with fine-scale observations. Rare events with small sample sizes are especially challenging for statistical methods.

Some users of environmental NWP, such as the U.S. military running NWP models on laptops or modest mobile computing platforms, may not know where the modeling system will be operating tomorrow or the next day. Many of these environmental model applications use high-resolution deterministic models and more practical solutions for model post-processing such as the unified post-processing system (UPPS) that creates on-the-fly adjustments to the model to reduce model biases (Stauffer et al. 2007a). These many new and exciting challenges for environmental NWP will likely keep us modelers busy for a long time.

References

Arakawa, A. 2004. The cumulus parameterization problem: Past, present, and future. *Journal of Climate* 7: 2493–2525.

Casati, B., L.J. Wilson, D.B. Stephenson et al. 2008. Forecast verification: Current status and future directions. *Meteorological Applications* 15: 3–18.

Deng, A., N.L. Seaman, G.K. Hunter, and D.R. Stauffer. 2004. Evaluation of interregional transport using the MM5-SCIPUFF system. *Journal of Applied Meteorology* 43: 1864–1886.

Deng, A. and D.R. Stauffer. 2006. On improving 4-km mesoscale model simulations. *Journal of Applied Meteorology and Climatology* 45: 361–381.

Deng, A., D. Stauffer, B. Gaudet, and G. Hunter. 2012. A rapidly relocatable high-resolution WRF system for military-defense, aviation and wind energy. 2.6. *13th Annual WRF Users' Workshop*, Boulder, CO, June 25–29. Available at https://www.regonline.com/AttendeeDocuments/1077122/43387256/43387256_1045166.pdf

Dixon, M., Z. Li, H. Lean, N. Roberts, and S. Ballard. 2009. Impact of data assimilation on forecasting convection over the United Kingdom using a high-resolution version of the met office unified model. *Monthly Weather Review* 3: 1562–1584.

Eckel, F.A. and C.F. Mass. 2005. Aspects of effective mesoscale, short-range ensemble forecasting. *Weather and Forecasting* 20: 328–350.

Fujita, T., D.J. Stensrud, and D.C. Dowell. 2007. Surface data assimilation using an ensemble Kalman filter approach with initial condition and model physics uncertainties. *Monthly Weather Review* 3: 1846–1868.

Grell, G.A., J. Dudhia, and D.R. Stauffer. 1995. A description of the fifth-generation Penn State/NCAR Mesoscale Model (MM5). NCAR Tech. Note NCAR/TN-398+STR. Available at http://www.mmm.ucar.edu/mm5/documents/mm5-desc-doc.html (accessed on July 15, 2012).

Hanna, S.R. and R. Yang. 2001. Evaluations of mesoscale models/simulations of near-surface winds, temperature gradients, and mixing depths. *Journal of Applied Meteorology* 0: 1095–1104.

Holt, T.R., D. Niyogi, F. Chen, K. Manning, M.A. LeMone, and A. Qureshi. 2006. Effect of land-atmosphere interactions on the IHOP 24–25 May 2002 convection case. *Monthly Weather Review* 4: 113–133.

Kalnay, E. 2003. *Atmospheric Modeling, Data Assimilation and Predictability*. Cambridge, MA: Cambridge University Press.

Kolczynski, Jr. W.C., D.R. Stauffer, S.E. Haupt, and A. Deng. 2009. Ensemble variance calibration for representing meteorological uncertainty for atmospheric transport and dispersion modeling. *Journal of Applied Meteorology and Climatology* 8: 2001–2021.

Kolczynski, W.C., D.R. Stauffer, S.E. Haupt, N.S. Altman, and A. Deng. 2011. Investigation of ensemble variance as a measure of true forecast variance. *Monthly Weather Review* 9: 3954–3963.

Lee, T.-Y. and S.-Y. Hong. 2005. Physical parameterization in next-generation NWP models. *Bulletin of the American Meteorological Society* 6: 1615–1618.

Lei L, D.R. Stauffer and A. Deng. 2012. A hybrid nudging-ensemble Kalman filter approach to data assimilation in WRF/DART. *Quart. J. Roy. Meteor. Soc.* doi:10.1002/qj.1939.

Leutbecher, M. and T.N. Palmer. 2008. Ensemble forecasting. *Journal of Computational Physics* 2: 3515–3539.

Li, X., W.K. Tao, A.P. Khain, J. Simpson, and D.E. Johnson. 2009. Sensitivity of a cloud-resolving model to bulk and explicit bin microphysical schemes. Part I: Comparisons. *Journal of the Atmospheric Sciences* 6 : 3–21.

Mahrt, L., S. Richardson, N. Seaman and D. Stauffer. 2010. Non-stationary drainage flows and motions in the cold pool, *Tellus* 6 : 698–705.

Orlanski, I. 1975. A rational subdivision of scales for atmospheric processes. *Bulletin of the American Meteorological Society* 6 : 527–530.

Reen, B.P., D.R. Stauffer, K.J. Davis, and A.R. Desai. 2006. A case study on the effects of heterogeneous soil moisture on mesoscale boundary layer structure in the southern Great Plains, USA. Part II: Mesoscale modelling, *Boundary Layer Meteorology* 1 : 275–314.

Reen, B.P. and D.R. Stauffer. 2010. Data assimilation strategies in the planetary boundary layer, *Boundary Layer Meteorology* 3 , doi:10.1007/s10546-010-9528-6.

Roebber, P.J., D.M. Schultz, B.A. Colle, and D.J. Stensrud. 2004. Toward improved prediction: High-resolution and ensemble modeling systems in operations. *Weather and Forecasting* 9 : 936–949.

Saleeby, S.M. and W.R. Cotton, 2005. A large-droplet mode and prognostic number concentration of cloud droplets in the Colorado State University Regional Atmospheric Modeling System (RAMS). Part II: Sensitivity to a Colorado winter snowfall event. *Journal of Applied Meteorology* 4 : 1912–1929.

Schroeder, A.J., D.R. Stauffer, N.L. Seaman et al. 2006. An automated high-resolution, rapidly relocatable meteorological nowcasting and prediction system. *Monthly Weather Review* 4 : 1237–1265.

Seaman, N.L. 2000. Meteorological modeling for air-quality assessments. *Atmospheric Environment* 4 : 2231–2259.

Seaman, N.L., B.J. Gaudet, D.R. Stauffer et al. 2012. Numerical prediction of sub-mesoscale flow in the nocturnal stable boundary layer over complex terrain, *Monthly Weather Review* 1 : 956–977.

Skamarock, W.C. 2004. Evaluating mesoscale NWP models using kinetic energy spectra. *Monthly Weather Review* 3 : 3019–3032.

Skamarock, W.C., J.B. Klemp, J. Dudhia et al. 2008. A description of the Advanced Research WRF Version 3. NCAR Tech. Note NCAR/TN-475+STR. Available at http://www.mmm.ucar.edu/wrf/users/docs/arw_v3.pdf (accessed on July 15, 2012).

Stauffer, D.R., A. Deng, G.K. Hunter et al. 2007a. NWP goes to wark J4.1. *22nd Conference on Weather Analysis and Forecasting/18th Conference on Numerical Weather Prediction*, Park City, UT, June 25–29. Available at http://ams.confex.com/ams/pdfpapers/124777.pdf

Stauffer, D., G. Hunter, A. Deng et al. 2009a. Realtime high-resolution mesoscale modeling for the Defense Threat Reduction Agency. 1A.3. *23rd Conference on Weather Analysis and Forecsating/19th Conference on Numerical Weather Prediction*, Omaha, NE, June 1–5. Available at http://ams.confex.com/ams/pdfpapers/154237.pdf

Stauffer, D.R., B.J. Gaudet, N.L. Seaman, J.C. Wyngaard, L. Mahrt, and S. Richardson. 2009b. Sub-kilometer numerical predictions in the nocturnal stable boundary layer. 18B.4. *23rd Conference on Weather Analysis and Forecasting/19th Conference on Numerical Weather Prediction*, Omaha, NE, June 1–5. Available at http://ams.confex.com/ams/pdfpapers/154288.pdf

Stauffer, D.R., G.K. Hunter, A. Deng et al. 2007b. On the role of atmospheric data assimilation and model resolution on model forecast accuracy for the Torino Winter Olympics. 11A.6. *22nd Conference on Weather Analysis and Forecasting/18th Conference on Numerical Weather Prediction*, Park City, UT, June 25–29. Available at http://ams.confex.com/ams/pdfpapers/124791.pdf

Stauffer, D.R. and N.L. Seaman. 1994. Multiscale four-dimensional data assimilation. *Journal of Applied Meteorology* 3 : 416–434.

Steppeler, J., R. Hess, U. Schättler, and L. Bonaventura. 2003. Review of numerical methods for nonhydrostatic weather prediction models. *Meteorology and Atmospheric Physics* 8 : 287–301.

Tanrikulu, S., D.R. Stauffer, N.L. Seaman, and A.J. Ranzieri. 2000. A field-coherence technique for meteorological field-program design for air-quality studies. Part II: Evaluation in the San Joaquin Valley. *Journal of Applied Meteorology* 39: 317–334.

Teixeira, J., C.A. Reynolds, and K. Judd. 2007. Time step sensitivity of nonlinear atmospheric models: Numerical convergence, truncation error growth, and ensemble design. *Journal of the Atmospheric Sciences* 6 : 175–189.

Thuburn, J. 2008. Some conservation issues for the dynamical cores of NWP and climate models. *Journal of Computational Physics* 2 : 3715–3730.

Wang, W., W.J. Shaw, T.E. Seiple, J.P. Rishel, and Y. Xie. 2008. An evaluation of a diagnostic wind model (CALMET). *Journal of Applied Meteorology and Climatology* 4 : 1739–1756.

Warner, T.T. and H.-M. Hsu. 2000. Nested-model simulation of moist convection: The impact of coarse-grid parameterized convection on fine-grid resolved convection. *Monthly Weather Review* 8 : 2211–2231.

Warner, T. T. 2011. *Numerical Weather and Climate Prediction.* Cambridge, MA: Cambridge University Press.

Wyngaard, J.C. 2004. Toward numerical modeling in the "terra incognita." *Journal of the Atmospheric Sciences* 6 : 1816–1826.

Yoden, S. 2007. Atmospheric predictability. *Journal of the Meteorological Society of Japan* 8 : 77–102.

Zhang, D.-L. and W.-Z. Zheng. 2004. Diurnal cycles of surface winds and temperatures as simulated by five boundary layer parameterizations. *Journal of Applied Meteorology* 4 : 157–169.

Laboratory Modeling of Environmental Flows

30

Physical Modeling of Hydraulics and Sediment Transport

Olivier Cazaillet
SOGREAH

Sultan Alam
SOGREAH

Clinton S. Willson
Louisiana State University

30.1 Introduction

Hydraulics and sediment transport in most natural and built systems are spatially and temporally complex, requiring some form of study to improve understanding of these systems and provide economical and reliable designs. Site-specific studies are critical for determining the various regimes that must be accounted for to optimize the safety, costs, and operation and minimize the environmental impact of the system. Prior to the advent of computers, all studies were either field- or lab-based. While field studies provide the most realistic data, the inability to actually control and measure some or all of the forcings and boundary/initial conditions make it diDcult to interpret the data, and past and current conditions may not be useful in predicting planned modifications or how structures may interact with the natural system. In addition, field-based studies can be expensive; particularly if the system domain is large and/or if one would like to measure the important processes and phenomena over long time scales, which is essential for an optimized and safe design.

The basic premise of physical modeling is to build a model of a natural or man-made prototype at a reduced scale that adequately reproduces all the main physical processes involved, such as fluid discharge, water surface elevations, flow patterns, sediments dynamics, and forces on structures. The sophistication of the physical model will vary with the design or study objective.

The law of similitude for the open-channel flow was published by Ferdinand Reech in 1852 and discovered again in 1872 by William Froude. The history of hydraulic modeling dated back to the end of the nineteenth century (O. Reynolds with the Mersey estuary model, 1885) and the first hydraulic laboratories to study hydraulic works were built in the 1920s and 1930s.

30.1.1 When to Use Hydraulic Models Today?

It is important for engineers, project managers, and owner/operators to understand the type of problems for which a physical scale model is a useful and, sometimes, unique tool. The quite exhaustive list that follows has been made from the studies carried out over the last 30 years in many prominent laboratories primarily in Europe and North America.

30.1.1.1 Hydraulic Structures Studies

   ft *River engineering and hydraulic structures*: flood protection, water intake, lateral diversion, culverts, drop structures, crossing of a river and its flood plain
   ft *High dams and run-of-river dams, weirs, and related structures*: spillways, downstream stilling basin, gated bottom outlets, flip buckets, plunge pools, flap gates, and radial gates
   ft *High lis height and complex locks (including saving basins)*

Handbook of Environmental Fluid Dynamics, Volume Two, edited by Harindra Joseph Shermal Fernando. © 2013 CRC Press/Taylor & Francis Group, LLC.
ISBN: 978-1-4665-5601-0.

ft *Storm water hydraulic structures*: complex head losses, drop shafts, and vortex drop shafts, lateral weir diversion, ed uent dispersion in rivers

ft *Pumping stations*: uniform discharge distribution, anti-vortex devices, pre-rotation requirements;

ft *Coastal and harbor engineering: breakwater stability*: rock or artificial unit block structures, caissons structures, forces on caissons and walls, harbor wave actions, and ship forces on mooring lines

30.1.1.2 Sedimentological Studies (Morphodynamics and Sediment Transport)

ft *Torrents' training*: bed-load sediment transport and mudflow flood protection, water intake with or without sediment diversion, sediment retaining barrage schemes, culverts and tunnels with bed-load sediment transport

ft *River studies*: erosion, morphodynamics and sediment transport modification caused by a new river scheme—set of groins for bank protection, lateral diversion of significant discharge for flood protection

ft *Run-of-river and high dam reservoir studies*: deposition area, flushing operations, plunge pool erosion

ft *Estuaries and deltas*: design and impact assessment of the development of an estuary: longitudinal dykes, dredging, impact on flood and ebb tide channels, sand bank evolution, possible depth for the navigation channels, sediment control

ft *Coastal studies*: impact of harbors on the coast line and on littoral sediment transport, design of beach protections—set of groins, offshore breakwaters, T-shape groins, beach nourishment, permanent river mouth design, tidal bay development

30.1.1.3 Environmental Studies

ft *Density models*: salt intrusion studies—anti-salt lock

ft *Specirc environmental studies*: air pollution dispersion problems scaled with water, sand dune progression, ship funnel design

30.1.1.4 Navigation Training

ft Ship handling training using manned models

30.1.2 Why Use a Scale Model?

To begin with, a physical model is a three-dimensional tool that is very helpful for engineers and scientists to understand the relationships between the various components of a particular design, how various structures of a whole scheme interact, or to imagine new solutions. However, in a more broad sense, the usefulness of a physical model can be seen in four areas: (1) *technical*: to ensure an appropriate and eDcient design, (2) *economical*: improve project optimization and cost savings, (3) *safety*: lower risk for both the hydraulic structures and the potentially impacted populations, and (4) *communication*: convey concepts and designs with technicians, residents, nongovernmental and governmental organizations and others, as necessary to aid in the understanding and acceptance of the project and of its hydraulic impacts.

Organizations have developed categories for physical models that include structure types, study/design objective or the degree to which certain physical processes can be modeled. Understanding the applications and limitations of a particular model design and study plan is critical for development of an appropriate model and presentation and defense of study results. For example, the U.S. Army Corps of Engineers (USACE) Committee on Channel Stabilization (CSS) has developed five categories for movable bed models that include, in general terms, (1) demonstration, education, and communication; (2) screening tool for alternatives to reduce maintenance or dredging; (3) screening tool for alternative channel and navigation alignments; and (4) screening tool for environmental evaluation of river modifications. Note that alternatives 2, 3, and 4 carry the caveat that improper predictions of system performance or response would not cause any potential project damage or danger to human life. Finally, category (5) is for a screening tool to study major navigation problems, or around structures and bridge approaches where the failure to perform as predicted in the model studies may endanger human life or be damaging to the project.

30.1.3 Physical Model and Mathematical Model: How to Choose, How to Mix?

When dealing with a hydraulic problem, two main questions must be asked in order to assess the interest and even the necessity of a scale model: (1) Is the hydraulic system suDciently complex to design (sizing and setting), with typically 3D flows (sometimes 2D), but not too large a domain of interest? (2) Is there a need to model sediment transport (bed load and/or suspended sediment transport) or at least scouring?

For many complex hydraulic structures, which are relatively small in size (spillways, energy dissipation basins, breakwaters, locks, water intakes or diversion structures, vortex shafts, flip buckets, plunge pools, etc.), 3D mathematical models may not be precise enough and are costly and computer time intensive. On the other hand, in the design of larger systems such as lateral embankments of a long river reach or the study of 2D flow in an estuary, mathematical models are obviously more adequate in terms of suDcient precision, cost-effectiveness, and rapidity of tests. However, in spite of advances in computational capabilities, the time scales that can be simulated in 3D numerical models may still be a limitation. Most physical modeling studies can be easily justified from an economic standpoint. In most cases, the cost of physical model design, construction, and tests is typically only a small percentage of the total project costs and the benefits (e.g., simplifying and optimizing the design, improving safety, reducing total project costs) are typically very high, particularly when one factors in the ability for project managers, decision-makers and the public to "see" how the system will work.

When dealing with sediment transport and erosion/deposition problems in mountainous torrents as well as in estuarine areas or in rivers with bed load or sediment load transport, the hydraulic scale model is also of particular interest. For studies of deltas, estuaries, reservoirs, or large river reaches, even if the model domain is large, the hydraulic scale model may be used due to the application of scale and sediment distortion principles. In these cases, the scale model is typically designed to primarily address the sediment problem and therefore, not all of the processes related to the hydrodynamics and hydraulic design are simulated. The hydrodynamics of these systems can be simulated using 2D mathematical models and detailed design questions can be addressed via larger-scale (but smaller in domain) physical models. For example, in France the Seine estuary embankments and the dredging schemes for navigation improvement and environment compensation and the Mont Saint Michel Bay study for the restoration of the maritime character of the site have been carried out with both 2D mathematical models for the hydraulic analysis of the flow patterns and with hydraulic sedimentological scale models for the sediment transport, morphodynamics and deposition/erosion studies.

In many cases, there is a close and complementary link between mathematical and the hydraulic scale modeling efforts—the two modeling approaches should not be thought of as competing. The mathematical model of a larger domain can give the boundary conditions for the hydraulic scale model focused on a more complex detailed local structure. For example, the downstream stage-discharge law of a dam spillway study can be obtained from a 1D mathematical model of a long reach downstream from the dam, or the upstream discharge distribution in a flood plain where a new highway with bridges is planned can be obtained from a 2D mathematical model of the whole valley. In that latter case, the scale model can provide the head loss details at the bridges and culverts under the highway that are crucial in the 2D mathematical model and, depending on the scale, can provide details on the smaller-scale hydraulics and scour around the bridge components.

30.2 Principles

30.2.1 The Main Principles

We can identify four main principles to ensure useful and reliable similarity between a model and its prototype: the principle of reality (i.e., the possible size of the model), also known as geometrical similarity; adequate reproduction of the hydrodynamics; accurate reproduction of the hydrodynamic forcings or dynamics similarity; and the good reproduction of the action of the water on its boundaries. Full similitude occurs when all are satisfied.

30.2.1.1 Principle of Reality

In order to ensure satisfactory working conditions, to protect and control for wind and other meteorological conditions and/or to better control the model system, it is often necessary to build a model inside a laboratory. While this often limits the physical size of the model, the ability to control the conditions is critical for ensuring quality results. Note that even an outdoor model faces limitations with regard to the size of the physical model. The physical size limitations imply the necessity of a practical and economical "geometrical similarity" that allows for the construction of a model at a reasonable cost, within the available space, but is perfectly rigorous in scientific terms and maintains

suDcient precision for the measured parameters consistent with the study objectives. Geometrical similarity means that the corresponding lengths L_m in the model and L_p in the prototype are given by the ratio $E(L) = L_m/L_p$, and that the corresponding angles are unchanged. In the case where space may be limited (particularly, for studying sediment transport in large systems), the model may be distorted; i.e., the vertical scale $E(H)$ is different from the horizontal scale $E(L)$. The model distortion is then given by $\Delta = E(H)/E(L)$. Distorted models will be discussed in a later section.

30.2.1.2 Reproduction of the Hydrodynamics (Flow and Wave)

In an undistorted model, the three-dimensional water motion (e.g., horizontal current patterns, vertical velocity profiles, secondary currents, height, and period of waves) must be accurately reproduced. In a distorted model, the vertical component of the flow field is typically not accurately reproduced. However, in a distorted model it is critical that the 2D horizontal water motion (e.g., horizontal current pattern, water surface elevations) be validated. Reproduction of the hydrodynamics is ensured through "kinematic similarity." If V is the velocity of a particle covering a small distance L, then the velocity can be found from $V = L/T$ and the velocity scale is given by $E(V) = E(L)/E(T)$ where $E(T)$ is the time scale.

30.2.1.3 Reproduction of the Forcings

Appropriate and relevant forcings must be modeled correctly to ensure the capacity of the model to predict the impact on or due to hydraulic structures. Scaling of the forces is achieved through "dynamic similarity"; i.e., the important forces impacting the hydrodynamics are in one to one ratio at the corresponding points of the prototype and the model, with corresponding angles unchanged. Forces that are considered include: inertial forces, F_i; gravity forces, F_g; pressure forces, F_p; viscosity forces, F_v; turbulence forces, F_i; and surface tension forces, F_a. Theoretically, the similarity in the patterns of forces should lead, in vector terms, to the scale ratio for any of the forces, F_j, as

$$E(F_j) = \frac{F_{jm}}{F_{jp}} = E(F_g) = E(F_p) = E(F_v) = E(F_i) = E(F_a)$$

To gain further insight into how the forces are scaled, consider a small cube-shaped volume of water Ω, of section dS, of mass m moving in the fluid along a horizontal curved path of radius r, at a depth h under water surface. The flow is assumed turbulent as is usually the case in a large majority of natural flows and we call u', v', and w' the turbulent velocity fluctuations at the particle location along the main direction and perpendicular to fluid flow. As the model is subjected to the gravity forces and typically uses the same fluid (i.e., water) as the prototype, the scale factors for gravitational acceleration, g, the fluid density, ρ, the dynamics viscosity, μ, and the surface tension, σ, are equal to 1; i.e., $E(g) = E(\rho) = E(\mu) = E(\sigma) = 1$. We can then express the forces and their scaling (for undistorted models using the same fluid) as follows:

$$F_i = \rho \Omega \frac{V^2}{r}; \quad E(F_i) = E(L)^4 E(T)^{-2}$$

where

$F_g = \rho \Omega g$; $E(F_g) = E(L)^3$

$F_t = \rho dS[u'v']$, where $[u'v']$ is the average of the product of the fluctuations along longitudinal and vertical direction through dS; $E(F_t) = E(L)^4 E(T)^{-2}$

$F_v = \mu dS \, dV/dn$ which is a result of shear stresses acting on dS where dV/dn is the velocity gradient perpendicular to dS; $E(F_v) = E(L)^2 E(T)^{-1}$

$F_p = \rho ghdS$, i.e., the pressure force acting on dS by the water column h; $E(F_p) = E(L)^3$

$F_\sigma \, \sigma L$, which only acts on particles at the water/air interface; $E(F_\sigma) = E(L)$

Note that for maritime studies, the use of saltwater in the model is usually avoided due to such practical reasons as corrosion problems, the large quantities of salt needed to match the natural salinity levels and the high salinity values in the wastewater. In these cases, corrections are typically made to the density of the solid element(s) to ensure adequate scaling of the forces acting on the structures and natural or artificial unit blocks.

30.2.1.4 Scaling of Forces

In order to ensure dynamic similarity, the *important* forces for the problem being studied must scale identically between the model and prototype (Chapter 1). Thus, if F_1 and F_2 are the dominant forces, then $E(F_1)/E(F_2) = 1$. The application of this concept to a problem where the inertial and gravity forces are dominant leads to $E(T) = E(L)^{1/2}$ and $E(V) = E(T) = E(L)^{1/2}$. If the flow is turbulent in the same way as the prototype, the turbulent forces are in the same ratio and $E(F_i) = E(F_g) = E(F_p) = E(F_t) = E(L)^3$.

For most open channel (free surface) flow problems, the inertial and gravity forces are most important and the ratio of these two forces is called the Froude number and is defined as $Fr = V/\sqrt{gH}$ where H is the hydraulic depth and $E(Fr) = 1$ for dynamic similarity.

In most prototype systems, the viscous and surface tension forces are typically negligible compared to the other forces. Provided this assumption holds in the model, it is not necessary to reproduce the same scale factor for these forces. However, during model design, it is very important to understand and study the relative importance of these two forces since the reduced scale may be such that they are no longer insignificant or are no longer in the same regime as in the prototype system. For open channel flow problems, the relative importance of the inertial and viscous forces is given by the Reynolds number, $Re = 4VR_h/\nu$, where R_h is the hydraulic radius of the wetted area of the cross-section in the flow, V, the mean velocity, and $\nu = \mu/\rho$ is the kinematic viscosity. Re is also used to characterize the turbulent regime.

30.2.1.5 Reproduction of the Action of the Water on Its Boundaries

Flows and waves cause the forces on solid boundaries; e.g., incipient motion and transport of sediment, forces on the obstacles such as gates, breakwaters, ships, etc. The flow dynamics in cases of submerged objects are mostly governed by inertial and viscous forces and, therefore, characterized by some form of the local Reynolds numbers, Re, depending upon the object size and shape. Other important forces that might need to be scaled include gravity and pressure forces, turbulent forces.

30.2.2 Limiting Conditions

30.2.2.1 Turbulence, Friction, and Reynolds Number

Typically, the most important limiting condition that must be satisfied for good modeling is the turbulence regime. For flows in pipes, lined canals, and culverts and tunnels, the friction head losses, J, are usually characterized by the dimensionless friction factor, λ, of the Darcy–Weisbach equation:

$$J = V^2 \frac{(\lambda/4R_h)}{2g}$$

λ is given by the Moody diagram as a function of the relative roughness of the pipe and the flow Reynolds number. Appropriate scaling requires that the dimensionless number λ should be the same in the model and in the prototype. This is the case if the flow in the model is rough turbulent since, from the Moody diagram it is obvious that in the rough turbulent regime, λ is not dependent on the Reynolds number value (i.e., Reynolds number similarity). This is fortunate in that for most hydraulics applications, Re is not equal in the model and in the prototype. However, if the geometric scale factor is too low, the flow regime in the model may be smooth turbulent instead of rough turbulent resulting in greater friction head losses in the model than in the prototype. Consequences include greater model water depths (compared to the prototype) for a given discharge or lower model discharge values for the same available head. Thus, if the purpose of the model study is to size a hydraulic structure, it is obvious that the results will be completely erroneous, or at least biased.

For natural flows, the rough turbulent friction head losses are usually characterized by the Keulegan equation for rough channels:

$$\frac{V}{u^*} = 6.25 + 5.75 \log\left(\frac{R_h}{k_s}\right) \quad \text{with} \quad u^* = \sqrt{ghI}, \, Re > 2000$$

$$\text{and} \quad \frac{u^* k_s}{\nu} > 70$$

where

k_s is the bottom roughness
u^* is the friction velocity
I is the water surface gradient

The condition, $E(V/u^*) = 1$ implies that $E(k_s) = E(R_h) = E(L)$. Therefore, the model roughness should be made at the geometrical scale. Note, that the conditions of rough turbulence still have to be checked through the Reynolds number.

30.2.2.2 Surface Tension Effects

The surface tension forces, resulting from molecular forces and acting at the liquid-gas interface, are typically very weak compared to other forces. The possible impact of the surface tension in the model is characterized by the Weber number, defined as $We = V/\sqrt{\sigma/\rho L}$. However, surface tension effects may become problematic if the scale of the model is too small. Some important cases where the influence of surface tension may be important include the discharge coeDcient of a model weir if the water depth over the crest is not suDcient; the minimal dimension of a bottom outlet for the good modeling of the discharge coeDcient and of the downstream jet shape; the vortex formation and vortex type reproduction (hydropower plant, pumping station); and the free-falling jet over a spillway crest and the consequences of the jet impact. Typically, scale models are used to design hydraulic structures at high discharges; in these cases, the scale effects of surface tension at low discharges can be neglected. However, if the study of low discharges is important or if the large size of the work necessitates a small scale, the risks of scale effects must be taken into consideration. This is a critical problem in geophysical modeling where the aspect ratios (horizontal/vertical scales) are large. Some examples are given next.

When studying the flow over a sharp-crested sill and/or a broad-crested weir, the influence of surface tension on the discharge coeDcient becomes negligible if the water height above the crest is greater than 15 mm, thus setting a minimum value for valid measurements. Studies have shown that the minimum value for the proper contraction coeDcient reproduction of flow through a sluice is 6 cm. Comparisons of a free-falling jet of a high arch dam overflow at different scales shows a different behavior of the downstream nappe and, consequently, of the peak dynamic pressures in the plunge pool.

Vortex formation and characteristics are often important processes addressed using scale models. However, a problem arises when a vertical flow toward an orifice is superimposed on a horizontal flow (which is typically a free-surface flow, but a vortex may be observed also in pressure flows). The reproduction in the model of the precise vortex shape, including air entrainment from the free surface, is usually limited by the surface tension, thus causing the model to underestimate the risk of vortex formation. Two approaches are now discussed. If good reproduction of vortex shape and air entrainment is important, as is the case for pumping stations studies, particular conditions concerning the Reynolds number ($R = Q/\upsilon S$) and Weber numbers ($W1 = \rho V^2 D/\sigma$, $W1 = \rho V^2 S/\sigma$) have to be checked and compared to values found in specialized literature. Here, S is the pump submergence and V the velocity at the intake mouth. As the unit pump discharge will usually range from 0.5 to 5 m³/s, the model scale will have to be between 1/5 and 1/25 to fulfill the Reynolds and Weber conditions. In the important case of the water intake of a low-head hydropower plant, where vortex formation is to be avoided, the scheme size typically prohibits the use of such large model scales. Because precise reproduction of the core development of the vortex is necessary, it is much more important to thoroughly reproduce all the geometrical approach conditions (which are the cause of the vortex formation by the vorticity induced in the approach current pattern). Usual scales for these studies are between 1/30 and 1/60.

30.2.3 Distortion

Model distortion is often used for studying large domains or for 2D problems with the understanding that local 3D effects will not be reproduced well. The model distortion, Δ, is defined as the ratio of the horizontal model scale, $E(L)$, to the vertical model scale, $E(H)$; i.e., $\Delta = E(L)/E(L)$. One of the primary concerns with distorted models is maintaining the Reynolds conditions that provide adequate friction head losses in the model (i.e., in most cases, the rough turbulence conditions must be fulfilled); this is not as easy as for undistorted models.

Froude scaling with the different horizontal and vertical scales leads to the following scales for distorted models:

$$E(V) = E(H)^{1/2}, \quad \text{where } V \text{ is the horizontal velocity}$$

$$E(Q) = E(H)^{3/2} E(L), \quad \text{where } Q \text{ is the discharge}$$

$$E(I) = \frac{E(H)}{E(L)} = \Delta, \text{ where } I \text{ is the water surface longitudinal slope}$$

The hydraulic radius scale, $E(R_h)$, is not a constant but will vary with the depth of water h. Examination of the Darcy–Weisbach equation shows that the friction head loss scale is a function of $E(R_h)$ and will also vary with h, and so does the roughness, k_s, scale.

$$E(k_s) = f\left(E(R_g(h)), \frac{k_{sp}}{R_{hp}}, E(h), \Delta \right)$$

Obviously, it is not possible to change the roughness in the model for each hydraulic condition. In practice, if we study the above function for typical river water depths, widths, and natural roughness, it is usually observed that for distortions less than 2.5–3, the use of a unique roughness is possible. However, it is important that this roughness be computed before model construction, possibly tested in flume, and varied as the result of model tests during a calibration phase. For these distorted hydraulic models of large size, typical horizontal scales are 1/100 to 1/300 and vertical scales are 1/50 to 1/100.

30.2.4 Sedimentological Models

There are two kinds of sedimentological models:

1. Models designed to study detailed problems such as scouring around an obstacle or downstream from a weir or spillway. In many of these studies, sediment transport is usually not represented since the objectives are the maximal scour depth, the shape of the scour hole or the

necessary structural protection. This class of models are never distorted as secondary 3D currents and vortices are the cause of the scouring and features such as the lateral slopes of the scour hole must be well reproduced.

2. Sediment transport models used to study the impact of one or more hydraulic structures on the site morphodynamics (e.g., torrent or river reach, whole estuary or portion of delta, beach section).

In sediment studies, one must be aware of potential issues regarding the sediment/water mixture and the kind of sediment used in the study:

1. Two-phase Newtonian flow: water and sediment. If the model sediment is not too fine at the geometrical scale ($d_{50} > 0.5$ or 0.6 mm), natural sand may be used in the model to represent boulder, gravel, or coarse sand of the prototype. The limit of 0.6 mm is typically used in the model design to avoid the formation of ripples that could completely change the friction head losses and drastically alter the sediment transport. If at the geometrical scale the model sediment is too fine (potentially creating problems due to sediment cohesion), artificially distorted sediment shall be used.

2. One-phase non-Newtonian flow: homogeneous mixture of highly concentrated sediment in water. This approach is typically used to study systems such as the mudflows observed in mountainous regions (Andes, Alps, Volcanoes like Pinatubo, etc.) that exhibit very high discharges and energy.

While Froude similarity is applied to the model for the hydrodynamics, another similarity law must be added concerning the inception of the sediment motion. This relation, called "Shields similarity," is based on the Shields curve that characterizes the beginning of the sediment motion according to two dimensionless parameters (Chapter 34): the critical sediment Reynolds number $Re_{cr}^* = u_{cr}^* D / \upsilon$ and the Shields parameter $\tau_{cr}^* = \tau_{ocr}/(\rho_s - \rho)gD$, where u_{cr}^* is the critical friction velocity of the sediment grain, $u_{cr}^* = (\tau_{ocr}/\rho)^{1/2}$, D is the sediment grain diameter, τ_{ocr} is the critical shear stress for which the sediment grain is lifted or dragged by the flow on the bottom of the river (or by the waves on the sea bottom), ρ is the water density and ρ_s is the grain density.

Examination of the Shields curve shows that for high values of Re^* (i.e., $Re^*>100$ for full turbulence around the grain), τ_c^* does not depend on the value of Re_c^*, so the similarity of the sediment incipient motion will be obtained under the condition $E(\tau_c^*) = 1$. In this case, the following scale models may be defined by applying the Shields similarity:

ft Undistorted model, undistorted sediment. The size of the model and the size of the natural sediment do not require any kind of distortion. Natural sand or gravel is used at the geometrical scale, $E(D) = E(L)$. Torrents are always studied under this condition.

ft Undistorted model, distorted sediment. The size of the study site does not need distortion, but the fine sediment size needs a distortion of the density to allow for suDcient model grain size to maintain $Re^*>100$ and to avoid undesirable ripples. Here, the grain size scale is $E(D) = f(E(L), E(\rho_s - \rho))$. This could be the case for a river reach or a storm water canal.

ft Distorted model, distorted sediment. The site size requires a geometric distortion Δ and the requirements to maintain a $Re^*>100$ and a suDciently large grain size dimension lead to a sediment distortion of density, $E(D) = f(E(\rho_s - \rho), E(H), \Delta)$. This is typically the case when studying estuaries, deltas, large reservoirs, or large river reaches.

If it is not possible to reach a suDciently high value of Re^* in the model, the similarity condition still needs to imply the conservation of the two dimensionless numbers, $E(Re^*) = 1$ and $E(\tau_c^*) = 1$. In that case, it is not always possible to find a solution and the choice of the distortion will be dependent on the sediment density or the sediment density will be chosen according to the model distortion. In practice, it is advisable to choose and test the artificial sediment properties (i.e., material and size) by preliminary measurements in a flume. In addition, it is highly recommended to use complete grain size curves in order to represent important sediment transport behaviors such as bed armoring, sediment sorting, bed friction, and sediment transport.

Once the sediment properties are chosen, the sediment discharge has to be correctly adjusted. The sedimentological time scale can be defined as the time scale required to fill in a known volume (e.g., pit). If the model and the sediment are not distorted, the sediment transport law will be well reproduced and the sedimentological time scale (t_s) will simply equal the hydraulic time scale; i.e., $E(t_s) = E(t)$. In the case of a distorted model, the sedimentological time scale will be different from the hydraulic time scale, mainly because of density distortion and because the model sediment transport law is different from the prototype; however, in all cases, $E(t_s) < E(t)$. This property is very interesting and proves to be very useful to test or study long-term morphodynamic evolution. In this case, it is necessary to calibrate the sedimentological time scale based on observed bed evolutions. This sedimentological time distortion applies for river as well as estuary or beach models. In that last case, the time scale reduction for the sediments will dictate the number of the tides reproduced in model, knowing that each tide is reproduced at the hydraulic time scale. For example, the following sedimentological time scales have been used for these studies:

ft Seine Estuary in France ($E(L) = 1/1000$, $E(H) = 1/100$): 1 year of prototype is represented by 5.6 h in the model and 45 model tides represent the annual cycle

ft Mont St Michel Bay in France ($E(L) = 1/400$, $E(H) = 1/65$): 1 year of prototype is represented by 6.25 h in the model

ft Lower Mississippi River in the United States ($E(L) = 1/12,000$, $E(H) = 1/500$): 1 year of prototype equals 30 min in model.

ft Rio Madeira in Brazil ($E(L) = 1/1000$, $E(H) = 1/100$): 1 year of prototype equals 2 h and 45 min in model.

30.3 Case Studies

This section contains case studies that contain some details that illustrate how proper model design along with a carefully planned testing program is able to provide insights or improve the performance and economics of a particular system or design.

30.3.1 Mardyck Inland Navigation Lock 1/12.5 Scale Hydraulic Model Studies

Context:

The Mardyck Inland Navigation Lock in Northern France near the Port on Dunkirk differs from most locks that are designed and operated for overcoming major difference in water levels between two canals or river reaches. The Mardyck lock is used for separating a saltwater canal connecting the Port of Dunkirk to a freshwater inland canal that is also used for irrigation and industrial use. Standard practice for dealing with this situation involves separation of the two canals by two locks each emptying into an intermediate low-level canal full of brackish water, which in turn flows into the sea at low tide. This arrangement has several drawbacks such as (1) the need for two navigation locks, (2) higher construction costs, (3) increased locking time, and (4) intrusion of a salt wedge into the freshwater canal.

Based on prototype tests in Holland at the Terneuzen navigation lock, the Dunkirk Port Authority asked SOGREAH to investigate the possibility of replacing the existing two-lock arrangement with a single lock. In the prototype, Dutch engineers were using an air bubble curtain to prevent the salt wedge from penetrating into a freshwater canal. After assisting in the prototype tests and discussion with the engineers at Delft Laboratory, it became evident that the efficiency of the air bubble curtain was limited and could not be applied at Mardyck, where any accidental increase in salt wedge penetration into the freshwater canal would be unacceptable. SOGREAH hydraulic engineers came up with a lock concept that satisfied the basic requirements of a solution using only one lock (Figure 30.1) and the navigation authorities agreed to further studies.

Two Physical Model Experiments:

Two models were built to test this design. First, a general model scaled at 1:12.5 (84 m long) was designed to reproduce the entire lock structure, filling system, and freshwater inflow and outflow systems. In that model, the water density used was the same as the density of the natural saltwater system. Second, a partial model was designed to look at a single jet mixing with freshwater for analysis of the interface thickness. To assess the model scale effect on the mixing zones, special comparative tests using a 1/1 scale model and a 1/10 scale model of saltwater jet injected into a column of freshwater were carried out. Additional details concerning the model design and testing can be found in Blanchet et al. (1963).

The Conceptual Design and Operating Modes of the Proposed Mardyck Navigation Lock:

The design concept basically relies upon the density difference between saltwater and freshwater. Replacement of freshwater from the lock is accomplished by injecting saltwater (1035 g/l) from the bottom of the lock by pumping through an elaborate hydraulic conduit and perforated slabs system. Injection of the denser saltwater from the bottom ensures a uniform distribution of saltwater over the entire bottom surface of the lock. The jet velocities, diameters, and their distributions are designed to minimize the mixing at the interface between the freshwater and the saltwater and tested using the 1/1 scale model and a 1/10 scale model of saltwater jet injected into a column of freshwater. These tests confirmed that it would be possible to keep the mixing to a minimum and a reasonable interface depth. With increasing depth of saltwater at the lock bottom, the pumping discharge can be increased and the saltwater acts as a liquid piston and forces freshwater out of the lock through the siphon ports all along the two large lateral aqueducts. Once all the freshwater within the lock chamber is replaced, the pumping is stopped and the lock is opened to the saltwater canal. To replace the saltwater in the lock with freshwater, the pumping is reversed and the salt water is pumped out of the lock from the bottom and lowering of the water surface in the lock makes the freshwater flow in

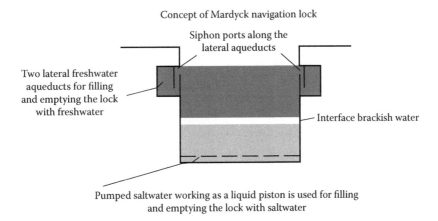

Concept of Mardyck navigation lock

Siphon ports along the lateral aqueducts

Two lateral freshwater aqueducts for filling and emptying the lock with freshwater

Interface brackish water

Pumped saltwater working as a liquid piston is used for filling and emptying the lock with saltwater

FIGURE 30.1 Schematic cross section of the Mardyck navigation lock and its mode of operation.

FIGURE 30.2 General view of the Mardyck navigation lock; freshwater inland canal is to the left.

through the siphon portals along the lateral aqueducts. As in the previous stage, the surface turbulent mixing between the saltwater in the lock and the freshwater surface jets flowing in is very critical and is achieved by controlled pumping discharge emptying saltwater.

Conclusion:
State-of-the-art hydraulic modeling allowed for successful testing and verification of the performance of an innovative navigation lock design connecting a seawater canal to a freshwater canal. This enabled (1) replacement of two locks by only one, (2) significant reduction in construction costs, (3) reduction in total locking time, and (4) elimination of any salt wedge intrusion into the freshwater canal. The Mardyck navigation lock (Figure 30.2) has been in operation since 1965 and is performing satisfactorily.

30.3.2 Sidney A. Murray Jr. Hydroelectric Project 1/100 Scale Hydraulic Model Studies

Context:
The Sidney A. Murray Jr. Hydroelectric Project is located at the Old River Control Complex of the Lower Mississippi River (Louisiana, USA) and takes advantage of the proximity of the Atchafalaya River to the Mississippi River. While both rivers flow to the Gulf of Mexico, there is a head differential between the two rivers at the project site due to the difference in the river lengths to the Gulf. The Mississippi is about 250 km longer, and although the average water surface slope is small, the resulting head differential is significant and varies between 2 and 6 m. The flow, and accompanying sediment, diversion from the Mississippi into the Atchafalaya River creates a complex flow and sediment transport pattern through the diversion project and created the following concerns related to sediment transport, siltation, and erosion:

1. Sediment transport patterns from the Mississippi River to the Atchafalaya River through the intake channel and the power plant.
2. Siltation of the intake channel during exceptionally high Mississippi water level.
3. Erosion of the tailrace channel with low tail-water level and high plant discharge.

Physical Model:
Answers to all three aforementioned aspects were addressed using a 1/100 undistorted comprehensive movable bed model. Sawdust was employed as a distorted model sediment, not very precise but did reproduce the sand transport. Model results were satisfactorily verified in the prototype since 1991 when the project went into operation. Additional details about the model design and testing can be found in Alam et al. (1987).

Intake Channel Layout and Design:
With the initial intake channel layout and cross section (Figure 30.3), it was impossible to obtain a uniform approach flow distribution at the power plant. The combined effect of the Mississippi River flow velocity, large width, and the off-take angle of the intake channel created a concentrated flow along the left bank. The resulting flow velocity distribution at a distance 457 m upstream of the power plant intake is as shown in Figure 30.4. Apart from the asymmetrical approach flow at the power intake that significantly increases the trash rack head loss, the long-term effect of such flow distribution increased aggradation in the model along the right bank and produced deep scour along the left bank. With time, these geomorphologic changes would further increase the asymmetry of the flow distribution and result in increased friction and trash rack head losses.

After testing of various alternative channel layout configurations it was found that, to improve the flow distribution at the power plant intake, the best arrangement consisted of an island at the channel entrance, which imposes a desired flow distribution at the upstream end of the channel. Figure 30.5 shows the channel entrance layout and the cross section that proved satisfactory. With this configuration, the flow distribution in the channel downstream of the island became fairly uniform at a distance 457 m upstream of the power plant intake as shown in Figure 30.6.

The improved velocity distribution completely changes the channel aggradation and erosion patterns observed with the initial channel configuration. With the reduced intensity of the maximum velocity and its location along the center of the channel, the erosive forces at the toe of the left bank side slope was reduced to a minimum. In addition, the improved flow distribution at the power plant intake resulted in a comparative decrease in the channel friction head loss.

Power Plant Tailrace Head Recovery:
In low head power projects, it is worthwhile to try to reduce to a minimum all sources of head loss. Total residual energy head at the draft tube outlet of the power plant is only 0.2–0.3 m. However, compared to the total head of 2–6 m, this is not insignificant. Systematic tests with low-head projects using bulb units indicated that it is possible to reduce tailrace head loss by proper hydraulic design of the tailrace structure, i.e., by reducing as much as practicable sudden changes in the flow velocities between the draft tube outlet and the tailrace channel. Although it is diDcult on a scale model to assess precisely the equivalent prototype head loss, it is possible to compare the head losses between alternative arrangements and use the one with

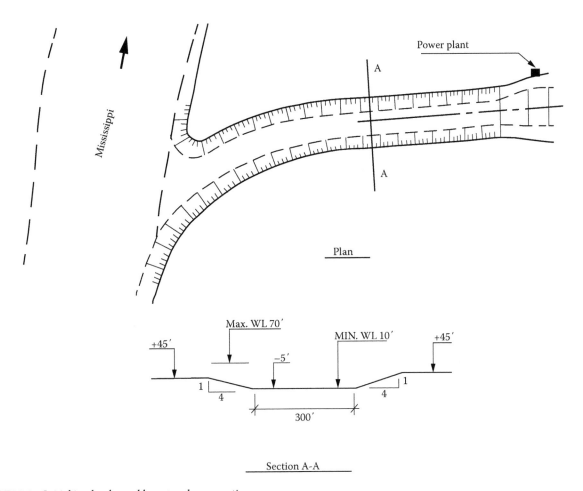

FIGURE3 0.3 Initial intake channel layout and cross section.

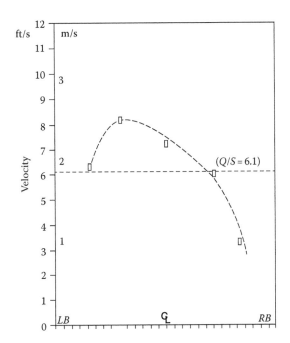

FIGURE 30.4 Typical flow velocity distribution in the intake channel at a distance of 457 m upstream of the power plant (Q_{pp} = 150,000 cfs, HWEL 45.7 ft.)

minimum head loss. Figure 30.7 shows the longitudinal section of the tailrace structure; with all the eight units in operation this arrangement recovered 80% of the velocity head with design plant discharge.

Sediment Issues:

The physical model predicted several phenomena that have since been verified through field data. First, model results showed sedimentation of an abundant part of the outflow channel. Second, during very high Mississippi River stages, the intake channel would be subjected to 1.5–3 m of sedimentation; however, subsequent lowering of the river stage would increase the flow velocities and these deposits would be washed away. Third, the model results indicated eDcient diversion of sediment to the Atchafalaya River. Daily sediment sampling at U.S. Army Corps flow diversion structures and at the hydro station (Figure 30.8) since 1991 has confirmed that on the average 30,000,000 ton of total sediment and 7,000,000 ton of sand is being diverted annually from the Mississippi to the Atchafalaya River; i.e., annually about 24% of water, 21% of total sediment load, and 30% of sand load is being diverted. Finally, the scale model also indicated that an initial tailrace channel invert at −15 ft would be scoured to about −55 to −60 ft; this has been verified in the prototype and has saved about $7 million of channel dredging costs.

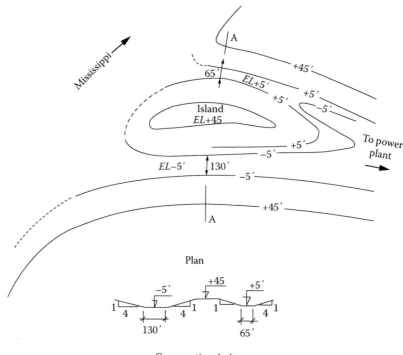

FIGURE3 0.5 Channel entrance layout and cross section with an island.

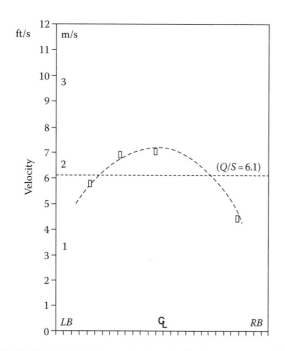

FIGURE 30.6 Typical flow velocity distribution upstream of the power plant with an island at the channel entrance.

Conclusion:

Hydraulic model studies of the Sidney A. Murray Jr. Hydro Station (Figure 30.9) solved diDcult hydraulic and sediment management problems at the project design stage and significantly enhanced its performance; plant power output exceeded that guaranteed by the equipment manufacturers. The model

studies enabled solutions of the intake channel flow velocity distribution at the power plant intake comprising eight 8.2 m Bulb turbines and the velocity head recovery at the powerhouse tailrace. A final note about the scale model costs: the total model construction and testing cost was about $1 million, which worked out to be about 0.2% of the total project cost.

30.3.3 Hydraulic Modeling of Sediment and Fish Egg Passage at Jirau, Brazil

Context:

A scale model was used for a 3750 MW Madeira River run-of-river hydropower project near Jirau, Brazil, for studying the flow conditions, sediment transport, and fish egg movement patterns in pre- and post-dam conditions. The hydraulic model also allows simulation of debris transport and provides an opportunity to optimize an eDcient debris removal system.

Description of the Model:

The model represents 20 km of the Madeira River upstream of the project site and 10 km downstream. The riverbed, with its existing cross sections, is represented as a fixed bed and will simulate the prototype stage discharge conditions and corresponding flow velocities. The model is geometrically distorted at 10 with a horizontal scale of 1/1000 and a vertical scale of 1/100. The velocity, time, and water discharge scales were determined using FroudeR Similarity Laws and the hydraulic conditions necessary for establishing initial movement of the sediment particles were obtained by using Shields incipient motion curve and through flume tests. The precise determination of the

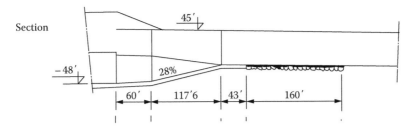

FIGURE3 0.7   Longitudinal section of the power plant tailrace arrangement.

FIGURE 30.8   General view of the old river control complex structures and the hydro station.

FIGURE 30.9   Sidney A. Murray Jr. Hydro Station, intake channel and the island at its entrance.

sediment material characteristics such as sediment material density (1.05) and grain size distribution curve (0.1–0.8 mm) is essential for carrying out tests reproducing sediment transport patterns similar to those observed in the natural system.

This physical model is equipped with the following measuring and control devices:

- ft Automatic systems to generate any kind of hydrograph and to record the discharge in the model continuously
- ft A set of eight echo-sounding water level recording systems to automatically and continuously monitor the water levels at eight locations along the Madeira river
- ft Two manual point gauges to measure and control the model water level upstream and downstream of the model at each discharge step
- ft An automatic laser scanner to survey the model surface area after a test and to survey the river bed allowing for comparison of sedimentation patterns between tests
- ft Seven thin plate weirs to adjust and measure the discharges passing through the power plants
- ft The required number of gates to control and adjust the discharge passing through the spillway
- ft A system to inject sediments according to the natural sediment-discharge laws observed at the upstream end of the model

Model Calibration with Natural Flow Conditions:

Additional details on model design and testing can be found in Cazaillet et al. (2011 and 2012). The model was calibrated first using several constant river discharges and corresponding known water levels at existing gauging stations. Satisfactory reproduction of the river water profiles with clear water required adjustment of local head losses due to the presence of major rock outcrops along the river (Figure 30.10).

Sediment transport tests required reproduction of the annual water discharge hydrograph and corresponding variations in sediment discharges and concentrations. Field measurements indicate that the sediment concentrations go up very quickly during the early stage of the rising discharge hydrograph and by the time the discharge reaches peak flow, the concentrations go down to minimum values and remain there for the rest of the water year. It was noticed on the model that some of the coarser fractions of the sediment materials that were deposited during the receding flood were gradually picked up during the rising discharge hydrographs, very similar to the process observed in nature.

FIGURE 30.10 General view of the Madeira River model in the project area during water profile adjustment tests.

FIGURE 30.11 Photo showing sedimentation in the low flow velocity areas along the right and left bank of the upper pool in the physical model.

Simulation of Flow and Sediment Transport Conditions with the Structures:

After satisfactory verification of the model with sediment transport under natural conditions, the complete project structural infrastructure was installed on the model and testing with accurate simulation of the powerhouse and spillway operations was possible. Operating conditions such as variations of the upper pool level as a function of river discharge hydrographs corresponding to average, high, and low flow water years were reproduced. Due to the increase in average water level by about 15 m and large increase in flow sections, the flow velocities decreased significantly in certain areas of the upper pool. Segregation of the coarser sediment particles and their deposition in the low flow velocity areas was immediate, and the increase in the extent of the deposit surface areas and their height increased with each hydrological cycle. After initial generalized sedimentation in the widest part of the upper pool, a new channel configuration started to develop. Based on the dominant river discharge, the river channel width should be about 900–1000 m, so where the upper pool width is about 3000 m wide, it will gradually fill up to the water surface elevation and the permanent stable channel configuration with its width of about 1000 m will gradually be created. This tendency started to appear after passing five annual hydrographs.

Simulation of Sediment Materials and Their Transport and Deposition Patterns:

The composition of the Madeira River sediment material is as follows: silt and clay 85%, fine sands (e.g., 65–250 μm) 14%, and sands (>250 μm) 1%. The model is designed to reproduce the transport of the sand fraction of the sediment material by the river flow and simulate permanent sedimentation and land building in the relatively low flow velocity areas and stagnant zones (Figure 30.11). However, as the model sediment material did not include silt and clay, the sedimentation process on the model was slower than it would naturally be. Surveys of sediment deposition in the stagnant zone upstream of the cofferdam between 2008 and 2010 (Figure 30.12) indicated very fast

sedimentation rates (about 15–20 m), leading us to conclude that the areas on the model where massive sand deposition has occurred would reach the water surface level of the upper pool faster than that observed on the model. In addition, during the annual lowering of the pool level from 90.00 to 82.50 m, the growth of vegetation and consolidation of the deposited silt and clay would further transition of these areas into firm land. It is expected that annual repetition of sedimentation, drying, and vegetation growth will lead to firm and fertile land formation up to the maximum pool level very quickly and this will also help to accelerate the creation of a deeper stable river channel formation process with an average width of about 900–1000 m even faster than that observed on the model.

Calibration of Fish-Egg Migration Patterns through the Upper Pool from Upstream to Downstream Under Natural Conditions and With the Structures in Place:

The Brazilian Environmental agency, IBAMA, is very concerned about the impact of the project on the fish egg and fish larvae migration patterns due to the modified channel configuration and flow conditions that are expected to occur after localized sedimentation in the upper pool. The Jirau HPP concessionaire "Energia Sustantavel do Brasil" contacted a specialized aqua culture firm to carry out field survey to obtain precise information on the early stage and evolution on fish eggs and fish larvae so that their transport patterns in the river under existing conditions and through the upper pool after the construction of the project could be simulated within the limit of the Jirau physical model. Data were collected on different fish species (*Jaú, Cachara, Curimbatá, Matrinxã, Tambaqui, Pirarara, Jurupensem*) over the seven different times of their life cycle: before fertilization; 1, 4, and 16 h after fertilization; after hatching; 12 h after hatching; and at first feeding. Specially prepared light-density synthetic material with specific gravities varying between 1.02 and 1.03 and diameters between 0.2 and 2 mm was used for simulating fish eggs and larvae. The color of this

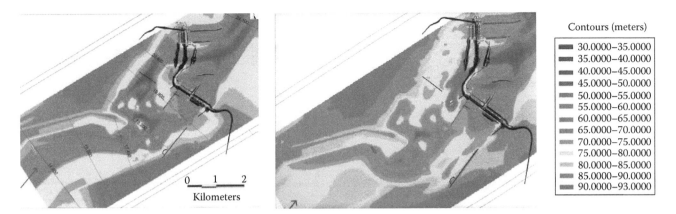

30.0000–35.0000
35.0000–40.0000
40.0000–45.0000
45.0000–50.0000
50.0000–55.0000
55.0000–60.0000
60.0000–65.0000
65.0000–70.0000
70.0000–75.0000
75.0000–80.0000
80.0000–85.0000
85.0000–90.0000
90.0000–93.0000

FIGURE 30.12 **(See color insert.)** Upper pool bed configuration as measured by the scanner: initial (left); after 10 years of operation (middle); and on the left is contour color code indicating sedimentation of the left arm to powerhouse two up to elevations 70–75 m.

material was orange so that they were easily detectable against the sediment material, which was black (Figure 30.13). At the beginning, large quantities of these materials were added to the water discharge at the upstream of the model and it was evident that most of it moved through the upper pool and then downstream of the structures. However, due to the mixing with the sediment, it was very diDcult to follow their migration patterns all the time. The approach chosen to avoid this problem was to run a special test with clear water and inject locally small quantities of fish egg and observe their migration pattern all along the upper pool. Model results indicated that the fish eggs and larvae move through the upper pool and the various structures in practically the same manner as they did in the past under natural conditions. A small percentage of the fish egg and larvae could be lost in stagnant zones where they may be trapped or beached and will never be able to join the main stream and move downstream; this is similar to conditions sometimes observed under natural conditions, during sand bar formation during the receding floods, and dry season.

FIGURE 30.13 Aspects of generalized fish egg transport on the model float down the upper pool upstream of the Jirau Hydro project structures.

Conclusions:

The state-of-the-art hydraulic model studies for the Jirau run-of-river hydropower project on the Madeira River in Brazil provided a clear understanding of the flow conditions, sediment transport and deposition patterns, and fish and fish egg movement patterns under natural conditions. This was accomplished using a relatively small model with distortion of geometric and density scales that was able to capture a large, complex system. The automatic laser scanning system enabled accurate surveying of the model bed surface allowing for precise evaluation of the upper pool sedimentation process and patterns as a function of years of operation. As expected, a large portion of the upper pool area started to fill up rapidly with sediments and gradually a stable channel started to form. The flow distribution became well defined and there were increasingly fewer low-velocity zones. Thus it may be assumed that the fish and fish egg migration process would gradually regain their original patterns in the pre-dam river flow conditions both upstream and downstream of the structures. Other important results of the study were to give recommendations to improve the sediment transport near the dam, to maintain the access channel to the powerhouses and to operate alternatively or simultaneously the two powerhouses (left bank and right bank powerhouse) according to the river discharge over the year.

30.3.4 Study of Water and Sediment Diversions from the Lower Mississippi River Using a Distorted Scale Physical Hydraulic Model

Context:

The Mississippi river one of the largest river in the world, forms one of the worldß largest watersheds, draining approximately 41% of the continental United States. For thousands of years, the Mississippi River was free to meander across what is now the Louisiana coast as it searched for more hydraulically eDcient routes to the Gulf of Mexico and responded to sea level rise. In the past 100 years or so, most coastal wetlands in the Lower Mississippi River Delta have been hydrologically isolated due to construction of levees on the river for flood control and navigation. These measures have prevented the historical flooding and introduction of sediments and nutrients into coastal wetlands, contributing to the high rates of marsh deterioration and subsidence. One of the strategies proposed for reversing this wetland loss is to build diversion structures designed to eDciently deliver river water and sediments into adjacent coastal wetlands, thereby reproducing the historic distributary processes that built the coast and slowing down the rate of wetland loss.

While numerical models have been developed by federal, state, and university scientists and engineers, the complexity of the lower Mississippi River and the long time scales, necessary for making decisions concerning optimal location, impacts on the navigation, and the potential for land building over decades, led the state of Louisiana to invest in a physical model study. The main requirement of such a model was to provide a clear understanding of comparative performance of sediment diversion at various locations along the Lower Mississippi River and, based on the results, choose the best location(s) for further detailed studies and final design. In the past, engineers around the world have used table-top models for relative simple hydraulic problems. But in the case of Mississippi River, the model would have to be fairly complex and would require a state-of-the- art physical hydraulic modeling.

Model Similitude Criteria and Scales:
The Louisiana Department of Natural Resources funded the construction of a small-scale physical hydraulic model of the Lower Mississippi River Delta reproducing about 9132 km² of the lower delta and surrounding Gulf of Mexico and ~100 km of the river above the Head of Passes and capable of simulating, qualitatively, the sand transport patters as a function of river discharge. The distorted model was designed and built in accordance with the Froude similarity law. With 1/12,000 horizontal and 1/500 vertical scales, the application of the distorted scale, distorted sediment equation given previously yielded a sediment material density of 1.05 and to satisfy the particle diameter Reynolds number required a model sediment diameter to be 3.2 times bigger than the prototype sand. Three sand sizes (medium, fine, and very fine) are modeled simultaneously. One of the most interesting features of this model is its capacity to reproduce the Mississippi River sand transport with a sedimentation time scale of ½ an hour equal to 1 yearß prototype time. These time scales allow for the economical and practical long-term simulations necessary to study the land building processes, which might take decades; e.g., 100 years of prototype delta building takes 50 h of model time. More details about the model can be found in SOGREAH (2003) and BCG (2004).

The model was built and verified in SOGREAH Hydraulic Laboratory in France (Figure 30.14), before it was transferred permanently to the Vincent A. Forte River and Coastal Research Laboratory located on the Louisiana State University Campus in Baton Rouge, Louisiana (SOGREAH, 2003). The model was verified hydraulically by comparison of the river stages along the river and discharges through the various passes in the lower river. Sediment transport was verified two ways. First, it was also

FIGURE 30.14   (See color insert.) General view of the model during verification tests at SOGREAH.

observed that the bulk of the sediment material moved before the discharge hydrograph peaked (at around 14,000–17,000 cm), which is very similar to field studies of sand transport in the prototype, and that the corresponding model turbulence was satisfied with a Reynolds number of about 6500. Second, model testing related to the annual sediment deposition and dredging along the Mississippi River and the corresponding dredging volume on the SSPM gave relatively comparable figures of both the volume and the area of dredging. This may be considered as a good verification of the model reproduction of the prototype sediment transport and deposition patterns as well as the sediment time scale.

Verir cation of Comparative E ciency of Sediment Diversion Structures:
In wide rivers like the Mississippi, the impact of secondary currents on the eDciency of sediment diversion may be significant and this aspect was verified by using two identical diversion structures: one at Myrtle Grove (RK 95 Above Head of Passes [AHP]) where the secondary currents are moving away from the river bank and the other at Magnolia (RK 76 AHP) where the secondary currents are moving toward the river bank. The tests were run using identical flow and sediment hydrographs that mimic that annual Mississippi River conditions for a prototype time of 100 years along with a ~0.9 m increase in sea level over that time (Alam and Willson, 2011). Figure 30.15 clearly shows the difference in the quantity of sediment (sand) diverted at Magnolia versus Myrtle Grove. Post experiment collection and quantification of the diverted sediment revealed that the quantity of sediment diverted at Magnolia is six times more than that at Myrtle Grove and that the particle size distribution of the diverted sediment was slightly coarser than the sediment material injected. Finally, the flow and sediment transport patterns in the stretch of the river have been compared to recent field data and found to be fairly representative of the natural conditions. All these results and observations confirm that the secondary currents are indeed entraining the larger sediment particles, which are near the bottom of the river channel.

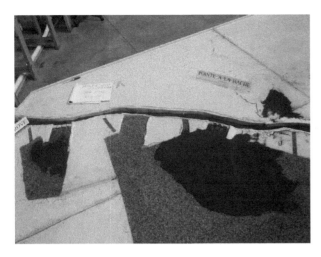

FIGURE 30.15 Comparison of sediment (in black) diversion using identical diversion structure, discharge, and total diversion time at Myrtle Grove on the left and Magnolia at the right.

Conclusion:

The complexity of water and sediment diversion from a river like the Mississippi is such that to date no satisfactory numerical model that could simulate the relevant processes reasonably well has been developed. Because of the large area to be reproduced, the very large quantities of water and sediment to be diverted, and the long time scales necessary for land building, physical hydraulic modeling of such a problem is diDcult. The state-of-the-art small-scale physical hydraulic model developed by SOGREAH Hydraulic Laboratory has, however, demonstrated that for comparison and preliminary decision making purposes, such models may be used with a reasonable degree of confidence and as a valuable tool for guiding further detailed studies and design.

30.4 Challenges and Conclusions

In spite of the increase in computational power and access to fast processing speeds, physical modeling is still an extremely valuable and robust tool. In certain types of situations, computer simulations can be faster and cheaper. There are, however, many cases where the complexity of the geometry, the prototype time scales required for the study, or the need for testing multiple alternatives give physical modeling a distinct advantage. Quite often nowadays mathematical and physical models are used simultaneously as complementary, the physical model being in size a part of a largest mathematical model. But, the ability for someone to personally observe a physical modeling experiment has tremendous value compared to seeing the results of a numerical model shown on a screen.

While hydraulic and, to a large extent, sediment transport physical modeling have proven to be valuable tools, increasing level of detail required in site studies and design (e.g., ecological and environmental impacts) are driving the need for more complex models. For physical models to continue to play a valuable and significant role, the testing methods and data acquisition technologies must also keep pace. This requires a long-term

investment and commitment in laboratory and human resources, oftentimes diDcult due to design budgets and a question about the usefulness of physical models. Thus, is it incumbent upon the engineers and scientists who use physical models to ensure that clients, stakeholders, and the public clearly understand the capabilities and limitations of physical modeling and are encouraged to either passively or actively participate in the project study.

References

Alam, S., C. Meyer, M. David, and G. Pinel, 1987, Sidney A. Murray Jr. Hydroelectric project—Hydraulic model studies, SOGREAH Report, July 1987.

Alam, S. and C.S. Willson, 2011, Progress report on sediment diversion testing on lower Mississippi River, Report submitted to LA Coastal Protection Authority, September 2011.

Blanchet, C., S. Alam, P. A. Kolkman, M. David, and L. Lieutaud, 1963, Ecluse de Mardyck—Etude sur modèle réduit au 1/12,5 des dispositifs spécieux de remplissage et vidange protégeant la branche fluviale contre lflntrusion des eux salées de la branche maritime. Sogreah Report, Mai 1963.

Brown, Cunningham, and Gannuch (BCG) Engineering Consultants, 2004, Report of the feasibility of small scale physical model of the lower Mississippi River delta for the testing Water and Sediment Diversion Projects. State of Louisiana Department of Natural Resources DNR, Contract No. 2503-03-04, December 2004.

Cazaillet, O., S. Alam, G. ExcoDer, 2003, Small-scale physical model of the Mississippi River delta, Louisiana Department of Natural Resources, Sogreah Report, September 2003.

Cazaillet, O., P. E. Loisel, A. L. F. A. Jorge, J. L.Q. Machado, T. F. Soares, and S. Alam, 2011, Modelo reduzido tridimesional do AHE Jirau, Transporte de sedimentos, material flutuante e ovos, larvas e juvenis de ictiofauna, Energia Sustentàvel do Brasil, Sogreah Report, January 2011.

Cazaillet, O., Validité des modèles sédimentologiques fluviaux, Etude de synthèse, Centre dfEtudes Techniques Maritimes et Fluviales (CETMEF), France, Sogreah Report, February 2012.

Cazaillet, O. and J. Vermeulen, 2009, Les problématiques du modèle physique hydraulique et sédimentologique: échelles, effets dflchelle et modes opératoires, Société Hydrotechnique de France, Colloque Modèles physiques Hydrauliques Outils indispensables du 21ième siècle, Lyon, November 2009.

Knauss, J., Swirling flow problems at intakes, Hydraulic structures design manual N°1, IAHR, 1987.

Rui, M. et al., Recent advances in hydraulic physical modelling, NATO ASI Series, Vol. 165, 1989.

Willson, C. S., N. Dill, S. Danchuk, W. Barlett, and R. Waldron, 2007, Physical and numerical modeling of river and sediment diversions, *Proceedings of the Sixth International Symposium on Coastal Engineering and Science of Coastal Sediment Processes*, May 13–17, 2007, New Orleans, LA. Eds. N.C. Kraus and J.D. Rosati, pp. 749–761.

31

Laboratory Modeling

Stuart B. Dalziel
University of Cambridge

31.1 Introduction

Experiments have always played an important role in advancing scientific understanding. While many aspects of environmental flows are "observable," the researcher lacks the ability to separate different elements and processes, which is often necessary to gain the key understanding, test theoretical and numerical models, or make predictions based on "what if" scenarios. Often we have a good idea of the equations describing the physics, but these are too complex to solve analytically and we are forced to utilize simplified systems of "model" equations. Numerical simulations entail approximate solutions to approximate sets of equations: we cannot model directly all the scales of turbulence in any real flow, or even begin to represent accurately the geometry of a porous medium. These limitations do not mean that theoretical and numerical activities have no value, but rather that at least some of their predictions and conclusions need to be tested against observations and against the more controlled conditions found in the laboratory.

Laboratory experiments have their own limitations. While a laboratory experiment is, in a sense, an *exact* simulation using an *exact* analogue computer, our ability to control the experiment or measure the behavior is far from exact. Some of these limitations come from technological restrictions, and some from the simple need to undertake the work at "laboratory scale" (i.e., at much smaller size than the natural phenomenon).

Experiments have only limited value in isolation. It is the interaction between experimenters and theoreticians and/or numerical modelers that imparts value to both approaches. Where possible, this interaction should be stronger than reading each other's papers, as much of the information potentially available from an experiment is effectively filtered out before it appears in print, and the weaknesses of a theoretical or numerical model may not be discussed in an accessible form. The role that is emerging for experiments, particularly to fulfill the needs of the numerical modeler, demands that experiments become more controlled and more quantitative. A place still exists for simple experiments with diagnostics provided by a stopwatch and ruler, but the value of and need for whole-field measurements is increasing.

Verification and validation of other approaches are not the only reason for doing experiments. Sometimes we do them to help train our intuition, or out of simple curiosity: to see what happens. Experiments can yield images of great, if transient, beauty that inevitably proves far more diﬃcult to capture in a photograph than the experimenter imagines. Experiments can reveal hidden depths to a flow that have escaped the artificial world of analytical modeling, or the idealized world of numerical modeling.

In this chapter, we explore laboratory modeling of environmental flows, discussing the mathematics allowing us to obtain useful results at the laboratory scale, providing an introduction

Handbook of Environmental Fluid Dynamics, Volume Two, edited by Harindra Joseph Shermal Fernando. © 2013 CRC Press/Taylor & Francis Group, LLC. ISBN: 978-1-4665-5601-0.

to a range of useful laboratory techniques and technologies, and considering some of the challenges and limitations with this approach to science. Examples of experimental results and the interaction between experiments and other scientific approaches may be found throughout both volumes of this Handbook.

31.2 Nature of Experiments

As is seen elsewhere in this Handbook, environmental fluid dynamics covers a very broad range of flows. Such flows typically include density variations, boundaries, turbulence, and the capacity to affect the environment in which we live. This brief kernel is often extended by free or mobile boundaries, multiple phases (whether solid, liquid, or gaseous), fluids with multiple constituents with different diffusivities, thermodynamical and physical properties, nonlinear mixing, chemical reactions, and complex equations of state. Some level of simplification is almost always essential when designing a laboratory experiment. This simplification is, on the whole, a good thing as it forces the experimenter to concentrate on determining what is important, an activity that inevitably ties the experiments more closely to related theory.

Even with simplified experiments, there remain two extremes of approaches: experiments can be "physical models" that appear like scaled replicas of the environment, or they can be drastically simplified versions of canonical flows that have little visual similarity with the target problem. Both approaches have their place, but it should be recognized that the physical model may well represent the cruder approximation to the motivating environmental flow and will almost certainly be more diDcultC to control, measure, and understand. It may, however, be much easier to "sell" a physical model to sponsors and the general public!

Performing a series of laboratory experiments can be broken down (somewhat arbitrarily) into a number of steps:

1. Reduce the problem into a simpler one that captures the essence
2. Select the control parameters
3. Design an experiment and apparatus
4. Determine the necessary experimental diagnostics
5. Set the control parameters and perform the experiment
6. Analyze the measurements/observations

In general, there will be some iteration through these steps, and the steps are not always executed in order. Although the steps are interrelated, it is worth considering issues associated with each in turn. Here we concentrate on steps 1–4 and discuss issues that need to be considered for a broad range of problems.

31.3 Step 1: Reduction of the Problem

The reductionist approach is central to science and plays a key role in laboratory modeling of environmental flows. Rather than attempting to model the whole of the environment, or even transplanting a part of it to the laboratory, we attempt to divide it into its simplest components so that we may understand them.

Once enlightenment is achieved (or progress has been made), complexity can start to be added back in.

If modeling a volcanic plume, for example, we might begin by considering the buoyant plume rising from a point source in a homogeneous environment. The source of buoyancy might be heat or (more likely) we might design the experiment "upside-down" to use water as the working fluid and add salt to make a dense plume that descends from the source. Establishing that a steady plume forms (in a time-average way) a conical structure, we might investigate how the source conditions (e.g., size, flow rate, density contrast) affect the flow. Each of these parameters is a "control parameter" that we expect to have some effect on the flow. The set of control parameters can also include other properties, such as the viscosity of the fluid or the diffusivity of the stratifying agent, although sometimes it can be diDcultC to change these.

Satisfied that we understand the basic behavior of plumes, we may then wish change the design of their experiment, adding further physical processes to the laboratory model. For example, we might add a stable ambient stratification and observe that the descent of the plume slows with the plume now spreading out at a finite depth. The presence of an ambient stratification adds further control parameters to the problem: the functional form of the stratification as well as its strength can be changed, although we are likely to restrict ourselves initially to changing only the strength. Further control parameters are added as we begin to explore the impact of the composition of the plume, or other features of the atmospheric environment. To mimic the ash from a volcano, we might add particles to the plume and determine how they settle. Atmospheric changes might include a mean flow and turbulence imparted by a grid inserted upstream of the plume. In each case, relatively simple experiments need to be devised to probe the additional physics, first as it impacts the basic conical plume, then how this additional physics interacts with more complex scenarios. It is seldom feasible to undertake experiments throughout the multidimensional space spanned by the experiment's control parameters. In some cases, the observed changes will vary continuously with changes in a control parameter, while in other cases there will be a distinct jump or bifurcation as the flow moves from one regime to another. We might be interested in the behavior within a particular regime, or the location (in parameter space) of the regime boundaries: our objective will influence the manner and order in which to probe the control parameter space, and the manner in which experimental diagnostics are obtained and used.

There is no universal rule of how to reduce a problem. Indeed, the answer is not unique and different strategies will have different merits and provide us with different insights. For a given problem, the choice will depend on many things. The aim of the research and the researcher's background and expertise are clearly important, as are the availability of apparatus, diagnostic equipment, and an assessment of which of the control parameters can really be "controlled." Fundamental differences in physics can also be important. For example, moving from three to two dimensions is often attractive and for some classes of flows can be very effective. However, for others, especially those involving turbulence, the dynamics of a quasi-two-dimensional experiment can differ fundamentally from that of a three-dimensional flow due to the (near) absence of vortex stretching and the associated cascade of energy from large to small scales.

31.4 Step 2: Selecting Control Parameters

Fundamental to all forms of modeling fluid flows is the idea that the flow variables—quantities such as the velocity field $\mathbf{u}(\mathbf{x},t)$, the density field $\rho(\mathbf{x},t)$ and temperature field $T(\mathbf{x},t)$—depend not only on space $\mathbf{x}$ and time t (the independent variables), but also on a set of *control parameters*. These control parameters typically include the geometry of the domain, the nature of any forcing, fluid properties, and some elements of the initial conditions and/or boundary conditions. While a numerical or analytical model might use the values of the control parameters appropriate for the "real" environmental flow, this is seldom possible for a laboratory experiment. Instead, we use the Buckingham π theorem to relate conditions that we can achieve in the laboratory, with conditions relevant to the environmental flows we are modeling.

The Buckingham π theorem (Buckingham, 1914) is based on the ideas of dimensional analysis and similarity developed during the nineteenth century by the likes of Stokes, Bélanger, Reynolds, and Rayleigh. For the purposes of this discussion, we can divide the complete set of dimensional variables into three categories: the m independent space ($\mathbf{x}$) and time (t) variables, n flow variables ($\mathbf{v}$, which might include the velocity, density, and temperature fields), and p control parameters ($\mathbf{P}$, such as channel width, imposed volume flow rate, background stratification, and fluid properties). For the present, we will assume that the set of control parameters is linearly independent and complete, and that the flow is repeatable in the sense that a given set of control parameters will always result in the same flow. In particular, this means that this we can assert (at least in principle) that the flow obeys an equation of the following form:

$$f(\mathbf{x},t,\mathbf{v};\mathbf{P})=0. \tag{31.1}$$

If between these $q = m + n + p$ variables there are r independent dimensions (e.g., length, time, mass), then the Buckingham π theorem shows us there will be precisely $q - r$ independent dimensionless quantities. The theorem does not distinguish between the independent variables, the flow variables, and the control parameters (all are, in some sense, *variables* for the equations describing the underlying physical laws), nor does it fix the choice of dimensionless quantities. However, generally we will choose to *nondimensionalize* the m independent variables ($\mathbf{x} \rightarrow \mathbf{X}$, $t \rightarrow T$) and n flow variables ($\mathbf{v} \rightarrow \mathbf{V}$) in terms of the dimensional control parameters, and thus end up with $p - r$

dimensionless control parameters $\mathbf{\Pi}$. We can thus rewrite (31.1) in dimensionless form as

$$\hat{f}(\mathbf{X},T,\mathbf{V};\mathbf{\Pi})=0.$$

The π theorem shows that two flows that share the same $q - r$ dimensionless variables are identical, except for scale. As we have insisted on a causal relationship between the control parameters and the flow variables, we can make the stronger statement that any two flows will be identical (except for scale) provided the $p - r$ dimensionless control parameters $\mathbf{\Pi}$ are the same, regardless of the value of the dimensional control parameters $\mathbf{P}$. This is the idea of "dynamical similarity."

In practice, we have to broaden the idea of dynamical similarity and relax the insistence that the $p - r$ dimensionless control parameters are all identical. Indeed, when devising an experiment, it is exceedingly rare (if not impossible) for us to establish or consider a complete set of control parameters. For example, the precise texture of a "smooth" flat surface is unlikely to be fully characterized, let alone matched, but ultimately might play some (hopefully insignificant) role in a turbulent flow. Instead, we concentrate our efforts on "matching" those control parameters that we believe are important and, especially for turbulent flows, making our comparisons on the basis of filtered, integrated, or statistical quantities. It is this restricted set of control parameters that will make an appearance in our analytical and numerical models, and that we will use to control our experiments.

Unfortunately, the need to perform the experiment at the laboratory scale, combined with the availability of suitable fluids (and equipment), means that we will actually be able to match very few, if any, of the dimensionless control parameters; so, a knowledge of how important a parameter is and over what range it is important either needs to be known a priori, or we must devise a series of experiments to determine the sensitivity to the control parameters. Indeed, although it is seldom necessary for us to probe the entire $p - r$ dimensions of the dimensionless control parameter space, some variation of a subset of the parameters is generally desirable. For those parameters we are unable to match in the laboratory, we normally hope that the flow will display asymptotic behavior that becomes relatively insensitive to the value of those parameters.

Although the Buckingham π theorem tells us how many dimensionless groups there should be, and provides us with a way of finding a suitable set, the groups themselves are not unique. Convention provides some guidance when selecting the groups, and allows us to make informed choices using previously published material. Some dimensionless parameters (e.g., the Reynolds number) are of nearly universal interest, while others (e.g., the critical Shields parameter, which tells us how readily particles are resuspended) will only play a role in a small but important subset of problems. A complete description of all the dimensionless groups that might arise for environmental flows would require a large volume all of its own, but it is valuable to discuss briefly some of those that are more common and central to environmental flows.

31.4.1 Reynolds Number

The Reynolds number, which represents the ratio of inertial to viscous forces, is one of the best-known dimensionless groupings (see Rott, 1990 for a discussion of the origins). Generically, it can be expressed as

$$Re = \frac{UL}{\nu},$$

where U is a velocity scale, L is a typical length scale of the flow and ν is the kinematic viscosity. At low Reynolds numbers, the flow is laminar, with viscous stresses balancing pressure and body forces. At high Reynolds number, the flow becomes turbulent, with nonlinearities and inertia dominating the dynamics at all but the smallest scales. Turbulent processes lead to a flux of energy from large scales to small scales (the so-called energy cascade), eventually reaching the "Kolmogorov length scale" $\eta = (\nu^3/\varepsilon)^{1/4}$ where viscosity once again becomes important to finally dissipate the energy as heat. (Here, ε is the rate of dissipation.) Not only does the Reynolds number indicate whether such a cascade could exist, but it also provides a measure for the range of scales (from L to η) over which inertia dominates the motion, although it is important to note that the range over which scalar fluctuations occur can be much larger than this as the "Batchelor scale" is smaller than η by a factor of $(\nu/\kappa)^{-1/2}$, where κ is the diffusivity of the scalar.

Although we often talk about *the* Reynolds number, it is quite common for there to be more than one for a given problem. For example, if we consider a particle-laden sewage outfall into a river, we might need to consider three distinct Reynolds numbers as possible control parameters:

$$Re_r = \frac{UH}{\nu}, \quad Re_p = \frac{V_s a}{\nu}, \quad Re_e = \frac{Q}{\pi d \nu_e}. \quad (31.2)$$

Here, U is the mean velocity and H the depth of the river*; V_s is the settling velocity of the particles of size a contained in the ed uent; Q is the volume flow rate through the outfall pipe of diameter d. The kinematic viscosity ν of the river water is likely to be important in the first two, while that of the ed uent ν_e may be important in the pipe and at the outfall. (Indeed, the ed uent may be non-Newtonian in character, and so even ascribing a *constant* viscosity may be misleading.)

The Reynolds number for the river, Re_r, is likely to be large. Mathematically, we might expect asymptotic behavior when $Re_r \ll 1$, and when $Re_r \gg 1$, although what is meant by "$\ll$" (much less than) and "$\gg$" (much greater than) needs some clarification. A more meaningful statement would be that for $Re_r > O(10^3)$, the river flow will probably be turbulent, with a vortical component to the flow that depends sensitively on small details of the upstream conditions and boundaries. Although any ability to match (or control) the instantaneous flow variable fields will be lost at high Reynolds numbers, as the Reynolds number is increased, there is frequently a convergence of statistics averaged across an ensemble. For a (statistically) steady flow, we would also expect convergence of temporal averages, and there may also be some convergence of suitable spatial averages. However, $Re_r \sim 10^3$ is the transition region where we might expect to see quite a strong dependence on the Reynolds number. Indeed, if we had smooth boundaries, the flow may still be laminar until beyond $Re_r \sim 10^4$.

In contrast, if Re_r is "small," then the flow will have a more laminar appearance, although quite how small Re_r has to be will depend in detail on the geometry. We can be fairly certain, however, of a relatively simple flow for $Re_r \lesssim 10$, but few rivers fall in this regime!

Simply matching Re_r for a large river in the laboratory setting by having the flow much faster will not work, as Re_r is not the only important dimensionless quantity that depends on the flow velocity U and depth H. For example, the (external) Froude number is given by

$$F_r = \frac{U}{\sqrt{gH}}, \quad (31.3)$$

where g is the acceleration due to gravity, which might need to be small, imposing additional constraints. On the assumption that we cannot change the kinematic viscosity of water,[K] and that g is fixed, then any attempt to simultaneously match both Re_r and F_r leads to a single solution,

$$U = \left(g\nu Re_r F_r^2 \right)^{1/3}, \quad H = \left(\frac{Re_r^2 \nu^2}{F_r^2 g} \right)^{1/3},$$

demonstrating that the only way in which we can match both Re_r and F_r simultaneously is by performing the experiment at the same scale as the river.

As performing experiments at full scale is seldom possible, it will be necessary to compromise, balancing the desire to match the Reynolds number with desire to match the Froude number, and in both cases asymptotic behaviors can provide us with guidance. For example, if we have a large but sluggish river of depth $H_{river} = 5\,m$ and velocity $U_{river} = 0.1\,m\,s^{-1}$, then the Reynolds number $Re_r = 5 \times 10^5$ is large and we expect the river to be turbulent, while the Froude number $F_r = 0.014$ is small, so we expect waves to be able to propagate easily both upstream and downstream. Exact matching of either of these parameters is unimportant, provided we maintain Re_r high enough for the flow to be turbulent and F_r small enough for the information carried by the waves to behave in a similar way. Suppose the laboratory channel is only $H_{lab} = 0.2\,m$ deep.

---

* For an open channel flow, it is often more appropriate to use the "hydraulic radius," the ratio of the area carrying the fluid to the wet perimeter, rather than simply the "depth" of the river.

[K] While increasing the viscosity of water is not too diDcult, significantly decreasing its viscosity is not possible.

We might choose to reduce the velocity to $U_{lab} = 0.05\,\mathrm{m\,s^{-1}}$, giving $Re_r = 10^4$ and $F_r = 0.036$. Although the Reynolds number is slightly on the low side, we could still ensure we had turbulent flow in the channel by either having a grid upstream to generate an appropriate level of turbulence, or by simply exaggerating the roughness of the channel. (In this way, we treat the upstream "relative turbulence intensity," u'/U, as an additional control parameter.) The Froude number, although a factor of two higher than that for the river, is still very small compared with unity, and so we would expect similar behavior for any surface waves. If there were no other limiting factors (e.g., the capacity of the pump driving the laboratory "river" or some other dimensionless control parameter), we would probably try varying U_{lab} by a factor of two or more in either direction to confirm our initial assessment that the flow is relatively insensitive to Re_r and F_r. (We shall discuss the Froude number further shortly, but for the present, turn to the other Reynolds numbers in this problem.)

The second Reynolds number defined in (31.2), Re_p, is frequently referred to as the "particle Reynolds number" and characterizes the settling of ed uent particles within the river. While the composition of the ed uent may vary tremendously from location to location, the particles will often be small with a density close to that of water, leading to relatively low settling velocities and the possibility the particles may be in a Stokes flow regime with $Re_p \ll 1$. If this is the case, then the exact particle Reynolds number is much less likely to be important than other dimensionless groups such as the ratio of the settling velocity V_s to the river velocity U, or perhaps the ratio of the settling velocity to the intensity of the turbulence u'. If $u' \gg V_s$, then the particles can become well mixed through the depth of the river, with only a gradual decay in concentration downstream of the ed uent outlet. Conversely, if $u' \ll V_s$, then the ratio V_s/U controls the distance a particle can travel downstream before settling from the flow. Of course, this picture is further complicated if we consider a "polydisperse" system with range of particle sizes and settling velocities, and we may choose to restrict our attention to mono- or bi-disperse suspensions. (Although we shall not discuss it here, a need to consider resuspension greatly complicates the matching of experiments with real river flows.)

Our final Reynolds number in (31.2) Re_e, characterizes the flow within the ed uent pipe. Simply applying the same 1/25 scale to the pipe as we have suggested for the river would lead to ridiculously high velocities if we were to attempt to match Re_e, or even if we wished to keep the flow within the pipe turbulent. However, we need not be too concerned about a low value for Re_e if we recognize that the coupling between pipe and river flow is essentially one way, and that details of the pipe flow are unimportant except for their influence on the formation of the ed uent plume. Maintaining the ratio between jet speed and river speed may be more appropriate, requiring $Q_{lab} = Q_{river}\left(d_{lab}^2 U_{lab}\right)/\left(d_{river}^2 U_{river}\right) \sim 2.4\times10^{-5}\,\mathrm{m^3\,s^{-1}}$, giving a Reynolds number Re_e of just under 10^3 in the pipe, suggesting a laminar flow. In practice, we often work with low Reynolds numbers as the sources of turbulent jets and plumes, designing the nozzles

through which they are released to trip turbulence in what would otherwise be a laminar outflow. Expected self-similar behavior of a turbulent plume near the nozzle also reduces the need for geometric similarity, broadening our experimental options.

Although we have illustrated Reynolds number-related issues with reference to a hypothetical river, many of the same considerations apply across the entire range of environmental flows. If we are looking at an airflow, whether on the scale of a room or the atmosphere, then often we will choose to model this in the laboratory using water. There are a number of reasons for this, but one of the most common is that the kinematic viscosity of water ($\nu_{water} \approx 10^{-6}\,\mathrm{m^2\,s^{-1}}$) is an order of magnitude smaller than that of air ($\nu_{air} \approx 10^{-5}\,\mathrm{m^2\,s^{-1}}$), allowing us to achieve higher Reynolds numbers at smaller scales through the use of water. There are other advantages to using water (e.g., simpler experimental diagnostics), but note here that there are, of course, also disadvantages in some cases, while in others substituting a liquid for a gas just will not work.

31.4.2 Froude Number

In our discussion on Reynolds numbers, we introduced the idea of a Froude number (see (31.3)) for a river. The Froude number is the ratio of the velocity of the river to the speed of long waves relative to the water. Here "long waves" means the wave length λ is much greater than the river depth H. Such waves are "nondispersive" (the wave speed does not depend on the wavelength), so that the phase and group velocities are identical at $c = c_p = c_g = (gH)^{1/2}$. These waves also travel faster than short gravity waves on the surface. In practice, we require $\lambda/H \gtrsim 10$ to approach this long-wave limit. If the speed of the waves relative to the water is greater than the speed with which the water is moving relative to an observer, then any waves created by a disturbance are able to propagate both upstream (with speed $c_1 = U - c$) and downstream (with speed $c_2 = U + c$). Such a flow is said to be "subcritical" and has a Froude number $F_r < 1$. If the Froude number is much less than unity, then the observer will see little difference in upstream- and downstream-propagating waves created by a localized disturbance. However, as F_r increases, the observer will notice a Doppler shift in the waves, with the wavelength of the upstream-propagating disturbance becoming shorter, while the length of the downstream-propagating disturbance increases. The upstream-propagating wave comes to rest when $F_r = 1$, and both left- and right-propagating waves being swept downstream for $F_r > 1$. This idea is made clear if we rewrite the Froude number in terms of $c_1 = U - c$ and $c_2 = U + c$:

$$F_r^2 = \frac{U^2}{gH} = \frac{U^2}{c^2} = 1 + \frac{c_1 c_2}{c^2}.$$

The bifurcation in the flow as it passes through $F_r = 1$ forms the basis of "hydraulic control" (see Vol. 1, Chapter 22 (Stratified hydraulics), a phenomenon that is very important not only for river flows, but also (in a slightly modified form) for exchange flow through doorways, flow through oceanic straits, and sometimes over mountain passes (see Vol. 1, Chapter 22).

Maintaining limits of $F_r \ll 1 = 1$ (c_1, c_2 large with opposite sign) or $F_r \gg 1$ (c_1, c_2 large and of same sign) is important for flows with a free surface, but the precise value of F_r may not matter much. However, for $F_r \sim O(1)$, it may well be more important to match the Froude number than (almost) any other control parameter: a flow with F_r slightly less than unity is qualitatively different from one with F_r slightly greater than unity.

Internal density differences in environmental flows can support wave motions even in the absence of a free surface. Indeed, even when there is a free surface, the internal "baroclinic" density variations may be more important. We can again define a Froude number as the ratio of some velocity scale to the wave speed, but its form will differ from F_r given by (31.3). Suppose, for example, we have a two-layer stratification with a lower layer of depth H_1, velocity U_1, and density ρ_1 beneath an upper layer of depth H_2, velocity U_2, and density ρ_2. Here, we define the "composite Froude number" as

$$F_c^2 = 1 + \frac{H_1 + H_2}{H_1 H_2 g'} c_1 c_2 = F_1^2 + F_2^2,$$

where c_1 and c_2 are again the speed of the upstream- and downstream-propagating waves relative to the observer. Here

$$g' = \frac{\rho_1 - \rho_2}{\bar{\rho}} g$$

is the "reduced gravity" (with $\bar{\rho}$ the mean density) and

$$F_1 = \frac{|U_1|}{\sqrt{gH_1}}, \quad F_2 = \frac{|U_2|}{\sqrt{gH_2}}$$

the densimetric layer Froude numbers. As with the free surface case, matching F_c is likely to be important if it is around unity, but unlike the free surface case (where g is fixed), here we may have some additional freedom to change g' (and hence the wave speed) by changing the density contrast between the two layers. Indeed, for laboratory experiments, it is frequently convenient to work with larger density differences (and hence larger g') than found in real environmental flows; the corresponding increase in wave speed allows us to use larger velocities and hence maintain higher Reynolds numbers than would otherwise be the case. If we make the density difference too large, however, the experiment may be contaminated by non-Boussinesq effects.

The structure of the waves in a continuous stratification is more complex again, with the energy propagating at an angle to the vertical, with the phase propagating perpendicular to the energy. A key control parameter here is the "buoyancy frequency" (also known as the "Brunt Väisälä frequency"):

$$N = \sqrt{-\frac{g}{\rho_0} \frac{d\rho_0}{dz}},$$

where $\rho_0(z)$ describes the stable background density stratification. Using the idea that a typical wave speed will scale as a length multiplied by N leads immediately to three distinct Froude numbers for a uniform flow U over an isolated topographic feature of length L and height h in a linearly stratified flow of depth H:

$$F_L = \frac{U}{NL}, \quad F_h = \frac{U}{Nh}, \quad F_H = \frac{U}{NH}.$$

Which one of these is the most important to match will depend on the regime we are in and what aspects of the flow we wish to consider, and the reader is referred to Vol. 1, Chapter 1 for a more complete discussion. (Of course, these three Froude numbers are related through the aspect ratios $A_h = h/L$ and $A_H = H/L$, and so we can consider only three of the set $\{F_L, F_h, F_H, A_h, A_H\}$ as independent control parameters.) We note here, however, that the dispersion relation for these waves, $\omega = N \cos \theta$, limits the wave frequency to $\omega \leq N$, which in turn limits the wave component of the fluid velocity to $u_w \leq N\eta_0$ (where η_0 is the wave amplitude), and so experimentally, we normally use stratifications that are much stronger than those found in the environment in an attempt to maintain an adequate Reynolds number.

31.4.3 Richardson Number

We normally refer to the Richardson number in the context of instability and mixing in density stratified flows (see Vol. 1, Chapters 1 and 20). It represents the ratio of the stabilizing effects of buoyancy to the destabilizing effects of inertia. As with so many other dimensionless control parameters, there is no unique definition. For a continuously stratified flow with a continuous shear (or at least a region of the flow where this is the case), the "gradient Richardson number"

$$Ri_g = N^2 \left(\frac{dU}{dz} \right)^{-2}$$

is arguably one of the most important parameters. Here $U = U(z)$ tells us the vertical structure of the nominally horizontal velocity field, and $N = N(z)$ is the buoyancy frequency of the nominally horizontal density surfaces. While it is well known that a necessary condition for shear instability to develop within such a region is that $Ri_g < \frac{1}{4}$, this is not always useful when designing an experiment. For many flows, the gradient Richardson number ends up as an internal parameter describing the flow that develops, rather than an external parameter under our direct control. More often we seek to control an "overall" or "bulk" Richardson number, which might take the form

$$Ri_b = \frac{\Delta \rho g H}{\rho \Delta U^2},$$

where $\Delta \rho$ is the density difference across a height H over which the velocity changes by ΔU. This form of Ri_b appears a little like a finite-difference approximation to Ri_g, but whereas Ri_g will typically vary with space and time, Ri_b attempts to represent the flow overall. It is important to note, however, that having $Ri_b \gg 1$ does not preclude instability or mixing as it is readily seen that Ri_g averaged over the depth is less than Ri_b for any nonlinearity in the shear, and at a given location, it is possible for Ri_g to be much less (the presence of layers and interfaces in a flow will often create interleaved regions of $Ri_g < \frac{1}{4}$ and $Ri_b \gg 1$).

The expression for Ri_b also looks a little like an inverse-square Froude number. While there are important distinctions, this close relationship can hamper any desire to set the two independently. Indeed, we often try to control either the Froude number or the Richardson number, but not both.

31.4.4 Rayleigh Number

Buoyancy-driven convection often plays an important role in environmental flows, whether due to an isolated source of buoyancy (see Vol. 1, Chapter 25 of *Handbook of Environmental Fluid Dynamics, Volume One*) or a distributed source (see Vol. 1, Chapter 37). Although natural convection from a distributed source is most frequently discussed in terms of thermal convection, other sources of buoyancy (e.g., salinity, suspended particles or bubbles, or compositional differences) have many similarities and can be of value in the laboratory. In each case, the ratio between the destabilizing effect of buoyancy and the stabilizing effects of viscosity ν and diffusivity (of density differences) κ can be described in terms of a Rayleigh number. For thermal convection due to a vertical temperature difference ΔT over a fluid layer of depth H, the Rayleigh number takes the form

$$Ra = \frac{gH^3 \alpha \Delta T}{\nu \kappa_T},$$

where κ_T is the thermal diffusivity and α is the thermal expansion coeDcient. Here, $\Delta \rho = \rho_0 \alpha \Delta T$ is a linear approximation to the change in density due to the temperature difference; this formulation is not appropriate for water near the 4°C density maximum where the density is strongly nonlinear in temperature.

Turbulent convection, which is most relevant for environmental flows, occurs for $Ra \gtrsim 10^5$. As Ra increases beyond this point, the turbulent interior changes from an unstable mean profile toward a weakly stable one (e.g., Castaing et al., 1989), bounded top and bottom by unstable boundary layers giving rise to thermal plumes (Malkus, 1954). Although the Rayleigh number in the laboratory may be much lower than that of the corresponding environmental flow, it is generally possible to reach the turbulent regime, perhaps through the imposition of substantially larger temperature differences. However, when using exaggerated temperature differences, we should be aware that basic fluid properties such as viscosity and heat capacity are also functions of temperature, which may lead to additional complexities when using exaggerated temperature differences, and that the laboratory flow will exhibit a much smaller range of scales.

Often, we will choose to use salt and water in place of heat and air to facilitate reaching the high Ra regime, but we should be aware that there will be differences. The real flow is likely to have boundaries with a finite thermal conductivity and heat capacity, whereas a laboratory tank is likely to be impervious to salt, acting as a perfect insulator. Moreover, the Prandtl number, representing the ratio of viscosity to thermal diffusivity, is $Pr = \nu/\kappa_T = 0.7$ for air, but the analogous Schmidt number for salt and water $Sc = \nu/\kappa_S$, where κ_S is diffusivity of the salt, is a factor of approximately 10^3 larger. For strongly turbulent thermal convection in air, the Batchelor scale is comparable with the Kolmogorov scale, but for salt in water, the Batchelor scale is approximately a factor of 30 smaller than the Kolmogorov scale.

The Rayleigh number may not always be the best choice for a dimensionless control parameter. If working with forced convection, in which there is an imposed flow, we might prefer to work with the Peclet number, $Pe = UL/\kappa_T = RePr$ (ratio of advection to diffusion), Grashof number, $Gr = gH^3 \alpha \Delta T/\nu^2 = Ra/Pr$ (ratio of buoyancy to viscosity), and perhaps the Prandtl number, rather than Re, Ra, and Pr. For other flows, we might prefer to set the Nusselt number, the ratio between the convective heat flux and the flux that could be achieved by conduction alone, rather than a Rayleigh number. This could be achieved by prescribing the heat fluxes rather than temperatures. (Often, however, the Nusselt number is a consequence of the flow rather than a control parameter.)

31.4.5 Effects of Rotation

As noted in the introduction to this Handbook, environmental fluid dynamics spans a broad range of scales. At the upper end the processes start to be influenced by the effects of the Earth⁵ rotation. This occurs in a number of different ways. For example, for a flow driven by internal density differences, we may wish to express one of our dimensionless control parameters as a Burger number,

$$B_{cont} = \frac{NH}{\Omega L} = \frac{N}{\Omega} A_H \quad \text{or} \quad B_{layer} = \frac{(g'H)^{1/2}}{\Omega L}$$

for a continuous (B_{cont}) or two-layer (B_{layer}) stratification, respectively. (Some authors define the Burger number as the square of these quantities.) Here, Ω is the angular frequency of the vertical component of the rotation and L is a horizontal length scale (such as the width of a channel) and related to the vertical scale (H) through an aspect ratio $A_H = H/L$. As the rotation rate Ω has not appeared in any of our earlier control parameters, it is clearly possible to set this independently, at least provided a suitable rotating table is available. We shall assume for this discussion that we can treat Ω as constant with $\mathbf{x}$ and t, even though in reality it is the vertical component that is primarily of interest and this varies with latitude.

The typical value of Ω for midlatitudes is $\sim 5 \times 10^{-5}$ rad s^{-1}, but to model processes in which rotation is important at a 1 m laboratory scale, we will often need to work with a rotation rate of $\Omega \sim O(1)$ rad s^{-1}, introducing a whole extra layer of complexity to the setup. The equilibrium shape of density surfaces is parabolic rather than flat, and we must be concerned with the Ekman number $Ek = \nu/(\Omega H^2)$ and Rossby number $Ro = U/(2\Omega L)$ of the flow. It should be stressed that only one new dimensionless control parameter is introduced by rotation, although the change in the physics might change our preferred set of dimensionless parameters. For example, we can write the Ekman number in terms of aspect ratio, Reynolds and Rossby numbers as $Ek = 2Ro/(A_H Re_r)$, or write $B_{cont} = 2Ro/F_H$. We shall not dwell on the additional complexities but instead refer the reader to Vol. 1, Chapter 29.

31.5 Step 3: Design and Construction of Experimental Apparatus

Laboratory experiments of environmental flows can be very simple. Indeed, often the simplest apparatus that can do the job is the best. This may be as uncomplicated as a pipe providing salt water to form a buoyant plume, or a channel with a removable barrier to initiate a gravity current. Such "appropriate technology" will often be reliable and cheap, but does impose its own limitations.

Basic laboratory procedures can impose limits to what can be achieved readily, and impose constraints on the design and construction of the experimental apparatus. Here we begin our discussion with the formation of density stratifications before considering issues of optical access, the construction of equipment, and health and safety considerations. For further discussion on general laboratory techniques, the reader is referred to Vol. 1, Chapter 32.

31.5.1 Forming Stratifications

Background density stratification plays such a central role in many environmental flows, but forming the stratification can be a time-consuming part of the experimental procedure and impose limits on what can be tackled. The simplest stratification comprises two homogeneous layers of different densities (with the dense layer beneath the light layer, unless we are interested in Rayleigh–Taylor instability!). The procedure is only slightly more complex than "put one layer in and then the other." Assuming we want to maintain a sharp interface between the two layers, then we need to be careful when adding the second layer to minimize the mixing. The order in which we add the two layers depends partly on our equipment and partly on personal preference.

Often the simplest strategy is to add the dense layer, then "float" the light layer on top, adding it through a floating sponge (typically supported by a collar of highly buoyant material such as expanded polystyrene; see Figure 31.1). The sponge allows the fluid being added (typically pumped through a flexible tube) to be dispersed over a broad area, minimizing the mixing between the layers. (It can be useful to place some filter paper or a piece of paper towel directly below the filling tube to prevent the flow penetrating the foam at this point.) It is best to start adding the fluid *very* slowly as this is when the most mixing will occur since the float will be sitting right on (or slightly below) the interface. While the flow rate can be increased as the layer thickness increases, the float must be allowed to rise freely and must remain beneath the filling tube.

Adding the denser layer after the lighter layer is very similar, except that some form of diffuser needs to be installed on the bottom of the tank. Filling from the bottom is particularly attractive if we need to automate the procedure, but filling from the top often allows us to use simpler apparatus. When the filling of the apparatus is complete, it is often possible to sharpen up an interface through careful siphoning of fluid from the interfacial region. A phenomenon known as "selective withdrawal" (see Vol. 1, Chapter 22 of *Handbook of Environmental Fluid Dynamics, Volume One*) means very slow siphoning allows us to remove the mixed fluid within the interface rather than the homogeneous layers above and below.

Filling a tank with a continuous stratification also requires the use of a float or diffuser to control the amount of mixing, although it is not necessary to completely eliminate the mixing. The trick here is to continuously change the density of the fluid being added through the float. If we want to create a complex stratification profile, or we have an odd-shaped tank, then the best strategy is to use a pair of computer-controlled peristaltic pumps—one supplying freshwater and the other salt water—to give precise control of $\rho(t)$ (and hence $\rho(V)$, where $V = V(t)$ is the

(a)

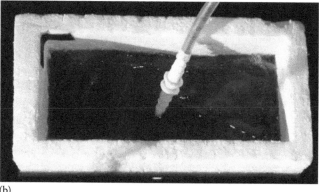
(b)

FIGURE 31.1 Simple foam float for creating a stratification by adding light fluid above dense. (a) Side view near beginning of creating a two-layer stratification. (b) Top view.

cumulative volume that has been supplied by pumps). However, if we are after a linear stratification in a tank with a cross-sectional area that does not vary with height, then Fortuin's (1960) "double bucket" arrangement can be employed. Water to fill the experimental tank is supplied from two identical buckets, connected at their base. One bucket is initially filled with freshwater and the other with salt water. If we are filling through a float, then we gradually draw off water from the salt water bucket. For every liter of water we pump into the tank, we have half a liter of freshwater flow into the salt water bucket; provided this bucket is mixed vigorously, then its density will decrease linearly as $V(t)$ increases. Although this arrangement guarantees $\rho(V)$ in the salt water bucket decreases linearly, we need to be careful to ensure any mixing in the tank is constant if we want to produce a linear stratification. This means we should avoid changing the flow rate through the float if at all possible. (Slight imperfections can be removed through some gentle stirring once filling is complete, or simply waiting some hours for molecular diffusion to smooth the density profile.)

Those new to experiments are sometimes devastated to find that bubbles have formed on the inside of the tank containing their carefully formed stratification. These bubbles roughen the boundary, can induce unwanted flow if they break free, and cause problems with image-based diagnostics. Wiping the bubbles away with a ruler or by hand before the start of the experiment is obvious for a homogeneous fluid, but the experimenter may be reluctant to do this with a stratified tank. Continuous stratifications, however, are surprisingly robust. Using a thin ruler or rod to scrape the bubbles from the walls of the tank is unlikely to lead to a noticeable change in the stratification. With care, the same can be done for stratifications with sharp interfaces, although there will inevitably be some thickening of the interface as a result. The formation of bubbles can be reduced by ensuring that the water has either been deaerated or at least that it has been allowed to reach laboratory temperature and sit for some time *after* the salt has been added but *before* filling the tank. Indeed, although not very environmentally friendly, it is sometimes worth starting with hot water (which will have a lower concentration of dissolved gases) and allow this to become undersaturated as it cools rather than waiting for the gas to diffuse out of cold water as it warms. In either case, if the water has not reached laboratory temperature before filling the tank with a continuous stratification, then layers might begin to form through a side-wall heating (cooling) mechanism (see Vol. 1, Chapters 35, 39), ruining an otherwise good stratification.

31.5.2 Optical Access

Many experimental diagnostics are based on imaging techniques requiring good, unobscured, and undistorted optical access to the flow. The lensing effect of curved boundaries can be greatly reduced by embedding a curved inner tank in a rectangular outer one and filling the space between with water or some other medium to (nearly) match refractive indices. Stray ambient light needs to be controlled both at the level of the laboratory, and within the tank. For many experiments, we only

need to see through some of the transparent walls of our tank, and it will generally help if light cannot enter through or reflect from the other sides. One of the best removable solutions is to use self-adhesive matt-black film (sold for covering books in some markets, and as a DIY finish in others) on the inside of a tank. This not only prevents light entering (or leaving), but also prevents the tank wall acting like a mirror (due to total internal reflection) and provides a cheap protective covering against inadvertent damage. While the film is slightly fiddly to apply, blacking the tank out in this way can lead to an order of magnitude improvement in the quality of the results. Giving other key components a black rather than a silver finish can also improve things dramatically.

31.5.3 The Apparatus

The range of experimental apparatus used for environmental flows is very broad. There are some common features, such as transparent, watertight tanks (typically made of glass, some type of acrylic or maybe polycarbonate), but beyond that it really depends on the nature of the experiments being envisaged. Designing and constructing experimental apparatus is as much of an art form as a science. The golden rules of "appropriate technology" and "keep it simple" are necessarily tempered by what might already be available, the technical resources, advice and purchasing power at our disposal, and also by what we are trying to achieve. We need to be aware of obvious material limitations of transparency, robustness, thermal properties, and resistance to corrosion (as and where appropriate). Before undertaking a major construction project, it can be worth doing some trials with a much simpler rig using existing equipment, even if it is not possible to achieve the desired parameter regime. The experience gained will combine with the ability to demonstrate to others what is needed and can prove invaluable when designing and constructing the final apparatus, much of which cannot be simply bought "off the shelf."

Experienced technicians are an invaluable asset, capable of converting a vague idea into a working solution that exceeds expectations. However, often researchers resort to "chewing gum and string" solutions to immediate problems, wanting a quick fix, no matter how fragile or imperfect. Duct tape (gaffer tape), blu-tack, and G-clamps are incredibly valuable and have their place, but are seldom long-term solutions when pursuing high-quality quantitative data, and should never be used in situations that compromise health and safety. Those of us lucky enough to have technicians and a workshop at our disposal soon come to realize that it is better to discuss what it is that we wish to achieve rather than simply describing the apparatus we have envisaged, as the technicians are likely to come up with a far better and more practical solution, even if their understanding of the science is incomplete.

31.5.4 Health and Safety

"Health and safety" may not have concerned research in the nineteenth century but has to be a consideration for the twenty-first-century scientist. Protecting personnel, or at least minimizing the risks they are exposed to, has to take priority over any experimental need. Where practicable, such health and safety considerations should be engineered into the experimental setup rather than simply relying on procedures or personal protective equipment (e.g., ear defenders, dusk masks, goggles, gloves, k). For experiments on environmental flows, it is likely that the biggest risks are going to be electrical, optical, and chemical.

Sensible location of electrical outlets, the use of low-voltage equipment where possible and residual current detectors (devices that cut the power if there is an imbalance in the current between the live and neutral wires) where not, along with routine testing of equipment, will help tremendously but could be defeated by operating equipment with wet hands. If the worst happens, strategically located emergency knock-down switches could save your life, or at least allow you to stop an experiment before serious damage occurs.

The desire to have optical access for diagnostics also introduces potential health and safety issues, especially if lasers or other high-power-density light sources are used. Depending on local rules, this can force us to completely enclose the experiment and install interlocks, effectively converting an experiment using a Class IV laser into a Class I device. Although some of the details can often be sorted out later, at least some consideration must be given at the design and construction stage. It is also frequently possible (although not necessarily desirable) to substitute other lower risk light sources: standard slide projectors are often all that is required.

Even chemicals that are harmless to most of us can cause anaphylaxis (acute hypersensitive allergic reaction) for others, or gradual sensitization with prolonged exposure. Indeed, basic safety aids such as latex gloves can cause serious problems for some. The choice of chemicals, as well as the procedures adopted for their use, needs careful consideration, as do issues such as storage and disposal. Similarly, the handling of particles should be considered carefully, especially if there is a risk of inhaling their dust.

It is not appropriate here to itemize all the risks and control measures that influence safety. It is critical, however, that the experimenter takes his or her responsibility seriously and undertakes some form of risk assessment to provide a focus and record of the various issues. As a rule, health and safety should not prevent us pursuing a line of research, but it may impact our experimental design and procedures. The cost of achieving a safe working environment and environmentally acceptable procedures should be considered at the initial planning stage, not only once the research has commenced.

31.6 Step 4: Experimental Diagnostics

Substantial insight can often be gained using simple "low-tech" experimental diagnostics. Direct observations with a stop watch, felt pen to mark the side of the tank, and ruler to subsequently take measurements are sometimes all that is necessary. Such measurements are only ever going to provide order one answers, but there are still many order one questions to be answered.

However, we are seldom satisfied with only the order one answer, forcing us to use a range of more sophisticated diagnostic techniques. Describing each of the technologies of value for laboratory experiments on environmental flows requires an entire volume of its own (e.g., Tropea et al., 2007), and even then coverage would be incomplete. Here we only have space for a very limited overview of a small subset of the key approaches. The reader is also referred to the other chapters in Part IV of this volume for additional discussion.

31.6.1 Velocity Measurements

Measuring the velocity field of a flow was once one of the hardest things to do, but is now commonplace. Here we shall give a brief overview of some of the techniques, although begin by noting that simple visualizations such as particle streaks (see Figure 31.2) can give a powerful qualitative impression of a flow.

31.6.1.1 Point Measurements

There are many technologies available for obtaining point measurements in a laboratory. Traditional ones such as impellors are now largely confined to flows along pipes (and even there ultrasonic measurements are often preferable). In contrast, hot-wire anemometry and Laser Doppler Velocimetry (LDV) are widely used due to their accuracy and high-frequency response. Although the former requires insertion of a probe into the flow, LDV can offer one, two, or three velocity components obtained in a nonintrusive way through the scattering of light as very small particles (often naturally occurring in the flow) are advected across fringes created by a pairs of intersecting laser beams. Although less widely used, Acoustic Doppler Velocimetry (ADV) can also be of value, especially when considering optically opaque particle-laden flows. The biggest limitation of these techniques, however, is that the velocity information is limited to a single point (actually an average over a small volume) at once, providing insuDcient spatial coverage for many researchers.

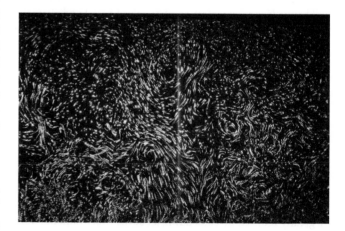

FIGURE 31.2 Particle streaks showing structure of turbulence above an oscillating grid. Here the streaks are generated digitally from a sequence of images, with a slowly fading memory of the previous locations of the particles.

31.6.1.2 Whole-Field Measurements

Particle Image Velocimetry (PIV) has become the almost universal solution to obtaining velocity measurements in fluid flows, and has spawned a massive literature in its own right (e.g., Adrian, 2005; Raffel et al., 2007). This is effectively a pattern-matching technique. The fluid is seeded with small, neutrally buoyant particles that (hopefully) faithfully follow the flow. (Pliolite™ particles, made from a resin used in the manufacture of solvent-based paints, are widely used for environmental flows.) By having a relatively high seeding density and illuminating the flow with a sheet of light (a nearly two-dimensional volume), a camera is used to capture a sequence of digital images of the random patterns of particles located within the light sheet. For each pair of images, software is then used to divide one image into "interrogation windows" (each window containing many particles while remaining smaller than the dominant scales of the flow) and determine how the pattern in each window has been moved and altered by matching it with the other image. This process requires that the patterns have not changed too much (e.g., by particles entering or leaving the light sheet, or by strongly distorting the pattern), thus requiring the interval between capturing the two images is kept relatively small. It is then a simple matter of relating the velocity to the distance moved in the interval between the pair of images in an Eulerian framework. There are many variants on this basic idea, some utilizing a single camera and yielding two components of the velocity in a plane, while others use multiple cameras and perhaps scanning light sheets to recover three-dimensional velocity fields in a plane or a volume of fluid. It must be stressed, however, that PIV does *not* require the use of lasers. Indeed, a slide projector and a standard video camera are often all that is required for laboratory work on environmental flows.

PIV is not the only approach to extracting velocity information from an extended part of the domain. Particle Tracking Velocimetry (PTV), like PIV, relies on images of particles, but rather than considering the flow in an Eulerian framework and relying on the spatial correlation of the velocity field to maintain coherence in the patterns in consecutive images, PTV relies on temporal correlations to be able to identify individual particles and the Lagrangian paths they follow over extended periods of time. Although not as widely used, PTV should not be seen as in competition with PIV, but rather as complementing PIV due to the Lagrangian nature of the data PTV recovers.

Optical velocity measurements in flows with varying density are often hampered by refractive index changes either blurring particles beyond recognition, or causing them to appear to move or be located somewhere other than where they actually are. In such cases, matching refractive indices can provide significant benefits. For example, 3 cm³ of propan2ol (isopropyl alcohol) per unit volume has about the same effect on the refractive index of water as 1 g salt in the same volume of water, but moves the density in the opposite direction. Hence, by forming a stratification that moves from saltwater at the bottom to a propan2ol–water solution at the top, we can remove all initial refractive index variations while being stably stratified in both salt and alcohol. Somewhat surprisingly, glycerol (at low concentrations) can be used in a similar way, although has the disadvantage of increasing the density, making it necessary to add even more salt. There are limits, however, to how strong a stratification can be refractive index matched in this way. Moreover, the diffusivities of salt, alcohol, and glycerol differ, leading to localized mismatch in refractive indices where mixing occurs, and potentially to double-diffusive effects (see Vol. 1, Chapter 39).

The problem of determining a velocity field from a sequence of images is often referred to as "optical flow," the underlying assumption of which is that light (image intensity) is conserved through a simple advection equation, $\partial I/\partial t + (\mathbf{u} \cdot \nabla)I = 0$, where I is the image intensity. PIV and PTV represent just two of the approaches to this problem, geared toward discrete, localized sources of light (i.e., the particles). The ideas behind optical flow can be applied to tracers other than particles (an example with dye is shown in Figure 31.3), but the resulting fields must be interpreted with care. While ∇I averaged over a particle does not yield a preferred direction due to its small size and approximately

(a)

(b)

FIGURE 31.3 (a) Dye pattern being advected by approximately uniform flow along a channel. (b) Optical flow measurement (using the Lucus–Kanade least squares method) of instantaneous velocity field.

circular image, for dye, concentration gradients at the scale of the interrogation windows limit our ability to determine **u** to the component in the direction of the mean gradient, potentially loosing the most significant parts of the velocity field.

The whole-field velocity techniques mentioned so far rely on optical access, but this is not always possible: multiphase flows rapidly become opaque as the concentration of the dispersed phase is increased; refractive index variations can blur particles beyond recognition; and the experimental design may simply not permit unhindered optical access. Magnetic Resonance Imagery (MRI) and Acoustic Tomography are just some of the technologies that can help, for a price.

31.6.2 Density Measurements

In the laboratory, density differences tend to come from one of three sources: fluid composition (e.g., the concentration of NaCl, common salt); temperature; or suspended material (whether particles, droplets, or buoyant bubbles). Although it is possible to create solutions of a known density through the use of tables (e.g., Lide, 2010), the ability to measure the density in different parts of a flow is generally desirable.

31.6.2.1 External Measurements

There are many ways in which density can be measured externally to the experiment from samples taken. Simply weighing a "density bottle" (containing a precisely known volume of a sample), or floating a hydrometer in a sample are simple and direct techniques, but ones that require accurate scales and relatively large sample sizes to achieve good accuracy. Indirect measures are often more convenient. The simple (essentially linear) relationship between density and refractive index for salt solutions allows us to use a handheld "refractometer" (determines the angle for total internal reflection) to obtain density measurement of small samples. Handheld conductivity probes also prove convenient when considering very small salinities, although these tend to require larger sample volumes. Where greater accuracy is required, many laboratories employ an oscillating u-tube density meter, capable of determining density to within 1 part in 10^5, as their prime reference. Such devices are expensive, but provide a reliable, high-precision measurement of the density of a small quantity (typically 1 mL) of any fluid.

31.6.2.2 Point Measurements

A conductivity probe (for salinity) or temperature probe traversed through the depth of a tank is often the most convenient way of determining the background stratification, and its evolution during an experiment. Such probes can also, of course, be located at fixed points, but the presence of a probe will almost inevitably disturb the flow to some degree. Conductivity probes (comprising immersed electrodes excited by an AC signal to prevent electrolysis) can have very high spatial resolutions and fast time responses, but some forms require

frequent calibration. Moreover, conductivity is a function not only of salinity, but also of temperature, complicating the measurement process where both salinity and heat are present. Temperature probes are often based on thermistors or thermocouples. These will typically have a slower time response than conductivity probes due to their own thermal mass and the need to encapsulate them to prevent electrolysis and corrosion from exposure to the working fluid. In general, thermistors are more sensitive than thermocouples, but have a more limited working range.

31.6.2.3 Whole-Field Measurements

Qualitative whole-field visualizations of the density field have been around for a long time. The addition of dyes to tag different sources or densities of fluid allows them to be followed readily. Shadowgraph, where approximately collimated light (e.g., from a slide projector) is projected through a density-stratified flow and onto a screen, highlights regions of high curvature in the refractive index (density) field. Shadowgraph provides a particularly clear visualization of interfaces and mixing events, but it is diDcult and often infeasible to obtain quantitative measurements from it. Conventional schlieren (see Settles, 2001) is sensitive to density gradients (making it more suitable than shadowgraph for continuous stratifications), but does not directly provide quantitative measurements and requires careful optical setups.

However, advances in imaging hardware and image processing techniques have allowed quantitative data to be extracted from many traditional measurement techniques. The absorption of light by simple dyes such as food coloring or potassium permanganate can be readily calibrated and used to determine the line-of-sight integral of concentration, giving either the density as a function of position, or the thickness of a layer (e.g., Holford and Dalziel, 1996). Fluorescent dyes such as disodium fluorescein or the rhodamine family of dyes (some of which are considered harmful in some jurisdictions), in combination with an intense light sheet with the correct spectral mix can provide stunning images of the density field on the plane of the light sheet. This technique, often referred to as planar laser-induced fluorescence (PLIF), can in fact work reasonably well with white light sources, although the high color temperature of an arc lamp produces much clearer images than the yellow light of a halogen lamp.

Whereas PLIF is of primary use in layered systems, or where fluids from discrete sources are interacting, other approaches are more amenable to continuous density variations. Synthetic schlieren (Dalziel et al., 2000, 2007) provides a straightforward method of obtaining line-of-sight-integrated whole-field density measurements by viewing a highly textured pattern placed on the far side of the flow. Refractive index variations due to density variations cause small apparent movements of the pattern; these apparent movements can be determined using algorithms very similar to those employed for PIV to obtain the gradients in the density field, which can subsequently be inverted to recover the density field itself.

31.6.3 Other Visualizations

Natural pearlescence is one of the most dramatically beautiful laboratory tracers. Pearlescence, extracted from fish scales, comprises small plate-like reflective crystals that act like mirrors and spend most of their time aligned with the flow. Most commonly, a small quantity of pearlescence (e.g., Mearlmaid AA) is mixed with water and illuminated with a sheet of light to observe the flow on a single plane. Unfortunately, the relationship between the details of the flow and the patterns formed is not trivial, making it diDcult to get quantitative information (Savao, 1985). However, the sheer beauty of the qualitative image makes this amongst the most impressive things to do for a live audience, but sadly one that often proves diDcult to capture adequately in a photograph (see Figure 31.4). Although there are relatively cheap man-made iridescent pigments that look similar, their particles are dense and settle rapidly, whereas the (expensive) natural variety manages to remain in suspension for extended periods of time.

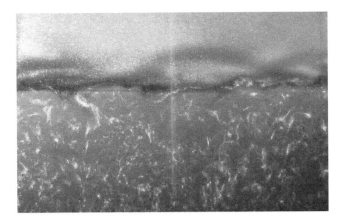

FIGURE 31.4 Stable density stratification is mixed from below by an oscillating grid (not visible), generating a well-mixed layer beneath a density interface. The turbulence from the grid drives an entrainment from the almost-quiescent upper layer into the turbulent mixed layer. Here, a fluorescent dye (disodium fluorescein) shows the extent of the mixed layer while pearlescence throughout the depth contrasts the structure of the turbulence in the mixed layer with the relatively quiescent fluid toward the top of the tank. The pearlescence also highlights the presence of internal gravity waves radiating from the disturbed interface, and their bending due to strong nonlinearities in the stratification above the interface. The flow is illuminated by a sheet of white light from an arc lamp.

The value of simply dripping some dye into the experiment, dropping some potassium permanganate crystals to generate streaks, or periodically adding dye to the float while creating a continuous stratification should not be underestimated (see Figure 31.5). While the results will generally be qualitative or semiquantitative, the insight that can be gained by watching a patch of dye being drawn out by the flow can be great indeed.

31.7 Step 5: Setting the Control Parameters and Performing the Experiment

Watching the first experiment can be exciting, but it can also be very dull or disappointing. Some experiments simply take a long time, with gradual erosion of a density interface over hours or days, while for others the flow remains invisible to the naked eye until extensive (and slow) computer-based analysis has been completed. Moreover, not all experiments will work first time, or every time; we should not simply give up, but try to work out what has gone wrong with the experiment, or with the thinking that lead to it.

Even once an experiment is working, it may be necessary to repeat it with nominally identical conditions, developing an ensemble of runs that provide the necessary statistics and prove (or otherwise) how robust and repeatable the phenomenon is. The time taken to perform each run will often decrease as experience is gained, but it is important not to become impatient and cut corners to get the experiments finished more quickly. It can be worth automating repetitive experiments, not only to remove the tedium, but to ensure more consistent procedures with few mistakes. Sometimes automation can significantly improve the quality of the experiments by allowing them to be done more slowly and carefully, or at times of day when the laboratory is otherwise inactive and free from disturbances of building ventilation and other researchers.

31.8 Step 6: Analyzing the Measurements

A picture will gradually emerge as the measurements collected during each run of an experiment are analyzed. Sometimes it is necessary to hold off the analysis until all the runs have been

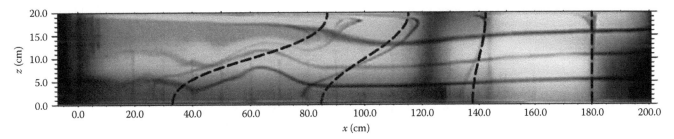

FIGURE 31.5 Nonequilibrium density intrusion propagating into a linear stratification. Dye was occasionally added to the float during filling to mark density surfaces, and potassium permanganate crystals were used to generate vertical dye lines. The intrusion itself was also dyed. The superimposed dashed lines indicate a columnar wave mode prediction of the disturbance propagating in front of the intrusion. (Courtesy of F. de Rooij.)

completed, but in general, this is a bad idea. Only once the analysis is started, will we know if the data collected is good enough for its intended purpose. It may be necessary to remove an imperfection in the apparatus or experimental procedure, or it may become clear that a higher sampling rate is necessary in order to capture the key dynamics in a video sequence. It is much better to spend a little time to establish this at an early stage than to have to redo the entire sequence of experiments; it is better again if experiments can be analyzed during what would otherwise be unproductive time between experiments. Indeed, the results of the analysis will usually guide how we explore parameter space, and how large an ensemble is required to obtain the necessary confidence in the results.

31.9 Major Challenges

Improvements in experimental diagnostics have brought experiments a long way over the last few decades. Accurate quantitative diagnostics in two dimensions are now commonplace, and three-dimensional diagnostics are becoming more widespread. Although there has been a tendency to concentrate on the velocity field, we should strive to get high-quality information for other dynamically important fields. Emerging technologies such as tomographic synthetic schlieren (see Figure 31.6) and the affordability of high-speed cameras in conjunction with rapidly scanned light sheets are paving the way. This opens new challenges, not the least of which are how to store and analyze the vast data sets produced, and how to convey adequately three-dimensional time-dependent information.

While technological advancements have helped with the experimental diagnostics, one of the major challenges for laboratory experiments remains the diDculty—or impossibility—of actually achieving dynamical similarity at the laboratory scale for most environmental flows. This is a severe limitation if we want to view the experiment as a miniaturization of reality, but it is much less problematic if we employ the experiment as a tool for understanding and exploring the key dynamics. The creative use of simplified experiments is a skill worth fostering, and one that can often make startling progress with relatively simple (and cheap) setups and diagnostics.

References

Adrian, R.J. 2005 Twenty years of particle image velocimetry. *Experiments in Fluids* **9** , 159–169.

Buckingham, E. 1914 On physically similar systems; illustrations of the use of dimensional analysis. *Physical Review* 4, 345–376.

Castaing, B., Gunaratne, G., Heslot, F., Kadanoff, L., Libchaber, A., Thomae, S., Wu, X.-Z., Zaleski, S., and Zanetti, G. 1989 Scaling of hard thermal turbulence in Rayleigh-Benard convection. *Journal of Fluid Mechanics* **Q** , 1–30.

Dalziel, S.B., Carr, M., Sveen, K.J., and Davies, P.A. 2007 Simultaneous synthetic Schlieren and PIV measurements for internal solitary waves. *Measurement Science and Technology* **8** , 533–547.

Dalziel, S.B., Hughes, G.O., and Sutherland, B.R. 2000 Whole field density measurements by synthetic schlierenfl*Experiments in Fluids* **8** , 322–335.

Fortuin, J.M.H. 1960 Theory and application of two supplementary methods of constructing density gradient columns. *Journal of Polymer Science* 4 , 505–515.

Hazewinkel, J., Maas, L.R.M., and Dalziel, S.B. 2011 Tomographic reconstruction of internal wave patterns in a paraboloid. *Experiments in Fluids* **θ** , 247–258.

Holford, J.M. and Dalziel, S.B. 1996 Measurements of layer depth in a two-layer flow. *Applied Science Research* **6** , 191–207.

Lide, D.R. (ed.) 2010 *CRC Handbook of Chemistry and Physics*, 90th Edition (Internet Version), Taylor & Francis Group, Boca Raton, FL (http://www.hbcpnetbase.com/)

Malkus, W.V.R. 1954 Discrete transitions in turbulent convection. Proceedings of the Royal Society of London, A. Vol. 225, No. 1161. 185–195.

Raffel, M., Willert, C.E., Wereley, S.T., and Kompenhans, J. 2007: *Particle Image Velocimetry: A Practical Guide* (2nd Edn), Springer-Verlag, Berlin, Germany. 448pp.

Rott, N. 1990 Note on the history of the Reynolds number. *Annual Review of Fluid Mechanics* **2** , 1–12 (doi:10.1146/annurev. fl.22.010190.000245).

Savaọ, O. 1985 On flow visualisation using reflective flakes. *Journal of Fluid Mechanics* **1** , 235–248.

Settles, G.S. 2001 *Schlieren and Shadowgraph Techniques: Visualizing Phenomena in Transparent Media*, Springer-Verlag, Berlin, Germany, 376pp.

Tropea, C., Yarin, A., and Foss, J.F. (eds.) 2007 *Springer Handbook of Experimental Fluid Mechanics*. Springer-Verlag, Berlin, Germany. 1557pp.

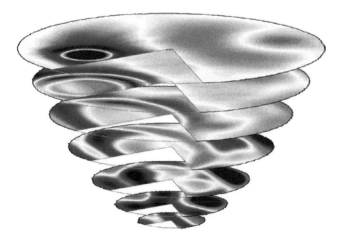

FIGURE 31.6 (See color insert.) Density perturbations due to internal gravity waves in a paraboloidal tank. The full three-dimensional density field was reconstructed using tomographic synthetic schlieren. (From Hazewinkel, J. et al., *Exp. Fluids*, 2011.)

32

General Laboratory Techniques

Thomas Peacock
*Massachusetts Institute
of Technology*

32.1 Introduction

Since the flows that characterize environmental systems typically present a significant challenge to analysis and numerical simulation, it is often the case that laboratory experiments are required to provide insight on environmental issues ranging from the fundamental to the applied. Some notable examples from the last 30 years include fundamental investigations of gravity currents advancing over a horizontal surface (Simpson and Britter 1979), the elegant study of lee-wave generation by flow past topography (Maxworthy 1979), research of filling boxes for the purpose of guiding the design of natural ventilation systems (Linden et al. 1990), and experiments on entrainment and mixing in stratified flows (Strang and Fernando 2001).

The rapid advancement of technology means that these are now exciting times for laboratory experimental research. Experiments that were not feasible a decade ago are now commonplace, thanks to high-speed, high-resolution cameras and powerful computers, among other things. The result is high-quality spatiotemporal data that is on par with the output from large-scale numerical simulations.

Since the field of environmental fluid dynamics is so diverse, it is not plausible in this chapter to present a comprehensive review of all the laboratory experimental methods that have evolved to study environmental flows. As such, this chapter provides an overview of the principal methods currently used, and the reader should consider this to be a summary of the most practical ways to make measurements. The focus of this chapter is on water-based experiments, which allow for relatively-high ($>O(10^2)$) Reynolds numbers due to the low kinematic viscosity of water. For more details on the approaches presented here, and other techniques, the reader is referred to three comprehensive reference books that should satisfy all needs (Goldstein 1996, Tavoularis 2005, Tropea et al. 2007).

32.2 Principles

Before conducting a laboratory experiment, it is important to clearly establish both its scope and goals. It must be determined which methods have the best chance of success, and common sense suggests using the least involved method that can achieve the desired result. The sequence of procedures to carry out the experiment must then be planned in advance and a careful record of the experiment maintained in a laboratory notebook. These records should be suDciently detailed that they could be used to repeat the experiment from scratch (a good rule of thumb is that no detail is too minor).

It is a challenge to ensure that the dynamic regime under investigation is relevant to the environmental problem in hand. This amounts to ensuring a clear understanding of the dimensionless quantities that govern the physical processes being studied and that the geophysical values of these quantities are reasonably reproduced in the laboratory experiment. It is not always possible to match all dimensionless numbers, however, a classic example being the inability to reproduce geophysical Reynolds numbers due to the much smaller length scales in the lab. Nevertheless, it is often still possible to achieve suDciently high Reynolds number that viscous effects are negligible in the laboratory experiment. Issues such as this are discussed in Chapter 31.

The experimentalist must be critical in assessing approximations and sources of error in a laboratory experiment. Other important principles to adhere to are the principles of repeatability and consistency. An experiment must be repeatable to confirm the validity of its result, and consistency checks must be performed to ensure that the data is correct and physically reasonable. A strong example of a consistency check is to make two forms of independent measurement (e.g., measure velocity by two different methods) and confirm agreement between the

Handbook of Environmental Fluid Dynamics, Volume Two, edited by Harindra Joseph Shermal Fernando. © 2013 CRC Press/Taylor & Francis Group, LLC.
ISBN: 978-1-4665-5601-0.

two resulting data sets. Another example of a consistency check is to process the experimental data and ensure it obeys known conservation laws (e.g., conservation of momentum, conservation of mass). Finally, an experimentalist must be brutally honest at all stages throughout an experiment; in many cases this will ultimately save time.

32.3 Methods

32.3.1 Flow Facilities

The heart of any laboratory experiment is the apparatus used to contain the fluid and generate the phenomenon to be studied. This apparatus can range from a simple rectangular tank to a sophisticated rotating turntable, and the flow phenomena can be as diverse as turbulent plumes and inertia-gravity waves. Since the possibilities are far too numerous to all be addressed here, this section focuses on an overview of a scenario that is of widespread interest in environmental fluid mechanics: a stratified flow facility.

32.3.1.1 Two-Tank Method

Density stratification is achieved using the two-tank method (Hill 2002), a schematic of which is presented in Figure 32.1. Two storage tanks are filled with water, with the density of the water in tank 2 (the mixing tank) being lower than that in tank 1 (the storage tank) due to differences in the concentration of a dissolved substance, such as salt. The water from tank 2 is pumped into the base of the lab tank, through some kind of flow conditioning device, such as sponge, to enforce laminar flow that suppresses mixing during the filling process. Water from tank 1 is simultaneously pumped into tank 2, where it is strongly mixed with the resident water. As a result of this arrangement, water pumped into the lab tank has an increasing density as a function of time and, furthermore, this water enters the lab tank as a laminar flow beneath the less-dense water already pumped into the tank, below which it naturally wants to reside.

If the pumping rate out of tank 2 is twice that out of tank 1, a linear density stratification results. More sophisticated two-tank systems can use computer control of the pumping rates to reliably produce nonlinear density stratifications (Hill 2002). It is also possible to operate the process in reverse, pumping from

tank 2 to tank 1 and from tank 1 to the lab tank; in this case the lab tank is filled from above, with progressively less dense water being introduced at the rising fluid surface via a float. The two-tank method is robust and works well for both stationary and rotating experimental facilities. If available, peristaltic pumps provide an accurate means of pumping that is little affected by increases in back pressure as the experimental tank fills up.

In the case that a two-layer stratification is required, the two-tank method is not necessary since one layer can simply be pumped carefully into the experimental apparatus after the other. Unless the two fluids are immiscible, however, some mixing inevitably occurs at the interface between the two fluid layers during the filling process, and furthermore diffusion acts over time to broaden the interface between the two layers. A practical way to counter this is to use a syringe to selectively withdraw fluid in the vicinity of the interface after filling, in which case interfaces as fine as a few mm can be achieved.

32.3.1.2 Forcing Mechanisms

Driving a stratified flow in a laboratory experiment is a significant challenge, which has resulted in the invention of several ingenious mechanisms. Classic techniques include lock-releases for producing gravity currents and simulating dam breaks (Rottman and Simpson 1983) and splitter plates for generating shear between two fluid layers of different density (Koop and Browand 1979). For continuously stratified fluids, a standard technique for studying stratified flow past obstacles is to use a towing tank, in which case the obstacle is moved relative to the fluid, rather than vice-versa. Alternatively, an Odell–Kovasznay pump (Odell and Kovasznay 1971) can be used to produce either uniform or shear flow in a stratified fluid. The pump, illustrated in Figure 32.2, comprises a stack of horizontal rotating discs that drive horizontal layers of the stratified fluid. This disc mechanism inevitably produces some turbulence and nonuniform spatial variations in horizontal flow velocities, but provided the fluid is suﬃciently strongly-stratified and there is a diffuser/settling chamber downstream of the pump, these inhomogeneities will be suppressed. Another possibility is to use a pair of pivoted end walls to produce a high-quality stratified shear flow for a limited period of time (McEwan and Baines 1974).

There are several ways to produce time-periodic uniform or shear flow in stratified fluids, and some of the approaches

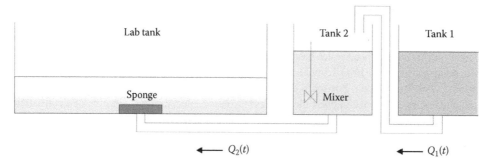

FIGURE 32.1 The two-tank method. Fluid is pumped at rate $Q_2(t)$ from tank 2 (mixing tank) to the base of the experimental tank. Simultaneously, fluid is pumped at rate $Q_1(t)$ from tank 1 (storage tank) to tank 2, where there is strong mixing. (From Hill, D.F., *Exp. Fluids*, 32, 434, 2002. With permission.)

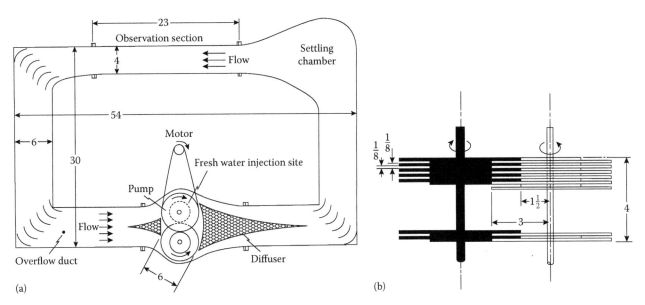

FIGURE 32.2 (a) Overhead and (b) side view of the Odell–Kovasznay pump in a recirculating channel. The dimensions are in inches. (From Odell, G.M. and Kovasznay, L.S.G., *J. Fluid Mech.*, 50(3), 535, 1971. With permission.)

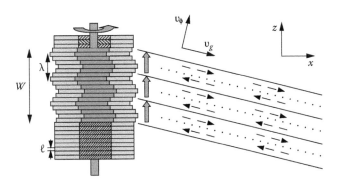

FIGURE 32.3 A schematic showing a series of stacked plates of thickness l on an eccentric camshaft for the purpose of providing horizontal forcing to a stratified fluid. In this example, the system in configured with a surface wavelength λ over a distance W to excite a downward-propagating internal wave field of group velocity v_g and phase velocity v_ϕ.

mentioned in the previous paragraph might even be adapted for this purpose. Alternatively, one can employ a vertically-plunging triangle-shaped wedge at one end of an experimental tank (Baines and Xin-Hua 1985); if the face of the plunger is inclined at 45° to the vertical, then the horizontal velocity forced in the fluid matches the vertical velocity of the plunger. Another novel technique is to force the fluid using a series of horizontal plates driven by eccentric camshafts (Gostiaux et al. 2007), as illustrated in Figure 32.3. The plates effectively set a boundary condition for the horizontal velocity field and may be used to drive uniform or vertically-varying time-periodic flow.

32.3.2 Fundamental Fluid Properties

In order to establish the dynamical regime of an experiment in terms of the governing dimensionless parameters, the physical properties of the working fluid must be accurately known.

To determine the Reynolds number $Re = \mu UL/\rho$, for example, requires that in addition to a characteristic length scale, L, and velocity scale, U, one accurately knows the fluid density, ρ, and the dynamic viscosity, μ. These properties vary with environmental conditions such as temperature, T, which therefore also require careful measurement. More complex, non-Newtonian fluids, used to model phenomena such as avalanches and landslides, for example, require several parameters for their characterization, in which case the reader is referred to Chapter 8 of Goldstein (1996) and Chapter 9 of Tropea et al. (2007).

Here, practical and reliable techniques for measuring fluid density, viscosity, and temperature of Newtonian fluids are summarized, under the caveat that a good experiment will require suDcient environmental control to ensure that these properties do not change throughout the course of an experiment. Indeed, it is good practice to make measurements of these quantities both before and after an experiment.

32.3.2.1 Density

The density of a fluid, ρ, is the mass per unit volume, which is measured in SI units of kg/m³. The most fundamental way to determine the fluid density is to accurately measure the mass and volume of a fluid sample, from which determination of the density naturally follows. Sophisticated methods for doing just this exist, but they are not necessarily the most practical means of density measurement. Other common approaches include hydrometers, which utilize buoyancy, and refractometers, which exploit a relation between refractive index and density.

A technique that provides a good combination of accuracy and practicality is a vibrating-tube densitometer. In a typical measurement system a small, hollow tube is filled with a sample of the fluid of interest and then excited by piezoelectric forcing, for example. The mass of the fluid contained within the tube influences the natural oscillation frequency, f, of the system according to the relation (Albert and Wood 1984):

$$\rho - \rho_0 = K\left(\frac{1}{f^2} - \frac{1}{f_0^2}\right),$$

where

    f_0 and ρ_0 are values for a reference fluid
    K is a calibration constant obtained from measurements using two fluid standards that will have been determined for any commercial product

Thus, the problem of accurately determining fluid density is converted to the problem of accurately determining the natural frequency of a vibrating tube system. Since both temperature and pressure influence the fluid density, for accurate measurement the system is subject to thermal control and is filled at the desired pressure. Commercial products using this measurement principle are capable of density measurements with an accuracy of $0.005\,kg/m^3$ for sample sizes as small as 1 mL, taking around 30 s to make a measurement.

32.3.2.2 Viscosity

The most common way of measuring fluid viscosity is via a capillary viscometer, in which case the kinematic viscosity, ν, in SI units of centi-Stokes (mm²/s) is determined by measuring the time taken, t, for the meniscus of a fluid sample to descend a prescribed height in a narrow glass tube under gravitational forcing alone. Assuming Hagen-Poiseuille flow within the narrow tube, to a good approximation the kinematic viscosity is given by (Einfeldt and Schmelzer 1982):

$$\nu = at - \frac{b}{t},$$

where the constants a and b are set by the construction of the viscometer and are typically determined using fluids of known viscosity (capillary viscometers come with calibration constants). Manufacturers quote an accuracy of better than 1% if the tests are performed under careful conditions (e.g., constant temperature, clean viscometer, vertical orientation of viscometer). There are also many other techniques to measure viscosity, including falling sphere viscometers, vibrational viscometers, and rotational viscometers; in some cases the technique measures the dynamic rather than the kinematic viscosity.

32.3.2.3 Temperature

A widely used commercial technology for measuring temperature is the thermistor, which is a type of resistor whose resistance varies with temperature. Thermistors have high sensitivity ($\sim0.01°C$) and stability, but can be fragile and are somewhat limited in the temperature ranges they can cover, which is typically not an issue for a fluid experiment. An alternative method is to use a thermocouple, which is a junction between two different metals that produces a voltage related to a temperature difference. This technology has

the benefit of being robust and cheap, and possessing an operational range of several hundred degrees, which is not so important for a typical fluid experiment. Typical pitfalls include a lack of stability and low accuracy ($\sim1°C$).

32.3.3 Measurements in Fluids

A laboratory experiment studies the evolution of quantities such as the velocity field and/or a scalar field such as density, temperature, or concentration of a tracer. In principle, one would like to obtain as detailed spatiotemporal data as possible for the physical variables of interest. The most basic form of measurement is a point measurement, in which a sensor provides the time series of a physical variable at a fixed location. In many cases, the experimental geometry, the cost of equipment and the nature of the flow field dictate that this is the only type of measurement that is feasible. Often several such sensors are used simultaneously to obtain data at multiple locations, or alternatively a single sensor is traversed through the fluid to make synchronized measurements at several locations. At the other end of the scale, modern technology is fast improving the ability to obtain fully three-dimensional, time-dependent data, and this type of measurement is becoming increasingly common.

This section reviews the principal techniques for measuring velocity, density, temperature, and concentration in fluid flows. In many cases, it is possible to employ more than one of these techniques simultaneously, allowing for synchronized measurement of multiple flow variables.

32.3.3.1 Velocity

Probably the most sought after physical variable is the velocity field. As such, there is a variety of sophisticated techniques for making measurements of this quantity.

32.3.3.1.1 Laser-Doppler Anemometry/Velocimetry

One of the best established techniques for making high-fidelity, nonintrusive point measurements of velocity is laser-Doppler anemometry (LDA) (Durst et al. 1981), which is also referred to as laser-Doppler velocimetry (LDV). The fundamental principle of operation is to measure the Doppler-shift in the frequency of laser light scattered by small particles travelling with a fluid. To reliably measure one component of velocity requires either a pair of laser beams and a single sensor or a single laser beam and dual sensing.

In the limit $2\pi d/\lambda \gg 1$, where d is the particle diameter and λ is the wavelength of the laser light, and provided that light is collected in forward scatter by a large light-collecting aperture, there is a simple explanation of the measurement technique that is often used as an alternative to the full scattering theory. As illustrated in Figure 32.4, an intersecting pair of laser beams forms a measurement volume in which the coherent beams produce interference fringes with spacing

$$d_f = \frac{\lambda}{2\sin(\theta/2)},$$

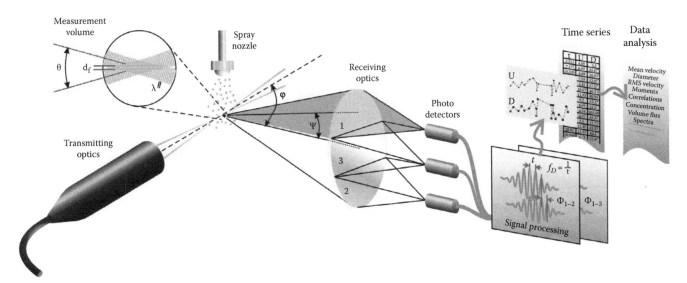

FIGURE 32.4 Schematic of the principle of operation laser-Doppler anemometry, with combined capability for phase-Doppler anemometry. Particles that travel through the intersection region of the two laser beams scatter light with a frequency that is Doppler-shifted in a manner that is linearly related to the particle velocity. Furthermore, multiple detectors enable the determination of phase information of the scattered light, which can be used to determine the size of a particle. (Courtesy of Dantec Dynamics, Skovlunde, Denmark.)

where θ is the interaction angle of the two beams. Tracer particles passing with velocity v perpendicular to the interference fringes through this measurement volume scatter light with a modulation frequency:

$$f = \frac{v}{d_f}.$$

Thus, by measuring f the component of the flow velocity perpendicular to the interference fringes can be determined. There is, however, still ambiguity in the direction of motion, so typically the frequency of one laser beam is shifted by a small amount, f^*, relative to the other using a Bragg cell, resulting in a moving fringe pattern. Particles moving in one direction therefore produce frequencies greater than f^* and particles moving in the other direction have frequencies less than f^*. Further improvements to this technique make use of two or three lasers of different colors to allow for simultaneous measurement of two or three velocity components, respectively.

A typical measurement volume takes the form of an ellipsoid with a minimum dimension of around 50 μm and maximum dimension of around 500 μm, enabling point-like measurements in macro-scale flows. A major advantage of the LDA technique is that it contains no undetermined coeDcients, eliminating the need for calibration, and is linear in response; it does, however, require a transparent fluid.

An extension of this approach that allows for simultaneous measurement of a particleß velocity and size is phase-Doppler anemometry (PDA). Mie scattering theory requires that light reflected off the surface of a particle, or refracted through it, acquires a phase shift that depends on the size of the particle. Multiple detectors positioned at different locations can be used to detect this phase shift and thus determine particle size, as also illustrated in Figure 32.4. This technique, however, is limited to low particle concentrations and spherical particles. For experiments in which the particle size and concentration is the quantity of primary interest then alternative techniques such as shadow-Doppler velocimetry (Hardalupas et al. 1994) and interferometric laser imaging (Dehaeck and van Beeck 2007) should be considered.

32.3.3.1.2 Thermal Anemometry

Thermal anemometry determines flow velocity from the convective cooling of a heated, fine metallic sensor or thin film positioned in the flow (Bruun 1995). The sensors are mm scale or smaller, and are constructed to have excellent thermal properties, to enable detection of small flow variations, and good mechanical strength, so as to withstand forcing by the flow. There are many practical challenges that arise in operating thermal anemometers, including a need for careful calibration, and there is also the fact that they are intrusive. Nevertheless, modern day anemometers use feedback electronics to enable a very-high-frequency response, making them widely used for turbulent flow measurements.

32.3.3.1.3 Particle-Image Velocimetry/Particle-Tracking Velocimetry

Perhaps the most widely used technique for obtaining velocity-field data is particle-image velocimetry (PIV) (Raffel et al. 2007). As illustrated in Figure 32.5, in its simplest (planar) form, the flow is seeded with small tracer particles, such as 10 μm glass spheres, and a plane of the flow field is illuminated using a fine laser sheet. A common laser source is a pulsed Nd: Yag laser, and the fine sheet is produced by passing the collimated laser beam through a cylindrical lens. By obtaining carefully timed images of the particles in the illuminated

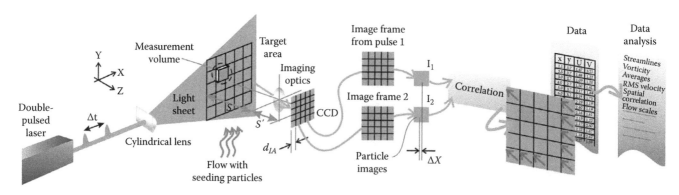

FIGURE 32.5 Schematic of the principle of operation of planar particle-image velocimetry. A cylindrical lens is used to produce a laser light sheet. The motion of particles in the plane of the laser sheet is recorded and then analyzed using cross-correlation methods. (Courtesy of Dantec Dynamics, Skovlunde, Denmark.)

plane, recording an accurate length scale for the images, and assuming that the particles follow the flow, the in-plane motion of the particles can be tracked via computer processing of image sequences.

For PIV analysis, each image is decomposed into an array of search windows, which may overlap. Ideally, within each search window there are several particles in a unique formation that is readily identifiable, and the particles are considerably brighter than the relatively-dark background. For a given search window in one image, the computer software searches the subsequent image to determine the new location of this particle formation using a cross-correlation function. For a well-configured experiment, a displacement $\vec{\delta}$ of several pixels occurs between subsequent images, and high-quality processing software can resolve this displacement to an accuracy of 0.1 pixels. Since the time Δt between images is known, the local velocity $\vec{v}$ in the plane of the laser sheet associated with a search window is as follows:

$$\vec{v} = \frac{\vec{\delta}}{\Delta t}.$$

There are several practical issues to address with planar PIV, a significant one being that it requires the dominant fluid motion to be in the plane of the laser sheet, otherwise particles constantly appear and disappear from view, which confuses the computer processing. Also, the method only works for optically transparent fluids.

An important consideration for stratified fluids is the impact of a spatially varying refractive index of the working fluid on the apparent position of the particles, which can be calculated using the ray tracing analysis for the Synthetic Schlieren technique described later in Section 32.3.3.2.2. This effect can be countered by refractive index matching (Daviero et al. 2001) or may practically be determined to be insignificant by demonstrating that there are no significant distortions of a pattern placed in the plane of the laser sheet and viewed through a typical flow field.

Ever more sophisticated PIV technology is being developed. A now standard technique is stereoscopic PIV, in which two cameras simultaneously image the motion of particles in a planar

laser sheet from slightly different perspectives, enabling the determination of all three velocity-field components in the plane of the laser sheet. An extension of this approach that provides three-dimensional velocity field data in a pseudo three-dimensional volume is to use a mechanical system (e.g., a traverse or an oscillating mirror) to scan the laser sheet through a volume, providing data in several neighboring planes; this requires either a large depth of field for the camera or refocusing of the camera for each plane of viewing, along with individual length-scale calibrations for each plane of view. Tomographic PIV employs three or four cameras and more sophisticated image processing routines to provide truly time-resolved, three-dimensional velocity field data in a three-dimensional volume; several commercial systems with this capability are available.

An alternative to PIV is particle-tracking velocimetry (PTV), which seeks to identify individual particles in subsequent images. An advantage of this method is that the velocity is ascribed to a particle rather than a cluster of particles, but the drawback is that a relatively low seeding-density is required in order for the algorithms to identify particles, leading to relatively sparse data that is randomly distributed over the flow domain. Overall this technique is currently not as widely used as PIV. There are hybrid techniques that use PIV to make a first calculation of the velocity field and then use this information to assist the PTV processing, with the ultimate goal of getting the highest quality data.

Another variant of this type of approach worth mentioning is the fluorescein dye-line method (Hopfinger et al. 1991). In this case a set of equidistant horizontal dye planes are created and illuminated by a vertical laser sheet. The dye lines deform with the constant-density surfaces (isopycnals) of the disturbed fluid and from their vertical displacements over a known time interval the vertical velocity field in the plane of the laser sheet can be calculated.

32.3.3.2 Density, Temperature, and Concentration

Laboratory studies of environmental flows are often concerned with passive or active scalar fields in a fluid flow. A passive dye tracer, for example, will give insight into the transport properties of an environmental flow. Important examples of active scalar fields are temperature or concentration of a dissolved salt, both of which influence the fluid density. Detailed experimental measurements of the density field are surprisingly challenging to obtain, and it is factors such as temperature and concentration that are measured experimentally, serving as a proxy for fluid density via a constitutive relation or an experimental calibration.

32.3.3.2.1 Fast Conductivity and Temperature Probes

One of the most common factors that influence fluid density is salinity, in which case a widely used technique for in situ measurement is to use an electrical conductivity probe. Such a probe utilizes a multiple-electrode arrangement to measure the electrical conductivity of a fluid, the sensor providing an output voltage that is proportional to the electrical conductivity, from which the fluid density can be determined. Both the electrical conductivity and the fluid density are influenced by fluid temperature, however, and so unless there is confidence that a laboratory experiment is truly isothermal, it is prudent to use a combined conductivity-temperature (CT) probe incorporating a fast-response thermistor (Head 1983). The measurement volume of such a sensor is small ($\sim 1 \, \mathrm{mm^3}$) and it can respond to fluctuations of several hundred Hz. The measurement technique is intrusive, however, and works best if the sensor is oriented into an oncoming flow. For ocean measurements, a pressure sensor is also included in the arrangement to form a conductivity-temperature-depth (CTD) probe, since the fluid density in deeper waters is also greatly influenced by pressure; this is typically not an issue for a laboratory experiment.

Good practice for the operation of a CT probe is to prepare two sets of identical fluid samples that cover the range of salinities in an experiment. Place the two sets of samples in a pair of thermal baths that are maintained at the minimum and maximum temperatures expected in the experiment. Then, measure the sample densities and create a calibration surface which relates the conductivity and temperature output voltages from the CT probe to the fluid density. The probe is subject to some drift and the calibration surface is expected to remain accurate to within 1% for several hours.

32.3.3.2.2 Synthetic Schlieren

A classical method for visualizing the density and/or concentration fields in fluid flows is the Schlieren technique, which produces optical flow visualizations by exploiting variations in refractive index of the fluid due to compressibility or spatial variations in concentration of dissolved substance. There is a very rich research history for this method, about which a great deal has already been written (Settles 2001). Here, we crudely summarize that the Schlieren method provides high-resolution images that have provided fundamental insight into many flow phenomena, but it is challenging to obtain quantitative rather than qualitative data using this approach.

An advancement of the Schlieren approach is the Synthetic Schlieren (SS) technique (Dalziel et al. 2007). This technique was initially developed for two-dimensional systems, and an outline of the SS method is presented in Figure 32.6. A random pattern (textured mask) is placed a distance B behind an experimental tank and imaged through the tank by a camera. A reference image of the pattern is obtained before the experiment begins, corresponding to a known reference state of the fluid (e.g., quiescent density-stratified fluid). When the experiment commences, flow that distorts the refractive index field compared to the base state results in apparent distortions of the random pattern viewed though the tank. If the system is two dimensional and the perturbations are small, ray-tracing arguments reveal that the apparent horizontal and vertical displacements, Δx and Δz,

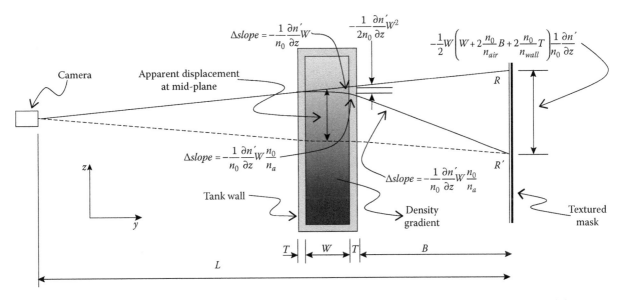

FIGURE 32.6 Schematic of the principle of operation of synthetic Schlieren. A textured mask is placed a distance B behind the experimental tank. The image is viewed through the tank using a camera positioned a large distance L away from the mask. Disturbances to the density field in the tank cause apparent displacements of features in the mask. (From Dalziel, S.B. et al., *Meas. Sci. Technol.*, 18, 533, 2007. With permission.)

of features in the random pattern are related to density-gradient perturbations, $\epsilon\rho'/\epsilon x$ and $\epsilon\rho'/\epsilon z$, via

$$(\Delta x, \Delta z) = -\frac{1}{2}\left[\frac{L - B - \left(1 - \dfrac{n_{air}}{2n_0}\right)W - 2\left(1 - \dfrac{n_{air}}{2n_{wall}}\right)T}{L - \left(1 - \dfrac{n_{air}}{2n_0}\right)W - 2\left(1 - \dfrac{n_{air}}{2n_{wall}}\right)T}\right]$$
$$\times W\left(W + 2\frac{n_{air}}{n_0}B + 2\frac{n_{air}}{n_{wall}}T\right)\frac{\beta}{\rho_0}\left(\frac{\partial\rho'}{\partial x}, \frac{\partial\rho'}{\partial z}\right),$$

where

  L is the distance between the random pattern and the camera
  W is the internal width of the experimental tank
  T is the thickness of the tank walls
  n_0 is the nominal refractive index of the working fluid
  n_{air} and n_{wall} are the refractive indices of the surrounding air and the tank walls, respectively

The displacements, Δx and Δz, are determined using the same cross-correlation algorithms utilized by the PIV method.

For two dimensional systems, SS provides nonintrusive, quantitative data on the density-gradient field. This data is typically of suDciently high quality that it may be spatially integrated to obtain the density field. The SS technique has also been applied to axisymmetric flows, where the spatial symmetry of the system again allows access to quantitative data. Most recently there have been promising advances using multiple viewpoints and tomographic reconstruction to determine three-dimensional density-gradient fields.

32.3.3.2.3 Planar Laser-Induced Fluorescence

A widely-used experimental technique for making two-dimensional measurements of scalar fields is planar laser-induced fluorescence (PLIF) (Houcine et al. 1996), a schematic of which is presented in Figure 32.7. PLIF uses a laser sheet to excite the fluorescence of a tracer dye whose concentration

may itself be the quantity of interest (e.g., when measuring transport and mixing within a fluid) or may serve as a proxy for a desired physical quantity (e.g., temperature, density).

While PLIF provides high-fidelity qualitative data, it requires careful calibration and experimental technique to produce reliable quantitative data. Several practical challenges include natural bleaching of the dye over time causing the fluorescence of the molecule to diminish, spatial and temporal variations in intensity of the laser sheet, and the fact that for high dye concentrations the intensity of light that reaches any given location is a function of the path that it has traversed, which has time varying properties due to the flow field; variations in temperature also affect fluorescence, but this can be exploited to make temperature measurements. These issues are well known, however, and there are established approaches to counter them. One example is the two-color approach, in which a second dye absorbs at the same wave length as the first dye but emits at a different wavelength, allowing the second dye to serve as a reference in the face of spatiotemporally varying illumination (Sakaibara and Adrian 1999).

32.4 Major Challenges

Progress in laboratory experimental techniques looks set to continue apace. One major challenge within reach is to establish methods for obtaining high-quality, three-dimensional spatiotemporal data from experimental fluid flows. Commercial systems are now available to do this for velocity fields and there is demand to develop this capability for other physical quantities. Another significant challenge is to combine techniques to make simultaneous measurements of a number of different physical quantities; simultaneous PIV/PLIF (Figure 32.7) and simultaneous PIV/SS (Dalziel et al. 2007) have already been realized, for example.

A challenge that comes along with rapid technological progress is to ensure that high-quality experiments and data analysis are performed. It is readily possible to generate "pretty pictures," but it is important that this data be validated

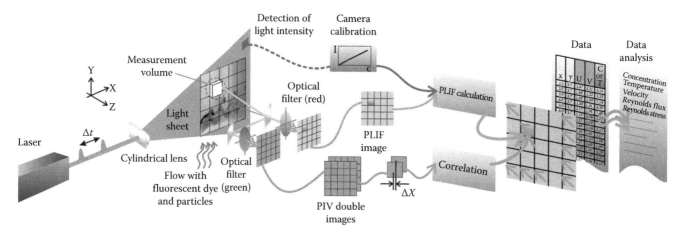

FIGURE 32.7 Schematic of the principle of operation of planar laser induced fluorescence (with simultaneous PIV). Dye concentrations in the plane of a laser sheet are recorded and may be used as a proxy for temperature or density. (Courtesy of Dantec Dynamics, Skovlunde, Denmark.)

and also be useful. Furthermore, with significant computational power vast amounts of data can be generated by even modest experiments. This data needs to be well managed and carefully interpreted, since it is only worthwhile performing such sophisticated experiments if the results provide reliable new insight.

References

Albert, H.J. and Wood, R.H. 1984. High-precision densimeter for fluids at temperatures to 700 K and pressures to 40 MPa. *Review of Scientir c Instruments* 55(4): 589–593.

Baines, P.G. and Xin-Hua, F. 1985. Internal tide generation at a continental shelf/slope: A comparison between theory and laboratory experiment. *Dynamics of Atmospheres and Oceans* 9: 297–314.

Bruun, H.H. 1995. *Hotwire Anemometry: Principles and Signal Analysis.* Oxford University Press, Oxford, U.K.

Dalziel, S.B., Carr, M., Svenn, J.K., and Davies, P.A. 2007. Simultaneous synthetic Schlieren and PIV measurements for internal solitary waves. *Measurement Science and Technology* 18: 533–547.

Daviero, G.J., Roberts, P.J.W., and Maile, K. 2001. Refractive index matching in large-scale stratified experiments. *Experiments in Fluids* 21(2): 119–126.

Dehaeck, S. and van Beeck, J.P.A.J. 2007. Desigiing a maximum precision interferometric particle imaging set-up. *Experiments in Fluids* 42(5): 767–781.

Durst, F., Melling, A., and Whitelaw, J.H. 1981. *Principles and Practice of Laser-Doppler Anemometry.* Academic Press, London, U.K.

Einfeldt, J. and Schmelzer, N. 1982. Theory of capillary viscometers taking into account surface tension effects. *Rheological Acta* 21: 95–102.

Goldstein, R.J. 1996. *Fluid Mechanics Measurements.* Taylor & Francis Group, Boca Raton, FL.

Gostiaux, L., Didelle, H., Mercier, S., and Dauxois, T. 2007. A novel internal waves generator. *Experiments in Fluids* 41(1): 123–130.

Hardalupas, Y., Hishida, K., Maeda, M., Morikita, H., Taylor, A.K.M.P., and Whitelaw, J.H. 1984. Shadow Doppler technique for sizing particles of arbitrary shape. *Applied Optics* 33(36): 8417–8426.

Head, M.J. 1983. The use of miniature four-electrode conductivity probes for high-resolution measurement of turbulent density or temperature variations in salt-stratified water flows. PhD thesis, University of California, Berkeley, CA.

Hill, D.F. 2002. General density gradients in general domains: the "two-tank" method revisited. *Experiments in Fluids* 32: 434–440.

Hopfinger, E.J., Flor, J.B., Chomaz, J.M., and Bonneton, P. 1991. Internal waves generated by a moving sphere and its wake in a stratified fluid. *Experiments in Fluids* 11: 255–261.

Houcine, I., Vivier, H., Plasari, E., David, R., and Villermaux, J. 1996. Planar laser induce fluorescence technique for measurements of concentration fields in continuous stirred tank reactors. *Experiments in Fluids* 22(2): 95–182.

Koop, G. and Browand, F.K. 1979. Instability and turbulence in a stratified fluid with shear. *Journal of Fluid Mechanics* 93(1): 135–159.

Linden, P.F., Lane-Serff, G.F., and Smeed, D.A. 1990. Emptying filling boxes: The fluid mechanics of natural ventilation. *Journal of Fluid Mechanics* 212: 309–335.

Maxworthy, T. 1979. A note on the internal solitary waves produced by tidal flow over a three-dimensional ridge. *Journal of Geophysical Research* 84(C1): 338–346.

McEwan, A.D. and Baines, P.G. 1974. Shear fronts and an experimental stratified shear flow. *Journal of Fluid Mechanics* 63(2): 257–272.

Odell, G.M. and Kovasznay, L.S.G. 1971. A new type of water channel with density stratification. *Journal of Fluid Mechanics* 50(3): 535–543.

Raffel, M., Willert, C., Wereley, S., and Kompenhans, J. 2007. *Particle Image Velocimetry: A Practical Guide.* Springer-Verlag, Berlin, Germany.

Rottman, J.W. and Simpson, J.E. 1983. Gravity currents produced by instantaneous release of a heavy fluid in a rectangular channel. *Journal of Fluid Mechanics* 135: 95–110.

Sakaibara, J. and Adrian, R.J. 1999. Whole field measurement of temperature in water using two-color laser induced fluorescence. *Experiments in Fluids* 26(1–2): 7–15.

Settles, G.S. 2001. *Schlieren and Shadowgraph Techniques: Visualizing Phenomena in Transparent Media.* Springer-Verlag, Berlin, Germany.

Simpson, J.E. and Britter, R.E. 1979. The dynamics of the head of a gravity current advancing over a horizontal slope. *Journal of Fluid Mechanics* 94(3): 477–495.

Strang, E.J. and Fernando, H.J.S. 2001. Entrainment and mixing in stratified shear flows. *Journal of Fluid Mechanics* 428: 349–386.

Tavoularis, S. 2005. *Measurement in Fluid Mechanics.* Cambridge University Press, Cambridge, MA.

Tropea, C., Yarin, A.L., and Foss, J.F. 2007. *Springer Handbook of Experimental Fluid Mechanics.* Springer-Verlag, Berlin, Germany.

Hot-Wire Anemometry in Turbulence Research

Eliezer Kit
Tel Aviv University

33.1 Introduction

This and the next chapter deal with different measurement techniques of turbulence and associated phenomena, such as two-phase flows or surface and internal wave generations. During the last 40 years, the main efforts were focused on the study of coherent structures in turbulence that are of medium or large size (big or intermediate eddies). To reach this goal one should have the ability to conduct simultaneous measurements of spatial flow patterns at a great number of points, on the one hand. On the other hand, there is a necessity to focus on the study of turbulence microstructures, enabling measurements of velocity derivativesﬁtensor, vorticity, dissipation, and enstrophy generation that require very high spatial resolution. New technologies emerging all the time have helped to develop new instruments, such as multi-arrays assembled of conventional single sensors (hot-wires/films) possessing very small volumes in space and providing a very high time response and comprehensive optical devices featuring the entire flow pattern in plane, or even in volume, by tracking solid particles or air bubbles: particle tracking velocimetry (PTV), particle image velocimetry (PIV), laser Doppler anemometry (LDA), fiber optic sensors for detection of void fraction in two-phase flow, and planar laser-induced fluorescence (PLIF) for concentration field measurements.

The number of high-quality studies is exploding. For this reason, we decided to base our presentation on the work in which we or our closest colleagues have been personally involved. We hope that in so doing we will be able to better convey our messages to the reader. Our intention is neither to imply that these are the most representative researches nor that they cover the entire spectrum of instruments and approaches. On the contrary, we acknowledge that many topics have not been adequately addressed in this presentation and we apologize to those who may think that the results of their study should have been referred to in this chapter.

In the 1960s starting with the seminal work of Kline et al. (1967), the interest in coherent structures in turbulent flow has grown immensely. Single hot-wire/film anemometry, which was the main instrument for the point-wise turbulent velocity measurements in pipe flow, or in free shear flows (boundary layer, mixing layer, or jet flow), appeared to be inappropriate for studying coherent structures. The natural extension of the use of the single hot-wire sensor was the employment of multiple probes assembled from many single sensors. These compound probes can disturb the flow, and attempts were made to decrease the disturbances by designing the traverses used to move these probes and the probe holders with better hydrodynamic shape.

During the four decades starting from the 1970s such probes were intensively used by Professor Wygnanski, his students, and colleagues, to study steady and pulsating pipe flows (Wygnanski and Champagne 1973, Wygnanski et al. 1976, Shemer et al. 1985), spot formation in boundary layers (Wygnanski et al. 1976, 1979, Glezer et al. 1989, Katz et al. 1990), jet and wall jet flows (Gutmark and Wygnanski 1976, Cohen and Wygnanski 1987, Likhachev et al. 2001), mixing layer (Wygnanski et al. 1979, Oster and Wygnanski 1982, Gaster et al. 1985, Weisbrot

Handbook of Environmental Fluid Dynamics, Volume Two, edited by Harindra Joseph Shermal Fernando. © 2013 CRC Press/Taylor & Francis Group, LLC.
ISBN: 978-1-4665-5601-0.

and Wygnanski 1988, Nygaard and Glezer 1991, Kit et al. 2005, 2007), and wakes (Wygnanski et al. 1986).

The measurement of vorticity and velocity gradients is always a challenge and in the continuation of this chapter we will present a method of approximate vorticity determination by finite differencesflcomputations of all velocity components measured by three- or four-hot-wire arrays located at fixed positions. The group led by Professor Wallace was the first that started to use multiple arrays (three arrays containing three inclined wires welded to four prongs with one of them common to all three wires and hence leading to undesirable cross-talk between the wires). In 1987 the results of the development and first successful use of a multi-sensor hot-wire probe for simultaneous measurements of all the components of the velocity gradient tensor in a turbulent boundary layer were published by Balint et al. (1987), see also Wallace and Vukoslavcevic (2010). Professor Tsinober, his colleagues, and students contributed significantly to further development and improvement of the probe based on the use of multiple arrays (five arrays with four inclined wires welded to eight separate prongs in each array, used for velocity vector measurements and with additional cold wire for simultaneous temperature measurements). The multi-sensor was successfully used for measurements in grid flow (Tsinober et al. 1992), in jet flow (using calibration unit as a jet generating facility), and in atmospheric flow (Kholmyansky et al. 2001, Gulitski et al. 2007a,b). Since then several experimentalists have used multi-sensor hot-wire probes of increasing complexity in turbulent boundary layers, wakes, jets, mixing layers, and grid flows.

The multiple arrays probes enable to assess the velocity gradient tensor and consequently strain tensor and vorticity by computing finite differences of the appropriate velocities at neighboring points, measured by various arrays placed at different span-wise and lateral locations. This chapter briefly describes and highlights these developments and points out some of the most important things about the turbulence measurements and the discovery of important features of coherent structures, by employing novel approaches for multi-sensor, multi-array hot-wire/film probe design and calibration.

Particle image velocimetry and other optical methods, which have been rapidly developed, have provided other means of access to the fundamental properties of turbulence that include assessment of the velocity gradient tensor (full or some components) and determination of the vorticity field (full vector or some of the components). Detailed presentation will be given in a following separate chapter.

Numerous computationalists have employed direct numerical simulations (DNS) in a wide variety of turbulent flows at ever increasing Reynolds numbers. There is an interaction and a mutual contribution between new advanced experimental methods and rapidly developing DNS. The former provide the new insight to comprehensive physical phenomena resulted in the development of various coherent structures such as hairpin vortices in boundary layers, two-dimensional (2D) Kelvin–Helmholtz span-wise vortices interacting with rib vortices representing the stream-wise legs of vortical structures

resembling hairpin vortices developing in free shear layers (jets and mixing layers). The later can be used for better interpretation of experimental results and for improving calibration procedures. One of the most striking examples is a study related to the use of Taylor hypothesis for evaluation of velocity derivatives in stream-wise direction. This study has its roots to a more fundamental physical problem related to determination of acceleration, local (defined as temporal derivative of velocity in Eulerian coordinates), convective and total (material).

By constructing a specially designed multi-sensor probe (five arrays) with the central array shifted out in the stream-wise direction, so-called NTH ("non-Taylor hypothesis") probe, Professor Tsinober and his group were able not only to assess the stream-wise velocity derivatives without invoking the Taylor hypothesis (Kholmyansky et al. 2001, Gulitski et al. 2007b) but also to estimate the convective acceleration a_c. Together with the local acceleration a_l, which is as usual evaluated from the acquired time series, it enabled them to compute the total acceleration a and to compare to the computed accelerations obtained in DNS simulations at relatively low Reynolds numbers (Tsinober et al. 2001). The strong antialignment of a_l and a_c is a manifestation of their strong mutual cancellation, so that total a appears to be much smaller than both a_l and a_c, which represents the main diDculty in reliable determination, e.g., of the variance of a concerning the noise involved in the measurements of convective and local counterparts of acceleration. Therefore, it is of utmost importance to use the DNS results (Tsinober et al. 2001) for statistical description of various components of acceleration and for validity study of Taylor random and conventional hypotheses.

33.2 Basic Data Processing and Treatment of Coherent Structures

33.2.1 Mean and Root Mean Square of Velocity Components

These are the most basic features of turbulent flow. In the experiments with array of multiple sensors used for instantaneous velocity measurements, comparison between root mean square (RMS) of different velocity components enables one to conclude about the level of anisotropy. Such results were presented in numerous papers. In the measurements in the grid and boundary layer flows with multiple arrays by Tsinober et al. (1992), the comparison of mean and RMS of similar velocity components at different arrays was used to estimate the quality of calibration procedure.

The mean value (first moment) of the ensemble of N sample functions of velocity component u_k (k is the number of sample function in ensemble) at time t_1 is then the arithmetic mean over the instantaneous values of the sample functions at time t_1. The variance (second moment) of the ensemble of N sample functions of velocity component u_k at time t_1 is then the arithmetic mean over the square of instantaneous values of the sample functions at time t_1 with the mean removed:

$$\mu_u(t_1) = \lim_{N \to \infty} \frac{1}{N} \sum_{k=1}^{N} u_k(t_1) \tag{33.1}$$

$$\sigma_u^2(t_1) = \lim_{N \to \infty} \frac{1}{N} \sum_{k=1}^{N} (u_k(t_1) - \mu_u(t_1))^2 \tag{33.2}$$

σ_u is the RMS of velocity $u(t)$. When the mean value is independent of t_1, the process is *stationary*. Otherwise, the process is *nonstationary*. Generally, however, statistics of a stationary random process are computed over a time average:

$$\mu_u = \bar{u} = \lim_{T \to \infty} \frac{1}{T} \int_0^T u(t)dt \tag{33.3}$$

$$\sigma_u^2 = RMS(u) = \lim_{T \to \infty} \frac{1}{T} \int_0^T (u(t) - \bar{u})^2 dt \tag{33.4}$$

If these values do not differ from those in (33.1 and 33.2) for a stationary process, then the process is said to be *ergodic*. All stationary processes encountered in fluid mechanics can be considered ergodic.

33.2.2 Correlation and Covariance Functions

A correlation or joint moment of the process at two different times can be computed by taking the ensemble average of the product of instantaneous values at two times t_1 and $t_1 + \tau$. These values can be written as

$$R_{uu}(t_1, \tau) = \lim_{N \to \infty} \frac{1}{N} \sum_{k=1}^{N} u_k(t_1) u_k(t_1 + \tau) \tag{33.5}$$

where R_{uu} is known as the autocorrelation function. When the process is stationary and the autocorrelation is only a function of τ

$$R_{uu}(\tau) = \lim_{T \to \infty} \frac{1}{T} \int_0^T u(t)u(t + \tau)dt \tag{33.6}$$

The covariance function is the autocorrelation function of single velocity component u with the mean removed and the cross-covariance function is the cross-correlation function of two velocity components u and v with the product of the means removed:

$$C_{uu}(\tau) = R_u(\tau) - \bar{u}^2$$
$$C_{uv}(\tau) = R_{uv}(\tau) - \overline{uv} \tag{33.7}$$

33.2.3 Estimators

If the time duration of the measurements of the random process is *limited*, only *estimators* can be determined from the given experiment. In many cases the expectation and variance of an estimator can be derived analytically, in particular, when the estimator for the mean is considered. The most common sample mean estimator is given by

$$\hat{\mu}_u = \frac{1}{N} \sum_{i=1}^{N} u_i \tag{33.8}$$

where u_i are individual samples of the process u. The estimator of the mean value given in (33.25) is nonbiased random number, $E(\hat{\mu}_u) = \mu_u$. The mean square error, or variance, of this estimator is then given by

$$\sigma_{\hat{\mu}_u}^2 = E((\hat{\mu}_u - \mu_u)^2) \tag{33.9}$$

Substituting (33.8) into (33.9) leads to

$$\sigma_{\hat{\mu}_u}^2 = E\left(\left(\frac{1}{N} \sum_{i=1}^{N} u_i - \mu_u\right)^2\right) = \frac{1}{N^2} E\left(\left(\sum_{i=1}^{N} (u_i - \mu_u)\right)^2\right) \tag{33.10}$$

If the condition $E(u_i u_j) = 0$ is satisfied, i.e., consecutive samples are uncorrelated or statistically independent, (33.27) can be further reduced to

$$\sigma_{\hat{\mu}_u}^2 = \frac{1}{N^2} E\left(\sum_{i=1}^{N} (u_i - \mu_u)^2\right) = \frac{\sigma_u^2}{N} \tag{33.11}$$

which states that the variance of the mean estimator decreases with increasing number of samples. This analysis has been performed for an estimator based on *discrete samples* u_i; however, a similar analysis could be made for a mean estimator based on the *continuous* stationary signal $u(t)$:

$$\hat{\mu}_u = \hat{u} = \frac{1}{T} \int_0^T u(t)dt \tag{33.12}$$

which differs from the true mean μ_u, since the integral is performed only over a finite time T. The variance of this estimator becomes

$$\sigma_{\hat{\mu}_u}^2 = E((\hat{u} - \bar{u})^2) = E(\hat{u}^2) - \bar{u}^2 \tag{33.13}$$

which can be evaluated using definition of autocorrelation function (33.6) as

$$\sigma_{\hat{\mu}_u}^2 = \frac{2\sigma_u^2 T_u}{T} \tag{33.14}$$

with the integral timescale $T_u = (1/\sigma_u^2) \int_0^\infty C_{uu}(\tau)d\tau$.

If the results given by (33.11) and (33.14) are equated, the condition for statistically independent samples can be obtained as $N = T/2T_u$.

This is an important result since it enables one to select a proper time of experiment, provided the $RMS(u)$ requested accuracy determined by N and the integral time-scale T_u are known.

33.2.4 Variance of Spatial Velocity Derivatives, Vorticity, Skewness, Dissipation

As explained earlier, the use of Taylor hypothesis enables one to evaluate the velocity derivative in a stream-wise (x) direction even from measurements with a single hot wire. Then, by utilizing the local isotropy approach, which states that in developed turbulent flow at small scales the turbulence is isotropic, the estimates of turbulent dissipation based on variance of stream-wise velocity derivative and enstrophy production based on third moment (skewness) of the same velocity derivative can be evaluated. Numerous studies were conducted using this approach. One of the most detailed studies is the work published by Champagne (1978).

Tsinober et al. (1992) used the isotropic relations to carry out performance tests for quality assessment of turbulence measurements. In particular, the following relations for spatial velocity derivatives were checked:

$$\overline{\left(\frac{\partial u_i}{\partial x_i}\right)^2} = 0.5 \overline{\left(\frac{\partial u_i}{\partial x_j}\right)^2}, \quad j \neq i, \text{ no summation over } i \quad (33.15)$$

The results were quite satisfactory and these subtle relations were satisfied within a 15% error.

In their work, the three or five multiple arrays enabled to compute all nine velocity derivatives using finite differences in spanwise and lateral directions and applying the "frozen" Taylor hypothesis in stream-wise direction. This approach was further developed in Gulitsky et al. (2007a) who shifted the central array in five-array arrangement by 1 mm that allowed computing the stream-wise velocity derivative without using Taylor hypothesis. Such approach enabled the authors, besides the checking of Taylor conventional and random hypotheses, to conduct a comprehensive study on statistics of acceleration vector (see more details in the section on hot-wire (film) anemometry). The general expression for the mean value of the dissipation is

$$\varepsilon = \nu \overline{\left(\frac{\partial u_i}{\partial x_j}\right)\left(\frac{\partial u_i}{\partial x_j}\right)} \quad (33.16)$$

that can be essentially simplified if local isotropy is assumed and the isotropic relations for velocity derivatives in (33.16) are adopted:

$$\varepsilon = 15\nu \overline{\left(\frac{\partial u_i}{\partial x_i}\right)^2}_{i=1, 2, \text{ or } 3, \text{ no summation over index } i} \quad (33.17)$$

This last expression is widely used for determination of kinetic energy dissipation. As shown by Tsinober et al. (1992), that even in case both relations (33.16) and (33.17) provide very similar results when the mean dissipation is considered, probability density distributions are essentially different (see Fig. 10 in Tsinober et al. 1992).

The full expression for the enstrophy generation $\overline{\omega_i \omega_j (\partial u_i/\partial x_j)}$ can be replaced by the skewness of stream-wise velocity derivative component (third moment) $\overline{\left(\partial u_i/\partial x_i\right)^3}$ multiplied by appropriate coefficient (−17.5). Tsinober et al. (1992) showed that the trends of both relations, e.g. their behavior versus distance from the grid, are similar but there is no quantitative correspondence. The probability density functions for both expressions differ significantly. The computed value of dimensionless skewness (33.18) is in good agreement with those obtained in various flows such as boundary layer, jet, mixing layer, grid flow, etc. (Tsinober et al., 1992, and references therein).

$$s = \overline{\left(\frac{\partial u_i}{\partial x_i}\right)^3} \Bigg/ \left(\overline{\left(\frac{\partial u_i}{\partial x_i}\right)^2}\right)^{3/2}, \quad i = 1,2,3, \text{ no summation over index } i$$

$$(33.18)$$

It is noteworthy that, in case of Gaussian stationary process (noise), the skewness of longitudinal derivative of any velocity component $e u_i/e x_i$ should be 0. It differs from zero due to the real physics (nonlinearity) involved into the enstrophy generation that determines vorticity dynamics. Therefore, the correspondence of this value to those obtained in numerous experiments may be used for validation of experimental methodology. For example, it can be used to eliminate the experimental runs with a very high level of noise. Again, the shape of density probability function for the enstrophy generating term differs essentially of that for skewness of stream-wise velocity derivative multiplied by the theoretically defined coefficient. For more details see Tsinober et al. (1992) and Champagne (1978).

33.2.4.1 Treatment of Coherent Structures: Triple Decomposition

The treatment follows the approach presented in Hussein (1983).

The triple decomposition of each instantaneous velocity component u_i means that, in any point of the flow in the presence of coherent structures, it can be represented as

$$u_i(\vec{r},t) = U_{iav}(\vec{r}) + u_{icoh}(\vec{r},\varphi_k) + u_{itur}(\vec{r},t) \quad (33.19)$$

Here u_i denotes any of the velocity components $u_1(u)$, $u_2(v)$, $u_3(w)$. U_{iav} is the time-averaged value of velocity in point $\vec{r}$. u_{icoh} is the phase-averaged value of any of the velocity components in point $\vec{r}$ for a given phase shift $\varphi_k = t_k\omega = 2\pi t_k/T$, the "coherent velocity." In particular, in experiments conducted in a mixing layer facility with externally imposed flapper oscillations (Kit et al. 2005) at frequency ω, u_{icoh} reflects the structure of K–H billows and is determined over the period of flapper oscillations, $0 < \varphi_k < 2\pi$.

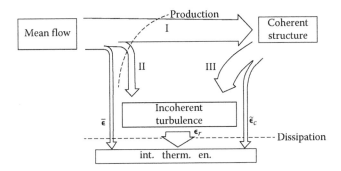

FIGURE 33.1 Schematic of the energy fluxes and kinetic energy dissipation between the mean, coherent, and incoherent velocity fields. (Reproduced from Hussain, A.K.M.F., *Phys. Fluids*, 26, 2816, 1983. With permission.)

u_{itur} is the incoherent turbulent part of velocity at point $\vec{r}$. $1/T$ is the forcing frequency applied to the flapper, $\omega = 2\pi/T$.

The energy exchange between these three components and kinetic energy dissipation of each component are nicely illustrated by the sketch presented in Hussein (1983).

Various expressions for energy fluxes and dissipations of turbulent kinetic energy are developed in Hussein (1983) and in great bulk of other papers.

In the most general case, the coherent structures in a turbulent shear flow occur at random phases and the treatment must involve phase-dependent information (Figure 33.1).

Thus, similar to the time average as presented earlier (33.3), the phase-average concept needs to be introduced. This is the ensemble average of any property, in particular, velocity components, at a phase of the structure as given in (33.20):

$$< f(x,y,z,t) > = \lim_{N \to \infty} \frac{1}{N} \sum_{i=1}^{N} f_i(x,y,z,t+t_i) \qquad (33.20)$$

where

 t is the time corresponding to the reference

 t_i denotes the random instants of occurrence of successive structures of the selected phase

When these structures occur at regular intervals (a situation that essentially occurs in vortex shedding or when structures are induced via controlled excitation as in Kit et al. 2005), this definition reduces to the periodic-phase average:

$$< f(x,t) > = \frac{1}{N} \lim_{N \to \infty} \sum_{i=1}^{N} f_i(x,t+iT) \qquad (33.21)$$

33.2.5 Use of Controlled Excitations: Two-, and Three-Dimensional Perturbation of Mixing Layer

The controlled excitation was used very successfully in numerous works to study development of instabilities in boundary layers using oscillating ribbons and in free shear flows to investigate

evolution of 2D Kelvin–Helmholtz rolls and generation of secondary instabilities such as stream-wise or "rib" vortices.

In Kit et al. (2005), the detection of coherent structures based on phase averaging of the velocity and vorticity data measured using two methodologies, hot-film anemometry (HFA) and PIV, enabled one to reconstruct a full three-dimensional (3D) pattern of the disturbed Kelvin–Helmholtz rolls.

The experiments reported in Kit et al. (2005) are conducted in a specially constructed facility combining two independent wind tunnels with two independent fans separated from each other by a horizontal splitter plate and a common work section. The mixing layer in the test section is obtained by merging the parallel streams generated in the two wind tunnels. A more detailed description of this tunnel is given in Oster and Wygnanski (1982). Experiments are carried out with the 5 m/s velocity of the lower and 2 m/s velocity of the upper streams.

In order to introduce artificial periodic oscillations into the mixing layer, a thin flapper is used, which spans the entire test section along the trailing edge of the splitter plate, and which is driven by an electromagnetic shaker actuated by a control sinusoidal signal. Two flappers are used. One had a straight trailing edge for generating 2D oscillations in the flow (so-called 2D flapper) and the second had a saw tooth curve trailing edge with a span-wise wavelength of 75 mm for introducing 3D perturbations into the mixing layer (3D flapper). The forcing frequency of the flapper was 20 Hz, the amplitude of the oscillations, measured under stroboscopic illumination, varied from 0.9 to 1.6 mm. For details see Kit et al. (2005).

Two kinds of experimental techniques were used in this work. The use of stereo particle image velocimetry (SPIV) enables to measure all three components of the velocity vector $\boldsymbol{u}$: u_1 in a stream-wise direction (along the x-coordinate), u_2 in a vertical direction (along the y-coordinate), and u_3 in a span-wise direction (along the z-coordinate) in the span-wise (zy) plane.

The second technique was constant temperature anemometer (CTA), by means of which the stream-wise component u_1 and the lateral component u_2 were obtained in a sequence of points in spanwise (zy) and stream-wise (xy) planes.

The acquisition of images obtained using SPIV system was performed in two different modes. In the first random-phase mode, acquisition of images was carried out under internal software signal with frequencies of 5 and 10 Hz. These images were therefore obtained in an *arbitrary* way relative to the control sinusoidal signal that was supplied to the flapper, and which, consequently, led to a random, relatively illuminated by the laser sheet plane of measurements, spatial placement of K–H structures in the mixing layer. In the second phase-locked mode, measured images were obtained with time shifts which were fixed relatively to the control sinusoidal signal supplied to the flapper. Consequently, all these images were strictly related to the phase of the K–H structures. More details in regard to SPIV system can be found in Kit et al. (2005, 2007).

Two x-hot-wire probes were used for velocity measurements: one had two degrees of freedom and was able to move in x and y directions, while the other had three degrees of freedom and could move in x, y, and z directions, with a resolution distance of 0.01 mm. The signals from the sensors were sampled at rates of 2000 or 4000 samples/s per channel and were converted to velocity values using polynomials with calibration coeDcients obtained with a standard calibration procedure, the description of which is following in the next section, and were stored for further processing. The calibration procedure included "in situ" measurements at 11 pitch angles varying in the range (∓45°); both wind tunnels have been running at the same velocity, thus providing the potential core, in which the calibration of the probe was performed flow with very low level of turbulence (<0.1%). Seven velocities varying in the range 1 ÷ 6 m/s were employed for calibration purposes.

U_{iav} is obtained from random-phase measurements of the velocity using ensemble averaging of SPIV images:

$$U_{iav}(\vec{r}) = \frac{1}{N_{seq}} \sum_{n=1}^{N_{seq}} u_i(\vec{r}, n) \qquad (33.22)$$

where N_{seq} is the number of random-phase measurements of the flow under investigation.

In the HW experiments

$$U_{iav}(\vec{r}) = 1/N \sum_{n=1}^{N} u_i(\vec{r}, t_n) \qquad (33.23)$$

where N is all velocity samples acquired at one spatial point.

u_{icoh} is obtained from phase-locked measurements using ensemble averaging of SPIV images

$$u_{icoh}(\vec{r}, \varphi_k) = 1/N_{seq}^k \sum_{n=1}^{N_{seq}^k} u(\vec{r}, n) - U_{iav}(\vec{r}) \qquad (33.24)$$

where N_{seq}^k is a number of measurements with the selected phase φ_k of the control signal applied to the flapper. In the HW experiments $u_{icoh}(r, \varphi_k)$ is obtained from a time series acquired at one spatial point by phase averaging the data for each phase φ_k of the control signal at the flapper:

$$u_{icoh}(\vec{r}, \varphi_k) = 1/N_{periods} \sum_{n=1}^{N_{periods}} u_i(\vec{r}, \varphi_{kn}) - U_{iav}(\vec{r}) \qquad (33.25)$$

Here $N_{periods}$ is the number of periods of flapper oscillations measured at each spatial point; φ_k is defined in the same manner as in the SPIV experiments relative to the flapper signal.

The turbulent component of the velocity u_{itur} is calculated as follows:

$$u_{itur}(\vec{r}, t) = u_i(\vec{r}, t) - U_{iav}(\vec{r}) - u_{icoh}(\vec{r}, \varphi_k) \qquad (33.26)$$

33.2.6 Reconstruction of the Coherent Velocity Field from SPIV Experiments

Measurements using HW allow one to obtain 100 or more values of u_{icoh} during the period of flapper oscillations at each spatial point, while the SPIV method enables to obtain only a few phase-locked points under the same flow conditions. So, the question arises concerning the minimum number of phase-locked experiments using SPIV necessary to obtain the reasonable behavior of u_{icoh}.

In the case of the forced mixing layer three harmonics are believed to be enough to represent $u_{icoh}(r, \varphi_k)$. This is due to an essentially stronger amplification of the fundamental mode relative to other harmonics in most of the flow domain except the region in the vicinity of the splitter plate. The ability to represent the signal with just three harmonics is generally satisfied in mixing layers, even when the second harmonic is of the order of the fundamental:

$$\begin{aligned} u_{icoh}(\vec{r}, \varphi_k) &= A_{i1}(\vec{r})\cos(\omega t + \xi_{i1}(\vec{r})) \\ &\quad + A_{i2}(\vec{r})\cos(2\omega t + \xi_{i2}(\vec{r})) + A_{i3}(\vec{r})\cos(3\omega t + \xi_{i3}(\vec{r})) \end{aligned}$$
$$(33.27)$$

By utilizing the hot-wire measurements, Kit et al. (2005) showed that seven time points distributed through the entire period are enough to compute all the unknown coeDcients and phase shifts. Least square method (LSM) has been developed that allows finding the coeDcients and phase shifts even in the case when the points are not equidistant. The advantage of the method is its ability to compute Fourier Transform (FT) coeDcients for unevenly distributed time points without employing any interpolation procedure. The details can be found in Kit et al. (2005).

The goal of the development of the LSM method was to obtain a comprehensive description of the time variation of the coherent structures (the primary K–H billows/rolls and their secondary instabilities) based on a very limited number of SPIV measurements.

The spatial structure of primary K–H billows may be determined by the structure of the span-wise component of vorticity $\omega_z = \partial v/\partial x - \partial u/\partial y$. The development of secondary instabilities and streaks may be characterized by the behavior of the stream-wise component of vorticity $\omega_x = \partial w/\partial y - \partial v/\partial z$. Using the velocity field obtained from the SPIV data processing in the zy-plane, the derivatives of velocity components in span-wise (z) and lateral (y) directions are computed and, applying the Taylor hypothesis, the velocity derivative in a stream-wise (x) direction $\partial u_i/\partial x = -1/U_{1av} \cdot \partial u_i/\partial t$ is obtained where U_{1av} is the local averaged stream-wise component. This enables to compute the time variations of *all* coherent vorticity components in the plane of measurements. The distributions of span-wise and stream-wise components of vorticity are given in Kit et al. (2005) and their detailed time evolution allowed the authors to follow the complicated dynamics of 2- and 3D instabilities in a plane mixing layer.

33.3 Thermal Anemometry

33.3.1 Brief Description of the Method

This method for turbulent velocity measurements exists almost for 100 years (King 1916) with many improvements over the years. Accordingly, the literature is very rich and superior handbooks and scientific papers can be easily found at the shelf. Some comprehensive chapter of books, monographs, or reviews on the subject have been contributed by Hinze (1975), Perry (1982), Comte-Bellot (1998), and many others. Just recently, a comprehensive Springer Handbook of Experimental Fluid Mechanics (Tropea et al. 2007) was published and it contains a very detailed chapter dealing with thermo-anemometry.

In this section, the main emphasis will be given to the various approaches of probe calibration especially in outdoor environment where the implementation of specially designed calibration units become problematic, to the generation of velocity data sets from row data using calibration procedures, to the express data analysis for the determination of data quality and to the more comprehensive analysis employed for coherent structure discrimination in turbulent flow.

Hot-wire anemometry is a versatile technique that can be used for the measurement of velocity and temperature fluctuations in the time domain for investigations in turbulent flows. With appropriate designs, hot-wire anemometry is useful over wide speed ranges from low subsonic to high supersonic flows. The method is capable of detecting turbulent perturbations with a large dynamic response because of the very small hot-wire thermal inertia and its correction in the anemometer. The hot wire/film can be operated in three popular modes: constant current, constant temperature, and the recently emerged—constant voltage:

- The current intensity in the sensor is maintained constant, I_w = constant, which leads to the constant current anemometer (CCA). Any change of velocity U creates a change across resistance R_w of the sensor and the measurable signal is directly this resistance change. It suffers from relatively high thermal inertia (Hinze 1975). However, being virtually a thermometer it is widely used to measure temperature fluctuations with a very fine wire (less than 1 μm in diameter), so-called cold wire, fed by very low electrical current (few mA).
- The resistance of the sensor is maintained constant, R_w = constant, which leads to the constant resistance anemometer, usually called a constant temperature anemometer (CTA). The measurable signal when a change in U occurs is then the change in current intensity I_w to be fed to the sensor to fulfill the imposed condition. In principle, since R_w is maintained constant, the thermal lag of the wire is automatically suppressed. Since its invention in 1960s, CTA became the most frequently used in turbulent velocity measurements.
- The voltage across the sensor is maintained constant, $V_w = R_w I_w$ = constant, which leads to the constant voltage

anemometer (CVA). The measurable signal is then the change in current intensity I_w to be fed to the sensor under that new imposed condition. Weiss and Comte-Bellot (2004), who investigated the signal-to-noise ratio (SNR) in a CVA in terms of the main constitutive elements of the circuit, arrived at a conclusion that concerning SNR, CVA mode of operation is preferable in comparison to two other modes.

33.3.1.1 Multiple Hot-Wires/Films Probes

The combination of inclined (slanted) sensors "feeling" differently the velocity vector arriving from different directions enables measurements of different velocity components simultaneously. Use of these sensor positions, or combinations of these positions, leads to the following possibilities:

- Single hot wires placed normal to stream-wise velocity U_1 mainly respond to the instantaneous longitudinal velocity U_1 when the velocity fluctuations are of relatively small amplitude.
- Cylindrical hot films work similarly to single hot wires, but they are more robust when experimenting in water, oils, or liquid metals, and even in slightly dirty gases. A thin quartz coating deposited on sensors prevents hydrolysis of the neighboring liquid.
- X wires are made of two inclined wires lying in a plane that includes mean velocity vector and the wires are often placed symmetrically relative to U_1, as in the letter X.
- Multi-sensor array, e.g., four-wire array used by Tsinober et al. (1992), shown in Figure 33.2 includes four inclined wires that enable to measure the entire velocity vector within a resolved cone. The size of the cone is limited and therefore it is highly desirable to keep the array axis as close as possible to be parallel to the mean velocity vector.

The relation between the voltages at individual sensors and the velocity components is essentially nonlinear and therefore calibration procedure is necessary to reconstruct the velocity field from the instantaneous measured voltages across the wires. These procedures will be described in more detail in the following for the most complicated multi-sensor arrays. Procedures for simpler probe configurations, x-wires, and single wires probes, can be readily derived from this more general case.

33.3.1.2 Calibration of Multi-Sensor Probes

33.3.1.2.1 Effective Cooling Velocity Approach

A formulation can be advanced based on the effective cooling velocity approach (following van Dijk and Nieuwstadt 2004). It is assumed that for a given absolute flow speed, the relation between an "effective cooling velocity" and the wires orientation relative to the flow is the same, and for implementing the cooling velocity concept a combination of the laws of King (1916) and Jørgensen (1971) can be used. These describe the response of

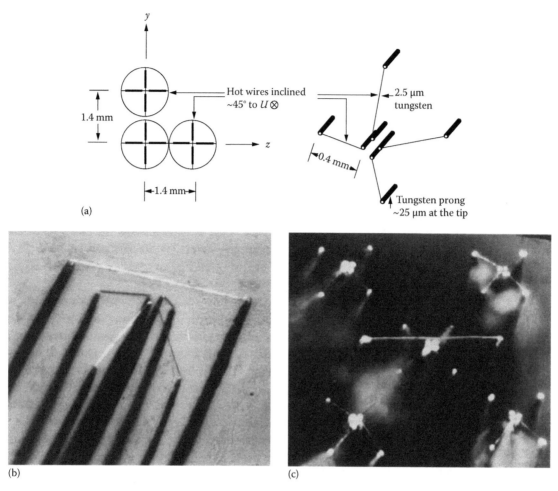

FIGURE 33.2 (a) Schematic of a 4-wire array and a 12-hot-wire probe, (b) close-up of an array, and (c) a 20-hot-wire probe with a cold wire. (Reproduced with permission from Tsinober, A. et al., *J. Fluid Mech.*, 242, 169. Copyright 1992, Cambridge University Press.)

hot wires with only a few parameters (five sensitivity parameters plus two angles per wire). The (modified) King's law for the absolute velocity dependence of a hot-wire signal E is

$$E^2 = A + BU_{eff}^n \qquad (33.28)$$

$$U_{eff} = |U| f(\text{orientation between the wire and the flow}) \qquad (33.29)$$

where f is a function. The Jørgensen's law for the directional sensitivity is

$$U_{eff}^2 = U_n^2 + k^2U_t^2 + h^2U_b^2 = U_{nn}^2 + k^2U_t^2 + (h^2-1)U_b^2; \quad U_{nn}^2 = U_n^2 + U_b^2 \qquad (33.30)$$

where

h is the bi-normal cooling coeDcient
k is the parallel cooling coeDcient
U_{nn} is the absolute value of velocity component normal to the wire

When the difference between normal and bi-normal cooling ($h = 1$) as well as the sensitivity of the wires for parallel flow ($k = 0$) are neglected, then (33.5) reduces to the "cosine law" is as follows:

$$U_{eff}^2 = U_{nn}^2 = U_n^2 + U_b^2 \qquad (33.31)$$

Usually (33.28 through 33.30) describes the single hot-wire/film response to an arbitrary selected external turbulent velocity field and is used in calibrations. Directional calibration by yawing and pitching of the probe enables the generation of calibration data sets that can be used to determine the calibration coeDcients and each wire's angles relatively to the coordinate system linked to the stem of the multi-probe. This type of calibration has been frequently used (see van Dijk and Nieuwstadt 2004, and references therein). The advantage of this approach that it is based on phenomenological relations involving clear physical insights and a very limited number of parameters necessitated for probe description. The later might also represent the drawback of the calibration procedure since it uses too few parameters to represent such complicated phenomena.

However, these equations can be also used to evaluate time series of voltages at each one of the wires of the multi-hot-film probe providing the time series of velocity components is given (simulated or measured) and all the coefficients in the King and Jørgensen relations for a given probe are known. These artificially obtained time series of velocities and voltages comprise a data set that can be used for analysis of calibration procedures.

van Dijk and Nieuwstadt (2004) applied a similar approach to "generate turbulence" and then used their generated data set to compare different calibration procedures. Unfortunately, they arrive at an incorrect conjecture in regards to the use of direct polynomial method (see below). Their statement "that direct models should not be used to analyze turbulence measurements" is misleading. Their artificial data sets of "turbulent velocity field" differ significantly from any real data set obtained in experiments with grid flow (Tsinober et al. 1992), jet flow, and atmospheric flow (Gulitski et al. 2007a,b). All the results from these experiments were thoroughly tested and confirmed against theoretical relations, which involve not just *velocity* fields but much more subtle flow characteristics such as high order momentum statistics, various accelerations, etc. based on temporal and spatial velocity *derivatives*.

33.3.1.2.2 Direct Polynomial Model

With the King/Jørgensen model, one calibrates a set of coeDcients in relations (33.28 through 33.30) that express the response of the probe as a function of the velocity vector. To interpret any sample of measured responses in terms of a velocity vector, one must invert the set of indirect response relations by solving a system of nonlinear equations at any time point. This can be numerically expensive when large data sets are to be processed. The inversion procedure can be avoided by adopting "direct" calibration relations, which express the velocity components as a function of the measured responses:

$$U_i = f_i(E_1, E_2, E_3, E_4) \qquad (33.32)$$

$$f_i(E_1, E_2, E_3, E_4) = \sum_{klmn} c_{iklmn} P_k(E_1) P_l(E_2) P_m(E_1) P_n(E_1),$$

$$\text{where } P_k(E) = E^k, 0 \le k, l, m, n \le 4, k+l+m+n \le 4 \qquad (33.33)$$

The 70 polynomial coeDcients c_{iklmn} for each velocity component can be found by solving a linear system of equations obtained applying a least square fit procedure to Equations 33.32 and 33.33 and by using appropriate velocities and corresponding voltages across each wire in array from measured calibration data set. Any power of polynomials can be used in the least square procedure but the selection of four as a power of polynomial has its justifications coming from the Kingß Law. In Tsinober et al. (1992) slightly different approach was used. Since three wires are suDcient to determine the full velocity vector, for each combination of three wires out of four in each array, a calibrating

function is produced, which gives the relation between the three velocity components voltages obtained in each of the three wires in array. The calibration functions for every velocity component in each array were constructed as 3D polynomials of the fourth order using Chebyshev orthogonal polynomials. Each such calibration function requires determination of 35 polynomial coeDcients c_{iklm}, which were found from 35 equations obtained by a least-square method using the information from measurements at 81 space points (9 yaw and 9 pitch angle positions) and 7 velocities, total 567 data points in calibration set. As indicated by van Dijk and Nieuwstadt (2004), the matrix equation that is involved in the linear least squares procedure for estimating coeDcients c can be optimized by adopting polynomials with a certain orthogonality relation, e.g., Legendre or Chebyshev polynomials. In Tsinober et al. (1992) the later have been used. When suDcient numerical precision is used, all types of polynomials should, however, give the same function f in (33.8), since all types of polynomials fill the same functional space.

The functional relation (33.33) and appropriate least square procedure can be readily simplified to be applied to x- or single-wires. There are numerous papers reporting successful use of the direct polynomial models with x- and single wires. One of these studies was recently reported in Kit et al. (2007), where a least square fit by fourth order polynomials was used to compute 15 polynomial coeDcients for each velocity component in the x-wires plane. An example of successful application of the method for a 10-element rake of single hot-wire probes that was used to survey the shear layer is given in Gaster et al. (1985). In both experiments the external excitation of the mixing layer was employed that enabled discrimination of the coherent K–H vortices by using the external harmonic signal as a phase reference.

Lemonis and Dracos (1995) suggested using polynomials with coeDcients determined through least square approximation to express the relation between voltages across the wires of multisensor array and velocities. Hence, in this indirect model (van Dijk and Nieuwstadt 2004) the King/Jorgensen relations are replaced by polynomial ones. Then, similarly to the approach adopted in the effective velocity model, one must invert the set of indirect response relations by solving a system of nonlinear equations at any time point. It complicates significantly the analysis in comparison to the direct method without justification of using the effective velocity model that has a stronger physical support.

33.3.1.2.3 Neural Network Calibration Model

Hot films/wires have been recently used in field experiments together with sonic anemometers, and attempts have been made to use suitably processed data from the latter to build a calibration set for the former. During the Cooperative Atmospheric Surface Exchange Study (CASES-99) in October 1999, Skelly et al. (2002) deployed two levels of triple-hot-film and sonic anemometers on a 5.5 m tower. Each hot film was colocated 5 cm from the sonic sensing path on a common boom. A similar setup to that of Skelly et al. (2002) has been deployed in the *Terrain-induced Rotors Experiment* (T-REX) in Owens Valley, California, during March-April 2006. The system was dubbed the *Outdoor*

Three-dimensional *In* situ calibrated *Hot*-film anemometry System (OTIHS). An in situ calibration method was suggested for OTIHS based on the first-use data obtained from simultaneous measurement of colocated sonic and hot-film anemometers, with the former providing velocity data for calibration of the latter (Poulos et al. 2006). The hot film was aligned with the prevailing wind direction measured by the sonic anemometer using an automated platform. This in situ method for hot-film multi-sensor probesflcalibration is based on a random data set from sonic anemometers. These data are unevenly spaced in the data domain, for which the applicability of standard interpolation techniques such as the Polynomial least square Fitting (PF) for producing calibration approximations is in doubt.

Considering the versatility of the OTIHS concept, in that it opens up the possibility of using hot-films/wires in the field without cumbersome calibration devises, Kit et al. (2010) decided to investigate possible approximation techniques for accurate interpolation of velocity–voltage data even in the presence of unevenly distributed calibration data. During this research, the Neural Network (NN) approach emerged as a strikingly successful method. A NN is a computational device that produces appropriate outputs from inputs, based on a selected architecture and subsequent training to perform intended tasks. If the general relationships are mapped as functions, say from Rn to Rm, then a NN with smooth activation functions can approximate continuous functions with compact support (i.e., all continuous functions whose domains are closed and bounded in Rn). This property is a result of the Stone–Weierstrass theorem: *all continuous functions with compact support can be approximated to any degree of accuracy by a neural network of one hidden layer with a sigmoid or hyperbolic tangent activation function* (Kit et al. and references therein). It is this property that was exploited for

calibrating hot-films/wires with uneven distributions of data sets. The following methodology was adopted:

1. A laboratory jet devise (calibration unit or calibrator) was used to obtain an *evenly distributed* calibration data set (henceforth called calibrator based set—CBS) by placing the hot film in the *potential core of the jet*, with a miniature Pitot tube alongside for velocity measurement. In the study by Kit et al. (2010), a combo assembled of two *x*–hot films was used (Figure 33.3). Therefore, calibration unit has to provide only yaw rotation of the probe, which is also enough for probe alignment. The calibration data set consists of 77 points, 7 variations of velocity and 11 even yaw angles varying within ∓30°. The upper *x*–hot film was rotated around its own axis by 90° from horizontal to vertical plane to enable measurements of vertical velocity component. The voltage to velocity calibration was realized using (a) PF and (b) training of a NN. The jet device was used both in the laboratory and field experiments.

2. The hot film was then placed within the probe volume of a sonic anemometer, and the combo was used for the measurements at different *far-r eld sections* of the laboratory jet where the *Vöw is turbulent* as well as in the field to measure atmospheric turbulence. The utility of these data sets are twofold: first to generate calibration data set (called sonic-based calibration data set—SBS) from low resolution sonic data (velocities) combined with low-pass filtered high resolution data (voltages) measured by hot films and to use the same data set and various calibration relations to extract turbulent velocity fields. The SBS was used to determine the coeDcients of a PF and to train the NN, as in CBS.

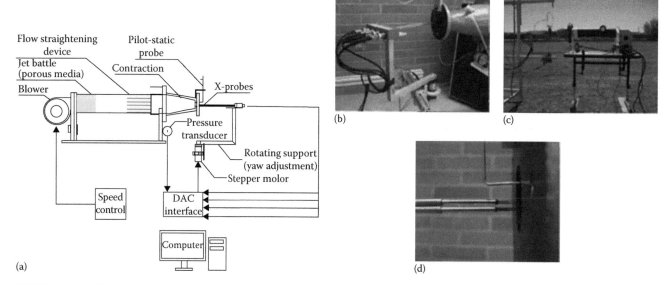

(a) (b) (c) (d)

FIGURE 33.3 Calibration setup: (a) calibration facility, (b) schematic of calibration unit, (c) calibration unit deployed in the field, and (d) a close-up of miniature Pitot tube and two *x*-hot-film probes. (Reproduced with permission from Kit, E. et al., *J. Atmos. Oceanic Technol.*, 27, 23. Copyright 2010, American Meteorological Society.)

TABLE 33.1 A List of Data Sets and Calibration Procedures Used and Nomenclature

Calibration Data Sets/ Approximations	Polynomial Fit (PF)	Neural Network (NN)
CBS (Calibrator-based data Set)	CBS-PF (1)	CBS-NN (2)
SBS (Sonic-based data Set)	SBS-PF (3)	SBS-NN (4)

Source: Reproduced with permission from Kit, E. and B. Gritz, *In-situ* calibration of hot-film probes using a co-located sonic anemometer: Angular probability distribution properties, *J. Atmos. Oceanic Technol.*, 28, 104–110. Copyright 2011, American Meteorological Society.

For the CBS, the interpolations based on NN and PF were found to be essentially identical, implying that both NN and PF work equally well. It is a most striking result indicating a very high potential of NN approach: it does not use any a priori information regarding the physics as King/Jorgensen relations do and yet works as successful as Polynomial Fit and it is in fact much more flexible than Polynomial least square Fit. Table 33.1 lists the data sets and calibration procedures used for all four cases.

33.3.1.3 Angular Density Probability Distributions

In the recent study Kit and Gritz (2011) tried to find out the reason for the failure of the Polynomial Fit algorithm with a sonic based calibration data set (SBS-PF). They developed theoretical expressions for the angular density probability distributions based on the following assumptions: (a) axisymmetric turbulent velocity field, (b) Gaussian density probability distribution for velocity components, (c) weak correlations between the velocity components, i.e., the probability density distribution of the entire velocity vector is a product of probabilities of its components. To investigate angular distribution of $\vec{v}$, it is better to switch to the spherical coordinate system (φ, θ, v) as shown in Figure 33.4.

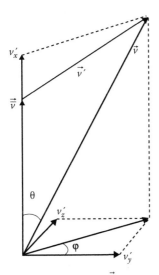

FIGURE 33.4 Velocities at a given point: $\vec{v}$ – mean velocity, $\vec{v}'$ – fluctuating part, $\vec{v}$ – full velocity; c – the deviation angle of $\vec{v}$ from $\vec{\bar{v}}$, h – the azimuth angle. (Reproduced with permission from Kit, E. and B. Gritz, 2011 *In-situ* calibration of hot-film probes using a co-located sonic anemometer: Angular probability distribution properties. *J. Atmos. Oceanic Technol.*, **28**, 104–110. Copyright 2011, American Meteorological Society.)

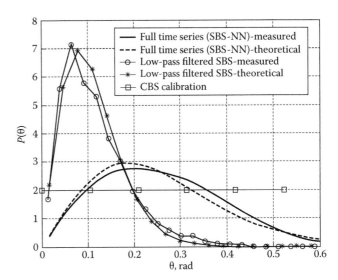

FIGURE 33.5 $P(\theta)$ for sonic and calibrator based calibration data sets and for full data set obtained using SBS-NN. The parameters σ_n, k_v, and k_w are computed from the corresponding data sets. (Reproduced with permission from Kit, E. and B. Gritz, 2011 *In-situ* calibration of hot-film probes using a co-located sonic anemometer: Angular probability distribution properties, *J. Atmos. Oceanic Technol.*, **28**, 104–110. Copyright 2011, American Meteorological Society.)

Θ is the deviation angle of $\vec{v}$ from $\vec{\bar{v}}$. φ is counted in counterclockwise direction from y axis, directed along v_y, $v = |\vec{v}|$. Finally, for an axisymmetric turbulent field, normalizing the velocity v and the standard deviation σ_x as $x = v/\bar{v}$, $\sigma_n = \sigma_x/\bar{v}$, the 2D density probability $P(\theta, x)$ yields

$$P(\theta, x) = \frac{x^2 \sin\theta}{(2\pi)^{1/2} k \cdot \sigma_n^3} \cdot \exp\left(-\frac{\left(x\cos\theta - 1\right)^2 + x^2 \sin^2\theta/k}{2\sigma_n^2}\right)$$

(33.34)

Kit and Gritz (2011) derived one-dimensional (1D) angular density probability function by integrating expression (33.34) over all x and computed the probability densities of velocity angles for the calibration data set as well as for a full velocity data set obtained using the Neural Network approach. It is noteworthy to outline that 1D probability function of velocity modulus can be similarly derived by integrating expression (33.34) over θ. The computed and experimental results are presented in Figure 33.5. The agreement between measured and theoretical angular probability distributions is good. The results indicate that the angular density probability of the low-pass filtered calibration data set is twice as narrow as that of the full velocity time series. This result can explain the failure of the Polynomial Fit to reconstruct satisfactorily the full velocity time series, due to the intrinsic property of this algorithm to ascribe a large weight to the highly concentrated points and a light weight to the thinly concentrated points while performing fitting. Brief description of the NN used in calibrations and more details including example of the matrix of activation functions and vectors of activation threshold for the hidden and output layers are presented in Kit et al. (2010).

References

Balint, J. L., Vukoslavcevic, P., and J. M. Wallace, 1987 A study of the vortical structure of the turbulent boundary layer. In *Advances in Turbulence* (eds. G. Comte-Bellot and J. Mathieu), Springer, Berlin, Germany, pp. 450–464.

Champagne, F. 1978 On the fine-scale structure of the turbulent velocity field. *J. Fluid Mech.*, **6**, 67–108.

Cohen, J. and I. Wygnanski, 1987 The evolution of instabilities in the axisymmetric jet. Part 1. The linear growth of disturbances near the nozzle. *J. Fluid Mech.*, **5**, 191–219.

Comte-Bellot, G. 1998 Hot-wire anemometry. In: *Handbook of Fluid Dynamics* (ed. R. W. Johnson), CRC, Boca Raton, FL, pp. 34.1–34.29.

van Dijk, A. and F. T. M. Nieuwstadt, 2004 The calibration of (multi-)hot-wire probes. 2. Velocity-calibration. *Exp. Fluids*, **6**, 550–564.

Gaster, M., Kit, E., and I. Wygnanski, 1985 Large-scale structures in a forced turbulent mixing layer. *J. Fluid Mech.*, **150**, 23–39.

Glezer A., Katz Y., and I. Wygnanski, 1989 On the breakdown of a wave packet trailing the turbulent spot in a laminar boundary layer. *J. Fluid Mech.*, **8**, 1–26.

Gulitski, G., Kholmyansky, M., Kinzelbach, W., Luthi, B., Tsinober, A., and S. Yorish, 2007a Velocity and temperature derivatives in high-Reynolds-number turbulent flows in the atmospheric surface layer. Part 1. Facilities, methods and some general results. *J. Fluid Mech.*, **589**, 57–81.

Gulitski, G., Kholmyansky, M., Kinzelbach, W., Luthi, B., Tsinober, A., and S. Yorish, 2007b Velocity and temperature derivatives in high-Reynolds-number turbulent flows in the atmospheric surface layer. Part 2. Accelerations and related matters. *J. Fluid Mech.*, **589**, 83–102.

Gutmark, E. and I. Wygnanski, 1976 Planar turbulent jet. *J. Fluid Mech.*, **3**, 465–495.

Hinze, J. O. 1975 *Turbulence*, McGraw Hill, New York.

Hussain, A. K. M. F. 1983 Coherent structures-reality and myth. *Phys. Fluids*, **26**, 2816–2850.

Jorgensen, F. 1971 Directional sensitivity of wire and fiber-film probes. *DISA Info.*, **1**, 31–37.

Katz, Y., Seifert, A., and I. Wygnanski, 1990 On the evolution of the turbulent spot in a laminar boundary layer with a favorable pressure gradient. *J. Fluid Mech.*, **2**, 1–22.

King, L. V. 1916 The linear hot-wire anemometer and its application in technical physics. *J. Franklin I.*, **8**, 1–25.

Kit, E., Cherkassky, A., Sant, T., and H. J. S. Fernando, 2010 *In-situ* calibration of hot-film probes using a co-located sonic anemometer: Implementation of a neural network. *J. Atmos. Oceanic Technol.*, **2**, 23–41.

Kit, E. and B. Gritz, 2011 *In-situ* calibration of hot-film probes using a co-located sonic anemometer: Angular probability distribution properties. *J. Atmos. Oceanic Technol.*, **8**, 104–110.

Kit, E., Krivonosova, O., Zhilenko, D., and D. Friedman, 2005 Reconstruction of large coherent structures from SPIV measurements in a forced turbulent mixing layer. *Exp. Fluids*, **9**, 761–770.

Kit, E., Wygnanski, I., Krivonosova, O., Zhilenko, D., and D. Friedman, 2007 On the periodically excited, plane turbulent mixing layer, emanating from a jagged partition. *J. Fluid Mech.*, **8**, 479–507.

Kholmyansky, M., Tsinober, A., and S. Yorish, 2001 Velocity derivatives in the atmospheric surface layer at $Re_\lambda = 104$. *Phys. Fluids*, **3**, 311–314.

Kline, S. J., Reynolds, W. C., Schraub, F. A., and P. W. Runstadler, 1967 The structure of turbulent boundary layers. *J. Fluid Mech.*, **0**, 741–773.

Lemonis G. and T. Dracos, 1995 A new calibration and data reduction method for turbulence measurement by multi-hotwire probes. *Exp Fluids*, **8**, 319–328.

Likhachev, O., Neuendorf, R., and I. Wygnanski, 2001 On streamwise vortices in a turbulent wall jet that flows over a convex surface. *Phys. Fluids*, **3**, 1822–1825.

Nygaard, K. J. and A. Glezer, 1991 Evolution of streamwise vortices and generation of small-scale motion in a plane mixing layer. *J. Fluid Mech.*, **3**, 257–301.

Oster, D. and I. Wygnanski, 1982 The forced mixing layer between parallel streams. *J. Fluid Mech.*, **3**, 91–130.

Perry, A. E. 1982 *Hot-Wire Anemometry*, Clarendon, Oxford, U.K.

Poulos, G. S., Semmer, S., Militzer, J., and G. Maclean, 2006 A novel method for the study of near-surface turbulence using 3-d hot-film anemometry: OTIHS. Preprints, *17th Symposium on Boundary Layers and Turbulence*, San Diego, CA, American Meteorological Society.

Shemer, L., Wygnanski, I., and E. Kit, 1985 Pulsating flow in a pipe. *J. Fluid Mech.*, **5**, 313–337.

Skelly, B. T., Miller, D. R., and T. H. Meyer, 2002 Triple-hot-film anemometer performance In CASES-99 and a comparison to sonic anemometer measurements. *Boundary-Layer Meteorol.*, **0**, 275–304.

Tropea, C., Yarin, A. L., and J. F. Foss (eds.), 2007 *Springer Handbook of Experimental Fluid Mechanics*, Springer-Verlag, Berlin, Germany.

Tsinober, A., Kit, E., and T. Dracos, 1992 Experimental investigation of the field of velocity gradients in turbulent flows. *J. Fluid Mech.*, **4**, 169–192.

Tsinober, A., Vedula, P., and P. K. Yeung, 2001 Random Taylor hypothesis and the behavior of local and convective accelerations in isotropic turbulence. *Phys. Fluids*, **13**, 1974–1984.

Wallace, J. M. and P. Vukoslavcevic, 2010 Measurement of the velocity gradient tensor in turbulent flows. *Annu. Rev. Fluid Mech.*, **4**, 157–181.

Weisbrot, I. and I. Wygnanski, 1988 On coherent structures in a highly excited mixing layer. *J. Fluid Mech.*, **9**, 137–160.

Weiss, J. and G. Comte-Bellot, 2004 Electronic noise in a constant voltage anemometer. *Rev. Sci. Instrum.*, **75**, 1290–1296.

Wygnanski, I. J. and F. H. Champagne, 1973 On transition in a pipe. Part I. The origin of puffs and slugs and the flow in a turbulent slug. *J Fluid Mech.*, **9** , 281–335.

Wygnanski, I., Champagne, F., and B. Marasli, 1986 On the large-scale structures in two-dimensional, small-deficit, turbulent wakes. *J. Fluid Mech.*, **6** , 31–71.

Wygnanski, I., Haritonidis, J. H., and R. E. Kaplan, 1979 On a Tollmien-Schlichting wave packet produced by a turbulent spot. *J. Fluid Mech.*, **9** , 505–528.

Wygnanski, I., Sokolov, M., and D. Friedman, 1976 On a turbulent spot in a laminar boundary layer. *J. Fluid Mech.*, **8** , 785–819.

34

Optical Methods and Unconventional Experimental Setups in Turbulence Research

Roi Gurka
Ben Gurion University of the Negev

Eliezer Kit
Tel Aviv University

34.1 Introduction

Particle image velocimetry (PIV) and other optical methods have been rapidly developed and advanced during the last two decades (1990s and 2000s), which have provided other means of access to the fundamental properties of turbulence that include assessment of the velocity gradient tensor (full or some components) and determination of the vorticity field (full vector or some of the components). This chapter briefly describes and highlights these remarkable developments and points out some of the most important things about the emerging methodology of turbulence measurements and the discovery of important features of coherent structures that develop in anisotropic turbulence due to shear, stratification, rotation, or electromotive forces. The essence of this methodology may be best expressed by the following citation:

> And this experiment you will make with a square glass vessel, keeping your eye at about the center of one of these walls; and in the boiling water with small movement you may drop a few grains of panic-grass because by means of the movement of these grains you can quickly know the movement of the water that carries them with it. And from this experiment you will be able to proceed to investigate many beautiful movements which result from one element penetrating into another (i.e., airflinto water). (Leonardo Da Vinci)

One of the most intriguing problems in measuring turbulent flows is determination of vorticity. Besides the multiple arrays probes that enable to assess the velocity gradient tensor and consequently strain tensor and vorticity by computing finite differences of the appropriate velocities at neighboring points measured by various arrays placed at different span-wise and lateral locations as explained in Chapter 33, other methods have been suggested. In particular, specially constructed Potential Difference Probe allows determining the vorticity component directed along the magnetic field. This approach has its roots in the field of Magnetohydrodynamics (MHD) and can be used in water when the latter is slightly salted.

In this section we present few nonconventional setups and instruments that were developed to study some special features of turbulent flow, e.g., (1) water tunnels to generate a water flow with strong magnetic field imparted to measure vorticity and helicity with an aid of potential difference probe based on MHD principles; (2) tanks with oscillating grids to measure parameters of Eulerian spectra in mean-zero velocity turbulence using two-component LDA and flying x-hot-film; (3) setups for shear generation at the density interface for studying turbulent mixing across the interface employing a specially developed *dual-probe-volume*, laser Doppler velocimeter, x-hot-films and various type of miniature conductivity probes for salinity measurements; (4) experimental rig that assembles wind tunnel and wave flume designed to study the wind-wave generation using various instrumentation: particle tracking velocimetry, x-hot-films, static and dynamic pressure tubes, wave gauges; and (5) two-phase facilities equipped with fast response sensors for determination of void fractions.

34.2 Optical Methods

34.2.1 Particle Image Velocimetry

34.2.1.1 Setup Description

Particle image velocimetry (PIV) is a nonintrusive state-of-the-art technique for flow measurements (e.g., Adrian, 1991; Raffel et al., 2007). The PIV technique is based on image recording of the illuminated flow field using seed particles. The experimental setup of a PIV system typically consists of several subsystems (see Figure 34.1). In most applications tracer particles are added to the investigated flow. A light sheet formed from a coherent light source illuminates the particles. The light scattered by the particles is recorded either on a single frame or on a sequence of frames. The displacement of the particle images between two consecutive light pulses is determined through evaluation of the PIV recordings and by applying a spatial cross-correlation function. Advanced postprocessing tools are used to handle the amount of acquired data. Figure 34.1 describes a typical setup for PIV recording. Small tracer particles are introduced to the flow. A plane within the flow is illuminated twice by means of a laser light sheet. The time delay between the pulses depends on the mean velocity and the image magnification. It is assumed that the tracer particles follow the local flow velocity between the two consecutive illuminations. The light scattered from the tracer particles is imaged via an optical lens on a digital camera.

The images once recorded are transferred to the computer disks directly through a frame grabber or indirectly through a video streaming card.

The images, acquired as pairs (frame A and frame B) corresponding to the two laser pulses, are then correlated using a cross-correlation function and image-processing tools in order to provide the velocity field. The velocity field obtained from the "classical" PIV technique is limited to two components only within the plane of imaging and provides high accuracy measurements. The distinction of the term "classical" is made to emphasize the add-ons that were added to the PIV technique along the years which enabled to increase the number of parameters that can be measured, as will be explained briefly in paragraph on "small-scale measurements using PIV."

The effectiveness of the measurement results strongly depends on a large number of parameters such as particles concentration, size distribution and shape, illumination source, recording device, and synchronization between the illumination, acquisition and recording systems (Huang et al., 1993). An appropriate choice of the different calculation parameters of the cross correlation analysis (e.g., interrogation area, time between pulses, scaling) will influence the results accuracy. Detailed descriptions of various implementations of the PIV method are given in Adrian (1991). The PIV components are described herein.

34.2.1.2 Seeding

The tracer particles are the information carriers. The introduction of particles into the flow is required to mark the flow field and to follow adequately the fluid motion. The accuracy of the velocity field is ultimately limited by the ability of the tracer particles to follow the instantaneous motion of the continuous phase. A compromise between reducing the particle size, to improve flow tracking, and increasing the particle size to improve light scattering is a necessity (Melling, 1997). To assure suDcient details throughout the flow field a higher spatial concentration of particles is generally needed. A uniform particle seeding size is desired in order to avoid excessive intensity from larger particles and background noise, decreasing the accuracy, from small particles. The choice of optimal diameter for seeding particles is a compromise between an adequate tracer response of the particles in the fluid, requiring small diameters and a high signal-to-noise ratio (SNR) of the scattered light signal, necessitating large diameters. Additional conditions have to be satisfied; the tracer particles should be naturally buoyant, chemical inert, of high index refraction, smaller than Kolmogorov scale (for PLIF, smaller than Batchelor scale), and of spherical shape so that drag on the particles will be minimized. The importance of using the appropriate particles, both type and size is crucial for the accuracy of the velocity measurements, especially of the higher moments of the velocity field and its associated derivatives (Hadad et al., 2010). Melling (1997) proposed basic guidelines to estimate the mean particle diameter to be chosen for a given flow field through simple calculation of the turbulent properties of the flow.

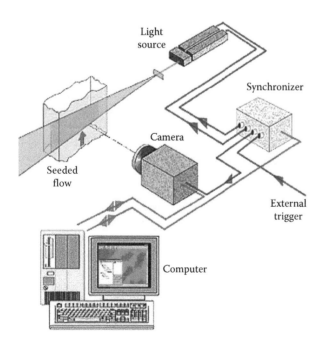

FIGURE3 4.1 Typical PIV setup.

34.2.1.3 Illumination

The illumination source is required to illuminate the tracer particles. The goal is to hit the particles with suDcient amount of energy, in order for diffraction to occur and light to be scattered from the particles. Therefore, the source of light should be collimated and monochromatic beam that can be easily shaped and maneuvered toward the test section as a thin light sheet. The most common light source used in PIV applications is a solid-state double-head Nd: YAG pulsed laser that emits light with a wavelength of 532 nm. The standard architecture of a PIV laser consists of two lasers fired independently at the required pulse separation. Therefore, the time separation can be adjusted to match the mean speed of the flow and the imaging magnification. The repetition rate of the YAG lasers is low; therefore, temporal resolution is not available. Recent developments in laser technology improved the performance of diode pumped Nd: YLF lasers. Their repetition rate can reach up to 10 kHz with pulses of energy ranging between 10 and 30 mJ/pulse at 527 nm. Once light is emitted from the laser, it will go through a set of optical lenses to form a thin sheet ("slice") of light, which will illuminate the flow field. The sheet is created using cylindrical lens, which generate a line of light from the collimated beam by expanding the light in one axis only. The thickness of the light sheet has a great importance in the PIV procedure; the goal is to increase the laser intensity and to avoid false vectors due to outliers. Therefore, a spherical lens is placed following the cylindrical lens (Adrian, 1991). A different light source that can be applicable for PIV applications is LED illumination. LED features similar properties to laser except for the lack of coherency. The drawback of using LED is that it is hard to manipulate and form a thin coherent light sheet, yet, the advantage is the low cost and maintenance. Recently, few works utilized LED in PIV measurements (Willert et al., 2010).

34.2.1.4 Imaging

The scattered light from the flow tracing particles is recorded on a digital camera (as shown in Figure 34.2; circular jet flow imaged by PIV, jet ext velocity is equal to 40 m/s).

Digital cameras can be either based on CCD technology or CMOS (Hain et al., 2007). An image of the tracer particles is recorded on the surface of the image sensor through an optical lens mounted on the camera. Both, the optical lens and the image sensor size determine the flow scales resolution and the dynamic range of velocities to be measured (Adrian, 1997).

Digital cameras are characterized by their image size (in pixel), frame rate, and dynamic range. Typical values for CCD cameras range from 1K × 1K up to 3 K × 4 K operating at a rate of about 10 frames per second, while CMOS cameras are capable of operating at much higher frame rates (up to 10 kHz) with similar image size. These cameras are based on different technologies (Hain et al., 2007). The CCD cameras have a higher quantum eDciency compare to CMOS. In addition, CMOS are noisier and require a precalibration process, which reduce the accuracy level of the acquired images. However, CMOS cameras with fast lasers enable to resolve the temporal scales of moderate turbulent

FIGURE 34.2 PIV image of a circular jet, exit velocity is 40 m/s.

flows in PIV mode. This feature provides a two-dimensional (2D) velocity field in time; therefore, the derivative *du/dt* can be calculated without any further assumption. For both imaging devices, the dynamic range is an important factor in the PIV technique. The dynamic range defines the gray scale level for each pixel. This means that each pixel will obtain a number representing its gray value. The higher the range means that the gray level distribution will be more homogeneous and later on will result in an accurate correlation value. The trade-off is time and computer resources since the image size is a function of its dynamic range and the image correlation to obtain the velocity fields consumes time, which is proportional to the image size and its dynamic range (Taylor et al., 2010).

34.2.1.5 Synchronization

In a PIV system, synchronization between the lasers and cameras has to be set. Figure 34.3 shows a schematic configuration of the time delays in a PIV system. The main principle is to trigger the laser to emit light once the camera shutter is open; a first image is acquired, than after a short time delay, a second laser

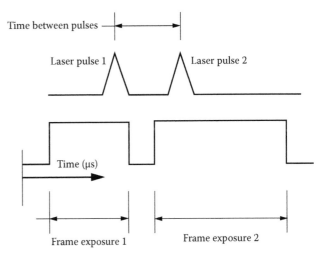

FIGURE 34.3 Temporal design of the PIV system.

will fire while the camera opens its shutter again; this is the so-called double-exposure mode.

The time delay between the laser pulses, *dt*, is set based on the mean velocity, *U*, of the flow and a predetermined pixel displacement, *X*, and the calibration ratio, *R*:

$$dt = \frac{X[\text{pixel}]*R[\text{m/pixel}]}{U[\text{m/s}]} \qquad (34.1)$$

This is motivated by guidelines proposed by Adrian (1991) to obtain values of pixel displacement larger than 6–8. Once two pairs of images have been acquired, the cycle repeats itself for the second pair and so on. Since the nature of the flow is turbulent, statistical analysis will be performed; therefore, a sufficient amount of images should be collected. The amount of data is limited by the convergence of the mean and rms properties and the capacity of the hard drives on the computer. For high-speed system, where fast cameras are used, an additional bottleneck emerges: data transfer rate. The transfer rates of high-speed cameras exceed 1 Gbyte/s. For this purposed, other type of data transfer has to be used instead of standard image cards. One of the available solutions is the use of video card, which stream the data into RAID disk arrays. This allows the acquisition of thousands of images per second directly to the disk (Taylor et al., 2010).

34.2.1.6 Control and Processing

The acquired images are analyzed using image processing tools and cross-correlation functions in order to obtain the 2D, two-component velocity field. The main steps are as follows:

1. Each PIV image is divided into interrogation areas over a square or rectangular grid.
2. For each window the local displacement is calculated based on a cross-correlation function.
3. Subpixel displacement is obtained using 2D Gaussian interpolation.
4. Erroneous vectors are removed by means of filters (e.g., global and local median filter).
5. Interpolation is applied at the positions of the missing data points.

For local velocity vector evaluation, the PIV images are divided into small sub areas called "interrogation areas." The basic principle is that the particles captured in a given interrogation window at the first illumination (corresponds to the first image) moved within the window frame for the second illumination (corresponds to the second image). A velocity vector is obtained for every chosen interrogation window; therefore, the number of vectors obtained for one map is the number of interrogation windows in an image. The velocity for each interrogation window equals to the mean pixel displacement of the particles over the time delay between the two laser pulses. The mean pixel displacement is obtained through correlation analysis and it corresponds to the distance from the center of the interrogation window where the correlation function obtains its maximum

value (Keane and Adrian, 1992). The cross-correlation function is applied to the gray level intensity for each pixel in a chosen interrogation window. It can be calculated directly (Roth and Katz, 2001) or indirectly using digital FFT to speed up the analysis time. The most common solver is the indirect operation (Raffel et al., 2007).

The pixel displacement is an integer value due to the discretization and using digital imaging. Therefore, the velocities exhibit a range of integer values, which limits the accuracy of the results. To overcome this problem, a subpixel interpolation is utilized. There are few interpolation methods that increase the accuracy of the results (Huang et al., 1997). The interpolation is based on fitting a known function (parabolic, Gaussian, etc.) to the 2D correlation values as shown in Figure 34.4. The Gaussian curve is based on fitting the curve to the three highest correlation values. The maximum value of the curve is the subpixel location, and, consequently, the velocity varies as a floating number rather than an integer. The use of subpixel interpolation increase the accuracy of the results, yet, it introduces an additional source of error: peak locking. The problem of peak locking is inherent in the process and to resolve it, one needs to either not use the subpixel interpolation or bypass it using other techniques (Chen and Katz, 2005).

The process is repeated for all interrogation areas of the PIV images. The quality of the data is determined by the quality of the images acquired. However, even with high-quality images that exhibit good contrast and a high concentration of seed particles, there is a finite probability that the outcome of the correlation will be erroneous or unrelated to the flow field being investigated. This can arise due to insuDcient particle pairs, out-of-plane velocity, strong velocity gradients, etc. For this reason, it is generally necessary to perform postprocessing on PIV data sets, removing the erroneous vectors with interpolation and applying smoothing techniques to reconstruct areas depleted by validation (Nogueira et al., 1997). Analysis and postprocessing software are essential for accurate interpretation of the data. Numerous works have been performed to provide additional mathematical tools in order to enhance the accuracy level of the correlation process (Raffel et al., 2007; Tropea et al., 2007 amongst many others including the *International Symposium on*

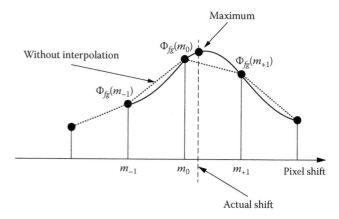

FIGURE 34.4 Subpixel interpolation.

PIV which is being held every 2 years from 1999), the image processing, and the filtering functions.

Recent developments are aimed towards resolving an additional bottleneck problem, which is the data size, acquired to be processed. Especially, for high-speed systems and multi-camera systems using large sensor size, the amount of data per experiment can reach more than few TB. Therefore, standard PIV correlation methods are not suitable. New developments are in the direction of graphical processing units (GPU), multi-core parallel implementations and real-time image processing on FPGA.

34.2.1.7 PIV Measurements of Small-Scale Turbulence

There is a special interest to obtain spatial distribution of the velocity field (Figure 34.5) and its gradients for the small scale turbulence. This is motivated by the fundamental aspects of turbulence such as dissipation and enstrophy (velocity gradients). From an environmental fluid mechanics point of view, the characterization of the small scales is crucial for the understanding of buoyancy flux related to cross gradients of salinity/temperature, sediment transport, turbidity, and dispersion transport such as contaminants or in a more general sense: turbulent diffusion. Therefore, it is essential to obtain high accurate measurements of the spatial distribution of the velocity fields in the smallest scales of the flow. In order to estimate the diffusion coeDcient and the dissipation constants, one needs to estimate the velocity gradients. Furthermore, most of the environmental flows are neither homogeneous nor isotropic; this requires instantaneous and spatial measurements of the velocity fields. Using PIV, one can estimate the velocity gradients in a given plane, e.g., five velocity gradients (du/dx, du/dy, dv/dx, dv/dy, dw/dz). For this purpose

the measurement resolution, accuracy, and suDcient statistics are crucial (Saarenrinne and Piirto, 2000).

For PIV, the resolution of the measurement is determined by the interrogation window size. This should be smaller or of the same order of magnitude as Kolmogorov half-length scale. The choice of the interrogation size is well defined; it is based on the image magnification, the time duration between the laser pulses and the desired pixel displacement. The pixel displacement should be as large as possible to minimize peak-locking effects and additional noise resulting from the correlation procedure. The obtained velocity field is proportional to the pixel displacement and therefore its accuracy relies on the aforementioned parameters and appropriate seeding. The spatial resolution is the grid spacing between the interrogation windows (usually taken as half of the interrogation window).

The accuracy of the velocity gradients obtained from PIV data depends not only on the velocity measurements but also on the numerical gradient scheme. The error resulting from PIV for the velocity gradients is relatively high (Westerweel, 2008) and should be handle with care. There are three distinct approaches to calculate dissipation/enstrophy through numerical derivation. The direct approach is to apply forward/backward/central difference scheme to the PIV velocity realizations. Several schemes of increasing order are listed in Raffel et al. (2007). For example, Liberzon et al. (2001) have showed that Richardson interpolator of fifth order (least square) is the most robust and accurate for the typical noise generated by PIV data. It is noteworthy that the error using standard derivation schemes increases linearly with the number of realizations whilst the least square technique exhibits a somewhat constant error. Another approach (see Raffel et al., 2007) is to apply "direct" estimate of dissipation using Kernel differentiation, for each point in the grid using n closest neighbors. This method is typically more accurate but not valid close to the boundaries. The third approach proposed by Lourenco and Kartopalii (1995) where the PIV vectors are extracted on a non-structured grid. This approach enables to overcome one of the main obstacles in gradient calculation using structured grid where ΔX and ΔY are fixed. Additional consideration has to be applied for the nature of the window size and spacing when performing spectral analysis (Hackett et al., 2009) to estimate dissipation where the window size serves as a box filter and attenuates the spectral distribution.

Since PIV is a planar technique, the dissipation will mostly be under/over the estimate of the real values. In order to fully resolve the dissipation constants, one needs to measure the velocity filed in three dimensions. There are several extensions of PIV that provides three-dimensional (3D) measurement of the velocity field (see review by Arroyo and Hirsch, 2008), such as tomography, holography, multi-plane, etc. Usually, these extensions have two major drawbacks: complexity and limitation either in the velocity or in the volume. These techniques enable one to measure the 3D velocity fields and in consequence to estimate the nine spatial velocity gradients using similar scheme as in 2D PIV.

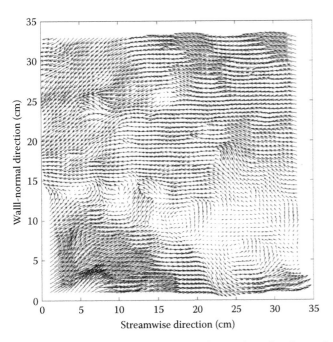

FIGURE 34.5 Velocity vector map of the bottom boundary layer of the Atlantic Ocean.

However, there are physical processes that the transport phenomena require to follow the movement of the fluid and information of dissipation and diffusion as the flow evolves, meaning to obtain Lagrangian information. For these cases, one might find useful to use a different approach to obtain the 3D velocity gradients in a field abbreviated as PTV (Particle Tracking Velocimetry) (Luthi et al., 2005). The technique is based on tracking of multiple tracer particles in a 3D volume based on photogrammetric principles. It is noteworthy that this method is based on Lagrangian principles, which are directly linked to the physics of the flow. Yet, the main drawback is the limitation by Reynolds number and the size of the field of view.

34.2.2 Planar Laser-Induced Fluorescence

34.2.2.1 General Description

The use of planar laser-induced fluorescence (PLIF) has become widespread in fluid mechanics experiments. PLIF is a nonintrusive technique for measuring the spatial scalar concentration field in continuous flow systems. PLIF is conceptually based on flow visualization, yet it provides a quantitative description of the investigated scalar. This has become feasible as laser and digital imaging had become in use in fluid mechanics measurements. Essentially, PLIF instrumentation is similar to PIV and the user does not have to change the PIV setup except for using a different type of particles and adjust the optical path. The underlying physical principle of this technique is based on the absorption and subsequent reemission of photons by fluorescent dye tracers. This allows, detecting small changes in the concentration accurately. The rapidity of the fluorescent emission compared to velocities in the flow allows following rapid concentration fluctuations. In the case of the passive (nonreacting) tracer, the effect on the existing flow condition such as pollutant dissolution and dispersion can be studied.

To be able to utilize the properties of a fluorescent dye in the flow field, information relating intensity to a known parameter, which characterizes the flow, is desired. In order to calibrate the images to obtain quantitative results, all parameters have to be characterized: light intensity, concentration, optical system, and experimental setup. The Beer–Lambert law is used to relate the absorption to the concentration of the medium. Considering all the parameters into Beer–Lambert law allows predicting the concentration variation from intensity variation within the measuring volume (Gaskey et al., 1990).

Some popular tracers in water are fluorescein and disodium salt, Rhodamine B, and Rhodamine 6G. Tracers with high absorption coeDcient and high quantum eDciency are generally favorable. The choice of the particles is based on the type of the flow medium: gas or liquid, and the scalar to be measured: concentration, temperature, pH, etc. PLIF was originally used in 1D form (Walker, 1987; Gaskey et al., 1990) by utilizing a laser source, optical components, receiving optics,

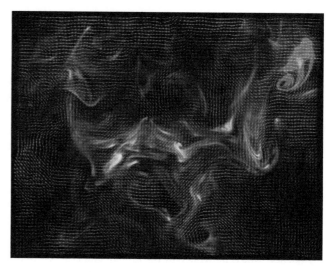

FIGURE 34.6 (See color insert.) Simultaneous measurement of velocity and scalar using PLIF and PIV.

photomultiplier, and signal processing unit. The evolution of the LIF technique is directly connected to the development of the digital solid-state sensor cameras (CCD cameras). The choice of imaging the flow replaced the photomultiplier and receiving optics making possible to measure simultaneously concentration not only in a single point but also along the line and later along the 2D plane. An optical filter is required with a cutoff wavelength similar to the wavelength of the incident light beam. The filter passes only the fluoresced light and hence reduces the noise to signal ratio of the image recording. The laser used in planar laser induced fluorescence must have part of its power band within the absorption band of the dye being used (Crimaldi, 2008). This makes laser selection important, as it is desired to utilize the peak absorption spectrum of the fluorescent dye. For example, acetone has an absorption band that lies between 225 and 320 nm, making it suitable for excitation by a XeCl excimer (λ = 308 nm), KrF (λ = 248 nm), or a frequency quadruple Nd: YAG (λ = 266 nm) laser. The laser beam emitted from a laser source is passed through a set of cylindrical and spherical lenses to form a thin laser sheet which will illuminate the investigated flow field similar to PIV. The light sheet created with a cylindrical lens has a Gaussian intensity distribution and therefore, corresponding gray value corrections are necessary van Cruyningen et al. (1991). Figure 34.6 demonstrates the utilization of 2D-PLIF in applied research. It shows a contaminant plume captured using PLIF in a water tunnel. PIV yields simultaneous velocities (arrows), revealing the interaction between the velocity field and the concentration of contaminants (Sarathi et al., 2010).

34.2.2.2 PLIF Image Corrections and Calibration Procedure

The PLIF technique relies on the relationship between the fluoresced light intensity, the concentration of tracer dye, and the intensity of the laser. At low concentrations levels (<120 μg/L) there is a linear relationship between the dye concentration and emitted light intensity. Calibration for basic concentration measurements is carried out using a similar method. A solution is prepared at a known concentration or temperature. The solution is isotropic such that it is calibrated for only each individual value of the scalar field. The fluorescent intensity of the solution is then measured and calibrated to the known value of the system. Since the image of the control volume is divided into pixels, each pixel must be calibrated based on the gray scale value of the intensity related to the scalar measurement, where local intensity of the laser is disregarded. The background emission could be found by averaging the intensity level over number of images. Each of the calibration images should to be corrected for laser attenuation due to the presence of the dye.

34.2.3 Laser Doppler Technique

The laser Doppler technique, most frequently referred to as laser Doppler anemometer (LDA) or laser Doppler velocimeter (LDV) is a widespread technique for point-wise velocity measurements. Some decisive advantages of the technique over more conventional probe-based techniques (hot-wire/film sensors) can be mentioned: nonintrusiveness, directional sensitivity, high spatial and temporal resolution, and high accuracy. The fringe model provides an alternative explanation of the laser Doppler principle, in which the wave fronts of two coherent beams interfere in the intersection volume, forming interference fringes (Figure 34.7). Tracers passing through the volume scatter modulated light yielding the Doppler frequency at the detector. The velocity can be readily assessed as the Doppler frequency f_D at the output of photomultiplier

$$u = \Delta x f_D = \frac{\lambda_b}{2\sin(\theta/2)} f_D \qquad (34.2)$$

where
  u is the particle velocity
  Δx is the fringes spacing
  λ_b is the beam length
  θ is the angle between beams
  f_D is the Doppler frequency

More details are in Tropea et al. (2007).

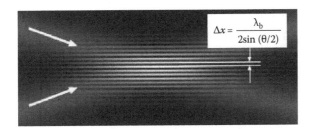

FIGURE 34.7 Interference pattern in the measurement volume-fringe module. (With kind permission from Springer Science + Business Media: *Springer Handbook of Experimental Fluid Mechanics*, 2007, Tropea, C., Yarin, A.L., and Foss, J.F. (Eds.).)

34.3 Experimental Setups

34.3.1 Potential Difference Probe for Turbulence Measurements, Vorticity and Velocity Statistics

It is very appealing to use Faraday's law relating the value of generated electromotive force to the velocity at a given magnetic field for measurements and analysis of turbulent velocity field. It seems to be very natural to use this methodology for measurements in liquid metals, where standard optical methods are completely inapplicable due to the lack of transparency and hot-film anemometry is questionable due to the high temperature and hostile environment. However, as explained in Tsinober et al. (1987) based on experiments performed in weakly salted water, the method has serious limitations that are discussed in the sequel.

The purpose of induction velometry is to deduce a velocity $\mathbf{u} = [u_i]$ ($i = 1, 2, 3$) from the measurements of an electrical field $\mathbf{e} = [e_i] = [\partial\varphi/\partial x_i]$ induced by the motion of an electrically conducting fluid in the presence of an external magnetic field $\mathbf{B}$. In the following it is assumed that the conductivity of the fluid is constant and large, the frequencies in the flow are not very large, and the magnetic Reynolds number is small. These conditions are usually well fulfilled for electrolytes and liquid metals up to frequencies at least as large as $10^5\,\mathrm{s}^{-1}$. Under these conditions Ohm's law in the form

$$\mathbf{j} = \sigma\,(\mathbf{e} + \mathbf{u} \times \mathbf{B}) \qquad (34.3)$$

is valid, the electrical field is a potential one, $\mathbf{e} = -\mathrm{grad}\ \varphi$ and the induced magnetic field is negligible. The external magnetic field is supposed to be uniform and stationary, though many of the results presented remain valid for a nonhomogeneous and/or time-dependent imposed magnetic field. The quantity $\mathbf{u} \times \mathbf{B}$ cannot be measured precisely unless the current $j = 0$ or is small in comparison with either $\mathbf{u} \times \mathbf{B}$ or $-\sigma$ grad φ. This occurs in some special cases, as for example in 2D flow with the magnetic field perpendicular to the plane of the flow (Kit, 1970). Thus, in general, the measured electric field—grad φ cannot be considered as $\mathbf{u} \times \mathbf{B}$.

In fact there is little hope of obtaining this relation in the general case of 3D nonhomogeneous nonisotopic turbulent flow. From the equation div $\mathbf{j} = 0$ and Ohm's law (34.3) follows the Poisson equation (34.4) for the electrical-field potential. The electrical field is thus uniquely defined by only one component of vorticity and the boundary conditions:

$$\nabla^2\varphi = \mathrm{div}\,(\mathbf{u} \times \mathbf{B}) = \mathbf{B} \cdot \boldsymbol{\omega} \qquad (34.4)$$

where $\boldsymbol{\omega} = \mathrm{rot}\ \mathbf{U}$. The electric field does not "feel" the two other vorticity components of the flow field, which are perpendicular to the magnetic field. Therefore, the only information about the flow field that could be obtained from the electrical field measurements is that which follows from the knowledge of $\boldsymbol{\omega}_B$.

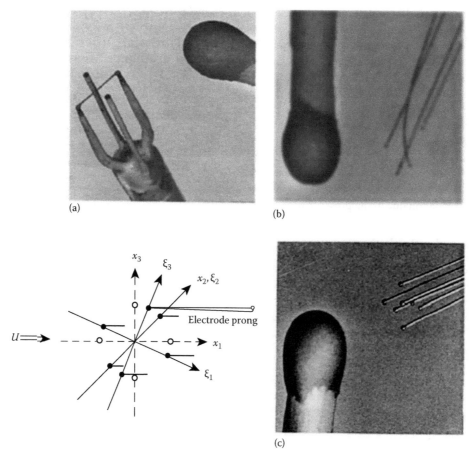

FIGURE 34.8 Probes used in the different experiments: (a) a two-electrode probe with a hot-film-type probe, (b) a four-electrode probe, (c) a schematic of the tip of a seven-electrode vorticity probe. The seven tips of the electrodes in the plane x_1x_3, are turned 22.5° relative to the x_1 in order to minimize the disturbances between them. Each one of the probes is photographed together with the head of a match. (From Tsinober, A. et al., *J. Fluid Mech.*, 175, 447, 1987. Reproduced with permission from Cambridge University Press.)

It is clear that in flows where ω_B is the only nonzero component of vorticity, measurements of the electrical field enable precise information on the velocity field to be obtained, as for the aforementioned example of 2D flow when $\omega \parallel B$, and $e_1 = Bu_3$, $e_3 = -Bu_1$. On the other hand, in the case of 2D flow such that $\omega \perp B$, grad φ = const and no local information about the flow field can be obtained. Nevertheless, in isotropic turbulence some interesting relations can be obtained. Probe details are shown in Figure 34.8.

As reported in Tsinober et al. (1987), most of the experiments exhibited good agreement with theoretical relations derived for isotropic turbulence despite a rather crude technique of simulation of isotropic homogeneous turbulence.

The seven electrode probe (Figure 34.8) was used to produce a central-difference approximation of the Laplacian $\nabla^2\varphi$, which as follows from Equation 34.4 allows *the vorticity component parallel to the imposed magnetic reld to be measured*. To minimize the mechanical contamination of the signal the approximating grid was turned 22.5° in the x_1x_3-plane (Figure 34.8c). This is possible owing to the invariance of a Laplace operator to rotation.

The method has the principal advantage *to be an absolute one, i.e., it does not require calibration at all*. The method was further advanced by Kholmyansky et al. (1991) to enable determination of one of ingredients $u_1\omega_1$ of full helicity $\mathbf{u}\cdot\mathbf{\omega}$. The experiments were performed in the big salt (1% concentration) water tunnel of the Laboratory for Turbulent Structures studies, Faculty of Engineering, Tel Aviv University. The magnetic field was about 1 T and was produced by a superconducting solenoid cooled by liquid helium. Special grid with circular mesh 32 mm in diameter and solidity 0.55 was used to perform experiments *with controlled sign of mean helicity of the turbulent fluctuations*. The results (Kholmyansky et al., 1991) confirmed the phenomenon of symmetry breaking and provided new results about properties of the vorticity field.

34.3.2 Oscillating Grid Experiments: "Flying" Hot Films, LDA Applications, Conductance Probes for Salinity Measurements

The oscillating grid experiments are widely used for studying of decaying turbulence in homogeneous and stratified fluids and of vertical mixing through the density interface. A great number of experiments using various modifications of the setups with

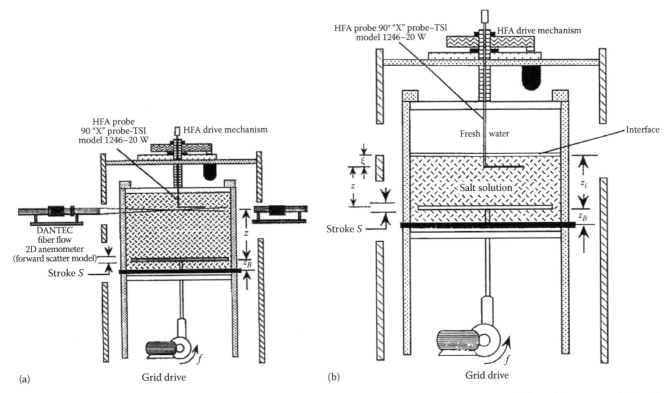

FIGURE 34.9 A schematic of the experimental facility: (a) with homogeneous fluid (From Kit, E. et al., *Phys. Fluids A*, 7(5), 1170, 1995.), (b) with stratified fluid. (From Kit, E. et al., *J. Fluid Mech.*, 334, 293, 1997. Reproduced with permission from Cambridge University Press.)

oscillating grids were conducted in Geophysical Fluid Dynamics Laboratory, Arizona State University, by Professor Fernando, his colleagues and students. Few examples that outline the applications of these setups, various instruments and procedures developed for turbulence studies are presented in this section. Kit et al. (1995, 1997) report on experiments that were conducted in a tank of a square cross section $47 \times 47\,cm^2$ and height 45 cm (see Figure 34.9).

With zero-mean-flow turbulence introduced in a water tank by an oscillating grid, the Eulerian frequency spectra of horizontal and vertical velocity components can be readily studied. The vertical (w') and horizontal (u') turbulent velocity fluctuations are measured using a two-component fiber optic laser-Doppler velocimeter in a forward scatter mode and the spectra are calculated accordingly. The theoretical form of the Eulerian frequency spectrum is proposed by Tennekes (1975):

$$E(\omega) = \alpha \varepsilon^{2/3} u^{2/3} \omega^{-5/3} \quad (34.5)$$

where ε is the rate of dissipation of turbulent kinetic energy. The evaluation of α from this expression is possible using the direct measurements of ε by utilizing x-type hot film that works in a mode of a "flying" wire. The hot film is mounted on a rotating arm driven by a special driving mechanism. The arm rotates and comes to a stop after one revolution.

The simplified effective velocity approach that assumes precise probe geometry is used for calibration of x-type hot film. The rotating arm has a short spin-up period and then achieves a constant speed. A record of 1024 velocity data points is acquired

before the probe comes back to its original position so as to avoid measurements in its own wake. The speed of the rotating probe tip is adjusted so that it is about 10–20 times the RMS velocity. This ensures the applicability of Taylor hypothesis for spatial measurements. The measurements are used to calculate the rate of dissipation ε of turbulent kinetic energy by *using isotropic strain rate relations*. For the calculations of ε the vertical component is mainly used because of its low sensitivity to the noise resulting from probe vibrations caused by the long arm. The normalized skewness $\overline{(\partial u_1/\partial x_1)^3}/\overline{\left((\partial u_1/\partial x_1)^2\right)}^{3/2}$ of stream-wise velocity derivative that is expected to be about −0.4 can serve to separate true turbulence from noise. Values of coeDcient α as appear in Equation 34.5 are reported to be ∼0.7–0.9 in Kit et al. (1995) and are in reasonable agreement with those known from other studies.

The same flow configuration (Figure 34.9b) has been used to study mixing across density interfaces by zero-mean-shear turbulence. In these experiments, stratified by salt fluid replaces the homogeneous one (Kit et al. 1997). The salinity is measured by the Fast Conductivity Sensor that is designed to make very rapid, high resolution measurements of electrical conductivity of water. The sensor consists of four electrodes (platinum spheres) supported by fused glass. Each electrode is electroplated with an amorphous platinum coating. The electronic circuit makes a four terminal measurement of the conductance of the water supplying an AC current between the inner electrodes of the sensor and measuring the AC voltage that develops across the outer electrodes. The spatial averaging volume of the sensor can be estimated to be a sphere of radius 0.5 mm. More details can be found in De Silva et al. (1999) and in Strang and Fernando (2001).

34.3.3 Setups for Shear Generation at the Density Interface

A closed-loop water facility with a dual stack, counter-rotating disk pump assembly designed after that of Odell and Kovasznay (1971) is used to study effects of shear on mixing and turbulence. A schematic of the apparatus is shown in Figure 34.10. The disk pump imparts mean momentum to the upper layer of a two-layer fluid and, hence, generates a velocity difference ΔU between the two layers. The pump consists of two sets of overlapping, counter-rotating disks that push forward the fluid mainly by the action of viscous forces. Stratification was established using salt and an aqueous solution of ethanol. Ethanol was introduced to create an optically homogeneous medium, enabling the use of optical measurement techniques in spite of stratification. For more details see De Silva et al. (1999). A *dual-probe-volume*, laser Doppler velocimeter operating in the forward scatter mode is developed and custom-built to measure the local fluid velocities at two points separated vertically by 0.27 cm. The LDV system is mounted on an adjustable traverse with 2 degrees of freedom, one parallel to the flow direction and the other in the vertical

direction, so that the measuring volumes could be positioned near the fluid interface. A *dual-tip*, four-electrode, conductivity probe (custom designed and manufactured by Precision Measuring Engineering), also having a vertical separation of 0.27 cm between the tips, is used to measure the buoyancy difference. The conductivity probe tips were bent 90° so that the tips face the oncoming fluid stream.

The tips of the probes are thin enough (0.5 mm) that they do not disturb the flow appreciably. The conductivity probe assembly is mounted on a vertical traverse and is positioned in such a way that the probe volumes of the LDV and the conductivity sensors lie in the same vertical plane, but horizontally displaced by about 0.2 cm with the conductivity probes located downstream of the LDV probe volumes. By using the local instantaneous horizontal velocity difference (Δu) and the buoyancy difference (Δb) measured by the probe assembly, it is possible to calculate the instantaneous local gradient Richardson number.

34.3.4 Water-Waves Generation in Flumes, Wave Gauges, and Static Pressure Measurements

Conducting field experiments in the open sea to study water-waves generation is, generally speaking, preferable as the collected data provide information on the measured parameters at

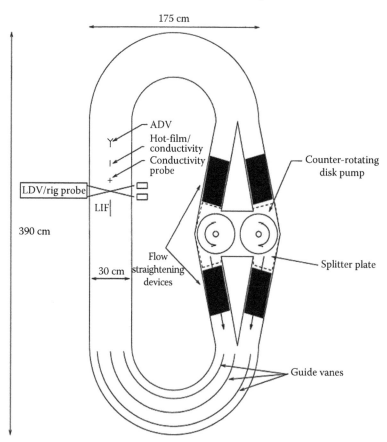

FIGURE 34.10 Schematic of closed-loop Odell/Kovasznay (recirculating) water facility. (From Strang E.J. and Fernando, H.J.S., *J. Fluid Mech.*, 428, 349, 2001. Reproduced with permission from Cambridge University Press.)

natural scales. The unstable nature of the wind and especially the limited ability to predict weather conditions are two major obstacles encountered in field measurements. An alternative to the field experiments is conducting the experiments in specially constructed wind-wave flumes. Laboratory experimental facilities offer the advantages of controlled environment at reasonable cost, as well as numerous additional advantages; among the most prominent is the possibility to create stable and steady airflow at a wide range of air velocities. Recently, a new setup equipped adequately with instruments enabling high quality measurements in air and water was developed in the Wave Research Laboratory, Tel Aviv University, by Professor Shemer and his students.

A small-scale wind-wave setup that allows generation of strong turbulent wind comparable with wind velocities measured in open sea under quite extreme conditions (Liberzon and Shemer, 2011) is shown in Figure 34.11. The experimental facility is suDciently long for study of initial stages of spatial water-waves evolution and enables an easy access to the test section for installation of instrumentation. The facility has a two-level configuration, with the wind tunnel mounted atop the wave tank, see Liberzon and Shemer (2011). Closed-loop airflow and the heat exchanger allowed achieving constant temperature and humidity of the airflow in the course of experimental runs.

The instruments carriage supports the measuring equipment, the power supplies and the sensors. The carriage position along the test section at the desired fetch is the only experimental parameter controlled manually. A conductivity sensor is used to attain accurate positioning of the airflow-sensing instruments relative to the constantly varying water-wave surface. The sensor consists of two parallel copper wires with exposed tips placed 5 mm below the air sensors and connected to a 1.5 V source through a 20 kΩ resistor. Instantaneous water surface elevation variations are measured by a capacitance-type wave gages. Anodized tantalum wire 0.5 mm in diameter is used. The vertical traverse is used to calibrate the gages statically at prescribed submerge depths in still water. Two pressure-sensing instruments, a static pressure probe, and a Pitot tube for mean velocity measurements, are used. To produce accurate static pressure variations measurements close to the water surface a commercially available static pressure probe (*A520, MAMAC Systems Inc*) is used. For detailed description of the static pressure probe characteristics and carefully performed validation tests see Liberzon and Shemer (2010). To measure the variations in both the vertical and the horizontal airflow velocity components, an x-hot-film (HF) probe (TSI *T-1241–20*) together with a commercially available multichannel anemometer (*AA Lab Systems*) is used. In situ calibration is performed using the sensor as well as the Pitot tube which are placed at identical elevation in the central part of the wind tunnel. Details can be found in Liberzon and Shemer (2011).

34.3.5 Detection of Air–Water Surfaces in Two-Phase Flows, Conductivity Probes, Fiber Optics Detectors

Two-phase flows are ubiquitous, however, they are exceptionally complicated and in most cases accurate quantitative description is not available. Thorough investigation of two-phase pipe flows is crucial for numerous industrial applications that require reliable predictive quantitative solutions for design and maintenance. Accurate experimental data on the instantaneous distribution of both phases within the pipe are necessary for understanding the governing physical mechanisms and essential for the construction of advanced theoretical models. A brief description of three methods, successfully used in the Two-Phase Laboratory of Tel Aviv University by Professors Barnea and Shemer, their colleagues and students, are presented.

The simplest method for detection of the local instantaneous void fraction is based on measuring the electrical conductance between a *tip probe* and a flat large electrode glued to the pipe wall. The probe consisted of Teflon-coated 0.2 mm in diameter stainless steel wire, which was accurately cut under a microscope at the tip. The probe has suDcient stiffness, while having a suDciently small measuring area. In the experiments (Barnea and Shemer, 1989), the tip electrode was located at the centerline of the pipe at a distance of 6 m from the entrance. Due to the low voltages employed, usually <2 V, a suDciently stable signal could be obtained. Due to the finite response time of the probe, voltages that differed from the extreme values are obtained. Software with adjustable thresholds was developed in order to obtain the binary signal from the raw data. The resulting binary signals can be further processed to extract meaningful statistical information and to enable detecting the intermittency of various flow regimes. More details are in Barnea and Shemer (1989).

Optical rber probes that are sensitive to the change in the refractive index of the surrounding medium are capable of detecting the local instantaneous phase (gas or liquid). The sensitive silica probe tip of commercially available optical fiber probe (Photonetics type FES) has a diameter of 0.14 mm. The return signals from the optical fiber sensors are amplified to yield an analog output of two levels (TTL signal) representing the instantaneous phase of the medium present at the fiber tip. The signal from the optical probe is very reliable especially when a multi-sensor probe is used due to the full independence of each

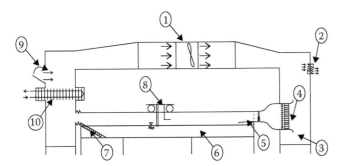

FIGURE 34.11 Wind-wave flume scheme: (1) Blower, (2) flow rate control hatch, (3) settling chamber, (4) contraction with honeycomb and nets, (5) flap, (6) test section, (7) beach, (8) instruments carriage, (9) maintenance hatch, and (10) heat exchanger.

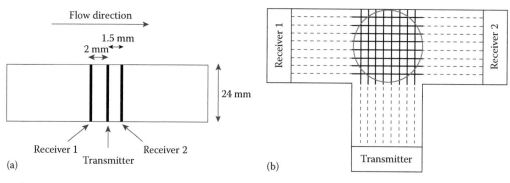

FIGURE 34.12 Schematic presentation of the wire-mesh sensor: (a) side view of the pipe and (b) cross-sectional pipe view. (Reproduced from *Chem. Eng. Sci.*, 63, Roitberg, E., Shemer, L., and Barnea, D. Hydrodynamic characteristics of gas-liquid slug flow in a downward inclined pipe, 3605–3613. Copyright 2008, with permission from Elsevier.)

sensor (no cross-talking) from the others. A study of the evolution of some hydrodynamic parameters of air-water slug flow in vertical pipe using four optical probes is presented in van Hout et al. (2001).

The wire-mesh sensor can be seen as an intrusive tomograph that enables measurements of the cross-sectional void fraction distribution. This instrument consists of layers of parallel wires, with the wire direction in each layer perpendicular to that in the neighboring layer (Figure 34.12).

The spacing between the first and the second layer is 1.5–2.0 mm. Each layer is built of eight parallel 0.125 mm diameter thin steel wires, with a constant spacing of 2.8 mm. The wires in the external layers have the same direction, while in the mid-layer the wires are perpendicular to those in the external layers. Therefore, the wires of any two successive layers form a rectangular mesh at which junction measurements of void fraction are actually performed. The operation principle of the wire-mesh sensor is based on the difference in the electrical conductivity of the two phases (water and air).

The mid-layer serves as a transmitter, while the external layers serve as receivers. Each wire in the transmitter plane is periodically activated by a multiplexer circuit by a short voltage pulse with duration ranging from 5.3 to 42.6 μs. An additional multiplexer circuit is used to connect consecutively each one of the wires in the receiver planes during the single pulse supplied to the transmitter wire. Each output pulse depends on the local instantaneous conductivity at the crossing point of the transmitter and the receiver wires. It is a remarkable development in comparison with a single tip conductance probe described earlier. The instrument is capable of sampling frequencies up to the maximum of 10 kHz for the whole cross section. The raw measurement results are recorded as a sequence of voltages at each junction and then further translated to the local instantaneous void fraction using the calibration coeDcients determined before the beginning of each experimental run. It is assumed that the void fraction at any junction of every mesh is determined by the phase distribution in the vicinity of the junction commensurable with the mesh size. More details about the use of the wire-mesh sensor that enables quantitative measurements of the cross-sectional void fraction distribution of a gas—liquid

slug that flows in a downward inclined pipe can be found in Roitberg et al. (2008).

References

Adrian, R.J. 1991 Particle-imaging techniques for experimental fluid mechanics. *Annual Review of Fluid Mechanics*, **3**:261–304.

Adrian, R.J. 1997 Dynamic ranges of velocity and spatial resolution of particle image velocimetry. *Measurements Science and Technology*, 8:1393–1398.

Arroyo, M.P. and Hinsch, K.D. 2008 Recent developments of PIV towards 3D measurements, 127–154, Topics Appl. Physics 112, A. Schroeder, C.E. Willert (Eds.): Particle Image Velocimetry.

Barnea, D. and Shemer, L. 1989 Void fraction measurements in vertical slug flow: Applications to slug characteristics and transition. *International Journal of Multiphase Flow*, **5**:495–504.

Chen, J. and Katz, J. 2005 Elimination of peak-locking error in PIV analysis using the correlation mapping method. *Measurement Science and Technology*, **6**:1605–1616.

Crimaldi, J.P. 2008 Planar laser induced fluorescence in aqueous flows. *Experiments in Fluids*, **4**:851–863.

De Silva, I.P.D., Brandt, A., Montenegro, L., and Fernando, H.J.S. 1999 Gradient Richardson number measurements in a stratified shear layer. *Dynamics of Atmospheres and Oceans*, **0**:47–63.

Gaskey, S., Vacus, P., David, R., and Villermaux, J. 1990 A method for the study of turbulent mixing using fluorescence spectroscopy. *Experiments in Fluids*, 9:137–147.

Hackett, E.E., Luznik, L., Katz, J., and Osborn, T.R. 2009 Effect of finite spatial resolution on the turbulent energy spectrum measured in the coastal ocean bottom boundary layer. *Journal of Atmospheric and Oceanic Technology*, 26:2610–2625.

Hadad, T., Ratner, D., Bernheim, A., Liberzon, A., and Gurka, R. 2010 Seeding particles for imaging techniques. *FEDSM-ICNMM2010*, Montreal, Canada.

Hain, R., Kahler, C.J., and Tropea, C. 2007 Comparison of CCD, CMOS and intensified cameras. *Experiments in Fluids*, **3**:403–411.

van Hout, R., Shemer, L., and Barnea, D. 2001 Evolution of hydrodynamic and statistical parameters of gas—Liquid slug flow along inclined pipes. *International Journal of Multiphase Flow* **2** :1579–1602.

Huang, H., Dabiri, D., and Gharib, M. 1997 On errors of digital particle image velocimetry. *Measurement Science and Technology*, 8:1427–1440.

Huang, H.T., Fiedler, H.E., and Wang, J.J. 1993 Limitation and improvement of PIV Part I: Limitation of conventional techniques due to deformation of particle image patterns. *Experiments in Fluids*, **5** :168–174.

Kaene, R.D. and Adrian, R.J. 1991 Theory of cross-correlation analysis of PIV images. *Applied Scientir c Research*, 49:191–215.

Kit, L.G. 1970 Turbulent velocity fluctuation measurements using a conduction anemometer with three-electrode probe. *Magnetohydrodynamics*, 6:480–487.

Kit, E., Fernando, H.J.S., and Brown, J.A. 1995 Experimental examination of Eulerian frequency spectra in zero-mean-shear turbulence. *Physics of Fluids A*, **7 5** :1170–1188.

Kit, E., Strang, E.J., and Fernando, H.J.S. 1997 Measurement of turbulence near shear-free density interfaces. *Journal of Fluid Mechanics*, **4** :293–314.

Kholmyansky, M., Kit, E., Teitel, M., and Tsinober, A. 1991 Some experimental results on velocity and vorticity measurements in turbulent grid flows with controlled sign of mean helicity. *Fluid Dynamics Research*, 7:65–75.

Liberzon, A., Gurka, R., and Hetsroni, G. 2001 Vorticity Characterization in a turbulent boundary layer using PIV and POD analysis. *4th International Symposium on Particle Image Velocimetry*, Gottingen, Germany.

Liberzon, D. and Shemer, L. 2010 An inexpensive method for measurements of static pressure fluctuations. *JTECH-A* **2** (4):776–784.

Liberzon, D. and Shemer, L. 2011 Experimental study of the initial stages of wind wavesflspatial evolution. *Journal of Fluid Mechanics*, **6** 1:462–498.

Lourenco, L. and Krothapalli, A. 1995 On the accuracy of velocity and vorticity measurements with PIV. *Experiments in Fluids*, **8** :421–428.

Luthi, B., Tsinober, A., and Kinzelblach, W. 2005 Lagrangian measurement of vorticity dynamics in turbulent flow. *Journal of Fluid Mechanics*, **8** :87–118.

Melling, A. 1997 Tracer particles and seeding for particle image velocimetry. *Measurement Science and Technology*, 8:1406–1416.

Nogueira, J., Lecuona, A., and Rodrrguez, P.A. 1997 Data validation, false vectors correction and derived magnitudes calculation on PIV data. *Measurement Science and Technology*, 8:1493–1501.

Odell, G.M. and Kovasznay, L.S.G. 1971 A new type of water channel with density stratification. *Journal of Fluid Mechanics*, **8** :535–543.

Raffel M., Willert, C.E., Wereley, S.T., and Kompenhans, J. 2007 *Particle Image Velocimetry: A Practical Guide*. Springer, New York.

Roitberg, E., Shemer, L., and Barnea, D. 2008 Hydrodynamic characteristics of gas-liquid slug flow in a downward inclined pipe. *Chemical Engineering Science*, 63: 3605–3613.

Roth, G. and Katz, J. 2001 Five techniques for increasing the speed and accuracy of PIV interrogation. *Measurement Science and Technology*, **2** :238–245.

Saarenrinne, P. and Piirto, M. 2000 Turbulent kinetic energy dissipation rate estimation from PIV vector fields, *Experiments in Fluids*, **Suppl**:S300–S307.

Sarathi, P., Gurka, R., Sullivan, P.J., and Kopp, G.A. 2010 Experimental measurements of expected mass fraction function in a contaminant plume. *Boundary Layer Meteorology*, **3** :10.1007/s10546-010-9526-8.

Strang E.J. and Fernando, H.J.S. 2001 Entrainment and mixing in stratified shear flows. *Journal of Fluid Mechanics*, **8** :349–386.

Taylor, Z., Gurka, R., Kopp, G.A., and Liberzon, A. 2010 Long duration, time-resolved PIV to study unsteady aerodynamics. *IEEE Transactions on Instrumentation and Measurement*, doi: 10.1109/TIM.2010.2047149, **59**(12):3262–3269.

Tennekes, H. 1975 Eulerian and Lagrangian time microscales in isotropic turbulence. *Journal of Fluid Mechanics*, **6** :561–567.

Tropea, C., Yarin, A.L., and Foss, J.F. (Eds.) 2007 *Springer Handbook of Experimental Fluid Mechanics*, Springer-Verlag, Berlin, Germany.

Tsinober, A., Kit, E., and Teitel, M. 1987 On the relevance of the potential difference method for turbulence measurements, *Journal of Fluid Mechanics*, **5** :447–461.

Van-Cruyningen, P., Lozano, A., and Hanson, R.K. 1991 Quantitative imaging of concentration by planar laser-induced fluorescence. *Experiments in Fluids*, **0** :41–49.

Walker, D.A. 1987 A fluorescence technique for measurement of concentration in mixing liquids. *Journal of Physics E: Scientir c Instrumentation*, **0** :217–224.

Westerweel, J. 2008 On velocity gradients in PIV interrogation. *Experiments in Fluids*, 4 :831–842.

Willert, C., Stasicki, B., Klinner, J., and Moessner, S. 2010 Pulsed operation of high-power light emitting diodes for imaging flow velocimetry. *Measurement Science and Technology*, **2** :075402.

Environmental
Measurements

35

Hydrophysical Measurements in Natural Waters

Elena Roget
University of Girona

35.1 Introduction

During the last 20 years great advances have been made in instrumentation to measure the different dynamical process of natural aquatic systems (internal waves, plumes, mixing processes including convection and double-diffusion, boundary layer dynamics, air–sea interactions, physical patchiness, sediment transport k). Knowledge obtained from the research associated with these measurements serves to better understand the ecology of these systems, contamination distribution, and to improve numerical models for better predictions. When long time series of data are available, global change can be traced.

The availability and the potential of modern instrumental technology provide unique, up-to-date support to engineers and managers for surveying and monitoring aquatic environments. This technology, together with the interdisciplinary knowledge of natural water systems gained over the years, helps to improve impact assessments of natural events or human activity, to make decisions in accidents, to assess aquatic transport and other water activities including rescue operations, to manage water resources, and to optimize infrastructure design.

Hydrophysical measurements are focused on the velocity and the thermodynamic variables of the basic equations of fluid dynamics and on turbulence variables which have expanded the range of temporal and spatial scales of measurements. The current technology for measuring hydrophysical variables is complex and does not always provide the desired variable directly. To process the data, users need a general understanding of the working principles of the instruments and the dynamics of the flow. In fact, depending on the background hydrodynamics, some measurements cannot even be taken, because the outputs cannot be interpreted in terms of well-identified physical variables. Here we review field instruments for mean flow and turbulence measurements and, to a lesser extent, the more classical approaches for measuring thermodynamic variables, with emphasis on small scales. We also introduce error sources, data analysis procedures, and the most relevant related physical concepts.

35.2 Principles

35.2.1 Mean Flow Measurements

35.2.1.1 Rotor Current Meters

Rotor velocimeters and vanes formed the first generation of commercial current meters. Although today traditional rotor current meters (RCMs) are being substituted in basic science by electronic current meters, they are still used for surveying and management tasks. RCMs use vanes to align the instrument with the flow and provide the local horizontal flow velocity based on the rotor speed calibration. RCMs can be classified according to the direction of their axis of rotation, which can be either vertical (with curved or straight blades) or horizontal (propeller type). The forward flow direction is measured by different types of compasses.

Handbook of Environmental Fluid Dynamics, Volume Two, edited by Harindra Joseph Shermal Fernando. © 2013 CRC Press/Taylor & Francis Group, LLC.
ISBN: 978-1-4665-5601-0.

Most RCMs contain a vector-averaging scheme and the mean velocity vector for the fixed sampling period is computed by adding and averaging the *xt y* components of the flow, which are measured at shorter subintervals. Propeller current meters can be set up to respond only to the flow component along their axes of rotation, so two perpendicular propellers can provide the flow velocity and its direction.

RCMs can be deployed in various ways: mounted on the bottom, moored with strings or hanging. RCMs are robust and reliable instruments, but objects in the water can interfere with their operation. In addition, when deployed for a long time, biological growth can also interfere with their functioning, and mechanical friction is a problem when measuring low velocities.

35.2.1.2 Electromagnetic Current Meters

Electromagnetic current meters (EMCMs) are the most traditional type of electronic current meter and their physical working principle involves electromagnetic induction to measure local flows. Figure 35.1 presents a typical EMCM sensor that creates a magnetic field B around a spherical head (the sensor). According to Faraday's law, a positive ion flowing through an upward field $\vec{B}$ at a velocity $\vec{V}$ experiences a magnetic force $\overrightarrow{F_m}$ that deviates it to the right (a negative ion will be deviated to its left). As a result of the cumulative deviation of ions in the sampling volume around the sensor, a voltage difference is produced

between a pair of electrodes located on either side of the sensor. The voltage measured between the two opposite electrodes is linearly proportional to the flow/ion velocity, and the velocity can be determined after appropriate calibration of the output voltage.

Most EMCMs contain two perpendicular pairs of electrodes within the same horizontal plane that measure the *xt y* flow components and do not need to be aligned with the flow either manually or with a vane. If a second magnetic field is created perpendicular to the first and two extra electrodes are added in the vertical direction, the third component of the velocity can also be measured. An internal compass is used to reference the velocity to earth coordinates. The typical diameter of a sensor ranges from about 5 to 50 cm, depending on the instrument. Sensors can have different shapes and if the instrument is measuring only one component of the flow, the sampling volume can be localized either below or above the sensor and may be smaller than the sensor.

Because EMCMs are sensitive to electronic interference and improper grounding can induce a drift of the measured velocity, the magnetic field is synchronically reversed at high frequencies so that static electric potential differences can be removed. For greater autonomy, most EMCMs use different sampling strategies, such as taking a specific number of samples per burst at a given sampling interval instead of at regular intervals.

EMCMs can be deployed like RCMs but they contain no moving parts so, in this sense, they are more robust. Biological growth can also modify the sampling volume. A minimum conductivity (~0.05 μs/cm) is required for the EMCM to work.

35.2.1.3 Acoustic Doppler Current Meters

Acoustic Doppler current meters (ADCMs) measure two or three components of the velocity based on the Doppler Effect. The Doppler Effect, the well-known effect on the frequency of the whistle on a passing train or the siren of a speeding ambulance, is outlined in Figure 35.2. When a particle moving at velocity

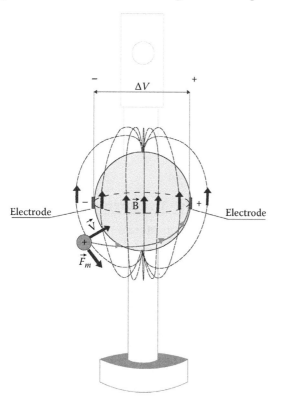

FIGURE 35.1 A positive ion deviating to its right when approaching a vertically upward magnetic field created by the sensor of an electromagnetic current meter. As a result of the accumulative deviation of ions (negatives in the opposite direction of the positives), a voltage difference is created between two electrodes.

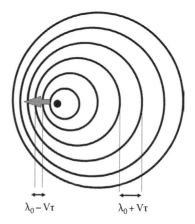

FIGURE 35.2 Instantaneous representation of several pulses emitted by a source at intervals of time τ when it is moving toward the left at constant velocity *V*. The observed distance between two consecutive pulses when the source is not moving, λ_0, appears to vary an amount Vτ when it moves, decreasing in front of the source path and increasing behind it.

V emits (or reflects) a sound peak after a time τ of emitting (or reflecting) a previous one, it will have already moved a distance $V\tau$. As a result the distance between consecutive peaks, λ_0, appears to be compressed in front of the particle (or its frequency increased) with respect to the case when particle has not moved.

Behind the particle, the second peak is delayed with respect to what would be expected if the emitting particle was not moving so, the signal appears to be stretched or its frequency decreased. Also when a particle moves away from the receiver, the next sound peak will take longer to reach the receiver than the previous one. Frequency shift and time dilation are alternative formulations of the Doppler effect. If the signal is slightly stretched or delayed, the phase the receiver measures will also be shifted.

The relative velocity between the source and a receiver is given by $V = c\Delta f_d/f_s$, where Δf_d is the Doppler shift frequency, f_s is the frequency of the source, and c the sound velocity. Motion perpendicular to the line connecting the source and the receiver does not introduce a Doppler shift.

The working principle of an ADCM is based on a sound generated by a transmitter and being reflected and Doppler shifted by the particulate matter moving with the water flow. ADCMs consist of three or four transductors—a transmitter and two or three acoustic receivers—depending on whether they provide two-dimensional (2D) or three-dimensional (3D) velocity data. As shown in Figure 35.3, the sensitive axes of the receivers are aligned to intersect with the transmitted beam in a common sampling volume located far from the probe. The sampling volume can be approximated by a cylinder with a diameter equal to the transmitted beam (usually 1 cm or less). The vertical extension of the sampling volume depends on the intersection of the transmitter and receiver beams, but also on the emitting signal (pulse length) and the receiver window (the period of time over which the return signal is sampled). These last variables are settled automatically after user specifications.

In 3D probes, the sensitive axis of each receiver is slanted at an angle 2α from the transmitter beam (see Figure 35.3) and the receivers are equally distributed within a circumference perpendicular to the beam transmitter. The imaginary axes between the sensitive axes of the receiver and the transmitted beam, and slanted at angle α from the transmitted beam, are called bistatic axes. The velocity of the particles along a bistatic axis within the sampling volume—the bistatic velocity—is proportional to the Doppler shift measured by the receiver. Bistatic velocities are converted to Cartesian velocities relative to the orientation of the probe with a transformation matrix to account for the probe geometry. ADCMs are equipped with internal compasses and tilt sensors to know the instrument orientation and allow velocity data to be reported within a local coordinate system.

Different types of ADCMs use different approaches to measure the Doppler Effect. Pulse-coherent ADCMs send a series of two pulses of sound separated by a time lag and the change in phase of the returning signal is measured. Techniques for measuring the phase shift are limited by the fact that if phase exceeds the range $[-\pi, +\pi]$, it will wrap around. That is known as an ambiguity jump. The problem can be solved to some extent by setting the lag between the two pulses considering the expected Doppler shift due to the water velocity. Other sampling strategies have also been implemented to overcome the problem. Users of commercial ADCMs have the option of introducing the expected maximum velocity and then, the corresponding operational settings for an optimal behavior of the instrument are fixed automatically.

The acoustic frequency of an ADCM is around of the order of 10 MHz. The simplest estimate of the velocity field that ADCMs can provide is referred to as a ping and the maximum sampling frequency can be set equal to the inverse of the time of a single ping, usually larger than 100 Hz. However, because increasing the number of points averaged per sample decreases the random noise, typical sampling rates of the order of 10 Hz are taken, depending on the application. The analysis of ADCM data to obtain mean flow and turbulence variables are reviewed in Sections 35.3.2.2 and 35.3.3.1.

35.2.1.4 Acoustic Doppler Current Meter Profilers

Acoustic Doppler current meter profilers (ADCMPs) can be considered an array of current meters for profiling based on the same physical principle as ADCMs. The acoustic frequency of ADCMPs is of the order of 10^2–10^3 kHz, and the profiling range decreases with higher frequency. Unlike ADCMs, individual transducers of ADCMPs act as transmitters and receivers (they are monostatic, not bistatic) and do not converge at a sampling volume but rather diverge at an angle α, sampling the water column (Figure 35.4). Because of the monostatic nature of the instrument, after a pulse has been emitted by the transducer, there is a short blanking period before it turns into a receiver and gives the time that the sound echoing from the closer sample volume starts reaching it. An overhead time is used to initialize the whole process for a ping.

As a general rule, ADCMPs need one beam for each component of the velocity, that is, one component of the velocity can be measured with a single-beam instrument. Nowadays, ADCMPs with four beams distributed according to the Janus configuration (Figure 35.4) are mostly used and will be discussed later. The benefit of an extra beam is to have complementary information to evaluate the quality of the output based on the difference

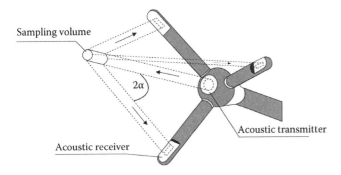

FIGURE 35.3 Scheme of an ADCM sensor showing the transmitter beam, the sampling volume and the sensitive axes of the three receivers.

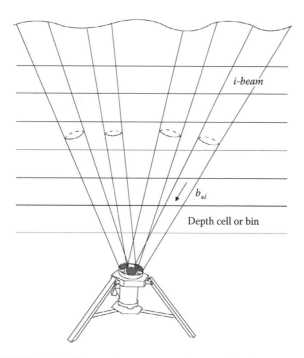

FIGURE 35.4 Scheme of an ADCMP showing the depth cells in which the sampling column of water has been divided. The along-beam velocity for one beam at a specific depth cell or bin is represented.

of two different estimations of the vertical velocity component (error velocity).

ADCMPs divide the water column into different segments, called bins or depth cells where, for each i-beam, they can provide a weighted average of the along-beam velocity for every ping (b_{*i}), the shortest period that it can average (see Figure 35.4). Consecutive measuring intervals of consecutive depth cells partially overlap, so data are partially correlated.

To obtain a single Cartesian velocity vector for a depth cell, ping-weighted average velocities along the different beams in the corresponding depth cell are assembled together, which only makes sense when the velocity field in the horizontal plane covered by the different beams is homogeneous. Note that obtained velocity vector for each depth cell correspond to a weighted average velocity over a depth range which can be of several meters. For turbulent flows, the homogeneity hypothesis can only be fulfilled statistically as commented in Section 35.2.2.3.

35.2.2 Turbulence Measurements

35.2.2.1 Relevant Turbulence Variables

The role of turbulence in the mean flow is captured within the single term $-\overline{u_i u_j}$ in the equation of motion:

$$\frac{\partial U_i}{\partial t} + U_j \frac{\partial U_i}{\partial x_j} = -\frac{1}{\rho} \frac{\partial P}{\partial x_i} - \frac{\rho'}{\rho} g_i + \frac{\partial}{\partial x_j}\left(\nu \frac{\partial U_i}{\partial x_j} - \overline{u'_i u'_j}\right)$$

(Kundu 1990). The term $-\rho \overline{u_i u_j}$, where ρ is the density, is known as the Reynolds stress tensor and specifies the flux due to turbulence of the j-momentum in the i-direction. Diagonal elements of the stress tensor determine the turbulent kinetic energy, $TKE = \rho\left(\overline{u_i^2}\right)/2 \equiv \rho q^2/2$, in which the Einstein summation convention is used and $i = 1 \div 3$. The lower-case notation in u_i, u_j indicates the (fluctuating) turbulent part of the velocities and the overbars indicate the average over many flow realizations (ensemble average). In many practical situations, the ensemble average is equivalent to time average (ergodic theorem). From the heat equation (Kundu 1990) the turbulent heat flux is given by $Q_i = -\rho C_p \overline{u_i T'}$, where Cp is the specific heat and $\overline{u_i T'}$ is the correlation between a component of the turbulent velocity and the turbulent temperature fluctuation.

In the equation of the turbulent kinetic energy for nondiffusive turbulence (Kundu 1990)

$$\frac{d(q^2/2)}{dt} = -\overline{u_i u_j}\frac{\partial U_i}{\partial x_j} + \frac{g}{\rho}\overline{\rho' w} - 2\nu\overline{e_{ij} e_{ij}} = 0,$$

the tensor $-\overline{u_i u_j}$ appears again together with the gradient of the mean velocity, $\partial U_i/\partial x_j$, a term that stands for the production of turbulence by the mean flow. The second term of the previous equation, where w is the vertical turbulent velocity, is the buoyancy flux and accounts for the effect of the stratification in the *TKE* balance. Depending on the sign, buoyancy generates turbulence (turbulence created by an unstable stratification, i.e., convection) or weakens it (due to the work against the buoyancy force).

The dissipation rate of the turbulent kinetic energy is given by $\varepsilon \equiv 2\nu\overline{e_{ij} e_{ij}}$, where e_{ij} is the fluctuating strain rate tensor $e_{ij} = (\partial u_i/\partial x_j + \partial u_j/\partial x_i)/2$ and ν is the molecular kinematic viscosity that in natural waters $\sim 10^{-6} m^2/s$ and substantially depends on the temperature. Analogously, the dissipation rate of the variance of scalars $\overline{\theta'^2}$ (i.e., temperature, salt, k) is $\varepsilon_\theta \equiv 2k_\theta\overline{\left(\partial\theta'/\partial x_i\right)^2}$, where k_θ is the molecular diffusivity of the scalar ($\sim 10^{-7}$ m²/s for temperature in natural waters and $\sim 10^{-9}$ m²/s for salt). When turbulence proprieties do not depend on the direction (isotropic turbulence), then $\varepsilon = 7.5\nu\overline{\left(\partial u/\partial z\right)^2}$ and $\varepsilon_\theta = 6k_\theta\overline{\left(\partial\theta'/\partial z\right)^2}$, where, in profiling measurements, z is the vertical direction and u one horizontal component of the turbulent flow.

In the spectral space (Kantha and Clayson 2000), $\varepsilon = 7.5\nu\int E_{\partial u/\partial z}(k)dk = 7.5\nu\int k^2 E_u(k)dk$ and $\varepsilon_\theta = 6k_\theta\int E_{\partial\theta'/\partial z}(k)dk = 6k_\theta\int k^2 E_{\theta'}(k)dk$, where $E_{\partial u/\partial z}$ and $E_{\partial\theta'/\partial z}$ are the one-dimensional (1D) spectra of the gradient of one component of the velocity (Section 35.3.3.3) and of the gradient of the scalar fluctuations. E_u and E_θ' are the turbulent velocity and scalar spectra (Section 35.3.3.4). Spectra can be obtained from the norm of the amplitude of the components of the Fourier transform of u, θ' $\partial u/\partial z$, $\partial\theta'/\partial z$, and k is the wave number.

All in all, the study of nondiffusive turbulence in natural systems requires measurements of two generic kinds of turbulent variables: the turbulent fluxes (including buoyancy and Reynolds stresses) and the dissipation rates, which, depending on the background, determine the eddy diffusivities and eddy viscosity (Wüest et al. 2005; Lozovatsky et al. 2006).

35.2.2.2 Turbulence Scales and the Turbulence Inertial Range

The smallest scales where energy is dissipated by the viscosity is $l_v = 2\pi(\nu^3/\varepsilon)^{1/4}$, where $\eta_k = (\nu^3/\varepsilon)^{1/4}$, is usually referred to as the Kolmogorov scale. Similarly to η_k, time and velocity microscales are defined as $\tau_k = (\nu/\varepsilon)^{1/2}$ and $u_k = (\nu\varepsilon)^{1/4}$. For the case of turbulent temperature fluctuations, $\theta' \equiv T'$, Batchelor defined the microscale $\eta_B = (k_T\nu^2/\varepsilon)^{1/4}$ where k_T is the thermal molecular diffusivity. In natural water systems ε ranges between 10^{-2} and 10^{-11} W/kg and in an active mixing layer of a lake under moderate wind it is $\sim 10^{-6}$–10^{-7} W/kg. Scales where dissipation dominates, $l_{v\,max}$, are about eight times larger than l_v (Section 35.3.3.3; for l_{Bmax} see Section 35.3.3.4). The smallest scales that turbulence measurements should resolve are summarized in Figure 35.5.

The range between the largest and smallest turbulence scales follows the relation $L = Re^{3/4}\eta_k$, where $Re = UL/\nu$ is the Reynolds number of the flow with U being its velocity, L its limiting scale, and ν the molecular kinematic viscosity. When stratification affects the turbulence, the largest scale that can overturn is approximately given by the Ozmidov scale $L_0 = \sqrt{\varepsilon/N^3}$ where $N = \sqrt{(g/\rho)(d\rho/dz)}$ is the buoyancy frequency. Turbulence

between $\sim L_0$ and the dissipative scales is isotropic and characterized by the velocity spectrum:

$$E_u = \alpha\varepsilon^{2/3}k^{-5/3}$$

where $\alpha = 0.52$ for the longitudinal (along-stream turbulent velocity) 1D spectra and $\alpha = 0.70$ for the transversal 1D spectra. This equation is usually called the $-5/3$ Kolmogorov law and the range where it holds is known as the inertial range.

35.2.2.3 Turbulence Measurements Using Electromagnetic and Acoustic Current Meters

Sampling frequencies of EMCMs, ADCMs, and ADCMPs allow, at least partially, resolving of the turbulence scales (Section 35.2.2.2). Pioneering work to estimate the Reynolds stresses in natural waters used EMCMs however, because the EMCMs sensor is located within the measured flow, for frequencies larger than $f = U/7d$, where d is the dimension of the sampling volume in the direction of the flow, there is a 10% decrease in the energy spectrum (Soulsby 1980). The nonintrusiveness of the sampling volume of ADCMs, together with their ability to sample at higher frequencies than EMCMs, has led to a wide acceptance of Doppler technology. ADCMs also have the advantage of measuring all three components of the velocity vector while most EMCMs provide only two.

ADCMPs can be used to measure turbulence provided that the turbulence field is statistically homogeneous across the whole horizontal section sampled by the i-beams ($i = 1 \div 4$); that is, the second-order moments of the turbulent velocity (u_*, v_*, w_*) for each depth cell must satisfy $\overline{u_{*1}^2} = \overline{u_{*2}^2} = \ldots$

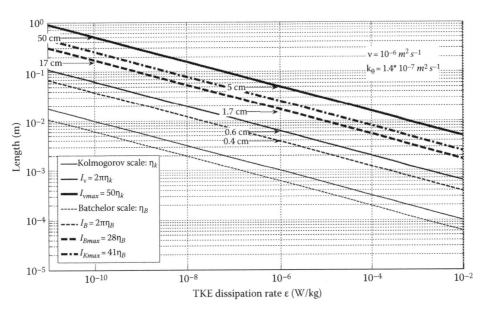

FIGURE 35.5 Representation of Kolmogorov and Batchelor microscales in logarithmic scale depending on ε, the rate of dissipation of the turbulent kinetic energy. Also depending on ε, the scale where the dissipation is higher according to the Panchev–Kesich model is shown.

and $\overline{u_{*_1} w_{*_1}} = \overline{u_{*_2} w_{*_2}} = \dots$. Obtaining turbulence variables from ADCMPs is reviewed in Section 35.3.3.2.

35.2.2.4 Airfoil Shear Sensors

Airfoil shear sensors (ASSs) have been specially designed to measure small-scale shear (i.e., $\partial u/\partial z$) and obtain the dissipation rate ε (Section 35.2.2.1). ASSs are sensitive to the component of the turbulent velocity, u, which is transverse to the path of the sensor (Figure 35.6). The transverse velocity component creates a lifting force at the airfoil which is sensed by a piezoceramic beam that generates an output voltage proportional to the velocity that has produced the force. Shear sensors must be mounted in moving vehicles or profilers, because the parameterization of the cross-flow force to obtain the velocity is based on the angle of attack, $\alpha < 10°$ (Figure 35.6). Sampling frequencies of the sensor, f_s, can be as high as ~1 kHz and their spatial resolution, Δz, depends on V, the velocity of the sensor, $\Delta z = V f_s^{-1}$. The lowest velocity of the profiler is usually fixed to allow the required minimum angle of attack and the highest velocity is set to avoid vehicle resonance. The working sinking velocity of standard profilers usually ranges between 0.3 and 1 m/s.

Other spatial resolution limitations are due to the size of the tip sensor ($\phi \sim 3$ mm) and the time response of the shear sensor τ (~2 ms), so that $\Delta z \geq V\tau$. The real spatial resolution of the probe is determined by the maximum of the limiting scales, $\Delta z_0 = \max[\phi, V f_s^{-1}, V\tau]$, according to which, output data series should be resampled or filtered.

Turbulence measurements with ASSs require accomplishment of the Taylor hypothesis: averaging over time is equivalent to averaging over the segments sampled with the moving sensor. This is the case when the rate of change of eddies is small compared to the time that it takes for the sensor to move across them. In these cases the turbulence field can be thought of as frozen. Because eddies will evolve faster in a more intense turbulent field, a rough test of Taylor's frozen hypothesis is that the ratio between the turbulent velocity variance and the sensor velocity $\overline{u^2}/V < 0.25$.

Regarding the large scales, there is also a limitation related to the length of the profiler, L. As a general rule, structures at wave numbers smaller than $L/2$ are contaminated by eddies generated by the profiler.

When profiling anchored at a fix station, time between consecutive cast should assure profiler-induced turbulence to be advected away. Airfoil shear sensors are sensitive to a single transversal direction, but if two sensors are mounted perpendicularly, the 2D shear within the plane perpendicular to the path of the sensor can be resolved.

35.2.3 Temperature and Salinity Measurements

For many years moored strings of temperature and conductivity sensors and profilers with CTD sensors have provided information about the water column and its dynamics. CTD stands for conductivity, temperature, and depth, although the set of the so-called CTD sensors do not measure depth but pressure. Conductivity is a measure of water's electrical properties, which are related to its ionic content and so depends on its salinity. In oceanographic campaigns, CTD sensors are often attached to a rosette wheel with bottles that can be closed by remote control at selected depths to collect water samples for more direct laboratory analysis of the physical and chemical properties of the water.

Standard CT sensors have a response time (Section 35.3.1.1) of the order of 100 ms so they have a lower time/spatial (if profiling) resolution compared to the so-called microstructure fast response sensors that have a response time of one order of magnitude less. Fast response sensors are however less accurate than standard precision sensors.

The microstructure temperature sensors are NTC (negative temperature coefficient) thermistors based on a semiconductor resistor that lowers its resistance with the increase of temperature. For natural aquatic system measurements, the NTC thermistor FP07 is widely used. The sensor is protected by a shock-resistant glass rod that becomes very thin at the top of the thermistor, allowing a response time of about 10 ms, depending on the velocity flow.

The most popular type of microstructure conductivity sensor for measuring natural waters is the four-electrode type. The conductivity cell is driven by a high frequency current applied to one pair of electrodes and the resulting voltage drop is measured between the other two to determine the conductance and, taking the geometry of the cell into account, the conductivity. A preemphasized signal magnifies the sensor output for rapidly changing conductivity and facilitates acquisition on the high frequency band, which conventional conductivity sensors cannot do.

In some commercial deployments fast conductivity and temperature sensors can be placed as close together as 1 mm to sample the same water volume and investigate turbulent mixing in stratified fluids. Fast conductivity and temperature sensors can also be incorporated in ADCMs to obtain the small-scale buoyancy flux (Section 35.2.2.1).

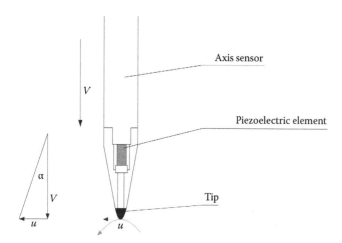

FIGURE 35.6 Scheme of an airfoil shear sensor falling at the speed V sampling a transverse turbulence velocity component u which leads to an angle of attack α.

35.3 Methods of Analysis

35.3.1 Measurement Errors and Data Segmentation

Instruments are limited by their own characteristics but also by flow and deployment conditions, that sometimes determine if measured data should be accepted or rejected. Some data are the result of a series of measurements (i.e., mean velocity computed from Doppler current meters), and turbulence variables (i.e., dissipation rates or Thorpe scales) are by definition statistical values so they must be computed from segments of data.

35.3.1.1 Accuracy, Precision, Resolution, and Aliasing

The accuracy of an instrument indicates the error with which it gives the value of the measured magnitude; it can be thought of as the bias that can be expected with respect to the real value. The precision indicates the differences that can be expected between two measurements of the same magnitude. So, precision is a measure of the dispersion around the mean value. If several measurements of the same stationary event are performed and the values are normally distributed, a dispersion error can be estimated as $e = \pm k\left(\sigma/\sqrt{N-1}\right)$, where $k = 1.96$ if $p < 0.05$ and $k = 2.58$ if $p < 0.01$, being p the lowest level of significance, N the number of measurements, and σ the standard deviation.

The instrument's resolution is the minimum difference provided by the measuring system. For discrete time series, the time resolution given by the sampling interval Δt must also be taken into account because it determines a frequency resolution of $\Delta f = 1/N\Delta t$, where N is the number of points of the data series. Analogously, when sampling along the space, at every Δx, the wave-number resolution is $\Delta k = 1/N\Delta x$.

In the case of discrete series, there is also a limit for the highest frequency (or wave-number) that can be resolved, known as the Nyquist frequency (wave-number), $f_N = 1/2\Delta t$ ($k_N = 1/2\Delta x$). If higher frequency components exist in the data, in the spectral analysis their energy is shifted to lower frequencies (aliasing effect), introducing an error that cannot be removed once data have been recorded.

35.3.1.2 Dynamic Errors

Dynamic errors are due to the different responses of the sensors to changing conditions and can have different sources. As an example, the time response of a sensor is a standard parameter that accounts for the time that a sensor needs to resolve 63% of the variation in the measurement in an evolving environment. The smaller the time response, the better is the measuring system. When a variable is computed by combining data recorded with sensors with different time responses, noise can be introduced in the derived variables. As an example, difference in the response time of the conductivity and temperature precision sensors generates artificial spikes in the salinity profiles (Section 35.3.4.1) and contaminates the density profiles (Section 35.3.4.2).

In this case, a standard approach to handling this problem is the use of a recursive filter:

$$C_i = C_{i-2}\exp\left(\frac{-\Delta t}{\tau}\right) + C_{i-1}\left(1 - \exp\left(\frac{-\Delta t}{\tau}\right)\right)$$

where
  Δt is the time within two consecutive data points
  τ is a time factor to be determined providing that the filtered conductivity, C_i, will give the smoothest possible salinity and density profiles

Generally, sensor time response depends on the flow velocity.

Another typical source of dynamic errors is the wake generated around a moving sensor, as is the case of the shed wakes created by decelerated CTD probes, or the eddies around a falling profiler. Other deployment issues, such as sensor misalignment and tilt or correlation between tilting angles and the measured turbulent velocity are also sources of dynamic errors.

35.3.1.3 Data Stationarity

As a first approach, mean values or other statistical variables are usually computed from temporal or spatial segments of data of equal length. However, a more accurate computation might require a segmentation based on stationarity. To assess mean stationarity (no trend) of a series of data, $\{X_n\}$ $n = 1 \div N$, a reverse arrangement test can be used (Stacey et al. 1999). The test is based on the number of times A that $X_i > X_j$ $i < j$ (number of reverse arrangements). If data are random the number of reverse arrangements is a random variable with the mean value of $N(N - 1)/4$ and a variance of $N(2N + 5)(N - 1)/72$, so that the z-score (the number of standard deviations for a particular realization) is given by

$$z = \left(A - \frac{N(N-1)}{4}\right)\left(\frac{2N^3 + 3N^2 - 5N}{72}\right)^{-1/2}$$

Then, for the segments of data that $-1.96 < z < 1.96$, the stationary hypothesis cannot be rejected at a confidence level of 95%.

To assess the statistical stationarity of fluctuating quantities around a mean value, the reverse arrangement test is not suitable, so other procedures like checking the variance level or comparing the spectra are used.

35.3.1.4 Noise and Meaningful Data

In geophysical data measurements, at high frequencies/wavenumbers the noise amplitude is often larger than real data and the so-called signal-to-noise ratio (SNR) is low. SNR is a measure of the intensity of the signal relative to the noise expressed in dB and defined as $10\log_{10}(I_s/I_n)$, where I_s and I_n are the signal and noise intensities. Data at temporal or spatial scales with low values of SNR are unreliable and should be disregarded, but sometimes it is difficult to determine the frequency or the wave-number cutoff, which can depend on the instrument or the measurement conditions. Because noise usually produces random variations around the mean, the probability of observing M consecutive values (run length) above and below the mean is expected to be the same and the probability distribution of the M run lengths is $P(M) = 2^{-M}$. The plot of the probability distribution of the run lengths of experimental data compared with the theoretical distribution shows the maximum M_{max}, where distributions are coincident, leading to the conclusion that noise dominates at these scales $< M_{max}\Delta$ (Stacey et al. 1999), where Δ is the interval of measurement.

The z-score (see Section 35.3.1.3) of the corresponding statistical test for randomness is given by

$$z = \frac{r + h - (2N_1N_2/N) - 1}{\sqrt{[2N_1N_2(2N_1N_2-1)]/[N^2(N-1)]]}}$$

where

    r is the number of times that consecutive data cross the mean value (number of sign inversions)
    N_1 is the number of positive data
    N_2 is the number of negative data
    N is the total number of data
    $h = 0.5$ when $(2N_1N_2/N + 1) > r$, and $h = -0.5$ when $(2N_1N_2/N + 1) < r$

For the segments of data that $-1.96 \leq z \leq 1.96$, the hypothesis that the series is random at a confidence level of 95% cannot be rejected.

Run length test are also used on segmentation algorithms to distinguish real data from noise segments.

35.3.2 Mean Flow Estimates

35.3.2.1 RCMs and EMCMs

The setup to obtain the mean flow from RCMs and EMCMs data (Sections 35.2.1.1 and 35.2.1.2) is rather simple; to a great extent it simply entails designing the sampling strategy and the mooring system. Both instruments are intrusive, but that has a minor effect on the mean horizontal velocity they determine directly.

Because EMCMs can bias (Section 35.2.1.2), data in low-velocity flows should be regarded with caution. Mechanical friction can also be a problem when RCMs measure low velocities. Horizontal mean flow velocities of EMCMs and ADCMs (Section 35.3.2.2) generally coincide pretty well.

35.3.2.2 Acoustic Doppler Current Meters

The sampling rate and the unambiguous velocity (Section 35.2.1.3) are the two major parameters given at the setup of the ADCM to obtain the three components of the velocity in the sampling volume. Based on these parameters, the number of

pings (the basic unit of measurement) averaged for each sample (the single output value) is internally optimized and fixed. Because time between pulses sets the correlation lags available so the velocity computation is not ambiguous, a larger pulse lag—a variable that users cannot modify—is needed for a higher velocity range so there will be less pings to average, diminishing the accuracy of the velocity estimation. Therefore, the chosen velocity range (maximum unambiguous velocity) should be as low as possible.

Errors in ADCM measurements are due to sampling failures, Doppler noise and velocity shear within the sampling volume (McLelland and Nicholas 2000). Doppler noise is inherent to the system and is produced by the variance of scattered velocities in the sample volume, which diminishes the resolution of the probe. Because Doppler noise is random, averaging multiple pings will reduce the scatter and the noise level will decrease by the square root of the number of samples averaged.

Because bistatic axes are slanted away from the beam transmitter at an angle $\alpha < 15°$ (Section 35.2.1.3), bistatic velocities are more sensitive to the motion parallel to the axis of the transmitter than to the transversal by approximately a factor of $\tan(\alpha)$. So, the component of the velocity parallel to the transmission beam is about $\tan(\alpha)$ times less noisy than the transversal components. In this sense, sensor geometry can be considered as a source of error which can be reduced by proper probe orientation.

Inaccurate sound speed is also a source of error that biases the data and decreases the accuracy of the instrument. The sound speed is taken into account in the internal processing of the instrument when the bistatic velocity is obtained from the Doppler shift. Sound speed (~1450 m/s in water) is mainly a function of temperature and salinity. Depending on the ADCM geometry, if the actual value of the sound speed is 5% larger than the given value, the vertical and horizontal components of the velocity could also be about 5% and 10% larger than the output given by the instrument (Sontek 2001).

The horizontal boundaries of the ADCM sampling volume are not well defined and nearby objects or boundaries (i.e., the bottom or the water surface) can interfere with velocity measurements when sidelobe energy reflects on them.

ADCMs process several quality parameters, such as the reflected signal strength (echo intensity), the signal-to-noise ratio (SNR; Section 35.3.1.4), and the correlation coeDcient between acoustic return signals for each beam. If there are not enough reflecting particles in the water, the return signal may not be stronger than the ambient electronic noise level. High resolution measurements (at higher sampling frequencies) require a higher SNR than mean flow measurements (at lower frequencies). The correlation coeDcients for each beam can range from zero to a maximum value corresponding to a perfect correlation. When correlation is low, the accuracy of the output decreases. Generally, the correlation should not be lower than 70% of the maximum value, although in high-turbulent environments a lower correlation could be accepted for mean velocity measurements. Neither the correlation coeDcients nor the SNR directly quantify the measurement error, but both are helpful tools for assessing data quality and, sometimes, can even be used as a basis for error parameterization.

35.3.2.3 Acoustic Doppler Current Meter Profilers

ADCMPs can only provide the mean velocity when the flow is locally horizontally homogeneous (Section 35.2.1.4), therefore, it is recommendable to average over the time, $\tau \gg L/U$, where L is the distance separating the beam pairs and U is the expected velocity flow (that is, sampling at frequencies $\ll U/L$).

In the case of a Janus configuration, for any single ping (indicated with a star, $*$) the Cartesian velocity for each bin cell (u_*, v_*, w_*) in the earth coordinates can be obtained from the along-beam velocities $(b_{*i}, i = 1 \div 4)$ following Lu and Luek (1999a):

$$u_* = \frac{b_{*2} - b_{*1}}{2 \sin \alpha} - \varphi_{*3} \frac{b_{*1} + b_{*2}}{2 \cos \alpha}$$

$$v_* = \frac{b_{*4} - b_{*3}}{2 \sin \alpha} + \varphi_{*2} \frac{b_{*3} + b_{*4}}{2 \cos \alpha}$$

$$w_* = -\frac{b_{*1} + b_{*2} + b_{*3} + b_{*4}}{4 \cos \alpha} - \varphi_{*3} \frac{b_{*2} - b_{*1}}{2 \sin \alpha} + \varphi_{*2} \frac{b_{*4} - b_{*3}}{2 \sin \alpha}$$

where φ_{*2} and φ_{*3} are the pitch and roll tilt angles of the instrument. The heading angle, φ_{*1}, can be added by an additional rotation. An error velocity equal to

$$e_* = -\frac{(b_{*1} + b_{*2}) - (b_{*3} + b_{*4})}{4 \cos \alpha}$$

is also obtained which is a measure of the error in the vertical component of the velocity due to either the flow not being horizontally homogeneous or the probe having been damaged.

Errors in individual ping estimates of the mean velocity are too high within the required standards of accuracy, so that they should be averaged by lowering the sampling frequency.

If the ADCMP is rigidly mounted and the tilt angles remain constant, the Cartesian mean velocity components at each bin depth can be obtained from the mean of the along-beam velocities, $\overline{b_{*i}}$, using equations formally equal to those given earlier. This is possible because for each ping, the Cartesian velocity components at each cell depth are linear combinations of the beam velocities; for example, the mean flow velocities along the x and y axes for each cell depth are given by $u = \left(\overline{b_{*2}} - \overline{b_{*1}} \right) / 2 \sin \alpha - \varphi_3 \left(\overline{b_{*1}} + \overline{b_{*2}} \right) / 2 \cos \alpha$ and $v = \left(\overline{b_{*4}} - \overline{b_{*3}} \right) / 2 \sin \alpha + \varphi_2 \left(\overline{b_{*3}} + \overline{b_{*4}} \right) / 2 \cos \alpha$ and analogously for w, the vertical component of the velocity vector.

In the case of nonrigid deployments, the tilt angles can change and the velocity in earth coordinates at each cell depth should therefore be computed ping by ping prior to averaging. That is, $u = \overline{\left(\left(b_{*2} - b_{*1} \right) / 2 \sin \alpha - \varphi_{*3} \left(b_{*1} + b_{*2} \right) / 2 \cos \alpha \right)}$ and analogously for the other two components of the velocity.

In the case of nonrigid deployments, tilt angles may correlate with the velocity induced by the movement of the instrument and bias the output velocity. Tilting, however, is not important when the scales of eddies are much larger than the distance between beams. Also, when they are much smaller, the tilt correlation between two different beams is small. However, when eddies and the distance between them are of comparable size, tilting is affecting weak vertical components of the flow, although it does not substantially influence the estimation of the horizontal velocity components (Lu and Lueck 1999b).

The sources of errors and the output quality parameters for ADCMPs are similar to those of ADCMs (Section 35.3.2.2). Also, the beam spreading, which is proportional to the squared distance along the beam, causes echo attenuation and decrease of SNR. Objects (i.e., mounting structures) and interfaces close to the sampling volume should be considered as possible sources of contamination. Considering the surface or bottom interfaces, the effective measurement range of an ADCMP can be estimated as $(1 - \cos\alpha)D$, where D is the distance from the center of the transducer to the interface and α is the transducer beam angle. That is, for an angle of 20° the farthest 6% of the D distance cannot be properly sampled, independent of the configuration setup. Also, near the head of ADCP there is a blank zone where measurements are not taken to avoid contamination of the return signal from the transmit energy.

ADCMPs usually provide a percent-good parameter based on SNR, the correlation values, and other quality parameters, which show the fraction of data which passed the set of rejection criteria. Depending on this parameter the mean flow velocity output is advisable to be rejected.

35.3.3 Turbulent Variables

35.3.3.1 EMCMs and ADCMs

EMCMs and ADCMs (Sections 35.2.1.2 and 35.2.1.3) measure two or three components of the velocity vector which can be decomposed into mean and turbulent components, so that the Reynolds stresses can be calculated directly from the covariance of two turbulent velocity components (Section 35.2.2.1).

The turbulent part of the velocity can be assessed for quality by computing its spectra and comparing it with the Kolmogorov $-5/3$ law in the inertial subrange (Section 35.2.2.2).

Due to their intrusive nature, EMCMs data are expected to diminish the true spectral energy for the scales comparable with the sensor size (Section 35.2.2.3).

The contribution of Doppler noise (Section 35.3.2.2) to shear stress measured by ADCMs can almost be suppressed for ideally constructed sensors if the noise in opposing beams is equal, but it cannot be suppressed when calculating normal stresses (Voulgaris and Trowbridge 1998). That is why ADCMs are more accurate when computing shear Reynolds stresses than normal Reynolds stresses and estimations of the turbulent kinetic energy are usually overestimated.

Performance of EMCMs and ADCMs for high turbulence flows—from 10% to 50% of the mean flow—are found to be in good agreement (MacVicar et al. 2007). ADCMs however usually measure the three components of the velocity at higher frequencies and within a smaller sampling volume.

35.3.3.2 Acoustic Doppler Current Meter Profilers

For rigidly mounted ADCMP, the shear Reynolds stresses (Section 35.2.2.1) can be determined ping by ping after Reynolds decomposition $u_* = U_* + u'_*$, $v_* = V_* + v'_*$ $w_* = W_* + w'_*$ and $e_* = E_* + e'_*$ if $E_* = 0$, so that for every depth cell (van Haren et al. 1994)

$$-\overline{uw} = -\overline{u_*'\left(w_*' - e'_*\right)}$$

$$-\overline{vw} = -\overline{v_*'\left(w_*' + e'_*\right)}$$

This approach, known as the correlation method, is also possible when the instrument is heading (Section 35.3.2.3) providing that rotation does not correlate with velocity fluctuations; in this case

$$-\overline{uw} = -\overline{u_*'\left(w_*' - e_*'\left(\cos 2\varphi_{*_1} - v_*'/u_*'\sin 2\varphi_{*_1}\right)\right)}$$

$$-\overline{vw} = -\overline{v_*'\left(w_*' + e'_*\left(\cos 2\varphi_{*_1} + v_*'/u_*'\sin 2\varphi_1\right)\right)}$$

Alternatively and for four-beam ADCMPs, the so-called variance method also provides the Reynolds stresses (Lohrmann et al. 1990). In this case the Reynolds decomposition is performed on the along-beam velocities (Section 35.3.2.3) so that $b_{*_i} = B_{*_i} + b'_{*_i}$ and averaging of the turbulent components must be done until the hypothesis of stationary turbulence is valid: $\overline{u_{*1}'^2} = \overline{u_{*2}'^2} = \cdots \equiv \overline{u_*'^2}$; $\overline{u_{*1}'w_{*1}'} = \overline{u_{*2}'w_{*2}'} = \cdots \equiv \overline{u_*'w_*'}$ (Section 35.2.2.3). Then, if the instrument is rigidly mounted, so that the pitch and roll angles, φ_2 and φ_3, are constant

$$-\overline{u'w'} = \frac{\overline{b_{*2}'^2} - \overline{b_{*1}'^2}}{2\sin 2\alpha} + \varphi_3\left(\overline{u_*'^2} - \overline{w_*'^2}\right) - \varphi_2\overline{u_*'v_*'}$$

and

$$-\overline{v'w'} = \frac{\overline{b_{*4}'^2} - \overline{b_{*3}'^2}}{2\sin 2\alpha} - \varphi_2\left(\overline{v_*'^2} - \overline{w_*'^2}\right) + \varphi_3\overline{u_*'v_*'}$$

where only the first-order terms of the tilting angles φ_2 and φ_3 are kept. The terms $\left(\overline{u_*'^2} - \overline{w_*'^2}\right)$, $\left(\overline{v_*'^2} - \overline{w_*'^2}\right)$, $\overline{u_*'v_*'}$ cannot be determined with the Janus configuration. Neglecting them introduces a bias to the estimates of the stresses but for isotropic turbulence and tilting angles of a few degrees bias is expected to be small (Lu and Lueck 1999b). The method also gives

$$S = \frac{1}{4\sin^2\alpha}\left(\overline{b_{*1}'^2} + \overline{b_{*2}'^2} + \overline{b_{*3}'^2} + \overline{b_{*4}'^2}\right)$$

$$-\left(\varphi_3\overline{u_*'w_*'} - \varphi_2\overline{v_*'w_*'}\right)\left(\frac{2}{\tan^2\alpha} - 1\right)$$

where S is related to q^2 (Section 35.2.2.1) by $S = (1/(1 + I))(1 + (2I/\tan^2\alpha))q^2/2$ and $I = \overline{w_*'^2}/\left(\overline{u_*'^2 + v_*'^2}\right)$. The last parameter is a measure of the isotropy and cannot be obtained with a Janus ADCMP but values of I from the literature can be used. Like in the case of ADCMs (Section 35.3.3.1), Doppler noise bias estimations of S but not the Reynolds stresses if noise in opposing beams is equal.

With the variance technique Reynolds stresses can be overestimated in more than 1 order of magnitude due to the increase of the velocity variance caused by wave motion, but techniques exist to address this problem (Whipple et al. 2006).

The along-beam velocity from high-resolution ADCMPs can be used to estimate ε by employing the along-beam velocity spectrum within a profiling range limited by attenuation of high frequency sound that allows cell sizes of a few centimeters (Lorke et al. 2005). If turbulence is homogeneous and locally isotropic, the Kolmogorov −5/3 law is used to estimate ε (Section 35.2.2.2). Because ADCMPs cannot resolve the along- and cross-stream components of the turbulent velocities, the value of α for the along-beam spectra ranges between the corresponding longitudinal and transversal values (Section 35.2.2.2), leading to an uncertainty of about 30%.

The ability of ADCMPs to measure turbulent fluxes and mean shear allow the computation of the turbulent shear production (Section 35.2.2.1).

35.3.3.3 Airfoil Shear Sensors

ASSs sense the relative velocity between the tip of the sensor and the water to measure small-scale shear (Section 35.2.2.4) and so vibrations of the vehicle where the sensor is mounted interfere with the turbulence signal if their frequencies are in the same range. Generally, noise affecting a narrow band of frequencies can be removed using Lanczos filters and classical high- and low-pass Butterworth filters can be used to remove the highest and lowest scales of the whole turbulent range. After that, the rate of dissipation of the turbulent kinetic energy, ε, can be obtained by computing the variance of the denoised small-scale shear and multiplying by 7.5V (Section 35.2.2.1). When relevant, variance loss due to denoising procedure has to be corrected (Prandke and Stips 1998).

The dissipation rate ε can also be evaluated by fitting the experimental shear spectra for a segment of data to a theoretical spectrum. Panchev and Kesich (1969) developed a model for homogeneous turbulence that has been corroborated (Gregg 1999) to satisfactorily reproduce the turbulence structure in natural aquatic systems for a wide range of energies. Roget et al. (2006) proposed an analytical approximation to the corresponding nondimensional transversal 1D shear spectrum $E_{(eu/ez)nd}$, which has the form

$$E_{(\partial u/\partial z)nd}(k_{nd}) = 0.9372 k_{nd}^{0.3748} \exp(-6.011 k_{nd}^{1.548})$$

where

$k_{nd} \equiv 2\pi k(\varepsilon/\nu^3)^{-1/4}$ is the nondimensional wave-number (k is given in cpm)

$E_{(eu/ez)nd} \equiv (2\pi)^{-1}(\varepsilon^3/\nu)^{-1/4}E$ (E is the dimensional shear spectra in s^{-2} cpm^{-1})

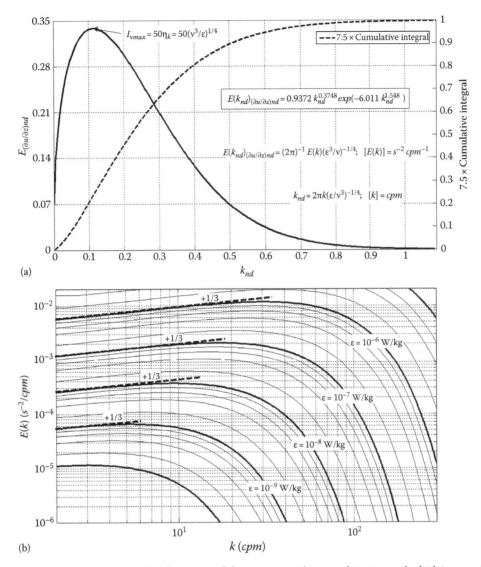

FIGURE 35.7 (a) Nondimensional 1D Panchev–Kesich transversal shear spectra and its cumulative integral, which is normalized in the way that $\varepsilon = 1$. (b) Several dimensional 1D Panchev–Kesich transversal shear spectra.

In Figure 35.7a, the nondimensional 1D Panchev–Kesich shear spectrum is plotted. It presents a maximum at $k_{nd} = 0.125$ that corresponds to 50 Kolmogorov scales ($50\eta_k$). The normalized cumulative integral of the dissipation rate is also represented in Figure 35.7a. As shown, 90% of ε is resolved at 0.5 k_{nd} ($12.5\eta_k$), and 95% at 0.6 k_{nd} ($10.5\eta_k$).

Given a value for the dissipation rate ε, the dimensional spectra, $E_{eu/ez}$, can be obtained from the formula for $E_{(eu/ez)nd}$. In Figure 35.7b, a family of dimensional spectra of longitudinal shear for different values of ε is presented in logarithmic scale. As observed there is a range at larger k where the spectral power increases as $k^{1/3}$. Within this range and according to the relation $E_{eu/ez}(k)dk = k^2 E_u(k)dk$ (Section 35.2.2.1), the velocity spectrum E_u is decreasing with $k^{-5/3}$ in correspondence to the Kolmogorov $-5/3$ law (Section 35.2.2.2).

Given the viscosity corresponding to the temperature of each segment of data, the Panchev-Kesich spectral model only

depends on ε, so the best fit of the experimental shear spectra to the theoretical model for the noncontaminated range of k also gives the estimates of ε.

35.3.3.4 Microstructure Temperature Data

Microstructure temperature measurements are less noisy than small-scale shear (Section 35.3.3.3) and provide a chance to estimate ε indirectly by fitting small-scale temperature data to the corresponding spectral turbulent temperature model which depends on both the dissipation rates of the velocity, ε, and the scalar, $\varepsilon_\theta = \varepsilon_n$. Like in the velocity case, the dissipation rate of temperature variance can be computed from $\varepsilon_T = 7.5\nu \, \text{var}(eT'/ez)$ (Section 35.2.2.1). However, because the smallest scales of the temperature turbulence field are smaller than those affected by shear (Section 35.2.2.2), they might not be resolved, leading to an erroneously lower value of ε_n. In this case, the corresponding loss of variance has to be corrected (Steinbuck et al. 2009).

Batchelor (1959) proposed a theoretical temperature spectrum that has the 1D nondimensional form:

$$E_{\theta nd}(k_{\theta nd}) = q\left[\frac{\exp(-q_B k_{\theta nd}{}^2)}{k_{\theta nd}} - \sqrt{q_B}\pi\left(1 - \mathrm{erf}\left(\sqrt{q_B}\,k_{\theta nd}\right)\right)\right]$$

where $E_{\theta nd} \equiv (2\pi)^{-1} E_\theta\, k_\theta^{-1/2}\, \varepsilon_\theta^{-1}\, (\varepsilon/\nu)^{3/4}$ and $k_{\theta nd} \equiv 2\pi k(\varepsilon/k_\theta\nu^2)^{-1/4} = 2\pi k\eta_B$ where η_B is the Batchelor scale (Section 35.2.2.2) and $\mathrm{erf}(y) = 2\pi^{-1}\int_0^y \exp(-x^2)dx$. q_B is a constant which is usually taken as 3.9 (Oakey 1982).

Kraichnan (1968) proposed an alternative model leading to

$$E_{\theta nd}(k_{\theta nd}) = q_K \frac{\exp\left(-\sqrt{6q_K}\,k_{\theta nd}\right)}{k_{\theta nd}}$$

where q_K is another constant taken as 7.5 (Smyth 1999; Sanchez et al. 2011). The relation between the temperature spectra and the temperature gradient spectra was given in Section 35.2.2.1 for dimensional spectra and also holds for nondimensional spectra. Both Batchelor and Kraichnan spectral models are reported to reproduce turbulence temperature spectra well in natural waters, although the corresponding temperature gradient spectra peak is slightly different, at $k_{\theta nd} = 0.22$ and 0.15, that is, at the scales of $28\eta_B$ and $41\eta_B$ (see Figure 35.5). To determine ε, several methods exist to optimize the empirical-theoretical fitting of the spectra (Luketina et al. 2001; Sanchez et al. 2011).

Microstructure measurements of the scalar components (i.e., temperature) are also useful in determining turbulent scales. Thorpe (1977) proposed reordering a measured vertical mixed density profile into a monotonic stable profile and computing the maximum of the root mean square of the vertical displacements associated with such a reordering to characterize the vertical size of overturns. When salinity is not relevant to density fluctuations, as it is often the case in lakes, Thorpe's scale can be obtained directly from microstructure temperature profiles.

35.3.4 Salinity and Density Calculations

35.3.4.1 Salinity

In the world's oceans the relative ratio of the ionic constituents can be considered nearly constant, so for practical reasons it was assumed that all samples with the same conductivity in relation to a reference solution and at the same temperature and pressure conditions also have the same salinity. Based on this ratio, the practical salinity scale (PSS 78) was defined and recommended by the UNESCO. PSS 78, salinity is nondimensional, although it is usually given in nondimensional practical scale units (psu) which approximately coincide with the part per thousand units (‰ or g/kg). However, in estuaries or in lake waters—where local geochemical processes can contribute to different types of ions—the relative ratio of predominant ions varies and the PSS 78 cannot be applied. In these cases, in situ $C(T,p)$ conductivity measurements are converted to the specific conductivity

at a reference temperature and pressure (i.e., 25°C and 0 Pa or $C(25,0)$) and then related to the salinity, S, so that $S = f(C(25,0))$. To obtain $C(25,0)$, the increase of conductivity due to pressure, $g_P(T,p)$ or to temperature, $g_T(T,p)$, should be determined either experimentally or analytically so that $C(25,0) = C(T,p)/g_p(T,p)$ $g_T(T,p)$. In general, g_T depends on the water composition and for limnological waters g_p can be approximated by the formula $g_P = 1 + (10^{-5}(1.856 - 0.05601)T + 0.0007\,T^2)P$, where p is in dbar and T in °C (Wüest et al. 1996). If some constituents are not ionic species, total salinity cannot be determined by CTD measurements alone. The role of nonionic components in the salinity can be determined from $S_{ni} = (1/\rho)\sum M_i C_i$, where M_i are the molar masses, C_i the molar concentrations, and ρ the density of water.

Recently, UNESCO has approved and recommended a new thermodynamic equation of seawater (TEOS-10) that requires absolute salinity in g/kg instead of the PSS 78. The range of salinity has increased compared to EOS 80 being from 0 up to 120 g/kg. This makes it applicable to estuaries and freshwater and salt lakes and accountable for horizontal density gradients due to the spatial variability of water composition. Absolute salinity S_A can be expressed in terms of practical salinity S_p as $S_A = (35.16504/35)S_p + \delta S_A$ where δS_A is the absolute salinity anomaly that depends on the earth coordinates and pressure (McDougall et al. 2009). For standard seawater composition $\delta S_A = 0$, but, in general, there is still no standardized procedure to account for absolute salinity.

35.3.4.2 Density

As a first approach, around a reference density state, density equation can be linearized as

$$\rho(T,S,p) = \rho(T_0,S_0,p_o)\left[1 - \alpha_T(T - T_0) + \beta_S(S - S_0) + K(p - p_0)\right]$$

where

    α_T is the thermal expansion coefficient
    β is the haline contraction coefficient
    K is the isothermal compressibility coefficient

However, more generally, density is approximated by $\rho(T,S,p) = \rho(T,S,p_o)/(1 - p/K(S,T,P))$, where K is the secant bulk modulus, which is the slope of the secant line cutting the pressure-compression curve (fractional change in volume) from a reference point to a close value on the curve. Based on this approach and considering the salinity in psu, UNESCO in 1981 recommended the seawater equation of state, known as EOS 80 but which at present is being substituted by TEOS-10 a new thermodynamic description of seawater based on Gibbs function and which requires salinity in g/kg (www.teos-10.org/).

35.4 Applications

The instruments that have been reviewed in previous sections can be deployed in different ways and can be configured differently depending on the applications and measuring approaches (real-time monitoring, surveys, long-term deployments). Doppler current meters can sample horizontal and vertical ranges (facing upward and downward), can be moored at the bottom, fixed at the side, mounted on ships, on drifters or on other surface platforms, and lowered. Microstructure profilers can be used either in profiling (rising or free-falling) or towing modes or incorporated into autonomous underwater vehicles. For surveys, CTD sensors are lowered through the water column, but they can also be moored on a buoyant line to record data in time. These sensors can be also mounted in combination with velocity current meters on different types of platforms. A wide variety of sensors are available to measure chemical and biological parameters (including chlorophyll, dissolved oxygen, pH, oxidation reduction potential, chloride, nitrate, and ammonia) and can also be mounted in combination with either velocity or CDT sensors to enhance interdisciplinary studies. Measurements of photosynthetically active radiation (PAR) as far down as 100 m are also possible giving information about the extent of the photic zone and also about radiative heat transmission.

Furthermore, in both limnological and ocean applications, different types of sensors are attached to surface and deep Lagrangian drifters, which are tracked by radar or global positioning systems to provide data about net displacements and dispersion rates. More recently, autonomous profiling floats and different types of autonomous underwater vehicles (including gliders, propelled by buoyancy control) have greatly supported data collection in natural waters.

Given the relevance of hydrodynamics in the state of the ecological system and in the quality of the water, and the large variety of uses of natural water systems, interdisciplinary approaches to present-day measurements in natural water (including physical measurements) are relevant not only for scientific activity but for the management of hydrological systems, water resources (including reservoirs), ports, harbors, fish farming, lakes, drilling platforms, waste water discharges, and for the study of global change when long-term data series are available.

35.5 Major Challenges

Last 20 years of proliferation of small-scale measurements in natural waters using commercial microstructure profilers and the optimization of Doppler technology have made indisputable strides toward a deeper knowledge of turbulence and related processes. At the same time, though, it has made evident the limitations of present approaches to determine turbulent fluxes, the need to standardize present and future procedures to obtain the turbulent variables, and the importance of quality parameters and error quantification. From a more fundamental point of view, further understanding of small-scale processes is still needed including numerical simulations. The identification of turbulence sources and a better understanding of the interaction between different hydrodynamic phenomena are other pending subjects for the near future. Interdisciplinary studies focused on small-scale dynamics, including the importance of biogenic mixing, should also shed more light on the global dynamics of water systems.

On the other hand, the large number of long-term deployments carried out during the last years and the use of autonomous vehicles has improved our understanding of large-scale dynamics. Remote sensing is also starting to provide valuable information about surface and subsurface hydrophysical and biochemical phenomena. The great efforts made during the last decade by the international community in data collection and assembling are noteworthy. An example is the portal of the Global Observing System Information Center (GOSIC), which was initially developed in 1997 at the University of Delaware and is now hosted at the NOAA's National Climatic Data Center in North Carolina, USA (http://gosic.org/). GOSIC links to different data bases such as the global array of temperature/salinity profiles recorded within the ARGO project (http://www.argo.net/) or the International Data Centre on the Hydrology of Lakes and Reservoirs, HYDROLARE, operated under the auspices of the World Meteorological Organization (http://www.hydrolare.ru/). To take full advantage of potential of recompilation and the use of multiple-data sources, data harmonization and standardization is a challenging task on which progress in the near future will be required. In the future, increasing computation capabilities may continue to offer powerful visualization techniques as support for data analysis and real-time monitoring.

The study of interactions between the atmosphere, the ocean, and the earth (including lakes) also need special measurements to obtain a better understanding of the global climate system and its components.

References

Batchelor G.K. 1959. Small scale variation of convected quantities like temperature in a turbulent fluid. *Journal of Fluid Mechanics* 5: 113–133.

Gregg M.C. 1999. Uncertainties and limitations in measuring ε and χ_T. *Journal of Atmospheric and Oceanic Technology* 16: 1483–1490.

Kantha L.H. and C.A. Clayson. 2000. *Small Scale Processes in Geophysical Fluid Flows*. International Geophysics Series 67. San Diego, CA: Academic Press.

Kraichnan R.H. 1968. Small-scale structure of a scalar field convected by turbulence. *Physics of Fluids* 11: 945–953.

Kundu P.K. 1990. *Fluid Mechanics*. San Diego, CA: Academic Press.

Lohrmann A., Hackett B., and L.P. Roed. 1990. High resolution measurements of turbulence, velocity and stress using a pulse-to-pulse coherent sonar. *Journal of Atmospheric and Oceanic Technology* 7: 19–37.

Lorke A. and A. Wüest. 2005. Application of coherent ADCP for turbulence measurements in the bottom boundary layer. *Journal of Atmospheric and Oceanic Technology* 22: 1821–1828.

Lozovatsky I., Roget E., Fernando H.J.S., Figueroa M., and S. Shapovalov. 2006. Sheared turbulence in a weakly stratified upper ocean. *Deep-Sea Research Part I—Oceanographic Research Papers* 53(2): 387–407.

Lu Y. and R.G. Lueck. 1999a. Using a broadband ADCP in a tidal channel. Part I: Mean flow and shear. *Journal of Atmospheric and Oceanic Technology* 16: 1556–1578.

Lu Y. and R.G. Lueck. 1999b. Using a broadband ADCP in a tidal channel. Part II: Turbulence. *Journal of Atmospheric and Oceanic Technology* 16: 1558–1567.

Luketina D.A. and J. Imberger. 2001. Determining turbulent kinetic energy dissipation from Batchelor curve fitting. *Journal of Atmospheric and Oceanic Technology* 18: 100–113.

MacVicar B.J., Beaulieu E., Champagne V., and A.G. Roy. 2007. Measuring water velocity in highly turbulent flows: field tests of an electromagnetic current meter (ECM) and an acoustic Doppler velocimeter (ADV). *Earth Surface Processes and Landforms* 32: 1412–1432. doi: 10.1002/esp.1497.

McDougall T.J., Jackett D.R., and F.J. Millero. 2009. An algorithm for estimating Absolute salinity in the global ocean. *Ocean Science Discussions* 6: 215–242.

McLelland S.J. and A.P. Nicholas. 2000. A new method for evaluating errors in high-frequency ADV measurements. *Hydrological Processes* 14: 351–366.

Oakey N.S. 1982. Determination of the rate of dissipation of turbulent energy from simultaneous temperature and velocity shear microstructure measurements. *Journal of Physical Oceanography* 12: 256–271.

Panchev S. and D. Kesich. 1969. Energy spectrum of isotropic turbulence at large wavenumbers. *Comptes rendus de l'académie Bulgare des sciences* 22: 627–630.

Prandke H. and A. Stips. 1998. Test measurements with an operational microstructure-turbulence profiler: detection limits of dissipation rates. *Aquatic Sciences* 60: 191–209.

Roget E., Lozovatsky I., Sanchez X., and M. Figueroa. 2006. Microstructure measurements in natural waters: methodology and applications. *Progress in Oceanography* 70: 126–148.

Sanchez X., Roget E., Planella J., and F. Forcat 2011. Small-scale spectrum of a scalar field in water: The Batchelor and Kraichnan models. *Journal of Physical Oceanography* 41: 2155–2167.

Smyth W.D. 1999. Dissipation range geometry and scalar spectra in sheared, stratified turbulence. *Journal of Fluid Mechanics* 401: 209–242.

Sontek 2001. SonTek/YSI Technical Notes. Accoustic Doppler Velocimeter Principles of Operation.

Soulsby R.L. 1980. Selecting record length and digitization rate for near-bed turbulence measurements. *Journal of Physical Oceanography* 10: 208–218.

Stacey M.T., Monismith S.G., and J.R. Burau. 1999. Measurements of Reynolds stress profiles in unstratified tidal flow. *Journal of Geophysical Research* 104: 10933–10949.

Steinbuck J.V., Stacey M.T., and S.G. Monismith. 2009. An evaluations of χ_T estimation techniques: Implications for Batchelor fitting and ε. *Journal of Atmospheric and Oceanic Technology* 26: 1652–1662.

Thorpe S.A. 1977. Turbulence and mixing in a Scottish loch. *Philosophical Transactions of the Royal Society of London* A286: 125–181.

Van Haren H., Oakey N., and C. Garrett. 1994. Measurements of internal wave band eddy fluxes above a sloping bottom. *Journal of Marine Research* 92: 909–946.

Voulgaris G. and J.H. Trowbridge. 1998. Evaluation of the Acoustic Doppler Velocimeter (ADV) for Turbulence Measurements. *Journal of Atmospheric and Oceanic Technology* 15: 272–289.

Whipple A.C., Luettich Jr. R.A., and H.E. Seim. 2006. Measurements of Reynolds stress in a wind-driven lagoonal estuary. *Ocean Dynamics* 56(3–4): 169–185. doi: 10.1007/s10236-005-0038-x.

Wüest A. and A. Lorke. 2005. Validation of microstructure-based diffusivity estimates using tracers in lakes and oceans. In: *Marine Turbulence: Theories, Observations and Models*. Cambridge, MA: Cambridge University Press.

Wüest A., Piepke G., and J.D. Halfman. 1996. Combined effects of dissolve solids and temperature on the density stratification of Lake Malawi. 1996. *The Limnology, Climatology and Paleoclimatology of the East African Lakes*. Eds. Johnson T. and E.O. Odada. New York: Gordon and Breach Scientific Publications, pp. 183–202.

36

Flow Measurements in the Atmosphere

Ronald J. Calhoun
Arizona State University

36.1 Introduction

The measurement of flow velocity is fundamental for under-standing the physics of air movement in the atmosphere. The scope of this chapter considers primarily measurements to obtain the velocity vector of the movement of air in the atmosphere. Therefore, the goal of this chapter is to describe the various commonly used methods to measure *winds*. Since a parcel of air deforms continuously when subjected to a shear stress, it may be treated as a *fluid* and its conservation of mass and momentum may be expressed in the form of partial differential equations. Under commonly accepted conditions for air, namely, that the fluid is *Newtonian* (i.e., applied shear is proportional to the velocity gradient), the governing mathematics are known as the *Navier–Stokes* equations.

Since velocity is a vector, complete measurement techniques would provide both the magnitude and the direction of the wind. The simplest example is a traditional cup-and-vane ane-mometer for which the spin rate is proportional to the wind speed and the vane points along the wind direction. However, many of the remote sensing approaches measure only a component of the wind, for example, along a given "look" direction. Indirect methods can be used to obtain remaining components. The indirect portion of the measurement may involve either supplementary measurements, additional assumptions, or theo-retical considerations. Since estimating nondirectly measured velocity components is an important and diﬃcult task in mod-ern remote sensing flow measurement, examples of frequently used algorithms and recent advances are given in the following.

36.2 Principles

Wind measuring instruments typically function based on sev-eral main principles: momentum transfer, pressure changes, heat transfer, and Doppler effects. Hot-wire anemometers, which use the rate of heat transfer from an electrically heat wire as a mea-sure of flow speed, are covered in Chapter 33. They are usually used in laboratory flows but have also been used in atmospheric flows to measure small-scale details of atmospheric turbulence. There are also several other types of instruments based on addi-tional principles, such as ion displacement and vortex shedding, which will not be covered here.

The operational principle of the cup anemometer is based on increased drag, or higher momentum transfer, when the open side of the cup faces the wind compared to the closed end. This causes a net torque on the rotor arm and the anemometer spins in proportion (approximately) to the wind speed. In contrast, sonic anemometers require no moving parts and measure, the time for emitted sound to travel between two ultrasonic transducers. Moving air changes the speed of sound. A tailwind speeds up sound propagation and a head wind slows it down. The length of time for sound to travel across the transducers is

$$t_1 = \frac{d}{c + u_{Air}}, \tag{36.1}$$

while the time for sound to travel in the other direction is

$$t_2 = \frac{d}{c - u_{Air}}, \tag{36.2}$$

Handbook of Environmental Fluid Dynamics, Volume Two, edited by Harindra Joseph Shermal Fernando. © 2013 CRC Press/Taylor & Francis Group, LLC. ISBN: 978-1-4665-5601-0.

where
- d is the distance between the transducers
- c is the speed of sound
- u_{Air} is the component of the wind speed along the transducer to transducer path (Calhoun et al. 2007a)

Then u_{Air} can be calculated from

$$u_{Air} = \frac{d}{2}\left[\frac{1}{t_1} - \frac{1}{t_2}\right]. \tag{36.3}$$

By measuring the components of wind speed in this way in several directions, i.e., sending and receiving sound pulses from and to different combinations of transducers, the two-dimensional (2D) or three-dimensional (3D) velocity vector can be measured.

Moving air, or particles moving with the wind, can be measured with a variety of devices which utilize the Doppler effect. An outgoing signal with a given frequency experiences a shift in frequency when backscattered. The amount of the frequency shift can be related to the speed of the movement of the backscattering object/medium in the signal beam direction. Sodars, radar profilers, and wind lidars operate on this principle. For coherent Doppler lidar, pulses are sent into the atmosphere through a turret usually on top of a box or room containing signal processing electronics (see Figure 36.1). Ambient aerosols scatter the laser light and a fraction of the light returns back to the receiver/emitter (backscatter). After pulse is emitted, the device pauses, waiting for backscatter returns for a time corresponding to the outer length of its range. For some lidars (coherent lidars normally use heterodyne detection), the outgoing laser light is precisely copied and mixed with the returned signal. (Formally, *heterodyning* means the creation of new frequencies by mixing two or more component frequencies.) The beat frequency is the difference between the two frequencies, outgoing and incoming.

36.3 Examples

36.3.1 Cup-and-Vane/Propeller Anemometers

Cup-and-vane anemometers are the standard in the wind power industry and in many meteorological applications. They can be frequently seen in airports, on wind turbines, and in use at weather stations. They are rugged in design, cost effective, and extensively tested. There are rigorous standards for error characterization and calibration based on wind tunnels. For example, many of today's high-end cup-and-vane anemometers are tested against the ASTM standard method D5096-90. A variety of companies manufacture these instruments, for example, *Met One*, *Vaisala*, and *WindSensor*. The P2546 Cup Anemometer is shown in Figure 36.2, originally developed at Risø National Laboratory, Denmark, and now marketed through *WindSensor* (see Pedersen 2003). Momentum is transferred from the wind to the cups causing the anemometer to spin. The rotational speed of the anemometer is counted typically via magnetic reed switch or a photocell. The rotational speed is related to the wind speed via a transfer function, which is usually roughly linear. An example of a transfer function is that for Vaisala's WAA252 anemometer: $U = 0.39 + 0.10 \times R$, where U is wind speed in m/s and R is output pulse rate in Hz. Early cup anemometers had four cups, but it was found that three cups gave faster response to changes in the wind speed. Frictional effects caused by the bearings are of greater relative importance at low wind speeds. Consequently, there is a finite minimum wind speed required in order for cup anemometers to function (~0.5 m/s). For the accuracy of cup anemometers, wind tunnels tests show that modern instruments can have standard deviations better than 0.2 m/s (0.17 m/s listed for the *Vaisala* WA25) in their normal range of operation, approximately between 0.5 and 60 m/s. Of course, this error gives a varying percent accuracy relative to wind speed. For example, 0.17 m/s

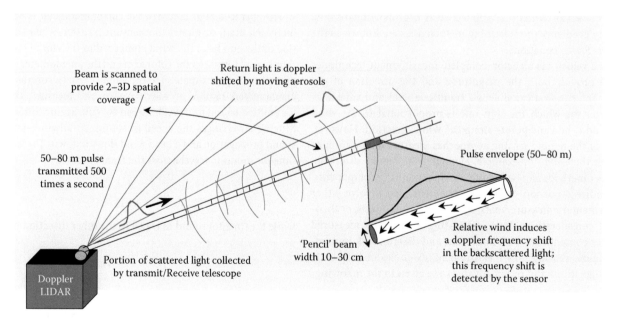

FIGURE 36.1 Coherent Doppler lidar principle of operation. (Courtesy of Lockheed Martin Coherent Technologies, Inc., Louisville, CO.)

FIGURE 36.2 Photo of P2546 cup anemometer. (Courtesy of WindSensor, Roskilde, Denmark.)

is equivalent to a 1% error at 17 m/s, but becomes a much larger percent near the minimum speed required for the anemometer to function. Note that there are a wide variety of accuracies reported in the technical specifications for cup anemometers. The accuracy for the DWS-V-DAC13 manufactured by *Carlo Gavazzi* is listed as " 0.5 m/s for winds ≤3 m/s and " 10% for winds ≥3 m/s.

One should be aware of various sources of bias when using cup anemometers. They are calibrated, typically, in wind tunnels under laminar-like flow conditions, but used most frequently in the atmosphere where the interaction of turbulence with the mechanical arrangement of the instrument can cause biases. The cups on the rotor arm respond more quickly to increases in wind than decreases. Therefore, there is a slight asymmetry—a tendency for the rotor arm to gain angular speed in response to an increase in wind velocity relatively quickly, but to respond to an equivalent decrease in wind more slowly. Asymmetric responses of cup anemometers in turbulent flows lead to the well known "overspeed" effect, where the mean velocity measured by the anemometer overestimates the real mean (Kristensen 1998, 2002). For the biases associated with turbulence effects, lateral variations in wind appear to be dominant. Other sources of bias are associated with sampling and nonlinearity in the calibration. However, Kristensen et al. (2003) suggest that in homogeneous and neutrally stratified atmospheric flow all of the biases together can be expected to be at or below 1% (for the cup anemometer they studied, i.e., the *WindSensor* P2546).

Propeller type anemometers generally have a spinning propeller on a vane which rotates into the wind (also called an *aerovane*). They also have a minimum wind required to obtain measurements and may be more fragile than cup anemometers, having two axes of motion. Resolution and accuracy are roughly

comparable with cup anemometers. For example, the *Young model 05103 Wind Sensor* has a useful range of 0–100 m/s, an accuracy of " 0.3 m/s for wind speed and " 3° for wind direction. The threshold propeller speed is 1.0 m/s and the vane requires at least 1.1 m/s. These sensors are most suited to measuring mean wind speeds and are of limited value measuring fluctuating components of the wind.

36.3.2 Sonic Anemometers

Sonic anemometers are state of the art for research-grade atmospheric flow measurements. Compared with cup anemometers, they have superior performance to measure the fluctuating components of atmospheric winds. Their accuracy can be ~0.01 m/s with up to 60 Hz data acquisition rates. They have no moving parts and have been ruggedized for harsh weather conditions. As described in Section 36.2 above, only the air speed along the direction of the transducers is measured, and the 2D or 3D velocity vector is obtained through sound measurements made between transducers in different directions.

Since sonics have higher sampling rates than cup anemometers they are often used to measure turbulence statistics. This is accomplished by first measuring mean wind speed components, often designated as $\overline{U}$, $\overline{V}$, $\overline{W}$ for the along-wind component, the lateral component, and the vertical component, respectively. Then the fluctuating components are found by calculating the differences between the original velocity component signals and the means, sometimes referred to as the perturbation velocities: u', v', w'. The bar or averaging is usually temporal, varying from seconds to hours, depending on quantities of interest. If the mean signal increases or decreases during the time period the ramping of the signal can be removed through detrending. Some commonly calculated statistics are variances $\overline{u'^2}$, $\overline{v'^2}$, $\overline{w'^2}$, and Reynolds stresses, $\overline{u'v'}$, $\overline{u'w'}$, $\overline{v'w'}$. Friction velocity is frequently calculated as $u_* = \sqrt{\overline{u'w'}^2 + \overline{v'w'}^2}$. The assumption for the calculation of the friction velocity is that the perturbed quantities are measured near the ground where fluxes are roughly constant with height.

There are a variety of companies which manufacture sonic anemometers. Figure 36.3 shows the Campbell Scientific CSAT3 probe. Care must be taken to orient the supporting structure relative to the mean wind to avoid measurements in its wake. Problems with measurements can be associated with birds, rain, and icing.

36.3.3 Pitot Tubes

A pitot tube (named after Henri Pitot, 1695–1771, who used the tubes to determine the flow speed of the river Seine) is a bent tube with a hole in the front and on the side. The hole in the front is placed directly into the wind and measures the stagnation pressure, which means that at the point of measurement (a stagnation point has no flow speed), all the kinetic energy of the flow has been converted into pressure. The hole on the side

FIGURE 36.3 Campbell scientific CSAT3 3-D sonic anemometer. (From Campbell Scientific product brochure at www.campbellsci.com/documents/product-brochures/b_csat3.pdf)

FIGURE3 6.4 Photo of heated pitot tube for aircraft.

measures static pressure which is the pressure in the free flowing wind without converting the kinetic energy of flow movement to additional pressure. The difference between these two pressures is the dynamic pressure which can be related to velocity. Apply the Bernoulli equation for precise relations between these quantities. The main uses of the pitot tube are in wind tunnels and on airplanes (see Figure 36.4). Malfunctions of airplane pitot tubes can have catastrophic results. Blocking of the pitot tube with ice or insect nests could cause inaccurate air speed readings. Pilots may incorrectly conclude that they are traveling too slowly, for example, and increase engine power to avoid a stall.

36.3.4 Sodars

Sodars (short for "Sound Detection and Ranging") function in a similar manner to radar except using acoustic energy. The frequencies of the sound used range from approximately 1000–4000 Hz. The principle of sodar is that emitted sound backscatters

from either wind velocity fluctuations or temperature variations in the atmosphere, depending on the type of sodar. Because humans can hear in the range from 20 to 20,000 Hz, these instruments are frequently bothersome and are best used in rural areas. Just as in Doppler lidar, the data is range gated, i.e., data is given in segments as a function of range from the device. The Doppler shift of the returning signal is measured to give the radial velocity. Multiple beams in different directions are needed to obtain the horizontal velocity vector. Beam steering is achieved through controlling the phases of the transducers. Sodars may be either monostatic (receiver and transmitter collocated) or bistatic (receiver and transmitter spatially separated). Because thermal inhomogeneities backscatter only at 180°, monostatic systems utilize only this type of backscatter. However, bistatic systems have a separation between transducer and receiver leading to angled beams, and consequently backscatter off both thermal variations and turbulence. The layout differences between monostatic and bistatic sodars also imply differences in sampling volume.

Limits of the vertical ranges for sodars are variable depending on atmospheric conditions and device design, but typical ranges are from 50 to 1000 m with 10–50 m range gates. Mini-sodars have more limited ranges of a few hundred meters. Ground clutter (nearby trees or power lines, for example) can be a significant problem for accuracy of sodars and care should be taken in site selection. With relatively high reliability and low operational costs, these devices can operate unattended for long periods.

36.3.5 Radar Profilers

Radar (RAdio Detection and Ranging) profilers are similar to Doppler lidars except for using radio waves instead of laser light. They use the Doppler effect, and therefore measure frequency shifts between their outgoing pulses of energy and the returns. The radio signal backscatters off of small-scale turbulent fluctuations. The radar profiler frequencies have been set so that they interact relatively strongly with turbulent fluctuations. For a UHF radar profiler, the size of eddies for which there is a strong interaction with the radar signal is approximately half the wavelength of the radar, or 16 cm. Similarly with the sodar, reconstruction of the velocity vector depends upon multiple-look directions and the assumption of local homogeneity of winds in the measured volume. Radar profiles have a larger vertical range than sodars, for example the

NOAA 915 MHz radar has a vertical range from 150 to 4000 with vertical resolutions from 10 to 50 m. Some profilers, called "tropospheric profilers" can reach up to 16 km vertically. Because different beam angles are needed for the retrieval algorithm, high elevations make stronger assumptions for the homogeneity of the winds. For example, measurements at 10 km in altitude assume homogeneity of the winds on the three kilometer scale. However, for the longer range, one must sacrifice resolution. Extra vertical range is usually achieved at the expense of vertical resolution. Because radar beams are not as focused as optical lasers, backscatter from side directions can be problematic if ground clutter is near. Effective experiment design often will couple a radar profiler with sodars. The radar profilers are well

FIGURE 36.5 A U.S. Department of Energy radar profiler. (From Wikimedia Commons, downloaded January 29, 2010.)

suited to obtain higher altitude measurements, and the sodar should measure well near the ground. Radar profilers are a well established technology and can operate effectively with minimal operator intervention (Figure 36.5).

36.3.6 Doppler Lidar

There are still only a handful of companies that make a modern, engineered coherent Doppler lidar with 3D scanning capability. For example, *Lockheed Martin Coherent Technologies* markets several generations of the *Windtracer*, which, because of its application in aircraft safety, has been placed in a number of airports around the world, as well as increasingly in research laboratories. Research groups in Arizona State University, the Meteorologische Institute in Karlsruhe, Germany, the Army Research Laboratory, and Curtin University in Western Australia have active research groups utilizing these types of lidars. Halo Photonics also markets a small Doppler lidar. Halo$ Doppler lidar has a relatively high pulse repetition frequency (PRF). The *WindTracer* emits 500–600 pulses per second in the current generation. Because returns are collected in discrete periods or "bins," a distance can be associated with each time "bin" according to the distance that light can travel in the time measured. Therefore, the Data is divided into range-gates, each of which is associated with a range corresponding to a given discrete range of time that it takes for signals to return. A typical Doppler lidar, currently, may utilize, for example, 100 range gates, each 65 m long, for a total range of 6.5 km. The laser pulses are cylindrically shaped, for the *WindTracer*, about 8 cm in diameter and the length of the range-gate (e.g., 60–150 m).

A simple lidar equation showing the relationships between range, power, backscatter, and extinction (the latter two being characteristics of the aerosol in the atmosphere) is given in Equation 36.3 (Calhoun et al. 2007b). It is valid for single elastic backscatter of the laser light off aerosols:

$$P(r) = \frac{C}{r^2}\beta(r)e^{-2\int_0^r k(r')dr'} \qquad (36.4)$$

where

r is the distance of a range gate from the lidar
$P(r)$ is the power received from the range r
$\beta(r)$ is the volumetric backscatter coeDcient of the atmosphere at range r
$k(r)$ is the corresponding extinction coeDcient at range r
C is a lidar constant (function of the physical characteristics of the lidar)

The equation shows that the received power, $P(r)$, is a function of the inverse square of the range, the backscatter, and also an exponential function of double the integrated extinction over the pathway.

36.4 Methods of Analysis

As described in the introduction, obtaining the velocity vector field may require indirect methods for some measurement techniques, especially for remote sensing of velocity. Therefore, in order to understand the suitability of various methods vis-à-vis different flow scenarios and to better assess error characteristics of the produced wind vectors, several algorithms are described subsequently. Vector retrieval algorithms have a wide range in complexity, from rapid and simple techniques with relatively strong assumptions to computationally intensive and high-fidelity techniques requiring their solutions to satisfy the governing partial differential equations.

For many remote sensors which directly measure the radial velocity, the *Velocity Azimuthal Display* (VAD) algorithm is commonly used to retrieve velocity vectors from measured radial velocities. One form of the VAD algorithm (defined and described in terms from Doppler lidar research) is given below (Krishnamurthy 2008). The basic idea is (1) to assume that over a given region of space, the velocity vectors can be treated as constant, and (2) by probing spatially separated portions of this region with the remote sensor, different components of the total velocity vector can be measured. A 2D velocity vector (u and v, below) can then be obtained using geometrical considerations. The velocity vector over the region of assumed homogeneity of the winds (the VAD sector) is estimated by minimizing the cost function:

$$L = \sum_n \left(\hat{u} \cdot \hat{r}_n - u_{r_n}\right)^2 \qquad (36.5)$$

where
ftis the vector dot product
$\mathbf{r}_n = \sin\phi\cos\theta\,\mathbf{x} + \cos\phi\cos\theta\,\mathbf{y}$ is the unit vector in the pointing direction of the laser beam, ϕ is the "azimuth angle" (defined as clockwise from North)
θ is the "elevation angle"
u_{r_n} is the "radial velocity" measured by the lidar
$\hat{u} = U\mathbf{x} + V\mathbf{y}$ (U and V components of wind vector)
n is the number of radial velocity measurements over the range ring

Minimizing L with respect to U and V, one obtains the following:

$$\frac{\partial L}{\partial U} = 2\sum_n \left(\hat{u} \bullet \hat{r}_n - u_{r_n} \right) \sin(az_n)\cos(el_n) \qquad (36.6)$$

$$\frac{\partial L}{\partial V} = 2\sum_n \left(\hat{u} \bullet \hat{r}_n - u_{r_n} \right) \cos(az_n)\sin(el_n) \qquad (36.7)$$

If one sets $eL/eU = 0$ and $eL/eV = 0$, a linear 2×2 system is obtained and can be solved to retrieve U and V. Notice that this method retrieves the 2D, horizontal velocity vector, i.e., the W component in the vertical is assumed close to zero. One might apply this algorithm, for example, to 360° Plan Position Indicator (PPI) scans of a Doppler lidar, where each range (corresponding to a "range ring") also corresponds to a given height. An example of the application of the VAD to PPI scans is shown in Figure 36.6.

It is also possible to apply the VAD technique to smaller *sectors*, where a sector might be defined, for example, as the region where $(r - \Delta r) < r < (r + \Delta r)$ and $(\phi - \Delta\phi) < \phi < (\phi + \Delta\phi)$. In this case, many vectors can be calculated for a PPI, one for each sector, but care must be taken to avoid mathematical instability by making the sectors too small. This may sometimes be referred to as *sector-VAD* and appears to function reasonably for some flow fields.

Optimal interpolation techniques have also been used to estimate the unmeasured portion of the velocity vector, both for radar data (Xu et al. 2006) and for Doppler lidar data (Choukulkar et al. 2012). These techniques are not expensive computationally and make use of background velocity and observations to obtain a maximum likelihood estimate of the flow field. The error statistics of both the background and observations are used in a forecast model to estimate the present state. The use of these approaches for Doppler lidar analysis is still developing.

In contrast, four-dimensional variational data assimilation (4DVAR), is algorithmically complex and computationally expensive. However, the resulting product satisfies the conservation of mass and momentum equations, and retrieved vector fields have suDcient fidelity to display coherent structures (Lin et al. 2008; Xia et al. 2008). For this work, Doppler lidar data was assimilated in a framework based on Large-Eddy Simulation (LES) with dynamical and consistency constraints on the set of possible interpretations of the radial velocity data. A useful consequence of the 4DVAR approach is the availability of the temperature field, since it is retrieved naturally as a consequence of the LES formulation. In Lin et al. (2008), wind and temperature fields were retrieved and the dominant coherent structures in the urban, the boundary layer, and their role in transport processes in convective and stably stratified conditions were studied. Their comparison of single and dual Doppler 4DVAR allowed the conclusion that single Doppler 4DVAR can be effective even for the tangential component of the vectors not directly measured by the single coherent Doppler lidar.

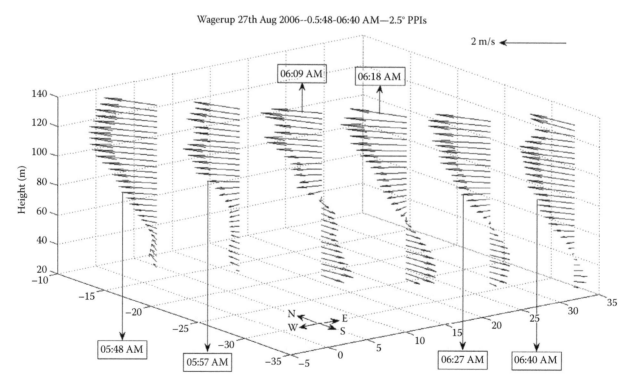

Wagerup 27th Aug 2006--0.5:48-06:40 AM—2.5° PPIs

FIGURE 36.6 VAD algorithm applied to 2.5° PPIs for August 27th during Wagerup 2006. (From Calhoun, R. et al., Meteorological mechanisms and pathways of pollution exposure: Coherent Doppler lidar deployment in Wagerup, Final report for Western Australian Department of Environment and Conservation, 2007b. With permission.)

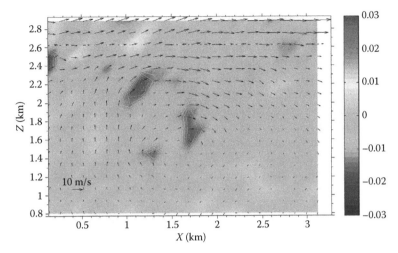

FIGURE 36.7 Two-dimensional vector retrieval for a co-scanned RHI plane during T-REX. Vorticity (background color) and swirling strength (contours without arrows) for a rotor at 11:04 am PST (1904 UTC), units are s⁻¹. (From Hill, M. et al., *J. Atmos. Sci.*, 67, 713–729, 2010. With permission.)

As described earlier, supplementary measurements can also be used to acquire unmeasured velocity components. Calhoun et al. (2006) presented algorithms for wind velocity profiles "virtual towers" from coordinated Doppler lidar scans (intersecting RHIs from different Doppler lidars) at a series of intersections upstream of a downtown urban area. Rather than using intersecting planes, co-scanning the same plane allows another form of vector retrieval especially valuable for tracking fluid mechanical structures on a 2D plane (Hill et al. 2010). Figure 36.7 in the following shows an example of educed velocities that captured a rotor during the T-REX campaign (Hill et al. 2010). This paper gives a first glimpse at the character and dynamics of clear air rotors and subrotors using 2D velocity vectors, vorticity, swirling strength, and velocity streamlines. Through the visualization of these flow parameters, the behavior and evolution of these vortical structures were observed.

In the Drechsel et al. (2009), 3D wind fields were reduced from single Doppler lidar and the continuity equation, as well as, with a dual Doppler approach. Both approaches were adapted from the radar community. Comparisons of wind fields obtained from this method were made with those from a wind profiler and the differences were found to be less than 1.1 m s⁻¹.

36.5 Major Challenges

Major challenges in modern wind measurements are to select instruments and methods appropriate for the investigation of interest, to understand the error characteristics, and to understand intercomparisons of results from inherently different measurement techniques. Measurement volumes, particularly for remote sensing instruments, may be complicated. For example, for angled beams emanating from a single point, the application of the VAD algorithm implies homogeneity of winds on increasingly large scales as the range increases. Further, the type of retrieval algorithm applied affects both the data resolution and error characteristics. Applying the assumption of local

homogeneity of winds in order to utilize the sector-VAD technique imposes a form of spatial averaging on the flow field. The basic output from a Doppler lidar is a spatially averaged radial velocity field requiring additional information of some form in order to resolve ambiguities. If high-frequency sonic or anemometer data is to be compared with Doppler lidar data, then a filter or averaging may be first applied, to mimic the spatial filter inherent in the Doppler lidar data. Error characterization in Doppler data analysis can be involved because of the propagation of uncertainty in the raw data (a function of range and Signal to Noise Ratio) through the vector retrieval algorithm. Secondly, in remote sensing measurements, "pointing error" should be estimated to understand the uncertainty in knowing which volume of space is sampled.

References

Calhoun, R., R. Heap, M. Princevac, R. Newsom, H. Fernando, and D. Ligon, 2006. Virtual towers using coherent Doppler lidar during the Joint Urban 2003 experiment, *Journal of Applied Meteorology*, 45, 1116–1126.

Calhoun, R., M. Princevac, and H. Fernando, 2007a. Atmospheric measurements, *Springer Handbook of Experimental Fluid Mechanics*, Eds. C. Tropea, A. Yarin, and J.F. Foss, p.1500,Springer Verlag, Berlin, Germany, ISBN: 3540251413.

Calhoun, R., C. Retallack, A. Christman, and H. Fernando, 2007a. Meteorological mechanisms and pathways of pollution exposure: Coherent Doppler lidar deployment in wagerup, Final Report for Western Australian Department of Environment and Conservation.

Choukulkar, A., R. Calhoun, B. Billings, and J. Doyle, 2012. A modified optimal interpolation technique for vector retrieval for coherent Doppler lidar, *IEEE Geoscience and Remote Sensing Letters*, (99), 1–5. DOI: 10.1109/LGRS.2012.2191762.

Drechsel, S., M. Chong, G. Mayr, M. Weissmann, R. Calhoun, and A. Dörnbrack, 2010. Three-dimensional wind retrieval: application of MUSCAT to dual Doppler lidar, *Journal of Atmospheric and Oceanic Technology*, 26(3), 635–646.

Hill, M., R. Calhoun, H. Fernando, A. Wieser, A. Dörnbrack, M. Weissmann, G. Mayr, and R. Newsom, 2010. Coplanar Doppler lidar retrieval of rotors from T-REX, *Journal of the Atmospheric Sciences*, 67(3), 713–729.

Krishnamurthy, R. 2008. Retrieval of the dissipation of turbulent kinetic energy from coherent Doppler lidar data, Master's thesis, Mechanical Engineering, Arizona State University, Phoenix, AZ.

Kristensen, L. 1998. Cup anemometer behavior in turbulent environments, *Journal of Atmospheric and Oceanic Technology*, 15, 5–17.

Kristensen, L. 2002. Can a cup anemometer "underspeed"? A heretical question, *Boundary-Layer Meteorology*, 103, 163–172.

Kristensen, L., O.F. Hansen, and J. Højstrup, 2003. Sampling bias on cup anemometer mean winds, *Wind Energy*, 6, 321–331.

Lin, C., Q. Xia, and R. Calhoun, 2008. Retrieval of urban boundary layer structures from Doppler lidar data. Part II: Proper orthogonal decomposition, *Journal of the Atmospheric Sciences*, 65(1), 21–42.

Pedersen, T. 2003. Characterization and classification of Risø P2546 cup anemometer, Risø-R-1364(EN), Risø National Laboratory, Roskilde, Denmark.

Xia, Q., C. Lin, R. Calhoun, and R. Newsom, 2008. Retrieval of urban boundary layer structures from Doppler lidar data. Part I: Accuracy assessment, *Journal of the Atmospheric Sciences*, 65(1), 3–20.

Xu, Q., S. Liu, and M. Xue, 2006. Background error covariance functions for velocity wind analysis using Doppler radar radial-velocity observations, *Q.J.R. Meteorol. Soc.*, 132(621C), 2887–2904.

37

Atmospheric Flux Measurements

Eric R. Pardyjak
University of Utah

Chad W. Higgins
Oregon State University

Marc B. Parlange
Ecole Polytechnique
Federal de Lausanne

37.1 Introduction

Quantification of near-surface atmospheric fluxes (hereafter AFs) is one of the oldest and most important aspects of environmental fluid dynamics (EFD), as these fluxes represent the transport of mass, momentum, and energy across boundaries of interest that are critical to human activities. In modeling, often, mean advective fluxes can be calculated directly using different (averaged/filtered) forms of the fundamental transport equations; however, the remaining turbulent flux terms represent the classic turbulence closure problem (Pope, 2000). The advent of a number of critical pieces of instrumentation has resulted in a greatly improved ability to probe atmospheric fluxes and understand the associated governing physical processes. Partly as a result of the increased availability of this instrumentation, there has also been a great increase in the number of projects and applications focusing on the measurement of atmospheric fluxes. With this increased activity, a need has developed for resources to aid students and researchers entering in this area. This chapter attempts to outline a practical methodology that can be utilized to aid researchers wishing to make reliable flux measurements in the atmospheric surface layer to test specific hypotheses related to EFD applications. Because of the depth of any individual aspect of the subject areas discussed later, it is impossible to provide a complete description. However, references are provided to useful resources where further details may be found. Finally, given the importance of understanding the linkages between atmospheric simulation and flux measurements, the chapter also provides guidance for integrating

modern flux modeling techniques and measurements with particular attention given to integration of measurements with large eddy simulations (LES).

37.2 Defining the Scope of the Problem

As experienced researchers know, developing a systematic experimental protocol to measure some specific quantity that is needed to test a hypothesis is iterative, requires critical thinking and is nonlinear in nature. To illustrate the process, consider Figure 37.1, which is essentially an application of the well-known "scientific method" to the measurement of atmospheric fluxes in the surface layer. The process begins with the development of a hypothesis based on existing theory and one⋔ current understanding of the problem. Ultimately, for the applications considered here, this step results in the identification of the atmospheric flux that we wish to measure. The next step in the processes represented by the three intersecting circles can be very diDcult and time consuming. This step represents the intersection of idealized theories, practical limitations associated with instrument constraints, and the realities of field deployment of instrumentation. It is during this portion of the processes that the viability of different methodologies is critically analyzed and evaluated to select the best measurement site and instruments to make the measurements. In this step, compromises to an ideal experiment are made. Estimates of the effects of these compromises are needed to ensure testability of the hypothesis. This process of selecting specific instrumentation, to be applied at a particular site, in a particular manner, usually

Handbook of Environmental Fluid Dynamics, Volume Two, edited by Harindra Joseph Shermal Fernando. © 2013 CRC Press/Taylor & Francis Group, LLC.
ISBN: 978-1-4665-5601-0.

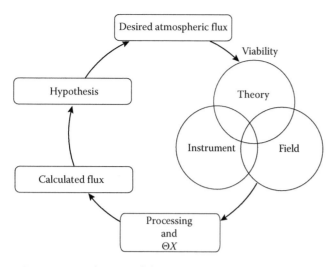

FIGURE 37.1 Schematic of the process for developing a successful atmospheric flux measurement.

fixes the type of data processing and quality control procedures that must be implemented. Hence, during the viability step, data processing methodologies, uncertainty analysis, and quality control must be considered. In the following sections, we suggest a methodology for developing the theoretical framework for defining the atmospheric fluxes to be measured and then discuss the processes associated with determining the viability of making a flux measurement.

37.3 Defining the Fluxes

In general, when considering the measurement of AFs and related hypotheses, one should begin with a general form of the basic transport equations in which the desired fluxes are embedded. For many applications, the following ensemble averaged equations are appropriate; however, other forms of the basic equations such as area averaged or filtered equations are also common and may give rise to additional flux terms. The ensemble averaged equations can be found in a number of resources (Stull, 1988), but are repeated here:

$$\frac{\partial \bar{u}_i}{\partial t} + \frac{\partial \overline{u_i u_j}}{\partial x_j} = -\delta_{i3} g \frac{\delta \bar{\theta}}{\theta_o} + f_c \varepsilon_{ij3} \bar{u}_j - \frac{1}{\rho} \frac{\partial \bar{p}}{\partial x_i} + v \frac{\partial^2 \bar{u}_i}{\partial x_j^2} - \frac{\partial \overline{u_i' u_j'}}{\partial x_j}$$

(37.1)

$$\frac{\partial \bar{\theta}}{\partial t} + \bar{u}_j \frac{\partial \bar{\theta}}{\partial x_j} = v_\theta \frac{\partial^2 \bar{\theta}}{\partial x_j^2} - \frac{\partial \overline{u_j' \theta'}}{\partial x_j} - \frac{1}{\rho C_p} \frac{\partial Q_j^*}{\partial x_j} - \frac{\lambda E}{\rho C_p}$$

(37.2)

$$\frac{\partial \bar{q}}{\partial t} + \bar{u}_j \frac{\partial \bar{q}}{\partial x_j} = v_q \frac{\partial^2 \bar{q}}{\partial x_j^2} - \frac{\partial \overline{u_j' q'}}{\partial x_j} + \frac{E}{\rho} + \frac{S_q}{\rho}$$

(37.3)

$$\frac{\partial \bar{c}}{\partial t} + \bar{u}_j \frac{\partial \bar{c}}{\partial x_j} = v_c \frac{\partial^2 \bar{c}}{\partial x_j^2} - \frac{\partial \overline{u_j' c'}}{\partial x_j} + S_c$$

(37.4)

Equations 37.1 through 37.4 are equations for momentum, enthalpy (energy), specific humidity, and mass (e.g. species concentration). While Equation 37.3 is simply a special case of Equation 37.4, we have written the two explicitly here so that the link between the two equations through the evaporation rate (E) is clear. To simplify the problem, it is usually convenient to rewrite the equations in dimensionless form. For example, Equation 37.1 may be rewritten as

$$\frac{\partial \bar{u}_i^*}{\partial t} + \bar{u}_j^* \frac{\partial \bar{u}_i^*}{\partial x_j^*} = -\delta_{i3} g \frac{\delta \bar{\theta}}{\theta_o} \frac{L}{U^2} + f_c \varepsilon_{ij3} \bar{u}_i^* \frac{L}{U^2} - \frac{\partial \bar{p}^*}{\partial x_j^*}$$

$$+ \frac{v}{LU} \frac{\partial^2 \bar{u}_i^*}{\partial x_j^{*2}} - \frac{\partial \overline{u_i'^* u_j'^*}}{\partial x_j^*}$$

(37.5)

where
  $Re = UL/v$ is the Reynolds number
  $R_O = U/f_c L$ is the Rossby number
  $Ri = g\delta\theta L/\theta_0 U^2$ a Richardson number

For most micrometeorological applications, the Reynolds number is very large and Coriolis force is small compared to inertial forces for the length scales under consideration. This allows terms 4 and 6 to be neglected. Well above the roughness elements in the inertial sublayer (or in the so-called constant flux layer), the horizontal pressure gradient can also be assumed to be negligible on the length scales considered (Monin and Obukhov, 1954). However, when individual wakes around trees and buildings are considered, this is certainly not the case. Figure 37.2 illustrates the complex behavior of momentum fluxes ($\overline{u'w'}$) that

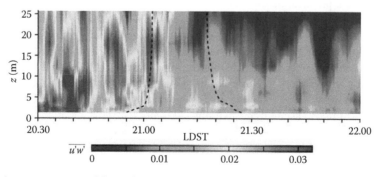

FIGURE 37.2 Contour plot of 3.33 min averages of the total momentum flux versus height and time of day from Metzger et al. (2007) during the evening transition of the ABL over the playa in Utah's west desert. (From Metzger M. et al., *Philos. Trans. R. Soc. A Math. Phys. Eng. Sci.*, 365, 859, 2007.)

exists even over relatively simple terrain such as the very flat and smooth playa in Utah's west desert.

A very typical application of these equations is the measurement of the flux of mass of a trace gas (e.g., CO_2) from a homogeneous surface (e.g., above a uniform agricultural crop). For such a case, Equation 37.4 is simplified and then integrated from the ground surface to the measurement height which is selected to be located in the constant flux layer. Following others (e.g., Baldocchi, 2003), if the flow is stationary, horizontally homogeneous, there is no mean vertical velocity component and molecular diffusion is much smaller than the turbulent diffusion, Equation 37.4 can be written as

$$\int_{z=0}^{z=z_m} \frac{\partial \overline{w'c'}}{\partial z}\, dz = -\int_{z=0}^{z=z_m} S_c\, dz$$

$$\overline{w'c'}\big|_{z=z_m} = \overline{w'c'}\big|_{z=0} - \int_{z=0}^{z=z_m} S_c\, dz \qquad (37.6)$$

According to Equation 37.6, the measured turbulent flux of c at height z_m, $\overline{w'c'}\big|_{z=z_m}$ represents the difference between the mass of c added (per unit area per unit time) at the canopy floor and that which emitted or absorbed within the canopy (per unit volume per unit time). The measurement of $\overline{w'c'}\big|_{z=z_m}$ is typically made using the eddy correlation technique described later. In this analysis, many assumptions have been made that should be checked, otherwise very erroneous conclusions may result.

In discussing idealized theory related to the measurement of atmospheric fluxes, it is necessary to introduce the concept of the so-called Monin–Obhukov similarity theory (MOST). This theory, developed by Monin and Obukhov (1954) and reviewed in many other places (e.g., Brutsaert, 2005; Foken, 2006), forms the fundamental basis for surface layer flux analysis. MOST has been extremely useful for the prediction of fluxes as well as the development of nondirect measurement techniques that rely on the theory to estimate atmospheric fluxes (discussed later). Essentially, the MOST hypothesis states that over an infinite rough surface which is horizontally uniform, there exists a region well above the height of the roughness elements where the fluxes are approximately constant; within this layer, all statistical quantities of the flow can be expressed in terms of the following variables: z, g/T_o, u_* and H. Where, z is the height above the ground, g/T_o is a buoyancy parameter (g is the acceleration of gravity and T_o the average temperature of the layer), $u_* = \sqrt{\tau_s/\rho}$ is the friction velocity characterized by τ_s the surface shear stress and ρ the fluid density, and $H = \rho c_p \overline{w'T'}$ is the sensible heat flux. Then, from dimensional analysis, the following length and temperature scales can be derived:

$$L = \frac{-u_*^3/\kappa}{(g/T_o)\overline{w'T'}} \qquad (37.7a)$$

$$T_* = \frac{-1}{\kappa u_*} \frac{H}{\rho c_p} \qquad (37.7b)$$

Within the constant flux layer, $\tau_s = \rho\overline{u'w'} = \text{constant}$ and $H/\rho c_p = \overline{w'T'} = \text{constant}$. Here w' is the vertical velocity fluctuation and u' the horizontal velocity fluctuation. As noted by Foken (2006), L, the Monin–Obukhov length scale, is historically interpreted as the thickness of a layer in which buoyancy forces are negligible, and more recently (if the denominator is interpreted as the buoyancy flux) as the ratio of mechanical generation of turbulent kinetic energy (tke) to buoyant production or destruction of tke. In the latter case, 37.7a is written (Stull, 1988) as

$$L = \frac{-u_*^3/\kappa}{(g/\Theta_{v,o})\overline{w'\theta_v'}} \qquad (37.8)$$

where

κ is the von Karman constant ($\sim$0.4)
θ_v' is the fluctuation of the virtual potential temperature
$\Theta_{v,o}$ is the virtual potential temperature of the layer

Again following others (Foken, 2006; Garratt, 1992), applying the Buckingham Pi theorem indicates that all dimensionless groups should be a function of z/L, using the mean wind, temperature, and moisture gradients leads to

$$\frac{\kappa z}{u_*} \frac{d\overline{u}}{dz} = \phi_m\left(\frac{z}{L}\right) \qquad (37.9a)$$

$$\frac{\kappa z}{\theta_{v*}} \frac{d\overline{\theta_v}}{dz} = \phi_h\left(\frac{z}{L}\right) \qquad (37.9b)$$

$$\frac{kz}{q_*} \frac{d\overline{q}}{dz} = \phi_w\left(\frac{z}{L}\right) \qquad (37.9c)$$

where

$\rho c_p \overline{w'\theta_v'} = \rho c_p u_* \theta_v$ is the buoyancy flux
q is the specific humidity
$\rho\overline{w'q'} = \rho u_* q_*$ is the water vapor flux

$\phi_m = (z/L)$, $\phi_h = (z/L)$ and $\phi_w = (z/L)$ are universal stability functions for momentum, buoyancy, and moisture fluxes that are given elsewhere (Arya, 2001; Foken, 2006; Garratt, 1992) and take on the value of unity for neutral conditions. When Equations 37.9a through 37.9c are integrated, they lead to the classical logarithmic profiles. These equations also show the basic relationship between the fluxes of momentum and heat and the gradients in velocity and temperature.

The basic theory has been tested and verified over many simple surfaces (see Foken, 2008 review) and appears to be surprisingly robust when the assumptions are weakly stretched. It is important to note that when the surface layer becomes very stable, it can become dominated by waves and intermittent turbulence, and the theory is not valid. MOST is also not valid in the roughness sublayer below the constant flux layer where there is both vertical and horizontal variability of the fluxes (Mahrt, 2000).

37.4 Evaluating the Viability of an Atmospheric Flux Experiment

Once the specific flux terms to be measured have been identified, instruments, a field site, and instrument placement/configuration must be determined. As indicated in Figure 37.1, these decisions represent a delicate balance between theory and practice. Generally, regardless of the application type, three issues must be considered in this phase: flux footprint, instrument bias and error, and the validity of theoretical assumptions. In the following sections, each of these issues is discussed in detail.

37.4.1 Flux Footprint

Sensor placement substantially affects the flux that is being measured. For example, flux-chamber measurements of CO_2 gas exchange at a point on the floor of a complex forest are likely to be very different from eddy covariance measurements made in the inertial layer above the forest canopy. The latter represents an integrated measurement of fluxes from a much larger area than the former. It is therefore very important to consider the "footprint" or region of influence of the flux measurement. A flux footprint is defined more precisely by Schmid (2002) as "the transfer function between the measured value and the set of forcings on the surface–atmosphere interface." For relatively simple topography and ranges of atmospheric stabilities, a number of analytical approaches for determining the footprint have been proposed (e.g., Horst and Weil, 1992; Hsieh et al., 2000). For scalars, the footprint function f is related to the measured flux on a tower of height z_m by considering the relative contribution of each upwind surface source ($S(x)$, typically in mass per unit mass per unit area) as follows (Hsieh et al., 2000):

$$F(x, z_m) = \int_{-\infty}^{x} S(x) f(x, z_m) dx \qquad (37.10)$$

For example, Hsieh et al. (2000) have proposed a model for a homogeneous velocity distribution in which the footprint function is dependent on atmospheric stability. In this model, S_o is the source strength or total surface flux (typically in mass per unit mass per unit area) and the fraction of the total surface flux can be expressed as

$$\frac{F(x, z_m)}{S_0} = \exp\left(\frac{-1}{k^2 x} D z_u^P |L|^{1-P}\right) \qquad (37.11)$$

where
  L is the Monin–Obukhov length scale
  z_m is the measurement height
  z_u is a combined length scale $z_u = z_m(\ln(z_m/z_o) - 1 + z_o/z_m)$
  x is the distance upwind of the measurement position or "fetch"
  D and P are constants that were determined using a stochastic Lagrangian dispersion model

D and P take on the following values depending atmospheric stability: neutral conditions ($L \sim 0$): $D = 0.91$ and $P = 1$, stable ($L > 0$): $D = 2.44$ and $P = 1.33$, and unstable ($L < 0$) $D = 0.28$, $P = 0.59$. Hence, the fetch $d = x$, which contributes to a specified fraction of the surface fluxes ($F^* = F/S_o$), can be obtained by solving Equation 37.11:

$$d = \frac{-D z_u^P |L|^{1-P}}{k^2 \ln(F^*)} \qquad (37.12)$$

Figure 37.3 shows an example of three different footprints (where $F^* = 0.9$) resulting from three atmospheric stabilities over simple terrain with the surface roughness (z_o) varying as shown with wind angle. From Figure 37.3 and Equation 37.12, it can be seen the footprint increases with increasing tower height and decreasing surface roughness. In addition, during the day when the atmosphere is very turbulent, the footprint is much smaller than typically found during the night under very stable atmospheric conditions.

This methodology is applicable for relatively simple configurations. For more complex topography or source configurations, more sophisticated modeling techniques may be needed to describe the footprint. Lagrangian stochastic (LS) methods and Backward Lagrangian Stochastic (BLS) are methods in which particle fluxes are calculated by simulating the stochastic motion of particles. With BLS models (Kljun et al., 2002). Particles are released at the measurement point and backward trajectories are calculated, typically using some variation of the Langevian equations (e.g., Thomson, 1987) run with backward velocities. The particles ultimately impact locations at their source, allowing one to understand the distribution of the source footprint even if it is quite nonhomogeneous.

37.4.2 Instrument Selection

Selection of instrumentation is extremely critical. Often times, the exact instrument required to make the desired flux measurement does not exist and may lead to the development of new instrumentation. In the following section, we survey a number of different types of methodologies for obtaining different types of fluxes within the atmospheric surface layer. More details of many of the methodologies have been reviewed elsewhere (e.g., Arya, 2001; Brutsaert, 2005; Foken, 2008; Garratt, 1992; Kaimal and Finnigan, 1994). In fact, Rana and Katerji (2000) review 10 of these methods in the context of evapotranspiration alone.

Eddy Covariance (*EC*)—Also referred to as the eddy correlation and eddy flux methods, this technique has been widely used and discussed in the literature and many excellent reviews exist (e.g., Baldocchi, 2003; Lee et al., 2004). EC has been applied to measure mass, momentum, and energy fluxes in many different environments such as over agricultural crops (Soegaard et al., 2003), cities (Rotach et al., 2005), forests (Baldocchi et al., 2001), and snow-covered regions. EC measurements of the trace gases

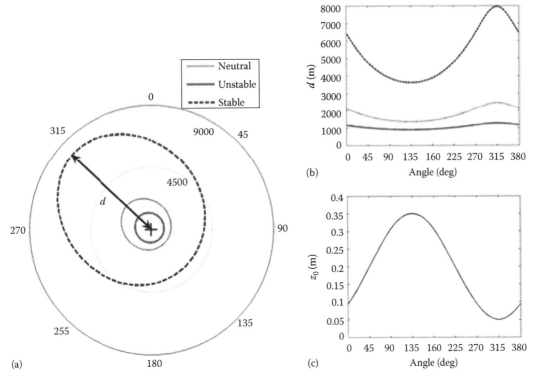

FIGURE 37.3 Schematic illustrating the effect of atmospheric stability and surface roughness on the flux footprint. (a) Polar plot indicating the length of the footprint d as a function of wind direction. Contours represent a boundary that contains 90% of the flux measured at the tower height z_m. Together (b) and (c) illustrate the relationship between roughness and footprint scale.

of CO_2 and H_2O (Baldocchi et al., 2001) have become extremely common and to a lesser extent, measurement of methane and N_2O fluxes (Smeets et al., 2009). Typically, these measurements are made at a relatively high sampling frequency (e.g., 10 Hz or greater) using close path or open path measurements for the gases. EC measurement of other more complex chemical species are less common, but fluxes of several volatile organic compounds (VOCs) have been made in forests (Westberg et al., 2001) and as part of urban field campaigns (Velasco et al., 2009). Particulate flux measurements are less common (Nemitz et al., 2008) than gas fluxes due to the inability of most instruments to capture the high-frequency portion of the turbulent spectrum.

EC is a direct measurement technique that provides correlations of sampled meteorological variables at point. The basic concept of the method is tied closely to the analysis discussed in the derivation of Equation 37.6 earlier. Fluxes that are measured are representative of a footprint surrounding the tower that varies depending on a number of parameters including tower height, surface roughness, and atmospheric stability. In the most simples form, the EC method is based on the measurement of the time average vertical flux of some scalar c and is written as $F_c = \overline{w'c'}$ which represents the correlation between the fluctuating turbulent vertical velocity component and the scalar as given in Equations 37.1 through 37.4. Velocities are typically based on (10–20 Hz) sonic anemometer measurements (Kaimal and Finnigan, 1994; Tropea et al., 2007). Because the technique utilizes instruments with high sampling frequencies, EC also allows the investigation of turbulence spectra. With respect to atmospheric fluxes, analysis of cospectrum and coherence spectrum (Biltoft and Pardyjak, 2009) can allow for a better understanding of the interaction of the two variables being measured at different scales.

Depending on the application (e.g., over homogeneous terrain versus within an urban street canyon) and the instruments selected (e.g., open path vs. closed path), different post-processing methodologies are needed and care must be exercised in interpreting EC data. Many of the details concerning these issues are presented elsewhere (e.g., Lee et al., 2004). For example, CO_2 and H_2O flux measurements require extreme care in the design of the experiment to ensure all theoretical assumptions are met. In addition, many corrections, to ensure confidence in the final flux measurements, usually need to be applied (e.g., Burba and Anderson, 2007). Examples of corrections/issues that need to be considered include sonic anemometer wake effects (Grelle and Lindroth, 1994), sonic anemometer angle of attack errors and coordinate rotation and sonic anemometer tilt correction errors (Wilczak et al., 2001), flux measurement errors associated with instruments that measure density rather than mixing ratio (Webb et al., 1980), and frequency response corrections. Additionally, consideration must be given to the averaging operators and methods used to assure statistical convergence (Lee et al., 2004). Over heterogeneous terrain, advection may be very important, yet still extremely diDcult to estimate (Feigenwinter et al., 2008).

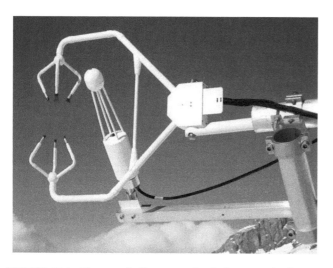

FIGURE 37.4 Photograph illustrating the deployment of a flux station with a 3D sonic anemometer/thermometer and an open-path infrared gas analyzer.

Note that additional details regarding processing of EC data are given in Appendix 37.A. Figure 37.4 shows a sample of typical mean fluxes obtained using EC. Shown are 30-min ensemble mean fluxes of CO_2, heat, and the friction velocity for a diurnal period at highly vegetated site in suburban area near Salt Lake City during late summer (September). Positive fluxes indicate the surface is acting as a source, while negative fluxes indicate the surface is a sink.

Drag Plate—A drag plate provides a direct measurement of the flux of momentum to the ground via the surface shear stress τ_s. With this method, the friction velocity is calculated directly from the definition $u_* = \sqrt{\tau_s/\rho}$. The technique is widely used in engineering wind tunnels where the flow is unidirectional, but is much more rare in the atmosphere due to many technical diDculties that must be overcome. A large hole is typically dug into a soil surface where the sensor is placed into the ground. It is extremely diDcult to make these measurements in the real atmosphere were the flow is omnidirectional; the sensor surface must have the same physical properties as the surrounding terrain and the instrument must be carefully leveled and care taken to minimize the width of the gap between the plate and surroundings. Typically, plates are circular to help overcome wind directionality problems. Sensor types vary, but Bradley (1969) successfully made drag plate measurements over grassland with a rigid mechanical drag plate, while Pasquill (1950) and Sadr and Klewicki (2000) made measurements using a drag plate floating on water over grassland and salt playa, respectively.

Hot Wire/Hot Film—This is a point measurement technique, traditionally used in laboratory wind tunnels, that can be used to obtain momentum fluxes ($u_i' u_j'$), and when combined with a so-called cold wire probe or thermocouple, can also be used to measure sensible heat fluxes $u_i' T'$ just as the EC technique does with sonic anemometers. Compared to sonic anemometers, hot-wire probes have the advantage of being much smaller (sampling

volumes are typically O (1 mm)) and are capable of resolving the high-frequency portion of the turbulence spectrum (up to approximately 100 kHz). This can be particularly important as the scales of turbulence get smaller, such as near surfaces and under stable conditions (Skelly et al., 2002). Unfortunately, the probes are extremely delicate (prone to wire breakage), require regular calibration, and are diDcult to deploy in the field. The use of the technique has been reviewed in great detail elsewhere (e.g., Tropea et al., 2007). While there are several types of hot-wire anemometers, the most common is the constant-current anemometer (CCA). The basic principle involves sending a constant, known current through a bridge circuit with one leg being the hot-wire sensor element (e.g., a 5 µm tungsten wire) bridge acting as a variable resister. A voltage is then measured across the probe. The voltage is proportional to the convective cooling of the wire. Typically, several wires are oriented at different angles with respect to each other to measure the different components of fluxes. Examples of applications in the atmospheric surface layer are the CASES-99 momentum and sensible heat flux measurements of Skelly et al. (2002) in Kansas over grassland and the momentum flux measurements of Priyadarshana and Klewicki (2004) over the salt playa in Utah\s west desert. A particularly novel flux measurement device is the so-called scalar transport probe of Metzger and Klewicki (2003) which combines hot-wire probes with fast-response concentration measurements to allow for flux (and flux gradient) measurements at a rate of over 300 Hz.

Particle Image Velocimetry (PIV)—Like hot-wire anemometry, this method that can be used to obtain mean velocities and turbulent momentum fluxes. However, since the technique involves imaging an entire plane, a field of data may be obtained and gradients of velocities are also relatively easily calculated. Because gradients of velocity components and fluxes are the actual terms found in the governing equations (see Equations 37.1 through 37.4), this technique is quite powerful. PIV is commonly used in the laboratory to study engineering flows, where it has been described in great detail (Tropea et al., 2007). Unfortunately, it is also diDcult to implement in the field due to the sensitivity of the equipment and technical diDculties. The basic principle behind PIV measurements is quite simple. The flow field is first seeded with neutrally buoyant particles to mark the flow. The particles are usually illuminated in a plane with a laser sheet pulsed at a known time increment and then imaged with digital cameras at two successive time steps. The technique determines a displacement vector from the successive images and velocities are calculated from the displacement vector and time step.

There are relatively few examples of momentum flux measurements that have been made using PIV in the surface layer. Figure 37.5 shows the experiment of Morris et al. (2007) that was made over the salt playa in Utah\s west desert. A more complex example is the measurements of van Hout et al. (2007), where momentum fluxes were measured above and within a corn canopy.

Flux-Prorle Measurement Methods—These techniques indirectly obtain surface fluxes under very ideal conditions. They are direct applications of the MOST flux-gradient similarity described earlier (e.g., Businger, 1971; Foken, 2006; Garratt, 1992).

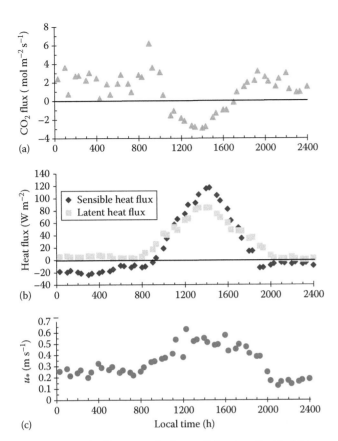

FIGURE 37.5 Plot illustrating the fluxes of (a) carbon dioxide, (b) latent and sensible heat, and (c) momentum flux at highly tree-covered urban site in Salt Lake City Utah. (Courtesy of P. Ramamurthy.)

If MOST assumptions are satisfied, then profile measurements of mean scalar quantities (e.g., temperature, specific humidity) and wind speed allow estimates of mean gradients and hence the surface fluxes described using Equation 37.9. *Bulk transfer methods* are type of profile method that utilizes the integrated forms of the flux–gradient relationships at two points. These methods are describes in many textbooks (e.g. Foken, 2006; Garrat, 1992; Stull, 1988).

Bowen Ratio Method—This is an important indirect method that is based on the ratio of fluxes and incorporates the use of the surface energy balance (SEB) to estimate fluxes of sensible and latent heat and does not directly use velocity measurements. The SEB for a thin horizontally homogeneous surface without advection effects is typically written: $Q^* = Q_E + Q_H + Q_G$ and the Bowen ratio is defined as the ratio of the sensible to latent heat fluxes: $B = Q_H/Q_E$ where Q_G is the ground heat flux and $Q^* = Q_{S\downarrow} + Q_{S\uparrow} + Q_{L\downarrow} + Q_{L\uparrow}$ is the net radiation flux composed of the individual components of incoming and outgoing short-wave and long-wave radiation. Here the sign convention is taken as positive for responsive fluxes pointing away from the surface and negative for radiative fluxes pointing toward the surface (e.g., Stull, 1988). Solving the SEB for the sensible and latent heat fluxes in terms of B yields

$$Q_H = B \frac{Q^* - Q_G}{1 + B}$$

and

$$Q_E = \frac{Q^* - Q_G}{1 + B}$$

The simplest application then involves utilizing the specification of the fluxes in a gradient form and with turbulent diffusion coeDcients for heat and heat and moisture (K_H and K_E):

$$B = \frac{Q_H}{Q_E} = \frac{\rho c_p \overline{w'T'}}{\rho L_v \overline{w'q'}} \approx \frac{\rho c_p K_H \, \partial \overline{T}/\partial z}{\rho L_v K_E \, \partial \overline{q}/\partial z}$$

If it is assumed that the diffusivities of heat and water vapor are the same and that the partial derivatives can be replaced by finite differences with temperature and specific humidity measured at the same two heights, then B is given by $B \approx (c_p/L_v)(\Delta \overline{T}/\Delta \overline{q})$ and can be measured using relatively inexpensive slow response pressure, temperature, humidity instruments (Foken, 2006).

The modified Bowen ratio method (Bussinger, 1986) utilizes some of these concepts, but relies on the fast-response measurement of at least the sensible heat flux using EC in conjunction with gradients of the mean quantities of two variables. Hence it assumes

$$\frac{F_c}{Q_H} \approx \frac{\Delta c}{\Delta T}$$

where the concentration and temperatures are measured at two heights. Note that model here has been used to estimate a wide variety of fluxes, including those of water vapor and CO_2 as well as the deposition of HNO_3, Hg, and others.

Surface Renewal (SR) Analysis—This is a technique that is used to estimate surface fluxes of scalars over vegetation without velocity measurements. It is based on the concept that particular coherent structures in flow over vegetation produce "ramp structures" in the time series of scalars that can be used to determine scalar fluxes (Paw et al., 1995). The model relates a Lagrangian conceptualization of transfer across a canopy interface to a Eulerian point measurement. In particular, a parcel of air rapidly descends to the surface, and for a period of time, the parcel is either enriched or depleted of a scalar. During this period, there is an associated change in concentration in the measured time series of the scalar. At some point, the parcel is rapidly ejected from the canopy, leading to a fast drop in the measured concentration. Using SR analysis, the sensible heat flux can be determined using the fast-response temperature measurement (e.g., fine wire thermocouple) as follows:

$$H = \rho c_p \frac{dT}{dt} \frac{V}{A}$$

where

dT/dt is the total temperature derivative
V/A is the volume of the air parcel over the horizontal area at the base of the parcel

A parcel of air with a characteristic height taken to be equal to the canopy height gives $h = V/A$. Several implementations of the method exist (Castellví et al., 2002). Recently, Castellví and Snyder (2010) compared SR to EC and lysimeter measurements over a grass field and found good agreement between all three methods.

Eddy Accumulation and REA—For the measurement of fluxes of many gases and particulates, instruments are often unavailable to measure concentration fluctuations fast enough to make EC practical. The eddy-accumulation (EA) technique (Desjardins, 1977) is essentially a conditional sampling method in which air samples are measured in "accumulating" reservoirs above and below an anemometer capable of measuring vertical velocity fluctuations. The upper reservoir accumulates the sample when the vertical velocity fluctuations (w') are positive and the lower instrument only samples when $w' < 0$ using a system that rapidly switches valves and ensures that the amount of air sampled is proportional to the strength of w'. As this method is diDcult to implement accurately, Businger and Oncley (1990) proposed a practical modification ("relaxation") of the original method where the measurement of c is made at a constant flow rate and the flux is calculated directly from the difference of the averaged concentrations made at the upper and lower samplers ($\overline{c^+}$ and $\overline{c^-}$). In their method, the flux is given by $\overline{w'c'} = \beta\sigma_w(\overline{c^+} - \overline{c^-})$, where σ_w is the standard deviation of the vertical velocity fluctuations and $\beta(z/L)$ is an empirical coeDcient (~ 0.6) that is relatively constant and given by $\beta(z/L) = r_{wc}(z/L)\sigma_c/(\overline{c^+} - \overline{c^-})$. Here, σ_c is the standard deviation of the concentration fluctuations and $r_{wc}(z/L)$ is a stability-dependant correlation coeDcient.

Chamber Techniques—These methods utilize enclosures that are placed over the ground (and vegetation) or water to make a local (small footprint), direct measurement of fluxes.

Depending on the type of chamber, the flux is obtained by measuring the change in concentration within the chamber or by measuring the difference in concentration entering and exiting the chamber. According to Livingston and Hutchinson (1995), closed static chambers are nonsteady-state devices without flow through that measured the change in concentration. Closed dynamic chambers are nonsteady devices that measure the change in concentration and open dynamic chambers are steady-state chambers that allow flow through the enclosure and measure the difference in concentration entering and exiting the chamber. These methods are often most appropriate for studying spatial scales ranging from organisms to leaves to plants, studying specific emission processes, and indentifying the variability sources within an ecosystem. Because there is a physical barrier between the ambient atmosphere and the surface being measured, they alter the local environment that is being measured, making the measurement unrepresentative. A review of both open- and closed-chamber methodologies can be found in Reicosky (1990a,b).

Raman LIDAR—A relatively new and exciting development in the measurement of AFs has been in the area of solar blind Raman LIDARs (Figure 37.6) to measure turbulent fields of moisture and temperature fluctuations and inferring latent and sensible heat fluxes in the atmospheric surface layer (Eichinger, 1994; Eichinger et al., 2006; Froidevaux et al., 2010). As shown in Figure 37.7, the methodology can resolve detailed turbulent moisture structures in the flow. Theoretically, current Raman LIDARs can measure at 1 Hz with 1.25 m spatial resolution over a path up to $\sim$700 m (Froidevaux et al., 2010). Here, we follow Froideveauxß (2010) description of the Raman LIDAR. The instrument operates by first emitting a beam of ultraviolet monochromatic light into the atmosphere and collecting

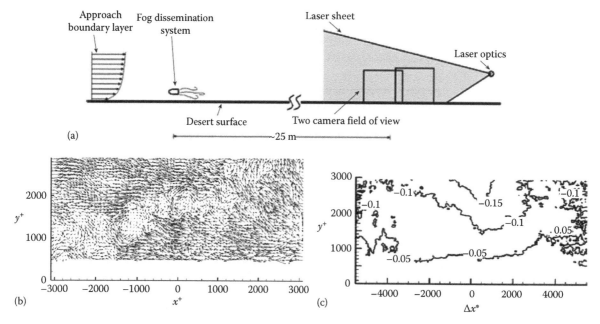

FIGURE 37.6 (a) Sketch of the PIV experiments of Morris et al. (2007), (b) snapshot of the instantaneous velocity field, and (c) cross-correlation coeDcient R_{uv}. Note that y is the vertical coordinate and x is the along-wind coordinate. (From Morris, S. et al., *J. Fluid Mech.* 580, 319, 2007.)

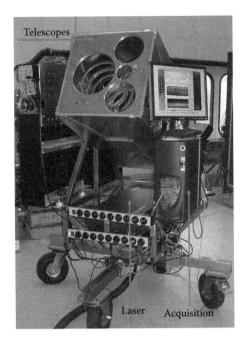

FIGURE 37.7 Photo of the four telescope solar blind EPFL Raman LIDAR.

the light fast enough to retrieve the position of the sounding volume using the known return time. The emitted radiation is in the solar blind region of the spectrum where wavelengths shorter than 300 nm are absorbed in the stratosphere by ozone. Telescopes are used to collect the returned backscattered radiation signal. Because of the Raman Effect, molecular scattering results in a shift in wavelength (Whiteman et al., 1992). The returned signal has a vibrational shift that can be used to measure water vapor mixing ratio and a purely rotational shift that can be used to measure temperature.

For these systems, the laser beam can be scanned to produce fields of mixing ratios or held in place to directly obtain variances of moisture fluctuations at different locations along the beam. Eichinger et al. (1994, 2003) have used the methodology to calculate fluxes from the vertical scans utilizing MOST flux-gradient profiles.

Scintillometry—This is a path-averaged technique based on the measurement of intensity fluctuations (scintillations) along a beam of electromagnetic radiation, which are associated with variations in index of refraction in the turbulent atmospheric surface layer. Since the refractive-index fluctuations are a function of humidity and temperature fluctuations, a relationship can be developed for the index-of-refraction structure function $\left(C_n^2\right)$ to provide an indirect measure of path-averaged turbulent fluxes of sensible and latent heat over distance ranging up to about 200 m (for small-aperture devices, Roth et al., 2006) and 10 km for large-aperture devices. The process of relating the temperature and humidity fluctuations to C_n^2 is done utilizing MOST. A number of different devices using different wavelengths, aperture sizes, and configurations exist, which can provide measurement of momentum fluxes. Several useful reviews of the methodology exist in the literature (e.g., Andrewas, 1990;

Hill, 1997), as well as device intercomparisons (Kleissl et al., 2009). Since the derived fluxes are based on the validity of the assumptions associated with MOST (homogeneous flat terrain), studies have investigated the applicability over slightly heterogeneous terrain and found if the violations of MOST are small, relatively good fluxes can still be obtained over more complicated terrain such as the roughness sublayer in urban areas. Roth et al. (2006) found agreement that was promising, but the results were particularly sensitive to the determination of the zero-plane displacement length.

Lysimeter—Many devices exist to estimate the flux of water vapor from the ground, including evaporation pans and atmometers (Brutsaert, 1982); however, lysimeters provide a very accurate way to directly measure the flux as well as the other components of the water balance (von Unold, 2008). A historical review and details of the measurement concepts of lysimeters are given in Brutsaert (1982). A lysimeter is essentially a large (often several meters in diameter and more than a meter deep) container of soil that is placed in the ground, on which vegetation may be grown, and is as representative as possible of the natural region surrounding the device. Wide ranges of surface covers have been studied, ranging from bare soil (Parlange and Katul, 1992) to crops (Pruitt, 1991) to trees (Edwards, 1986). Weighing lysimeters directly measure evapotranspiration through a mass balance, while nonweighing units utilize a volume balance. Deeper lysimeters usually use free drainage, while shallower units may utilize suction to ensure similar moisture profiles to the surrounding area. While lysimeters cannot be used to capture turbulent scale fluctuations, they can be used down to sub-hourly measurements (e.g., Parlange and Katul, 1992) depending on the accuracy of the device (Rana and Katerji, 2000). In addition to large traditional lysimeters, low-cost micro-lysimeters can also be used (Heusinkveld et al., 2006).

Sap Flow Measurements—This technique uses sap flow within individual trees to determine the tree transpiration component of water vapor fluxes from a forest. The method has the advantages of not requiring homogeneous terrain for the simplifications that are required for meteorological methods (Wilson et al., 2001) and is relatively inexpensive. When combined with EC, sap flux measurements can theoretically be used to separate tree transpiration from total water vapor flux (Granier et al., 1996; Oren et al., 1998). The measurements can be made on relatively short timescales (~30 min). Estimates of total forest fluxes require "up-scaling" from representative tree measurements (Oren et al., 1999).

The Granier sap flow sensor is a thermal dissipation probe (Granier, 1985) that is based on the measurement of heat diffusion in the sapwood xylem of a tree. In this technique, two cylindrical probes containing thermocouples are inserted into the sapwood of the bole of a tree, spaced vertically. The upper sensor is heated at constant power by passing a current through a constantan element. The xylem sap flux density (J_s in m s^{-1}) can then be estimated from Granier (1987) as $J_s = \alpha K^n$, where

$$K = \frac{T_o - T}{T - T_\infty}$$

Here, T is the temperature measured at the heated probe, T_y is the temperature of the unheated probe, and T_o is the temperature of the heated probe with zero flow ($J_S = 0$), and then, the total flow rate in a tree is given by $F = J_S A_S$, where A_s is the cross-sectional area of the sapwood at the heated probe.

37.5 Integrating Modern Flux Modeling Techniques and Measurements

37.5.1 Subgrid Scale Fluxes

In addition to fluxes of heat, water vapor, and momentum explained in the previous section, and the vast literature behind the theory and instrumentation (this can be conceptualized as a RANS approach), another recently emerging field is the measurement of subgrid scale (SGS) fluxes with arrays of fast-response instrumentation (sonic anemometers). These subgrid fluxes differ from the eddy fluxes in two unique ways: (1) they are described by a three-dimensional (3D) vector instead of a scalar vertical flux as in the eddy viscosity case; and (2) they are time local, so they can be calculated instantaneously at each measurement interval as opposed to a single flux value for a given averaging period. Subgrid, or subfilter fluxes are defined as fluxes that occur due to small, unresolved scales. This is an idea originating from the LES numerical technique. LES is a numerical technique in which the Navier–Stokes equations

$$\frac{\partial u_i}{\partial t} + u_j \frac{\partial u_i}{\partial x_j} = -\frac{1}{\rho}\frac{\partial p}{\partial x_i} + \nu \frac{\partial^2 u_i}{\partial x_j^2} \tag{37.13}$$

are treated with a high-pass filter which explicitly separates the resolved scales from the subgrid scales. The scale at which this divide takes place is the so-called filter scale, Δ. The filtering operation is carried out by convolving a filtering function (given as G in the following) with a known function or data stream. This filtering operation is expressed mathematically in a single dimension through the integral

$$\tilde{u}_i(x) = \int_{-\infty}^{\infty} G(\xi) u_i(x - \xi) d\xi \tag{37.14}$$

Applying the filtering operation 37.14 to the Navier–Stokes equations (Equation 37.13) and assuming the filtering operation commutes with the derivative will yield the LES prognostic equations. Note that here we have yet to define the shape of the filter function G, and it is unnecessary for the development of the LES equations:

$$\frac{\partial \tilde{u}_i}{\partial t} + u_i \frac{\partial \tilde{u}_i}{\partial x_j} = -\frac{1}{\rho}\frac{\partial \tilde{p}}{\partial x_i} + \nu \frac{\partial^2 \tilde{u}_i}{\partial x_j^2} - \frac{\partial \tau_{ij}}{\partial x_j} \tag{37.15}$$

Here, the SGS flux of momentum is given by τ_{ij}. This term has been given considerable attention in the literature as it must be parameterized within the LES. Verifying models for the SGS

stress and associated scalar fluxes is one of the main motivations for the experiments that measure the SGS stress:

$$\tau_{ij} = \widetilde{u_i u_j} - \tilde{u}_i \tilde{u}_j \tag{37.16}$$

And by analogy, the corresponding SGS heat flux:

$$q_j = \widetilde{\theta u_j} - \tilde{\theta}\tilde{u}_j \tag{37.17}$$

There are many choices of filter functions (G), and although it is unnecessary to prescribe the functional form of the filter to derive the LES equations, one must choose the form of the filter function to explicitly carry out the filtering operation (e.g., when treating data streams). The three most common filters are given as follows; however, all filters must be energy conserving, thus they must satisfy

$$1 = \int_{-\infty}^{\infty} G(\xi) d\xi \tag{37.18}$$

Top-hat, Gaussian, and spectral cutoff, are the three most common filtering functions, and are described by the following equations:

$$G(x)_{spectral} = \frac{\Delta^2}{\pi x} \sin\left(\frac{\pi x}{\Delta}\right) \tag{37.19}$$

The spectral filter has the advantage that turbulence spectra is cut sharply at a given wave number π/Δ. This property makes a clear separation in scales; however, it can be impractical to use on data as the filter has a wide range in real space, with important contributions to the filtered quantity reaching far beyond the filtering length scale; see Figure 37.8.

$$G(x) = \sqrt{\frac{6}{\pi \Delta^2}} \exp\left(-\frac{6x^2}{\Delta^2}\right) \tag{37.20}$$

The Gaussian filter is often chosen when a large spatial resolution is available, and is considered a good compromise between the spectral (mentioned earlier) and top-hat (later) filters since it has local and smooth properties in both real and spectral space. It will not cut the spectra as sharply as the spectral cutoff filter, but it will greatly diminish the energy of scales above the specified filter scale.

$$G(x)_{tophat} = \begin{cases} \dfrac{1}{\Delta} \text{ for } -\dfrac{\Delta}{2} \le x \le \dfrac{\Delta}{2} \\ 0 \text{ otherwise} \end{cases} \tag{37.21}$$

The top-hat filter is local in real space and is often the filter of choice when there is a limited spatial resolution of a data stream. However, it has the property that its Fourier transform contains all high wave number scales at low amplitudes. For this reason,

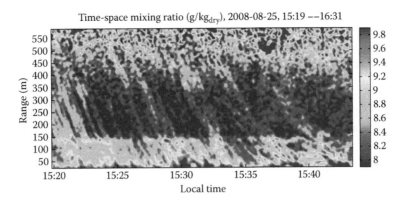

Time-space mixing ratio (g/kg$_{dry}$), 2008-08-25, 15:19 --16:31

FIGURE 37.8 **(See color insert.)** Contour plot of the mixing ratio of water vapor (g/kg) as a function of distance and time over a small lake in Switzerland using the EPFL Raman LIDAR.

it does not attenuate the turbulence spectra as sharply as the spectral cutoff or Gaussian filter functions. A plot of each filter is given in real space to demonstrate the relative features of each function in Figure 37.8.

37.5.2 Models of SGS fluxes

It is possible to completely resolve the full SGS fluxes through experimentation; however, these fluxes are not known explicitly within a simulation. Therefore, the SGS fluxes must be modeled. Typically, one of the following models is used. Note that the primary motivation for measuring the SGS flux through experimentation is to investigate the properties of these models.

37.5.3 Smagorinsky Model

Smagorinsky (1963) proposed a simple eddy-viscosity model based on local variables characterizing the motions at the length scale of the computational grid. In this model, the deviatoric part of the SGS stress tensor,

$$\tau_{ij} = \widetilde{u_i u_j} - \tilde{u}_i \tilde{u}_j,$$

is set proportional to the strain-rate tensor, $\tilde{S}_{ij} = (1/2)(e_i \tilde{u}_j + e_j \tilde{u}_i)$, characterizing the local deformation of the resolved velocity field. In these expressions, a tilde denotes spatial filtering at a length scale Δ. The model is written as

$$\tau_{ij} - \frac{1}{3}\tau_{kk}\delta_{ij} = -2\nu_T \tilde{S}_{ij}$$

The constant of proportionality is the eddy viscosity ν_T, which is written as $\nu_T = \lambda^2 |\tilde{S}|$, where $|\tilde{S}| = (2S_{ij}S_{ij})^{1/2}$. Here λ is a mixing length scale, while $\lambda |\tilde{S}|$ is a characteristic velocity scale estimated from the shear scale $|\tilde{S}|$ and the mixing length. The mixing length must be chosen judiciously. For locations far from boundaries and in the absence of buoyancy and rotation effects, the only length scale available to characterize the local turbulence structure of the simulated flow is the filter scale, Δ. Dimensionally it follows that $\lambda = c_s \Delta$, where c_s is a dimensionless model parameter. This parameter must be specified in LES, and has been the subject of much attention in the literature (Meneveau and Katz, 2000).

Lilly (1967) showed how c_s could be evaluated from basic knowledge of turbulence, and thus c_s is often referred to as the "Smagorinsky-Lilly" constant in the literature. Central to Lilly's development was the realization that the most important effect of the SGS model upon the dynamics of the large-scale structures is the amount of kinetic energy the model extracts. Hence, the energetics of the flow computed in an LES takes on a special role. Lilly (1967) derives the transport equation for the subgrid kinetic energy $E = (1/2)\tau_{kk}$ and obtains

$$\frac{\partial E}{\partial t} + \tilde{u}_k \frac{\partial E}{\partial x_k} - \nu \left[\frac{\partial^2 E}{\partial x_k^2} - \overline{\left(\frac{\partial u_i}{\partial x_k}\right)^2} + \left(\frac{\partial \tilde{u}_i}{\partial x_k}\right)^2 \right]$$

$$= \frac{1}{2}\tau_{ij}\tilde{S}_{ij} - \frac{\partial}{\partial x_k}\left(\frac{\widetilde{\tilde{u}_k u_i^2}}{2} - \frac{\tilde{u}_k \tilde{u}_i^2}{2} - \widetilde{\tilde{u}_i \tilde{u}_k u_i} + \widetilde{\tilde{u}_i}\ \tilde{u}_k + \frac{\widetilde{\tilde{u}_k p}}{\rho} - \frac{\tilde{u}_k \tilde{p}}{\rho} \right)$$

where $\tilde{S}_{ij} = \frac{1}{2}(\partial_j \tilde{u}_i + \partial_i \tilde{u}_j)$. Taking the ensemble average of this equation (denoted later by angled brackets), and assuming steady-state conditions, one obtains the equality of molecular dissipation of SGS kinetic energy and its rate of production:

$$\nu \left[\left\langle \overline{\left(\frac{\partial u_i}{\partial x_k}\right)^2} \right\rangle - \left\langle \left(\frac{\partial \tilde{u}_i}{\partial x_k}\right)^2 \right\rangle \right] = -\langle \tau_{ij}\tilde{S}_{ij} \rangle \quad (37.22)$$

The quantity $\langle \tau_{ij}\tilde{S}_{ij} \rangle$ is interpreted as the mean flux of kinetic energy from the range of resolved scales into the SGS range, and also appears as a sink in the equation for resolved kinetic energy, $\frac{1}{2}\tilde{u}_k \tilde{u}_k$. When Δ is in the inertial range, the first term on the left-hand side of Equation 37.22 dominates and equals the overall rate of dissipation by viscosity. Hence, we can write $\varepsilon = -\langle \tau_{ij}\tilde{S}_{ij} \rangle$.

Lilly then makes the next step in his derivation by replacing τ_{ij} with the Smagorinsky closure. One obtains the expression

$$\varepsilon = 2^{3/2}(c_s \Delta)^2 \left\langle \left(\tilde{S}_{ij}\tilde{S}_{ij}\right)^{3/2} \right\rangle \quad (37.23)$$

as a condition for the Smagorinsky model to extract kinetic energy from the resolved scales at the correct rate. Two more assumptions are required to complete Lilly's original argument: (1) that at the grid scale Δ, the turbulence exhibits a universal Kolmogorov spectrum $E(k) = c_k \varepsilon^{\frac{2}{3}} k^{-\frac{5}{3}}$ with turbulence statistics that are isotropic; this assumption is justified when Δ pertains to the inertial range of turbulence, and (2) the third-order statistics of the strain-rate magnitude may be approximated with its second-order moment as

$$\left\langle \left(\tilde{S}_{ij} \tilde{S}_{ij} \right)^{\frac{3}{2}} \right\rangle \approx \left\langle \tilde{S}_{ij} \tilde{S}_{ij} \right\rangle^{\frac{3}{2}}$$

The latter assumption is not explicitly stated in Lilly's paper since he did not elaborate on the nature of statistical averaging underlying the argument. The accuracy of this assumption was recently tested with direct numerical simulation (DNS) data and deviations on the order of 20% were observed in the inertial. Equation 37.23 still leaves the task of evaluating the second-order moment and strain-rate tensor contraction $\langle \tilde{S}_{ij} \tilde{S}_{ij} \rangle$. Using standard techniques from isotropic turbulence analysis, it is straightforward to show that

$$\left\langle \tilde{S}_{ij} \tilde{S}_{ij} \right\rangle = \int_0^{\pi/\Delta} k^2 E(k) dk$$

where, as in Lilly (1967), a spherical spectral sharp filter is used to cutoff the integration in wave number space at a wave number π/Δ. Substituting this into Equation 37.23 and using the Kolmogorov spectrum $E(k) = c_k \varepsilon^{\frac{2}{3}} k^{-\frac{5}{3}}$ and solving for the coefficient c_s, one obtains Lilly's result as follows:

$$c_s = \frac{1}{\pi} \left(\frac{2}{3c_k} \right)^{\frac{3}{4}} \sim 0.165 \quad (\text{for } c_k = 1.6)$$

The analysis has been reproduced in tutorial detail in Pope (2001).

37.5.4 Scale-Dependent Dynamic (Germano, 1996)

One way to specify the coefficient c_s is to obtain it directly through the simulation dynamically. If one filters the simulation results at a scale 2Δ, where Δ is the original filter scale, we obtain an equation for the SGS closure at the original filter scale Δ

$$\tau_{ij} - 2c_{s,\Delta}^2 \Delta^2 |\tilde{S}| \tilde{S}_{ij}$$

and a second SGS closure for the second filter scale 2Δ

$$T_{ij} = -2c_{s,2\Delta}^2 \Delta^2 |\hat{\tilde{S}}| \hat{\tilde{S}}_{ij}$$

Then, using the Germano identity

$$L_{ij} = T_{ij} - \hat{\tau}_{ij} = \widehat{\tilde{u}_i \tilde{u}_j} - \hat{\tilde{u}}_i \hat{\tilde{u}}_j$$

next we assume scale invariance

$$c_{s,\Delta}^2 = c_{s,2\Delta}^2 = \frac{\langle L_{ij} M_{ij} \rangle}{\langle M_{kl} M_{kl} \rangle}$$

where

$$M_{ij} = 2\Delta^2 \left[\widehat{|\tilde{S}| \tilde{S}_{ij}} - 2^2 |\hat{\tilde{S}}| \hat{\tilde{S}}_{ij} \right]$$

This form minimizes the error in the calculation of c_s, but assumes scale invariance. A similar procedure that relaxes this assumption filters the resolved field at a third scale: 4Δ, and assumes a power-law variation in scale of the quantity c_s:

$$c_{s,\Delta}^2 = \frac{c_{s,2\Delta}^4}{c_{s,4\Delta}^2} = c_{s,2\Delta}^2 / \beta$$

37.5.5 Nonlinear Model

The nonlinear (or tensor eddy viscosity) SGS parameterization (see Clark et al., 1979; Leonard, 1974; Liu et al., 1994)

$$\tau_{ij}^{nl} = C_{nl} \Delta^2 \left(\frac{\partial \tilde{u}_i}{\partial x_k} \frac{\partial \tilde{u}_j}{\partial x_k} - \frac{1}{3} \frac{\partial \tilde{u}_m}{\partial x_k} \frac{\partial \tilde{u}_m}{\partial x_k} \delta_{ij} \right)$$

yields significantly better results in correlation studies, and allows for negative dissipation of turbulent kinetic energy. Although this latter feature is thought to be physically more realistic, negative dissipation can cause numerical errors to grow, leading to numerical instability.

37.5.6 Mixed Model

A third formulation, the so-called mixed model, is a linear combination of the Smagorinsky and nonlinear model:

$$\tau_{ij}^{mix} = -2c_s^2 \Delta^2 |\tilde{S}| \tilde{S}_{ij} + C_{nl} \Delta^2 \left(\frac{\partial \tilde{u}_i}{\partial x_k} \frac{\partial \tilde{u}_j}{\partial x_k} - \frac{1}{3} \frac{\partial \tilde{u}_m}{\partial x_k} \frac{\partial \tilde{u}_m}{\partial x_k} \delta_{ij} \right)$$

and has been shown to perform well in a variety of flow situations when the dimensionless model coefficients are chosen to match the mean dissipation. (For a review of applications of the mixed model in LES, see Meneveau and Katz, 2000.)

37.5.7 Concept of A Priori Testing

It is often difficult to tease out the effect of an SGS model's performance on the LES output due to the integrated nature of model results (i.e., errors can arise through the mesh discretization, boundary conditions, numerical errors, and imperfect SGS closures). Therefore, it is often easier to compare real and modeled SGS stresses directly, in the absence of simulation results. This is the so-called *a priori* approach. Here we envision that our sensor array samples within a typical grid cell of a numerical simulation. This over-sampling of the grid cell allows for spatial filtering (see next section), and the calculation of the full SGS stress tensor and the filtered strain rate along with other relevant turbulent parameters, thus allowing for a direct comparison between the measured and modeled SGS stress tensor.

37.5.8 Experimental History

SGS fluxes can be attained from a single stationary anemometer. It is typical, however, to attain such measurements using horizontal arrays of sensors shown schematically in Figure 37.9 (used by Tong et al., 1999), through multiple horizontal arrays shown schematically in Figure 37.10 (Bou-Zeid et al., 2009; Higgins et al., 2003; Horst et al., 2004; Kleissl et al., 2004; Porté-Agel et al., 2000) or through a vertical grid array. In Figure 37.10, the imagined filter section is shown with the filter length scale Δ. The essential difference between the array geometry is on the filters that can be applied to the data and the other turbulence statistics that can be computed. While a single fast-response sensor is suDcient to capture full SGS fluxes (with assumptions

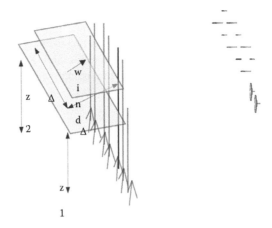

FIGURE 37.11 A schematic representation of a typical SGS flux experiment with two horizontal arrays of anemometers displaced by a vertical distance.

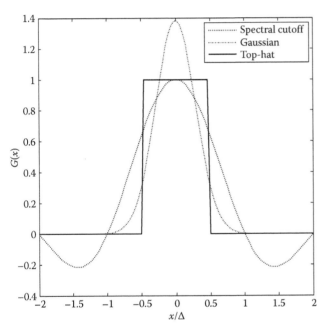

FIGURE 37.9 Illustration of the behavior of spectral, top-hat, and Gaussian filter functions in real space.

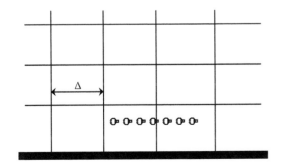

FIGURE 37.10 Illustration of a horizontal array of sonic anemometers used in an a priori field study. We imagine the atmosphere as a LES domain, while our experiment setup measures the fine-scale turbulence within a small number of grid cells that have a characteristic size Δ.

about the applicability of one-dimensional filtering), it is diDcult to give this measurement any meaning as an experiment involving one sensor is incapable of resolving cross-stream gradients of the scalar and velocity field.

Previous studies of this type contributed to the general understanding of sweeps and ejections (Porté-Agel et al., 2000), the properties of the filtered variables (Tong et al., 1999), the development of the scale-dependent dynamic procedure, the influence of surface waves (Sullivan et al., 2006), and strongly stable buoyant conditions (Bou-zeid et al., 2009).

Typical a priori experiments in the ABL have the geometry shown in Figure 37.11. This is the geometry of the so-called Horizontal Array Turbulence Study (HATS) approach, due to the fact that this configuration is the minimum requirement to fully resolve all of the velocity and scalar gradient terms. However, this configuration relies on 2D filtering.

If one wants an experiment in which all terms of the velocity gradient are computed with 3D filtered variables, an array geometry such as that of Higgins et al. (2007, 2009) is required.

37.5.9 Data Treatment

To evaluate the SGS flux (Equations 37.16 and 37.17), one must carry out the filtering operation in Equation 37.18 while choosing an appropriate filter form for the data available. When spatial data resolution is sparse, the top-hat filter may be the only choice. When enough data is present to produce meaningful turbulent spectra, any filter may be chosen. It is easiest to carry out the mathematical operation in Fourier space to take advantage of the properties of the convolution. This operation is also available in many commercial software packages and libraries. For example, the commercial software MATLAB® (www.mathworks.com) is capable of performing this operation with the conv(U,G, same) command. Where G is a vector containing the filter function, U is the unfiltered data of interest, and "same" is an option for the convolution function. Similar routines exist in LINPAC (www.linpac.com) and other programming libraries.

37.5.10 Data Availability

Approximately 50% of the data used in the literature are available freely through the National Center for Atmospheric Research. If such data are of interest, they can be found on the NCAR webpage www.ncar.ucar.edu.

Appendix 37.A: Practical Information for Sonic Data Analysis and Eddy Correlation

Basic anemometer operating principle: A 3D ultrasonic anemometer measures wind speed in three dimensions using three pairs of non-orthogonally orientated transducers. Each pair of transducers transmits and receives the ultrasonic signal. The time-of-flight is directly related to the wind speed along the sonic transducer axis, and the speed of sound in air is related to the temperature and humidity through the air density. For more details, see, e.g., Tropea et al. (2007).

Tilt corrections: When the sonic is initially set up, one will try to orient the vertical axis of the anemometer with the "true" vertical direction, but there will always be small deviations from this ideal orientation. Thus, the data are virtually aligned to flow-defined coordinates through rotation maps characterized by the matrix $\overline{A}$:

$$\vec{x} \rightarrow \overline{A} \cdot \vec{x}$$

For the Yaw correction, $\overline{A}$ is given as follows:

$$\overline{A} = \begin{pmatrix} \dfrac{\overline{u}}{\sqrt{\overline{u}^2 + \overline{v}^2}} & \dfrac{\overline{v}}{\sqrt{\overline{u}^2 + \overline{v}^2}} & 0 \\ \dfrac{-\overline{v}}{\sqrt{\overline{u}^2 + \overline{v}^2}} & \dfrac{\overline{u}}{\sqrt{\overline{u}^2 + \overline{v}^2}} & 0 \\ 0 & 0 & 1 \end{pmatrix}$$

After the Yaw has been corrected, the pitch can be corrected by defining

$$\overline{A} = \begin{pmatrix} \dfrac{\overline{u}}{\sqrt{\overline{u}^2 + \overline{v}^2}} & 0 & \dfrac{\overline{w}}{\sqrt{\overline{u}^2 + \overline{v}^2}} \\ 0 & 1 & 0 \\ \dfrac{-\overline{w}}{\sqrt{\overline{u}^2 + \overline{v}^2}} & 0 & \dfrac{\overline{u}}{\sqrt{\overline{u}^2 + \overline{v}^2}} \end{pmatrix}$$

If the mean wind direction does not change with height, there is no lateral velocity correlation. Therefore, all correlation of v' and w' is attributed to misalignment of the sonic anemometer with the mean wind direction. To realign the measurements to the proper coordinate frame, one must roll the coordinate frame around the mean wind direction. In practice, this is achieved through the transform matrix:

$$\overline{A} = \begin{pmatrix} 1 & 0 & 0 \\ 0 & \cos\beta & \sin\beta \\ 0 & -\sin\beta & \cos\beta \end{pmatrix} \quad \text{with } \beta = \frac{1}{2}\tan^{-1}\frac{2\overline{v'w'}}{\overline{v'^2} - \overline{w'^2}}$$

References

Arya S.P.: 2001, *Introduction to Micrometeorology*, 2nd edn., Academic Press, San Diego, CA.

Baldocchi D.D.: 2003, Assessing the eddy covariance technique for evaluating carbon dioxide exchange rates of ecosystems: Past, present and future, *Global Change Biol.* 9: 479–492.

Baldocchi D., Falgae E., Gu L., Olson R., Hollinger D., Running S., Anthoni P. et al.: 2001, Fluxnet: A new tool to study the temporal and spatial variability of ecosystem-scale carbon dioxide, water vapor, and energy flux densities. *Bull. Am. Meteorol. Soc.* 82: 2415–2434.

Biltoft C.B. and Pardyjak E.R.: 2009, A note on calculating coherence statistical significance. *J. Atmos. Oceanic Technol.* 26: 403–410.

Brutsaert W.: 2005, *Hydrology: An Introduction*, Cambridge University Press, Cambridge, U.K.

Castellví F., Perez P.J., and Ibanez M.: 2002, A method based on high-frequency temperature measurements to estimate the sensible heat flux avoiding the height dependence. *Water Resour. Res.* 38: 1084–1104.

Castellví F. and Snyder R.L.: 2010, A comparison between latent heat fluxes over grass using a weighing lysimeter and surface renewal analysis. *J. Hydrol.* 381: 213–220.

Edwards W.R.N.: 1986, Precision weighing lysimetry for trees, using a simplified tared-balance design. *Tree Physiol.* 1: 127–144.

Foken T.: 2008, The energy balance closure problem: An overview. *Ecol. Appl.* 18: 1351–1367.

Froidevaux M.: 2010, Land-atmosphere interactions measured with Raman lidar. PhD thesis. École Polytechnique Fédérale de Lausanne, Lausanne, Switzerland.

Froidevaux M., Higgins C., Simeonov V., Serikov I., van den Bergh H., Calhoun R., Ristori P., Pardyjak E., and Parlange M.: 2010, Turbulent atmospheric boundary layer evaporation (TABLE) experiment: Preliminary results. *25theInternational Laser Radar Conference*, July 5–9, 2010, St. Petersburg, Russia.

Garratt J.R.: 1992, *The Atmospheric Boundary Layer*, Cambridge University Press, Cambridge, U.K.

Grelle A. and Lindroth A.: 1994, Flow distortion by a solent sonic anemometer: Wind tunnel calibration and its assessment for flux measurements over forest and field. *J. Atmos. Ocean. Technol.* 11: 1529–1542.

Heusinkveld B.G., Berkowicz S.M., Jacobs A.F.G., Holtslag A.A.M., and Hillen W.C.A.M.: 2006, An automated microlysimeter to study dew formation and evaporation in arid and semiarid regions. *J. Hydrometeorol.* 7: 825–832.

Higgins C.W., Parlange M.B., and Meneveau C.: 2003, Alignment trends of velocity gradients and subgrid-scale fluxes in the turbulent atmospheric boundary layer, *Boundary-Layer Meterol.*, 109: 59–68.

Horst T.W., Kleissl J., Lenschow D.H., Meneveau C., Moeng C.-H., Parlange M.B., Sullivan P.P., and Weil J.C.: 2004, HATS, field observations to obtain spatially-filtered turbulence fields from transverse arrays of sonic anemometers in the atmospheric surface layer, *J. Atmos. Sci.* 61: 1566–1581.

Horst T.W. and Weil J.C.: 1992, Footprint estimation for scalar flux measurements in the atmospheric surface layer. *Boundary-Layer Meteorol.* 59: 279–296.

van Hout R., Zhu W., Luznik L., Katz J., Kleissl J., and Parlange M.B.: 2007, PIV measurements in the atmospheric boundary layer within and above a mature corn canopy. Part I: Statistics and energy flux. *J. Atmos. Sci.* 64: 2805–2824.

Hsieh C.-I., Katul G., and Tze-wen C.: 2000, An approximate analytical model for footprint estimation of scalar fluxes in thermally stratified atmospheric flows. *Adv. Water Resour.* 23:765–772.

Kaimal J.C. and Finnigan J.J.: 1994, *Atmospheric Boundary Layer Flows: Their Structure and Measurement*, Oxford University Press, Oxford, U.K.

Kleissl J., Parlange M.B., and Meneveau C.: 2004, Field experimental study of dynamic Smagorinsky models in the atmospheric surface layer. *J. Atmos. Sci.* 61 (18): 2296–2307.

Kleissl J., Watts C.J., Rodriguez J.C., Naif S., and Vivoni E.R.: 2009, Scintillometer intercomparison study—Continued. *Boundary-Layer Meteorol.* 130: 437–443.

Kljun N., Rotach M.W., and Schmid H.P.: 2002, A three-dimensional backward Lagrangian footprint model for a wide range of boundary-layer stratifications. *Boundary-Layer Meteorol.* 103: 205–226.

Lee X., Massman W., and Law B.: 2004, *Handbook of Micrometeorology: A Guide to Surface Flux Measurement and Analysis*, Kluwer Academic Publishers, Dordrecht, the Netherlands.

Leonard A.: 1974, Energy cascade in large-eddy simulations of turbulent fluid flows. *Adv. Geophys.* 18: 237.

Livingston G.P. and Hutchinson G.L.: 1995, Enclosure-based measurement of trace gas exchange: Applications and sources of error. In Matson P.A. and Harriss R.C. (Eds.), *Biogenic Trace Gases: Measuring Emissions from Soil and Water*. Blackwell Science, Cambridge, U.K., pp. 14–50.

Mahrt L.: 2000, Surface heterogeneity and vertical structure of the boundary layer. *Boundary-Layer Meteorol.* 96: 33–62.

Metzger M. and Klewicki J.C.: 2003, Development and characterization of a probe to measure scalar transport. *Meas. Sci. Technol.* 14: 1437–1448.

Metzger M., McKeon B.J., and Holmes H.: 2007, The near-neutral atmospheric surface layer: Turbulence and non-stationarity. *Philos. Trans. R. Soc. A Math. Phys. Eng. Sci.* 365: 859–876.

Monin A.S. and Obukhov A.M.: 1954, Basic laws of turbulent mixing in the surface layer of the atmosphere. *Tr. Akad. Nauk SSSR Geophiz. Inst.* 24: 163–187.

Morris S., Stolpa S., Slaboch P., and Klewicki J.C.: 2007, Near-surface particle image velocimetry measurements in a transitionally rough-wall atmospheric boundary layer. *J. Fluid Mech.* 580: 319–338.

Parlange M.B. and Katul G.: 1992, Estimation of the diurnal variation of potential evaporation from a wet bare soil surface. *J. Hydrol.* 132: 71–89.

Paw U.K.T., Qiu J., Su H.B., Watanabe T., and Brunet Y.: 1995, Surface renewal analysis: A new method to obtain scalar fluxes. *Agric. For. Meteorol.* 74: 119–137.

Porté-Agel F., Meneveau C., and Parlange M.B.: 2000, A scale dependent dynamic model for large eddy simulation: Application to a neutral atmospheric boundary layer, *J. Fluid Mech.* 415: 261–284.

Pruitt W.O.: 1991, Development of crop coeDcients using lysimeters. In Allen, R.G., Howell, T.A., Pruitt, W.O, Walter, L.A., and Jensen, M.E. (Eds.), *Lysimeters for Evapotranspiration and Environmental Measurements. Proc. ASCE Int. Symp. Lysimetry*, Honolulu, HI, ASCE, New York, p. 444.

Rana G. and Katerji N.: 2000, Measurement and estimation of actual evapotranspiration in the field under Mediterranean climate: A review. *Eur. J. Agron.* 13: 125–153.

Roth M., Salmond A., and Satyanarayana A.N.V.: 2006, Methodological considerations regarding the measurement of turbulent fluxes in the urban roughness sublayer: The role of scintillometery. *Boundary-Layer Meteorol.* 121: 351–375.

Schmid H.P.: 2002, Footprint modeling for vegetation atmosphere exchange studies: A review and perspective. *Agric. For. Meteorol.* 113: 159–183.

Smagorinsky J.: 1963, General circulation experiments with primitive equations. *Mon. Wea. Rev.* 91: 99–164.

Smeets C., Holzinger R., Vigano I., Goldstein A.H., Röckmann T.: 2009, Eddy covariance methane measurements at a Ponderosa pine plantation in California. *Atmos. Chem. Phys.* 9: 8365–8375.

Soegaard H., Jensen N.O., Boegh E., Hasager C.B., Schelde K., and Thomsen A.: 2003, Carbon dioxide exchange over agricultural landscape using eddy correlation and footprint modelling. *Agric. For. Meteorol.* 114: 153–173.

Stull R. B.: 1988, *An Introduction to Boundary Layer Meteorology*, Kluwer Academic Publishers, Dordrecht, the Netherlands.

Thomson D.J.: 1987, Criteria for the selection of stochastic models of particle trajectories in turbulent flows. *J. Fluid Mech.* 180: 529–556.

Tong C.N., Wyngaard J.C., and Brasseur J.G.: 1999, Experimental study of the subgrid-scale stress in the atmospheric surface layer. *J. Atmos. Sci.* 56: 2277–2292.

Tropea C., Yarin A.L., and Foss J.F. (Eds.): 2007, *Springer Handbook of Experimental Fluid Mechanics*, Springer-Verlag, Berlin, Germany.

Webb E.K., Pearman G.I., and Leuning R.: 1980, Correction of flux measurements for density effects due to heat and water vapour transfer. *Q. J. R. Meteorol. Soc.* 106: 85–100.

Westberg H., Lamb B., Hafer R., Hills A., Shepson P., and Vogel C.: 2001, Measurement of isoprene fluxes at the PROPHET site. *J. Geophys. Res.* 106: 24347–24358.

Wilczak J., Oncley S., and Stage S.: 2001, Sonic anemometer tilt correction algorithms. *Boundary-Layer Meteorol.* 99: 127–150.

Wilson K.B., Hanson P.J., Mulholland P.J., Baldocchi D.D., and Wullschleger S.D.: 2001, A comparison of methods for determining forest evapotranspiration and its components: Sap-flow, soil water budget, eddy covariance and catchment water balance. *Agric. For. Meteorol.* 106: 153–168.

38

Clear-Air Radar Profiling of Wind and Turbulence in the Lower Atmosphere

Jean-Luc Caccia
University of the South, Toulon-Var

38.1 Introduction

In this chapter, a radar technique capable of measuring the temporal evolution of vertical profiles of the three components of the wind vector and turbulence parameters is presented. Although the utilization and interpretation of those radar-derived dynamic parameters for low-atmospheric dynamics studies constitute our main purpose, some measurement physics and a few technological aspects are also briefly summarized.

Clear-air radar refers to ground-based active remote-sensing instrumentation transmitting radio waves upward across the atmosphere at wavelengths allowing measurements of wind and atmospheric reflectivity at any time (day or night) and whatever the meteorological conditions. It is not to be confused with meteorological or weather radars devoted to precipitation and hydrometeor measurements and with airborne or satellite radars. The clear-air radar technique is also sometimes referred to as the MST (mesosphere, stratosphere, troposphere)

radar technique according to its range of measurements, which depends on configuration parameters such as the frequency, the transmitted power, and the antenna surface. The technique uses the weak backscattering from refractive-index fluctuations at radio wave frequencies due to atmospheric turbulence and temperature/humidity sheets. The atmospheric density, and thus the refractive-index fluctuations, quasi-exponentially decrease with increasing altitude, so that the maximum altitude range for detecting atmospheric echoes is 20–25 km. Above about 60 km, in the ionosphere, backscattering arises due to the presence of ionized particles and electrons that enhance the scattering from turbulence.

In the present review, we focus on wind and turbulence measurements in neutral atmosphere from the atmospheric boundary layer (ABL) to the lower stratosphere. In this altitude range radar, radio waves are backscattered from turbulent eddies having a scale of one half of the transmitted wavelength (Bragg backscattering condition). The most efficient frequency bands for Bragg scattering from clear-air turbulence

Handbook of Environmental Fluid Dynamics, Volume Two, edited by Harindra Joseph Shermal Fernando. © 2013 CRC Press/Taylor & Francis Group, LLC. ISBN: 978-1-4665-5601-0.

are the VHF (Very High Frequency, 30–300 MHz, 1–10 m) and the UHF (Ultra High Frequency, 0.3–3 GHz, 0.1–1 m) bands. Because of their transmitted power and wavelength, and their antenna dimensions, VHF radars are generally used for investigating dynamic processes from above the ABL (1000–1500 m) up to the upper troposphere and lower stratosphere (10–25 km). Due to their higher frequency and lower sensitivity, UHF radars can provide measurements within the ABL (about 100 m) and up to the free troposphere (3–6 km). Those two kinds of radar are therefore complementary tools for low-atmospheric dynamic studies.

A theoretical approach of the processes of interaction between turbulence and electromagnetic waves can be found in the book by Tatarskii (1961). The first measurements of wind and turbulence by a VHF radar were reported by Woodman and Guillèn (1974) with the Jicamarca radar (Peru). Since then, many comprehensive reviews about MST radar advances have been published (e.g., Gage and Balsley 1978, Röttger and Larsen 1990, Hocking 1997, Fukao 2007).

Although a large variety of techniques and methodologies can be applied for retrieving atmospheric parameters, the Doppler Beam Swinging (DBS) technique is the earliest and still the most commonly used. That is why we mainly focus on this technique in the next sections. However, a description of some alternative techniques is briefly presented in Sections 38.2 and 38.5.

38.2 Principles

38.2.1 Radar Antennas and Electromagnetic Beams

The antenna of UHF or VHF radars consists of an array of electrically fed dipoles transmitting successive pulses of electromagnetic waves toward the atmosphere. For the VHF case, when all the dipoles are fed in phase, the main lobe direction is vertical while it is oblique when they are appropriately phase shifted. For the UHF case, an antenna panel (one horizontal and the others slanted) transmitting a beam perpendicularly to its surface is used for each direction. For a wavelength λ and an array dimension D, the angular half-power beamwidth, given by the classical diffraction theory, is approximately λ/D. Since the radar beamwidths are comparable at UHF and VHF (i.e., typically from 3° to 8°), a simple calculation leads to an antenna array dimension of a few meters, in the UHF case, and of several tens of meters, in the VHF case. In DBS mode, the radar beam is sequentially switched among at least three beam directions. One is vertical and the others are steered from 10° to 20° off zenith at perpendicular azimuthal directions. Although three directions are required to obtain the three wind components, many radars use five beams in order to slightly reduce estimation errors and to study the azimuthal dependence of the radar echoes. Two radars, one UHF and one VHF, are showed on Figure 38.1a through c, respectively.

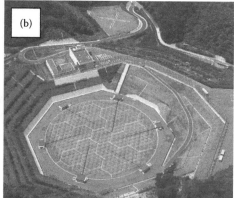

FIGURE 38.1   Examples of radars: (a) the Nice airport UHF radar, and (b and c) the MU VHF radar at Shigaraki (Japan). The length of a UHF antenna panel is 1 m, while the diameter of the VHF antenna array, shown in (b), is 100 m. Some details of the individual antennas are shown in (c).

38.2.2 Doppler Spectra, Wind, and Turbulence Parameters

Since the backscattering structures (turbulence eddies and sheets) are advected by the wind, the echoes received from a given altitude along a given beam direction are frequency shifted. This frequency shift is due to the Doppler Effect produced by the radial velocity, which is the projection of the wind vector along the beam axis. After a synchronous demodulation of the received signals, the power spectral density of time series of the signals is computed using an FFT (Fast Fourier Transform) algorithm. Such a Doppler spectrum mainly consists of a spectral peak superimposed to a constant positive offset. The radial velocity value, and the signal and the noise power are given by the frequency corresponding to the peak center, and the integrated value of the spectral peak and of the offset level, respectively. Assuming the wind field is homogeneous within the horizontal beam-steered area and stationary during the DBS sequence, the three wind components are retrieved by geometrical recombination of the three radial velocities. The radar-derived turbulent parameters are the refractive-index fluctuations structure constant, C_N^2, which is estimated from the classical radar equation using the signal-to-noise ratio (SNR) (e.g., Röttger 1980), the turbulent kinetic energy dissipation rate, ε, and the turbulent diffusivity, K, which are estimated from the spectral-peak width (e.g., Hocking 1983, Jacoby-Koaly et al. 2002). As a result, a series of Doppler spectra are obtained at different altitudes, leading to a vertical profile of radial velocities and turbulence parameters.

It is to be noticed that the wind estimates are directly derived from easy-to-determine frequency values and under very reasonable assumptions. It is not the case for turbulence parameters that are estimated from much more complex equations assuming additional assumptions (e.g., Hocking 1983). Therefore, for meteorological applications, and mainly in an operational context, only the wind data are used. Consequently, the VHF and UHF ST (Stratosphere, Troposphere) radars are often considered as, and called, "wind profiler radars" by the nonspecialist scientific community.

38.2.3 Resolutions and Accuracy

The temporal resolution is given by the duration necessary to obtain a reliable estimate of vertical profiles. A value of 3–5 min is currently necessary for both UHF and VHF radars. Because it is short compared to the time scale of the classical meteorological dynamic processes, it is generally considered that the radars continuously observe the temporal evolution of the vertical dynamic structure of the lower atmosphere.

The vertical resolution Δz is given by the pulse width τ ($\Delta z = c\tau/2$) and is typically $\Delta z = 150$–300 m, at VHF, and $\Delta z = 75$–150 m, at UHF.

Determining the accuracy of the radar-derived parameters is very diDcult because they are the result of different physical and statistical processes under different assumptions. Regarding the wind measurements, error bars of 0.5 and 0.05 m s^{-1} for the horizontal wind components and the vertical velocity, respectively, are currently considered.

38.2.4 Summary of Major Features

As previously mentioned, VHF and UHF ST radars are complementary tools for monitoring dynamic processes of the lower atmosphere. Measuring the temporal evolution of the vertical profiles of the three wind components obtained during all weather conditions constitutes the main feature of those instruments. Regarding the time and vertical resolutions, ST radars are very useful for mesoscale meteorology studies (1–100 km horizontal-scale) and more generally for mesoscale dynamic process studies where wind systems and/or atmospheric waves are to be investigated. Vertical profiles of turbulence parameters such as C_N^2, ε, and K can also be retrieved and used in specific studies where turbulent layers due to static and/or dynamic instabilities play a significant role.

Three main features are worth emphasizing: (1) Observations are made continuously in time above the radar site; (2) at a given time, the wind measurements over the altitude range are simultaneous and made along a vertical direction (Eulerian measurements), while balloon measurements provide values at various altitudes at different times and at different positions (Lagrangian measurements); and (3) ST radars are unique tools for measuring profiles of the air vertical velocity.

38.2.5 Brief Description of Other Techniques

In addition to the DBS mode, other observational modes based on the use of several receiving antennas have also been implemented early after the development of the MST/ST radars. These modes can be an alternative to the DBS mode for measuring winds ("Spaced Antenna" [SA] techniques) or for improving the angular resolution of the radars, often limited by their beamwidth ("Spatial Domain Interferometry" [SDI] or "Coherent Radar Imaging"). The SA technique is fundamentally based on signal delays between two antennas for retrieving the wind component along the antenna axis. A large number of estimation methods have been developed, assuming more or less sophisticated models of backscattering from turbulence. These methods also allow one to retrieve turbulence parameters. SDI is used for determining the direction of arrival of targets through beam forming. Beamforming consists of combining the signals from individual receivers into a set of narrow-focused beams within a relatively wide transmitted beam. The technique has been transposed in the frequency domain ("Range Imaging") by combining signals from closely spaced frequencies switched pulse to pulse for range imaging along the direction of observations (Luce et al. 2001). High-resolution methods such as maximum entropy method and Capon's method have been implemented for both improving resolutions and for rejecting outliers, clutter, or interferences. Range resolution improvement strongly depends on SNR.

For SNR larger than 0 dB, the range resolution can be improved by a factor 5–10 with respect to the initial resolution given by the pulse width.

38.3 Examples

38.3.1 National Wind Profiler Networks

The first national ST radar network was developed in the United States in the early 1990s: the NOAA profiler network (NPN) (http://www.profiler.noaa.gov/npn/). Thirty-five 404-MHz band profilers are operated and the data are used in numerical weather prediction (NWP) and operational forecasts. For the same purpose, since the mid-1990s, the CWINDE (COST Wind Initiative for a Network Demonstration in Europe) European wind profiler network has also been operated (http://www.met-oDce.gov.uk/research/interproj/cwinde/profiler). It consists of 25 45-MHz to 1290-MHz radars covering the whole Western Europe. The Japan Meteorological Agency (JMA) has also deployed a wind profiler network called WINDAS (WInd profiler Network and Data Acquisition System), and has been operating since 2001. It consists of 31 1.3-GHz radars covering the main islands of Japan (Ishihara et al. 2006). In general, the radar-derived wind profiles data are assimilated and used as initial values in mesoscale NWP models and taken into account in model reanalyses.

38.3.2 Observation of Troposphere–Stratosphere Exchanges

As already mentioned, VHF radars are suitable tools for studying troposphere–stratosphere exchange (TSE) at mesoscale. Many studies about the description and the interpretation of dynamic processes contributing to TSE have been published, for example, the passage and the evolution of an upper-level potential-vorticity (PV) streamer (Hoinka et al. 2003), an upper-level front/tropopause folding system (Caccia and Cammas 1998) or a cold vortex/tropopause funnel system (Fukao et al. 1989). In those studies, time–height cross sections of wind mesoscale systems, including the vertical velocity field, and associated turbulence obtained by one or several VHF radars are reported and analyzed.

Due to the strong horizontal stratification of the low stratosphere, the VHF echoes are enhanced around the vertical direction. As a result, the vertical to oblique echo power ratio, also called aspect ratio, increases at the tropopause altitude. The temporal evolution of the altitude where the vertical profiles of aspect ratio or vertical echo power are enhanced can be used for monitoring the tropopause altitude (e.g., Caccia and Cammas 1998).

Figure 38.2a shows the wind field observed by a VHF radar after an upper-level PV streamer has passed over the Alps (Hoinka et al. 2003). The corresponding simulation by the mesoscale non-hydrostatic numerical model Meso-NH (Lafore et al. 1998) is shown in Figure 38.2b. The main structures, such as the jet stream region and the associated tropopause variations are correctly reproduced by the model.

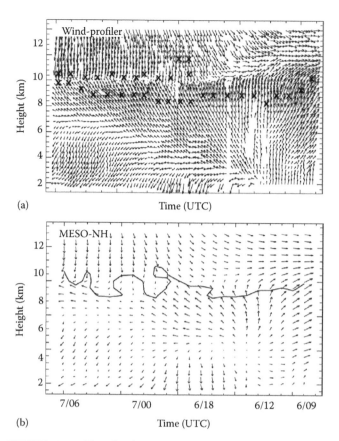

FIGURE 38.2 Time–height cross section of the horizontal wind observed by a VHF wind profiler (a) (Météo-France/CNRM) and simulated by meso-NH (b) above Lonate-Pozzolo (Piedmont, Italy) during an upper-level PV-streamer passage. The crosses on the top panel indicate the locations of the VHF vertical echo power enhancement; the lowermost ones correspond to tropopause heights. The full line on the bottom panel indicates the PV tropopause of 3.5 PV units. Since the structure propagates from west to east, time increases from right (6 November 0900 UTC) to left (7 November 0600 UTC). The wind vector is indicated by an arrow so that its length is proportional to the wind speed and an orientation toward the bottom indicates a northerly wind. (After Hoinka, K.P. et al., *Q. J. R. Meteorol. Soc.* 129, 609, 2003.)

38.3.3 Observation of Mesoscale Gravity Waves

The sources of gravity waves (GW) can be different dynamic processes in the lower atmosphere such as flow passage over mountains, shear instabilities, geostrophic adjustment near jet streams, cold/warm front passages, and convective systems. The associated mechanisms act as dynamic perturbations of a stratified atmosphere initially in equilibrium. After their generation, it is now well established that GWs are able to transport momentum and kinetic energy over large distances.

VHF radars make possible direct observations of GW activities in the free troposphere and the lower stratosphere such as those reported by Röttger (1979) during a frontal passage, by Caccia et al. (1997) during a mountain/lee wave event or Murayama et al. (1994), during a strong jet stream passage. In addition to the source identification and the description of the wind field perturbed by GW, radar observations also helped to characterize the

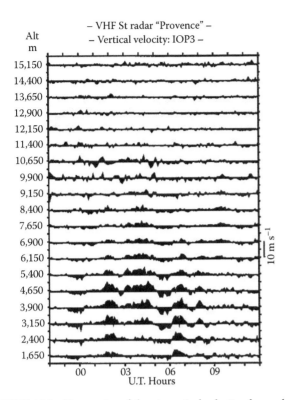

FIGURE 38.3 Time series of the air vertical velocity observed by a VHF radar ("Provence," CNRS/LSEET) from 1,650 to 15,150 m of altitude (ASL) every 750 m. Those observations are made downstream the Pyrénées during a mountain/lee wave event.

wave propagation mode (freely upward in the lower atmosphere or trapped in a horizontal wave duct) (Caccia et al. 1997), the parameterization of the wave–mean flow interactions (Palmer et al. 1986), the estimation of wave-induced momentum fluxes (Murayama et al. 1994), or the determination of GW spectra using vertical (Ecklund et al. 1985) or horizontal (Balsley and Carter 1982) velocities. The vertical velocity field perturbed by the mountain/lee waves observed by a VHF radar installed downstream the mountain chain (here the Pyrénées) and reported in Caccia et al. (1997) is shown in Figure 38.3. That figure clearly shows that the waves are trapped within the troposphere. During that event, a maximum value of 6 m s^{-1} for the vertical velocity has been recorded at 0645 UT and at 3900 m of altitude (ASL).

38.3.4 Observation of Convective Systems

Convective systems may have vertical extent from the top of the ABL up to the tropopause layer. Consequently, both UHF and VHF radars can be very useful tools in order to describe the wind field around and inside the convective cells. Besides, for such particular studies, the vertical velocity field is a crucial parameter and is directly measured by ST radars.

Many mesoscale convective systems in subtropical and tropical regions have been described by UHF and VHF radars. Those systems involve a multi-scale structure of vertical motions associated with a cyclonic circulation background and their

propagation velocity is determined by synoptic-scale conditions. Typical horizontal scales from several tens to a few hundreds of kilometers, vertical scales from several hundreds of meters to several kilometers, and time scales from several tens of minutes to several hours are classical features of those convective cells. Different types of mesoscale convective systems have been observed by radars, e.g., the passage of a system of multi-mesoscale strong updraft regions within cyclonic rotor circulation (Shigabaki et al. 1997), mesoscale squall line systems (Augustine and Zipser 1987), or tropical convective cells during the monsoon break (Cifelli and Rutledge 1994).

Convective systems with larger vertical extent are generally found in the equatorial atmosphere. Above equatorial regions, the tropopause is between 15 and 17 km, so that only powerful VHF radars can be suitable for studying high-tropospheric convective processes. For instance, the generation of GWs by deep cumulus convection was evidenced by Dhaka et al. (2006), with an MST radar installed in Sumatra (Indonesia). Many results obtained with the same equatorial radar have been reported by Fukao (2006).

38.3.5 Observation of Atmospheric Boundary Layer Processes

The observations of dynamic processes within the ABL are usually made using UHF radars for both wind and turbulence measurements. Radar data have been mainly useful in several studies, such as the monitoring of the ABL vertical structure, the low-level jets (LLJ), the breeze regimes, some orographic processes (flow splitting, flow braking, downslope windk), and the transport of aerosols and atmospheric constituents. All those issues and interactions between some of concerned processes have been addressed using UHF radar observations.

Details about the method used for monitoring the ABL height using C_N^2 and ε-measurements can be found in the paper by Jacoby-Koaly et al. (2002). A wind regime (mistral) passing from a high vertical extent to a LLJ-like structure due to variations of the upstream conditions has been reported by Guénard et al. (2005). UHF observations of that event are shown in Figure 38.4a and b together with the corresponding simulations by Meso-NH in Figure 38.4c and d. The model/observations comparison shows a good agreement for the onset of the mistral regime, but the model overestimates its duration and its vertical extent for both the sites which are 90 km apart. This illustrates the diDculty for a mesoscale model to correctly reproduce a parameter field with sharp gradients and rapid temporal variations as it is the case for the mistral wind field.

A case of mistral/sea breeze (two opposite flows) interactions giving rise to an ABL development up to 2 km of altitude and allowing aerosols to reach that level has been presented by Bastin et al. (2006). The ABL dynamic structure corresponding to this case, as observed by a UHF radar installed near the coast, is shown in Figure 38.5. The mistral braking due to the sea breeze is clearly visible within the ABL every day around noon in Figure 38.5a, as well as the development of the convective ABL using the kinetic energy, turbulent dissipation rate, ε, in Figure 38.5b.

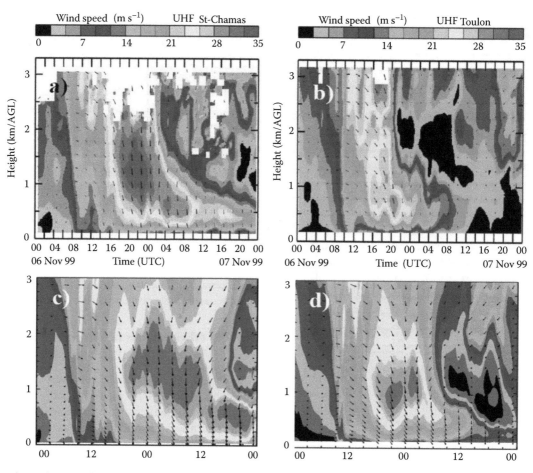

FIGURE 38.4 **(See color insert.)** Time–height cross sections of the horizontal wind field observed by two UHF radars installed 90 km apart (a) at St-Chamas (EDF/LA/CRA) and (b) at Toulon (Degréane/LSEET), and the corresponding Meso-NH simulations (c) and (d), respectively, during the MAP campaign. The wind vector is given in colored levels with superimposed arrows indicating the wind direction so that an arrow toward the bottom corresponds to a northerly wind. The same wind speed scale, i.e., 0–35 m s^{-1}, timescale, and altitude range, i.e., 0–3 km, are used in (a), (b), (c), and (d). (After Caccia, J.-L. et al., *Metorol. Zeit.* 10, 469, 2001.)

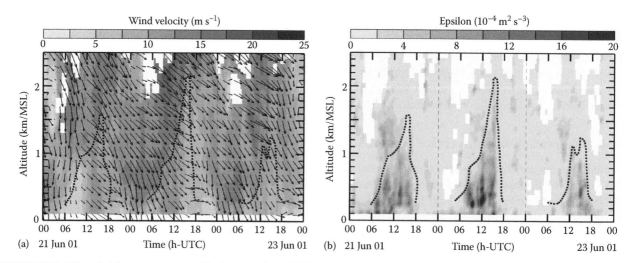

FIGURE 38.5 Time–height cross sections of the horizontal wind field (a) and the kinetic-energy turbulent dissipation rate, ε, (b) observed by a UHF radar installed at St-Chamas, near the Mediterranean coast, during the ESCOMPTE campaign. The wind speed and the ε values are given in gray levels with superimposed arrows in (a) indicating the wind direction as for the previous figures. The dotted line, both in (a) and (b), shows the temporal evolution of the ABL top following the C$_N^2$-method.

38.4 Methods of Analysis

38.4.1 Campaigns and Databases

Numerous scientific results about mesoscale meteorology have been obtained in the context of coordinated and multi-instrumented campaigns, including networks of UHF and VHF radars adapted to the scientific objectives. Their primary use is to produce databases for the scientific community involved in the campaign and, then, for the entire scientific community according to a data diffusion policy. In this context, the radar data are analyzed with complementary data from other instruments.

During the MAP (Mesoscale Alpine Program in 1999, http://www.map.ethz.ch) campaign, Hoinka et al. (2003) completely described the propagation and the evolution of an upper-level PV anomaly from the Atlantic atmosphere to the Alps using airborne in situ and lidar measurements, satellite images, mesoscale model simulation, and VHF radar data. During PYREX (PYRenean EXperiment in 1990), Caccia et al. (1997) made the identification and the scaling (the horizontal and vertical extent and the life-time) of a nonstationary mountain lee wave trapped in the troposphere using airborne (aircraft and constant volume balloons), radiosounding, and two VHF radar data. During ESCOMPTE (Expérience sur Site pour COntraindre les Modèles de Pollution et de Transport dﬀEmission in 2001, http://medias.obs-mip.fr/escompte), Bastin et al. (2006) described in detail the mistral/sea breeze interactions using meteorological surface stations, airborne lidar and UHF radar data, and non-hydrostatic mesoscale model simulation.

38.4.2 Validation of Data Assimilation in Numerical Weather Prediction Models

Another very important role played by the radar wind profiler is its contribution to improve the reliability of the mesoscale numerical atmospheric models. This can be done using two ways: (1) research approach: analyzing the differences between radar data and the same parameters produced by numerical models and (2) operational approach: using the radar data as constraints for the model through an assimilation operation. In the Section 38.3.1, the role of the wind profilers in national and operational meteorological networks has been already presented.

In a research program devoted to improve the knowledge of various mesoscale dynamic processes, a set of instruments is generally deployed and the collected data are directly compared to mesoscale numerical model simulations. The purpose is clearly to improve the reliability of the model. In such a context, the experimental data are processed and are subject to quality control by an expert before being used in the comparison procedure. For instance, Hoinka et al. (2003) and Bastin et al. (2006) (see Section 38.4.1) obtained a quite good agreement between parameters estimated from radar data and those provided by Meso-NH (mesoscale non-hydrostatic) model simulations (e.g., see Figures 38.2 and 38.4).

The assimilation of radar data in operational models is much more diﬀcult. It is mainly because the determination of a reliable data quality criterion is a very complex task and makes very dangerous an automatic assimilation of (routine) radar data by operational models. Indeed, despite the space and time consensus, and the automatic rejection methods applied to the raw data, it is impossible to assess the total reliability of the processed data. However, assimilation of radar data is not impossible and can be useful. For example, Ishihara et al. (2006) presented an impact experiment conducted with a numerical mesoscale model of the JMA and showed that the assimilation of WINDAS data (see Section 38.3.1) improved the accuracy of the forecasting of a mesoscale weather system, evolution associated with heavy rainfall events.

38.5 Major Challenge

38.5.1 Radar Data Assimilation in Operational Mesoscale Models

As already mentioned, the assimilation of UHF or VHF radar data in mesoscale numerical models using an automatic and routine procedure is a very diﬀcult task. In the case of specific campaigns or research programs, the processed data corresponding to the period on study are carefully reprocessed by experts prior to be sent to the assimilation procedure, and thus taken into account in the model forecast or simulation. Successful results have been already obtained by using this specific approach (e.g., Ishihara et al. 2006), but, to our knowledge, a routine assimilation procedure in operational mesoscale models has never been implemented yet.

In order to minimize error sources in the wind component measurements, making diﬀcult the assimilation, it is currently proposed to assimilate radial velocities directly obtained by the Doppler spectra method (see Section 38.2.2). By this way, the hypotheses of stationarity and homogeneity of the wind field is no longer assumed, since the assimilation procedure takes account for the exact location and time of the measurements. If necessary, an estimated wind profile is then retrieved using model equations. That procedure, still under testing, requires an assimilation-module-specific interface between the radar measurements and the model.

38.5.2 Fine-Scale Observations

Due to their limited resolutions in time and range, standard ST radars in DBS mode cannot resolve the vertical distribution of turbulent layers and sheets. Even though recent technologies can provide suﬀcient bandwidths for shorter pulses and then higher-range resolution, e.g., 37.5 m with the updated VHF MPI-Sousy radar (Woodman et al. 2007), frequency-domain interferometry techniques are still promising alternatives. As an illustration of range imaging (RIM) using frequency diversity, Figure 38.6 shows a time–height intensity plot of SNR with the vertical beam before (a) and after (b) the Capon RIM

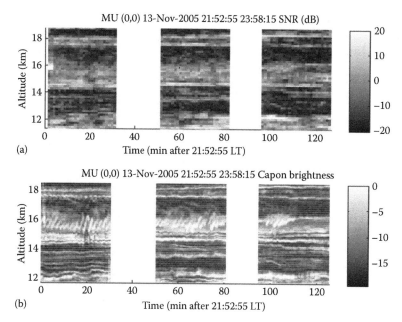

FIGURE 38.6 Time–height intensity plot of SNR with the vertical beam before (a) and after (b) the Capon RIM processing with the MU radar (Shigaraki, Japan). The vertical and time resolutions are significantly increased from (a) to (b).

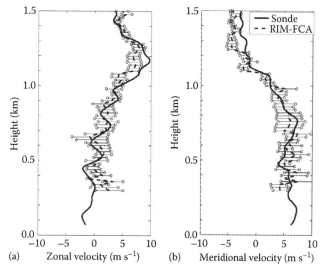

FIGURE 38.7 Vertical profiles of the zonal (a) and meridional (b) component of the wind using the SA/RIM method applied to radar data (dotted line) and obtained by balloon (solid line). The vertical resolution for the radar profiles is 20 m. The superimposed horizontal lines with small circles indicate the estimation errors. (After Yu, T.-Y. and Brown, W.O.J., *Radio Sci.*, 39, RS1011, 2004.)

processing with the MU radar (Shigaraki, Japan) in the tropopause region. This illustrates the vertical and time-resolution improvement using that processing technique. The «S-shaped» structures resulting from a Kelvin–Helmholtz instability are clearly visible after RIM.

The first estimations of horizontal winds with a high vertical resolution (here 20 m) using combined SA/RIM and compared

with in situ observations are shown in Figure 38.7. Those results are promising for demonstrating the usefulness of the FII/RIM technique for small-scale studies of the horizontal winds.

References

Augustine, J.A. and E.J. Zipser 1987. The use of wind profilers in a mesoscale experiment. *Bull. Am. Meteorol. Soc.* 68: 4–17.

Balsley, B.B. and D.A. Carter 1982. The spectrum of atmospheric velocity fluctuations at 8 km and 86 km. *Geophys. Res. Lett.* 9: 465–468.

Caccia, J.-L., J.-P. Aubagnac, G. Béthenod et al. 2001. The French ST-radar network during MAP: Observational and scientific aspects. *Metorol. Zeit.* 10: 469–478.

Caccia, J.-L., B. Benech and V. Klaus 1997. Space-time description of nonstationary trapped lee waves using ST radars, aircraft, and constant volume balloons during the PYREX experiment. *J. Atmos. Sci.* 54: 1821–1833.

Caccia, J.-L. and J.-P. Cammas 1998. VHF-ST radar observations of an upper-level front using vertical and oblique-beam CN2 measurements. *Mon. Wea. Rev.* 126: 483–501.

Cifelli, R. and S.A. Rutledge 1994. Vertical motion structure in maritime continent mesoscale convective systems: Results from a 50-MHz profiler. *J. Atmos. Sci.* 51: 2631–2652.

Dhaka, S.K., M.K. Yamamoto, Y. Shibagaki, H. Hashiguchi, S. Fukao and H.-Y. Chun 2006. Equatorial atmosphere radar observations of short vertical wavelength gravity waves in the upper troposphere and lower stratosphere region induced by localized convection. *Geophys. Res. Lett.* 33: L19805, doi:10.1029/2006GL027026.

Ecklund, W.L., B.B. Balsley, D.A. Carter, A.C. Riddle, M. Crochet and R. Garello 1985. Observations of vertical motions in the troposphere and lower stratosphere using three closely spaced ST radars. *Radio Sci.* 20: 1196–1206.

Fukao, S. 2006. Coupling processes in the equatorial atmosphere (CPEA): A project overview. *J. Meteorol. Soc. Jpn.* 84A: 1–18.

Fukao, S. 2007. Recent advances in atmospheric radar study. *J. Meteorol. Soc. Jpn.* 85B: 215–239.

Fukao, S., M.D. Yamanaka, H. Matsumoto, T. Sato, T. Tsuda and S. Kato 1989. Wind fluctuations near a cold vortex-tropopause funnel system observed by the MU radar. *Pure Appl. Geophys.* 130: 463–479.

Gage, K.S. and B.B. Balsley 1978. Doppler radar probing of the clear atmosphere. *Bull. Am. Meteorol. Soc.* 59: 1074–1093.

Hocking, W.K. 1983. On the extraction of atmospheric turbulence parameters from radar backscatter Doppler spectra, Part I: Theory. *J. Atmos. Terr. Phys.* 45: 89–102.

Hocking, W.K. 1997. Recent advances in radar instrumentation and techniques for studies of the mesosphere, stratosphere and troposphere. *Radio Sci.* 32: 2241–2270.

Hoinka, K.P., E. Richard, G. Poberaj, R. Busen, J.-L. Caccia, A. Fix and H. Mannstein 2003. Analysis of a potential-vorticity streamer crossing the Alps during MAP IOP 15 on 6 November 1999. *Q. J. R. Meteorol. Soc.* 129: 609–632.

Ishihara, M., Y. Kato, T. Abo, K. Kobayashi and Y. Izumikawa 2006. Characteristics and performance of the operational wind profiler network of the Japan Meteorological Agency. *J. Meteorol. Soc. Jpn.* 84: 1085–1096.

Jacoby-Koaly S., B. Campistron, S. Bernard, B. Bénech, F. Ardhuin-Girard, J. Dessens, E. Dupont and B. Carissimo 2002. Turbulent dissipation rate in the boundary layer via UHF wind profiler Doppler spectral width measurements. *Boundary-Layer Meteorol.* 103: 361–389.

Luce, H., M. Yamamoto, S. Fukao, D. Hélal, and M. Crochet 2001. A frequency radar interferometric imaging applied with high resolution methods. *J. Atmos. Sol. Terr. Phys.* 63: 221–234.

Murayama, Y., T. Tsuda and S. Fukao 1994. Seasonal variation of gravity wave activity in the lower atmosphere observed with the MU radar. *J. Geophys. Res.* 99: 23057–23069.

Palmer, T.N., G.J. Shutts and R. Swinbank 1986. Alleviation of a systematic westerly bias in general circulation and numerical weather prediction models through an orographic gravity wave drag parametrization. *Q. J. R. Meteorol. Soc.* 112: 1001–1039.

Röttger, J. 1979. VHF radar observations of a frontal passage. *J. Appl. Meteorol.* 18: 85–91.

Röttger, J. 1980. Structure and dynamics of the stratosphere and mesosphere revealed by VHF radar investigations. *Pure Appl. Geophys.* 118: 494–527.

Röttger, J. and M.F. Larsen 1990. UHF/VHF radar techniques for atmospheric research and wind profile applications. In *Radar in Meteorology*, D. Atlas, Ed., American Meteorological Society, Boston, MA, pp. 235–281.

Shigabaki, Y., M.D. Yamanaka, H. Hashiguchi, A. Watanabe, H. Uyeda, Y. Maekawa and S. Fukao 1997. Hierarchical structures of vertical velocity variations and precipitating clouds near the Baiu frontal cyclone center observed by the MU and meteorological radars. *J. Meteorol. Soc. Jpn.* 75: 569–595.

Tatarskii, V.I. 1961. *Wave Propagation in a Turbulent Medium*, Dover, New York.

Woodman, R.F. and A. Guillèn 1974. Radar observations of winds and turbulence in the stratosphere and mesosphere. *J. Atmos. Sci.* 31: 493–505.

Woodman R.F., G. Michhue, J. Röttger and O. Castillo 2007. The MPI-Sousy-VHF radar at Jicamarca: High altitude resolution capabilities. In *Proceedings of MST11*, Gadanki, India, pp. 334–337.

Yu, T.-Y. and W.O.J. Brown 2004. High-resolution atmospheric profiling using combined spaced antenna and range imaging techniques. *Radio Sci.* 39: RS1011, doi:10.1029/2003RS002907.

Index

545

T - #0722 - 101024 - C12 - 279/216/27 - PB - 9781138374744 - Gloss Lamination